FREE ELECTRON LASERS 1998

FREE ELECTRON LASERS 1998

Proceedings of the Twentieth International
Free Electron Laser Conference,
Williamsburg, Virginia, USA, August 16-21, 1998

Editors

G.R. NEIL

Thomas Jefferson National Accelerator Facility MS 89
Newport News, Virginia, USA

S.V. BENSON

Thomas Jefferson Laboratory - Free Electron Laser Deputy Program Manager
Newport News, Virginia, USA

1999

ELSEVIER

Amsterdam - Lausanne - New York - Oxford - Shannon - Singapore - Tokyo

ELSEVIER SCIENCE B.V.
Sara Burgerhartstraat 25
P.O. Box 211, 1000 AE Amsterdam, The Netherlands

First edition 1999

Library of Congress Cataloging in Publication Data
A catalog record from the Library of Congress has been applied for.

ISBN: 0 444 50248 3

♾The paper used in this publication meets the requirements of ANSI/NISO Z39.48-1992 (Permanence of Paper).
Printed in The Netherlands.

Preface

The 20th International Free Electron Laser Conference and 5th FEL User Workshop was held August 16–22, 1998 in historic Williamsburg, Virginia, USA. The conference was hosted by the Thomas Jefferson National Accelerator Facility (Jefferson Lab), a U.S. Department of Energy nuclear physics research laboratory in Newport News, Viginia. The close coupling between the Users Workshop and the Conference allowed excellent interactions between the FEL physics community and those who utilize the light output of FELs for research in a wide variety of scientific fields.

In addition to the technical sessions, the 275 attendees of both sessions from 20 countries had an opportunity to tour Jefferson Lab and see the recently commissioned high-average-power FEL User Facility. There they also toured the associated Applied Research Center with exhibits from five local universities as well as an industrial applications poster session. The conference banquet was held in the Mariner's Museum featuring nautical exhibits and artifacts from the Titanic.

The distribution of attendees included: US, 158; Japan, 34; Germany, 13; Netherlands, 9; France, 8; Russia, 7; Israel, 5; China, 4; England, 4; Italy, 4; Sweden, 2; Korea, 3; Scotland, 3; Taiwan, 2; India, 1; Canada, 1. Thanks to the generous sponsorship of the U.S. Department of Navy Office of Naval Research and the U.S. Department of Energy, we were able to provide support for 31 students and scholars from 11 countries. In addition, we are grateful for the support from the other organizations listed on our sponsor page which allowed us to provide a much more enjoyable conference for all attendees.

The approach to the scientific program utilized a select group of international scientists from the program committee to choose general topics for the various sessions. They appointed session chairs for each of those areas to choose contributed and invited talks in the field from the most current research activities in progress. Sessions highlighting research in the following topics were: New Lasing, FEL Theory, SASE FELs, Accelerator Technology, FEL Technology, Linac-Based FELS, Storage-Ring-Based FELs, UV and X-ray Sources, and New Concepts. In addition, the breadth of interest in applying FELs is evident from the session topics held at the Users Workshop: Physics and Chemistry Research, Applications Research, Industrial Applications, Biomedical, Biophysical, Biochemical, and Photochemical Applications, Applications in Condensed Matter, Materials, Plasmas, Semiconductors, and Surface Physics.

The growth in the FEL field in the past year was visible in the number of papers presented: 224. Of that number we have selected (with the help of reviews by referees comprising the Program Committee and others) 89 of these for publication in NIMA. Following the standards of NIMA only those papers presented at the conference and which meet the normal criteria for scientific publications (accuracy, scientific merit, and novelty) were considered for publication. The editors would like to express their deep appreciation to those who reviewed the submissions and helped us select the papers you see.

The FEL field is quite active in a few specific areas at this time. Several groups are investigating self-amplified spontaneous emission (SASE) as a route to a 0.1 nm FEL. Although the technical challenge are large, a growing portion of the community believes this is a feasible goal and have begun planning 4th generation light sources based on this technique. Already, demonstrations of SASE by many orders of magnitude in an unguided (by external means) optical mode have been achieved in the IR with extension into the UV soon to follow. Other groups are extending the

PII: S0168-9002(99)00506-9

applications of FELs by evolutionary changes in the capabilities of user facilities around the world. Many of these utilize other sources of radiation synchronously with tunable FEL beams. An emerging trend is the use of Thompson scattered photons from the electron beam. Because of the Doppler shift involved, the photons can be up-scattered into the X-ray (keV) or even gamma ray (MeV) regime forming a useful picosecond probe for analysis of materials or nuclear structure. Other groups continue to extend the range of FEL operation. A new record was set this year for short wavelength lasing (210 nm) as well as production of the highest CW average power yet for an FEL (311 W). Millimeter wave FELs are close to producing powers that greatly exceed even this high level. A number of facilities began operation. Several exciting applications of FEL light were cited in the user workshop involving laser ablation of corneas, clinical use of FEL beams, and several interesting pump probe experiments using FEL radiation with synchrotron emissions or other lasers in semiconductors. FEL beams have been coupled with near-field microscopy for bond specific studies of impurities and defects in materials such as diamond crystallites.

We hope this volume gives the reader an appreciation of the state of FEL technology and coveys the sense of excitement and interest in the field. Despite the fact that it has been 22 years since the first demonstration of lasing in an FEL oscillator, the field continues to provide interesting areas for study and application. We look forward to the conferences in the years to follow for the results of initiatives already underway, for the new ideas that are certain to appear, and for the useful application of FEL photons in ways currently unimagined.

George R. Neil
Chair

Stephen V. Benson
Program chair

Conference Chair

G.R. Neil

Conference Program Chair

S. Benson

User Workshop Chair

G. Edwards

International Executive Committee

I. Ben-Zvi, Brookhaven National Laboratory, USA
W.B. Colson, Naval Postgraduate School, USA
A. Gover, Tel-Aviv University, Israel
H. Hama, Institute for Molecular Science, Japan
K.J. Kim, Lawrence Berkeley Laboratory, USA
J.M.J. Madey, Duke University, USA
K. Mima, Osaka University, USA
G.R. Neil, Thomas Jefferson National Accelerator Facility, USA
J.M. Ortega, LURE, France
M.W. Poole, EPRSC, Daresbury, UK
A. Renieri, ENEA, Frascati, Italy
C.W. Roberson, Office of Naval Research, USA
J. Rossbach, DESY, Germany
T. Smith, Stanford University, USA
P. Sprangle, Naval Research Laboratory,USA
M. Van Der Wiel, FOM, The Netherlands
N.A. Vinokurov, Budker Institute, Russia
T.L. Yang, Institute of Atomic Energy, China

Program Committee

M.E. Couprie, CEA/LURE, France
G. Dattoli, ENEA, Italy
H. Freund, Science Applications International, USA
J. Goldstein, Los Alamos National Laboratory, USA
E. Jerby, Tel-Aviv University, Israel
V. Litvinenko, Duke University, USA
A. Lumpkin, Argonne National Laboratory, USA
M. Mendenhall, Vanderbilt University, USA
E.J. Minehara, Japan Atomic Energy Res. Inst., Japan
D. Nolle, Univ. of Dortmund, Germany
D. Oepts, FOM, The Netherlands
S. Okuda, Osaka University, Japan
G. Ramian, University of California at Santa Barbara, USA
J. Rossbach, DESY, Germany
T. Tomimasu, Free Electron Laser Institute, Japan
P.J.M. Van der Slot, Nederlands Centrum voor Laser Research, The Netherlands
E. Szarmes, Duke University, USA
A. Varfolomeev, Kurchatov Institute, Russia
R. Walker, Sincrotrone Trieste, Italy
M. Xie, Lawrence Berkeley Laboratory, USA
T. Yang, CIAE, China
L. Yu, Brookhaven National Laboratory, USA

PII: S0168-9002(99)00507-0

20th international **FEL98** **conference** Williamsburg, VA
August 16-21, 1998

SPONSORS

The Organizers of FEL98 would like to express their gratitude to the following sponsors for the support provided:

Eruditio et Religio
Duke University

Office of Naval Research

DUPONT
REG. U.S. PAT. & TM. OFF.

SAIC

Department of Energy · United States of America

Jefferson Lab
Thomas Jefferson National Accelerator Facility

Professor John Walsh wins the 1998 International Free-Electron Laser Prize

The International Free-Electron Laser Prize is awarded each year to recognize individuals for pioneering contributions to the field of free-electron lasers. At the 20th International Free-Electron Laser Conference, the Prize for 1998 was awarded to Professor John Walsh, of Dartmouth College, in recognition of his outstanding leadership in research and teaching in the field of free-electron lasers.

Professor Walsh's long and distinguished career in free-electron lasers and microwave devices at Dartmouth has inspired many students, both graduate and undergraduate. His research epitomizes the pioneering spirit, characterized always by innovation and imagination. He is one of the few in our field who is equally at home in discussions of both long-wavelength, low-energy devices and short-wavelength, high-energy machines, and as often as not his research has been outside the mainstream of either community. For example, he has been a pioneer in the development of Cherenkov free-electron lasers, and recently demonstrated the first evidence of gain on a new device called a Smith–Purcell free-electron laser. Indeed, it is fitting that this new concept should be reported for the first time at this Conference. In addition, Professor Walsh has been ahead of the rest of us in the development of compact free-electron lasers. At a time when ever-larger free-electron lasers are the focus of our attention, he has had the imagination to conceive of a "pocket" free-electron laser, and the perseverance to build a tabletop free-electron laser out of a cast-off scanning electron microscope. He realized early the importance of "brightness", rather than current, in electron beams for small devices, and recognized the high brightness available from the guns and optics of electron microscopes. For the first time, it now seems possible for anyone to have a free-electron laser in a corner of her or his laboratory.

Never one to feel restricted to follow the scientific paths taken by others, he has also branched out from the mountains of New Hampshire to form international collaborations, notably Oxford and Frascati. It is no surprise that support for his nomination came from around the world, and we are delighted to have the opportunity to honour his achievements with the 1998 Free-Electron Laser Prize.

Charles Brau, Chair
Giuseppi Dattoli
Richard Pantel
Nikolai Vinokurov

PII: S0168-9002(99)00509-4

TITANIC
Fortune & Fate

ELSEVIER

Nuclear Instruments and Methods in Physics Research A 429 (1999) xv–xxiii

NUCLEAR INSTRUMENTS & METHODS IN PHYSICS RESEARCH
Section A

www.elsevier.nl/locate/nima

Contents

Section II. FEL theory

Section III. Long wavelength FEL experiments

Section IV. Linac based FEL experiments

Section VII. Accelerator theory and technology

Section X. FEL applications

Part II

ELSEVIER

Nuclear Instruments and Methods in Physics Research A 429 (1999) 1–8

NUCLEAR INSTRUMENTS & METHODS IN PHYSICS RESEARCH
Section A

www.elsevier.nl/locate/nima

The equation of motion of an electron: a debate in classical and quantum physics[☆]

Kwang-Je Kim

Argonne National Laboratory, Argonne, IL 60439 and The University of Chicago, Chicago, IL 60637, USA

Abstract

The current status of understanding of the equation of motion of an electron is summarized. Classically, a consistent, linearized theory exists for an electron of finite extent, as long as the size of the electron is larger than the classical electron radius. Nonrelativistic quantum mechanics seems to offer a fine theory even in the point-particle limit.

1. Introduction

For almost 100 years there has been consideration of the proper equation of motion of an electron. Many fine physicists, beginning with Abraham and Lorentz, have worked on this subject, and there are hundreds of papers in the literature. Here we summarize the present state of understanding, with some historical background, giving the major contributions through the years. This paper is a shortened version of a recent review article written by the author and Sessler [1].

In contrast with what most physicists believe, it is seen that the linearized classical theory, when it is applied in the appropriate regime (nonquantum), is in fine shape: it is finite, has no contradiction with relativity, has no runaway solutions, and has no acausal behavior.

The Abraham–Lorentz equation [2,3] for a point electron involving a third derivative in time suffers from two major problems: contradiction with relativity and runaway and acausal behavior. The work of Poincaré [4] and Dirac [5] solves the problem with relativity. The runaway problem is solved by going to an extended model of an electron described by a difference-differential equation. The equation for the nonrelativistic case was derived by Sommerfeld [6] and Page [7] and was generalized to the relativistic case by Caldirola [8]. The extended model is finite and causal if the electron size a is larger than the classical electron radius $r_e = e^2/mc^2 = 2.7 \times 10^{-13}$ cm.

Classical theory is clearly not appropriate for examining behavior at a distance less than the Compton wavelength of an electron, $\lambda = \hbar/mc = 4 \times 10^{-11}$ cm. The work by Moniz and Sharp [9,10] indicates that in nonrelativistic quantum mechanics an electron behaves as an extended particle with the size of the Compton wavelength: the equation of motion is finite and causal even in the point particle limit as long as the Compton

☆Work supported by the U.S. Department of Energy, Office of Basic Energy Sciences, under Contract No. W-31-109-ENG-38.

0168-9002/99/$ - see front matter
PII: S0168-9002(99)00047-9

wavelength is larger than the classical electron radius. Furthermore, the mass correction δm is not only finite but actually vanishes in quantum theory. We present a new quantum mechanical derivation of this interesting result [11]. In quantum electrodynamic (QED) analysis, recent work by Low [12] shows that the electron motion is finite in perturbation theory in $\alpha = e^2/\hbar c$. However, a proper QED analysis has not yet been obtained (and maybe, those do *not* even exist).

There are several excellent textbooks [13–15] and review articles [16–18] on the classical electron theory.

2. Abraham–Lorentz and other classical equations of motion

We start by writing various forms of equation of motion of an electron in nonrelativistic classical mechanics. Let the electron trajectory be $\boldsymbol{y}(t)$. The electron is really a charge distribution centered at $\boldsymbol{y}(t)$ represented by

$$ef(\boldsymbol{x} - \boldsymbol{y}(t)), \tag{1}$$

where e is the electron charge, and $f(\boldsymbol{x})$ is a spherically symmetric function, normalized so that

$$\int f(\boldsymbol{x})\, \mathrm{d}^3x = 1. \tag{2}$$

The nonrelativistic equation of motion is determined by the Lorentz force law:

$$m_0\, \ddot{\boldsymbol{y}}(t) = \boldsymbol{F}_{\mathrm{ext}} + e\int f(\boldsymbol{x} - \boldsymbol{y}(t))\,(\boldsymbol{E}(\boldsymbol{x}) + \frac{1}{c}\dot{\boldsymbol{y}}(t)\times \boldsymbol{B}(\boldsymbol{x}))\, \mathrm{d}^3x, \tag{3}$$

where m_0 is the bare mass, the dot represents the time derivative, $\boldsymbol{F}_{\mathrm{ext}}$ is the external force, and $\boldsymbol{E}$ and $\boldsymbol{B}$ are, respectively, the electric and magnet field in Gaussian units. The second term in the above equation is the electromagnetic self force.

Solving Maxwell's equation to determine the electromagnetic field [12,1], the self force arising from the electric field is found to be

$$m_0\, \ddot{\boldsymbol{y}}(t) = \boldsymbol{F}_{\mathrm{ext}}(t) - \frac{2}{3}e^2\frac{4\pi}{(2\pi)^3}\int d^3k\, |\tilde{f}(\boldsymbol{k})|^2 \times \int_0^\infty d\tau\, \boldsymbol{Y}(t,\tau)\cos kc\tau, \tag{4}$$

where

$$\tilde{f}(k) = \int d^3k e^{-ik\cdot x} f(\boldsymbol{x}),$$

and

$$\boldsymbol{Y}(t,\tau) = e^{ik\cdot y(t)} e^{ik\cdot y(t-\tau)}\dot{\boldsymbol{y}}(t-\tau). \tag{5}$$

The force arising from the magnetic field is nonlinear in $\boldsymbol{y}$ (and its derivative), which we neglect. In linear approximation, the exponential factors in Eq. (5) are replaced by 1. Therefore,

$$\boldsymbol{Y}(t,\tau) = \dot{\boldsymbol{y}}(t-\tau). \tag{6}$$

Eqs. (4) and (6) are the desired expressions for the classical equation of motion and can be shown to be identical to the power series expression derived by Lorentz for linearized radiation reaction in the nonrelativistic approximation [3,13].

For the case of a spherical shell of radius a,

$$f(\boldsymbol{x}) = \frac{\delta(|\boldsymbol{x}| - a)}{4\pi a^2}, \quad \tilde{f}(\boldsymbol{k}) = \frac{\sin ka}{ka}. \tag{7}$$

Eqs. (4) and (6) become

$$m_0\ddot{\boldsymbol{y}}(t) = \boldsymbol{F}_{\mathrm{ext}}(t) + \frac{e^2}{3a^2c}[\dot{\boldsymbol{y}}(t - 2a/c) - \dot{\boldsymbol{y}}(t)]. \tag{8}$$

This differential-difference equation will be referred to as the Sommerfeld–Page equation because they derived it [6,7].

Expanding $\dot{\boldsymbol{y}}(t - 2a/c)$ in Eq. (8) in a Taylor series and neglecting terms that vanish as $a \to 0$,

$$m_0\, \ddot{\boldsymbol{y}}(t) = \boldsymbol{F}_{\mathrm{ext}}(t) - \delta m\ddot{\boldsymbol{y}}(t) + \frac{2}{3}\frac{e^2}{c^3}\dddot{\boldsymbol{y}}(t), \tag{9}$$

where

$$\delta m = \frac{4}{3}\frac{e^2}{2ac^2}. \tag{10}$$

Eq. (9) will be referred to as the Abraham–Lorentz equation in the following. It is the simplest form of the equation of motion, taking into account the electromagnetic self force in a nonrelativistic linear approximation and in the point-particle limit.

3. Problems with relativity and Poincaré stress

The second term on the RHS of the Abraham–Lorentz equation can be interpreted as the inertia due to the electromagnetic mass δm. However, it is in contradiction with the notion of relativity because δm is different from the electrostatic mass $e^2/2ac^2$ by a factor of 4/3.

To see the meaning of the third derivative term in the Abraham–Lorentz equation, we move the δm term to the LHS, multiply both sides by $\dot{y}$, and integrate over a finite interval of time. The LHS becomes the increase in the electron's kinetic energy. The RHS is

$$-\frac{2}{3}\frac{e^2}{c^3}\int_{t_1}^{t_2}\ddot{y}^2\,\mathrm{d}t+\frac{2}{3}\frac{e^2}{c^3}\dot{y}\cdot\ddot{y}|_{t_1}^{t_2}. \tag{11}$$

As long as the second term is negligible, this is the negative of the well-known dipole radiation. The last term in Eq. (9) is therefore reasonable. However, we will see later that this term gives rise to runaway or acausal behavior.

Eq. (9) therefore contains two major difficulties: (i) a contradiction with relativity, and (ii) runaway and preacceleration behavior. We shall see in the following how these troubles are avoided. The bottom line is that there is no problem if the shell radius a is larger than the classical electron radius r_e.

In 1903, before special relativity was established, Abraham derived [2] the force equation, which is the relativistic generalization of Eq. (9). Abraham also found the rate of work done on an electron, the power equation. Although these equations were derived before relativity was fully established, he used the correct relativistic model of an electron in which the spherical charge distribution in the rest frame is contracted to a spheroid for a moving electron. The derivations of these equations are difficult and confirmed by Schott [19] who carried out a very rigorous and complicated calculation.

It is not necessary to write down these equations here. However, we remark that Abraham's power equation for nonrelativistic speed becomes

$$\frac{\mathrm{d}E}{\mathrm{d}t}=\boldsymbol{F}_{\mathrm{ext}}\cdot\dot{\boldsymbol{y}}-\frac{5}{6}\frac{e^2}{ac^2}\dot{\boldsymbol{y}}\cdot\ddot{\boldsymbol{y}}+\frac{2}{3}\frac{e^2}{c^3}\dot{\boldsymbol{y}}\cdot\dddot{\boldsymbol{y}}. \tag{12}$$

Note that $\dot{\boldsymbol{y}}\cdot$ force (Eq. (9)) does not yield the power (Eq. (12)). This is a fundamental difficulty. The problem was solved by Poincaré [4] with a nonelectromagnetic stress and Dirac by invoking covariance [5].

Poincaré, in his paper submitted in 1905 [4] (without knowledge of Einstein's work on special relativity), observed that a purely electromagnetic model of the electron, such as the charged sphere, is not internally consistent because it will fly apart due to the electrostatic repulsion. To counteract the repulsive force, he imagined that the inside of the sphere provides a uniform negative pressure (or stress) $p=e^2/8\pi a^4$.

Relative contributions of Poincaré and Einstein to special relativity is a subject of some debate and considerable historical interest (see Longunov in Ref. [4]).

To see how this Poincaré stress solves the problem, consider the charged shell in motion. Due to the Lorentz contraction, the sphere becomes an oblate spheroid with the minor axis in the direction of the motion reduced by $\gamma=1/\sqrt{1-\dot{\boldsymbol{y}}^2/c^2}$. The work done by the mechanical force is given by the pressure times the volume change. (The pressure is relativistically invariant since the force and the area element transform the same way.) Thus the mechanical system must lose energy at the rate

$$\begin{aligned}\frac{\mathrm{d}E}{\mathrm{d}t}&=-\frac{e^2}{8\pi a^4}\frac{\mathrm{d}}{\mathrm{d}t}\left(\frac{1}{\gamma}\frac{4}{3}\pi a^3\right)\\&=-\frac{2}{3}\frac{e^2}{a}\frac{\mathrm{d}}{\mathrm{d}t}\frac{1}{4\gamma}.\end{aligned} \tag{13}$$

Here the expression $4\pi a^3/3\gamma$ is the volume of the spheroid. In the nonrelativistic approximation Eq. (13) becomes $\mathrm{d}E/\mathrm{d}t=(1/6)(e^2/a)\dot{\boldsymbol{y}}\cdot\ddot{\boldsymbol{y}}/c^2$. If this is added to Eq. (12), the discrepancy between the force and the power equation is removed. It can be shown that the relativistic loss of mechanical energy, Eq. (13), removes a similar difficulty in relativistic force and power equations [16,1].

However, the problem of the mass – the fact that the electromagnetic mass is 4/3 times the electrostatic mass – is not solved yet. One way of solving the problem is to say that the bare mass contains a term $-(1/3)$ times electrostatic mass. For a more formal approach [20], we can introduce a stress tensor representing the Poincaré stress by

$$\Theta^{\mu\nu}_{\text{Poincaré}} = \frac{e^2}{8\pi a^3}(g^{\mu\nu} - qu^\mu u^\nu)\Theta(a - r) \tag{14}$$

where $g^{\mu\nu} = (1, -1, -1, -1)$, q is an arbitrary constant, u^μ is the velocity four-vector, Θ is the step function, and r is the radial coordinate in the rest frame of the shell. The quantity $\Theta^{\mu\nu}_{\text{Poincaré}}$ is constructed so that

$$\partial_\mu \Theta^{\mu\nu} = \partial_\mu(\Theta^{\mu\nu}_{\text{EM}} + \Theta^{\mu\nu}_{\text{Poincaré}}) = 0, \tag{15}$$

where $\Theta^{\mu\nu}_{\text{EM}}$ is the electromagnetic stress tensor associated with the Coulomb field of a spherical shell. Eq. (15) assures that the total momentum

$$p^\mu = \int d^3x \Theta^{\mu 0} \tag{16}$$

is a four vector [12]. The momentum associated with the Poincaré stress alone is

$$p^\mu_{\text{Poincaré}} = \int a^3 x \Theta^{\mu 0}_{\text{Poincaré}} = +\frac{e^2}{2a}\frac{1}{3}\left[-\frac{1}{\gamma} + q\gamma, q\gamma \boldsymbol{u}\right]. \tag{17}$$

We need to add $\mathrm{d}p^\mu_{\text{Poincaré}}/\mathrm{d}t$ to the force and the power equations. Our previous result (Eq. (13)) corresponds to the case $q = 0$. If, on the other hand, we choose $q = 1$, we remove the so-called 4/3 problem.

The above method of solving the 4/3 problem is clearly rather formal and arbitrary. There are also more intuitive approaches, for example by Boyer [21] who notes that the Poincaré stress may not act at the same time in all parts of the moving electron. A review of different solutions of the 4/3 problem is given by Rohrlich [18].

The saga of the 4/3 problem is in some sense a story of how special relativity proved itself as a theory of internal consistency and beauty. The investigation of the electron's equation of motion started while special relativity was still evolving. Therefore, there was doubt whether the electron theory was consistent with relativity, which lingered even after relativity was fully established. Einstein, never doubting relativity, wasted no time in checking covariance of the electron theory. He was too busy working out general relativity!

With this choice of q, the force-power equation in the relativistic case becomes

$$m_0 c^2 \frac{\mathrm{d}}{\mathrm{d}s} u^\mu = f^\mu_{\text{ext}} + \Gamma^\mu - \frac{e^2}{2a}\frac{\mathrm{d}u^\mu}{\mathrm{d}s}, \tag{18}$$

where

$$\Gamma^\mu = \frac{2}{3}e^2\left(\frac{\mathrm{d}^2 u^\mu}{\mathrm{d}s^2} + \left(\frac{\mathrm{d}u^\alpha}{\mathrm{d}s}\frac{\mathrm{d}u_\alpha}{\mathrm{d}s}\right)u^\mu\right). \tag{19}$$

We now have a relativistically covariant equation of motion of an electron with the observed mass $m = m_0 + \delta m$ and $\delta m = e^2/2ac^2$.

Before closing this section, we need to mention Dirac's contribution [5]. He understood that the problem with the Abraham–Lorentz equation was in trying to approach too near an electron. He therefore devised an ingenious way to avoid the difficulty. Stay a finite distance away from an electron and demand relativistic covariance. He then obtained Eq. (18), which is therefore referred to as the Abraham–Lorentz–Dirac equation.

Actually, Dirac's derivation of Eq. (18) is valid only up to a term of the form $\mathrm{d}B^\mu/\mathrm{d}s$, where B^μ is a four vector with the restriction $u^\mu \mathrm{d}B_\mu/\mathrm{d}s = 0$. The simplest choice, $B^\mu = ku^\mu$, gives rise to a term that can be incorporated into the inertial term ($mc\mathrm{d}u^\mu/\mathrm{d}s$) in Eq. (18). The next order term involving $\mathrm{d}u^\mu/\mathrm{d}s$ is complicated, which is disposed of by saying that *a simple thing like an electron cannot possibly have such a complication.*

If we are willing to make a few very plausible assumptions, the Abraham–Lorentz–Dirac equation can be derived very easily as follows [18]:

$$m_0 c^2 \frac{\mathrm{d}u^\mu}{\mathrm{d}s} = f^\mu_{\text{ext}} + X^\mu, \tag{20}$$

where X^μ includes radiative effects. Both sides of Eq. (20) are orthogonal to u^μ. Thus we may write $X^\mu = (g^{\mu\nu} - u^\mu u^\nu)Y_\nu$. Try Y^μ of the form $a\, u^\mu + b\, \mathrm{d}u^\mu/\mathrm{d}s + c\, \mathrm{d}^2u^\mu/\mathrm{d}s^2$. The first term does not contribute, and the second term is of the same form as the mass term, which can be incorporated into m. With the third term, it is easy to show that X^μ is proportional to Γ^μ. (Here we are using $(\mathrm{d}u^\alpha/\mathrm{d}s)\mathrm{d}u_\alpha/\mathrm{d}s + \mathrm{d}^2u_\alpha/\mathrm{d}s^2 \cdot u^\alpha = 0$.) The coefficient is determined by demanding consistency with the nonrelativistic Abraham–Lorentz equation, leading to $\Gamma^\mu = X^\mu$. End of the proof.

4. Runaway and acausality

The general solution of the Abraham–Lorentz equation is

$$m\ddot{y}(t) = e^{t/t_e}[m\,\ddot{y}(0) - 1/t_e \int_0^t dt' e^{-t/t_e} F_{ext}(t')], \quad (21)$$

where

$$t_e = \frac{2}{3}\frac{r_e}{c} = \frac{2}{3}\frac{e^2}{mc^3}. \quad (22)$$

The solution in general exhibits exponential growth, i.e., runaway behavior. Dirac noted that the runaway can be avoided if we choose the initial condition

$$m\ddot{y}(0) = 1/t_e \int_0^\infty dt' e^{-t'/t_e} F_{ext}(t'). \quad (23)$$

Then

$$m\ddot{y}(t) = \int_0^\infty d\alpha e^{-\alpha} F_{ext}(t + t_e\alpha). \quad (24)$$

This solution now exhibits preacceleration: the particle starts to move before the force is applied and the initial condition depends upon the entire future path.

Nevertheless, it satisfies the Rohrlich criteria of "*the unobservability of very small charges*" [15]. That is, as the charge becomes very small we should not have a solution widely different from that of an uncharged particle.

The Abraham–Lorentz equation exhibits runaway and preacceleration because, as the electron radius a vanishes, the bare mass m_0 becomes negative to keep the observed mass m finite. A particle with a negative mass can clearly supply an infinite amount of energy.

The difficulty can be avoided in an extended electron model. To see this, we return to the Sommerfeld–Page Eq. (8), which can be written in the following form [9,10,17]

$$\ddot{y}(t) = \frac{F_{ext}(t)}{m(1 - ct_e/a)} + \left(\frac{c}{2a}\right)\left(\frac{ct_e}{a}\right)\frac{1}{(1 - ct_e/a)} \times [\dot{y}(t - 2a/c) - \dot{y}(t)]. \quad (25)$$

In the above expression, the combination $m(1 - ct_e/a)$ is simply the bare mass m_0. Assume that there are no external forces, and there exists a runaway solution of the form $y(t) = y(0)e^{\alpha t}$ with a positive real part of α. From Eq. (25) it is easy to see that this is possible only when $1 - ct_e/a$ is negative. Thus the runaway solution is possible if and only if the electron radius a is less than $(2/3)\,r_e$. For $a > (2/3)r_e$ we get damped oscillatory solutions.

It is sometimes stated that the runaway behavior is due to the infinite energy associated with a point electron. This is false because the runaway occurs for even a finite a as long as it is less than $(2/3)\,r_e$. Runaway occurs if and only if the bare mass is negative (and, therefore, the Hamiltonian is no longer positive definite).

We can also show that if $a > ct_e$, the motion is causal with no preacceleration. This can be seen most easily if Eq. (25) is turned into an integral equation with a Green's function [9,10,17]:

$$\dot{y}(t) = \int_{-\infty}^{\infty} G(t - t')\, F_{ext}(t')\, dt', \quad (26)$$

where

$$G(t - t') = \frac{1}{2\pi m}\int_{-\infty}^{\infty} \frac{e^{-i\omega(t-t')}d\omega}{B}, \quad (27)$$

and

$$B = 1 - \frac{i}{\omega}\left(\frac{c}{2a}\right)\left(\frac{ct_e}{a}\right)\left[e^{(i\omega 2a/c)} - \frac{i\omega 2a}{c} - 1\right]. \quad (28)$$

It can be shown that all poles of $1/B(\omega)$ occur in the lower half ω-plane if

$$a > \tfrac{2}{3} r_e. \quad (29)$$

Then the Green's function $G(\tau)$ vanishes for $\tau < 0$. Therefore, the motion is causal. On the other hand, if $a < \frac{2}{3}r_e$, then there is in general acausal behavior.

A relativistic generalization of the Sommerfeld–Page equation was conjectured by Caldirola [8] and proved by Yaghjian [16]:

$$m_0 c^2 \dot{u}^\mu = f^\mu_{ext} + \frac{2}{3}\frac{e^2}{a}\frac{1}{2a} \times [u^\mu(s - 2a) - u^\mu(s)u_\alpha(s)u^\alpha(s - 2a)]. \quad (30)$$

The Caldirola equation is fine: no runaways and no causality problems. However, when we take $a \to 0$, then there are problems.

Yaghjian [16] attempted to produce a causal equation with no runaways for a point electron. He modified the Abraham–Lorentz Eq. (9) by multiplying the third derivative term by a function $\eta(t)$, which changes smoothly from 0 for $t < 0$ to 1 for $t > 2a/c$ (or by multiplying the Γ^{μ} term by a similar function $\eta(s)$), arguing that the Taylor series expansion breaks down when the force changes abruptly. However, the solution of the modified equation still exhibits acausal behavior.

5. Quantum mechanical analysis

It had been hoped that the difficulty of the classical theory in taking a point particle limit might be solved by quantum mechanics. The observation that the mass renormalization δm is less singular in quantum electrodynamics than in classical theory seemed to reinforce this hope. However, a real quantum mechanical analysis of the electron equation has not been carried out until modern time. The reason behind this lack of activity appears to be the success of the renormalization theory in quantum electrodynamics, which established that all observable phenomena can be calculated to a finite answer order by order in perturbation theory in terms of the observed mass and charge.

In 1977, Moniz and Sharp presented a very interesting quantum mechanical calculation [9,10], according to which the equation of motion in nonrelativistic quantum mechanics is finite even in the point particle limit. The derivation is based on the elaborate manipulations of infinite series.

Here another derivation will be summarized [11].

The derivation of the quantum mechanical equation of motion for Heisenberg operators proceeds similarly to the derivation in the classical case given in Section 2. Again neglecting the magnetic term, one arrives at an equation identical in form to Eq. (4), but the operator $\boldsymbol{Y}$ is given by the symmetrized version of the classical expression, Eq. (5), as follows:

$$\begin{aligned}\boldsymbol{Y}(t,\tau) &= \tfrac{1}{4}\{\mathrm{e}^{\mathrm{i}\boldsymbol{k}\cdot\boldsymbol{y}_2}, \{\mathrm{e}^{-\mathrm{i}\boldsymbol{k}\cdot\boldsymbol{y}_1}, \dot{\boldsymbol{y}}_1\}_+\}_+ \\ &= \tfrac{1}{4}(\mathrm{e}^{\mathrm{i}\boldsymbol{k}\cdot\boldsymbol{y}_2}\mathrm{e}^{-\mathrm{i}\boldsymbol{k}\cdot\boldsymbol{y}_1})(\dot{\boldsymbol{y}}_1 + \mathrm{e}^{\mathrm{i}\boldsymbol{k}\cdot\boldsymbol{y}_1}\dot{\boldsymbol{y}}_1\mathrm{e}^{-\mathrm{i}\boldsymbol{k}\cdot\boldsymbol{y}_1}) \\ &\quad + [\boldsymbol{k} \to -\boldsymbol{k}]^+.\end{aligned} \tag{31}$$

Here $\boldsymbol{y}_2 = \boldsymbol{y}(t)$, $\boldsymbol{y}_1 = \boldsymbol{y}(t-\tau)$, $\{\}_+$ is the anticommutator, and $[\boldsymbol{k} \to -\boldsymbol{k}]^+$ indicates terms obtained by changing the sign of $\boldsymbol{k}$ and taking the Hermitian conjugate.

Noting that $\boldsymbol{p} = \boldsymbol{P} - e\boldsymbol{A}/c = m_0\dot{\boldsymbol{y}}$ is the kinetic momentum operator ($\boldsymbol{P}$ = canonical momentum), we obtain

$$\mathrm{e}^{\mathrm{i}\boldsymbol{k}\cdot\boldsymbol{y}_1}\dot{\boldsymbol{y}}_1\mathrm{e}^{-\mathrm{i}\boldsymbol{k}\cdot\boldsymbol{y}_1} = \dot{\boldsymbol{y}}_1 - c\boldsymbol{k}\lambda, \tag{32}$$

where $\lambda = \hbar/m_0c$ is the Compton wavelength for the bare mass m_0. In classical mechanics, we would have replaced the factor $\exp(\mathrm{i}\boldsymbol{k}\cdot\boldsymbol{y}_1)$ by 1 since it would at most contribute to nonlinear terms. Such a procedure is not justified in quantum mechanics as is clear from Eq. (32). Similarly, the factor $\exp(\mathrm{i}\boldsymbol{k}\cdot\boldsymbol{y}_2)\cdot\exp(-\mathrm{i}\boldsymbol{k}\cdot\boldsymbol{y}_1)$ in Eq. (31) cannot simply be replaced by 1. Instead, we reduce the product of exponential operators using a technique similar to that used by Baier and Katkov [22] in their calculation of quantum synchrotron radiation. The steps used here follow closely a simplified formulation of the Baier-Katkov reduction by Cahn and Jackson [23]. We write

$$\begin{aligned}&\exp(\mathrm{i}\boldsymbol{k}\cdot\boldsymbol{y}_2)\exp(-\mathrm{i}\boldsymbol{k}\cdot\boldsymbol{y}_1) \\ &\quad= \exp(\mathrm{i}\tau H/\hbar)\exp(\mathrm{i}\boldsymbol{k}\cdot\boldsymbol{y}_1)\exp(-\mathrm{i}\tau H/\hbar) \\ &\qquad\times\exp(-\mathrm{i}\boldsymbol{k}\cdot\boldsymbol{y}_1) \\ &\quad= \exp(\mathrm{i}\tau H(\boldsymbol{p})/\hbar) \\ &\qquad\times\exp(-\mathrm{i}\tau H(\boldsymbol{p}-\hbar k)/\hbar)\,(\tau = t - t').\end{aligned} \tag{33}$$

Here H is the Hamiltonian. Since H is a sum of the kinetic energy $\boldsymbol{p}^2/2m_0$ and the electromagnetic energy, it follows that

$$H(\boldsymbol{p} - \hbar\boldsymbol{k}) = \frac{\hbar^2k^2}{2m_0} + H(\boldsymbol{p}) - \frac{\hbar\boldsymbol{p}\cdot\boldsymbol{k}}{m_0}. \tag{34}$$

Now, using the Campbell–Baker–Hausdorff formula [24] to first order in $\boldsymbol{p}$, we can show that

$$\exp(\mathrm{i}\boldsymbol{k}\cdot\boldsymbol{y}_2)\exp(-\mathrm{i}\boldsymbol{k}\cdot\boldsymbol{y}_1)=\exp\left(-\frac{\mathrm{i}\tau\hbar k^2}{2m_0}\right)\times[1+\mathrm{i}\boldsymbol{k}\cdot(\boldsymbol{y}_2-\boldsymbol{y}_1)]. \quad (35)$$

Thus

$$\boldsymbol{Y}=\frac{1}{2}\left\{\exp\left(-\frac{\mathrm{i}\tau\hbar k^2}{2m_0}\right)(1+\mathrm{i}\boldsymbol{k}\cdot(\boldsymbol{y}_2-\boldsymbol{y}_1))\times\left(\dot{\boldsymbol{y}}_1-\frac{c\boldsymbol{k}\lambda}{2}\right)\right\}+[\boldsymbol{k}\to-\boldsymbol{k}]^{+}. \quad (36)$$

Noting that odd power in $\boldsymbol{k}$ vanishes after d^3k-integration and $\langle k_x^2\rangle=\frac{1}{3}k^2$, one obtains (keeping only terms linear in $\boldsymbol{y}$)

$$\boldsymbol{Y}(t,\tau)=\dot{\boldsymbol{y}}(t-\tau)\cos(\lambda c\tau k^2/2)-\frac{k^2c\lambda}{6}(\boldsymbol{y}(t)-\boldsymbol{y}(t-\tau))\sin(\lambda c\tau k^2/2). \quad (37)$$

The linearized electron's equation of motion in nonrelativistic quantum mechanics is then given by inserting Eq. (37) into Eq. (4). Expanding the operator $\boldsymbol{y}(t-\tau)$ in Eq. (37) in a Taylor series around $\tau=0$ and inserting it in Eq. (4), an equation involving a sum of derivatives of $\boldsymbol{y}(t)$ is obtained. It can be shown that the coefficients of these derivatives are exactly those derived in Moniz and Sharp.

In particular, the coefficient of $\ddot{\boldsymbol{y}}(t)$, the quantum mechanical self-mass δm, is found to be

$$\delta m=\left(\frac{2}{3}\frac{e^2}{c^2}\right)\left(1+\frac{\lambda}{6}\frac{\partial}{\partial\lambda}\right)\left(1+\lambda\frac{\partial}{\partial\lambda}\right)\Omega_0. \quad (38)$$

Here

$$\Omega_0=\frac{2}{\pi}P\int_0^\infty\frac{|\tilde{f}(k)|^2}{(1-\lambda^2k^2/4)}\,\mathrm{d}k, \quad (39)$$

where P denotes the principal value integration. Eq. (39) is remarkable in that the self-mass remains finite in the point particle limit $\tilde{f}(\boldsymbol{k})\to 1$. *In fact, the self-mass vanishes in this limit!* $\delta m\to 0$ ($a\to 0$ with λ fixed in quantum mechanics).

Therefore, λ (bare mass) $=\lambda$ (observed mass). This is why we used the same notation for these two Compton wavelengths.

In another limit, $\lambda\to 0$,

$$\Omega_0=\frac{2}{\pi}P\int_0^\infty|\tilde{f}(k)|^2\,\mathrm{d}k, \quad (40)$$

and the theory reproduces the classical result, as it should.

Whether the quantum mechanical equation of motion exhibits runaway or acausal behavior can again be studied by writing the quantum mechanical equation of motion in a Green's function form, similar to that at the end of Section 4. Such an analysis in the point particle limit was carried out by Moniz and Sharp [9,10], who found that the motion is causal with no runaways if

$$\alpha=\frac{e^2}{\hbar c}=\frac{r_e}{\lambda}<1.75. \quad (41)$$

In quantum mechanics, an electron is spread over a Compton wavelength. Thus, the above condition is reasonable in view of the classical causality condition Eq. (29), which can be written as $r_e/a<1.5$.

Nonrelativistic quantum mechanics is not valid for rapid motion with frequency $\hbar\omega\gtrsim mc^2$ or if we approach a distance λ near the electron, and a full QED analysis must be performed. The only calculation by QED was reported by Low [12]. He has not derived an equation of motion, but he has shown that the motion is finite in each order of α. This is reasonable in view of the well-established renormalization theory that gives a finite answer for any physical process in perturbation expansion in α. However, the perturbation theory is not suitable for taking the classical limit, and it is unclear whether QED can actually produce an equation of motion.

6. Concluding remarks

The impression one gets from reading text books is that the classical electron theory is in trouble due to runaway solutions and acausal behavior. However, we have seen that classical theory is actually fine if the electron is taken to be a spherical object of not too small a radius, greater than the classical electron radius r_e. The restriction is reasonable since r_e is about 100 times smaller than the Compton wavelength λ, and we cannot consider distances less than λ without considering quantum mechanics.

The nonrelativistic quantum theory also looks fine: In fact, the quantum theory is better behaved than the classical theory because it is finite and causal irrespective of the size of the electron, as long as $\alpha = r_e/\lambda < 1.75$. The inequality is certainly satisfied in the real world where $\alpha = 1/137$. The fact that the quantum theory is better behaved is also reasonable because the electron is smeared out due to the uncertainty principle. The quantum theory as reviewed here has the appropriate feature of having the correct classical limit. However, the limit of validity of the nonrelativistic quantum analysis is not really understood. The vanishing of the self-mass δm in the point-particle limit must be accordingly interpreted with care.

In fact, we know that the nonrelativistic treatment cannot strictly be valid in the point-particle limit because of vacuum polarization. It is reasonable that a full QED calculation will be at least as well-behaved as is the case in nonrelativistic quantum mechanics. Unfortunately, there exists as yet no real calculation in QED to confirm these conjectures.

Acknowledgements

I would like to thank Andy Sessler for collaboration with Ref. [1], on which this paper is based. I would also like to thank Dave Jackson, Francis Low, Robert Mills, Fritz Rohrlich, and David Sharp for useful remarks.

References

[1] K.-J. Kim, A.M. Sessler, The Equation of Motion of an Electron, ANL preprint LS-270 (1998), Proceedings of the Eighth Workshop on Advanced Accelerator Concepts, July 5-11, 1998, Baltimore, MD.

[2] M. Abraham, Theorie der Elektrizitat, vol II: Elektromagnetische Theorie der Strahlung, Teubner, Leipzig, 1905.

[3] H.A. Lorentz, The Theory of Electrons, (second ed. 1916), Leipzig, Teubner 1909.

[4] H. Poincaré, On the dynamics of the electron, Rendiconti del Circolo Matematico di Palermo 21 (1906) 129; translated by Scientific Translation Service, Ann Arbor, MI. The article is also translated into English with comments, in: A.A. Logunov (Ed.), On the articles by Henri Poincaré, Publishing Department of the Joint Institute of Nuclear Research, Dubna, 1995.

[5] P.A.M. Dirac, Proc. Roy. Soc. London A 167 (1938) 148.

[6] A. Sommerfeld, Simplified deduction of the field and the forces of an electron moving in any given way, Akad. van Wetensch. te Amsterdam 13 (1904) (English translation 7 (1905)) 346.

[7] L. Page, Phys. Rev. 11 (1918) 376.

[8] P. Caldirola, Nuovo Cimento 3 (Supplemento 2) (1956) 297.

[9] E.J. Moniz, D.H. Sharp, Phys. Rev. D 15 (1977) 2850.

[10] H. Grotch, E. Kazes, F. Rohrlich, D.H. Sharp, Acta Physica Austriaca 54 (1982) 31.

[11] K.-J. Kim, submitted for publication.

[12] F. Low, Run-away electrons in relativistic spin 1/2 quantum electrodynamics, Preprint MIT–CTP-2522, 1997.

[13] W.K.H. Panofsky, M. Phillips, in: Classical Electricity and Magnetism, Addison-Wesley, Reading, MA, 1962.

[14] J.D. Jackson, Classical Electrodynamics, third ed., Wiley, New York, 1998 (second ed., 1975). (Chapter 16 of third edition contains a review of the equation of motion of an electron, much more extensive than in previous editions.)

[15] F. Rohrlich, Classical Charged Particles, Addison-Wesley, Reading, MA, 1965 (second ed. 1990).

[16] A.D. Yaghijian, in: Relativistic Dynamics of a Charged Sphere, Lecture Notes in Physics 11, Springer, Berlin, 1992.

[17] P. Pearle, Classical Electron Models, in: D. Teplitz (Ed.), Electromagnetism: Path to Research, Plenum, New York, 1982.

[18] F. Rohrlich, Am. J. Phys. 65 (1997) 105.

[19] G.A. Schott, in: Electromagnetic Radiation, Cambridge University Press, Cambridge, 1912.

[20] J. Schwinger, Found. Phy. 13 (1983) 373.

[21] T.H. Boyer, Phys. Rev. D 25 (1982) 3246.

[22] V.N. Baier, V.M. Katkov, Sov. Phys. JETP 26 (1968) 854.

[23] R.N. Cahn, J.D. Jackson, unpublished Lawrence Berkeley National Laboratory note.

[24] A.J. Dragt, J.M. Finn, J. Math. Phys. 17 (1976) 2215.

ELSEVIER

Nuclear Instruments and Methods in Physics Research A 429 (1999) 9–11

NUCLEAR INSTRUMENTS & METHODS IN PHYSICS RESEARCH Section A

www.elsevier.nl/locate/nima

A 0.1 kW operation of the JAERI superconducting RF linac-based FEL

E.J. Minehara*, M. Sugimoto, M. Sawamura, R. Nagai, N. Kikuzawa, T. Yamanouchi, N. Nishimori

Free Electron Laser Laboratory, Advanced Photon Research Center, Kansai Research Establishment, Japan Atomic Energy Research Institute, 2-4 Shirakata-shirane, Tokai-mura, Naka-gun, Ibaraki-ken, 319-1195 Japan

Abstract

First and stable laser oscillation has been obtained around a wavelength of 24 μm using the Japan Atomic Energy Research Institute, Tokai (JAERI) superconducting RF-linac-based FEL driver and far infrared FEL device. The electron beam energy and spread are 15.8 MeV and 0.6%, respectively. The beam current varies between 2 and 4 mA, and the pulse width between 0.1 and 0.9 ms, respectively. The near-concentric optical resonator is 14.4 m long, and uses gold-coated copper mirrors 120 mm in diameter. The hybrid, planar undulator has 52 periods, 33 mm in length and $K = 0.7$. Remote-controlled actuators precisely adjust the optical axes and distance of the mirrors in order to coincide with the electron beam and micro pulse repetition rate, respectively, before oscillation. The power has been measured and is scattered from 10^7 to 10^8 times higher than that of the spontaneous emission. During the first successful operation, the highest average FEL power was measured to be about a hundred watts. The FWHM of the FEL spectrum is around the Fourier-transform limited value, and less than 0.09 μm, which corresponds to $\Delta\lambda/\lambda = 0.37\%$. The detuning curve of the cavity is asymmetric, and spans about 15 μm.

1. Introduction

A prototype for a quasi-CW or very long-pulse, and high-average power free electron laser (FEL) driven by a 15 MeV superconducting RF linac has been developed, and constructed at Tokai, Japan Atomic Energy Research Institute (JAERI) starting in 1989 [1–4]. High performance in both cryogenic stand-by loss (< 3.5 W at 4.5 K) and accelerating fields ($E_{acc} < 8.3$ MV/m and $Q < 2\times10^9$) in the JAERI superconducting RF linac modules have been successfully realized. A so-called zero-boil-off and world-first transportable capabilities without any serious vibrational problem as achieved in the JAERI FEL accelerator vault.

Since modification and related maintenance of the cryogenic refrigerator system for the driver were completed in the middle of October 1995, the system has run with no trouble, and the driver has been continuously run very successfully up to now. The optical resonator system and related electron

*Corresponding author. Tel.: + 81-29-282-5464; + 81-29-282-6752; + 81-29-282-5455; fax: + 81-29-270-5923; + 81-29-282-5939.

E-mail address: minehara@felwu0.tokai.jaeri.go.jp, minehara@olive.ocn.ne.jp (E.J. Minehara)

0168-9002/99/$ - see front matter
PII: S0168-9002(99)00050-9

beam transport system were modified to realize larger acceptance than previously for both the undulator radiation and energetic electron beam. An alignment and distance measurement system was newly developed, and successfully applied to actual preparatory measurements for lasing in the JAERI FEL. A far-infrared light transport line and detector room was built to realize a low-loss and low-noise measurement near the accelerator vault in April 1996.

In order to realize the quasi-CW and very long-pulse operation, we have improved the electron gun grid-pulsar, the high-voltage power supply, and the RF amplitude and phase control systems for the JAERI superconducting RF linac. The improvements in the electron gun and those related to the RF system are still under way.

Beam tests and commissioning of the JAERI superconducting RF linac as an FEL driver were successfully performed to get an electron beam ranging from 10 to 23 MeV with a nearly full transmission and a full current, and relatively short macro pulse of 0.01 ms. Strong and stable oscillation in the wavelength of around 24 μm have been observed by using the Ge(Cu) detectors with a home-made, fast current amplifier system, commercially available MCT detectors, and thermopiles. As shown in Fig. 1, a sudden increase of the light signal was observed a few tens of microseconds after the onset of the beam current pulse.

2. Experiment

Fig. 1 shows the first and stable laser oscillation with the JAERI superconducting RF linac based FEL. The average current for a macro pulse is measured to be around 4 mA. The lower trace of the figure is a far-infrared light signal waveform of a Ge–Cu detector. The total cavity loss and the FEL gain are scattered between 1.1% and 1.9%, and estimated to be from 10% to 30%, respectively. They were obtained from decay and rising times of the output pulse. Fig. 2 shows the e-beam current signal in the lower trace and FEL light output signal of the upper trace over a half millisecond.

As shown in Fig. 3, the FEL spectrum was measured with a grating spectrometer and a commercially available pyroelectric line-sensor of 64 channels located at the focal plane during the operation. The FWHM of the FEL spectrum is less than 0.09 μm, which corresponds to $\Delta\lambda/\lambda = 0.37\%$. The tuning range of the cavity spans about 15 μm. The detuning curve was measured reliably to be asymmetric, long-tailed at shorter cavity lengths, and short-tailed or abruptly cut off at longer cavity lengths.

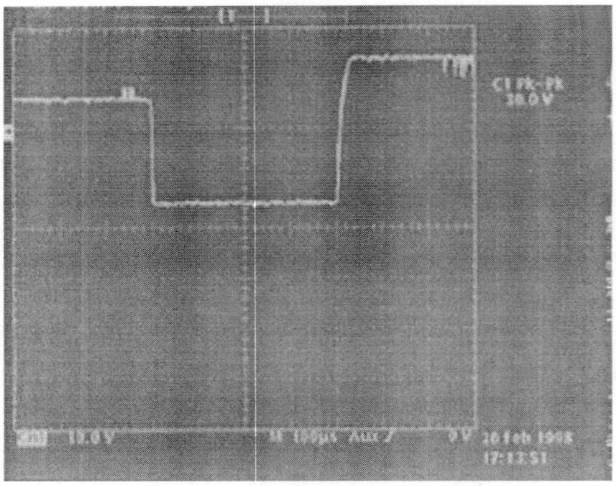

Fig. 1. Stable oscillation at a wavelength of around 24 μm. The pulse duration is over 0.4 ms, the beam current ranging from 2 to 4 mA, and the maximum laser power averages about a hundred watts.

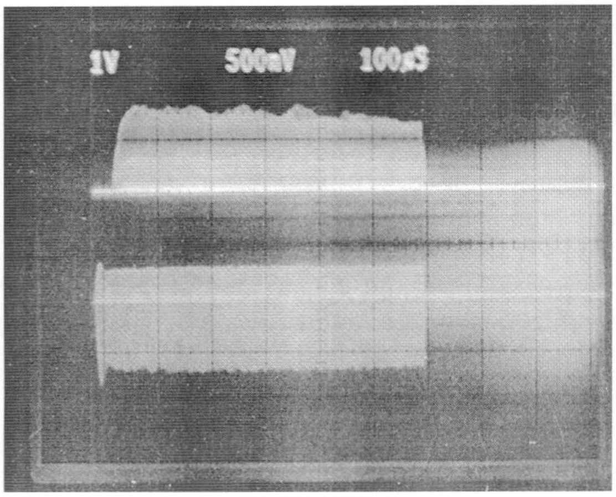

Fig. 2. Electron-beam current signal in the lower trace and FEL light output signal in the upper trace over a half millisecond.

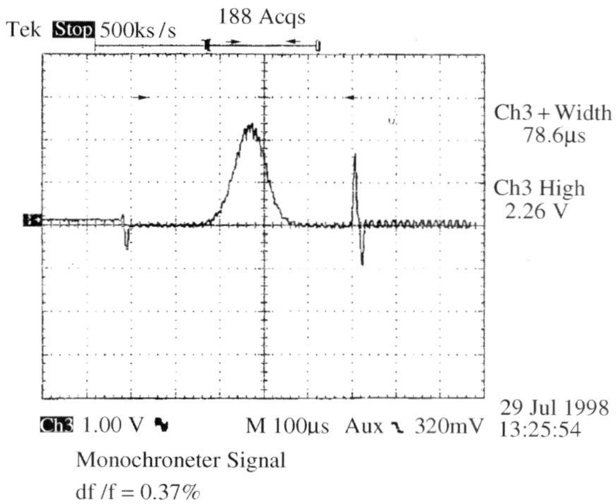

Fig. 3. The FEL spectrum was measured with a grating spectrometer and a commercially available pyroelectric line-sensor of 64 channels fixed at the spectrometer focal plane during the operation. The FWHM of the FEL spectrum is less than 0.09 μm, which corresponds to $\Delta\lambda/\lambda = 0.37\%$.

Stable laser oscillation has been obtained at a wavelength of 24 μm using the JAERI superconducting RF linac-based FEL driver. During the first successful operation, the highest average FEL power was measured to be about a hundred watts. The JAERI superconducting RF linac operational parameters are summarized as the following: (1) electron beam energy of 15.8 MeV from 0.1 to 0.9 ms, (2) beam current ranging from 2 to 4 mA, (3) energy spread around 0.6%, and (4) repetition rate 10.4125 MHz.

The near-concentric optical resonator is 14.4 m long, and uses gold-coated copper mirrors 120 mm in diameter. The hybrid, planar undulator has 52 periods, 33 mm in length and $K = 0.7$. Remotely controlled actuators adjust the optical axes and the distance between the mirrors in order to coincide with the electron beam and micropulse repetition rate, respectively, before oscillation. The power has been measured and is scattered from 10^7 to 10^8 times higher than that of the spontaneous emission.

References

[1] E.J. Minehara et al., Nucl. Instr. and Meth. A 331 (1993) 276.

[2] M. Sawamura et al., Nucl. Instr. and Meth. A 318 (1992) 127.

[3] R. Nagai et al., Nucl. Instr. and Meth. A 358 (1995) 403.

[4] E.J. Minehara, et al., proceedings of the Particle Accelerator Conference, Dallas, 1995, pp. 159–161.

ELSEVIER

Nuclear Instruments and Methods in Physics Research A 429 (1999) 12–16

NUCLEAR INSTRUMENTS & METHODS IN PHYSICS RESEARCH Section A

www.elsevier.nl/locate/nima

First lasing of the Dutch Fusion-FEM: 730 kW, 200 GHz

W.H. Urbanus[a,*], W.A. Bongers[a], C.A.J. van der Geer[a], P. Manintveld[a], J. Plomp[a], J. Pluygers[a], A.J. Poelman[a], P.H.M. Smeets[a], F.C. Schüller[a], A.G.A. Verhoeven[a], V.L. Bratman[b], G.G. Denisov[b], A.V. Savilov[b], M.Yu. Shmelyov[b], M. Caplan[c], A.A. Varfolomeev[d]

[a]*FOM Instituut voor Plasmafysica Rijnhuizen, Association EURATOM-FOM, P.O. Box 1207, 3430 BE Nieuwegein, The Netherlands*
[b]*Institute of Applied Physics, Russian Academy of Sciences, 46 Ulyanov Street, Nizhny Novgorod 603600, Russia*
[c]*Lawrence Livermore National Laboratories, P.O. Box 808, Livermore, CA 94551, USA*
[d]*Russian Research Center "Kurchatov Institute", Moscow 123182, Russia*

Abstract

A high-power electrostatic free-electron maser is operated at various frequencies. An output power of 730 kW at 206 GHz is generated with a 7.2 A, 1.77 MeV electron beam, and 360 kW at 167 GHz is generated with a 7.0 A, 1.61 MeV electron beam. It is shown experimentally and by simulations that, depending on the electron beam energy, the FEM can operate in single-frequency regime. First experiments were done without electron beam energy recovery system, and the pulse length was limited to 12 μs. Nevertheless, many aspects of generation of mm-wave power have been explored, such as the dependency on the electron beam energy and beam current and cavity settings such as the feedback coefficient. The achieved parameters and the FEM dynamics are in good accordance with simulations.

PACS: 41.75.H; 41.60.C; 29.17

Keywords: Maser; FEM; Electron beam mm-waves

1. Introduction

The principal target of present free-electron maser (FEM) research is the realization of a source of mm-wave radiation of high average power, high system efficiency, and broad tunability. The achievement of these targets may culminate in the use of FEMs as power sources for electron cyclotron applications on magnetically-confined plasmas in future fusion research devices, such as ITER. For such applications, power sources of at least 1 MW in the frequency range from 140 to 200 GHz at a system efficiency of 50% are required. Fast tunability over a small range and a Gaussian output beam would be an advantage.

A promising approach to realize long-pulse operation is the use of electrostatic beam acceleration

* Corresponding author. Tel.: + 31-30-609-6999; fax: + 31-30-603-1204.
E-mail address: urbanus@rijnh.nl (W.H. Urbanus)

0168-9002/99/$ - see front matter
PII: S0168-9002(99)00052-2

and deceleration. In this scheme the electron beam is accelerated to the interaction region, and afterwards it is decelerated and collected in a multistage depressed collector. The main beam power is supplied at the collector side, i.e., at low voltage, while the accelerator voltage generator has to supply only the beam *loss* current. Consequently, to reach long-pulse operation, the loss current has to be extremely low, in our case of the order of 0.2%. This target was reached successfully [1].

2. Layout of the Fusion-FEM

A description of the Fusion-FEM experiment is given in Ref. [2]. The system basically consists of an 80-kV triode electron gun, a DC accelerator, and a mm-wave cavity with the undulator [3]. The undulator is step-tapered to enhance the electronic efficiency; the first 20 periods have a field amplitude of 0.2 T and the latter 16 periods have a field amplitude of 0.16 T [4]. In a later stage the DC decelerator and the depressed collector will be added. In the experiments described here, the beam recovery system was not installed. Consequently, the beam current is fed from the capacitance of the HV terminal (1 nF), and thus the beam energy (terminal voltage) drops by 1 kV/A_{beam}/μs.

In order to limit electron beam losses a simple, straight electron beam line is used [1]. This requires the mm-wave beam to be directed off-axis for outcoupling and feedback. The mm-wave cavity consists of the operating waveguide inside the undulator – a rectangular corrugated waveguide with a cross section of 15×20 mm^2 – and two stepped waveguides at either side, see Fig. 1 [5]. Corrugated waveguides are used to reduce Ohmic losses. In the stepped waveguides the mm-wave beam emerging from the operating waveguide splits into two identical off-axis beams. At the position of full separation, mirrors are placed. The backward propagating beams merge back into one beam. This way, a 100% reflector is realized at the upstream side of the undulator. At the downstream side a similar system is used, but here one of the mirrors can be shifted in longitudinal direction (direction of beam propagation). Upon propagating backwards, the beams have a phase difference, which results into one on-axis beam and two off-axis beams. The on-axis beam is the feedback power, the two off-axis beams are coupled out. By adjusting the position of the mirror, the feedback

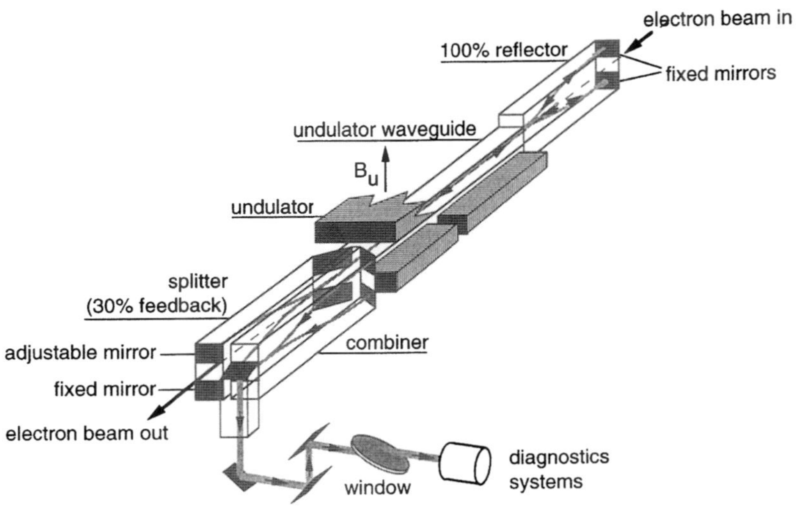

Fig. 1. Layout of the mm-wave cavity, for details see text. The diagnostics system consists of a fast semi-conductor detector (to measure the time structure) coupled to a mixer (to measure the frequency), a calorimeter (to measure the mm-wave power) and an absorption sheet viewed by an IR-camera (to measure the beam profile).

can be varied between 0% and 100%. Frequency tuning of the stepped waveguides requires the upper and lower walls of the waveguides to be adjusted. The operating waveguide is wide band and does not need any adjustment. The electron beam enters and leaves the cavity through an opening between the mirrors and follows a straight path.

3. mm-wave power

Since the beam energy drops during the pulse, the amplification band shifts across the frequency band of the cavity. This changes the characteristics of the interaction and results in variations of the output power during the pulse.

A net output power of 730 kW was generated by a 1.77 MeV, 7.2 A electron beam, see Fig. 2a. For this beam energy single-frequency operation was reached, see Fig. 2b. For a slightly lower electron beam energy the mm-wave output power showed to be more chaotic, see Fig. 2c. As seen in Fig. 2d, several frequencies are generated, and the output power strongly fluctuates on a short-time scale. Note that in both cases the frequency locks to specific values, i.e., the frequency does not follow the electron beam energy.

The FEM was also operated at 161 GHz for an electron beam energy around 1.65. Fig. 3a and b show the mm-wave output power for a 1.61 MeV, 7.1 A electron beam (single frequency), and for a 1.59 MeV, 7.0 A electron beam (multi frequency), respectively. At these frequencies, around 160 GHz, in principle a similar behavior is observed with single or multi-frequency beams, depending on the electron beam energy with respect to the cavity

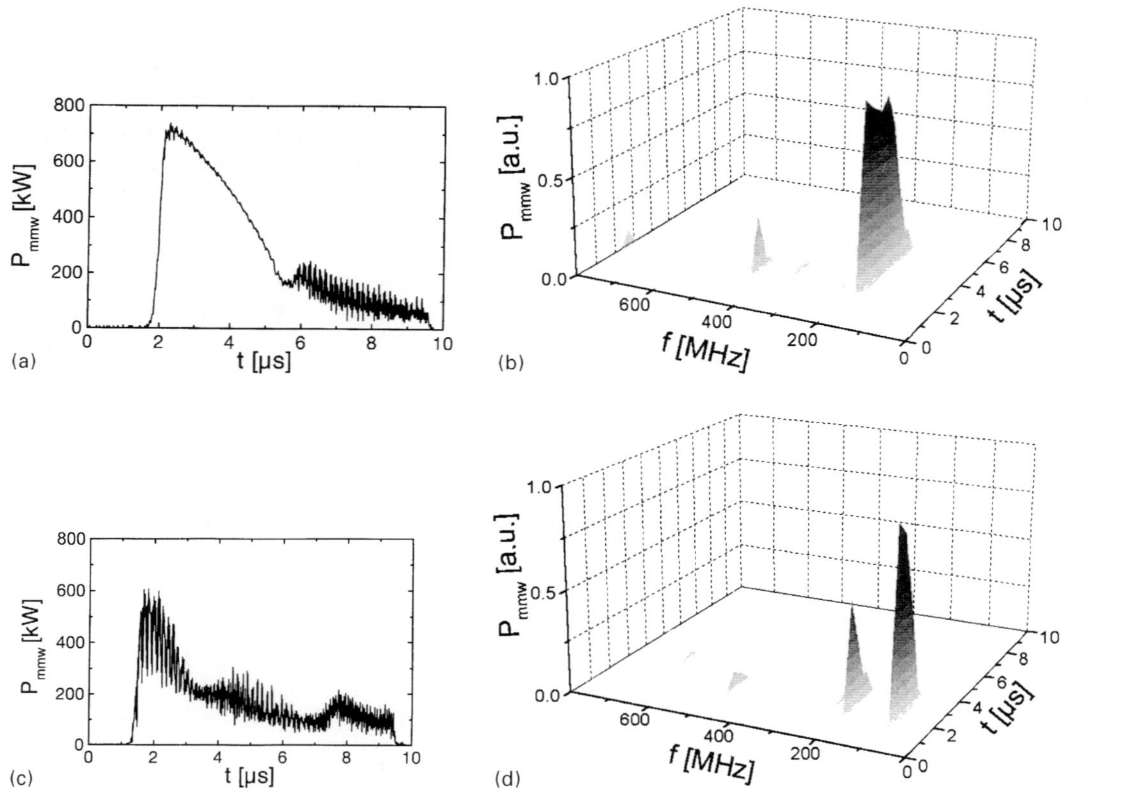

Fig. 2. Single-frequency mm-wave output power, P_{mmw}, for a 1.77 MeV, 7.2 A electron beam (a), and the corresponding frequency spectrum, shifted over 205.7 GHz (b). Multi-frequency mm-wave output power for a 1.76 MeV, 7.2 A electron beam (c), and the corresponding frequency spectrum, shifted over 205.7 GHz (d). The electron beam starts at $t = 0$.

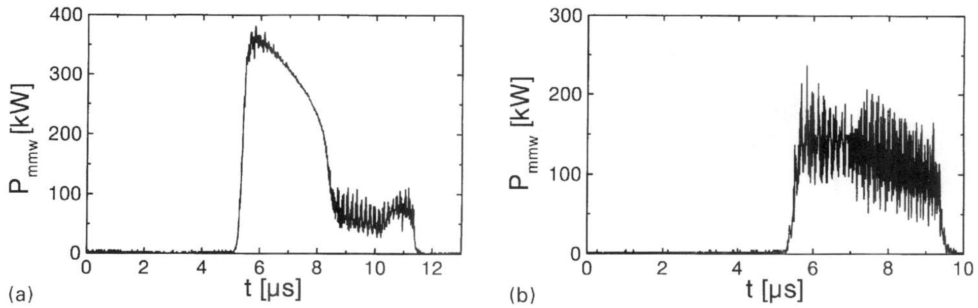

Fig. 3. mm-wave output power around 160 GHz, P_{mmw}, for a 1.61 MeV, 7.1 A electron beam (single frequency) (a), and for a 1.59 MeV, 7.0 A electron beam (multi frequency) (b).

frequency band. In this case the output power is lower, which is due to high losses in the present undulator waveguide. Presently, a new waveguide is under construction.

4. Comparison between experiments and simulations

The experiments show that the temporal dynamics of the mm-wave output beam strongly depend on the electron beam energy. For specific beam energies a single-frequency output beam is generated while for only slightly different electron beam energies, of the order of a few keV lower, mm-wave beams are generated which contain various frequencies and show strong power fluctuations on a short-time scale.

These phenomena are connected with the overlap between the amplification band, which depends on the electron beam energy, and the frequency band of the cavity. The width of the amplification band is of the order of 5 GHz while the width of the cavity frequency band is of the order of 15 GHz. When the amplification band overlaps with the higher frequency part of the cavity frequency band, the following process takes place. Firstly, a number of longitudinal modes (frequencies) grow from noise to a certain level in the linear amplification regime. In this regime there is no interaction between the various longitudinal modes. During the pulse the electron energy drops and thus the amplification band shifts towards lower frequencies. If the amplification band is still in the cavity frequency band, the mm-wave power grows further and reaches the non-linear regime. Then mode competition takes place and the mode with the highest gain grows on the expense of other modes. After some time, typically a few μs, only one mode survives. As shown in Fig. 2b and d the output beam locks to a specific frequency and does not follow the drop of the electron beam energy. Consequently, when the amplification band shifts further towards lower frequencies, the power decreases since the amplification for this mode gets lower. The mode dies out and a new mode can be excited, at a lower frequency, which again grows to high power and than dies out. The frequency step from one mode to the next one is of the order of the width of the amplification band. A simulation result of this process is shown in Fig. 4a.

Another situation occurs when the amplification band overlaps with the lower frequencies of the cavity band. In this case the interaction time is much shorter, and the feedback, which is frequency-dependent, is lower. Consequently, the mm-wave power does not reach the non-linear regime and a number of modes occur in the output beam. Since there is no mode competition, there is no mode which suppresses other modes. The output beam contains several frequencies and the power fluctuates strongly. A simulation result of this situation is shown in Fig. 4b. A more detailed explanation on mode competition and power built-up can be found in an accompanying paper [6].

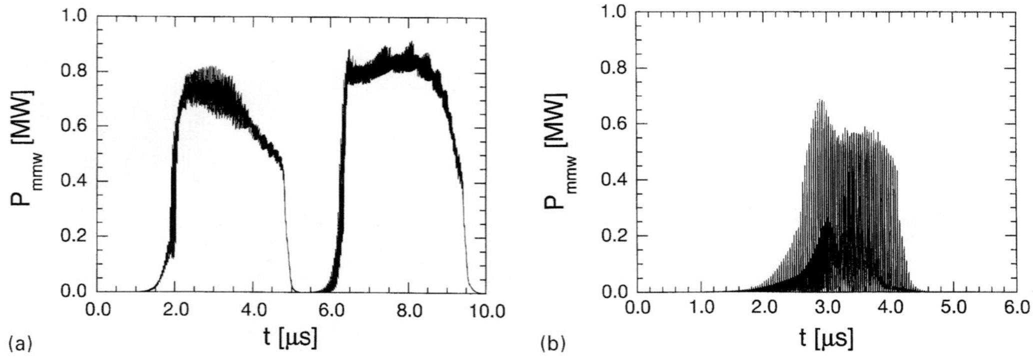

Fig. 4. Simulation results for two different values of the electron beam energy at the start of the pulse. Other parameters are identical: beam current of 7.2 A and a feedback coefficient of 0.55. In (a) the situation is shown for a high electron beam energy and the amplification band overlaps with the higher frequencies of the cavity frequency band. In (b) the amplification band overlaps with the lower frequencies of the cavity frequency band.

5. Conclusions

The Fusion-FEM generates 730 kW of mm-wave power at 200 GHz, with a 7.2 A, 1.77 MeV electron beam. The start-up time of the mm-wave beam agrees well with simulations. For this frequency the electronic efficiency is 5.7%, which is slightly higher than expected. These results were obtained with low cavity losses. For another setting, with higher cavity losses, the output power is lower. Around 160 GHz 380 kW of output power is generated by a 7.2 A electron beam. Although the cavity losses are frequency dependent, for all cases the process of power built-up and mode competition is similar; depending on the electron beam energy at the start of the pulse, either a high power, single-frequency output beam is generated, or a strongly fluctuating multi-frequency beam is generated.

Acknowledgements

This work was performed as part of the research program of the association agreement of EURATOM and the “Stichting voor Fundamenteel Onderzoek der Materie” (FOM) with financial support of the “Nederlandse Organisatie voor Wetenschappelijk Onderzoek” (NWO) and EURATOM.

References

[1] M. Valentini, C.A.J. van der Geer, A.G.A. Verhoeven, M.J. van der Wiel, W.H. Urbanus, Nucl. Instr. and Meth. A 390 (1997) 409.

[2] W.H. Urbanus et al., Nucl. Instr. and Meth. A 331 (1993) 235.

[3] A.A. Varfolomeev et al., Nucl. Instr. and Meth. A 341 (1994) 462.

[4] M. Caplan, R.W.B. Best, A.G.A. Verhoeven, M.J. van der Wiel, W.H. Urbanus, V.L. Bratman, G.G. Denisov, Nucl. Instr. and Meth. A 331 (1993) 243.

[5] G.G. Denisov, M. Yu. Shmelyov, in: M. von Ortenberg, H.U. Müller (Eds.), Proceedings of the 21st International Conference on IR and MM-waves, IR and MM-waves, vol. 1, 1996, p. BF3.

[6] A.V. Savilov, V.L. Bratman, G.G. Denisov, Nucl. Instr. and Meth. A 429 (1999) 46.

ELSEVIER

Nuclear Instruments and Methods in Physics Research A 429 (1999) 17–20

NUCLEAR INSTRUMENTS & METHODS IN PHYSICS RESEARCH Section A

www.elsevier.nl/locate/nima

First lasing of a Cherenkov free-electron laser with annular electron beam☆

J. Wieland[a], J. Couperus[b], P.J.M. van der Slot[b,*], W.J. Witteman[a,b]

[a]*Department of Applied Physics, University of Twente, P.O. Box 217, 7500 AE Enschede, The Netherlands*
[b]*Netherlands Centrum voor Laser Research B.V., P.O. Box 2662, 7500 CR Enschede, The Netherlands*

Abstract

A Cherenkov FEL with an annular electron beam has been succesfully operated. During the first lasing experiments output was found for accelerating voltages ranging from 85 to 150 kV and currents from 1 to 10 A. At a particular accelerating voltage lasing was observed for several distinct settings of the magnetic field (0.7–1.6 T). This field is used to compress and guide the electron beam along the dielectric liner. Depending on the magnetic field setting, the output was reproducible or showed large fluctuations within a single laser pulse as well as from pulse to pulse. Due to a misalignment of the cathode the total output power is only of the order of 100 W. From theoretical considerations the spectrum is expected to center around 19 GHz.

PACS: 41.60.Cr; 41.60.Bq; 07.57.Hm

Keywords: Cherenkov FEL; Waveguide; Simulations; Dielectric liner

1. Introduction

A Cherenkov FEL (CFEL) is capable of generating coherent radiation in the millimeter and submillimeter wavelength region at moderate electron beam energies (100–500 kV). The Cherenkov FEL is a slow-wave device as the phase velocity of the radiation field is reduced by a dielectric medium to below the speed of light in vacuum. Though the radiation field is basically confined to the dielectric medium, it has an evanescent part outside the dielectric which allows an electron beam to couple with the radiation field. Though the CFEL has been investigated in different configurations [1–5], this is the first Cherenkov FEL with the aim to achieve a high average power (~ 1 kW at the longer wavelengths). In this CFEL a cylindrical waveguide with a dielectric liner is used. As the radiation field decays approximately exponentially on the scale of the radiation wavelength the electron beam has to move closely to the liner and an annular instead of a solid electron beam results in a better coupling between radiation field and electron beam.

☆This work is supported in part by the Netherlands Technology Foundation (STW).

* Corresponding author. Tel.: + 31-53-489-1110; fax: + 31-53-489-1102.

E-mail address: p.j.m.vanderslot@tn.utwente.nl (P.J.M. van der Slot)

0168-9002/99/$ - see front matter
PII: S0168-9002(99)00057-1

2. Experimental setup

The experimental layout of the CFEL has been described previously [6]. The magnetic field required for beam compression and guidance is produced by a superconducting solenoid [7]. The current setup uses a cylindrical waveguide with an inner diameter of 14 mm and is lined with Al_2O_3 ($\varepsilon_r \approx 10$) of 1.5 mm thickness and a length of 40 cm. The rest of the design parameters are summarized in Table 1 together with the values used to obtain first lasing. The waveguide is tapered from a diameter of 14 mm to a diameter of 30 mm just after the liner. The taper is designed to have a reflectivity of less than 0.1% and a parasitic mode suppression of better than 20 dB [8]. This allows the electron beam to expand in the diverging magnetic field before it is lost on the beam dump while the mode of the radiation field is preserved. Expansion of the electron beam is required to limit the energy deposition on the beam dump. A thin mylar foil is used as a vacuum window and the open end of a cylindrical waveguide (30 mm diameter) is used as an outcouple horn in these preliminary experiments. Reflections at the vacuum window and/or outcouple horn are very small and therefore the laser is configured as an amplifier starting from noise. The gap between electron beam and dielectric liner, i.e., $R_{\text{diel.}} - R_{\text{beam}}$, is only 0.5 mm. Therefore the waveguide, superconducting magnet and cathode should be carefully aligned. Especially, the cathode proved to be difficult to align. When the complete system was considered to be aligned only 80% of the emitted current was transported along the dielectric liner and collected at the beam dump.

Table 1
Main system parameters, the symbols have the usual meaning

	Design	First lasing
V_{beam} (kV)	200–500	100–150
I_{beam} (A)	30–100	1–15
R_{beam} (mm)	5	≤ 5.5
T_{beam} (μs)	10	10
Rep. rate (Hz)	≤ 10	3
$R_{\text{diel.}}$ (mm)	$R_b + 0.5$	5.5
$d_{\text{diel.}}$ (mm)	0.25–1.5	1.5
B (T)	≤ 2.5 T	08–1.75 T

The remaining 20% of the beam was lost in the waveguide wall before the dielectric liner. It turned out that this was due to a small misalignment of the cathode with respect to the waveguide. The mounting of the cathode will be modified to provide better alignment capabilities. The operating frequency is determined by the beam accelerating voltage and the waveguide dispersion characteristics [9]. With the current parameters (and a perfectly centered electron beam), numerical simulations show that the TM_{01} mode at about 19 GHz is the dominant mode.

3. Results and discussion

Despite the fact that part of the beam was lost in the waveguide wall before the liner, the quality of the remaining part of the beam was sufficient to observe output for different voltages (85–150 kV) and different settings of the longitudinal magnetic field strength. As the CFEL mechanism is based on a longitudinal interaction, the strength of the magnetic guide field is not expected to have a substantial influence on the process. In this configuration however, the spread in longitudinal momentum that the electron beam acquires during compression is greatly affected by the magnetic field. For example at $V_{\text{beam}} = 126$ kV output was found for B around 1.55 (± 5%), 1.28 (± 5%) (see Fig. 1) and

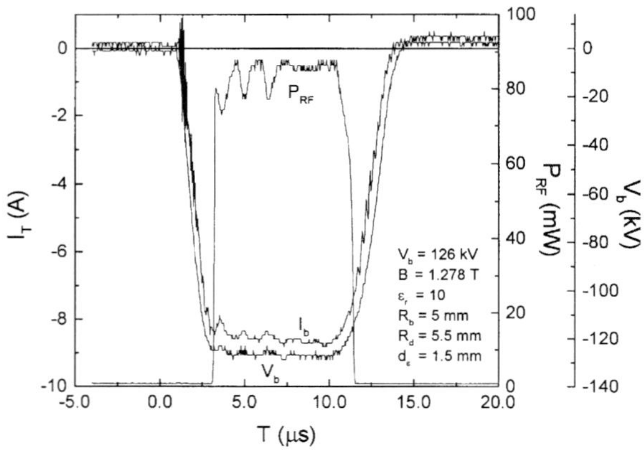

Fig. 1. Example of stable output where output power (as measured by the detector) follows oscillations in current. The total output power amounts to approximately 40 W.

0.75 T (≥ 5%). For this voltage setting the optimal magnetic field as obtained from orbit calculations, should be around 1.28 T. At a field strength of 1.28 T the radial mode structure is measured to be corresponding to the expected TM_{01} mode (see Fig. 4) for a Cherenkov interaction. Not enough data is available for the other two field strengths to decide on the type of interaction, which may include cyclotron-like interactions. This will be a point of further investigation. The output was measured using a standard gain horn mounted on a waveguide diode detector operating in the frequency range 18–26.5 GHz. The spectral distribution of the radiation will be measured in future experiments. In general, the transmitted current is reduced (about 60% instead of around 80%) and less stable for low values of the magnetic field as can be seen in Fig 2. Also the laser output was rather unstable. For values in between, no output could be observed. The transmitted current was about 8.5 A at the higher magnetic field settings (80% transmission). The cathode was operated in the space-charge limited regime. An example of unstable output at a lower magnetic field setting is shown in Fig. 2 for an accelerating voltage of 146 kV. As shown by Fig. 3 only a small current of about 1 A was required in order to distinguish the laser output from the background noise (V_{beam} = 135 kV and B = 0.746 T). Here the cathode was operated in the temperature limited regime. In all mentioned cases emitted radiation was

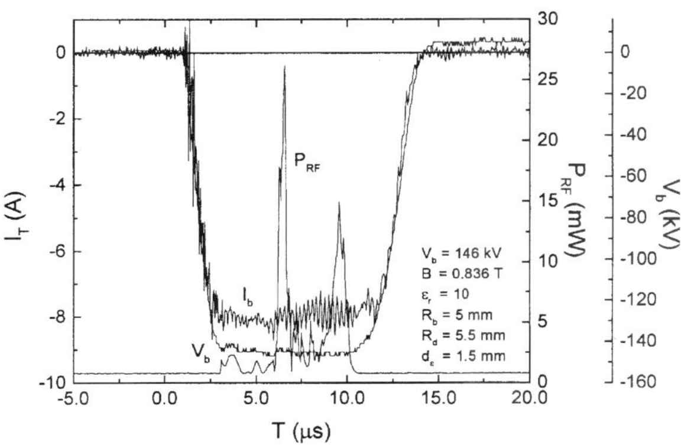

Fig. 2. Example of unstable detector signal where large fluctuations in power are observed within a single pulse as well as from pulse to pulse.

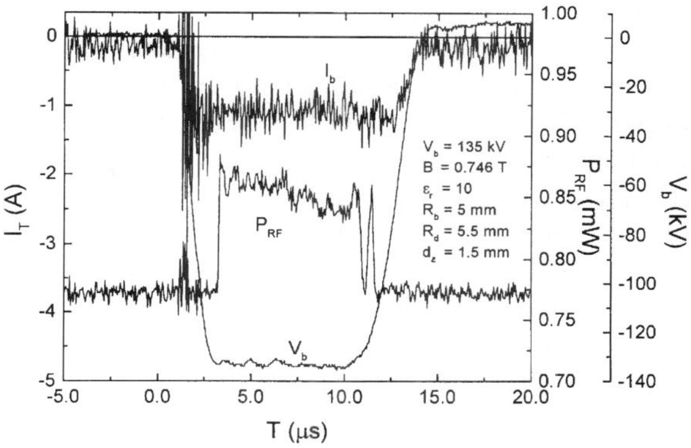

Fig. 3. Measured detector signal at low current ($I_T \approx 1$ A).

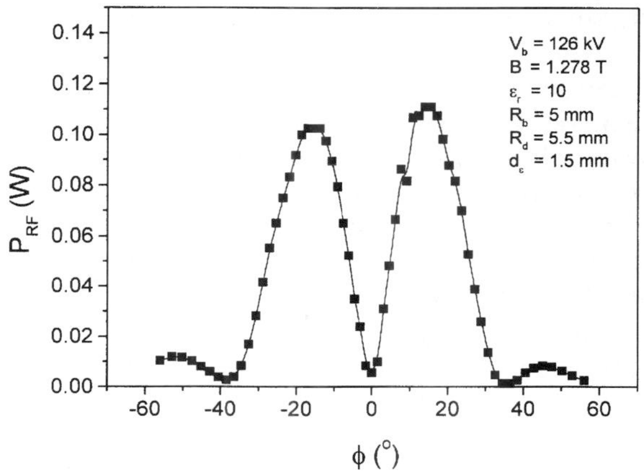

Fig. 4. Angular distribution of emitted radiation. Measurement conditions are the same as in Fig. 1.

observed over the full flat part of the accelerating voltage (which is about 8 μs). However, by tuning the magnetic field shorter pulses could also be found. The emitted radiation pattern has been measured as well in the horizontal plane and is similar to the expected mode (TM_{01}) of operation (see Fig. 4). Here ϕ is the angle with respect to the axis of the waveguide in the horizontal plane. The receiving horn was positioned at a distance of 0.942 m from the outcouple horn and has an aperture of 19.24 cm^2. So far measurements were only performed in the horizontal plane. Assuming cylindrical symmetry this pattern can be integrated to obtain the total emitted power. The total power thus obtained is about 40 W. This level was confirmed by a preliminary energy measurement using a Joule meter [10]. For this value of the beam current, a numerical model shows that the length of the interaction region is not long enough to reach saturation. However, the measured power is much lower than the level of about 30 kW that this model predicts at 19 GHz using an interaction length of 40 cm. This is to be expected as the longitudinal position of the superconducting magnet has not been optimised yet, and coupling of the electron beam to the radiation field is reduced due to the fact that, at this moment, the electron beam is not coaxial with the waveguide.

4. Summary

To summarize, a small current of approximately 1 A was already sufficient to observe output radiation above the background noise. This current was produced by temperature-limited operation of the cathode. By switching to space-charge limited emission, the transmitted current increased to about 8.5 A at 126 kV accelerating voltage. A total emitted power of 40 W was observed in a donut like mode.

References

[1] J. Walsh, B. Johnson, G. Dattoli, A. Renieri, Phys. Rev. Lett. 53 (1984) 779.
[2] W. Main, R. Cherry, E. Garate, IEEE Trans. Plasma Sci. 18 (1990) 507.
[3] F. Ciocci, A. Doria, G.P. Gallerano, I. Giabbai, M.F. Kimmit, G. Messina, A. Renieri, J.E. Walsh, Phys. Rev. Lett. 66 (6) (1991) 699.
[4] E.E. Fisch, J.E. Walsh, Appl. Phys. Lett. 60 (11) (1992) 1298.
[5] T.B. Zhang, T.C. Marshall, J.L. Hirshfield, IEEE Trans. Plasma Sci. 26 (3) (1998) 787.
[6] P.J.M. van der Slot, J. Couperus, W.J. Witteman, A.N. Lebedev, E.G. Krastelev, A.V. Agafonov, V.S. Voronin, V.A. Krasnapolsky, Nucl. Instr. and Meth. A 358 (1995) 100.
[7] W.A.J. Wessel, A. den Ouden, H.J.G. Krooshoop, H.H.J. ten Kate, J. Wieland, P.J.M. van der Slot, IEEE Trans. Appl. Superconduct. 8 (1999) accepted.
[8] M. Thumm, Forschungszentrum Karlsruhe, Karlsruhe, Germany, Private communication.
[9] R.A. Waldron, Theory of Guided Electromagnetic Waves, Van Nostrand Reinhold Company Ltd., New York, 1970.
[10] P.J.M. van der Slot, Ph.D. Thesis, University of Twente, The Netherlands, 1992.

ELSEVIER

Nuclear Instruments and Methods in Physics Research A 429 (1999) 21–26

NUCLEAR INSTRUMENTS & METHODS IN PHYSICS RESEARCH Section A

www.elsevier.nl/locate/nima

First lasing of the Darmstadt cw free electron laser

M. Brunken*, S. Döbert, R. Eichhorn, H. Genz, H.-D. Gräf, H. Loos, A. Richter, B. Schweizer, A. Stascheck, T. Wesp

Institut für Kernphysik, Technische Universität Darmstadt, D-64289 Darmstadt, Germany

Abstract

The Darmstadt cw FEL designed for wavelengths between 3 and 10 μm driven by the superconducting electron accelerator S-DALINAC first lased on December 1st, 1996 and has operated thereafter successfully in the wavelength region between 6.6 and 7.8 μm. The pulsed electron beam employed had a micro pulse length of about 2 ps, with a repetition rate of 10 MHz and a peak current of 2.7 A while its energy was varied between 29.6 and 31.5 MeV. A wedged pole hybrid undulator, with 80 periods each of 0.032 m length and a magnetic field strength of 0.15–0.4 T, was located in between a 15.01 m long optical cavity equipped with two high reflectivity (99.8%) mirrors of 0.05 m diameter. Due to the low beam current special care with respect to the electron and optical beam properties was necessary to meet the stringent conditions in order to reach a minute small signal gain of at least a few percent resulting in amplification. Saturation was obtained after about 2000 repetitions of the photon pulse inside the cavity. The Darmstadt FEL experiment is unique with respect to its variable electron beam macro pulse structure and it is thus the first linac-based FEL providing a cw photon beam. First experiments using the FEL were performed for characterizing laser light parameters as well as for tissue ablation. Macropulse lengths of 4–8 ms at an average outcoupled power of 100 mW corresponding to a peak power density of 100 MW/cm^2 at a beam spot of 150 μm were employed to investigate the thermal ablation of bovine cornea. © 1999 Elsevier Science B.V. All rights reserved.

1. Introduction

The Darmstadt free electron laser experiment is tailored around the superconducting electron accelerator S-DALINAC that has gone into operation in 1991. It was designed especially to serve the needs of nuclear physics experiments and it thus provides a cw electron beam up to 130 MeV with an average beam current of 60 μA at a micro pulse repetition rate of 3 GHz. It is its interesting cw feature that initiated the design and construction of the free electron laser providing accordingly a cw photon beam.

The existing accelerator concept was modified with respect to the electron gun and the room temperature injection in order to provide pulses with a length of 2 ps and a peak current of 2.7 A. With electron energies between 25 and 50 MeV provided immediately behind the main linac a hybrid undulator and a nearly concentric optical cavity with dielectric mirrors the photon wavelength region falls between 3 and 10 μm. The Darmstadt FEL experiment is after Stanford [1] the second one using a superconducting accelerator and the first in Europe.

In the present paper we report on the design and commissioning of the FEL. In Section 2 the

* Corresponding author.
E-mail address: brunken@ikp.tu-darmstadt.de (M. Brunken)

PII: S0168-9002(99)00060-1

accelerator and the FEL are described but we mention only the main aspects referring to earlier publications for details [2,3]. The commissioning of the FEL is subject of Section 3 while a number of performance tests are summarized in Section 4, which is followed by the conclusion given in Section 5.

2. Layout of the FEL

The layout of the accelerator (Fig. 1) illustrates its principle of operation. The electron gun (1) is located on a high voltage terminal of 250 kV and emits bunches with a width of 1 ns at a repetition rate of 10 MHz (the 300th subharmonic of the accelerator's RF frequency of 3 GHz) for FEL operation. In the chopper/prebuncher section (2) the electron bunches are first compressed to 370 ps by the 600 MHz subharmonic chopper and then to a width of 5 ps by the subharmonic prebuncher. Acceleration of the beam up to 10 MeV is achieved by a 2-cell ($\beta = 0.85$) and a 5-cell ($\beta = 1$) superconducting capture cavity and two 20-cell superconducting acceleration cavities (3). After passing the 180° arc (4) compressing the bunches to a width of 2 ps magnetically the beam enters the 40 MeV main linac (5) consisting of eight 20-cell superconducting cavities. When leaving the main linac the beam can either be extracted to serve a number of nuclear physics experiments or it can be recirculated and reinjected to the main linac once or twice by the appropriated beam transport system.

The FEL (7–10) is located at the bypass of the first recirculation and is shown in detail in Fig. 2. The electron beam enters the FEL area (Fig. 2) traveling down (from right to left) the straight section of the first recirculating beamline of the S-DALINAC. An achromatic beam transport system (1) consisting of two dipole and three quadrupole magnets translates the electron beam onto the axis of the optical cavity. A fourth quadrupole behind the second dipole is needed for a final matching of the electron beam to the undulator (2). After passing the undulator the electron beam is guided by the refocusing system (3) to the beam dump (4). For the determination of the electron bunch length a diagnostics station (5) is installed behind the undulator. The measurement of the electron-beam energy spread using optical transition radiation (OTR) is performed behind the third dipole (6). The spontaneous radiation emitted by the electrons hits the downstream cavity mirror (7) where it is reflected back and focussed through the undulator to hit the upstream cavity mirror (8).

The undulator was designed in close collaboration with the manufacturer[1] and resulted in a hybrid configuration [4] with cobalt–samarium permanent magnets and wedged pole pieces made from vanadium permendur.

The concept of the optical cavity was changed with respect to our earlier work [2–7] in order to have a strong overlap of the optical mode and the electron beam in the front and end region of the undulator as well as in its central part. The radii of curvature of the cavity mirrors were therefore increased to $R_1 = R_2 = 8.50$ m leading to a less divergent optical mode resulting in losses of 0.9% per round trip only using mirrors of 99.8% reflectivity.

The dielectric resonator mirrors[2] are capable to reflect within the wavelength range from about $\pm$ 25% around the central wavelength. During the experiment two different mirror sets were used, one with two high reflecting mirrors having a reflectivity of 99.8% and a second with one high reflector and an outcoupling mirror with a reflectivity of 99%. Table 1 contains the main parameters of the Darmstadt FEL experiment.

3. Commissioning of the FEL

The electron beam diagnosis was performed using transition radiation that originates from the interaction of electrons with thin aluminium foils placed at appropriate positions within the beam line. Thus, parameters as size, emittance, energy spread and bunch length would be deduced. The obtained electron distribution is shown in Fig. 3. It yields in this particular case a bunch length of (3.3 $\pm$ 0.3) ps and a peak current of (1.3 $\pm$ 0.3) A is

[1] Spectra Technology Inc., Bellevue, WA, USA.

[2] Delivered by II–VI Inc., Saxonburg, PA, USA.

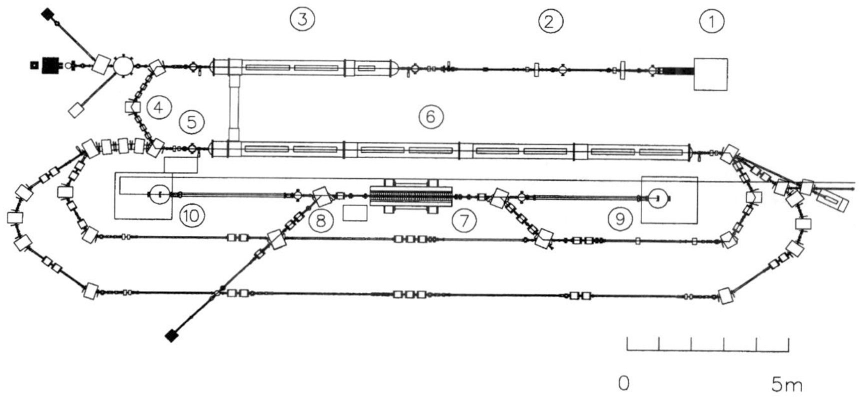

Fig. 1. Layout of the accelerator and the FEL.The numbers stand for: (1) electron gun, (2) chopper/prebuncher section, (3) injector linac, (4) 180° arc, (5) diagnostics station, (6) main linac, (7) undulator, (8) diagnostics station, (9) and (10) optical tables.

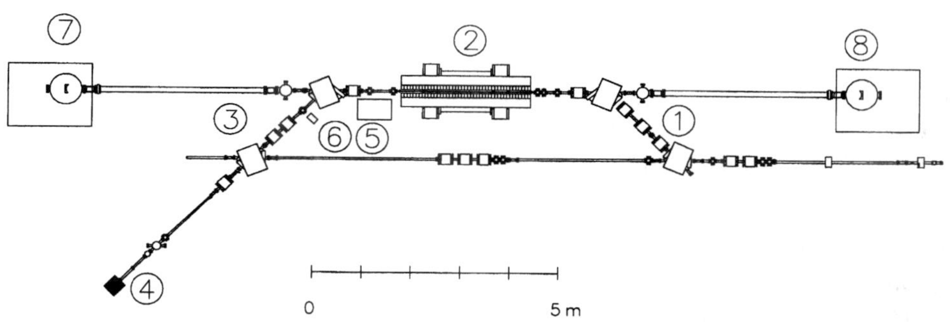

Fig. 2. Beam transport system of the FEL. Numbers denote: (1) achromatic bypass, (2) undulator, (3) refocusing system, (4) beam dump, (5) diagnostic station, (6) OTR station, (7) downstream optical mirror and (8) upstream optical mirror.

Table 1
S-DALINAC and FEL properties

Electron beam		
Electron energy	25–50	MeV
Average current	60	μA
Peak current	2.7	A
Repetition rate	10	MHz
Operation mode	cw	
Undulator and cavity		
Number of periods	80	
Wavelength λ_U	3.2	cm
Undulator strength	0.45–1.12	
Cavity length	15	m
Q-value	120	
Laser		
Wavelength	3–10	μm
Small signal gain	3–5	%
Pulse energy	300	nJ

determined behind the undulator. It should be mentioned that this refers only to the lower limit of the peak current, since possible additional power losses in the power spectrum not noticed in the efficiency calculation would result in a smaller bunch length and thus a larger peak current.

The Q-value of the optical cavity was determined by measuring the decay of the spontaneous emission intensity stored in the optical cavity. This was accomplished using the pulsed operation mode of the S-DALINAC. It enables to switch off the electron beam within two successive electron bunches. Fig. 4 exhibits that this procedure leads to an exponential decay of the intensity of the stored spontaneous emission. It was achieved by triggering an oscilloscope on the falling edge of the electron macropulse and observing the total intensity of the

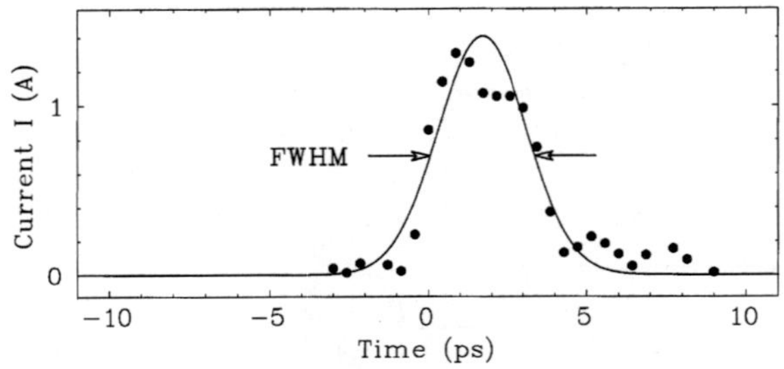

Fig. 3. Reconstructed electron distribution. Values below − 3 ps and above 9 ps are not displayed. The solid line represents a fit of a parabolic distribution to the data points which yields a FWHM of 3.3 ps.

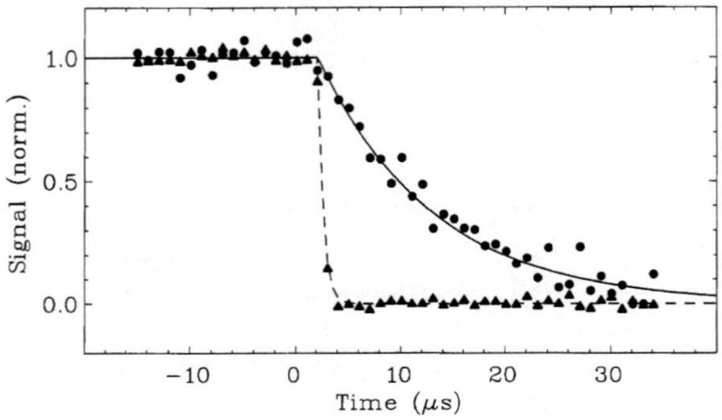

Fig. 4. Decay of the spontaneous emission after switching off the electron macro pulse (triangles) and similar for the case in which the photon intensity is stored inside the optical cavity (circles). The experimental resolution is indicated by the dashed curve. The solid line represents a fit of an exponential to the data points which allows to extract the Q value of the optical cavity as described in the text.

spontaneous emission for the aligned optical cavity simultaneously. The time constant of the electronic data collection system of the setup was determined by bypassing the DS resonator mirror with a tilting mirror inside the vacuum chamber and performing the same measurement. For the calculation of the Q value this time constant was taken into account. Using two high reflecting resonator mirrors a Q value of 120 was obtained, the combination of a high and a low reflector yields a Q value of 60.

4. First lasing

The first startup of the S-DALINAC FEL was observed on December 1st 1996 for electrons with

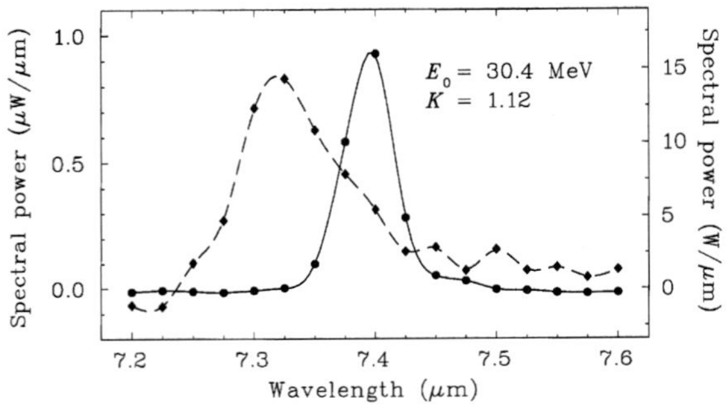

Fig. 5. Spontaneous emission spectrum (dashed line, left scale) and first laser startup of the S-DALINAC FEL (solid line, right scale). Note the spectrum of spontaneous emission is multiplied by a factor of 10^7.

an energy E_0 of 30.4 MeV and an undulator K of 1.12 resulting in a laser wavelength of 7.4 μm. After matching the electron beam to the undulator and optimizing the bunch length the cavity was finally adjusted by slightly tilting the cavity mirrors. With a $Q > 100$ the length of the optical resonator was adjusted. The build up of the light power by some orders of magnitude indicated the correct cavity length. The net gain, however, was only of the order of 0.5% which clearly indicates the difficulties for commissioning of the S-DALINAC FEL since its parameter set allows only very small variances from the optimum value. By slightly optimizing the electron beam parameters the expected gain of a factor 10^7 in the power level could be achieved. A more comfortable net gain between 3% and 5% was established this way also at a later attempt.

The spectra of spontaneous and stimulated emission (Fig. 5) reveal the power gain. The power calibration was performed directly behind the monchromator but at the total beam intensity using a HgCdTe detector for spontaneous emission and a thermopile laser power meter for the laser beam.

The saturation power of the FEL was determined using a power meter situated on the optical tables at both ends of the resonator. In the resonator setup with one outcoupling mirror ($R = 99.0\%$) and one high reflective mirror ($R = 99.8\%$) we measured a laser power of 2.0 W through the outcoupling mirror and 0.24 W through the high reflective mirror during the macropulse with an electron beam of 30 MeV and 60 μA.

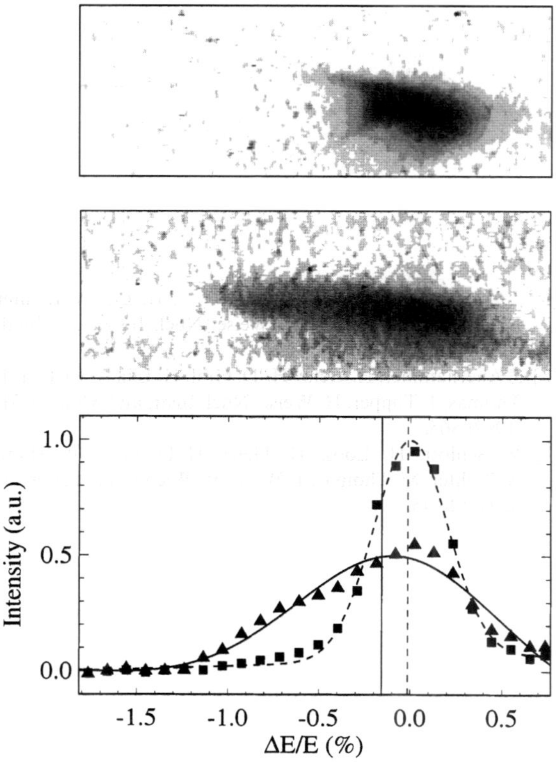

Fig. 6. CCD image of the electron beam image recorded behind the undulator in case of an ordinary electron beam without lasing (upper photo) and in the case of lasing (center part). The intensity deduced by means of a frame grabber is displayed in the lower part for the nonlasing (squares) and the lasing (circles) case.

To detect the energy loss of the electron beam the energy distribution behind the undulator with and without laser operation was determined. Two different beam spots are shown in the upper and middle part of Fig. 6. Electrons with higher energies can be found on the right side, those with lower energies on the left side of the foil. The upper image displays the ordinary beam spot with no lasing resulting in the symmetric energy distribution shown in the lower part.

When the laser is turned on, the electron beam shape changes: some electrons gain energy during their pass through the undulator, but most of the electrons lose power which they supply to the laser field. Fig. 6 shows at the bottom the projection of the electron distribution onto the energy axis. The center of mass of the energy spectrum during laser operation is shifted to lower energies compared to the spectrum without laser operation. For the electron beam with 60 μA current and 30 MeV electron energy the total beam power is 1.8 kW. Since electron energy and beam power are proportional the observed relative energy loss of 0.14% corresponds to a power loss of 2.5 W in the electron beam. This is in fair agreement with the measured laser power.

Wavelengths between 6.6 and 7.8 μm were achieved by using undulator K-values between 0.97 and 1.12 and electron beam energies between 29.6 and 31.4 MeV. The tunability is limited by several machine parameters: The electron energy generated by the accelerator was varied only in a range of ± 1 MeV. Larger changes would demand a completely new steering of the electron beam through the accelerator. Furthermore, it should be mentioned that the use of dielectric mirrors causes a limitation to wavelengths between 6.5 and 8.5 μm due to the reflectivity characteristics of these mirrors. Finally, changes of the undulator gap also affect the small signal gain. This determines the lower edge of the undulator K.

5. Conclusion

The Darmstadt FEL project has succeeded in providing a cw laser beam in the wavelength region between 6.6 and 7.8 μm. Due to the special feature of the superconducting accelerator delivering an electron pulse of arbitrary selectable macro pulse length a laser beam with the according time structure has become available. The output power amounts to 3 W. Using dielectric mirrors with different reflectivity regions the Darmstadt FEL will provide photons in the wavelength region between 3 and 10 μm in the future. First experiments dealing with the ablation of various tissue have successfully been performed.

Acknowledgements

We gratefully remember the team of the first hours Klaus Alrutz–Ziemssen, Jutta Auerhammer,

Jürgen Töpper and Rainer Hahn who provided the essential layout. In this context we would like to thank Andreas Gaupp (BESSY) for his assistance in the conceptional state of the project. Furthermore, we owe special thanks to Wim van Ammersfort, Alan Gillespie, Maurice Kimmit, Dino Jaroszynski, Lex van der Meer, Jean–Michel Ortega, Günther Renz, Gert Spindler, Todd Smith, Alan Schwettman and Marnix van der Wiel for numerous stimulating discussions and encouraging remarks that cheered us up during the frustrating periods of the endeavour. Finally, we gratefully acknowledge the financial support provided by the BMFT/BMBF under contract No. 05 345 EAI, the European Network contract number SCI-0471C(A), the DFG Graduiertenkolleg "Physics and Technology of accelerators" and DESY Hamburg.

References

[1] R. Rohatgi, T.I. Smith, H.A. Schwettman, R.L. Swent, Nucl. Instr. and Meth. A 272 (1988) 32.

[2] V. Aab, K. Alrutz-Ziemssen, H.-D. Gräf, A. Richter, A. Gaupp, Nucl. Instr. and Meth. A 272 (1988) 53.

[3] A. Richter, in: S. Myers (Ed.), Proceedings EPAC96, Barcelona, 1996, p. 110.

[4] K. Alrutz-Ziemssen, J. Auerhammer, H. Genz, H.-D. Gräf, A. Richter, J. Töpper, H. Weise, Nucl. Instr. and Meth. A 304 (1991) 159.

[5] J. Auerhammer, H. Genz, H.-D. Gräf, H. Liu, A. Richter, F. Thomas, J. Töpper, H. Weise, Nucl. Instr. and Meth. A 318 (1992) 184.

[6] J. Auerhammer, H. Genz, H.-D. Gräf, H. Liu, A. Richter, F. Thomas, J. Töpper, H. Weise, Nucl. Instr. and Meth. A 318 (1992) 865.

[7] V. Schlott, H. Loos, H. Genz, H.-D. Gräf, R. Hahn, A. Richter, M. Thomas, T. Wesp, M. Wiencken, Part. Accel. 52 (1996) 45.

ELSEVIER

Nuclear Instruments and Methods in Physics Research A 429 (1999) 27–32

NUCLEAR INSTRUMENTS & METHODS IN PHYSICS RESEARCH Section A

www.elsevier.nl/locate/nima

First lasing of the Jefferson Lab IR Demo FEL

S. Benson*, G. Biallas, C. Bohn, D. Douglas, H.F. Dylla, R. Evans, J. Fugitt, R. Hill, K. Jordan, G. Krafft, R. Legg[1], R. Li, L. Merminga, G.R. Neil, D. Oepts, P. Piot, J. Preble, M. Shinn, T. Siggins, R. Walker, B. Yunn

Thomas Jefferson National Accelerator Facility, 12000 Jefferson Avenue, Newport News, VA 23606, USA

Abstract

As reported in previous work (Neil, Proceedings of the 1998 European Particle Accelerator Conference, Stockholm, Sweden, 1998, p. 88; Shinn, in: Free-Electron Laser Challenges, Proceedings of SPIE, vol. 2988, 1997, p. 170), Jefferson Lab is building a free-electron laser capable of generating a continuous wave kilowatt laser beam. The driver-accelerator consists of a superconducting, energy-recovery accelerator. The initial stage of the program was to produce over 100 W of average power with no recirculation. In order to provide maximum gain the initial wavelength was chosen to be 5 μm and the initial beam energy was chosen to be 38.5 MeV. On June 17, 1998, the laser produced 155 W cw power at the laser output with a 98% reflective output coupler. On July 28th, 311 W cw power was obtained using a 90% reflective output coupler. A summary of the commissioning activities to date as well as some novel lasing results will be summarized in this paper. Present work is concentrated on optimizing lasing at 5 μm, obtaining lasing at 3 μm, and commissioning the recirculation transport in preparation for kilowatt lasing this fall.

Keywords: Lasing; Free-electron laser; Recirculation

1. Introduction

At the 18th International FEL in Rome we introduced the design of a free-electron laser (FEL) driven by a recirculating, energy-recovered continuous electron beam accelerator to demonstrate scalability to higher powers for industrial applications [1,2]. The design of this accelerator is such that the full current of 5 mA can only be produced with energy recovery. Without energy recovery only 1.1 mA of current can be produced due to limits on available rf power. Since the repetition rate of the accelerator can be reduced by a factor of two, and since the electron beam quality improves as the charge is lowered, the gain is not very much lower than that at full beam current when the laser is operated with 1.1 mA. This makes it possible to lase in "first light" mode without energy recovery and optimize the laser before energy recovery is attempted. Simulations of the beam using

* Corresponding author. Tel.: + 1-757-269-5026; fax: + 1-757-269-7352.

E-mail address: felman@jlab.org (S. Benson)

[1] Presently at General Atomics, San Diego, CA, USA.

0168-9002/99/$ - see front matter
PII: S0168-9002(99)00061-3

PARMELA indicated that we should be able to achieve an emittance of 5 mm mrad with an energy spread of 50 keV and a bunch length of 370 fs (all quantities rms). Using these numbers the predicted gain is well over 100% indicating a good margin for lasing. This paper is a description of the first lasing process and some of the interesting results we have gained since achieving first light.

2. Driver accelerator

The driver accelerator is shown in Fig. 1. The design and measured performance are shown in Table 1. Microbunches with an rms bunch length of 20 ps are produced in a DC photocathode gun and accelerated to 350 keV [3–5]. The bunches are compressed by a copper buncher cavity operating

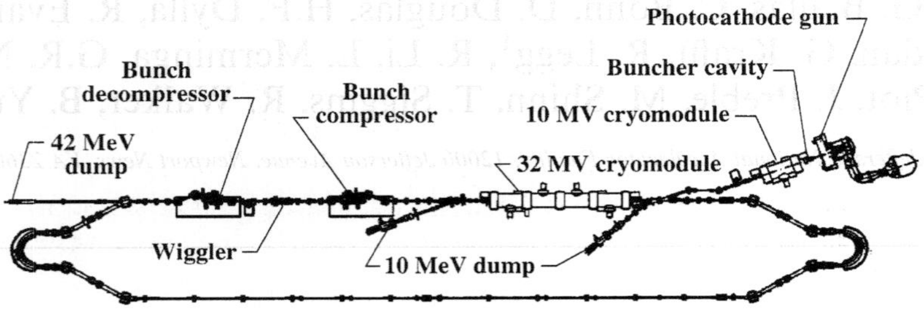

Fig. 1. IR demo layout. See text for explanation of operation.

Table 1
Measured parameters for the IR Demo driver accelerator and laser

Property	Design	Achieved	Unit
Pulse length	1	0.4 ± 0.1	ps rms
Charge per bunch	60	60 ± 2	pC
Peak current	22	60 ± 15	A
Average current	1.1	1.2	mA
Transverse emittance	< 8.7	7.5 ± 1.5	mm-mrad rms
Energy spread	< 0.22	0.3 ± 0.2	% rms
Kinetic energy	38	48	MeV
Repetition rate	18.7	18.7	MHz
Position stability	50	< 100	μm
Angular stability	100	< 200	μrad
Wiggler wavelength	2.7	2.7	cm
Number of periods	40	40.5	–
K_{rms}	1.0	0.98	–
Wiggler rms phase error	< 5	2.6	°
Trajectory wander	< 100	< 100	μm p-p
Optical cavity length	8.0105	8.0105	m
Rayleigh range	40	40	cm
Mirror radii	2.54	2.54	cm
Mirror radius of curvature	4.045	4.050	m
Mirror losses	< 0.001	< 0.001	
Mirror tilt tolerance[a]	5	~ 5	μrad
Center wavelength	4.8	4.8	μm
Output coupler refl.	98, 90	97.6, 90.5	%
HR reflectivity	> 99.5	99.85	%

[a]This is the mirror tilt necessary to rotate the optical mode by $\frac{1}{3}$ of the mode divergence. This much tilt has been shown to reduce the power by around 10% when the 90% output coupler is used and the laser is in pulsed mode.

at the fundamental accelerating frequency of 1.497 GHz. They are then injected into a high performance superconducting rf (SRF) cavity pair operating at a mean gradient of 10 MV/m. The output beam from this is transported through an achromat, injected into an eight cavity SRF cryomodule, and accelerated up to $\sim$ 38 MeV. The electron beam is then bent around the two optical cavity mirrors passing through the wiggler along the way and captured in a water-cooled copper dump.

The program schedule was very aggressive with first-light hardware installed 15 months after first funding. Commissioning of the injector was started with the wiggler removed so that the radiation dose to the NdFeB permanent magnets could be minimized. The design current of 1.1 mA cw was achieved 21 months after first funding. The beam quality at 60 pC (as shown in Table 1) was found to be more than sufficient for lasing. In fact, the bunch length and vertical emittance are very close to the values predicted by PARMELA. The horizontal emittance is larger than PARMELA either due to wakefield effects or CSR induced emittance growth [6]. Sensitivity to the phase of the cryomodule cavities has prevented good measurements of energy spread when the laser is optimized.

Once first light was achieved the emphasis in the accelerator development shifted to commissioning the recirculation system [7]. To date both pulsed and cw beams have been transported with essentially no loss to the energy recovery dump. The aperture of the system is sufficient that the exhaust beam of the laser can also be propagated with no losses through the recirculation loop back to the accelerating module. There are losses in the final transport to the energy recovery dump at 9.5 MeV, however, leading to a limit on average power. This loss is being investigated so that higher laser power can be achieved with energy recovery.

3. Laser commissioning

Once it was determined that the beam quality was sufficient for lasing [8,9], the wiggler and optical cavity mirrors were installed [10]. For first light we chose to use a high reflector and a 97.6% reflective output coupler. Using macropulses 200 μs in duration the electron beam was aligned with the wiggler viewers and the cavity length was scanned. The laser power output immediately rose to saturation. Lasing at 4.9 μm was achieved over a 10 μm range of cavity length with no optimization. The laser was then operated cw. Using a power meter at the exit of the laser, 155 W of average power was outcoupled and over 110 W of power delivered to the optical diagnostic room.

After changing to a 90% output coupler, 311 W of cw laser power was measured on a power meter at the laser exit. Lasing weakly with energy recovery was possible but, as noted above, limited due to beam loss at 9.5 MeV. Once the recovered beam is better matched into the accelerating structure we expect to lase at high power with energy recovery.

4. Lasing results

In general the IR Demo laser has been very predictable and easy to run. When the electron beam and optical cavity are set up carefully the laser lases very strongly before any optimization. If the laser power is subsequently optimized, the resulting electron beam configuration is not much different from the original setup. Using measured values for the beam parameters, one ideally expects a small signal gain of approximately 90%. When setting up the system the gain should not be reduced by more than a factor of 3 due to electron beam and optical misalignments. This still allows sufficient gain to lase to saturation. Variation of the electron bunch parameters indicates that the highest power occurs for the shortest electron bunch length and the smallest longitudinal emittance as expected. Sensitivities to variation in steering, focussing, cavity phases, and average current are qualitatively similar to expectations. Note that the mirror sensitivity in Table 1 is for pulsed or low power operation with large cavity losses. When the mirrors heat up or when the output coupling is small, we have found that the sensitivity to mirror tilt is much smaller.

In Fig. 2a we show the power as a function of cavity length as the micropulse repetition rate is varied [11]. To our knowledge this is the first time

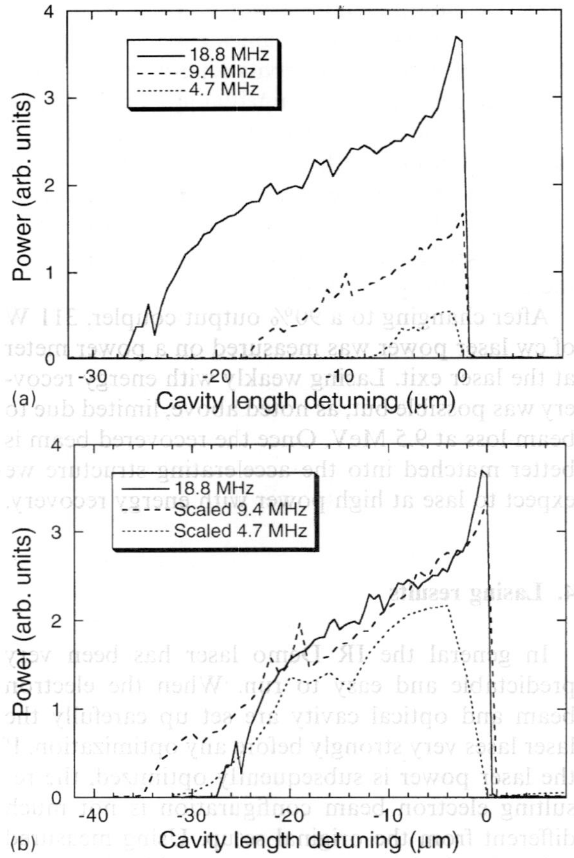

Fig. 2. Laser power vs. cavity length for three different micropulse repetition rate. The round trip frequency is 18.813 MHz so the three curves correspond to one, two and four round trips per gain pass. In (a) we show the raw data. In (b) we have scaled the data by a factor of two and four in both axes to study the variation of the extraction efficiency with the repetition rate.

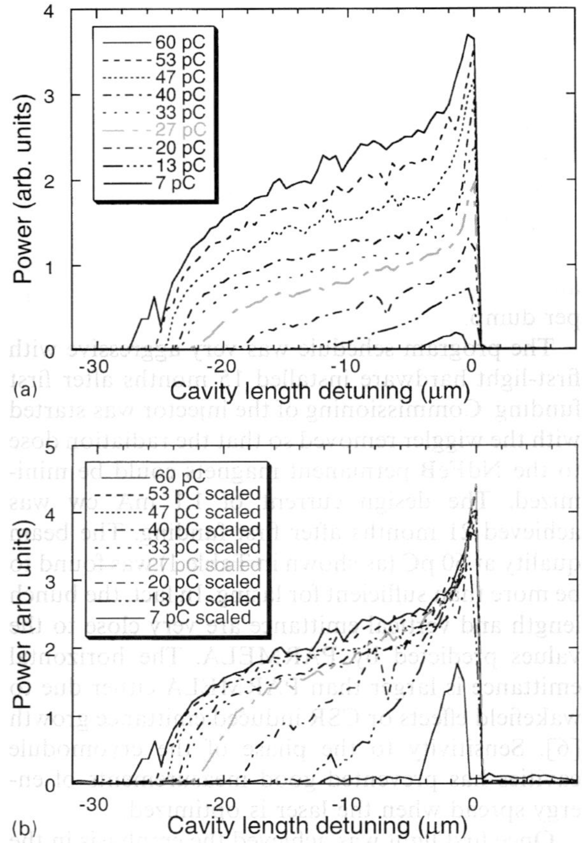

Fig. 3. Laser power vs. bunch charge. In (a) is the raw data. In (b) we have scaled the power to the maximum charge so that the extraction vs. charge is evident.

this has been reported for an FEL. The total cavity loss is 11% per round trip so the threshold gain (defined as $G_{th} = (1-\Gamma)^{-N} - 1$ where Γ is the round trip cavity loss) is 12.4% for 18.8 MHz repetition rate, 26.3% for 9.4 MHz, and 59.4% for 4.7 MHz. We see from the detuning width in Fig. 2a that the gain must be well in excess of 59.4%. The electron beam in this case was pulsed with a 1.2% duty cycle. The electron gun was run at 327 kV for this data so the emittance is not as small as in Table 1. The other parameters are similar. Mirror heating effects should have been negligible. In Fig. 2b we have scaled both the power and the cavity length detuning by the reduction in the frequency to give an indication of the extraction efficiency as a function of the optical delay from synchronism. Note that the optical delay scales linearly with the number of round trips per gain pass. The curves in Fig. 2b are remarkably similar to each other. Note, however, that the scaled cavity length detuning curve is actually shorter for a smaller threshold gain (9.4 MHz case compared to 18.8 MHz). This is a very puzzling feature, which may be due to optical guiding effects [12].

In Fig. 3a we show the power vs. cavity length as a function of bunch charge. In Fig. 3b the power is scaled to the bunch charge. As expected the laser power is approximately proportional to the bunch charge. The length of the cavity detuning curve is a very non-linear function of the charge, however. This is not entirely unexpected since the emittance and energy spread are smaller when the charge is smaller. It is a bit surprising, though, since the bunch length and transverse match are not optimized for the low charge operation. If we assume that the bunch length is unchanged, the predicted gain for 13 pC is around 25% and the length of the detuning curve according to super-mode theory [13] is 13 μm, very close to what is observed.

Note how the shape of the curves also changes from convex to concave as the charge is reduced. Also note that the predicted detuning curve length for even 60% gain (and the gain must be at least this large as shown in Fig. 2) is 31 μm. As noted above, the gain calculated from measured parameters is approximately 90% which leads to an expectation of a 46 μm long cavity length detuning curve. The length of the curve seems to be shorter than possible assuming a one-dimensional super-mode theory.

Spectral measurements have also confirmed that we have quite high gain. When lasing near the synchronous detuning a strong sideband is observed. This should not be the case unless the gain-to-loss ratio is greater than around 4 which implies at least 50% small signal gain. As expected the spectra become much narrower as the cavity length is shortened. The spectral brightness varies little over the central part of the detuning curve.

With pulsed operation at 60 Hz and a duty cycle of 1.2%, 4.1 W of laser power was detected in the optical diagnostic lab. The losses in the transport are at least 10% so this means that the laser power was at least 380 W during the macropulse and may have been as much as 480 W. During cw operation with a similar electron beam setup, the power did not exceed 311 W despite repeated optimization. This indicates that mirror heating reduced the laser output by 18–35%. Calculations of mirror distortion [10] indicated that the mirror absorption had to be less than 0.1% to achieve over 300 W of laser power. Measurements of the mirrors at the Naval Advanced Warfare Center in China Lake CA put an upper limit of 0.1% on the absorption.

5. Conclusions

In general this laser behaves according to expectations. One exception to this observation is the cavity length detuning curve. There seems to be some phenomenon that is keeping the detuning curves short when the gain is high. One possibility is guiding. Simulations have hinted that guiding effects can shorten a cavity length detuning curve even though the gain is increased. This might explain the very slow change in the cavity length detuning curve length with pulse charge.

With high gain and a 40 period wiggler it is not unreasonable to expect that one might extract over 1% of the electron beam power. Since the electron beam power is in excess of 40 kW, this would indicate that over 400 W of laser power is reasonable to achieve. Mirror distortion has limited operation to only 311 W but operation at 3 μm with sapphire mirrors should allow us to increase the power in “first light” mode to over 400 W. Recirculation can raise the electron beam power to 240 kW but may limit the extraction efficiency to no more than 0.75%. This setup should allow much higher power (close to 2 kW) if the quality of the sapphire mirror coatings is similar to those on the CaF_2 mirrors now in use.

Acknowledgements

The authors would like to thank all those at Jefferson Lab whose diligent and effective work made this project successful. Work supported by the U.S. Department of Energy under contract DE-AC05-84-ER40150, the Office of Naval Research, the Commonwealth of Virginia, and the Laser Processing Consortium.

References

[1] G.R. Neil, Thomas Jefferson national accelerator facility FEL industrial applications, Proceedings of the 1998 European Particle Accelerator Conference, Stockholm, Sweden, 1998, p. 88.

[2] M.D. Shinn, Jefferson laboratory IR demo project, in: P.G. O'Shea, H.E. Bennett (Eds.), Free-Electron Laser Challenges, Proceedings of SPIE, vol. 2988, 1997, p. 170.

[3] C. Sinclair, Nucl. Instr. and Meth. A 318 (1992) 410.
[4] S. Benson, M. Shinn, Development of an accelerator-ready photocathode drive laser at CEBAF, Proceedings of the 1995 Particle Accelerator Conference, Washington D.C, 1995, p. 1052.
[5] P. Piot et al., Experimental results from an injector for an IR FEL, Proceedings of the 1998 European Particle Accelerator Conference, Stockholm, Sweden, 1998, p. 1447.
[6] R. Li, C.L. Bohn, J. Bisognano, Nucl. Instr. and Meth. A 429 (1999) 310.
[7] D.R. Douglas, Lattice design for a high power infrared FEL, Proceedings of the 1997 Particle Accelerator Conference, Vancouver, Can., 1997, p. 1351.
[8] G. Krafft, Electron beam diagnostics for Jefferson lab's high power free-electron laser, Proceedings of the 1997 Particle Accelerator Conference, Vancouver, Can., 1997, p. 912.
[9] G. Krafft et al., Measuring and characterizing ultrashort bunches in the Jefferson lab free-electron laser, Proceedings of the 1998 European Particle Accelerator Conference, Stockholm, Sweden, 1998, p. 1580.
[10] S.V. Benson et al., Nucl. Instr. and Meth. A 407 (1998) 401.
[11] K. Jordan et al., Pulse selection control for the IR FEL photocathode drive laser, Proceedings of the 1997 Particle Accelerator Conference, Vancouver, Can., 1997, p. 2529.
[12] S.V. Benson, J.M.J. Madey, Opt. Commun. 56 (1985) 212.
[13] G. Dattoli et al., Nucl. Instr. and Meth. A 285 (1989) 108.

ELSEVIER

Nuclear Instruments and Methods in Physics Research A 429 (1999) 33–36

NUCLEAR INSTRUMENTS & METHODS IN PHYSICS RESEARCH Section A

www.elsevier.nl/locate/nima

Long wavelength free-electron lasers in 1998

H.P. Freund[a,*], V.L. Granatstein[b]

[a]*Naval Research Laboratory, Washington, DC 20375, USA*
[b]*University of Maryland, College Park, MD 20742, USA*

Abstract

A summary of the current status and most important future directions for long wavelength (≥ 0.5 mm) free-electron lasers is presented. The distinction between long and short wavelengths is a natural one. For example, space-charge effects may be important for the high currents typically employed at long wavelengths, and the dominant interaction mechanism is often coherent Raman scattering. In addition, dispersion due to the dielectric effects and finite transverse dimensions in the drift tubes and cavities are important at longer wavelengths. The principal goals at long wavelengths are to achieve much higher average powers with good overall efficiency in a compact design.

PACS: 41.60.Cr; 52.75.M

Keywords: Free-electron lasers; Space-charge effect; Wavelength

This is the fifth paper dealing with the status of long wavelength free-electron lasers (FEL) and our purpose, as in the first four papers [1–4], is to provide a summary of long wavelength (⩾ 0.5 mm) FELs which are currently either in operation or under construction. Also, as before, a companion paper appears in this volume dealing with "short wavelength" FELs [5]. Our search in compiling this list relies upon (1) contributions to the general literature and the proceedings of the annual Free-Electron Laser Conferences, and (2) personal contacts. The latter was an important source of information concerning FEL programs that are either in the early operational stages or still under construction and have not as yet been described in archival literature. Despite our best efforts, however, inadvertent omissions are still possible.

The distinction between long and short wavelength FELs is natural because higher current and lower energy beams are typically employed in this regime and space-charge effects are more important. In particular, the dominant interaction mechanism is often coherent Raman scattering. Also, while short wavelength FELs excite optical modes, dispersion due to the beam dielectric effects and finite transverse dimensions in the drift tubes and cavities are important effects at longer wavelengths.

The long wavelength FEL table lists parameters describing: wavelength/frequency, peak power, pulse time, repetition rate, beam voltage and current, wiggler period and field strength, and the type

* Corresponding author.

0168-9002/99/$ - see front matter
PII: S0168-9002(99)00062-5

Table 1

FEL [Ref.]	λ (mm)/f (/GHz)	P_{peak}(MW)	τ_p (μs)	ν_{rep} (Hz)	V_b/I_b (MV/A)	λw(cm)/K(type)	Type/Acc
UCSB [14]	≤ 2.5/120	≤ 0.015	6	2	6/2	7.14/1. (P)	O/ES
UCSB [14]	1/300	UC	CW		2/2	3/0.73 (H)	O/ES
SU/IAP [15]	9.4/32	0.75	0.1	SS	0.3/50	2.3/0.64	O/PL
CESTA [16]	8/35 (100/3)	90 (300)	0.03	SS	2.2/800	12/1.2 (H)	A/IL
CESTA [16]	8/35	40	0.03	SS	2.2/800	12/1.2 (H)	SASE/IL
CESTA [17]	8/35	80	0.03	SS	2.2/800	12/1.2 (H)	A/IL
FOM [13]	1.5/206	0.73	12	1	1.77/7.2	2/0.67 (P)	O/ES
NRL [18]	8/35	0.027	1	6	0.1/10	0.64/0.2(CH)	A/M
KAERI [19]	12/26	0.001	10–30	SS	0.4/2	3.2/0.39(H)	O/ES
ENEA [20]	2–3.5/85–150	0.0015	5.5	40	2.3/0.35	2.5/1.4(P)	O/MI
ENEA [21]	0.4–0.8/375–750	UC	5.5	20	5.5/0.25	2.5/1.7 (P)	O/MI
CU [22]	12.5/24	0.2	0.15	SS	0.6/100	1.85/0.2 (H)	A/PL
TAU/WI [23]	3/100	0.012	70	SS	1.4/1.4	4.4/0.82 (P)	O/ES
TAU [24]	68/4.48	0.0035	5	1	0.07/0.8	4.4/0.12 (P)	O/ES
KEK/JAERI [25,26]	3/9.4	150	0.005	0.05	2/700	16/3.3 (P)	A/IL
JINR/IAP [27]	8/31	48	0.2	2	0.8/170	6/0.84 (H)	O/IL
IAP	10.3/45	7	0.0025	SS	0.5/120	2.4/0.07 (P)	O/PL
INP/IAP [28]	4/75	200	1	SS	1/2000	4/0.3 (P)	O/PL

A–Amplifier
CA-Coaxial
CESTA–Centre d'Etudes Scientifique e Techniques d'Aquitaine
CH - CHI Wiggler
CU - Columbia University, USA
DS - Design study
DU- Duke University
ES - Electrostatic accelerator
H - Helical wiggler
IAP - Institute of Applied Physics, Russia
INP - Institute of Nuclear Physics, Russia
IL - Induction linac
JAERI - Japan Atomic Energy Research Institute
JINR - Joint Institute for Nuclear Research, Russia
KAERI - Korean Atomic Energy Research Institute
KEK - National Laboratory for High Energy Physics, Japan
UCD/SLAC - University of California at Davis/Stanford Linear Accelerator Center
NRL - Naval Research Laboratory
M - Modulator
MI - Microtron
O - Oscillator
P - Planar wiggler
PL - Pulse line accelerator
PS - Power supply
RFL - Radio frequency linac
SASE - Self-amplified spontaneous emission
SS - Single shot
TAU - Tel Aviv University
UC - Under construction
UM - University of Maryland, USA
SU - University of Stratthclyde
UT - University of Twente, Netherlands
WI - Weizmann Institute, Israel.

of device and accelerator employed. We have chosen to denote the wiggler parameter by K ($= 0.0934B_w\lambda_w$, where B_w is the wiggler amplitude in kG and λ_w is the wiggler period in cm); however, this parameter is often called a_w in the literature. Note that in the case of planar wiggler designs this does denote the peak and not the rms amplitude. The wiggler type shown refers to helical (H), planar (P), circular (C), and coaxial hybrid iron (CH) wigglers [6]. Note that an axial guide field is often used in conjunction with helical wigglers in intense beam FELs, but that the brevity of this format does not permit inclusion of this information in Table 1.

Long wavelength FELs employ many types of accelerator. However, most of the experiments employ pulse line accelerators which operate in a single-shot (SS) mode with high peak powers. Historically, the pioneering work on long wavelength FELs was performed by Phillips [7] prior to 1964 using modulators. Long wavelength FEL research began again at various laboratories in the early 1970s with the intention of extending Phillips' work using intense relativistic electron beams. Much of this work was directed at exploring the basic physics of the Raman regime [8], and high average power was not an essential goal. Hence, pulse line accelerators are often used since they produce beam currents of many kiloamperes. In order to achieve the necessary beam quality from pulse line accelerators, careful design of the diode is essential [9] and a large fraction ($\approx 90\%$) of the beam is typically "scraped" off so that only a central core with a low velocity spread is injected into the wiggler. As a result, collective FELs built using pulse line accelerators suffer from the disadvantages of a low wall-plug efficiency and single-shot operation.

At present, the principal thrusts of long wavelength FEL research are the achievements of high average powers, broad bandwidths, and compact systems. Hence, the development of practical long wavelength FELs depends upon accelerator technologies other than pulse line accelerators.

Induction linacs have the advantage that they can be repetitively pulsed and it is possible, at least in principle, to produce high average powers and high efficiencies. However, there are two practical difficulties. The first is the achievement of high repetition rates. The ELF experiment, [10] which produced a peak power of ≈ 1 GW at 35 GHz operated with a pulse time of 10–20 ns and a repetition rate of 0.5 Hz for an average power of only ≈ 7 W. The second difficulty is emittance growth resulting in unacceptably high thermal spreads, and the necessity of beam scraping ($\approx 80\%$ of the beam was scraped off in the ELF experiment). Thus, while the extraction efficiency was about 34% from the beam that passed through the wiggler, the overall efficiency was closer to 7%. Recently, an induction linac-based FEL oscillator in Russia [11] demonstrated that high efficiency operation corresponds to excitation of a single mode in a Bragg resonator.

Other promising approaches for achieving high average powers involve long pulse modulators and electrostatic accelerators. In the case of electrostatic accelerators, full CW operation is possible if a high degree of beam energy recovery can be achieved by using depressed collectors. This has been accomplished at the University of California at Santa Barbara [12]. However, the degree of beam energy recovery which can be achieved decreases with increasing extraction efficiency of the FEL, and this may prove to be a limiting factor in the technology. A significant experiment using electrostatic accelerators is currently under construction at FOM in The Netherlands [13]. The goal of this experiment is to build a 1 MW CW FEL operating in the frequency range 130–260 GHz to heat plasmas in magnetic fusion reactors. At the present time, they have employed a 1.77 MeV/7.2 A electron beam to achieve an output power 730 kW at a frequency of 206 GHz for an extraction efficiency of 5.7%. The energy recovery system is not presently operational; hence, the pulse time was 12 μs. The overall wallplug efficiency is expected to be much larger than 5.7% once the energy recovery system is functional.

In summary, the ultimate goals of long wavelength FEL research are to achieve much higher average powers with good overall efficiency and more compact designs than have yet been achieved. At the present time, high voltage/low current electrostatic accelerators and long pulse modulators remain the prime candidates for drivers for these systems.

This work was supported in part by the Naval Research Laboratory and the Office of Naval Research.

References

[1] H.P. Freund, V.L. Granatstein, Nucl. Instr. and Meth. A 358 (1995) 551.

[2] H.P. Freund, V.L. Granatstein, Nucl. Instr. and Meth. A 375 (1996) 665.

[3] H.P. Freund, V.L. Granatstein, Nucl. Instr. and Meth. A 393 (1997) 9.

[4] H.P. Freund, V.L. Granatstein, Nucl. Instr. and Meth. A 407 (1997) 30.

[5] W.B. Colson, Nucl. Instr. and Meth. A 429 (1999) 310.

[6] R.H. Jackson, H.P. Freund, D.E. Pershing, J.M. Taccetti, Nucl. Instr. and Meth. A 341 (1994) 454.

[7] R.M. Phillips, IRE Trans. Electron Dev. ED- 7 (1960) 231.

[8] H.P. Freund, T.M. Antonsen, Jr., Principles of Free-electron Lasers, second ed., Chapman & Hall, London, 1996.
[9] R.H. Jackson, S.H. Gold, R.K. Parker, H.P. Freund, P.C. Efthimion, V.L. Granatstein, M. Herndon, A.K. Kinkead, J.E. Kosakowski, T.J.T. Kwan, J. Quantum Electron. QE-19 (1983) 346.
[10] T.J. Orzechowski, B.R. Anderson, J.C. Clark, W.M. Fawley, A.C. Paul, D. Prosnitz, E.T. Scharlemann, S.M. Yarema, D.B. Hopkins, A.M. Sessler, J.S. Wurtele, Phys. Rev. Lett. 57 (1986) 2172.
[11] N.S. Ginzburg, personal communication.
[12] G. Ramian, Nucl. Instr. and Meth. A 318 (1992) 225.
[13] W.H. Urbanus, W.A. Bongers, C.A.J. van der Geer, P. Manintveld, J. Plomp, J. Pluygers, A.J. Poelman, P.H.M. Smeets, A.J.A. Verhoeven, V.L. Bratman, G.G. Denisov, A.V. Savilov, M. Yu. Shmelyov, M. Caplan, A.A. Varfolomeev, S.V. Tolmachev, S.N. Ivanchenkov, Phys. Rev. Lett., in press, 1999.
[14] G. Ramian, personal communication.
[15] A.D.R. Phelps, A.W. Cross, D.A. Jaroszynski, C.G. Whyte, W. He, N.S. Ginzburg, N.Yu. Peskov, Paper Presented at the 22nd Internaitonal Conference on Infrared and Millimeter Waves, Wintergreen, VA 1997.
[16] J. Gardelle, T. Lefevre, G. Marchese, M. Padois, J.L. Rullier, Nucl. Instr. and Meth. A 429 (1999) 202.
[17] J. Gardelle, T. Lefevre, G. Marchese, M. Padois, J.L. Rullier, J.T. Donohue, Nucl. Instr. and Meth. A 429 (1999) 111.
[18] J.M. Taccetti, R.H. Jackson, H.P. Freund, D.E. Pershing, V.L. Granatstein, Nucl. Instr. and Meth. A 429 (1999) 116.
[19] B.C. Lee, personal communication.
[20] A. Doria, G.P. Gallerano, E. Giovenale, M.F. Kimmitt, G. Messina, Nucl. Instr. and Meth. A 375 (1996) ABS–11.
[21] A. Doria, M. Faraone, G.P. Gallerano, E. Giovenale, S. Letarde, G. Messina, Nucl. Instr. and Meth. A 393 (1997) II–69.
[22] Y.H. Liu, T.C. Marshall, Phys. Rev. E 56 (1997) 2161.
[23] A. Abramovich, A. Arensburg, D. Chairman, A. Eichenbaum, M. Draiznin, A. Gover, H. Kleinman, I. Merhasin, Y. Pinhasi, J.S. Sokolowski, I.M. Yakover, Nucl. Instr. and Meth. A 393 (1997) II–41.
[24] A. Abramowich, Y. Pinhasi, M. Arbel, L. Gilutin, H. Kleinman, A. Eichenbaum, I.M. Yakover, A. Gover, Nucl. Instr. and Meth. A 407 (1998) 87.
[25] K. Takayama, J. Kishiro .Oza, T. Ozaki, S. Hiramatsu, H. Katoh, J. Appl. Phys. 77 (1995) 5467.
[26] K. Saito, K. Takeyama, T. Ozaki, J. Kishiro, K. Ebihara, S. Hiramatsu, Nucl. Instr. and Meth. A 375 (1996) 237.
[27] N.S. Ginzburg, personal communication.
[28] N.Yu. Peskov, personal communication.

ELSEVIER

Nuclear Instruments and Methods in Physics Research A 429 (1999) 37–40

NUCLEAR INSTRUMENTS & METHODS IN PHYSICS RESEARCH Section A

www.elsevier.nl/locate/nima

Short wavelength free electron lasers in 1998

W.B. Colson*

Physics Department, Naval Postgraduate School, Monterey, CA 93943, USA

Abstract

A table is presented that describes the characteristics of 58 short wavelength free electron lasers, operating and proposed, around the world. 1999 Published by Elsevier Science B.V.

Table 1 lists existing and proposed relativistic free electron lasers (FELs) in 1998. The top part of the table lists existing FELs. These are substantially complete experiments that may not be operating at the present time. The bottom part of the table lists proposed FELs. Each FEL, existing or proposed, is identified by a location, or institution, followed by the FEL's name in parentheses. The table can be found at `http://www.physics.nps.navy.mil/fel.html`. Additions and corrections can be transmitted to us for inclusion on the table in the future.

The first column of the table lists the operating wavelength λ, or wavelength range, in micrometers (μm). The large range of operating wavelengths, six orders of magnitude, indicates the flexible design characteristics of the FEL mechanism. In the second column, σ_z is the electron pulse length divided by the speed of light c, and ranges from CW to short sub-picosecond pulse time scales. The expected optical pulse length can be 3–5 times shorter or longer than the electron pulse depending the optical cavity Q, the FEL desynchronism, and the FEL gain. Most FEL oscillators produce an optical spectrum that is Fourier transform limited by the optical pulse length.

The electron beam energy E and peak current I provided by the accelerator are listed in the third and fourth columns in units of MeV and A. The accelerator type is listed as the first entry in the last column with a code such as RF for the radio-frequency linac. While there are a variety of accelerators used, most are RF with some electron storage rings, microtrons, and electrostatic accelerators. Storage rings tend to be used for the short wavelength applications, while the electrostatic accelerators provide longer wavelengths.

The next three columns list the number of undulator periods N, the undulator wavelength λ_0 in centimeters, and the undulator parameter $K = e\bar{B}\lambda_0/2\pi mc^2$ where e is the electron charge magnitude, $\bar{B}$ is the rms undulator field strength, and m is the electron mass. For an FEL klystron undulator, there are multiple undulator sections as listed in the N-column. Note that the range of values for N,λ_0, and K are much smaller than for the other parameters indicating that most undulators are similar. Only a few of the FELs use the klystron undulator at present, and the rest use

* Tel.: + 1-408-656-2765; fax: + 1-408-656-2834.
E-mail address: colson@physics.nps.navy.mil (W.B. Colson)

0168-9002/99/$ - see front matter 1999 Published by Elsevier Science B.V.
PII: S0168-9002(99)00063-7

Table 1
Relativistic short wavelength free electron lasers (1998)

FELs	λ (μm)	σ_z	E (MeV)	I (A)	N	λ_0 (cm)	K (rms)	Acc., type [Ref.]
EXISTING								
UCSB (mm FEL)	340	25 μs	6	2	42	7.1	0.7	EA,O [1]
Florida (CREOL)	355	8 μs	1.3	0.13	185	0.8	0.1	EA,O [25]
Dartmouth (FEL)	200	CW	0.04	0.001	50	300	–	SP,O [2]
Himeji (LEENA)	65–75	10 ps	5.4	10	50	1.6	0.5	RF,O [3,4]
UCSB (FIR FEL)	60	25 μs	6	2	150	2	0.1	EA,O [1]
Osaka (ILE/ILT)	47	3 ps	8	50	50	2	0.5	RF,O [5]
Tokyo (UT-FEL)	43	10 ps	13	20	40	4	0.7	RF,O [6]
Nieuwegein (FELIX)	4–200	1 ps	50	50	38	6.5	1.8	RF,O [7]
Osaka (ISIR)	40	30 ps	17	50	32	6	1	RF,O [8]
Bruyeres (ELSA)	20	30 ps	18	100	30	3	0.8	RF,O [9]
Osaka (FELI4)	18–40	10 ps	33	40	30	8	1.3–1.7	RF,O [52]
Stanford (FIREFLY)	15–65	1–5 ps	15–32	14	25	6	1	RF,O [10]
UCLA-Kurchatov	16	3 ps	13.5	80	40	1.5	1	RF,A [33]
Frascati (LISA)	15	7 ps	25	5	50	4.4	1	RF,O [11]
UCLA-Kurchatov-LANL	12	5 ps	18	170	100	2	0.7	RF,A [53]
AES (CIRFEL)	12–21	5 ps	9–14	100	73	1.4	0.2	RF,O [12]
Beijing (IHEP)	10	4 ps	30	14	50	3	1	RF,O [13]
Orsay (CLIO)	3–53	0.1–3 ps	21–50	80	38	5	1.4	RF,O [14]
LANL (RAFEL)	15.5	15 ps	17	300	200	2	0.9	RF,O [15]
Osaka (FELI1)	5.5	10 ps	33.2	42	58	3.4	1	RF,O [16]
Darmstadt (IR-FEL)	5	2 ps	40	2.7	80	3.2	1	RF,O [17]
TJNAF (FEL)	5	0.4 ps	38	60	40	2.7	0.96	RF,O [56]
Stanford (SCAFEL)	3–13	0.5–12 ps	22–45	10	72	3.1	0.8	RF,O [18]
Vanderbilt (FELI)	2.0–9.8	0.7 ps	43	50	52	2.3	1	RF,O [19]
Duke (MarkIII)	3	3 ps	44	20	47	2.3	1	RF,O [20]
Osaka (FELI)	2–6	2 ps	170	100	50	6	1.3	RF,O [5]
Osaka (FELI2)	1.88	10 ps	68	42	78	3.8	1	RF,O [16]
BNL (ATF)	0.5	6 ps	50	100	70	0.88	0.4	RF,O [21]
Tsukuba (NIJI-IV)	0.35	160 ps	300	5	2 × 42	7.2	2	SR,O [22]
Orsay (Super-ACO)	0.3–0.6	15 ps	800	0.1	2 × 10	13	4.5	SR,O [23]
Osaka (FELI3)	0.3–0.7	5 ps	155	60	67	4	0.7–1.4	RF,O [54]
Okazaki (UVSOR)	0.24	6 ps	500	5	2 × 8	11	2	SR,O [24]
Duke (OK-4)	0.2–0.4	2.5 ps	500	12	2 × 33	10	2.4	SR,O [42]
PROPOSED								
Netherlands (TEUFEL)	180	20 ps	6	350	50	2.5	1	RF,O [26]
Rutgers (IRFEL)	140	25 ps	38	1.4	50	20	1	MA,O [27]
Moscow (Lebedev)	100	20 ps	30	0.25	35	3.2	0.8	MA,O [28]
Osaka (ILE/ILT)	95	3 ps	9	50	10	6	3	RF,O [49]
Tokai (SCARLET)	24–28	10 ps	20	10	52	3.3	0.7	RF,O [5]
LBL (IRFEL)	3–50	30 ps	55	60	40	5	1	RF,O [29]
TJNAF (IRFEL)	2.5–25	1.5 ps	200	36	2 × 12	20	4	RF,O [30]
Boeing (kWFEL)	0.2–4	9 ps	120	300	220	2.18	1.31	RF,O [31]
Osaka (ILE/ILT)	12	3 ps	9	50	30	0.79	0.22	RF,O [50]
Stanford (FEL)	10	4 ps	24	25	52	2.6	0.9	RF,O [32]
Novosibirsk (RTM)	2–11	20 ps	98	100	4 × 36	9	1.6	RF,O [34,35]
BNL (HGHG)	3.4	10 ps	30	110	83	1.8	1.4	RF,A [36]
TJNAF (UVFEL)	0.16–1	0.2 ps	200	270	72	3.3	1.3	RF,O [30]
Rocketdyne (FEL)	0.84	3 ps	90	500	160	2.4	1.4	RF,MOPA [37]
Dortmund (DELTA)	0.4	50 ps	500	90	17	25	2	SR,O [38,39]
Harima (SUBARU)	0.2–10	26 ps	1500	50	33,65	16,32	8	SR,O [45]

(Cont.).

Table 1 *Continued*

FELs	λ (μm)	σ_z	E (MeV)	I (A)	N	λ_0 (cm)	K (rms)	Acc., type [Ref.]
Italy (ELETTRA)	0.2–0.3	30 ps	1000	230	42	10	3	SR,O [55]
BNL (DUVFEL)	0.1	6 ps	230	1000	256	2.89	1.2	RF,A [40]
ANL (APSFEL)	0.12	1 ps	440	150	12 × 72	3.3	3.1	RF,A [51]
Frascati (COSA)	0.08	10 ps	215	200	400	1.416	1	RF,O [41]
DESY (TTF1)	0.042	0.8 ps	390	500	490	2.73	0.9	RF,A [48]
Duke (VUV)	0.05–2	10 ps	1000	350	2 × 33	10	2	SR,O [46]
DESY (TTF2)	0.006	0.17 ps	1000	2500	981	2.73	0.9	RF,A [43]
SLAC (LCLS)	0.00015	0.07 ps	14 350	3400	3328	3	3.7	RF,A [44]
DESY (TESLA)	0.0001	0.08 ps	35 000	5000	1200	5	4.2	RF,A [47]

RF – RF Linac Accelerator; MA – Microtron Accelerator; SR – Electron Storage Ring; EA – Electrostatic Accelerator; A - FEL Amplifier; O – FEL Oscillator; MOPA – Master-Oscillator Power-Amplifier; SP – Smith–Purcell Oscillator.

the conventional periodic undulator. The FEL resonance condition,

$$\lambda = \frac{\lambda_0(1 + K^2)}{2\gamma^2}$$

where γ is the relativistic Lorentz factor $\gamma = E/mc^2$, provides a relationship that can be used to derive K from λ, E, and λ_0. The middle entry of the last column lists the FEL type: "O" for oscillator, "A" for amplifier, etc. Most of the FELs are oscillators, but recent interest in short wavelength FELs has produced several proposals for amplifiers that avoid the need for mirrors. A reference describing the FEL is provided at the end of each line entry.

For the conventional undulator, the peak optical power can be estimated by the fraction of the electron beam peak power that spans the undulator spectral bandwidth, 1/4 N, or $P \approx EI/4eN$. For the FEL using a storage ring, the optical power causing saturation is substantially less than this estimate and depends on ring properties. For the high-gain FEL amplifier, the optical power at saturation can be substantially more. The average FEL power is determined by the duty cycle, or spacing between electron pulses, and is generally many orders of magnitude lower than the peak power.

In the FEL oscillator, the optical mode that best couples to the electron beam in an undulator of length $L = N\lambda_0$ has Rayleigh length $z_0 \approx L/\sqrt{12}$ and has a mode waist radius of $w_0 \approx \sqrt{N\gamma\lambda/\pi}$. The FEL optical mode typically has more than 90% of the power in the fundamental mode described by these parameters.

Acknowledgements

The author is grateful for support of this work by the Naval Postgraduate School and the Directed Energy Weapons Group, SPAWAR, U.S. Navy.

References

[1] G. Ramian, Nucl. Instr. and Meth. A 318 (1992) 225.
[2] J. Urata et al., Phys. Rev. Lett. 80 (1998) 516.
[3] T. Mochizuki et al., Nucl. Instr. and Meth. A 393 (1997) II-47.
[4] S. Miyamoto, T. Mochizuki, J. Japan Soc. Infrared Science and Technology 7 (1997) 73.
[5] M. Asakawa et al., Proceedings of the Fifteenth Annual Meeting of the Laser Society of Japan, Osaka, Japan, January 19–20, 1995.
[6] E. Nishimura et al., Nucl. Instr. and Meth. A 341 (1994) 39.
[7] D. Oepts et al., Infrared Phys. Technol. 36 (1995) 297.
[8] S. Okuda et al., Nucl. Instr. and Meth. A 341 (1994) 59.
[9] P. Guimbal et al., Nucl. Instr. and Meth. A 341 (1994) 43.
[10] K.W. Berryman, T.I. Smith, Nucl. Instr. and Meth. A 375 (1996) 6.
[11] M. Castellano et al., Nucl. Instr. and Meth. A 304 (1991) 204.
[12] I.S. Lehrman et al., Nucl. Instr. and Meth. A 393 (1997) 178.
[13] J. Xie et al., Nucl. Instr. and Meth. A 341 (1994) 34.

[14] J.M. Ortega et al., Nucl. Instr. and Meth. A 375 (1996) 618.
[15] R.L. Sheffield et al., Proc. SPIE 2988 (1997) 28.
[16] A. Kobayashi et al., Nucl. Instr. and Meth. A 375 (1996) 317.
[17] J. Auerhammer et al., Nucl. Instr. and Meth. A 341 (1994) 63.
[18] H.A. Schwettman et al., Nucl. Instr. and Meth. A 375 (1996) 662.
[19] C. Brau, Nucl. Instr. and Meth. A 318 (1992) 38.
[20] S.V. Benson et al., Nucl. Instr. and Meth. A 250 (1986) 39.
[21] K. Batchelor et al., Nucl. Instr. and Meth. A 318 (1992) 159.
[22] T. Yamazaki et al., Nucl. Instr. and Meth. A 341 (1994) ABS3.
[23] M.E. Couprie et al., Nucl. Instr. and Meth. A 407 (1998) 215.
[24] H. Hama et al., Nucl. Instr. and Meth. A 341 (1994) 12.
[25] M. Tecimer et al., Nucl. Instr. and Meth. A 341 (1994) A126.
[26] J.I.M. Botman et al., Nucl. Instr. and Meth. A 341 (1994) 402.
[27] E.D. Shaw et al., Nucl. Instr. and Meth. A 318 (1992) 47.
[28] K.A. Belovintsev et al., Nucl. Instr. and Meth. A 341 (1994) ABS45.
[29] K.J. Kim et al., Nucl. Instr. and Meth. A 341 (1994) 280.
[30] S. Benson et al., Nucl. Instr. and Meth. A 429 (1999) 27.
[31] J.L. Adamski et al., Proceedings of the SPIE: LASE'97, High-Power Lasers and Applications, San Jose, CA, February 13, 1997.
[32] J.F. Schmerge, R. Pantell, Nucl. Instr. and Meth. A 341 (1994) 335.
[33] M. Hogan et al., Phys. Rev. Lett. A 80 (1998) 289.
[34] N.G. Gavrilov et al., SPIE Proc. 2988 (1997) 23.
[35] N.A. Vinokurov et al., Nucl. Instr. and Meth. A 331 (1993) 3.
[36] I. Ben-Zvi et al., Nucl. Instr. and Meth. A 318 (1992) 208.
[37] R.J. Burke et al., Proceedings of SPIE: Laser Power Beaming, Los Angeles, vol. 2121, January 27–28, 1994.
[38] D. Nolle et al., Nucl. Instr. and Meth. A 341 (1994) ABS7.
[39] Schmidt et al., Nucl. Instr. and Meth. A 341 (1994) ABS9.
[40] E.D. Johnson, Nucl. Instr. and Meth. A 393 (1997) II-12.
[41] F. Ciocci et al., A. Torre, IEEE J. Quantum Electron. 31 (1995) 1242.
[42] V.N. Litvinenko et al., Nucl. Instr. and Meth. A 407 (1998) 8.
[43] W. Brefeld et al., Nucl. Instr. and Meth. A 375 (1996) 295.
[44] M. Cornacchia, Proc. SPIE 2998 (1997) 2.
[45] S. Miyamoto et al., Report of the Spring-Eight International Workshop on 30 m Long Straight Sections, Kobe, Japan, August 9, 1997.
[46] V.N. Litvinenko et al., Nucl. Instr. and Meth. A 358 (1995) 369.
[47] R. Brinkmann et al., Nucl. Instr. and Meth. A 393 (1997) 86.
[48] W. Brefeld et al., Nucl. Instr. and Meth. A 393 (1997) 119.
[49] M. Asakawa et al., Nucl. Instr. and Meth. A 375 (1996) 416.
[50] N. Ohigashi et al., Nucl. Instr. and Meth. A 375 (1996) 469.
[51] S.V. Milton et al., Nucl. Instr. and Meth. A 407 (1998) 210.
[52] T. Takii et al., Nucl. Instr. and Meth. A 407 (1998) 21.
[53] M. Hogan et al., Phys. Rev. Lett., 80 (1998) 4867.
[54] T. Tomimasu et al., Nucl. Instr. and Meth. A 393 (1997) 188.
[55] R.P. Walker et al., Nucl. Instr. and Meth. A 429 (1999) 179.
[56] G. Neil et al., Proceedings 20th FEL Conference, Williamsburg, VA, 1998 Elsevier Science, Amsterdam 1999; p. II. 29.

Nuclear Instruments and Methods in Physics Research A 429 (1999) 41–45

NUCLEAR INSTRUMENTS & METHODS IN PHYSICS RESEARCH Section A

www.elsevier.nl/locate/nima

Growth of the energy spread due to the radiative interaction in a short electron bunch moving in an undulator

E.L. Saldin[a], E.A. Schneidmiller[a], M.V. Yurkov[b,*]

[a]*Automatic Systems Corporation, 443050 Samara, Russia*

[b]*Joint Institute for Nuclear Research, Particle Physics Laboratory (LSVE), Dubna 141980, Moscow Region, Russia*

Abstract

This paper presents investigations of the longitudinal radiative force in an electron bunch moving in an undulator (wiggler). An analytical solution is obtained for a Gaussian longitudinal bunch profile. Radiative interaction of the particles in an intense microbunch induces a correlated energy spread in the electron beam. Numerical estimates presented in this paper show that this effect can be important for free electron lasers.

PACS: 41.60; 41.85.L; 52.75.Ms

Keywords: Radiative interaction; undulator; Wiggler

1. Introduction

The theory of radiative interactions of the electrons in an intense microbunch traversing a curved trajectory is intensely studied nowadays. This is motivated by the practical importance of radiative effects for beam dynamics in linear colliders and short-wavelength free electron lasers. When an intense electron bunch passes bending magnets, bunch compressors, wigglers, etc., the radiative interaction induces an energy spread in the electron beam and can lead to transverse emittance dilution in dispersive regions.

In this paper we present the results of analytic calculation of the radiative interaction of particles in a bunch with a line-charge distribution moving in an undulator (a wiggler) [1]. Transient effects (when the bunch enters and leaves the undulator) are not considered. Shielding effects the (influence of a vacuum chamber on the radiative process) are neglected as well. Practical application of the results obtained is illustrated with numerical examples for proposed SASE FELs. Also, benchmarking of the numerical simulation codes would be difficult without rigorous analytical results.

2. General solution for a bunch moving in an undulator

Let us consider an electron bunch with a linear density of particles $\lambda(s)$ moving along the z direction in an undulator with magnetic field

$$H_x = H_w \cos(k_w z).$$

* Corresponding author. Tel. + 7-09621-62154; fax: + 7-09621-65767.

E-mail address: yurkov@sunse.jinr.ru (M.V. Yurkov)

0168-9002/99/$ - see front matter
PII: S0168-9002(99)00064-9

The transverse and the longitudinal velocities of an electron can be approximated by

$$\beta_y = \frac{K}{\gamma}\sin(k_w z), \quad \beta_z = \beta - \frac{K^2}{2\gamma^2}\sin^2(k_w z)$$

where $K = eH_w/k_w mc^2$ is the undulator parameter, $\gamma = \varepsilon/mc^2$ is the relativistic factor, $(1 + K^2/2)/\gamma^2 \ll 1$ and $\beta \simeq 1 - \frac{1}{2}\gamma^2$. The transverse coordinate of the electron oscillates as

$$y = -\frac{K}{\gamma k_w}\cos(k_w z).$$

The bunch length is assumed to be much shorter than the undulator period $2\pi/k_w$. We calculate the longitudinal (along the particle's velocity) radiative force assuming the motion of the particles to be given. Only the interaction connected with the curvature is considered and we omit the trivial longitudinal space-charge forces in a bunch moving on a straight line, since they can be calculated separately. The calculations are based on the general algorithm for an arbitrary small-angle trajectory described in Ref. [1]. Leaving out the details of the calculations, we present the final result for the rate of the energy change as a function of the positions of the electron in the bunch and in the undulator, s and z, respectively [1]:

$$\frac{d\varepsilon}{c\,dt} = e^2 k_w \int_{-\infty}^{s} ds' D(\hat{s} - \hat{s}', K, \hat{z}) \frac{d\lambda(s')}{ds'} \tag{1}$$

where $\hat{s} = \gamma^2 k_w s$ and $\hat{z} = k_w z$,

$$D(\hat{s},K,\hat{z}) = \frac{1}{\hat{s}} - 2\frac{\Delta - K^2 B(\Delta,\hat{z})}{\Delta^2 + K^2 B^2(\Delta,\hat{z})}[\sin\Delta\cos\hat{z} + (1 - \cos\Delta)\sin\hat{z}] \tag{2}$$

$$B(\Delta,\hat{z}) = (1 - \cos\Delta - \Delta\sin\Delta)\cos\hat{z} + (\Delta\cos\Delta - \sin\Delta)\sin\hat{z} \tag{3}$$

and Δ is the solution of the transcendental equation:

$$\hat{s} = \frac{\Delta}{2}\left(1 + \frac{K^2}{2}\right) + \frac{K^2}{4\Delta}\{[2(1 - \cos\Delta) - \Delta\sin\Delta] \times (\cos\Delta\cos 2\hat{z} + \sin\Delta\sin 2\hat{z}) - 2(1 - \cos\Delta)\}. \tag{4}$$

It follows from the geometry of the problem and from Eqs. (2)–(4) that the function D is periodic in z with the period equal to half of the undulator period π/k_w (or, to π in terms of the normalized position $\hat{z}$). In the following, we will study the rate of the energy change averaged over the z coordinate:

$$\frac{d\bar{\varepsilon}}{c\,dt} = e^2 k_w \int_{-\infty}^{s} ds' \bar{D}(\hat{s} - \hat{s}', K) \frac{d\lambda(s')}{ds'} \tag{5}$$

where

$$\bar{D}(\hat{s},K) = \frac{1}{\pi}\int_0^{\pi} d\hat{z} D(\hat{s},K,\hat{z}). \tag{6}$$

In the case of small values of the undulator parameter, $K \ll 1$, function $\bar{D}(\hat{s},K)$ takes the simple form:

$$\bar{D}(\hat{s},K) = -K^2\left(\frac{\sin^2\hat{s}}{\hat{s}} + \frac{\sin 2\hat{s}}{2\hat{s}^2} - \frac{\sin^2\hat{s}}{\hat{s}^3}\right). \tag{7}$$

3. Averaged solution for a Gaussian bunch

In this section we consider a bunch with a Gaussian distribution of linear density:

$$\lambda(s) = \frac{N}{\sqrt{2\pi}\sigma}\exp\left[-\frac{s^2}{2\sigma^2}\right]. \tag{8}$$

The averaged solution for the Gaussian bunch can be written in the form:

$$\frac{d\bar{\varepsilon}}{c\,dt} = \frac{e^2 N K^2}{\sqrt{2\pi}\sigma^2\gamma^2}\bar{G}(p,K,x) \tag{9}$$

where $x = s/\sigma$ and p is the bunch length parameter:

$$p = \frac{\gamma^2 k_w \sigma}{1 + K^2/2}.$$

In the general case, the function $\bar{G}$ should be calculated by means of numerical integration of Eq. (5). Nevertheless, in some parameter regions it can be expressed analytically. Let us study the practically important case of a long bunch, $p \gg 1$. First, we consider the case of small K. Under these conditions the function $\bar{G}$ can be calculated analytically using Eqs. (5) and (7)

$$\bar{G}(p,x) = \frac{x}{2}\exp\left(-\frac{x^2}{2}\right)\ln p + F(x). \qquad (10)$$

Here the parameter p is reduced to $p \simeq \gamma^2 k_w \sigma$, and function $F(x)$ has the form:

$$F(x) = \frac{1}{4}(C + 3\ln 2 - 2)x\exp\left(-\frac{x^2}{2}\right) - \sqrt{\frac{\pi}{8}}\left[1 + \mathrm{erf}\left(\frac{x}{\sqrt{2}}\right) - x\exp\left(-\frac{x^2}{2}\right) \times \int_0^x \mathrm{d}x' \exp\left(\frac{(x')^2}{2}\right)\left(1 + \mathrm{erf}\left(\frac{x'}{\sqrt{2}}\right)\right)\right] \qquad (11)$$

where $C = 0.577\ldots$ is Euler's constant and erf(…) is the error function [2]. The plot of the function $F(x)$ is presented in Fig. 1. Fig. 2 presents plots of the function $\bar{G}$ calculated for different values of parameter p.

In the case of an arbitrary value of the undulator parameter K it is difficult to find an explicit analytical solution. Nevertheless, using the results of numerical integration of Eq. (5), we can write function $\bar{G}$ in the following form ($p \gg 1$):

$$\bar{G}(p,K,x) = \frac{x}{2}\exp\left(-\frac{x^2}{2}\right)[\ln p + g(K)] + F(x) \qquad (12)$$

where function $g(K)$ changes from 0 to 1 when K changes from small to large values. The plot of this function is presented in Fig. 3.

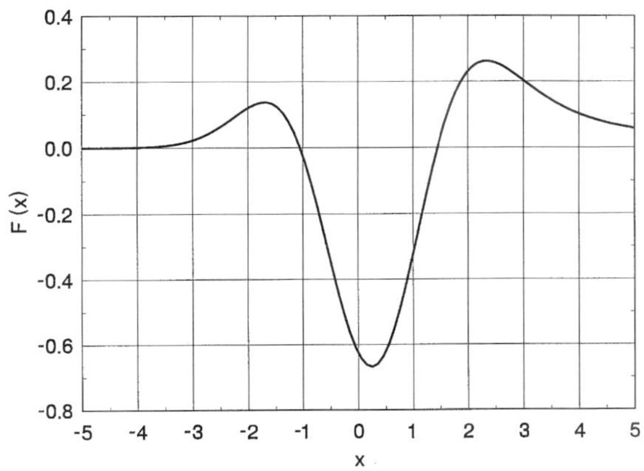

Fig. 1. Function $F(x)$ given by Eq. (11).

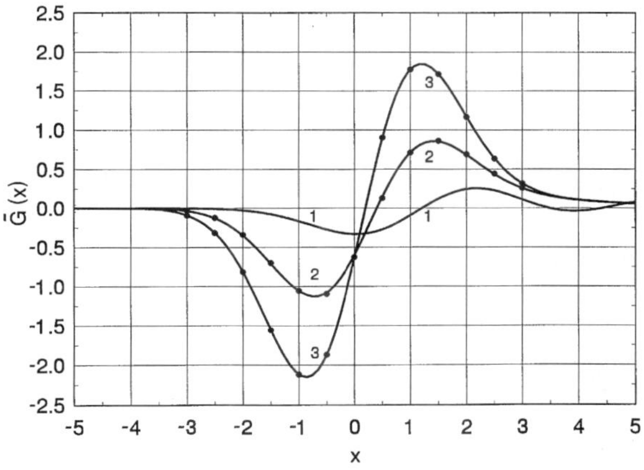

Fig. 2. Function $\bar{G}$ for small value of the undulator parameter K and different values of the bunch length parameter p. Curve (1): $p = 1$, curve (2): $p = 30$, and curve (3): $p = 1000$. The curves are the results of numerical integration of Eq. (5) and the circles are calculated with the help of analytical formula (10) for large values of parameter p.

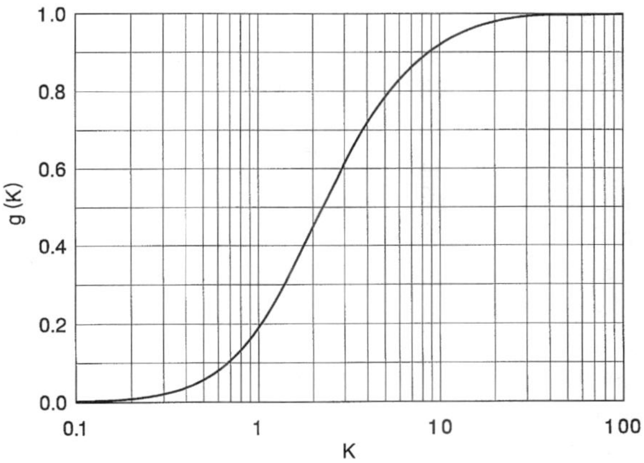

Fig. 3. Function $g(K)$ entering Eq. (12).

In the conclusion of this section we present the formula for the induced correlated energy spread in the Gaussian bunch due to radiative interaction. Using expression (12) we write down this formula in the form convenient for practical calculations ($p \gg 1$):

$$\sigma_\gamma = 0.219\frac{IK^2L_w}{I_A\sigma\gamma^2}\sqrt{[\ln p + g(K)]^2 - 0.933[\ln p + g(K)] + 1.54} \qquad (13)$$

where $I = \mathrm{Nec}/\sqrt{2\pi}\sigma$ is the peak current, $I_A = 17$ kA is the Alfven current, L_w is the undulator length and

$$mc^2\sigma_\gamma = \sqrt{\langle\bar{\varepsilon}^2\rangle - \langle\bar{\varepsilon}\rangle^2}.$$

4. Discussion

Let us perform estimations of the applicability region of the results obtained for the practically important case of a long bunch, $\sigma \gg (1 + K^2/2)/\gamma^2 k_w$. First, we consider the transient effects. When the behaviour of the radiative forces after the bunch leaves the undulator is not important from a practical point of view (which is true for FELs), only the entrance transient effect is of interest. In this case we can write down the following limitation on the undulator (wiggler) length L_w, allowing us to neglect transient effects:

$$L_w \gg \sigma\gamma_z^2 \qquad (14)$$

where $\sigma\gamma_z^2 = \sigma\gamma^2/(1 + K^2/2)$ is the typical formation length of the radiation.

Second, we estimate the region of parameters where we can neglect the influence of the bunch transverse size and of the vacuum pipe. It follows from simple geometrical consideration that a characteristic measure distinguishing these effects is the mean geometric value of the bunch length and the formation length of the radiation. Thus, we can roughly estimate the region where the considered effects can be neglected

$$\sigma_\perp \ll \sigma\gamma_z \ll b. \qquad (15)$$

Here $\sigma_\perp$ and b are transverse dimensions of the bunch and of the vacuum chamber, respectively.

When the above-mentioned limitations are not satisfied, the effects considered become important leading to suppression of the radiative interaction. In other words, the model considered in this paper describes the worst-case approximation, which makes it useful for quick estimations of the radiative interaction effects. Also, analytical results presented in the paper can serve as primary standards for testing numerical simulation codes.

In conclusion, let us illustrate practical application of the results obtained with two numerical examples. The first one is the 6 nm SASE FEL under construction at the TESLA test facility at DESY [3]. Parameters of the project are: an energy of 1 GeV, an rms bunch length of 50 μm, z peak current of 2.5 kA, an undulator period of 2.73 cm, K is 1.27 and the undulator length is 27 m. Substituting these values into formula (13), we find that the induced correlated energy spread is equal to $\sigma_\gamma/\gamma = 4 \times 10^{-5}$, which is negligible. Besides, conditions (14) and(15) for the shielding are not satisfied, since the diameter of the vacuum chamber is equal to 1 cm. This will lead to further reduction of the effect. It should be noticed that such a situation is typical for VUV and X-ray FEL projects.

The second example is the proposal by the Duke University [4] to construct a 1.4 μm SASE FEL using the PALADIN wiggler. The energy is 200 MeV, the rms bunch length is 50 μm, the bunch radius is 0.125 cm, the peak current is 2.5 kA, the wiggler period is 8 cm, K is 3 and the wiggler length is 15 m. Assuming the size of the vacuum chamber to be about 2 cm, we find that conditions (14) and (15) are met and that our simple model provides a correct estimation of the effect. According to formula (13), we obtain $\sigma_\gamma/\gamma = 8.4 \times 10^{-3}$. Effective operation of the free electron laser requires $\sigma_\gamma/\gamma \ll \rho$, where ρ is the FEL parameter [5]. For the Duke SASE FEL $\rho = 4.8 \times 10^{-3}$, and the latter condition is strongly violated. Besides, the SASE coherence length [6] is of the order of the bunch length. Thus, the large energy spread will be induced within the coherence length and the FEL process will be destroyed.

Acknowledgements

We wish to thank R. Brinkmann, C. Bohn, Ya. Derbenev, M. Dohlus, P. Emma, D. Jaroszynski, J. Krzywinski, R. Li, T. Limberg, J. Rossbach and V. Shiltsev for useful discussions on the radiative interaction effects.

References

[1] E.L. Saldin, E.A. Schneidmiller, M.V. Yurkov, Radiative interaction of electrons in a bunch moving in an undulator, DESY Print, TESLA-FEL 97-08, Hamburg, 1997.

[2] M. Abramowitz, I.A. Stegun, Handbook of Mathematical Functions, National Bureau of Standards, 1964.
[3] J. Rossbach, Nucl. Instr. and Meth. A 375 (1996) 269.
[4] P.G. O'Shea, C.P. Neumann, J.M.J. Madey, H.P. Freund, Nucl. Instr. and Meth. A 393 (1997) 129.
[5] R. Bonifacio, C. Pellegrini, L. Narducci, Opt. Commun. 50 (1984) 373.
[6] E.L. Saldin, E.A. Schneidmiller, M.V. Yurkov, Opt. Commun. 148 (1998) 383.

ELSEVIER

Nuclear Instruments and Methods in Physics Research A 429 (1999) 46–51

NUCLEAR INSTRUMENTS & METHODS IN PHYSICS RESEARCH Section A

www.elsevier.nl/locate/nima

Temporal dynamics of fusion-FEM oscillations: comparison of experiment and simulations

A.V. Savilov[a,*], V.L. Bratman[a], G.G. Denisov[a], W.A. Bongers[b], C.A.J. van der Geer[b], P. Manintveld[b], A.G.A. Verhoeven[b], W.H. Urbanus[b]

[a]*Institute of Applied Physics, Russian Academy of Sciences, 46 Ulyanov Street, Nizhny Novgorod 603600, Russia*

[b]*FOM Institut for Plasmafysica "Rijnhuizen", Association EURATOM-FOM P.O. Box 1207, 3430 BE Nieuwegein, The Netherlands*

Abstract

At the first stage of the Fusion-FEM experiment, a high-power electrostatic free-electron laser operating in the mm-wave regime was tested without the energy recovery system. In this situation, a slow temporal drop of the acceleration voltage complicated significantly the dynamics of the FEM operation. In this paper, experimental results are compared to simulations. It is demonstrated that the decrease of the frequency of the output radiation, caused by the voltage drop, has a stepped character; the value of this step is as large as the width of the amplification band.

Keywords: Fusion-FEM oscillators; Electron-beam pulse; Electron-wave resonance

1. Introduction

A free-electron maser (FEM) is under development at the FOM-Institute for Plasma Physics (Netherlands) for ECW applications on future fusion research devices such as ITER [1–4]. It is designed to generate coherent RF radiation with the power 1 MW in the frequency range 130–260 GHz with the fast frequency tuning over 10%. The needed level of the power is generated by using a step-profiled undulator, which results in an extraction efficiency of over 5% within the entire frequency range. According to the design, the electron-energy recovery system should provide a total system efficiency of over 60%.

The microwave system of the FEM is a relatively low-Q cavity, consisting of a rectangular operating waveguide with corrugated walls (needed to decrease ohmic losses), and of two newly-developed quasi-optical stepped reflectors [5,3] formed by segments of a wide waveguide, each terminated by conventional mirrors (Fig. 1).The operation of the reflectors is based on the effect of the multiplication of wave beams. For the input reflector, the reflection coefficient is nearly 100%, while the output reflector can be controlled within almost the entire range of 0–100%. In order to provide fast frequency tuning, the frequency band of the reflectors is correspondingly wide. This leads to a significant complication of the temporal dynamics of oscillations due to interaction of many longitudinal modes.

*Corresponding author. Tel.: + 7-8312-384575; fax: + 7-8312-362061.

E-mail address: savilov@appl.sci-nnov.ru (A.V. Savilov)

0168-9002/99/$ - see front matter
PII: S0168-9002(99)00065-0

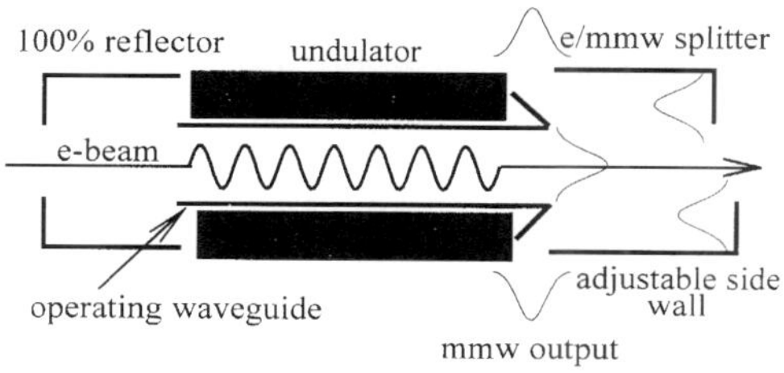

Fig. 1. Schematic of the Fusion-FEM microwave system.

For the planned regimes of the Fusion-FEM with energy recovery and a long electron-beam pulse, the dynamics of the mode interaction was preliminarily studied in detail [6–8]. This work is devoted to the first stage of the experiment, when the FEM was tested without recovery system (inverse set-up) [4]. Due to this fact, in this experiment there was a drop of the accelerating voltage in time. Evidently, this effect significantly complicates the process of the mode interaction and, simultaneously, makes it significantly more interesting. During the electron-beam pulse, the frequency of the electron-wave resonance decreases and, therefore, the amplification band shifts to lower frequencies. Thus, different modes are in the resonance with electrons in different temporal intervals of the electron-beam pulse. As it is observed in the experiment and predicted by simulations, the dynamics of the FEM significantly depends on the initial value of the voltage (the voltage at the beginning of the electron pulse). At some initial voltages, the voltage drop leads to the decrease of the frequency of the output radiation. It is important to note that though the voltage drops slowly (typically by 6–8 kV/μs), the change in frequency has a stepped character.

2. Equations of maser excitation

In a FEM-oscillator, a large number of longitudinal modes (of the order of the square of the electron Lorentz factor, γ^2) usually exists in the amplification band. At the initial (small-signal) stage of the excitation all these modes arise from noise, and afterwards, at the nonlinear stage, they take part in the complicated nonstationary process of the interaction. For a low-Q FEM-oscillator, this process can be simulated within the framework of the spatio-temporal approach [9–11]. The simplest 1D set of equations includes the equations for the complex amplitudes of forward and backward waves,

$$\frac{\partial a_{\rightarrow}}{\partial \zeta} + \frac{c}{v_{\rm gr}}\frac{\partial a_{\rightarrow}}{\partial \tau} = 2{\rm i}\langle {\rm e}^{-{\rm i}\theta}\rangle_{\theta_0}$$

$$\frac{\partial a_{\leftarrow}}{\partial \zeta} - \frac{c}{v_{\rm gr}}\frac{\partial a_{\leftarrow}}{\partial \tau} = 0 \quad (1)$$

and the equations of the electron coupling with the forward wave,

$$\frac{\partial u}{\partial \zeta} + \frac{c}{v_{\rm e}}\frac{\partial u}{\partial \tau} = {\rm Im}(a_{\rightarrow} + q^2\rho){\rm e}^{{\rm i}\theta},$$

$$\frac{\partial \theta}{\partial \zeta} + \frac{c}{v_{\rm e}}\frac{\partial \theta}{\partial \tau} = u - \delta(\zeta). \quad (2)$$

Here $u = \mu C^{-1}(1 - \mathscr{E}/\mathscr{E}_0)$ is the normalized energy loss of a particle, $\zeta = h_0 C_z$ is the normalized coordinate, q is the space charge parameter, $\theta = \omega_0 t - h_0 z$ is the electron phase respect to the resonant combination wave, $\tau = ch_0 Ct$ is the normalized time, $\omega_0 = v_{\rm e} h_0(\omega_0)$ is the resonance frequency, $h = k_{\parallel} + h_{\rm u}$ is the combination wavenumber, δ is the mismatch of the electron-wave combination resonance, $v_{\rm e}$ and $v_{\rm gr}$ are the unperturbed longitudinal electron velocity and the group velocity of the wave, c is the speed of light, $C = \sqrt[3]{eI\mu\kappa^2/mc^3 N\gamma_0}$ is the Pierce amplification parameter, $\mu = \gamma_0^{-2}$ and $\kappa = eA_u/mc^2\gamma_0$ are the parameter of the inertial bunching of particles and the coefficient of the electron-wave coupling. The initial conditions for the electron beam, being unmodulated at the input, are written in the following way

$$u(\zeta = 0, \tau) = u_0(\tau), \qquad \theta(\zeta = 0, \tau) = \theta_0 \quad (3)$$

where the initial phases are distributed homogeneously over the interval $[0, 2\pi)$. A slow drop of the accelerating voltage is described by the change of the initial electron energy with time, $u_0(\tau)$. In Eq. (2), the dependence of the mismatch on the coordinate, $\delta(\zeta)$, describes the tapering of the undulator field.

The simplest boundary condition for the RF field (not taking into account the feedback frequency dispersion) can be represented in the following form [11]

$$a_{\leftarrow}(L,\tau) = R_{\text{out}} a_{\rightarrow}(L,\tau - T),$$

$$a_{\rightarrow}(0,\tau) = R_{\text{inp}} a_{\leftarrow}(0,\tau - T) \tag{4}$$

where R_{inp} and R_{out} are the amplitude reflection coefficients of the input and the output reflectors, T is the reflection time, L is the normalized length of the interaction region. However, in the considered FEM the dependence of the reflection coefficients on the frequency is very important [6–8]. Within the framework of the spatio-temporal approach Eqs. (1)–(3), one can take into account the dispersive properties of the feedback by simulation of the motion of the wave beams through the reflectors. This results in the following modification of the boundary condition Eq. (4) for the output reflector [12]:

$$a_{\leftarrow}(L,\tau) = \sum_{\text{m,p}} R_{\text{m,p}} a_{\rightarrow}(L,\tau - T_{\text{m}} - T_{\text{p}}). \tag{5}$$

Here the coefficients $R_{\text{m,p}}$, T_{m} and T_{p} are defined by the transverse and longitudinal dimensions of the reflectors. A similar boundary condition takes place for the input reflector.

Thus, the set of equations Eqs. (1)–(3), and (5) describes the basic features of the nonstationary dynamics of the FEM with taking into account the most important effects: the dependence of the reflection coefficients on the frequency, the tapered undulator, RF space-charge field, and the voltage drop in time.

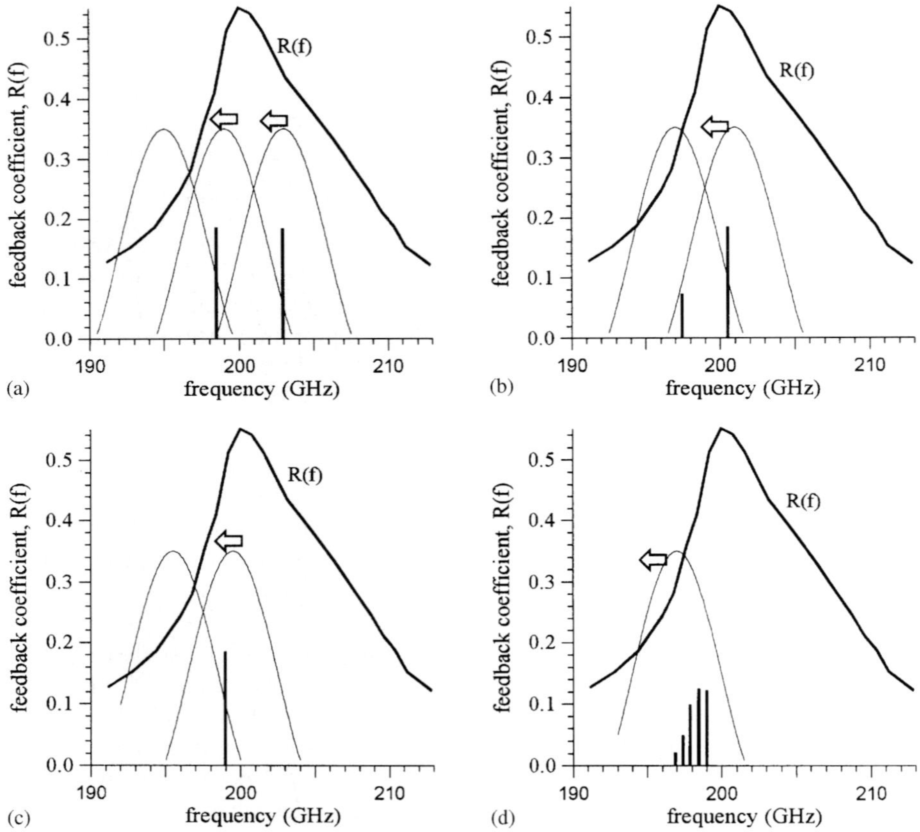

Fig. 2. Illustration of the mode excitation for different initial position of the amplification band with respect to the feedback frequency band, $R(f)$, (different initial voltages).

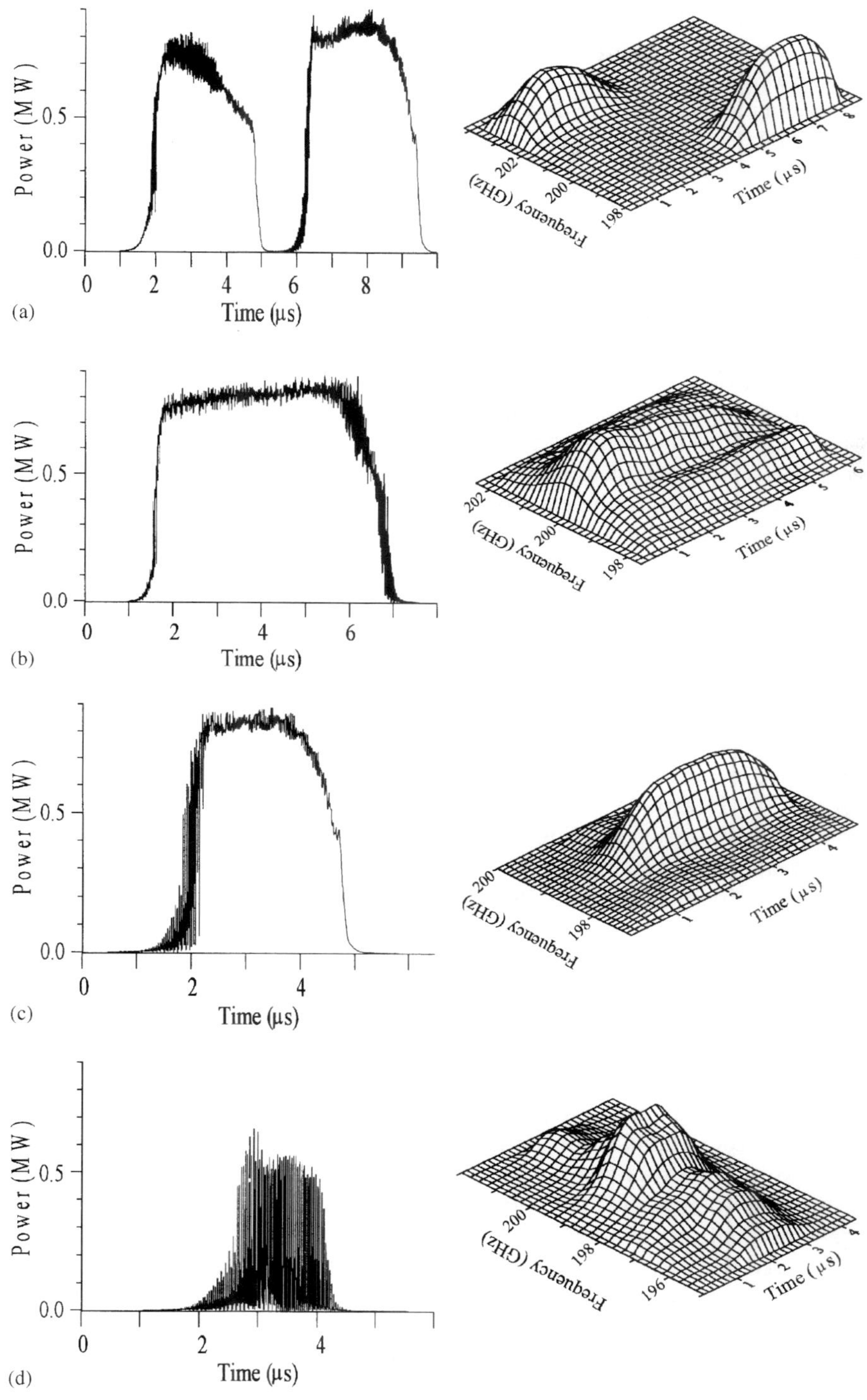

Fig. 3. Output RF power and spectrum of longitudinal modes versus the time. As compared with the case (a), the initial voltage is lower by 8.6 kV (b), 29.3 kV (c), and 30.7 kV (d).

3. FEM dynamics during the voltage drop

In the simulations, we consider the FEM operation at the electron-beam current of 7.2 A and the frequency around 200 GHz. Simulations show that the dynamics of the FEM significantly depends on the value of the initial electron energy (accelerating voltage). Namely, it depends on the initial position of the amplification band with respect to the frequency band of the feedback. The most interesting case is when at the beginning of the electron pulse the center of the amplification band is at the higher frequency border of the feedback band (Fig. 2a). In this case, at the beginning of the electron pulse a mode, placed at this border of the feedback band (203 GHz), is excited after a some transient process. Then, due to the voltage drop during the electron pulse, the position in frequency of the center of the amplification band decreases in time. Thus, the amplification band shifts towards lower frequencies and passes through the entire frequency band of the feedback, being quite broad as compared with the amplification band (which amounts about 10 GHz). During this process, the 203 GHz mode gradually runs out of the amplification band. Correspondingly, its power decreases in time almost down to zero (Fig. 3a). Once the 203 GHz mode is out of the amplification band and, therefore, its coupling with electrons becomes weak, another mode with a lower frequency (198.5 GHz) is excited. With a further drop of the voltage, this mode also runs out of the amplification band, but no new modes are excited because no feedback is provided for lower frequencies. Thus, while the voltage drops slowly, the change in the frequency of the output radiation is not a continuous but a stepped character. Evidently, the value of the frequency step is of the order of the width of the amplification band. This result is confirmed by the experiment, where a stepped change in the frequency with the corresponding temporal modulation of the output power have been observed (Fig. 4).

A different situation takes place at a slightly lower initial electron energy. In this case, the amplification band is at lower frequencies (Fig. 2b) at the beginning of the pulse, as compared to the previous case. Correspondingly, a mode at lower frequency (200.5 GHz) is excited, and simulations predict the single-frequency operation of this mode during almost the entire RF wave pulse (Fig. 3b). Only at the final stage of the FEM operation another modes with lower frequencies (near 198.5 GHz) are excited, but their share is small due to a weak feedback. According to simulations, if the initial voltage is even lower, then a mode with a lower frequency (199 GHz) is excited at the beginning of the maser operation, and the output radiation is single-moded during the entire RF pulse (Fig. 3c). In addition, the duration of the RF pulse decreases.

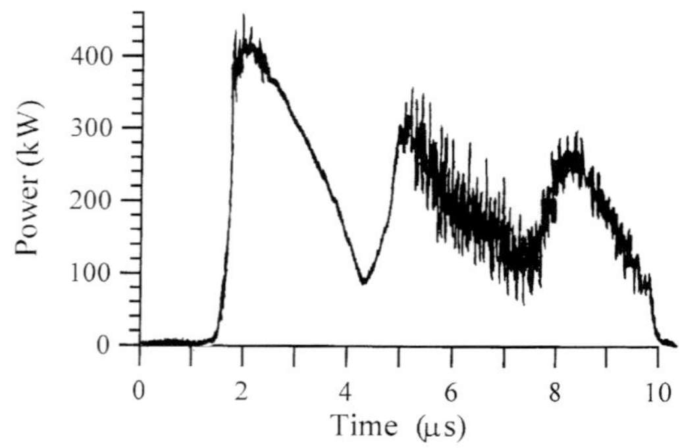

Fig. 4. Experimental demonstration of the stepped change of the operating frequency due to the voltage drop. The output power versus time for the case of feedback coefficient of 0.50 and initial voltage of 1.773 MV.

The situation is significantly changed when the initial electron energy is so low that at the beginning of the FEM operation the amplification band is close to the lower frequency border of the feedback band (Fig. 2d). In this case, due to a weak feedback for all modes of the amplification band, the output power is low. This means that the regime of the FEM operation has almost the small-signal character. So, the interaction between different modes of the amplification band is weak and, therefore, there is no mode competition (mode suppression). Thus, many modes of the amplification band with relatively low frequencies (199–197 GHz) are excited (Fig. 3d). It is important that in the case of relatively low values of the initial electron energy, the dependence of the feedback coefficient on the frequency causes a strong sensitivity of the FEM dynamics to the initial electron energy. Actually, as compared to the case shown in Fig. 3c, in the case of Fig. 3d the initial voltage is only 1.4 kV

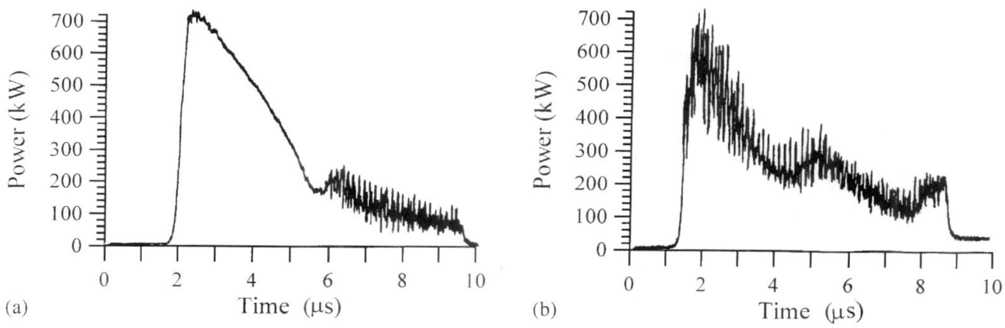

Fig. 5. The output power versus time. Experimental results for the case of feedback coefficient of 0.55, and initial voltages of 1.770 MV (a) and 1.769 MV (b).

lower. This theoretical result is in a good accordance with the experiment. For example, in Figs. 5a and b the experimental results for initial voltages 1.770 and 1.769 MV are given. It is seen that a 1 kV decrease of the initial voltage leads to a decrease in the peak of the output power, as well as to significant temporal modulation of the power due to the multi-mode operation.

4. Conclusion

At the first stage of the experiment, the Fusion-FEM oscillator was tested without a recovery system (inverse set-up), which led to a drop of the accelerating voltage in time. According to simulations, which are in a good accordance with the experimental results, this effect significantly complicated the process of the interaction of different longitudinal modes. During the electron-beam pulse, the frequency of the electron-wave resonance decreases and, therefore, different modes are in the resonance with electrons in different temporal intervals of the electron-beam pulse. In this situation, the dynamics of the FEM significantly depends on the initial value of the voltage (at the beginning of the electron pulse). At some initial voltages, the voltage drop leads to a decrease of the frequency of the output radiation. It is important that although the voltage drops slowly, the change in frequency has a stepped character. The value of this frequency step is as large as the width of the amplification band of the FEM.

References

[1] A.G.A. Verhoeven et al., Proceedings of the 17th International Conference on IR and MM Waves, Passadena, USA, 1992.
[2] W.H. Urbanus et al., Proceedings of the 10th International Workshop on Strong Microwaves in Plasmas, Nizhny Novgorod, Russia, vol. 2., 1997, p. 762.
[3] W.H. Urbanus et al., Proceedings of the 10th Joint Workshop on EC Emission and EC Heating, Ameland, The Netherlands, 1997, p. 497.
[4] W.H. Urbanus et al., Nucl. Instr. and Meth. A 429 (1999) 12.
[5] G.G. Denisov, D.A. Lukovnikov, M.Yu. Shmelyov, Digest of 18th International Conference on the IR and MM Waves, Colchester, UK, 1993, p. 485.
[6] V.L. Bratman, A.V. Savilov, Nucl. Instr. and Meth. A 358 (1995) 182.
[7] M. Caplan, B. Levush, T.M. Antonsen Jr., A.V. Tulupov, W.H. Urbanus, Nucl. Instr. and Meth. A 358 (1995) 174.
[8] V.L. Bratman, G.G. Denisov, A.V. Savilov et al., Nucl. Instr. and Meth. A 407 (1997) 40.
[9] Ya.L. Bogomolov, V.L. Bratman, N.S. Ginzburg, M.I. Petelin, A.D. Yunakovsky, Opt. Commun. 36 (1981) 209.
[10] T.M. Antonsen Jr., B. Levush, Phys. Fluids. B. 1 (1989) 1097.
[11] N.S. Ginzburg, A.S. Sergeev, Zhurnal Tekhnicheskoi Fiziki 61 (1991) 133.
[12] A.V. Savilov, Opt. Commun. 123 (1996) 133.

ELSEVIER

Nuclear Instruments and Methods in Physics Research A 429 (1999) 52–57

NUCLEAR INSTRUMENTS & METHODS IN PHYSICS RESEARCH
Section A

www.elsevier.nl/locate/nima

Role of recovery pass beam phase error in RF system design for same cell energy recovery FELs[☆]

A.M. Vetter*

Boeing Information, Space & Defense Group, P.O. Box 3999 MC 2T-50, Seattle, WA 98124-2499, USA

Abstract

Recovery of residual energy in the electron beam leaving the FEL interaction region allows considerable improvement in two problem areas of particular concern in high average power designs: (1) the RF power required to generate a given average optical output power is reduced, and (2) the power and energy of the beam which must be dumped are reduced, with concomitant reductions in the amount of heat which must be removed and in the radiation shielding requirements. Recirculation of the beam for a second pass through the linac allows the residual beam power to be recovered in the same RF structure used for acceleration, minimizing the investment in structure and yielding a compact layout. If the energy recovered from the beam is adjusted so that the part which interacted with the FEL optical fields is reduced to the same energy as the part of the beam which did not ("differential" energy recovery), then a relationship between the RF power required, the power delivered to the FEL optical mode, the beam current, and the linac structure's external coupling coefficient is established. For a 100 kW optical output example using eight TESLA-type superconducting cavities, a minimum of 250 kW of RF power is required if Q_e is adjusted to 2×10^6. This would be less power than that required for beam loading in the injector linac.

1. Introduction

The wiggler extraction efficiency for near IR and visible FELs is typically less than 10%. Thus, at least 90% of the energy invested in the beam remains there after the beam exits the FEL interaction. In high average power FELs, this represents an enormous amount of wasted energy which not only was obtained at considerable expense in the form of RF power, but which will ultimately generate unpleasant radiation when the beam is stopped. Recovery of a large part of this residual beam energy, preferably in the form of RF energy usable in the beam acceleration process, is therefore strongly motivated, and has been the subject of theoretical and experimental study at several laboratories for over a decade [1–7]. Two basic architectures can accomplish this objective: (1) residual energy is extracted in a dedicated energy recovery structure through which the beam passes on its way between the FEL wiggler and the beam dump, the recovered energy being transmitted by waveguide back to the accelerating structure where it is combined with the output of the high-power RF amplifier, and (2) the beam is recirculated back through the accelerating structure for a second pass, for which it is phased to deliver energy to the RF fields.

☆Work supported by USA/SSDC under Contract DASG60-97-C-0105.

*Tel.: + 1-206-544-5986;.
E-mail address: arthur.m.vetter@boeing.com (A.M. Vetter)

0168-9002/99/$ - see front matter
PII: S0168-9002(99)00066-2

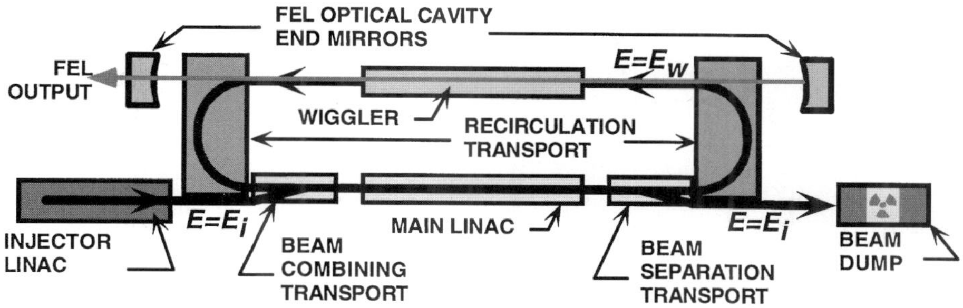

Fig. 1. Schematic layout for same cell energy recovery FEL.

Because the recovered beam energy would be converted to RF energy in the same structure where RF energy is converted to beam energy for the accelerated beam, this architecture is termed "same cell" energy recovery.

Contrasted with the first architecture option mentioned above, the same cell recovery approach (shown schematically in Fig. 1) eliminates the need for additional RF energy recovery structure, recovered power transmission waveguide, and power combining components, and enjoys a more compact layout. Because of significant reduction in the RF power which must be coupled into the accelerating structure to satisfy the net beam loading requirements, same cell energy recovery is arguably more amenable to the use of superconducting accelerator structure, which further reduces the total RF power requirements of the linac.

The price of these benefits includes, at a minimum: complication of beam transport problems, especially in connection with accommodating the wide energy spread of the recovery pass electron beam; accelerator mode instabilities involving interaction between the FEL and the RF system; and possible transverse beam breakup instabilities. It is our objective here to examine the RF system design implications of certain approaches to the beam transport problems in a same cell energy recovery FEL employing superconducting linac structure.

2. Efficiencies and trapping fraction

Consider the energy recovery FEL layout illustrated in Fig. 1, where the beam leaves the injector with energy E_i, gains energy E_a in the main linac, and enters the wiggler with energy $E_w = E_i + E_a$. We will adopt a simple, two-component model for the interaction of the beam with the optical fields in the wiggler: a fraction F_t of the beam is "trapped" in the FEL ponderomotive well and delivers power P_γ to the FEL photon field. The portion of the beam which was *not* trapped returns to the main linac with energy E_w, whereas the *trapped* portion loses energy ΔE and returns to the main linac with energy E_t for the energy recovery pass. Thus, the bimodal energy spectrum of the electron beam leaving the wiggler is approximated by two mono-energetic beams which are temporally superposed at the wiggler exit.

The overall wiggler extraction efficiency is η_w, so

$$P_\gamma = \eta_w P_b = \frac{F_t I_b}{e} \Delta E \tag{1}$$

where P_b is the total electron beam power at the wiggler entrance. The trapped beam energy loss is

$$\Delta E = E_w - E_t = \eta_t E_w \tag{2}$$

where

$$\eta_t = 1 - \frac{E_t}{E_w} \tag{3}$$

is the energy conversion efficiency for the trapped part of the beam. Substituting Eq. (2) in (1), we obtain

$$\eta_w P_b = \frac{F_t I_b}{e} \eta_t E_w = F_t \eta_t P_b \tag{4}$$

so

$$\eta_t = \frac{\eta_w}{F_t} \quad (5)$$

and

$$E_t = E_w(1 - \eta_t) = E_w\left(1 - \frac{\eta_w}{F_t}\right). \quad (6)$$

3. Differential energy recovery

If both the trapped and untrapped parts of the recovery pass beam are decelerated by the same amount, then the trapped part will reach the separation transport with an energy which is still ΔE lower than the energy of the untrapped beam, which is the injection energy E_i. The normalized energy separation $\Delta E/E_i$ may be inconveniently large for the separation transport to accommodate. *For the purposes of the present analysis, we will assume that both the trapped and untrapped components of the recovery pass beam arrive at the separation transport with the same energy* E_i, a situation which we refer to as "differential energy recovery".

The energy extracted from the trapped portion of the beam is controlled by shifting the phase of the trapped part of the beam δ_t relative to the untrapped part, which passes through the linac at maximum decelerating phase. The energy removed from the untrapped part is E_a, and from the trapped part is

$$\begin{aligned} E_a \cos\delta_t &= E_t - E_i = E_w(1 - \eta_t) - E_i \\ &= E_w - E_i - \eta_t E_w = E_a - \eta_t E_w \end{aligned} \quad (7)$$

so

$$\cos\delta_t = 1 - \eta_t \frac{E_w}{E_a}. \quad (8)$$

When the trapped portion of the beam arrives for the second pass through the linac at a phase which is different from that of maximum deceleration, it appears to the RF system as a phase "error" which tends to pull the phase of the RF fields in the structure. We can calculate how much effort will be required from the RF system to correct this reactive beam loading by drawing and analyzing the structure voltage phasor diagram.

4. Contributions to accelerator structure voltage

In the limit of very large coupling coefficient (i.e., when $Q_e \ll Q_o$) it is convenient to define a loaded shunt impedance

$$R_l = 2\left(\frac{R_s}{Q_o}\right)Q_e \quad (9)$$

where R_s is the shunt impedance conventionally defined by

$$P_c = \frac{V_c^2}{2R_s} \quad (10)$$

P_c being the dissipative losses of the structure when the amplitude of the accelerating voltage waveform is V_c. With this definition, the RF input from the high-power amplifier produces a voltage component

$$V_k = 2\sqrt{R_l P_f}\,e^{i\phi} \quad (11)$$

ϕ being the phase of the RF input of the cavity and P_f being the forward power. Each beam component induces a voltage contribution in phase with the beam current. Thus, the accelerating pass beam induces a voltage

$$V_1 = I_b R_l \quad (12)$$

the trapped part of the recovery pass beam induces a voltage

$$V_t = I_t R_l \quad (13)$$

and so on.

To illustrate the various contributions to the complex total cavity voltage V_c, let us assume the parameters introduced above have the numerical values shown in Table 1, which also gives derived values for a main linac consisting of eight structures. The resulting voltage phasors are shown in Fig. 2, where we choose the phase origin such that the accelerating pass beam is aligned with the negative real axis. It is assumed for simplicity that no quadrature component of cavity voltage is required; i.e., $V_c = V_I + 0i$.

Table 1
Numerical values for FEL system parameters used to determine cavity voltages shown in Fig. 2

Parameter		Value
[a]Beam energy at wiggler	E_w	100 MeV
[a]Injection energy	E_I	20 MeV
Main linac energy gain	E_a	80 MeV
[a]Wiggler extraction efficiency	η_w	5%
[a]Trapping fraction	F_t	0.5
Trapped beam energy conversion efficiency	η_t	10%
[a]Beam current	I_b	20 mA
Trapped beam current	I_t	10 mA
[a]Number of linac structures	N	8
Structure accelerating voltage	V_I	10 MV
[a]Loaded shunt impedance/structure	R_l	1 GΩ
Accelerated beam-induced voltage	V_1	− 20 MV
Untrapped beam-induced voltage	V_u	+ 10 MV
Trapped beam phase shift	δ_t	29°
Trapped beam-induced voltage	V_t	(8.75 + i4.84) MV

[a]Parameters (a) are assumed a priori, while other quantities are derived using equations in the text above.

5. RF power required

Examination of Fig. 2 shows that the triangle has these dimensions:

base $= V_t + V_I = I_t R_l + V_I$

left side $= V_t = I_t R_l$

right side $= V_k$.

The Law of Cosines allows us to find the square of the right side in terms of the base, the left side, and the angle δ_t between them.

$$\begin{aligned}4R_lP_f &= V_k^2\\ &= (I_tR_l + V_I)^2 + (I_tR_l)^2\\ &\quad - 2I_tR_l(I_tR_l + V_I)\cos\delta_t.\end{aligned} \tag{14}$$

Expanding the binomial products, collecting terms, and using Eq. (8), we arrive at

$$4R_lP_f = V_I^2 + 2\eta_t\left(\frac{E_w}{E_a}\right)I_tR_l(I_tR_l + V_I) \tag{15}$$

whence

$$P_f = \frac{V_I^2}{4R_l} + \frac{\eta_t I_t E_w}{2E_a}(I_tR_l + V_I). \tag{16}$$

Using Eqs. (1) and (4), we have

$$P_\gamma = \frac{F_t I_b}{e}\eta_t E_w = \frac{\eta_t I_t E_w}{e} \tag{17}$$

which, with the fact that $E_w = eV_a = NeV_I$ where N is the number of structures in the linac, allows us to express Eq. (16) as

$$P_f = \frac{V_I^2}{4R_l} + \frac{P_\gamma}{2N}\left(1 + \frac{I_tR_l}{V_I}\right). \tag{18}$$

Using the numerical example above,

$$\begin{aligned}P_f &= \frac{(10\ \text{MV})^2}{4\times 1\ \text{G}\Omega} + \frac{100\ \text{kW}}{2\times 8}\left(1 + \frac{0.01\ \text{A}\times 1\ \text{G}\Omega}{10\ \text{MV}}\right)\\ &= 25\ \text{kW} + 12.5\ \text{kW} = 37.5\ \text{kW}.\end{aligned} \tag{19}$$

6. Optimum external coupling

Is it possible to improve on this requirement by adjusting R_l? We differentiate Eq. (18) with respect

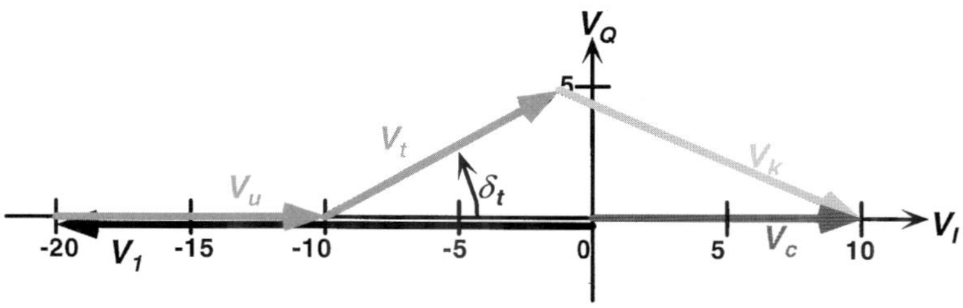

Fig. 2. Phasor diagram for one of the eight structures of the main linac in the FEL described in Table 1.

to R_l, obtaining

$$\frac{dP_f}{dR_l} = -\frac{V_I^2}{4R_l^2} + \frac{P_\gamma}{2N}\left(\frac{I_t}{V_I}\right). \tag{20}$$

For minimum P_f, the derivative will vanish so that

$$\frac{V_I^2}{4R_l} = \frac{P_\gamma}{2N}\left(\frac{R_l I_t}{V_I}\right) \tag{21}$$

or, alternatively,

$$R_{l,\text{opt}} = \sqrt{\frac{V_I I_t}{2P_\gamma/N}\frac{V_I}{I_t}} = \sqrt{V_a I t_{2P_\gamma}\frac{V_I}{I_t}}. \tag{22}$$

Substitution of Eq. (21) in Eq. (18) gives the minimum value for P_f.

$$P_{f,\min} = \frac{P_\gamma}{N}\left(\frac{1}{2} + \frac{I_t R_l}{V_I}\right) = \frac{P_\gamma}{N}\left(\frac{1}{2} + \sqrt{\frac{V_I I_t}{2(P_\gamma/N)}}\right)$$

$$= \frac{P_\gamma}{N}\left(\frac{1}{2} + \sqrt{\frac{eV_a}{2\Delta E}}\right). \tag{23}$$

Note that since $eV_a > \Delta E$, the quantity under the radical in Eq. (23) always exceeds 1/2, and the quantity inside the parentheses therefore always exceeds unity. The total RF power requirement for the main linac $NP_{f,\min}$ therefore always exceeds the power P_γ delivered to the photon field.

To see how far from optimum the value of R_l assumed in Table 1 is, we evaluate Eq. (22):

$$R_{l,\text{opt}} = \sqrt{\frac{V_I I_t}{2P_\gamma/N}\frac{V_I}{I_t}}$$

$$= \sqrt{\frac{10\ \text{MV} \times 0.01\ \text{A}}{2 \times (100\ \text{kW}/8)}\frac{10\ \text{MV}}{0.01\ \text{A}}}$$

$$= \sqrt{\frac{100\ \text{kW}}{25\ \text{kW}} \times 1\ \text{G}\Omega} = 2\ \text{G}\Omega. \tag{24}$$

The lowest possible RF power requirement would be

$$P_{f,\min} = \frac{P_\gamma}{N}\left(\frac{1}{2} + \sqrt{\frac{V_a}{2\Delta E}}\right)$$

$$= 12.5\ \text{kW}(\tfrac{1}{2} + \sqrt{4}) = 31.25\ \text{kW}. \tag{25}$$

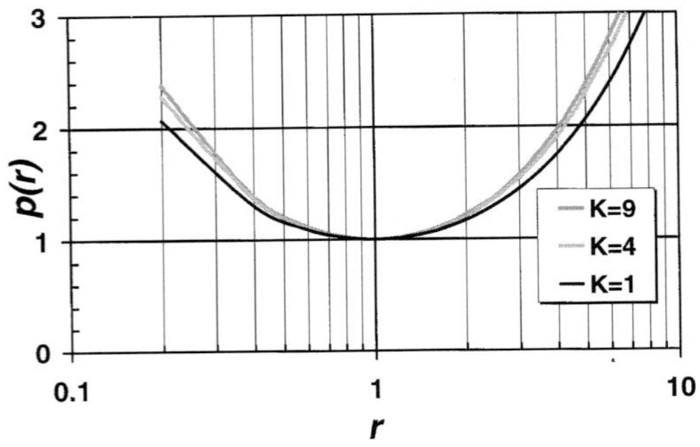

Fig. 3. Power requirement normalized to minimum value plotted against loaded shunt impedance normalized to optimum value.

Thus, by missing the optimum loaded shunt impedance by a factor of 2, we suffer a 20% increase in RF power required to deliver the nominal beam to the wiggler.

Defining the normalized loaded shunt impedance r and required forward power p by

$$r = \frac{R_l}{R_{l,\text{opt}}} \tag{26}$$

and

$$p = \frac{P_f}{P_{f,\min}} \tag{27}$$

we can express the required RF power in terms of the minimum requirement

$$p = \frac{1 + \sqrt{K}(\frac{1}{r} + r)}{1 + 2\sqrt{K}} \tag{28}$$

where $K = V_a I_t / 2P_\gamma = V_a / 2\eta_t V_w$. Eq. (28) is plotted for $K = 1, 4$, and 9 in Fig. 3.

7. Conclusions

We have seen that if differential, same cell energy recovery is implemented in a high average power (100 kW) FEL architecture incorporating TESLA type linac structures for 80% of the beam energy delivered to the wiggler, a minimum of 250 kW of RF power would be required by the main linac to overcome the reactive beam loading of the recovery pass beam (this is in addition to the RF power consumed by the injector linac, which would be at least 400 kW for beam loading alone in this example, where $P_b = E_I I_b = 20\ \text{MV} \times 20\ \text{mA}$). The required Q_e for the main linac structures falls within the nominal range for the high power coupler for that structure.

While the RF power required for the main linac could be further reduced by resorting to non-differential energy recovery, it would not reduce the injector linac power requirements which already dominate those of the main linac in differential energy recovery mode.

References

[1] D.W. Feldman et al., Proceedings of the IEEE Particle Accelerator Conference, Washington, DC, 1987, p. 221.

[2] D.W. Feldman et al., IEEE J. Quantum Electron. QE-23 (1987) 1476.

[3] R. Rohatgi et al., Proceedings of the 1987 IEEE Particle Accelerator Conference, Washington, DC, 1987, p. 230.

[4] N.S. Sereno, L.S. Cardman, Proceedings of the IEEE Particle Accelerator Conference, Washington, DC, 1993, p. 3247.

[5] G.R. Neil et al., Nucl. Instr. and Meth. A 358 (1995) 159.

[6] N.A. Vinokurov et al., Nucl. Instr. and Meth. A 375 (1996) 403.

[7] L. Merminga et al., Nucl. Instr. and Meth. A 375 (1996) ABS 39.

ELSEVIER

Nuclear Instruments and Methods in Physics Research A 429 (1999) 58–64

NUCLEAR INSTRUMENTS & METHODS IN PHYSICS RESEARCH
Section A

www.elsevier.nl/locate/nima

Analysis of the FEL-RF interaction in recirculating, energy-recovering linacs with an FEL

L. Merminga*, P. Alexeev, S. Benson, A. Bolshakov, L. Doolittle, G. Neil

Thomas Jefferson National Accelerator Facility, MS 58, 12000 Jefferson Avenue, Newport News, VA 23606, USA

Abstract

Recirculating, energy-recovering linacs can be used as driver accelerators for high power FELs. Instabilities which arise from fluctuations of the cavity fields are investigated. Energy changes can cause beam loss on apertures, phase oscillations and optical cavity detuning. These effects in turn cause changes in the laser output power through a time-varying FEL gain function. All three effects change the beam-induced voltage in the cavities and can lead to unstable variations of the accelerating field and output laser power. We have developed a model of the coupled system and solved it both analytically and numerically. It includes the beam-cavity interaction, low level RF feedback, and the electron–photon interaction. The latter includes the FEL gain function in terms of cavity detuning, energy offset, and is valid both in the small signal gain and in the saturated regimes. We have demonstrated that in the limit of small perturbations, the linear theory agrees with the numerical solutions and have performed numerical simulations for the IR FEL presently being commissioned at Jefferson Lab.

Keywords: FEL-RF interactions; Beam-cavity interaction; Electron–photon beam interaction

1. Introduction

Jefferson Lab's infrared free electron laser (IR-FEL) is presently being commissioned to produce continuous-wave, kW-level light at 3–6 μm wavelength. Output power of 311 W at 5 μm has been achieved to date with straight ahead operation [1,2], and commissioning of recirculation with energy recovery is currently underway. The IRFEL injector consists of a 350 kV photocathode gun, followed by a copper buncher cavity and a CEBAF-type 1497 MHz superconducting RF (SRF) cryounit to generate an average accelerating gradient of 10 MV/m, boosting the beam to 10 MeV. The linac uses a full CEBAF-type 1497 MHz SRF cryomodule to generate an average accelerating gradient of 8 MV/m, for a resulting beam energy of 42 MeV. The design average beam current is 5 mA. The beam is transported from the linac to the wiggler where the lasing process takes place. A transport lattice recirculates the spent beam back to the linac for deceleration and energy recovery where about 75% of the beam power is converted into RF power. The 10 MeV beam is then transported to a dump. Electron beam recirculation

* Corresponding author. Tel.: 757-269-6281; fax: 757-269-7658.

E-mail address: merminga@jlab.org (L. Merminga)

0168-9002/99/$ - see front matter
PII: S0168-9002(99)00067-4

reduces power consumption, RF equipment capital costs, and beam dump shielding requirements. It is arguably essential as FEL technology is scaled to higher average power levels.

Fluctuations of the cavity fields in the linac can cause beam loss on apertures, phase oscillations and optical cavity detuning. Beam loss can originate from an energy error which shifts the beam centroid off its central trajectory and leads to beam scraping on apertures. Phase shifts can originate from an energy error coupled to the finite compaction factor, M_{56}, of an non-isochronous arc. Compaction factor is defined as the transfer matrix element which multiplied by an energy offset $\Delta E/E$ equals the resulting deviation in the path length, Δl,

$$\Delta l = M_{56}\frac{\Delta E}{E}. \tag{1}$$

Furthermore, the time derivative of an energy error coupled to the M_{56} will result in a shift of the bunch arrival frequency at the wiggler, which is equivalent to optical cavity detuning. Both cavity detuning and electron beam energy error can cause changes in the FEL gain function, which also changes the laser output power. The total change of the laser power is the sum of two terms: one just described and a second one caused by the electron beam energy error directly. Moreover, changes in the laser power will change the energy of the recirculated beam, potentially leading to additional beam loss on apertures and phase shift of the decelerating beam.

All three effects – beam loss, phase shifts and laser power variations – change the beam-induced voltage in the cavities through the recirculating beam. Depending on the RF feedback characteristics this can lead to instabilities both of the accelerating field and the laser output power.

We have developed a system of equations which models the complete system of the RF and the FEL. We have obtained an analytical solution of the equations which is valid in the linear regime where perturbations from equilibrium are small, in order to: (a) perform stability analysis to determine stability boundaries, growth rates of instabilities and threshold currents; (b) obtain an insight on the stability conditions of the coupled system; (c) benchmark the numerical model. We have also developed a code which numerically integrates the equations of motion and has the capability of dealing with transient effects and response to large parameter variations, nonlinear effects, precise representation of RF control feedback system and noise sources such as microphonic.

In Section 2 we present the system of equations that describe our model. In Section 3 we show comparisons between analytical and numerical results and conclude with simulations of the complete system, including the FEL, and its expected behavior during normal operating conditions.

2. The model

The interaction of the beam with the cavity fields can be described, to a very good approximation, by the following first-order differential equation,

$$\frac{d\tilde{V}_c}{dt} + \frac{\omega_0}{2Q_L}(1 - i\tan\Psi)\tilde{V}_c = \frac{\omega_0 R_L}{2Q_L}(\tilde{I}_g - \tilde{I}_b) \tag{2}$$

where ω_0 is the cavity resonant frequency, Q_L is the loaded Q of the cavity and R_L is the loaded shunt impedance $R_L = (R/Q)Q_L$. The beam in the cavity is represented by a current generator. In arriving at Eq. (2) we assume that the cavity voltage, generator and beam current vary as $e^{i\omega t}$, where ω is the RF frequency, and $\tilde{V}_c, \tilde{I}_g$ and $\tilde{I}_b$ are the corresponding complex amplitudes (phasors) in the rotating frame of reference, varying slowly with time. In this equation I_b (absence of tilde denotes the magnitude of the corresponding quantity) is equal to the average beam current (in the limit of short bunches). Also Ψ is the tuning angle defined by $\tan\Psi = -2Q_L(\omega - \omega_0)/\omega_0$. In steady-state the generator power is given by

$$P_g = \frac{(1+\beta)}{4\beta}I_g^2 R_L$$

where β is the cavity coupling coefficient, and can be calculated from $Q_L = Q_0/(1+\beta)$.

The cavity voltage is perturbed in amplitude and phase, by $\hat{v}(t)$ and $\hat{\phi}(t)$, respectively,

$$\tilde{V}_c = [V_0 + \hat{v}(t)]e^{i\hat{\phi}(t)}$$

where V_0 is the steady-state cavity voltage. All phases are calculated with respect to the steady-state cavity voltage.

The beam current $\tilde{I}_b$ is equal to the sum of the accelerating and decelerating beam current,

$$\tilde{I}_b = \tilde{I}_1 + \tilde{I}_2. \tag{3}$$

Assuming that the accelerating beam remains unperturbed, $\tilde{I}_1$ can be written as

$$\tilde{I}_1 = I_0 e^{i\Psi_1}$$

where Ψ_1 is the accelerating beam phase. The decelerating beam however can, in principle, be perturbed both in magnitude and phase,

$$\tilde{I}_2 = I_0[1 + \mu(\varepsilon)]e^{i[\Psi_2 + \phi_2(t)]}$$

where Ψ_2 is the steady-state phase of the decelerating beam, ϕ_2 is the phase shift from stead-state and is given by

$$\phi_2(t) = h\varepsilon_1 + (h - h_1)\varepsilon_2 \tag{4}$$

where ε_1 is the energy error of the accelerating beam at the cryomodule exit, originating from cavity voltage fluctuations, and ε_2 is the energy error due to the interaction with the photon beam. Also h and h_1 are defined as

$$h = \frac{M_{56}\omega}{cE_b}, \qquad h_1 = \frac{M_{56}^1\omega}{cE_b}$$

where M_{56} is the compaction factor from the cryomodule exit to its entrance and M_{56}^1 is the compaction factor between the cryomodule exit and the wiggler. E_b is the electron beam energy at the cryomodule exit. $\mu(\varepsilon)$ is a function of beam losses and ε is the total energy error, $\varepsilon = \varepsilon_1 + \varepsilon_2$.

2.1. Beam losses

We have two ways of expressing the beam loss parameter μ as function of the energy error ε, linear and nonlinear. In the linear expression,

$$\mu(\varepsilon) = -\frac{\eta_x}{L_{\text{lin}}E_b}\varepsilon \tag{5}$$

η_x is the horizontal dispersion of the arc and L_{lin} is a loss coefficient which characterizes the amount of beam loss. $L_{\text{lin}} = 1$ implies that 1 mm offset produces beam loss at the 10^{-3} level.

While a linear beam loss term is relevant for small signal stability analysis, it is clearly inadequate for large signals. A more realistic representation of the beam loss as function of energy deviation is obtained assuming that μ is a nonlinear function determined by the shape of the energy distribution of the electron beam in the transverse dimension, horizontal dispersion of the arc, aperture of the vacuum chamber, and initial displacement of beam centroid with respect to the center of the vacuum chamber. If σ is the standard deviation of an assumed Gaussian transverse beam distribution, δ is the semi-aperture of the vacuum chamber in energy units, α is the steady-state shift of the unperturbed beam from the center of the vacuum chamber in energy units, and ε is the energy error, the expression for the beam loss function is [3]

$$\mu(\varepsilon) = \frac{1}{2}\left[\operatorname{erf}\left(\frac{\delta - \alpha - \varepsilon}{\sigma\sqrt{2}}\right) + \operatorname{erf}\left(\frac{\delta + \alpha + \varepsilon}{\sigma\sqrt{2}}\right) - \operatorname{erf}\left(\frac{\delta - \alpha}{\sigma\sqrt{2}}\right) - \operatorname{erf}\left(\frac{\delta + \alpha}{\sigma\sqrt{2}}\right)\right]. \tag{6}$$

2.2. Feedback

In the open loop analysis we assume that the generator current $\tilde{I}_g$ is constant and is expressed in polar form by

$$\tilde{I}_g = I_{g_0} e^{i\Psi_{go}}.$$

In the presence of feedback, the generator current $\tilde{I}_g$ is no longer constant, but it depends on the cavity voltage vector and its history. In the existing Jefferson Lab RF control module [4], it assumes the form

$$\tilde{I}_g = [I_{g_0} + \Delta I_g(t)]e^{i[\Psi_{go} + \Delta\Psi_g(t)]}$$

where $\Delta I_g(t)$ is the additional signal providing amplitude feedback, and $\Delta\Psi_g(t)$ is the additional signal providing phase feedback [5]. The transfer function in the feedback path is presently modeled as a low-pass filter with gain G and roll-off frequency

$(2\pi T)^{-1}$. Therefore the Laplace transforms of ΔI_g and $\Delta\Psi_g$ are

$$\frac{\Delta I_g(s)}{I_{g_0}} = -\frac{G}{1+sT}\frac{\hat{v}(s)}{V_0} \tag{7}$$

and

$$\Delta\Psi_g(s) = -\frac{G}{1+sT}\hat{\phi}(s) \tag{8}$$

where $\hat{v}(s)$ and $\hat{\phi}(s)$ are the errors in the amplitude and phase of the cavity field.

2.3. Laser power

Two basic processes are considered in deriving the equation for the laser power. These are (a) the interaction with the electron beam as it passes through the wiggler and (b) the power losses through the mirrors of the optical cavity. Using conservation of energy arguments the equation for the laser power can be written as

$$\frac{dP}{dt} = \frac{1}{\tau}\left[-1 + G_l\left(\frac{\tau}{T_{las}} - 1\right)\right]P \tag{9}$$

where τ is the decay time of the optical cavity, $T_{las} = 2L_w/c$ is the round trip time, G_l is the FEL gain function and P denotes the power stored in the optical cavity. Presently we use Dattoli's [6] FEL gain function given by

$$G_l = -0.85 g_0 \frac{\Theta}{\Theta_s}\frac{\ln[(\Theta/\Theta_s)(1-\mu_c/3)f(x)] - 1}{1-(\varepsilon_1/aE_b)^2}. \tag{10}$$

The function $f(x)$ is the saturation function

$$f(x) = \frac{bx}{1-e^{-bx}}$$

where $x = P/P_s$ is the ratio of laser power to saturation laser power, and $b = 1.045\pi/2$. The variable Θ is defined as

$$\Theta = \frac{4\delta L}{g_0\Delta}$$

where $\Delta = N\lambda_L$ is the slippage, N is the number of wiggler periods, λ_L the laser wavelength, and δL the cavity detuning. $\Theta_s = 0.456$ is a constant. g_0 is the gain coefficient and μ_c is ratio between slippage and electron bunch length, $\mu_c = \Delta/\sigma_z$. Finally $a \simeq 1/2N$ and ε_1 is the electron beam energy error.

The FEL gain function thus written allows us to calculate the effect of a slow shift in the arrival frequency of the electron bunches on the laser output power. The electron bunch frequency shift $\delta\omega$ is given by

$$\delta\omega = -h_1\frac{d\varepsilon_1}{dt}$$

and this is equivalent to cavity detuning,

$$\frac{\delta L}{L} = \frac{\delta\omega}{\omega_0}$$

which enters in the expression for Θ. Also notice in the denominator of the FEL gain function the direct dependency of the FEL gain on the electron beam energy error. The saturation function $f(x)$ allows the formalism to be valid both in the small signal gain regime ($x \ll 1$) and in the saturated regime ($x \geq 1$). Finally, the electron beam energy error due to the interaction with the photon beam is given by

$$\varepsilon_2 = -\frac{\delta(G_l P)}{I_0}$$

and contains contributions both from the gain function and the laser power.

In the analytical model we linearize the complete system of equations and take the Laplace transform. The results can be expressed as a system of three algebraic equations $MA = 0$, where M is a 3×3 matrix and A is the column vector with $\hat{v}$, $\hat{\phi}$ and ΔP as components. The determinant of M is then set to zero and the roots of s, the Laplace transform variable, are examined. The real parts of the roots will provide the damping or growth rates of perturbations. The imaginary parts of the roots will give the oscillation frequencies of $\hat{v}$, $\hat{\phi}$ and ΔP. If all roots have zero or negative real parts, the system is stable.

For the numerical solution of the system of equations the code "laser" was developed. The program is written in C++ using Linux. We have compared the analytical model with the numerical solution for the cases of phase oscillations and beam

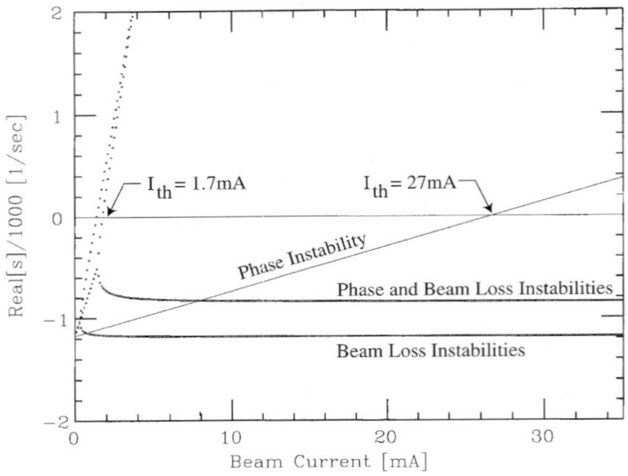

Fig. 1. Analytical solution of the real part of s as function of beam current for the cases of phase instabilities, beam loss instabilities and both combined.

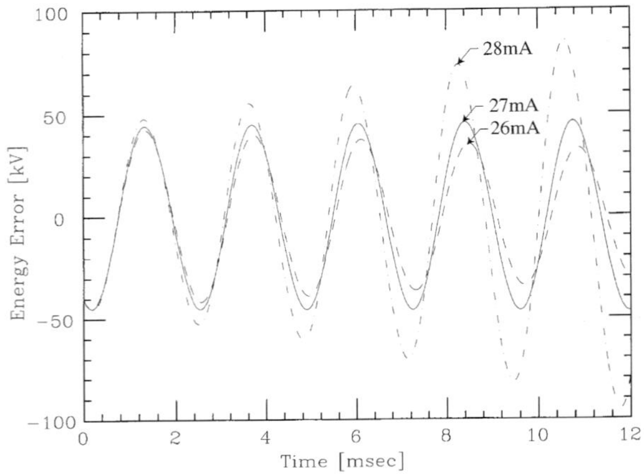

Fig. 2. Numerical solution of the time dependence of energy perturbation in the case of phase instabilities.

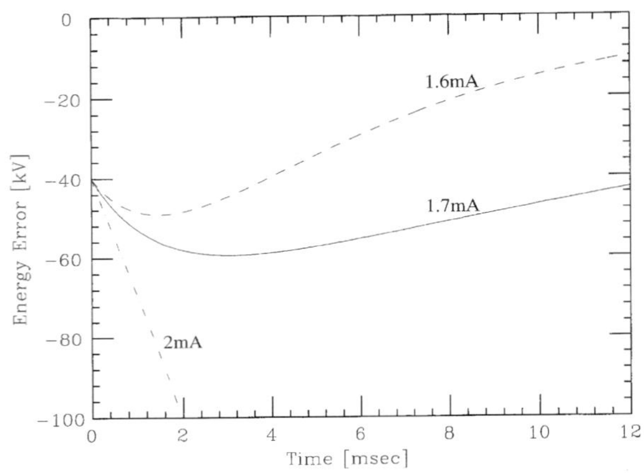

Fig. 3. Numerical solution of the time dependence of energy perturbation in the case of beam loss and phase instabilities.

loss instabilities open and closed loop and we present the results in the next section.

3. Numerical benchmarking and simulation results

Fig. 1 is the plot of the real part of s, the Laplace transform variable, as function of beam current, as predicted by the linear theory (closed form solution is given in Ref. [7]) for nominal operating parameters of Jefferson Lab's IRFEL, with the FEL turned off. For the design value of $M_{56} = -0.15$ m and assuming a linear scraping model with $L_{\text{lin}} = 1$, then in the absence of feedback the system is unstable at $I_0 = 5$ mA. The instability threshold is 1.7 mA, where the stability boundary curve intersects the $s = 0$ line. The threshold current for the phase instability alone (assuming an "infinite" energy aperture) is 27 mA. Figs. 2 and 3 display the time dependence of the energy perturbation as calculated from the code "laser". Growth/Damping in the oscillation amplitude signifies instability/stability of the system. In Fig. 2 only phase oscillations are present, while in Fig. 3 the linear beam loss mechanism has been added. The dashed curve of Fig. 2 corresponds to 26 mA, the dot–dash curve to 28 mA and the solid curve to 27 mA, where the stability threshold appears to be, in agreement with the analytical solution. Similarly in Fig. 3 the dashed curve which corresponds to 1.6 mA damps towards zero, whereas the dot–dash curve which corresponds to 2 mA diverges. The stability threshold appears to be close to 1.7 mA in agreement with the analytical model. Finally comparisons between analytical and numerical results with the feedback loops closed show very good agreement as well.

Next we present numerical results on the stability of the complete system, including the FEL. Fig. 4 displays the time dependence of the energy perturbation ε_1 with nominal M_{56}, 5 mA of

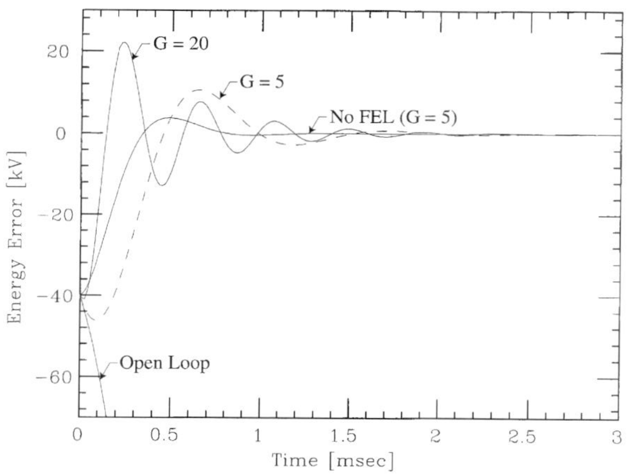

Fig. 4. Numerical solutions of the complete system: Energy perturbation as function of time at 5 mA beam current, feedback loops open and closed at different gains, and the FEL on and off.

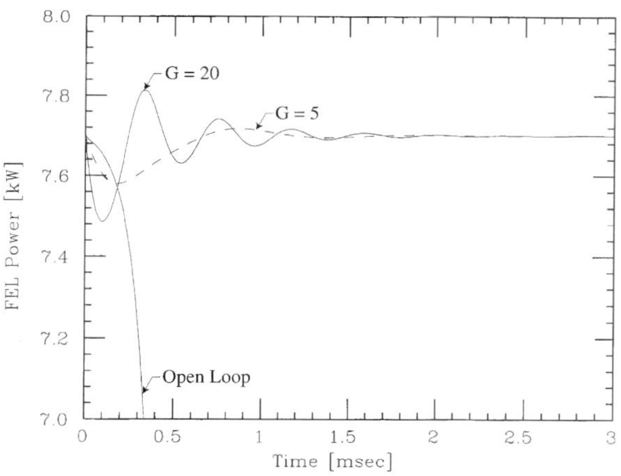

Fig. 5. Numerical solutions of the complete system: Intracavity laser power as function of time at 5 mA beam current, and feedback loops open and closed at different gains.

beam current, nonlinear scraping ($\sigma = 0.75\%$, $\delta = 2.5\%$, $\alpha = 1.0\%$), and the FEL turned on with output laser power approximately 1 kW, which corresponds to intracavity power of 7.7 kW. The different curves correspond to the open loop case, closed loop with zero-frequency gain G (as defined in Eqs. (7) and (8)) equal to 5 and 20 and the FEL turned off and gain $G = 5$. The roll-off frequency of the feedback is assumed to be 1.5 kHz. Fig. 5 diplays the time dependence of the intracavity laser power for the same parameters. These simulation results suggest first, that the interaction of the FEL with the electron beam is unstable open loop, and second, that modest gains and bandwidths well within the range of the low level RF control feedback are sufficient to ensure stability.

4. Conclusions

We developed a model of the interaction of the RF with an FEL which we solved analytically and numerically. The model includes beam-cavity interaction, low level RF feedback, and the electron–photon beam interaction through a time-varying FEL gain function. The gain function includes cavity detuning, energy offset and is valid in the small signal gain and saturated regimes. We are presently benchmarking the numerical model against the analytical one. We have shown that this interaction can be unstable open loop for the parameters of the Jefferson Lab IRFEL, however modest gains at reasonable frequencies are required to stabilize the system. Future plans include further comparisons between analytical and numerical solutions, expansion of the analysis to include a gain function more appropriate for our parameter regime, and direct comparisons with experimental data from Jefferson Lab's IRFEL.

Acknowledgements

The authors are grateful to Drs. Jean Delayen and Joe Bisognano for careful reading of the manuscript and useful comments. This work was supported by the Virginia Center for Innovative Technology and DOE Contract #DE-AC05-84ER40150.

References

[1] S.V. Benson et al., Nucl. Instr. and Meth. A 429 (1999) 27.

[2] C.L. Bohn et al., Performance of the Accelerator Driver of Jefferson Laboratory's Free Electron Laser, Proceedings 20th Free Electron Laser Conference, Williamsburg, VA, 1998, Elsevier Science, Amsterdam, 1999, p. II. 43.
[3] A. Bolshakov, Energy stability and light power stability in recirculating, energy-recovering linacs in presence of an FEL, Jefferson Lab Internal Rep. TN-97-039, 1997.
[4] S.N. Simrock, RF control system for CEBAF, Proceedings of the 1991 Particle Accelerator Conference, pp. 2515–2519.
[5] J.R. Delayen, Phase and amplitude stabilization of beam-loaded superconducting resonators, Proceedings of 1992 Linear Accelerator Conference, Ottawa, AECL Research Rep., AECL-10728, pp. 371–373.
[6] G. Dattoli et al., An introduction to the theory of free electron lasers, CERN 90-03, pp. 254–286.
[7] L. Merminga, J.J. Bisognano, J.R. Delayen, Energy stability in recirculating, energy-recovering linacs in the presence of an FEL, Proceedings of the EPAC 1996, Barcelona, Spain.

ELSEVIER

Nuclear Instruments and Methods in Physics Research A 429 (1999) 65–69

NUCLEAR INSTRUMENTS & METHODS IN PHYSICS RESEARCH Section A

www.elsevier.nl/locate/nima

Stabilization of spatio-temporal dynamics of free-electron laser operation under effect of spread in electron velocity

A.V. Savilov*

Institute of Applied Physics, Russian Academy of Sciences, 46 Ulyanov Street, Nizhny Novgorod 603600, Russia

Abstract

The effect of velocity spread on the dynamics of the interaction of longitudinal modes is studied for a FEL-oscillator with broadband feedback. It is shown that the spread shortens the process of mode competition and increases the electron-current threshold of stable single-mode operation. In addition, the spread leads to a shift in the frequency of the excited mode to the range of the most efficient electron-wave interaction. This results in a very weak sensitivity of the efficiency to the spread. © 1999 Elsevier Science B.V. All rights reserved.

Keywords: FEL-oscillator; Low-Q-cavity; Spatio-temporal dynamics

1. Introduction

In a free-electron laser (FEL) a large number of longitudinal modes (of the order of γ^2, where γ is the Lorentz-factor of electrons) usually exists in the amplification band. In FEL-oscillators, the dynamics of their interaction strongly depends on dispersive properties of the feedback. If its frequency band is significantly narrower than the amplification band, then the feedback determines the frequency of the excited mode. The opposite case of a very broadband feedback (which can be necessary, for instance, in order to provide frequency tuning) is more interesting. In this case, all modes of the amplification band arise from initial noise at the small-signal stage of excitation. Then, at the nonlinear stage of the transient process, they take part in the complicated process of competition. The result of this process and, correspondingly, the spectrum of the operating (build-up) regime are defined mainly by the value of the electron current [1–3]. If the current is close to the threshold value, then single-frequency operation of the so-called "main" mode, which is excited at the small-signal stage with the greatest gain, is achieved. However, if the current exceeds some threshold, then the single-mode operation is not stable, and the final stage represents the operation of a number of longitudinal modes.

An important problem in the theory of FEL-oscillators is the effect of electron-velocity spread on the dynamics of excitation, as well as on characteristics of the operating regime (mode spectrum and the value of output power). In oscillators with narrow frequency band of the feedback, where the

* Tel.: + 7-8312-384575; fax: + 7-8312-362061.
E-mail address: savilov@appl.sci-nnov.ru (A.V. Savilov)

0168-9002/99/$ - see front matter
PII: S0168-9002(99)00068-6

operation frequency is fixed, the regime of maximal electron efficiency can be achieved by optimizing the mismatch between the operating frequency and the frequency of the exact electron-wave synchronism. In this situation, the spread evidently decreases the output power (see, e.g. Refs. [4,5]). However, in the case of a broadband feedback, when the frequency is not fixed and, therefore, does not correspond to the regime of maximum efficiency, this problem is not so clear. The spread leads to a shift of the frequency of the main mode to the range of the most efficient electron-wave interaction. Therefore, though the spread decreases the maximum possible efficiency, the decrease in the efficiency of electron interaction with the "main" mode, which is excited in reality after the transient process, can be very small [6]. The effect of the spread on the dynamics of the transient process is also interesting. On the one hand, due to the fact that different fractions of the beam are in resonance with longitudinal modes having different frequencies, the spread increases the number of competing modes. This should complicate the transient process. On the other hand, the velocity spread increases the starting current of the oscillator. In principle, this should lead to the increase of the electron-current threshold of the stability of the single-frequency operation, as well as to a shortening of the transient process. In this work, the effect of the spread on the interaction dynamics of longitudinal modes is studied for a FEL-oscillator with a relatively low-Q cavity and with a broadband feedback.

2. Spatio-temporal approach

We consider a FEL-oscillator with the microwave system consisting of a waveguide segment with two (input and output) frequency-broadband mirrors (Fig. 1). For a number of important cases the dynamics of such an oscillator can be successfully studied within the framework of the spatio-temporal approach on the basis of the simplest 1D nonstationary equations [1–3] for complex amplitudes of forward (a_+) and backward (a_-) waves

$$\frac{\partial a_+}{\partial \zeta} + \frac{\partial a_+}{\partial \tau} = 2\mathrm{i}\rho, \quad \frac{\partial a_-}{\partial \zeta} - \varepsilon\frac{\partial a_-}{\partial \tau} = 0 \tag{1}$$

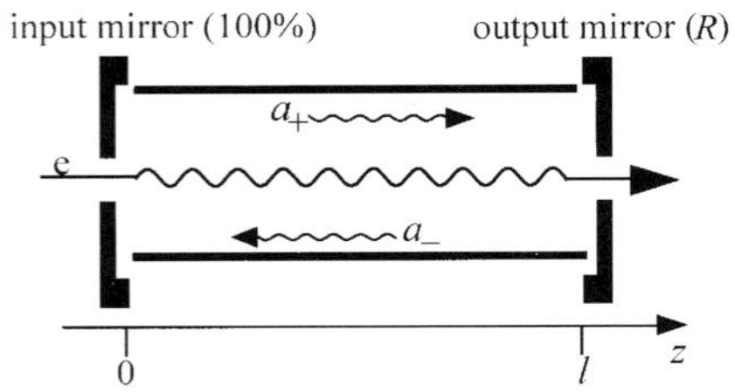

Fig. 1. Schematic diagram of a FEL-oscillator.

and motion equations for electrons coupling with the forward wave

$$\frac{\partial u}{\partial \zeta} = \mathrm{Im}\, a_+ \mathrm{e}^{\mathrm{i}\theta}, \quad \frac{\partial \theta}{\partial \zeta} = u - \delta. \tag{2}$$

Here $u = \mu C^{-1}(1 - \mathscr{E}/\mathscr{E}_0)$ is the normalized change in electron energy, $\zeta = h_s C z$ is the normalized coordinate, $\theta = \omega_s t - h_s z$ is the electron phase with respect to the synchronous combination wave, $\tau = h_s C(\hat{v}_e t - z)(\hat{v}_e/v_{gr} - 1)^{-1}$ is the normalized time, $\varepsilon = (\hat{v}_e + v_{gr})/(\hat{v}_e - v_{gr})$ is so-called slippage parameter, $\omega_s = \hat{v}_e h_s$ is the synchronous frequency, $h_s = k_{\|s} + k_u$ is the wavenumber of the synchronous combination wave, v_{gr} is the group wave velocity, $\hat{v}_e$ is the averaged longitudinal velocity of particles, $C^3 = eI\mu\kappa^2/mc^3N\gamma_0$ is the Pierce parameter, I is the electron current, and $\mu = \gamma_0^{-2}$ and $\kappa = eA_u/mc^2\gamma_0$ are the parameters of the inertial bunching of particles and the coefficient of the electron-wave coupling, respectively. The initial conditions for the unmodulated electron beam at the input can be written in the following way

$$u(\tau,\zeta = 0) = 0, \quad \theta(\tau,\zeta = 0) = \theta_0 \tag{3}$$

where the initial phases, θ_0, are distributed homogeneously over the interval $[0,2\pi)$.

In our model we assume that if the RF wave is absent, then the transverse electron motion represents mainly two types of oscillations. Along with "operating" oscillations in the undulator field, electrons perform oscillations affected by a some focusing system (for instance, cyclotron oscillations in a guiding magnetic field or betatron oscillations affected by the focusing strength of the undulator). The presence of the second type of oscillations leads to a spread in transverse electron velocity. This means that a spread in longitudinal velocity also takes place (all electrons have the same initial energy). In the equations of electron motion, Eqs. (2), the velocity spread is taken into account by introducing different parameters $\delta = (v_e/\hat{v}_e - 1)/C$ for different fractions of the electron beam.

In Eq. (1), excitation of the RF wave is defined by the first harmonic of the electron current

$$\rho = \frac{1}{2\pi}\int_0^{2\pi} d\theta_0 \int_{-\infty}^{+\infty} f(\delta)\, d\delta e^{-i\theta} \quad (4)$$

where $f(\delta)$ is the distribution function of electrons over the initial longitudinal velocity. The time dependence of the RF amplitudes describes the frequency spectrum of excited longitudinal modes:

$$a(\tau) = \int a_\Delta e^{i\Delta\tau}\, d\Delta.$$

Here the mismatch of a mode $\Delta = \frac{\omega - \omega_s}{\omega_s}(1 - \frac{v_{gr}}{\hat{v}_e})C^{-1}$ is the normalized difference between the frequency of this mode and the synchronous frequency. The simplest boundary condition for the RF field describes the reflection of the waves from the input and the output mirrors:

$$a_+(\tau,\zeta = 0) = a_-(\tau,\zeta = 0),$$

$$a_-(\tau,\zeta = L) = Ra_+(\tau,\zeta = L). \quad (5)$$

Here R is the amplitude reflection coefficient of the output mirror, which does not depend on the frequency (for the input mirror, 100% reflectivity is assumed), and $L = h_s Cl$ is the normalized length of the cavity.

The set of Eqs. (1)–(5) (with no velocity spread taken into account) was studied in work [3]. At a fixed R the type of the operating regime is defined by the electron-current parameter, L. The oscillator is excited if L is larger than the starting value, $L_{st}(R)$. If parameter L does not exceed the starting value significantly, then the spectrum of the operating regime contains only one excited mode, which has the greatest small-signal single-pass gain (the "main" mode). However, when L exceeds some threshold, the operation regime has the multi-mode character with fast temporal modulation of the output power due to "beats" of excited modes.

3. Results of simulations and their interpretation

The spatio-temporal system Eqs. (1)–(5) can be solved numerically using standard routines, which are based on the method of cubic Hermite polynomials [7]. As an example, we study the case of a moderately low-Q cavity, $R = 0.5$. The distribution function of electrons over their initial velocity, $f(\delta)$, is assumed uniform within the interval $-D \leq \delta \leq D$. In simulations, one assumes $\varepsilon = 3$. Results of simulations for various electron-current and spread parameters, L and D, are given in Figs. 2–4, where the dependencies of the normalized electron efficiency (output power)

$$\eta(\tau) = \frac{1}{2\pi}\int_0^{2\pi} d\theta_0 \int_{-\infty}^{+\infty} f(\delta)\, d\delta\, u(\tau,L,\theta_0,\delta)$$

as well as frequency spectra of operating stages, are shown.

We consider the three following typical cases: when the electron-current parameter, L, is smaller than the threshold of the stable single-mode operation (Fig. 2); when it slightly exceeds this threshold (Fig. 3) and when the electron-current threshold of the single-mode operation is significantly exceeded (Fig. 4). In the first case, for the ideal electron beam

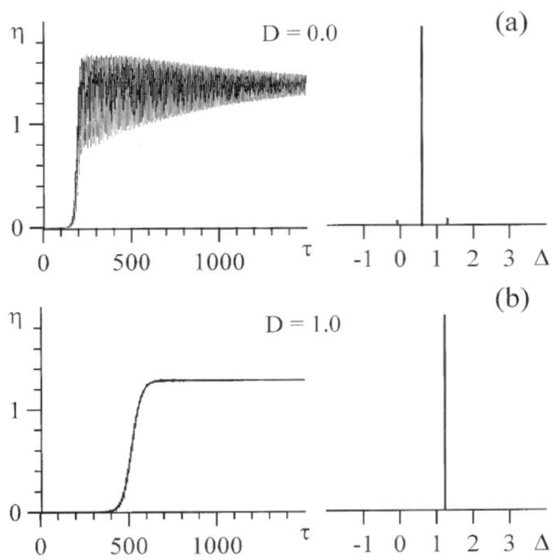

Fig. 2. Electron-current parameter $L = 2.5$. Normalized output power versus time, $\eta(\tau)$, and the frequency spectrum of the operating regime for the cases of the electron beam without (a) and with velocity spread $D = 1$ (b).

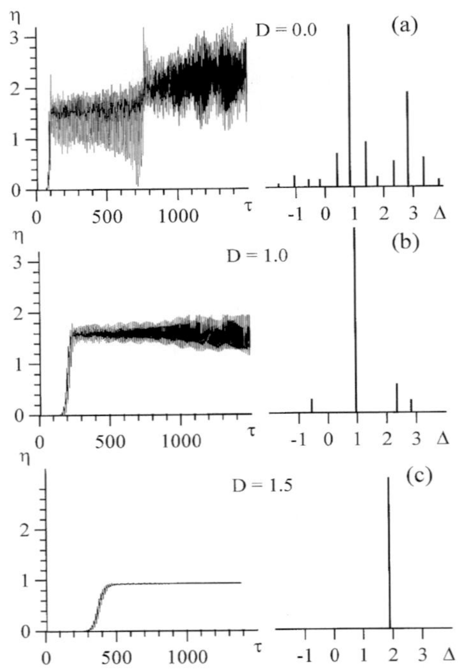

Fig. 3. Electron-current parameter $L = 3.3$. Normalized output power versus time, $\eta(\tau)$, and the frequency spectrum of the opening regime for the cases of the electron beam without (a), and with velocity spread $D = 1$ (b) and $D = 1.5$ (c).

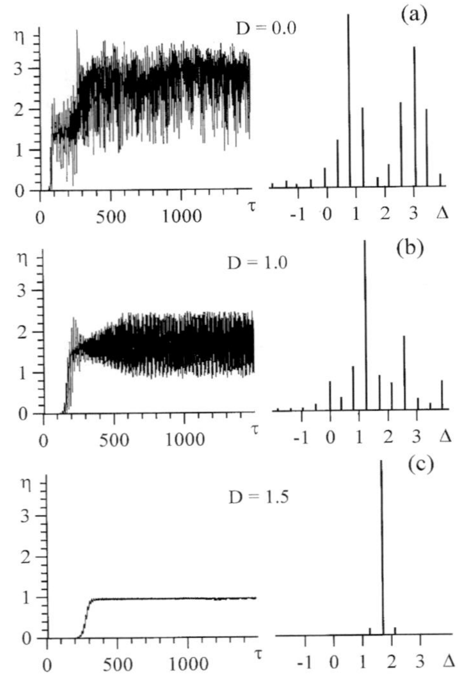

Fig. 4. Electron-current parameter $L = 3.6$. Normalized output power versus time, $\eta(\tau)$, and the frequency spectrum of the operating regime for the cases of the electron beam without (a), and with velocity spread $D = 1$ (b) and $D = 1.5$ (c).

(without the velocity spread, $D = 0$) the single-mode operation builds up after some transient time (Fig. 2a). For the beam with spread $D = 1$, the single-mode operation is also stable (Fig. 2b). Moreover, the spread decreases the time of the mode competition. At the same time, as compared with the case $D = 0$, the output power of the excited mode is practically the same, but its frequency is shifted. For the larger spread, the starting current becomes larger than the operating one, and the oscillator is not excited.

If the parameter L slightly exceeds the threshold of the single-mode operation, then for the ideal electron beam, $D = 0$, several modes are excited (Fig. 3a). In this case, a spread $D = 1$ improves the operating regime significantly (Fig. 3b). The spectrum of the output radiation becomes almost single-mode. As for the output power, it almost does not decrease. The larger spread, $D = 1.5$, results in very pure single-mode operation with a smaller output power (Fig. 3c). A similar situation takes place in the case, when the electron-current parameter significantly exceeds the single-mode threshold. In this case, for the ideal electron beam the output radiation contains a lot of modes (Fig. 4a). The increase of velocity spread improves the spectrum. For a spread $D = 1$, the output radiation is still multimode, but its spectrum is purer when compared with the case $D = 0$ (Fig. 4b). A spread $D = 1.5$ leads to a pure single-mode operation (Fig. 4c).

Thus, the velocity spread increases the starting current of the oscillator. Evidently, the electron-current threshold of the stable single-mode operation also increases. Due to this fact, the spread improves the mode spectrum of the output radiation and provides the single-mode operation at larger electron currents as compared with the case of the ideal electron beam. At the same time, for single-mode regimes, a quite large spread does not

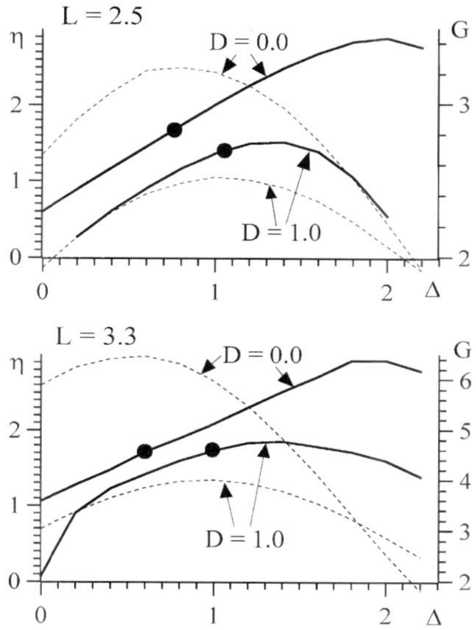

Fig. 5. Normalized single-mode electron efficiency, η, (solid curves) and small-signal gain, G, (dashed curves) as functions of the normalized frequency, Δ, for various parameters of the electron current, L, and of the spread, D. Efficiencies of corresponding "main" modes are indicated by black circles.

result in a significant decrease of the output power. This can be explained by the shift of the frequency of the excited mode due to the spread [6]. If the frequency band of the feedback is broad as compared with the amplification band, then the feedback does not fix the frequency of the excited mode. In this situation, the so-called "main" mode with a frequency, being close to the center of the amplification band (the maximum of the small-signal gain), is excited [3]. Fig. 5 illustrates the single-mode electron efficiency, $\eta(\Delta)$, and the small-signal single-pass gain of the modes, $G(\Delta) = |a_\Delta(L)/a_\Delta(0)|$, as functions of their normalized frequencies, Δ. As seen from this figure, the spread $D = 1$ is quite large: in the both cases ($L = 2.5$ and $L = 3.3$) it leads to the almost a factor of two decrease in the maximum efficiency. However, the spread causes a shift of the amplification band, $G(\Delta)$, to the frequency of the maximum efficiency. Due to this fact, the spread practically does not change the electron efficiency, corresponding to the "main" modes with maximal $G(\Delta)$, which are really excited.

4. Conclusion

The effect of the spread on the dynamics of the interaction of longitudinal modes is studied for a FEL-oscillator with broadband feedback and with a relatively low-Q cavity. Simulations show that even a large (at the limit of the hydrodynamic amplification stage) velocity spread causes a positive effect. Due to the spread, the duration of the transient process of the mode competition reduces, and the electron-current threshold of the single-frequency operation significantly increases. At the same time, for single-mode regimes, the output power decreases very slowly with the increase of the velocity spread. This is caused by a broad frequency band of the feedback. In the situation, when the operating frequency is not fixed by the feedback, the spread leads to a shift of the frequency of the excited mode to the "optimal" range of the most efficient electron-wave interaction. Thus, even a large velocity spread does not result in a significant decrease of the electron efficiency.

Acknowledgements

This work was supported by the Russian Foundation for Basic Research, Grant No. 98-02-17068.

References

[1] Ya.L. Bogomolov, V.L. Bratman, N.S. Ginzburg, M.I. Petelin, A.D. Yunakovsky, Opt. Commun. 36 (1981) 209.

[2] T.M. Antonsen, Jr., B. Levush, Phys. Fluids B 1 (1989) 1097.

[3] N.S. Ginzburg, A.S. Sergeev, Zhurnal Tekhnicheskoi Fiziki 61 (1991) 133 (in Russian).

[4] L.V. Rodygin, A.V. Smorgonsky, Zhurnal Tekhnicheskoi Fiziki 52 (1982) 2013 (in Russian).

[5] L.K. Gover, J. Feinstein, R.H. Pantell, IEEE J. Quant. Electron. QE- 21 (1985) 470.

[6] A.V. Savilov, Zhurnal Tekhnicheskoi Fiziki 65 (1995) 66 (in Russian).

[7] R.F. Sincovec, N.K. Madsen, ACM Trans. Math. Software 1 (1975) 232.

ELSEVIER

Nuclear Instruments and Methods in Physics Research A 429 (1999) 70–75

NUCLEAR INSTRUMENTS & METHODS IN PHYSICS RESEARCH Section A

www.elsevier.nl/locate/nima

Power and bandwidth studies for W-band free-electron masers

H.P. Freund*, R.H. Jackson[1], B.G. Danly, B. Levush

Naval Research Laboratory, Washington, DC 20375, USA

Abstract

Power and bandwidth studies for a high power W-band (≈ 94 GHz) FEM amplifier is presented for a helical wiggler/cylindrical waveguide configuration using 3D ARACHNE simulation code (H.P. Freund, T.M. Antonsen, Jr., Principles of Free-electron Lasers, 2nd edn., Chapman & Hall, London, 1996, Ch. 5). Using a 300 kV/20 A electron beam with a normalized emittance of 95 mm mrad, a 600 G wiggler field with a 0.88 cm period, and a strong guide field of 20 kG, efficiencies of greater than 8% are possible with a FWHM bandwidth of 4.5 GHz.

PACS: 41.60.Cr; 52 − 75.M

Keywords: Bandwidth; W-band; FEM amplifier; Gyro-klystron; Waveguide

1. Introduction

High-power W-band (75–110 GHz) amplifiers are under investigation for applications both in advanced accelerators and radar. Potential radar applications include high-resolution imaging, precision tracking and cloud physics studies. W-band coupled-cavity TWTs have been developed and deployed in several operating instrumentation radar systems. These millimeter wave amplifiers have delivered peak powers of approximately 5 kW with a 10% duty factor and 1 GHz bandwidth [1]. These power levels are nearing the limits of this technology, and other technologies are needed to obtain a significant increase in power at these frequencies. In particular, fast wave devices such as free-electron masers (FEM) and gyro-klystrons must be considered. Gyro-klystron experiments conducted in Russia and in the US have demonstrated high peak powers and high efficiencies. In W-band, a pulsed four-cavity gyro-klystron amplifier achieved 65 kW peak output power at 26% efficiency with 300 MHz bandwidth [2]. A CW version of this device demonstrated 2.5 kW average output power. Experiments at the Naval Research Laboratory (NRL) have further improved four cavity gyro-klystron performance and produced a 67 kW peak output power with a 28% efficiency and 460 MHz full-width at half-maximum (FWHM) bandwidth [3]. More recently, these experiments have produced a record peak output power of 84 kW at 34% efficiency, and significantly

* Corresponding author; Science Applications International Corporation, 1710 Goodridge Drive, McLean, VA 22102, USA; Tel.: + 1-202-767-0034; fax: + 1-202-767-0082.

E-mail address: freund@mmace.nrl.navy.mil (H.P. Freund)

[1] Permanent address: Lucent Technologies, Incorporation, Norcross, GA 30071, USA.

0168-9002/99/$ - see front matter
PII: S0168-9002(99)00069-8

enhanced the bandwidth to 640 MHz FWHM at the output power of 60 kW with a 25% efficiency.

In this paper, we consider the FEM to design a device that can produce both high powers and broad bandwidths. W-band FEM amplifier experiments have been conducted using a 450 kV/17 A sheet electron beam injected into a planar wiggler with a 9.6 cm period [4,5]. Maximum linear gains of 30 dB for 1 W (and 24 dB for 1 kW) of injected power were found with a peak efficiency of 3%. The basic configuration under consideration here employs an electron beam propagating through a cylindrical waveguide in the presence of a helical wiggler field and an axial guide field.

The analysis is based upon the 3D nonlinear simulation code ARACHNE [6] that has been benchmarked, with good agreement, against earlier FEM experiments at NRL [7,8] and MIT [9,10]. In addition, ARACHNE has been used in previous feasibility studies [11,12]. ARACHNE is a slow-time-scale formulation where the electromagnetic field is expanded as a superposition of the vacuum TE and TM modes of a cylindrical waveguide, and the space-charge field is expanded in terms of a superposition of the Gould-Trivelpiece modes. A set of slow-time-scale equations is derived for the evolution of the amplitude and phase of these modes together with the electron beam. The self-electric and self-magnetic fields generated by the bulk charge and current densities of the beam are also included [13]. These equations are integrated simultaneously with the 3D Lorentz force equations.

Issues that must be addressed are thermal damage at high power levels and the suppression of oscillations. Thermal damage can occur in high average power devices from two sources: beam loss to the circuit, and wall loading due to the intense electromagnetic fields. However, calculations indicate that wall loading in the TE_{11} mode is not a major problem for the W-band design presented here. In order to suppress oscillations, no single amplifier section should have gains in excess of about 20–26 dB depending upon the required load match. High gain amplifiers, therefore, often have multiple sections separated by severs in which the gain in each section does not exceed to oscillation threshold.

2. Beam propagation

Stable beam propagation though the device is critically important for high average power operation because even low levels of loss can result in problems. As such, we first consider the conditions needed for beam propagation with minimal wall losses.

The single-particle orbits in FEMs fall into three categories of steady-state helical trajectories. Group I and Group II orbits describe a parallel orientation of the wiggler helicity and axial field. Group I refers to weak axial fields where the Larmor period is longer than the wiggler period, and Group II refers to the opposite. The third category is the reversed-field case where the guide field is oriented anti-parallel to the wiggler, which is an extension of the Group I orbits.

We confine the discussion to the Group II orbits where the guide field is strong enough that the beam is confined over the entire interaction region. A W-band wave-particle resonance is obtained using a 300 kV electron beam and a wiggler with a 600 G field amplitude and a 0.88 cm period. The steady-state helical trajectories corresponding to this case are shown in Fig. 1 as a function of the axial guide field. Amplification at 94 GHz is found for axial magnetic fields in the neighborhood of 20 kG on the Group II branch, and this is the case we shall focus on.

Using ARACHNE to study beam propagation, we find that there is no beam loss corresponding to a 20 kG axial field unless the beam has an

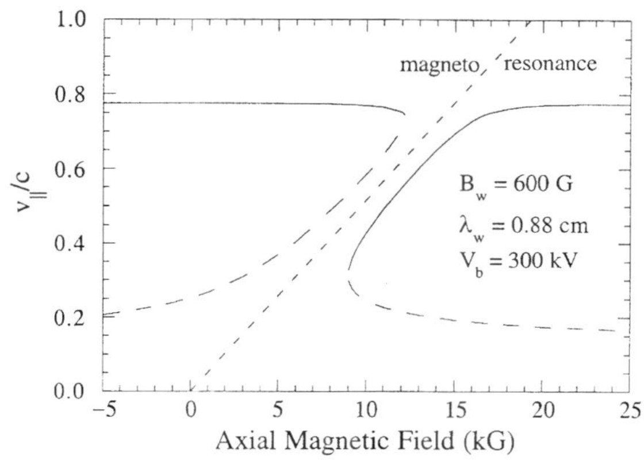

Fig. 1. An illustration of steady-state orbits.

extremely high emittance ($\approx$ 600 mm mrad. As such, beam loss to the drift tube wall during the flattop portion of the voltage pulse is unlikely to be a problem for this Group II orbit configuration.

3. Wave amplification

The gain band is centered at $\approx$ 94 GHz in a waveguide with a 0.3 cm radius for the beam and magnetic fields under study. Specifically, a wiggler amplitude and period $B_w = 600$ G and $\lambda_w = 0.88$ cm, an axial field of $B_0 = 20$ kG, a beam voltage and current of 300 kV and 20 A with a radius of 0.15 cm. We also assume an entry taper of $N_w = 10$ wiggler periods. The amplification of a 1 kW TE_{11} mode drive signal at 94 GHz is shown in Fig. 2 for a beam with a normalized emittance of 95 mm mrad. Saturation is after 61 cm at a power of 408 kW for an efficiency of 6.8%. The exponential growth is $\approx$ 0.68 dB/cm.

The effect of beam emittance on the interaction efficiency is shown in Fig. 3 for varying saturation lengths. A peak efficiency of about 7.0% is obtained for an ideal beam, and the efficiency falls off relatively slowly as the emittance increases up to 100 mm mrad and more rapidly thereafter.

In order to study the instantaneous bandwidth, the interaction length was fixed at 61 cm corresponding to the saturation length at 94 GHz of a 95 mm mrad beam (see Fig. 4). Power and effiiency are shown in Fig. 4 versus frequency. The parameters were chosen so that the center of the gain band and the peak growth rate occur near 94 GHz; however, the maximum efficiency of about 8.5% is found at a frequency between 92–93 GHz. The principal result is that the instantaneous bandwidth (FWHM) is about 4.0 GHz extending from 91.4 to 95.6 GHz with peak power levels of $\approx$ 500 kW. As a consequence, duty factors of from 1% to 10% result in average powers of up to 5–50 kW.

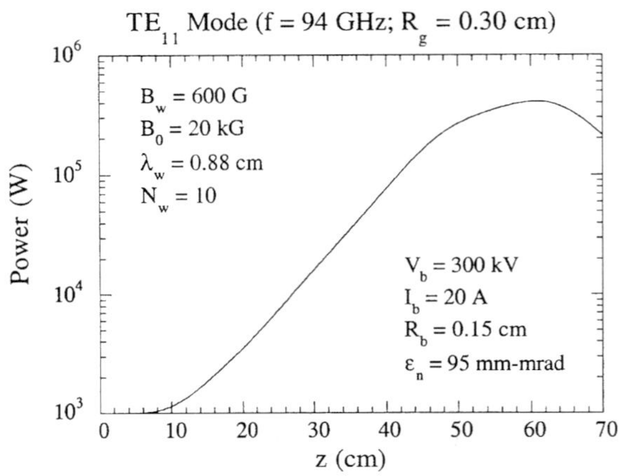

Fig. 2. Amplification of a 1 kW drive signal at 94 GHz for Group II orbit parameters.

4. Wiggler imperfections

Imperfections in the wiggler arise due to a variety of causes in the manufacturing process, and result

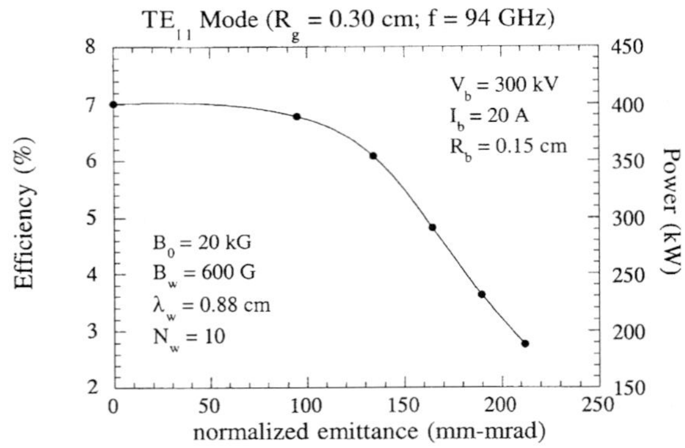

Fig. 3. Plot of the interaction efficiency and power versus normalized emittance.

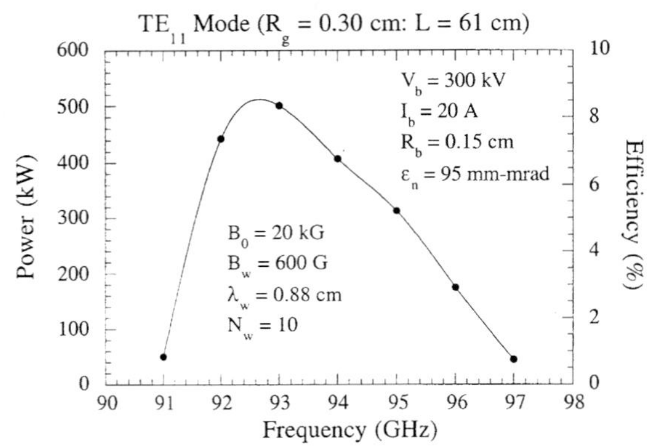

Fig. 4. Instantaneous bandwidth for a 95 mm mrad beam over a length of 61 cm.

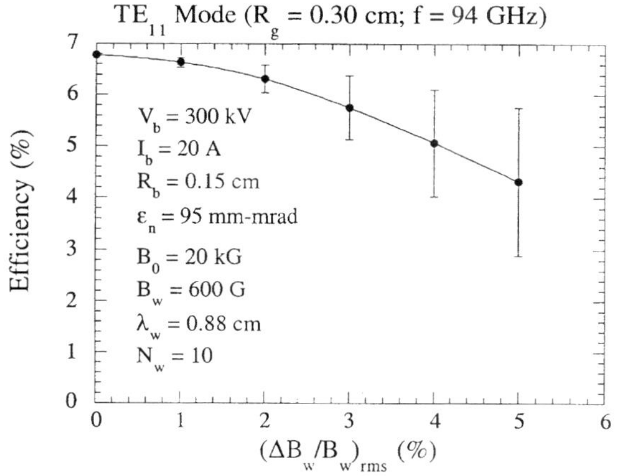

Fig. 5. The efficiency versus amplitude imperfections for the Group II example.

in random fluctuations in both the amplitude and phase of the wiggler. The associated fluctuations in the electron orbits can cause degradations in the gain and efficiency. Wiggler imperfections are modeled in ARACHNE by generating a random series of amplitude fluctuations at intervals of the wiggler period, and then smoothly tapering the amplitude between these points. An ensemble average can be found by running ARACHNE for many such choices of wiggler fluctuation distributions at a given value of the rms error. For the cases under consideration, convergence is found to within about 1% using 20 such randomly generated distributions.

The effect of amplitude imperfections on the design under study is shown in Fig. 5. It is seen that the efficiency begins to degrade for $(\Delta B_w/B_w)_{rms} \geq 1\%$, and drops by 30% for $(\Delta B_w/B_w)_{rms} \approx 5\%$. Because the current state of wiggler technology is easily able to construct wigglers with rms imperfections less than 1%, we conclude that wiggler imperfections will not be a limiting factor.

5. Attenuation and severs

In order to model the effect of wall losses and severs, ARACHNE has been modified to include a variable loss rate. In this context, severs are treated as very large damping rates over some extended region. Thus, a typical loss profile will include a background damping rate as determined by the wall losses as well as a region where the damping rate is extremely high to model a sever.

Resistive wall loading in high-intensity fields results in reduction of the amplification and in wall heating. Heating is important only near the downstream portion of the waveguide where the power is high. The lost rate for the TE_{1n} mode is given by [14]

$$\Gamma_{1n} = \frac{R_s}{R_g \eta_0}\left(1 - \frac{\omega_{co}^2}{\omega^2}\right)^{-1/2}\left[\frac{\omega_{co}^2}{\omega^2} + \frac{l^2}{x_{1n}^2 - l^2}\right] \quad (1)$$

where R_s denotes the surface resistivity, $\eta_0[=120\,\pi\,\Omega]$ is the free space impedance, R_g is the waveguide radius, ω_{co} is the cutoff frequency, and x_{1n} is the nth zero of $J_1'(x_{1n}) = 0$. The ideal surface resistivity of copper waveguide is $R_s = 2.61 \times 10^{-7} f^{1/2}\,\Omega$, where the frequency is in Hz. However, more detailed estimates of R_s have been made at SLAC [15] where, after careful surface preparation, an estimate of $R_s \approx 3.39 \times 10^{-7} f^{1/2}\,\Omega$ was obtained. The loss rate ranges from $\Gamma_{11} \approx$ 0.0033–0.00433 dB/cm for the ideal and estimated surface resistivities at 94 GHz. The power loading per unit area on the waveguide wall can also be calculated and is given in W/cm^2 by [14]

$$\frac{\Delta P}{\Delta A} = (4.07 \times 10^{-4} - 5.29 \times 10^{-4})P \quad (2)$$

at 94 GHz, where the power P is in units of W. If the peak power is ≈ 400 kW and the duty factor ranges from 1% to 10%, then the average wall heating is $\approx$ 1.6–21 W/cm^2. This is, in principle, relatively easy to cool.

A sever is needed to suppress oscillation. In general, if G denotes the gain pass in dB, ρ_o and ρ_i denote the power reflection coefficients in dB at the output, and L denotes the cold loss rate, then oscillation may occur when [16] $G - L - \rho_o - \rho_i > 0$. The cold circuit loss at 94 GHz is ≈ 0.0035 dB/cm; hence, L $\approx$ 0.18 dB over a length of 50 cm. It is difficult to design efficient input and output couplers over broad bandwidths; hence, we assume that 10% of the power is reflected at either end of the interaction region and $\rho_o \approx \rho_i \approx 10$ dB. As a result, G cannot exceed about 20 dB in

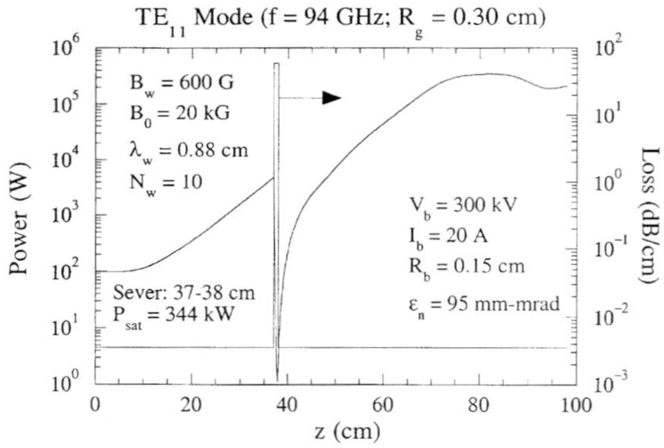

Fig. 6. The evolution of the power and the loss used to model the Group II configuration.

any single section. Assuming a drive power of 100–1000 W and a saturated power of 450 kW, we expect the total gain to $\approx$27–37 dB. This necessitates the use of a sever.

We now include both the cold circuit attenuation of 0.0035 dB/cm and a sever. To ensure that the gains in the input and output sections $\leq$ 20 dB, we model the sever as a region of 60 dB/cm loss extending from 37 to 38 cm beyond the start of the wiggler. In a real device, the sever region must be carefully designed to minimize reflections by the tapering of the loss rates on both sides of the sever. However, this model will provide some guidance for the performance of the device with a sever.

The loss model and the evolution of the power with axial position are shown in Fig. 6. Starting with a drive power of 100 W, a gain of 16.7 dB is obtained in the input section which is attenuated rapidly by the sever. Growth is rapid after the sever, and saturation is found after a total length of 76 cm at a power level of 344 kW. Note that the effect of the sever and attenuation is to cause a reduction in the overall efficiency from about 8.6% without attenuation to 5.9% with attenuation.

6. Summary and discussion

We have analyzed the FEM interaction for a wide bandwidth 94 GHz amplifier using the ARACHNE simulation code. A high axial field Group II configuration was studied using a short period wiggler ($\approx$ 0.88 cm) which requires beam energies of $\approx$ 300 keV to amplify W-band. The efficiencies found in simulation are $\approx$ 6–8% yielding peak powers of $\approx$ 300–450 kW. This yields average powers of $\approx$ 3–45 kW for duty factors of between 1% and 10%. The bandwidths observed ranged from 4% to 5%. No beam losses to the wall were observed for reasonable choices of the emittance, and average wall loading in the interaction region is estimated to range between 1.6 and 21 W/cm^2 for these duty factors.

Beam losses to the wall and wall loading (in all but the sever region) do not appear to be insurmountable problems. The implementation of a sever is challenging, however, and may necessitate breaking the wiggler into two segments allowing room to locate external loads and cooling channels. A second issue left for the future is the actual design of the sever to minimize reflections. The stability of an amplifier requires the sever to have a good match over the entire operating bandwidth. In practice, this is usually accomplished with the insertion of variable attenuation in the region before and after the sever. The simple sever model used here ignores both the difficulty in dissipating the large power levels and in the attenuation matching. An actual engineering design must address both these issues; nevertheless, the present model is a useful initial study of the effect of the sever on bandwidth an output in an FEM amplifier.

Other issues left pending are a detailed engineering design of the input/output couplers. A variety of coupler designs have been used in high-power millimeter-wave gyroklystrons and FEMs. Input/output couplers [17] are in use in a high average power W-Band gyroklystron [3] that operates at 94 GHz with a 600 MHz bandwidth. The output coupler is designed to handle peak powers of up to 120 kW. In addition, a high average power coupler has been implemented on the FOM Fusion FEM [18] which makes use of two quasi-optical stepped reflectors. This coupler operates at frequencies in excess of 160 GHz and handle powers in the MW range. Hence, we do not feel that the coupler design issues represent insurmountable obstacles.

Finally, wiggler imperfections do not appear to pose a problem as long as the rms level of wiggler imperfections is less than or of the order of 2%, which is within the state-of-the-art in wiggler manufacture. As a result, we conclude that the FEM is a feasible option for a high peak power, broad bandwidth W-band amplifier when design considerations do not rule out the use of high-voltage electron beams. The average power capability of such a device requires further evaluation of the issues mentioned above.

Acknowledgements

This work was supported by the Office of Naval Research. Computational work was supported by the DoD HPC Sheared Resource Centers CEWES and NAVO.

References

[1] B.G. James, M. Kreutzer, High Power Millitron TWTs in W Band, Conference Digest, in: R. Temkin (Ed.), 12th International Conference On Infrared and Millimeter Waves, Orlando, FL, 1995, pp. 13–14
[2] I. Antakov et al., Proc. SPIE 2104 (1993) 466.
[3] M. Blank et al., Phys. Rev. Lett. 79 (1997) 4485.
[4] W.W. Destler et al., Phys. Plasmas 1 (1994) 1708.
[5] S. Cheng et al., IEEE Trans. Plasma Sci. 24 (1996) 751.
[6] H.P. Freund, T.M. Antonsen Jr., Principles of Free-electron Lasers, second ed., Chapman & Hall, London, 1996 (Chapter 5).
[7] D.E. Pershing et al., Nucl. Instr. and Meth. A 304 (1991) 127.
[8] D.E. Pershing et al., Nucl. Instr. and Meth. A 355 (1995) 104.
[9] J. Fajans et al., Phys. Rev. Lett. 57 (1986) 579.
[10] D.A. Kirkpatrick et al., Phys. Fluids B 1 (1989) 1511.
[11] H.P. Freund et al., Nucl. Instr. and Meth. A 358 (1995) 163.
[12] M. Blank et al., Nucl. Instr. and Meth. A 375 (1996) 169.
[13] H.P. Freund et al., Phys. Fluids B 5 (1993) 2318.
[14] P. Bhartia, I.J. Bahl, Millimeter Wave Engineering and Applications, Wiley, New York, 1984.
[15] R. Fowkes, R.M. Phillips, personal communication.
[16] J.F. Gittins, Power Traveling Wave Tubes, American Elsevier, New York, 1965, p. 175.
[17] D. Pershing et al., Circuit Aspects of the NRL/Industrial 94 GHz Gyroklystron Amplifier, Conference Digest, in: H.P. Freund (Ed.), 22nd International Conference On Infrared and Millimeter Waves, Wintergreen, VA, 1997, pp. 231–232.
[18] V.L. Bratman et al., Nucl. Instr. and Meth. A 407 (1998) 40.

ELSEVIER

Nuclear Instruments and Methods in Physics Research A 429 (1999) 76–81

NUCLEAR INSTRUMENTS & METHODS IN PHYSICS RESEARCH Section A

www.elsevier.nl/locate/nima

Collective effects in short wavelength free-electron lasers

H.P. Freund[a,*], R. Tatchyn[b]

[a]Science Applications International Corporation, 1710, Goodridge Drive McLean, VA 22102, USA
[b]Stanford Synchrotron Radiation Laboratory, Stanford University, Stanford, CA 94309, USA

Abstract

The self-fields generated by the propagation of pulsed electron beams through the wiggler fields in free-electron lasers are calculated and shown to be diamagnetic. A one-dimensional theory for pulsed beams is applied to proposed designs for fourth generation light sources of high-intensity X-rays, which employ beams with current densities of several hundred MA/cm^2. The theory indicates that the diamagnetic effect is unlikely to affect either the growth length or the fundamental emission energy and bandwidth of these devices. © 1999 Elsevier Science B.V. All rights reserved.

Keywords: Free-electron lasers; One-dimensional theory

Free-electron lasers (FELs) operate by the axial bunching of the electron beam [1]. At lower current densities, this bunching is due to the ponderomotive force produced by the beating of the magnetostatic wiggler field and the radiation, and the interaction is the stimulated Compton scattering of the wiggler off the electron beam. Collective beam-plasma waves are excited in the process, however, and the space-charge forces can exceed the ponderomotive forces for sufficiently high current densities. When this occurs, the interaction proceeds by the stimulated Raman scattering of the beam-plasma waves off the wiggler field. Collective effects also arise from the self-fields generated by the bulk charge and current distributions of the electron beam and, depending upon the field configuration, the propagation of the beam through the wiggler can excite either a diamagnetic or paramagnetic response [1–3].

Collective effects are important in FELs driven by long pulse accelerators where the wavelengths exceed several millimeters and the electron beams have energies up to several MeV and current densities of several hundred A/cm^2. Collective effects have been unimportant in submillimeter wavelength FELs because such systems were, typically, driven by RF linacs with energies exceeding 10 MeV and peak currents less than several hundred Amperes. More recently, there is interest in FELs driven by multi-GeV RF linacs to replace synchrotrons as fourth generation light sources of high-intensity X-rays [4,5] with 1–60 Å wavelengths. Raman effects are unimportant at these wavelengths; however, as these experiments employ pulse-compressed beams with peak currents of several kA and current densities of several

* Corresponding author. Tel.: 202-767-0034; fax: 202-734-1280.
E-mail address: freund@mmace.nrl.navy.mil (H.P. Freund)

0168-9002/99/$ - see front matter
PII: S0168-9002(99)00073-X

hundred MA/cm^2, it is of interest to re-examine the effects of the bulk self-fields generated by such beams.

The original formulations [2,3] of the diamagnetic/paramagnetic effect were applied to long wavelength FELs and made use of a one-dimensional analysis. The beams in these devices were long pulse (i.e., longer than the wiggler) and employed a helically symmetric wiggler in conjunction with a solenoid. The paramagnetic effect arises only for a strong solenoidal field. Since the X-ray FELs under study do not employ solenoidal focusing, the diamagnetic field, $\boldsymbol{B}_{\rm d}$, generated by the propagation of an assumed DC (viz., infinitely long) beam through a helical wiggler is [2]

$$\boldsymbol{B}_{\rm d} = -\frac{\mu}{1+\mu}\boldsymbol{B}_{\rm w}, \tag{1}$$

where $\boldsymbol{B}_{\rm w}$ is the wiggler field, the diamagnetic field parameter $\mu \equiv \omega_{\rm b}^2/\gamma c^2 k_{\rm w}^2$, $\omega_{\rm b}$ is the beam plasma frequency, $k_{\rm w}(\equiv 2\pi/\lambda_{\rm w}$, where $\lambda_{\rm w}$ is the wiggler period) is the wiggler wavenumber, γ_0 is the beam relativistic factor, and c is the speed of light in *vacuo*. This diamagnetic field is generated by the transverse periodic motion of the beam due to the wiggler, and has the effect of partially screening the beam from the wiggler field.

The largest value for μ to date was achieved at the naval research laboratory [6] where $\mu \approx 0.25$ and this provides for an effective shielding of $\approx 20\%$ of the wiggler field. In contrast, μ exceeds unity for the fourth-generation light source designs. These include the linac coherent light source (LCLS) at the Stanford Linear Accelerator Center which, in one of its initial designs [4], employs a 15 GeV/5 kA beam with a 16 μm radius, and the TESLA test facility (TTF) at the Deutsches elektronen-synchrotron [5] using a 1 GeV/2.5 kA beam with a 55 μm radius. Since

$$\mu \cong 0.0059\frac{\lambda_{\rm w}^2 I_{\rm b}}{R_{\rm b}^2\gamma_0} \tag{2}$$

for a current $I_{\rm b}$ in kA (note that $\lambda_{\rm w}$ and $R_{\rm b}$ must have the same units), this means that $\mu \approx 3.6$ for the LCLS which uses a wiggler with $\lambda_{\rm w} = 3$ cm. For the TTF $\lambda_{\rm w} = 2.73$ cm, and $\mu \approx 2.0$. Since the theory for long pulse beams with these diamagnetic field parameters imply the large-scale screening of the wiggler field with a resultant catastrophic decrease in performance, it is important to revisit the analysis for the short pulse beams which are relevant to these fourth generation light source designs. It is our purpose in this paper to extend the previous analyses to short pulse beams.

The diamagnetic field can be found in closed form for arbitrary values of μ for a long pulse beam as in Eq. (1). However, for short pulse beams it is simpler to employ a perturbative approach, which is valid, as long as the self-field effect is not too large. In this approach, we first calculate the self-fields due to the lowest-order wiggler-induced motion, and then use those self-fields to calculate perturbed trajectories. This procedure can be repeated to obtain the self-fields to arbitrary accuracy.

We impose a one-dimensional approximation where the helical wiggler field can be written as

$$\boldsymbol{B}_{\rm w} = B_{\rm w}[\hat{\boldsymbol{e}}_x \cos k_{\rm w}z + \hat{\boldsymbol{e}}_y \sin k_{\rm w}z]. \tag{3}$$

This simplifies the analysis and is valid as long as the beam radius is much less than the wiggler period and $a_{\rm w}/\gamma_0 \ll 1$, where $a_{\rm w} = eB_{\rm w}/m_e c^2 k_{\rm w}$ is the wiggler strength parameter, which is the case in the fourth generation light sources designs. The steady-state orbits in this field are [1]

$$\boldsymbol{v}_0 = v_{\rm w}[\hat{\boldsymbol{e}}_x \cos k_{\rm w}z + \hat{\boldsymbol{e}}_y \sin k_{\rm w}z] + v_{\|}\hat{\boldsymbol{e}}_z \tag{4}$$

where $v_{\rm w} = -eB_{\rm w}/\gamma_0 m_e c k_{\rm w}$, e and m_e are the electronic charge and rest mass, and the axial and transverse components of the velocity satisfy $v_{\rm w}^2 + v_{\|}^2 = (1 - 1/\gamma_0^2)c^2$. As in the earlier cited analyses, we also employ a one-dimensional treatment of the electron beam and Maxwell's equations in order to derive a closed-form result and obtain initial conservative estimates of the self-fields. The pulse shape is described by a beam density of the form $n_0(z,t) = n_{\rm b}S(z - v_{\|}t)$, where $n_{\rm b}$ is a constant and the shape function $S(z - v_{\|}t)$ is unspecified at this point. Maxwell's equations for this lowest order state are

$$\nabla \times \boldsymbol{E} = -\frac{1}{c}\frac{\partial}{\partial t}\boldsymbol{B}, \tag{5}$$

$$\nabla\times\boldsymbol{B}=\frac{1}{c}\frac{\partial}{\partial t}\boldsymbol{E}+\left(\mu k_{\mathrm{w}}\boldsymbol{B}_{\mathrm{w}}-\frac{m_e}{e}\,\omega_{\mathrm{b}}^2\beta_{||}\hat{\boldsymbol{e}}_z\right)S(\zeta) \tag{6}$$

$$\nabla\cdot\boldsymbol{E}=-\frac{m_e}{e}\,\omega_{\mathrm{b}}^2\,S(\zeta) \tag{7}$$

and $\nabla\cdot B=0$, where $\beta_{||}\equiv v_{||}/c$, $\zeta=z-v_{||}t$, and $\omega_b^2\equiv 4\pi e^2 n_{\mathrm{b}}/m_{\mathrm{e}}$.

To lowest order in the wiggler field, Maxwell's equations are

$$\nabla\times\boldsymbol{E}_0=-\frac{1}{c}\frac{\partial}{\partial t}\boldsymbol{B}_0 \tag{8}$$

$$\nabla\times\boldsymbol{B}_0=\frac{1}{c}\frac{\partial}{\partial t}\boldsymbol{E}_0-\frac{m_e}{e}\,\omega_{\mathrm{b}}^2\beta_{||}\hat{\boldsymbol{e}}_z S(\zeta), \tag{9}$$

$$\nabla\cdot\boldsymbol{E}_0=-\frac{m_e}{e}\,\omega_{\mathrm{b}}^2 S(\zeta) \tag{10}$$

and $\nabla\cdot B_0=0$. It can be verified by substitution that

$$\boldsymbol{E}_0=-\frac{m_e}{e}\,\omega_{\mathrm{b}}^2 r\hat{\boldsymbol{e}}_r S(\zeta) \tag{11}$$

and

$$\boldsymbol{B}_0=-\frac{m_e}{e}\,\omega_{\mathrm{b}}^2\beta_{||} r\hat{\boldsymbol{e}}_\theta S(\zeta) \tag{12}$$

are solutions to the lowest order equations correct to within terms of $O(\gamma_{||}^{-2})$ where $\gamma_{||}^2=(1-\beta_{||}^2)^{-1}$. It should also be noted that these are exact solutions for a long pulse or CW beam for which $S(\zeta)=1$, and are the familiar radial-electric and azimuthal-magnetic fields due to a cylindrical beam [7]. As is known, these fields cancel to terms of order $\gamma_{||}^{-2}$ in the net Lorentz force on any bunch electron and exert a typically negligible effect on its trajectory for ultrarelativistic values of $\gamma_{||}$.

The diamagnetic/paramagnetic contributions are found to first order in the wiggler field, for which

$$\nabla\times E_1=-\frac{1}{c}\frac{\partial}{\partial t}\boldsymbol{B}_1 \tag{13}$$

$$\nabla\times\boldsymbol{B}_1=\frac{1}{c}\frac{\partial}{\partial t}\boldsymbol{E}_1+\mu k_{\mathrm{w}}\boldsymbol{B}_{\mathrm{w}}S(\zeta) \tag{14}$$

and $\nabla\cdot\boldsymbol{B}_1=\nabla\cdot\boldsymbol{E}_1=0$. The long pulse result (1) can be recovered when $S(\zeta)=1$ by observing that the temporal derivatives vanish when the sources are independent of time. Hence, $E_1=0$ and $\nabla\times\boldsymbol{B}_1=\mu k_{\mathrm{w}}\boldsymbol{B}_{\mathrm{w}}$. Noting that $\nabla\times\boldsymbol{B}_{\mathrm{w}}=-k_{\mathrm{w}}\boldsymbol{B}_{\mathrm{w}}$, we obtain $B_1=-\mu\boldsymbol{B}_{\mathrm{w}}$ which corresponds to Eq. (1) when $\mu\ll 1$. This diamagnetic field screens the beam from the wiggler, and results in a reduced transverse wiggler velocity $\Delta v_{\mathrm{w}}=-\mu v_{\mathrm{w}}$ and an enhanced axial velocity; hence,

$$\boldsymbol{v}\cong(1-\mu)v_{\mathrm{w}}[\hat{\boldsymbol{e}}_x\cos k_{\mathrm{w}}z+\hat{\boldsymbol{e}}_y\sin k_{\mathrm{w}}z]+v_{||}\left(1+\mu\frac{v_{\mathrm{w}}^2}{v_{||}^2}\right)^{1/2}\hat{\boldsymbol{e}}_z \tag{15}$$

and the growth rate of the interaction is reduced. In the absence of the diamagnetic correction to the trajectories, the resonant wavelength is $\lambda=\lambda_{\mathrm{w}}/2\gamma_{||}^2$. An additional effect of the diamagnetic field is to reduce the wavelength, and we find that

$$\lambda\cong\frac{\lambda_{\mathrm{w}}}{2\gamma_{||}^2}\left(1-2\frac{\Delta v_{\mathrm{w}}}{v_{\mathrm{w}}}\right)=\frac{\lambda_{\mathrm{w}}(1-2\mu)}{2\gamma_{||}^2}. \tag{16}$$

We turn now to the pulsed beam. For simplicity, we treat a square-top pulse in which the shape function is $S(\zeta)=H(\zeta)-H(\zeta-\Delta z)$, where Δz is the pulse width and H is the Heaviside function. We look for harmonic solutions in the limit where $\Delta z\leq\lambda_{\mathrm{w}}$, i.e., we do not incorporate the finite undulator length into the initial boundary conditions at this stage of our analysis. The solutions we find are of the form

$$\begin{pmatrix}\boldsymbol{B}_1\\ \boldsymbol{E}_1\end{pmatrix}=\sum_n\left[\begin{pmatrix}\boldsymbol{a}^{(n)}(t)\\ \boldsymbol{c}^{(n)}(t)\end{pmatrix}\sin nk_{\mathrm{w}}z+\begin{pmatrix}\boldsymbol{b}^{(n)}(t)\\ \boldsymbol{d}^{(n)}(t)\end{pmatrix}\cos nk_{\mathrm{w}}z\right] \tag{17}$$

where $\boldsymbol{a}^{(n)}=a_x^{(n)}\hat{\boldsymbol{e}}+a_y^{(n)}\hat{\boldsymbol{e}}_y$, and similarly for $\boldsymbol{b}^{(n)}$, $\boldsymbol{c}^{(n)}$, and $\boldsymbol{d}^{(n)}$. These coefficients are obtained by substitution into Ampere's and Faraday's laws, after which we find

$$\boldsymbol{B}_1=-\frac{\Delta z}{\lambda_{\mathrm{w}}}\mu\boldsymbol{B}_{\mathrm{w}}\times\left[1-2\sum_{n=1}^{\infty}\frac{[(1+\beta_{||}^2)n^2-1]F_n(\zeta-\Delta z,\zeta)}{4n^2-(1+n^2/\gamma_{||}^2)^2}\right]$$

$$-\frac{2\Delta z}{\lambda_w}\mu\,\hat{e}_z\times B_w\left[\frac{G_1(\zeta-\Delta z,\zeta)}{4-\beta_{||}^2}+\sum_{n=2}^{\infty}\frac{n(1-n^2/\gamma_{||}^2)G_n(\zeta-\Delta z,\zeta)}{4n^2-(1+n^2/\gamma_{||}^2)^2}\right] \tag{18}$$

$$E_1=\frac{\Delta z}{\lambda_w}\mu\beta_{||}B_w\left[\frac{8-3\beta_{||}^2}{\beta_{||}^2(4-\beta_{||}^2)}G_1(\zeta-\Delta z,\zeta)+2\sum_{n=2}^{\infty}\frac{n(1+n^2/\gamma_{||}^2)G_n(\zeta-\Delta z,\zeta)}{4n^2-(1+n^2/\gamma_{||}^2)^2}\right]-\frac{\Delta z}{\lambda_w}\mu\beta_{||}\hat{e}_z\times B_w\left[\frac{F_1(\zeta-\Delta z,\zeta)}{\beta_{||}^2}+\sum_{n=2}^{\infty}\frac{4n^2F_n(\zeta-\Delta z,\zeta)}{4n^2-(1+n^2/\gamma_{||}^2)^2}\right] \tag{19}$$

where

$$F_n(z_1,z_2)=\frac{\sin nk_wz_1-\sin nk_wz_2}{nk_w(z_2-z_1)} \tag{20}$$

$$G_n(z_1,z_2)=\frac{\cos nk_wz_1-\cos nk_wz_2}{nk_w(z_2-z_1)}. \tag{21}$$

There are two points that should be noted. Firstly, the self-fields scale as $(\mu\Delta z/\lambda_w)$ for a short pulsed beam; hence, the self-field effect can be small even if $\mu\geq 1$ for a pulsed beam when $\Delta z\ll\lambda_w$. Secondly, while B_1 is diamagnetic, the displacement current E_1 also contributes to the electron orbits and partially cancels the effect of B_1.

The equilibrium state can be determined by solution of the continuity and momentum transfer equations

$$\left(\frac{\partial}{\partial t}+v_0\cdot\nabla\right)n_1+n_0\nabla\cdot v_1=0 \tag{22}$$

$$\frac{d}{dt}v_1=\frac{e}{\gamma_0m_e}\frac{\Delta z}{\lambda_w}\mu\beta_{||}\hat{e}_z\times B_w\left[1+\frac{(4-3\beta_{||}^2)F_1}{\beta_{||}^2(4-\beta_{||}^2)}+2\sum_{n=2}^{\infty}\frac{(1+n^2/\gamma_{||}^2)F_n}{4n^2-(1+n^2/\gamma_{||}^2)^2}\right]+\frac{e}{\gamma_0m_e}\frac{\Delta z}{\lambda_w}\mu\beta_{||}B_w\left[\frac{(8-\beta_{||}^2)G_1}{\beta_{||}^2(4-\beta_{||}^2)}+\sum_{n=2}^{\infty}\frac{4nG_n}{4n^2-(1+n^2/\gamma_{||}^2)^2}\right] \tag{23}$$

to first order in E_1, B_1, and B_w. Since it is beyond the scope of the present work to present a stability analysis of this equilibrium state, we focus on the momentum transfer equation, which will provide a measure of the diamagnetic shift in the wavelength. When $\Delta z=\lambda_w$

$$\frac{d}{dt}v_1=\frac{e}{\gamma_0m_e}\mu\beta_{||}\hat{e}_z\times B_w \tag{24}$$

and we recover the long pulse result (15).

For conditions specific to the proposed fourth generation light sources, however, $\Delta z<\lambda_w$, $\beta_{||}\approx 1$, $\gamma_{||}\gg 1$. In addition, $|\zeta|\leqslant\Delta z\ll\lambda_w$ and we may assume that $\zeta=0$ to within a good approximation to obtain

$$\frac{d}{dt}v_1=\frac{e}{\gamma_0m_e}\frac{\Delta z}{\lambda_w}\mu\beta_{||}[\Phi(\Delta z)\hat{e}_z\times B_w+\Psi(\Delta z)B_w] \tag{25}$$

where,

$$\Phi(\Delta z)\equiv 1+\tfrac{1}{3}F_1(-\Delta z,0)+2\sum_{n=2}^{\infty}\frac{F_n(-\Delta z,0)}{4n^2-1} \tag{26}$$

$$\Psi(\Delta z)\equiv\tfrac{7}{3}G_1(-\Delta z,0)+4\sum_{n=2}^{\infty}\frac{nG_n(-\Delta z,0)}{4n^2-1} \tag{27}$$

In the long pulse limit ($\Delta z\to\lambda_w$), $\Phi\to 1$ and $\Psi\to 0$ and we recover Eq. (24). In contrast, $\Phi\to\frac{1}{3}$ and $\Psi\to 0$ as $\Delta z\to 0$. The general behavior of these functions is illustrated in Fig. 1. It is clear that $\Phi\approx\frac{1}{3}$ for $\Delta z/\lambda_w<0.01$, but that $\Psi\approx 0$ only when $\Delta z/\lambda_w=1$ or $\Delta z/\lambda_w<10^{-6}$. When $10^{-6}<\Delta z/\lambda_w<1$, the self-fields exert forces on the electrons which are proportional to both $\hat{e}_z\times B_w$ and B_w.

In the ultra-short beam limit in which $\Phi\approx 1/3$ and $\Psi\approx 0$, we have that

$$\frac{d}{dt}v_1\cong\frac{e}{\gamma_0m_e}\frac{\Delta z}{3\lambda_w}\mu\beta_{||}\hat{e}_z\times B_w \tag{28}$$

The solution follows immediately after noting that $v_{1,z}=\text{constant}$ and $\dot{B}_w=k_wv_z\hat{e}_z\times B_w$. Since $v_{1,z}\approx O(v_w^2/v_{||})$, the orbit is given approximately by

$$v\cong\left(1-\frac{\Delta z}{3\lambda_w}\mu\right)v_w[\hat{e}_x\cos k_wz+\hat{e}_y\sin k_wz]+v_{||}\left(1+\frac{\Delta z}{3\lambda_w}\mu\frac{v_w^2}{v_{||}^2}\right)\hat{e}_z. \tag{29}$$

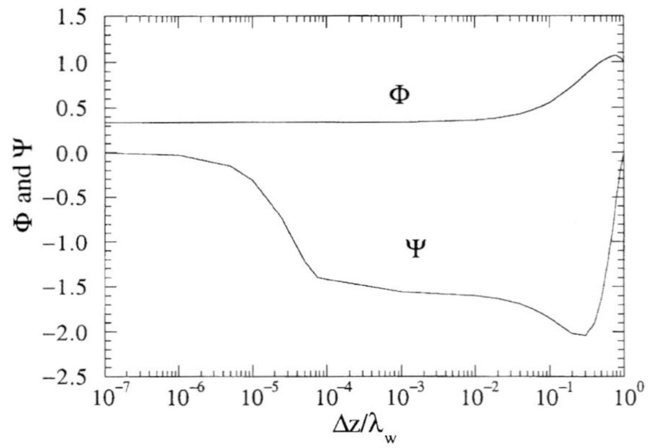

Fig. 1. Plots of Φ and Ψ versus the pulse length.

As a result, the reduction in the transverse velocity is $\Delta v_w/v_w = -\mu(\Delta z/3\lambda_w)$ corresponding to a resonant wavelength of

$$\lambda \cong \frac{\lambda_w}{2\gamma_{||}^2}\left(1 - \frac{2}{3}\frac{\Delta z}{\lambda_w}\mu\frac{a_w^2}{1+a_w^2}\right). \tag{30}$$

The diamagnetic effect, therefore, is substantially smaller than for a long pulse beam.

In the intermediate, but short pulse length, regime

$$\frac{d}{dt}\boldsymbol{v}_1 = \frac{e}{\gamma_0 m_e}\frac{\Delta z}{\lambda_w}\mu\beta_{||}[\Phi\hat{\boldsymbol{e}}_z \times \boldsymbol{B}_w + \Psi\boldsymbol{B}_w] \tag{31}$$

and the trajectory is given by

$$\boldsymbol{v} \cong \left(1 - \Phi\frac{\Delta z}{\lambda_w}\mu\right)v_w[\hat{\boldsymbol{e}}_x \cos k_w z + \hat{\boldsymbol{e}}_y \sin k_w z] + \Psi\frac{\Delta z}{\lambda_w}\mu v_w[-\hat{\boldsymbol{e}}_x \sin k_w z + \hat{\boldsymbol{e}}_y \cos k_w z] + v_{||}\left(1 + \Phi\frac{\Delta z}{\lambda_w}\mu\frac{v_w^2}{v_{||}^2}\right)\hat{\boldsymbol{e}}_z. \tag{32}$$

This results in a wavelength of

$$\lambda \cong \frac{\lambda_w}{2\gamma_{||}^2}\left(1 - 2\Phi\frac{\Delta z}{\lambda_w}\mu\frac{a_w^2}{1+a_w^2}\right). \tag{33}$$

We now explore the consequences of these results for selected FELs in the LCLS class. We first note that the critical dimension Δz depends implicitly on the assumed length of the insertion device. For a long undulator with length $L_u > 2\gamma_{||}^2 L_b$ (with L_b the FWHM bunch length), the fields generated by any part of the beam will propagate over the whole bunch, and Δz will consequently equal the bunch length. This is in fact the approximation under which the foregoing analysis has been carried out, resulting in the maximum possible estimate of the self-fields. For a shorter undulator, the fields from any transverse slice of the bunch will propagate over a distance $\Delta z = N_u\lambda$, where N_u is the number of undulator periods. For an FEL, the coherent fields will be generated predominantly within the last gain length of the undulator, viz., for a distance $\Delta z \approx 0.1N_u\lambda$. In all three cases, the self-field components (in the one-dimensional approximation) can be shown to assume the form of Eqs. (18) and (19), viz., to scale linearly with $\Delta z/\lambda_w$. For the LCLS, where $N_u = 3300$, $\lambda = 0.15$ nm, and $L_b \approx 80$ μm, these three cases considered lead to $\Delta z/\lambda_w$ values of 2.7×10^{-3}, 1.65×10^{-5}, and 1.65×10^{-6}, respectively. In addition, the wiggler design has planar symmetry with an on-axis field of 13 kG, which is equivalent to a helical wiggler with an on-axis field of 9.2 kG and $a_w \approx 2.6$. Using the values of Φ, Ψ, and $(\mu\Delta z/\lambda_w)$ associated with these three values of $\Delta z/\lambda_w$ leads, via Eq. (33), to wavelength shifts of 5.6×10^{-3}, 3.4×10^{-5}, and 3.4×10^{-6}, respectively. For the TTF, with $L_b \approx 54$ μm, a helical-equivalent $a_w 0.9$, $N_u = 915$ and $\lambda = 6.4$ nm, the corresponding wavelength shifts are 1.2×10^{-3}, 1.3×10^{-4}, and 1.3×10^{-5}. It is of interest to note that while the shortest-scale self-field shift for the LCLS insertion device appear to be negligible (even in the one-dimensional approximation), the corresponding wavelength shifts predicted for the TTF is approximately four times larger, conceivably at the threshold of measurability. In view of this, it is necessary to consider further the anticipated corrections resulting from three-dimensional effects.

In general, the extent to which the one-dimensional approximation is valid can be gauged from the ratio of the beam's transverse diameter, D_b, to the length over which the self-field interaction propagates in the frame of the electron bunch. If this ratio is approximately equal to or greater than unity, the transverse slices of the bunch contributing to the field at the point of observation can be approximated by transversely infinite charge sheets, and the one-dimensional approximation is justifiable (at least for self-fields along the longitudinal bunch axis). For ratios substantially smaller than unity most of the bunch slices contributing to the observed self-field will appear as sources of reduced transverse extent, with their radiation fields falling of as $(z-z')^{-1}$ (where z' is the point of observation and z the longitudinal bunch coordinate) in the limit $|z-z'| \to \infty$. For the smallest assumed lab-frame interaction length, viz., $\Delta z \approx 0.1 N_u \lambda_w$, the ratio in question is given by $\approx 10 D_b \gamma_{||}/N_u \lambda_w$. For both the LCLS and the TTF this works out to approximately 0.04, indicating a relatively strong departure from the assumed one-dimensional approximation. This suggests that a fuller three-dimensional analysis of the self-fields in finite-length wigglers should be undertaken for a more precise estimation of self-field effects, and that these are likely to be substantially reduced with respect to the self-fields derived in this paper.

This work was supported in part by the Department of Energy Office of Basic Energy Sciences and High-Energy and Nuclear Physics Department of Energy Contract DE-AC03-76SF00515.

References

[1] H.P. Freund, T.M. Antonsen Jr., Principles of Free-electron Lasers, second ed., Chapman&Hall, London, 1996.
[2] T.J.T. Kwan, J.M. Dawson, Phys. Fluids 22 (1979) 1089.
[3] N.S. Ginzburg, IEEE Trans. Plasma Sci. PS-15 (1987) 411.
[4] R. Tatchyn et al., Nucl. Instr. and Meth. A 375 (1996) 274.
[5] B. Faatz et al., Nucl. Instr. and Meth. A 375 (1996) 441.
[6] R.K. Parker et al., Phys. Rev. Lett. 48 (1982) 238.
[7] R.C. Davidson, Theory of Nonneutral Plasmas, Benjamin, Reading, MA, 1974.

ELSEVIER

Nuclear Instruments and Methods in Physics Research A 429 (1999) 82–87

NUCLEAR INSTRUMENTS & METHODS IN PHYSICS RESEARCH Section A

www.elsevier.nl/locate/nima

Feedback control of cavity detuning in short-pulse free-electron lasers

S.B. Song[a,*], Duk-In Choi[a], Sun Kook Kim[b], Sang June Hahn[c], Jeong Sik Choi[d]

[a]*Department of Physics, Korea Advanced Institute of Science and Technology, Taejon 305-701, South Korea*
[b]*Korea Atomic Energy Research Institute, Daejon, South Korea*
[c]*Chung-Ang University, Seoul, South Korea*
[d]*Dong-Shin University, Naju, South Korea*

Abstract

Feedback control of cavity detuning length in a short-pulse free-electron laser oscillator system is suggested and numerical simulation results are presented. By changing the detuning length with reference to the position of the optical pulse at the end of every pass, the system operation is very stable, and very high power and comparably good spectral characteristics have been achieved. Feasibility of this scheme is also discussed. © 1999 Elsevier Science B.V. All rights reserved.

1. Introduction

Many free-electron laser oscillators are driven by a train of electron bunches shorter than or equal to the slippage distance. This is because RF accelerators are widely used in short wavelength FEL oscillators [1]. So there has been growing interest in FELs of this type.

Short-pulse FEL oscillators are known as rich sources of various interesting phenomena, like superradiance, limit cycle oscillation, chaos, and so on. The short-pulse FEL oscillator as a nonlinear dynamical system has been analyzed numerically, and the superradiant phenomenon has been observed numerically and studied analytically [2–5]. Dynamic cavity desynchronization is also suggested and its realization is carried out for this type of oscillator, to get maximum small-signal gain at the amplifying stage, and to get maximum output power at saturation [6–8].

In this paper, we try to actively control the desynchronism between the optical pulse and the electron bunch in numerical simulations. Because of the lethargy effect, the centroid of the optical pulse travels slower than the speed of light in vacuum, especially at the startup. We check the position of the centroid of the optical pulse at the end of each round-trip, and decide the magnitude of cavity detuning for the next round-trip.

2. Feedback control

It is well known that matching between the electron beam micropulse and the optical pulse is very important. There is an optimal cavity detuning

* Corresponding author.

0168-9002/99/$ - see front matter
PII: S0168-9002(99)00076-5

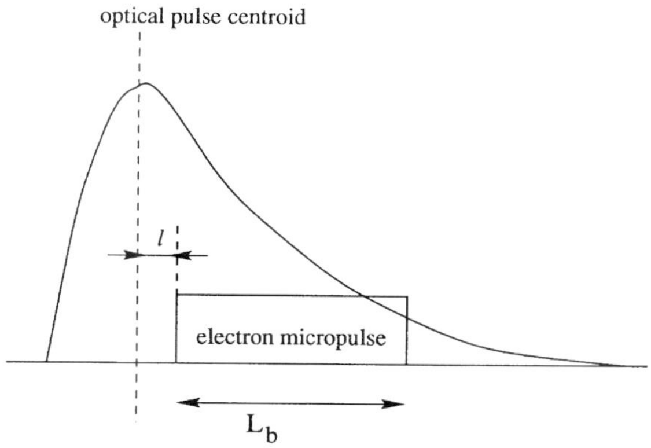

Fig. 1. Matching the new electron bunch of the optical pulse.

value for the maximum gain, while the maximum saturation power exists at a smaller cavity detuning. With decreasing (in absolute value) cavity detuning, the system changes from steady-state regime to limit-cycle oscillation regime, and further decreases of detuning causes the onset of chaos via periodic doubling. Near the perfect synchronism, the system is extinguished due to the lethargy effect.

All these phenomena are due to the mismatch between the optical pulse and the electron bunch. Because of the slippage effect, it is impossible to match the electron bunch with the centroid of the optical pulse all the time during the operation, but it would be possible to change the cavity detuning at the end of each pass.

The idea of feedback control is to change the cavity detuning value with respect to the position of the optical pulse, such that the new electron bunch can be launched to perfectly match the centroid of the optical pulse, as in Fig. 1. Here l is the electron micropulse position with respect to the optical centroid in units of wavelength.

To simulate light and electron pulse propagation, we used the usual one-dimensional model of equations in the Compton approximation. The governing equations are [9],

$$\left(\frac{\partial}{\partial z}+\frac{1}{v_{\|}}\frac{\partial}{\partial t}\right)\gamma_j = -\frac{k_s a_w a_s}{\gamma_j}\sin\psi_j \tag{1}$$

$$\left(\frac{\partial}{\partial z}+\frac{1}{v_{\|}}\frac{\partial}{\partial t}\right)\theta_j = k_w - k_s\frac{(1+a_w^2)}{2\gamma_j^2} \tag{2}$$

$$\left(\frac{\partial}{\partial z}+\frac{1}{c}\frac{\partial}{\partial t}\right)a_s e^{i\phi_s} = i\eta F(z-v_{\|}t)a_w\left\langle\frac{e^{-i\theta}}{\gamma}\right\rangle. \tag{3}$$

Here γ is the electron's relativistic Lorenz factor, $j=1,2,\ldots,N_p$ the electron index, $v_{\|}$ the electron velocity parallel to the direction of the light pulses, k_s the wave number of the light signal, a_w, a_s are normalized vector potentials of the wiggler field and the light signal, respectively. ψ_j is the ponderomotive phase of the jth particle defined as

$$\psi = \theta(z) + \phi_s = (k_w + k_s)z - \omega_s t(z) + \phi_s$$

where k_w is the wave number of the wiggler field, ω_s the light frequency, and ϕ_s the light field phase. In Eq. (3), $\eta = \omega_p^2/2k_s c^2$ (w_p is electron plasma frequency), $F(z-v_{\|}t)$ is the axial electron beam micropulse profile normalized to 1 at the peak ($0 \le F \le 1$), and the angular brackets indicate an average over the electrons. In these simulations, most of parameters are those of FEL-1 in FELIX, and the rectangular profile is assumed for the electron micropulse. Note that the slippage length in the device (38λ) is longer than the electron beam length ($\approx 25\lambda$).

How the feedback has acted is shown in Fig. 2. At the startup of the oscillation, the optical pulse retards to $\delta L \approx -6.9\lambda$, and at the 32the pass, the zero detuning first appeared. It means that the peak which has the largest amplitude in the optical pulse remained the strongest after one round trip. Between the 30th pass and the 35th pass the peak of the optical pulse wanders, as two peaks in the pulse compete, and after the 40th pass, the peak of the optical pulse almost recovered the speed of light in vacuum. However, to maintain the signal, the feedback control should not be stopped. (It is like making a rod stand on a finger by moving the finger in the direction of falling continuously.) The averaged output power is oscillating somewhat, but is damping out.

In Fig. 3, the electron beam length is set equal to the slippage length. The optical pulse retards to $\delta L \approx -2.75\lambda$ at the startup, and if we keep the electron bunch tracking the optical pulse position, at around the 50–100th pass the centroid of the optical pulse sometimes exceeds the speed of light

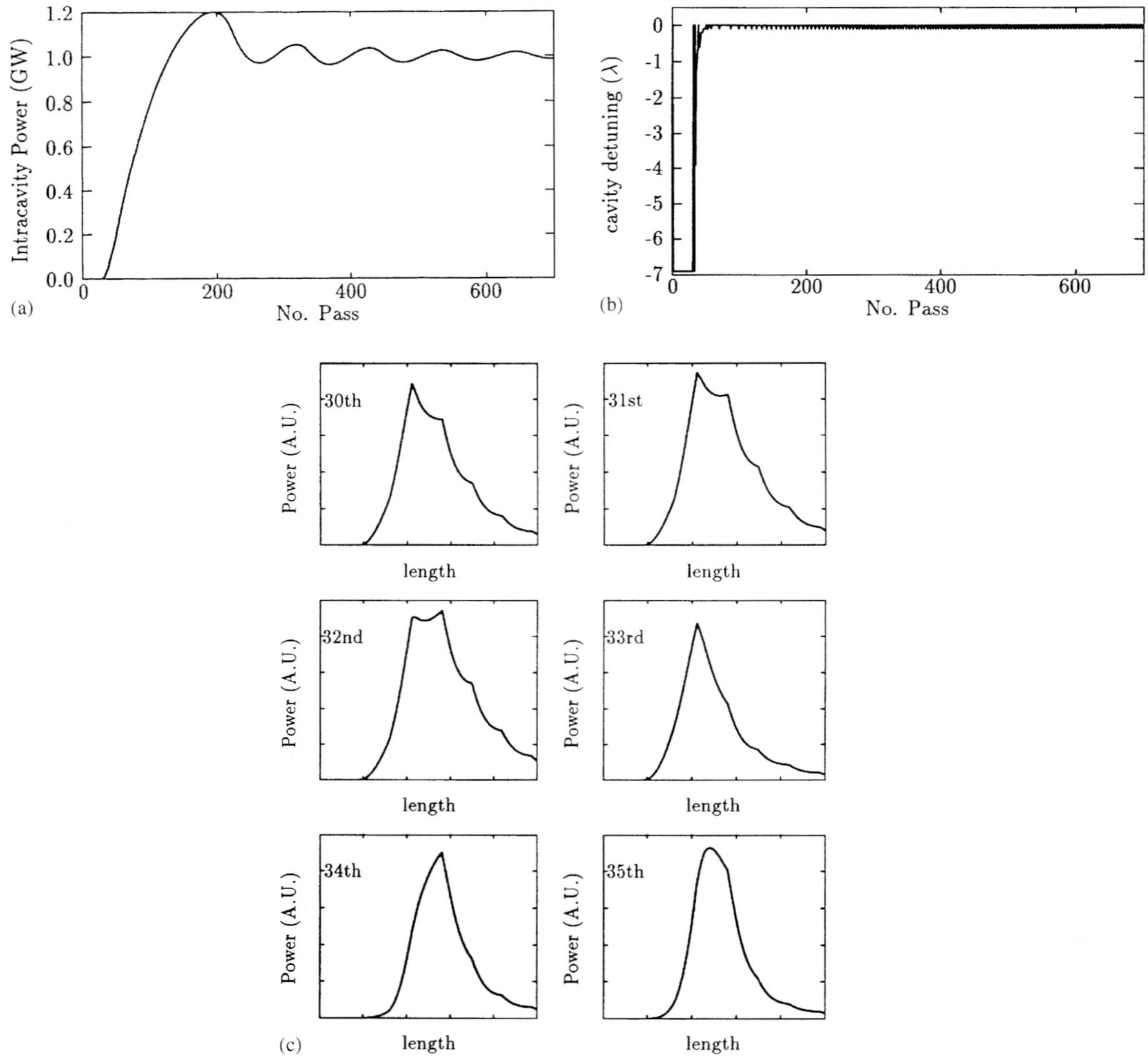

Fig. 2. Feedback control of the cavity detuning; (a) the micropulse averaged intracavity power of the optical pulse, and (b) the controlled cavity detuning value. (c) Optical pulse profiles at 30th–35th pass.

in vacuum. This is because we position the tail of the new electron bunch at the centroid of the optical pulse, making the electron bunch start slightly in advance of the optical pulse. By doing this, the electrons, while slipping back, effectively interact with the major part of the optical pulse.

3. Power and spectrum enhancement

Fig. 4 is the intracavity power of the optical micropulse; (a) is without the feedback control of the cavity detuning, and (b) is with feedback control after the 200th pass. In the case of feedback control, the "l value" is set to 0. The FEL parameters are the

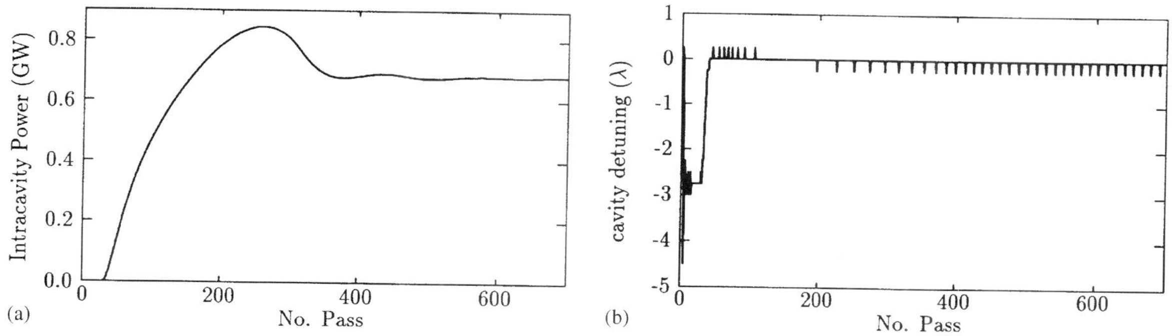

Fig. 3. Feedback control of the cavity detuning; $L_b = N\lambda$. (a) the micropulse averaged intracavity power of the optical pulse, and (b) the controlled cavity detuning value.

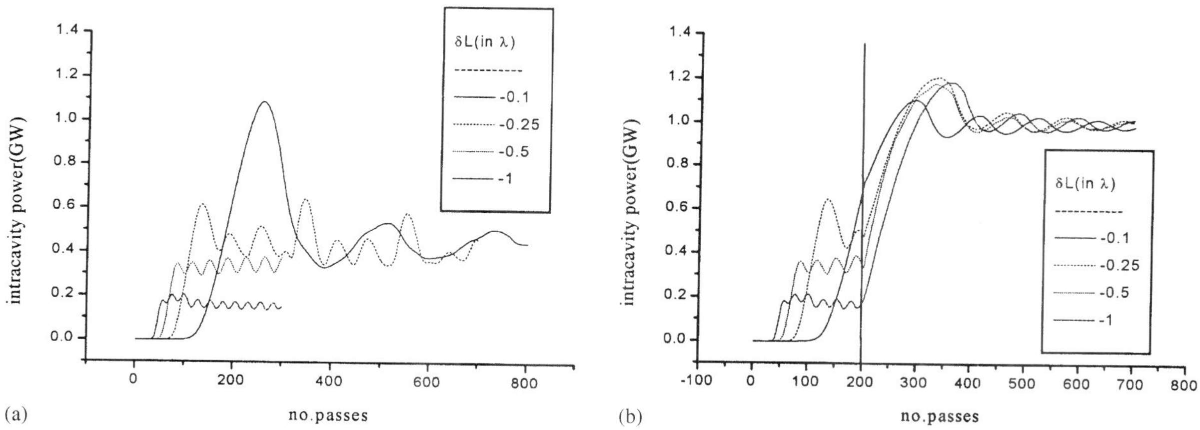

Fig. 4. Intracavity power of the optical micropulse; (a) before feedback control, (b) feedback control starting at the 200th pass.

same as those in Fig. 2. As we have expected, they all go to the same status, so the controlled status is quite stable and there seems to be no initial condition dependency. The output power in a feedback controlled system is very high and there is no chaotic fluctuations or stable oscillations. It is also clear that the system is not in a transient state, but in a steady state.

In Fig. 5, we compared the spectral profile of a feedback controlled signal with other statically detuned signals. The spectral bandwidth of the feedback-controlled signal in Fig. 5(d) is narrower than in case (a), where δL is -0.1λ, and in case (b), where δL is $-\lambda$ and the system is in the limit cycle oscillation regime. The FWHM is about $3\lambda_0$ in this simulation. Moreover, the spectrum is much smoother in the feedback-controlled case, and the power level is also much higher, as shown in Fig. 4. Though the sharpness of the spectrum is far better in case (c), where $\delta L = -3\lambda$, the power level of this region is much less than even that of (b). These results suggest that by feedback control of cavity detuning, a very high power level and comparably good spectrum profile may be possible at the same time in the short-pulse FEL oscillators.

The power level of FCS (feedback controlled signal) will depend on the parameter l, which is defined in Fig. 1. The l parameter determines the initial position of the electron bunch with respect to the optical pulse. Because of the slippage, there

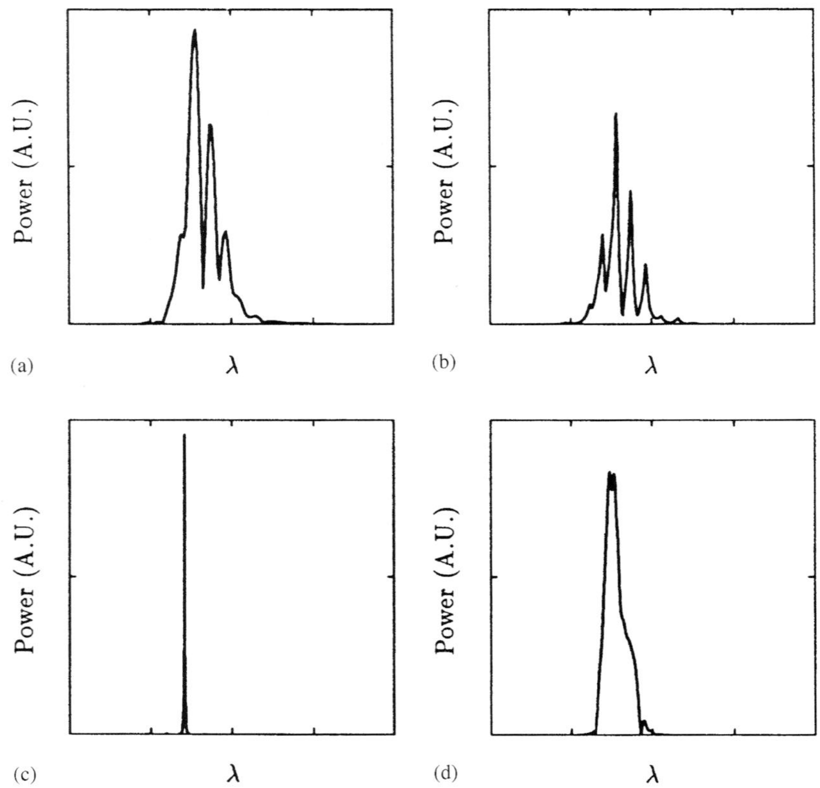

Fig. 5. Power spectrum; (a) $\delta L = -0.1\lambda$, (b) $\delta L = -\lambda$, (c) $\delta L = -3\lambda$, (d) feedback-controlled case.

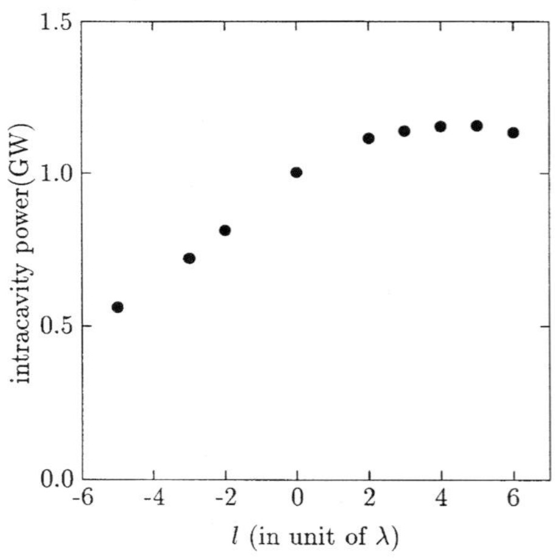

Fig. 6. Time-averaged power of the steady-state regime. The variable l is defined in Fig. 1.

must be an optimal $l > 0$, which maximizes the steady-state power. We computed the time-averaged steady-state power as a function of l, and the results are shown in Fig. 6. The optimal l value is 5λ in this case. From $l = 0$, the power decreases relatively fast as l becomes negative, while only slight changes are seen when l has positive values.

4. Conclusion and discussion

We have suggested the feedback control of cavity detuning for a short-pulse free-electron laser system to achieve high power, good spectrum, and stable operation. Using feedback control, we have found the steady state near the zero cavity detuning regime, and it does not depend on the initial condition of the system. The high power with a feedback-controlled signal may be understood as a superradiant behavior [5,10]. The transient regime near zero cavity detuning in an ideal case without loss has been studied previously [10] and it may be possible to describe the feedback control of cavity detuning as a stabilization of the transient regime. The analysis is under way.

Realization of this feedback control would be not so easy. The required feedback time is, for the FEL-I system in FOM, about 40 ns. The RF-signal modulation technique used in the DCD experiment may be used for this type of control [8]. At the steady state, the cavity detuning value changes only sightly, but at the amplification stage, the required change of cavity detuning per one round trip can be up to several wavelengths. Even if the required feedback time and scale is achievable, it would be very difficult to identify the position of the optical pulse at the end of each pass.

As an alternative, because feedback cavity detuning signals in these simulations are almost periodic at the steady-state stage, it may yield a similar result if we change the cavity detuning periodically. But in this case, there is no stabilizing force in the system, so it would work only for a limited time and only when the near synchronism state is not heavily unstable. To make the system reach its steady state is also a problem.

References

[1] W.B. Colson, Nucl. Instr. and Meth. A 393 (1997) 6.
[2] Sang June Hahn et al., Phys. Rev. E 48 (1993) 2162.
[3] Sang June Hahn et al., Phys. Lett. A 176 (1993) 339.
[4] N. Piovella et al., Phys. Rev. E 52 (1995) 5470.
[5] D.A. Jaroszynski et al., Phys. Rev. Lett. 78 (1997) 1699.
[6] P. Sprangle et al., Phys. Rev. A 28 (1983) 2300.
[7] D.A. Jaroszynski et al., Nucl. Instr. and Meth. A 296 (1990) 480.
[8] R.J. Bakker et al., Phys. Rev. E 48 (1993) R3256.
[9] N.M. Kroll, et al., J. Qu. Elec. QE-17 (1981) 1496.
[10] N. Piovella, Phys. Rev. E 51 (1995) 5147.

ELSEVIER

Nuclear Instruments and Methods in Physics Research A 429 (1999) 88–93

NUCLEAR INSTRUMENTS & METHODS IN PHYSICS RESEARCH Section A

www.elsevier.nl/locate/nima

Ginzburg–Landau model for a long-pulse low-gain free-electron laser oscillator

C.S. Ng*, A. Bhattacharjee

Department of Physics and Astronomy, The University of Iowa, Iowa City, IA 52242, USA

Abstract

The Ginzburg–Landau model for the radiation field of a free-electron laser (FEL) was originally derived for a high-gain amplifier (S.Y. Cai and A. Bhattacharjee, Phys. Rev. A 43 (1991) 6934). With a view to making precise comparisons with experimental data from the long-pulse FEL oscillator at the University of California at Santa Barbara (UCSB) (L.R. Elias, G. Ramian, J. Hu, A. Amir, Phys. Rev. Lett. 57 (1986) 424), we have developed a new formulation of the Ginzburg–Landau model starting from the low-gain oscillator equations. We implement a small-amplitude expansion of the radiation field, and derive the coefficients of the Ginzburg–Landau equation by analysis as well as by Mathematica. Stability analysis of the Ginzburg–Landau equation produces results similar to those obtained by T.M. Antonsen and B. Levush (Phys. Fluids B 1 (1989) 1097). These include the stability of the main mode (no Benjamin–Feir instability), phase-unstable off-centered modes (Eckhaus instability), as well as relaxation to the single mode which occurs much faster in amplitude than in phase. We obtain the saturated radiation amplitude a_0 as functions of the detuning parameter p_{inj} and cavity loss, and determine the phase instability boundary in the $a_0 - p_{\text{inj}}$ plane. The probability of realizing a single mode starting with random initial conditions is calculated and compared with spectral measurements from the UCSB FEL.

PACS: 41.60.Cr; 42.65.Tg; 52.35.Mw

Keywords: Ginzburg–Landau model; Free electron; Single-mode state

1. Introduction

Cai and Bhattacharjee [1] have proposed that the radiation field of a free-electron laser (FEL) amplifier can be modeled by the complex Ginzburg–Landau equation (GLE) of the general form

$$\frac{\partial A}{\partial z} = A + (1 + \mathrm{i}c_1)\frac{\partial^2 A}{\partial t^2} - (1 + \mathrm{i}c_2)|A|^2 A. \tag{1}$$

In Eq. (1), $A(z, t)$ is a complex scalar field, z and t are real independent variables, and c_1 and c_2 are real parameters. In previous work [2], motivated largely by the experimental results from the long-pulse FEL at the University of California at Santa Barbara (UCSB) [3,4], we have applied the GLE to discuss the issues of mode competition and single-mode operation. In particular, the GLE has enabled us to resolve a long-standing controversy [5–11] on whether single-mode operation was truly attained in the UCSB FEL. Our results tend to support the conclusion [8–11] that a single mode was most probably not realized in the experiment

* Corresponding author.

0168-9002/99/$ - see front matter
PII: S0168-9002(99)00077-7

[3,4]. The early disappearance of temporal variations in the radiation power, seen in the experiment [3,4], is due to the much faster decay rate of amplitude perturbations than phase perturbations. However, the mere absence of temporal variations in the power is not sufficient to ensure that a single mode is realized because it conceals the presence of multiple modes rapidly varying in phase.

Our earlier derivation of the GLE [1,2] was based on the equations for a FEL amplifier in the high-gain, Compton regime. Since our qualitative conclusions on a single-mode operation [2] are similar to those of Antonsen and Levush [8–11] who rely on the low-gain oscillator equations, we have conjectured [2] that the underlying physics of mode-competition and single-mode operation is similar in amplifiers and oscillators. In this paper, we prove the conjecture by deriving the GLE from the low-gain oscillator equations. In so doing, we make a strong case for the universal applicability of the GLE to a long-pulse FEL, independent of whether the FEL is configured as an oscillator or an amplifier.

The following is a plan of this paper. In Section 2, we present a derivation of the GLE from the low-gain oscillator equations. In Section 3, we discuss the stability of a single-mode solution of the GLE. With a view to comparing theoretical predictions with direct spectral measurements from the UCSB FEL [12], we report in Section 4 the results of a large number of high-resolution simulations of the GLE with random initial conditions. We conclude in Section 5 with a summary.

2. Derivation of the GLE

We derive the GLE from the equations for a low-gain oscillator. In particular, it is convenient to use the formulation used by Antonsen and Levush [8–11] who represent the radiation field by the Fourier series

$$a(\tau_s,\tau_0) = \sum_{n=-\infty}^{\infty} a_n(\tau_s)\exp(-in\pi\tau_0). \tag{2}$$

In Eq. (2), a is the normalized signal amplitude, related to the vector potential A for the radiation field by the relation

$$a = \frac{qA}{mc^2}\left(\frac{\omega L}{c}\right)^2 \frac{K(1+K^2/2)}{[\gamma_R^2-(1+K^2/2)]^2} \tag{3}$$

where q, m, γ_R, and c is the electron charge, mass, relativistic factor and speed of light, ω is the reference frequency of the radiation, $K = qA_w/mc^2\gamma_R$ is the wiggler parameter, L is the length of the interaction region, and A_w is the magnetic potential of the wiggler. The time dependence of a is separated into a fast time τ_0 associated with the time of transit of the radiation through the empty cavity, and a slower time τ_s associated with the decay time of the radiation in the empty cavity. Specifically, we write $\tau_0 = tv_g/L_c$, where t is the physical time, L_c is the cavity length, v_g is the axial group velocity and $\tau_s = tvv_g/2L_c$, where v is the fraction of the power lost from the radiation field per round trip. Note that a is a periodic function of τ_0, with a period of 2. The evolution of a in slow time is given by the equation

$$\frac{da_n}{d\tau_s} + \frac{1}{2}a_n = -\frac{i\hat{I}}{4v}\int_0^2 \frac{d\tau_0}{2}\int_0^1 d\xi \langle \exp\{-i[\psi - n\pi(\varepsilon\xi+\tau_0)]\}\rangle \tag{4}$$

where

$$\hat{I} = \frac{4\pi j}{I_A}\frac{L^3\omega}{v_g}\frac{K(1+K^2/2)}{[\gamma_R^2-(1+K^2/2)]^2}$$

is the normalized current. Here $I_A = \gamma_R\beta_z mc^3/q$ is the so-called Alfvén limiting current, j is the effective beam current density, $\beta_z = v_z/c$ is the normalized electron axial velocity, $\xi = z/L$ is the normalized axial coordinate,

$$\varepsilon = \frac{L}{L_c}\left(\frac{v_g}{v_z} - 1\right) \tag{5}$$

is the slippage parameter, and $\psi = (k_w + k_z)z - \omega t$ is the relative phase of an electron, with k_w, k_z denoting the wave number of the wiggler and the radiation field. Note that the angle bracket in Eq. (4) represents an ensemble average over entrance phases ψ_0, assumed to be uniformly distributed between 0 and 2π. The phase ψ satisfies the one-dimensional pendulum equation,

$$\frac{dp}{d\xi} \equiv \frac{d^2\psi}{d\xi^2} = \mathrm{Im}\left(\sum_{n=-\infty}^{\infty} a_n \exp\{i[\psi - n\pi(\varepsilon\xi + \tau_0)]\} \right) \quad (6)$$

subject to the boundary condition

$$\left.\frac{d\psi}{d\xi}\right|_{\xi=0} = p_0 = \text{constant}. \quad (7)$$

Eqs. (4), (6) and (7) are the basic equations underlying the numerical simulations discussed in Refs. [8–11].

Now, we define $Q \equiv \langle \exp(-i\psi) \rangle$ and observe that Q has the following functional dependence:

$$Q = Q[p_0, \xi, a(\tau_s, \tau_0), \varepsilon\partial a(\tau_s, \tau_0)/\partial\tau_0, \varepsilon^2\partial^2 a(\tau_s, \tau_0)/\partial\tau_0^2, \ldots]. \quad (8)$$

By Eq. (2), we can rewrite Eq. (4) as

$$\frac{\partial a}{\partial \tau_s} + \frac{1}{2}a = -\frac{i\hat{I}}{4v}\int_0^1 d\xi Q[p_0, \xi, a(\tau_s, \tau_0 - \varepsilon\xi), \varepsilon\partial a(\tau_s, \tau_0 - \varepsilon\xi)/\partial\tau_0, \ldots]. \quad (9)$$

Expressing ψ in a Taylor series

$$\psi(\xi) = \sum_{n=0}^{\infty} \frac{1}{n!} \left.\frac{d^n\psi}{d\xi^n}\right|_{\xi=0} \xi^n \quad (10)$$

we can write, in the low-gain approximation,

$$\exp[-i\psi(\xi)] = \exp[-i(\psi_0 + p_0\xi)] \sum_{m=0}^{\infty} \frac{1}{m!}\left[-i\sum_{n=2}^{\infty} \frac{\xi^n}{n!} \left.\frac{d^n\psi}{d\xi^n}\right|_{\xi=0} \right]^m. \quad (11)$$

Note that by Eq. (6), the mth term in the above series is $O(a^m)$. We will restrict ourselves to the case of small a, and so keep only terms up to $O(a^3)$. Also, it can be easily seen that terms of a raised to odd power are the only ones to survive the ensemble average. So we obtain

$$Q \approx \left\langle e^{-i(\psi_0 + p_0\xi)} \left\{ -i\sum_{n=2}^{\infty} \frac{\xi^n}{n!}\frac{d^n\psi_0}{d\xi^n} + \frac{i}{6}\left[\sum_{n=2}^{\infty} \frac{\xi^n}{n!}\frac{d^n\psi_0}{d\xi^n} \right]^3 \right\} \right\rangle. \quad (12)$$

Let us first consider the linear term of Eq. (12). By Eqs. (2), (6), (9) and (12), it can be shown analytically that

$$\frac{\partial a}{\partial \tau_s} + \frac{1}{2}a = \frac{\hat{I}}{4v}\left[G_0(p_0)a - iG_1(p_0)\varepsilon\frac{\partial a}{\partial \tau_0} - \frac{1}{2}G_2(p_0)\varepsilon^2\frac{\partial^2 a}{\partial \tau_0^2} \right] \quad (13)$$

where

$$G_0(p_0) = \frac{1}{2}\frac{d}{dp_0}\left(\frac{\cos p_0 - 1}{p_0^2} \right) + \frac{i}{2}\frac{d}{dp_0}\left(\frac{p_0 - \sin p_0}{p_0^2} \right) \quad (14)$$

$$G_1(p_0) = \frac{dG_0}{dp_0}, \quad G_2(p_0) = \frac{d^2G_0}{dp_0^2}. \quad (15)$$

Note that the real part G_{0r} is just the standard small gain function of an oscillator which is positive for $0 < p_0 < 6.28$ with a maximum at $p_0 \approx 2.6$.

The third-order term in Eq. (12) is more complicated. If we neglect all contributions due to $\varepsilon\partial/\partial\tau_0$ as well as its higher powers, which is justified if $|a|$ and ε are sufficiently small, it can be seen that there will be an additional term proportional to $|a|^2 a$ in Eq. (13). The final equation is of the form

$$\frac{\partial a}{\partial \tau_s} + \frac{1}{2}a = \frac{\hat{I}}{4v}\left[G_0(p_0)a - iG_1(p_0)\varepsilon\frac{\partial a}{\partial \tau_0} - \frac{1}{2}G_2(p_0)\varepsilon^2\frac{\partial^2 a}{\partial \tau_0^2} + \beta(p_0)|a|^2 a \right]. \quad (16)$$

Although the coefficient β cannot be written in closed form, we can express it as a power series in p_0, that is,

$$\beta(p_0) = \sum_{n=0}^{\infty} \beta_n p_0^n \quad (17)$$

and calculate the coefficient β_n using Mathematica up to $n = 25$. We also find that $\beta(p_0)$ is well-approximated by the relation

$$\beta(p_0) \approx -2.3|G_0(p_0)|^2 G_0(p_0) \quad (18)$$

for $|p_0| < 6$.

Eq. (16) is already in the form of the GLE. We can transform it into the standard form (1) by using the transformations given in Ref. [2].

3. Stability of a single mode

Single mode solutions of Eq. (1) and their stability properties have been discussed in detail in Ref. [2]. The single mode with the largest gain is stable, that is, there is no Benjamin–Feir instability of the GLE if $1 + c_1 c_2 > 0$. Using the transformations given in Ref. [2], we can show easily that $c_1 = G_{2i}/G_{2r}$, $c_2 = \beta_i/\beta_r$, and furthermore,

$$1 + c_1 c_2 = 1 + G_{2i}\beta_i/G_{2r}\beta_r > 0 \tag{19}$$

for $|p_0| < 4.632$. However, away from the gain maximum, the system is unstable to the Eckhaus instability. Following the method discussed in Ref. [2], we can obtain analytically the following stability condition for the slippage parameter $\varepsilon \to 0$:

$$\left(1 + \frac{G_{2i}\beta_i}{G_{2r}\beta_r}\right)|a_0|^2 > \frac{G_{1r}^2}{G_{2r}\beta_r}\left(1 + \frac{\beta_i^2}{\beta_r^2}\right). \tag{20}$$

Here

$$|a_0|^2 = \frac{1}{\beta_r}\left[\frac{G_m}{\chi} - G_{0r}\right] \tag{21}$$

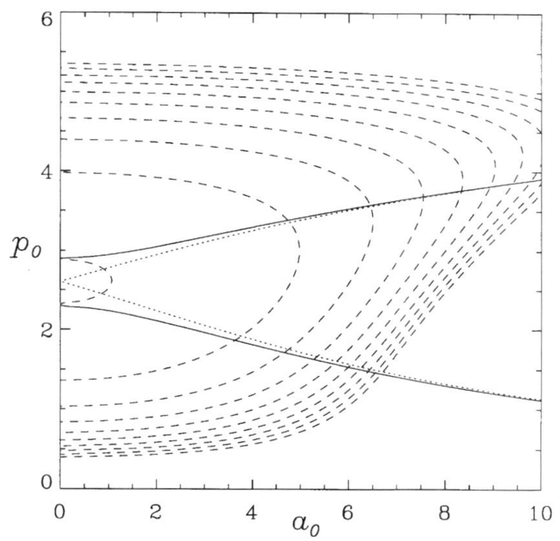

Fig. 1. Contours of constant χ (dashed curves), from 1 (most inner curve) to 4 (most outer curve) with a constant increment of $\frac{1}{3}$ using $G_m = \frac{1}{15}$ calculated from Eq. (21), plotted on the (a_0, p_0) space. The phase stability boundary for $\varepsilon = 0.2$ (solid curve), and for $\varepsilon \to 0$ (dotted curve) is calculated from Eq. (20). The central region is stable.

is the square of amplitude of the complex saturated radiation field, G_m is a constant chosen to be close to the maximum value of G_{0r} and $\chi \equiv G_m\hat{I}/4v$. Note that it is always possible to choose $|a_0|^2$ so that the Eckhaus stability condition (20) is violated. As shown in Ref. [2], the Eckhaus instability is a phase instability because the phase perturbation grows much faster than the amplitude perturbation in the linear stage of the dynamics.

In Fig. 1 we plot contours of constant χ (dashed lines), calculated from Eq. (21) in (a_0, p_0) space. They are qualitatively similar to those reported in Refs. [8–11]. When $\varepsilon > 0$, the analytical condition (20) is not accurate for small $|a_0|$, and so we determine the stable domain numerically. The solid lines in Fig. 1 define the stability boundaries for $\varepsilon = 0.2$, with the stable region lying in-between the two lines. These curves also are qualitatively similar to those reported in Refs. [8–11]. The dotted line in Fig. 1 represents the stability boundaries according to the analytical condition (20) which is valid in the limit $\varepsilon \to 0$. For small values of ε (for example, $\varepsilon \approx 3.3 \times 10^{-3}$ for the UCSB FEL), the numerical stability boundaries cannot be distinguished from the $\varepsilon \to 0$ boundaries when plotted on the scale of Fig. 1.

4. Spectrum and statistics

In Ref. [2], we have already presented numerical simulations for the GLE showing that it takes a rather long time, compared with the electron pulse length, for the radiation field to relax to a single mode. While the amplitude relaxes quickly to a nearly constant value, multiple modes persist in the phase. This can be easily explained by the fact that the decay rate of the amplitude perturbation is much larger than that of the phase perturbation. The spectral half-width of the radiation field is found to decay as $\tau_s^{-1/2}$, in agreement with Refs. [8–11].

In the experiment reported in Ref. [12], the spectrum is directly observed using a spectrometer which is capable of resolving 20 modes in a given pulse. About 90 spectra were taken and about 29% of them were found to be a "single mode" state, defined in Ref. [12] as a state in which one mode has power at least twice as large as any other mode. Note that this is a much less stringent definition of a single mode state than we have adopted so far. In the simulations discussed in Refs. [8–11,13], based on numerical solutions of Eqs. (4)–(7), many modes were found within the spectral half-width. It was estimated that only about 10–18% cases fall in the "single-mode" category, according to the less stringent definition adopted in Ref. [13].

Here, we present our own simulations of the GLE [Eq. (16)]. In all runs, we use parameters characteristics of the UCSB experiment with $\varepsilon = 3.3 \times 10^{-3}$, $p_0 = 2.606$ which is near maximum gain and $\chi = 3$, with $G_m = G_{0r}$ (2.606). Note that a value of χ larger than approximately 4 will excite the sideband instability [8–11] which will invalidate the GLE. For ease of comparison with Refs. [8–11,13], we take $\chi = 3$. We use periodic boundary conditions on τ_0 with a period of 2, and a total of 1024 Fourier modes including 390 modes under the positive gain curve. Initially, random small amplitudes are assigned to all the Fourier modes. Then the evolution of the radiation field as a function of τ_s is obtained by solving Eq. (16) by a pseudo-spectral method. Note that the slow time $\tau_s = t\nu v_g/2L_c = t\nu/0.05\,\mu s$ for the parameters of the UCSB experiment. Since ν is the fractional power loss per round trip, its maximum value is one. The maximum time t is determined by the pulse length of the electron, which is about 50 μs. Therefore, the maximum required value of τ_s is about 1000. In reality, since ν is smaller than unity, τ_s is of the order of a few hundred.

We now present the statistical results from 10^4 runs, each with a random initial condition. In Fig. 2, we plot the probability P_{20} (solid line) of finding a "single mode" within the central 20-mode range, defined in Refs. [12,13], as a function of τ_s. We see that the probability grows very rapidly for $\tau_s < 50$, but grows much more slowly after that at a level around 10%. This agrees with the estimates given in Ref. [13] but is only about a third of the experimental finding [12]. It should be borne in mind, however, that the simulation sample is much larger than the experimental sample (90 shots), and it is possible that the experimentally calculated probabilities on the basis of a small number of

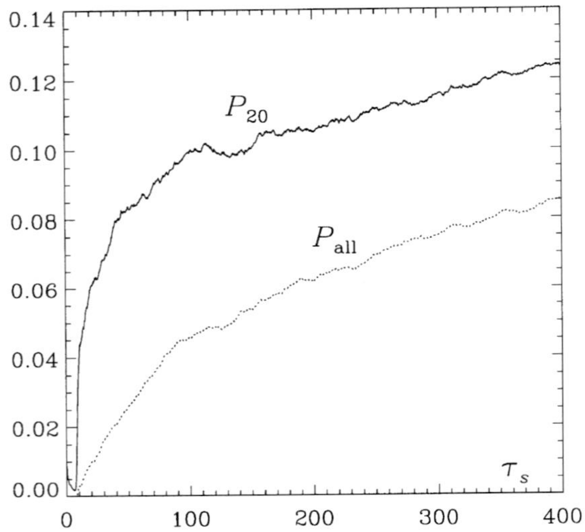

Fig. 2. Solid curve: the probability P_{20} of finding a "single mode" using the definition of Refs. [12,13], within the central 20-mode range as a function of τ_s based on 10^4 random runs. Dotted curve: the probability P_{all} of a "single mode" for the whole spectrum.

shots overestimate P_{20} due to the presence of large fluctuations.

In Fig. 2, the dotted line represents P_{all}, the probability of a "single mode" for the entire spectrum. Note that $P_{all} \leqslant P_{20}$ by definition. We see that there exits a substantial difference (about 0.04) between the two curves for τ_s as large as 400. This implies that about a third of the so-called "single-mode" cases, inferred by calculating P_{20}, are not, in fact, single modes even by the less stringent definition adopted in Ref. [12] because it is possible to find a mode with a larger amplitude outside of the central 20-mode range. We now show that all these so-called "single-mode" cases do not truly represent single modes at all if we determine single-mode states by the more stringent and accurate condition $\Delta n < 1$, where

$$\Delta n = \langle (n - \langle n \rangle)^2 \rangle^{1/2} \tag{22}$$

with

$$\langle n \rangle = \sum_n n|a_n| / \sum_n |a_n|. \tag{23}$$

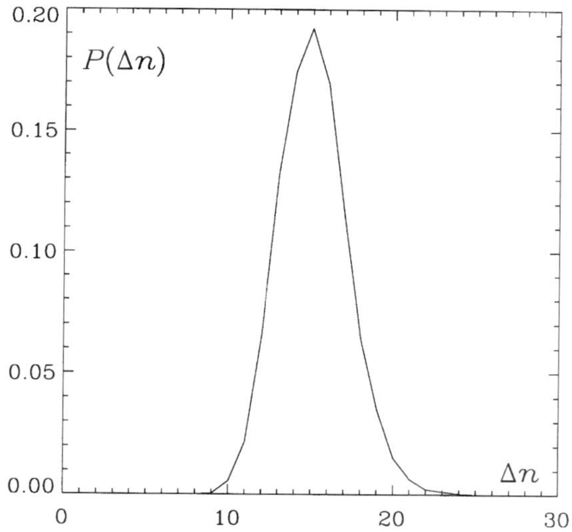

Fig. 3. Probability density $P(\Delta n)$ at $\tau_s = 400$ based on the same set of runs as in Fig. 2. Note that the probability of obtaining $\Delta n < 1$, a true single-mode state, is negligible.

Fig. 3 shows the probability density $P(\Delta n)$ for different Δn values at $\tau_s = 400$. We see that the most probable value is about $\Delta n \approx 15$. The probability for Δn to be near one, representing a true single-mode state, is negligibly small.

5. Conclusion

We have presented a new derivation of the GLE for a long-pulse, low-gain FEL oscillator. The saturation level contours and phase stability boundary curves are found to be qualitatively similar to those of Refs. [8–11]. Based on a large number of numerical simulations of the GLE, we have determined that the most probable state realized in a typical electron pulse-length consists of about 15 modes. The probability of realizing a true single-mode state is found to be negligibly small.

Acknowledgements

This research is supported by the Air Force Office of Scientific Research Grant No. F49620-96-1-0068 and the Department of Energy Grant No. DE-FG02-91ER40669.

References

[1] S.Y. Cai, A. Bhattacharjee, Phys. Rev. A 43 (1991) 6934.

[2] C.S. Ng, A. Bhattacharjee, Nucl. Instr. and Meth. A 407 (1988) 34; Phys. Rev. E. 58 (1998) 3826.

[3] L.R. Elias, G. Ramian, J. Hu, A. Amir, Phys. Rev. Lett. 57 (1986) 424.

[4] L.R. Elias, I. Kimel, Nucl. Instr. and Meth. A 296 (1990) 144.

[5] I. Kimel, L.R. Elias, Phys. Rev. A 38 (1988) 2889.

[6] I. Kimel, L.R. Elias, Nucl. Instr. and Meth. A 296 (1990) 528.

[7] I. Kimel, L.R. Elias, Nucl. Instr. and Meth. A 341 (1994) 191.

[8] B. Levush, T.M. Antonsen Jr., Nucl. Instr. and Meth. A 272 (1988) 375.

[9] B. Levush, T.M. Antonsen Jr., Nucl. Instr. and Meth. A 285 (1989) 136.

[10] T.M. Antonsen Jr., B. Levush, Phys. Rev. Lett. 62 (1989) 1488.

[11] T.M. Antonsen Jr., B. Levush, Phys. Fluids B 1 (1989) 1097.

[12] B.G. Danly, S.G. Evangelides, T.S. Chu, R.J. Temkin, G. Ramian, J. Hu, Phys. Rev. Lett. 65 (1990) 2251.

[13] T.M. Antonsen Jr., B. Levush, Phys. Fluids B 2 (1990) 2791.

ELSEVIER

Nuclear Instruments and Methods in Physics Research A 429 (1999) 94–100

NUCLEAR INSTRUMENTS & METHODS IN PHYSICS RESEARCH Section A

www.elsevier.nl/locate/nima

Experimental observation of wiggler superradiance under group synchronism condition

N.S. Ginzburg[a,*], N.Yu. Peskov[a], I.V. Zotova[a], A.S. Sergeev[a], A.D.R. Phelps[b], A.W. Cross[b], W. He[b], K. Ronald[b], V.G. Shpak[c], M.I. Yalandin[c], S.A. Shunailov[c], M.R. Ulmaskulov[c]

[a]*Institute of Applied Physics, Russian Academy of Sciences, 46 Ulyanov Str. 603600 Nizhny Novgorod, Russia*
[b]*Department of Physics and Applied Physics, University of Strathclyde, Glasgow, G4 0NG, UK*
[c]*Institute of Electrophysics, Russian Academy of Science, 620049 Ekaterinburg, Russia*

Abstract

The first results of the observation of superradiance from a single, subnanosecond, high current, electron bunch passing through a wiggler immersed in a guide magnetic field are presented. The 300–500 ps microwave pulses were generated in the high gain regime for both the conventional and reverse directions of the guide magnetic field. The dependence of the radiation power on the interaction length as well as the absolute value of the power, 100–200 kW, were related with the development of self-bunching and consequently with coherent emission. © 1999 Elsevier Science B.V. All rights reserved.

1. Introduction

In recent years much attention has been paid to theoretical consideration of superradiance (SR) from space-localized ensembles of electrons [1–11]. This phenomenon includes features present in both stimulated (self-bunching and coherence) as well as spontaneous processes (absence of threshold). It is reasonable to consider SR in a specific situation when the electron pulse duration essentially exceeds the wavelength (otherwise it is effectively traditional spontaneous emission without self-bunching) while at the same time is less than the interaction length (in contrast to the mechanism of stimulated emission of quasi-continuous electron beams which are used extensively in microwave electronics – travelling wave tubes, backward wave oscillators, cyclotron resonance masers, free-electron lasers (FELs), etc. [12,13]). Thus superradiance may be regarded as pulse stimulated emission from a single short electron bunch which includes particle self-bunching as well as slippage and the escape of radiation from the electron pulse. Normally slippage of radiation is associated with the difference between the electromagnetic wave group velocity and the electron bunch axial velocity [1–4]. However, in the special case of group synchronism conditions when the wave group velocity and the electron bunch axial velocity coincides

* Corresponding author. Fax: + 7-831-2-384-552.
E-mail address: ginzburg@appl.sci-nnov.ru (N.S. Ginzburg)

0168-9002/99/$ - see front matter
PII: S0168-9002(99)00078-9

some slippage will arise from the wave packet expansion caused by dispersion [11,14].

Superradiance can be associated with several different mechanisms of stimulated emission: bremsstrahlung, cyclotron, Cherenkov, etc. Experimental observations of cyclotron and Cherenkov superradiance in the microwave waveband have been studied and reported in Refs. [14–16]. A RADAN 303 accelerator with a subnanosecond slicer was used as a driver of high current subnanosecond electron bunches. Subnanosecond microwave pulses with peak power up to several hundreds of kilowatts in the case of cyclotron emission and up to several megawatts in the case of Cherenkov emission have been recorded. The microwave signal was amplified from the noise level up to the level mentioned above during the single passage of an electron bunch through the interaction region which typically was within the range 10–30 cm. Therefore, the radiation was associated with the high gain regime. In Ref. [17] results of the observation of magnetic Bremsstrahlung SR from a series of picosecond electron bunches passing through a wiggler have been reported. In this experiment a superradiation-like regime was obtained in the FELIX infrared FEL when the detuning parameter tended to zero. The maximum power of the SR pulse in these experiments achieved a level of several hundred MW but pulse amplification occurred in the low-gain regime in which the electromagnetic pulse travelling between the mirrors interacts with a whole train of electron pulses.

In the present paper we describe results of the first experimental observation of microwave magnetic bremsstrahlung SR in the high gain regime. Radiation was emitted by a single, high current, electron bunch moving in a combined wiggler and homogeneous guide magnetic field (see Fig. 1). The electrons radiated under the synchronism condition (see Fig. 2).

$$\omega - k_{\parallel}v_{\parallel} = \Omega_{\mathrm{B}}. \tag{1}$$

Here frequency ω and longitudinal wave number $k_{\parallel}$ are related by the dispersion characteristic of an operating waveguide mode, $\Omega_{\mathrm{B}} = 2\pi v_{\parallel}/\lambda_{\mathrm{W}}$ is the bounce frequency, $v_{\parallel}$ is the electron drift velocity, and λ_{W} is the wiggler period.

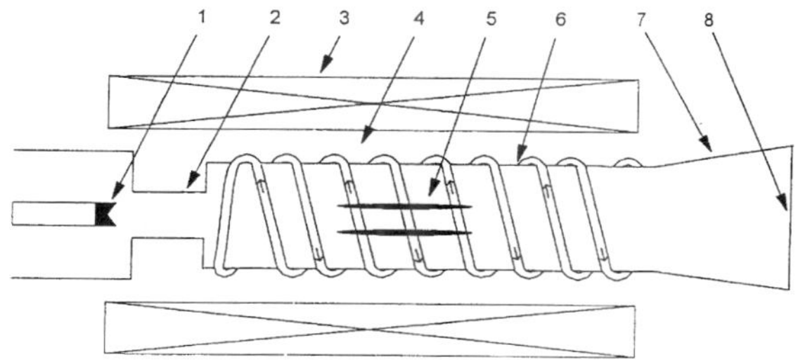

Fig. 1. Experimental configuration of the interaction region (1 – cathode; 2 – anode; 3 – guiding field solenoid; 4 – wiggler coil; 5 – electron pulse; 6 – drift chamber; 7 – horn; 8 – microwave window.)

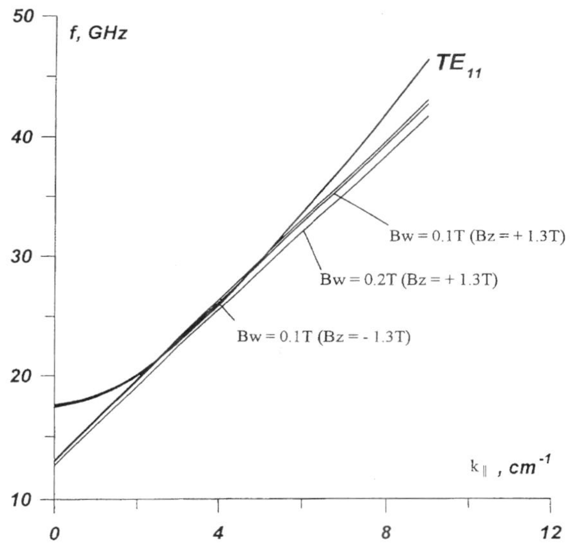

Fig. 2. Dispersion diagram showing the waveguide mode and the electron beam grazing condition.

2. Experimental set-up

A RADAN 303 accelerator with a subnanosecond slicer was used to inject 0.3–0.5 ns, 0.1–1 kA, 220 keV electron pulses [18,19]. A short duration electron pulse was generated when this sub-nanosecond voltage pulse was applied across a magnetically insulated coaxial diode which utilized a cold, explosive-emission cathode. The fast electron bunch current and accelerating voltage pulse were measured using a strip line current probe and an in-line capacitive probe, respectively. Typical oscillograms of the electron current before the collimator and voltage across the cathode are presented in Fig. 3. These sub-nanosecond signals were recorded using a Tektronix 7250 digitizing oscilloscope with a single-event 7 GHz analog bandwidth. For measurement of the radiation a hot-carrier germanium detector with a rise time of 200 ps was used.

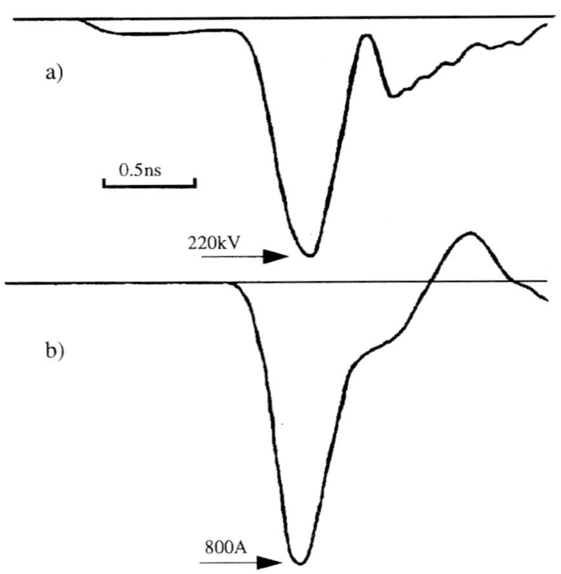

Fig. 3. Typical waveforms of (a) the accelerating voltage and (b) pulsed electron current before the collimator.

High-current electron pulses were transported through the interaction space of total length up to 30 cm in a longitudinal guide magnetic field of up to 1.4 T. A collimator was used to improve electron beam quality. The collimated electron beam diameter was 4 mm. The peak electron current after the collimator was reduced from 800 to 250 A. Transverse momentum was imparted to the electrons by the helical wiggler. Adiabatic tapering of the wiggler entrance over 6 periods was used to provide smooth excitation of the bounce oscillations. In the regular section of the wiggler with total length 20 cm and period of 1.6 cm electrons moved along helical trajectories with constant longitudinal velocity. These trajectories correspond to the stationary orbits of type groups I and II [20]. For these orbits the calculated dependencies of the transverse electron velocity on the strength of the guide magnetic field are presented in Fig. 4 for different amplitudes of the wiggler field. Solid lines correspond to an ideal wiggler while the dashed lines were found from simulations of electron trajectories in the real wiggler (taking into account the transverse inhomogeneity of the wiggler field). It was possible to switch the direction of the guide magnetic field. To facilitate the transportation of the high-current electron bunch through the interaction space the magnitude of the uniform magnetic field needs to be rather high, in excess of 1 T. In such a guide field Group I steady orbits can be realized only for the so-called reverse direction of

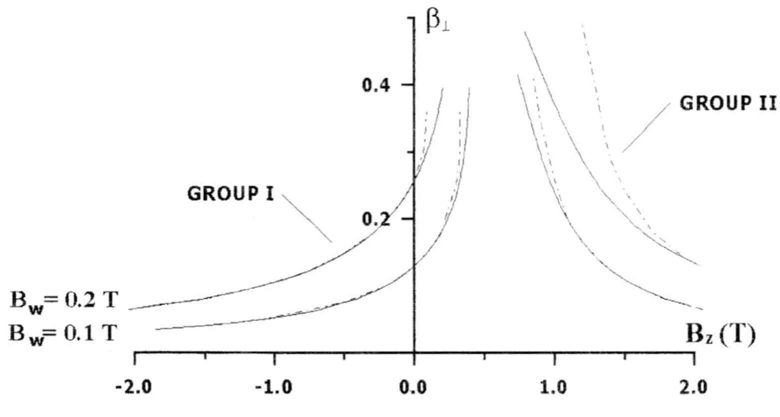

Fig. 4. Transverse electron velocity as a function of both guide and wiggler magnetic fields.

the guide magnetic field [21,22]. At the same time electron motion over the group II orbits occurred for the conventional direction of the guide field. In this case the strength of the magnetic field exceeded the cyclotron resonance value. It should be noted that for the group II orbits near magneto-resonance, increasing the amplitude of the wiggler field results in a drastic increase of the amplitude of the bounce oscillation and full interruption of the bunch transport through the wiggler (Fig. 7).

3. Results

The superradiant pulses with typical duration 300–500 ps have been observed for both (conventional and reverse) directions of the guide magnetic field and correspondingly for both groups of steady-state orbits. The oscillograms of microwave pulses under different amplitudes of wiggler field and different directions of guide field are shown in Fig. 5. The maximum radiation power was obtained for a conventionally directed magnetic field of 1.3 T. This result can be explained by the fact that under the experimental parameters (see Fig. 4) the transverse electron velocity for the conventional magnetic field direction exceeds the similar value under reverse field. The optimal amplitude of the wiggler field was about 0.2 T. For higher amplitude magnetic fields, as is easily seen from Fig. 6, considerable losses of electron current passing through the wiggler occurred. As follows from Fig. 4 such losses can be explained by an increase of the electrons transverse momentum and a decrease of the longitudinal momentum. According to simulations for a guide magnetic field of 1.3 T, when the wiggler field exceeds 0.3 T, the electron drift velocity tends to zero and electrons reflect from the wiggler adiabatic entrance.

According to measurements of the mode pattern the radiation was associated with excitation of the

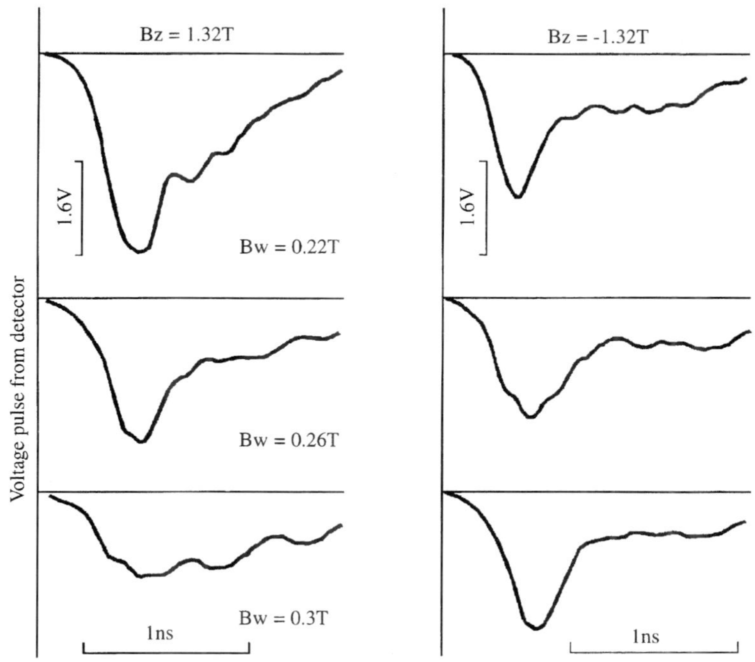

Fig. 5. Examples of detected millimetre-wave pulses for both the conventional and the reversed guide field regimes and for a range of wiggler magnetic field values.

operating mode TE_{11}. The spectrum measurements using cut-off waveguides and metallic grids show good agreement with the frequency which is predicted from the dispersion diagram (Fig. 2). The radiation frequency was approximately 28 GHz. Note that the waveguide diameter 9.8 mm and wiggler period were chosen to match the grazing condition for the operating TE_{11} mode. We believe that space charge effects were not very significant. Taking into account the reduction of the space charge force caused by the metallic waveguide walls we estimate a longitudinal plasma frequency $f_p < 1$ GHz. Therefore, a rather small shift of the beam line on the dispersion diagram can be associated with space charge. As this shift was relatively small the influence of space charge on the experiment can be easily compensated by adjustment of the electron longitudinal velocity through the variation of the wiggler field.

As follows from the theoretical considerations for waveguide propagation of radiation there are some advantages (like increased gain) in generation of SR pulses under grazing conditions when the electron bunch velocity coincides with the electromagnetic wave group velocity. It should be noted that even under exact grazing conditions there are some factors which enable the electromagnetic radiation to escape from the electron bunch. These include the effects of electronic frequency detuning caused by finite electron beam density as well as waveguide dispersion. For cyclotron SR the benefit of operating close to grazing incidence was proven experimentally (see details in Refs. [14,15]), because in that case it was possible to vary the operating frequency around the point corresponding to grazing incidence by changing the magnetic field and hence the gyrofrequency. In the case of emission in a wiggler the bounce frequency was in fact determined by the wiggler period and it was practically impossible to mismatch the system from the grazing condition without changing the electron beam energy. Therefore, we have not been able to confirm experimentally that this grazing condition is the preferred one for the production of wiggler SR.

According to theoretical analysis [23], under the grazing condition for a short electron bunch, the temporal growth rate of the SR instability in the reference frame moving with the bunch is given by

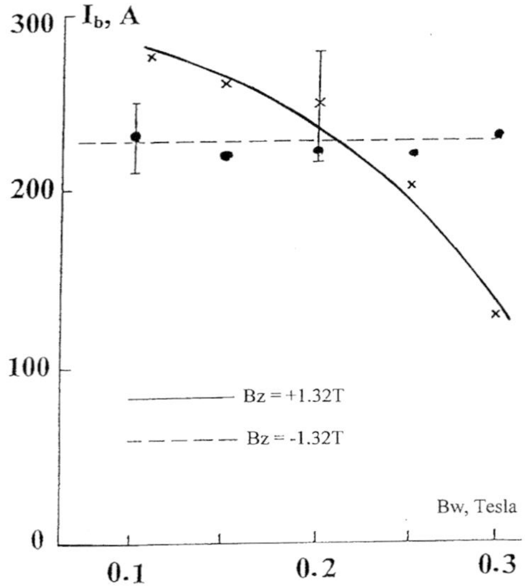

Fig. 6. Transported electron beam current as a function of the magnitude of the wiggler magnetic field.

the relation

$$|\mathrm{Im}\,\omega'| = \omega_c \sin\left(\frac{2\pi}{5}\right)\left[\frac{1}{4\sqrt{2\pi}}\left(\frac{eI_0}{m_e c^3}\right)\frac{b\lambda_W \chi_{||}^2 \beta_\perp^2}{R^2 \quad \beta_{||}^2} \times \frac{J_{m-1}^2(\nu_n R_0/R)}{J_m^2(\nu_n)(1-m^2/\nu_n^2)}\right]^{2/5} \tag{2}$$

where I_0 is the peak current, b the bunch length, $\beta_\perp$ the transverse velocity in the wiggler field (see, Fig. 4), $\chi = (1-(v_{||}/c)^2)^{-1/2}$, J_m the mth-order Bessel function, m is the azimuthal mode index, ν_n the nth root of the equation $J'_m(\nu) = 0$, R_0 is the electron injection radius and R is the waveguide radius. For excitation of the TE_{11} mode by a bunch of length 5 cm and current 200 A, from Eq. (2) we estimate the growth rate to be $7\times10^9\,\mathrm{s}^{-1}$. For the growth time $T' = 1/\mathrm{Im}\,\omega'$ the electron bunch will travel a distance $L = T' v_{||}\chi_{||}$. So in the lab frame the spatial gain for the amplitude of the radiation field will be defined by the relation $\Gamma = (\mathrm{Im}\,\omega')/v_{||}\chi_{||} \sim 0.14\,\mathrm{cm}^{-1}$.

Fig. 7 illustrates the experimental dependence of the peak power on the length of the interaction of the electron pulse with the wiggler field. It is obvious that the radiation power is quite small for short interaction lengths and increases exponentially as

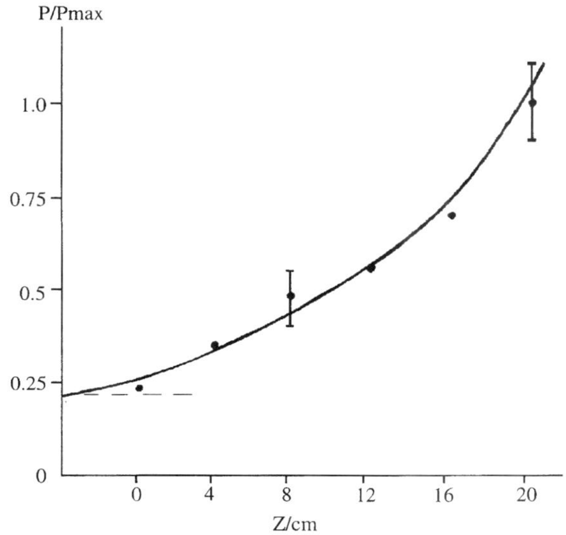

Fig. 7. Peak radiated millimetre-wave power as a function of the length of the wiggler's uniform interaction region for the conditions $B_z = +1.3$ T and $B_w = 0.22$ T.

this length increases. This can be interpreted as the natural level of spontaneous emission caused by the electron density fluctuations as well as the radiation from the front edges of the pulse being relatively small, while the observed electromagnetic pulse behaviour is related with the development of longitudinal self-bunching and coherent emission from the bunched electron pulse. However, the experimental gain is 2–3 times smaller than the theoretical prediction. This discrepancy can be explained by velocity spread as well as space charge effects. The observation of some radiation even when the length of the homogeneous section of the wiggler tends to zero is considered to be due to two factors. Firstly, there is some magnetic bremsstrahlung produced from the tapered sections of the wiggler and secondly, even in the absence of a wiggler field there is some background cyclotron radiation from the electron motion in the uniform guide field. Fig. 7 also shows that the maximum length of the homogeneous part of the wiggler was not sufficient to observe the nonlinear saturation regime. For the present experimental configuration the maximum power of the SR pulses was estimated as being in the range 100–200 kW.

4. Conclusion

Note in conclusion that further experiments using longer wigglers are planned which give the possibility of achieving saturation, and according to simulations will result in the emission of megawatt, sub-nanosecond, microwave pulses. It is also interesting to compare the generation of superradiant pulses in wigglers with different periods to investigate the dependence when operating near and far from grazing incidence. Another important problem is the verification of scaling laws such as the dependence of peak power and electromagnetic pulse duration on electron bunch current in the high gain regime. At present the experimental equipment does not allow the electron bunch current to be changed without variation of the electron energy, which is typical for explosive emission cold cathodes. Verification of scaling laws is also complicated due to the pulse-to-pulse instability of the short high current electron pulse generated from such a cathode when operating in the substantially nonstationary sub-nanosecond regime.

Meanwhile the significance of the new experimental results reported here are that single electron bunches with length 5–7 cm passing through a wiggler with total length 30 cm radiated coherent, high power 300–500 ps pulses in the K_a band. Because generation of these unique short duration pulses was shown to be associated with self-bunching we believe these results should be interpreted as superradiance. This conclusion is supported by theoretical considerations [23] based on a similar model as described in Refs. [11,14].

Acknowledgements

This work was supported by the Russian Fund for Fundamental Research, grant 98-02-17308 and by the United Kingdom Defense and Evaluation Research Agency and Engineering and Physical Sciences Research Council. The authors wish to acknowledge help with the wiggler construction provided by A. Kaminsky's group at the Joint Institute of Nuclear Research, Dubna.

References

[1] R.H. Bonifacio, C. Maroli, N. Piovella, Opt. Commun. 68 (1988) 369.

[2] R.H. Bonifacio, N. Piovella, B.W.J. McNeil, Phys. Rev. A 44 (1991) 3441.

[3] R.H. Bonifacio, F. Casagrande, G. Cerchiony, L.D. Souza, P. Pierini, N. Piovella, Riv. Nuovo Cimento 13 (1990) N9.

[4] R.H. Bonifacio, R. Corsini, L. De Salvo, P. Pierini, N. Piovella, Riv. Nuovo Cimento 15 (1992) N11.

[5] N.S. Ginzburg, Pis'ma v ZhTeor. Fiz. 14 (1988) 440.

[6] N.S. Ginzburg, A.S. Sergeev, Sov. Phys. JETP Lett. 60 (1994) 513.

[7] G.R.M. Robb, N.S. Ginzburg, A.D.R. Phelps, A.S. Sergeev, Phys. Rev. Lett. 77 (1996) 1492.

[8] N. Piovella, P. Chaix, G. Svets, D. Jaroszynski, Phys. Rev. E 53 (1995) 5470.

[9] V.V. Zheleznyakov, V.V. Kocharovsky, Vl.V. Kocharovsky, Usp. Fiz. Nauk. 159 (1989) 194.

[10] N.S. Ginzburg, A.S. Sergeev, Opt. Commun. 91 (1992) 140.

[11] N.S. Ginzburg, I.V. Zotova, A.S. Sergeev, Sov. Phys. JETP Lett. 60 (1994) 513.

[12] V. Granatstein, I. Alexef (Eds.), High-Power Microwave Sources, Artech House, Boston, 1987, p. 421.

[13] T.C. Marshall, Free-Electron Lasers, Macmillan, New York, 1985.

[14] N.S. Ginzburg, I.V. Zotova, A.S. Sergeev et al., Phys. Rev. Lett. 78 (1997) 2365.

[15] N.S. Ginzburg, I.V. Zotova, A.S. Sergeev, I.V. Konoplev, V.G. Shpak, M.I. Yalandin, S.A. Shunailov, M.R. Ulmaskulov, Sov. Phys. JETP Lett. 63 (1996) 331.

[16] N.S. Ginzburg, I.V. Zotova, A.S. Sergeev et al., Nucl. Instr. and Meth. A 393 (1997) 352.

[17] D.A. Jaroszynski, P. Chaix, N. Piovella et al., Phys. Rev. Lett. 78 (1997) 1699.

[18] G.A. Mesyats, V.G. Shpak, M.I. Yalandin, S.A. Shunailov, Proceedings of the eighth International Pulsed Power Conference San-Diego, USA, 1991, p. 73.

[19] G.A. Mesyats, V.G. Shpak, M.I. Yalandin, S.A. Shunailov, Proceedings of SPIE International Symposium: Intense Microwave Pulses, San-Diego, USA 2154, 1994, p. 262.

[20] H.P. Freund, P. Sprangle, Phys. Rev. A 24 (1981) 1965.

[21] M.E. Conde, G. Bekefi, Phys. Rev. Lett. 67 (1991) 3082.

[22] A.K. Kaminsky, A.A. Kaminsky, V.P. Sarantsev et al., Nucl. Instr. and Meth. A 375 (1996) 215.

[23] N.S. Ginzburg, I.V. Zotova, A.S. Sergeev, Pis'ma v ZhTeor. Fiz., submitted for publication.

ELSEVIER

Nuclear Instruments and Methods in Physics Research A 429 (1999) 101–106

NUCLEAR INSTRUMENTS & METHODS IN PHYSICS RESEARCH Section A

www.elsevier.nl/locate/nima

Study of radiation build-up and spectral evolution in the Israeli electrostatic accelerator free-electron laser oscillator

A. Abramovich[a,*], Y. Pinhasi[b], Y. Yakover[a], A. Gover[a], J.S. Sokolowski[a], M. Canter[a]

[a]*Faculty of Engineering, Tel-Aviv University, Physical Electronics, 69978 Ramat-Aviv, Israel*
[b]*Faculty of Engineering, Department of Electrical and Electronic Engineering, The College of Judea and Samaria, Ariel, Israel*

Abstract

Time–frequency study of oscillation build-up in an electrostatic accelerator free-electron laser (EA-FEL) oscillator is presented. The unique features of EA-FEL and its capacity to operate in a long pulse mode, enable observation and study of linear and non-linear processes taking place in the evolution of radiation in the EA-FEL oscillator. The experimental data recorded with the aid of a digital 4 GS/s fast sampling oscilloscope, was analyzed using various signal processing techniques to obtain time–frequency phase space presentation of the power spectral density evolution. This presentation make it possible to follow the radiation field coherence evolution from noise to saturation including intermediate stages of mode build-up from noise and competition between longitudinal modes until single-mode steady-state operation is established. The experimental results were also compared with the results of a FEL oscillator simulation code. The code predicts well the behavior of the oscillator in its various regimes of operation.

Keywords: Electrostatic accelerator-free electron laser; RF detector

1. Introduction

We report single mode and multimode operation of an electrostatic accelerator free electron laser (EA-FEL) oscillator using an internal cavity.

The FEL is based on the 6 MeV Tandem Van de Graaff accelerator of the Weizman Institute. It was converted into a high-current electron beam accelerator, and was modified so as to enable the insertion of a magnetostatic wiggler containing a mm-wave resonator and electron-optical focusing elements in the positively charged HV terminal (see Fig. 1) [1,2]. The FEL is presently operating in a pulsed mode at 100 GHz. Table 1 summarizes the parameters of the EA-FEL.

Since the first lasing of the EA-FEL [3], we significantly improved the current control of the e-beam and the experimental instrumentation. These improvements enable us to increase the e-beam current through the wiggler to 1 A and to analyze the spectral components of the FEL radiation as function of time.

* Corresponding author. Tel.: + 972-3-640-8271; fax: + 972-3-642-3508.
E-mail address: amir007@eng.tau.ac.il (A. Abramovich)

0168-9002/99/$ - see front matter
PII: S0168-9002(99)00080-7

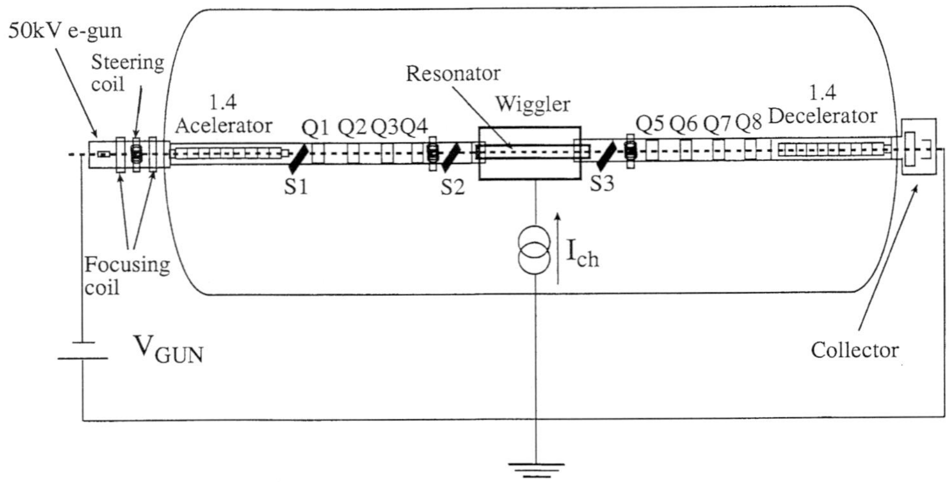

Fig. 1. Electron beam line of the Israeli EA-FEL.

Table 1
Parameters of the tandem FEL

Accelerator:	
Electron beam energy	$E_k = 1.4$ MeV
Beam current	$I_0 = 1.5$ A
Wiggler:	
Magnetic induction	$B = 2000$ G
Period length	$\lambda_w = 4.4$ cm
Number of periods	$N_w = 20$
Waveguide resonator:	
Curved parallel plate resonator	
Mode	TE_{01}
Total reflectivity	$R = 91\%$
Resonator length	$L_c = 130.8$ cm
Quality factor	$A = 30{,}000$

2. Experimental setup

Fig. 1 shows the layout of the electron beam line along the FEL. The electron optics of this FEL is based on a straight line geometry and a positive HV terminal. The electron injection system is based on a 43 kV Pierce-type e-gun and employs focusing and steering coils after the cathode. The parallel-flow electron gun produces a high-quality electron beam (with emittance of $\varepsilon_n = 10\pi$ mm mrad [2]), which is injected to the acceleration section. The following four quadrupole lenses (Q1–Q4) are used to control the e-beam and determine optimal entrance conditions to the wiggler [4]. The wiggler and the mm-wave resonator, forming the FEL interaction region, are installed inside the positively charged high voltage terminal, located at the center of the tank along the symmetry axis. The quadrupoles Q5, Q6, Q7 and Q8 collimate the e-beam into the deceleration section prior to collection by a depressed collector, which provides e-beam energy recuperation. Three diagnostic screens S1, S2 and S3 were used for monitoring of the e-beam transport.

A permanent magnet wiggler, arranged in a Halbach planar configuration [5] is employed. Two long magnets were used, one on each side of the wiggler, to focus the e-beam in the lateral (wiggling) plane by means of a lateral magnetic gradient which they produce on that axis [6]. The RF resonator utilizes curved parallel plates as a waveguide structure and has two quasioptical Talbot effect reflectors (wave splitters) one at each resonator end, which enable e-beam passage into and out of the resonator [7,8]. This type of resonator is characterized by very small ohmic and total radiation losses of about 9%, as was determined in Ref. [8] by loss measurements prior to the installation of the resonator in the HV terminal.

3. Experimental results

The EA-FEL was operated in a pulsed mode, using the current control electrode of the e-gun, to produce 20 μs duration e-beam pulses. The power and spectral evolution of the radiation build-up process were measured with the aid of the diagnostic setup shown in Fig. 2. Power measurements were made using the RF detector. Simultaneously, measurements of the IF signal were made, using a mixer and a local oscillator. Oscilloscope traces of the e-beam currents, RF power, and high voltage droop are shown in Fig. 3. Trace 1 is the voltage droop at the terminal during the e-beam pulse, trace 2 is the cathode current pulse, trace 3 is the power measured by the RF detector, and trace 4 is the e-beam current measured at the wiggler exit. The measurements show that the high voltage droop during an e-beam 20 μs pulse is about 22 kV. The e-beam current transiting the wiggler is about 1 A, and the RF output power is about 8 kW. The IF signal was recorded by a 4 Gsample/s digital oscilloscope and stored and processed in a PC computer.

4. Analysis and simulation

The time–frequency diagram shown in Fig. 4 includes frequencies obtained by mixing the local oscillator with the exited radiation of the longitudinal modes. Therefore, the intermediate frequency (IF) signal consists of down converted

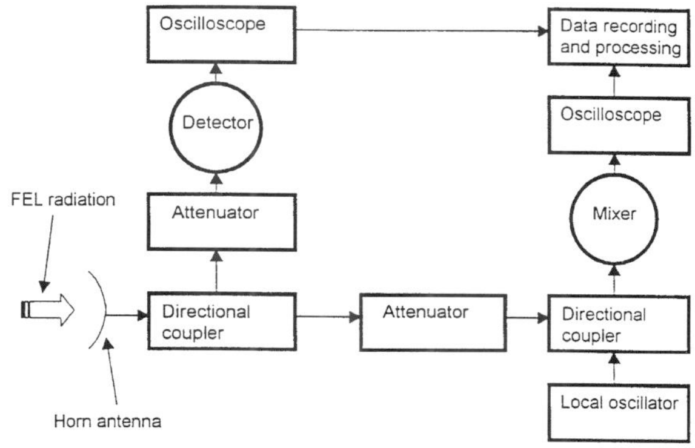

Fig. 2. Experimental setup for measuring the FEL RF power and spectral components of the radiation.

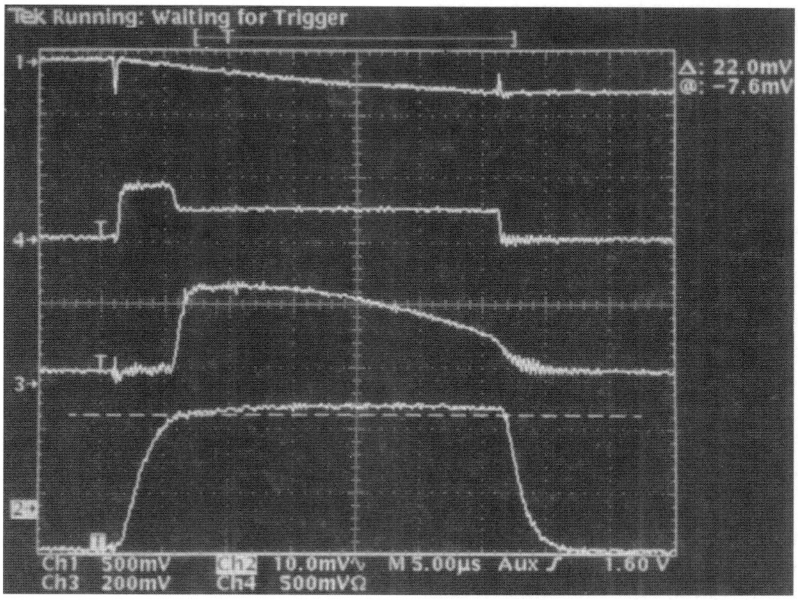

Fig. 3. Experimental measurements of high voltage droop (trace 1), E-beam current at the cathode (trace 2), output radiation power of the EA-FEL (trace 3) and e-beam current after interaction region (trace 4).

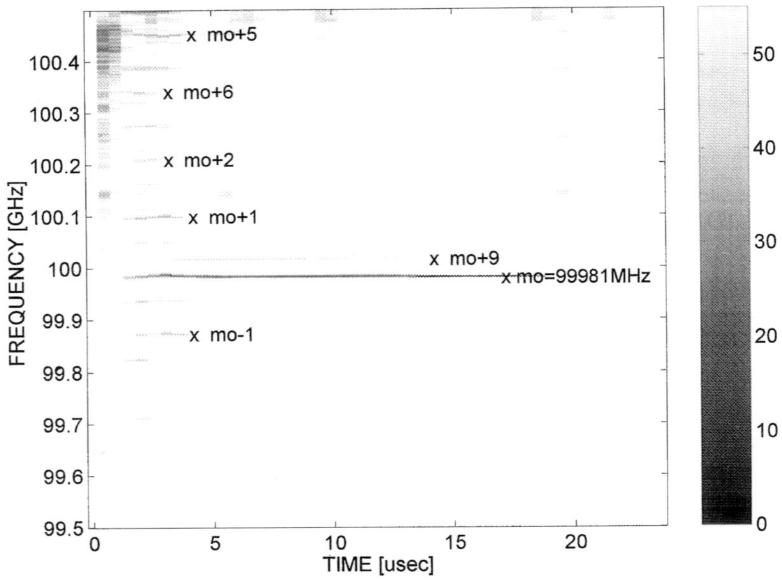

Fig. 4. Spectral analysis of the IF signal for single-mode operation at 99.981 GHz (where the gray scale is attenuation in dB and the maximum peak is normalized to 0 dB).

longitudinal mode frequencies;

$$f_{\mathrm{IF_m}} = |f_{\mathrm{m}} - f_{\mathrm{LO}}| \tag{1}$$

where f_{LO} is the local oscillator frequency, f_{m} is the eigen-frequency of the FEL oscillator longitudinal mode m excited, and $f_{\mathrm{IF_m}}$ is the down converted IF signal. The gray scale is the attenuation in dB (the maximum peak was normalized to 0 dB). The spectral analysis shows that we obtain single-mode operation at an eigenfrequency of 99.981 GHz (marked m_0 in Fig. 4). Furthermore, we can see evidence of mode competition between longitudinal modes of the resonator at the beginning of the radiation build-up process. The free-spectral range (FSR) between the eigen-frequencies of the resonator is given by

$$\Delta f_{\mathrm{FSR}} = v_{\mathrm{g}}/2L_{\mathrm{c}} \approx 112\ \mathrm{MHz} \tag{2}$$

where v_{g} is the group velocity of the electromagnetic wave and L_{c} is the resonator length [8]. The spectrum shown in Fig. 4 is obtained by subtracting the local oscillator frequency from all of the IF components. Therefore, some of the components are image frequencies of the exited modes. Using experimental data obtained with differing local oscillator frequencies and with the calculated value of Δf_{FSR} we prove that the steady-state single frequency, which evolves at 99.981 GHz (marked as m_0 at Fig. 4), is the main mode eigenfrequency of the FEL oscillator. The components at 99.873 GHz is the $m_0 - 1$ mode, at 100.099 GHz is the $m_0 + 1$ mode and at 100.210 GHz is the $m_0 + 2$ mode. Other frequencies shown in Fig. 4 are the image frequencies resulting from mixing of the $m_0 + 5$, $m_0 + 6$ and $m_0 + 9$ eigenfrequencies of the FEL oscillator which are at 100.547 GHz, 100.658 GHz and 100.983 GHz, respectively.

During the e-beam pulse the voltage droops at the terminal by 22 kV (see Fig. 3), which causes the small signal gain curve to move towards lower frequencies. Using our 3D simulation code FEL 3D [9,10], the small signal gain of the FEL is calculated. Fig. 5 shows FEL 3D simulation code results of small signal gain of the EA-FEL for two values of the accelerating voltage; $V = 1.400$ MV and 22 kV lower ($V = 1.378$ MV), which correspond to the voltatge droop in our experiment. The simulation results show that the frequency component m_0 at 99.981 GHz is still under the gain curve even after the voltage drops by 22 kV.

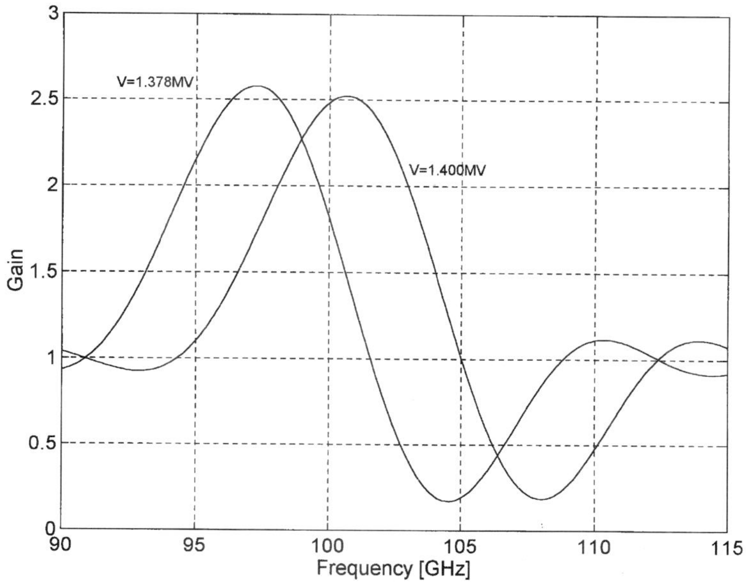

Fig. 5. FEL 3D simulation code results for small signal gain in two cases of accelerating voltages 1.4 and 1.378 MV which corresponds to the high voltage droop.

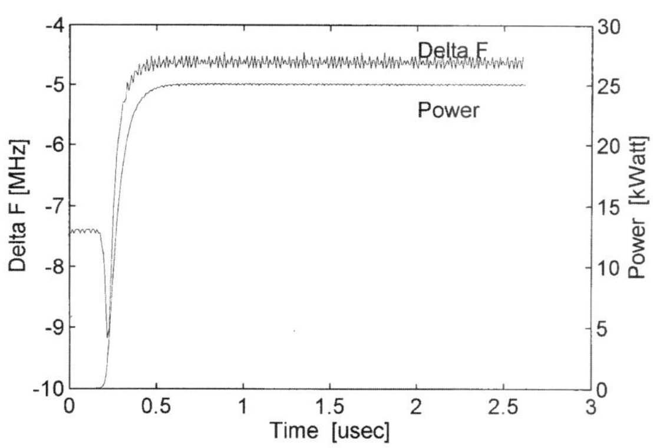

Fig. 6. Simulation results of radiation power buildup and frequency deviation.

The FEL 3D code was also used to simulate main mode radiation build-up from noise (spontaneous emission) level up to saturation. The initial spontaneous emission power (of the main mode) that was used in the simulation was $P_{sp}^{mode} = 1\,\mu W$. This was calculated from analytical expression [11] using the parameters of Table 1.

Fig. 6 shows simulation results of power evolution and instantaneous frequency deviation (from the main mode frequency as a function of time). Inspection of Fig. 6 reveals that a transient frequency instability occurs as the oscillator reaches the non-linear regime and enters to saturation. This effect was also observed experimentally as shown in Fig. 7 and can be interpreted as a relaxation oscillation phenomenon [12].

Acknowledgements

This work was supported by the Israeli ministries of Energy and Science Foundation, and the Meyer Fund.

References

[1] E. Jerby et al., Nucl. Instr. and Meth. A 259 (1987) 263.

[2] A. Arensburg et al., Nucl. Instr. and Meth. A 375 ABS (1996) 1

[3] A. Abramovich et al., Appl. Phys. Lett. 71 (1997) 26.

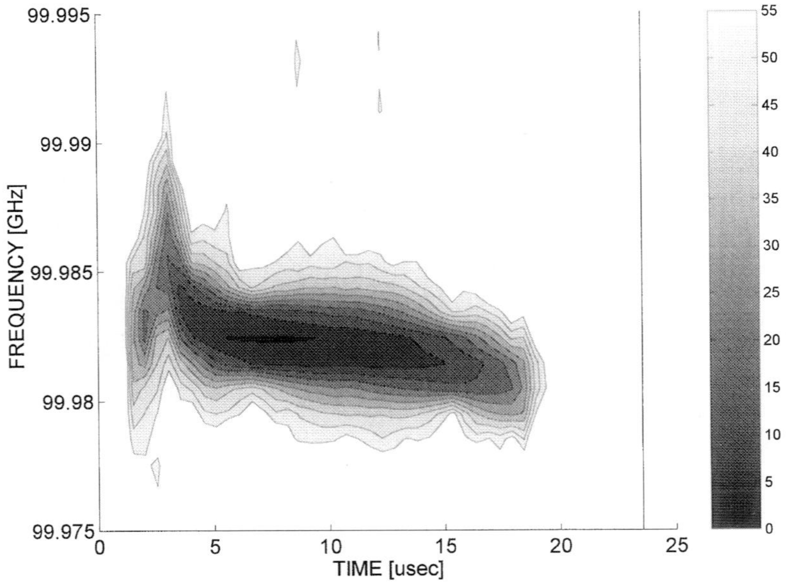

Fig. 7. Expansion of the time–frequency analysis of the recorded IF signal near the dominant mode.

[4] I. Merhasine et al., Nucl. Instr. and Meth., 1998, submitted for publication.
[5] K. Halback, Nucl. Instr. and Meth. A 169 (1980) 1.
[6] A. Abramovich et al., Nucl. Instr. and Meth. A 407 (1998) 81.
[7] B.C. Lee et al., Nucl. Instr. and Meth. A 375 (1996) 28.
[8] I.M. Yakover, Y. Pinhasi, A. Gover, Nucl. Instr. and Meth. A 358 (1995) 323.
[9] Y. Pinhasi et al., Int. J. Electron. 78 (1995) 581.
[10] Y. Pinhasi et al., Phys. Rev. E 54 (1996) 6774.
[11] A. Gover et al., in: K. Buttom (Ed.), Design Consideration for free electron lasers, Academic Press, New York, 1984.
[12] A. Yariv, Quantum Electronics, third ed., Wiley, New York, 1989.

ELSEVIER Nuclear Instruments and Methods in Physics Research A 429 (1999) 107–110

NUCLEAR INSTRUMENTS & METHODS IN PHYSICS RESEARCH Section A

www.elsevier.nl/locate/nima

Efficiency enhancement of a pre-bunched free-electron maser oscillator by locking to a single eigen frequency of the resonator

A. Abramovich[a,*], Y. Pinhasi[b], H. Kleinman[a], A. Eichenbaum[a], Y.M. Yakover[a], A. Gover[a]

[a]*Tel-Aviv University, Faculty of Engineering, Physical Electronics, 69978 Ramat-Aviv, Israel*
[b]*The College of Judea and Samaria, Faculty of Engineering, Department of Electrical and Electronic Engineering, Ariel, Israel*

Abstract

We present simulations results and experimental evidence of radiation energy extraction efficiency enhancement by use of pre-bunching in a free electron maser (FEM). This enhancement is attained by locking the oscillator to operate at a distinct resonator eigen frequency (longitudinal mode) which is different from the maximum-gain frequency of the free running oscillator. The mode selection is accomplished by means of pre-bunching at the desired eigen frequency. It was shown that maximum extraction efficiency is obtained if the dominant resonator mode (obtained by pre-bunching at that frequency) is the one having the lowest eigen frequency under the net gain larger than unity curve (which is below the maximum gain eigen-frequency mode that would normally dominate in a free running oscillator). The numerical simulations predict that it should be possible to obtain a basic extraction efficiency enhancement of the oscillator of 50%. Experimental results so far have shown an improvement of about 30%.

Keywords: Eigen frequency; Maser; e-beam

1. Introduction

Selection and operation of oscillators at a desired frequency by using seed radiation or modulation of the e-beam is a well-known technique for enhancing the build-up process and for avoiding multi-mode operation in microwave tubes and in FEMs [1–4]. Mode locking by pre-bunching is different from seed radiation since no RF radiation is introduced into the FEM cavity. Pre-bunching of the e-beam is preferable over seed radiation injection into the cavity because pre-bunching is unidirectional while insertion of RF seed radiation may cause RF outcoupling from the cavity and thus increases total resonator losses. In a FEM with a resonator there are several longitudinal modes (eigen frequencies) under the net gain larger than one curve competing with each other and deriving their energy from the e-beam, until the mode having the highest gain wins and the growth of other modes is suppressed. Pre-bunching of the e-beam provides a unique handle for interfering in the mode competition process by injecting an e-beam prebunched at a desired eigen frequency of the resonator. This interference can shorten the mode competition process time, in particular if the

* Corresponding author. Tel.: 972-3-6408271; fax: + 972-3-6423508.

E-mail address: amir007@eng.tau.ac.il (A. Abramovich)

0168-9002/99/$ - see front matter
PII: S0168-9002(99)00081-9

pre-bunching frequency is equal to one of the eigen frequencies of the FEM resonator that satisfy the oscillation condition. On the other hand, the mode competition process can become longer if the pre-bunching frequency is between two eigen frequencies of the FEM resonator that satisfy the oscillation condition [5]. In this work we present a new method of efficiency enhancement by locking the FEM oscillator to an eigen frequency having a large detuning parameter θ, which enables attainment of greater energy extraction from the e-beam [6,7], than that which would be attained if the largest gain eigen frequency prevailed at saturation (as in a free running FEM).

2. Experimental setup

The experimental setup is shown in Fig. 1. The e-beam passes a pre-bunching region which is a traveling wave tube interaction section and enters the FEM interaction region, which is a waveguide resonator located inside of a planar magnetic wiggler. As we reported in a previous publication [3] it is possible to select and make dominant in the resonator, at will, any one of the 5 eigenmodes of the FEL oscillator that have a net gain greater than unity. This is accomplished by choosing an RF-buncher frequency close to the desired resonator eigen frequency, with a bunching RF power input high enough to give the desired mode a sufficient head start in the mode competition process. Several milliwatts of RF input to the buncher were sufficient to select a desired mode and win the mode competition against the other (possibly higher gain) modes. The process always resulted in single mode operation. A radiation diagnostic system described in Fig. 1 was used to sample the power that was obtained in each selected mode and in order to verify its frequency and spectral purity. These measurements were repeated for each mode using various reflectivities of the resonator output coupler. The radiation was coupled out through a hole in a copper plate placed beyond a Teflon window at the end of the resonator waveguide. By using reflectors with different holes we were able to change their reflectivity from 60% to 100%. By changing the reflectivity we controle the I/I_{trh} parameter where, I is the e-beam current and I_{trh} is the oscillator threshold current of the FEM [6,7]. An RF sampler prior to the load provides information on power evolution with time for all modes existing in the resonator.

A power splitter divides the sampled power; part of it is fed to a diode detector and another part to a mixer. The detector output and the IF output of the mixer are monitored and recorded with the aid

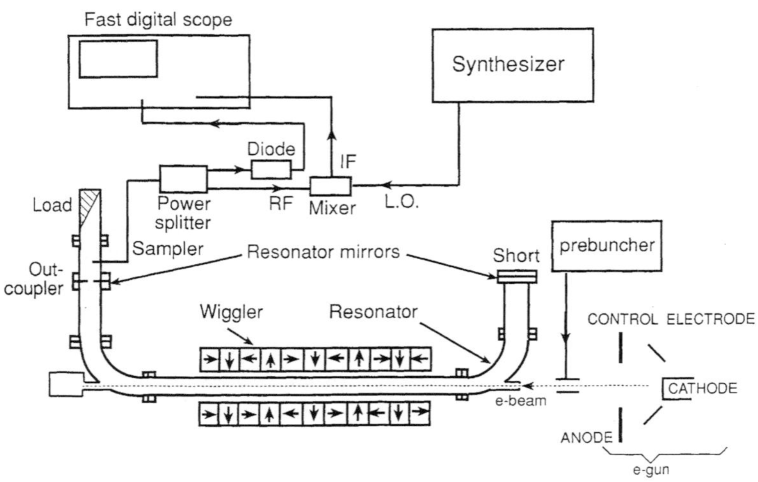

Fig. 1. Experimental setup of the FEM.

Table 1
Parameters of the Mini FEM

Accelerator:	
Electron beam energy	$E_k = 70$ keV
Beam current	$I_0 = 0.9$ A
Wiggler:	
Magnetic induction	$B = 320$ G
Period length	$\lambda_w = 4.44$ cm
Number of periods	$N_w = 17$
Waveguide resonator:	
Rectangular waveguide cavity	2.21 cm × 4.75 cm
Mode	TE_{01}
Frequency	4.5 GHz
Resonator length	$L_c = 138$ cm
Output power	$P = 3.5$ KW

of a fast digital scope. The fast digital scope has a 1000 MHz bandwidth and a sample rate of 4 Gsample/s. These scope parameters enable simultaneous recording of 5 longitudinal modes (which are the modes that satisfy the oscillation condition [3]); the frequency difference between adjacent modes is about 80 MHz and the net gain greater than unity bandwidth is about 400 MHz [3]. The main operation parameters of the FEM are given in Table 1.

3. Simulations, experimental results and discussion

Simulations of FEM operation were made using the FEL3D code developed in our research group [8,9]. Fig. 2 shows simulation results of efficiency for each of our FEM for the five resonator modes having a net gain greater than one as a function of the reflectivity of the rear reflector. The simulations predict an efficiency of about 10% for an eigen frequency of 4.302 GHz, while the highest gain eigen frequency of 4.466 GHz (which is normally dominant in a free running operation of the FEM) has an expected efficiency below 7%. These results were obtained using the FEL3D code, which was used independently for each of the net gain eigen frequencies of the resonator.

Pre-bunching of the e-beam (see Fig. 1) enables interference in the mode competition process and determines which of the five longitudinal modes of the resonator will become dominant in FEM operation. Fig. 3 shows experimental results of efficiency enhancement due to mode selection by

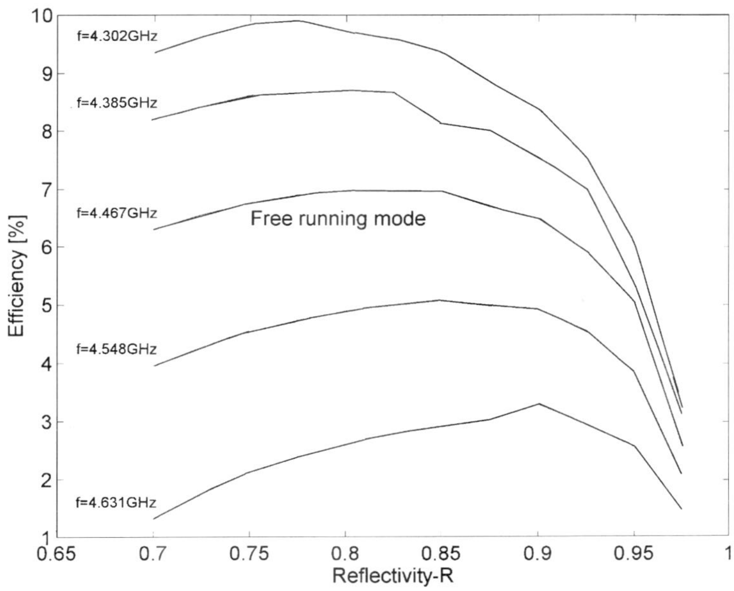

Fig. 2. Simulation prediction of efficiency for the five net gain greater than one eigen frequencies of FEM.

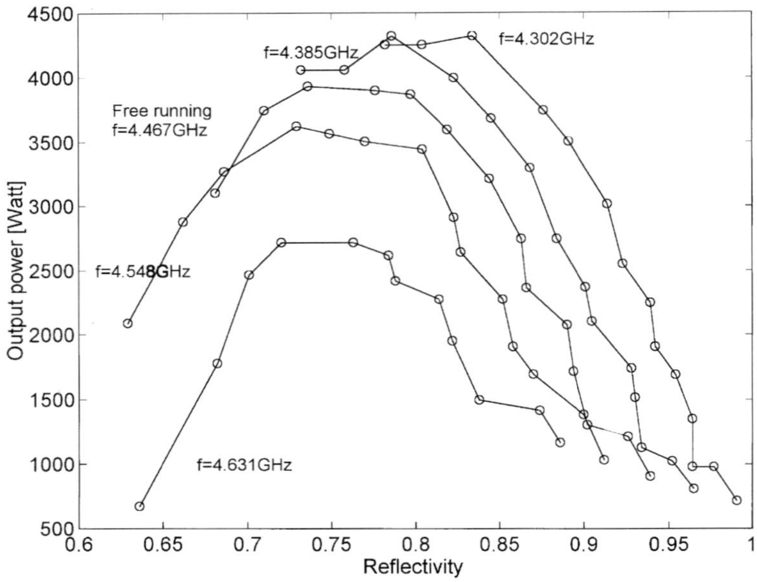

Fig. 3. Experimental obtained efficiency for each of the five eigen frequencies of the FEM having a gain greater than one (modes selected by pre-bunching).

pre-bunching in FEM operation. The experimental results show clearly a higher efficiency for the lowest eigen-frequency mode ($f = 4.302$ GHz) for reflectivities above $R = 0.8$ in good agreement with the simulation results. Under these conditions the efficiency is about 8% for the lowest eigen frequency of 4.302 GHz; it is about 5.5% for the free running mode of 4.467 GHz. For reflectivities below $R = 0.8$ the lowest eigen frequency did not build-up sufficiently, probably due to greater out-coupling loss and a low small signal gain.

In this work we have demonstrated a new method of significant radiation efficiency enhancement of FEMs by mode selection via pre-bunching of the e-beam. The experimental set-up enables us to select the mode with the largest detuning parameter θ (out of the 5 modes having net small signal gain greater than one). Our simulation code FEL3D predicts that this mode will be the most efficient. This prediction was confirmed by our experimental results as given in Fig. 3.

Acknowledgements

This work was supported by the Israel National Science Foundation and the Ministry of Science.

References

[1] V.L. Granastein, I. Alexeff, High Power Microwave Sources, Artch House, Boston, 1987 (Chapter 5).
[2] R.P. Fischer et al., Phys. Rev. Lett. 72 (1994) 2395.
[3] M. Cohen et al., Phys. Rev. Lett. 74 (1995) 3812.
[4] A. Eichenbaum et al., Nucl. Instr. and Meth. A 393 (1997) 361.
[5] A. Abramovich et al., Nucl. Instr. and Meth. A 407 (1998) 87.
[6] T.M. Antonsen, B. Levush, Phys. Rev. Lett. 62 (1989) 1488.
[7] T.M. Antonsen, B. Levush, Phys. Fluids B 1 (1989) 1097.
[8] Y. Pinhasi et al., Int. J. Electron. 78 (1995) 581.
[9] Y. Pinhasi et al., Phys. Rev. E 54 (1996) 6774.

ELSEVIER

Nuclear Instruments and Methods in Physics Research A 429 (1999) 111–115

NUCLEAR INSTRUMENTS & METHODS IN PHYSICS RESEARCH
Section A

www.elsevier.nl/locate/nima

Production of high power microwaves for particle acceleration with an FEL bunched electron beam

J. Gardelle[a,*], T. Lefevre[a], G. Marchese[a], M. Padois[a], J.L. Rullier[a], J.T. Donohue[b]

[a]*CEA/Centre d'Etudes Scientifiques et Techniques d'Aquitaine, B.P. 2, 33114 Le Barp, France*
[b]*Centre d'Etudes Nucléaires de Bordeaux-Gradignan, B.P. 120, 33175 Gradignan, France*

Abstract

Among the studies in the framework of high gradient linear electron–positron collider research, the Two-Beam Accelerator (TBA) is a very promising concept, and two projects are in progress, the Compact Linear Collider project at CERN (W. Schnell, Report no. CERN SL/92-51 and CLIC note 184; K. Hübner, CERN/PS 92-43, CLIC note 176; S. Van der Meer, CERN/PS 89-50, CLIC note 97.) and the Relativistic Klystron-TBA project at LBNL (Technical Review Committee, International Linear Collider Technical Review Committee Report 1995, SLAC-R-95-471, 1995). In a TBA an extremely intense low-energy electron beam, called the drive beam, is bunched at the desired operating frequency, and upon passing through resonant cavities generates radio-frequency power for accelerating the main beam. Among the different approaches to the production of a suitable drive beam, the use of an FEL has been proposed and is under active study at CEA/CESTA.

1. Introduction

Among the means under study for generating the intense bunched drive beam needed for a Two-Beam Accelerator (TBA) design, the use of a Free-Electron Laser (FEL) has been proposed by Shay and co-workers at Lawrence Livermore National Laboratory (LLNL) [1]. In an FEL, the electron beam, initially uniform along the axis, forms small bunches in the troughs of the co-propagating electromagnetic wave. The generation of high output power in a microwave FEL indicates that considerable bunching of the electrons has indeed taken place. Direct optical evidence for this bunching has been obtained at CEA/CESTA in the amplifier mode at 35 GHz [2–4] and also in the SASE mode [5]. In Section 2, we describe the FEL experiment which uses the induction accelerator LELIA (2.2 MeV, 800 A). Measurements of output power and frequency have been performed and a comparison with a numerical calculation is given. Section 3 deals with electron beam bunching analysis. Optical diagnostics allow us to study the bunching as a function of both time in the beam current pulse and interaction length. A suitable adiabatic exit has been developed to extract the bunched beam from the wiggler in order to perform the experiment described in Section 4 to test resonant RF-cavities at 35 GHz. The main issues are beam extraction from the wiggler, beam focusing at a smaller radius than that of the cavity and debunching due to the combination of the effects of longitudinal space

* Corresponding author.

0168-9002/99/$ - see front matter
PII: S0168-9002(99)00082-0

charge and solenoidal focusing. The last part of the paper is devoted to the proposal which extends the present work to a higher energy by using the induction accelerator PIVAIR (7 MeV, 3.5 kA), as a test stand for high-power RF structure experiments. With greater electron beam energy the beam remains bunched over long distances, allowing us to test various RF components of the two TBA projects.

2. The FEL experiment

Intense emission from the far infrared to the millimetre range is due to the strong electron bunching which appears in high-gain FELs which utilise low energy but very intense electron beams (≈ 1 MeV, ≈ 1000 A). This bunching mechanism occurs at the FEL wavelength so that electrons are able to emit radiation in phase leading to higher peak power levels [6]. At CESTA we have observed this bunching in an FEL amplifier. The electron beam is produced by the induction linac LELIA which delivers a 800 A, 2.2 MeV, 60 ns electron beam in single-shot-operation. After exiting LELIA, the beam is transported and matched into a 12-cm period helical wiggler where it amplifies a co-propagating 35 GHz, 10 kW electromagnetic wave (EM) generated by a magnetron. The wiggler is a 26-period pulsed bifilar helix. The first six periods as well as the last five periods are strapped with metallic wires in order to obtain adiabatic entry and exit of the electron beam. Three solenoids are used after the accelerator exit to guide and match the beam in the wiggler (i.e. to minimise betatron oscillations). The experimental setup is shown in Fig. 1. The FEL resonance condition at 35 GHz is obtained when the wiggler magnetic field amplitude is 1100 G. In that case we obtain the best amplification, the gain is exponential and the FEL saturates at 80 MW after 20 periods of interaction. The circles in Fig. 2 give the measured power as a function of the interaction length. To obtain these data we used a kicker magnet movable along the wiggler in order to stop the interaction by deflecting the beam into the pipe. The solid line corresponds to numerical results obtained with the SOLITUDE code in the TE_{11} mode [7]. Good

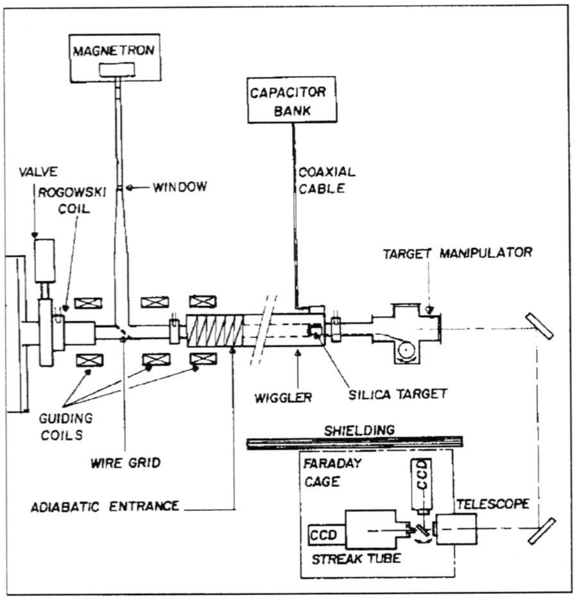

Fig. 1. Experimental setup showing the optical diagnostic for bunching experiment.

agreement with the experiment is observed because we used the measured electron beam initial conditions in the code. The FEL frequency is equal to the magnetron frequency (35.04 GHz). It was measured by a heterodyne technique in which the FEL output signal is mixed with the radiation from a variable frequency of 34–36 GHz local oscillator [8].

3. Bunching analysis

In order to measure electron beam bunching, we use an optical diagnostic based on the Cherenkov emission produced by the electrons when they are stopped in a movable fused-silica target. A mechanical system under vacuum allows us to place the target at any desired position in the 60 mm-diameter beam pipe. A gated camera and a fast streak camera (2 ps resolution) are used to collect and analyse a small part of the visible Cherenkov light. The gated camera gives the beam transverse position so that we are able to precisely place the slit of the streak camera in the beam spot. Bunching quality can be described by the bunching parameter b,

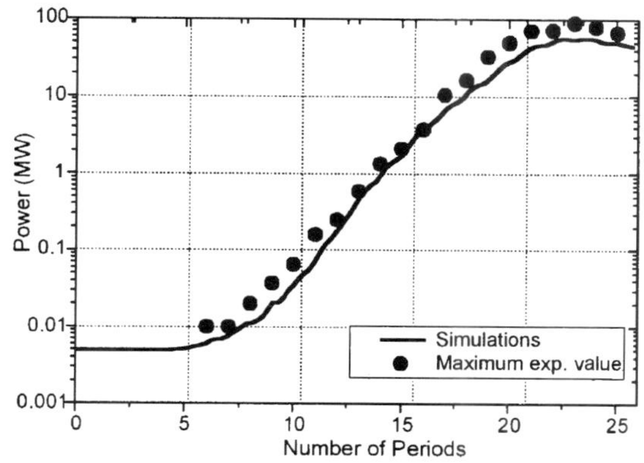

Fig. 2. Comparison between experimental and numerical power growth in the wiggler.

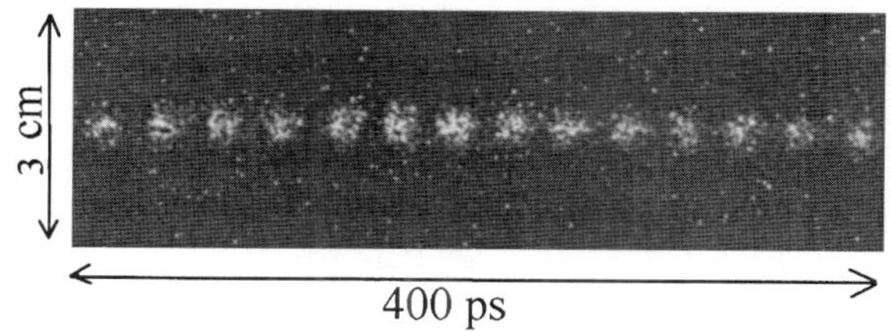

Fig. 3. Example of a streak camera image showing 14 bunches at 35 GHz.

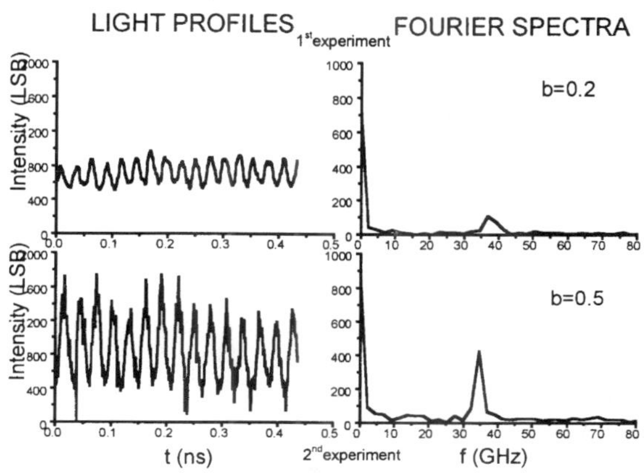

Fig. 4. Comparison between bunching measurements in the two amplifier experiments.

which is defined by the ratio of the Fourier component at 35 GHz of the current profile divided by the value of the average. Fig. 3 shows the best photograph of the bunches we have obtained. They are more pronounced than in the first experiment [1] and b has been increased from 0.2 to 0.5 as shown in Fig. 4, thanks to a higher current propagating inside the wiggler. Measurements as a function of distance inside the wiggler have shown bunching is maximum just before the FEL saturation, as expected from numerical calculations [9]. The adiabatic exit, where the wiggler magnetic field decreases from 1100 G to zero, begins at this location and the beam leaves the wiggler close to the propagation axis.

4. Test of RF-cavities

The transverse momentum acquired by electrons in the wiggler is lost and propagation on axis is possible behind the wiggler if an external force is applied to compensate the defocusing effects of emittance and space charge. In our case we require the beam to traverse a small radio-frequency (RF) cavity located as close as possible to the wiggler exit to minimise the longitudinal space charge effects which rapidly debunch the beam as it propagates downstream. Fig. 5 gives the experimental configuration we have adopted where a solenoid has been inserted to produce a small waist at the cavity location. A PARMELA[1] calculation has shown the beam remains bunched while passing through the cavity.

Two kinds of cavities are considered. The first has no output port (see Fig. 6) and acts as a rebuncher if it is slightly detuned from the driving 35 GHz frequency. Detuning can be obtained and adjusted by placing rings of different thickness inside the cavity resulting in a variation of its internal radius. Optical measurements will be performed to analyse the effects of the beam-cavity interaction on the longitudinal profile of the electron beam. A major challenge is to propagate and to focus the beam in such a small cavity inside a beam pipe whose diameter decreases from 60 to only 4 mm at the

[1] PARMELA is a transport code written and maintained by the Los Alamos National Laboratory Accelerator Code Group.

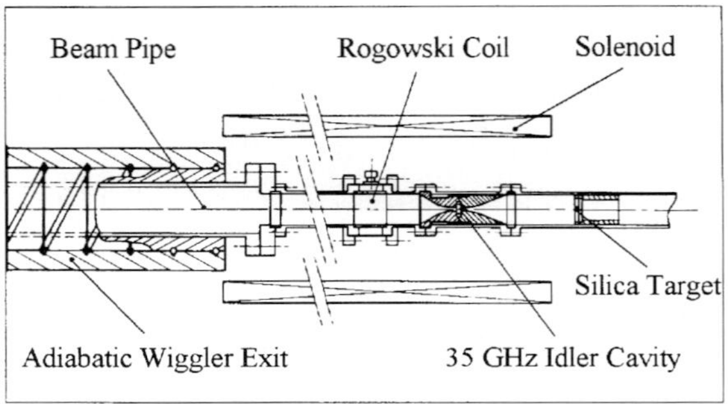

Fig. 5. Detail of the end of the experimental line which shows the end of the wiggler at left, the focusing solenoid, the idler cavity and the movable silica target.

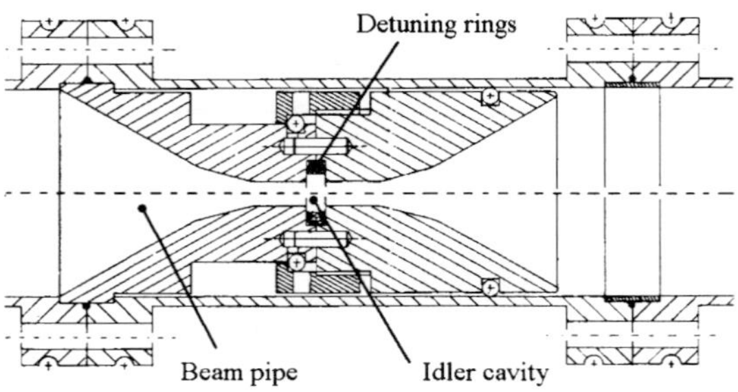

Fig. 6. Drawing of the idler cavity showing in detail the mechanism used to change internal rings.

cavity position. The second kind of cavity has one output port to a standard Ka-band waveguide in order to extract electromagnetic power. We plan to test both a low-Q cavity and a medium-Q cavity. They differ mainly by the size of the aperture between the cavity itself and the output waveguide. Calculations have been made with the GDFIDL code [10] and an example of the mesh used is given in Fig. 7. About 200 MW of output power is expected for an 800 A electron beam current with a bunching parameter $b = 0.5$. Measurements of frequency and phase of the output signal will be performed and compared with those of the FEL.

5. Future experiment

To improve the experiment, we need a higher-energy electron beam. The higher the energy, the lower the effects of longitudinal space charge defocusing. We have recently proposed [11] to use the PIVAIR facility at CESTA as a microwave FEL-buncher. PIVAIR is a single shot induction linac developed for radiographic applications. A 3.5 kA electron beam is extracted from a velvet cathode and is accelerated up to 7.2 MeV. The energy spread of the beam is much less than the LELIA machine (0.5% for 50 ns) and the normalised emittance is 1000 π mm mrad. The FEL calculations done with the code SOLITUDE have shown that for a 1 kA beam, about 1 GW of RF power at 35 GHz is expected with a 6 m-long pulsed helical wiggler with a period of 20 cm and a wiggler field of 1900 G. The shot-to-shot reproducibility of PIVAIR is excellent and it will allow us to perform FEL experiments at high power levels not only in the present application but also for FEL physics studies in SASE [12].

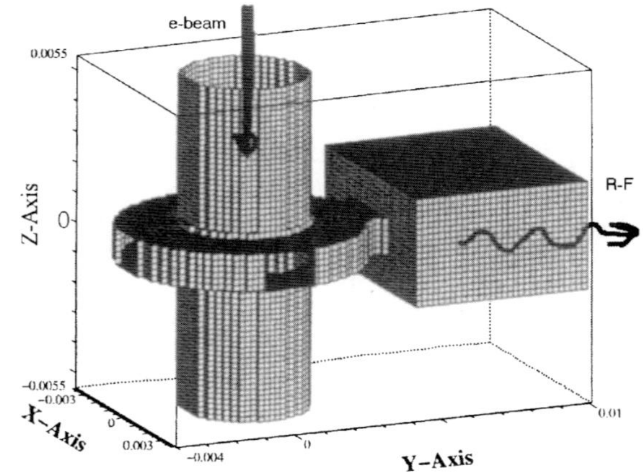

Fig. 7. Example of the mesh used for the open cavity calculations with the GDFIDL code.

6. Conclusion

We have presented the recent results of the CESTA-FEL experiment. The main application concerns TBA research where our induction linac-based FEL is used to produce an intense bunched electron beam to test high-frequency RF cavities. We are now optimising beam extraction from the wiggler and beam transport to the cavity location.

This experiment will be upgraded in the future by using the more powerful electron beam produced by the PIVAIR induction linac. Simulations have shown that the high quality bunched beam produced by a new wiggler will propagate far enough to provide a test bench for various components of the TBA projects. It is important both for the CLIC and the Relativistic Klystron-TBA (RKTBA) projects where the production of power at high frequency as well as knowledge about breakdown limits are of prime importance.

Acknowledgements

We wish to thank our collaborators in the CLIC group at CERN and the RK-TBA group at LBNL, in particular Steven Lidia for performing the calculations of the cavities.

References

[1] H.D. Shay et al., Nucl. Instr. and Meth. A 304 (1991) 262.
[2] J. Gardelle, J. Labrouche, J.L. Rullier, Phys. Rev. Lett. 76 (1996) 4532.
[3] J. Gardelle, T. Lefevre, G. Marchese, J.L. Rullier, J.T. Donohue, Phys. Rev. Lett. 79 (1997) 3905.
[4] J. Gardelle, J. Labrouche, G. Marchese, J.L. Rullier, D. Villate, J.T. Donohue, Phys. Plasmas 3 (1996) 4197.
[5] T. Lefevre, J. Gardelle, G. Marchese, J.L. Rullier, J.T. Donohue, Phys. Rev. Lett. 82 (1999) 323.
[6] H.P. Freund, T.M. Antonsen Jr., Principles of Free-electron-Lasers, second edition, Chapman & Hall, 1996, London.
[7] J. Gardelle, J. Labrouche, P. Le Taillandier, P. Gouard, Phys. Rev. E 50 (1994) 4973.
[8] J. Gardelle, T. Lefevre, G. Marchese, J.L. Rullier, J.T. Donohue, IEEE Trans. Plasma Sci. 25 (1997) 1419.
[9] R. Bonifacio et al., Rivista del nouvo cimento 13 (9) (1990) 1.
[10] W. Bruns, IEEE Trans. Magn. 32 (3) (1996) 1453.
[11] J. Gardelle, J.L. Rullier, CEA Internal Report, 1998.
[12] J.T. Donohue et al., Nucl. Instr. and Meth. A 429 (1999) 202.

ELSEVIER

Nuclear Instruments and Methods in Physics Research A 429 (1999) 116–120

NUCLEAR INSTRUMENTS & METHODS IN PHYSICS RESEARCH Section A

www.elsevier.nl/locate/nima

A Ka-band CHI-wiggler free-electron maser: experimental results

J.M. Taccetti[a,*], R.H. Jackson[b], H.P. Freund[c], D.E. Pershing[d], V.L. Granatstein[e]

[a]*Los Alamos National Laboratory, MS E526, Los Alamos, NM 87545, USA*
[b]*Naval Research Laboratory, Washington, DC 20375-5347, USA*
[c]*SAIC, McLean, VA 22102, USA*
[d]*MRC, Newington, VA 22122, USA*
[e]*University of Maryland, College Park, MD 20742, USA*

Abstract

Due to the nature of the Coaxial Hybrid Iron wiggler, a resonance between the transverse wiggler and cyclotron motions of the beam exists which can enhance the Free-Electron Maser interaction. The difference in behavior on either side of this resonance is presented. Amplification with a bandwidth on the order of a few percent was achieved. In spite of beam propagation problems, the gain was about 5 dB for operation below gyroresonance, and 7 dB above it, when wideband grazing intersection operation was achieved. A gain as high as 13.5 dB was obtained for cutting intersection between the beam line and the TE_{01} dispersion curve. © 1999 Elsevier Science B.V. All rights reserved.

PACS: 41.60.Cr

Keywords: Coaxial; Wiggler; Microwaves; Free-electron maser; Small-period wigglers

1. Introduction

The free-electron laser interaction provides the possibility for high-power, high-frequency radiation sources, although almost always coupled with the need for high-voltage operation. The Coaxial Hybrid Iron (CHI) wiggler [1–5] configuration overcomes this obstacle by allowing the use of short periods and gap spacings while maintaining excellent beam focusing properties and high magnetic fields. This reduces the overall system size while maintaining the ability to operate at high frequencies and power levels. The Ka-band Free-Electron Maser (FEM), with $g/\lambda_w > 1$ (g is the coaxial wiggler gap width and λ_w the period), demonstrates the operation of this wiggler configuration and its potential for applications to higher-frequency devices.

2. Experimental setup

The FEM amplifier consists of three main subsystems: the interaction area, the electron beam and the rf circuit. The interaction area is made up of a CHI wiggler with $\lambda_w = 6.4$ mm, a 60 λ_w uniform

* Corresponding author.
E-mail address: taccetti@lanl.gov (J.M. Taccetti)

0168-9002/99/$ - see front matter
PII: S0168-9002(99)00083-2

section and a coaxial gap width of 1.09 λ_w. The electron beam is formed in a thermionic hollow-cathode diode, with design voltage and current of 100 kV and 10 A, respectively. This beam is collected immediately after traversing the wiggler, on the outer walls of a coaxial drift tube. The rf is coupled in at a point between the electron gun and the wiggler, using a novel wideband TE_{01} outer coaxial to TE_{01} inner coaxial mode coupler. A similar design is used to monitor approximately 25% of the power after the interaction, while the rest of the rf is dumped into a water load/calorimeter. The inner coaxial section of the wiggler is held concentrically (the device is mounted horizontally) by three sets of three tungsten 'spokes' each, at different axial positions along the FEM.

Two DC breaks with accompanying current monitors were used to determine the fraction of the beam which makes it from the gun to the wiggler and from there to the collector. A misalignment in the magnetic axis of the electron gun and that of the interaction region is believed to be the cause of serious beam propagation problems experienced during the experiment. Only about 10% of the beam making it to the wiggler reached the collector in a typical measurement. RF amplification was nevertheless achieved. (A measurement of the beam quality from the gun alone was not performed, and therefore cathode emission problems cannot be completely ruled out.)

The voltage pulse applied has a FWHM of 2.4 μs. The 1 μs input rf pulse was timed such that it coincided with the flatter section of the voltage pulse. Our setup allowed the detailed study of the dependence of the rf output pulse shape on the variations in the voltage pulse.

3. Experimental results

A transition (or gyroresonant) field B_T can be defined in terms of the original energy of the beam, γ, the axial velocity of the beam, β_z, and $k_w = 2\pi/\lambda_w$. This field,

$$B_T = \frac{mc^2}{e}\gamma\beta_z k_w \tag{1}$$

is that for which the wiggler motion and the cyclotron motion of the beam are resonant. Operation below and above this value is termed Group I and Group II operation, respectively. The dependence of B_T on β_z demonstrates how its value is not constant but rather varies dynamically during the interaction.

RF amplification of Group I orbits was achieved at 0.76 B_T. The output rf pulse shape varied with voltage, as expected, but was not overly sensitive. Varying the input power from 0 to 1.2 kW resulted in a gain of approximately 4 dB for all these values, as shown in Fig. 1. The uncertainty in the measured gain values is on the order of the size of the symbols used in the figure. The beam voltage in this case was 113.1 kV, the current reaching the collector, I_c, was 1.9 A, and the estimated $\alpha = v_{transverse}/v_{axial}$ was 0.39. This value of α was obtained by assuming a grazing intersection of the uncoupled TE_{01} coaxial mode dispersion curve and the CHI beam line, a safe assumption due to the mode selectivity of the output rf monitor. This high α indicates a large amount of axial energy of the beam is being coupled to transverse motion before entering the wiggler. This is supported by comparison of measurements of gain as a function of frequency against a simulation (performed on CHIFEL [3]), as shown in Fig. 2, where the simulation is shown as a solid line. Both agree fairly well on the low-frequency side. The high-frequency discrepancy is

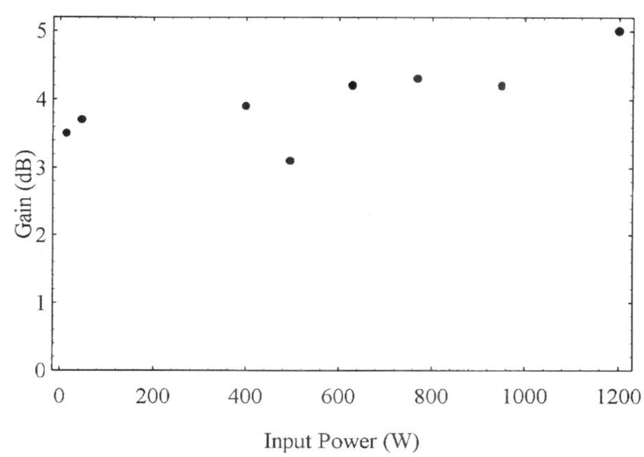

Fig. 1. Gain as a function of input power for $B_0 = 8.5$ kG and $V_b = 113.1 \pm 0.9$ kV ($I_c = 1.9 \pm 0.1$ A).

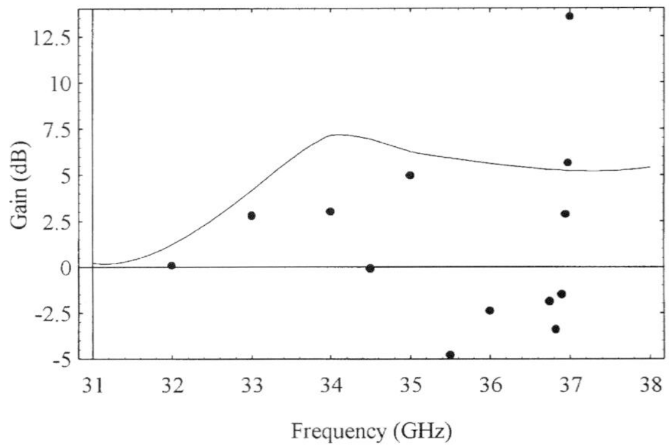

Fig. 2. Gain as a function of drive frequency in Group I ($B_0 = 8.5$ kG, $V_b = 116.3 \pm 0.7$ kV, $I_c = 2.05 \pm 0.06$ A). Points are measured values, and the solid line is from CHIFEL simulations (details in text).

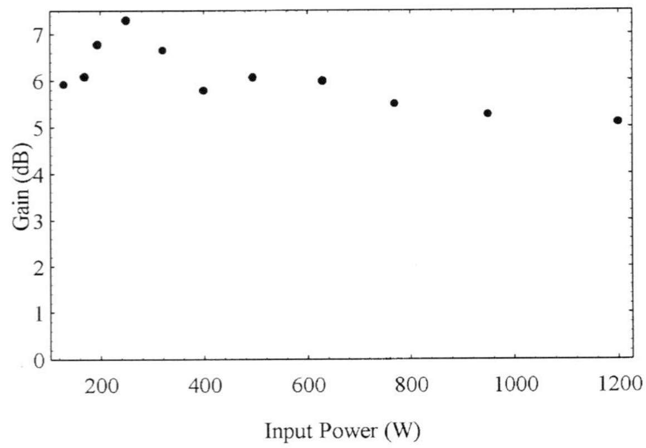

Fig. 3. Gain as a function of input power for $B_0 = 12.5$ kG and $V_b = 119.7 \pm 0.4$ kV ($I_c = 1.69 \pm 0.07$ A).

believed to be due to reflections in the rf circuit caused by limitations of the output rf monitor. In the simulation, the maximum beam width was chosen such that all of the 2.1 A (as in the experiment) makes it through the interaction region. For interaction to occur in the simulation, the voltage had to be lowered to 102 kV, from the 116.3 kV used in the experiment, indicating again that a large fraction of the beam energy is turned into transverse motion prior to entering the wiggler. This is believed to be a result of the magnetic axis misalignment mentioned previously.

Group II operation was also achieved at $1.1B_T$. The output rf pulse was in this case much more sensitive to slight variations in voltage. This is due to the proximity to B_T, as a variation in axial velocity causes a variation in B_T, which in turn causes a quite drastic change in α. The gain measured as a function of input power in this case was from 5 to 7 dB in the range from 0 to 1.2 kW input power and is shown in Fig. 3.The bandwidth for this case was about 5% and centered about 35 GHz. This is shown in Fig. 4. A simulation such as the one performed for Group I orbits was not obtained for operation above gyroresonance. Preliminary calculations did not show the same kind of agreement, but further studies are necessary as far as the cause for this disagreement is concerned.

A map of the observed behavior of the amplified rf pulse as a function of voltage and applied field is

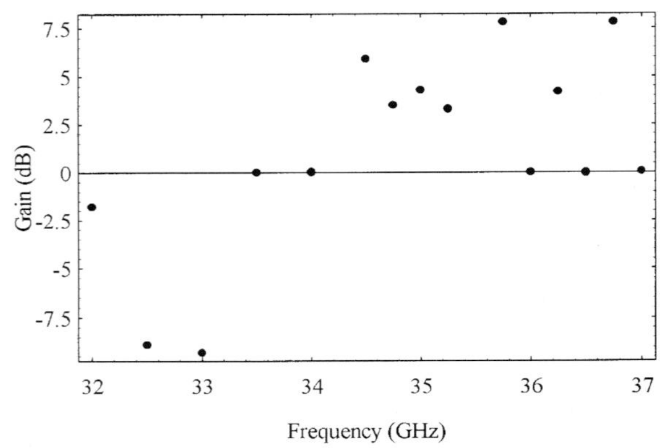

Fig. 4. Gain as a function of drive frequency in Group II ($B_0 = 12.0$ kG, $V_b = 114.5 \pm 0.5$ kV, $I_c = 2.17 \pm 0.08$ A).

shown in Fig. 5. The beam voltage was swept between 67 and 140 kV for various values of the applied solenoidal fields (from 8.5 to 10.5 kG). The gyroresonant field, B_T, varies with axial velocity, and therefore depends on the beam voltage and the beam α. As seen in Eq. (1), it increases with increasing axial velocity. As one moves towards the upper left-hand corner of the figure, with low fields and high voltages, one moves below gyroresonance. The opposite is the case towards the lower right-hand corner, with high fields and low voltages. The behavior of the output rf pulse as a function of

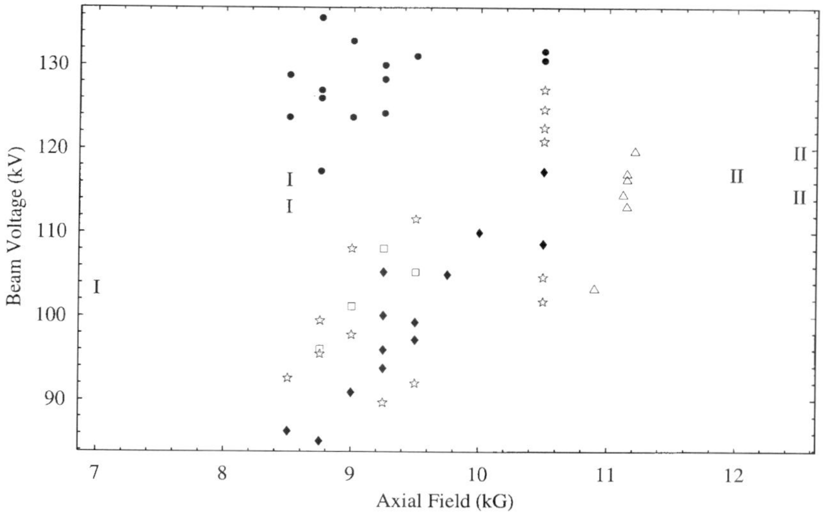

Fig. 5. Map of amplified rf pulse behavior when varying voltage at various solenoidal fields. Shown are points where 'rabbit ears' appear (boxes), points where they coalesce and take over flat top of pulse (solid diamonds), and other spurious oscillations (stars and solid dots). Operating points where amplification was achieved are represented by the text 'I' and 'II' for Groups I and II, respectively. Triangles represent the B_T values calculated for these cases.

voltage was similar for different fields, although shifted in voltage. The different types of behavior encountered are distinguished in Fig. 5 by different symbols. The points where stable amplification was achieved are indicated on the figure by the labels 'I' and 'II', for Group I and II operation, respectively. The triangles represent B_T values for these operating points. The solid diamonds represent points where a major oscillation dominated the rf pulse, beginning usually with 'rabbit ears' appearing on the outside of the pulse at higher voltages and moving inward toward the center of the pulse as the voltage is lowered. The boxes represent the voltages at which these rabbit ears appeared. The stars and solid dots instead represent oscillations which only occur on the flat part of the pulse and exist only for a certain voltage. There is a central region where the oscillations die down and amplification is possible (Group I), though this region does seem to get narrower (in voltage) as the interaction field is increased. The stars indicate cyclotron resonance oscillation points.

The rabbit ears can be understood as follows. During a voltage pulse, the beam goes through gyroresonance twice if the parameters are set for Group I orbits. At low voltages, during the rise of the pulse, the beam starts out in Group II. As the voltage increases, the transition field moves up to and through the operating point, the orbits become Group I. The reverse process takes place during the fall of the voltage pulse. The rabbit ears become more noticeable as the magnetic field is increased because they occur at a higher point on the pulse where there is more current present.

A linear gyrotron oscillator start current code [6] was used to determine the validity of the values for α and beam axial spread chosen for a current within the possible bounds. Many of the parameters used are unknown, such as the interaction length, the guiding center radius, and the velocity spread, but it was found that it is possible to start gyrotron oscillation at values of α and magnetic fields as expected from the uncoupled dispersion curves. Varying the cavity length in the simulations causes oscillation at various frequencies, which agrees with the observed spectrum (oscillations throughout the band).

The higher voltage oscillations (the solid dots in the figure) are not completely understood at this time, though it is believed they may represent an interaction with the TE_{12} mode, which somehow is either getting coupled out or affecting the TE_{01} mode coupling.

4. Conclusion

Operation of the first FEM amplifier utilizing a CHI-wiggler configuration has been demonstrated, with a gap to wiggler period ratio in excess of one. The dependence of the output rf on the voltage has been studied on either side of gyroresonance, approaching to within 10% of it on the Group II side. It was found that operation is limited by cyclotron oscillations on Group I, and by high sensitivity to voltage on Group II. In conclusion, in spite of beam propagation problems, the CHI wiggler displays exceptional potential for higher frequency operation.

Acknowledgements

This work was supported in part by the Office of Naval Research and the Naval Research Laboratory through a contract with the University of Maryland.

References

[1] R.H. Jackson, H.P. Freund, D.E. Pershing, J.M. Taccetti, Nucl. Instr. and Meth. A 341 (1994) 454.
[2] J.M. Taccetti, R.H. Jackson, H.P. Freund, D.E. Pershing, M. Blank, V.L. Granatstein, Nucl. Instr. and Meth. A 358 (1995) 193.
[3] H.P. Freund, R.H. Jackson, D.E. Pershing, J.M. Taccetti, Phys. Plasmas 1 (1994) 1046.
[4] J. Feinstein, U.S. Patent No. 4,876,687.
[5] R.H. Jackson, H.P. Freund, D.E. Pershing, J.M. Taccetti, U.S. Patent No. 5,499,255.
[6] P.E. Latham, S.M. Miller, C.D. Striffler, Phys. Rev. A 45 (1992) 1197.

ELSEVIER

Nuclear Instruments and Methods in Physics Research A 429 (1999) 121–124

NUCLEAR INSTRUMENTS & METHODS IN PHYSICS RESEARCH Section A

www.elsevier.nl/locate/nima

Project of a large orbit FEM-oscillator operated at the second harmonic

N.S. Ginzburg[a], N.Yu. Peskov[a,*], I.E. Kamenetsky[a], A.K. Kaminsky[b], A.A. Kaminsky[b], S.N. Sedykh[b], A.P. Sergeev[b]

[a]*Institute of Applied Physics, Russian Academy of Sciences, 603600 Nizhny Novgorod, Russia*
[b]*Joint Institute for Nuclear Research, 141980 Dubna, Russia*

Abstract

To operate at harmonics of the bounce-frequency in the JINR-IAP FEM-oscillator driven by the LIU-3000 linac (JINR, Dubna) we propose using a large electron orbit regime. In this regime a thin axial beam moving in a helical wiggler with an axially guided magnetic field excites at the *n*th bounce harmonic only the waveguide mode with the same azimuthal index. For effective FEM operation in this regime, the radius of electron orbit should be comparable with the transverse inhomogeneity of the resonant wave. A 75 GHz FEM project for operation at the second harmonic of the bounce-frequency is presented.

Keywords: Free-electron lasers and masers; Powerful millimeter and submillimeter radiation; High harmonics operation

1. Introduction

An FEM-oscillator with a Bragg resonator and a reversed guide magnetic field was studied both theoretically and experimentally in collaboration between JINR (Dubna) and IAP RAS (N. Novgorod) during the last few years [1,2]. In these experiments a thin solid electron beam was generated by the LIU-3000 linac (1 MV/200 A/200 ns) (JINR, Dubna). This magnetically guided beam was axially injected into a helical wiggler, where the operating transverse velocity was produced in a tapered wiggler section. The interaction between the electrons, oscillating in regular wiggler section with the bounce-frequency $\Omega_b = 2\pi v_{\|}/d$, and $TE_{1,1}$ mode of a circular waveguide was provided at the fundamental harmonic ($n = 1$) of the bounce-frequency under the resonance condition

$$\omega - hv_{\|} = n\Omega_b. \quad (1)$$

As a result of the experiments in the wiggler of period $d = 6$ cm at the frequency of 31 GHz, an output power of 37 MW and an efficiency of 26% were obtained [1,2].

The radiation frequency in the FEM can be increased, in principal, using a wiggler with a shorter period. However, in practice, such a possibility is rather restricted because of wiggler overheating, probability of breakdown, etc. Note also that a decrease in the wiggler period leads to an increase in

* Corresponding author. Tel.: + 7-8312-384-575; fax + 7-8312-362-061;
E-mail address: peskov@appl.sci-nnov.ru (N. Yu. Peskov)

0168-9002/99/$-see front matter
PII: S0168-9002(99)00084-4

the transverse inhomogeneity of the wiggler field. This results in an increase in the velocity spread that the beam electrons obtain in the wiggler and, thus, a drop in the FEM efficiency.

An attractive way to increase the radiation frequency while maintaining the geometrical parameters of the electron-optical system is to use FEM operation at harmonics of the bounce-frequency [3,4]. This allows operation at a radiation frequency n time higher than the fundamental harmonic frequency. Moreover, a thin axially injected beam will excite at the nth harmonic only that waveguide mode with the azimuthal index $m = n$. Thus, a discrimination of excitation of "parasitic" waveguide modes takes place. The suggested selective mechanism is similar to the one used in the so-called Large Orbit Gyrotrons [5,6].

2. Main equations

Let us investigate the energy extraction from a thin solid electron beam moving in a regular wiggler section along a helical trajectory with a transverse velocity $v_\perp$ and an orbit radius $a_\perp = v_\perp/\Omega_b$. The interaction between the beam and the $TE_{m,k}$ mode of a circular waveguide at the resonance condition (1) may be described by the following system of average equations:

$$\frac{dE}{dz} = \mathrm{Re}\left\{(-1)^m eA_0 \frac{v_\perp}{v_\parallel} J_{m-n}(\kappa_\perp R) e^{i(n-m)\Psi} J'_m(\kappa_\perp a_\perp) e^{i\Theta}\right\} \quad \text{(2a)}$$

$$\frac{d\Theta}{dz} = \frac{1}{v_\parallel}(\omega - n\Omega_b) - h \quad \text{(2b)}$$

where $E = \gamma mc^2$ is the electron energy, $\Theta = \omega t - hz - n\Omega_b t$ is the phase with respect to the resonant wave, A_0 is the amplitude of the resonant wave, $\kappa_\perp = \sqrt{\omega^2/c^2 - h^2}$ is the transverse wavenumber of the resonant wave, $(R\,;\,\Psi)$ are the polar coordinate of the beam injection, $J_s(x)$ is the Bessel function of order s and $J'_s = dJ_s/dx$. The boundary conditions for a monoenergetic, unmodulated beam have the form

$$E|_{z=0} = E_0, \quad \Theta|_{z=0} = \Theta_0 \in [0, 2\pi). \quad (3)$$

The term $J_{m-n}(\kappa_\perp R)$ in Eq. (2a) describes the mode selection mechanism discussed above. Clearly, in the case of axial beam injection $R = 0$ only an interaction with the waveguide mode having the azimuthal index equal to the harmonic number (i.e. $m = n$) is possible. For this case we reduce Eqs. ((2a) and b) to the form

$$\frac{du}{dZ} = \sin\Theta, \quad \frac{d\Theta}{dZ} = \Delta + u\,\mathit{sign}\,\mu. \quad (4)$$

Here the following dimensionless variables are used: $Z = hz\sqrt{\alpha_s\rho\mu}$, $u = (1 - \gamma/\gamma_0)\sqrt{\mu/\alpha_s\rho}$, $\mu = d\beta_\parallel^{-1}/d\gamma$ is the inertial bunching parameter [7–9], $\Delta = [\beta_{ph}\beta_{\parallel 0}^{-1}(1 - n\Omega_b/\omega) - 1]/\sqrt{\alpha_s\rho\mu}$ is the initial mismatch from the synchronism, $\alpha_s = eA_0/\gamma_0 hmc^2$, $\rho = J'_n(\kappa_\perp a_\perp)\beta_\perp/\beta_\parallel$ is the electron-wave coupling coefficient. Electron efficiency in the new variables is given as

$$\eta = \sqrt{\frac{\alpha_s\rho}{\mu}}\frac{\hat{\eta}}{1 - \gamma_0^{-1}} \quad \hat{\eta} = \frac{1}{2\pi}\int_0^{2\pi} u\, d\Theta_0. \quad (5)$$

The maximum in the normalized efficiency $\hat{\eta}_{max} = 1.37$ is achieved with $\Delta_{opt} = -1.3\,\mathrm{sign}\,\mu$ and $L_{opt} = 3.9$. The starting current of the oscillation is

$$I_{st} = \frac{mc^3}{e}\frac{\gamma_0 SN}{\pi l^2 Q\mu\rho^2} \quad (6)$$

where $S = \pi D^2/4$ is the waveguide cross-section, l is the interaction length, $N = (1 - 4m^2/\kappa_\perp^2 D^2) \times J_m^2(\kappa_\perp D/2)$ is the wave norm and Q is the Q-factor of the resonator.

Under the condition $\kappa_\perp a_\perp \ll 1$ the electron-wave coupling coefficients $\rho \approx (\kappa_\perp a_\perp)^{n-1}\beta_\perp/2^n\beta_\parallel(n-1)!$ decrease very rapidly with an increase of the harmonic number. This results in a significant increase in the starting current (6) and a decrease in the efficiency (5). At the same time, an effective interaction is possible under the condition $\kappa_\perp a_\perp \sim 1$, i.e. when the radius of electron orbit is comparable with the transverse inhomogeneity of the resonant wave. Thus, for FEM operation at high harmonics, an increase in the radius of the electron orbit is needed.

The radius of the electron orbit in a helical wiggler versus the strength of the guide magnetic field is presented in Fig. 1. For effective operation at the bounce-frequency harmonics with the parameters of the experiments [1,2] an electron orbit radius $a_\perp \approx 0.2$–0.3 is required. It can be achieved with a so-called reversed guide magnetic field (marked by the "-" sign in Fig. 1) [10,11], where the cyclotron rotation of the beam electrons and the rotation imposed by the helical wiggler have opposite directions. In this case, the radius of the electron orbit in the wiggler may be approximated as [7–9]

$$a_\perp = \frac{eH_w d}{2\pi\gamma mc(\omega_H + \Omega_b)}. \tag{7}$$

A theoretical analysis shows [2,12] that FEM operating regimes far from the cyclotron resonance, including the reversed guide field regime, provide a higher quality helical electron beam formed in a slowly up-tapered wiggler entrance and possess a rather small sensitivity to the initial velocity spread in the beam. This is corroborated by the results of the FEM-amplifier [10,11] and oscillator [1,2] experiments, where a maximum for millimeter wavelength FEM efficiency of 25–30% was obtained in the mentioned regime. The high efficiency achieved in the experiments with reversed field configuration at fundamental bounce-frequency harmonic permit us to expect a rather high efficiency for FEM operation at higher harmonics as well.

3. Project of 4 mm FEM operated at second harmonic

Using the theoretical consideration carried out above let us estimate the possibility of constructing a 75 GHz FEL operating at the second bounce-frequency harmonic using the LIU-3000 linac. The helical wiggler of period 6 cm and wiggler field amplitude of 0.1–0.15 T will drive the operating transverse velocity $\beta_\perp \approx 0.2$–0.3. For the thin axially injected beam the radius of the electron orbit will be about 0.2–0.3 cm under a reversed guide field of strength ~ 0.2 T. A $TE_{2,1}$ mode of a 2-cm diameter circular waveguide is chosen as the operating mode. A two-mirror Bragg resonator with $Q \approx 2000$ should provide the selective feedback for the operating mode at the designed frequency. Theoretical analysis based on the average equations of motion (4) as well as computer simulations using a 3-D code, which solves the full (non averaged) equations, demonstrates that an efficiency of about 15% and an output power of

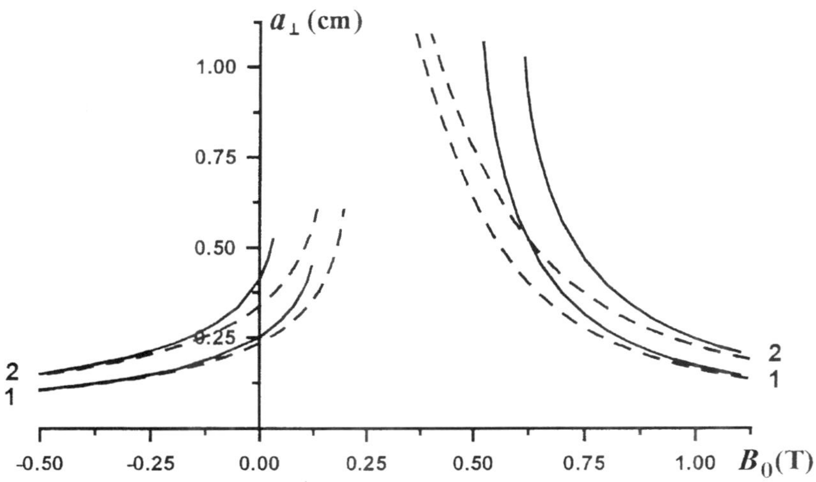

Fig. 1. Dependence of the radius of the electron orbit in a slowly up-tapered helical wiggler vs. the strength of the guide magnetic field: 1 – $B_W = 0.1$ T, 2 – $B_W = 0.14$ T. Dashed lines correspond to approximations by Eq. (7). ($d = 6$ cm, $\gamma = 2.6$).

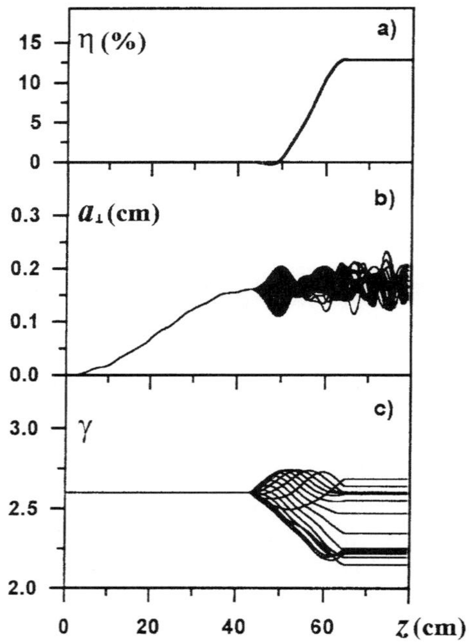

Fig. 2. Dependence on longitudinal coordinate of electron efficiency (a), energy (b) and radius of rotation (c) of electrons with different initial phase under excitation of the $TE_{2,1}$ mode. $\lambda = 4$ mm, $d = 6$ cm, $\gamma = 2.6$, $B_W = 0.1$ T, $B_0 = -0.2$ T, $A_0 = 350$ kV/cm, $D = 2$ cm.

15–20 MW may be obtained for parameters of this experiment (Fig. 2). The starting current for the oscillator is about 20 A. An experimental study of this FEM scheme is currently in progress.

Acknowledgements

This work is supported by Grants 97-02-16643 and 7-02-17379 of Russian Foundation for Basic Research.

References

[1] N.S. Ginzburg, A.K. Kaminsky, A.A. Kaminsky, N.Yu. Peskov, S.N. Sedykh, A.P. Sergeev, A.S. Sergeev, Proceedings of the Seventeenth International FEL Conference, New York, USA, 1995; p.215.
[2] N.S. Ginzburg, A.A. Kaminsky, A.K. Kaminsky, N.Yu. Peskov, S.N. Sedykh, A.P. Sergeev, A.S. Sergeev, IEEE Trans. Plasma Sci. 26 (3) (1998) 541.
[3] N.S. Ginzburg, ZhTF, 51 (4) (1981) 764 (in Russian).
[4] W.B. Colson, IEEE J. Quantum. Electron QE-17 (8) (1981) 1417.
[5] D.B. McDermott, N.C. Luhmann, A. Kupiszewski, H.P. Jory, Phys. Fluids 26 (1983) 1936.
[6] V.L. Bratman, Yu.K. Kalynov, N.G. Kolganov, Proceedings of the Third International Workshop on Strong microwaves in plasmas, vol 2, Nizhny Novgorod, Russia, 1996, p. 745.
[7] H.P. Freund, S. Jounston, P. Sprangle, IEEE J. Quantum. Electron QE-19 (3) (1983) 322.
[8] H.P. Freund, P. Sprangle, D. e.a. Dillenberg, Phys. Rev. A 24 (4) (1981) 1965.
[9] N.S. Ginzburg, N.Yu. Peskov, ZhTF 58 (5) (1998) 859 (in Russian).
[10] A.A. Kaminsky, A.K. Kaminsky, S.B. Rubin, Particle Accelerators 33 (1990) 189.
[11] M.E. Conde, G. Bekefi, Phys. Rev. Lett. 67 (1991) 3082.
[12] N.Yu. Peskov, S.V. Samsonov, N.S. Ginzburg, V.L. Bratman, Proceedings of the Nineteenth International FEL Conference, Beijing, China, 1997, p.114.

ELSEVIER

Nuclear Instruments and Methods in Physics Research A 429 (1999) 125–130

NUCLEAR INSTRUMENTS & METHODS IN PHYSICS RESEARCH Section A

www.elsevier.nl/locate/nima

First lasing of the regenerative amplifier FEL

Dinh C. Nguyen*, Richard L. Sheffield, Clifford M. Fortgang, John C. Goldstein, John M. Kinross-Wright, Nizar A. Ebrahim

Los Alamos National Laboratory, MS H851, Los Alamos, NM 87545, USA

Abstract

The regenerative amplifier free-electron laser (RAFEL) is a high-gain RF-linac FEL capable of producing high optical power from a compact design. The combination of a high-gain and small optical feedback enables the FEL to reach saturation and produce a high optical power and high extraction efficiency without the risk of optical damage to the mirrors. This paper summarizes the first lasing of the regenerative amplifier FEL and describes recent experimental results. The highest optical energy achieved thus far at 16.3 μm is 1.7 J over a train of 900 micropulses. We infer pulse energy of 1.9 mJ in each 16 ps micropulse, corresponding to a peak power of 120 MW. © 1999 Published by Elsevier Science B.V. All rights reserved.

Keywords: Regenerative amplifier; High-gain; SASE; High power; RF-linac FEL; Cs_2Te; Photocathode

1. Regenerative amplifier FEL

High-gain RF-linac FELs operating in the self-amplified spontaneous emission (SASE) regime have recently emerged as a potentially viable technology for generating short-wavelength coherent radiation [1]. High-power FELs, prone to optical damage in the resonator mirrors because of high intracavity power, can also benefit from the single-pass nature of SASE. A variation of SASE, which we call the regenerative amplifier FEL, uses mirrors to provide a small amount of optical feedback to restart the amplification process in a high-gain wiggler [2]. The large single-pass gain allows the optical intensity to build-up in a few passes to a sufficiently high level for efficient energy extraction. Using a large outcoupling allows most of the optical power to exit the cavity, thereby reducing the risk of optical damage and increasing the amount of light that exits the cavity as useful power.

The regenerative amplifier FEL offers a number of unique attributes: very large cavity detuning length (on the order of millimeters), fast build-up and ring-down (a factor of 3 in successive passes), broad output spectra (> 5%), a high extraction efficiency (> 2%), and a very high peak power (hundreds of MW). This paper summarizes the experimental conditions for realizing the RAFEL concept and describes the first lasing results as well as some recent accomplishments.

2. Experimental setup

The experimental setup of the regenerative amplifier FEL has been described in detail elsewhere

* Corresponding author. Tel.: 505-667-9385; fax: 505-667-8207.

E-mail address: dcnguyen@lanl.gov (D.C. Nguyen)

0168-9002/99/$ - see front matter © 1999 Published by Elsevier Science B.V. All rights reserved.
PII: S0168-9002(99)00090-X

[3]. Table 1 summarizes the RAFEL experimental conditions. The electron beam is generated by a compact, 1.2-m-long, 1300-MHz photoinjector/linac capable of producing a maximum beam energy of 20 MeV. In operation, the cesium telluride (Cs_2Te) photocathode was illuminated with the 8-ps ultraviolet pulses from a frequency-quadrupled modelocked Nd:YLF laser. The photocathodes for this experiment have been enlarged so that they can be uniformly illuminated with a 7-mm-radius drive laser spot. The large emission radius reduces space-charge effects and results in a higher peak current [3]. Following a report of successful application of protective coating to reduce the sensitivity of Cs_2KSb and Cs_3Sb cathodes [4], we applied a thin film (~40 Å) of CsBr on the Cs_2Te photocathodes. The coated Cs_2Te cathodes are much more rugged than uncoated cathodes. The quantum efficiency for Cs_2Te cathodes, approximately 10% for freshly made cathodes, is reduced to 5–6% after application of the coating (Fig. 1). The QE remains unchanged for more than two months of operation without any sign of degradation.

Table 1
Summary of parameters for standard RAFEL operating conditions

Beam energy	E_b	16.7 MeV
Peak current	I	270 A
Micropulse charge	Q	4.5 nC
Bunch length	τ	16 ps
Normalized emittance[a]	ε_n	7 π-mm-mrad
Energy spread	$\Delta\gamma/\gamma$	0.5%
Rms beam radius inside wiggler	r_b	0.2 mm
Wiggler period (fixed)	λ_w	2 cm
Wiggler length	L_w	2 m (1 m uniform)
On-axis field	B_w	(0.7–0.5) T
Wiggler parameter	a_w	0.92–0.65
Wiggler gap	g_w	(5.9–9.5) mm
Betatron period	λ_β	1 m
Wavelength	λ	16.3 μm
Pierce parameter	ρ	0.02
Gain length (measured)	L_G	15 cm

[a] Inferred from the measured matched beam radius.

The generated electron beam is focused to the entrance of the wiggler by two solenoids, one around the photocathode and the other at a location 0.4 m in front of the wiggler. The wiggler is designed to provide two-plane sextupole focusing to maintain the same electron beam radius through

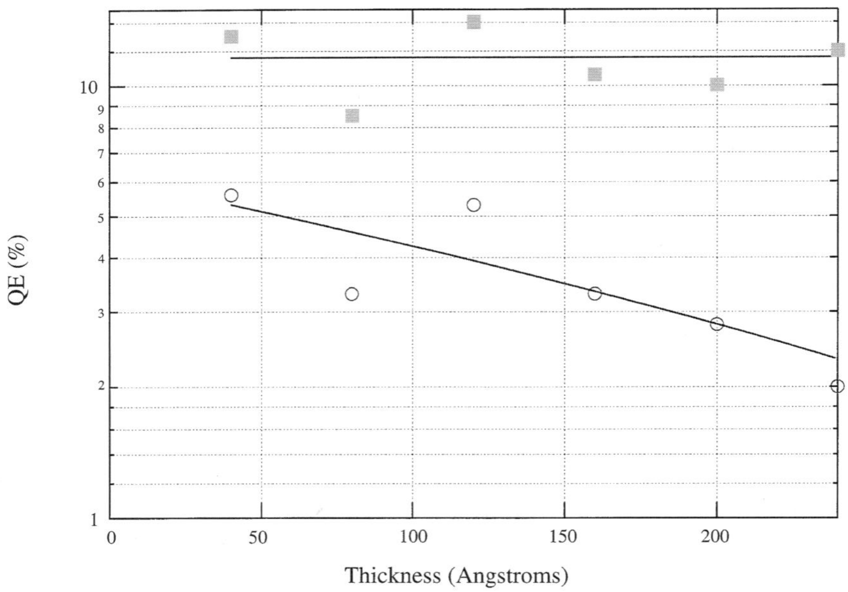

Fig. 1. Quantum efficiencies of uncoated and CsBr-coated CsTe photocathodes versus CsBr film thickness.

the wiggler. The design of the RAFEL wiggler has been described in detail previously [5]. The wiggler has a constant period of 2 cm, with a 1-m uniform section (peak field = 0.7 T) and a 1-m tapered section (peak field = 0.7–0.5 T). The wiggler is bracketed with two annular mirrors that form a part of the feedback cavity. The electron beam converges through a 5-mm-diameter hole in the first annular mirror to a 0.2-mm-radius spot at the wiggler entrance. Both the electron and the FEL beams go through a 12-mm-diameter hole in the downstream annular mirror. Part of the optical beam is reflected off the downstream annular mirror, collimated by a pair of spherical and cylindrical mirrors (approximating a 90° paraboloid), and then refocused to the wiggler entrance by the second paraboloid. Most of the high-power optical beam exits through the large hole in the downstream annular and expands to a 2-in-diameter beam at the KBr vacuum window. The optical beam is measured with a Molectron J50 energy meter, a slow HgCdTe detector, or a fast Cu : Ge detector. After traversing the second annular mirror, the electron beam enters a spectrometer dipole, turns 120°, and terminates in the beam dump located in the ground.

3. Experimental results

The RAFEL small-signal gain was measured in an SASE experiment that is described in detail elsewhere [6]. From the best fit to the data, we obtained a value of the gain coefficient that indicates there was 8 power gain lengths in our wiggler. This corresponds to a single-pass gain of 330 (33 000%). This small-signal gain is too large to be measured by conventional optical build-up methods. The large-signal gain as measured by the optical build-up is as high as 300%. Immediately after installing the feedback optics, we observed an optical power that exceeded the SASE power by more than six orders of magnitude (Fig. 2). The measured HgCdTe signal in the large signal regime follows the transmitted current macropulse closely and exhibits much more fluctuations than the single-pass signal. By adjusting the feedback cavity length, we measured a detuning length FWHM

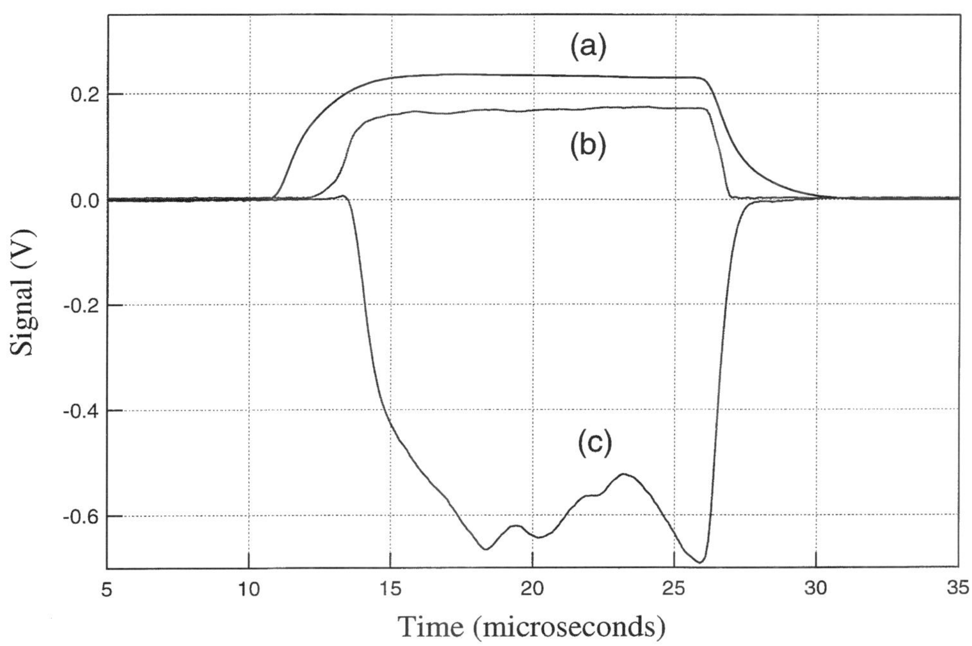

Fig. 2. Oscilloscope traces of (a) cavity RF field, (b) beam current, (c) HgCdTe voltage in the large-signal (lasing) regime.

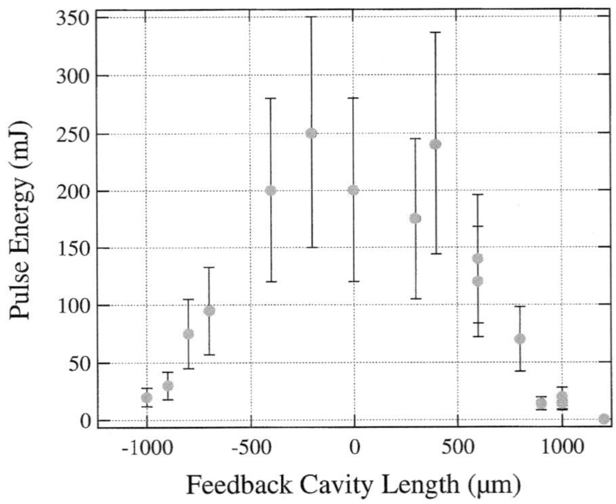

Fig. 3. RAFEL cavity detuning length.

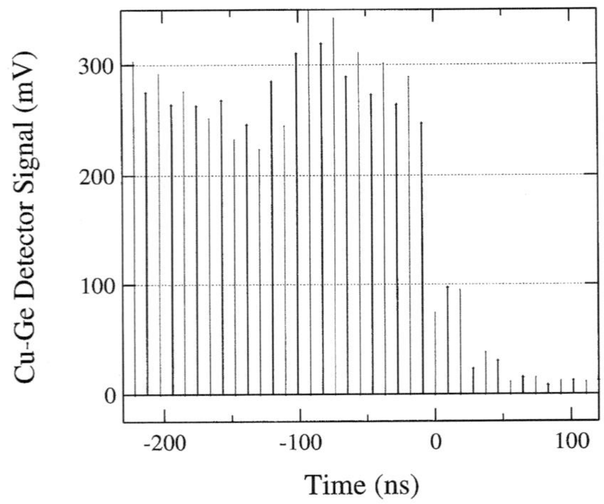

Fig. 5. Cu : Ge detector signals of individual micropulses showing fast ring-down.

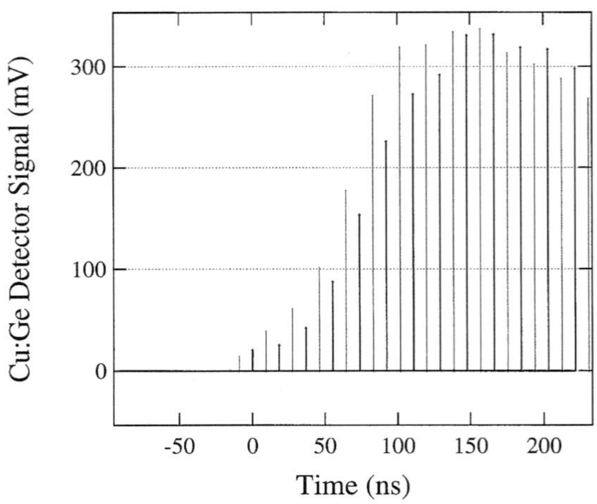

Fig. 4. Cu : Ge detector signals of individual micropulses showing fast optical build-up near saturation.

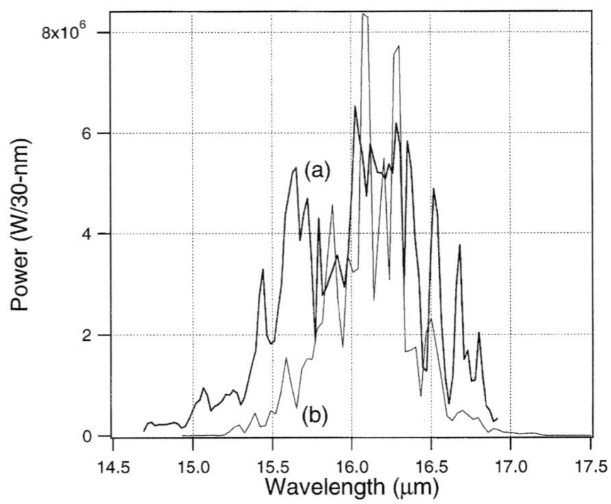

Fig. 6. Output optical spectra: measured (solid) and calculated with FELEX (dashed).

greater than 1 mm (Fig. 3), which greatly relaxes the mechanical stability requirements for the feedback cavity.

The optical build up to saturation was recorded with a high-speed copper-doped germanium detector that integrated the optical energy over each micropulse and yielded the pulse energies of the individual micropulses. As there are two optical pulses in the feedback cavity, two sets of optical micropulses build up from intrinsically different gain conditions and achieve saturation at different times (Fig. 4). Regardless of the gain conditions, both sets of micropulses achieve the same saturation level and when the electrons are turned off, they decay together as a pair of micropulses (Fig. 5). Because of the large outcoupling, the cavity ring-down is fast: the FEL power drops by a factor of 3 in successive passes. From the ring-down

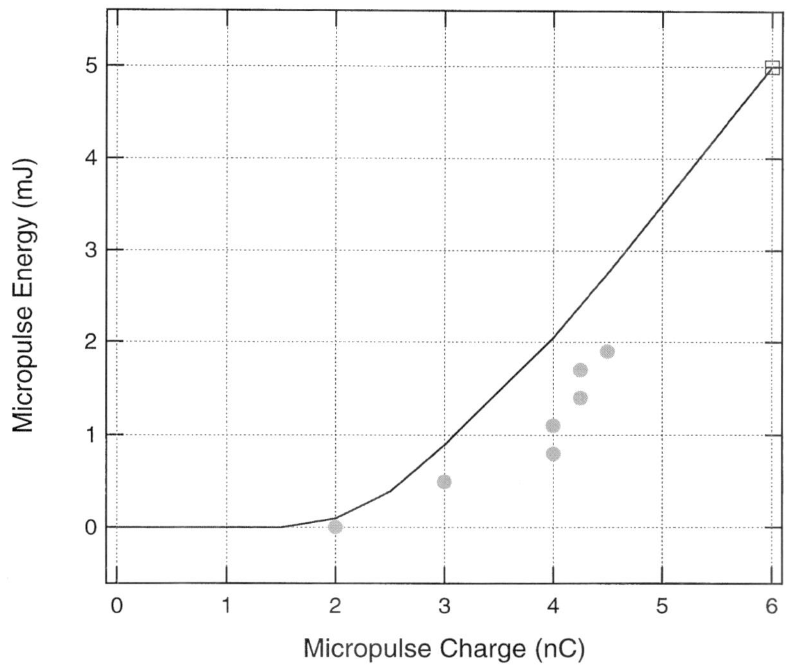

Fig. 7. Output micropulse energy versus micropulse charge; experimental measurements (solid circles), FELEX prediction (square) and extrapolation to lower charge based on FELEX (line).

measurements, we estimate the feedback cavity has an outcoupling of 66%. Only 25% of the total power reflected back into the feedback cavity gets injected into the wiggler. The feedback fraction is thus only 8%.

The output spectrum was measured with a Jarrell–Ash monochromator and a fast HgCdTe detector. The signal was integrated over the portion of the macropulse where the cavity field is stable to within 1%. The spiky nature of the spectrum is similar to those predicted by FELEX simulations at the 9th and 10th passes (Fig. 6). The breadth of the measured spectrum compared to prediction may be due to the fact that the measurement was integrated over several passes.

During the first lasing experiment, the micropulse charge was limited to 3 nC or less, and the measured energy integrated over a 9-μs macropulse (~1000 micropulses) was 0.5 J, yielding a micropulse energy of 0.5 mJ in each 10-ps micropulse. More recently, with a higher micropulse charge of 4.5 nC, we achieved a macropulse energy of 1.7 J over a 8-μs macropulse (~900 micropulses), corresponding to a micropulse energy of 1.9 mJ in approximately 16 ps. The corresponding average power during the macropulse is 200 kW and the peak power during a micropulse is estimated at 120 MW. Since these results were obtained with 4.5 nC of charge at 16.7 MeV, corresponding to 75 mJ of energy in each electron micropulse, we deduced that 2.3% of the beam power was converted to FEL light exiting the cavity. The measured micropulse energy is plotted versus micropulse charge in Fig. 7. For comparison, the FELEX prediction at 6 nC is included, together with our estimates of the maximum micropulse energy that can be obtained (solid line).

4. Conclusion

We have achieved very efficient lasing with the regenerative amplifier free-electron laser. This experiment demonstrates the utility of optical feedback to achieve saturation in a high-gain SASE FEL. The highest optical energy achieved thus far is 1.7 J over a train of 900 micropulses. We deduced a pulse energy of 1.9 mJ in each 16-ps micropulse, corresponding to a peak power of 120 MW. This new FEL has operated at high peak power without optical damage to the cavity mirrors. The RAFEL also exhibits a very large feedback cavity detuning length. The FEL output efficiency, which is the efficiency of conversion from electron beam energy to the light outside the FEL cavity, for this initial demonstration is 2.3%. Work is in progress to improve the RAFEL efficiency and to explore the capability of this high-gain free-electron laser.

Acknowledgements

We thank R.F. Harrison, R.W. Brown, H.J. Martinez and S.K. Volz for their assistance. This work was performed under the auspices of the U.S. Department of Energy and supported (in part) by funds provided by the University of California for the conduct of discretionary research by Los Alamos National Laboratory.

References

[1] For example, see R. Tatchyn et al., Nucl. Instr. and Meth. A 375 (1996) 275.
[2] D.C. Nguyen et al., Nucl. Instr. and Meth. A 393 (1997) 252.
[3] D.C. Nguyen et al., Initial results of the infrared regenerative amplifier FEL, in: Coherent Electron-beam X-ray Sources, Proceedings SPIE, vol. 3154, 1997, pp. 39–50.
[4] A. Buzulutskov et al., Nucl. Instr. and Meth. A 400 (1997) 173.
[5] C.M. Fortgang, Nucl. Instr. and Meth. A 393 (1997) 385.
[6] D.C. Nguyen et al., Phys. Rev. Lett. 81 (1998) 810.

ELSEVIER

Nuclear Instruments and Methods in Physics Research A 429 (1999) 131–135

NUCLEAR INSTRUMENTS & METHODS IN PHYSICS RESEARCH Section A

www.elsevier.nl/locate/nima

Enhancement of harmonic generation using a two section undulator

R. Prazeres[a], F. Glotin[a], D.A. Jaroszynski[b,*], J.M. Ortega[a], C. Rippon[a]

[a] *L.U.R.E., Centre Universitaire Paris-Sud, bât 209d, BP34, 91898 Orsay cedex, France*

[b] *University of Strathclyde, Dept. of Phys. and Appl. Phys., John Anderson Building, 107, Rottenrow, Glasgow-Scotland, G4ONG, UK*

Abstract

Enhancement of the 2nd and 3rd harmonic of the wavelength of a Free-Electron Laser (FEL) has been measured when a single electron beam is crossing a two-section undulator. To produce the harmonic radiation enhancement, the undulator is arranged so that the resonance wavelength of the 2nd undulator (downstream) matches a harmonic of the 1st undulator (upstream). Both the fundamental and the harmonic optical fields evolve in the same optical cavity and are coupled out with different extraction efficiency, through a hole in one of the cavity mirrors. We present measurements that show that the optical power at the 2nd and 3rd harmonic can be enhanced, by about one order of magnitude, in two configurations: when the resonance wavelength of the 2nd undulator matches the harmonic of 1st one (harmonic configuation), or when the gap of the 2nd undulator is slightly larger than first one (step-tapered configuration). We examine the dependence of the harmonic power on the gap of the 2nd undulator. This fundamental/harmonic mode of operation of the FEL may have useful applications in the production of coherent X-ray and VUV radiation, a spectral range where high reflectivity optical cavity mirrors are difficult or impossible to manufacture.

1. Introduction

In a Compton free-energy laser (FEL), the motion of the electrons in a planar undulator is mainly dominated by a transverse oscillation arising from the periodic magnet field of the undulator. The field radiated by the electrons is spiky with many harmonics when the rms undulator deflection parameter, K, is larger than one. On-axis, only odd harmonics are predicted, and in the saturated FEL oscillator, harmonic radiation is usually observed [1], amounting to not more than 1% of the total radiation. As far as the user of FEL radiation is concerned, this is a nuisance and needs to be suppressed or filtered out. When $K > 1$ the laser can have enough gain to oscillate independently and saturate on a harmonic [2,3]. In the presence of the large intracavity fields of the saturated FEL oscillator, the bunching of the electron beam contains many higher harmonics of the fundamental frequency of the laser. This fact is taken advantage of in the "CLIO" Free-Electron Laser, which uses an undulator with two independant sections. The first section of the FEL acts as a buncher and the second section as a radiator [4,5]. Generating harmonics by frequency multiplication in an optical klystron [6] also relies on anharmonic bunching although the bunching optical field is usually derived from an external laser source.

* Corresponding author. Tel.: + 44-141-548-3057; fax: + 44-141-552-2891.

E-mail address: d.a.jaroszynski@strath.ac.uk (D.A. Jaroszynski)

0168-9002/99/$ - see front matter
PII: S0168-9002(99)00092-3

The "CLIO" FEL is designed as a user facility [7] and operates in the wavelength range 3–50 μm, producing picosecond to subpicosecond optical radiation with peak powers as high as 100 MW. The FEL is powered by 25–50 MeV electron bunches, each with a charge of 1 nC and a peak current of 70 A, derived from an S-band linac. The undulator consists of two planar sections, each with $N_u = 19$ magnetic periods of $\lambda_u = 50.5$ mm. Each section has independently adjustable gaps g_1 and g_2, corresponding, respectively, to the deflection parameters K_1 and K_2. A single undulator section ($N_u = 19$) has a gain $G_o > 20\%$ for $K \approx 1$, allowing the oscillator to saturate with one section only, when the optical cavity detuning is set for optimum gain. By adjusting the step in the undulator, $\Delta K = K_2 - K_1$, the laser may operate in a variety of configurations: (i) with a straight undulator, $\Delta K = 0$, (ii) with a "reverse-step taper", $\Delta K \approx 0.03$, leading to an enhancement of efficiency and spectral brightness [8], and (iii) on two colour operations, when $|\Delta K| > 0.1$, allowing production of two independent and simultaneous colours in the laser spectrum [9].

To observe enhanced harmonic radiation from the FEL with the harmonic $n = 2$ and $n = 3$, a series of band-pass filters and a monochromator have been used to analyse the radiation spectrum at the fundamental and harmonics. Fast infrared HgCdTe detectors, and a calorimetric power meter have been used to measure the relative and absolute powers, respectively.

2. Measurements

To observe enhanced coherent harmonic emission on "CLIO" the gap g_1 of the 1st undulator was set close to its minimum value of $g_1 = 17$ mm, giving $K_1 \approx 2.3$ and a resonance wavelength of $\lambda_1 \approx 10$ μm for an electron beam of $\gamma mc^2 =$ 50 MeV. The fundamental wavelength of the laser radiation, produced in the 1st undulator, is measured at $\lambda \approx 10$ μm. Fig. 1 displays a curve representing the 3rd-harmonic intensity as a function of the gap g_2 of the 2nd undulator. A grating monochromator allows a selection of the 3rd harmonic. The gap g_2 is scanned continuously from 17 mm

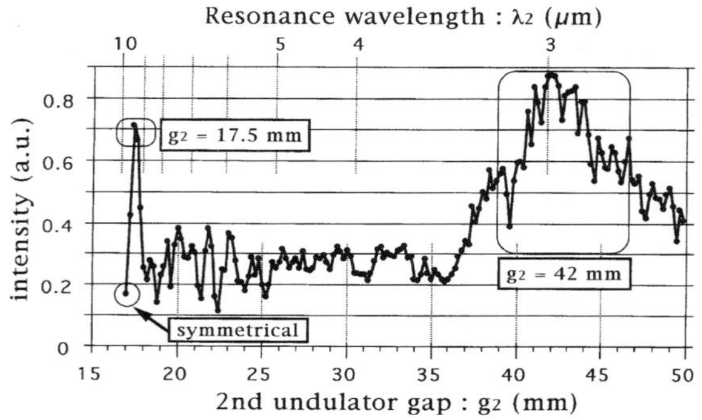

Fig. 1. Intensity of the 3rd harmonic as a function of the gap g_2 of the 2nd undulator. The gap g_1 of the 1st undulator is set at 17 mm. The resonance wavelength λ_2 corresponding to the gap g_2 is displayed on the axis above the curve.

(flat undulator) up to 50 mm. This is equivalent to a scanning of the resonant wavelength λ_2 from 10 to 2.5 μm, which the corresponding axis is displayed in Fig. 1 above the curve. The enhancement of the 3rd-harmonic radiation occurs for two different areas of the resonance: $g_2 = 17.5$ mm and $g_2 \approx 42$ mm. The gap $g_2 \approx 42$ mm giving $k_2 = 0.48$, corresponds to the resonance of the 3rd harmonic of 10 μm, at $\lambda_2 = 3.14$ μm. This resonance extends from a gap $g_2 = 38$ mm to 47 mm, corresponding to a width $\Delta K_2 = 0.25$ in FWHM, and to a width of resonance wavelength of $\Delta\lambda_2 = 0.35$ μm, representing 10% of the harmonic wavelength $\lambda_2 = 3.14$ μm. The resonance at the gap $g_2 =$ 17.5 mm occurs with the undulator in a "step-tapered" configuration, where the K_2 of the second undulator is slightly smaller than the K_1 of first one, with a difference of $K_1 - K_2 = 0.07$ for the case of Fig. 1. This peak is rather sharp, but it is always present in the series of scans that have been done within the same experimental conditions. Much theoretical work has been done in the past on the "tapered" configuration of the FELs, to allow, in principle, a larger efficiency of the FEL. Nevertheless, at present, there are no FELs working with tapered undulators, and a detailed theoretical analysis has shown that the chaotic behaviour of the laser spectrum at saturation does not allow this principle to work [10]. The "step-tapered" undulator is certainly different than the linear "tapered", however, the measurements displayed in

Fig. 1 show that it is efficient for harmonic enhancement. In these conditions, the power at the harmonic wavelength can be enhanced by about one order of magnitude as compared with the power obtained with flat undulator configuration ($g_1 = g_2$). The resulting macropulses, for a cavity detuning of $\delta L \approx \lambda_1$, are more than 6 μs long (out of a 10 μs long electron pulse train), implying that the gain is high enough to achieve saturation with a single section.

The enhancement of the 2nd harmonic has also clearly been observed. Fig. 2 displays a curve representing the 2nd-harmonic intensity as a function of the gap g_2 of the 2nd undulator. A grating monochromator allows a selection of the 2nd harmonic. The 1st undulator was still set to the minimum value of $g_1 = 17\,\text{mm}$, giving $K_1 \approx 2.3$ and $\lambda_1 \approx 10$ μm for a beam energy of $\gamma mc^2 = 50$ MeV. The axis of the corresponding resonance wavelength is displayed in Fig. 2 above the curve. It can be seen from the figure that there is an enhancement of the 2nd harmonic in the region of $g_2 \approx 27\,\text{mm}$, i.e. $K_2 \approx 1.2$, corresponding to the resonance at $\lambda_2 = 4.75$ μm of the 2nd harmonic. Enhancement occurs also for the "step-tapered" configuration, with $g_2 \approx 17.5\,\text{mm}$ and $K_1 - K_2 \approx 0.07$. Note that the harmonic enhancement occurs, in spite of a decreasing of the FEL average power $\langle P \rangle$, at the fundamental wavelength. Indeed, the maximum of power $\langle P \rangle = 350$ mW is obtained with the flat undulator ($g_1 = g_2$), and the enhancement of harmonic occurs with the gaps $g_1 = 17\,\text{mm}$ and $g_2 \approx 27\,\text{mm}$, when the power is only $\langle P \rangle = 225\,\text{mW}$. Taking into account the ratio between the harmonic intensity, in Fig. 2, and the FEL power, the 2nd harmonic is enhanced by a factor of 10. The production of 2nd harmonic is surprising at first glance as the conventional theory of a planar undulator, which predicts zero gain and power on axis at harmonic two. However, as the FEL optical cavity can support antisymmetric radiation fields, such as the TEM11 (Hermite–Gauss) mode, the gain and power are non-negligible [11,12]. In addition, the radiated 2nd-harmonic on-axis power is large if there is harmonic bunching of the electrons close to the 2nd-harmonic frequency as they pass through the undulator, as will be shown below. These are perhaps the main reasons why, in most FELs, a significant 2nd-harmonic power is always produced. Another intriguing feature of Fig. 2 is the presence of two peaks of resonance of the 2nd harmonic, at $g_2 = 26$ and $28\,\text{mm}$, i.e. $K_2 = 1.3$ and 1.15, respectively. The separation between these two peaks is of $\Delta K_2 = 0.15$ representing a separation of the resonance wavelengths of $\Delta\lambda_2 = 0.5$ μm, representing 10% of the harmonic wavelength $\lambda_2 = 4.75$ μm. The enhancement of 3rd harmonic, shown in Fig. 1, exhibits a single large peak, with 10% of linewidth, i.e. a linewidth equivalent to the separation of the two peaks in Fig. 2. The spectrum of the harmonic radiation has been measured, both with the flat undulator ($g_1 = g_2$) and with the undulators in "harmonic resonance configuration". No difference was observed between these spectra. For example, for both cases, the 2nd harmonic is peaking at $\lambda_2 = 4.78$ μm with a linewidth of $\Delta\lambda_2 = 0.07$ μm FWHM corresponding to 15%.

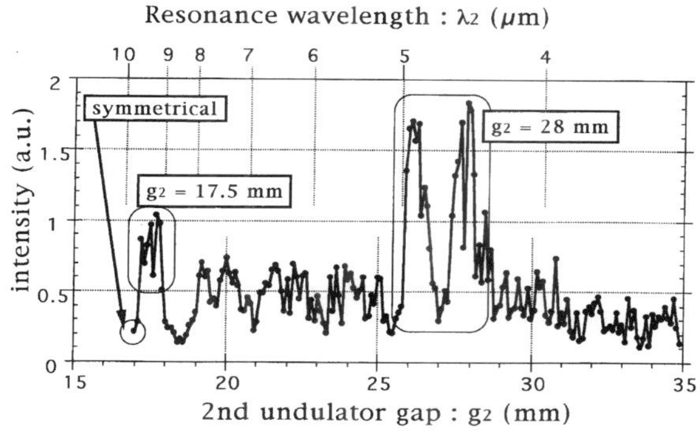

Fig. 2. Intensity of the 2nd harmonic as a function of the gap g_2 of the 2nd undulator. The gap g_1 of the 1st undulator is set at 17 mm. The reasonance wavelength λ_2 corresponding to the gap g_2 is displayed on the axis above the curve.

In order to observe in more detail the effect of the "step-taper" configuration on the harmonics, a measurement of the 1st- and 2nd-harmonics intensities as been done with the gap g_2 scanning close to the value of g_1. The 1st undulator was set to a value of $g_1 = 18$ mm, giving $K_1 \approx 2.15$ and $\lambda_1 \approx 8.8$ μm for a beam energy of $\gamma mc^2 = 50$ MeV. This measurement is shown in Fig. 3. The fundamental wavelength is peaking at the gap $g_2 = 18$ mm, i.e. with the flat undulator. The 2nd harmonic is peaking at $g_2 = 18.3$ mm, i.e. $K_2 = 2.11$, where the harmonic is enhanced by a factor of about 10 as compared with the flat undulator. Note that the laser vanishes for the gap $g_2 = 17.6$, i.e. $K_2 = 2.2$, because the laser power produced at fundamental wavelength in the 1st undulator is reabsorbed in the 2nd one. This effect has already been observed when the "CLIO" FEL is operating in two-colour [13]. In this case, two different wavelengths are produced by the FEL when the two undulators are adjusted at different gaps, $g_1 \neq g_2$. The harmonic experiment uses the same configuration, but in addition, the 2nd resonance wavelength is adjusted to be a harmonic of the 1st one.

We have verified experimentally that the enhancement of the intensity of the harmonics is not a simple effect of laser in the 2nd undulator, independantly of the 1st undulator, as in the two colour operation. This also would give an enhancement of the signal at the harmonic resonance, because of the monochromator selection of the measured signal. The intensity, displayed in vertical axis of Figs. 1 and 2, has been measured through a monochromator in order to only select the harmonic component. As a consequence, the peak which is observed in the scan of Figs. 1 and 2, at harmonic resonance, also could be due only to the fact that the laser power of the 2nd undulator is passing through the monochromator slits when its wavelength is fitting. In this case, the intensity obtained by scaning g_2 would also exhibit a peak, as it is observed in Figs. 1 and 2. In order to show that the enhancement is not due to such effect, the monochromator has been adjusted to 4.75 μm, corresponding to the fundamental wavelength of the 2nd undulator at $g_2 = 27$ mm, and the 1st undulator has been scanned. In these conditions, a strong signal is observed only if the gaps $[g_1, g_2]$ are in harmonic configuration, i.e., respectively, at 17 and 27 mm. If the signal was only due to simple laser effect of the 2nd undulator, it would be more or less independant of the 1st undulator gap. In conclusion, this shows that the enhancement of the signal in Figs. 1 and 2 corresponds to a real effect of harmonic prebunching in the 1st undulator, followed by amplification of the harmonic in the second one. Note that the harmonic energy is also stored in the mirror cavity of the FEL, like the fundamental component of the laser. As a consequence, the harmonic enhancement observed here is not a simple effect of Coherent Harmonic Generation (CHG), which is a single-pass process [6]. The process observed here involves several consequences of the FEL process: the CHG for laser feeding, the gain on the harmonic for amplification, and the harmonic pre-bunching as a booster.

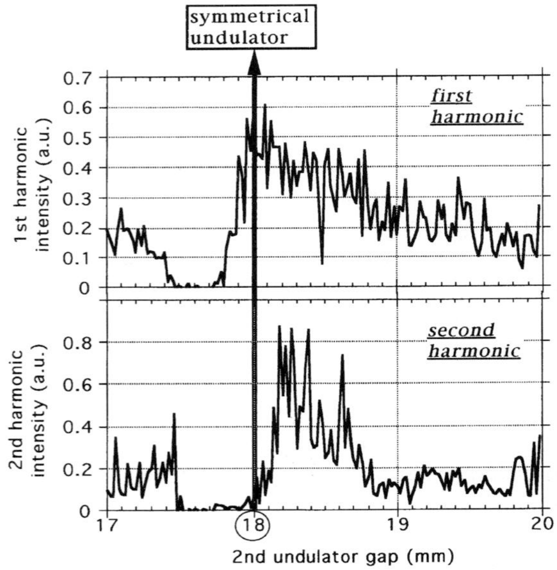

Fig. 3. Intensity of the 1st and 2nd harmonics as a function of the gap g_2 of the 2nd undulator. The gap g_1 of the 1st undulator is set at 18 mm.

The measurements shown in Figs. 1 and 2 have also been carried out with an electron beam of $\gamma mc^2 = 38$ MeV. The result was similar to 50 MeV: the enhancement at resonance of 3rd harmonic exhibits one large peak at $g_2 = 40$ mm, and the resonance of 2nd harmonic exhibits two peaks at $g_2 = 26.5$ and 28.5 mm. An enhancement also occurs with a "step-tapered" configuration ($g_1 =$ 17 mm and $g_2 = 17.5$ mm), though less evidently as for the energy of $\gamma mc^2 = 50$ MeV.

3. Conclusion

The operation of FELs in the VUV spectral range is difficult because the gain drops strongly at short wavelengths, and high reflectivity mirrors are difficult or impossible to manufacture. Much work is done now with self-amplified spontaneous emission (SASE) in order to reach the short wavelengths [14,16]. Nevertheless, the lasing on the harmonics may also be a good candidate to reach the VUV range. An enhancement of the harmonic radiation is necessary to allow sufficient power emission, and the measurements described here are showing an enhancement of one order of magnitude with the 2nd and the 3rd harmonics. The double section of the undulator of CLIO allowed harmonic enhancement, by adjusting the two undulator sections, respectively, in fundamental and harmonic resonance. Also, equivalent quantity of enhancement of harmonic has been measured with the "step-taper" configuration of the undulator, i.e. with the gap of the 2nd section slightly larger as the 1st section. Note that enhancement of the harmonics has already been observed, with the CLIO FEL [15], but only with the 2nd harmonic, not the 3rd one, and the harmonic enhancement with "step-taper" configuration was not observed. The enhancement process with the undulator in the fundamental/harmonic configuration requires a large gap between both undulators, whereas the "step-taper" configuration needs only about $\frac{1}{2}$mm of gap between sections, and undulators with several "step-tapered" sections is realistic.

References

[1] D. Bamford, D. Deacon, Nucl. Instr. and Meth. A 296 (1980) 89.

[2] S. Benson, J. Madey, Phys. Rev. A 39 (1989) 1579.

[3] R. Warren et al., Nucl. Instr. and Meth. A 296 (1990) 84.

[4] R. Bonifacio, L.D. Slavo, P. Pierini, Nucl. Instr. and Meth. A 293 (1990) 627.

[5] L. Yu, Nucl. Instr. and Meth. A 358 (1995) 74.

[6] R. Prazeres et al., IEEE J. Quantum Electron. QE-27 (1991) 1061.

[7] J.M. Ortega et al., Nucl. Instr. and Meth. A 375 (1996) 618.

[8] D. Jaroszynski, R. Prazeres, F. Glotin, J.M. Ortega, Phys. Rev. Lett. 74 (1995) 2224.

[9] D. Jaroszynski, R. Prazeres, F. Glotin, J.M. Ortega, Phys. Rev. Lett. 72 (1994) 2387.

[10] P. Chaix et al., Nucl. Instr. and Meth. A 331 (1993) 379.

[11] H. Bluem et al., Phys. Rev. Lett. 67 (1991) 824.

[12] M. Schmitt, C. Elliot, Phys. Rev. A 34 (1986) 4843.

[13] R. Prazeres, F. Glotin, C. Insa, D.A. Jaroszynski, J.M. Ortega, Nucl. Instr. and Meth. A 407 (1998) 464.

[14] R. Prazeres, J.M. Ortega, F. Glotin, D.A. Jaroszynski, O. Marcouillé, Phys. Rev. Lett. 78 (1997) 2124.

[15] D.A. Jaroszynski et al., Nucl. Instr. and Meth. A 375 (1996) 456.

[16] D.A. Jaroszynski et al., Nucl. Instr. and Meth. A 358 (1995) 228.

ELSEVIER Nuclear Instruments and Methods in Physics Research A 429 (1999) 136–140

NUCLEAR INSTRUMENTS & METHODS IN PHYSICS RESEARCH Section A

www.elsevier.nl/locate/nima

Simultaneous two-color lasing in the mid-IR and far-IR region with two undulators and one RF linac at the FELI

A. Zako*, Y. Kanazawa, Y. Konishi, S. Yamaguchi, A. Nagai, T. Tomimasu

Free Electron Research Institute, Inc. (FELI), 2-9-5 Tsuda-yamate, Hirakata, Osaka 573-0128, Japan

Abstract

Simultaneous two-color FEL lasing on the same electron macropulse has been demonstrated at 4.65 and 19.4 μm using a 33 MeV electron beam and two FEL facilities (FEL-1 and FEL-4) of the FELI. The average powers are 11 mW at 4.65 μm and 2 mW at 19.4 μm. Their spectral widths (FWHM) are 0.85% and 1.06%. They are delivered to users through two evacuated optical pipes. FEL-1 achieved lasing in the mid-infrared region with the undulator-1 (UND-1; $\lambda_u = 34$ mm, $K = 0.5$–1.5, $N_u = 34$) in 1994 and FEL-4 achieved lasing at far-infrared region with the undulator-4 (UND-4; $\lambda_u = 80$ mm, $K = 1.26$–3.37, $N_u = 30$) in 1996. UND-4 is installed at the downstream of the UND-1 and an electron beam is transported through an S-type electron beam transport line including two 45° bending magnets.

1. Introduction

Since first lasing at the mid-infrared (MIR) region in 1994 at the FELI [1], we have succeeded in lasing in the near-infrared, visible and UV regions [2]. We have also succeeded in lasing in the far-infrared (FIR) region ($\lambda = 18.6$–40 μm) with a Halbach-type undulator in December 1996 [3]. Our FEL system has mainly been used for FEL application research in the 5–13 μm region so we have no machine time to lase at wavelengths longer than 40 μm and shorter than 278 nm.

Simultaneous lasing in the MIR region with two undulators has been achieved at CLIO [4] where a single beam and a common optical cavity is used. Therefore, two wavelengths of FEL light are delivered to the user simultaneously, and their color difference is small. Quasi-simultaneous FEL operation of the FIR and MIR regions has also been demonstrated at the Stanford FEL Center [5]. At Stanford, interleaved electron beams of different energies are generated by a single linac, separated and then used to independently drive the FIR-FEL facility and the MIR-FEL facility. They lase on alternate macropulses with the optical beam delivered to two users at once.

FELI has also achieved two-color oscillations using the FELI linac, beam transport systems and the two FEL facilities (FEL-1 and FEL-4). This paper describes the results of the first simultaneous two-color lasing.

2. Experimental setup

Fig. 1 shows a schematic layout of the simultaneous two-color FEL experiment. The electron injector generates a 22.3125-MHz electron micropulse train lasting 24 μs. These pulse trains repeat

* Corresponding author.

0168-9002/99/$ - see front matter
PII: S0168-9002(99)00093-5

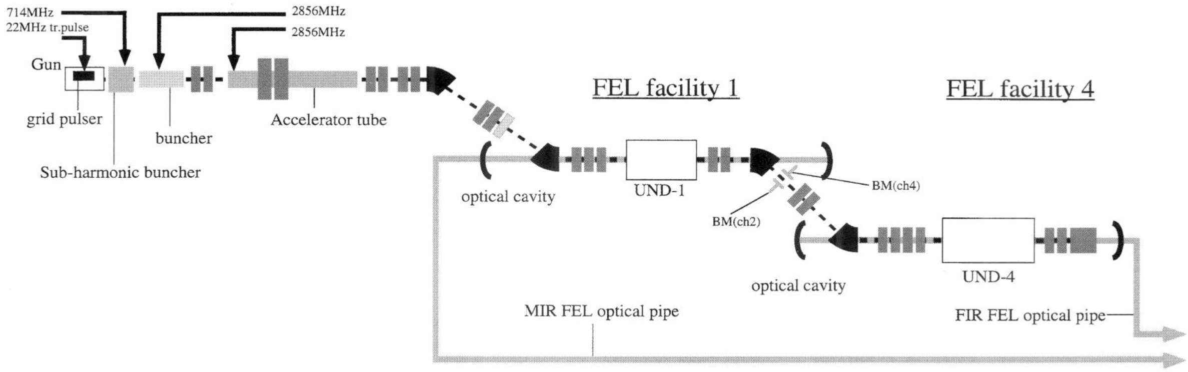

Fig. 1. A schematic layout of the simultaneous two-color FEL experiment.

Table 1
Characteristics of each FEL oscillators

	Mid-IR	Far-IR
Undulator type	Halbach	Halbach
Undulator period (mm)	34	80
Period number	58	30
Gap length (mm)	15–25	30–55
K value	0.56–1.42	3.37–1.26
Material of permanent magnet	Nd–Fe–B	Sm–Co
Cavity type	Fabry–Perot	Fabry–Perot
Cavity length (m)	6.72	6.72
Rayleigh length (m)	1.0	0.54
g-parameter	− 0.93 − 0.76	− 0.89 − 1.01
Mirror curvature (m)	3.490	3.555
Material of the mirrors	Au/Cu	Au/Cu
Extraction	Hole	Hole
Aperture dia. of an outer coupler (mm)	1.5	2.5
Window for FEL guide	KRS-5 ZnSe	Diamond KRS-5, polystyrene

at 10 pulses per second. The electron micropluses are bunched and accelerated with a 714-MHz sub-harmonic buncher, a 2856-MHz buncher and a 2.93-m long accelerator tube. Accelerated electron beams are transported to the MIR undulator (UND-1) through two (22.5° and − 22.5°) bending electro-magnets and focusing quadrupole magnets. Once passed through UND-1, they are transported to the FIR undulator (UND-4) through two (45° and − 45°) bending electromagnets and focusing quadrupole magnets. Table 1 shows the parameter lists of these two FEL oscillators. The two FEL wavelengths are guided through two evacuated optical pipes. The entrance windows of the optical pipes are KRS-5 for MIR-FEL and diamond for FIR-FEL and the exit windows at the monitor room and the user rooms are ZnSe for MIR and KRS-5 or polystyrene for FIR.

3. Simultaneous two-color lasing experiment

Fig. 2 shows a layout of the FEL measurement systems installed at the monitor room. They are set on three separate platforms. The first is the optical reflection platform where the FEL light is switched to the other two platforms which are the FEL measurement platforms. The MIR-FEL and the FIR-FEL pass through two individual optical guides and so we provide two measurement systems. The FIR-FEL measurement platform, however, is usually used for FEL application research. FEL wavelengths are measured using a monochromator and a 128-element pyroelectric linear array detector. A grating in the monochromator must be scanned to measure a wavelength so that it is effective to use the linear array detector at the exit of the monochromator to measure the light from pulsed lasers-like FELs. HgCdTe detectors are used to measure FEL macropulse signals. They are very sensitive and so care must be taken to guard

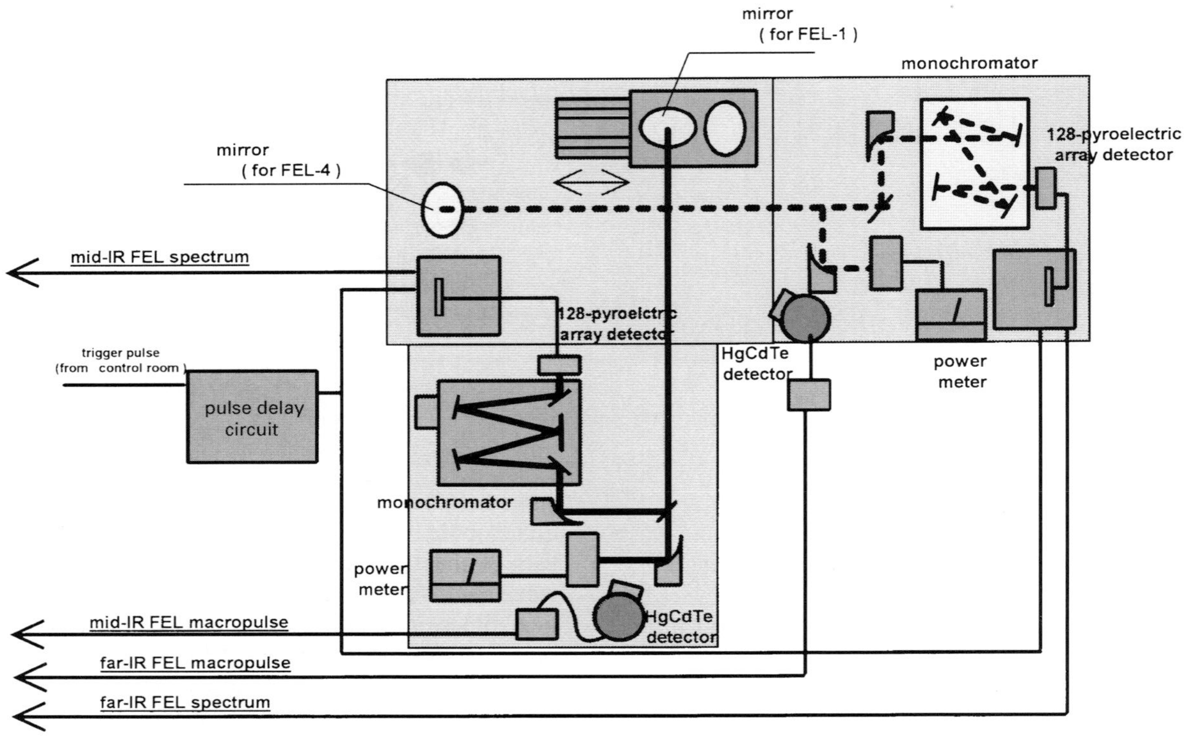

Fig. 2. A layout of the two-color FEL measurement systems.

against measuring scattered FEL light with a partition. FEL average powers are measured with powermeters.

In these experiments, widening the undulator gap length of UND-1 to 50 mm (K-value = 0.06), we tried to lase at the FIR range with UND-4 and fixed the oscillating condition. Next, gradually narrowing the UND-1 gap length to 25 mm (K-value = 0.56), we tried to lase also in the MIR region while maintaining the lasing condition in the FIR region. Finally, we achieved simultaneous two-color FEL lasing with FEL-1 and FEL-4. Fig. 3 shows a spectra of MIR- and FIR-FELs. The lasing wavelengths are 4.65 and 19.4 μm and their spectral spread (FWHM) are 0.85% and 1.06%, respectively. Fig. 4 shows the FEL macropulse signals of the MIR and FIR lasers measured with HgCdTe detectors and the electron beam macropulse measured with an electrostatic probe (button monitors [6]) which was set at the entrance of UND-1. The macropulse of the FIR-FEL should rise more quickly than that of the MIR-FEL because the small signal gains of the FIR- and MIR-FELs are 30% and 20%, respectively. However, it is not so. A part of the electron beam loses its energy to the MIR-FEL at UND-1 and the beam energy spread becomes broader at UND-4 and so the gain of the FIR-FEL with UND-4 decreases. The beam trajectory also moves.

The button monitor signals set between two 45° bending magnets of the downstream of UND-1 clearly show this phenomenon. Two button monitors (ch2, ch4) are set on the electron beam duct horizontally. Thus, the difference between both the signals should be zero if the electron beam passes through the center of the beam duct. The signals in Fig. 5 show the differences between two button monitors with mid-infrared lasing and when a shutter is inserted into the optical cavity line of FEL-1 to stop lasing. The ch2 signal rises about 5 μs later than the electron macropulse and its shape is similar to that of the MIR-FEL macropulse. This is

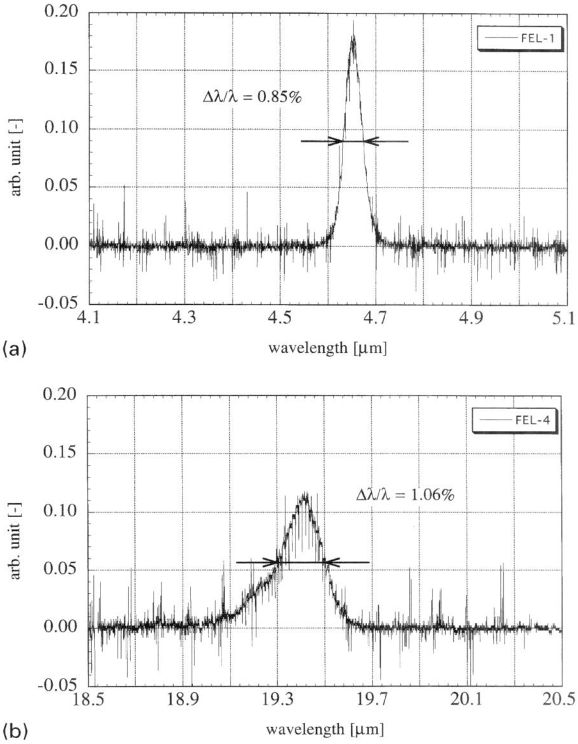

Fig. 3. The FEL spectra of (a) MIR region with UND-1 ($\lambda = 4.65\,\mu m$, $\Delta\lambda/\lambda = 0.85\%$) and (b) FIR region with UND-4 ($\lambda = 19.4\,\mu m$, $\Delta\lambda/\lambda = 1.06\%$).

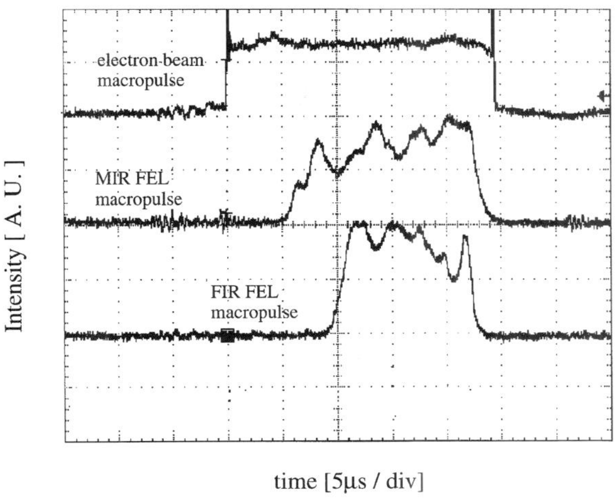

Fig. 4. The FEL macropulse signals of MIR and FIR region from HgCdTe detectors and the electron beam macropulse from button monitors set in the entrance of UND-1.

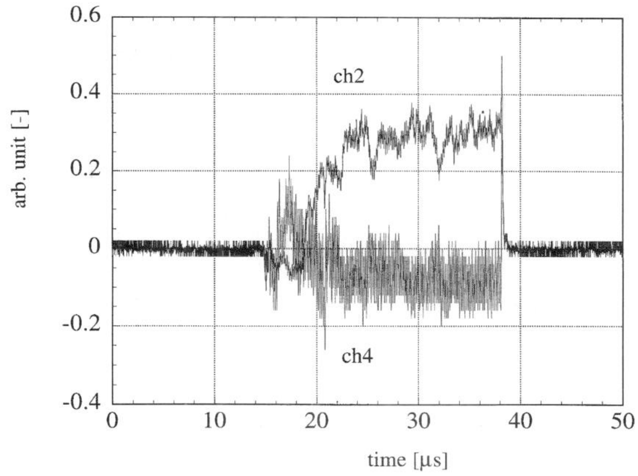

Fig. 5. The signals show the difference between two button monitors with MIR lasing and when a shutter is inserted into the optical cavity of the MIR-FEL to stop lasing.

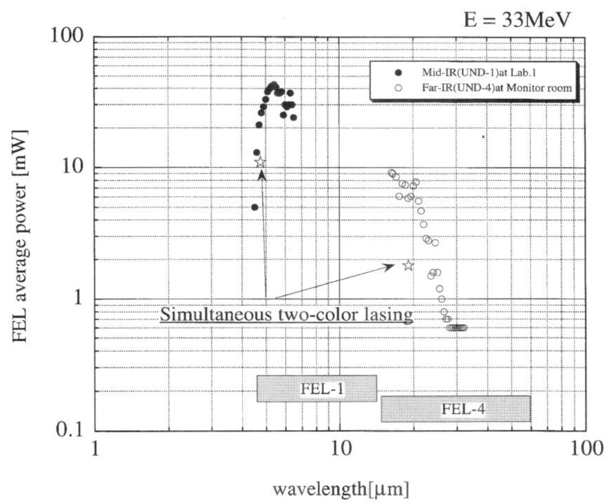

Fig. 6. The average FEL powers at mono-color lasing and at simultaneous two-color lasing.

caused by a decrease in the electron energy. Fig. 6 shows the average FEL powers with single color lasing and at simultaneous two-color lasing. The average powers are 11 mW at 4.65 μm and 2 mW at 19.4 μm for the two-color lasing.

4. Conclusion

We have achieved simultaneous two-color lasing with one linac and two undulators of the FELI. The lasing wavelengths are 4.65 and 19.4 μm and the spectral widths are 0.85% and 1.06%, respectively. The average powers are 11 mW at 4.65 μm and 2 mW at 19.4 μm. It is not easy to expand the range of tunability, normally an advantage of the FELs, because the electron beam loses energy to the MIR-FEL at UND-1 and its trajectory moves at UND-4 according to its energy loss.

References

[1] T. Tomimasu et al., IEEE Proceedings PAC'95, Dallas, May 1–5, 1995, p. 257.

[2] T. Tomimasu et al., Nucl. Instr. and Meth. A 383 (1996) 337.

[3] T. Takii et al., Nucl. Instr. and Meth. A 407 (1998) 21.

[4] D. Jaroszynski et al., Phys. Rev. Lett. 74 (1995) 2224.

[5] K.W. Berryman, T.I. Smith, Nucl. Instr. and Meth. A 375 (1996) 6.

[6] A. Zako et al., Proceedings of the Second Asian Symposium on Free Electron Lasers, Novosibirsk, June 13–16, 1995, p. 57.

ELSEVIER

Nuclear Instruments and Methods in Physics Research A 429 (1999) 141–145

NUCLEAR INSTRUMENTS & METHODS IN PHYSICS RESEARCH Section A

www.elsevier.nl/locate/nima

Linac-based UV-FEL macropulse shape and gain estimate at the FELI

T. Tomimasu*, E. Oshita, Y. Kanazawa, A. Zako

Free Electron Laser Research Institute, Inc. (FELI) 2-9-5, Tsuda-Yamate, Hirakata, Osaka 573-0128, Japan

Abstract

Linac-based UV-FEL macropulses were measured with an Si-photodetector using a 144-MeV electron beam from the FELI linac at the UV-FEL facility (FEL-3) with a 6.72-m long optical cavity and a 2.68-m ($\lambda_u = 4$ cm) long undulator. The minimum rise-time of a 350-nm FEL macropulse shape is about 8 μs, while the electron beam macropulse continues for 17 μs. A net gain is estimated from the rise-time of the 350-nm FEL macropulse and a gain reduction due to the filling factor is also estimated. The estimated gain is 19.0% for a 144-MeV, 132-A electron beam of $\varepsilon_n = 26$ mm mrad and $\Delta E/E$ (FWHM) = 0.5%. The UV-FEL facility has broken the world record for the shortest wavelength oscillation of linac-based FELs with a thermionic gun by reaching 278 nm. These parameters can be used to find a saturation length of a SASE device in the VUV range.

Keywords: Ultraviolet free electron laser (UV-FEL); FEL macropulse shape; Gain estimate; Filling factor; SASE

1. Introduction

The first lasings at 278–370 nm of linac-based free electron lasers (FELs) have been achieved by using the FELI linac with a thermionic gun and the ultraviolet (UV) FEL facility (FEL-3) since December 1995 [1–3]. The FELI keeps a world record of the shortest wavelength oscillation of linac-based FELs. A gain estimate from these UV-FEL macropulse shapes can be used to find a scheme of a self-amplified spontaneous emission (SASE) device in the VUV range [4]. It seems to be useful to check the reliability of existing simulation codes. This paper describes (1) the measurement of FEL macropulses at the UV-range, (2) a gain estimate from the rise-time of the 350-nm FEL macropulse, and (3) a gain reduction due to the filling factor considering the dimensionless Rayleigh length and the dimensionless electron beam waist radius [5].

2. FELI 165-MeV linac and UV-FEL facility (FEL-3)

Fig. 1 shows the layout of the FELI 165-MeV linac, four S-type BT lines and four FEL facilities (FEL-1, FEL-2, FEL-3 and FEL-4). FEL-3 consists of a 6.72-m long optical cavity and a 2.68-m (λ_u = 4 cm, $N = 67$, $K_{max} = 1.9$) long undulator (undulator-3) [6]. The optical cavity 3 is of the Fabry–Perot type which consists of two mirror chambers and a vacuum duct connecting them. Fig. 2 shows (a) a bird's eye view of the optical cavity 3 and

* Corresponding author. Tel.: 81-720-96-0414; fax: 81-720-96-0421.

E-mail address: tomimasu@feli.co.jp (T. Tomimasu)

0168-9002/99/$ - see front matter
PII: S0168-9002(99)00094-7

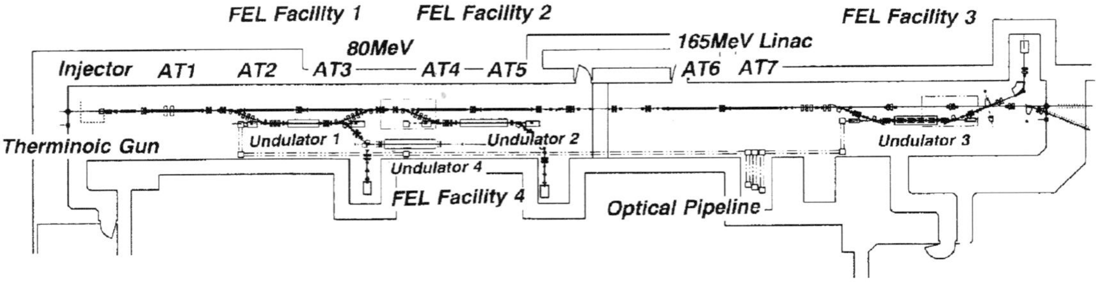

Fig. 1. Layout of the FELI 165-MeV linac, S-type BT lines and four FEL facilities.

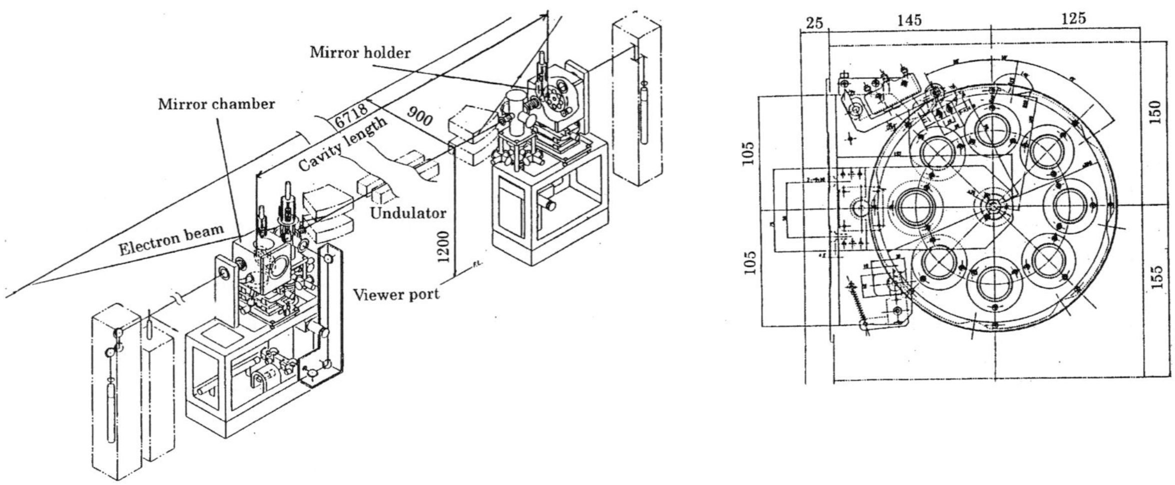

Fig. 2. (a) Bird's eye view of optical cavity 3 and (b) rotating-type mirror holder.

a rotating type mirror holder. The same method used for the case of the MIR-FEL and IR-FEL facilities was used to align the cavity mirrors [7]. The rotating-type mirror holder of each mirror chamber can accommodate eight mirrors. Each cavity mirror is a spherical and dielectric multilayer mirror of Ta_2O_5/SiO_2 or HfO_2/SiO_2. The cavity loss per pass is 0.5% (mirror loss is 0.1% and extraction efficiency is 0.2% per mirror and 0.4% per pass) [8]. Table 1 shows the main parameters of the optical cavity 3.

The FELI 144–155-MeV electron beam consists of a train of several pico-second (ps), 0.6-nC electron bunches (micropulses) repeating at 22.3125 MHz. The electron micropulses usually continue for 24 μs (macropulse) and the macropulse repeats at 10 Hz.

The FELI 6-MeV injector with a thermionic gun was carefully designed to reduce the emittance growth in the bunching process from the thermionic gun (EIMAC Y646B) to a 714-MHz prebuncher and a 2856-MHz buncher. The gun is driven by a 22.3125-MHz, 0.5-ns wide, 150-V grid pulser (Kentech Instruments, Ltd.). The short-bunching and strong-focusing system of the injector [9] and a stable 24-μs RF pulse system [10] allow lasing at 278 nm in spite of a small signal gain of 6%. The gun injection voltage should be above 110 kV and the allowable focusing field range capable of lasing is 0.11–0.12 T near the inlet of the buncher. The

Table 1
Main parameters of optical cavity for FEL facility 3

Type	Optical mode
Length	6.718 m
Rayleigh length	2.0 m
g parameter	− 0.44
	− 0.49
Mirror curvature	4.656 m
	4.5 m
Mirror type	Multilayer of Ta_2O_5/SiO_2 or HfO_2/SiO_2
Reflectivity	99.6%
Transmission	0.4% (0.2% per mirror)
Intra-cavity peak power	0.9 GW at 350 nm

following seven accelerating waveguides with a length of 2.93 m are of the linearly narrowed iris type to prevent beam blow up (BBU) effects at high peak current acceleration [11].

The electron beam size and position are always monitored and controlled to pass through the center of the accelerating waveguides, the quadrupole magnets, the S-type BT line and the UV-FEL facility. The beam emittance is measured with three Al-foil OTR beam profile monitors installed in a 2.7-m long undulator vacuum chamber. Each Al-foil has a 1-mm diameter aperture. The S-type BT line can focus nearly 90% of the electron beam to pass through the 1-mm diameter aperture. A two profile monitor emittance measurement method is used for its simplicity and the short time needed for data acquisition. The normalized emittance of the 144–155-MeV (0.6-nC micropulse) beam is estimated to be 26 mm mrad.

3. UV-FEL beam quality measurement

A diagnostic system for UV-FEL beam quality measurement is shown in Fig. 3. The macropulse shapes are measured with an Si-photodiode (Si-APD C5331-04, Hamamatsu Photonics, K.K.) and the spectra are measured with a Czerny–Turner-type monochromater equipped with an ICCD camera (Oriel Instruments) with an optical fiber and a linear array sensor (Intra Spec IV). The macropulse power is also measured with an array of

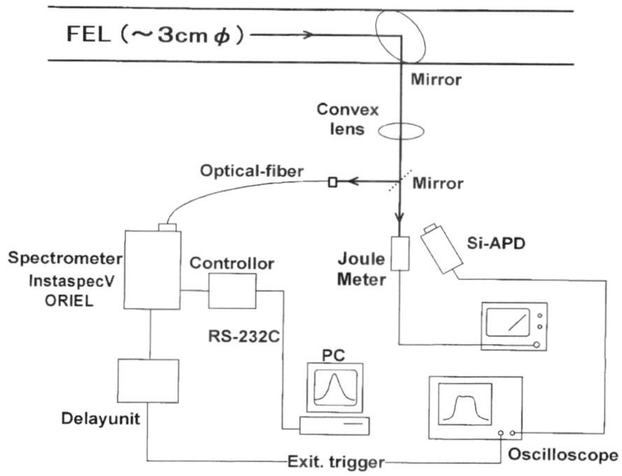

Fig. 3. Visible- and UV-FEL beam diagnostic system.

pyro-electric detectors (Type ED100, Gentec). The UV-FEL micropulse length is measured with a streak camera (Hamamatsu Photonics, K.K.). The micropulses are usually 1.7 ps in length and 0.8 ps in length at a cavity de-tuning condition [12].

The usual 350-nm FEL micropulse energy is 3.1 μJ and the micropulse peak power is 1.8 MW at the outlet of the multi-layer mirror (Ta_2O_5/SiO_2). The intra-cavity power is 0.9 GW and the peak power density at the mirror inner surface is near 6 GW/cm^2 (10 mJ/cm^2). The 278-nm FEL micropulse energy is 0.35 μJ and the peak power is 0.2 MW at the outlet of the multi-layer mirror (HfO_2/SiO_2). The peak power density at the mirror inner surface is one-tenth of the 350-nm FEL beam. Fig. 4 shows the micropulse FEL peak powers and the intra-cavity peak powers of FEL-3 in the 278–600-nm range. Peak power densities at the dielectric multi-layer mirrors exceed the several GW level as shown in Fig. 4 so that the mirror of Ta_2O_5/SiO_2 was damaged by X-ray assisted-FEL ablation within 50 h of operation [8].

Fig. 5 shows a 350-nm FEL macropulse shape with a 17 μs electron macropulse. The rise-time of the 350-nm FEL macropulse shape is nearly 8 μs, while the electron beam macropulse continues for 17 μs. This means that it takes 8 μs before an intra-cavity optical pulse reaches a saturated value and that a saturation length of a SASE device in the UV range is 480 m (= 2.68 m multiplied by 179)

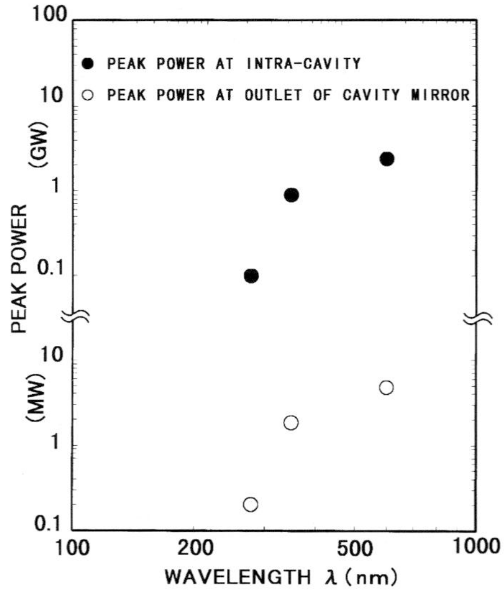

Fig. 4. Micropulse FEL peak powers and intra-cavity powers of FEL-3.

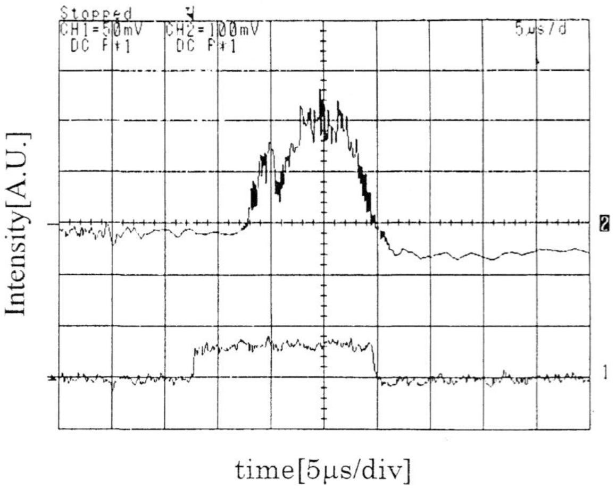

Fig. 5. 350-nm FEL macropulse and 17-μs electron macropulse.

for a 144-MeV, 132-A electron beam (0.6-nC micropulse train) with a normalized emittance of 26 mm mrad. The electron beam macropulse is measured with button monitors [13] at the entrance of the undulator-3. The UV-FEL macropulse length is about 9 μs and the UV-FEL macropulse consists of two hundred saturated FEL micropulses, since the micropulse separation is 44.8 ns.

4. Gain estimate

The rise-time τ_R is defined as the time needed to reach a saturated value $I_s = I(\tau_R)$ of an FEL intra-cavity power and is given by

$$\tau_R\ (\mu s) = 0.67 \times 10^{-2} \ln(I_s/I_i) L_c\ (m)/g_n \tag{1}$$

assuming that the intra-cavity power grows with time according to the exponential law: $I(t) = I_i \exp(g_n t/T_c)$ with I_i being the initial spontaneously generated power, g_n the net gain, T_c the round-trip period ($T_c = 2L_c/c$) and L_c the cavity length.

For all FELI FEL facilities, L_c is 6.72 m long and in the case of the MIR-FEL facility the usual macropulse length is 18 μs and then the rise-time is 6 μs, since the net gain is near 20% [14]. Thus, the relation between the rise-time τ_R and the net gain g_n is simplified as

$$\tau_R\ (\mu s) = 1.2/g_n, \tag{2}$$

and from Eqs. (1) and (2), we obtain $\ln(I_s/I_i) = 27$ and $I_s/I_i = 5.3 \times 10^{11}$.

From Eq. (2) and $\tau_R = 8$ μs of the 350-nm FEL macropulse, we obtain the net gain $g_{n\,350\,nm} = 15\%$.

The small signal gain is reviewed for an electron beam passing through a planar undulator with the period length λ_u, the period number N and the magnetic field B_0, and the beam is the practical pulse beam with a normalized beam emittance ε_n and an rms energy spread σ_e.

The small signal gain of a planar undulator for the practical beam is estimated by the following equation:

$$G_P^{PU} = C\{J_0(\xi) - J_1(\xi)\}^2 (I_0/377)(\lambda\lambda_u)^{1/2} N^2 \times (K^2/2)/(1 + K^2/2)^{3/2} \tag{3}$$

where I_0 is the peak current of micropulses, λ is the FEL wavelength, and other parameters are defined as follows: $\xi = (\frac{1}{4})(K^2/2)/(1 + K^2/2)$, $K = 93.4\ B_0\ (T)\ \lambda_u\ (m)$. A gain reduction factor $C = C_c C_e C_x C_y$ was estimated for the pulse length effect C_c, for energy spread C_e and for beam emittance $C_x C_y$, by many workers [6,15,16].

However, as these references do not describe a gain reduction due to the filling factor, it is calculated according to the report of Small et al. [5].

The dimensionless Rayleigh length $Z_0 = Z_R/L_u$ of the UV-FEL facility is 0.75 from Table 1. This length gives a 90% of the maximum normalized gain, though the average filling factor over the undulator drops to half of its peak value at $Z_0 = 0.75$. The dimensionless electron beam waist radius $\sigma_x = r_b(\pi/L_u\lambda)^{1/2}$ of this facility is 0.74 for the 350-nm FEL, where $r_b(=0.4$ mm) is the electron beam waist radius. This radius also gives 90% of the maximum normalized gain for a normalized emittance beam of 16 mm mrad. The good agreement of these gain reduction factors show that our parameters on the undulator length and the beam waist radius have been reasonably selected. Thus, an estimated gain reduction is 81%. From the net gain of 15%, the gain reduction of 81% and the cavity loss per pass of 0.5%, we obtain an estimated gain of 19.0%. The small signal gain for the FELI linac beam with a relative energy spread of 0.5% (FWHM) and a normalized emittance of 26 mm mrad is given as a function of electron energy (100–180 MeV) and gap length (15–35 mm) in Ref. [6]. From these data we can estimate that the small signal gain of 19.0% can be given by a 144-MeV, 132-A electron beam of a relative energy spread of 0.5% and a normalized emittance of 26 mm mrad.

5. Conclusions

Linac-based UV-FEL macropulse shapes have been measured with a silicon avalanche photodiode. The net gain g_n is estimated from the 8-μs rise-time of the 350-nm FEL macropulse and is given by $1.2/\tau_R$ (μs). From this relation we can understand that the ratio I_s/I_i of the saturated intra-cavity power I_s to the initial spontaneously generated power I_i reaches 5.3×10^{11} at lasing at the FELI FEL facilities.

References

[1] T. Tomimasu et al., Nucl. Instr. and Meth. A 383 (1996) 337.

[2] T. Tomimasu et al., Nucl. Instr. and Meth. A 393 (1997) 188.

[3] T. Tomimasu et al., CP413, in: R. Bnifacio, W.A. Barletta (Eds.), Towards X-Ray Free Electron Lasers, AIP, New York, pp. 127–141.

[4] W. Brefeld et al., Nucl. Instr. and Meth. A 375 (1996) 295.

[5] D.W. Small et al., Nucl. Instr. and Meth. A 375 (1996) ABS61.

[6] Y. Miyauchi et al., Nucl. Instr. and Meth. A 375 (1996) ABS42.

[7] K. Saeki et al., Nucl. Instr. and Meth. A 375 (1996) 12.

[8] S. Okuma et al., Proceedings AFEL '97, Hirakata, January 21–24, 1997, pp. 95–98.

[9] T. Tomimasu et al., Nucl. Instr. and Meth. A 407 (1998) 370.

[10] E. Oshita et al., IEEE Proceedings PAC '95, Dallas, May 1–5, 1995, pp. 1608–1610.

[11] T. Tomimasu et al., IEEE Trans. Nucl. Sci. NS-28(3) (1981) 3523.

[12] K. Wakita et al., Proceedings AFEL '97, Hirakata, January 21–24, 1997, pp. 87–90.

[13] A. Zako et al., Proceedings second Asian Symposium on Free Electron Lasers, Novosibirsk, June 13–16, 1995, pp. 57–60.

[14] T. Tomimasu et al., Proceedings AFEL '97, Hirakata, January 21–24, 1997, pp. 65–74.

[15] G. Dattoli et al., IEEE Quantum Electron. 20 (1984) 637.

[16] P.W. van Amersfoort et al., The FELIX Project Status Report, FOM, April 1988, pp. 3–19.

ELSEVIER

Nuclear Instruments and Methods in Physics Research A 429 (1999) 146–150

NUCLEAR INSTRUMENTS & METHODS IN PHYSICS RESEARCH Section A

www.elsevier.nl/locate/nima

Upgrade of the ISIR-FEL at Osaka University and oscillation experiments in the sub-millimeter wavelength region

R. Kato, S. Okuda, G. Kondo, S. Kondo, H. Kobayashi, T. Igo, S. Suemine, G. Isoyama*

Institute of Scientific and Industrial Research, Osaka University, 8-1 Mihogaoka, Ibaraki, Osaka 567-0047, Japan

Abstract

The far-infrared free electron laser (FEL) at the Institute of Scientific and Industrial Research (ISIR), Osaka University has been upgraded to extend the wavelength region further to the long wavelength side. The present modification includes replacement of the optical resonator and installation of an evacuated optical transport line from the FEL to the measurement room as well as installation of a far-infrared monochromator for measuring the wavelength of FEL light. We recently started experiments using the new FEL system. The loss of light in the optical resonator was measured and it is experimentally confirmed that the diffraction loss is reduced considerably after the upgrade of the optical resonator.

PACS: 41.60.Cr; 41.85.Lc; 42.25.Fx; 29.17. + w

Keywords: Free-electron laser(s); RF linac; Optical resonator; Diffraction loss; Far-infrared; Sub-millimeter

1. Introduction

We have been developing the far-infrared free-electron laser (FEL) based on the L-band linac at the Institute of Scientific and Industrial Research, Osaka University since 1989. After we obtained the first lasing in 1994 at wavelengths around 40 μm [1–3], we began modifying the FEL system to make it suitable for user experiments and to extend the wavelength region to the longer wavelength side [4]. Modifications of the FEL system are being made step by step so as to maximize the FEL gain and to reduce the diffraction loss of light in the optical resonator in the long wavelength region. They include (i) remodeling of the wiggler from a fixed magnet gap type to a variable one, (ii) modification of the bending magnets on both sides of the wiggler and replacement of the vacuum chambers for them to enlarge the aperture of these parts in the optical resonator. After these modifications, we conducted FEL experiments and observed lasing at wavelengths from 21 up to 126 μm [5].

We continue to upgrade the FEL system. In order to extend the wavelength region further, we have replaced the mirrors and the mirror mount stages for the optical resonator with new ones in order to reduce the diffraction loss in the long wavelength region. In the previous experiments, it

* Corresonding author. Tel.: + 81-6-6879-8485; fax: + 81-6-6879-8489.

E-mail address: isoyama@sanken.osaka-u.ac.jp (G. Isoyama)

0168-9002/99/$ - see front matter
PII: S0168-9002(99)00095-9

was found that FEL light could not be detected at many wavelengths due to absorption mainly by water vapor in the air in the optical transport line from the FEL to a measurement room. We, therefore, have installed an evacuated optical transport line from the FEL to the measurement room. In the previous experiments, the wavelength of FEL light was mainly determined by calculation using the magnet gap and the electron energy, and the wavelength was measured with a small monochromator once in a while to calibrate the electron energy. We have also installed a far-infrared spectrometer which can be evacuated. These modifications of the FEL were completed recently and we began experimenting with the new system.

In this paper, we will report the upgraded FEL system and preliminary results of experiments using the new system.

2. New FEL system

Fig. 1 schematically shows the L-band linac and the new FEL system as well as the measurement system. The L-band linac has the sub-harmonic buncher (SHB) system composed of two 12th and one 6th SHBs in order to produce an intense single bunch beam. In FEL experiments, the electron beam with a peak current of 0.6 A is injected from a thermoionic gun with a cathode area of 3 cm^2 (EIMAC, YU-156) into the SHB system, and the second 12th and the 6th SHBs are powered to make an electron beam having the macropulse duration of 1.8 μs and micropulse intervals of 9.2 ns. Then the electron beam is accelerated to 10–19 MeV. The main characteristics of the electron beam are listed in Table 1. The electron beam is transported through the achromatic beam transport line to the FEL system. The main parameters of the FEL system are listed in Table 2. The wiggler is a planar type. The magnet gap of the wiggler can be varied from 30 to 120 mm, for which $K = 0.013–1.472$. By changing the electron energy from 10 to 19 MeV and the magnet gap from 30 to 120 mm, it is possible to cover the wavelength region from 20 to 200 μm with the fundamental peak.

2.1. Optical resonator

The original mirror mount stages for the optical resonator had some technical problems; the angular stability of the mount stages was not good enough and only the cavity length could be remotely controlled. The latter problem restricts freedom to conduct experiments over the wide range of the undulator gap and the electron energy. For example, the axis of the optical resonator has to be horizontally shifted as the magnet gap of the

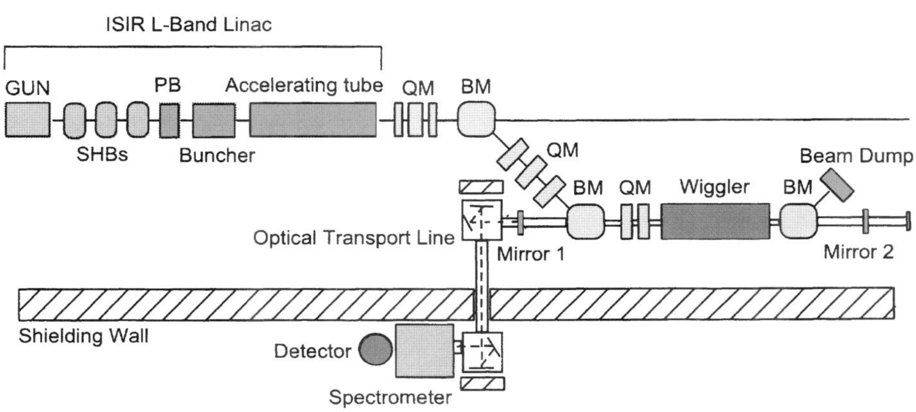

Fig. 1. Schematic layout of the ISIR L-band linac and the upgraded FEL system. The mirrors and the mirror mount stages in the optical resonator were replaced with new ones with larger aperture. The evacuated optical transport line from the FEL to the measurement room and the evacuated far-infrared spectrometer were newly installed.

Table 1
Main parameters of the electron beam

Energy	10–19 MeV
Accelerating frequency	1300 MHz
Bunch spacing	9.2 ns
Charge/bunch	2 nC
Peak current/bunch	50 A
Bunch length	20–30 ps
Macropulse length	1.8 μs
Normalized emittance	
YU-156	200π mm mrad

Table 2
Main parameters of the wiggler and the optical resonator

Wiggler	
Magnet	Nd–Fe–B
Length	1920 mm
Number of periods	32
Magnet gap	30–120 mm
K-value	0.013–1.472
Optical resonator	
Cavity length	5531 mm
Radii of mirrors	
M1	3385 mm
M2	2877 mm
Rayleigh range	1000 mm
Waist radius	5.6 mm (at 100 μm)

wiggler is varied so that the electron beam moves along the axis of the optical resonator. We, therefore, have replaced the mirror mount stages with new ones which have higher angular stability and can be fully remotely controlled. At this opportunity, the vacuum chambers near the mirror mount stages and the mirrors have been also replaced with new ones in order to reduce the diffraction loss further.

The new mirror mount stages made by Micro-Controle have three degrees of freedom to position the mirror and two degrees of freedom for adjustment of the tilts of the mirror, and all the movements are remotely controllable. The effective diameter of the mirrors is increased from 60 to 80 mm. A result of a simulation based on an open resonator model shows that the diffraction loss at 150 μm will be reduced from 22% to 15% by the change of mirrors. The vacuum chambers near the mirror mount stages and outside of the bending magnets shown in Fig. 1 have been replaced with new ones with the inner diameter of 95 mm.

2.2. Optical transport line

In the original system, laser light was taken from the optical resonator through a 3 mm hole in the downstream mirror and transported in the air to the measurement room. In order to make the light transport line simple in the new system, laser light is taken out through a 3 mm hole in the upstream mirror as shown in Fig. 1. The optical transport system consists of two mirror chambers and a straight pipe with an inner diameter of 100 mm which goes through the concrete shield wall of 3 m thick between the linac and the measurement rooms. The light transport system is evacuated with a rotary pump. Laser light coming through the hole in the mirror goes into one of the mirror chamber through a window at the entrance of the chamber, separating the higher vacuum in the optical resonator and the lower vacuum in the transport system. The window is a synthesized diamond plate of 20 mm in diameter and 0.2 mm thick, which has constant transmittance of approximately 70% in all the spectral regions longer than the ultraviolet except for small absorption around 5 μm. In the mirror chambers, there is a pair of plane and concave mirrors coated with gold. One of the mirrors can be adjusted from the outside of the mirror chamber by mechanical coupling to steer light after evacuation. The concave mirror in the mirror chamber of the measurement room focuses light onto the entrance slit of the spectrometer, which is connected to the optical transport system with no window.

2.3. Spectrometer and detectors

A grating-type far-infrared spectrometer is used to monochromatize laser light. It is a cross Czerny–Turner type monochromator with an effective aperture ratio of *f*/4.0. Two spherical mirrors with a focal length of 500 mm are used as collimating and focusing mirrors. They are square mirrors of 120×120 mm^2. A plane reflective grating with

7.9 grooves/mm (Milton Roy) is used, which has a square shape of $64 \times 64\,mm^2$ and the blaze wavelength of 112.5 μm. The monochromator can be used in the wavelength region from 70 to 160 μm with the grating. Spectral resolution is dependent on the measured wavelength and the slit width, and it is $\sim 0.6\,\mu m$ in the present experiments. A turntable for scanning of wavelength is driven with a pulse motor, which is remotely controlled with a personal computer. The monochromator is evacuated with the same rotary pump through the optical transport line. The monochromatized light is taken out through a single crystal quartz window of 2 mm thick to the air, and detected with a Ge:Ga photoconductive detector cooled with liquid helium, which is placed closed to the window. Teflon sheets are inserted in front of the detector as an attenuator to avoid saturation of the output signal when necessary.

The Ge:Ga detector has the highest sensitivity around 105 μm and the sensitivity drops steeply as the wavelength increases. It is approximately 10% of the peak value at 140 μm. The detector has the short wavelength cut-off filter below 50 μm. Therefore the sensitivity of the detector ranges from 50 to 140 μm. We used two kinds of Ge:Ga detectors. One is a so-called slow detector, which has the high detection sensitivity but has the slow time response. We used this detector for measuring the spectrum of FEL. The other is a fast detector for measuring the time evolution of FEL light.

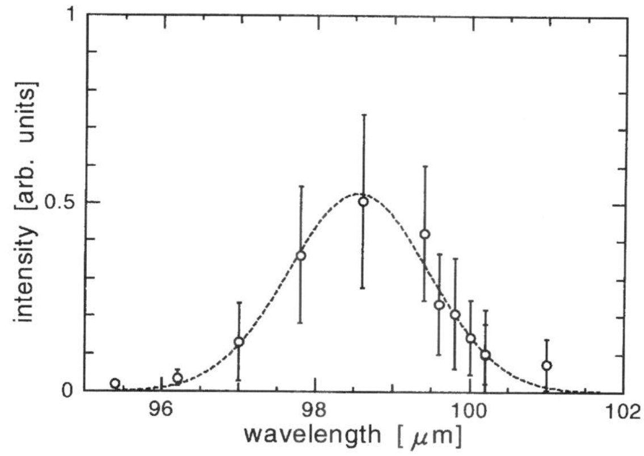

Fig. 2. FEL spectrum as a function of the wavelength. The dashed line is a fitted Gaussian distribution. The peak wavelength is 98.5 μm and the spectral width is 2.1 μm (FWHM). The error bars are due to fluctuation of the light intensity.

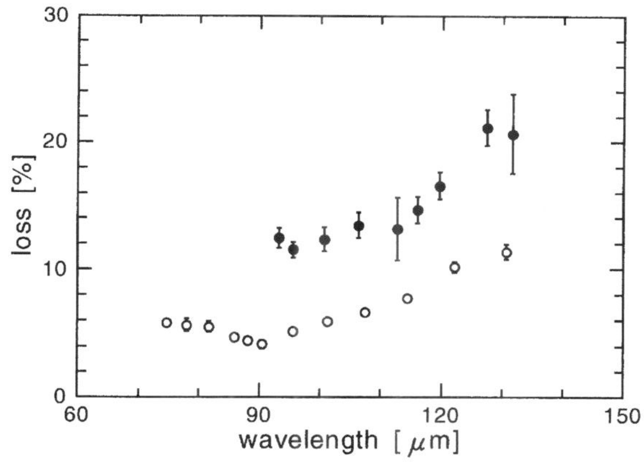

Fig. 3. The loss of light in the optical resonator as a function of the wavelength. The open and the solid circles are the loss after and before the upgrade of the mirror system in the optical resonator, respectively. The diffraction loss is reduced considerably after the upgrade.

3. Experimental results and discussions

After completing the upgrade of the FEL system, we conducted FEL experiments in the sub-millimeter wavelength region. The electron energy was decreased to approximately 12 MeV. Laser light coming through the monochromator was detected with the slow detector. An example of the measured FEL spectrum is shown in Fig. 2. A time-integrated value of the output signal was taken as the intensity of the light. The error bars shown in Fig. 2 are mostly due to fluctuation of the light intensity and not by measuring errors. The peak wavelength in Fig. 2 is 98.5 μm and the spectral width is 2.1 μm (FWHM). The measured relative bandwidth of 2.2% is smaller than the homogeneous bandwidth of the spontaneous emission, $\sim 1/N = 3.2\%$, where N is the number of wiggler period, due to the FEL oscillation.

The time evolution of FEL light was measured with the fast detector, and the gain and the loss were derived from the time spectrum. The measured loss of light in the optical resonator is shown as a function of the wavelength in Fig. 3. The solid circles show the loss measured before the present upgrade and the open circles after the upgrade. Since the monochromator was not used in these measurements, the wavelength of light was derived from the electron energy and the K-value of the wiggler. The electron energy measured with a momentum analyzer using a dipole magnet was calibrated with wavelength measurements with the monochromator at some wavelengths. It was experimentally confirmed that owing to the present upgrade of the optical resonator the loss was considerably reduced.

Acknowledgements

This work was supported in part by a Grant-In-Aid for Scientific Research from the Ministry of Education, Science, Sports and Culture of Japan.

References

[1] S. Okuda, Y. Honda, N. Kimura, J. Ohkuma, T. Yamamoto, S. Suemine, T. Okada, S. Ishida, T. Yamamoto, S. Takeda, K. Tsumori, T. Hori, Nucl. Instr. and Meth. A 358 (1995) 244.

[2] S. Okuda, Y. Honda, N. Kimura, J. Ohkuma, T. Yamamoto, S. Suemine, T. Okada, S. Takeda, K. Tsumori, T. Hori, Nucl. Instr. and Meth. A 358 (1995) 248.

[3] S. Okuda, S. Ishida, G. Isoyama, Y. Honda, R. Kato, Nucl. Instr. and Meth. A 375 (1996) 329.

[4] S. Okuda, R. Kato, Y. Nakajima, G. Kondo, Y. Iwase, H. Kobayashi, G. Isoyama, Free Electron Lasers 1996 (1997) II–103.

[5] R. Kato, S. Okuda, Y. Nakajima, G. Kondo, Y. Iwase, H. Kobayashi, S. Suemine, G. Isoyama, Nucl. Instr. and Meth. A 407 (1998) 157.

ELSEVIER

Nuclear Instruments and Methods in Physics Research A 429 (1999) 151–158

NUCLEAR INSTRUMENTS & METHODS IN PHYSICS RESEARCH Section A

www.elsevier.nl/locate/nima

OK-4/Duke storage ring FEL lasing in the deep-UV☆

V.N. Litvinenko*, S.H. Park, I.V. Pinayev, Y. Wu, M. Emamian, N. Hower, P. Morcombe, O. Oakeley, G. Swift, P. Wang

Free Electron Laser Laboratory, Box 90319, Department of Physics, Duke University, Durham, NC 27708-0319, USA

Abstract

In this paper we present the result of the deep-UV lasing with the OK-4/Duke storage ring FEL. The OK-4 FEL was initially commissioned at Duke Free Electron Laboratory in November 1996 followed by a yearlong shutdown. After re-establishing the storage ring operation in early 1998, the OK-4 FEL lasing was demonstrated in the wavelength range of 226–256 nm in April 1998. The OK-4 FEL successfully lased from 217 nm to 235 nm in August 1998. A number of lasing modes with electron energies ranging from 300 to 750 MeV have been established to accommodate application researches. Starting September 1998, the coherent deep-UV radiation is used for applications in surgery, cell biology, surface physics, and nuclear physics.

Keywords: Deep-UV Lasing; Storage ring; Lasing modes

1. Introduction

The OK-4/Duke storage ring FEL has been in development since April 1992 when the Duke FEL Laboratory (DFELL) and Budker Institute of Nuclear Physics (BINP, Novosibirsk Russia) signed a Memorandum of Understanding on collaborative research in development of the XUV FEL [1]. The OK-4 FEL, previously employed for the visible and the UV lasing at the VEPP-3 storage ring [2,3], was moved to the DFELL in May of 1995. In 1998, the ownership of the OK-4 FEL was transferred to the Duke University.

The OK-4/Duke storage ring FEL demonstrated the first lasing in the near UV and the visible (from 345 to 413 nm) during its commissioning in November, 1996 [4]. Following the initial lasing, the coherent radiation of the OK-4 FEL was used for production of nearly monochromatic γ-rays via Compton back-scattering [5]. By the end of 1996, the Duke storage ring was shut down for reconditioning one of the klystrons used for its 250 MeV linac-injector. The year of 1997 saw just a few sporadic operations of the Duke storage ring without serious attempts to lase at short wavelengths. The first quarter of 1998 was spent on the demonstration experiments in surface physics using tunable spontaneous radiation [13] and on the demonstration of full tunability of the OK-4

☆ Work supported by Office of Naval Research grant #N00014-941-0818 and Air Force Office of Scientific Research grants #F49620-93-1-0590 and F49620-95-1-4076.

* Corresponding author.

0168-9002/99/$-see front matter
PII: S0168-9002(99)00096-0

spontaneous radiation as required by our funding agency [6]. To extend the capabilities of the application research, a two-story user facility – the Keck Life Science Laboratory was scheduled to start construction in the second quarter of 1998. The construction of this facility adjacent to the main FEL building forced a three-month accelerator shutdown due to the modification of the radiation shielding in the initial phase of the construction. Taking advantage of a two-day operation window in April 1998 prior to the construction shutdown, we demonstrated the OK-4 FEL tunability from 226 to 256 nm and also tested the operation of the newly built gain modulator. We produced giant pulses with peak power sufficient to drill tiny holes with diameter of approximately 50 μm through a piece of plastic.

After the three-month construction shutdown, efforts were devoted to improving the setup of the linac-injector as well as the reliability of this system. A more powerful N_2-laser with time jitter less than 1 ns was installed to allow the single-bunch mode of injection. An attempt in late July 1998 to lase around 193 nm yielded no success.

A set of mirrors with the central wavelength of 225 nm was installed on 10 August 1998. During the following two operation shifts, the OK-4 FEL lased and was tuned between 217 and 235 nm. This set of mirrors degraded rather quickly in a couple of weeks and became unusable. Since the 245 nm mirrors with HfO_2/SiO_2 dialectic coating [2] showed little degradation of reflectivity during their initial use, they have been installed and currently provide the laser beam for the application research.

In Section 2 we briefly describe the recent modifications of the OK-4/Duke storage ring FEL and parameters used for the above experiments. In Section 3 we present the results of the OK-4 lasing around 245 and 225 nm as well as a summary of an unsuccessful attempt to lase around 193 nm. In Section 4 we discuss the strategy for further advance towards the VUV lasing using the Duke storage ring.

2. The OK-4/Duke storage ring FEL

The layout of the Duke storage ring with the OK-4 FEL as well as most of the parameters has been presented in our previous publications [7–9]. The recent developments of the Duke storage ring involved the installation of 34 units of Bergoz BPM electronics, the new timing system, the gain modulator [10], the NIST wiggler [11], and a new nitrogen laser for single-bunch injection. We have also acquired a UV Hamamatsu streak camera with a 2 ps resolution for both the storage ring and the OK-4 FEL diagnostics.

2.1. Duke storage ring

The installation of the NIST undulator [11] and its vacuum chamber had affected both the broadband impedance and the dynamic aperture of the Duke storage ring. The impedance of the vacuum chamber increased by about $\Delta Zn/n \sim 0.5\,\Omega$. Because of strong uncompensated fields, the NIST undulator when in operation reduced the dynamic aperture significantly at injection energy and prohibited any good injection. In order to minimize these impacts, we keep the "jaws" of the NIST undulator opened to the maximum during regular operations.

In July 1998 we installed 34 Bergoz electronic units for beam position monitoring with 2–10 μm resolution. The diagnostics for closed orbit measurements in the Duke storage ring are still under development. A set of software tools will have to be developed to utilize the full potential of the new orbit measurement system. Using a program currently available we were able to visualize the orbit, therefore, to speed-up the alignment of the electron beam in the OK-4 FEL. This system finally ended an era during which the OK-4 FEL lasing was achieved blindfolded.

In July 1998 we improved the performance of the 250 MeV linac-injector, especially by installing a new nitrogen laser with a larger energy per shot and low timing jitter (less than 1 ns) for single-bunch injection. At present, this system is capable of injecting 0.5 mA per shot into the desirable RF bucket of storage ring with no spills. The main problem with the previous nitrogen laser was a low injection current (~ 50–100 μA/shot) and spills into nearby RF buckets due to a large time jitter. A reasonable injection rate allowed us to study the single-bunch current limitation in the Duke storage

Table 1
Typical Duke storage ring parameters for the deep-UV lasing

		Comments
Electron energy (GeV)	0.3–0.75	depending on user demand
RF voltage (178 MHz), kV	200	for April 1998 run[a]
	200-550	for July/August 1998 runs
Number of electron bunches	1	standard users
	2	γ-rays production
Beam current (mA), per bunch	4-8	
Broad band impedance, $Z_n/n(\Omega)$	3.25 ± 0.25	
Typical lifetime with laser on (hs)[b]	2	at 0.7 GeV

[a]A temporary set-up without the RF circulator limited the maximum RF voltage to 200 kV.
[b]Lifetime is presently limited by a poor vacuum in the arcs. We plan to address this problem next year.

ring. At present, the maximum stored current in a single-bunch mode is limited to 12 mA. The maximum stored current in the single-bunch varies daily and can be as low as 5 mA/bunch. Above this threshold, the electron beam develops vertical instability causing the beam loss. This threshold value does not exhibit strong dependence on the global parameters such as chromaticity or betatron tunes. It depends mostly on the beam orbit and on the positions of high-order mode tuners in the RF cavity. We are currently studying the nature of this single-bunch instability so that more current can be stored in the single-bunch.

The OK-4 deep-UV lasing has been realized at a set of different beam energies. Generally, lasing is easier to achieve and is more stable when operating at higher e-beam energies (i.e. from 500 to 750 MeV). We attribute this effect to a better stability of the electron beam and power supplies. A set of typical Duke storage ring parameters used for lasing is listed in the Table 1.

2.2. OK-4 FEL magnetic system and its optical cavity

The OK-4 FEL comprises of two electromagnetic wigglers (each 3.5 m long), an electromagnetic buncher located between wigglers and a 53.7-m long optical cavity. The general description of this system can be found in our earlier publications [4,7–9], while the feedback and control system for the OK-4 FEL optical cavity is described in detail in Ref. [12]. The main change to the OK-4/Duke FEL since initial operation in 1996 was the installation and commissioning of the gain modulator for production of giant pulses. The gain modulator displaces the electron beam orbit from the axis of the optical cavity to stop lasing. The absence of the lasing provides for electron beam cooling towards its natural energy spread. After cooling for a few damping times, the gain modulator returns the cooled electron beam back onto the optical axis adiabatically to generate a giant pulse. With a beam current of 4–8 mA/bunch, the giant pulse has a typical macropulse duration of 100–200 μs. This process is repeated with a rate of 1–30 Hz, depending on the damping time.

Rather large value of broadband impedance and a low value of compaction factor ($\alpha_c = 0.0086$) of the Duke storage set a threshold of microwave instability at about 0.1 mA/bunch at 700 MeV. With typical electron beam parameters used for our experiments, the start-up gain is defined by well-developed microwave instability: gain $\sim I_b^{1/3} V_{rf}^{1/6}$. As shown in Fig. 1, the gain depends weakly on the beam current and the typical value of the gain is 5–10% per pass in the deep-UV range for the set of e-beam parameters listed above.

We heavily utilized the OK-4 diagnostics system for optimization and attainment of the maximum OK-4 gain. In order to optimize the gain, we matched the strength of the OK-4 buncher with the e-beam energy spread. We relied on a fully automated spectrum measurement system to verify both the optimal setting of the OK-4 magnetic system as well as the quality of the electron beam.

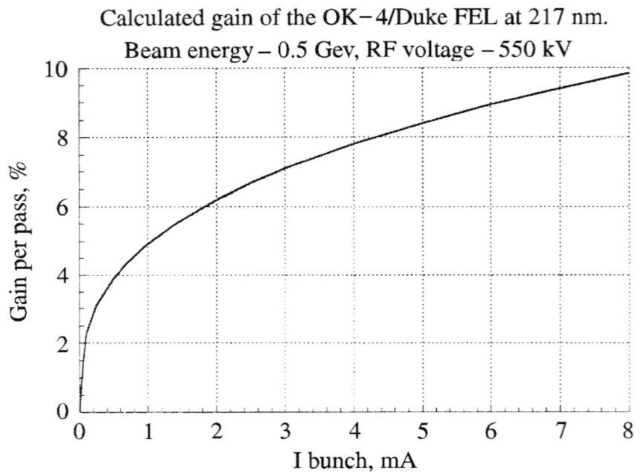

Fig. 1. Typical calculated dependence of the OK-4/Duke FEL gain on the electron current per bunch. We used broad-band impedance of $Z_n/n = 3.25\Omega$ for this calculations.

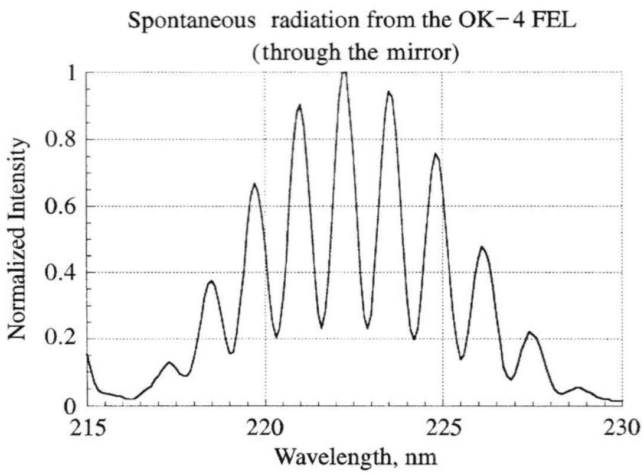

Fig. 2. A typical measured spontaneous radiation spectrum of the OK-4/Duke FEL. The depth of the fine structure (neighboring min/max ~ 0.245) is optimized for the maximum OK-4 FEL gain. The OK-4 wigglers were tuned for the wavelength, $\lambda = 222.47$ nm, the buncher was tuned to a slippage $N_d\lambda$ with $N_d = 141.25$ and the total slippage was $(N_w + N_d)\lambda$ where $N_w = 33.5$ is the number of periods in each of the OK-4 wigglers.

Fig. 2 shows a typical spectrum of spontaneous radiation with optimal tuning of the OK-4 for a maximum gain at 222 nm. We calibrated the UV monochromator using those well-known lines in the Hg lamp spectrum to ensure a measurement accuracy of better than ± 0.1 nm. These procedures are proved to be very fruitful – the measured gain of the OK-4/Duke FEL was very close to predicted values.

A Hamamatsu deep-UV streak camera with a 2 ps resolution has been used for optimization of the RF cavity operation via the proper setting selections for the RF high-order mode tuners. It has also been used for synchronization of the electron beam with the optical beam, i.e. matching of their round trip times. Changing the revolution frequency provided means for fine-tuning of the lasing conditions.

2.3. Mirrors

We have used three types of mirrors with central wavelengths of 245, 225 and 193 nm, respectively. The mirror substrates are made of UV fused silica with 27.46 m radii of curvature. The substrates were custom made by Lumonics Optics Group (Ontario, Canada). Lumonics produced the multilayer coatings for 245 nm (HfO_2/SiO_2) and 225 nm (Al_2O_3/MgF_2) mirrors, while Acton Research Corporation produced coatings for 193 nm.

We had expected the OK-4 FEL to lase in all three wavelength ranges since the measured initial losses of these mirrors were all below the OK-4 gain. However, the losses of 193 nm mirrors increased dramatically due to exposure to the OK-4 spontaneous radiation. So far, we have failed to lase around 193 nm. The degradation of the 225 nm mirrors, due to the same effect, has limited their useful lifetime to a couple of weeks.

To monitor the degradation of the mirrors, we measure the cavity losses routinely by measuring the cavity ring-down time. We used this method to measure dependence of the round-trip losses on the wavelength. By measuring the cavity losses at the threshold of lasing, we are able to obtain the OK-4 gain (see Figs. 3 and 4 for details).

With HfO_2 top layer coating, the 245 nm mirrors hold quite well after more than 4 weeks of lasing operations with about 0.5 amp-h of exposure. Our experiments with 245 nm mirrors confirmed our expectations [2] that the use of HfO_2 as a top layer provides for good radiation resistance.

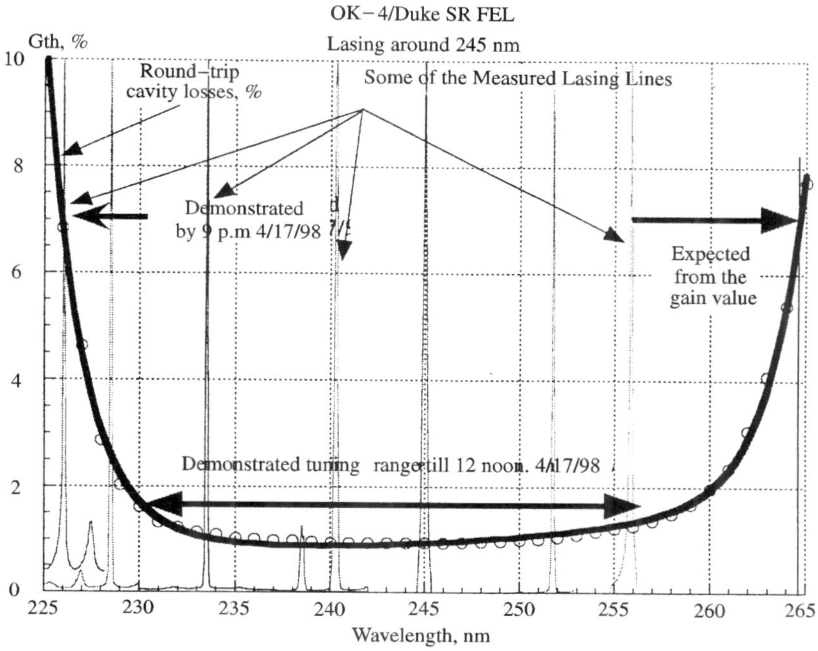

Fig. 3. Tunability of the OK-4/Duke FEL measured in April 1998: 226–265 nm. The thick-solid curve shows measured round trip losses and thin curves show samples of measured spectra with a 1.98 mA single-bunch electron beam at 500 MeV and 200 kV RF voltage. The stop-band, i.e. the shortest or the longest attainable lasing wavelength defines the level of the FEL gain at this wavelength. This graph indicates that the OK-4 FEL gain is larger than 6.9% per pass with the above beam parameters.

3. Deep-UV operations with the OK-4/Duke storage ring FEL

During 1998, we had attempted to lase with OK-4 FEL in three distinctive deep-UV spectrum ranges:

- Successful lasing within 226–256 nm range, 16–17 April 1998;
- Unsuccessful lasing attempt around 193 nm, 20 July–28 July, 1998;
- Successful lasing within 217–232 nm range, 11–12 August 1998.

It is worth noting that lasing at both 226 and 217 nm in April and August 1998 established new shortest wavelength records for free-electron lasers.

3.1. Lasing around 245 nm

Initial lasing was achieved rather easily at both 500 and 700 MeV due to small round-trip losses of just above 1% at 245 nm. Focused on obtaining the shortest lasing wavelength, we demonstrated a tuning range of 226–256 nm. The available gain of the OK-4 FEL would have allowed a full tunability of 226–265 nm (16%) with this set of mirrors (Fig. 3). Transparency of these mirrors grows drastically and makes lasing far beyond this range unrealistic.

Since the 245 nm mirrors are rather stable, we are using them for our pilot user program [8], which supports the photoemission electron microscopy [14], cell biology, UV surgery and nuclear physics experiments. The OK-4 FEL lasing has been achieved at different energies from 300 to 800 MeV. Operating the OK-4 at different energies allows us to optimize the user-requested parameters such as the average out-coupled power, peak power, out-coupled energy in macropulse, or energies of γ-rays. For example, by operating the storage ring between 300 and 750 MeV,

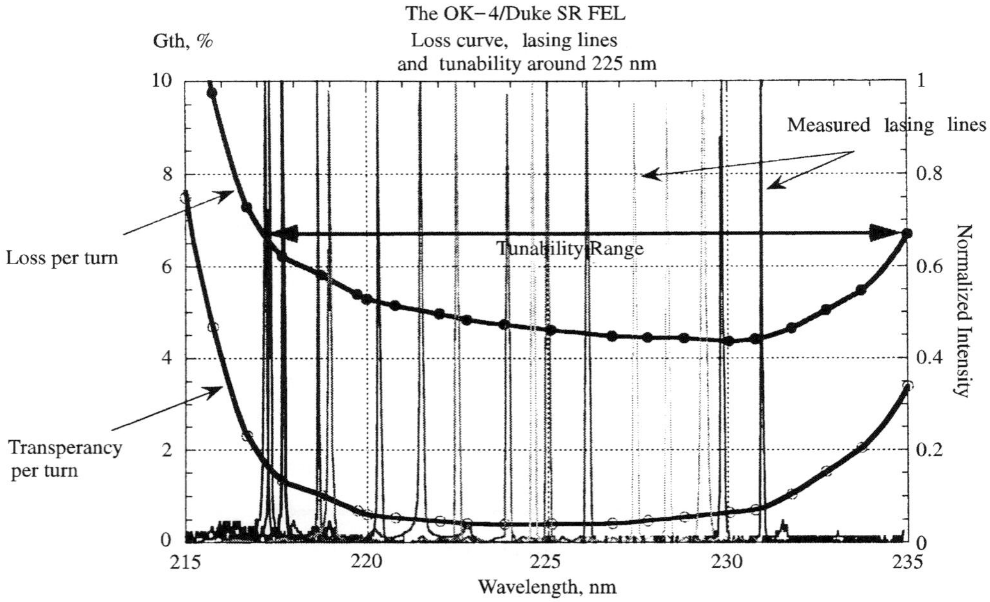

Fig. 4. Tunability of the OK-4/Duke FEL measured in August 1998: 217–235 nm. The thick-solid curve shows measured round trip losses and transparencies of the mirrors; thin curves show samples of measured spectra with a 1.50 mA single-bunch electron beam at 500 MeV and 200 kV RF voltage. This graph indicates that the OK-4 FEL gain is larger than 6.5% per pass with the above beam parameters.

monochromatic γ-ray beams in the energy range of 6–45 MeV were produced.

In September 1998 a 100 mW of average laser power per mirror was out-coupled from the 245 nm mirrors using 700 MeV electron beam with 16 mA average current in two bunches. Turning on the gain modulator, giant pulses with 50–200 μs duration were generated at a maximum rep rate of 30 Hz with 0.25–0.4 mJ per macropulse per mirror and a peak out-coupled power of about 0.3 MW. The transparency of these mirrors is too low for the optimized out-coupling. At the central wavelength of 245 nm, the mirror transparency of 0.035% and initial loss of 0.47% provide an outcoupling efficiency less than 8%. Even though the 245 nm mirrors coated with HfO_2/SiO_2 are very stable and radiation resistive, in the long run their losses tend to grow to about 1% per mirror [2], which will further reduce the extraction efficiency. We have recently ordered a new set of mirrors with transparency of 0.5–0.75% at the central wavelength for better extraction efficiency.

The evaluation of the OK-4/Duke FEL performance in this range of spectra will be completed by the end of 1998 and we plan to publish a detailed description of OK-4 FEL parameters in a future paper.

3.2. Lasing around 225 nm

Mirrors with central wavelength of 225 nm (Al_2O_3/MgF_2) were stable enough to support the OK-4 FEL operation for six days. Losses increased almost immediately to 4–4.5% per pass at the central wavelength and then increased gradually to about 20% per pass after six days of exposure. Attainment of the lasing in the OK-4 was difficult because of rather high initial losses. First lasing was achieved after 8 h of tuning both the OK-4 and Duke storage ring at 500 MeV. We demonstrated lasing and tunability within the 217–231 nm wavelength range on 11 August 1998 (see Fig. 4) with 2–3 mA/bunch beam current at 500 MeV. Later we also demonstrated lasing at 700 MeV,

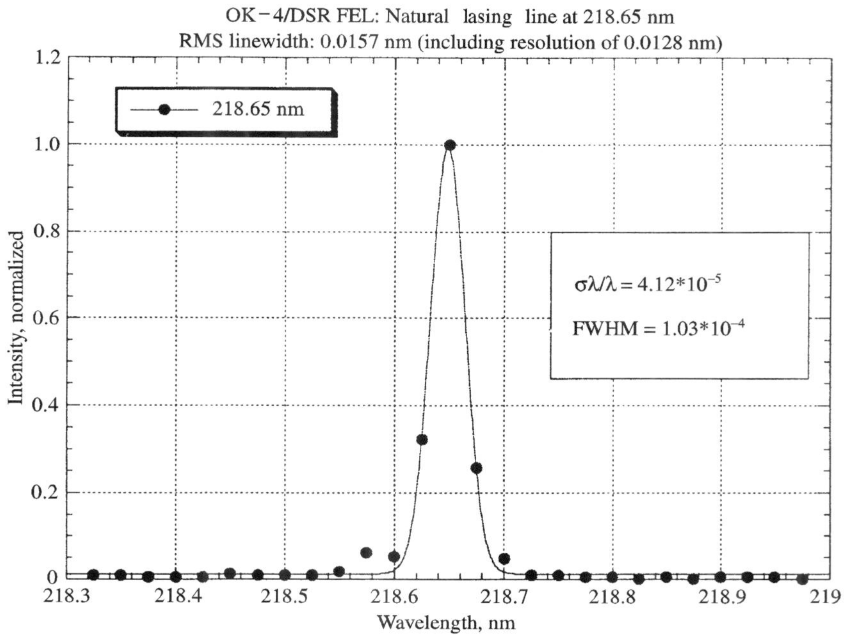

Fig. 5. One of the measured time-averaged spectra of the OK-4/Duke FEL lasing in the deep-UV. The linewidth depends on the setting of the OK-4 buncher as well as on the de-tuning from synchronism.

which sped up the degradation of the mirror. In spite of high cavity losses, we were capable of lasing with narrow lasing lines typical for the OK-4 FEL operation. One of measured lasing lines with a FWHM of 0.01% is shown in Fig. 5.

3.3. Lasing attempt 193 nm

We tried to lase around 193 nm using high reflective coatings made by Acton Research Corporation. Initial reflectivity of the 193 nm mirrors was about 98%, which set initial threshold gain at 4% per pass. Unfortunately, the 193 nm mirrors degraded faster than we could manage to tune the OK-4 FEL for lasing. After three shifts, the round trip losses increased to 32% per pass at 193 nm, which forced us to abandon the lasing attempt.

4. Future plans to advance below 200 nm

It becomes rather transparent from our experience that standard dielectric coatings below 240 nm deteriorate very quickly in the presence of strong spontaneous radiation from a high K planar wiggler. We plan to test a number of un-obvious dielectric coatings for mirrors in the range of 150–255 nm for their stability in these conditions. We also plan to test a Brewster plate as well as grazing incidence Al mirrors to protect the downstream mirror from the exposure to the OK-4 spontaneous radiation.

A more radical solution has also been planned to install on the Duke storage ring a new OK-5 "Blue Devil" FEL; a 19-m long distributed optical klystron with four helical electromagnetic wigglers and three bunchers. The helical wigglers could be switched between the left and right circular polarizations as well as provide all intermediate polarizations [14]. This feature of the OK-5 FEL is of great importance for user applications related to dichroic systems as well the study of CP-violating processes in nuclear physics.

The use of helical wigglers will boost the FEL gain by a factor of four compared with a planar one. It will also eliminate on-axis radiation of high order harmonics [15], which are responsible for the degradation of the mirrors. We expect such a system to be installed on the Duke storage ring by the end of 1999 and to be operational in early 2000.

5. Conclusions

The OK-4/Duke storage ring FEL has demonstrated lasing as well as generated usable laser beams in the deep-UV range of spectrum. We expect that the OK-4 FEL will be operated reliably to deliver coherent UV radiation from 226 to 265 nm and monochromatic γ-rays from 6 to 45 MeV to the pilot applications. The exploration of the shorter wavelength range with the OK-4/Duke storage ring FEL will continue in parallel with the user applications.

Acknowledgements

Authors are thankful to T. Brinkley, J. Faircloth, R. Guenther, J. Gustavsson, M. Johnson, L. Kennard, S. Lowther, H. Mercardo, J. Murphy, J. Patterson, C. Ratcliff, G. Ridout from Duke University Free Electron Laboratory, M.G. Fedotov, N.G. Gavrilov, V.M. Popik, V.N. Repkov, L.G. Isaeva, G.N. Kulipanov, G.Ya. Kurkin, S.F. Mikhailov, A.N. Skrinsky, N.A. Vinokurov, P.D. Vobly from Budker Institute of Nuclear Physics and A. Lumpkin from Argonne National laboratory for their important contributions to the success of the OK-4/Duke storage ring FEL. Authors would like to acknowledge the Office of Naval Research for its continuous support of the Duke OK-4 UV FEL and its applications.

References

[1] V.N. Litvinenko, J.M.J. Madey, N.A. Vinokurov, Proceedings of 1993 IEEE Particle Accelerator Conference, Washington, DC, 17–20 May 1993, pp. 1442–1445.

[2] V.N. Litvinenko, The optical klystron on VEPP-3 storage ring bypass – lasing in the visible and the ultraviolet, Thesis, Novosibirsk, 1989.

[3] G.N. Kulipanov, V.N. Litvinenko, I.V. Pinaev, V.M. Popik, A.N. Skrinsky, S.A. Sokolov, N.A. Vinokurov, Nucl. Instr. and Meth. A 296 (1990) 1.

[4] V.N. Litvinenko et al., Nucl. Instr. Meth. A 407 (1998) 8.

[5] V.N. Litvinenko et al., Phys. Rev. Lett. 78(4) (1997) 16, pp. 4569–1472.

[6] V.N. Litvinenko et al., Capabilities of the Duke storage ring FEL facility for applications research in spectral range from the IR to γ-rays, Presented at the 20th Int. Free Electron Laser Conference, Williamsburg, VA, 16–21 August, 1998.

[7] Y. Wu, V.N. Litvinenko, B. Burnham, S.H. Park, J.M.J. Madey, IEEE Trans. Nucl. Sci. NS-44 (5) (1997) 1.

[8] Y. Wu, V.N. Litvinenko, I.V. Pinayev, S.H. Park, B. Burnham, Nucl. Instr. and Meth. B 144 (1998) 90.

[9] V.N. Litvinenko, S.H. Park, I.V. Pinayev, Y. Wu, The Duke XUV FEL storage ring facility, Proceedings of the first APAC Conference, 23–27 March, 1998, KEK, Tsukuba, Japan, 1998, p. 834. V.N. Litvinenko, B. Burnham, J.M.J Madey, Y. Wu, Nucl. Instr. and Meth. A 375 (1996) 46–52.

[10] I. Pinayev et al., Critical issues for deep UV lasing with the OK-4/Duke storage ring FEL, Presented at the 20th Int. Free Electron Laser Conference, Williamsburg, VA, 16–21 August, 1998.

[11] L. Johnson, private communication.

[12] System for control and stabilizing of OK-4/Duke FEL optical cavity, I.V. Pinayev, M. Emamian, V.N. Litvinenko, S.H. Park, Y. Wu, AIP Proceedings of eighth Beam Instrumentation Workshop, 4–7 May 1998, Stanford, Palo Alto, CA, CP 451, 1998, p. 545.

[13] H. Ade, Y. Wang, S.L. English, J. Hartman, R.F. Davis, R.J. Nemanich, V.N. Litvinenko, I.V. Pinayev, Y. Wu, J.M.J. Madey, A free electron laser – photoemisson electron microscope system (FEL-PEEM), Surf. Rev. Lett. 5 (6) (1998) 1257.

[14] N.A. Vinokurov, private communication.

[15] N.A. Vinokurov, FELs with storage rings, in: Relativistic High Frequency Electronics, Gorky, 1990.

ELSEVIER

Nuclear Instruments and Methods in Physics Research A 429 (1999) 159–164

NUCLEAR INSTRUMENTS & METHODS IN PHYSICS RESEARCH Section A

www.elsevier.nl/locate/nima

Lasing down to the deep UV in the NIJI-IV FEL

K. Yamada[a,*], N. Sei[a], T. Yamazaki[b], H. Ohgaki[a], V.N. Litvinenko[c], T. Mikado[a], S. Sugiyama[a], M. Kawai[d], M. Yokoyama[d]

[a]*Electrotechnical Laboratory, 1-1-4 Umezono, Tsukuba, Ibaraki 3058568, Japan*
[b]*Institute of Advanced Energy, Kyoto University, Gokasho, Uji, Kyoto 611-0011, Japan*
[c]*FEL Laboratory, Physics Department, Duke University, Durham NC 27708, USA*
[d]*Kawasaki Heavy Industries, Ltd., 118 Futatsuzuka, Noda, Chiba 278, Japan*

Abstract

Lasing has been obtained at the wavelengths down to the deep UV on the NIJI-IV free electron laser (FEL) system. The shortest wavelength obtained was 228 nm and the laser peak power was estimated to be $\sim$ 200 and $\sim$ 40 mW at 300 and 240 nm, respectively. Suppression of the fast head–tail instability in the electron beam led to a drastic enhancement in the laser gain which enabled us to achieve such short-wavelength lasing even in a compact storage ring such as NIJI-IV. Low-loss cavity mirrors having Al_2O_3/SiO_2 dielectric layers are being prepared to obtain FEL operation at much shorter wavelengths.

1. Introduction

Free electron lasers (FELs) have an excellent potential to generate coherent and high-power light in a very wide wavelength range, compared with other lasers. The vacuum ultraviolet (VUV), shorter than $\sim$ 200 nm, is especially attractive for FELs, because it is not easy to extract high-power coherent light with a wide wavelength tunability in the VUV, even if frequency up-conversion of conventional-type lasers in nonlinear crystals, metal vapors and rare gases is used. Efforts to shorten the FEL wavelength down to the deep UV have been made intensively using storage rings[1] [1,2]. To shorten the wavelength toward the VUV, a large-scale accelerator is usually required, because of the necessity of high-energy and high-peak-current electron beam with a very good beam quality. However, it is also very interesting to study the feasibility of a compact accelerator in short wavelength FELs from a laser application point of view.

At the Electrotechnical Laboratory (ETL), an FEL experiment has been carried out on a compact storage ring NIJI-IV [3,4], at wavelengths from the visible (595 nm) to the near UV (349 nm), after the first lasing and a series of FEL experiments in the visible on another storage ring TERAS [5,6]. However, the FEL gain obtainable in the NIJI-IV has been restricted to 1% or less due to its compactness and simple structure, which has been making it difficult to achieve lasing below 300 nm, where the laser cavity composed of state-of-the-art low-loss dielectric multilayer mirrors still has a loss larger by about one order of magnitude, compared

* Corresponding author. Tel.: + 81-298-54-5679; fax: + 81-298-54-5683.

E-mail address: kyamada@etl.go.jp (K. Yamada)

[1] Recently, FEL has been achieved in the deep UV at 217 nm also on a 1-GeV storage ring in the Duke university.

0168-9002/99/$ - see front matter
PII: S0168-9002(99)00097-2

with that for 350 nm. In the NIJI-IV system, the gain limitation comes from several kinds of instabilities [7] in the stored beam, such as coupled-bunch instability, fast head–tail instability and longitudinal microwave instability, which have resulted in an insufficient peak current and a relatively large beam-energy spread. In order to increase the peak current and enhance the laser gain, efforts have been made to suppress such beam instabilities [8–10]. Degradation characteristics of the laser-cavity mirrors were also investigated at the wavelengths shorter than 300 nm [11,12]. These efforts enabled us to shorten the laser wavelength down to the deep UV. On March 20, 1998, lasing was obtained between 315 and 287 nm by using dielectric multilayer mirrors optimized at 300 nm. Lasing was also achieved from 253 to 228 nm on May 12, 1998, by changing the mirrors to those optimized at 240 nm. Here we report the FEL characteristics down to the deep UV and a brief prospect toward VUV FEL in the NIJI-IV.

2. NIJI-IV FEL system

The NIJI-IV is a compact and simple racetrack-type storage ring dedicated to the FEL research. The ring is usually operated at ~310 MeV for experiments in the UV range. The ring lattice is basically composed of only six bending dipole magnets and 12 focusing quadrupole magnets. Such a simple structure enabled us to take a 7.25-m dispersion-free straight section within a circumference of only ~29.6 m. In this long straight section, a 6.3-m optical klystron (OK) [13] is installed. However, such a structure made it difficult to suppress beam instabilities under higher beam current condition, which has been limited the peak current to be ~1 A and the laser gain to be ~1% or less.

Since the NIJI-IV is operated with only one electron bunch in the FEL experiment [4,9], single-bunch instabilities, such as fast head–tail instability and longitudinal microwave instability, should be mainly taken into account. The former has been the most serious factor to limit the peak current in the NIJI-IV. It is well known that the fast head–tail instability can be suppressed by the correction of the ring chromaticity [7], induced mainly from an energy aberration of the quadrupole magnets. We designed compact sextupole–quadrupole–sextupole (SQS) combination magnets [9,10] and replaced four old quadrupoles, QF_2 (see Ref. [4]), with them. By optimizing the excitation current in the SQS, the horizontal and vertical chromaticities, ξ_x and ξ_y, were successfully reduced respectively to $+0.1$ and $+0.2$ from their natural values of -2.97 and -4.17 [10]. This resulted in an increase of the peak beam current up to ~5 A (at an average beam current of 30 mA per bunch) and has produced sufficient enhancement of the FEL gain.

The laser cavity was composed of two low-loss dielectric multilayer mirrors manufactured with the ion beam sputtering technique and its length was finely adjusted around 14.8 m with an accuracy of 0.2 μm so that the roundtrip of the light pulse in the cavity is tuned to the revolution of the electron bunch in the ring. In the present experiment, two sets of mirrors having alternating HfO_2/SiO_2 dielectric layers optimized at ~300 and ~240 nm were used. Their initial loss was 1300 and 2700 ppm (per mirror) at 300 and 240 nm, respectively. The evolution of cavity degradation for 240 nm during the exposure to the undulator radiation is shown in Fig. 1. It is found in Fig. 1 that the cavity indicates both surface- and volume-type degradations [12]

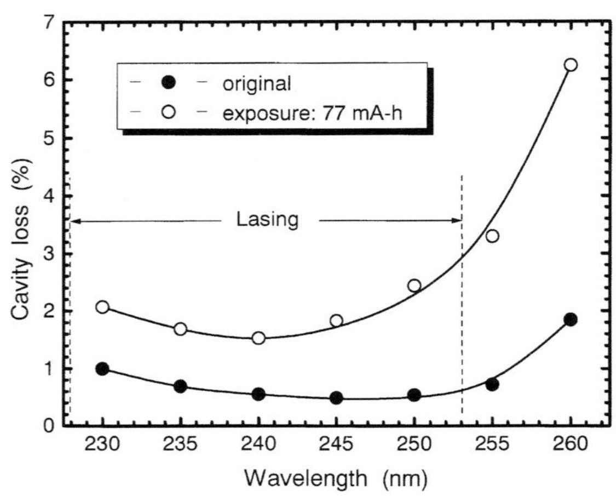

Fig. 1. Evolution of degradation in the laser cavity optimized at 240 nm. Arrows indicate the range where the lasing was obtained.

and the cavity loss (per mirrors pair) increases up to 1.53% at 240 nm after 77-mA h exposure to the undulator radiation. As shown in the previous section, we can obtain the lasing between 253 and 228 nm under relatively good cavity condition. In addition, it was also found that we obtained lasing in a wide range around 240 nm even after 77-mA h exposure. Considering the cavity degradation shown in Fig. 1, these facts suggest that the FEL gain obtained is 2–3% around 240 nm which is in good agreement with the simple gain calculation [10].

3. Lasing characteristics

Figs. 2(a) and (b) show typical FEL spectra observed with a monochromator equipped with a photodiode array around 300 and 240 nm, respectively. Each line in Fig. 2 was obtained for different magnetic flux density in the OK, by changing the gap between two rows of OK magnets. The envelope of the lines reflects the dependence of mirror reflectance on the wavelength [12] and the line intensity is found to be rather small at both long and short wavelength fringes. Fig. 3 shows the dependence of the laser-line intensity, proportional to the average FEL power, on the detuning length between a light pulse and an electron bunch in the laser cavity at 300 nm. It is found that the laser output can be obtained within the cavity detuning of ± 60 μm. Temporal laser waveforms observed with a photomultiplier for typical detuning conditions are also indicated in the figure. In the storage-ring-based FELs, the laser should be emitted as a continuous "micro-temporal" pulse train with a time interval of successive electron bunches, $\sim$ 100 ns in the NIJI-IV. This corresponds to a CW lasing in FELs. However, in some cases where the gain is sufficiently large and the laser buildup is very fast, the laser output indicates an irregular burst of "macro-temporal" pulses of a few ms in duration, due to sudden laser build up and resultant gain depression through beam heating. Such a burst lasing is seen when the cavity detune is nearly zero, as shown in Fig. 3, where the duration of macropulses and the period between them are

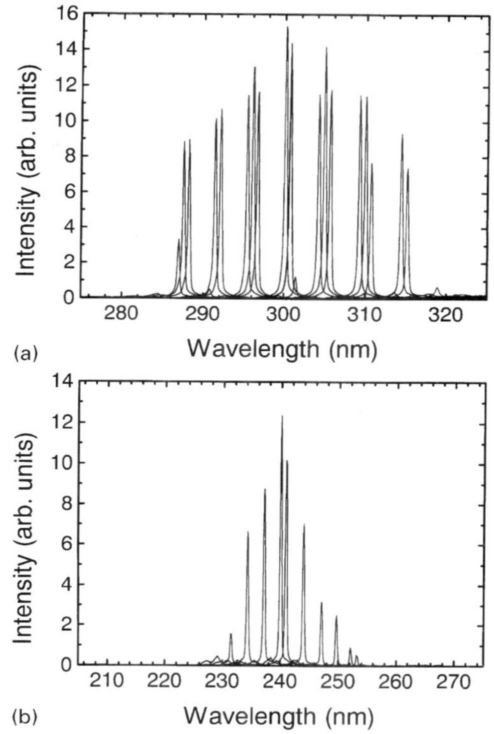

Fig. 2. Typical FEL spectra around 300 nm (a) and 240 nm (b). Each line was obtained by changing the gap of the OK magnets. The deflection parameter K was 2.02 an 1.70 at 300 and at 240 nm, respectively.

observed to be ~ 1 and ~ 5 ms, respectively. A CW-like lasing is also observed with a 20- and 40-μm detuning, though the laser power becomes smaller. The "micro-temporal" pulse was observed with a streak camera. Fig. 4 is a typical example at the wavelength and the average beam current of 300 nm and ~ 4 mA, respectively. By decomposing the observed waveform into two Gaussian components, the laser-pulse width and the electron-bunch length are measured to be 92 and 280 ps in FWHM. It should be noted that the signal height for the electron-bunch component is emphasized, because a lot of off-axis undulator radiation was focused onto the entrance slit of the streak camera with a collecting lens. The laser power at the micro-pulse peak in the burst lasing was estimated to be ~ 200 and ~ 40 mW at 300

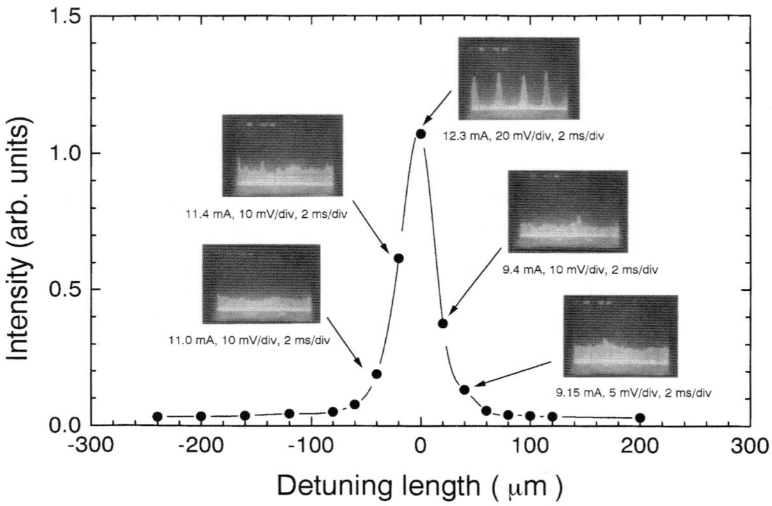

Fig. 3. Dependence of the laser-line intensity on the detuning length between a light pulse and an electron bunch. Oscilloscope traces of the laser waveforms for typical detuning conditions are also indicated in the figure. The average beam current and the oscilloscope scales are indicated under each trace.

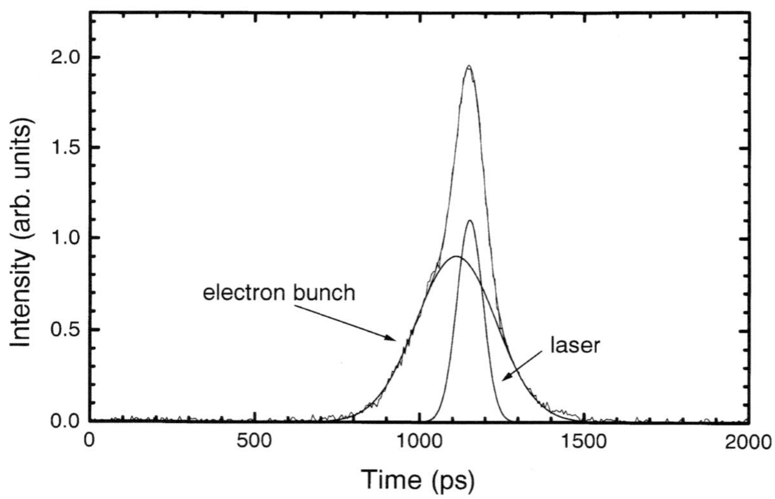

Fig. 4. Typical micro-temporal pulse observed at 240 nm with a streak camera.

and 240 nm, respectively. In order to obtain a higher laser power, it will be necessary to increase both the electron-beam energy and the transmittance of the cavity mirrors as well as peak beam current. The duration of FEL operation per beam injection was about 1.5 and 1 h for 300 and 240 nm, respectively.

The lateral intensity profile of the laser spot near the cavity-mirror surface was always monitored with a CCD camera during the experiment to check the cavity alignment. Fig. 5 shows an example at 240 nm. Though the CCD image was rather noisy due to its low photosensitivity, we found that a TEM_{00} mode can be obtained within a cavity misalignment of ± 40 μrad. Fig. 6 shows the FEL line at the shortest wavelength (228 nm) obtained in this experiment. Although the intensity of the line is not large due to a small reflectance (a large loss) of the cavity mirrors at 228 nm, obvious FEL amplification can be seen in the figure. Lasing with higher power at shorter wavelength is expected by optimizing the working wavelength of the cavity mirrors. We are preparing new mirrors having Al_2O_3/SiO_2 dielectric layers whose light-absorption loss can be sufficiently small even below 200 nm. It was already confirmed that we could obtain a laser cavity whose loss is 5700 ppm (per two mirrors) at 220 nm by using Al_2O_3/SiO_2 mirrors.

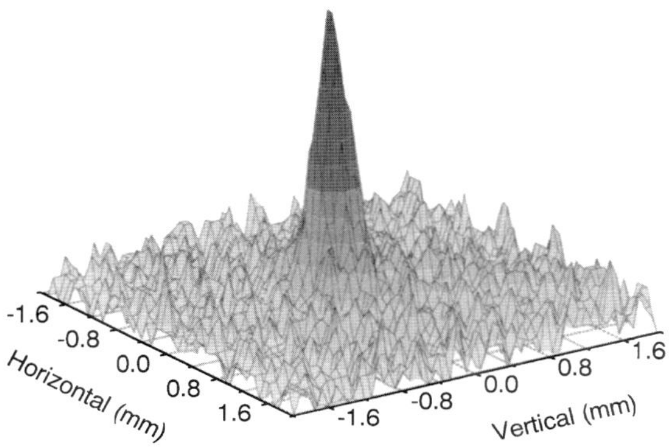

Fig. 5. Lateral intensity profile of the laser spot observed with a CCD camera.

4. Conclusion

The FEL based on a compact storage ring NIJI-IV was operated at the wavelengths down to the deep UV to study the feasibility of a compact accelerator in short wavelength FELs. By suppressing the fast head–tail beam instability, the FEL gain was drastically enhanced and the lasing wavelength has reached 228 nm. The shortest wavelength obtained in this experiment was determined by the wavelength-dependent loss of the cavity mirrors. Now we are preparing Al_2O_3/SiO_2 multilayer mirrors optimized below 220 nm. We expect to obtain lasing down to ~ 200 nm with the new mirrors. To achieve FELs in the VUV range, a large-scale accelerator is usually required, because of the necessity

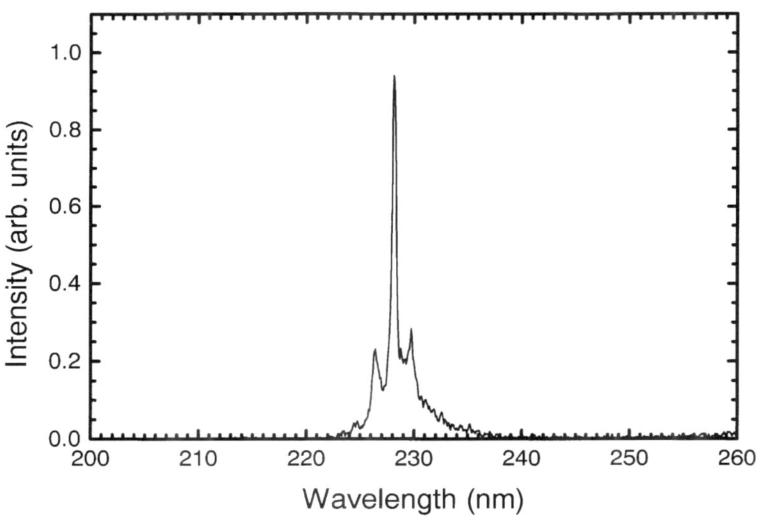

Fig. 6. FEL line spectrum at the shortest wavelength (228 nm) obtained in this experiment.

of high-peak-current electron beams with very good beam quality. Extrapolating from the present results, however, we can expect FEL gain of more than 5% in the NIJI-IV, if a further increase of the peak beam current to more than 10 A is achieved. This will be realized by shortening the electron-bunch length. In this condition, lasing in the VUV can be also expected even in such a compact storage ring NIJI-IV.

Acknowledgements

The authors would like to thank Mr. K. Ito and Mr. K. Eto of Japan Aviation Electronics Industry for manufacturing the test samples of low-loss mirrors. This work was supported by Peaceful Utilization Technology of Nuclear Energy from the Science and Technology Agency of Japan.

References

[1] G.N. Kulipanov, V.N. Litvinenko, I.V. Pinaev, V.M. Popik, A.N. Skrinsky, A.S. Sokolov, N.A. Vinokurov, Nucl. Instr. and Meth. A 296 (1990) 1.

[2] H. Hama, K. Kimura, M. Hosaka, J. Yamazaki, T. Kinoshita, Proceedings of the Third Asian Symposium on Free Electron Lasers and Fifth Symposium on FEL Applications (AFEL'97), Hirakata, 1997, Ionics Publishing, Tokyo, 1997, p. 17.

[3] T. Yamazaki et al., Nucl. Instr. and Meth. A 331 (1993) 27.

[4] T. Yamazaki, K. Yamada, N. Sei, H. Ohgaki, M. Kawai, M. Yokoyama, S. Hamada, S. Sugiyama, T. Mikado, R. Suzuki, T. Noguchi, M. Chiwaki, A. Iwata, Jpn. J. Appl. Phys. 33 (1994) L1224.

[5] T. Yamazaki, K. Yamada, S. Sugiyama, H. Ohgaki, T. Tomimasu, T. Noguchi, T. Mikado, M. Chiwaki, R. Suzuki, Nucl. Instr. and Meth. A 309 (1991) 343.

[6] K. Yamada, T. Yamazaki, S. Sugiyama, T. Tomimasu, H. Ohgaki, T. Noguchi, T. Mikado, M. Chiwaki, R. Suzuki, Nucl. Instr. and Meth. A 331 (1993) 103.

[7] H. Wiedemann, Particle Accelerator Physics II, Springer, Berlin, 1995, pp. 382–405 (Chapter 10).

[8] N. Sei, H. Ohgaki, K. Yamada, T. Yamazaki, Nucl. Instr. and Meth. A 393 (1997) 38.

[9] K. Yamada, T. Yamazaki, N. Sei, H. Ohgaki, T. Mikado, T. Noguchi, S. Sugiyama, R. Suzuki, T. Ohdaira, M. Kawai, M. Yokoyama, in: P.G. O'shea, H.E. Bennet (Eds.), Proceedings SPIE, San Jose, 1997, SPIE, Bellingham, vol. 2988, 1997, p. 134.

[10] N. Sei, K. Yamada, H. Ohgaki, V.N. Litvinenko, T. Mikado, T. Yamazaki, Nucl. Instr. and Meth. A 429 (1999) 185.

[11] K. Yamada et al., Nucl. Instr. and Meth. A 358 (1995) 392.

[12] K. Yamada, T. Yamazaki, N. Sei, R. Suzuki, T. Ohdaira, T. Shimizu, M. Kawai, M. Yokoyama, T. Mikado, T. Noguchi, S. Sugiyamma, H. Ohgaki, Nucl. Instr. and Meth. A 393 (1997) 44.

[13] T. Yamazaki, K. Yamada, S. Sugiyama, H. Ohgaki, T. Tomimasu, M. Kawai, Nucl. Instr. and Meth. A 318 (1992) 142.

ELSEVIER

Nuclear Instruments and Methods in Physics Research A 429 (1999) 165–171

NUCLEAR INSTRUMENTS & METHODS IN PHYSICS RESEARCH Section A

www.elsevier.nl/locate/nima

Inter-dependence of the electron beam excitations with the free electron laser stability on the super-ACO storage ring

M.-E. Couprie[a,b,*], R. Roux[a,b], D. Nutarelli[a,b], E. Renault[a,b], M. Billardon[b,c]

[a]*Service Photons, Atomes et Molécules; CEA/DSM/DRECAM Bat.522, 91191 Gif sur Yvette, France*

[b]*Laboratoire pour L'Utilisation du Rayonnement Electromagnétique (LURE), Bât. 209D, Université Paris-Sud, BP34, 91 898 Orsay Cédex, France*

[c]*Ecole Supérieure de Physique et Chimie Industrielle (ESPCI), 10 rue Vauquelin, 75231 Cédex, France*

Abstract

Storage ring free electron lasers have a complex dynamics as compared to the LINAC driven FEL sources since both the laser and the recirculating electron beam behaviours are involved. Electron beam perturbations can strongly affect the FEL operation (start-up, stability) whereas the FEL can stabilize beam instabilities. Experimental analysis together with simulations are reported here. Improvements of the Super-ACO FEL for users is discussed, and consequences are given in terms of electron beam tolerances for a source development for users.

PACS: 41.60.Cr

Keywords: Electron beam excitations; Super-ACO storage ring

1. Introduction

The Super-ACO free electron laser has first been operated for users in 1993 [1,2]. Generally, each data acquisition performed by the users requires the FEL to remain stable during roughly half an hour. The different FEL characteristics were followed versus time with accurate means of measurements: a double sweep streak camera for the evolution of the laser and electron beam temporal distribution [3], a scanning Fabry-Perot for the laser spectral line [4], and a photomultiplier for the average intensity. It then clearly appeared that the FEL drifts could be related to some beam instabilities, such as line excitation at 50 Hz and its harmonics, coherent or complex synchrotron oscillations, beam orbit deviations, etc. Even though the FEL itself generally stabilizes the beam excitations, its own stability is strongly affected, and is sometimes not sufficient for performing user applications. The source of the beam excitations have to be fought: for example, a feedback damping the quadrupolar modes of synchrotron oscillations has been developed on Super-ACO [5], a feedback correcting the orbit transverse deviations is under way. In addition, the FEL is also stabilized [6]: its temporal drift is compensated with a longitudinal feedback system allowing the jitter to be reduced to

* Corresponding author. Tel.: + 33-1-64-46-80-44; fax: + 33-1-64-46-41-48.

E-mail address: couprie@lure.u-psud.fr (M.E. Couprie)

0168-9002/99/$ - see front matter
PII: S0168-9002(99)00098-4

10 ps, the spectral drift to 0.05 Å and the intensity fluctuations to a few %, owing to the up-grade of the electronics. The longitudinal feedback on the FEL is also very efficient with the operation with a harmonic RF cavity at 500 MHz, with a bunch length reduced by a factor of 2 [7], with relatively low-vertical and -longitudinal excitations. Under these conditions, the FEL has been used for applications, in combination with synchrotron radiation for dynamical studies on excited species. Because of the higher gain achieved under these conditions on Super-ACO, more than 300 mW of average power can be outcoupled by transmission for users, setting the highest average power obtained with a UV FEL. In addition, shorter wavelength operation for users can be expected in the future.

2. Characteristic times for the electron beam and for the FEL

2.1. Characteristic times of the electron beam

According to a simple analysis, the electrons in a storage ring adopt a temporal Gaussian repartition around the position of the synchronous electron for which the RF electric field perfectly compensates the energy radiated by synchrotron radiation. Their energy repartition is also Gaussian, around the synchronous electron. The bunch length σ_l and the energy spread σ_γ are defined as the rms values of the distributions [8]. The electrons obey a second-order differential equation, whose solution corresponds to a damped oscillator, with synchrotron frequency f_S (μs range) and synchrotron damping time τ_S (ms range). Bunch length and energy spread generally grow when the current increases, corresponding to the microwave instability regime [9], depending on the impedance of the vacuum chamber. Potential well distortion, corresponds to an increase of the bunch length without modification of the energy spread, defines the stationary solution. Finally, coherent or complex synchrotron oscillations can govern the beam, with a limited amplitude preventing beam blow-up.

In the transverse plane, the electrons perform a few betatrons oscillations per round trip, resulting from the fact that electrons are differently focused and defocussed according to their energy. A consequent change of the path ΔL in reference to the circumference of the ring L is proportional to the energy difference ΔE, relative to the reference energy E, according to $\Delta L = \alpha L(\Delta E/E)$ with α the momentum compaction factor; introducing longitudinal-to-transverse coupling.

2.2. Characteristic times of the FEL

The FEL rising time is defined by Eq. (1) and corresponds to a few tens of μs.

$$\tau_r \sim T_{ph}/(P \ln(G/P)). \tag{1}$$

T_{ph} being the round trip time in the optical resonator, G the gain per pass and P the cavity losses. The FEL grows to the detriment of the kinetic energy of the electron bunches, resulting in a modification of the energy spread of the electron bunch accumulated over many passes. The FEL-induced "heating" of the electron bunch corresponds to the simple saturation scheme for the saturation of a storage ring FEL [10]. It is fully valid when the electron bunch is not submitted to bunch lengthening due to microwave instability or potential well distortion, or more generally to any kind of instabilities [11]. It corresponds experimentally to a small current stored in the ring. Small perturbations to the FEL equilibrium state can be analysed with the dimensionless laser intensity and energy spread. It leads to a system of equations, which can be linearized in the case of small movements by a differential equation of second order, whose general solution is a damped oscillator, of damping time τ_s and frequency $F_r = (1/\pi\sqrt{2})\sqrt{1/\tau_r\tau_s}$ [12]. This macro-temporal structure is systematically observed for specific detuning conditions (synchronization between the optical pulse stored in the optical resonator and the electron bunch circulating in the storage ring). The behaviour versus detuning has been explained by a more elaborate model, in the temporal space [13] and in the frequency domain [14]. Nevertheless, a rather high current should be stored in order to perform pump-probe, two-color experiments, leading to a more complex dynamics of the FEL interaction, and competing with the phenomena responsible for the bunch anomalous lengthening. New saturation schemes are observed [13]. Different situations are analyzed in the following sections, considering different time scales for the electron beam excitations.

3. Stability in the microsecond range

3.1. Experimental results

The microsecond range is characteristic of synchrotron oscillations. With the operation of a single RF cavity at 100 MHz, the dipolar modes (corresponding to a displacement of the whole bunch distribution) are damped using a feedback system, analogous to the one developed by Pedersen in CERN [15]. Then current thresholds can be distinguished for the quadrupolar modes at 28 kHz and 55–60 mA in two equidistant bunches, for the sextupolar modes at 40 kHz and 85 mA and for the octupolar modes above 120 mA. The Super-ACO FEL naturally operates in a stable regime below the threshold of quadrupolar modes of synchrotron oscillations. In presence of the quadrupolar modes, the FEL gain follows the temporal evolution of the bunch distribution, through the modulation of electronic density term at 35 μs. With rather low gain and high cavity losses, the FEL rising time is of the order of 100 μs, and it sees a beam distribution average over the coherent synchrotron oscillations. The characteristic times of the FEL at equilibrium (in the ms range) are long compared to the synchrotron period, and the FEL sees an average of the synchrotron motion over several ms.

When the FEL is established, the FEL heating competes with the coherent synchrotron motion (due to the deformation of bunch distribution from a sharp peaked to a flattened profile) and the quadrupolar modes of synchrotron oscillations are damped [5] as currently observed with a spectrum analyser or with a streak camera. Such a phenomenon also occurs with the sextupolar modes of synchrotron oscillations [16].

More complex situations are observed when Super-ACO is operated with two RF cavities at 100 and 500 MHz, in a bunch shortening mode [17]. A higher electronic density interacting with the impedance of the storage ring can lead to complex synchrotron oscillations, as shown in Fig. 1a. The streak camera image sometimes exhibits a so-called "fish bone" instability, instead of the pure quadrupolar or sextupolar modes. The instability gets stronger with higher voltages applied on the 500 MHz cavity. Indeed, the spectrum analyser can show two close lines around the frequency of the dipolar mode, of the order of 28 kHz for 90 kV on the 500 MHz cavity. The electron bunch profile can be extremely distorted, evolving even more rapidly at a ns scale. When the beam excitations are not too important, the average gain is not strongly reduced and the FEL can start. When established, the FEL suppresses both the "fish-bone" instability and the sideband content on the spectrum analyser, as shown in Fig. 1b.

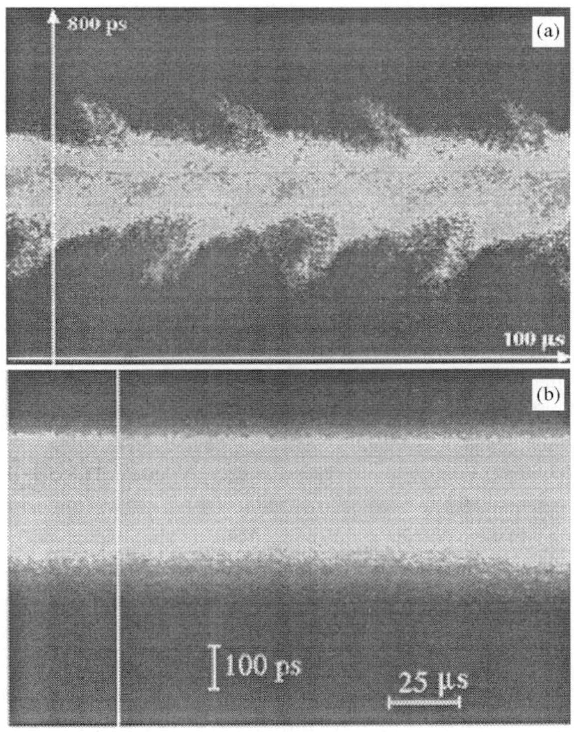

Fig. 1. (a) Complex coherent synchrotron oscillations observed on Super-ACO with two RF cavities with a Hamamatsu double sweep streak camera. (b) Instability damped with the FEL establishment.

3.2. Theoretical analysis

Theoretical analyses have been developed [18], taking into account the FEL interaction and the influence of parasitic fields located in the vacuum chamber, using a pass to pass model based on modified equations of the longitudinal movement of the electrons [8].

$$\tau_{n+1} = \tau_n - \alpha T_o \varepsilon_n, \tag{2}$$

$$\begin{aligned}\varepsilon_{n+1} = {} & \varepsilon_n + a\sin(\omega_{RF}\cdot\tau_{n+1} + \phi_1) \\ & + b\sin(5\omega_{RF}\cdot\tau_{n+1} + \phi_5) - U_{rad} - 2T_o\cdot\varepsilon_n/\tau_S \\ & + \delta u_{random} + \mathrm{Gex}\cdot\rho(\tau) + c\sqrt{I}\sin(\omega_{las}\tau_n + \phi),\end{aligned} \tag{3}$$

where τ_n is the relative position of the electron at pass n with respect to the synchronous electron, ε_n its relative energy, α the momentum compaction factor, T_o the revolution time, ω_{RF} the pulsation of the main RF cavity with associated phase ϕ_1, ϕ_5 the phase of the fifth harmonic cavity, U_{rad} the energy radiated by synchrotron radiation, δU_{random} is associated with the stochastic process of photon emission, Gex represents a coupling force corresponding to a model of resistive impedance, ω_{las} is the FEL optical pulsation and ϕ its relative phase, a, b, c being constants. The modelling of the FEL interaction represents the microbunching induced on the electron bunch, while the laser position is locked at the maximum of the electronic density, whereas the laser light quantum emission is neglected.

The model reproduces easily the Gaussian distributions of the bunch in the simple cases, when the beam is not perturbed by external fields (Gex = 0) without FEL interaction ($c = 0$). It shows as well the bunch length reduction due to the use of additional harmonic RF cavity. Considering the interaction with the wakefield in the vacuum chamber (Gex $\neq 0$), the model simulates the coherent modes of synchrotron oscillations associated with an incoherent bunch lengthening for intermediate values of Gex. When Gex is increased furthermore, the model represents situations similar to the "fish-bone" instability, as shown in Fig. 2, and informs one about the growth of the instability. The analysis of the electrons trajectory allows the shape of the streak camera images to be better understood: the electron in the center of the distribution remains stable whereas the external electrons perform giant synchrotron oscillation, whose amplitude is dependant on the phase relative to the synchronous electron. Greater values of Gex lead to beam blow-up. Then, the evolution of the longitudinal instabilities has been followed when the FEL interaction is set up. The model clearly shows a suppression of the synchrotron oscillations induced by the FEL. Indeed, the disappearance of the "fish bone" instability by the FEL is also represented by the model. Actually, the FEL interaction creating a bunching with a very small spatial period, destroys the coherence between the electrons in the bunch, preventing them from performing their coherent or complex motion. The results concerning the damping of the coherent synchrotron oscillations by the FEL are in agreement with the previous model developed by Renieri [19], relying on the same principles.

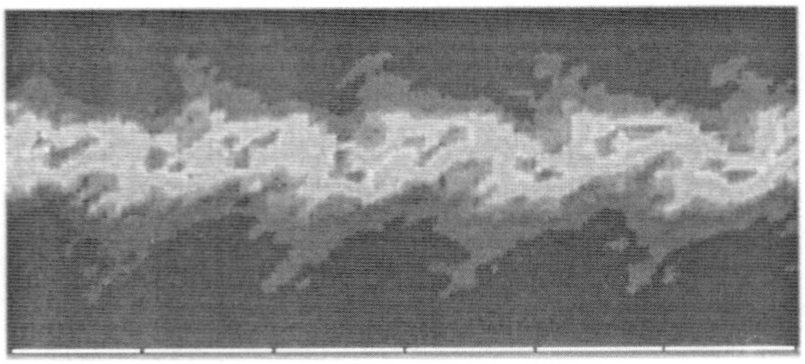

Fig. 2. Simulation of a "fish-bone" instability (V scale = 800 ps, H scale = 300 μs). The synchrotron period of 75 μs (compared to 25 μs in Fig. 1) obtained in the simulation corresponding to a setting of the voltages of the RF cavities different from that employed for the experiment in Fig. 1a.

3.3. Consequences for the FEL stability

Under the presence of quadrupolar modes of coherent synchrotron oscillations, the FEL operates without good stability. Temporal jitter of the FEL micropulse could reach 100–200 ps, so that the FEL was driven in a different detuning zone (see Fig. 3), from "cw" to pulsed. The spectral drift could be as great as 10 Å, compared to 1 Å for normal operation without coherent synchrotron oscillations, and the macrotemporal structure is

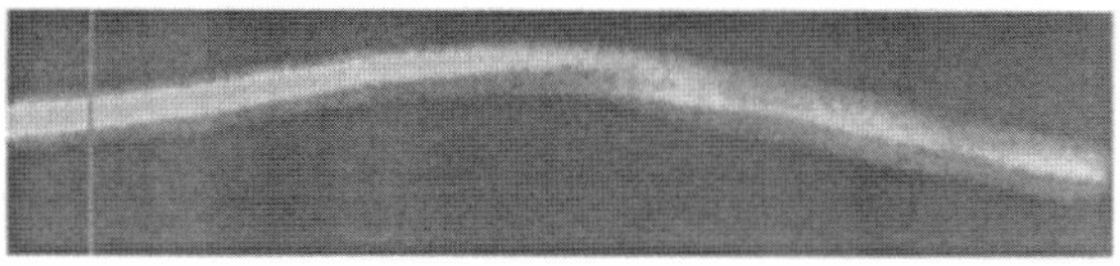

Fig. 3. Temporal drift of the FEL micropulse driving it from the 'cw' operation to a pulsed regime (V scale: 300 ps, H scale: 18 ms).

mainly pulsed at the ms range, with complex modulations [5]. The longitudinal feedback system, developed for maintaining the FEL micropulse position at perfect synchronization by readjustment of the RF revolution frequency [6], cannot be generally switched on, because the instabilities are too strong. However, high-current operation corresponding to the range of current in which coherent modes of coherent synchrotron oscillations are present is desired because of the high-photon flux delivered to the users exploiting synchrotron radiation for pump-probe two-color experiments. For that purpose, a feedback system, damping these modes of synchrotron oscillations, by a modification of the phase of the RF cavity, has been developed [5]. Consequently, the FEL could be operated with very good stability, provided the feedback on the electron beam and the longitudinal feedback on the FEL were present.

Under the presence of the complex coherent synchrotron oscillations observed with the 500 MHz operation, the FEL has been operated for users with good stability, in presence of its longitudinal feedback, for intermediate RF voltages (up to 180 kV recently). When the FEL is established, the RF voltage can be further increased because the FEL has already stabilized the beam. In these conditions, the FEL is very sensitive to any perturbations, which can stop it. In that case, the FEL can be re-established only if the RF voltage is lowered, in order to reduce the complex synchrotron oscillations.

4. Stability in the millisecond range

Once coherent synchrotron oscillations are permanently established, a long time range measurement with the streak camera only provides the bunch distribution average over time. Nevertheless, the "fish-bone instability", apparently erratically occurs when it is observed on few hundred μs scale, can follow a more reproducible behaviour at a longer time scale. Saw-tooth regimes ruling the beam evolution with the two RF cavities on Super-ACO were observed in a ms time scale with the double sweep streak camera, the dissector in the phase oscillation mode [20] or even by looking at the energy evolution with the spectrum of the optical klystron [21] as shown in Fig. 4a. Such a regime could even be seen with the 100 MHz cavity operated alone, but with a very low level of intensity.

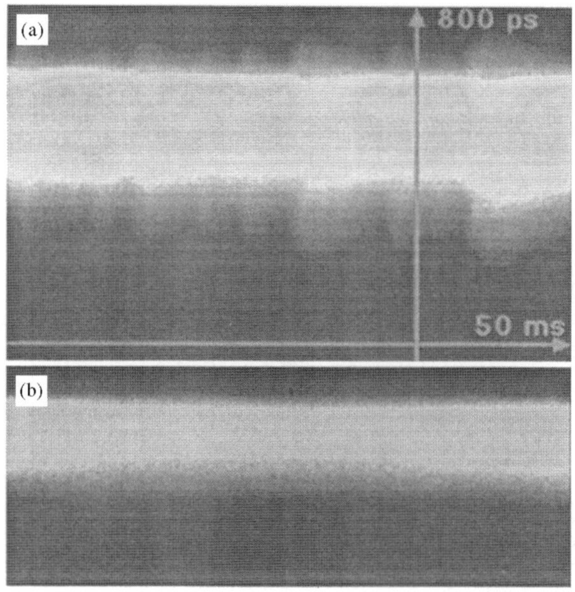

Fig. 4. (a) Sawtooth instability observed on Super-ACO with the 500 MHz cavity (FEL off) with the double sweep streak camera; (b) damped by the FEL establishment (V scale = 1 ns, H scale = 20 ms).

When the FEL is established, these "saw-tooth" instabilities disappear, as can be seen in the streak camera image in Fig. 4b, to be compared to Fig. 4a. The analysis of the bunch distribution in both cases, using a data treatment by the moments method, clearly shows a bunch stabilization and a bunch shortening at a 20 ms scale (see Fig. 5). The observed bunch shortening is in disagreement with the FEL saturation model by increase of energy spread, since the energy spread and the bunch

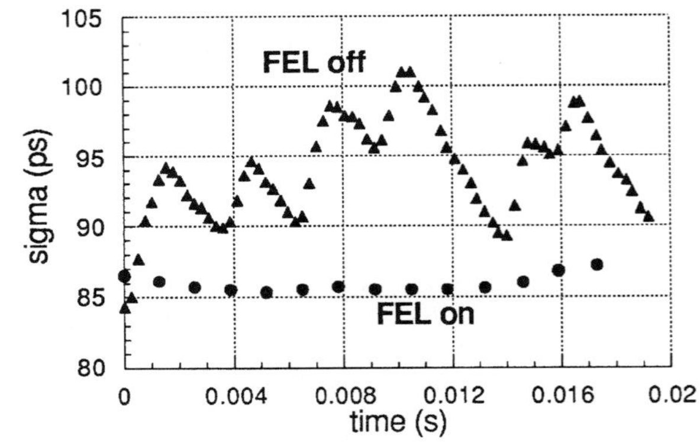

Fig. 5. RMS value of the distribution versus time with and without the FEL.

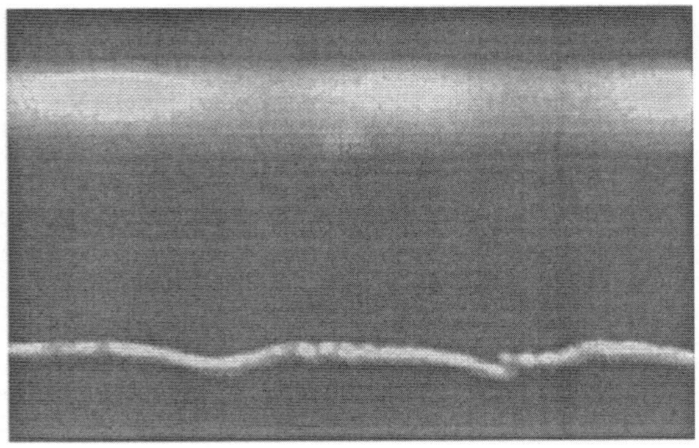

Fig. 6. Transverse instability at 50 Hz, seen on the electron beam (upper image) and on the FEL (lower image). V scale: 1.7 ns, H scale = 50 ms.

length are linked by the following relationship $\sigma_l = (\alpha/\omega_s)\sigma_\gamma$. Once again, as in the case of coherent synchrotron oscillations, the FEL interaction through the microbunching, competes with the growth of the saw-tooth instability, and stabilizes the beam. A further analysis shows that the bunch shortening occurs only on the steep risetime of the longitudinal bunch distribution, whereas the falltime remains unchanged, except from the reduction of the saw-tooth fluctuations.

Indeed, even when the longitudinal instabilities are damped, the FEL stability can also be affected by transverse instabilities. The most current situation observed on Super-ACO is an orbit modulation at 50 Hz, the line frequency. These transverse instabilities on the beam orbit can easily be observed with the double sweep streak camera, when synchrotron radiation is focused onto the iris. A change of the orbit leads to a modification of the intensity seen by the CCD, as illustrated in Fig. 6. The origin of the instabilities could be related to line modulation on a power supply of a magnetic element or to some parasitic field in the RF components. Clear line modulation on the FEL was also seen with vibrations of a primary pump transmitted by the floor to the mirror holder. Sometimes, transverse instabilities around 66–70 Hz can be currently seen on Super-ACO. A transverse feedback on the beam orbit under development was set up for the FEL operation, but the improvement of stability is not really clear since the feedback is limited to a few Hz and leaves the line modulation practically unmodified.

In addition, with the operation of the 500 MHz cavity, strong vertical excitations appear on the beam, that were damped by the FEL operation [17]. These instabilities could be reduced by an important increase of the chromaticity (or hexapolar fields), and even though, under these conditions, the FEL could induce a slight reduction of the beam transverse size. These observations have been first interpreted in terms of head tail instability [22].

Besides, strong modulations close to the laser natural frequency can lead to a modulation at the period of the modulation, at twice the period and even to chaotic behaviour [23]. The FEL is particularly sensitive to a very small perturbation, close to its natural frequency.

Longer time duration can also be considered, with the influence of mechanical vibrations and mirror heating.

5. Consequences for the Super-ACO present FEL performances

From a dynamical point of view, in presence of the FEL longitudinal feedback system, the FEL has been able to operate very close to the Fourier limit, by a factor 1.3 above compared to the previous values of 4–5 both on Super-ACO and UVSOR [24]. The measurements show a pulse duration of 15 ps FWHM (see Fig. 7) and a laser line of 0.3 Å. It

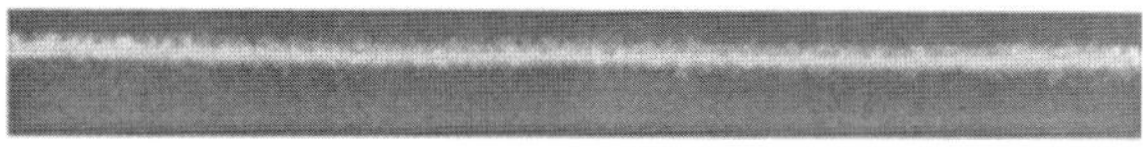

Fig. 7. FEL pulse duration: FWHM = 15 ps, H scale = 500 μs.

should be pointed out that the FEL line width is 1 Å without feedback. Clearly, the feedback system allows the FEL to properly proceed in its temporal and spectral narrowing, avoiding perturbations. With the slightly higher gain achieved with the 500 MHz cavity, the Fourier limit can be obtained in a shorter period of time.

Besides, the gain enhancement associated with an enhancement of the FEL stability led to a slight increase of the FEL output power, and with a set of more transparent mirrors, more than 300 mW could be extracted from one mirror for the users. It represents the highest FEL average power delivered so far from a UV FEL.

6. Conclusion

Storage ring free electron lasers represent a very complex system, involving both the FEL interaction and the electron beam behaviour accumulated over many phases. New phenomena occur for high-current storage, or for operation with two RF cavities. The understanding and the control of the FEL performances, considering the various processes, is a main issue for the FEL sources devoted to users. From such analysis, tolerances for the electron beam can be derived in order to provide a very stable FEL.

References

[1] M.E. Couprie, P. Tauc, F. Merola, A. Delboulbé, D. Garzella, T. Hara, M. Billardon, Rev. Scient. Inst. 65 (5) (1994) 1485.

[2] M. Marsi, M.E. Couprie, L. Nahon, D. Garzella, A. Delboulbé, T. Hara, R. Bakker, G. Indlekofer, M. Billardon, A. Taleb-Ibrahimi, Appl. Phys. Lett. 70 (7) (1997) 895.

[3] T. Hara, M.E. Couprie, A. Delboulbé, P. Troussel, D. Gontier, M. Billardon, Nucl. Instr. and Meth. A 341 (1994) 21.

[4] T. Hara, M.E. Couprie, A. Delboulbé, D. Garzella, L. Nahon, M. Billardon, Nucl. Instr. and Meth. A. 358 (1995) 341.

[5] R. Roux, M.E. Couprie, R.J. Bakker, D. Garzella, D. Nutarelli, L. Nahon, M. Billardon, Phys. Rev. E 58 (1998) 6584.

[6] M.E. Couprie, D. Garzella, T. Hara, J.H. Codarbox, M. Billardon, Nucl. Instr. and Meth. A 358 (1995) 374.

[7] M.E. Couprie, D. Garzella, A. Delboulbé, M. Velghe, M. Billardon, Nucl. Instr. and Meth A 331 (1993) 84.

[8] M. Sands, The Physics of Electron Storage Rings: an introduction, SLAC Rep. 121, 1970.

[9] J.L. Laclare, Bunch beam coherent instabilities, 264-326 CAS Cern Acc. School, CERN 87-03/21 Ap 87.

[10] G. Dattoli, A. Renieri, Nuovo Cimento, B 59, pp 1–39, 1980.

[11] H. Hama, J. Yamazaki, T. Kinoshita, K. Kimura, G. Isoyama, Nucl. Instr. and Meth. A 358 (1995) 365.

[12] P. Elleaume, J. Phys. 45 (1984) 997.

[13] M. Billardon, D. Garzella, M.E. Couprie, Phys. Rev. Lett. 69 (16) 19 (1992) 2368.

[14] T. Hara, Etudes du laser à électrons libres de courtes longueurs d'onde et de ses applications, Ph.D. Thesis, Université de Paris-Sud, July 12, 1995.

[15] F. Pedersen, F. Sacherer, IEEE, Trans. Nucl. Sci. NS-24 (3) (1977) 1396.

[16] M.E. Couprie, M. Velghe, D. Jaroszynski, M. Billardon, Nucl. Instr. and Meth. A 304 (1991) 58.

[17] M. Billardon, M.E. Couprie, L. Nahon, D. Nutarelli, B. Visentin, A. Delboulbé, G. Flynn, R. Roux, Proceedings of the sixth European Particule Accelerator Conference, Stockholm, 22–26 June, Institute of Physics Publishing Bristol and Philadelphia, 1998.

[18] R. Roux, M. Billardon, Modelization of the Super-ACO longitudinal instabilities with the 500 MHz cavity, Workshop on Non linear Problems in charged beam transport in linear and recirculated accelerators, analysis of transverse and longitudinal instabilities, Frascati, 13-15/05/98, Nuovo Cimento A 112 (1999).

[19] G. Dattoli, L. Giannessi, A. Renieri, Nucl. Instr. and Meth. A 358 (1995) 338.

[20] M.E. Couprie, T. Hara, D. Gontier, P. Troussel, D. Garzella, A. Delboulbé, M. Billardon, Phys. Rev. E 53 (2) (1996) 1871.

[21] M.E. Couprie, M. Billardon, IEEE, J. Quantum Electron. 30 (3) (1994) 781.

[22] G. Dattoli, M. Migliorati, L. Mezi, A. Renieri, M. Billardon, M.E. Couprie, D. Naturelli, R. Roux, M. Billardon, Phys. Rev. E 58 (1998) 6584.

[23] M. Billardon et al., Phys. Rev. Lett. 65 (1990) 713.

[24] K. Kimura et al., Nucl. Instr. and Meth. A 375 (1996) 62.

ELSEVIER Nuclear Instruments and Methods in Physics Research A 429 (1999) 172–178

NUCLEAR INSTRUMENTS & METHODS IN PHYSICS RESEARCH Section A

www.elsevier.nl/locate/nima

Longitudinal beam dynamics and FEL interaction on a negative momentum compaction storage ring

H. Hama*, M. Hosaka

UVSOR Facility, Institute for Molecular Science, Myodaiji, Okazaki 444 Japan

Abstract

A simulation study for longitudinal phase space of the electron bunch including effects of potential-well distortion and FEL interaction has been performed for both positive and negative momentum compaction factors on a storage ring. The trend of the bunch lengthening is well interpreted by the wake field due to inductive impedance. In spite of large energy spread in the case of the negative momentum compaction, FEL oscillation may be possible. The peak power of Q-switched lasing is higher than that in positive momentum compaction. There was, however, no significant difference in FEL power and particle distribution in the longitudinal phase space at the equilibrated state.

Keywords: Storage ring; Potential-well distortion; Inductive impedance; Momentum compaction factor

1. Introduction

Single-bunch instability on storage rings is normally dominated by the wake field induced by interaction between the electron bunch and the impedance of the environment such as the vacuum chamber. Particularly, the broad band-inductive impedance is a major source of bunch lengthening and growth in energy spread. The bunch is lengthened as the inductive wake field increases on conventional storage rings whose momentum compaction factor (α) is positive. In this case the bunch shape becomes much more parabolic and then the peak current saturates at even higher beam current, which results from the so-called *potential-well distortion* [1]. On the other hand, in the case when the ring is operated with a negative momentum compaction factor, potential-well distortion due to the inductive impedance acts to shorten the bunch length. Below the threshold for negative-mass instability (*longitudinal microwave instability*) [2], the bunch length is getting shorter as the beam current increases. As one easily expects from the Boussard phase-stability criterion [3], the threshold current for longitudinal microwave instability on the negative α ring is lower than that on a positive α ring because of the higher peak current due to the bunch-shortening effect. However, the bunch lengthening (shortening) is dominated by potential-well distortion on storage rings except for very old rings whose ring-impedances are pretty high, and one normally has not observed clear evidence of the longitudinal mode coupling effect [4].

*Corresponding author. Tel.: + 81-564-55-7206; fax: + 81-564-54-7079.

E-mail address: hama@ims.ac.jp (H. Hama)

0168-9002/99/$-see front matter
PII: S0168-9002(99)00099-6

The concept of the longitudinal microwave instability being suppressed by the faster growth rate of the energy spread due to the FEL interaction has been already introduced by Dattoli et al. [5]. In this article, we have, however, employed more practical effects of the ring impedance to interpret experimental data of the bunch lengthening (shortening) and the increase of the energy spread by potential-well distortion, and numerical simulations of the longitudinal phase space of the electron beam and the FEL power evolution have been performed for both the positive and the negative momentum compaction rings.

2. Potential-well distortion

2.1. Haissinski equation

The single-particle Hamiltonian in the longitudinal phase space at a normalized energy deviation $\varepsilon\,(\equiv \delta e/\sigma_{\varepsilon 0})$ and a normalized time deviation $\tau\,(\equiv \delta t/\sigma_{\tau 0})$ is in general written as

$$H = -\frac{1}{2}\varepsilon^2 - \frac{1}{2}\tau^2 - \frac{e}{\alpha\sigma_{\varepsilon 0}^2 E_0 T_0}\int_0^\tau V_w(\bar{\tau})\,\mathrm{d}\bar{\tau} \tag{1}$$

where $\sigma_{\varepsilon 0}$ and $\sigma_{\tau 0}$ denote the natural energy spread and the natural bunch length, respectively. The last term in Eq. (1) represents the wake potential created ahead by particles in the same bunch, where E_0 and T_0 are the central energy and the revolution time, respectively. A static solution of the Vlasov equation for a longitudinal distribution function $\rho(\tau)$ results from the Hamiltonian of Eq. (1) and has been derived to be

$$\rho(\tau) = K\exp\left(-\frac{1}{2}\tau^2 - \frac{e}{\alpha\sigma_{\varepsilon 0}^2 E_0 T_0}\int_0^\tau V_w(\bar{\tau})\,\mathrm{d}\bar{\tau}\right) \tag{2}$$

where a normalization factor K is determined such that the complete integral of $\rho(\tau)$ is equal to unity. The wake field is normally interpreted by assuming a broad-band (low quality factor) resonator with a very high resonant frequency. Here we introduce a more practical and simple model to create the wake field. The wake field in the broad-band (BB) resonator is assumed to be divided into inductive and resistive parts, that is

$$V_{\mathrm{w}} = -L\frac{\mathrm{d}I}{\mathrm{d}t} - RI \tag{3}$$

where L and R are the inductance and the resistance, and I denotes the current in the bunch. Inserting Eq. (3) into Eq. (2) and after differentiation with respect to τ, the Haissinski equation [6]

$$\frac{\mathrm{d}f}{\mathrm{d}t} = -\frac{tf \pm \xi f^2}{1 \pm f}\begin{pmatrix} + & \text{for} & \alpha > 0 & \text{and } t \equiv \tau \\ - & & \alpha < 0 & \end{pmatrix} \tag{4}$$

is obtained, where $f \equiv aLN_p\rho$, and $\xi \equiv R\sigma_{\tau 0}/L$ is a normalized R–L ratio. Defining a normalized total charge

$$\Gamma \equiv \int_{-\infty}^{+\infty} f\,\mathrm{d}t = \frac{aLN_p}{\sigma_{\tau 0}} = \frac{\tilde{I}L}{|\alpha|(E_0/e)\sigma_{\varepsilon 0}^2\sigma_{\tau 0}} \tag{5}$$

where $\tilde{I}$ denotes an average beam current, Eq. (4) can be numerically solved as a function of Γ. Examples of calculated bunch shapes for $\xi = 0$ (pure inductive), 0.2 and 0.5 are shown in Fig. 1. It can be noted that the bunch lengthening or shortening due to potential-well distortion is essentially governed by the inductive impedance in the case of the low ξ, and the resistive impedance causes an asymmetric shape, which is generally detected as a loss factor. As far as our recent measurements of the loss factor and the bunch lengthening on the UVSOR storage ring at the nominal momentum compaction factor ($\alpha \approx 0.035$), ξ seemed to be less than 0.2. Accordingly we have mainly employed the pure inductive impedance in further calculation to save cpu time.

2.2. Simulation of bunch lengthening

We should note that the validity of the solution of the Haissinski equation is in a frame of the static Vlasov equation. In other words, a constant rms energy spread $\sigma_{\varepsilon 0}$ supports intentional solutions. Consequently, we have performed a simulation study to examine whether the natural energy spread would fluctuate. Details of the simulation will be reported elsewhere, and basic formulae are presented.

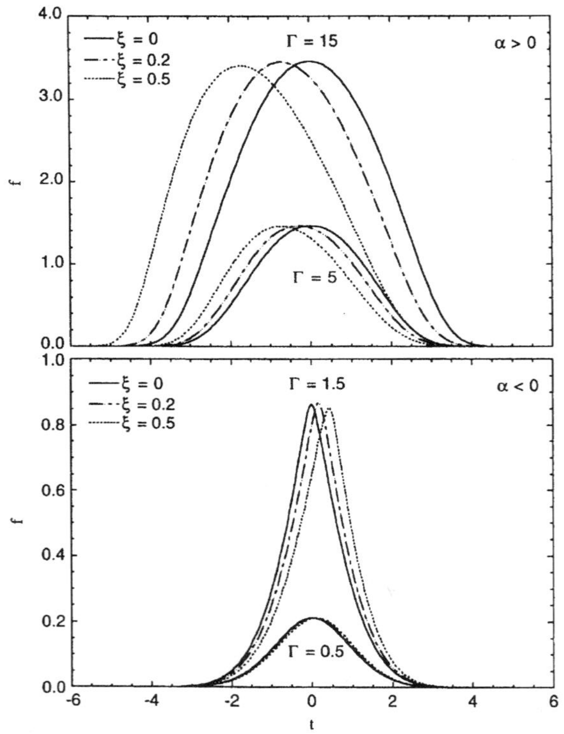

Fig. 1. Bunch shapes evaluated from Hassinski equation for the positive α (upper) and the negative α (lower).

An ith electron which is characterized by the energy deviation and the time deviation from the synchronous phase as p_i (δt_i, δe_i) changes its position and energy turn by turn. From the definition of the first-order momentum compaction factor, the arrival time of the electron changes after one turn as

$$\Delta t_i = \alpha \delta e_i (T_0/E_0). \tag{6}$$

Change of the energy in one turn is expressed by

$$\Delta e_i = \delta e_i e^{-U_0/E_0} - U_0 - g(\delta U_0) - V_{\mathrm{wake}}(\delta t_i) + V_{\mathrm{RF}} \cos(\phi_0 + \omega_{\mathrm{RF}} \delta t_i) \tag{7}$$

where the first term expresses the synchrotron damping and the last term is the acceleration in an RF cavity. Energy loss due to synchrotron radiation U_0 should fluctuate due to the quantum effect, so that the Gaussian random number g is added to produce a finite energy spread and the bunch length, which should correspond to the natural ones in a case of no wake field. Since the wake field results from the collective effect of the bunch, the particles are put into bins in the simulation to create the wake field (*particle-in-cell* method). Results of the simulation for the pure inductive impedance are shown in Fig. 2. A very good agreement with the solution of the Haissinski equation for the bunch lengthening can be seen for the positive α. In case of the negative α, the bunch shortening occurs in a low-charge region, but then the bunch length increases in a region above the upper bound of the Haissinski equation, which is made up by an enhancement of the energy spread.

Experimental data are also shown in Fig. 2 [7], where an inductance of $L = 130$ nH and calculated $\sigma_{\tau 0}$ and $\sigma_{\varepsilon 0}$ are used to normalize to the charge Γ of Eq. (5). One can see not only an agreement of qualitative behavior of the bunch lengthening but also that the absolute values agree well with experimental data. The growth of the energy spread in the negative α operation is not reproduced well by the simulation, and an inductance of $L = 70$ nH is suitable to fit the experimental data. The model of the wake field expressed in Eq. (3) is generally able to be applied for the low-frequency regime of the BB resonator, and frequency dependence of the impedance has been neglected in the simulation. In the high-frequency regime, the resistive part steeply increases; meanwhile the inductive part gradually decreases. Because the bunch is shortened due to the inductive impedance in the negative α operation, which means the high-frequency component gets larger in the spectrum, the simulation employing the constant inductive impedance would not reproduce all of the experimental data. Moreover, the experimentally deduced energy spread from modulation factors of the spontaneous radiation must be affected by the transverse beam emittance [8], so that there have been some ambiguities in the data. However, it may be concluded that the overall trend of the bunch lengthening for both the positive and the negative α operation has been well interpreted by potential-well distortion due to the inductive impedance based on the BB resonator model.

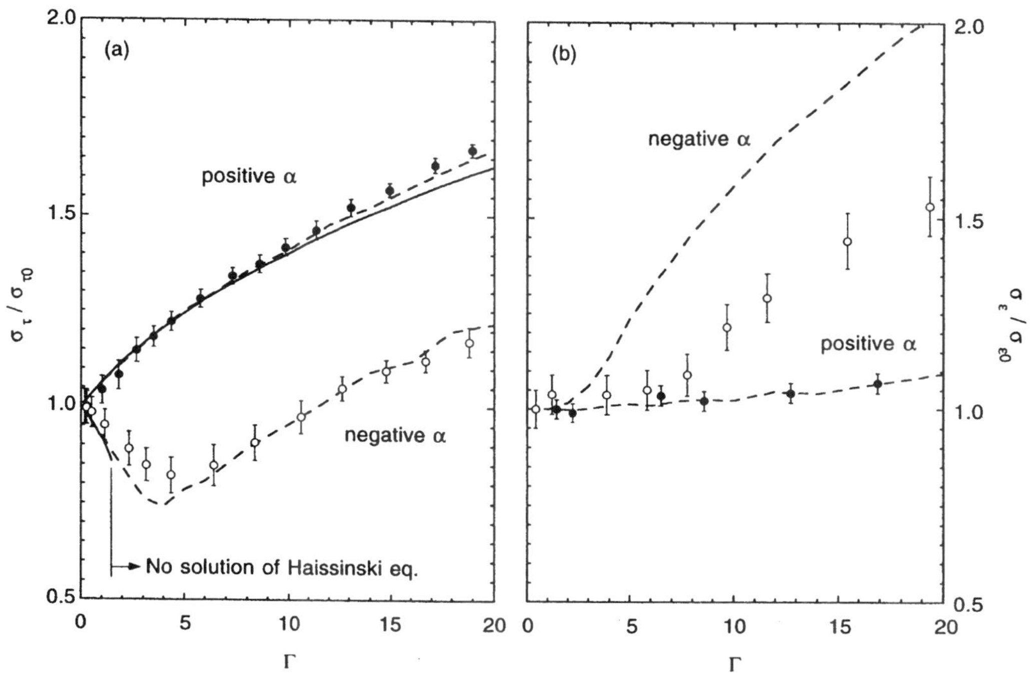

Fig. 2. Normalized charge dependences of the bunch length (a) and the energy spread (b). The solid lines in (a) indicate the numerical solution of the Haissinski equation with the pure inductive impedance, and the broken lines in both figures are results of the simulation of the bunch lengthening. Measured bunch length and energy spread for the positive α and the negative α operations are shown by closed and open circles, respectively, where the inductive impedance of 130 nH is used to normalized the average beam current to the charge Γ.

3. Simulation of the FEL oscillation

3.1. Q-switched lasing

Q-switched lasing can provide much higher peak power than that of cw-lasing because the laser develops from a completely damped state of the electron bunch [9]. To discuss evolution of the longitudinal phase space, Q-switched lasing is quite complicated because (1) the FEL power density may reach a strong saturation regime, and then (2) the final momentum distribution of the electrons is quite non-linear versus the FEL power, particularly in the case when an optical klystron is used. Furthermore (3) it seems to be affected by the synchrotron oscillation even during a short period of the lasing. Although these issues have been investigated by simulations and analytical methods [10,11], here we focused on differences in the Q-switched lasing coming from the sign of the momentum compaction factor and potential-well distortion.

The FEL interaction has been added to Eqs. (6) and (7) for the simulation, where the equation of motion of the electrons including the ponderomotive force is numerically evaluated for the individual particle during propagating in an optical klystron. To simplify the model, the longitudinal FEL distribution is also divided into the identical cells of particles. Parameters of the storage ring and the FEL devices such as the optical klystron have been chosen to be close to those of the UVSOR-FEL, which are shown in Table 1. A total charge of $\Gamma = 10$ is chosen as a typical beam current (~ 25 mA on the UVSOR ring), and the result of the bunch length simulation is used as an initial phase-space distribution of the electrons. The electrons always get into random phases of the optical wave because of the large magnitude of α, but the perfect synchronism between the optical bunch and

Table 1
Employed parameters for the simulation

Storage ring parameters		Optical parameters	
Beam energy	$E_0 = 600$ MeV	Polarity	Complete helical
RF frequency	$f_{RF} = 90.108$ MHz	Period length	$\lambda_u = 0.11$ m
RF voltage	$V_{RF} = 46$ kV	Period number	$N = 9 \times 2$
Natural energy spread	$\sigma_{\varepsilon 0} = 203$ keV	Dispersive section	$N_d = 100$
Natural bunch length	$\sigma_{\tau 0} = 132$ ps	FEL wavelength	$\lambda_{FEL} = 270$ nm
Normalized charge	$\Gamma = 10$	Peak gain	$g_0 = 1.6\%$ at $\hat{I} = 10$ A
Inductive impedance	$L = 130$ nH	Cavity loss	0.5%

the electron bunch is kept at the entrance of the optical klystron at each turn.

One of representative results is summarized in Fig. 3, where a cavity loss of 0.5% is employed which is a typical realistic value. In the case of the negative α, the maximum Q-switched laser power is much higher than that of the positive α, which is considered to result from the high peak current of the shorter bunch. However, a phase-space oscillation with a large amplitude begins from the early stage of lasing, and meanwhile the evolution of the phase space is more moderate in the positive α case. Those details can be seen in the two-dimensional particle distributions shown in Fig. 4. Although the initial particle distribution is completely collapsed at the start-up of lasing, the electron population is still centered in the bunch and the FEL can grow. After the FEL interaction, the phase space becomes far from the equilibrium state of the inductive wake field; then the particle distribution goes into a violent oscillation. The evolution of the particle distribution of the positive α case is in good agreement with our previous observation by using a dual-sweep streak camera [12].

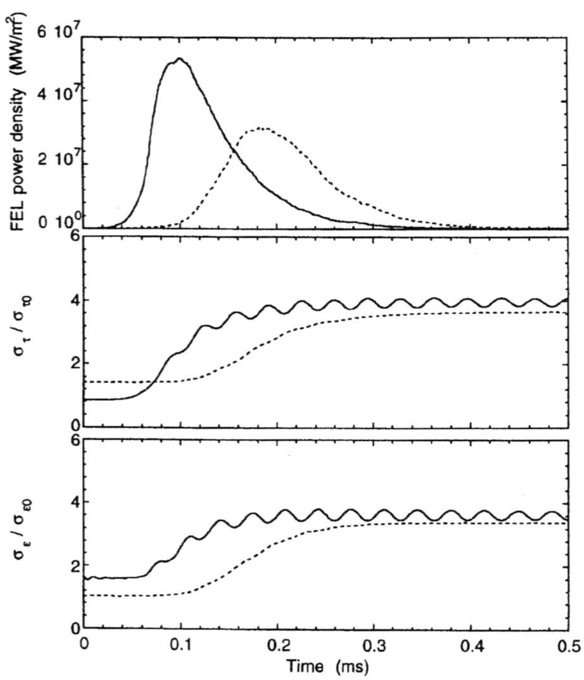

Fig. 3. Results of simulation of the FEL oscillation for the first macropulse. The solid line and the dotted line indicate the negative α and the positive α operations, respectively.

3.2. Equilibrium state

Because of a very slow synchrotron damping time, the simulation is unlikely to reach an equilibrium state of FEL power and phase space. To reach an equilibrium state in a reasonable cpu time, the damping time was reduced in the following calculations. However, in this case the natural bunch length is also reduced because it comes from an equilibration between the quantum excitation and the synchrotron damping. A change of the rms energy spread in one turn can be expressed as

$$\Delta\sigma_\varepsilon = [\sqrt{\sigma_\varepsilon^2 + \delta U_0^2} - \sigma_\varepsilon] - \frac{T_0}{\tau_\varepsilon}\sigma_\varepsilon \tag{8}$$

where δU_0 is the rms energy width of synchrotron radiation used for the Gaussian random number in Eq. (7) and τ_ε is the damping time. The equilibrium energy spread ($\Delta\sigma_\varepsilon = 0$) is $\sigma_\varepsilon = \delta U_0\sqrt{\tau_\varepsilon/(2T_0)}$. If

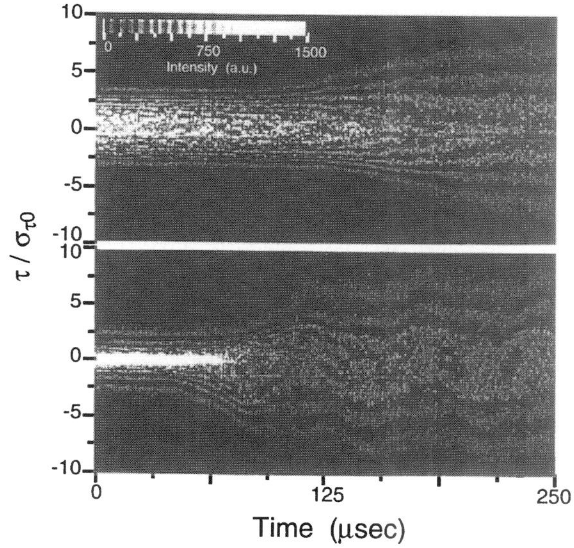

Fig. 4. Simulated electron distributions during a lasing of the first macropulse for the positive α (upper) and the negative α (lower) operations.

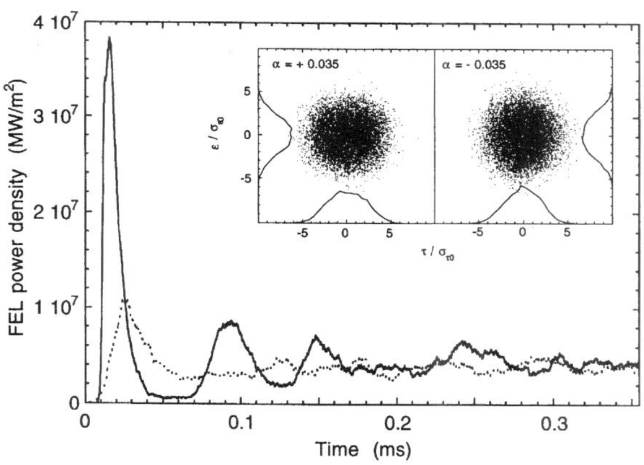

Fig. 5. Simulated FEL power evolutions for the negative α and the positive α operations indicated by solid and dotted lines, respectively, where the synchrotron damping time is reduced by a factor of 100. The insert shows the final longitudinal phase spaces of the electrons.

the damping time is shortened to be τ_ε/χ, the width of the Gaussian random number must be increased to $\sqrt{\chi}\delta U_0$ to create the same natural bunch length and the energy spread. A result of simulated FEL power evolution employing a fast damping time ($\chi = 100$) is shown in Fig. 5, where the cavity loss is rather large (1%). Since the saturated FEL power is evaluated to be inversely proportional to the damping time from the first-order approximation [13], the simulation is obviously overestimated, but they may be relatively comparable with each other. The first macropulse in the negative α is still much larger than that in the positive α and following macropulses quickly come back. However, one cannot see any significant difference in the saturation levels of the FEL power. At a glance, there is also no apparent difference in the Poincare plots of phase spaces shown in the insert. It can be noted that a beam heating effect due to the FEL interaction is much stronger than the effect of the wake field and dominates the longitudinal dynamics.

4. Concluding remarks

Above the upper bound of the Haissinski equation for the negative α operation, the simulation shows the bunch is lengthened by increasing the energy spread, which is in good agreement with the experimental data. For the positive α operation, the trend of the bunch lengthening mostly corresponded to the solution of the Haissinski equation and to the simulation as well. Since the Boussard criterion for the microwave instability took only the natural energy spread and the bunch length into account, the bunch lengthening effect due to the inductive wake field should be included; consequently the threshold current for microwave instability may be rather high.

Even though the energy spread is increased at the negative α, the FEL oscillation is possible and there particularly seems an advantage for the Q-switched macropulse because the electron population is centered in the bunch. Because of no transverse effect included in the simulation, the result may not always represent the actual phenomena. However, the longitudinal beam dynamics normally plays an important role in the FEL oscillation, so that these results will be valuable for further experimental study.

References

[1] B. Zotter, Proceedings of the CERN Accelerator School, Paris 1984, CERN Rept. 85-19, CERN, Geneva, 1985.

[2] E. Keil, W. Schnell, CERN Rept. ISR-TH-RF/69-48, CERN, Geneva, 1969.

[3] D. Boussard, Int. CERN Rept. Lab II/RF/Int 75-2, CERN, Geneva, 1975.

[4] A.W. Chao, AIP Proceedings No. 105, American Institute of Physics, New York, 1983.

[5] G. Dattoli, L. Mezi, A. Renieri, G.K. Voykov, Nucl. Instr. and Meth. A 393 (1997) 70.

[6] J. Haissinski, Nouvo Cimento 18B (1) (1973) 72.

[7] M. Hosoka, H. Hama, J. Yamazaki, T. Kinoshita, Nucl. Instr. and Meth. A 407 (1998) 234.

[8] P. Elleaume, J. Phys. C 44 (1983) C1-333.

[9] P. Elleaume, Nucl. Instr. and Meth. A 304 (1985) 28.

[10] V.N. Litvinenko, B. Burnham, J.M.J. Madey, Y. Wu, Nucl. Instr. and Meth. A 358 (1995) 334.

[11] G. Dattoli, L. Giannessi, A. Renieri, Nucl. Instr. and Meth. A 358 (1995) 338.

[12] H. Hama, J. Yamazaki, T. Kinoshita, K. Kimura, G. Isoyama, Nucl. Instr. and Meth. A 358 (1995) 365.

[13] H. Hama, K. Kimura, J. Yamazaki, S. Takano, T. Kinoshita, M.-E. Couprie, Nucl. Instr. and Meth. A 375 (1996) 32.

ELSEVIER

Nuclear Instruments and Methods in Physics Research A 429 (1999) 179–184

NUCLEAR INSTRUMENTS & METHODS IN PHYSICS RESEARCH Section A

www.elsevier.nl/locate/nima

European project to develop a UV/VUV free-electron laser facility on the ELETTRA storage ring☆

R.P. Walker[a,*], B. Diviacco[a], C. Fava[a], A. Gambitta[a], M. Marsi[a], F. Mazzolini[a], M.E. Couprie[b], L. Nahon[b], D. Nutarelli[b], E. Renault[b], R. Roux[b], M.W. Poole[c], N. Bliss[c], A. Chesworth[c], J.A. Clarke[c], D. Nölle[d], H. Quick[d], G. Dattoli[e], L. Giannessi[e], L. Mezi[e], P.L. Ottaviani[e], A. Torre[e], M. Eriksson[f], S. Werin[f]

[a]*Sincrotrone Trieste, Padriciano 99, 34012 Trieste, Italy*
[b]*SPAM-CEA and LURE, Orsay, France*
[c]*CLRC-Daresbury Laboratory, Warrington, UK*
[d]*University of Dortmund, Dortmund, Germany*
[e]*ENEA-Frascati, Frascati, Italy*
[f]*MAX-lab, Lund, Sweden*

Abstract

The main features and novel technical aspects of a new European project to integrate a free-electron laser on an existing "third generation" synchrotron radiation user facility are described, including the design of the optical cavity and undulator, the electron beam characteristics and a first assessment of the predicted laser performance. © 1999 Elsevier Science B.V. All rights reserved.

Keywords: Synchrotron radiation; ELETTRA; Storage ring; Free-electron laser facility

1. Introduction

Sincrotrone Trieste, in collaboration with SPAM/LURE, CLRC-Daresbury Laboratory, the University of Dortmund (DELTA Project), ENEA-Frascati and Max-lab, and with partial EU funding, has recently begun a 3 year project to develop a UV/VUV free-electron laser on ELETTRA, Europe's first "third generation" high-brightness synchrotron radiation source for the VUV/soft X-ray region.

The goals of the project are orientated towards its future use as a *user facility*, rather than in FEL development per se, and in general towards the advancement of SRFELs as part of future European synchrotron radiation facilities (e.g. the proposed SOLEIL (France) and DIAMOND (UK) projects). The aims of the project are therefore to demonstrate:

- Lasing over a broad spectral range from 350 nm to below 200 nm; although setting lasing wavelength records is not the primary goal of the project, extending operation below 200 nm is

☆Partly funded under EC Contract ERBFMGECT980102.

*Corresponding author. Tel.: + 39-40-3758225; fax: + 39-40-226338.

E-mail address: r.walker@elettra.trieste.it (R.P. Walker)

PII: S0168-9002(99)00100-X

nevertheless of very great interest in order to extend the range of scientific possibilities for the utilization of the radiation.

- Operation of the laser while continuing to provide stable synchrotron radiation (SR) beams for other users; even though the operational energy and current will be lower than that in routine operation, the FEL mode is still of potential interest for a significant number of beamlines that do not require the highest photon energies and which in particular want to perform experiments that make use of the time structure of the radiation.
- Use of the laser to perform pilot experiments both by itself and also in synchronism with SR from another beamline.

We hope to meet these objectives, not only by means of an active collaboration between all of the European laboratories presently interested in SRFEL development, but through a combination of the following technical features:

- Use of a high-quality electron beam from a "3rd generation" synchrotron radiation source providing a low emittance and short bunch length and therefore a higher gain than in presently operating experiments.
- Operation at higher energy than existing experiments, providing the potential of higher power output and better compatibility with regular SR operation.
- Use of a helical undulator/optical klystron to reduce the power density of the spontaneous radiation and hence the degradation of the mirrors, as successfully demonstrated at UVSOR [1], as well as for increased gain compared to a linearly polarized device.
- Use of a sophisticated mirror chamber that will contain three remotely interchangeable mirrors to make the best use of the limited commissioning time, and also to allow a greater wavelength range to be covered for future experiments.

The combination of higher gain and lower spontaneous radiation power should alleviate the main difficulty that has been associated with UV storage ring FELs, namely mirror deterioration, and therefore may allow more flexibility in the type and quality of mirrors that can be used, permitting for example a trade-off between reflectivity and wavelength range. However, mirror quality will still remain a very important factor, particularly for extending to the shortest wavelengths and also for higher power operation [2]. A programme of mirror development will therefore be pursued, and partly in connection with this project the mirror characterization system developed by the Super-ACO group for the UV [2] will be extended to shorter wavelengths.

2. Choice of main parameters

2.1. Energy

The advantage of FEL operation at lower energy is that it provides higher gain and reduced spontaneous power emission, whereas higher energy offers higher laser power, generally better beam stability and lifetime and better compatibility with normal SR utilization. As a reasonable compromise, and also for convenience, initial operation will be at 1 GeV, since this is the normal injection energy of ELETTRA. Nevertheless, since ELETTRA normally operates at 2 GeV for SR users, higher energy operation would clearly be beneficial. This is also of great importance for the integration of FELs on future SR facilities, and so we also plan to explore this aspect at a later phase of the present project.

2.2. Undulator/optical klystron parameters

The choice of undulator period generally involves a compromise between higher gain (favouring a shorter period length with increased number of periods) and lower spontaneous power emission (favouring a longer period and therefore reduced field strength at the same operating wavelength). In addition, a longer period allows a device with a higher K value to be built that allows more scope for a later increase in operating energy. In the present case the spontaneous radiation power is of much less concern because of the choice of a helical undulator in which the power density is minimum on-axis. In fact, the integrated power over practical mirror sizes depends only weakly on undulator period, and marginally favours *shorter* period lengths since the on-axis power density decreases with increasing K value.

Another important factor that has to be taken into account is the fact that the undulator will also be used as a source of radiation with both linear and circular polarization for a photoemission electron microscopy (PEEM) beamline, designed to operate in the region 30 eV–1 keV at 2 GeV. Optimization of the undulator for this purpose also involves a compromise between obtaining the highest flux at high energy (favouring a shorter period) and reducing the radiation power density to a level that does not deteriorate the beamline optical elements (favouring longer periods).

After examining both aspects in detail it was concluded that a period of 100 mm is the best choice for both applications. The corresponding minimum K value for operation at 350 nm is then 5.08 (circular mode, 1 GeV).

Having thus chosen the period length, in order to provide more flexibility to increase the operating energy at a later stage of the project, it is clearly beneficial to increase the field strength to the maximum reasonably achievable value, taking economic factors also into account. The K value for the proposed design (see below) is in the range 5.9–6.3, depending on the actual permanent magnet performance and final operating gap that will be achieved. In this way operation at 350 nm should be possible up to about 1.2 GeV, and 250 nm up to 1.4 GeV.

2.3. Optical cavity parameters

The practical lengths of cavities that allow synchronism with possible bunch patterns in ELETTRA (harmonic number = 432) are 21.6, 32.4 and 43.2 m, corresponding to 6, 4 and 3 bunch operations, respectively. The length chosen is 32.4 m which is the shortest that allows both cavity mirror chambers to be located outside the ring shielding wall, a necessary condition because of the limited periods of accessibility to the ring tunnel. Fig. 1 shows the planned layout of the optical cavity in the corresponding ELETTRA straight section. For radiation safety reasons the front mirror will be located in a separately shielded area to which access will only be permitted with the beamline shutter closed.

Table 1 lists the main parameters of the optical cavity. The present choice of 4 m for the Rayleigh range is a compromise between increased gain and reduced laser power density on the mirrors (lower ZR) or better stability (larger ZR). Centring the mode at the undulator centre then leads to the mirror radii in the table.

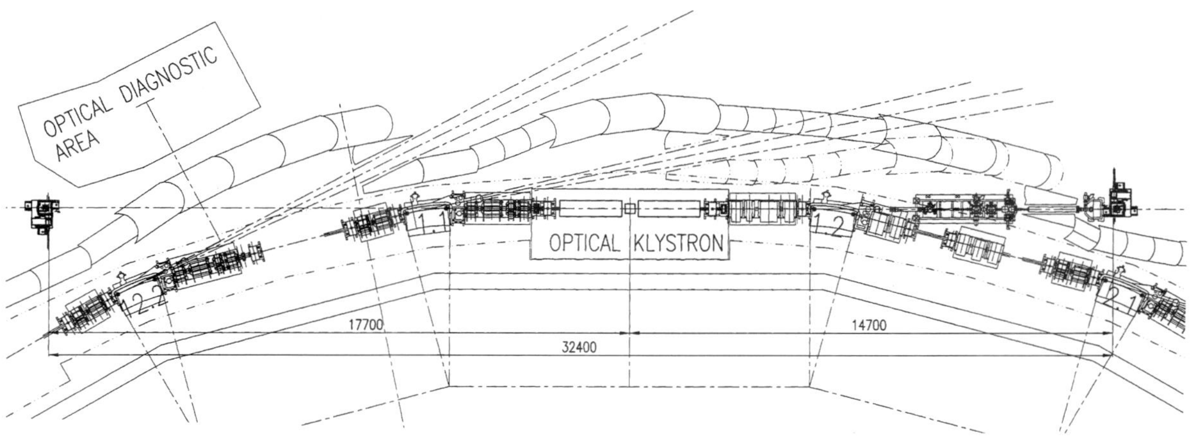

Fig. 1. Layout of the undulator (optical klystron) and optical cavity on ELETTRA.

Table 1
Main parameters of the ELETTRA FEL project

FEL beam		
Wavelength range	350 nm – < 200 nm	
Initial value	350 nm	
Average power	≤ 10 W	
Peak power	≤ 150 kW	
Pulse energy	≤ 1 μJ	
Pulse length (rms)	∼ 1–5 ps	
Line width	$\sim 0.3\text{–}1.5\times10^{-3}$	
Repetition rate	4.6 MHz	
Electron beam		
Energy	1 GeV	
No. bunches	4	
Max. total current	100 mA	
Natural rms emittance	1.7 nm rad	
Natural rms energy spread	0.04%	
Natural rms bunch length	6.3 ps	
Current/bunch	10 mA	20 mA
Bunch length	25 ps	30 ps
Peak current	140 A	230 A
Energy spread	0.16%	0.19%
Optical cavity		
Cavity length	32.4 m	
Distances from mirrors to undulator centre	17.7 m, 14.7 m	
Waist position	Centre of undulator	
Rayleigh length	4 m	
Mirror radii	18.6, 15.8 m	
Stability parameter	0.78 m	
Waist size (w, 350 nm)	0.67 nm	
Undulator		
Type	Permanent magnet, APPLE-2, optical klystron	
Period length	100 mm	
No. of periods	2 × 20	
Total length (including modulator)	4.7 m	
Minimum gap	18 mm	
Permanent magnet block sizes	35 mm × 35 mm × 25 mm	
Maximum K value (circular mode)	5.9–6.3	
Nominal K value (350 nm)	5.08	

3. Mirror chambers

Fig. 2 shows the present design of the mirror chambers which contain three remotely interchangeable mirrors of up to 40 mm diameter, as well as permitting a straight-through direction for use of the spontaneous undulator radiation. Positioning of the mirrors is achieved by eight remotely controllable systems: three-axis positioning external to the vacuum, in-vacuum vertical motion for mirror exchange, in-vacuum coarse (motor) and fine (piezo) pitch and yaw motions. Inserting and removing mirrors will be performed via a separately pumped side chamber, so as not to disturb the vacuum of the main mirror chamber. The units are designed to reach UHV conditions ($\sim 10^{-10}$ mbar) and so all components are bakeable. The chambers will be pumped by a 400 l/s ion pump and two 70 l/s NEG cartridges. The support will be constructed of synthetic granite in order to be rigid, but minimizing thermal expansion.

4. Undulator/optical klystron

A Sasaki APPLE-2 design has been chosen for the undulator construction in order to give the required flexibility for producing synchrotron radiation with both circular (right and left handed) and linear polarization. This is also a convenient choice since a number of other devices of this type will be built for ELETTRA in the next 2 years [3]: two other straight sections will each accommodate two devices of different period length each with a total length of 2.0–2.1 m. The first device (36 periods of 6 cm) is presently under final assembly, and a 7-period prototype has already been successfully tested [3]. The optical klystron configuration will therefore be implemented by means of two separate undulators of standard length with a suitably designed modulator in between.

Such an arrangement offers a further flexibility – the possibility of generating linear polarization by interference of two circularly polarized beams that are polarized in opposite directions [4]. The advantage is that the radiation contains only a single harmonic on-axis and has therefore the same low power density characteristic of a helical device.

Having decided the period length (see above), and given the minimum magnet gap of 18 mm, only the permanent magnet block sizes remains to be decided. Assuming the same block clamping arrangement used on the first device [3], a square

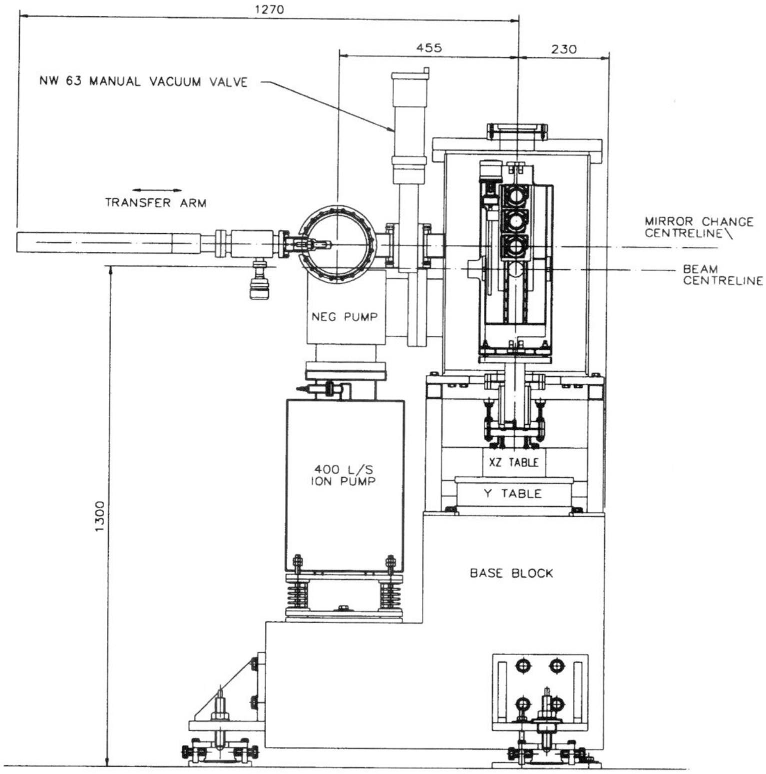

Fig. 2. Optical cavity mirror chamber design.

cross section is appropriate in order to reduce the number of different block types and increase the options available for block sorting. The proposed choice of $35 \times 35\ mm^2$ blocks is a reasonable compromise between performance and increasing cost and magnetic forces. The final parameters are summarized in Table 1.

5. Electron beam performance

Table 1 lists the relevant parameters of ELETTRA in the FEL-mode. Currents in excess of 50 mA have been obtained in the single bunch mode, however recent tests in 4-bunch mode have so far been restricted to 25 mA/bunch in order not to risk excessive parasitic mode loss heating until temperature monitoring of critical components (e.g. bellows and other objects with RF shielding) can be installed. The maximum permitted total current of 100 mA was able to accumulate easily in the 4-bunch mode. The lifetime was of the order of 1.5 h, dominated by the scattering of electrons within the same bunch (Touschek effect). The measured value is consistent with the calculated Touschek lifetime using the measured bunch lengths (see below) and assuming reasonable values for the emittance ($2 \times$ the natural value to account for the expected degree of intra-beam scattering) and coupling ($\sim$ 3%).

Earlier measurements of bunch length in single bunch mode using both a streak camera and fast photodiode gave consistent results, namely 25 ps (rms) at 10 mA/bunch and 30 ps at 20 mA/bunch, i.e. a factor of 4 and 4.8 times larger than the theoretical zero-current value (6.3. ps), respectively. Applying the same factor to the energy spread results in the values given in Table 1.

Beam stability is generally more difficult to achieve at 1 GeV compared to the normal operating energy of 2 GeV, since the thresholds for coupled bunch instabilities are much lower, because of the much longer radiation damping times. With a complete filling of the ring therefore the beam is unstable, however in the four-bunch case the situation is radically different because there are many fewer coupled bunch modes. In preliminary experiments it has been verified that by appropriate tuning of the RF cavity temperatures, which control the higher-order mode frequencies, it is possible to find a situation in which all modes are stable with a maximum residual rms amplitude of the order of 0.5° of RF phase, corresponding to 3 ps, significantly smaller therefore than the measured bunch lengths. The transverse stability also appears to be very good, with transverse betatron sidebands only appearing at the highest currents.

Further studies of bunch length and beam stability will be made over the forthcoming months using a new dual-sweep streak camera due to be commissioned in October. Attempts will also be made to measure the energy spread and also to increase the vertical beam size in order to improve the beam lifetime, by moving closer to the coupling resonance. The development of a digital transverse feedback system for normal multibunch operation is also proceeding in parallel, which could be applied to the FEL mode if required.

6. FEL performance

Using the electron beam, undulator and optical cavity parameters given in Table 1 results in a calculated small signal gain of 40% per pass, in the case of a standard helical undulator at 350 nm with 20 mA/bunch. In the optical klystron case, a maximum of 60% is obtained, with an optimized modulator delay parameter of Nd $\approx$ 20. Because of the high K value the gain does not change significantly between 350 and 200 nm.

Preliminary calculations of the average power have also been carried out. Depending on the model used to take into account the combined effect of FEL interaction and the microwave instability on the longitudinal distribution of the electron beam, the resulting average output power lies in the range 1–10 W.

Concerning the characteristics of the laser pulse, theory predicts at the Fourier limit a laser pulse length given by $\sigma_l = \sqrt{\sigma_z N \lambda}$, where σ_z is the electron bunch length, and corresponding linewidth $\Delta\lambda/\lambda = \sqrt{(\lambda/N\sigma_z)}/\pi$ [5]. In the present case this predicts a laser pulse length of about 1 ps and linewidth of 3×10^{-4}. In practice larger values than those corresponding to the Fourier limit have been observed depending on the stability and detuning [6,7]. Based on this experience we expect the FEL pulse in the present case to be in the range 1–5 ps with a linewidth of 0.03–0.15%. To take a specific example, 5 W average power corresponds to a pulse energy of 1 μJ and with 3 ps (rms) pulse width corresponds to a peak power of 150 kW.

7. Conclusion

An active project to develop the first storage ring FEL on an operating third-generation synchrotron radiation user facility has been described. Preliminary calculations suggest that the combined effects of a higher quality, higher energy electron beam, together with the use of a helical undulator, should allow FEL performance levels that significantly exceed present day values to be achieved.

References

[1] H. Hama et al., Nucl. Instr. and Meth. A 393 (1997) 23.
[2] D. Nutarelli, Proceedings 20th Free Electron Laser Conference, Williamsburg, VA, 1998, Elsevier Science, Amsterdam, 1999, p. II.63.
[3] B. Diviacco et al., Proceedings of the 1998 European Particle Accelerator Conference, p. 2216.
[4] S. Sasaki, Proceedings of the 1997 Particle Accelerator Conference, p. 802.
[5] G. Dattoli, A. Renieri, Nuovo Cimento 59 B (1980) 1.
[6] K. Kimura, Nucl. Instr. and Meth. A 375 (1996) 62.
[7] M.E. Couprie, Nucl. Instr. and Meth. A 429 (1999) 165.

ELSEVIER

Nuclear Instruments and Methods in Physics Research A 429 (1999) 185–190

NUCLEAR INSTRUMENTS & METHODS IN PHYSICS RESEARCH Section A

www.elsevier.nl/locate/nima

Improvement of the beam quality by chromaticity correction for wavelength shortening in the NIJI-IV FEL

N. Sei[a,*], K. Yamada[a], H. Ohgaki[a], V.N. Litvinenko[b], T. Mikado[a], T. Yamazaki[c]

[a] *Electrotechnical Laboratory, 1-1-4 Umezono, Tsukuba, Ibaraki 3058568, Japan*

[b] *FEL Laboratory, Physics Department, Duke University, Durham, NC 27708, USA*

[c] *Institute of Advanced Energy, Kyoto University, Gokasho, Uji, Kyoto 611, Japan*

Abstract

Electron-beam qualities improved by chromaticity correction in the storage ring NIJI-IV were investigated at the beam energy of 309 MeV. Sextupole–quadrupole–sextupole (SQS) magnets, which were installed in all of the short-straight sections in NIJI-IV, perfectly corrected a horizontal and a vertical chromaticity. This improvement suppressed a head–tail instability, so that higher beam current ($\sim$ 30 mA) and higher peak-electron density ($\sim 6\times10^{16}\ \mathrm{m}^{-3}$) were available for FEL experiments. The maximum FEL gain was estimated to be about 2.5% at a wavelength of 240 nm. The lasing of an FEL around 300 nm was achieved in March 1998, and the lasing of an FEL at around 240 nm was successfully observed in May 1998. The shortest wavelength of FELs with the NIJI-IV FEL system was 228 nm.

Keywords: Beam quality; NIJI-IV; Chromaticity correction

1. Introduction

We had already achieved the lasing of FELs between 595 and 349 nm in 1994 [1], but it was difficult to shorten the lasing wavelength due to a limitation on the peak-electron density in a bunch. Various electron-beam instabilities were observed in the NIJI-IV FEL system that disturbed the electron bunch and reduced the density. A head–tail instability, which caused a limitation on the single-bunch current, was the most troublesome one. It was hard to obtain a large amount of beam current ($\sim$10 mA) in the single-bunch operation. Then we designed SQS magnets [2] and installed them in the restricted spaces between the bending magnets where the dispersion function was large in the operating mode.

The experiments with these SQS magnets started in January 1998. It was found that SQS magnets perfectly compensated both horizontal and vertical chromaticities. Unusual bunch lengthening which was observed before this improvement decreased, and a higher beam current ($\sim$ 30 mA) in single-bunch operation was available for FEL experiments. The beam cross section did not increase as

* Corresponding author. Tel: + 81-298-54-5680; fax: + 81-298-54-5683.

E-mail address: sei@etl.go.jp (N. Sei)

0168-9002/99/$ - see front matter
PII: S0168-9002(99)00101-1

the beam current increased. The peak-electron density in a bunch was thus not limited and the head–tail instability was adequately suppressed by the chromaticity correction. The peak-electron density in a bunch was over $6\times10^{16}\ m^{-3}$. This maximum attainable density is 5 times higher than the density before the improvement [3].

As a result, we achieved lasing at a wavelength of around 300 nm in March 1998 [4] and around 240 nm in May 1998. The electron energy during the lasing was 309 MeV, and the deflection parameter of an optical klystron was set to 2.02 (300 nm) and 1.70 (240 nm). The maximum FEL gain at a wavelength of 240 nm was 2.5%. In this article, we discuss the improvement of the beam qualities and its effect for the wavelength shortening in the NIJI-IV FEL.

2. Chromaticity correction

Though several beam instabilities were observed in the NIJI-IV FEL system, we concluded that the limitation of the beam current was caused by a head–tail instability [2]. The other instabilities disturbed the beam quality but were not associated with beam loss directly. There were not enough sextupole magnets in NIJI-IV at that time. The proper location for sextupole magnets where the dispersion function was large enough was on each side of the QF2 magnet in the short-straight sections, but there was only a small space for two sextupoles and one quadrupole magnet. We therefore developed a compact SQS magnet that had a small pole length by using a small bore of 45 mm [2]. Both sextupole magnets (SF and SD) were designed to correct the horizontal and vertical chromaticities with a magnet current of 5.00 A at an electron energy of 310 MeV. An SQS magnet was installed between bending magnet 1 and 2 in 1996, and the rest of the SQS magnets were installed in each of the short-straight sections in 1997. At the same time, an RF electrode was replaced because the old one caused vacuum leaks.

Electron injection into NIJI-IV restarted in January 1998. The operation mode of the ring was changed slightly because of the injection efficiency. The present machine function of NIJI-IV is shown in Fig. 1. The natural horizontal and vertical chromaticities calculated with the code TRACY2 [5] were -2.969 and -4.173, respectively. The maximum stored current in the ring ranged up to 260 mA, which almost equaled that before this improvement.

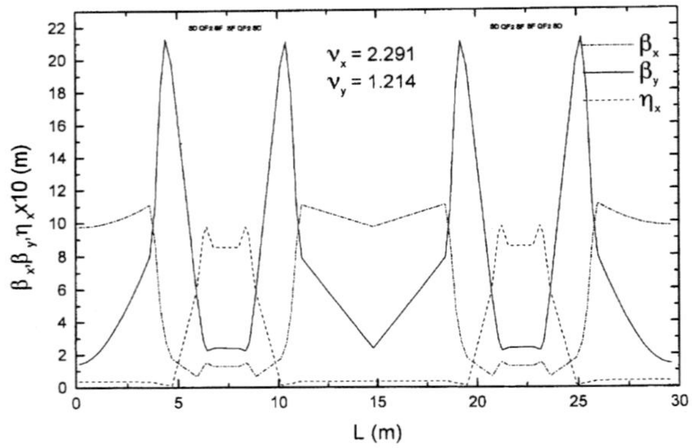

Fig. 1. Present machine function of NIJI-IV.

The chromaticities were measured by observing tune shifts with change of revolution frequency. Fig. 2 shows the experimental data of the chromaticities. We note that both chromaticities were corrected to be zero in the case of SF = 5.09 A and SD = 4.99 A. In order to suppress the head-tail instability, the chromaticities should be corrected to be small positive values [6]. This correction can effectively damp dipole and quadrupole modes of the head–tail oscillations with the help of a radiation dumping. Therefore, SF and SD were set to 5.29 and 5.18 A, respectively. The horizontal chromaticity was 0.08 and the vertical chromaticity was 0.15 in the present operation mode.

3. Improvement of the beam quality

In order to improve the beam quality, an RF-KO method [7] was applied to the full 16-bunch electron beam to realize a single-bunch mode. In the present operation mode with the chromaticity correction, higher beam current ($\sim$30 mA) in single-bunch operation was available for FEL experiments. Fig. 3 shows the dependence of the bunch length on the beam current, which was

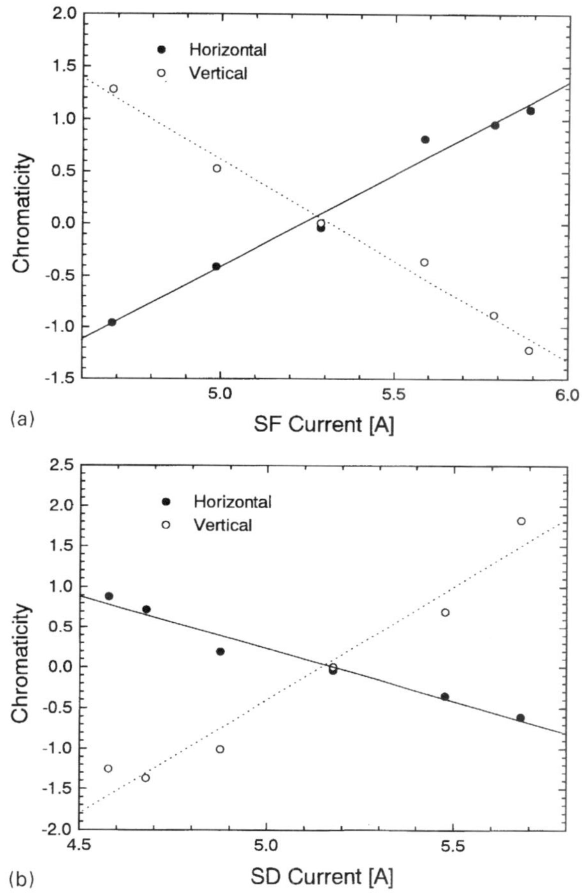

Fig. 2. Measured chromaticities with changing the SF current (a) and changing the SD current (b). The SD current is fixed to be 5.176 A in Fig. 1(a) and the SF current is fixed to be 5.287 A in Fig. 1(b). The solid and break lines mean a fitting line for the horizontal and the vertical chromaticity, respectively.

measured by a streak camera with a resolution of 18 ps. We note that the bunch length after the improvement was 30% shorter than that before the improvement. The main reason for this is that twice as much RF power became available for the experiments due to a new RF electrode. In the case of multi-bunch operation, a bunch lengthening occurred even at low current due to a strong coupled-bunch instability. A bunch lengthening in single-bunch operation became large above a current of 2 mA. Though Fig. 3 does not show it, a bump which was reported before [2] often appeared from

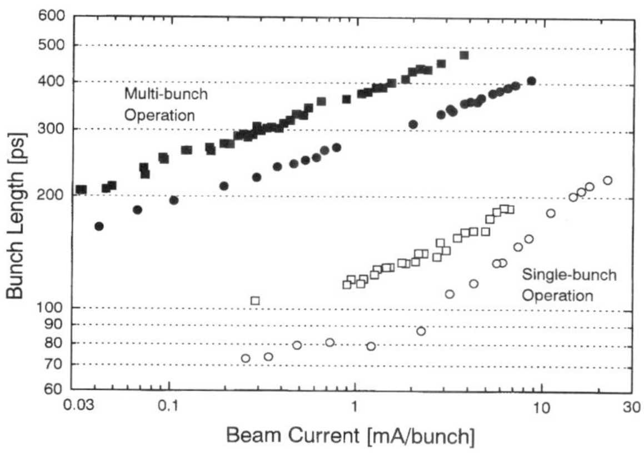

Fig. 3. Dependence of the bunch length on the beam current at the beam energy of about 309 MeV. The rectangular and the circle symbols represent the bunch length before and after the improvement, respectively.

2–4 mA in single-bunch operation. This phenomenon could be a kind of mode transition.

The dependence of the energy spread on the beam current was also measured. The energy spread was measured using a modulation factor f of the spontaneous radiation spectra from the optical klystron. It is known that the measured f is the simple product of the individual f factors, that is to say, $f = f_m f_\varepsilon f_\gamma$ [8]. Here f_m is due to the monochromator, f_ε is due to angular spread of the electron beam and f_γ is due to the energy spread. The energy spread is given as a coefficient by the following equation:

$$-\log\frac{f}{f_m} = 8\pi\left(\frac{\sigma_\gamma}{\gamma}\right)^2 (N + N_d)^2 \qquad (1)$$

with

$$f_m \approx \left[\frac{\sin(\pi(N + N_d)\Delta\lambda/\lambda)}{\pi(N + N_d)\Delta\lambda/\lambda}\right]^2 \qquad (2)$$

where N is the period number of one undulator, N_d is the interference order due to the dispersive section, $\Delta\lambda$ is a resolution of the monochromator and λ is the observing wavelength. We can control N_d by changing the gap of the dispersive section, so that the relation between f/f_m and $(N + N_d)^2$ is found. Fig. 4 illustrates the measured energy-spread data. An increase in the energy spread as well as the

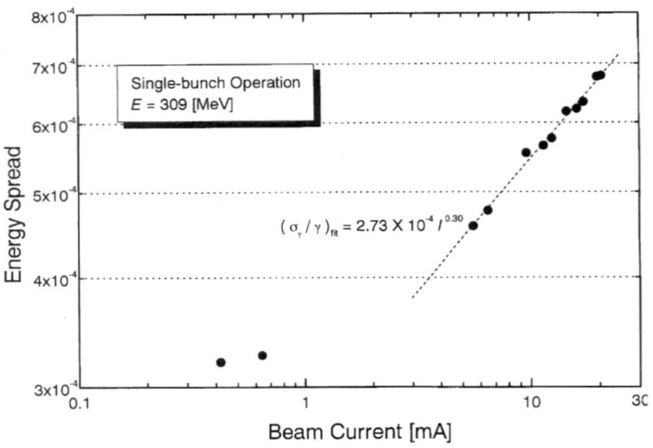

Fig. 4. Dependence of the energy spread on the beam current.

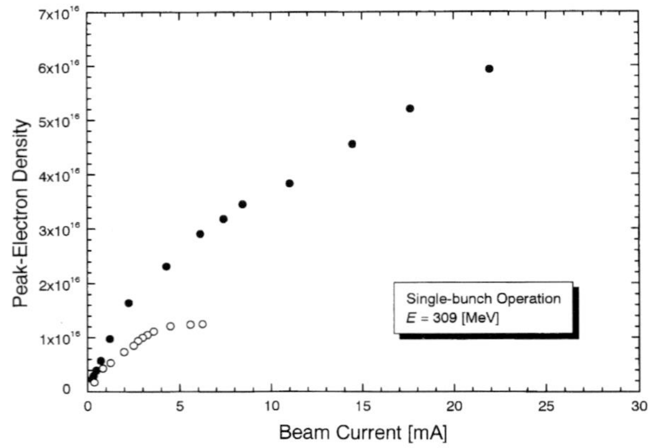

Fig. 5. Experimental measurement of the peak-electron density. The open and the solid symbols represent the peak-electron density before and after the improvement, respectively.

bunch lengthening was also confirmed above a beam current of 2 mA. This suggests that the bunch lengthening would be caused mainly by a microwave instability, and that the threshold of this instability is about 2 mA.

A CCD camera was used to measure beam sizes at the center of the dispersive section. The beam sizes were almost constant in single-bunch operation up to 30 mA. The horizontal size σ_x was about 0.77 mm and the vertical one σ_y was about 0.26 mm. This large value of σ_y indicates that the vertical emittance was rather large ($>10^{-8}$ mrad). The reason for the large vertical emittance could be the closed orbit distortion (COD) in NIJI-IV. However, the dependence of the beam cross section on the beam current which was observed at an electron energy of 263 MeV before [3] did not occur. The larger σ_y and the higher electron energy should suppress the multiple Touschek effect. Though the theoretical Touschek lifetime in single-bunch operation is about 60 min at the bunch current of 10 mA, the practical lifetime is about 40 min at the same bunch current. Scattering between residual gas and the electron beam leads to this difference in the lifetime.

The peak-electron density of a bunch in m^{-3}, ρ_{peak} is experimentally determined with the beam parameters above:

$$\rho_{peak} = \frac{I_e}{(2\pi)^{2/3} e f_{rev} \sigma_x \sigma_y \sigma_l} \tag{3}$$

where I_e is the beam current, e is the elementary charge and f_{rev} is the revolution frequency. The dependence of ρ_{peak} on the beam current before and after the improvement is shown in Fig. 5. The peak electron density after the improvement was about 5 times as high as that before the improvement, and it was over $6 \times 10^{16}\ m^{-3}$. The peak-electron density was not limited in the measured region of the beam current. This fact suggests that the head–tail instability, which limited the peak-electron density before the improvement, was adequately suppressed by the chromaticity correction. If we can store higher beam current in single-bunch operation, higher electron density will be available for FEL experiments.

It is known that FEL gain G_0 for an optical klystron is expressed by the following equations [9].

$$G_0 = 1.12 \times 10^{-13} \lambda_u^2 N^2 (N + N_d) K^2 \times [J_1(\xi) - J_0(\xi)]^2 f \rho_{peak} F_f \gamma^{-3}$$

$$\xi = K^2(4 + 2K^2)^{-2} \tag{4}$$

where γ is the relativistic electron energy, λ_u is the period length of the undulator and F_f is the filling factor. The gain for the NIJI-IV FEL system was estimated using Eqs. (4) and plotted in Fig. 6. The deflection parameter of the optical klystron was set to 2.02 for an FEL at 300 nm and 1.70 for an FEL

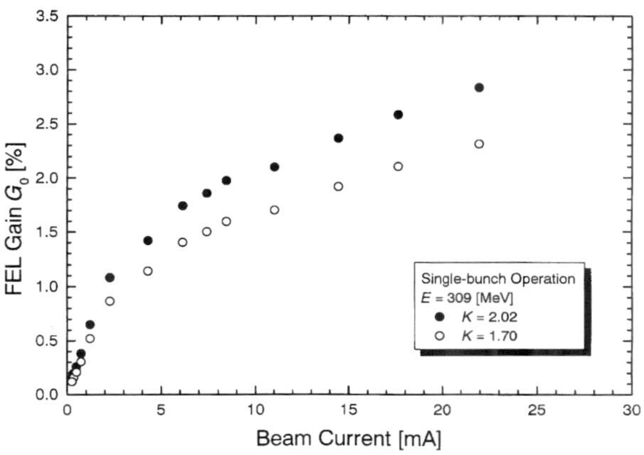

Fig. 6. Calculated FEL gain at wavelength of 300 nm (solid circle) and 240 nm (open circle). The parameter N_d is set to be 65 because the gain is at its maximum in the region of 20–25 mA. FEL experiments usually start from this current region.

at 240 nm, respectively. As Fig. 6 shows, an FEL gain of ~3.2% at 300 nm was available for the present NIJI-IV FEL system. We can conclude that the FEL gain increases by a factor of 5 due to the chromaticity correction [2]. The FEL gain at 240 nm was about 20% lower than that at 300 nm because of the difference of the deflection parameter. Though the initial cavity loss of the optical cavity in NIJI-IV was 0.3% at 300 nm and 0.5% at 240 nm, the loss increased to about 0.5% at 300 nm and 1.5% at 240 nm due to radiation damage of the cavity mirrors. Thus it is noted that the NIJI-IV FEL system should have enough gain to obtain lasing at both wavelengths. FEL experiments were carried out starting in March 1998, and we achieved the lasing at around 300 nm on March 20 and around 240 nm on May 7. Specifically, the laser wavelength with mirrors optimized around 240 nm could be varied from 253 to 228 nm. The characteristics of the FEL under 300 nm are reported in another paper [10] in detail.

4. Conclusions

We have shown that the electron-beam quality was improved by the chromaticity correction with the SQS magnets. We confirmed that the SQS magnets almost corrected both chromaticities with the design current. Anomalous bunch lengthening was decreased and higher beam current (~30 mA) in single-bunch operation was obtained. Current-dependent energy spread was also observed in the region of the high beam current ($\geqslant$ 2 mA). Therefore the bunch lengthening would come from a microwave instability. It was confirmed that the beam cross section hardly depended on the electron-beam current in single-bunch operation. The peak-electron density in a bunch that was available for the FEL experiments is 5 times as high as that before this improvement. The current limit at which the peak-electron density saturates has not been observed in single-bunch operation yet. This fact suggests that the head–tail instability, which limited the peak-electron density before this improvement, is adequately suppressed by the chromaticity correction. At present, the single-bunch current is limited by a kind of multi-bunch instability before the RF-KO method reduces the number of bunches. If we improve the single-bunch injection into NIJI-IV as reported previously [11], the peak-electron density will be higher. We will replace an RF cavity in this Autumn because the present one is damaged on the inside surface by electric discharge. This improvement might reduce the multi-bunch instabilities and allow higher peak- electron density.

The improvement of the beam quality due to the chromaticity correction greatly contributed to the lasing of the FEL at the shorter wavelength. We accomplished FEL lasing at around 300 and 240 nm, and shortened the wavelength down to 228 nm. The available FEL gain was estimated to be about 2.5% at 240 nm. Therefore, it should be possible to obtain an FEL gain of 2.0% even at 200 nm. It will be a challenge to achieve FEL lasing at even shorter wavelength with the NIJI-IV FEL system.

References

[1] T. Yamazaki et al., Nucl. Instr. and Meth. A 358 (1995) 353.

[2] N. Sei et al., Nucl. Instr. and Meth. A 393 (1997) 38.

[3] N. Sei et al., Nucl. Instr. and Meth. A 407 (1998) 187.

[4] N. Sei et al., Proceeding of the First Asian Particle Accelerator Conference, 1998.
[5] Rewritten from TRACY by J. Bengson. For original reference, see H. Nishimura, LBL Report 25236, ESG-4, 1988.
[6] H. Winick, Synchrotron Radiation Sources, Ch. 12, World Science, 1994.
[7] M. Yokoyama et al., Nucl. Instr. and Meth. A 331 (1993) ABS 34.
[8] D.A.G. Deacon et al., Appl. Phys. B 34 (1984) 207.
[9] M. Billardon et al., IEEE J. QE-21 (1985) 805.
[10] K. Yamada et al., The Twentieth International Free Electron Laser Conference, Williamsburg, USA, 1998.
[11] M. Yokoyama et al., Nucl. Instr. and Meth. A 341 (1994) 367.

ELSEVIER

Nuclear Instruments and Methods in Physics Research A 429 (1999) 191–196

NUCLEAR INSTRUMENTS & METHODS IN PHYSICS RESEARCH
Section A

www.elsevier.nl/locate/nima

Influences of electron beam properties on spontaneous radiation from an optical klystron

M. Hosaka*, J. Yamazaki, H. Hama

UVSOR Facility, Institute for Molecular Science, Myodaiji Okazaki 444, Japan

Abstract

Spectrum of the radiation from an optical klystron is very sensitive to the beam energy spread which is an important factor for the storage ring free electron laser. In order to make use of the spectrum for a measurement of the relative energy spread, we have investigated influences of the transverse beam property on the spectrum using a helical optical klystron on the UVSOR storage ring. From analysis of the experimental results, it has been found that the most important effect comes from inhomogeneous magnetic fields and a horizontal beam size.

Keywords: Optical klystron; Modulation factor

1. Introduction

Information about the energy spread of the electron beam is frequently crucial to understand the longitudinal dynamics of free electron laser oscillation on storage ring [1,2]. Conventionally, the beam energy spread in a storage ring has been evaluated by measuring the size of the beam cross section at a position where the dispersion function is large. However, precise measurement is generally very difficult because ambiguity in the Twiss parameters leads to large errors. One of the most effective methods to deduce the beam energy spread is to observe spontaneous radiation from an optical klystron. Intensity modulation of the spectrum which is a result of interference of radiation from two undulators is sensitive to the energy spread. However, the spectrum might be also affected by the transverse characteristics of the beam such as the beam size or the angular spread.

In this article, we present an experimental study on the influences of the electron beam transverse property on the spectrum of the helical optical klystron and discuss the so-called modulation factor by comparison with calculations based on an ideal model.

2. Radiation from helical optical klystron

Since the radiation from an optical klystron can be expressed as a superposition of radiation fields from two identical undulators [3], the radiation

* Corresponding author. Tel.: + 81-564-55-7402; fax: + 81-564-54-7079.

E-mail address: hosaka@ims.ac.jp (M. Hosaka)

0168-9002/99/$ - see front matter
PII: S0168-9002(99)00102-3

power from an electron, per unit solid angle and per frequency is written as

$$\frac{d^2W}{d\omega\, d\Omega} = 2\frac{d^2W}{d\omega\, d\Omega}\bigg|_{1\,\text{undulator}}(1+\cos\alpha) \tag{1}$$

with

$$\alpha = \alpha_u + \alpha_d \tag{2}$$

$$\alpha_u = \omega(T_u - l/c) \tag{3}$$

$$\alpha_d = \omega(T_d - d/c) \tag{4}$$

where T_u and T_d are time durations for the electron to pass through the undulator section with a total length of l and the dispersive section with a total length of d, respectively. Phase shifts α_u, α_d can be approximately calculated as

$$\alpha_u = \frac{\pi l}{\lambda\gamma^2}\left[1 + \frac{e^2}{lm^2c^2}\int_0^l\left\{\left(\int_0^s B_x(z)\,dz\right)^2 + \left(\int_0^s B_y(z)\,dz\right)^2\right\}ds + \gamma^2\theta^2\right] \tag{5}$$

$$\alpha_d = \frac{\pi d}{\lambda\gamma^2}\left[1 + \frac{e^2}{dm^2c^2}\int_l^{l+d}\left\{\left(\int_l^s B_x(z)\,dz\right)^2 + \left(\int_l^s B_y(z)\,dz\right)^2\right\}ds + \gamma^2\theta^2\right] \tag{6}$$

where e is the electron charge, m the mass, γ the relativistic factor and $\lambda(=2\pi/\omega)$ is the light wavelength, and $B_x(z)$ and $B_y(z)$ are magnetic fields in the horizontal and the vertical direction at the longitudinal coordinate z, respectively. Integration in Eqs. (5) and (6) are along the undulator and the dispersive section, respectively. In the case of a helical (elliptical) optical klystron, $B_x(z)$ and $B_y(z)$ are given as

$$B_x(z) = B_{u(d)x}\sin(2\pi/\lambda_{u(d)}),$$

$$B_y(z) = B_{u(d)y}\cos(2\pi/\lambda_{u(d)}) \tag{7}$$

where $B_{u(d)x}$ and $B_{u(d)y}$ are the peak magnetic field in the undulator (dispersive) section for the horizontal and the vertical direction, respectively and $\lambda_{u(d)}$ is the period of the undulator (dispersive) section. Eq. (5) can be also written using deflection parameters K_x, K_y as

$$\alpha_u = \frac{\pi l}{\lambda\gamma^2}[1 + (K_x^2 + K_y^2)/2 + \gamma^2\theta^2] \tag{8}$$

with

$$K_x = eB_{ux}\lambda_u/2\pi mc, \quad K_y = eB_{uy}\lambda_u/2\pi mc. \tag{9}$$

Supposing the electron beam has distributions in its energy, angle and horizontal position around central values γ_m, θ_m and x_m, respectively, the radiation spectrum may be given by integrating Eq. (1) incoherently over the distributions as

$$\frac{d^2W}{d\omega\, d\Omega} = 2\frac{d^2W}{d\omega\, d\Omega}\bigg|_{1\,\text{undulator}}\int d\gamma\int d\chi\int d\theta\rho(\gamma,\theta,\chi) \times\{1+\cos\alpha(\gamma,\theta,\chi,\lambda)\} \approx 2\frac{d^2W}{d\omega\, d\Omega}\bigg|_{1\,\text{undulator}} \times\{1 + f_{mod}\cos\alpha(\gamma_m,\theta_m,\chi_m,\lambda)\} \tag{10}$$

where $\rho(\gamma,\theta,\chi)$ represents the normalized density of the electrons and $f_{mod}(<1)$ is the so-called modulation factor. We neglect the contribution of the vertical beam size in Eq. (10), because it is much smaller than the horizontal one in an ordinary storage ring with a small transverse coupling. Assuming these contributions to be independent, the modulation factor can be written as

$$f_{mod} = f_\gamma f_\theta f_\chi \tag{11}$$

where f_γ, f_θ and f_χ are contributions from the energy, the angular and the position spreads, respectively.

3. Experiment

In order to investigate the effects of the transverse beam property to the modulation factor, we changed the horizontal beam emittance of the UVSOR storage ring. Because of restriction of an injection scheme of the electron beam from the booster synchrotron to the storage ring, the horizontal betatron tune cannot be changed much. Therefore, we maintained the horizontal and the vertical tune at constant values ($\upsilon_x = 3.17$, $\upsilon_y = 1.38$) and varied the beam emittance by controlling the shape of the dispersion function in the bending magnet. In Fig. 1,

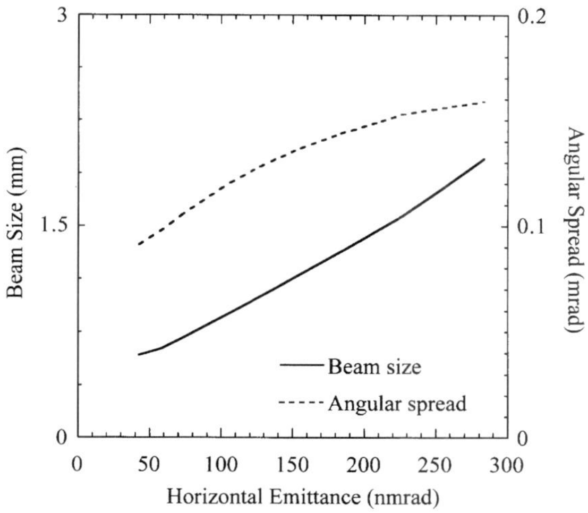

Fig. 1. Beam size and angular spread at the center of the helical optical klystron as a function of the beam emittance.

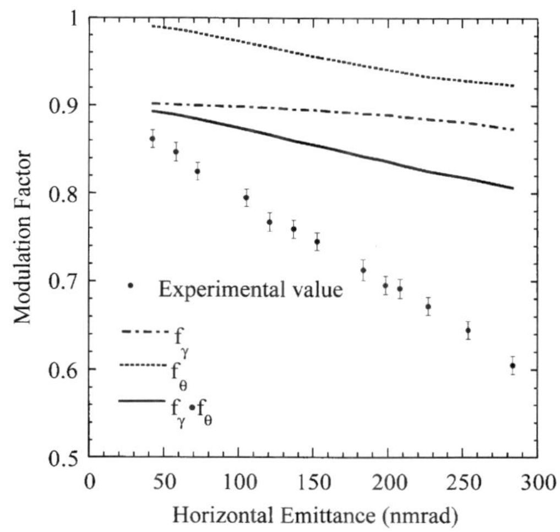

Fig. 2. Measured modulation factor as a function of the beam emittance. Solid and broken lines are the contributions of the natural energy spread and angular spread, respectively.

the rms beam size and the angular spread at the center of the helical klystron are shown as a function of the transverse emittance at operation points used for the experiment.

At each operating point, the radiation from the helical optical klystron (UNKO-3) was measured using a monochrometer, and the resonant wavelength of 350 nm was chosen. The storage ring was operated at an energy of 600 MeV and a single-bunch mode with a very low beam current (~ 0.3 mA) was employed in order to minimize effects of potential-well distortion [4]. Basic parameters of the experiment are shown in Table 1. The modulation factors deduced from observed spectra are plotted as a function of the beam emittance in Fig. 2.

4. Effect of longitudinal beam property

Our concern is to examine the influence of the transverse beam property to the modulation factor in this study, therefore we have carefully eliminated contributions of the longitudinal effect of the beam.

As mentioned earlier, the experiment was performed with a very low beam current. Nevertheless,

Table 1
Basic parameters of the helical optical klystron (UNKO-3) in the experiment

		Undulator section	Dispersive section
Number of periods:		9	1
Period length:		11.0 cm	33.0 cm
Peak magnetic field ($\lambda = 350$ nm)		Inhomogeneous parameter (horizontal)	
B_{ux}	0.271 T	a_{ux}	− 0.41
B_{uy}	0.271 T	b_{ux}	− 1.13
B_{dx}	0.291 T	a_{dx}	− 3.77
B_{dy}	0.715 T	b_{dx}	− 0.72

a weak longitudinal collective synchrotron oscillation was observed as sidebands of the beam frequency spectrum. This collective oscillation may be explained by a slight instability of the rf cavity of the storage ring. At a beam current of 1.0 mA, the energy amplitude of the collective oscillation estimated from the intensity of the frequency spectrum is around 4 eV. The value is about 2 orders of magnitude smaller than the natural energy spread of the electron beam and the contribution to the modulation factor can be neglected. We also measured bunch shape of the electron beam using a streak camera, because the bunch length and the energy spread is directly connected with each other at the very low beam current. Measured bunch length agreed well with one deduced from calculated natural energy spread. Accordingly, we treat the energy distribution of the electrons in the bunch as the Gaussian with a width of the natural energy spread σ_γ/γ, then the contribution to the modulation factor can be calculated using Eq. (10) as

$$f_\gamma = \exp\left[-8\pi^2(N+N_d)^2\left(\frac{\sigma_\gamma}{\gamma}\right)^2\right] \quad (12)$$

where $N+N_d$ corresponds to $(\alpha_u+\alpha_d)/2\pi$ at no energy deviation. The values of the natural energy spread slightly depend on the lattice parameters and the modulation factor as well. In Fig. 2, f_γ is plotted at each operating point.

5. Effect of transverse beam property

As seen in Fig. 2, it is clear that the modulation factor is considerably affected by the transverse beam emittance. Effect of the emittance is differentiated into the angular spread and the beam size.

Because no collective transverse oscillation was observed in the frequency spectrum of the beam signal, the Gaussian distribution with a width of σ_θ calculated from the natural emittance and the lattice function was used as the angular spread. Using Eq. (9), the contribution to the modulation factor is given as

$$f_\theta = \left[1+4\left\{\frac{\pi(l+d)}{\lambda}\right\}^2\sigma_\theta^4\right]^{-1/4}. \quad (13)$$

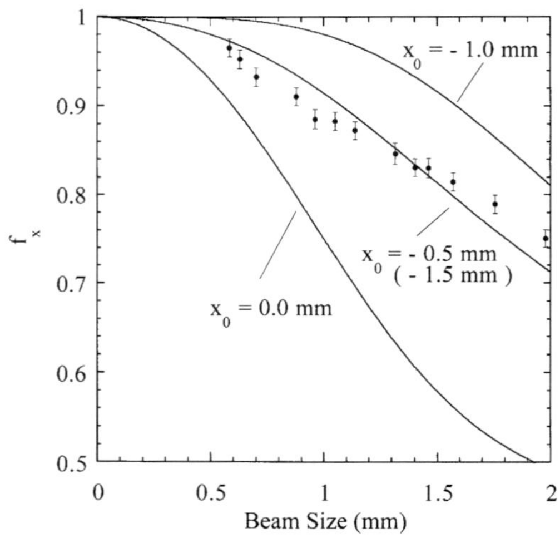

Fig. 3. Contribution of the beam size to the modulation factor f_x as a function of the beam size. The lines are calculations for $x_0 = 0.0$, -0.5 (-1.5), -1.0 mm.

The calculated values are plotted in Fig. 2 and it is clear the contribution is not so large.

Since no other significant effects to the modulation factor is expected, the contribution of the beam size to the modulation factor f_x can be deduced by dividing the experimental values by the calculated contribution of the energy and the angular spread. The deduced f_x are plotted as a function of the rms beam size in Fig. 3. As seen in the figure, f_x rapidly decreases with the beam size, and it is apparent that the most important transverse effect is the horizontal beam size.

In order to evaluate the experimental result, we performed a model calculation based on a consideration that the decrease of the modulation factor due to the beam size is closely related to the uniformity of the magnetic field. In a conventional planar optical klystron with sufficiently large magnets, the magnetic field is almost uniform in the horizontal direction, so that the effect of the horizontal beam size may be small. On the other hand, in case of the helical optical klystron, the magnetic field is not uniform in both the horizontal and the vertical directions. The transverse distribution of magnetic field is approximately given by

$$B_x(x,y,z) = (1+\tfrac{1}{2}a_xk^2x^2+\tfrac{1}{2}a_yk^2y^2)B_x(0,0,z) \quad (14)$$

$$B_y(x,y,z) = (1+\tfrac{1}{2}b_xk^2x^2+\tfrac{1}{2}b_yk^2y^2)B_y(0,0,z) \quad (15)$$

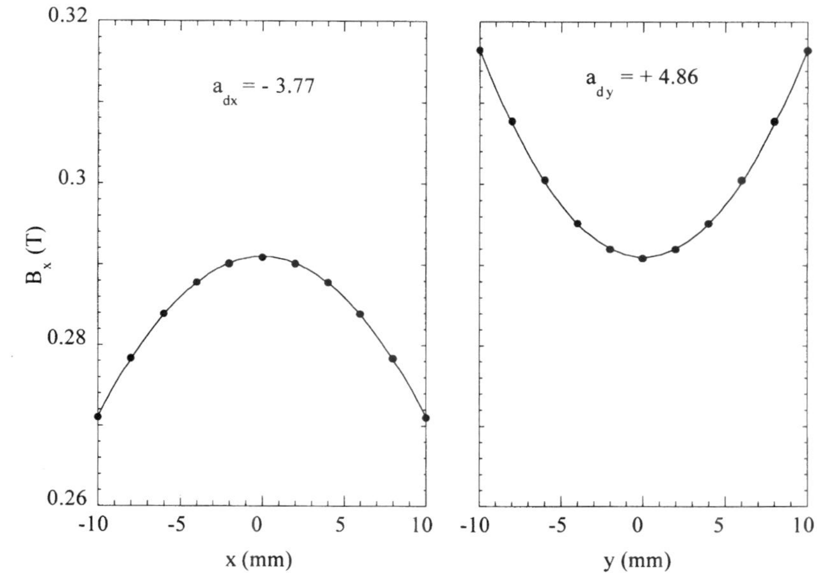

Fig. 4. Calculated horizontal and vertical distributions of the horizontal magnetic field in the dispersive section of the UNKO-3.

where $k = 2\pi/\lambda_0$ and λ_0 is a period length of trigonometric wiggler field, and inhomogeneous parameters a_x, a_y, b_x and b_y should satisfy relations, $a_x + a_y = 1$, $b_x + b_y = 1$ from Maxwell's equations. The inhomogeneous parameters are deduced from a numerical calculation for the magnetic field based on the equivalent current method and an example is shown in Fig. 4.

When the electron beam is transversely distributed around a position x_0 from the center of the magnetic field with a width of σ_x, one can calculate the modulation factor using Eqs. (5), (6), (10), (14) and (15) as

$$f_x = \{1 + (s_0\sigma_x^2)^2\}^{-1/4} \exp\left(\frac{-\frac{1}{2}s_1^2\sigma_x^2}{1 + (s_0\sigma_x^2)^4}\right) \tag{16}$$

with

$$s_1 = s_0 x_0 + s_d \tag{17}$$

$$s_0 = \frac{\pi}{\lambda}\left(\frac{e}{mc\gamma}\right)^2 \{(a_{dx}B_{dx}^2 + b_{dx}B_{dy}^2)d + (a_{ux}B_{ux}^2 + b_{ux}B_{uy}^2)l\} \tag{18}$$

$$s_d = \frac{\pi}{\lambda}\left(\frac{e}{mc\gamma}\right)^2 (a_{dx}B_{dx}^2 + 2b_{dy}B_{dy}^2)\frac{eB_{dy}}{mc\gamma}\left(\frac{d}{2\pi}\right)^2 \tag{19}$$

where subscripts d and u represent the dispersive and the undulator section, respectively. The parameter s_d in Eq. (17) describes the combined effect of the inhomogeneous magnetic field and the wiggled beam trajectory in the horizontal direction at the dispersive section. We neglect the vertical beam size because the trajectory of the electron beam in the dispersive section is wiggled largely only in the horizontal direction [5]. One can notice that the beam size effect depend on the horizontal trajectory of the beam from Eq. (16). When σ_x is small, Eq. (16) can be approximately rewritten as

$$f_x \propto \exp\left\{-\frac{1}{2}\sigma_x^2 s_0^2\left(x_0 + \frac{s_d}{s_0}\right)^2\right\}. \tag{20}$$

It should be noted that the horizontal beam position that compensates the beam-size effect is not on the center of the magnetic field and a calculated value of x_0 to minimize f_x is -1.0 mm.

Because it is very difficult to know the exact beam position relative to the magnetic center of the UNKO-3, the experimental data are fitted with the function given in Eq. (16) with x_0 as a fitting parameter. As seen in Fig. 3, the fitting quality is not so bad and the obtained value of x_0 is -0.5 mm

(− 1.5 mm is another possibility). It can be concluded that the default orbit used in the experiment so far has been not well optimized for the modulation factor.

6. Summary

We have studied the effect of the beam transverse property on the modulation factor of the radiation spectrum from the helical optical klystron. The agreement between the experimental results and the model calculation shows that the horizontal beam size and position are the main contribution to decrease the modulation factor.

Although the optical klystron seems to be a suitable diagnostic tool to observe the longitudinal beam property, one should carefully take account of these effects especially when the beam dimension is large. Even in a planar optical klystron, the magnetic field is not completely uniform and the inhomogeneous parameters have finite values (at the planar mode of UNKO-3, b_{dx} is $\sim \frac{1}{3}$ relative to that at the helical mode).

References

[1] M. Hosaka, H. Hama, J. Yamazaki, T. Kinoshita, Nucl. Instr. and Meth. A 407 (1998) 234.

[2] H. Hama, M. Hosaka, J. Yamazaki, T. Kinoshita, Nucl. Instr. and Meth. B 144 (1998) 75.

[3] P. Elleaume, J. Phys. Colloq. 44 (1983) C1-333.

[4] J.L. Laclare. Proceedings of the CERN Accelerator School, Advance Accelerator Physics, Oxford, 1985, CERN, Geneva, CERN 87-03, 1987, p. 264.

[5] H. Hama, Nucl. Instr. and Meth. A 375 (1996) 57.

ELSEVIER

Nuclear Instruments and Methods in Physics Research A 429 (1999) 197–201

NUCLEAR INSTRUMENTS & METHODS IN PHYSICS RESEARCH Section A

www.elsevier.nl/locate/nima

Numerical simulations of the UCLA/LANL/RRCKI/SLAC experiment on a high-gain SASE FEL

E.L. Saldin[a], E.A. Schneidmiller[a], M.V. Yurkov[b,*]

[a]Automatic Systems Corporation, 443050 Samara, Russia
[b]Particle Physics Laboratory (LSVE), Joint Institute for Nuclear Research, 141980 Dubna, Moscow Region, Russia

Abstract

In this paper we present a theoretical analysis of a recent SASE FEL experiment performed by a UCLA/LANL/RRCKI/SLAC team reporting on a high power gain of about 10^5 at the wavelength of 12 μm. The region of physical parameters of this experiment (as well as of future X-ray FELs) does not allow one to apply available analytical techniques for a quantitative description of the results obtained. The analysis presented in this paper is based on the results produced by the three-dimensional, time-dependent FEL simulation code FAST. It is shown that, within the limit of accuracy of the experiment, the data obtained agree fully with the results of numerical simulations.

PACS: 41.60. Cr; 52.75. Ms

Keywords: High-gain SASE FEL; Numerical simulations

1. Introduction

Recently, the operation of a high-gain FEL amplifier starting from noise has been reported [1] (see Table 1). A power gain of more than 10^5 has been obtained. Parameters of the output radiation have been measured with high accuracy. In particular, it has been demonstrated that the fluctuations of the energy in the radiation pulse follow the gamma distribution predicted in Refs. [2,3]. A thorough analysis shows that even though this experiment was performed at a relatively long wavelength, the physical processes in this SASE FEL are similar to those expected to occur in a short-wavelength SASE FEL. This is connected with the fact that the aperture of the vacuum chamber is significantly larger than the transverse size of the beam radiation mode. As a result, waveguide effects do not influence the amplification process. Also, the effects of coherent synchrotron radiation are not important with the experimental parameters. A helpful factor simplifying analysis of this experiment is the careful choice of experimental conditions. That is, the undulator field has been tuned precisely and distortions of the electron beam trajectory are

*Corresponding author. Tel.: + 7-09621-62154; fax: + 7-09621-65767.

E-mail address: yurkov@sunse.jinr.ru (M.V. Yurkov)

0168-9002/99/$ - see front matter
PII: S0168-9002(99)00103-5

Table 1
Parameters of the UCLA/LANL SASE FEL [1]

Electron beam	
Energy (MeV)	18
Charge per micropulse (nC)	0.3–2.2
Transverse spot size (σ) (μm)	115–145
Energy spread (rms) (%)	~ 0.3
Pulse length (FWHM) (ps)	7–13
Peak current (A)	40–170
Undulator	
Period (cm)	2.05
Number of periods	98
Undulator parameter K	1
Betatron wavelength (m)	1.2
FEL	
Radiation wavelength (μm)	12
Power gain length (at 2.2 nC) (cm)	~ 14

significantly less than the typical transverse size of the beam radiation mode. There is also a significant safety margin with respect to the energy spread and the emittance of the electron beam.

In this paper we perform a theoretical analysis of the high-gain SASE FEL experiment [1]. The analysis is based on the results produced by the three-dimensional, time-dependent FEL simulation code FAST [4]. Parameters of the electron beam and of the undulator [1] have been used as the input parameters for the numerical simulation code without making any corrections, as is described in Section 3. It is shown that the experimental results are in good agreement with the results of numerical calculations. Statistical simulations of the energy fluctuations in the radiation pulse performed over several thousand shots give a result identical to the experimental one. That is, fluctuations of the energy in the radiation pulse follow a gamma distribution with the value of the parameter $M \simeq 8$.

2. Region of physical parameters

We begin our analysis within the framework of the steady-state approximation in order to estimate the power of different physical effects influencing the operation of the FEL amplifier. A numerical solution of the corresponding eigenvalue equation [5–7] shows that the main physical effects defining the operation of the UCLA/LANL SASE FEL are

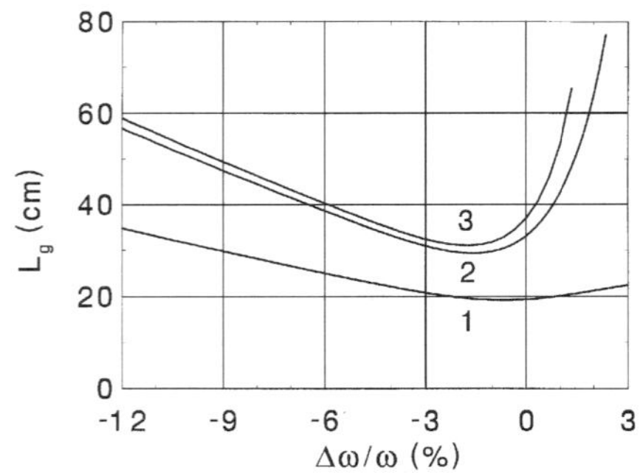

Fig. 1. The field gain length versus the frequency deviation from the resonance value. Calculations have been performed in the steady-state approximation. Curve 1 is calculated with only the diffraction effects taken into account, curve 2 includes also the space charge effects, and curve 3 is calculated taking into account all the effects (diffraction, space charge and energy spread).

diffraction effects and space charge effects. It is seen in Fig. 1 that the field gain length should be about 30 cm. Also, an analysis of the imaginary part of the eigenvalue shows that the slippage effect will be suppressed by a factor of four with respect to kinematic slippage due to the fact that the group velocity of the amplified wave, $\partial\omega/\partial k$, is less than the velocity of light c.

3. Numerical analysis of the experiment

Numerical simulations have been performed with the three-dimensional, time-dependent FEL simulation code FAST [4]. The equations of motion (kinetic equation or equations for macroparticle motion) and Maxwell's equations are solved simultaneously taking into account the slippage effect. Radiation fields are calculated using an integral solution of Maxwell's equations. The code allows one to simulate the radiation from the electron bunch of any transverse and longitudinal bunch shape; to simulate simultaneously an external seed with superimposed shot noise in the electron beam; to take into account energy spread in the electron beam and the space charge fields; and to simulate a high-gain, high-efficiency FEL amplifier with a tapered undulator. The code is extremely fast thus allowing precise statistical calculations to be performed.

To calculate averaged characteristics of the FEL amplifier we performed several thousand statistically independent runs. Input data for the numerical simulation code are the value of the undulator period and the peak field, and the bunch charge, length and radius. When performing simulations we used two models of the axial profile of the bunch current, a Gaussian one

$$I(z) = \frac{Qc}{\sqrt{2\pi}\sigma_z}\exp\left[-\frac{z^2}{2\sigma_z^2}\right]$$

with $\sigma_z = \sigma_z^{\mathrm{HWHM}}/\sqrt{2\ln 2}$, and a parabolic one:

$$I(z) = \frac{3Qc}{4\sigma_z}\left[1-\frac{z^2}{\sigma_z^2}\right], \quad |z| < \sigma_z$$

with $\sigma_z = \sqrt{2}\sigma_z^{\mathrm{HWHM}}$. The transverse distribution of the beam current density was assumed to be Gaussian:

$$j(z,r) = \frac{I(z)}{2\pi\sigma_r^2}\exp\left[-\frac{r^2}{2\sigma_r^2}\right]$$

with $\sigma_r = \sigma_r^{\mathrm{HWHM}}/\sqrt{2\ln 2}$. HWHM values are defined by fitting formulae presented in Ref. [1]: $\sigma_{z,r}^{\mathrm{HWHM}} = \sqrt{a^2 + (bQ)^2}$. Parameters for the spot size are $a = 120$ μm, $b = 38$ μm/nC, and for the pulse length $a = 3$ ps, $b = 2.2$ ps/nC.

In Fig. 2 we present the typical time structure of the radiation pulse at the undulator exit with a bunch charge of 2.2 nC. Averaging over 1000 independent shots gives the value of the averaged radiation pulse shape (curve 2 in Fig. 2). Analysis of the latter plot allows one to obtain the value of the "effective" shot noise power at the undulator entrance to be used in the steady-state codes for obtaining the approximate value of the average output power [8,9]. The value of the "effective" power of shot noise is about 20 mW in the case under study. It is seen also from Fig. 2 that the slippage of the radiation with respect to the electron bunch is significantly smaller than the kinematic one. This is connected with the fact that the group velocity of the spikes, $\partial\omega/\partial k$, is less than the velocity of light c as mentioned in Section 2.

In Fig. 3 we present a comparison of experimental and simulation results for different values of the charge. Fig. 4 presents the dependence of the

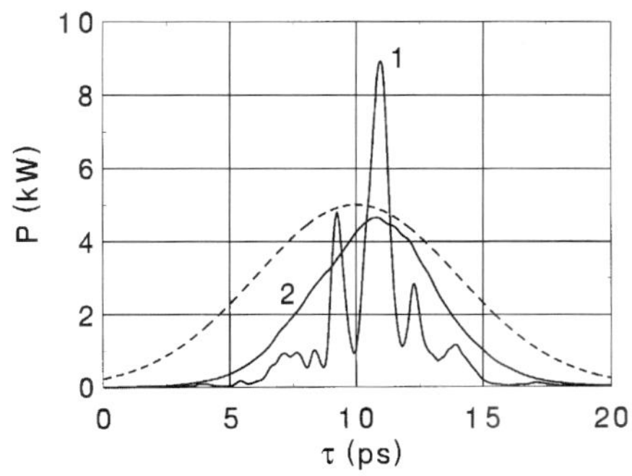

Fig. 2. Typical time structure of the radiation pulse at the undulator exit (curve 1) and time structure of the radiation pulse averaged over 1000 statistically independent runs (curve 2). The dashed line presents an axial profile of the beam current. Charge in the electron bunch is 2.2 nC.

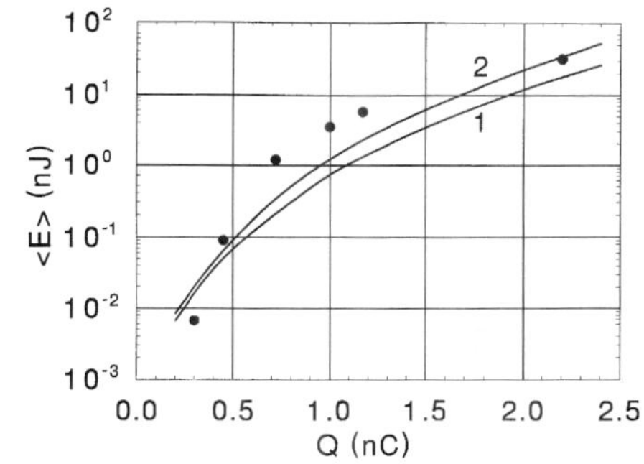

Fig. 3. Dependence of the averaged energy in the radiation pulse versus the bunch charge. Curves 1 and 2 correspond to the Gaussian and the parabolic axial beam profiles. The circles are experimental results [1].

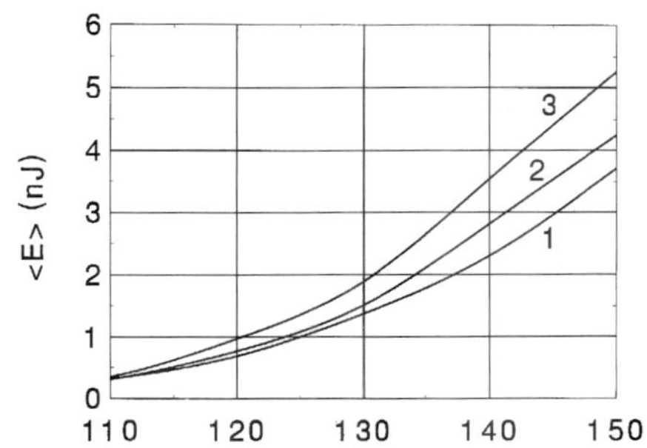

Fig. 4. Dependence of the averaged energy in the radiation pulse on the transverse bunch size. Curves 1, 2 and 3 correspond to values of the longitudinal HWHM bunch size σ_z^{HWHM} of 3.9, 3.7 and 3.5 ps, respectively. The bunch charge is 1 nC.

energy in the radiation pulse on the transverse and longitudinal beam sizes at a fixed value of the bunch charge of 1 nC. It is seen that within the uncertainties in measuring these values [1,10,11], the numerical and experimental results agree rather well. One can also obtain an unusual behaviour of the energy in the radiation pulse versus the transverse beam size. That is, the radiation energy increases as the transverse beam size increases. This effect is connected with the fact mentioned in Section 2 that the UCLA/LANL FEL amplifier operates in a regime of strong space charge effects. In the case under study the diffraction parameter B is much less than unity, and increasing the beam size results only in logarithmic decrease of the field gain due to diffractive effects [5,6]. On the other hand, there is strong influence of the space charge fields. Increasing the transverse beam size leads to a quadratic decrease of the space charge parameter which results in the increase of the field gain. In the region of parameters traced in Fig. 4 this effect dominates above the diffraction effects. As a result, the field gain and the energy in the radiation pulse grow with the increase of the transverse size of the electron beam. Calculations show that such a tendency will take place up to the value of diffraction parameter $B \sim 0.3$. Above this point the space charge effect becomes a small perturbation to the FEL process and increase of the transverse beam size will lead to the decrease of the field gain (see, e.g. Ref. [12]).

So, we see that there is good agreement between measurements of the energy in the radiation pulse and the simulation results. Indeed, over the entire range of electron beam charge (see Fig. 3), the relative difference expressed in the units of the gain length is less than one power gain length. Taking into account good agreement of the energy in the radiation pulse, we can expect much better agreement with the probability distribution of the energy in the radiation pulse. We performed the corresponding simulations at the value of the bunch charge of 2.2 nC. Simulations show that the probability distribution of the energy in the radiation pulse is quite close to the gamma distribution:

$$p(E) = \frac{M^M}{\Gamma(M)}\left(\frac{E}{\langle E\rangle}\right)^{M-1}\frac{1}{\langle E\rangle}\exp\left(-M\frac{E}{\langle E\rangle}\right)$$

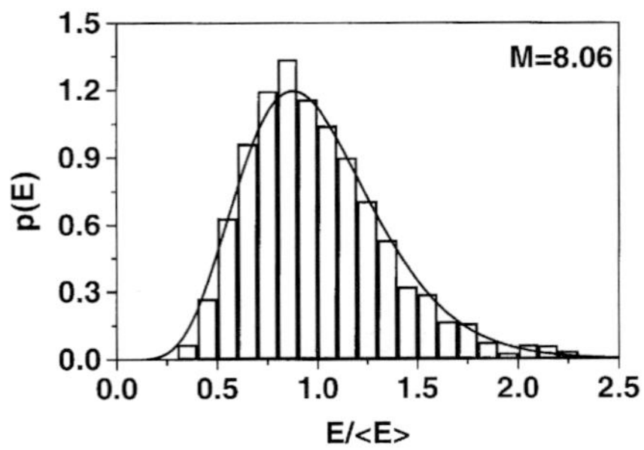

Fig. 5. Probability distribution of the energy in the radiation pulse at a bunch charge of 2.2 nC calculated over 2400 statistically independent runs. The solid curve represents a gamma distribution with $M = 8.06$.

where $\Gamma(M)$ is the gamma function of argument M. The parameter of the distribution is equal to $M = 1/\sigma^2$, where $\sigma^2 = \langle E^2 - \langle E\rangle^2\rangle/\langle E\rangle^2$ is the normalized dispersion of the energy distribution. The first set of runs has been performed at fixed values of the bunch charge, transverse beam size, and longitudinal beam size. Simulations with the Gaussian and the parabolic bunch profiles give very close results, namely that $M \simeq 11.5$ and $\sigma \simeq 30\%$. On the other hand, the experiment yields fluctuations of about $\sigma \simeq 37\%$. Such a visible difference indicates that shot-to-shot fluctuations of the beam parameters contribute to the fluctuations of the radiation energy. To take these fluctuations into account we performed 2400 statistically independent runs. During each run the fluctuations have been introduced in the following limits: $\pm 0.75\%$ for the bunch charge, $\pm 5.5\%$ for the transverse beam size and $\pm 6\%$ for the bunch length [1]. The results of the simulations are shown in Fig. 5. The probability distribution of the radiation energy follows a gamma distribution with $\sigma \simeq 35\%$ and $M \simeq 8$. This result is in good agreement with the experiment.

4. Conclusion

In conclusion, we should like to point out that there is no doubt that the UCLA/LANL experiment [1] is a proof-of-principle of a high-gain SASE FEL. Even though it was performed at a relatively long wavelength, the physics of its operation is described with the same equations as future VUV and X-ray SASE FELs. All the simulations presented in this paper have been performed with the simulation code developed for simulation of short-wavelength SASE FELs. It is seen that there is good agreement between theoretical predictions and experimental results, which forms a reliable base for the future design of short-wavelength SASE FELs.

Acknowledgements

We are extremely grateful to C. Pellegrini and A. Varfolomeev for providing us with experimental results and fruitful discussions. We wish to thank B. Faatz, J. Feldhaus, J. Krzywinski, G. Materlik, T. Möller, C. Pagani, J. Pflüger, S. Reiche, J. Roßbach and J.R. Schneider for many useful discussions.

References

[1] M. Hogan, C. Pellegrini, J. Rosenzweig, A. Anderson, P. Frigola, A. Tremaine, C. Fortgang, D. Nguyen, R. Sheffield, J. Kinross-Wright, A. Varfolomeev, A.A. Varfolomeev, S. Tolmachev, R. Carr, Phys. Rev. Lett. 81 (1998) 4867.

[2] E.L. Saldin, E.A. Schneidmiller, M.V. Yurkov, DESY Print TESLA-FEL 97-02, April, 1997.

[3] E.L. Saldin, E.A. Schneidmiller, M.V. Yurkov, Opt. Commun. 148 (1998) 383.

[4] E.L. Saldin, E.A. Schneidmiller, M.V. Yurkov, Nucl. Instr. and Meth. A 429 (1999) 233.

[5] E.L. Saldin, E.A. Schneidmiller, M.V. Yurkov, Phys. Rep. 260 (1995) 187.

[6] E.L. Saldin, E.A. Schneidmiller, M.V. Yurkov, Opt. Commun. 97 (1993) 272.

[7] E.L. Saldin, E.A. Schneidmiller, M.V. Yurkov, DESY Print May 1995, TESLA-FEL 95-02, Hamburg, DESY, 1995.

[8] K.J. Kim, Nucl. Instr. and Meth. A 250 (1986) 396.

[9] J.M. Wang, L.H. Yu, Nucl. Instr. and Meth. A 250 (1986) 484.

[10] R.L. Sheffield et al., Nucl. Instr. and Meth. A 341 (1994) 371.

[11] S.H. Kong et al., Nucl. Instr. and Meth. A 358 (1995) 284.

[12] E.L. Saldin, E.A. Schneidmiller, M.V. Yurkov, Proceedings of the Fifth European Particle Accelerator Conference, vol. 1, Institute of Physics Publishing, Bristol, 1996, pp. 471–473.

ELSEVIER

Nuclear Instruments and Methods in Physics Research A 429 (1999) 202–208

NUCLEAR INSTRUMENTS & METHODS IN PHYSICS RESEARCH Section A

www.elsevier.nl/locate/nima

Comparison of beam bunching in amplifier and SASE modes at the CEA-CESTA free-electron laser

J.T. Donohue[a,*], J. Gardelle[b], T. Lefevre[b], G. Marchese[b], M. Padois[b], J.L. Rullier[b]

[a]*Centre d'Etudes Nucléaires de Bordeaux Gradignan, 33175 Gradignan, France*

[b]*CEA/Centre d'Etudes Scientifiques et Techniques d'Aquitaine, 33114 Le Barp, France*

Abstract

In previous studies of bunching we operated our 2.2 MeV, 800 A single shot FEL in the amplifier mode, and observed output power and bunching not only at the injected frequency of 35 GHz but also at the second-FEL resonant frequency near 3 GHz. In order to study the mechanism of the low-frequency bunching we have performed a series of measurements in SASE mode. Here we compare and contrast results obtained in these two modes. For both modes a detailed study of the formation of the low-frequency bunches during the pulse has been performed using a streak camera, as a function of both position in the wiggler and time in the pulse. Analysis of the images shows that despite the great difference in output power, the strong bunching we see is quite similar. In the amplifier mode the coexistence of the bunching at high frequency masks some of the details of the low-frequency bunch formation, whereas in the SASE mode the temporal and spatial behavior of bunching at 3 GHz is clearly seen. © 1999 Elsevier Science B.V. All rights reserved.

PACS: 41.60.Cr; 41.75.Ht; 41.75.Lx

Keywords: Low-frequency bunching; SASE measurements; Free-electron laser

1. Introduction

The free electron laser (FEL) community has recently become interested in the phenomenon called self-amplified spontaneous emission (SASE) [1], and several groups have reported new experimental results [2–4]. One reason for this interest is the possibility of obtaining intense sources of far ultraviolet or X-rays by using GeV energy electron beams in a SASE FEL [5,6]. Theoretical work on SASE may be traced back to the early days of FEL [7–10], and a thorough discussion of superradiance in high-gain FELs is given in the review article of Bonifacio and co-workers [11]. These authors point out that a key parameter in superradiance is slippage, which may be defined as $v_g/v_z - 1$, where v_g and v_z denote the group velocity of the radiation and the mean axial velocity of the electrons, respectively. Our results were obtained in a waveguide which has two FEL resonant frequencies and they concern mainly the lower frequency for which the group velocity is much less than the electron axial speed. The corresponding slippage parameter is approximately -0.7, and previous experiments have not operated in such a regime.

* Corresponding author.

0168-9002/99/$ - see front matter
PII: S0168-9002(99)00104-7

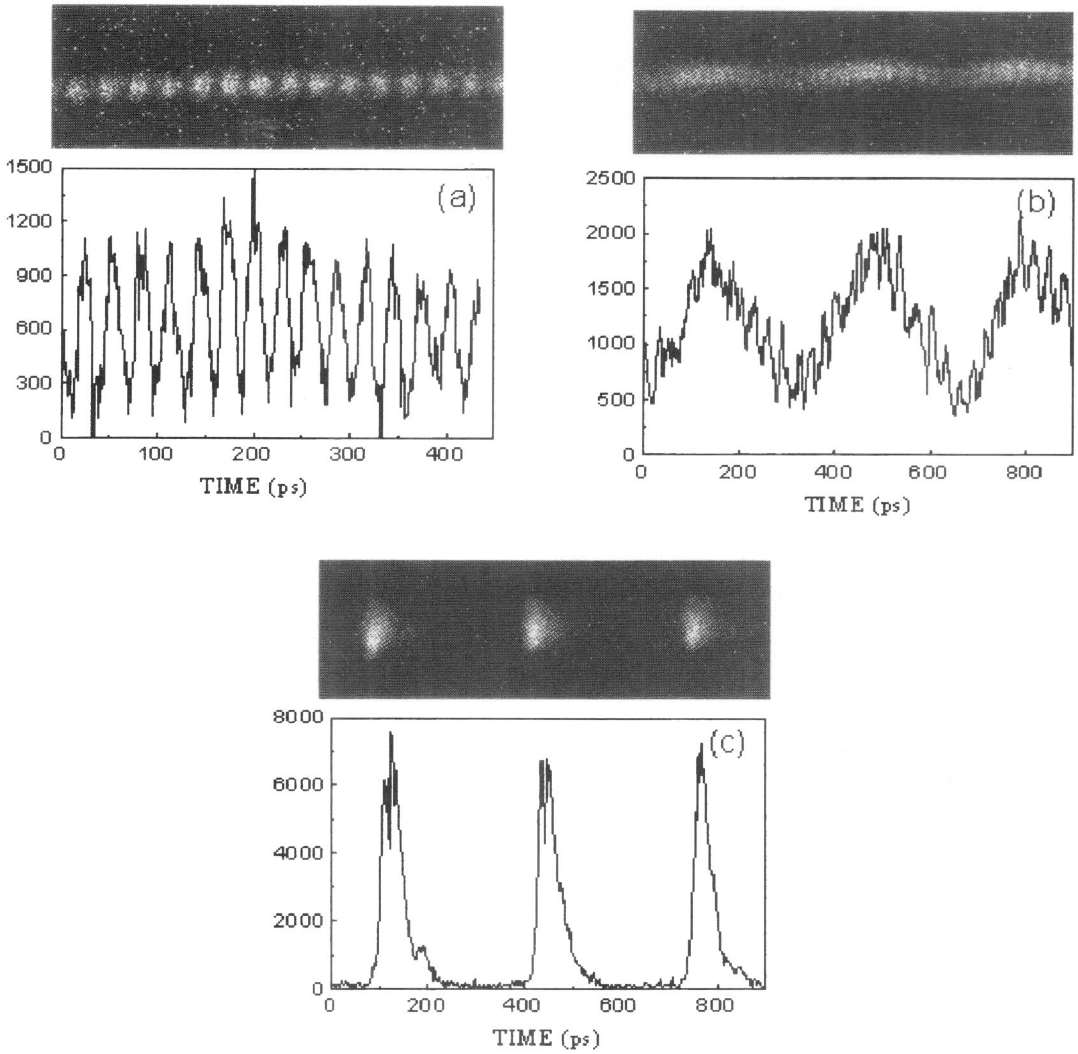

Fig. 1. Streak camera pictures and their digitized representations, obtained when running in the amplifier mode at 35 GHz. Strong electron bunching at the lower resonant frequency is apparent at the end of the pulse.

2. Bunching in amplifier mode

Although the observation of high-power output from a FEL is convincing evidence that bunching has occurred, optical measurement of the bunching itself has been performed only recently. With the CEA-CESTA FEL operating in the amplifier mode, a direct optical observation of electron bunching at 35 GHz was carried out [12,13]. After improving the experiment, we obtained the streak camera pictures and their digitised representations that are shown in Fig. 1 [14]. Early in the beam pulse the electrons were bunched at the injection frequency, but as the pulse continued, very strong bunching at a frequency near 3 GHz emerged and finally dominated the high-frequency bunching. Since the strong signal at 3 GHz, which corresponds to the second-FEL resonance in our 61 mm

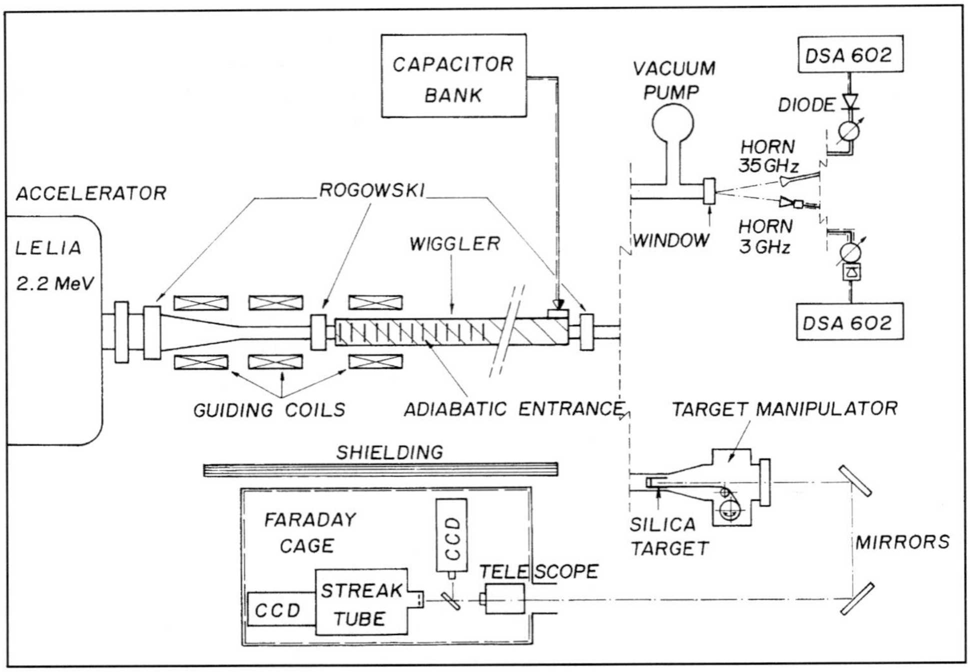

Fig. 2. The SASE experimental lay-out, with at right, two distinct detector configurations. At the top are shown the horns and lines for measuring output power and frequency, while below are shown the beam position camera and the streak camera for observing bunching.

diameter waveguide, was not injected, it may be considered as evidence for SASE. However, the presence of the strongly amplified 35 GHz signal makes such a straightforward interpretation risky, and it was decided to repeat the experiment in the SASE mode, with no input signal at either resonant frequency.

3. SASE experiment

A schematic of our experiment is shown in Fig. 2. The induction linac LELIA [15] delivers a 800 A electron beam of energy 2.2 MeV. This beam is transported by three solenoidal magnets and then injected into a helical wiggler consisting of 26 periods of 12 cm. No axial guide field is employed in the wiggler itself, and a six period adiabatic entrance gradually increases the wiggler field from zero to its desired value.

In the right-hand side of the figure we show two distinct set-ups, the upper for observing RF power at both frequencies, the lower for performing optical measurements of beam position (with the gated camera) and bunching (with the streak camera). Frequency measurements were made using heterodyne methods for the upper frequency and direct signal analysis with a fast oscilloscope for the lower. In order to study output power as a function of axial distance, a movable permanent magnet was used to deflect the electron beam into the beam-pipe at any desired position. The bunching of the electrons was observed by causing the beam to strike a 2 mm-thick fused silica target, producing thereby Cerenkov radiation at visible wavelengths. The target could be moved under vacuum, allowing us to observe the bunching anywhere along the axis. By varying the trigger delays of the cameras we could study the bunching as a function of time in the beam pulse.

In Fig. 3a we show the beam current, as measured at the wiggler entry (solid curve), and the light intensities (circles) as observed with the gated camera when the target is placed at period 15 of the

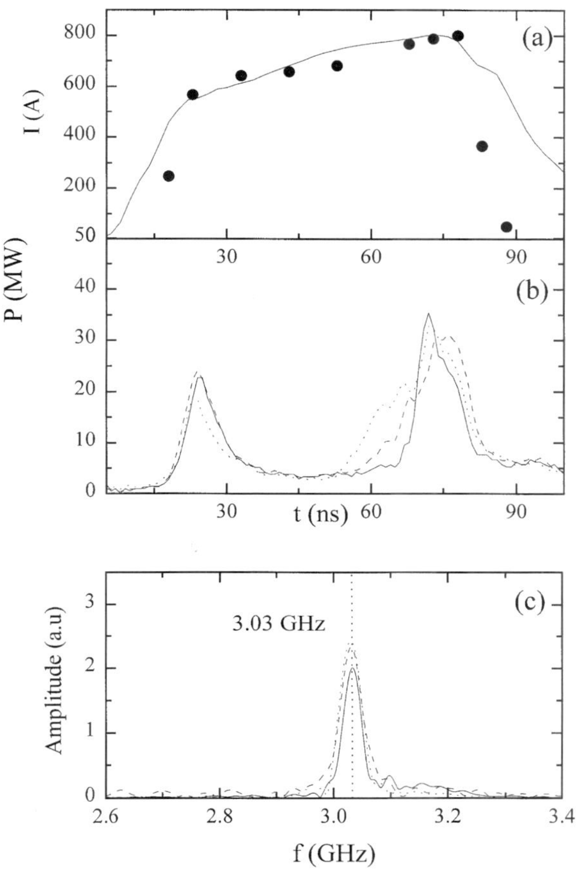

Fig. 3. (a) The beam current at the wiggler entrance (solid curve) vs. time together with integrated beam intensities seen by the camera at period 15 using a 5 ns exposure time (squares). (b) Reproducibility of the output power in the 3 GHz horn as a function of time in the pulse. (c) Fourier analysis of the 3 GHz horn signal for three different shots.

wiggler, both as a function of time. We have arbitrarily normalized the latter to agree with the maximum of the former, and we observe fair agreement between these two measurements, both of which indicate that the bulk of the current pulse lasts 60 ns. Fig. 3b shows output power as a function of time as detected with the low-frequency horn, and using the full length of the wiggler. The maximum RF power at 3 GHz was 40 ± 10 MW, which is somewhat less than what we observed in the amplifier mode in previous operation. The signal always had a two-peaked structure as shown.

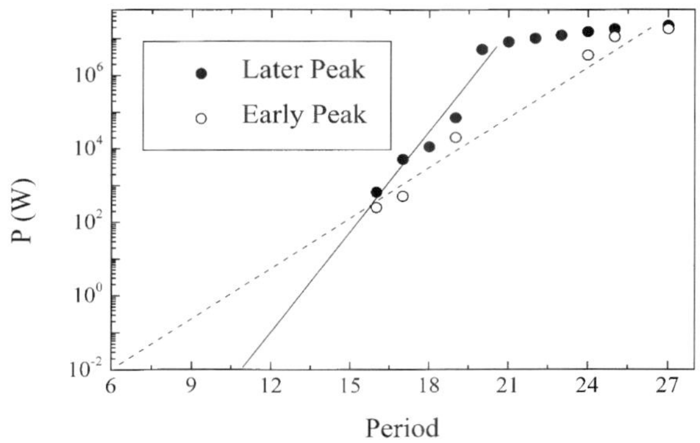

Fig. 4. Output power at 3 GHz as a function of the interaction length, for the first and second peaks.

At the upper frequency, the output power was limited to a few MW. Finally in Fig. 3c we show the Fourier analysis of the low-frequency horn signal for three shots, in the region near 3 GHz. Again a high degree of reproducibility is seen. On the basis of this stability, we feel justified in treating on the same footing data from different shots.

The output power at 3 GHz as a function of the interaction length is shown in Fig. 4, treating the early and later peak powers separately. For the former, the signal displays slow exponential growth throughout the wiggler, while the latter, after showing rapid growth early in the wiggler, displays a slower growth rate near the end of the wiggler, although saturation has not yet been reached. Since the time and space dependence of the SASE mechanism appears to play a major role in our experiment, we have mapped out the two-dimensional region spanned by axial distance and time in the pulse for which the output power and the bunching at 3 GHz are important. The output power map is shown in Fig. 5a, while the map showing the charge per bunch above a smooth background is displayed in Fig. 5b. In the latter, the region above the dashed line is characterized by increasingly sharp bunching, accompanied by substantial loss of beam, until in the upper right-hand corner there are no electrons. The maps are mutually consistent, and show that the strongest bunching and highest power occur near the end of the wiggler. However, the output power appears to last somewhat longer than the beam pulse, which is consistent with

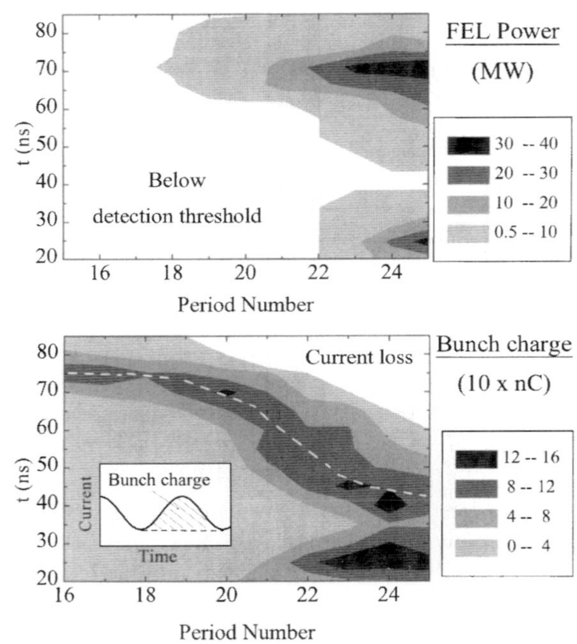

Fig. 5. Output power at 3 GHz (a) and charge per bunch above background (b) as a function of interaction length and time in pulse.

our situation of a slow wave and strong negative slippage.

4. Details of bunching

In Fig. 6 we show a comparison between our previous measurements of bunching in the amplifier mode (at 35 GHz, top) and the SASE measurements (bottom). The time interval between two adjacent photographs is 6 ns. In the amplifier mode, and for early times, the bunching at 35 GHz is clear, and the bunching at 3 GHz is barely visible under the high-frequency bunching. Towards the end of the amplifier mode pulse, the bunching at 35 GHz disappears, and the signal strongly resembles the corresponding SASE signal seen below it. In fact, if one ignores the presence of the 35 GHz bunching in the upper photographs, the 3 GHz bunching is quite similar in both cases. On the basis of these images, we suggest that the mechanism of generation of the 3 GHz bunching is independent of the presence of an appreciable signal at 35 GHz. However, the difference in output powers suggests that there may still be some cooperative effects between the two frequencies.

A sequence of streak camera images, taken at period 23 in the wiggler, and for various time slices in the beam pulse is displayed in Fig. 7. Below each photograph appears the corresponding beam current profile as a function of time. These pictures were taken using the nominal 100 ps/mm sweep speed of the streak camera, and each exposure lasts 1.66 ns. As can be seen by following the frames, the low-frequency bunching, weak in the first frame, grows with time until a large fraction of the current is confined to a short time interval. An important aspect of these images is the substantial transverse widening of the beam from early to late times, easily seen by comparing the first and last frames. We also see a significant decrease in the total current with time, which implies that electrons are being lost as the bunching sharpens. We have performed similar but less thorough analyses at several different target positions, and the time dependence of bunching we observe is similar to that shown here, even for such early periods as 17, where the output power is only of order 10 kW. However, the bunching occurs at increasingly later times as one moves upstream. From these observations we conclude that at fixed position, the transition from a uniform to a bunched beam takes place in a short time, 2–3 ns, and is then followed by a slow but steady loss of current.

In these photographs and the current profiles one begins with a relatively unbunched beam in frame 1, next one observes bunching at the fundamental frequency (i.e., 3 GHz) in frame 2, then one sees second and higher harmonics as the frame numbers increase. In the final frames, where the current has decreased substantially, the bunches appear almost as a sequence of Dirac Delta functions. However, these sharp peaks are caused by the loss from the beam of electrons situated between adjacent peaks. In our experiment, the slippage is both large and negative, and the frequency is very close to cutoff. Some combination of these effects might be the cause of this loss, and more experiments are necessary to understand the phenomena. It is interesting to note that Barletta and collaborators [16] have generated current distributions resembling

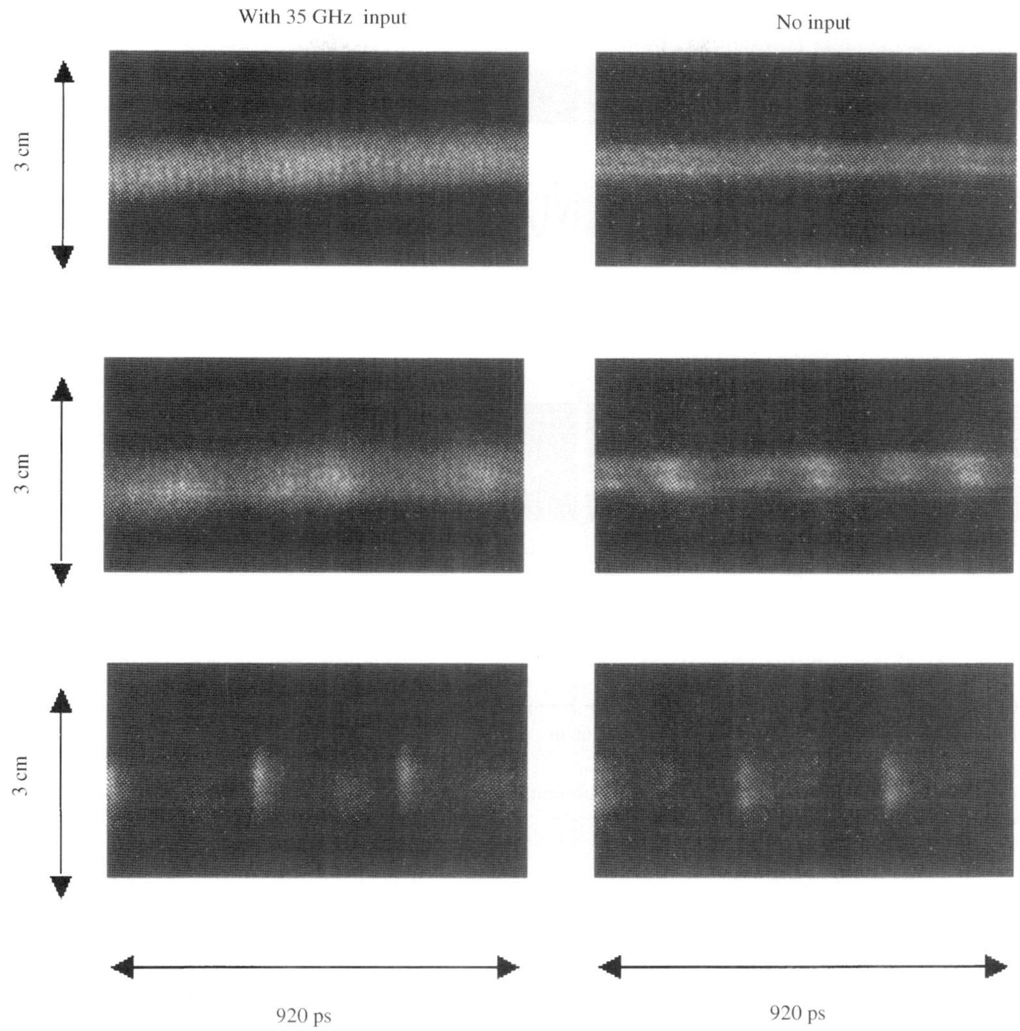

Fig. 6. Streak camera photographs of bunching when the FEL is running in the amplifier mode (top) and SASE mode (bottom), for three different values of time in the beam pulse.

ours, although the beam energy and output power level are quite different.

5. Conclusions

In conclusion, we have observed a highly reproducible SASE mechanism, responsible for high microwave power output (≈ 40 MW) and sharp electron bunching at the lower resonant frequency of our pulsed FEL. In comparison with our previous observation of bunching at 3 GHz when running in the amplifier mode at 35 GHz, the bunching we observe here is uncontaminated by appreciable interference from the upper frequency, and a very detailed study of its time dependence has been presented. We observed that significant bunching can occur without high output power, and that extremely sharp bunching is accompanied by an important loss of electrons.

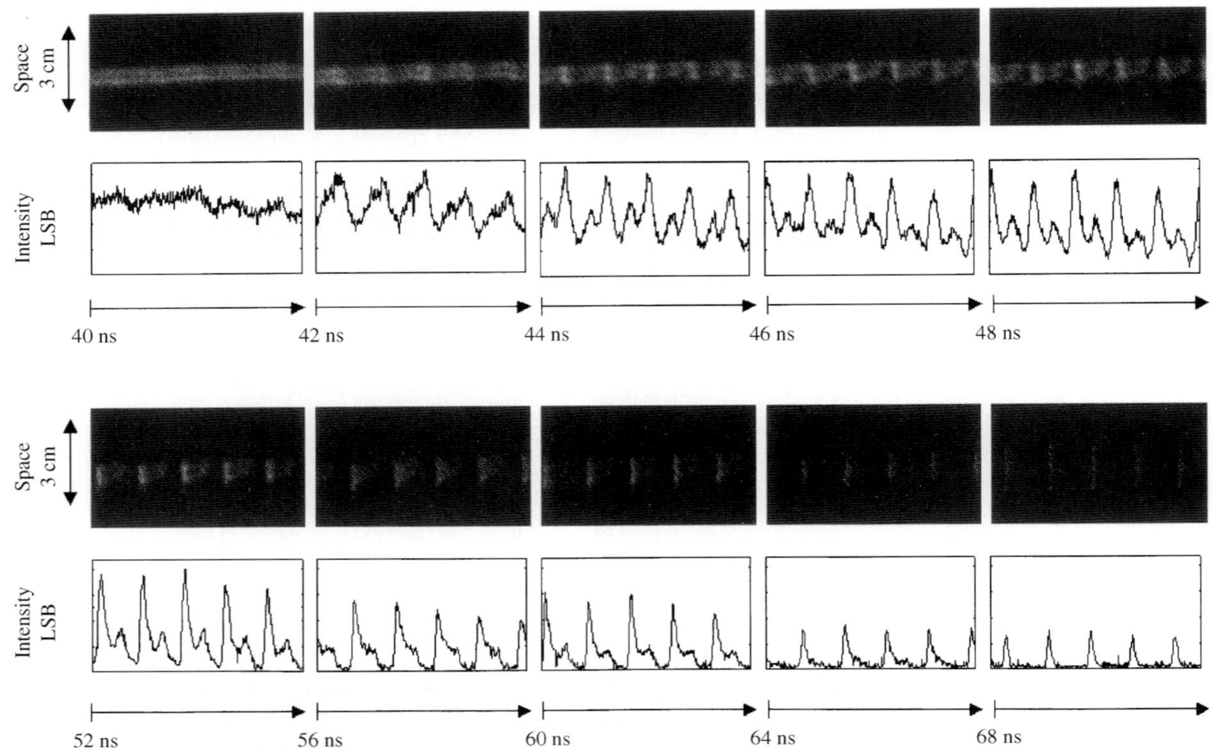

Fig. 7. Streak camera photographs and their digitized representations, as observed at wiggler period 23, for ten different times in the beam pulse and a 1.66 ns time interval.

Acknowledgements

We thank D. Gogny for his advice and encouragement, M. Lavergne, C. Vermare, and D. Villate for their help with the optical measurements.

References

[1] K.-J. Kim, Nucl. Instr. and Meth. A 393 (1997) 147.
[2] R. Prazères et al., Phys. Rev. Lett. 78 (1997) 2124.
[3] M. Hogan et al., Phys. Rev. Lett. 80 (1998) 289.
[4] D. Bocek et al., Nucl. Instr. and Meth. A 375 (1996) 13.
[5] R. Thatchin et al., Nucl. Instr. and Meth. A 375 (27) (1996) 4.
[6] J. Rossbach et al., Nucl. Instr. and Meth. A 375 (1996) 269.
[7] H.P. Freund, T.M. Antonsen, Principles of Free-electrons Lasers, Second ed., Chapman & Hall, London, 1996.
[8] R. Bonifacio, C. Pellegrini, L. Narducci, Opt. Commun. 50 (1984) 373.
[9] Ya.S. Derbenev, A.M. Kondratenko, E.L. Saldin, Nucl. Instr. and Meth. A 193 (1982) 415.
[10] K.-J. Kim, Phys. Rev. Lett. 57 (1986) 1871.
[11] R. Bonifacio et al., Riv. Nuovo Cimento 13 (9) (1990).
[12] J. Gardelle, J. Labrouche, J.L. Rullier, Phys. Rev. Lett. 76 (1996) 4532.
[13] J. Gardelle et al., Phys. Plasma 3 (1996) 4197.
[14] J. Gardelle et al., Phys. Rev. Lett. 79 (1997) 3905.
[15] J. Bardy et al., Nucl. Instr. and Meth. A 304 (1991) 311.
[16] W. Barletta et al., Nucl. Instr. and Meth. A 329 (1993) 348.

ELSEVIER

Nuclear Instruments and Methods in Physics Research A 429 (1999) 209–212

NUCLEAR INSTRUMENTS & METHODS IN PHYSICS RESEARCH Section A

www.elsevier.nl/locate/nima

Measured free-electron laser microbunching using coherent transition radiation

A. Tremaine[a,*], J. Rosenzweig[a], S. Anderson[a], P. Frigola[a], M. Hogan[a], A. Murokh[a], C. Pellegrini[a], D. Nguyen[b], R. Sheffield[b]

[a]*University of California, Los Angeles, 405 Higard Avenue, Los Angeles, CA 90095, USA*
[b]*Los Alamos National Laboratory, Los Alamos, NM 87545, USA*

Abstract

The microbunch distribution of an electron beam exiting a SASE free-electron laser has been measured using the emitted coherent transition radiation (CTR) produced from a thin aluminum foil placed at the end of the undulator. The wavelength of the coherent transition radiation is shown to be the same as the FEL wavelength, and thus a measure of the beam microbunch spacing. Also, the study of the CTR linewidth and angular acceptance of the radiation captured are shown to be derived from this coherent radiative process. Scattering effects on the forward emitted transition radiation from the electron beam traversing an aluminum foil are also considered. © 1999 Elsevier Science B.V. All rights reserved.

Keywords: CTR; FEL wavelength; Microbunching; Scattering effects

1. Introduction

Longitudinal modulation of electron beams is now generated with many types of devices such as the free-electron laser (FEL) [1], inverse FELs [2], and plasma and structure acceleration [3]. The periodic beam modulation formed from these devices is at time scales where usual diagnostic methods like streak cameras [4] and RF sweeping [5] cannot be used. Instead of these time-domain methods, frequency-domain measurements using coherent transition radiation (CTR) from metallic foils have shown promise in the measurement of very short electron pulses [6–8]. The high-gain process of FELs is directly related to beam micobunching and gives an opportunity to use CTR to measure microbunch spacing down to several femtoseconds.

2. Theoretical background

Describing the beam distribution exiting an FEL has been done in Ref. [9], but in the present experiment, asymmetries in transverse beam dimensions were present at the exit of the wiggler and at the CTR foil and must be included. Thus, the electron

* Corresponding author.

PII: S0168-9002(99)00105-9

beam distribution now looks like

$$f(r,z)=\frac{N_b}{(2\pi)^{3/2}\sigma_x\sigma_y\sigma_z}\exp\left(-\frac{x^2}{2\sigma_x^2}-\frac{y^2}{2\sigma_y^2}-\frac{z^2}{2\sigma_z^2}\right)\times\left[1+\sum_{n=1}^{\infty}b_n\sin(nk_r,z)\right] \quad (1)$$

where N_b is the number of electrons, k_r is the FEL radiation wavenumber and thus the modulated micro-bunch beam wavenumber, and b_n is the bunching factor. Analysis of the CTR follows Ref. [8] where the energy spectrum is given by

$$\frac{d^2U}{d\omega\, d\Omega}\cong N_b^2F_L(\omega)F_T(\omega,\theta)\chi(\theta)\left.\frac{d^2U}{d\omega\, d\Omega}\right|_{\text{single } e-}. \quad (2)$$

Here $F_L(\omega)$ and $F_T(\omega,\theta)$ are the square of the longitudinal and transverse beam profile Fourier transforms. $\chi(\theta)$ is a divergence factor which is usually ignored in analysis, but will be shown to be very important for forward emitted CTR. Putting Eq. (1) into Eq. (2) and integrating, the CTR photon number angular dependence is found to be

$$\frac{dN_{\text{CTR}}}{d\theta}=\frac{\alpha(N_bb_N)^2}{4\sqrt{\pi}nk_r\sigma_z}\left(\frac{\sin^3(\theta)}{(1-\beta\cos(\theta))^2}*\chi(\theta)\right)\times\exp[-(nk_r\sin(\theta))^2(\sigma_x^2\sin^2(\phi)+\sigma_y^2\cos^2(\phi))] \quad (3)$$

where θ and ϕ are the polar and azimuthal angles and α is the fine structure constant. Most of the CTR light is found in a small annular cone with a maximum, using an axisymmetric beam of size σ_r, near $\theta\approx 1/\sqrt{2nk_r\sigma_r}$. Ignoring the divergence factor and integrating Eq. (2), the number of photons can be found by the straightforward relation

$$N_{\text{CTR}}=\frac{\alpha(N_bb_n)^2}{4\sqrt{\pi}k_r\sigma_z}\left(\frac{\gamma}{nk_r}\right)^4\left(\frac{\sigma_x^2+\sigma_y^2}{\sigma_x^3\sigma_y^3}\right). \quad (4)$$

In order to best maximize the number of photons, Eq. (4) shows the electron beam must be very dense at the foil. It should also be noted that the above relation is for normal incidence of the electron beam on the foil.

3. Experimental setup

The present experiment was performed at Los Alamos National Laboratory. A 1300 MHz RF photoinjector produces a 100 bunch train of low-emittance high-current bunches. Relevant beam parameters are given in Table 1. The wiggler used is the permanent magnet 2 m Kurchatov undulator with a period of 2.01 cm and is the same one used in the high-gain SASE FEL in Ref. [10].[1] For the CTR experiment, a 6 μm foil was put on an insertable mount which when inserted was 2 cm away from the last period of the wiggler and normal to the electron beam. Also, the foil covered the entire exit aperture of the wiggler so when inserted, all the FEL radiation would be reflected back and the only light to propagate down the optical beamline would be the forward emitted CTR. The close proximity of the foil to the end of the wiggler is very important. First, there would be very little electron beam debunching [11] from space-charge effects and second, the source points and the optical beamline for the FEL (foil extracted) and CTR (foil inserted) radiation would be the same. A calibrated HgCdTe detector was installed 3.5 m away from the source point causing the angular acceptance of the optical beamline to be about 12 mrad. Mainly coherent radiation would be collected, since the incoherent spectrum is peaked at a much larger angle, $\theta_{\text{incoh}}\approx 1/\gamma\approx 30$ mrad.

For the present experiment, the system was run at a charge of 1.5 nC and the conditions for the high-gain SASE FEL in Ref. [9] were reestablished. Once the maximum SASE signal was obtained and the foil inserted, it was found that a minor RF phase adjustment of 2° was necessary in order to maximize the CTR signal (and thus the micro-bunching of the beam) emanating from the foil. In this case, adjustments to minimize the spot size through RF focusing would enhance the CTR radiation as predicted by Eq. (4) and also negligibly change the final beam energy. Simulations using the 3-D FEL code Ginger were done for a series of experimental parameters giving a bunching factor

[1] SASE FEL gain in excess of 10^5 was first seen in this experiment.

Table 1
Electron beam and undulator experimental parameters

Parameter	Value
Beam energy (E)	17.5 MeV
Peak current (I)	140 A
Charge/bunch (Q)	1.5 nC
Bunch length (τ)	11 ps
(FWHM)	
Energy spread ($\Delta\gamma/\gamma$)	0.5%
Wiggler period (λ_W)	2.01 cm
On axis field (B_0)	7.4 kG
FEL wavelength (λ_r)	13 μm
FEL parameter (ρ)	0.008
Rms beam sizes (σ_x, σ_y)	210, 160 μm

between 0.008 and 0.01 for the first harmonic and negligible bunching for the higher harmonics.

4. Results

In order to compare the measured photon number with that predicted in Eq. (4), the attenuation factor $\chi(\theta)$ introduced in Ref. [7] must now be examined. When the radial scattering of the beam through the foil is much less than the angular spread of the incoherent radiation, $\sigma_r' \approx \theta_{scat} \ll 1/\gamma$, the attenuation can be ignored. When this condition is violated, $\chi(\theta)$ becomes a complicated integral and must be done numerically and quickly becomes less than unity. Defining a degradation factor, η, it is found the forward emitted CTR signal through the 6 μm foil is reduced by a factor of 0.61. It should be noted that this experiment was first attempted with a 50 μm foil in which the CTR was greatly smaller than expected ($\eta = 0.11$) leading to the examination of scattering issues in the foil. Using the predictions from Ginger and numerically integrating Eq. (2), the range of expected photons is $N_\gamma = 2.8 \times 10^8$–$4.4 \times 10^8$. The calibrated measured photon number per pulse from the HgCdTe detector was 3.5×10^8. To within experimental and simulation uncertainty, the numbers agree quite well.

Next the SASE and CTR signals were sent through a Jerrel Ash monochrometer. The results are shown in Fig. 1. The SASE signal is attenuated by a factor of 3 and the CTR signal is multiplied by 12 to give it the same scale as SASE and the resolution for the monochrometer was found to be 0.177 μm. As seen in the figure, the signals are nearly centered around the same wavelength (13 μm) as expected. This shows that the microbunching of the beam is at the SASE radiation wavelength agreeing with theory.

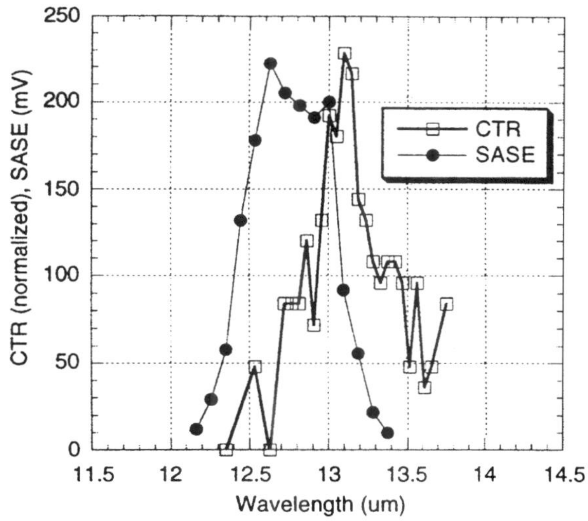

Fig. 1. CTR/SASE through monochromator versus wavelength.

5. Conclusion

The microbunching of an electron beam due to the FEL gain process has been shown to have a spacing equivalent to that of the radiation wavelength. Also, the narrow angular spectrum and form of the spectral linewidth were measured and shown to correspond with the theoretical predictions. When doing forward scattered CTR experiments, foil thickness and scattering effects must be carefully considered. This method of diagnosing a microbunched electron beam is being planned for the Visible FEL experiment at Brookehaven National Lab in which the microbunch spacing and radiation wavelength will be near 800 nm.

References

[1] R. Bonifacio, C. Pellegrini, L. Narducci, Opt. Comm. 50 (1984) 373.

[2] A. Van Steenburgen et al., Phys. Rev. Lett. 80 (1998) 289.
[3] J. Rosenzweig et al., Phys. Rev. Lett. 74 (1995) 2467.
[4] J. Gardelle et al., Phys. Rev. Lett. 79 (1996) 3905.
[5] B. Carlsten, S.J. Russell, Phys. Rev. E 53 (1996) R2072.
[6] U. Happek, A.J. Sievers, E.B. Blum, Phys. Rev. Lett. 67 (1991) 2962.
[7] A. Murokh et al., Nucl. Instr. and Meth. A, submitted for publication.
[8] Yukio Shibata et al., Phys. Rev. E 50 (1994) 1479.
[9] J.B. Rosenzweig, G. Travish, A. Tremaine, Nucl. Instr. Meth. A 365 (1995) 255.
[10] M. Hogan et al., Phys. Rev. Lett., submitted for publication.
[11] J. Rosenzweig et al., Nucl. Instr. Meth. A 393 (1997) 376.

ELSEVIER

Nuclear Instruments and Methods in Physics Research A 429 (1999) 213–218

NUCLEAR INSTRUMENTS & METHODS IN PHYSICS RESEARCH Section A

www.elsevier.nl/locate/nima

Power spectrum calculation for the Cornell Wiggler A SASE experiment at BNL

Li Hua Yu

National Synchrotron Light Source, Brookhaven National Laboratory, Upton, NY 11973, USA

Abstract

Recently, we showed (Yu, Phys. Rev. E 58 (1998) 4991) that the widely used simulation code TDA3D, even though a single-frequency code, can be used to determine the power spectrum in the SASE process with excellent approximation in the exponential growth regime. In this paper, we apply this method to the BNL Cornell Wiggler A SASE experiment as an example. When the gain is not very high, there are many modes in the radiation, which seems to make the analytical calculation very difficult. However, we show that the increment of the radiation due to SASE over the spontaneous radiation can be expanded in terms of guided modes with rapid convergence. Thus when the spontaneous radiation is subtracted from the SASE power during the calculation, there is a good agreement between the analytical theory and the numerical simulation. © 1999 Elsevier Science B.V. All rights reserved.

Keywords: Power spectrum; Simulation code; SASE; Coupling coefficient

1. Introduction

Recently, we showed [1] that the TDA3D code, which has been modified to include harmonic generation calculation, can be used to calculate the power spectrum. One reason that made this possible is that in the linear regime there is a very simple scaling relation between the number of simulation particles and the output power: the output power is inversely proportional to the number of simulation particles. Hence, the number of simulation particles can be made much smaller than the actual number of electrons in the beam, making the simulation practical.

This method uses an entirely different approach to reduce the number of simulation particles as compared with the simulation codes such as GINGER. We shall briefly compare these two approaches. The codes such as GINGER use an artificial initial distribution to simulate the SASE start-up process. To suppress the increased shot noise due to the limited number of simulation particles, the codes are based on a distribution with equally spaced particles. To introduce a controlled noise, they generate a random deviation from the equally spaced distribution with a controlled rms value of the displacements. The rms displacement is chosen to reproduce the same mean and variance of the bunching parameter. While the mean and variance of the bunching parameter simulate the initial status of the system, it is not evident that the higher moments of this quantity would not affect the high-gain process, it is also not evident that the mean and variance of the relevant quantity would remain to be the same as the realistic distribution during the high-gain process, even though the

PII: S0168-9002(99)00106-0

simulations did show an agreement with the linear high-gain theory.

As compared with these codes, our method uses a realistic distribution instead of an artificial evenly spaced distribution. We do not attempt to suppress the noise due to the limited number of simulation particles. Rather, use the scaling relation to go from the simulation case with increased start-up noise (due to the reduced number of simulation particles) to the realistic case. In this manner, we obtain the correct radiation power in the linear regime.

Another reason that we can use a single-frequency code such as TDA3D to calculate the intrinsically broad band SASE process is the following. In the original TDA code, all the simulation particles are limited to within one optical wavelength, or, one cell. During a later modification of the code for harmonic generation calculation, we extended the code such that the simulation particles could be in an arbitrary number of wavelengths. It is easy to see that if the number of cells is n_ℓ, then the code is describing a fictitious electron beam distribution with longitudinal periodic structure of n_ℓ optical wavelengths. That is, we artificially set a periodic boundary condition on the electron beam with period equal to n_ℓ optical wavelengths. In this case, the radiation spectrum has a line structure with frequency spacing ω_s/n_ℓ, where ω_s is the optical frequency.

We denote the slippage distance by $l_s = N_w\lambda_s$, and the distribution period by $l = n_\ell\lambda_s$, then when $l \gg l_s$, the line spacing ω_s/n_ℓ is much narrower than the spontaneous radiation width ω_s/N_w and the dense line structure gives a profile of the spontaneous spectrum. When we choose the period to be equal to the slippage distance $l = l_s$, i.e., when $n_\ell = N_w$, the line spacing is equal to the radiation spectrum width, and hence there is only one line. The slippage $N_w\lambda_s$ is equal to the spacing between the periodic boundaries of the electron beam. Hence there is no interaction between any two of the idealized periods of the electron distribution. The calculated output energy within one idealized period of the electron distribution is the same as it would be from a non-periodic structure in the electron beam, i.e., the same as for the realistic case for SASE process. The output power is shown to be $1/N_w(\partial P_{spon.}/\partial\omega/\omega)$.

Now from the 1D analytic theory of SASE, we know that the full bandwidth of the SASE spectrum is [1] $(1/N_w)\sqrt{L_w 3/4\pi L_G} \cong (1/N_w)\sqrt{L_w/4L_G}$. This width is narrower than $2/N_w$ as long as $L_w < 16L_G$. So when the wiggler length is much shorter than 16 power gain lengths, if we choose n_ℓ to be equal to the number of periods N_w, to good approximation there is only one line within the bandwidth centered around the resonant frequency. Therefore, when the wiggler length is much shorter than 16 power gain lengths TDA3D serves as a good approximation to the output power even though it handles only one single spectral line. In this paper we assume the electron bunch is much longer than the slippage and the bunch shape is sufficiently smooth.

In Section 2, we shall apply the new method to simulate the spontaneous radiation power spectrum in the BNL Cornell Wiggler A experiment as an example and a check of the calculation. In Section 3, we apply the method to the SASE calculation of the same experiment, and describe the analytical calculation.

2. The calculation of the spontaneous radiation spectrum

We consider the parameters for the BNL Cornell Wiggler A SASE experiment: the radiation resonant wavelength is $\lambda_s = 5$ μm, the wiggler period is $\lambda_w = 3.3$ cm, the wiggler length is $L_w = 1.98$ m, the number of period is $N_w = 60$, with wiggler parameter $K_{max} = 1.44$, and the e-beam energy is $\gamma = 82$. We take a small current $I_0 = 10$ A to calculate spontaneous radiation. Our analytical calculation based on the well-known spontaneous radiation theory shows that the power spectrum, integrated over the full solid angle of the radiation, is as shown by the solid curve in Fig. 1. Because when the radiation angle is deviating from the forward direction, the wavelength is always shifted to longer, even though for a very small solid angle the radiation spectrum is a simple sinc function of width $1/N_w$, the integrated spectrum over all angle is more like a step function with rising width $1/N_w$ near the resonant wavelength λ_s. We can show that

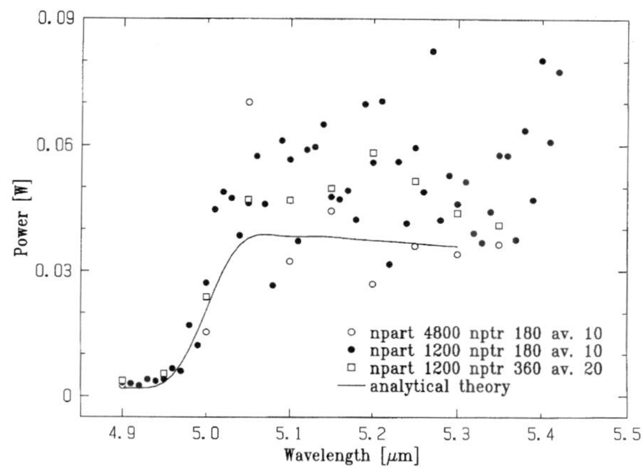

Fig. 1. Output power spectrum for spontaneous radiation as a function of wavelength.

the peak of the power spectrum, integrated over the full solid angle, is given by

$$\left(\frac{\partial P_{\text{spon.}}}{\partial\omega/\omega}\right)_{\text{peak}} \cong \pi\theta_{\text{w}}^2 B_0 \tag{1}$$

where $\theta_{\text{w}} \equiv \sqrt{2\lambda_{\text{s}}/L_{\text{w}}}$ is defined as the opening angle of the spontaneous radiation within a sufficiently narrow bandwidth, and B_0 is the brightness in the forward direction and at the resonant frequency, given by

$$B_0 = \left(\frac{\partial^2 P_{\text{spon.}}}{(\partial\omega/\omega)\partial\Omega}\right)_{\theta=0,\omega=\omega_{\text{s}}}$$

$$= \frac{eZ_0 I_0}{4\pi} N_{\text{w}}^2 \gamma^2 \frac{K_{\max}^2}{(1+K_{\max}^2/2)^2}[JJ]^2\omega_{\text{s}} \tag{2}$$

with $Z_0 = 377\ \Omega$ the vacuum impedance, and $[JJ]$ the Bessel factor. We can also show that at the resonant wavelength λ_{s}, the power spectrum is half of the peak height, and the peak is positioned at a longer wavelength, away from λ_{s} by a space of order but less than $\lambda_{\text{s}}/N_{\text{w}}$. For our example, these formulas give $\theta_w = 2.2$ mrad and $1/N_{\text{w}}(\partial P_{\text{spon.}}/(\partial\omega/\omega))_{\text{peak}} = 0.043$ W.

As explained before, to calculate the power spectrum, we choose the number of wavelength cells in the TDA3D calculation to be $N_{\text{w}} = 60$. The number of electrons in one wavelength is then $N = I_0\lambda_{\text{s}}/ec = 1.1\times10^6$. Because the algorithm used by TDA3D [2], we must always specify an input power to normalize the calculation, we choose the input power to be 10^{-10} W, which is small enough that the output is entirely determined by the shot noise of the simulation particle, and not affected by this number. When the number of simulation particles is $N' = 1200$, for the given parameters, we found that at $\lambda = 5.05$ μm, the radiation power reaches the peak value of $P' = 40$ W, after averaging over many runs. Thus, using the scaling relation, we find that the real radiation power spectrum is

$$\frac{1}{N_{\text{w}}}\left(\frac{\partial P_{\text{spon.}}}{\partial\omega/\omega}\right)_{\text{peak}} = \frac{N'}{N}\langle P'\rangle$$

$$= \frac{1200}{1.1\times10^6}\times 40 = 0.04\ \text{W}. \tag{3}$$

To test the convergence of the simulation, Fig. 1 gives the results for several sets of simulation of the power spectrum as a function of the wavelength λ and compare them with the analytical theory. The figure also indicates the number of runs of TDA3D for each of the average points. The figure shows that when the number of radial mesh points NPTR = 360 and average over 20 runs, the simulation is converging to the analytical theory.

3. The simulation of SASE

To simulate SASE, we assume an idealized distribution used in Ref. [3]. The transverse distribution is a step function profile with a constant current density within a radius of $2\sigma_x$, where $\sigma_x = 170$ μm is the rms radius, the current density is zero outside this radius. We assume all the electron momentum is parallel to the wiggler axis. We assume both the horizontal and vertical focusing is zero. Use the method described in Section 2, we plot in Fig. 2 the power spectrum at the resonant frequency ω_{s} as a function of current, varied from zero to 110 A, every point is an average over 10 run with different initial random number seeds. In this calculation we used five azimuthal modes with $m = 0, \pm 1, \pm 2$.

For this idealized model, the SASE power in the guided modes is explicitly solved [3,4]. The power spectrum in a mode $n = \{j, m\}$ (j is the radial node number, m is the axial node number) is

$$\left(\frac{\mathrm{d}P_{\mathrm{nn}}}{\mathrm{d}\omega}\right)_{\mathrm{SASE}} = \frac{1}{9}e^{L_{\mathrm{w}}/L_{\mathrm{G}_n}}C_n(\tilde{a})\left[\frac{2L_{\mathrm{G}_n}}{L_{\mathrm{w}}}\left(\frac{\mathrm{d}P}{\mathrm{d}\omega}\right)^{L_{\mathrm{w}}}_{\mathrm{spon.}}\right] \tag{4}$$

where L_{G_n} is the power gain length, and $\tilde{a}$ is the scaled beam size defined by $\tilde{a}^2 = 16k_{\mathrm{s}}k_{\mathrm{w}}\rho\sigma_x^2$, with k_{s}, k_{w} the wavenumber for the radiation and the wiggler, respectively. ρ is the Pierce parameter [5] given by

$$(2\rho)^3 = \frac{n_0 Z_0 e^2 K_{\mathrm{rms}}^2 [JJ]^2}{2m\gamma^3 k_{\mathrm{w}}^2 c} \tag{5}$$

where n_0 is the peak current density, in our case it is just the current density within the edge radius $2\sigma_x$, since it is a constant. K_{rms} is the rms wiggler parameter. The gain length is given by $L_{\mathrm{G}_n} = \lambda_{\mathrm{w}}/8\pi\rho\mathrm{Im}(\lambda_n(\tilde{a}))$. Thus the power spectrum Eq. (4) is completely determined by the scaled beam size $\tilde{a}$ through two functions: the coupling coefficient C_n and scaled growth rate λ_{n}. The physical meaning of Eq. (4) is clear now: the SASE input noise is the spontaneous radiation power spectrum within two power gain lengths $[(2L_{\mathrm{G}_n}/L_{\mathrm{w}})(\mathrm{d}P/\mathrm{d}\omega)^{L_{\mathrm{w}}}_{\mathrm{spon.}}]$, and this input noise is coupled by the coupling C_n, and then amplified by a factor $\frac{1}{9}\mathrm{e}^{L_{\mathrm{w}}/L_{\mathrm{G}_n}}$ to give the output power spectrum.

The two functions C_n and λ_n are calculated and given in detail by Ref. [1], and to a good approximation when $\tilde{a} \geq 0.25$, the calculated results are fit with

$$\mathrm{Im}(\lambda_n) \cong \frac{\sqrt{3}}{2}\mathrm{e}^{-(1/\tilde{a}\sqrt{1+\tilde{a}^2})(\alpha_0+\alpha_1(1/\tilde{a}^2))} \tag{6}$$

and

$$C_n(\tilde{a}) \cong \frac{\sqrt{3}}{\pi\tilde{a}^2}\mathrm{e}^{-(1/\tilde{a}\sqrt{1+\tilde{a}^2})(\beta_0+\beta_1(1/\tilde{a}^2))} \tag{7}$$

where for the mode $\{1,0\}$, we have $\alpha_0 = 0.397$, $\alpha_1 = -0.0067$, $\beta_0 = 1.093$, $\beta_1 = -0.02$; while for the mode $\{1, \pm 1\}$ we have $\alpha_0 = 1.2625$, $\alpha_1 = -0.1494$, $\beta_0 = 5.082$, $\beta_1 = -0.5707$.

As an example, let us take $I_0 = 110$ A. Using Eq. (5), we find that the Pierce parameter $\rho = 8 \times 10^{-3}$, and the scaled beam size $\tilde{a} = 0.95$. Applying these to Eq. (6), and Eq. (7), we find the power gain lengths and coupling coefficients for the mode $n = \{1,0\}$ to be $L_{\mathrm{G}} = 0.26$ m, $C = 0.24$, and for mode $m = \{1, \pm 1\}$, $L_{\mathrm{G}} = 0.43$ m, $C = 0.021$.

Using Eq. (4), we can calculate the power spectrum in each mode, and sum over modes. However, empirically, we find it is convenient to calculate the increment of SASE power over spontaneous radiation power ratio by summing over the corresponding ratio increments for all the modes. Thus we have

$$\frac{(\mathrm{d}P/\mathrm{d}\omega)_{\mathrm{SASE}}}{(\mathrm{d}P/\mathrm{d}\omega)^{L_{\mathrm{w}}}_{\mathrm{Spon.}}} - 1 = \left[\frac{1}{9}\mathrm{e}^{1.98/0.26} - 1\right] \times 0.24 \times \frac{2\times 0.26}{1.98}$$
$$+ 2\times\left[\frac{1}{9}\mathrm{e}^{1.98/0.43} - 1\right] \times 0.021 \times \frac{2\times 0.43}{1.98} + \cdots$$
$$= 13.6 + 0.2 + \cdots \approx 14. \tag{8}$$

The 1 in each term in the square parentheses is the subtraction of the contribution from the spontaneous radiation. The extra factor 2 in the second term is due to the two modes with $m = \pm 1$. For higher modes, the gain factor $\frac{1}{9}\mathrm{e}^{L_{\mathrm{w}}/L_{\mathrm{G}_n}}$ is rapidly reduced to nearly one or even smaller than one, and the formula Eq. (4) is not valid. However, the gain for these higher modes are negligible, this means that they only contribute to the spontaneous radiation, so $(\frac{1}{9}\mathrm{e}^{L_{\mathrm{w}}/L_{\mathrm{G}_n}} - 1)$ for these modes can simply be replaced by 0 as an approximation. The SASE over spontaneous radiation ratio is then $14 + 1 = 15$. The SASE power spectrum calculated this way is plotted against the simulation results in Fig. 2, showing a good agreement.

Naturally, one familiar with one-dimensional high-gain FEL theory would raise a question. When the gain is not very high there are three longitudinal terms, i.e., in addition to the growth term, there are other two terms comparable with the growth term: one is exponentially decaying, the other is oscillating, and the formula Eq. (4) is not valid. Why we can still use this formula even the total SASE over spontaneous radiation ratio is rather small, as shown in Fig. 3.

The answer is that, for the two modes we wrote down in Eq. (8), the gain factor is indeed much larger than one. In three-dimensional theory, the corresponding decaying mode and oscillating mode should be calculated as other transverse modes.

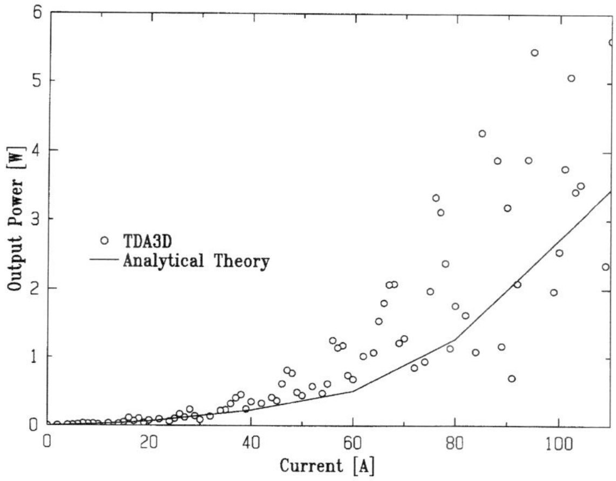

Fig. 2. Output power versus current for the step function model.

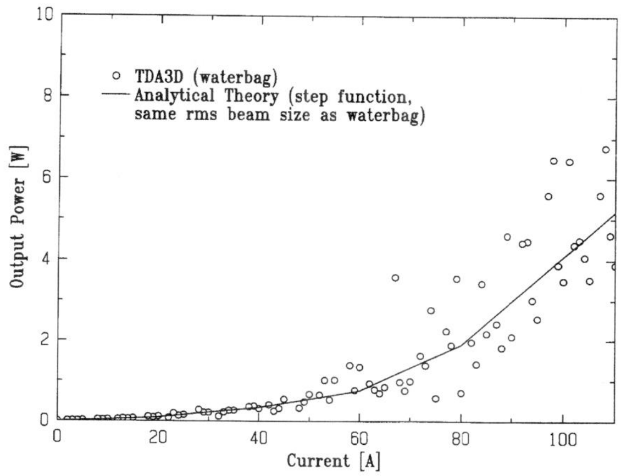

Fig. 3. Output power versus current for the waterbag model.

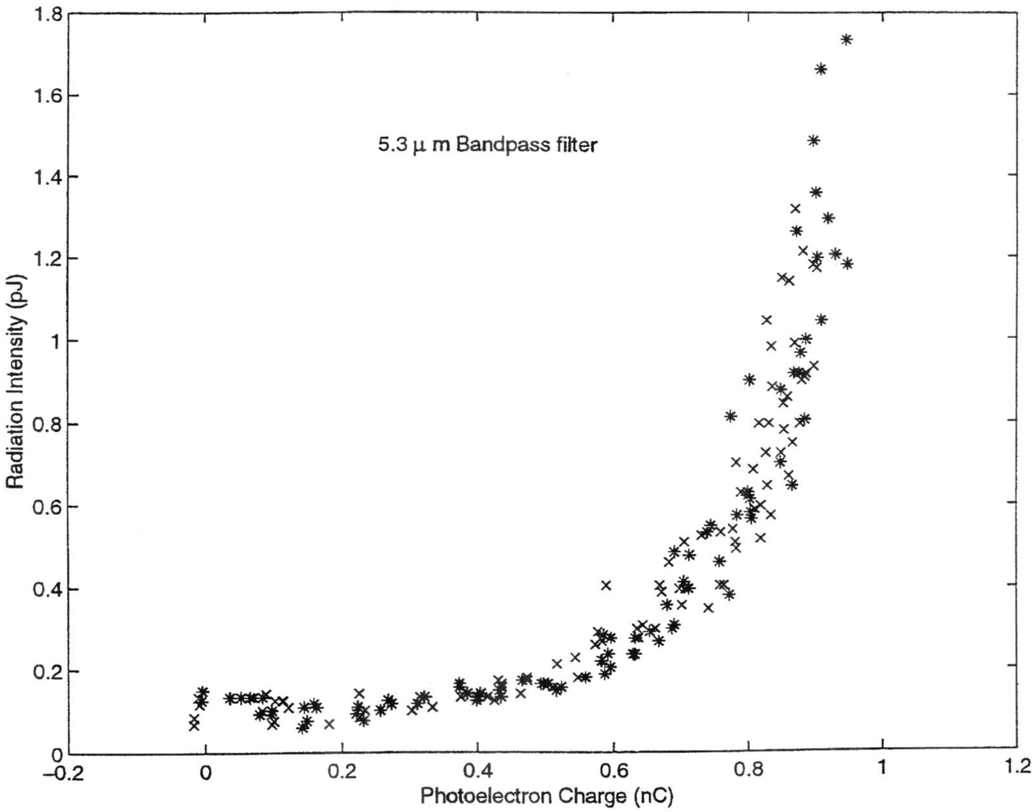

Fig. 4. Radiation energy versus e-beam charge in the Cornell Wiggler A SASE experiment at BNL.

They are neglected because we are only calculating gain, and these modes only contribute to the spontaneous radiation, which is subtracted from SASE calculation. If we do not subtract the spontaneous radiation from the SASE, the series would converge very slowly. In addition, for the higher modes, the growth term is not large enough to dominate over the other decaying and oscillating terms, so the calculation becomes very difficult. In short, using guided modes to calculate spontaneous radiation is very difficult and unnecessary. By subtracting the spontaneous radiation, and only calculating the gain, we avoid this difficulty, and obtain an empirically excellent approximation.

Up to now, we used an idealized step function beam profile to test the calculation. To compare with the experiment, we use a more realistic waterbag model. We choose the same rms beam size $\sigma_x = 170\ \mu\text{m}$. This corresponds a normalized emittance $\varepsilon_n = 4$ mm mrad for our case with a focusing betatron wavelength $\lambda_\beta = 3.7$ m. We take the local energy spread to be $(\Delta\gamma/\gamma)_{\text{rms}} = 4 \times 10^{-4}$. These parameters lead to a power gain length with waterbag model $L_G = 0.26$ m, same as the step function model case before.

The calculated results are shown in Fig. 3. The dots are the simulation with the waterbag model, and the solid line is the calculation using the analytical method Eq. (8) for the step function model with the same rms beam size. The agreement is very good considering the crude nature of the approximation.

A recent SASE experimental result is plotted in Fig. 4. We did not have enough time to characterize the e-beam before this conference. However, the plot clearly shows that the beam quality is better than our originally designed parameters used for our calculation in this paper (110 A, 4 mm mrad).

References

[1] L.H. Yu, Phys. Rev. E 58 (1998) 4991.

[2] T.M. Tran, J.S. Wurtele, Comput. Phys. Commun. 54 (1989) 263.

[3] S. Krinsky, L.H. Yu, Phys. Rev. A 35 (8) (1987) 3406.

[4] L.H. Yu, S. Krinsky, Nucl. Instr. and Meth. A 285 (1989) 119.

[5] R. Bonifacio, C. Pellegrini, L.M. Narducci, Opt. Commun. 50 (1984) 373.

ELSEVIER

Nuclear Instruments and Methods in Physics Research A 429 (1999) 219–224

NUCLEAR INSTRUMENTS & METHODS IN PHYSICS RESEARCH Section A

www.elsevier.nl/locate/nima

Electron beam matching and its influence on the performance of high-gain free-electron lasers

H.P. Freund[a,*], P.G. O'Shea[b]

[a]Science Applications International Corporation, 1710 Goodridge Drive, McLean, VA 22102, USA

[b]Free-Electron Laser Laboratory, Duke University, Durham, NC 27708, USA

Abstract

We consider the influence of electron beam on the performance of high-gain FELs. We test this hypothesis for a SASE configuration using a 3-D multimode simulation. The gain length predicted for a matched beam agrees well with analytic theory, and the simulation indicates that while the gain length is optimized for a matched beam the saturated power may not be optimized. In previous work we have considered a flat-top beam model while we also consider a Gaussian beam model, and conclude the two are in substantial agreement. © 1999 Elsevier Science B.V. All rights reserved.

PACS: 41.60.Cr; 52.75.M

Keywords: Free-electron lasers; SASE; FEL; Gaussian beam model; Simulation

High-gain free-electron lasers (FELs) are proposed as *fourth generation* light sources [1–4]. Most analytic and numerical studies assume optimal performance for a *matched* electron beam [5]. While this assumption has been verified by simulation for low-gain oscillators [6], the dynamics of a high-gain self-amplified spontaneous emission (SASE) FEL are different. Our purpose is to study the effect of matched and unmatched beams of FEL performance.

High-gain FELs differ substantially from low-gain oscillators. Because the optical cavity acts as a mode filter, the radiation pattern is governed largely by the aperture and mirror geometry, and optical guiding is not dominant [7]. Hence, the optical power is concentrated in a single-TEM_{00} Gaussian mode [8]. In contrast, the gain length is much less than the wiggler length and betatron oscillation period in a high-gain amplifier and diffraction is limited only by optical guiding. Even when the beam envelope performs betatron oscillations, the optical mode follows the beam envelope [9]. Therefore, the optical mode content of a high-gain FEL is more complex than that of a low-gain oscillator.

To analyze high-gain FEL amplifiers, we employ a previously described 3-D, nonlinear formalism which simulates a planar wiggler with an ensemble of Gauss–Hermite optical modes [10,11]. We assume that the electron beam pulse length exceeds the slippage distance through the wiggler; hence, finite pulse-length effects are negligible. In addition, a monochromatic assumption is made which permits an average over the fast time scale. The dynamical equations for the mode amplitudes and phases are integrated simultaneously with the 3-D Lorentz force equations for the electrons. No

* Corresponding author. Tel.: + 1-202-767-0034; fax: + 1-202-734-1280.

E-mail address: freund@mmace.nrl.navy.mil (H.P. Freund)

PII: S0168-9002(99)00107-2

wiggler average is imposed on the orbit equations, which are integrated subject to detailed wiggler and radiation fields.

The simulation self-consistently models beam injection into the wiggler over an adiabatic entrance region of N_w wiggler periods, and the initial state of the electron beam is described at the wiggler entrance. Two distributions are used: one with a flat-top density profile, and one which models a matched Gaussian beam. The advantage of the flat-top model is that it is not restricted to the case of a matched beam which permits us to study the detailed effects of beam matching.

The flat-top distribution describes a cylindrical beam with a momentum space distribution which is monoenergetic but with a pitch-angle spread corresponding to a nonzero emittance, i.e.

$$F_b(\boldsymbol{p}_0) = \frac{\exp[-(p_{z0}-p_0)^2/\Delta p_z^2]\delta(p_0^2-p_{\perp 0}^2-p_{z0}^2)H(p_{z0})}{\pi\int_0^{p_0}\mathrm{d}p_{z0}\exp[-(p_{z0}-p_0)^2/\Delta p_z^2]} \tag{1}$$

where p_0 and Δp_z denote the bulk momentum and the axial momentum spread, and H is the Heaviside function. The axial energy spread associated with the distribution is

$$\frac{\Delta\gamma_z}{\gamma_0} = 1 - \frac{1}{\sqrt{1+2(\gamma_0^2-1)\Delta p_z/p_0}} \tag{2}$$

where $\gamma_0^2 = 1 + p_0^2/m_e^2c^2$. The transverse emittance of this distribution is given by $\varepsilon \approx R_b\theta_p$, where R_b is the beam radius prior to injection, and θ_p is the maximum pitch angle given by $\theta_p = \sin^{-1}(p_{\perp,\max}/p_0)$ where $p_{\perp,\max} \cong \sqrt{2p_0\Delta p_z}$ in the limit where $\Delta p_z \ll p_0$. Further, since $\Delta p_z/p_0 \approx \Delta\gamma_z/\gamma_0^3$ for $\gamma_0 \gg 1$ this corresponds to a normalized emittance of

$$\varepsilon_n \cong R_b\sqrt{\frac{2\Delta\gamma_z}{\gamma_0}}. \tag{3}$$

Here R_b is the beam radius at the start of the adiabatic region. The beam radius through the wiggler exhibits betatron oscillations, except for a matched beam. We fix ε_n and vary R_b over a range that includes the matched beam. This permits a determination of the variation in the exponentiation [i.e., gain] length and saturation efficiency with R_b.

The Gaussian beam distribution we use is also monoenergetic and is given by

$$F_b(x_0, y_0, \boldsymbol{p}_0) = \frac{\exp[-x_0^2/\sigma_x^2 - y_0^2/\sigma_y^2 - p_{x0}^2/\sigma_{px}^2 - p_{y0}^2/\sigma_{py}^2]}{(2\pi)^2\sigma_x\sigma_y\sigma_{px}\sigma_{py}} \times \delta[p_{z0} - \sqrt{p_0^2 - p_{\perp 0}^2}] \tag{4}$$

where $\sigma_x, \sigma_y, \sigma_{px}, \sigma_{py}$ denote the spread in coordinates and momentum. This distribution can, in general, treat elliptic beams; however, we confine ourselves in this work to cases where $\sigma_x = \sigma_y$ and $\sigma_{px} = \sigma_{py}$. For a matched beam these parameters are given in terms of the normalized emittance by $\sigma_x = \sigma_y = \sqrt{\hat\beta\varepsilon_n/\gamma}$ and $\sigma_{px} = \sigma_{py} = m_ec\sqrt{\gamma\varepsilon_n/\hat\beta}$.

We study a proposed SASE demonstration experiment [12] using the PALADIN wiggler [13]. The electron beam has an energy of 200 MeV, a current of 2500 A, and $\varepsilon_n \approx$ 20–40 mm mrad. The PALADIN wiggler is a hybrid design with parabolic pole faces and an 8-cm period, and the field for each period is individually controllable with maximum on-axis fields of $\approx$ 4 kG, and a variable entry taper region. The parabolic-pole-faces are modeled in the simulation, and we choose an amplitude of 4.0 kG with an entry taper region $N_w = 5$.

The first consideration is to study the variation in the matched beam radius with emittance using the flat-top distribution. The matched beam radius is $R_M = \sqrt{\hat\beta\varepsilon_n/\gamma}$, where $\hat\beta = \gamma\lambda_w/\delta\pi a_w$, λ_w is the wiggler period, a_w is the normalized rms wiggler parameter, and δ is a number which depends on the type of focusing. Since we consider a wiggler with equal two-plane focusing $\delta = \sqrt{2}$. The evolution of the beam radius for R_b = 0.3–0.8 mm is shown in Fig. 1 for our choice of beam and wiggler parameters and ε_n = 30 mm mrad. The oscillation in R_b is minimized (denoting a matched beam) for $R_b \approx 0.5$ mm. Corresponding choices for R_b which yield a matched beam $R_b = 0.42$ mm for $\varepsilon_n = 20$ mm mrad, and $R_b = 0.6$ mm for $\varepsilon_n = 40$ mm mrad. Note also that the betatron period in the wiggler is $\lambda_\beta \approx 12$ m.

We compare the simulation (using both distributions) with a linearized analytic theory (i.e.,

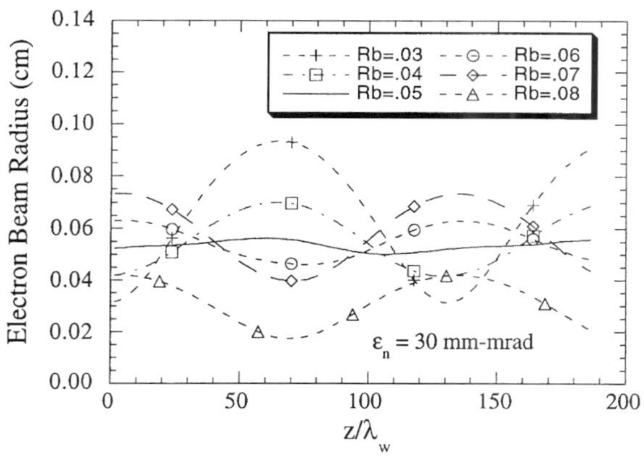

Fig. 1. Evolution of the electron beam radius for various choices of the initial beam radius and a normalized emittance of 30 mm mrad.

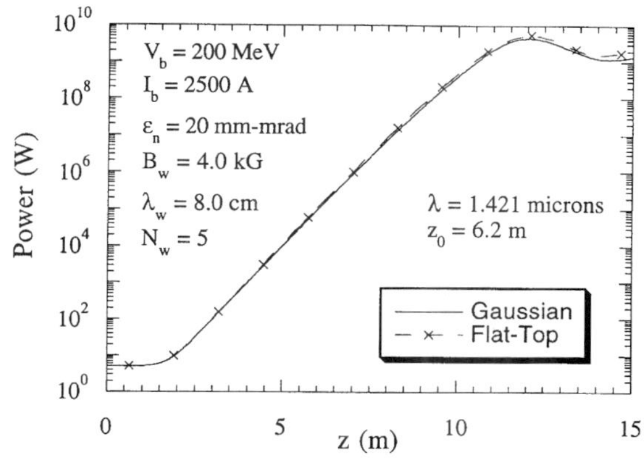

Fig. 2. Evolution of the power versus axial position for the Gaussian and flat-top beam models.

exponential gain phase) [14,15] for a matched beam. The linear theory was originally derived for a waterbag distribution [14] and subsequently extended to deal with a Gaussian beam in both momentum space and radial profile [15]. The matched beam radius in this model refers to the rms radius of the Gaussian beam profile, and the theory yields estimates of 0.41, 0.51, and 0.58 mm, respectively, for $\varepsilon_n = 20$, 30, and 40 mm mrad. These estimates are in close agreement with the simulation subject to the identification of R_b with the rms beam radius in the linearized analytic theory for a Gaussian radial profile.

For the parameters of interest, peak amplification is found for $\lambda \approx 1.41$–1.44 μm. We first compare the two beam models at the wavelength corresponding to peak growth ($\lambda = 1.421$ μm) for a matched beam and a normalized emittance of 20 mm mrad. The start is chosen to be 0.5 W based upon standard approximations for the spontaneous emission [2,15]. We consider an initial beam waist corresponding to a Rayleigh length of $z_0\ (= \pi w_0^2/\lambda) \approx 6.2$ m. A comparison of the evolution of the power versus axial position for the two-beam models is shown in Fig. 2. Note that 39 transverse modes were used in the simulation, and that the total initial power was in the TEM_{00} mode. It is clear from the figure that the two-beam models are is substantial agreement for the simulation of a matched beam, and we now proceed to consider the agreement between the simulations and the linear analytic theory.

Some discussion is required about the monochromatic model for a SASE FEL where the entire spontaneous noise spectrum is amplified. Monochromatic simulations are in use in designing X-ray SASE FELs at the Stanford Linear Accelerator Center [4] for the linac coherent light source (LCLS) at $\lambda \approx 1.5$ Å and at the Deutsches Elektronen-Synchrotron [17] for the TESLA test facility (TTF) for $\lambda \approx 60$ Å. No experimental validation of these simulations exists at these wavelengths; however, an earlier SASE FEL experiment at $\lambda \approx 600$ μm has shown the utility of such models in describing several important spectral characteristics [18]. The monochromatic formulation can explain the bulk exponential growth in a SASE FEL, and can also reproduce the bulk features of the unsaturated spectrum. However, it is unable to treat spectral noise, the evolution of the saturated state, and finite pulse-length effects.

The variation in the gain length, L_g, versus λ found in simulation is shown in Fig. 3 for the flat-top model and a matched beam with $\varepsilon_n = 20$, 30, and 40 mm mrad. Otherwise the beam and wiggler parameters were as used previously. The peak growth rate occurs for $\lambda = 1.426$ μm regardless of emittance, and the minimum $L_g = 0.45$, 0.49, and 0.53 m, respectively for $\varepsilon_n = 20$, 30, and 40 mm mrad. The linear analytic theory [15] yields

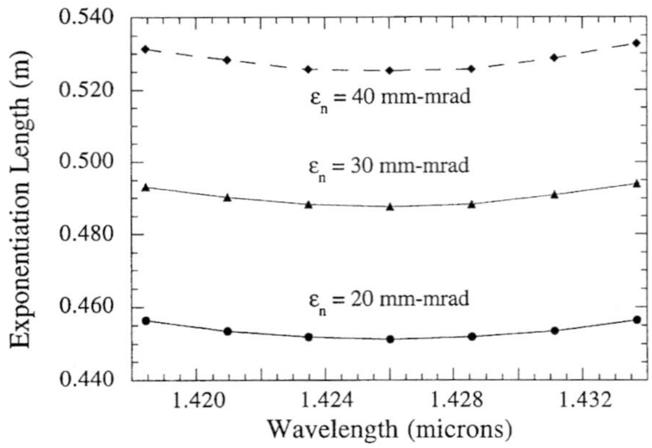

Fig. 3. Variation of the exponentiation length with wavelength.

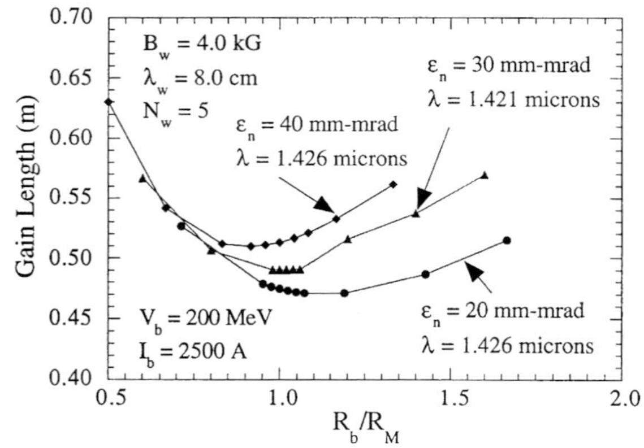

Fig. 4. Variation of the exponentiation length with initial beam radius.

$L_g = 0.45, 0.51$, and 0.57 m for these emittances and Rayleigh lengths subject to the identification of the Gaussian beam rms radius with R_b in the nonlinear formulation. The Gaussian beam model also is in substantial agreement with both the flat-top beam and the linear analytic theory, and predicts growth lengths of 0.45, 0.48, and 0.52 m for $\varepsilon_n = 20$, 30, and 40 mm mrad. For a matched beam, the simulations and the linear theory are in good agreement. A summary of the comparison is given in Table 1.

The parameters of the proposed Duke/PALADIN FEL are in the general category covering the LCLS and TTF. These include two LCLS designs [4] at 5 and 15 GeV, and the TTF design [17] at 1 GeV. The design parameters vary widely; nevertheless, these designs exhibit similar scaling with regard to many essential features of the FEL physics as shown by a number of dimensionless parameters characterizing the interaction. Firstly, $L_g/\lambda_\beta \approx 0.05$ and $L_s/\lambda_\beta \approx 1$ which is characteristic for all of these high-gain systems. The diffraction parameter $B\,(\equiv 4\pi R_b^2/\lambda L_g)$ is a measure of the relative importance of diffraction over a gain length [2,19], and diffraction is relatively unimportant when $B > 1$, as satisfied for all the examples. Hence, the present study is relevant to a wide range of practical SASE designs.

On the matched beam effects, we plot L_g versus R_b from simulation in Fig. 4 for $\varepsilon_n = 20$, 30, and 40 mm mrad. The simulations for $\varepsilon_n = 20$ and 40 mm mrad were for $\lambda = 1.426$ μm corresponding to the optimal L_g, and L_g was minimized for a matched beam. The 30 mm mrad case was for $\lambda = 1.421$ μm to determine if the behavior differed for wavelengths removed from the maximum growth rate, but the matched beam optimized L_g even here. In all cases, L_g varied weakly with R_b. For example, at $\varepsilon_n = 30$ mm mrad, L_g increased

Table 1
Comparison of the simulation and the linear theory for the matched beam radius and gain length. The gain length is found at $\lambda = 1.426$ μm, $z_0 = 8.8$ m

		Matched beam radius (mm)		Gain length (m)		
ε_n (mm mrad)	z_0 (m)	Flat-top beam	Linear theory	Flat-top beam	Gaussian beam	Linear theory
20	6.2	0.42	0.41	0.45	0.45	0.45
30	8.8	0.50	0.51	0.49	0.48	0.51
40	12.7	0.60	0.58	0.53	0.52	0.57

from a minimum of 0.47 m at $R_b \approx 0.45$ mm to a maximum of 0.57 m for $R_b \approx 0.8$ mm. This is an increase of only 21% for a change in R_b of 78%. It is important to observe (see Fig. 1) that such a large change in R_b from the matched beam condition results in a relatively large oscillation in the beam envelope. Optical guiding is described in simulation by the inclusion of higher-order modes. The start power is assumed to be only in the TEM_{00} mode, and optical guiding is described by the growth of higher-order modes. The saturated state for the present examples shows substantial power in the higher-order modes; in particular, while the total optical power is ≈ 3.8 GW, the TEM_{00} mode power is only 0.9 GW; hence, we conclude that optical guiding mitigates much of the effect of the betatron oscillations. A discussion of the emittance requirements for optical guiding and related issues for short-λ SASE FELs can be found in Ref. [16].

Examination of the saturated power versus R_b, however, leads to a more equivocal conclusion regarding the value of beam matching. In Fig. 5 we plot the saturated power (for whatever wiggler lengths are required to reach saturation) for the examples shown in Fig. 4. The saturated power for $\varepsilon_n = 40$ mm mrad is higher than for 30 mm mrad due to the different wavelengths used. Evidently, while a matched beam corresponds to the peak saturated power for $\varepsilon_n = 20$ mm mrad, this is not so for either $\varepsilon_n = 30$ or 40 mm mrad. In the latter cases, the saturated power is relatively constant over a broad range of R_b, and may be higher for unmatched beams. Of course, since L_g is longer for the unmatched beams, a longer wiggler is required.

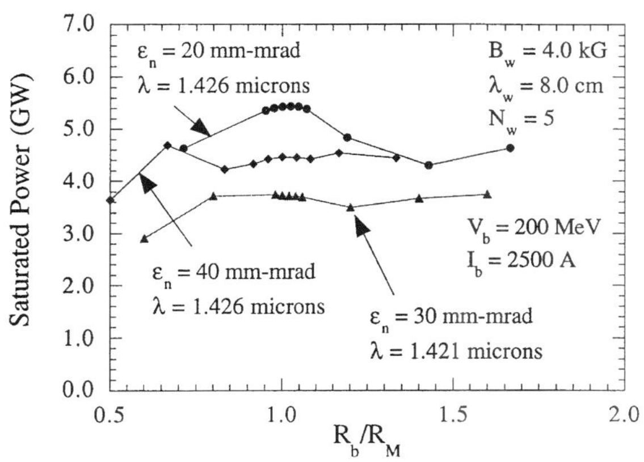

Fig. 5. Variation of the saturated power with initial beam radius.

In summary, we employed a nonlinear FEL simulation to determine if a matched beam optimizes the performance of a high-gain FEL amplifier. Comparison of L_g from simulation using both the flat-top and Gaussian beam models and linear theory for a matched beam shows good agreement. Using the simulation to investigate the effect of unmatched beams, we find that a matched beam yields an optimal growth rate. However, the increase in L_g with increases in the oscillation of the beam envelope is relatively mild. Further, a matched beam does not necessarily yield the greatest efficiency. Note that for the SASE FELs considered, the gain lengths are substantially less than both the Rayleigh ranges and the betatron periods.

For a long enough wiggler to reach saturation, the power is relatively independent of (while L_s is weakly dependent on) the degree of beam mismatch. We also attribute this weak dependence on the beam mismatch to optical guiding since the bulk of the saturated power is in the higher-order modes (the TEM_{00} mode constitutes $\approx 20\%$ of the total power) for all cases considered. We have shown that a variation in R_b/R_M in the range 0.6–1.6 results in an increase in L_g of up to 21%, while the saturated power can either increase or decrease depending upon the detailed parameters. Thus, the principal result of a matched beam is a small reduction in wiggler length with no certain increase in efficiency. This is particularly significant for nonthermalized beams where there is a sub-microbunch phase-space correlation (i.e., the phase space Courant–Snyder parameters vary along the bunch [20]), and it is impossible to match the entire bunch into the wiggler with conventional magnetic optics. This greatly simplifies the design of high-gain SASE FELs since it implies that performance is weakly affected by matching the beam into the wiggler. The agreement between Gaussian and flat-top beam models for the matched beam case suggests that FEL performance is weakly dependent upon the detailed transverse beam model.

This work was supported by the Office of Naval Research MFEL program. Computational work was supported by the North Carolina Super Computing Center.

References

[1] K.J. Kim, Nucl. Instr. and Meth. A 358 (1995) 31.
[2] R. Bonifacio, C. Pellegrini, L.M. Narducci, Opt. Commun. 50 (1984) 373.
[3] W.B. Colson, Nucl. Instr. and Meth. A 393 (1996) 82.
[4] M. Cornacchia et al., in: P.G. O'Shea, H. Bennet (Eds.), Free-Electron Laser Challenges, Proc. SPIE 2988 (1997) 5.
[5] W.B. Colson, in: W.B. Colson, C. Pellegrini, A. Renieri (Eds.), Laser Handbook, Vol. 6, North-Holland, Amsterdam, 1990 (Chapter 5).
[6] J.C. Goldstein, personal communication.
[7] C.A. Brau, Free-Electron Lasers, Academic Press, San Diego, 1990 (Chapter 4).
[8] B. Newnam et al., Nucl. Instr. and Meth. A 237 (1985) 187.
[9] E.T. Scharlemann, in: W.B. Colson, C. Pellegrini, A. Renieri (Eds.), Laser Handbook, Vol. 6, North-Holland, Amsterdam, 1990 (Chapter 9).
[10] H.P. Freund, Phys. Rev. E 52 (1995) 5401.
[11] H.P. Freund, T.M. Antonsen Jr., Principles of Free-electron Lasers, Second ed., Chapman&Hall, London, 1996 (Chapter 12).
[12] P.G. O'Shea et al., Nucl. Instr. and Meth. A 393 (1997) 129.
[13] G.A. Deis et al., IEEE Trans. Magn. 34 (1988) 1019.
[14] L.H. Yu et al., Phys. Rev. Lett. 64 (1990) 3011.
[15] M. Xie, Proceedings of the IEEE 1995 Particle Accelerator Conference, IEEE Cat. No. 95CH35843, 1995, p. 183.
[16] K.J. Kim, M. Xie, Nucl. Instr. and Meth. A 331 (1993) 359.
[17] B. Faatz et al., Nucl. Instr. and Meth. A 375 (1996) 441.
[18] D.A. Kirkpatrick et al., Phys. Fluids B 1 (1989) 1511.
[19] J. Rossbach et al., Nucl. Instr. and Meth. A 374 (1996) 401.
[20] P.G. O'Shea, Phys. Rev. E 57 (1998) 1081.

ELSEVIER

Nuclear Instruments and Methods in Physics Research A 429 (1999) 225–228

NUCLEAR INSTRUMENTS & METHODS IN PHYSICS RESEARCH Section A

www.elsevier.nl/locate/nima

An integral equation based computer code for high-gain free-electron lasers

Roger J. Dejus[a,*], Oleg A. Shevchenko[b], Nikolai A. Vinokurov[b]

[a]*Advanced Photon Source, Argonne National Laboratory, 9700 S. Cass Avenue, Argonne, IL 60439, USA*
[b]*Budker Institute of Nuclear Physics, 11 Ac. Lavrentyev Prosp., 630090 Novosibirsk, Russia*

Abstract

A computer code for gain optimization of high-gain free-electron lasers (FELs) is described. The electron motion is along precalculated period-averaged trajectories, and the finite-emittance electron beam is represented by a set of thin "partial" beams. The radiation field amplitudes are calculated at these thin beams only. The system of linear integral equations for these field amplitudes and the Fourier harmonics of the current of each thin beam is solved numerically. The code is aimed for design optimization of high-gain short-wavelength FELs with nonideal magnetic systems (breaks between undulators with quadrupoles and magnetic bunchers; field and steering errors). Both self-amplified spontaneous emission (SASE) and external input signal options can be treated. A typical run for a UV FEL, several gain lengths long, takes only 1 min on a Pentium II personal computer (333 MHz), which makes it possible to run the code in optimization loops. Results for the Advanced Photon Source FEL project are presented. © 1999 Elsevier Science B.V. All rights reserved.

Keywords: Free-electron laser; Simulation

1. Introduction

The short-wavelength high-gain free-electron lasers [1,2] offer the possibility to extend the free-electron laser (FEL) operation to the X-ray energy range. The advantage of such a device is that mirrors are not needed for the operation; however, tight requirements for the quality of the electron beam and the undulator magnetic fields are essential. An electron beam with high peak current, low transverse emittance and small energy spread is necessary for successful operation. The undulator needs to be very long (typically tens of meters) and carefully aligned with respect to the magnetic elements and to the beam. To aid in the technical design, it is useful to have a fast and versatile computer code that calculates the signal growth for a multisectional undulator with breaks between the sections, taking into account quadrupoles, magnetic bunchers, steering coils, undulator magnetic field errors, and beam-steering errors at the entrance.

However there are few working codes, especially in the 3-D case. This article describes a computer code that calculates the linear time-independent growth rate of radiation in a single pass FEL for a multisegmented system. It was applied to the

* Corresponding author.

0168-9002/99/$ - see front matter
PII: S0168-9002(99)00108-4

optimization of design parameters of the FEL under construction at Argonne National Laboratory [3,4].

2. Basic equations

We use a mathematical model based on solving a set of integral equations to describe the process of coherent radiation of the beam in the undulator field [5]. In this model, the internal structure of the beam is represented by a set of N thin beams with different initial conditions $x_q(0), \dot{x}_q(0), y_q(0), \dot{y}_q(0)$. (The point is used to denote a derivative with respect to the longitudinal coordinate z, and q is the number of the beam.) Using these initial conditions we calculate the trajectories $x_q(z)$, $y_q(z)$, neglecting the influence of the radiation field. (We consider the motion averaged over the undulator period.) Thus, one can consider the motion of the electrons in the transverse direction as if they were small beans strung on a thin rigid wire (trajectory). The problem is therefore reduced to one-dimensional motion along precalculated trajectories. A planar undulator with a vertical magnetic field given by $B_0(z)\sin[k_w z + \varphi(z)]$ is considered. It is convenient to use the z coordinate as the independent variable, the relative energy deviation Δ as the canonical momentum ($E = \gamma mc^2(1 + \Delta)$ is the electron energy, m is the mass of the electron and c is the speed of light), and the time delay $\tau = t - t_1$ with respect to the equilibrium unperturbed particle time

$$t_1 = \int_0^z \left[1 + \frac{1}{2\gamma_{\|}^2} + \frac{\dot{x}_q^2 + \dot{y}_q^2 + k_x^2 x_q^2 + k_y^2 y_q^2}{2}\right]\frac{dz'}{c}$$

as the canonical coordinate, where

$$\gamma_{\|}^2 = \frac{\gamma^2}{1 + K^2/2}, \qquad K = \frac{eB_0}{k_w mc^2}$$

is the deflection parameter, and k_x^2 and k_y^2 the rigidities of horizontal and vertical undulator focusing, respectively. In these variables, the projection of the "velocity" in the phase space to the energy axis is simply the projection of the force (caused by the radiation electric field) on the particle velocity. The longitudinal distribution function $F^q(z, \Delta, \tau)$ of each beam obeys the Liouville equation

$$\frac{\partial F^q}{\partial z} - \frac{\Delta}{\gamma_{\|}^2 c}\frac{\partial F^q}{\partial \tau} + \frac{eK(JJ)}{\gamma^2 mc^2}$$
$$\times \mathrm{Im}\left[A^q(z,t)\mathrm{e}^{\mathrm{i}k_w z + \mathrm{i}\varphi - \mathrm{i}k_0\int_0^z\left(\frac{\dot{x}_q^2 + \dot{y}_q^2 + k_x^2 x_q^2 + k_y^2 y_q^2}{2} + \frac{1}{2\gamma_{\|}^2}\right)dz' - \mathrm{i}k_0 c\tau}\right]$$
$$\times \frac{\partial F^q}{\partial \Delta} = 0 \tag{1}$$

where the electric field of the radiation on the qth beam is represented by $E_x^q = A^q(z,t)\mathrm{e}^{\mathrm{i}k_0(z-ct)}$ + *complex conjugate*, $2\pi/k_0$ is the wavelength near the fundamental harmonic of the undulator radiation, (JJ) is the standard combination of Bessel functions, and e is the charge of the electron.

In our case, the radiation field may be calculated in the paraxial approximation. As the relativistic electron emits in the forward direction, one can neglect the backward radiation. This allows us to rewrite the system of N partial differential equations as integral equations in which the integration is carried out from the undulator entrance to the current longitudinal point in z, like in the Volterra equations. This feature allows us to construct a simple explicit numerical algorithm. The obtained equations are nonlinear as the radiation field depends on the beam current, which is derived from integration of the distribution function. To simplify the calculations, we consider only the linear regime when the distribution function can be written as a sum of a time-independent part and a small perturbation. By making a Fourier transformation and carrying out integration by energy, we obtain the equations for the beam current harmonics $j_\omega^q(z) = \int_{-\infty}^{\infty}\int_{-\infty}^{\infty} F^q \mathrm{e}^{\mathrm{i}(k_0 c + \omega)\tau}\, d\Delta\, d\tau$ for all N thin beams. The final system of $2N$ equations can be written as

$$j_\omega^q(z) = \int_{-\infty}^{\infty} f_\omega^q(\Delta, 0)\mathrm{e}^{-\mathrm{i}k_0\Delta\int_0^z dz'/\gamma_{\|}^2}\, d\Delta$$
$$+ \frac{ie}{2\gamma^2 mc^2}\int_0^z \phi^q\left(k_0\int_{z'}^z \frac{dz''}{\gamma_{\|}^2}\right)A^q(z')$$
$$\times \mathrm{e}^{-\mathrm{i}k_0\int_0^{z'}\frac{\dot{x}_q^2 + \dot{y}_q^2 + k_x^2 x_q^2 + k_y^2 y_q^2}{2}dz''}\mathrm{e}^{\mathrm{i}k_w z' - \mathrm{i}(k_0 + (\omega/c))\int_0^{z'} dz''/2\gamma_{\|}^2}$$
$$\times K(JJ)\mathrm{e}^{\mathrm{i}\varphi}\, dz'$$

$$A^q(z') = A^q_{0\omega}(z')\mathrm{e}^{-\mathrm{i}(\omega/c)z'}$$

$$-\frac{k_0 I}{2\gamma c}\int_0^{z'}\sum_p \frac{1}{z'-z''}\,\mathrm{e}^{-\mathrm{i}k_w z'' + \mathrm{i}(k_0+(\omega/c))\int_0^{z''}\mathrm{d}z'''/2\gamma_{\|}^2}$$

$$\times \mathrm{e}^{\mathrm{i}k_0\frac{[x_q(z')-x_p(z'')]^2+[y_q(z')-y_p(z'')]^2}{2(z'-z'')}+\mathrm{i}k_0\int_0^{z''}\frac{\dot{x}_p^2+\dot{y}_p^2+k_x^2x_p^2+k_y^2y_p^2}{2}\,\mathrm{d}z'''}$$

$$\times j^p_\omega(z'')K(z'')(JJ)\,\mathrm{e}^{-\mathrm{i}\varphi(z'')}\,\mathrm{d}z'' \quad (2)$$

where $f^q_\omega(\Delta, 0)$ is the harmonic of the initial distribution function, I is the beam current, $A^q_{0\omega}(z) = \int_{-\infty}^{\infty} A^q_0(z,t)\mathrm{e}^{\mathrm{i}\omega t}\,\mathrm{d}t$ describes the external wave (input signal), and $\phi^q(x) = \int_{-\infty}^{\infty}\frac{\partial F^q_0}{\partial \Delta}\mathrm{e}^{-\mathrm{i}x\Delta}\,\mathrm{d}\Delta$, where $F^q_0(\Delta)$ is the unperturbed distribution function.

3. Description of the code

To solve the system of integral equations, we use a trapezoidal estimation for the integrals. This leads to the following set of equations for the discrete values of $J(q,n)$ and $A(q,n)$ at $z = \sum_{m=1}^{n-1} h(m)$, where h is the step length,

$$A(q,n) = \sum_{m=1}^{n}\sum_p G_{qpnm}\cdot J(p,m) + A_0(q,n)$$

$$J(q,n) = \sum_{m=1}^{n-1} H_{qnm}\cdot A(q,m) + J_0(q,n) \quad (3)$$

where A_0 and J_0 correspond to the inhomogeneous terms of Eq. (2). To calculate $J(q,n)$, we need only the preceding values of $A(q,m)$ $(1 \le m < n)$. Thus, solving this system becomes trivial.

The magnetic system is described by the following input parameters: the deflection parameter K, the phase φ of the undulator field, the sextupole focusing parameter along each step, the optical strength of the thin quadrupole lens, and the vertical and horizontal angle deflections (kicks) at the entrance of each step.

We use a unit monoenergetic excitation of one of the thin beams to calculate the self-amplified spontaneous emission case, which corresponds to the situation of a single particle moving into the beam without initial density fluctuations. The output intensity will be the sum of the contributions from the different particles because we consider only the linear regime and the initial noise is random.

4. Results

The code has been used for optimization of the UV FEL at Argonne National Laboratory. The dashed curve in Fig. 1 shows the results for a homogeneous undulator with sextupole focusing. The normalized growth rate is

$$F = \frac{1}{2k_\mathrm{w}D}\frac{\mathrm{d}}{\mathrm{d}z}\ln\left|\sum_{q=1}^{N}\mathrm{e}^{\mathrm{i}k_0\int_0^z\frac{\dot{x}_q^2+\dot{y}_q^2+k_x^2x_q^2+k_y^2y_q^2}{2}\,\mathrm{d}z'}j^q_\omega(z)\right|^2$$

where $D = 2\sqrt{\frac{eI}{\gamma mc^3(1/2+K^{-2})}}(JJ)$.

It is clear that after approximately two gain lengths L_g the growth rate becomes constant indicating that only one eigenmode is prevalent. The normalization was chosen such that $F(\infty) = 1/(2k_\mathrm{w}DL_\mathrm{g})$ will be equal to the scaling factor for this eigenmode, which was calculated earlier (see e.g., Refs. [2,5,6]). It was seen from many different runs, that this steady-state growth rate did not depend on the initial noise.

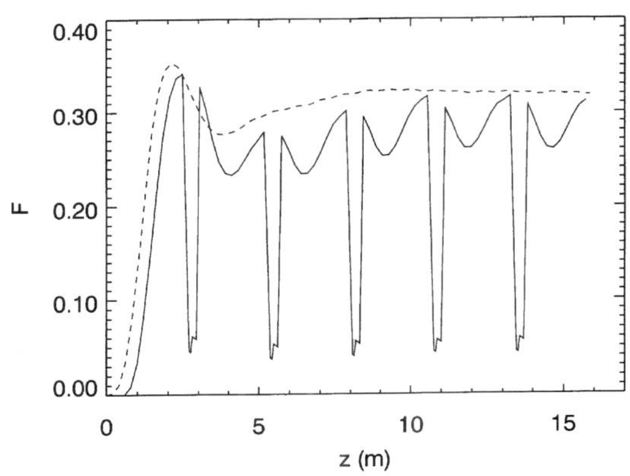

Fig. 1. The dimensionless scaling factor versus distance along the undulator for a homogeneous undulator (dashed curve) and for an inhomogeneous undulator (solid curve).

The solid curve in Fig. 1 shows the results for the real project with breaks and horizontally focusing quadrupoles [7]. The lower but nonzero growth rate in the breaks corresponds to the rudimental bunching that take place there. The small growth rate reduction in undulators in comparison with the homogeneous case is, probably, due to beating of beta-functions, caused by the inhomogeneity of the focusing.

5. Discussion

According to Eq. (3), the calculation time is proportional to the square of the number of the partial beams N and to the square of the longitudinal number of steps n, which makes it feasible to quickly study very long systems (a typical run for the APS UV FEL takes only one minute on a 333 MHz Pentium II personal computer). Commonly, it is sufficient to use about 10 steps per undulator and three steps for the breaks with the quadrupole inside, however, for simulations using magnetic field errors from magnetic measurements the number of steps tends to be larger – typically one step per period.

The typical number of the partial beams for the APS runs was 49, which proved to be sufficient. We also found that the results for the APS project parameters were insensitive to different initial distributions of the partial beams. For shorter wavelengths (X-rays), when the diffraction is less significant, the number of partial beams needs to be increased, and more attention to the actual initial beam distributions must be given. The code was used to define tolerances for the undulator alignments and to optimize the break length and focusing strength for the APS project [7,8].

In spite of the fact that the general integral equation [5] is nonlinear and time-dependent, the corresponding extension of this code for nonlinear phenomena seems unrealistic because it uses linearity and stationarity in several places. However, since the major part of the length of a self-amplified spontaneous emission (SASE) FEL operates with a small signal, the code is a convenient design tool to make an optimal system for beam bunching. The last section of the superradiant FEL has to be considered separately from both theoretical and engineering points of view.

The angle distribution of the radiation intensity was added recently but has not yet been extensively tested. The calculation of the spectrum of the radiation intensity is underway and will be tested in the near future.

Acknowledgements

This work was supported by the U.S. Department of Energy, Office of Basic Energy Sciences, under Contract No. W-31-109-ENG-38.

References

[1] A.M. Kondratenko, E.L. Saldin, Particle Accelerators, vol. 10, Gordon and Breach Science Publishers, Inc., 1980, p. 207.
[2] K.-J. Kim, M. Xie, Nucl. Instr. and Meth. A 331 (1993) 359.
[3] S.V. Milton et al., Nucl. Instr. and Meth. A 407 (1998) 210.
[4] N.A. Vinokurov, R.J. Dejus, H. Friedsam, E.S. Gluskin, J. Maines, S.V. Milton, E.R. Moog, E.M. Trakhtenberg, I.B. Vasserman, Proc. SPIE 2988 (1997) 64.
[5] N.A. Vinokurov, The Integral Equation for a High Gain FEL, Argonne National Laboratory Report, ANL/APS/TB-27, 1996.
[6] M. Xie, Proceedings of the 1995 Particle Accelerator Conference, Dallas, TX IEEE, Piscataway, NJ, 1995, p. 183.
[7] E. Gluskin, R.J. Dejus, P.K. Den Hartog, B.N. Deriy, O.A. Makarov, S.V. Milton, E.R. Moog, V.I. Ogurtsov, E.M. Trakhtenberg, K.E. Robinson, I.B. Vasserman, N.A. Vinokurov, Nucl. Instr. and Meth. A 429 (1999) 358.
[8] R.J. Dejus, I.B. Vasserman, Short Note on Undulator Alignments and Beam Tolerances for the APS FEL at 220 MeV, Argonne National Laboratory Internal Report, APS/IN/LEUTL/98-1, 1998.

ELSEVIER

Nuclear Instruments and Methods in Physics Research A 429 (1999) 229–232

NUCLEAR INSTRUMENTS & METHODS IN PHYSICS RESEARCH Section A

www.elsevier.nl/locate/nima

The formation of transverse coherence in SASE FELs

E.L. Saldin[a], E.A. Schneidmiller[a], M.V. Yurkov[b,*]

[a]*Automatic Systems Corporation, 443050 Samara, Russia*

[b]*Particle Physics Laboratory (LSVE), Joint Institute for Nuclear Research, Moscow Region, 141980 Dubna, Russia*

Abstract

This report presents a fully three-dimensional study of the amplification process in the self amplified spontaneous emission (SASE) free electron laser. Investigations are based on the data obtained with the three-dimensional, time-dependent FEL simulation code FAST. Analysis of the data shows that the statistical properties of the radiation can be described with Gaussian statistics. In particular, fluctuations of the instantaneous radiation intensity at one space point follow the negative exponential law, while the finite-time integrals of the radiation intensity (both in space and in time) follow a gamma-distribution. Numerical examples presented in the paper correspond to the 70 nm SASE FEL under construction at the TESLA test facility at DESY.

PACS: 41.60.Cr; 52.75.Ms; 42.25.Kb

Keywords: SASE; Transverse coherence; Amplification process

1. Introduction

The process of amplification in a self-amplified spontaneous emission free electron laser (SASE FEL) starts from the shot noise in the electron beam. Since the FEL amplifier is a narrow bandwidth device, it selects from the shot noise and begins to amplify only a narrow fraction of the fluctuation spectrum. During the amplification process the longitudinal coherence of the radiation increases and the process of formation of transverse coherence takes place. The process of formation of the longitudinal coherence can be studied in the framework of the one-dimensional theory, since it is mainly connected with the slippage effect. Reliable analytical and numerical methods allow a detailed study of this effect [1]. As for the investigation of transverse coherence formation, there are only qualitative predictions based on the results of the steady-state approximation. Namely, in the linear high-gain limit, the radiation of the electron beam in the undulator may be represented as a set of modes having different growth rates. For a sufficient undulator length the fundamental mode should survive, having the most gain. On the other hand, the saturation gain in an SASE FEL is limited and one should perform a quantitative study if such a single-mode asymptote is to be realized in a practical device. At present there are no reliable analytical approaches giving these quantitative results, and the information on the process of transverse coherence formation can be obtained only

* Corresponding author. Tel.: + 7-09621-62154; fax: + 7-09621-65767.

E-mail address: yurkov@sunse.jinr.ru (M.V. Yurkov)

0168-9002/99/$ - see front matter
PII: S0168-9002(99)00109-6

Table 1
Parameters of the 70 nm SASE FEL at DESY

Electron beam	
Energy, ε_0	300 MeV
Peak current, I_0	500 A
rms bunch length, σ_z	250 μm
Normalized rms emittance, ε_n	2π mm mrad
External β-function,	100 cm
Number of bunches per train	7200
Repetition rate	10 Hz
Undulator	
Type	Planar
Length of undulator, L_w	13 m
Period, λ_w	2.73 cm
Peak magnetic field, H_w	4.97 kGs
Radiation	
Wavelength, λ	71.4 nm
Power averaged over pulse	300 MW
Flash energy	0.5 mJ
Average power	45 W

with the fully three-dimensional, time-dependent simulation codes. Recently, it has been reported that two such codes, FAST and GENESIS, have been produced [2,3].

In this paper we present a numerical study of the process of amplification in the SASE FEL using the numerical simulation code FAST [2]. To be specific, we present numerical examples for the 70 nm SASE FEL under construction at the TESLA test facility at DESY (see Table 1) [4]. The results obtained are used to plan experiments on the characterization of the photon beam from this FEL [5].

2. Results of numerical simulations

Numerical simulations have been performed with the three-dimensional, time-dependent FEL simulation code FAST [2]. Parameters of the simulations have been chosen to achieve a simulation accuracy of about 1%. The size of the longitudinal slice is equal to four radiation wavelengths. The numbers of the radial and azimuthal divisions are equal to 16 and 50, respectively, and the number of azimuthal modes for the radiation field calculations has been set to nine. The numerical simulation code produces the matrices for the field values in the Fresnel diffraction zone. Typical output results for the instantaneous power flux density inside the undulator are presented in Fig. 1. Using the data for the radiation field in the near field, the post-processor code can calculate the distributions of the radiation intensity in the far field (see Fig. 2).

The amplification process starts from the shot noise. At the beginning of the amplification process there are many transverse radiation modes contributing to the FEL power. With increasing undulator length the higher modes begin to contribute less and the field distribution tends asymptotically to that predicted by the steady-state theory [6,7].

Integration of the radiation flux density across the electron beam allows one to obtain the time structure of the radiation pulse (see Fig. 3). In Fig. 4 we present the corresponding probability distribution of the radiation power obtained from a large number of statistical runs. It is seen that the probability distribution follows the gamma distribution

$$p(P) = \frac{M^M}{\Gamma(M)}\left(\frac{P}{\langle P\rangle}\right)^{M-1}\frac{1}{\langle P\rangle}\exp\left(-M\frac{P}{\langle P\rangle}\right)$$

where $\Gamma(M)$ is the gamma function of argument M. The parameter of the distribution is equal to $M = 1/\sigma^2$, where $\sigma^2 = \langle P^2 - \langle P\rangle^2\rangle/\langle P\rangle^2$ is the normalized dispersion of the distribution of the instantaneous radiation power. In the case under study, the physical sense of the parameter M is the average number of transverse radiation modes.

Statistical simulations allow us to develop a quantitative description of this transverse coherence formation process in terms of the average number of the transverse radiation modes, M (see Fig. 5). With increasing undulator length the number of the transverse radiation modes asymptotically approaches one. Analysis of this plot shows that we can expect almost full transverse coherence of the radiation at the exit of the 70 nm SASE FEL at DESY.

Values of practical interest are the saturation length and the saturation power. Since the temporal structure of the radiation is spiky (see Fig. 3), it is reasonable to calculate the value of the

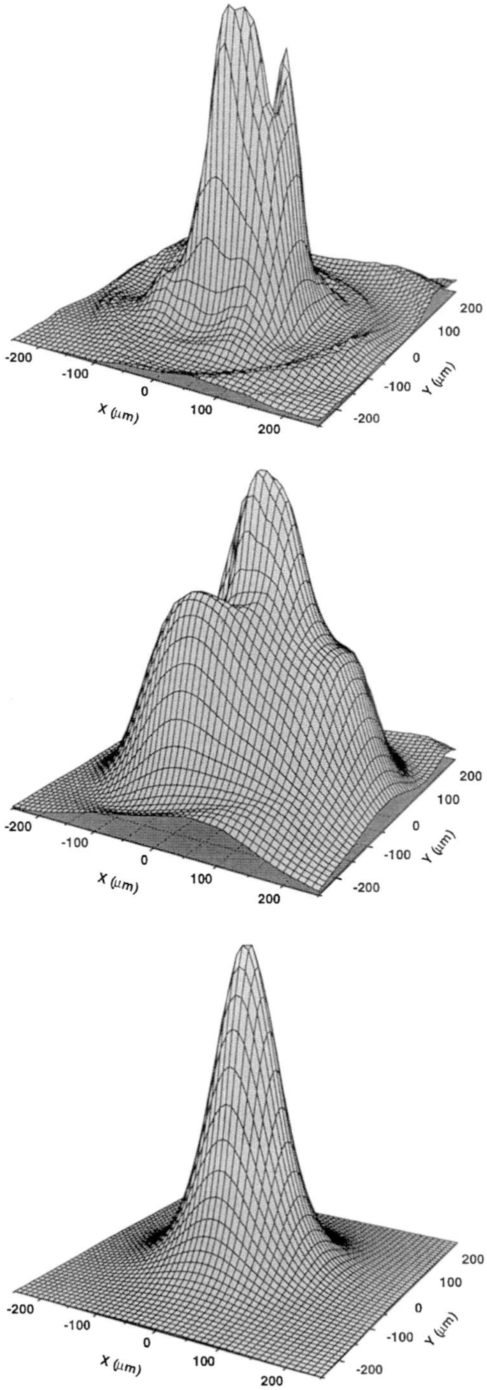

Fig. 1. Distribution of the instantaneous power flux density in the near zone at the undulator length of 2, 4.5 and 8 m (upper, middle and lower plot, respectively).

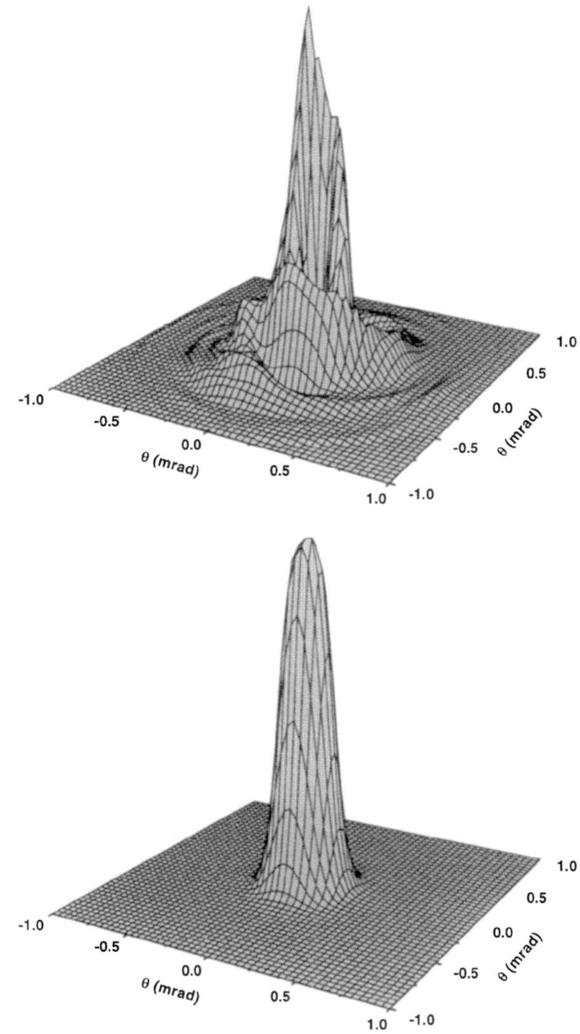

Fig. 2. Directivity of the instantaneous radiation intensity in the far field at the undulator length of 2 and 8 m (upper and lower plot, respectively).

radiation power averaged over many pulses. In Fig. 6 we present the dependency of the averaged radiation power on the undulator length. The calculations have been performed with the nonlinear simulation code for negligibly small energy spread. It is seen that with such an optimistic scenario one can expect to reach saturation in the 70 nm SASE FEL at DESY. Analysis of the plot in Fig. 6 also allows one to calculate that the value of the "effective" shot noise power to be used in the steady-state

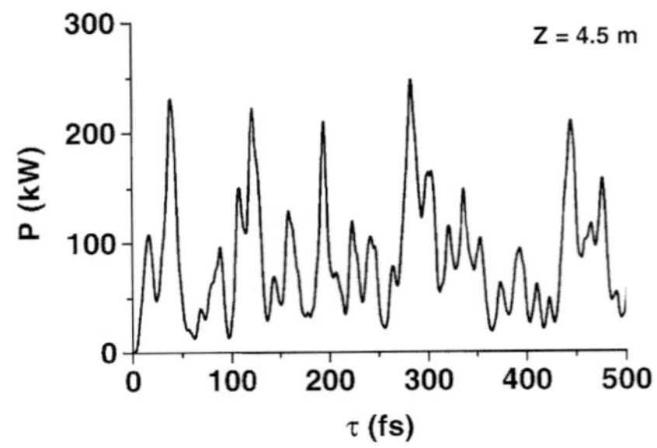

Fig. 3. Temporal structure of the radiation pulse at the undulator length of 4.5 m.

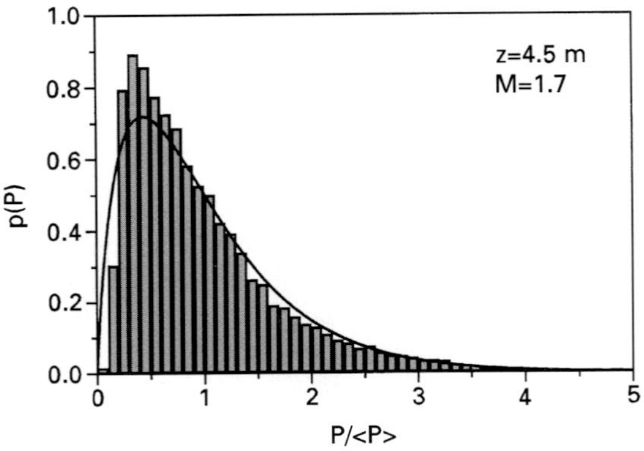

Fig. 4. The probability distribution of the instantaneous radiation power at the undulator length of 4.5 m.

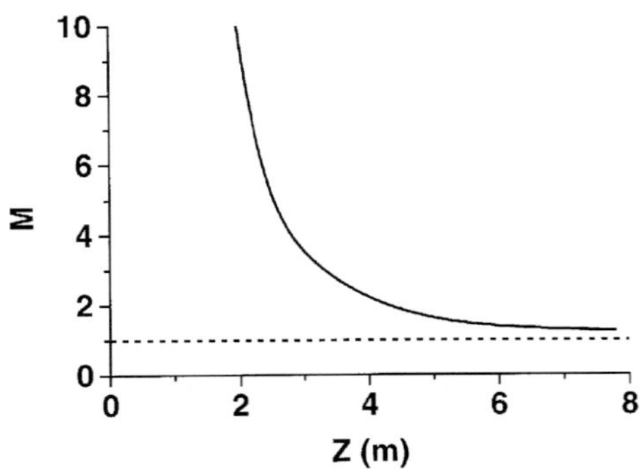

Fig. 5. The dependence of the number of transverse modes, M, versus the undulator length.

simulation code is about 10 W. The dashed curve in Fig. 6 presents the results of the steady-state simulations. It is seen that the steady-state simulations give a reliable estimation for the values of the

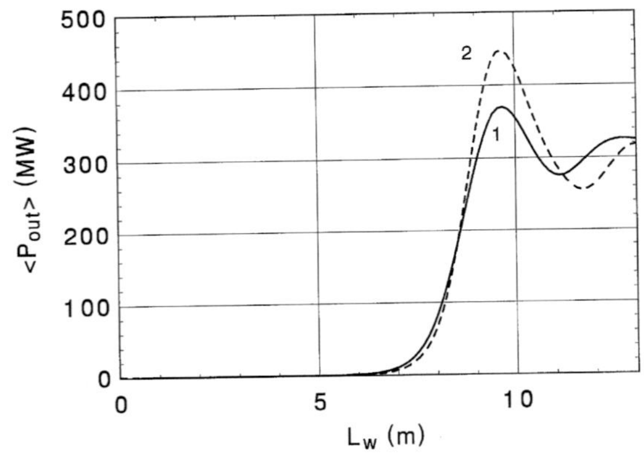

Fig. 6. The dependence of the radiation power averaged over the pulse length on the undulator length. The dashed curve presents the results of the steady-state simulation with an input power of 10 W.

saturation length and the average saturation power for the parameters of the 70 nm SASE FEL at DESY.

Acknowledgements

We are grateful to B. Faatz, J. Feldhaus, G. Materlik, J. Pflüger, S. Reiche, J. Rossbach and J.R. Schneider for many useful discussions.

References

[1] E.L. Saldin, E.A. Schneidmiller, M.V. Yurkov, Opt. Commun. 148 (1998) 383.

[2] E.L. Saldin, E.A. Schneidmiller, M.V. Yurkov, FAST: three-dimensional time-dependent FEL simulation code, Presentation at the 20th FEL Conference, Williamsburg, 1998.

[3] S. Reiche, GENESIS 1.3 – A Fully 3D time dependent FEL simulation code, Presentation at the 20th FEL Conference, Williamsburg, 1998.

[4] W. Brefeld et al., Nucl. Instr. and Meth. A 393 (1997) 119.

[5] Proceedings of the Workshop on the Characterization of the Photon Beam Emitted by the VUV SASE FEL at DESY, Hamburg, April 2–3, 1998.

[6] E.L. Saldin, E.A. Schneidmiller, M.V. Yurkov, Opt. Commun. 97 (1993) 272.

[7] E.L. Saldin, E.A. Schneidmiller, M.V. Yurkov, Phys. Rep. 260 (1995) 187.

ELSEVIER

Nuclear Instruments and Methods in Physics Research A 429 (1999) 233–237

NUCLEAR INSTRUMENTS & METHODS IN PHYSICS RESEARCH Section A

www.elsevier.nl/locate/nima

FAST: a three-dimensional time-dependent FEL simulation code

E.L. Saldin[a], E.A. Schneidmiller[a], M.V. Yurkov[b,*]

[a]*Automatic Systems Corporation, 443050 Samara, Russia*

[b]*Joint Institute for Nuclear Research, Particle Physics Laboratory (LSVE), Dubna, 141980 Moscow Region, Russia*

Abstract

In this report we briefly describe the three-dimensional, time-dependent FEL simulation code FAST. The equations of motion of the particles and Maxwell's equations are solved simultaneously taking into account the slippage effect. Radiation fields are calculated using an integral solution of Maxwell's equations. A special technique has been developed for fast calculations of the radiation field, drastically reducing the required CPU time. As a result, the developed code allows one to use a personal computer for time-dependent simulations. The code allows one to simulate the radiation from the electron bunch of any transverse and longitudinal bunch shape; to simulate simultaneously an external seed with superimposed noise in the electron beam; to take into account energy spread in the electron beam and the space charge fields; and to simulate a high-gain, high-efficiency FEL amplifier with a tapered undulator. It is important to note that there are no significant memory limitations in the developed code and an electron bunch of any length can be simulated.

PACS: 41.60.Cr; 07.05.T

Keywords: FEL amplifier; SASE FEL; Simulation code

1. Introduction

Complete calculation of the parameters of an FEL amplifier can be performed only with numerical simulation codes. At present there are several numerical simulation codes calculating the amplification process in the FEL amplifier using the steady-state approximation. Such an approximation describes rather well the case when the FEL amplifier is seeded by monochromatic external radiation and when the slippage effect can be neglected. Such a simplification allows one to simulate the electron beam with one slice equal to the radiation wavelength, thus significantly reducing the requirements for computer resources. Nevertheless, the steady-state simulations do not provide a correct result when the slippage of the radiation is comparable with the length of the electron bunch. The steady-state code cannot be used in principle

* Corresponding author. Tel.: + 7-09621-62154; fax: + 7-09621-65767.

E-mail address: Yurkov@sunse.jinr.ru (M.V. Yurkov)

0168-9002/99/$ - see front matter
PII: S0168-9002(99)00110-2

for the simulations of the FEL amplifier starting from the shot noise (SASE FELs), and a complete simulation of SASE FELs can be done only with a fully three-dimensional, time-dependent simulation code.

Recently, the development has been reported of one dimensional, time-dependent simulation codes which revealed the possibility to perform detailed investigation of the process of formation of the longitudinal coherence in the SASE FELs. Nevertheless, the one-dimensional approximation omits an essential effect of the diffraction of the radiation giving only a rough estimate of the SASE FEL parameters. Complete simulation of the physical process in the SASE FEL can be done only with a three dimensional, time-dependent simulation code. Unfortunately, the progress in this field is rather limited, and the main reason for this is due to limited possibilities of the computers.

In this paper we report on the development of a fast three dimensional, time-dependent simulation code FAST. The ideas implemented during the construction of this code allowed a significant reduction of the requirements on the computer, and the simulations of actual devices can be performed using a conventional personal computer.

2. General approach

Time-dependent simulations of the FEL amplifier should be performed by simultaneous solutions of Maxwell's equations and the equations of motion of the electrons. Nevertheless, the problem formulated in such a general form cannot be implemented in the simulation code and some physical approximation should be made. Here it is reasonable to remember that the free electron laser is a resonance device amplifying the radiation within a narrow bandwidth. Using this resonance approximation we present the transverse beam current density as $j_1(\boldsymbol{r},t) = \tilde{j}_1(\boldsymbol{r},t)\exp(\mathrm{i}\omega(z/c - t)) + \mathrm{C.C.}$, where ω corresponds to the resonance FEL frequency and $\tilde{j}_1(\boldsymbol{r},t)$ is the slowly varying complex amplitude. The radiation field is also presented in the resonance approximation, $E(\boldsymbol{r},t) = \tilde{E}(\boldsymbol{r},t)\exp(\mathrm{i}\omega(z/c - t)) + \mathrm{C.C.}$ with the slowly varying complex amplitude $\tilde{E}(\boldsymbol{r},t)$. Using paraxial approximation one can obtain the following expression for $\tilde{E}(\boldsymbol{r},t)$:

$$\tilde{E}(z,\boldsymbol{r}_\perp,t) = \mathrm{i}\frac{\omega}{c^2}\int_0^z \frac{\mathrm{d}z'}{z-z'}\int \mathrm{d}\boldsymbol{r}_\perp \tilde{j}_1\left(z',\boldsymbol{r}'_\perp,t-\frac{z-z'}{c}\right) \times \exp\left[\frac{\mathrm{i}\omega|\boldsymbol{r}_\perp - \boldsymbol{r}'_\perp|^2}{2c(z-z')}\right]. \quad (1)$$

The simulation code is organized as follows. We divide the electron beam in a large number of elementary volumes. The longitudinal size of each volume is equal to (or a multiple of) the radiation wavelength. Also, the electron beam is divided into a large number of divisions in the transverse direction. The FEL equations for the particle motion in each elementary volume are solved at each integration step. Then the radiation fields are calculated for each elementary volume using an integral solution (1). At the next integration step these fields are substituted into the FEL equations, etc. As a result, one can trace the evolution of the radiation field and the particle distribution when the electron beam passes the undulator. One can obtain from the integrals (1) that the radiation field at each point is defined only by the sources located closer than the slippage distance and it is not necessary to keep in the memory all the current sources. The procedure of the simulations begins from the tail slice of the electron bunch and the procedure of integration is performed over the whole undulator length. Then the equations of motion for the second slice are integrated taking into account the radiation field from the first slice, etc. As a result, the self-consistent FEL equations can be integrated for an electron bunch of any length. The memory requirements for the code are rather moderate. Our experience shows that a few tens of megabytes is sufficient to simulate most practical devices with sufficient accuracy.

The code is realized in two versions: linear and nonlinear. The linear simulation code is based on solution of the kinetic equation describing the evolution of the distribution function of the electron beam, and the nonlinear simulation code uses the traditional technique of macroparticles for the simulation of the distribution function of the electron beam.

3. Self-consistent equations

To be specific, in this section we present explicit formulae implemented in the code for the simulation of the FEL amplifier with an axially symmetric electron beam. The transverse distribution of the beam current density is assumed to be Gaussian; $j(r,z) = I(z)\exp(-r^2/2\sigma_r^2)/\sqrt{(2\pi\sigma_r^2)}$ with $\sigma_r = \sqrt{\varepsilon_n\beta/\gamma}$ where ε_n is the rms normalized emittance, β is the focusing beta function and $\gamma = \mathscr{E}_0/m_e c^2$ is the relativistic factor. The electron beam is divided into $L = l_b/k\lambda$ slices in the longitudinal direction (k is an integer), in M slices over the azimuthal angle ϕ, and in N divisions in the radial direction (we use the polar coordinate system (r,ϕ,z) here). As a result, we have $L \times M \times N$ elementary volumes. The self-consistent equations for the linear simulation code are as follows [2,3]:

$$\frac{\mathrm{d}^2 b_1}{\mathrm{d}\hat{z}^2} + 2\mathrm{i}\hat{C}\frac{\mathrm{d}b_1}{\mathrm{d}\hat{z}} + [\hat{\Lambda}_p^2\alpha(\hat{r},\hat{z},t) - \hat{C}^2]b_1$$

$$= \sum_n \exp(\mathrm{i}n\phi)U_r^{(n)}(\hat{r},\hat{z},t) \quad (2)$$

where $b_1(\hat{r},\hat{z},\phi,t)$ is the beam bunching in the elementary volume, $\alpha(\hat{r},\hat{z},t) = I(\hat{z},t)\exp(-\hat{r}^2)/I_{max}$ and $U_r^{(n)}(\hat{z},\hat{r})$ is the nth azimuthal harmonic of the effective potential of interaction of the particle with the electromagnetic radiation

$$U_r^{(n)} = U_{ext}^{(n)}(\hat{z},\hat{r},t)$$

$$+ \mathrm{e}^{\mathrm{i}n\pi/2}\int_0^{\hat{z}} \frac{\mathrm{d}\hat{z}'}{\hat{z}-\hat{z}'}\int \hat{r}'\mathrm{d}\hat{r}'\alpha(\hat{r}',\hat{z}',t-(z-z')/c)$$

$$\times b_1^{(n)}(\hat{z}',\hat{r}',t-(z-z')/c))$$

$$\times J_n\left(\frac{B\hat{r}\hat{r}'}{\hat{z}-\hat{z}'}\right)\exp\left\{\frac{\mathrm{i}B(\hat{r}^2+\hat{r}'^2)}{2(\hat{z}-\hat{z}')}\right\}. \quad (3)$$

Here term $U_{ext}^{(n)}(\hat{z},\hat{r})$ corresponds to the external electromagnetic field, $b_1^{(n)}(\hat{r},\hat{z},t)$ are azimuthal harmonics of the beam bunching calculated using the values of $b_1(\hat{r},\hat{z},\phi,t)$ and J_n are the Bessel functions.

When writing down the normalized equations we used the following notations. The transverse coordinate is normalized as $\hat{r} = r/\sqrt{2\sigma_r^2}$. The corresponding reduced variables are as follows: $\hat{z} = \Gamma z$ is the reduced longitudinal coordinate, $\hat{C} = (2\pi/\lambda_w - \omega/(2c\gamma_z^2))/\Gamma$ is the detuning parameter, $\hat{\Lambda}_p^2 = \Lambda_p^2/\Gamma^2 = 4c^2(\theta_s\sigma_r\omega A_{JJ})^{-2}$ is the space charge parameter, $B = 2\Gamma\sigma_r^2\omega/c$ is the diffraction parameter, and the gain parameter Γ is

$$\Gamma = [I_{max}A_{JJ}^2\omega^2\theta_s^2/(2I_A c^2\gamma_z^2\gamma)]^{1/2} \quad (4)$$

where $\omega = 2\pi c/\lambda$ is the frequency of the radiation field and $I_A = m_e c^3/e$. The undulator is assumed to be planar with magnetic field amplitude H_w and period λ_w. The undulator parameter K, the angle of electron oscillations θ_s, the longitudinal relativistic factor γ_z and the factor A_{JJ} are defined as follows: $K = eH_w\lambda_w/2\pi m_e c^2$, $\theta_s = K/\gamma$, $\gamma_z^2 = \gamma^2/(1+K^2/2)$ and $A_{JJ} = J_0(K^2/(4+2K^2)) - J_1(K^2/(4+2K^2))$, where J_0 and J_1 are the Bessel functions.

In the nonlinear simulation code, the electron beam distribution function is simulated with N_p macroparticles in each elementary volume. The equations of motion for the macroparticle written down in the "energy-phase" variables are as follows (for the low efficiency approximation):

$$\frac{\mathrm{d}\hat{P}}{\mathrm{d}\hat{z}} = \mathrm{Re}\left[2\mathrm{i}\mathrm{e}^{\mathrm{i}\psi}\sum_n \exp(\mathrm{i}n\phi)U_r^{(n)}(\hat{r},\hat{z})\right] + U_c$$

$$\frac{\mathrm{d}\psi}{\mathrm{d}\hat{z}} = \hat{C} + \hat{P} \quad (5)$$

where $\hat{P} = (\mathscr{E} - \mathscr{E}_0)/\rho\mathscr{E}_0$ is the reduced energy deviation, ψ is the phase of the macroparticle within the longitudinal slice and $\rho = c\gamma_z^2\Gamma/\omega$ is the efficiency parameter. The initial energy spread is simulated with the additional distribution of the particles according to the Gaussian law

$$\mathrm{d}w = \frac{1}{\sqrt{2\pi\hat{\Lambda}_T^2}}\exp\left[-\frac{\hat{P}^2}{2\hat{\Lambda}_T^2}\right]\mathrm{d}\hat{P}$$

where $\hat{\Lambda}_T^2 = \langle(\Delta\mathscr{E})^2\rangle/(\rho^2\mathscr{E}_0^2)$ is the energy spread parameter. An algorithm for the calculation of the space charge contribution, U_c, can be found in Ref. [3]. The complex amplitude of the beam bunching is calculated by averaging the macroparticle ensemble in the elementary volume, $b_1 = \langle\exp(\mathrm{i}\psi_k)\rangle$, and is used for the calculations of the azimuthal harmonics of $b_1^{(n)}$. Eqs. (5) can be simply extended to the case of a high efficiency approximation and undulator tapering [3,4].

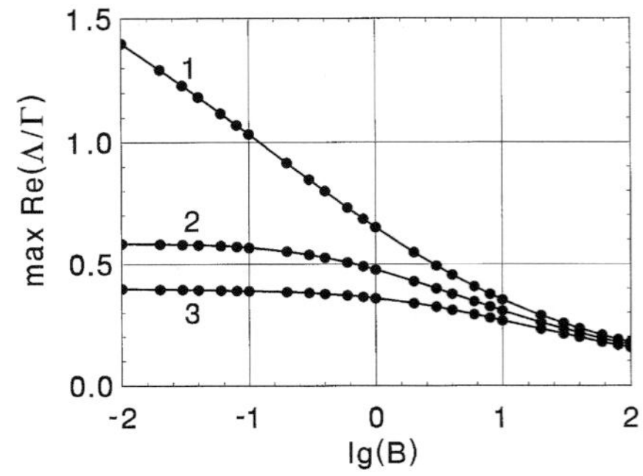

Fig. 1. The dependence of the maximal reduced field gain max $\mathrm{Re}(\Lambda/\Gamma)$ on the diffraction parameter B. Here $\hat{\Lambda}_{\mathrm{p}}^2 \to 0$, $\hat{\Lambda}_{\mathrm{T}}^2 = 0$. Curve (1): TEM_{00} mode, curve (2): TEM_{10} mode and curve (3): TEM_{01} mode.

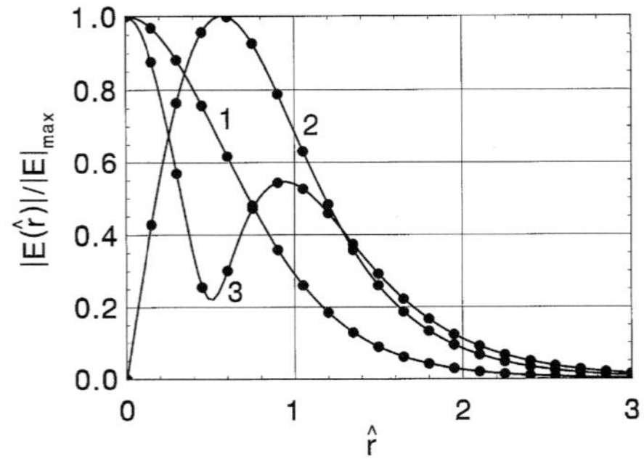

Fig. 2. Transverse distribution of the radiation field for TEM_{00}, TEM_{10} and TEM_{01}. Here $B = 10$, $\hat{\Lambda}_{\mathrm{p}}^2 \to 0$ and $\hat{\Lambda}_{\mathrm{T}}^2 = 0$. Curve (1): TEM_{00}, curve (2): TEM_{10} mode and curve (3): TEM_{01} mode. Solid curves are analytical results and the circles are the results of the numerical simulation code.

A complication for the case of the betatron oscillations and the undulator field errors to be taken into account is also straightforward [5].

Analysis of the integrals (3) shows that at the chosen radial mesh and at a fixed integration step Δz, calculations can be simplified significantly, since there is a finite set of the combination of the values $z - z'$, r and r'. The integrals over Δz for each of the combinations are calculated only once for the "unit" source term b_1. Then these data are transferred into the simulation code solving the self-consistent equations. Such a trick allows one to drastically reduce the required CPU time for the field calculations, since at each integration step the computer calculates simple sums weighted by the current sources.

The procedure for the solution of the self-consistent Eq. (5) has been described in the previous section. The accuracy of the calculations is controlled by changing the number of the axial, radial and azimuthal divisions, and the number of the azimuthal modes for the calculations of the radiation field. The criterium is that the final result is independent of the details of the simulations. Figs. 1 and 2 present the test results of the simulation code operating in a high gain, steady-state limit. It is seen that there is a good agreement between analytical [1] and simulation results. The relative accuracy for the calculation of the radiation field and the gain is about 0.1%.

At the exit the simulation code produces the matrices for the field values in the Fresnel diffraction zone. Then the post-processor programs are used to extract additional information for the field distribution in the far diffraction zone, for the spectrum, for the time, space and spectral correlation functions, and for the probability distributions of the radiation power and the radiation energy (see, e.g. Ref. [6]).

4. Initial conditions for the start-up from noise

The initial shot noise in the electron beam is simulated according to the algorithm presented in Ref. [6]. The number of particles per elementary volume N_{v} is large, so the bunching in each box is the sum of a large number of random phasors with fixed amplitudes and phases uniformly distributed on $(0, 2\pi)$. Using the central limit theorem, we can conclude that the phases of the bunching parameters are also distributed uniformly and the squared moduli of amplitudes, $|b_1|^2$, are distributed in accordance with the negative exponential distribution:

$$p(|b_1|^2) = \frac{1}{\langle |b_1|^2 \rangle} \exp\left(- \frac{|b_1|^2}{\langle |b_1|^2 \rangle} \right) \tag{6}$$

where $\langle |b_1|^2 \rangle = 1/N_{\mathrm{v}}$. The distribution of the modules, $|b_1|$, is the Rayleigh probability density function. So, we use the negative exponential random generator setting $1/N_{\mathrm{v}}$ as a mean value to extract the values of $|b_1|^2$ for each box and then we extract the square root to find the values of $|b_1|$. The phases of b_1 are produced by a random generator of a uniform distribution from 0 to 2π. These values are directly used as input parameters for the linear simulation code. In the nonlinear simulation code the macroparticles are distributed in such a way that the resulting bunching corresponds to the target value of b_1 in each elementary volume.

5. Conclusion

In conclusion, we should notice that the speed of calculations is an essential parameter for the FEL code calculating the start-up from noise. The reason for this is that the most important characteristics of the SASE FEL (the spectrum, time, space and spectral correlation functions and the probability distributions of the radiation power and energy) can be calculated only with statistical analysis of a large number of simulation runs [6]. For instance, the number of the simulation runs required for the calculation of the probability distribution of the radiation energy in the pulse is about several thousands. The presented code allows one to calculate all the above-mentioned statistical parameters of the SASE FEL within a reasonable time. For instance, a typical simulation run with the linear simulation code for the parameters of the UCLA/LANL/RRCKI/SLAC SASE FEL [7] takes about 1 min on a VAX processor [8]. This is about two orders of magnitude less than the time required by other time-dependent codes (such as GINGER or GENESIS) to obtain the same physical result [9,10].

Acknowledgements

We are grateful to B. Faatz, J. Feldhaus, G. Materlik, J. Pflüger, S. Reiche, J. Roßbach and J.R. Schneider for many useful discussions.

References

[1] E.L. Saldin, E.A. Schneidmiller, M.V. Yurkov, Opt. Commun. 97 (1993) 272.
[2] E.L. Saldin, E.A. Schneidmiller, M.V. Yurkov, Opt. Commun. 95 (1993) 141.
[3] E.L. Saldin, E.A. Schneidmiller, M.V. Yurkov, Phys. Rep. 260 (1995) 187.
[4] C. Pagani, E.L. Saldin, E.A. Schneidmiller, M.V. Yurkov, Nucl. Instr. and Meth. A 423 (1999) 190.
[5] T.M. Tran, J.S. Wurtele, Comput. Phys. Commun. 54 (1989) 263.
[6] E.L. Saldin, E.A. Schneidmiller, M.V. Yurkov, Opt. Commun. 148 (1998) 383.
[7] M. Hogan, C. Pellegrini, J. Rosenzweig, A. Anderson, P. Frigola, A. Tremaine, C. Fortgang, D. Nguyen, R. Sheffield, J. Kinross-Wright, A. Varfolomeev, A.A. Varfolomeev, S. Tolmachev, R. Carr, Phys. Rev. Lett. 81 (1998) 4867.
[8] E.L. Saldin, E.A. Schneidmiller, M.V. Yurkov, Numerical simulations of the UCLA/LANL/RRCKI/SLAC experiment on a high gain SASE FEL, Presentation at the 20th FEL Conference, Williamsburg, 1998.
[9] P. Pierini, W. Fawley, Nucl. Instr. and Meth. A 375 (1996) 332.
[10] S. Reiche, GENESIS 1.3 – A Fully 3D Time Dependent FEL Simulation Code, Presentation at the 20th FEL Conference, Williamsburg, 1998.

ELSEVIER

Nuclear Instruments and Methods in Physics Research A 429 (1999) 238–242

NUCLEAR INSTRUMENTS & METHODS IN PHYSICS RESEARCH
Section A

www.elsevier.nl/locate/nima

Influence of electron beam halos on the FEL performance

B. Faatz*, S. Reiche

Deutsches Elektronen-Synchrotron DESY, Notkestraße 85, 22603 Hamburg, Germany

Abstract

For single-pass free-electron lasers (FEL), such as amplifiers and SASE devices, saturation of the radiation power has to be reached within the length of the undulator. Therefore, detailed knowledge of electron beam parameters is crucial. So far, simulations have been performed with a given rms emittance and energy spread. At short radiation wavelengths, bunch compressors are used to compress the electron beam to achieve the desired high peak currents. In addition, external focusing along the entire undulator is used to maintain a constant small radius. The rotation of phase space due to compression might lead to a significant part of the bunch in tails that could increase the gain length. Furthermore, it is in general not possible to match both the beam core and the tail to the focusing structure. In this contribution, the influence of these tails, both transverse and in energy, on the FEL performance will be investigated. Simulations will be performed for beam parameters that have been assumed for the TESLA Test Facility FEL at DESY.

Keywords: Electron beam halos; FEL; TTF; SASE device

1. Introduction

In many existing free electron lasers, radiation power is amplified using a feedback system. In such a system, the most crucial constraint is that the gain exceeds the sum of all losses. As long as the number of passes needed to reach saturation is small compared to the total number of roundtrips of the field inside the cavity, the exact value of the single-pass gain is of less importance. For a single-pass device, more detailed knowledge of all parameters reducing the gain has to be obtained in order to determine the undulator length. In most studies thus far, the influence of beam emittance and energy spread has been investigated. In addition, a more exact estimate of the shotnoise power (in case of a SASE FEL [1]), the influence of magnetic errors and any misalignment [2] has been studied by several authors. In all these studies, however, the electron beam was assumed to be either Gaussian or parabolic. This is not necessarily the case. First simulation studies of the bunch compressor of the TESLA Test Facility [3] have shown that the electron beam can have either tails or spatially separated distributions, both in any transverse plane and in energy. In this paper, the influence of two possible distributions on the FEL performance is studied. Two identical Gaussian distributions, spatially separated in the x-direction, are used. Only their distance and relative fraction is varied. Independently, two Gaussian distributions with the same central energy, but with different rms-width provide the initial settings for the studies. The fraction of the

* Corresponding author: Tel.: + 49-40-8998-4513; fax: + 49-40-8998-2787.

E-mail address: faatz@desy.de (B. Faatz)

0168-9002/99/$ - see front matter
PII: S0168-9002(99)00111-4

two distributions has been changed in the same manner as for the transverse displacement.

2. Results

Simulations with GENESIS 1.3 [4] have been performed for the TTF-FEL parameters. This is a fully three-dimensional code, including time dependence. For these simulations, only the time-independent part has been used. Deviating from the TTF parameters the FODO-structure has been changed to get an almost constant β-function, with the average value equal to the value resulting from the real structure.

A first set of simulations has been performed assuming two identical Gaussian distributions, spatially separated by a certain amount in the x-direction. An example, with an offset of 250 μm and a charge of 70% in the large Gaussian and 30% in the smaller, is shown in Fig. 1. A similar displacement in y, or an initial kick in either x or y-direction would give similar results, the latter due to the FODO-structure, which makes the electron beam perform a betatron oscillation (see Table 1).

Results are shown in Fig. 2. The displacement in x varies from 50 to 250 μm, with the center of the total beam always on axis, e.g. $\langle x \rangle = 0$. This means that, depending on the fraction in both

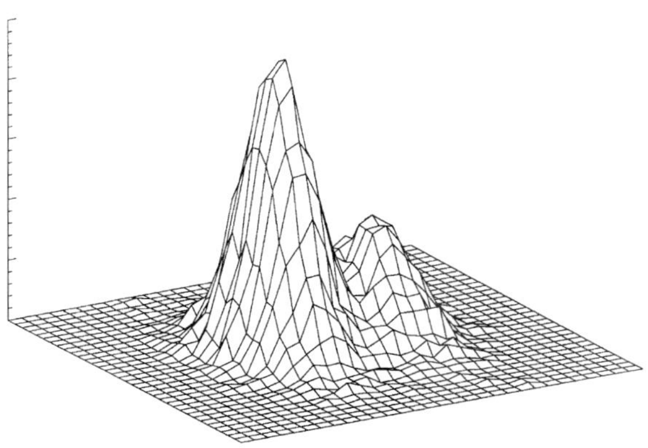

Fig. 1. Example of a transverse electron beam distribution, in this case separated by 250 μm. The smaller Gaussian contains 30% of the charge.

Table 1
Undulator and optics parameters for the TTF FEL (Phase I at 230 MeV)

Electron beam	
Peak current	500 A
Normalized rms emittance	2π mm mrad
rms energy spread	500 keV
Average beam size	77 μm
ρ-parameter	4×10^{-3}
Undulator	
Period length	27.3 mm
Undulator peak field	0.497 T

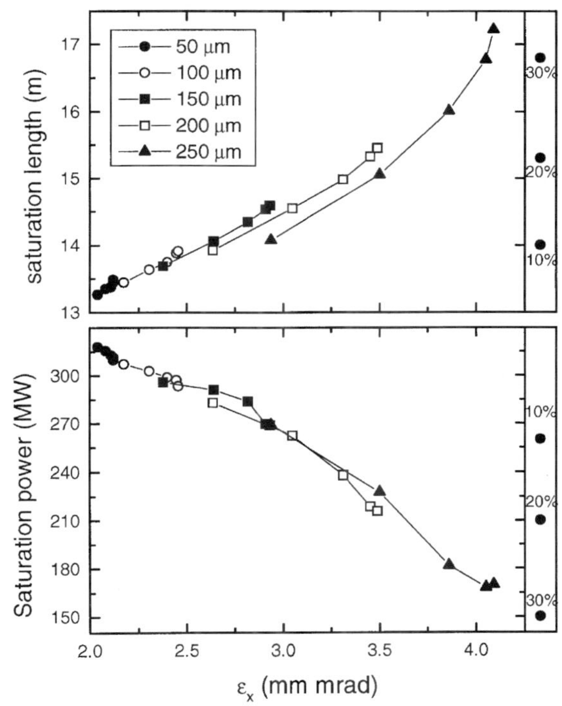

Fig. 2. Saturation length (top figure) and power (bottom figure) versus normalized emittance for an electron beam consisting of two Gaussians separated in the x-direction in the range 50–250 μm. The fraction of electrons in one distribution varies from 10% to 50%, keeping the total charge constant. The points on the right indicate the saturation length and power for a single Gaussian beam with a reduced current given by the indicated fraction.

distributions and on their relative offset, both are off-axis, performing a betatron oscillation. The relative fraction of the smaller Gaussian increases from 10% to 50% if saturation is reached within

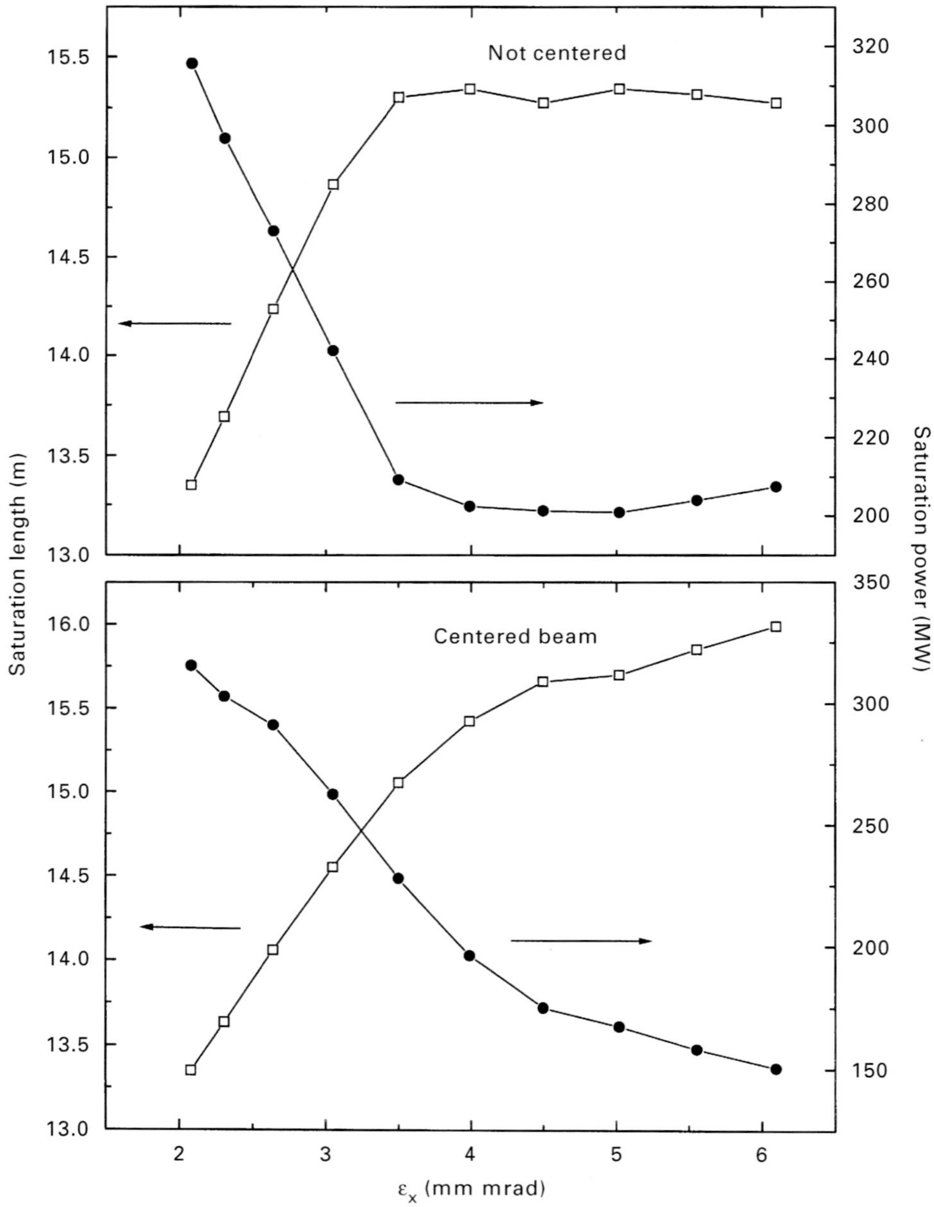

Fig. 3. Saturation length versus normalized emittance for an electron beam consisting of two Gaussians separated in the x-direction in the range 50–500 μm. The fraction of electrons in one distribution is 20%. Either the largest fraction is put on axis (top figure) or the center of the entire beam, as before (bottom figure).

17.5 m. Especially for larger offsets, saturation is only reached for smaller fractions. The normalized emittance given in the horizontal scale is calculated in the usual way, e.g. $e_x^2 = \langle \Delta x^2 \rangle \langle \Delta x'^2 \rangle - \langle (\Delta x\, \Delta x')^2 \rangle$. As it can be seen, independent of the offset of the beams or their relative fraction, both power and saturation length seem to be related to the emittance calculated this way. One can also calculate the saturation power and length for an electron beam without halo, but with the same normalized emittance and the same mismatch of the initial conditions (resulting in the same beam envelope along the longitudinal axis). The results are within a few percent of the results shown in Fig. 2.

More important is, however, whether it is beneficial to cut away part of the electron beam, place the remainder on axis and match it to the FODO-lattice. Results are shown to the right in Fig. 2. The reduction in current varies from 10% to 30%. As can be seen, within the parameter range studied, the saturation length for a given fraction in the halo tends towards the point where the current has been reduced by the same amount. If one would not place the center of the beam on axis, but the largest fraction of the beam (90–50%), reducing the current gives similar results.

For a fraction of 20% in the halo, the offset has been extended to 500 μm. Results are shown in Fig. 3. In this case, two different positions of the beam are studied. In the top figure, the largest fraction of the beam is put on axis. For smaller offsets, this gives a slightly larger saturation length because of the large betatron oscillation performed by the off-axis part. For very large offsets, however, the 20% fraction no longer participates in the interaction and the saturation length and power are equal to the values obtained by cutting away this part of the current. In the bottom figure, the saturation length is slightly shorter for small offsets. As the offset increases, the saturation length increases beyond the value obtained for a 20% reduced current, because the large betatron oscillation of both parts of the beam reduce the interaction with the field. Therefore, the saturation length still increases (and the power decreases), but at a smaller rate than before.

The next set of simulations shows the influence of two overlapping Gaussians in energy. Fig. 4 shows saturation power and length for an energy width of the halo compared to the core of the beam from 2 to 5 times. In case the halo has a two times larger energy width up to 50% halo, saturation is still reached within the undulator length studied here.

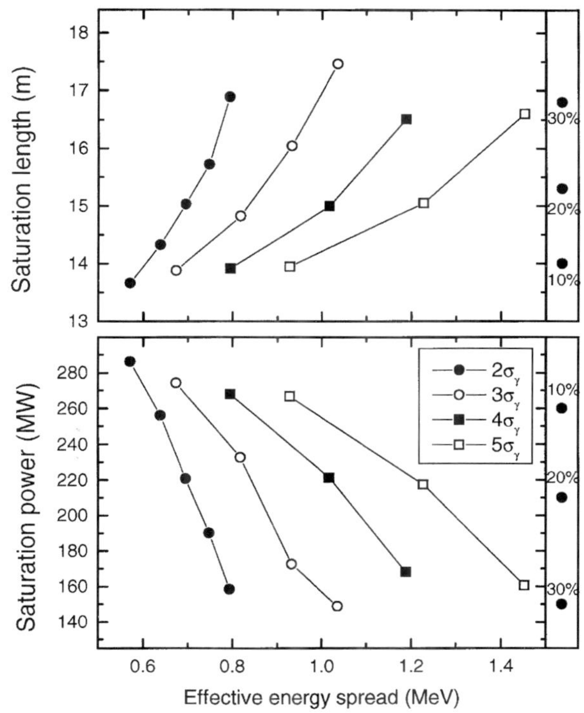

Fig. 4. Saturation length (top figure) and power (bottom figure) versus effective energy spread for an electron beam consisting of two Gaussians in energy. The core beam always has the same energy width, the halo has a width of up to 5 times the width of the core beam. The fraction in the two parts varies from 10% and 90% to 50% in both. The points on the right indicate the saturation length and power for a single Gaussian beam with a reduced current given by the indicated fraction.

For a larger width this number decreases to 30%. The horizontal scale, the effective energy spread (given in MeV), is simply $\langle \gamma^2 \rangle - \langle \gamma \rangle^2$, with γ the Lorentz factor. There is no obvious relation between this parameter and the gain. Simulations of a single Gaussian energy distribution with this same width give different results. As can be seen, the levels are almost constant for an energy spread of the halo exceeding $2\sigma_\gamma$. In turn, these values are almost identical to the results with the current reduced by the same fraction as it was in the halo. This can be understood as follows. The energy spread of the normal beam without halo is about half of the ρ-parameter. The bucket in longitudinal phase space in which the electrons are captured have a height of no more than ρ, a value which is reached close to saturation. All electrons outside of this region will not interact with the electron beam. Therefore, for a given fraction in the halo, increasing the energy spread will not change saturation power or length if it exceeds ρ significantly. Changing the fraction will, however, reduce the gain, because of a reduced part that remains captured inside the bucket.

3. Conclusions

Simulations of halos have shown that they can strongly influence the performance of the FEL. For transverse distribution consisting of two (partly) separated but otherwise equal distributions, one benefits from realigning the electron beam to get the center on axis and to minimize the betatron oscillations for small offsets. For larger offsets, the large betatron oscillation reduces the gain if the center is placed on axis. For two overlapping Gaussians in energy, both with the same central energy but with different width, the saturation length and power cannot be changed significantly. Cutting away the halo, if this would be possible, and thus reducing the current, does not give a better performance for the parameters studied here.

Acknowledgements

The authors would like to thank H.-D. Nuhn, C. Pagani and E.A. Schneidmiller for their useful suggestions.

References

[1] K.J. Kim, Nucl. Instr. and Meth., A 250 (1986) 396.
[2] B. Faatz, J. Pflüger, Y.M. Nikiting, Nucl. Instr. and Meth. A 393 (1997) 388.
[3] DESY Print TESLA-FEL 95-03, Hamburg, DESY, 1995.
[4] S. Reiche, Nucl. Instr. and Meth. A 429 (1999) 243.

ELSEVIER

Nuclear Instruments and Methods in Physics Research A 429 (1999) 243–248

NUCLEAR INSTRUMENTS & METHODS IN PHYSICS RESEARCH Section A

www.elsevier.nl/locate/nima

GENESIS 1.3: a fully 3D time-dependent FEL simulation code

S. Reiche*

Deutsches Elektronen Synchrotron, Notkestraße 85, 22603 Hamburg, Germany

Abstract

Numerical simulation codes are basic tools for designing Free Electron Lasers (FEL). They are used to study the impact of different parameters, e.g. wiggler errors and external focusing, which allow FEL users to optimize the performance. For faster execution some simulation codes assume radial symmetry or decompose the radiation field into a few azimuthal modes, although then this treatment does not include the full description of the FEL. This contribution describes the new FEL code GENESIS 1.3 which uses a fully three-dimensional representation of the FEL equations in the paraxial approximation for time-dependent and steady-state simulations of single-pass FEL. In particular this approach is suitable for cases where the radial symmetry is broken by the electron beam distribution as well as by wiggler errors, betatron motion and off axis injection of the electron beam. The results, presented here, are based on the parameters of the TESLA Test Facility FEL at DESY.

Keywords: Free-electron lasers; GENESIS 1.3; Numerical simulation

1. Introduction

With the design and construction of Free Electron Lasers (FEL), many codes have been developed [1–5] over the years in order to describe the physics taking place in different regimes. Recent research is done in the field of Self-Amplified Spontaneous Emission (SASE) FEL [6–8]. To investigate the properties of this radiation source as well as the extension of SASE FEL's wavelength to the VUV or X-ray regime, new codes or extensions of established codes are needed. To cover these aspects and others such as the influence of wakefields, a time-dependent code has been developed called GENESIS 1.3.

The algorithm to solve the FEL equation in the paraxial approximation is similar to TDA3D [2]. In fact GENESIS 1.3 is mainly based on TDA3D although major modifications have been made. One of the major improvement is the replacement of the radial mesh with a full Cartesian mesh using the Alternating Direction Implicit (ADI) integration scheme [9] to solve the field equation. This allows the user to study non-axi-symmetric cases such as undulator field errors or beam halos.

The basic idea of the extended algorithm for time-dependent simulation is to solve the equation for a given slice of the electron bunch and a certain integration length significantly smaller than the gain length before advancing the radiation field to the next slice and replacing it with the radiation field of the trailing slices. This allows only a fractional part of the electron beam and radiation field to be kept in memory.

The source code of GENESIS 1.3 is written in standard ANSI Fortran 77 and can be compiled

* Tel.: + 49-40-8998-3448; fax: + 49-40-8998-4305.

0168-9002/99/$ - see front matter
PII: S0168-9002(99)00114-X

and linked with common Fortran compilers on any platform. The two restrictions for a successful execution are the support of double complex precision numbers and enough memory for time-dependent simulation. During execution the code will read from or write to files using only standard Fortran formats such as sequential or direct access and formatted or unformatted in- and output. GENESIS 1.3 does not support any graphics as output. For visualization of the output data the postprocessor XGENESIS can be used running under the IDL environment [10].

So far the tests which have been made do not show any significant deviation from other well-tested codes or analytic results. The CPU time consumption is moderate and steady-state simulation, even on Personal Computers, will run in the range of minutes. For time-dependent simulation the CPU time scales linearly with the number of macroparticle slices.

To illustrate the application regime of GENESIS 1.3, only few important examples will be shown in this paper. All simulations are based on the parameters of the TESLA Test Facility Free Electron Laser (TTF-FEL) which are listed in Table 1.

Table 1
Parameters of TTF-FEL used for the simulations. If two values are given the first one corresponds to Phase I of the TTF-FEL, the second to Phase II

Electron beam	
Energy	250/1000 MeV
Energy spread	0.5/1.0 MeV
Normalized emittance	2π mm mrad
Ave. beam size	60/50 μm
Peak current	500/2500 A
Bunch length	250/50 μm
Undulator	
Number of modules	3/5
Length of modules	4.5 m
Period length	2.73 cm
Undulator peak field	0.497 T

2. Time-dependent simulation

One of the most important applications of time-dependent codes is the simulation of SASE FELs. Following the algorithm described in Ref. [11] spontaneous emission seeds the FEL amplifier instead of an external seeding field commonly used in steady-state simulations. As an example a compact

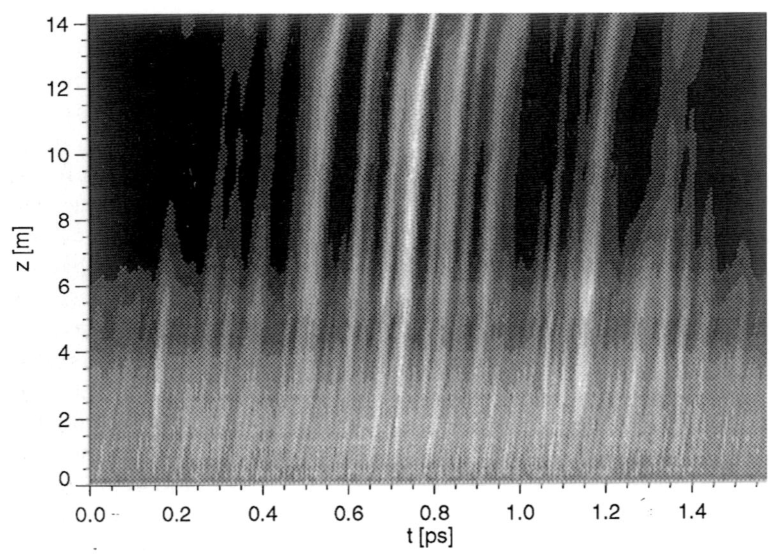

Fig. 1. Radiation power of TTF-FEL (Phase I) versus longitudinal position in the electron bunch (horizontal axis) and undulator (vertical axis). The radiation pulse is normalized to unity for all undulator positions to exclude the dominant exponential growth of the radiation field.

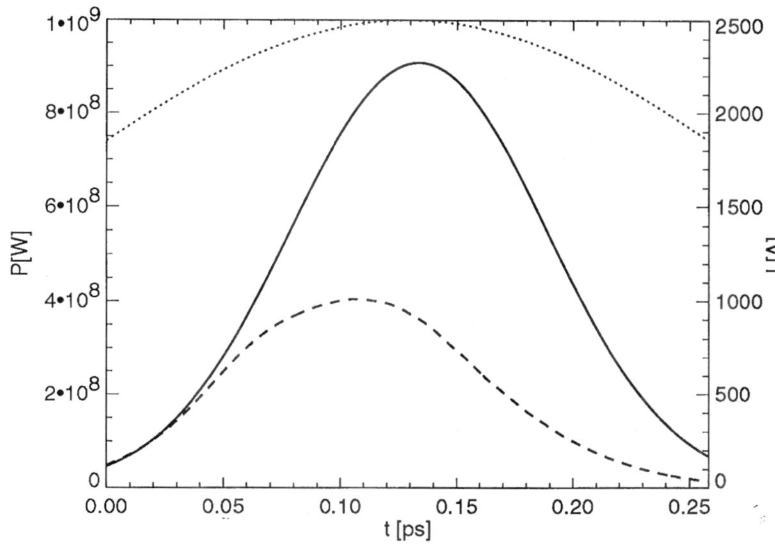

Fig. 2. FEL radiation of TTF-FEL (Phase II) including energy losses of the electron beam due to wakefields (dashed line). The power of the radiation pulse is significantly decreased in comparison to the FEL radiation excluding the effect of wakefields (solid line). Both simulations used the same set of parameters such as the same longitudinal beam current profile (dotted line) and same seeding field.

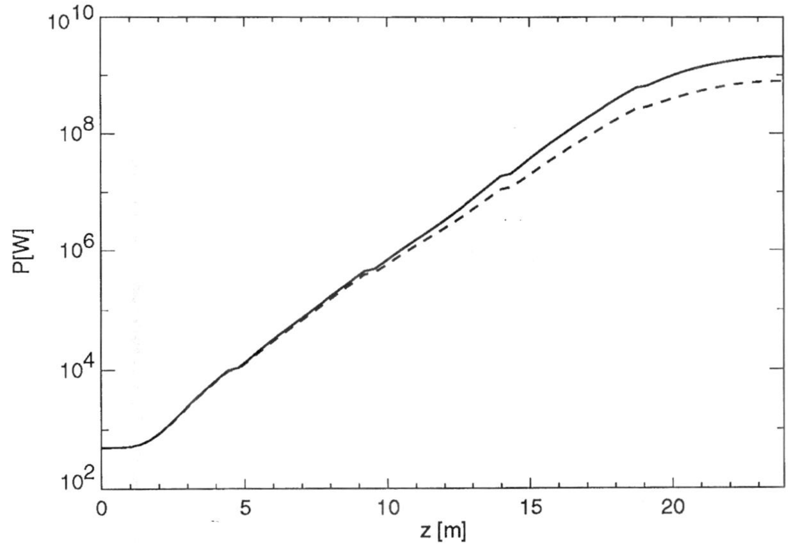

Fig. 3. Radiation power versus undulator position including (dashed line) and excluding (solid line) wakefields.

representation of the result for a single run is shown in Fig. 1 using the parameters of the TTF-FEL Phase 1 (see Table 1). For each longitudinal position in the undulator (vertical axis) the radiation pulse (horizontal axis) is normalized in such a way that the maximum always has the same value. In this way the structure of the pulse is clearly visible for any position in the undulator. Otherwise the

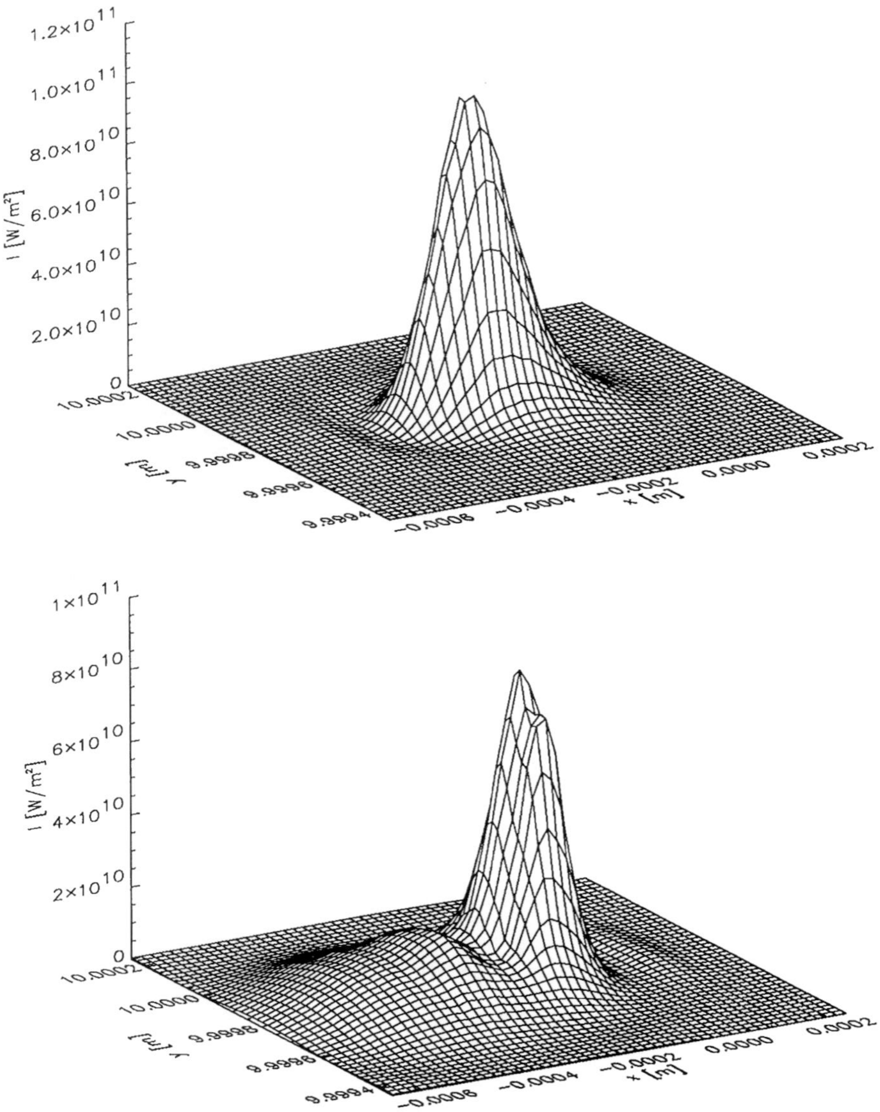

Fig. 4. Intensity of radiation field for FEL simulation including magnetic field errors. The data shows a typical case where the electron beam is kicked towards one direction in the horizontal plane (upper plot) and bent back towards the undulator axis (lower plot).

exponential growth of the SASE FEL amplification process will dominate. It is seen that the number of spikes is decreasing along the undulator and the contrast is enhanced. Both effects arise due to growing longitudinal and transverse coherence.

To demonstrate another application regime for time-dependent simulation, the influence of wakefields on the performance of the TTF-FEL is investigated. This is of particular concern for Phase II with its short rms bunch length of 50 μm. The geometric and electromagnetic properties of the

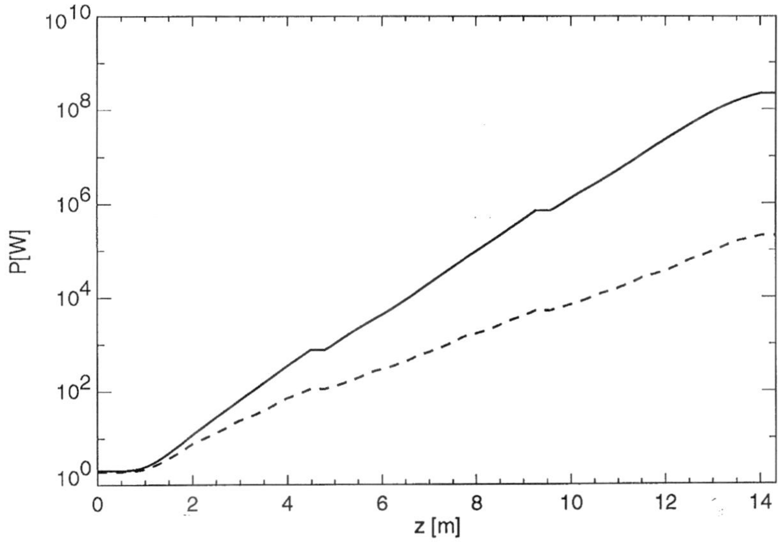

Fig. 5. Radiation power versus undulator position for the TTF-FEL (Phase I) including magnetic field errors (dashed line). For the calculation 400 independent runs are averaged with a relative rms field error of 0.3%. The undisturbed FEL performance is shown by the solid line.

beam pipe provide three major sources for wakefields: conductivity, surface roughness and geometric changes of the beam pipe along the undulator. For the parameters of the TTF-FEL all three wake potentials have nearly the same amplitude but different shapes [12]. For a time-dependent simulation covering a length of about 100 μm around the peak current, the gradient of the total wake potential varies between − 100 and 30 keV/m.

Because most electrons are shifted away from the FEL resonant condition due to the wakefields, the total gain is significantly reduced. For a position close to saturation the radiation pulse is plotted in Fig. 2. In contrast to the undisturbed FEL performance the maximum of the radiation pulse is trailing behind the maximum of the beam current distribution. The reason is that slippage of the radiation field has less influence than the detuning of the electron beam which is stronger in the front part of the electron beam.

In Fig. 3 the radiation power with and without wakefields taken into account is plotted as a function of the longitudinal position in the undulator. It is seen that the power reduction due to wakefields is larger in the last part of the undulator due to the accumulated energy losses of the electron beam. The saturation length remains nearly unchanged and is reached in this case when most of the beam is completely detuned.

3. Non-axi-symmetric simulation

As an example of broken axi-symmetry the case of undulator field errors for the TTF-FEL (Phase I) is simulated [13].

For a relative rms error of 0.3% for the field errors the radiation distribution at two positions within the undulator is presented in Fig. 4. At these positions the electron beam centroid was deflected from the undulator axis and bent backwards due to strong focusing. The radiation field, unable to follow the rapid motion of the electron beam, becomes distorted. The steeper edges of the radiation field distribution induce a larger diffraction in addition to a reduced gain guiding because the electron beam position has changed transversely. The lower plot of Fig. 4 shows such a large diffracted part of the radiation field.

The FEL gain is also reduced by other effects such as loss of the synchronization condition of the electron beam and radiation field. For this case the coherent transverse kick reduces the longitudinal velocity of the electron beam. The average power gain versus longitudinal position in the undulator is shown in Fig. 5 with a saturation power reduced by nearly 3 orders of magnitudes relative to the undisturbed motion of the electron beam.

4. Conclusion

Although several FEL codes already exist, there is an increasing demand to cover new aspects of FEL performance in order to keep up with the advance research. GENESIS 1.3 contains several new features and allows the user to describe the problem in a more flexible manner. Even for cases which are not covered by the standard features of GENESIS 1.3, the user has the option to modify the source code for his own purposes, because GENESIS 1.3 is distributed free. Thus GENESIS 1.3 provides a new and useful tool for FEL-physics.

Acknowledgements

The author would like to thank B. Faatz, E.L. Saldin, E.A. Schneidmiller, H. Schlarb and M.V. Yurkov for their useful discussions.

References

[1] G.A. Travish, Proceeding of the X-ray Workshop, June 1997, Gargnano, Italy.
[2] T.-M. Tran, J.S. Wurtele, Comput. Physics Commun. 54 (1989) 263.
[3] E.T. Scharlemann et al., Nucl. Instr. and Meth. A 250 (1986) 150.
[4] W.M. Fawley, CBP Tech Note-104, 1995.
[5] E.L. Saldin et al., Nucl. Instr. and Meth. A 429 (1999) 233.
[6] M. Hogan et al., Presented at the 10th Free Electron Laser Conference, Williamsburg, VA, 1998.
[7] DESY Print TESLA-FEL 95-03, Hamburg, DESY, 1995.
[8] R. Tatchyn et al., Nucl. Instr. and Meth. A 375 (1996) 274.
[9] W.H. Press et al., Numerical Recipes in Fortran 77, Cambridge University Press, Cambridge, 1992.
[10] IDL – Interactive Data Language, Research System Inc.
[11] C. Penman, B.W.J. McNeil, Opt. Commun. 90 (1992) 82.
[12] M. Dohlus et al., DESY Print TESLA-FEL 98-02, Hamburg, DESY, 1998.
[13] B. Faatz et al., Proceeding of the FEL 96 Conference, 1997, p. 277.

ELSEVIER

Nuclear Instruments and Methods in Physics Research A 429 (1999) 249–256

NUCLEAR INSTRUMENTS & METHODS IN PHYSICS RESEARCH Section A

www.elsevier.nl/locate/nima

FEL simulations for the LCLS☆

Heinz-Dieter Nuhn

Stanford Linear Accelerator Center, Stanford University, CA 94309-0210, USA

Abstract

A first design study report has recently been completed (The LCLS Design Study Group, LCLS Design Study Report, April 1998, SLAC-R-521) for the linac coherent light source (LCLS), a proposal to build an X-ray free electron laser (FEL) at the Stanford Linear Accelerator Center (SLAC) as a single pass self-amplified spontaneous emission (SASE) amplifier. The proposal includes the use of a very low emittance electron beam accelerated up to 15 GeV by the last third of the SLAC linac to produce sub-picosecond X-ray pulses with high brightness and full transverse coherence in a 112-m long undulator. Many aspects of the FEL design have been analyzed with FEL simulation codes. The paper discusses some of the results of these aspects, i.e. temporal X-ray pulse structure and power spectrum, trajectory errors and effects of undulator beam tube wakefields. © 1999 Elsevier Science B.V. All rights reserved.

Keywords: FEL simulations; LCLS; GINGER

1. Introduction

The Stanford Linear Accelerator Center (SLAC) and collaborating institutions are proposing to build the linac coherent light source (LCLS), a free electron laser (FEL) facility operating in the wavelength range 1.5–15 Å. Since optical cavities are not available for the wavelength range of interest, due to the lack of good reflecting surfaces to form the optical cavity mirrors, the LCLS is based on the process of self-amplified spontaneous emission (SASE) [2]. No mirrors are used. Lasing is achieved in a single pass of an electron bunch through a long undulator. The basic parameter set for the proposed LCLS FEL is displayed in Table 1 for the short wavelength limit. The design of the LCLS is based on a hybrid permanent magnet undulator that comprises 52 segments each 1.92 m long separated by 24 cm long gaps. The total length of the segments is 99.84 m or 3328 undulator periods. Between the segments are electron beam position monitors and permanent magnet quadrupole magnets for focusing and trajectory control. The discussion in this paper is focused on the short wavelength limit of the LCLS proposed operational range.

2. Time dependent simulation results from GINGER

The analysis of startup from noise, saturation, as well as the temporal and spectral structure of the radiation pulse for the LCLS design uses the 2D, time-dependent simulation code GINGER [3,4],

☆Work supported by the Department of Energy (Contract DE-AC03-76SF00515), Office of Basic Energy Sciences, Division of Material Sciences.

0168-9002/99/$ - see front matter
PII: S0168-9002(99)00116-3

Table 1
Basic LCLS FEL Parameters

Radiation wavelength λ_r	1.5 Å
Undulator period λ_u	3 cm
Peak magnetic field	1.32 T
a_w	2.62
Electron energy	14.35 GeV
Norm. electron beam emittance	1.5 mm mrad
Peak current	3400 A
RMS bunch length	20 μm
Uncorrelated energy spread	2.0×10^{-4}
FEL parameter (1D)	4.7×10^{-4}
Rayleigh length	40 m
Power gain length L_G	5.8 m
Error free saturation length	94 m

which models the interaction of the 3D motion of the electron beam with an axisymmetric, multi-frequency radiation field. SASE startup from shot noise is modeled by adding random fluctuations to the macro-particles' longitudinal and transverse coordinates [5]. Segment separations or lumped focusing are not implemented in GINGER. It has been shown that for the LCLS, and particularly at the short wavelength limit, they have little effect on output power and on the amount of magnet material needed [6,7]. Output power increase will, of course, pause between segments. Figures that depend on the longitudinal coordinate, z, would show the effect.

Fig. 1 shows the development of the pulse averaged peak power for the error-free LCLS undulator. Exponential growth starts after roughly 15 m. Saturation occurs before the end of the undulator.

In order to reduce the simulation time for the analysis, GINGER is run in a mode in which both the longitudinal electron beam distribution and the radiation field are assumed to be periodic, with a period much larger than the total slippage length. The simulation is only done for one periodicity window. Electron beam slices that slip outside the window on one side will enter it again on the other side. The use of periodic boundary conditions is not believed to lead to significant, unphysical effects. Periodic boundary conditions cannot be used to study the effect of correlated energy spread and bunch emittance or electron bunch end effects.

Fig. 2 shows the time structure of the radiation field for a periodicity window of 12 fs at the end of a 100 m long undulator. The optical pulse is composed of a number of superradiant spikes spaced at a distance related to the cooperation length $2\pi L_c = 4\pi L_G \lambda_r/\lambda_u = 1.2\ \mathrm{fs}\times c$ as predicted from 1D theory [8].

Fig. 3 shows the normalized intensity of the radiation pulse along the window (abscissa) as function of the windows position within the undulator (ordinate). After the initial random distribution in the lethargy regime, and until saturation, a superradiant spike structure forms with a group velocity different from the phase velocity of the radiation. The window position is coupled to the phase of the radiation. The group velocity of the spikes was derived in Ref. [8] to be $v_s = 3v_{||}/2 + v_{||}/c$. Using $v_{||} = c(1-\lambda_r/\lambda_u)$ for the average longitudinal speed of the electron beam and the fact that $\lambda_u \ll \lambda_r$, the spike velocity v_s is thus predicted to $v_s = c(1-\frac{2}{3}\times\lambda_r/\lambda_u)$ or $1 - v_s/c = \frac{2}{3}\times\lambda_r/\lambda_u = 0.33\times10^{-10}$, i.e., in the exponential gain regime spikes fall behind the radiation field by $\frac{1}{3}\times\lambda_r$ every undulator period or they move ahead of the electron beam by $\frac{2}{3}\times\lambda_r$ during the same time. The simulation predicts a value of $1 - v_s/c \approx 0.3\times10^{-10}$, slightly smaller than the prediction of the 1D theory.

The analysis of beam transport from the gun to the undulator through linac and bunch compressors predicts that a correlation between the average energy and the transverse position of a slice of the bunch and its longitudinal position within the bunch is to be expected. At the entrance to the undulator, the electron bunch is expected to be homogeneous in intensity over most of its core longitudinal position (see Ref. [1], p. 7–10). Over the same range, the average value of the energy distribution follows a slowly varying function (covering an rms width of 0.1%) while the width of the distribution stays roughly constant ($\Delta\gamma_{\mathrm{corr}} < 0.02\%$). At both ends of the bunch is a strong increase in energy spread, which will suppress the production of micro-bunching and radiation enhancement. The energy correlation will widen the spectral distribution of the total radiation pulse to about $2\Delta\gamma_{\mathrm{corr}} \approx 0.02\%$).

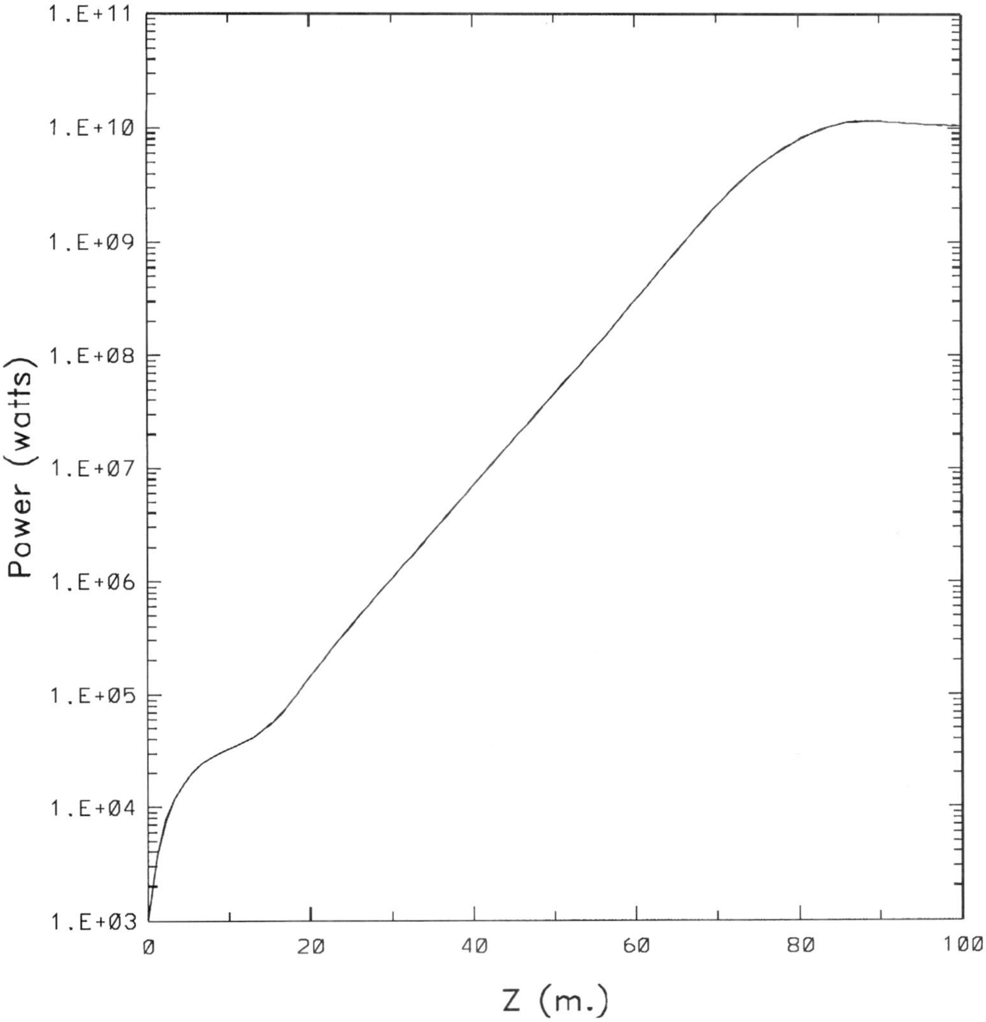

Fig. 1. Predicted development of the peak radiation power. The power is averaged over a window.

Fig. 4 shows the predicted spectrum after saturation for a slice from the center region of the radiation pulse. The power is binned into wavelengths intervals of 0.0003 Å. Besides the main spectral line, two other peaks at slightly longer wavelength and lower power level are visible.

Fig. 5 shows how the bandwidth of the radiation decreases until saturation is reached. At saturation the rms bandwidth is about $1/N_{sat} \approx 0.003$ as predicted in Ref. [9].

3. Trajectory analysis

Deviations of the electron trajectory from the ideal straight line inside the undulator can be caused by on-axis magnetic field errors, quadrupole misalignments and the limited precision of beam position detection and correction. Trajectory errors can become a major source of gain reduction if they are not kept small. The quadrupoles along the LCLS undulator are made from permanent magnet

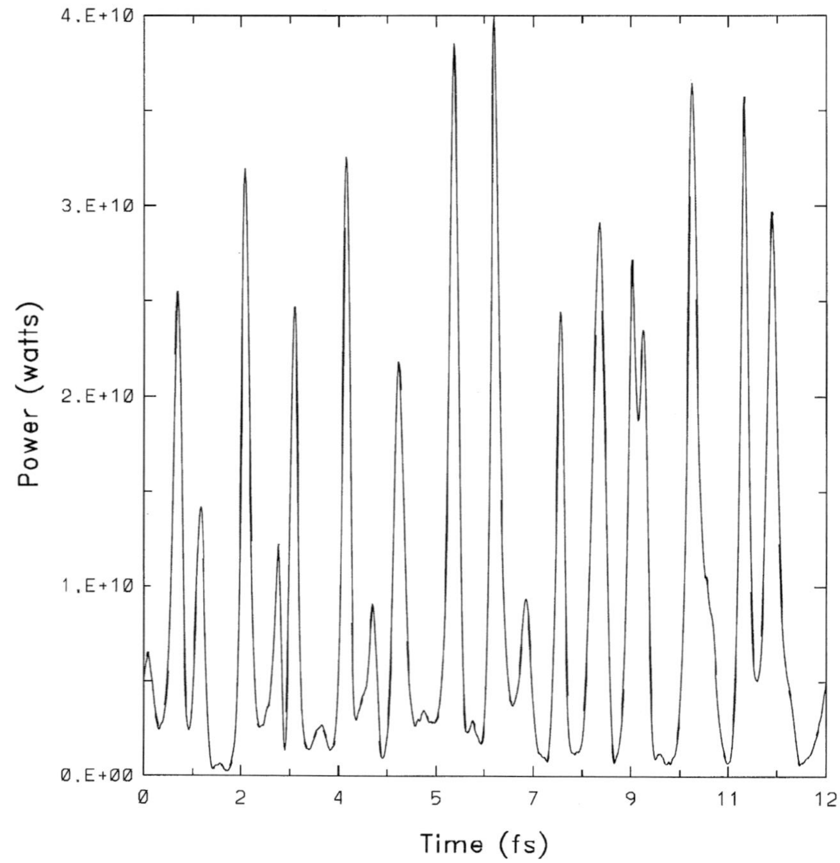

Fig. 2. Predicted temporal structure of a typical LCLS X-ray pulse after saturation.

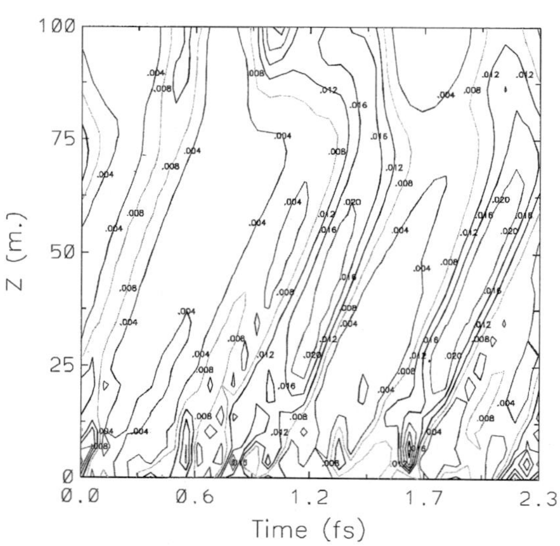

Fig. 3. Prediction of the development of the temporal structure of a typical LCLS X-ray pulse along the undulator. A short periodicity window is used for clarity.

material and are transversely movable to correct the trajectory in both the horizontal and vertical planes. Thus, the calibration and resolution of the beam position monitor system, not the quadrupole alignment, is the primary source for orbit errors. The choice of high-resolution cavity BPMs and the development of a beam-based-alignment procedure [10] for BPM calibration allow to produce trajectories close to the ideal case. What remains is the random walk trajectory between two beam position monitors caused by magnetic field errors from imperfections

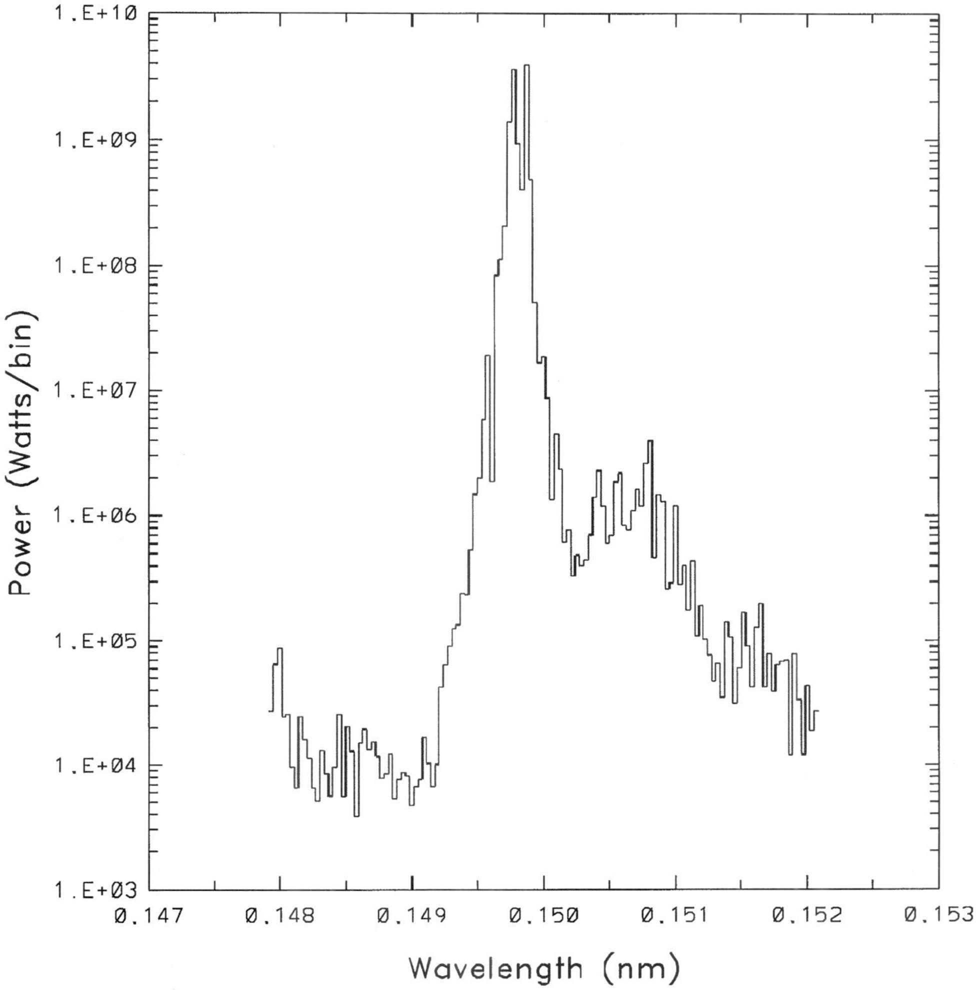

Fig. 4. LCLS spectrum of a longitudinal pulse slice of a typical X-ray pulse around saturation.

of the 128 undulator poles that lie between two LCLS BPM/corrector pairs and from external fields. Shimming in combination with sophisticated field error measurement methods, such as the stretched wire technique, can be used to reduce the random walk trajectory to a very high degree. Extensive FEL simulations have been used to determine the degree of field and trajectory errors that can be allowed before an increase in saturation length occurs.

A quantity that is of interest for undulator designers is the rms deviation of the on-axis magnetic field from the ideal case. This quantity can be used as input to the monochromatic 3D simulation code, FRED3D [11]. The specific LCLS undulator field will cause one particular random-walk trajectory, which can only be known after the undulator is built. The simulations use randomly generated Gaussian error distributions with specified rms values, truncated to 3 sigma.

Fig. 6 shows the distribution of maximum random walk amplitudes that can be generated by a given rms magnetic field error, assuming that correction occurs every 2.16 m. The distribution

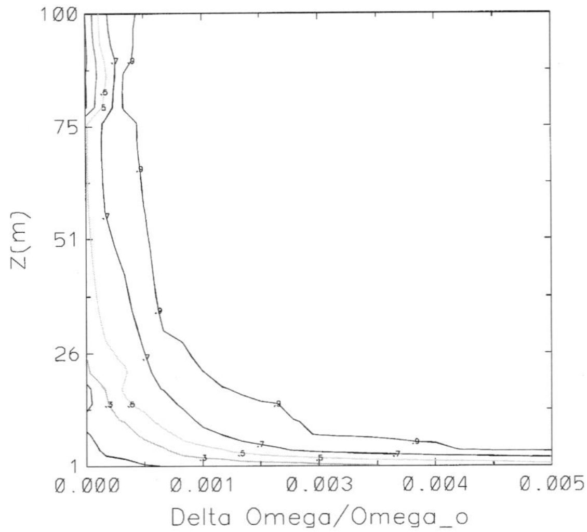

Fig. 5. Width of the power spectrum of a longitudinal slice of the radiation pulse vs. Z.

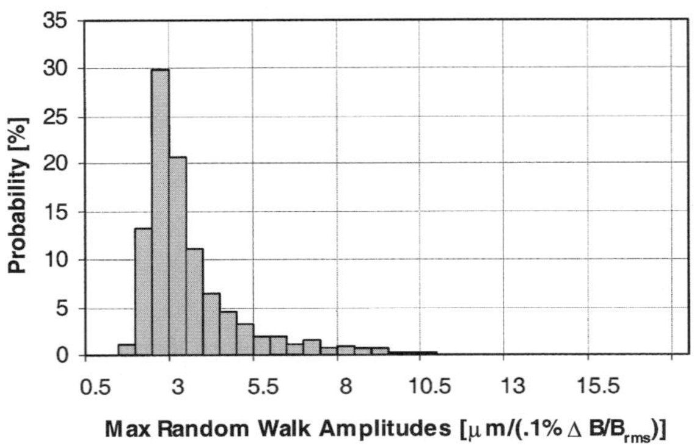

Fig. 6. Distributions of the maximum random walk amplitude per rms magnetic field error for a perfectly steered LCLS orbit. Parameters are $\gamma = 28077$, $\langle\beta\rangle = 18$ m/rad, $\Delta z_{corr} = 2.16$ m.

peaks below 3 μm/0.1% rms error amplitude. Simulations show that up to rms errors of 0.1% trajectories at or below the peak of the distribution, i.e. maximum trajectory deviations below 3 μm, will not increase saturation length.

4. Wall roughness and resistive wall impedance

When the electron beam moves through the undulator it will excite longitudinal and transverse wakefields due to the resistance and the discontinuities of the beam tube wall. The forces due to the wakefields are correlated with longitudinal position, z_1, within the bunch. For a longitudinal bunch slice of length dz, the longitudinal wakefield will generate a change in energy, dγ, at a rate dγ/dz. If the average beam tube properties do not depend on the position within the undulator, a slice's average energy will linearly increase or decrease as it moves along the undulator, depending on its position within the bunch. The transverse electron distribution within a slice is not affected.

If the rate at which the slices' energy change occurs becomes too large, FEL dynamics will be negatively effected due to de-trapping. Only the average energy change from this effect can be corrected by tapering. To ensure FEL gain, tolerances for wall roughness and wall resistance need to be established.

In order to simulate the effect with the GINGER code, a position-dependent energy loss term $\mathrm{d}\gamma_{wake}/\mathrm{d}z(z_1) = a\sin(2\pi z_1/b)$ was added to the FEL equations by the author of GINGER [12]. The parameter b specifies the periodicity window width, z_1 is the relative position of a slice within the window, and the amplitude parameter, a, is varied during the study.

Fig. 7 shows the result of a typical run with the parameter values $a = 0.5\,\mathrm{m}^{-1}$ and $b = 18.56$ fs. The ordinate shows the slices' peak power change in units of watts over the length of the undulator, while the abscissa indicates the temporal position of the electron and photon beam slices relative to the beginning of the window at the entrance to the undulator. Note that the electron slices fall behind (i.e. move to the right hand side in the graph) by 1.668 fs with respect to the photon slices during propagation through the undulator, due to slippage. The gain of the radiation spikes is reduced when the energy loss, $\Delta\gamma$, grows outside of the range $-30 \leq \Delta\gamma \leq 20$. For the example shown in the figure, the final reduction in peak power is about one order of magnitude at bunch locations at which the energy loss $\Delta\gamma$ is above this range. The reduction is less at bunch locations where slices gain energy, due to compensation of the electrons' energy loss from radiation. A significant increase in output power occurs at the position where the

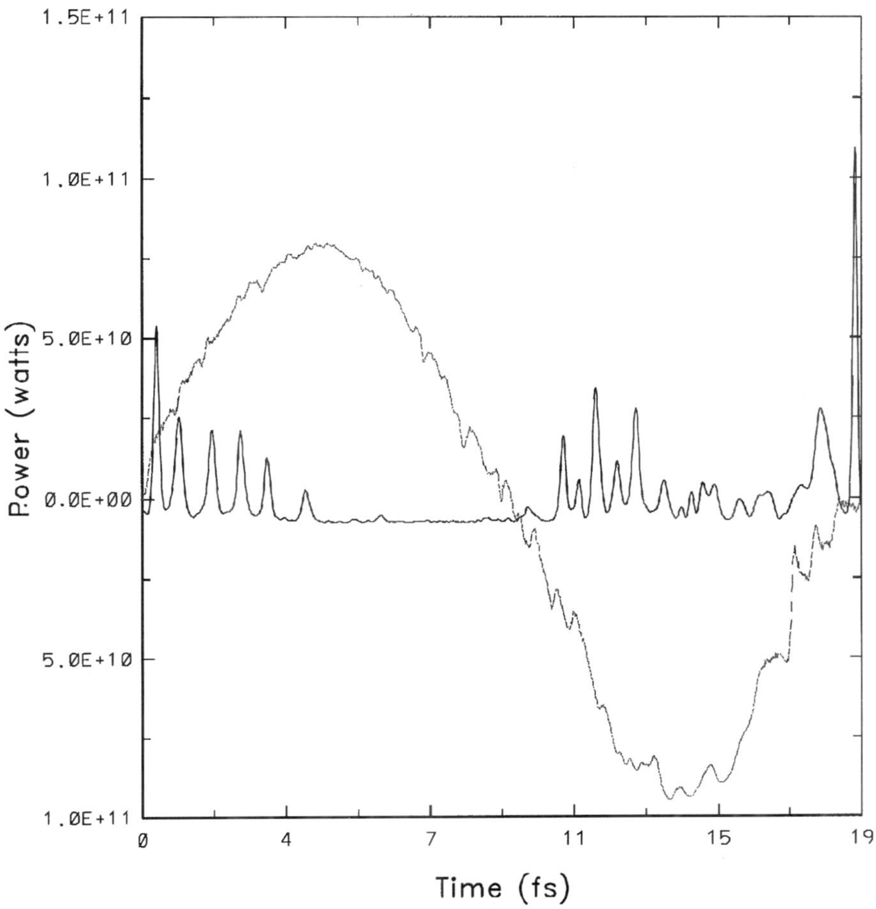

Fig. 7. Effect of beam tube wakefields on the temporal structure of the LCLS optical beam. The sinusoidal function corresponds to energy loss of the electron beam the other function to energy gain of the radiation field.

combination of energy increase from wakefields and the slippage work together to keep the slice in resonance with the ponderomotive potential. For those spikes, growth continues.

The tolerance for the total rms energy change relative to the average bunch energy for a given bunch has been set to 0.001 for the LCLS, which corresponds to a $\Delta\gamma$ of 28. At the present design only a part of the bunch is expected to satisfy this tight tolerance even with the wall roughness tolerance as low as 100 nm.

References

[1] The LCLS Design Study Group, LCLS – Design Study Report, April 1998. SLAC-R-521.

[2] R. Bonifacio, C. Pellegrini, L. Narducci, Opt. Commun. 50 (6) (1985).

[3] R.A. Jong, W.M. Fawley, E.T. Scharlemann, SPIE 1045 (1989) 18.

[4] P. Pierini, W.M. Fawley, Shot noise startup of the 6 nm SASE FEL at the TESLA test facility, in Proceedings of the Seventeenth FEL Conference (FEL95), New York, NY, USA, August 21–25, 1995, pp. 332–335.

[5] W.M. Fawley, A.M. Sessler, E.T. Scharlemann, Coherence and linewidth studies of a 4-nm high power FEL, Proceedings of the 1993 Particle Accelerator Conference, Washington DC, 1993.
[6] N.A. Vinokurov, A.A. Zholents, W.M. Fawley, K.J. Kim, Critical issues for high power FEL based on micron recuperator/electron out coupling scheme, Talk given at Photonic West 97 Conference, San Jose, CA, 8–14 Feb 1997 LBL-40081.
[7] K.-J. Kim, Undulator interruption in high gain free electron lasers, Nucl. Instr. and Meth. A 407 (1998) 126–129. Presented at the 19th International FEL Conference (FEL97), Beijing, China.
[8] R. Bonifacio et al., Phys. Rev. Lett. 73 (1) (1994) 70.
[9] K.-J. Kim, Phys. Rev. Lett. 57 (15) (1986) 1871.
[10] P. Emma, R. Carr, H.-D. Nuhn, Nucl. Instr. and Meth. A 429 (1999) 407.
[11] E.T. Scharlemann, J. Appl. Phys. 58 (6) (1985) 2154.
[12] W.M. Fawley, Private communication, 1997.

ELSEVIER

Nuclear Instruments and Methods in Physics Research A 429 (1999) 257–263

NUCLEAR INSTRUMENTS & METHODS IN PHYSICS RESEARCH Section A

www.elsevier.nl/locate/nima

Photoelectric field emission from needle cathodes

C. Hernandez-Garcia*, C.A. Brau

Department of Physics and Astronomy, Vanderbilt University, Nashville, TN 37235, USA

Abstract

Experiments have been carried out at laser intensities of the order of 10^8 W/m^2 using a cw argon laser and tungsten needles with roughly 1-μm tip radius operated at voltages up to 50 kV. The results show a nonlinear dependence of the photocurrent on the laser intensity and a strong dependence of the photocurrent on the electric field. Comparison with theoretical calculations based on the Fowler–Nordheim tunneling theory of field emission indicates that the photoexcited electrons rapidly relax to energy levels just above the Fermi level, from which they tunnel out.

Keywords: Electron beam; Field emission; Photo-field emission

1. Introduction

The emission of electrons from the surfaces of metals in vacuum occurs via three mechanisms. In thermionic emission the electrons accumulate sufficient thermal energy to pass over the classical barrier presented by the work function of the metal, which is reduced by the Schottky effect when the field is sufficiently strong. In photoelectric emission the electrons receive enough energy from the photon to pass over the barrier at the surface of the metal. In field emission the electrons tunnel out through the barrier when it is thinned by a strong electric field. To have a barrier thin enough for significant tunneling to occur, the electric field at the surface must be of the order of 10^9 V/m, which can be achieved near the tips of sharp needles even for modest voltages.

Field emission was first observed around the turn of the century, before tunneling was understood. A correct (quantum-mechanical) description of the phenomenon was proposed by Fowler and Nordheim in 1929. According to this model, the electrons in the metal possess a Fermi–Dirac energy distribution, which at low temperature cuts off sharply at the Fermi energy. Electrons near the top of the occupied levels tunnel through the barrier and account for the predicted current density [1]

$$J_e = \frac{q^2}{16\pi^2\hbar}\frac{E^2}{\varphi_W}\exp\left[-\frac{4\sqrt{2mq}}{3\hbar}\frac{\varphi_W^{3/2}}{E}f(y)\right] \quad (1)$$

where m is the electron mass, q the elementary charge, $\hbar$ the Planck's constant, ϕ_W the work function, E the electric field, and

$$y = \sqrt{\frac{q}{4\pi\varepsilon_0}}\frac{\sqrt{E}}{\phi_W} \quad (2)$$

* Corresponding author.

0168-9002/99/$ - see front matter
PII: S0168-9002(99)00118-7

in which ε_0 is the permittivity of free space. The function $f(y)$ is a dimensionless elliptical function introduced to account for the image forces near the surface, and is actually closely approximated by $f(y) \approx \cos(\pi y/2)$. Formula (1), known as the Fowler–Nordheim equation, predicts a strong dependence of the current density on the electric field. Experimental results confirm the theoretical predictions in remarkable detail. The current density has been observed to follow the Fowler–Nordheim relation over more than six orders of magnitude, up to a current density of 10^{11} A/m^2 [2]. Above this value, space-charge effects reduce the field at the surface. The longitudinal energy distribution has been carefully measured, and confirms both the predicted shape and the predicted width of about 1 eV [3].

At elevated needle temperatures the emission increases due to faster tunneling by electrons thermally excited to higher energy [4]. The current density is computed from the integral

$$J_e = \int_0^{E_F + \phi_W} dE\, f_T(E) P(E) \tag{3}$$

where $f_T(E)$ is the Fermi–Dirac density of electrons in an infinitely deep, three-dimensional rectangular potential well at the energy E and temperature T. $P(E)$ is the tunneling probability for an electron at the energy E [5]. This simple model is in excellent agreement with previous experimental results, including the distribution of electron energy in beams emitted from needles at elevated temperatures [6]. The voltage dependence of thermionic emission is much weaker than that of field emission, so at sufficiently high temperature and low electric field the emission changes from field emission to thermionic emission.

Shining a laser on the tip of a needle turns on field emission at surface electric fields well below those otherwise required for significant emission. In experiments conducted at low laser intensities, most of the attention has been directed at the emission of electrons at energies $\hbar\omega$ above the Fermi energy, where $\hbar\omega$ is the photon energy. For laser intensities below about 10^8 W/m^2 the emission is observed to be linear in the laser intensity, with a quantum efficiency of the order of 10^{-6} for the emission of electrons at the photoexcited energy [7]. In more recent experiments using higher-laser intensities, the attention has been focused on the total photocurrent. At laser intensities of the order of 10^{10}–10^{14} W/m^2, the quantum efficiency is observed to be of the order of unity, with only a weak dependence on the surface electric field and the laser wavelength [8,9]. The largest total current that has been observed so far is 2 A, extracted in nanosecond pulses from needles with a tip radius of 50 nm [8]. The corresponding current density is 10^{14} A/m^2 which is near the space-charge limit for a needle of this size. The current pulse was observed to follow the laser pulse. In all cases the photon energy was between 1 and 3 eV, which is less than the work function of the metal, even after the Schottky correction. The effect has been interpreted as single-photon photoemission in which the electron is excited to an energy level above the Fermi level from which the electron can easily tunnel through the thinner part of the surface barrier. However, in the experiments by Boussakaya et al. [8] the tip of the needle was heated to an extremely high temperature and thermionic emission could contribute to the current. In fact, in some of the measurements the current seems to have exceeded the space-charge limit, which suggests that the emission occurred over a larger area or a longer time than the laser pulse on the tip of the needle. However, in the experiments of Ramian and Garate [9], the estimated temperature rise of the surface of the needle was about 300°C, which is not enough to account for the observed current. The experiments described below were undertaken to examine the mechanism for the observed photoemission.

2. Experimental techniques

The experimental apparatus is illustrated schematically in Fig. 1. The tungsten needle was held in a pin vice supported on a ceramic insulator inside the cathode structure. The outer structure of the cathode could be adjusted with a micrometer actuator to expose more or less of the needle. The outer structure of the anode was likewise insulated from ground, and could be adjusted with a micrometer actuator to vary the distance from the tip of the

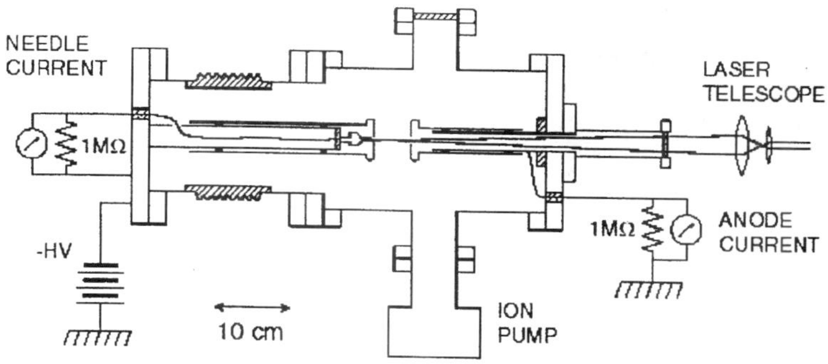

Fig. 1. Schematic diagram of the experimental apparatus.

needle to the face of the anode. A 2-mm hole in the anode was used to pass the laser beam to the needle tip for photoelectric current measurements, and will be used later for measurements of the electron-beam properties.

The cathode bias was provided by an unregulated 50-kV power supply. The electron-beam current was measured by a digital 200-mV meter across a resistor in series with the needle. This eliminates spurious current from the surfaces of the cathode structure. The anode current was measured with a voltmeter across a 1-MΩ shunt. Differences between the anode and cathode currents could be accounted for by the current transmitted through the hole in the anode.

The final vacuum was maintained by a 20-l/s ion pump. The system was generally baked out overnight at about 200°C to outgas the walls of the chamber. Most of the measurements were made at pressures below 10^{-7} Torr, but the effects of contamination by the residual gas (mostly water vapor) were observed, as discussed below.

The tip of the needle was illuminated by an argon-ion laser. This laser operates on a number of lines in the green part of the spectrum, near 500 nm. The maximum total power (all lines) was about 1 W. The laser beam was focused on the tip of the needle at normal incidence through the hole in the anode. The measured beam profile was nearly Gaussian, and the radius at the focus was about 35 μm at the $1/e$ (intensity) point. At maximum power the intensity on the tip of the needle was 2.5×10^8 W/m^2.

The needles were fabricated from 150-μm diameter, 99.9 + % pure, polycrystalline tungsten wire. To form a point the wire was inserted 1 mm into a 1-M NaOH solution and etched using 10 V AC. The etching was continued until the current stopped. This was observed to give reproducible tips with a radius of about 1 μm and a cone half-angle of 15°. An electron micrograph of the needle used to generate the data reported here is shown in Fig. 2.

After the needles were installed in the vacuum system, they were heated overnight at 2200–2400°C by electron bombardment to clean and smooth the surface. The temperatures were measured using an optical pyrometer, and are believed to be good to ± 50°C. Processing at temperatures above 2400°C caused the needle to evaporate and become thinner and sharper. After several hours of operation at low current or accidental operation at high current (over 100 μA) the current would become erratic and the processing had to be repeated. While doing measurements of the current it was observed that after a few minutes of operation at the ambient pressures used in these experiments the effects of contamination by the residual gas could be observed. The effects took the form of a weaker dependence of the current on the voltage, and sometimes an increased total current. These effects can be accounted for by a monolayer of hydrogen partly covering the surface, which is known to reduce the work function [10]. In this case the surface could be cleaned by illuminating the tip with the laser and

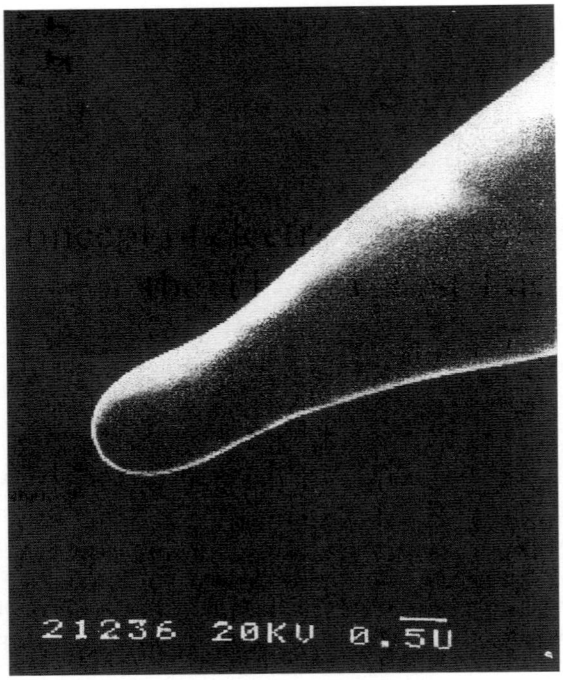

Fig. 2. Electron micrograph of the tip of a tungsten needle.

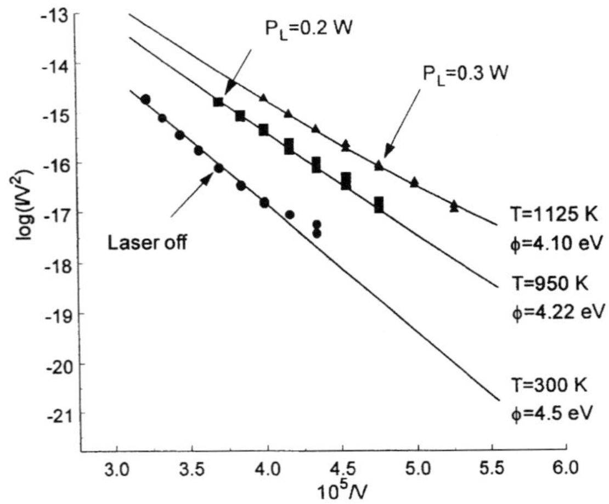

Fig. 3. Fowler–Nordheim plot of observed field-emission current.

heating it to about 1500°C for a period of a few minutes.

3. Experimental results

When precautions were taken to smooth the needle by electron-bombardment heating and clean the needle by laser heating, results like those shown in the Fowler–Nordheim plot in Fig. 3 were obtained. With the laser off, the current density should obey Eq. (1), which predicts nearly a straight line on this plot. The theoretical curve was computed using Eq. (1) and integrating the current density over the surface of the tip. To calculate the electric field at the surface of the needle, the anode–cathode–needle system was approximated by a conducting ellipsoid in a uniform electric field [11]. The only unknown parameter in the calculation was the radius of the needle tip. In cases when the radius could be checked it usually was close to that determined from the simulations. As shown by the electron micrograph in Fig. 2, the observed tip radius of the needle used in the present experiments was about 0.9 μm, but the shape was not an ellipsoid. The best fit to the data was obtained using a tip radius of 0.82 μm. When the needles became contaminated after a few minutes of exposure to the residual gas in the vacuum chamber, the current became erratic and showed a weaker dependence of current on voltage. This effect is evident in the lowest points of the "laser off" curve in Fig. 3, and could generally be corrected by a few minutes of heating to 1500°C using the laser at full power. Eventually, the needles became so contaminated or otherwise damaged that good performance could be restored only by electron-beam bombardment heating to about 2300°C, and ultimately the needles had to be replaced. When the needles were damaged by excessive current or many hours of use and were removed from the vacuum chamber, the tip of the needle was observed to be rough, with the appearance of bumps and lumps on the surface.

When the laser was turned on the current increased by as much as two orders of magnitude, as shown in Fig. 3. The photocurrent increased with laser power in a monotonic but nonlinear fashion, and was characterized by a steep voltage dependence. The highest observed quantum efficiency was of the order of 3%, obtained at 50% of the maximum laser power (about 0.5 W total). At the highest laser powers, the needles were observed to become incandescent. Using the optical pyrometer, the temperature was found to be as high as 1500°C. Although it is not shown in Fig. 3, at sufficiently low voltages the current at these high temperatures was observed to be almost independent of the voltage. This is attributed to thermionic emission from the tip and sides of the needle over the incandescent portion, which was as much as a few millimeters long. At a fixed voltage, the quantum efficiency was observed to be an exponentially increasing function of the laser power. If we extrapolate the results to higher power it appears that at laser powers in excess of a few Watts, depending on the voltage, the quantum efficiency will approach unity, in accord with previous measurements by Boussoukaya et al. [8], and by Ramian and Garate [9].

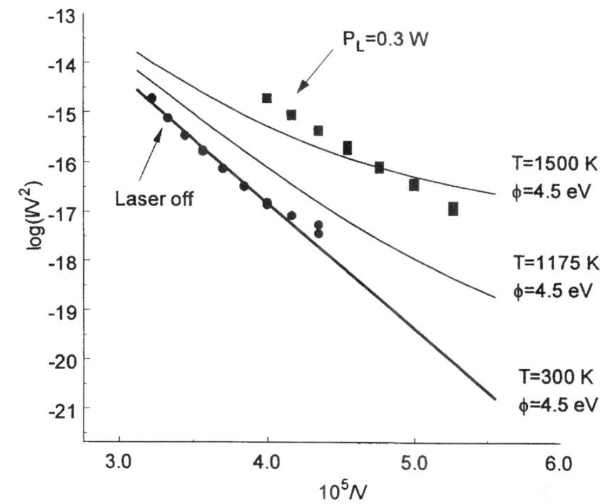

Fig. 4. Fowler–Nordheim plot of field emission at elevated electron temperature.

4. Interpretation

Since the needles were observed to become incandescent at high incident laser power, it is tempting to attribute the increased electron emission to an elevated electron temperature. The Fowler–Nordheim theory can be extended to higher temperatures by integrating the tunneling probability over a Fermi distribution corresponding to the higher temperature [4]. Comparison of the present experimental data with the extended Fowler–Nordheim theory, as illustrated in Fig. 4, shows that this interpretation cannot explain the observed photocurrent. The emission calculated for the observed temperature of the needle is too small, and if the electron temperature is increased above the measured temperature to account for the large enhancement of the emission, then the dependence of the current on voltage is too weak. Therefore, the observed current is evidently some sort of nonthermal photoemission. In no case was the Schottky correction large enough to reduce the barrier below the energy of a single laser photon. Attempts to interpret the observed current as a multiphoton photoemission process also failed. When the current is plotted as a function of the laser power for fixed voltage, the slope on a log–log plot decreases from roughly 5 to about 3 as the voltage changes from 20 to 24 kV. Since the photon energy is 2.5 eV and the work function is 4.5 eV (or lower, when the Schottky effect is included), slopes greater than 2 are difficult to explain.

The steep voltage dependence of the observed photocurrent is a clear indication that the electrons tunnel through the surface barrier from electron energy levels not far above the Fermi level. It is not possible to explain the steep voltage dependence in any other way. At elevated electron temperatures, field emission is dominated by electrons thermally excited to levels above the Fermi level where the surface barrier is thinner, and this accounts for the weaker voltage dependence observed at high temperatures [6]. An excellent fit to the experimental data is obtained by using the Fowler–Nordheim theory with a reduced work function that depends on the laser intensity. The solid curves in Fig. 3 show the best fit of this model to the experimental data. Indicated next to each curve is the effective work function ϕ_W required to fit the data. At the higher laser powers the needles were heated above room temperature and became incandescent. The theoretical curves have been corrected to account for elevated needle temperature, as indicated in Fig. 3. In general, the best fit was obtained when the electron temperature was slightly (50–100°C) below the measured temperature of the needle tip.

Fig. 5 shows a plot of the effective work function ϕ_W as a function of the laser power. The decreased work function is equivalent to an elevated Fermi level. Within the scatter of the data, the work function is a linearly decreasing function of the laser power, indicating that the shift in the Fermi level increases linearly with laser power. A simple linear relaxation model can be used to account for this observation. We assume that the electrons are initially excited by the laser photons to energy levels well above the Fermi level, after which they relax rapidly to levels just above the Fermi energy, as illustrated in Fig. 6. They then reside in these levels for a time τ. As the laser power increases, the energy to which these levels are filled increases in proportion, at least near the surface, which lowers the effective work function of the metal. The levels into which the electrons relax could be some sort of surface states [12], or merely higher levels in the conduction band. Although tungsten has a complex conduction-band structure composed of s and d electrons, we assume here a simple model of a free-electron gas in an infinitely deep potential well, and place the two s electrons in the conduction band. The electron density is then $N_e = 1.26\times10^{29}$ /m^3, and the Fermi energy is $E_F = 9.19$ eV. Since the number of energy levels below the Fermi level is $\propto E_F^{3/2}$, the shift in the effective Fermi energy caused by placing ΔN_e electrons in levels above the nominal Fermi energy is $\Delta E_F/E_F = \frac{2}{3}\Delta N_e/N_e$, for small shifts of the Fermi level. If the probability of photoexcitation is P per unit time, then the shift in the effective Fermi energy is

$$\frac{\Delta E_F}{E_F} = \frac{2}{3}P\tau. \tag{4}$$

To estimate the excitation probability P we proceed as follows. The incident laser photons are absorbed by the electrons near the surface of the metal to a skin depth of the order of 15 nm. The excitation probability is then

$$P = \frac{S(1-R)}{\hbar\omega_L N_e d} \tag{5}$$

where S is the incident laser intensity, R the reflectance, d the skin depth, and ω_L the laser frequency. From the experimental results shown in Fig. 5, we

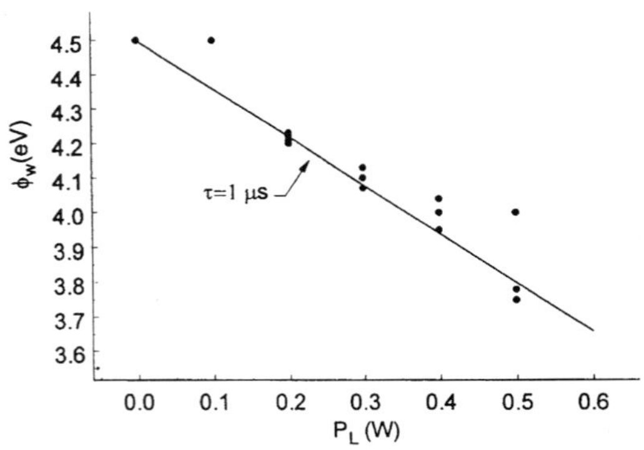

Fig. 5. Effective work function as a function of laser power.

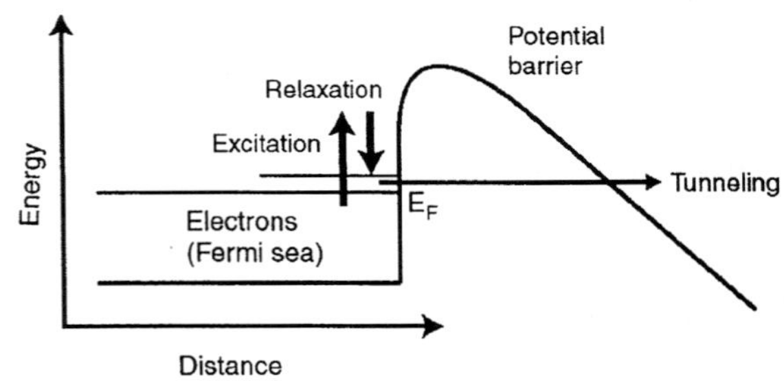

Fig. 6. Schematic diagram showing photoexcitation, followed by relaxation to levels just above the Fermi energy and tunneling through the energy barrier at the surface of the metal.

find that the effective residence time according to this model is $\tau = 1$ μs.

This effective relaxation time is extremely long compared with the time for the relaxation of the electron energy observed in other experiments, such as thermally-assisted multiphoton photoemission [13]. The cooling time observed in those experiments was of the order of 1 ps, due in part to the very small heat capacity of the conduction electrons. This supports the suggestion of rapid relaxation of the photoexcited electrons to levels near the Fermi level. However, while those experiments were sensitive to the relaxation of hot electrons to a Fermi distribution at the lattice temperature, they may not have been sensitive to a shift of the Fermi level, which would have been very small.

To estimate the energy-loss time on more fundamental grounds, we note that for electrons near the Fermi energy the rate of emission of phonons is of the order of 10^{11}/s [14]. But the ratio of the phonon energy to the phonon momentum is v_s, the speed of sound, so the energy-loss time is longer than the momentum-loss time (the phonon-emission time) by the factor v_e/v_s, where v_e is the speed of an electron near the Fermi energy. Thus, we would estimate that that the energy-loss time in tungsten is of the order of 1–10 ns at room temperature, and less at elevated temperature. This is much shorter than the observed residence time. The explanation might lie in the details of the band structure of tungsten. For example, electrons from lower levels, such as the 4f or 5d levels, might be excited into s levels in the conduction band above the Fermi energy, from which they could have difficulty relaxing back into the original levels. Another possibility is that the electrons are excited to, and become trapped in, some sort of surface states [12].

5. Conclusions

By operating at laser intensities of the order of 10^8 W/m^2, intermediate between those used in previous experiments, it has been possible to examine the transition from the small quantum efficiencies observed at low intensities to the unit quantum efficiencies observed at high laser intensities. It is clear that in this range of incident intensities the emission is nonlinear in the laser intensity. It is also clear that the emission is dominated by electrons tunneling through the surface barrier from states just above the Fermi energy. However, the nature of the emitting levels is not at all clear.

Acknowledgements

The authors gratefully acknowledge helpful and illuminating discussions with Drs. Marcus Mendenhall, Leonard Feldman, and Sokrates Pantelides of Vanderbilt University, and Peder Estrup of Brown University. This work was supported by the Office of Naval Research under Grant N000149411023.

References

[1] P.W. Hawkes, E. Kasper, Principles of Electron Optics, Vol. 2, Academic Press, San Diego, 1989, pp. 918–933.
[2] W.P. Dyke, W.W. Dolan, Field Emission, in: L. Marton (Ed.), Advances in Electronics and Electron Physics, Vol. II, Academic Press, San Diego, 1956, p. 106.
[3] E.W. Mueller, Z. Physik 120 (1943) 261.
[4] W.W. Dolan, W.P. Dyke, Phys. Rev. 95 (1954) 327.
[5] W.P. Dyke, W.W. Dolan, Field emission, in: L. Marton (Ed.), Advances in Electronics and Electron Physics, Vol. II, Academic Press, San Diego, 1956, p. 93.
[6] W.P. Dyke, W.W. Dolan, Field emission, in: L. Marton (Ed.), Advances in Electronics and Electron Physics, Vol. II, Academic Press, San Diego, 1956, p. 95.
[7] D. Venus, M.J.G. Lee, Surf. Sci. 116 (1982) 359.
[8] M. Boussoukaya et al., Nucl. Instr. and Meth. A 279 (1989) 405.
[9] G. Ramian, E. Garate, Presented at the Sixteenth International Free-Electron Laser Conference, Stanford, CA, August 23, 1994.
[10] L.D. Schmidt, in: R. Gomer (Ed.), Interactions on Metal Surfaces, Springer, Berlin, 1975, p. 63.
[11] L.D. Landau, E.M. Lifshitz, L.P. Pitaevskii (Eds.), Electrodynamics of Continuous Media, second ed., Pergamon Press, Oxford, 1984, pp. 19–29.
[12] D. Venus, M.J.G. Lee, Surf. Sci. 172 (1986) 477.
[13] J.G. Fujimoto, Phys. Rev. Lett. 53 (1984) 1837.
[14] C. Kittel, Introduction to Solid State Physics, seventh ed., Wiley, New York, 1996, p. 665.

ELSEVIER

Nuclear Instruments and Methods in Physics Research A 429 (1999) 264–268

NUCLEAR INSTRUMENTS & METHODS IN PHYSICS RESEARCH Section A

www.elsevier.nl/locate/nima

Self-interaction of subpico-second electron bunch traveling through a chicane-based bunch-compressor

Ryoichi Hajima[a,*], Koji Yoshii[b], Toru Ueda[b], Fumio Sakai[c], Hideyuki Kotaki[c], Shuji Kondoh[c], Masaki Kando[c], Kenichi Kinoshita[b], Hideki Harano[b], Takahiro Watanabe[b], Mitsuru Uesaka[b], Hideki Dewa[c], Kazuhisa Nakajima[c,d]

[a]*Department of Quantum Engineering and Systems Science, University of Tokyo, Hongo 7-3-1, Bunkyo-ku, Tokyo 113-8656, Japan*
[b]*Nuclear Engineering Research Laboratory, University of Tokyo, 2-22 Shirakata-Shirane, Tokai-mura, Ibaraki 319-1106, Japan*
[c]*Advanced Photon Research Center, Japan Atomic Energy Research Institute, 2-4 Shirakata-Shirane, Tokai-mura, Ibaraki 319-1100, Japan*
[d]*High Energy Accelerator Research Organization (KEK), 1-1 Oho, Tsukuba, Ibaraki 305-0801, Japan*

Abstract

A photo-cathode RF-gun and a chicane-based bunch-compressor are installed on an S-band linac which had been used for a UT-FEL experiment. Electron bunches extracted from the photo-cathode RF-gun are accelerated by an S-band structure up to 20 MeV and compressed by a chicane magnet. Since the bunch has very small longitudinal size and relatively low energy, coherent synchrotron radiation emitted from the bunch in the chicane creates a nonuniform energy loss in the bunch and degrades the performance of the bunch compressor. In the present paper, the performance of the bunch-compressor under the influence of coherent synchrotron radiation is studied. Preliminary experimental results are also presented.

Keywords: Chicane; Photo-cathode RF-gun; Bunch compressor; Coherent synchrotron radiation

1. Introduction

A photo-cathode RF-gun and a chicane-based bunch-compressor are installed on an S-band linac which had been used for a UT-FEL experiment [1]. Several new experiments such as femto-second X-ray generation, laser wakefield acceleration, and picosecond X-ray diffractometry are being carried out as new applications of the linac [2]. An electron bunch extracted from the photo-cathode RF-gun is accelerated by the S-band structure up to 20 MeV and can be compressed by a chicane magnet. Electron bunches as short as 400 fs have been obtained.

In recent years it has been reported that emittance growth caused by the self-field of electrons traveling though a circular path may become a severe problem in the design of bunch compressors for an extremely low emittance beam [3–5]. This self-field in circular motion, which is called the coherent synchrotron radiation (CSR) force or

* Corresponding author. Tel.: + 81-3-5689-7283; fax: + 81-3-3818-3455.

E-mail address: hajima@q.t.u-tokyo.ac.jp (R. Hajima)

0168-9002/99/$ - see front matter
PII: S0168-9002(99)00120-5

the noninertial space charge force, is the subject of the present study.

The emittance growth by the CSR force is not a major problem in typical storage rings, because the CSR is almost completely shielded by the conducting walls of the beam pipe. In the chicane magnet of our linac, however, the shielding of CSR is very weak because of small bending radius and short bunch length. The dimensionless shielding parameter $R\sigma^2/h^3$, which gives the shielding effect for two parallel plates, is less than 0.1 and the CSR power emitted from the bunch is the same order as an open structure [6]. Hence the CSR and noninertial space charge force must be taken into account to study the performance of the bunch compressor.

2. Analysis of bunch compression including CSR and noninertial space charge force

The performance of the chicane-based bunch compressor under the influence of CSR and the noninertial space charge force is studied in this section. There are two approaches to the study of the CSR force: a one-dimensional analytical method [3,7] and a three-dimensional numerical simulation [5]. In the one-dimensional analytical approach an electron bunch is assumed to be a line charge of Gaussian or uniform distribution and the CSR force applied on each electron in the bunch is calculated by integrating a Liénard–Wiechert potential of other electrons over the bunch. The three-dimensional simulation is based on particle tracking and calculates the self-field from the Liénart–Wiechert potential. In the present study we use the latter approach, because the line charge assumption in the one-dimensional analysis does not give a good approximation for pancake-shaped electron bunches in the bunch compressor. The particle-tracking simulation also has the advantage of taking the noninertial space charge force into account. This is the non-canceled force between two particles having different paths in a circular trajectory and never appears in a one-dimensional analysis.

Using a particle tracking code modified to include CSR and noninertial space charge forces [8], we calculate the performance of the bunch compressor and estimate the effect of the self-field applied on electron bunches in the chicane magnet. Since the shielding of CSR by metallic walls in our chicane is very weak as described above, we neglect the shielding in these simulations.

An ideal compression of an electron bunch having zero initial transverse emittance is studied to distinguish the effect of the self-field. We prepare, before the compressor, an electron bunch having zero transverse emittance, uniform radial distribution of $r = 3$ mm, Gaussian temporal profile of $\Delta\tau = 24$ ps (FWHM), charge of 1 nC and an average energy of 20 MeV. The parameters of the chicane are chosen so that the optimum bunch compression is obtained for a simulation without CSR and noninertial space charge forces, where the bending angle and path length in each magnet are 25° and 15.5 cm, respectively, all the gaps between rectangular magnets are 15 cm, and the initial energy modulation of the bunch is 50 keV/ps.

The longitudinal phase space and temporal profile of compressed bunch are shown in Figs. 1 and 2. If only the usual self-field is taken into account, the compressor shows excellent performance and the bunch is compressed to as little as 0.1 ps (FWHM). The simulation including the CSR and noninertial space charge forces shows degradation of the performance, and the bunch length is 0.5 ps (FWHM) after the chicane. This performance degradation is the result of an additional nonuniform energy modulation arising from CSR and noninertial space charge forces during circular motion in the chicane magnet.

Making precise control of four individual magnets with trim coils, we can obtain an electron bunch a little shorter than 0.5 ps even if CSR and noninertial space charge forces are included. This has only a minor effect for improving the performance, and electron bunches as short as 0.1 ps cannot be obtained, because the longitudinal phase space has been distorted by CSR and noninertial space charge forces.

Fig. 1 also shows the decrease in the average energy due to the CSR force, which corresponds to a total CSR power of 250 μJ emitted from the bunch as it goes through the chicane.

The calculated transverse emittance growth is listed in Table 1. It shows that CSR and noninertial

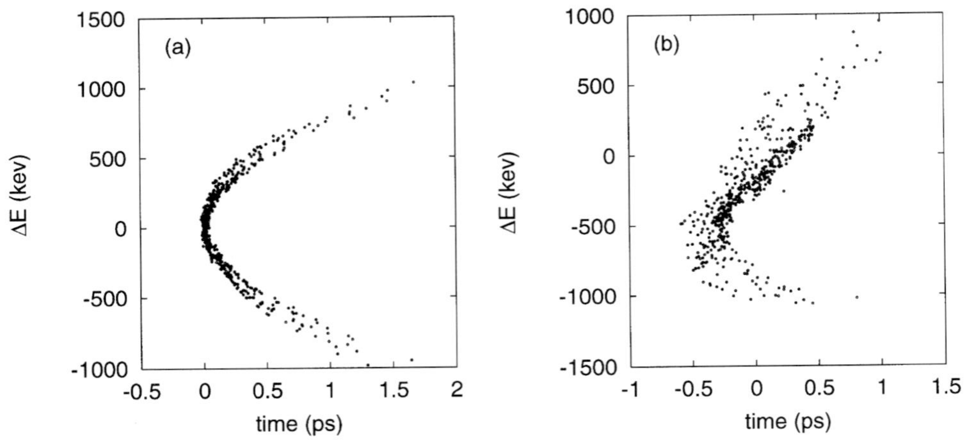

Fig. 1. Calculated longitudinal phase space of compressed bunches (a) without CSR and the noninertial space charge forces and (b) with CSR and noninertial space charge forces.

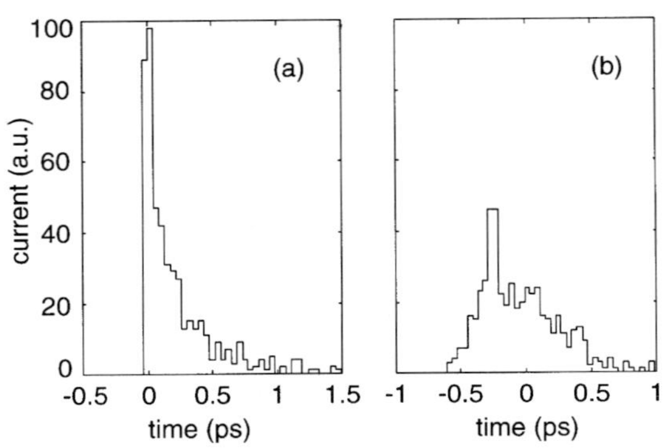

Fig. 2. Calculated temporal profile of compressed bunches (a) without noninertial space charge and CSR forces and (b) with noninertial space charge and CSR forces.

Table 1
Normalized transverse emittance growth

Self-field	$\Delta\varepsilon_x$(πmm mrad)
Usual	16
Usual + CSR	22
Usual + CSR + NI[a]	29

[a]NI is noninertial space charge force.

space charge forces cause additional emittance growth of the electron bunch, while emittance growth due to higher order chromatic aberration and fringe field of the magnets still exists in simulations without CSR and noninertial space charge forces.

3. Experiment for the direct observation of CSR force

An experiment was carried out for the direct observation of the CSR force and the noninertial space charge field affecting the electron motion in the chicane magnet. In the experiment, the energy spectrum of the electron bunch was measured downstream of the chicane at various chicane fields to detect the energy change of electrons arising from the CSR force.

Fig. 3 shows the experimental setup. The beam current at the outlet of an energy analyzer was measured with a co-axial Faraday cup and a current transformer (CM1) installed after the accelerating cells was used as a reference of beam current. Using a fluorescent screen just after the analyzer, we determined the beam optics carefully and kept the same position and injection angle of the electron beam at the entrance of the analyzer for precise energy measurement.

Fig. 4 shows the measured energy spectrum for two different chicane fields, $B = 0$ and 2320 G, where the bunch charge and length were 1.4 nC and 5 ps (FWHM), and the path length in each magnet was 16 cm. The accelerator phase was chosen so

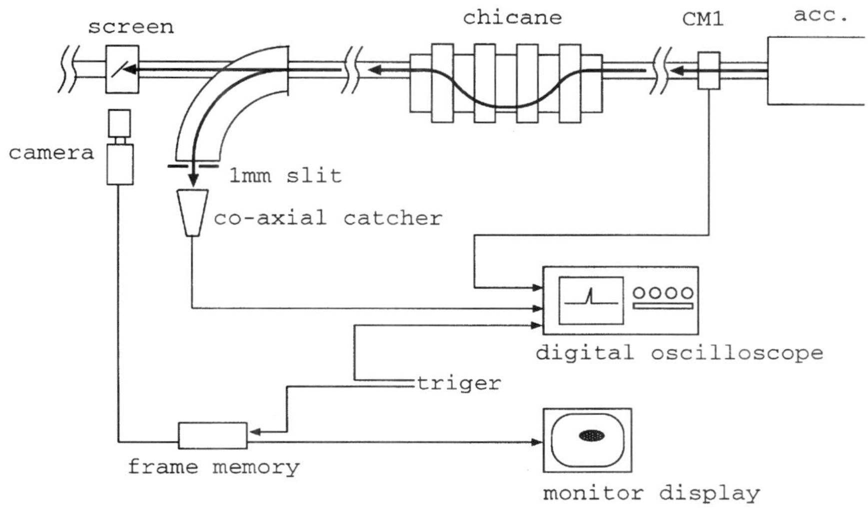

Fig. 3. Experimental setup.

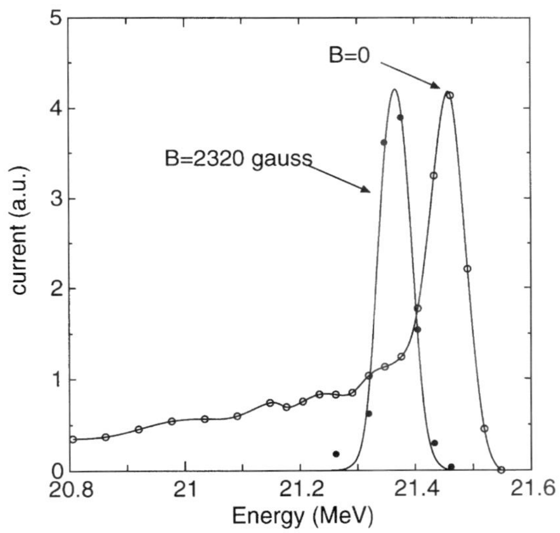

Fig. 4. Measured energy spectrum for chicane field of $B = 2320$ and 0 G.

that the electron bunch was on the crest of the accelerating field, where neither compression nor decompression occurs. The measured energy spectrum shows two distinctive features: one is the shift of the peak in energy and the other is the disappearance of the lower-energy electrons when a large field is applied to the chicane.

The wide low-energy tail appearing in the measurement without a chicane field is considered to be a tail of the electron bunch behind the phase necessary to stay on the crest of the RF wave. This tail experiences a non-zero transverse field in the RF-gun and has a relatively large slice emittance. The bunch measurement with a streak camera and the observation of the transverse profile at the screen support this assumption. According to the above consideration, the disappearance of the lower energy electrons with a large chicane field can be explained by electrons being scraped off at the outlet of the chicane, where the aperture diameter is only 15 mm and the beam acceptance becomes smaller for the larger chicane field.

The shift of peak in the energy spectrum is evidence for the CSR force, which decreases the total energy of the electron bunch. The measured energy shift in Fig. 4 is 94 keV, while a 3-D numerical simulation gives 230 keV as the energy shift of electrons for the same parameters as the experiment but with the injection of a uniform energy beam. If we assume a bunch charge of 0.6 nC, that is, the value after getting rid of the low-energy tail, the numerical simulation gives 105 keV as shown in Figs. 5 and 6. In the numerical simulation we can see that electrons around the center of the bunch, where current density is higher than in other parts,

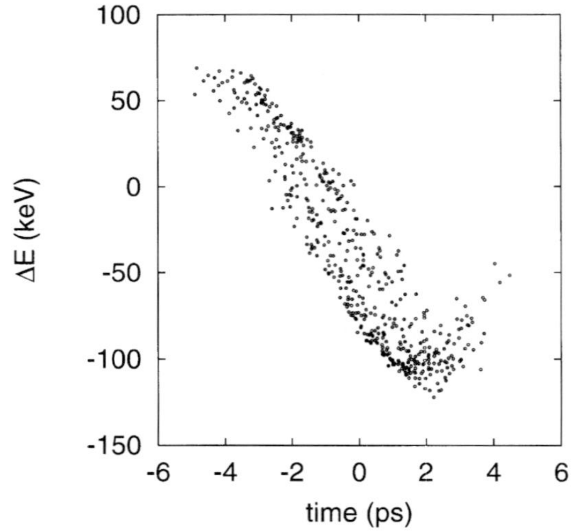

Fig. 5. Longitudinal phase space after the chicane obtained by numerical simulation.

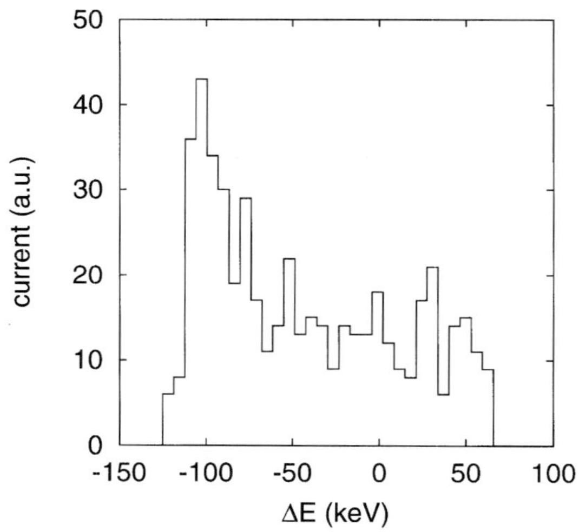

Fig. 6. Energy spectrum after the chicane obtained by numerical simulation.

lose more energy by emitting stronger CSR and slip backward due to the dispersion of the chicane. Electrons near the head of the bunch are, to the contrary, accelerated by CSR emitted from the tail. Consequently, the energy spread grows continuously through the chicane. In the experiment, however, the growth of the energy spread could not be observed. This was because that the energy resolution in the measurement was not sufficiently sensitive and the growth of the energy spread is concealed by the initial energy spread.

Further discussion including transient phenomena between a straight path and a circular orbit still remains as a future work.

4. Summary

We have studied the effect of coherent synchrotron radiation (CSR) and the noninertial space charge forces on the performance of a chicane-based bunch compressor with 3-D particle tracking. The lengthening of the temporal profile and the emittance growth of compressed bunches are observed as a result of nonuniform energy modulation caused by CSR and noninertial space charge forces during the circular motion in the chicane. An experimental result for the direct observation of the CSR force has also been presented. The observed energy shift is consistent with the energy loss of electrons due to coherent synchrotron radiation.

References

[1] E. Nishimura et al., Nucl. Instr. and Meth. A 341 (1994) 39.
[2] H. Harano et al., Proceedings of EPAC-98.
[3] Ya S. Derbenev et al., Michrobunch radiative tail–head interaction, unpublished.
[4] E.L. Saldin et al., On the coherent radiation of an electron bunch moving in an arc of a circle, unpublished.
[5] B.E. Carlsten, Phys. Rev. E 54 (1996) 838.
[6] R. Hajima, E. Ikeno, Nucl. Instr. and Meth. A 429 (1999) 315.
[7] J. Schwinger, On radiation by electrons in a betatron, unpublished.
[8] R. Hajima, Proceedings of the International Computational Accelerator Physics Conference (ICAP-98).

ELSEVIER

Nuclear Instruments and Methods in Physics Research A 429 (1999) 269–275

NUCLEAR INSTRUMENTS & METHODS IN PHYSICS RESEARCH Section A

www.elsevier.nl/locate/nima

Accelerator design of the KHI FEL device

M. Yokoyama*, F. Oda, A. Nakayama, K. Nomaru, M. Kawai

Kawasaki Heavy Industries, Ltd.,118, Futatsuzuka, Noda, Chiba 278-8585, Japan

Abstract

Kawasaki Heavy Industries, Ltd. (KHI) has designed a compact free electron laser device that can provide a powerful and tunable (5–16 μm) light source. The FEL device consists of a 32 MeV accelerator, a beam transport system with two 25-degree bending magnets, an undulator of 40 periods and 32 mm period length, and a cavity mirror system. Using some computer codes to optimize the beam optics of the accelerator could provide a high-quality beam for the FEL.

Keywords: Electron laser; Magnet; Accelerator

1. Introduction

Kawasaki Heavy Industries, Ltd. (KHI) has completed the design of a compact linac-based FEL device. The FEL device can provide a powerful and tunable (5–16 μm in the infrared region) light source. The device will be installed in the IR FEL Research Center at the Science University of Tokyo in 1999 [1]. The center is dedicated to the development of IR FEL and exploring applications in chemistry, biology and materials science.

There are many FEL centers being utilized in the world. However, only two types of injectors have been adopted. One consists of a DC electron gun, a sub-harmonic buncher, focusing coils, and a buncher. The other consists of an RF-gun, focusing magnets and an α-magnet which was adopted by the MARK-III (at Vanderbilt University and Duke University) and the BFEL at IHEP [2–5]. We selected the latter type as an injector for the KHI FEL in order to make a compact FEL device and generate low emittance beams easily. The performance of the KHI FEL device in comparison with MARK-III and BFEL is described first in this paper.

A new type of RF-gun with (on axis coupled structure) (OCS) was designed [6] and will be used for the KHI FEL. Higher FEL gain needs higher peak current in the undulator. The electron beam from the new RF-gun must be bunched, electrons with different energy from design removed, and the remaining electrons must be accelerated and transported to the undulator to produce a high-peak current in the undulator. These designs are important for an FEL device. The beam optics calculations of the KHI FEL accelerator are described next in this paper. They are carried out by using some computer codes such as ARUFA and TRANSPORT [7]. The beam optics was optimized so as to provide a high-quality beam with high peak current.

* Corresponding author.

0168-9002/99/$ - see front matter
PII: S0168-9002(99)00122-9

2. Performance of KHI FEL

The fundamental parameters of KHI FEL, MARK-III and BFEL are shown in Table 1 [2–5]. Those of the accelerator and the undulator for the KHI FEL were determined to provide a laser with a wavelength range of 5–16 μm at a fixed energy of 32 MeV, to make the total size (the accelerator and the undulator) smaller, and to obtain the required FEL gain. It is easier to operate the KHI FEL because of the fixed energy. The beam energy from the RF-gun of the KHI FEL is 1.9 MeV and higher. The beam emittance growth of KHI FEL in the injector can be reduced. The cavity length is 3.36 m which is 2^5 times 10.5 cm (one period length of 2856 MHz of the RF frequency). In the future, the capability of a beam burst mode of 89.3 MHz, which is 2856 MHz/32 will be obtained by using an RF gun with a grid. The grid pulse for the electron beam emission will be synchronized with the divided 89.3 MHz.

The rise time for FEL lasing must be significantly less enough at 7 μs, which is the length of the electron beam macro pulse. The KHI FEL was designed to have a rise time of 2 μs. An FEL gain of at least 17% is required to achieve this rise time. The FEL gain estimate was calculated by using the equation from Dattoli et al. [8].

3. Optics simulation

3.1. Beam parameter of the RF-gun

Fig. 1 shows the emittance data of the electron beam at the exit of the design RF-gun. The normalized emittance is 11 πmm mrad. An FEL gain of more than 17% requires a peak current of more than 30 A and an energy spread of less than 0.5%. At the injector, the energy spread should be as small as possible and definitely under 100 kV (0.5% of 32 MeV). The electrons from the RF-gun during 0–12 ps of the RF phase have an electron charge of 0.17 nC and an energy range of 1.95–1.89 MeV. The electron beam during 0–12 ps must therefore be bunched to 5 ps by an α-magnet and injected into an accelerator structure. The time delay Δt through the α-magnet is

$$\Delta t = 0.2681\sqrt{E}/(c\beta\sqrt{G}) \tag{1}$$

where E is the momentum of beam in MeV/c, c is velocity of light in m/s, β is the ratio of light velocity

Table 1
The fundamental parameters of KHI FEL, MARK-III, and BFEL devices

	KHI-FEL	MARK-III	BFEL	Unit
Accelerator				
Beam macropulse length	7	6	4.5	μs
Beam energy from the RF-gun	1.9	0.9	1.2	MeV
Beam energy in the undulator	32(Max. 40)	26–45	24–28	MeV
Micropulse peak current	> 30	20–40	20	A
Energy spread (FWHM)	~ 0.5	0.3–0.8	0.7	%
Normalized emittance	11	7	20	πmm mrad
Repetition rate	10	20	3.125	Hz
Undulator				
Number of periods	40	47	50	—
Undulator period	32	23	30	mm
Maximum magnetic field	0.83	0.47	0.42	T
Laser				
Wavelength	5–16	1.5–905	9–11	μm
Micro-pulse repetition rate	2856	2857	2856	MHz
Cavity length	3.36	2.064	2.519	m

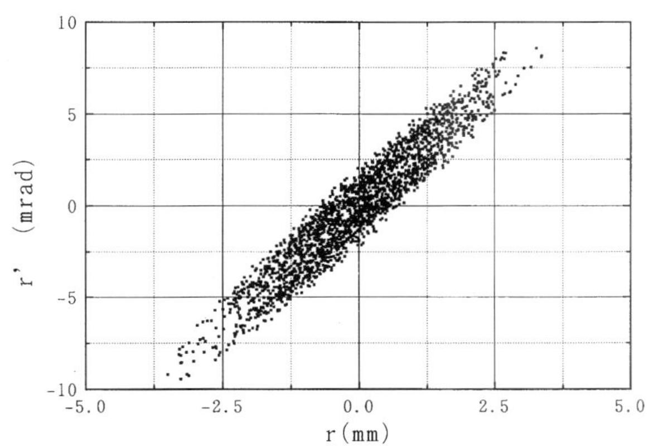

Fig. 1. The beam emittance at the exit of the RF-gun.

to electron velocity, and G is the gradient of the magnetic field of the α-magnet in T/m [9]. The lengths from the RF-gun to the α-magnet and from the α-magnet to the accelerating structure must be considered as a debunching section. The calculated bunch length in the case of the KHI linac is shown in Fig. 2. The gradient of the magnetic field of the α-magnet should therefore be tuned to 1.75 T/m.

The beam optics from the RF-gun to the accelerating structure was simulated and optimized by using ARUFA code. The code solves for the electrons momentum using the Runge–Kutta method in order to more easily include the special magnetic fields of the α-magnet and quadrupole magnets. The space charge effect was considered due to the low-energy beam. The code was used for the simulation from the exit of the RF-gun to the entrance of the accelerating structure.

3.2. Simulation from the RF-gun to the accelerator

Positions and strength of the magnetic field gradient of these magnets were optimized to satisfy the following requirements: (1) The horizontal beam size should be as small as possible at the turning point of the α-magnet. If it is larger than the beam size due to the energy distribution and dispersion, good electrons with an energy of more than 1.89 MeV are cut by the slit. (2) Beam sizes

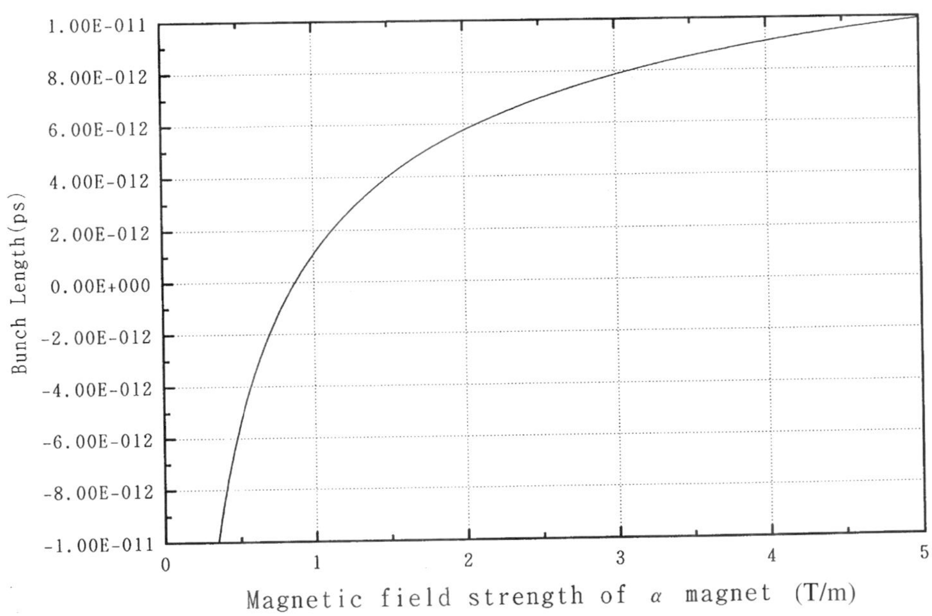

Fig. 2. The relation between the bunch length and the gradient of the magnetic field strength of the α-magnet.

should be made as small as possible and the Twiss parameter α should be positive at the entrance to the accelerating structure.

3.3. Simulation from the accelerator to the undulator

Beam optics from the accelerating structure to an undulator was optimized by using the TRANSPORT code. The accelerating structure of length 3 m is selected to accelerate the electron beam from 1.95 to 32 MeV. The double bend achromat lattice was adopted. A bending angle of 25° and a bending radius of 0.7 m were optimized to be without an interaction with the cavity mirror chamber. The following requirements were considered: (3) The Twiss parameters should be matched to the

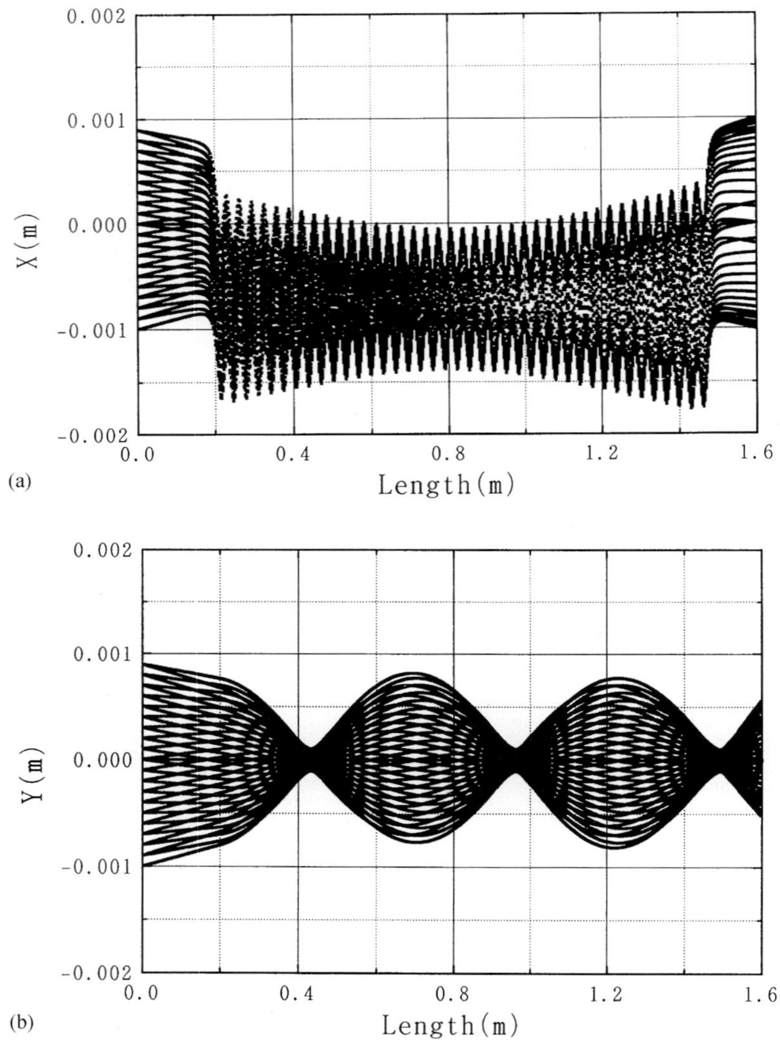

Fig. 3. The electron trajectories of (a) the horizontal direction (b) the vertical direction in the undulator. Each electron in the horizontal direction has a divergence and position on an ellipse with $\alpha_x = 2$, $\beta_x = 2$ m, emittance = 0.5 πmm mrad at the entrance point of the undulator. In the vertical direction it has a divergence and position on an ellipse with $\alpha_y = 2$, $\beta_y = 2$ m, emittance = 0.5 πmm mrad at the entrance point of the undulator, respectively.

optimized ones (this is discussed in Section 3.5) at the entrance of the undulator. (4) The dispersion at the entrance of the undulator should be as small as possible. The vertical beam size should be as small as possible to pass easily through the undulator at a minimum gap of 8 mm.

3.4. Simulation in the undulator

Beam optics in the undulator was evaluated by the same tracking code as ARUFA. The calculated magnetic field of the undulator was used for the code. Electrons with position and divergence on the emittance ellipse were input to the code. The ellipse is calculated from the Twiss parameters and emittance. Optimized values of Twiss parameters were $\alpha_x = 2$, $\beta_y = 2$ m, $\alpha_y = 2$, $\beta_x = 2$ m. The trajectories at the optimized values are shown in Fig. 3(a) and (b). If emittance is less than 0.5 πmm mrad, beam sizes can be less than 0.5 mm. Beams with sizes of 1 mm can easily go through an undulator gap of 8 mm which the FEL wavelength requires.

3.5. Result of the simulation

The optimized beam sizes from RF-gun to the exit of the undulator are shown in Fig. 4. The optimized gradients of each magnet are shown in Table 2. In requirement (1), though horizontal beam size is about 3 mm at the turning point of the α-magnet, it appeared that the beam size for the direction of the slit motion is less than 0.5 mm as shown in Fig. 5. In requirement (2), beam sizes and αs at the entrance of the accelerating structure are about 1 mm and nearly 0, respectively. In requirements (3) and (4), Twiss parameters and the dispersion at the entrance of the undulator were $\alpha_x = 2$, $\beta_y = 2\,m$, $\alpha_y = 2$, $\beta_x = 2$ m and approximately 0.1 m, respectively. The beam sizes in the undulator could be less than 0.5 mm. All

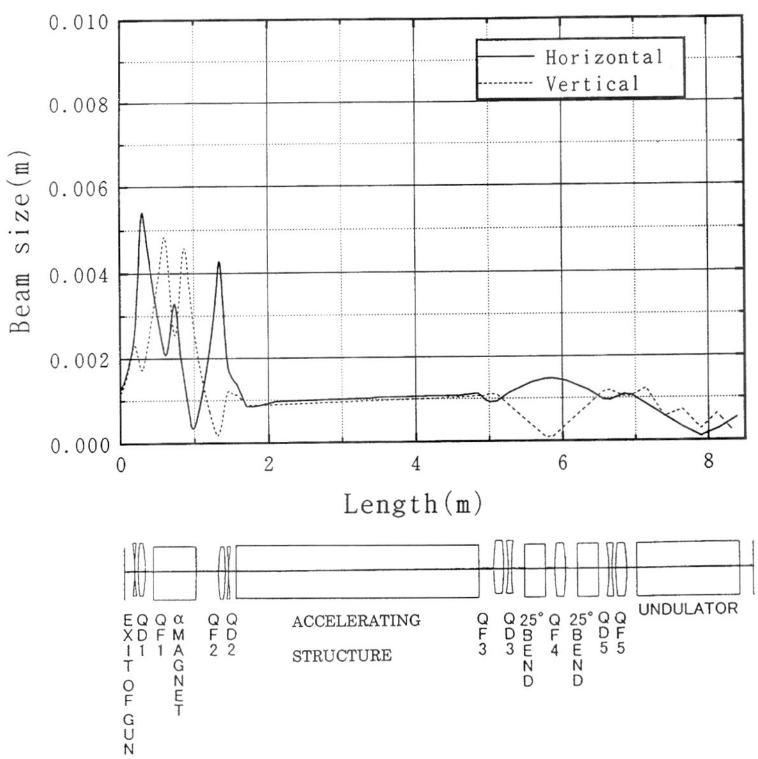

Fig. 4. Beam sizes from the exit of the RF-gun to the exit of the undulator.

Table 2
The optimized magnetic field gradients of each magnet

Name	Effective length (m)	Gradient of magnetic field (T/m)
QD1	0.05	− 1.3
QF1	0.05	1.5
α	0.316	− 1.75
QF2	0.05	2.3
QD2	0.05	− 2.1
QF3	0.1	0.97
QD3	0.1	− 2.4
QF4	0.1	0.52
QD4	0.1	− 2.08
QF5	0.1	1.76

requirements were sufficiently satisfied. A schematic view of the KHI FEL device with the optimized magnet positions is shown in Fig. 6.

4. Conclusion

The KHI FEL is a compact free electron laser device which can provide a powerful and tunable (5–16 μm) light source. Beam optics of the KHI accelerator were optimized by some codes such as ARUFA and TRANSPORT. Requirements from the FEL gain and the FEL wavelength were satisfied. The optimized beam sizes should be less than 0.5 mm in the undulator. These are small enough to

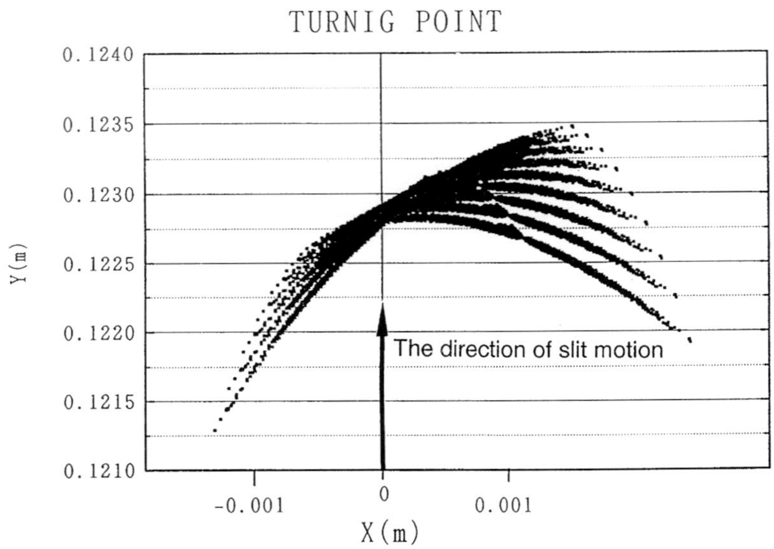

Fig. 5. The beam trajectories at the turning point of the α-magnet.

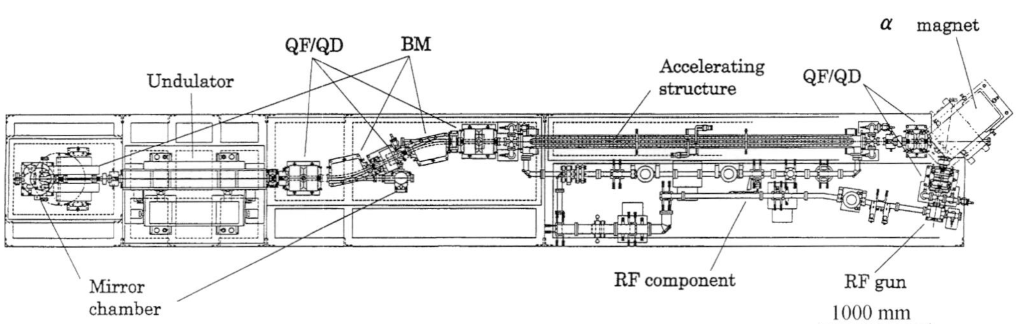

Fig. 6. The schematic view of the optimized KHI FEL device.

produce a high peak current density which is expected to produce high FEL gain and to allow the electron beam to pass easily through the undulator gap of 8 mm which the FEL wavelength requires. Therefore, FEL lasing of the KHI FEL device should be achieved easily.

The KHI-FEL will be constructed in April, 1999. Simulated and optimized beam optics will be measured and confirmed.

References

[1] H. Kuroda, A. Iwata, M. Kawai, The Proceedings of the 12th Russian Synchrotron Radiation Conference SR98, Novosibirsk, 1998.

[2] S.V. Benson, J.M.J. Madey, J. Schmaltz, M. Marc, W. Wadensweiler, G.A. Wesrenskow, Nucl. Instr. and Meth. A 250 (1986) 39.

[3] S.V. Benson, W.S. Fann, B.A. Hooper, J.M.J. Madey, E.B. Szarmes, B. Richman, L. Vintro, Nucl. Instr. and Meth. A 296 (1990) 110.

[4] G.A. Barnett, J.M.J. Madey, C.B. Mckee, K.D. Straub, E.B. Szarmes, Nucl. Instr. and Meth. A 375 (1996) 97.

[5] J.X.J. Zhuang, Y.H.S. Zhong, Y. Li, S. Lin, R. Ying, Y. Zhong, L. Zhang, G. Wu, Y. Zhang, C. Chao, L. Li, Z. Fu, J. Su, Y. Wang, G. Wang, Nucl. Instr. and Meth. A 358 (1995) 256.

[6] F. Oda, M. Yokoyama, M. Kawai, A. Nakayama, E. Tanabe, Nucl. Instr. and Meth., A 429 (1999) 332.

[7] K.L. Brown, F. Rothacker, D.C. Carey, C, Iselin, Å@CERN80-04, 1980.

[8] G. Dattoli, T. Letardi, J.M.J. Madey, A. Renieri, IEEE J. Quantum Electron. QE-20 (1984) 637.

[9] G. Baria, C. Brassard, P.F. Hinrichsen, J.P.Labrie, J.P. Martin, IEEE 3498 (1981).

ELSEVIER

Nuclear Instruments and Methods in Physics Research A 429 (1999) 276–280

NUCLEAR INSTRUMENTS & METHODS IN PHYSICS RESEARCH
Section A

www.elsevier.nl/locate/nima

Concept of electron beam diagnostic for the VUV SASE FEL at the TESLA Test Facility (TTF FEL) at DESY

U. Hahn, J. Pflüger, G. Schmidt*

Deutsches Elektronen Synchrotron, Notkestraße 85, 22603 Hamburg, Germany

Abstract

The electron beam trajectory inside an undulator with integrated strong focusing quadrupoles is disturbed by any kind of magnetic or alignment errors of the guiding field. The electron and photon beam must overlap over the entire undulator length to achieve an optimum output of the TTF FEL (A VUV Free Electron Laser at the TESLA Test Facility: Conceptual Design Rep., DESY Print TESLA-FEL 95-03, Hamburg, 1995). Therefore, it is necessary to measure and correct the electron beam trajectory. The orbit correction in the undulator is based on two principles of orbit measurement. The absolute position of the electron beam inside the undulator can be measured at 4 points of support with calibrated monitors. The second method of measuring the beam trajectory is a beam based alignment algorithm which uses relative orbit changes at 30 distributed beam position monitors along the undulator. A mismatching of the optic at the entrance of the undulator can be seen by measuring the beam size at different locations along the undulator. © 1999 Published by Elsevier Science B.V. All rights reserved.

1. Introduction

The TTF FEL uses an undulator with integrated strong focusing quadrupoles. The undulator consists of three undulator sections with integrated strong focusing quadrupoles. Each undulator section contains 10 quadrupole magnets building a FODO structure. At the entrance, the exit and in between the undulator sections diagnostic blocks are installed as shown in Fig. 1. Orbit deviations inside the undulator are introduced by unknown, random dipole kicks, due to the error in the planar undulator structure, and kicks due to misaligned quadrupoles. To correct the errors of the 30 quadrupoles, for each quadrupole a steerer and a beam position monitor (BPM) are installed.

The diagnostic inside the undulator is necessary to align the electron beam within 10 μm rms deviation to a straight line. This ensures the overlap of the electron and photon beam over the entire undulator length to keep the gain reduction of the FEL smaller than 15% in phase 1 of the TTF FEL project [1]. It is foreseen to use the undulator at electron beam energies between 200 and 500 MeV. Due to the fact that the magnet strength is fixed inside the undulator the optical functions inside the undulator change with the beam energy. To get a periodic beta function the optic is adapted to the beam energy at the entrance of the undulator. A mismatching of the optic at the entrance of the undulator can be detected by using four wire scanners along the undulator.

* Corresponding author. Tel. + 49-40-8998-3510; fax: + 49-40-8998-4305.

PII: S0168-9002(99)00125-4

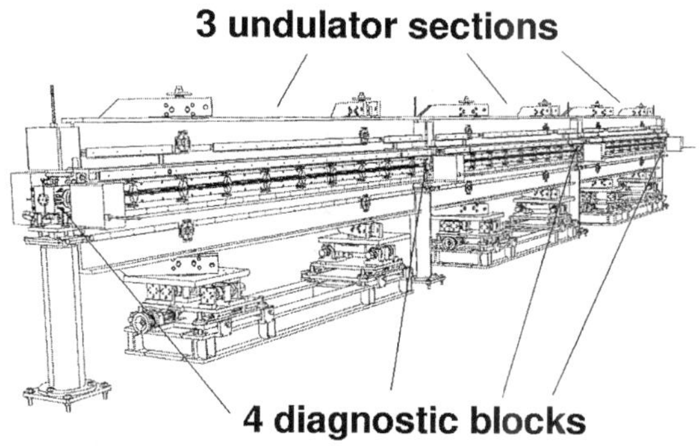

Fig. 1. Three undulator sections with four diagnostic blocks.

Table 1
Parameters of TTF FEL Phase *I* [1]

Beam energy	200–500 MeV
Normalized emittance	2 π mm mrad
Average beam size inside the undulator	0.08 mm
Charge per bunch	0.1–1 nC
Bunch repetition rate	1 MHz
Macropulse repetition rate	up to 10 Hz
Macropulse length	up to 800 μs
Chamber diameter inside the undulator	9.6 mm
Number of undulator sections	3
Undulator gap height	12 mm
Undulator section length	4.5 m
Undulator period length	27.3 mm
Gradient inside the undulator	12.5 T/m
Number of BPM's inside one section	10
Number of diagnostic blocks	4
Number of wire scanners	4 hor. and 4 ver.
Number of cavity monitors	4 hor. and 4 ver.

2. Different type of monitors for the undulator

To measure the orbit deviations due to quadrupole errors two different methods can be used.

1. The absolute beam position with respect to the undulator axis can be measured at several points of support and then be corrected. This is done by calibrated wire scanners [2–4], which allow the electron beam position to be measured in the reference frame of the undulator.
2. In a second step a beam based alignment procedure is used to reduce the offset over the entire undulator [5,6].

The absolute error of the electron beam orbit with respect to the undulator axis is measured by means of several wire scanners which are located before, after and between the undulator sections. The wire scanner uses a carbon or quartz wire of 5 μm diameter, which moves through the beam. The beam profile is obtained by measuring the bremsstrahlung emitted from electrons hitting the wire as a function of the wire position [2,3].

The same wire scanner can be used as a slow or a fast wire scanner. Slow means that for each macropulse, consisting of 800 single bunches with 1 MHz repetition rate (see Table 1) only one wire position inside the beam pipe is measured without moving the wire.

Fast means that during one macropulse the full beam profile and beam position is measured. In both cases the wire scanner cannot give the beam position with only one single electron bunch.

The wire scanners define a reference axis through the entire undulator with an absolute accuracy better than 15 μm. This is achieved by calibrating the wire position with respect to the reference mark on the wire scanner front plate (see Fig. 4). This is done before the installation of the complete assembly. The reference mark is then used to define the position of the wire in the undulator coordinate system after the installation of the complete undulator system. The resolution of the beam position measurement with the wire scanner is of the order of 1 μm [3].

In a second step the wire scanner will be used to calibrate cavity monitors [7] which are located in the same diagnostic block (see Fig. 2). The cavity monitor is a non-destructive device and allows the position of a single electron bunch to be measured. This is the reason why after calibration the cavity monitor will be the standard device for orbit measurements.

To allow this calibration both monitors are located in a so-called diagnostic block. Each diagnostic block contains a pair of wire scanners for horizontal and vertical beam profile and position measurement and a pair of cavity monitors (see Fig. 4). A cut through the diagnostic block is given in Fig. 2.

Fig. 3 shows the undulator chamber [10] with integrated BPMs and correctors, which are used for

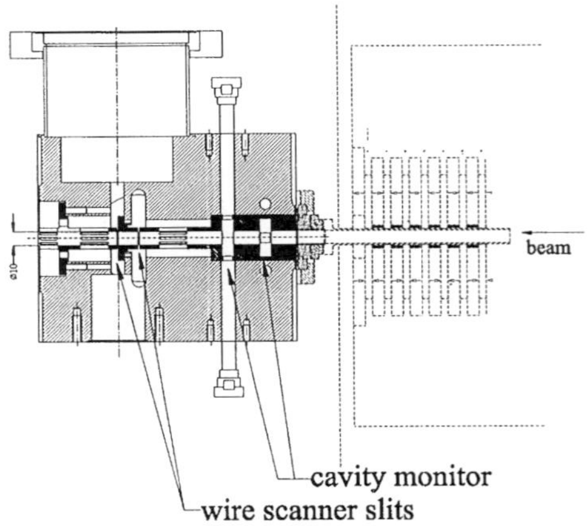

Fig. 2. Cut through the diagnostic block showing the end of an undulator section. The beam is coming from the right. On the left the slits in the beam pipe are visible which allow the wire to pass through the beam.

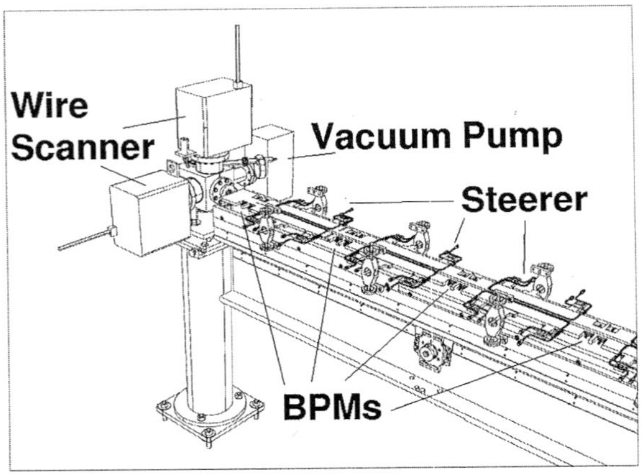

Fig. 4. Diagnostic block with part of the undulator chamber and the lower part of the undulator magnet structure. The BPMs and steerer are visible.

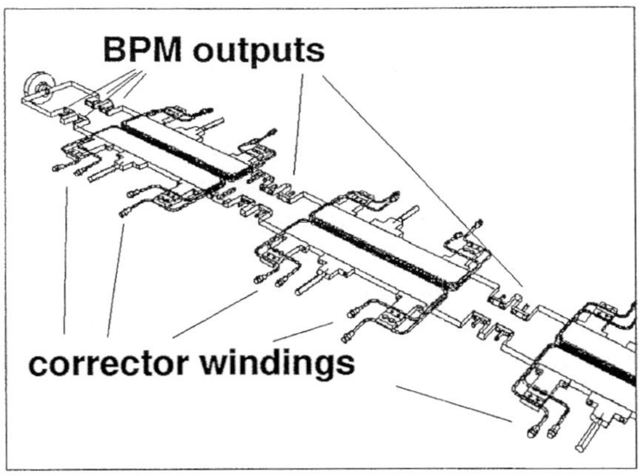

Fig. 3. Typical section of the undulator chamber [10]. Steerer windings and the 4 channels of the BPMs are visible.

the beam-based alignment. For each quadrupole inside the undulator a corrector and BPM is integrated into the chamber. Two types of BPMs are used, both optimized for short electron bunches and both allowing a single bunch position measurement with an accuracy of 1 μm [8,9]. This should allow the trajectory inside the undulator to be optimized to the desired rms value of less than 10 μm [5,6].

3. Procedure to align the electron beam

Three different phases can be distinguished.

For the startup the electron beam must be brought through the undulator. All BPMs can be used without any additional alignment or calibration. The accuracy due to mechanical and electrical tolerances of the monitors of a few hundred micro meter is sufficient.

To protect the undulator against radiation damage the startup will be done with only a few single bunches in the macropulse, a reduced macropulse repetition rate and a reduced charge per electron bunch (0.1 nC instead of 1 nC, see Table 1). All types of monitors are able to work in this mode.

The next step is to measure the electron beam position in the reference frame of the undulator with the wire scanners inside the diagnostic blocks. The electron trajectory is now fixed at 4 points with an accuracy better than 15 μm.

In a third step the orbit is optimized with a beam based alignment procedure. Using the BPMs and corrector coils inside the undulator sections (see Figs. 3 and 4) it should be possible to achieve a straight orbit within the needed 10 μm [5,6]. The BPMs and steerer are visible.

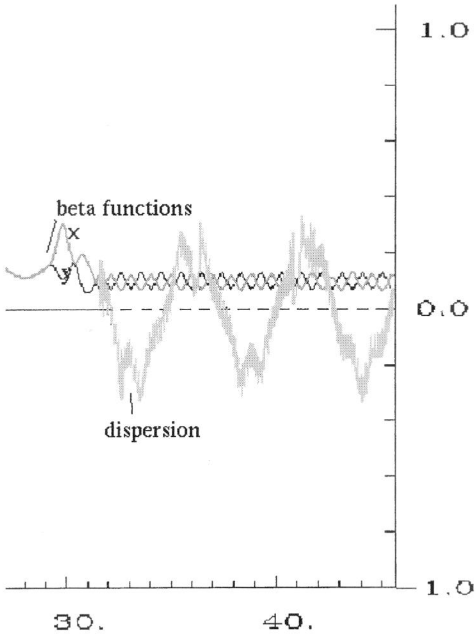

Fig. 5. Matched optic: The beta functions β and the horizontal dispersion $D(s)$ are shown for a matched optic at a beam energy of 200 MeV. (x-axis: longitudinal position in (m), undulator starts at 32 m.) The vertical dispersion is $\approx$ 0. (y-axis: the beta function is shown as $(\sqrt{\beta/m} \times 1/10)$ and the horizontal dispersion is shown as $(-D(s)/m \times 1/10000)$.)

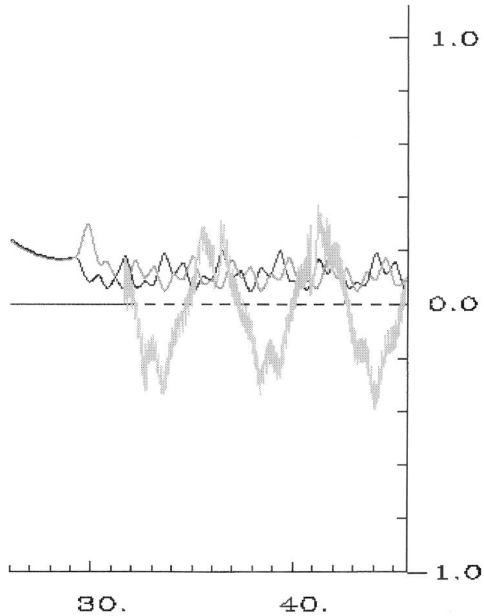

Fig. 6. Mismatched optic: the beta functions and horizontal dispersion are shown for a mismatched optic at a beam energy of 200 MeV. Same units as in Fig. 5 are used.

4. Measurement of optical mismatching

The beam optic before the undulator must be matched to the optic of the undulator to obtain a regular optic inside the undulator. Due to the constant gradient of the quadrupoles the matching depends on the beam energy. A good matching results into the same beam size at all wire scanner locations. Fig. 5 shows a matched optic. A mismatching causes a beta beat which results into different beam sizes at the locations of the wire scanners. This can be measured with the help of the wire scanners and used for optimizing the matching. Fig. 6 shows an example for a mismatched optic.

5. Conclusion

The electron beam diagnostic will allow an easy start up of the FEL undulator. In a second step the necessary precision of the beam trajectory can be achieved by a precise measurement of the electron beam orbit in the reference frame of the undulator and with a beam based alignment procedure. The wire scanners check the correct matching of the beam optics to the undulator.

References

[1] A VUV Free Electron Laser at the TESLA Test Facility: Conceptual Design Rep., DESY Print TESLA-FEL 95-03, Hamburg, 1995.
[2] K. Wittenburg, H. Schultz, DESY; A Proposal for using Wire Scanners at the LINAC Test Facilities; DESY-TESLA 94-15, 1994.
[3] G. Schmidt, S. Striganov, K. Wittenburg, Estimation of the TTF-FEL Wire Scanner Signal, Internal Report.
[4] Cern, With support from the Beam Instrumentation Group (BI).
[5] P. Castro, Orbit correction by dispersion minimization in an undulator with superimposed FODO lattice, Proceedings of the 1998 EPAC, Stockholm.

[6] K. Flöttmann, B. Faatz, E. Czuchry, J. Roßbach, Beam Based Alignment Procedure for an Undulator with Superimposed FODO Lattice, DESY Print TESLA-FEL 97-05, Hamburg, 1997.

[7] R. Lorenz, T. Kamps, M. Wendt, Beam position measurement inside the FEL undulator at the TESLA Test Facility linac, Proceedings of the DIPAC 97, Frascati, Italy, p. 73–75.

[8] T. Kamps et al., New microwave beam position monitors for the TESLA Test Facility FEL, eighth Beam Instrumentation Workshop, SLAC, May 4–7, 1998.

[9] M. Wendt, Coaxial line monitors, private communication.

[10] APS, Argonne, The chamber is designed and built in collaboration with the Experimental Facility Division of the Advanced Photon Source (APS).

ELSEVIER

Nuclear Instruments and Methods in Physics Research A 429 (1999) 281–286

NUCLEAR INSTRUMENTS & METHODS IN PHYSICS RESEARCH Section A

www.elsevier.nl/locate/nima

Space charge calculations of elliptical cross-section electron pulses in PARMELA☆

B.E.C. Koltenbah, C.G. Parazzoli*

Boeing Information, Space & Defense Group, P.O. Box 3999, MC 87-85, Seattle, WA 98124-2499, USA

Abstract

The Boeing version of the PARMELA code has been modified to compute the space charge effects for electron pulses with highly elliptical transverse cross-sections. A dynamic gridding routine has been added to allow good resolution for pulses as they evolve in time. The results from calculations for the chicane buncher in the 1 kW visible FEL beam line at Boeing indicate that the old circular algorithm of the SCHEFF subroutine overestimates the emittance growth in the bend plane by 30–40%.

Keywords: Space charge; Microbunch compression; Simulation; FEL

1. Introduction

The LANL-developed PARMELA code is a very useful, computationally efficient tool for electron beam line design, but it relies upon simplifying assumptions. One critical assumption in PARMELA is that the electron pulse cross-section is nearly circular through the beam line. This is strongly violated in the buncher of the 1 kW visible FEL beam line currently under construction at The Boeing Company in Seattle [1,2]. Modifications have been made to PARMELA to allow modeling pulses of elliptical cross section. These changes are more general than those presented in Ref. [3]. A dynamic grid has been added to improve resolution as the pulse evolves in time.

* Corresponding author. Tel.: + 1-253-773-8299.

E-mail address: claudio.g.parazzoli@boeing.com (C.G. Parazzoli)

☆Work supported by USA/SSDC under Contract DASG60-97-C-0105.

The modifications are described in Section 2. Section 3 presents test runs showing the differences between results produced by the old (circular) and new (elliptical) algorithms in the SCHEFF subroutine. PARMELA with the old version of SCHEFF consistently overestimates the emittance increase in the buncher by approximately 30–40%.

2. PARMELA modifications

Modifications have been made to PARMELA to enhance its ability to model electron bunches which, through the course of traversing various beam line elements such as the bends and quadrupoles, can become greatly elongated in the cross sectional plane (x, y). The PARMELA SCHEFF subroutine calculates the space charge impulse applied to a macro-particle at each time step. A detailed description of the method can be found in Ref. [4]. The modifications have been made to this section and related subroutines.

0168-9002/99/$ - see front matter
PII: S0168-9002(99)00126-6

2.1. SCHEFF circular method

The fundamental assumption made in the old SCHEFF subroutine is that the macroparticles within the electron bunch can be described by a set of concentric circular tori with rectangular cross-section. Fig. 1 shows a diagram of this assumption. Each torus is assumed to have a uniform distribution of charge, and the charge of each macroparticle within a given torus is spread uniformly throughout the torus. The circular tori of rectangular cross-section can be described by a rectangular grid (also called the mesh), where each cell within the mesh represents the intersection of a given torus with a radial plane ($\theta = \text{constant}$) as shown in Fig. 1(c) and (d). The SCHEFF subroutine performs all the calculations in the rest frame of the electrons and makes the assumption that all the particles are stationary in this frame. The mesh consists of a grid of points in two-dimensional (r, z) space which lie in between the rectangular cells of Fig. 1(d). These points are the observation points for the electric fields generated by the tori of uniform unit charge.

The electric field components in the electrons' rest frame are E_r and E_z. E_θ vanishes because of the assumed azimuthal symmetry. The components are computed at each observation point by performing a two-dimensional integration in (r, z) space, where each element of integration is actually a circular ring of charge that intersects the cell. The expressions for the electric field in the electrons' rest frame in MKS units at an observation point $\boldsymbol{r}$ are as follows (see Ref. [3] for a more detailed derivation):

$$E_r(r,z) = \frac{1}{4\pi\varepsilon_0}\frac{q}{2\pi a}\frac{2a}{r[z^2 + (r+a)^2]^{1/2}} \times \left[K(k) - \frac{z^2 - (r^2 - a^2)}{z^2 + (r+a)^2}E(k)\right],$$

$$E_r(r,z) = \frac{1}{4\pi\varepsilon_0}\frac{q}{2\pi a}\frac{4azE(k)}{[z^2 + (r+a)^2]^{1/2}[z^2 + (r-a)^2]}$$

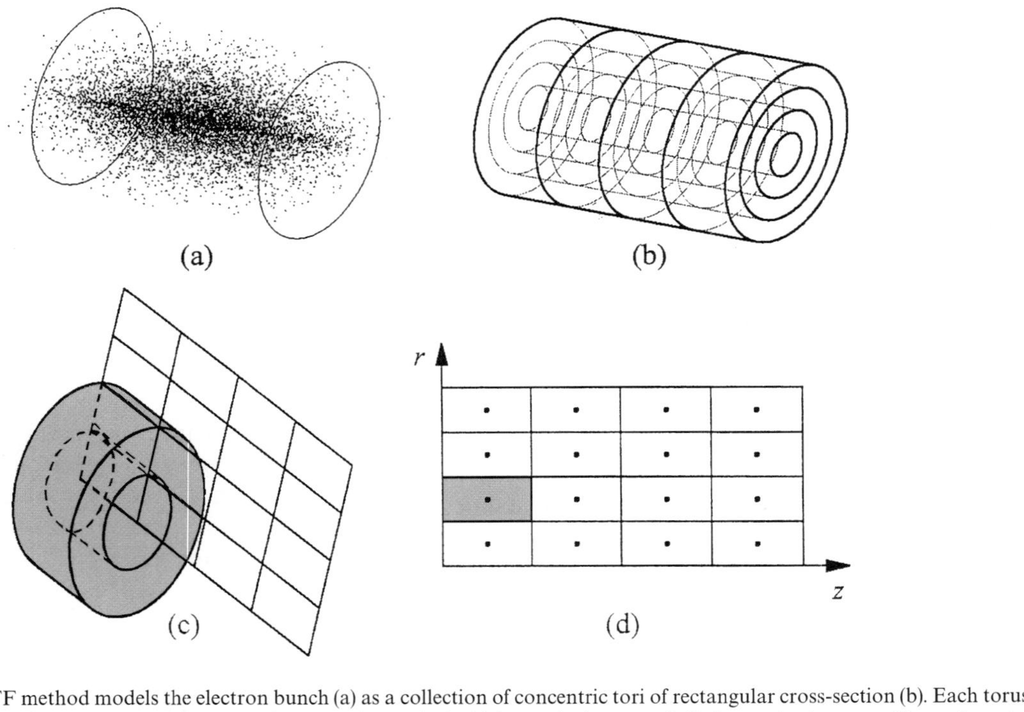

Fig. 1. The SCHEFF method models the electron bunch (a) as a collection of concentric tori of rectangular cross-section (b). Each torus in (b) is assumed to have a uniform distribution of charge, the total of which equals the sum of all charges in the torus. The tori can be described on a two-dimensional grid (d) where each rectangular cell is the intersection of a torus with a radial plane ($\theta = \text{constant}$) (c).

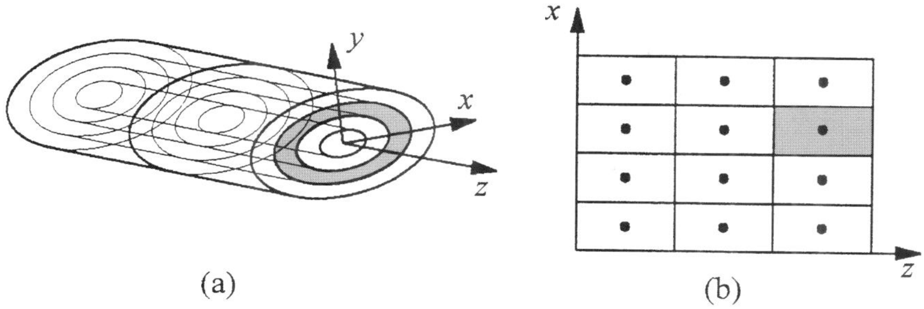

Fig. 2. The modifications in SCHEFF now model the electron bunch as a collection of concentric elliptical tori of rectangular cross-section as shown in (a). The two-dimensional grid shown in (b) is the cross-section of the tori along the x-axis, or $\theta = 0$.

where $k = [4ar/(z^2 + (r + a)^2)]^{1/2}$ and E and K are the complete elliptical integrals of second and first kind, respectively.

We note at this point that we use the original SCHEFF method, although we have not verified that the charge assignment and field interpolation functions are symmetric and thus assure particle momentum conservation [5].

2.2. The SCHEFF elliptical method

The critical modifications made to SCHEFF and related subroutines extend the circular modeling of the electron bunch to allow for bunches of elliptical cross-section. The process is essentially the same as with the circular torus method described above, except that the new algorithm relaxes the assumption of azimuthal symmetry.

The electron bunch is assumed to have elliptical–cylindrical symmetry, and it is modeled as a group of elliptical tori of rectangular cross-section as seen in Fig. 2. Note that, in the transverse plane, these ellipses are concentric, not confocal. That is, they share the same center as well as the same ellipticity e (ratio of semi-minor to semi-major axes) and orientation, but not the same foci. Once again, the elliptical tori may be characterized by a two-dimensional rectangular (r, z) grid. In the circular case, this grid could be arbitrarily defined as the intersection of the tori with ($\theta = \text{constant}$) radial plane. Now this new grid represents the intersection of the tori with the $\theta = 0$ radial plane. Note that the "radial width" of the cross-sections has a θ-dependence due to the change in perpendicular distances as a function of θ.

Again, the normalized electric field is computed at every observation point. Due to elliptical symmetry, only the electric field values for $0 < \theta < \pi/2$ need to be computed. The integral expressions for the electric fields, in the electron rest frame, at the location $\mathbf{r}(r, \theta, z)$, due to an elliptical torus of uniform unit charge and ellipticity e may be expressed as

$$E_r(\mathbf{r}) = \frac{1}{4\pi\varepsilon_0}\frac{q}{v}\int_{z_1}^{z_2}\int_{r_1}^{r_2} eu \int_0^{2\pi} \frac{r - r'\cos(\theta - \theta')}{[(z - z')^2 + r^2 + r'^2 - 2rr'\cos(\theta - \theta')]^{3/2}}\,\mathrm{d}v\,\mathrm{d}u\,\mathrm{d}z'$$

$$E_\theta(\mathbf{r}) = \frac{1}{4\pi\varepsilon_0}\frac{q}{v}\int_{z_1}^{z_2}\int_{r_1}^{r_2} eu \int_0^{2\pi} \frac{r'\sin(\theta - \theta')}{[(z - z')^2 + r^2 + r'^2 - 2rr'\cos(\theta - \theta')]^{3/2}}\,\mathrm{d}v\,\mathrm{d}u\,\mathrm{d}z'$$

$$E_z(\mathbf{r}) = \frac{1}{4\pi\varepsilon_0}\frac{q}{v}\int_{z_1}^{z_2}\int_{r_1}^{r_2} eu \int_0^{2\pi} \frac{z - z'}{[(z - z')^2 + r^2 + r'^2 - 2rr'\cos(\theta - \theta')]^{3/2}}\,\mathrm{d}v\,\mathrm{d}u\,\mathrm{d}z'$$

where $u = \sqrt{x'^2 + (y'/e)^2}$, $v = \arctan((y'/e)/x')$, $r' = u\sqrt{\cos^2 v + e^2 \sin^2 v}$, $\theta' = \arctan(e \tan v)$, and the volume of integration is over the torus: $v = 0$–2π (the analogous azimuthal variable); $u = r_1$ to r_2 (the analogous radial variable along the semi-major axis); $z = z_1 - z_2$ (the longitudinal variable). There are no known closed-form solutions for these integrals, so an additional numerical integration is required. Note that in the limit where the ellipses are actually concentric circles ($e = 1$), u and v become r and θ, and the proper limit for the circular case is retrieved.

2.3. Dynamic gridding

Another modification made to the SCHEFF subroutine involves the dynamic gridding of the $(x = r, z)$ space shown in Fig. 1. On each call to SCHEFF, the extents and characteristics of the bunch are checked, and if the parameters of the bunch have not changed by more than a prescribed amount, then the old grid is used. If a pertinent parameter has changed, then the grid is re-meshed and the program continues. This is repeated as often as necessary. The parameters that are checked are the bunch's radial and longitudinal extents and its ellipticity.

3. Comparison of circular and elliptical algorithms

Several test runs have been made to evaluate the differences arising from using the new elliptical algorithm versus the old circular algorithm. We will report here only on the calculations for the buncher of the Boeing visible FEL beam line [1,2]. The buncher is shown in Fig. 3 along with its physical attributes. The chicane buncher operates at about 20 MeV, where the three-dipole bending magnets (shaded regions) have field strength of 1340 G. With an energy slew of − 510 keV/cm, the electron pulse enters the buncher with length L and exits with an approximate length of $L/6$. The electron pulse is shown to emphasize its orientation through the buncher; its size is exaggerated for emphasis.

The middle of the electron pulse bends along the dashed line of Fig. 3. The head of the pulse, having a smaller energy than the middle, bends with a smaller radius of curvature, and the tail of the pulse, having a larger energy than the middle, bends with a larger radius of curvature. As a result, the pulse does not directly follow the path, but rather is twisted. While in the second dipole, the pulse comes back together, however the length of the pulse becomes compressed due to the difference in path lengths of the electrons from the tail to the head. During the third bend, the compression of the pulse is halted. The electron pulse parameters and the results of the calculations are shown in Fig. 4.

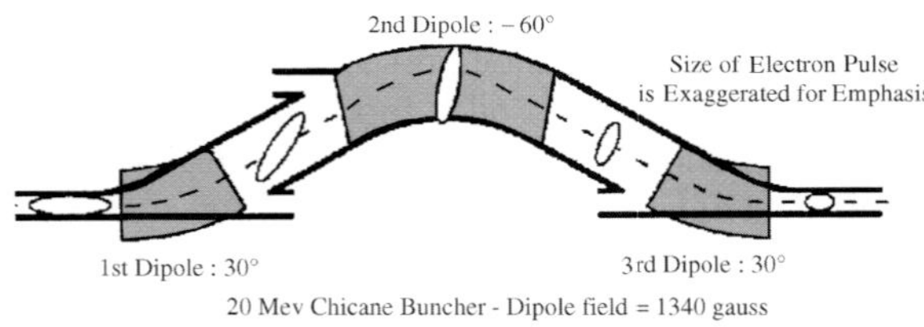

Fig. 3. Diagram of the 20 MeV chicane buncher.

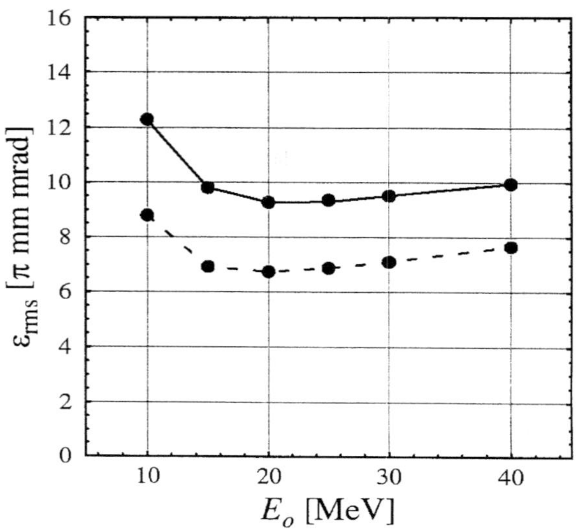

Fig. 4. Normalized x (bend) plane rms emittance for varying energies of the electron pulse at the buncher entrance from old (circular algorithm) calculations (solid line), and new (elliptical algorithm) calculations (dashed line). The normalized entrance emittance in the x and y plane is 1.0 π mm mrad, the pulse charge is 3.0 nC, the superimposed energy slew is − 510 keV/cm.

The results with the new elliptical algorithm predict that emittance will grow in the bend (x-) plane 30–40% less than the circular results indicate. Both algorithms show negligible emittance growth in the y-plane.

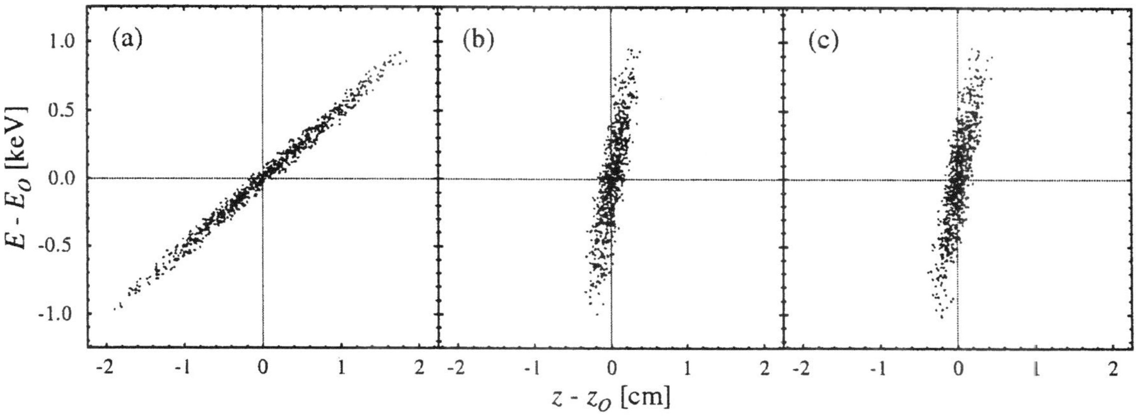

Fig. 5. Difference between particle energy and average pulse energy versus longitudinal particle position relative to the bunch centroid position, (a) at the entrance of the buncher, (b) at the exit of the buncher from the elliptical and (c) circular calculations. The electron bunch parameters are the same as in Fig. 4.

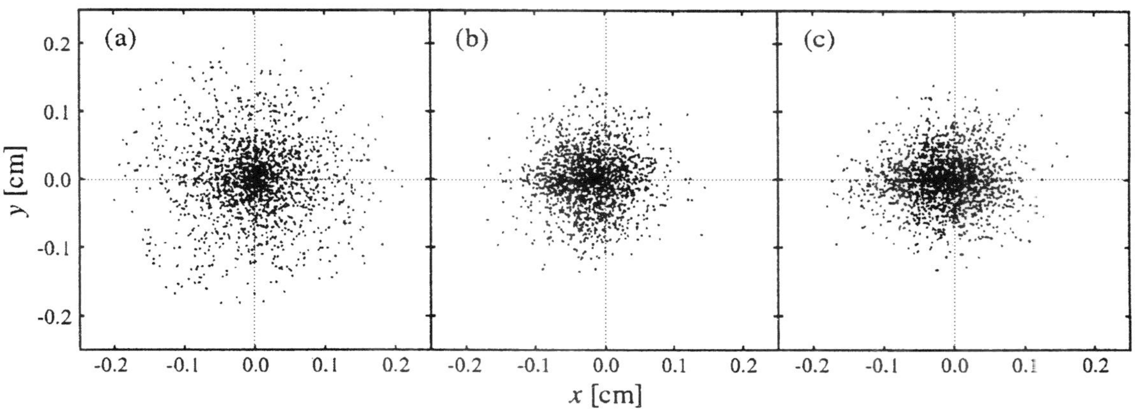

Fig. 6. Electron bunch y–x cross-section plots from the elliptical calculations at the (a) entrance and (b) exit of the buncher and (c) from the circular calculations at the exit of the buncher.

Through the course of the runs, the ellipticity was monitored and found to be on the order of $e = 0.01$ through the middle of the buncher. Clearly, modeling the electron bunch as a collection of circular tori with uniform charge distribution does not seem adequate for such a highly elliptic bunch.

A careful scrutiny of the phase spaces of the particles at the exit of the buncher shows that the spot size computed by the elliptical tori method is smaller than the one computed by the circular tori method. Phase space plots for both elliptical and circular runs for $E_0 = 20$ MeV are shown in the following figures. Fig. 5 shows the difference between the particle energy and the average energy versus the particles longitudinal position. The energy spread of the pulse at the exit is slightly larger in the circular case, however in both cases the particles are "bunched" by the same amount, roughly by 80% from a length of about 4–0.8 cm.

The current pulse shape at the buncher exit is nearly identical in both sets of calculations. The

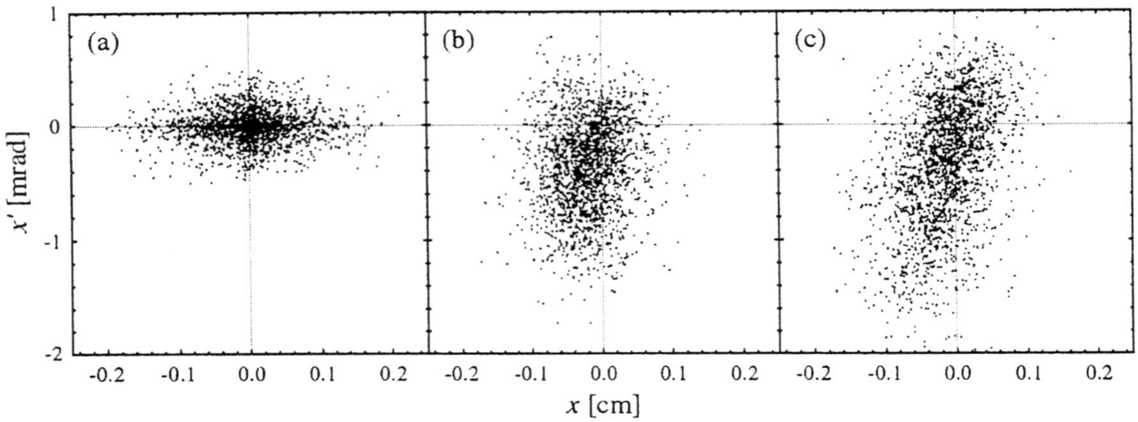

Fig. 7. x-(bend) plane transverse phase space from the elliptical calculations at the (a) entrance and (b) exit of the buncher and from the (c) circular calculations at the exit of the buncher.

differences between the results of the elliptical and circular calculations are most apparent when comparing the spot size and shape as well as the transverse phase space. Fig. 6 shows the (y, x) cross-section view of the particle bunch at the entrance and exit of the buncher. In the elliptical calculations, it is clear that the electron bunch spot size is smaller at the exit of the buncher, and it remains more symmetric than in the circular calculations.

Fig. 7 shows a comparison of the x-transverse phase space, the bend plane, from both calculations. Here the differences in results are most clear. The reduction in emittance from the circular to elliptical calculations can be surmised from the exit transverse phase space since emittance is related to the area of these plots. In both cases, the transverse phase space is not symmetrical, but the asymmetry and amount of divergence is more pronounced in the circular calculations than in the elliptical calculations.

4. Summary and conclusion

The Boeing version of the PARMELA code has been modified to compute more correctly the space charge effects for electron pulses with highly elliptical transverse cross-sections. A dynamic meshing routine has been added to allow good resolution for pulses that undergo significant changes in their longitudinal or transverse dimensions. The results from the calculations for the chicane buncher of the 1 kW visible FEL line at Boeing, indicate that the old circular algorithm of the SCHEFF subroutine overestimates the emittance growth in the bend plane by 30–40%.

References

[1] J.L. Adamski et al., Nucl. Instr. and Meth. A 375 (1996) 288.
[2] D.H. Dowell et al., Nucl. Instr. and Meth. A 393 (1997) 184.
[3] P. Lapostolle et al., Nucl. Instr. and Meth. A 379 (1996) 21.
[4] P. Zhou, R.G. Carter, A User's Guide to PARMELA, Lancaster University, Department of Engineering, May, 1989.
[5] R.W. Hockney, J.W. Eastwood, Computer Simulations using Particles, Hilger (Adam), Bristol, 1989.

ELSEVIER

Nuclear Instruments and Methods in Physics Research A 429 (1999) 287–292

NUCLEAR INSTRUMENTS & METHODS IN PHYSICS RESEARCH
Section A

www.elsevier.nl/locate/nima

A non-destructive electron beam diagnostic for a SASE FEL using coherent off-axis undulator radiation

C.P. Neuman*, M.L. Ponds, G.A. Barnett, J.M.J. Madey, P.G. O'Shea

Duke University Free Electron Laser Laboratory, Durham, NC 27708-0319, USA

Abstract

We show that by observing coherent off-axis undulator radiation (COUR) from a short diagnostic wiggler, it may be possible to determine the length and structure of a short electron bunch. Typically the on-axis undulator radiation is incoherent, but at angles of a few degrees, the wavelength of the emitted radiation may be comparable to the length of a short electron bunch, and thus coherence effects emerge. Due to such coherence effects, the intensity of the emitted radiation may change by up to a factor of 10^9 as the angle of observation is increased. The radiation becomes coherent in a way which depends on the length and structure of the electron bunch. Observing COUR disturbs the electron bunch negligibly. Thus, COUR can be used as a non-destructive diagnostic which would allow for optimization of FEL performance while an FEL is operating. Such a diagnostic could be used for proposed SASE FELs, which use short electron bunches. We present two methods to describe the theory for COUR, and we use these methods to calculate the expected outcome of a COUR experiment. We propose an experiment to demonstrate COUR effects and their applications to SASE FELs.

Keywords: SASE FELs; FEL; Beam diagnostic; Undulator radiation; Off-axis undulator radiation

1. Introduction

For decades there has been interest in obtaining coherent radiation from free electrons. Schwinger discussed the properties of radiation from accelerated electrons in 1949 [1], and the (unwanted) coherent radiation in a synchrotron was studied theoretically in the early 1950s [2]. Also in that decade, the idea existed to make use of coherent mm-wave radiation from bunched electrons [3]. In recent years, different methods have been used to obtain coherent radiation from free electrons [3]. These methods include passing electrons though thin metal foil to produce transition radiation [4,5], generating synchrotron radiation from bending magnets [6], passing electrons near a metal grating to produce Smith–Purcell radiation [7,8], and generating undulator radiation by passing electrons through periodic arrays of magnets [9,10]. In our study, we investigate the possibility of obtaining coherent, mm-wave radiation by observing undulator radiation at an angle to the beam axis in cases where the on-axis radiation is *incoherent* [11].

In addition, there is a continuing challenge to accurately characterize short electron bunches. There has been much interest in recent years in

* Corresponding author. Tel.: + 1-919-660-2667; fax: + 1-919-660-2671.

E-mail address: charles.neuman@duke.edu (C.P. Neuman)

0168-9002/99/$ - see front matter
PII: S0168-9002(99)00127-8

producing ultra-short electron bunches for SASE (self-amplified spontaneous emission), or single-pass, FELs [12–15] and for linear colliders. One planned SASE FEL experiment will use the PALADIN wiggler at Duke [14]. The quality of the electron beam is of utmost importance for these applications; and beam characteristics and their effects on FEL performance have been studied extensively [16,17]. It would be extremely useful to have a device which could characterize a short electron bunch without disturbing the electron beam. Then the electron beam could be used for an experiment while it is being diagnosed in real time. For example, a SASE FEL requires an accurately characterized short electron bunch. With a non-destructive beam diagnostic, the electron bunch length could be known, and the FEL output could be optimized while the FEL is operating.

Our study shows that it may be possible to use coherent, off-axis undulator radiation (COUR) to determine the length and structure of an electron bunch in real time. As the angle of observation θ changes from 0° to angles as large as 15°, the emitted undulator radiation changes from incoherent to coherent in a way that is dependent on the length and structure of the electron bunch. Furthermore, since no rearrangement of the electrons in a bunch (microbunching) is necessary in order to obtain COUR, a short wiggler may be used. Thus, the electron bunch is passed relatively undisturbed. The result is the possibility of a non-destructive electron beam diagnostic.

2. Theory

2.1. Introduction

Coherence effects may be present in undulator radiation, depending on the angle of observation. Typically an electron bunch is longer than the wavelength of light which is emitted on-axis from a wiggler. For example, an electron bunch may be a few hundred microns long, and the emitted on-axis radiation may have a wavelength of 1 μm. In this case, assuming a "smooth" longitudinal electron bunch distribution, the radiation is incoherent, and the intensity scales linearly with the number of electrons, which is generally on the order of 10^9. However, at angles to the axis, coherence effects may emerge. The wavelength of the radiation by an electron in a wiggler increases with the angle of observation:

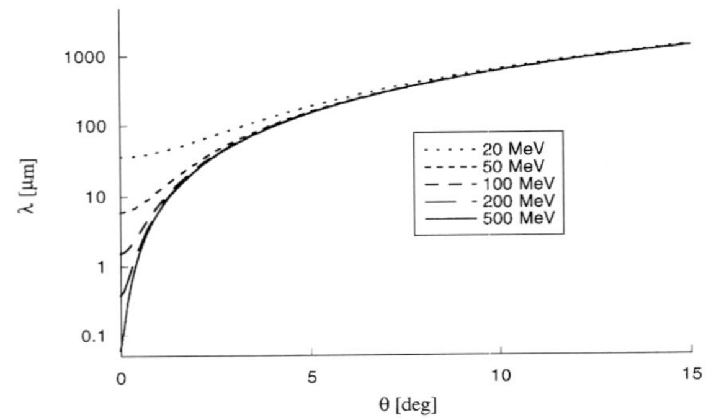

Fig. 1. $\lambda(\theta)$ for various beam energies. Note that $\theta > 5°$, $\lambda(\theta)$ is independent of electron beam energy.

$$\lambda(\theta) = \frac{\lambda_w}{2\gamma^2}[(1 + a_w^2)\cos\theta + 2\gamma^2(1 - \cos\theta)] \tag{1}$$

where λ_w is the wiggler period, a_w is the wiggler parameter, and γ is the relativistic factor for the electron beam [18] (see Fig. 1). When this wavelength is longer than the electron bunch, the radiation will be coherent. At $\theta = 15°$, for example, the radiation may have a wavelength of 1 mm or more which is much longer than a short electron bunch, say 200 μm. The electrons thus radiate coherently, and the intensity scales quadratically with the number of electrons. At intermediate angles, such as $\theta = 5°$, the wavelength of the emitted radiation may be comparable to the bunch length, and some coherence effects should be observed.

As the angle of observation is gradually changed from small to larger angles, the radiation observed will become more coherent, and the intensity will change by up to a factor of N, the number of electrons, where N is generally on the order of 10^9. The way this intensity changes with the angle of observation will depend on the dimensions and structure of the electron bunch. Thus by observing the radiation spectrum through a range of angles, one can potentially characterize an electron bunch.

To calculate the total power radiated at various angles, two methods are used: a multi-mode analysis and a $dP/d\Omega$ analysis. The multi-mode analysis involves expanding the vector potential of the radiation in a set of orthogonal Gaussian optical modes, and calculating the energy radiated into each mode. The $dP/d\Omega$ analysis allows for a calculation of the total energy radiated. The multi-mode analysis is a novel technique in electrodynamics [11], and for the single-electron case it is not limited to the far-field as is the $dP/d\Omega$ analysis.

2.2. Multi-mode analysis, single electron

In order to calculate the energy of the expected COUR for a given angle, we expand the vector potential in a set of orthogonal modes (see Fig. 2) and derive an expression for the evolution of each mode amplitude as the electron passes through the wiggler. From this mode amplitude, the energy deposited into the mode is obtained. The result simplifies for the single-electron case. For a bunch of N electrons, the result involves a bunch form factor, which accounts for the effects of the bunch dimensions.

The vector potential is expanded in Gauss–Laguerre optical modes and longitudinal cavity modes:

$$\boldsymbol{A}(\boldsymbol{r},t) = \sum_{k,p,q} a_{kpq}(t)u_{kpq}(\boldsymbol{r})\frac{e^{-i(kz-\omega t)}}{\sqrt{L_{\mathrm{cav}}}}\hat{x} \tag{2}$$

where

$$u_{kpq}(\boldsymbol{r}) = \sqrt{\frac{2p!}{\pi(p+q)!}}\frac{1}{w(z)}\left(\frac{\sqrt{2}r}{w(z)}\right)^{q} L_p^q\left(\frac{2r^2}{w(z)^2}\right) \times e^{-r^2/w(z)^2 + i[(2p+q+1)\tan^{-1}(z/z_R) - q\phi]} \tag{3}$$

$$w(z)^2 = w_0^2\left[1 + \left(\frac{z}{z_{\mathrm{R}}}\right)^2\right]$$

w_0 is the minimum spot size, $z_{\mathrm{R}} = kw_0^2/2$ is the Rayleigh range, and L_{cav} is the cavity length (the cavity is theoretical, and the final result does not depend on L_{cav}).

The modes $u_{kpq}(\boldsymbol{r})e^{-i(kz-\omega t)}/\sqrt{L_{\mathrm{cav}}}$ satisfy the wave equation

$$\left(\nabla^2 - \frac{1}{c^2}\frac{\partial^2}{\partial t^2}\right)\boldsymbol{A}(\boldsymbol{r},t) = 0 \tag{4}$$

taking into account the slowly varying envelope approximation, which assumes that the envelope of $\boldsymbol{A}(\boldsymbol{r},t)$ varies slowly compared to its oscillations in time. The modes also satisfy the orthogonality condition

$$\iiint d^3r\, u_{kpq}(\boldsymbol{r})u^*_{k'p'q'}(\boldsymbol{r})\frac{e^{-i(k-k')z}}{L_{\mathrm{cav}}} = \delta_{kk'}\delta_{pp'}\delta_{qq'}. \tag{5}$$

The expansion, Eq. (2), is inserted into the wave equation for the vector potential,

$$\left(\nabla^2 - \mu_0\varepsilon_0\frac{\partial^2}{\partial t^2}\right)\boldsymbol{A}(\boldsymbol{r},t) = -\mu_0\boldsymbol{J}(\boldsymbol{r},t) \tag{6}$$

and Eqs. (4) and (5) and the slowly varying envelope approximation are used to obtain the following result:

$$a_{kpq}(t) = a_{kpq}(0) + \frac{-1}{2i\omega\varepsilon_0}\int_0^t dt \iiint d^3r J_x(\boldsymbol{r},t)U^*_{kpq}(\boldsymbol{r},t) \tag{7}$$

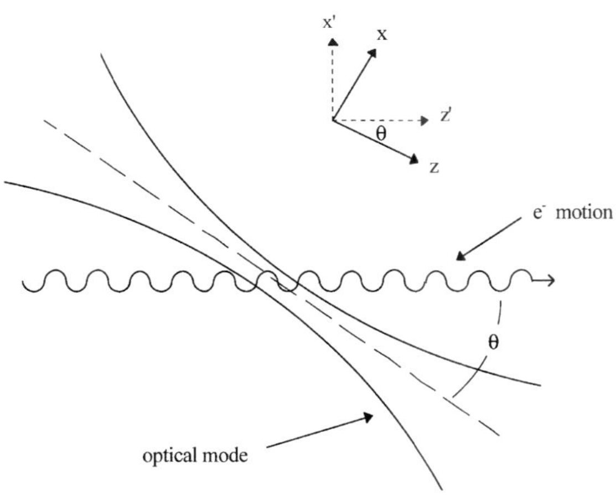

Fig. 2. Geometry for a multi-mode analysis of COUR showing a Gaussian optical mode at an angle θ to the electron beam axis. The electron motion is in the $\hat{x}' - \hat{z}'$ plane, and $\hat{z}$ is the optic axis.

which describes the time evolution of the mode amplitude. This is ultimately evaluated at the time when the electron leaves the wiggler. The analysis up to this point is known as an $\boldsymbol{A}\cdot\boldsymbol{J}$ (pronounced "A dot J") analysis [19].

The electron current density can be expressed as $\boldsymbol{J}(\boldsymbol{r},t) = -ec\boldsymbol{\beta}(t)\rho(\boldsymbol{r},t)$, where $\rho(\boldsymbol{r},t)$ is the electron charge density, and $\boldsymbol{\beta}(t)$ is the electron velocity normalized by c. For a single electron, a point particle, $\rho(\boldsymbol{r},t) = \delta(\boldsymbol{r} - \boldsymbol{r}_e(t))$, where $\boldsymbol{r}_e(t)$ is the electron trajectory, and Eq. (7) reduces to

$$a_{kpq}(t) = a_{kpq}(0) + \frac{ec}{2i\omega\varepsilon_0}\int_0^t dt\beta_x(t)U^*_{kpq}(\boldsymbol{r}_e(t),t). \qquad (8)$$

In this approximation, the electron bunch remains unmodified by the interaction with the optical field.

The electron velocity in the primed coordinates is calculated by applying the Lorentz force equation to the periodic wiggler field:

$$\beta_{x'} = \frac{\dot{x}'}{c} = \frac{a_w\sqrt{2}}{\gamma}\cos(k_w z')$$

$$\beta_{z'} = \frac{\dot{z}'}{c} = \left[1 - \frac{1}{2\gamma^2}\left(1 + \frac{a_w^2}{2}\right)\right] - \frac{a_w^2}{4\gamma^2}\cos(2k_w z'). \qquad (9)$$

Using the approximation $z' \approx \bar{\beta}_{z'}ct$ these equations of motion are integrated to give the electron trajectory in the primed coordinates:

$$x'_e(t) = \frac{a_w\sqrt{2}}{\gamma k_w\bar{\beta}_{z'}}\sin(k_w\bar{\beta}_{z'}ct)$$

$$z'_e(t) = \bar{\beta}_{z'}ct - \frac{a_w}{8\gamma^2 k_w\bar{\beta}_{z'}}\sin(2k_w\bar{\beta}_{z'}ct) - \frac{L}{2} \qquad (10)$$

where

$$\bar{\beta}_{z'} = 1 - \frac{1}{2\gamma^2}\left(1 + \frac{a_w^2}{2}\right)$$

and L is the wiggler length. At $t = 0$, $z' = -L/2$, and the particle is entering the wiggler.

The electron trajectory and velocity are expressed in the unprimed coordinates as follows:

$$x_e(t) = x'_e(t)\cos\theta + z'_e(t)\sin\theta$$

$$z_e(t) = z'_e(t)\cos\theta - x'_e(t)\sin\theta \qquad (11)$$

$$\beta_x(t) = \beta_{x'}(t)\cos\theta + \beta_{z'}(t)\sin\theta$$

where $x'_e(t)$ and $z'_e(t)$ are given by Eqs. (10), and $\beta_{x'}(t)$ and $\beta_{z'}(t)$ are given by Eqs. (9), with $z' \approx \bar{\beta}_{z'}ct$:

$$\beta_{x'}(t) = \frac{a_w\sqrt{2}}{\gamma}\cos(k_w\bar{\beta}_{z'}ct)$$

$$\beta_{z'}(t) = \bar{\beta}_{z'} - \frac{a_w^2}{4\gamma^2}\cos(2k_w\bar{\beta}_{z'}ct). \qquad (12)$$

The total energy deposited into the mode is calculated from the mode coefficient a_{kpq}. The spectral energy is given by

$$\frac{dW_{kpq}}{dk} = \frac{\varepsilon_0}{2}\iiint d^3r|E|^2 + \frac{1}{2\mu_0}\iiint d^3r|B|^2 = \varepsilon_0\omega^2 a_{kpq}a^*_{kpq} \qquad (13)$$

where $\boldsymbol{E} = -\partial\boldsymbol{A}/\partial t \approx i\omega\boldsymbol{A}$ follows from the slowly varying envelope approximation. The total energy W_{pq} for a single mode is given by

$$W_{pq} = \sum_k \omega^2\varepsilon_0\, a_{kpq}\, a^*_{kpq} = \frac{2L_{cav}\varepsilon_0}{\pi}\int_0^\infty dk\omega^2 a_{kpq}\, a^*_{kpq} \qquad (14)$$

and the total energy is given by $W = \sum_{pq} W_{kpq}$.

2.3. $\frac{dP}{d\Omega}$ analysis, single electron

The total power radiated by COUR is also calculated by using a $dP/d\Omega$ analysis, which gives the total power radiated in a solid angle $d\Omega$ by an accelerated electron. The result is valid for the far-field region which, for a 0.5 m wiggler, would ideally be a distance of about 5 m from the wiggler. Shorter distances simply scale the far-field result.

We begin with Jackson 14.38 (in SI units), which gives the power radiated per solid angle by an accelerating electron [22]:

$$\frac{dP}{d\Omega} = \frac{1}{4\pi\varepsilon_0}\frac{e^2}{4\pi}\frac{1}{c}\frac{|\hat{n}\times\{(\hat{n}-\boldsymbol{\beta})\times\dot{\boldsymbol{\beta}}\}|^2}{(1-\hat{n}\cdot\boldsymbol{\beta})^5} \qquad (15)$$

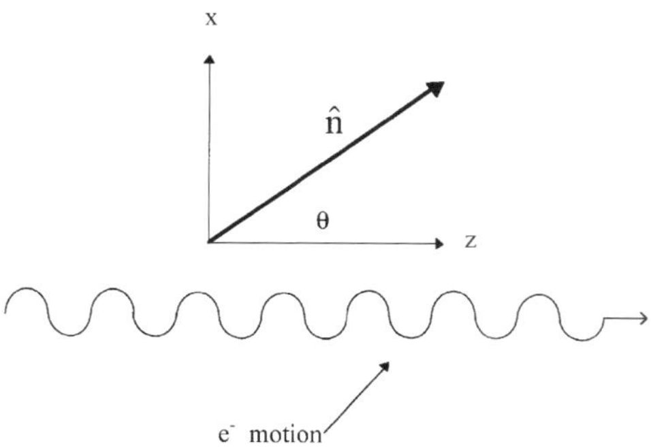

Fig. 3. Coordinate system for $dP/d\Omega$ analysis. The electron motion is in the $\hat{x}-\hat{z}$ plane, and $\hat{n}$ is the direction of observation.

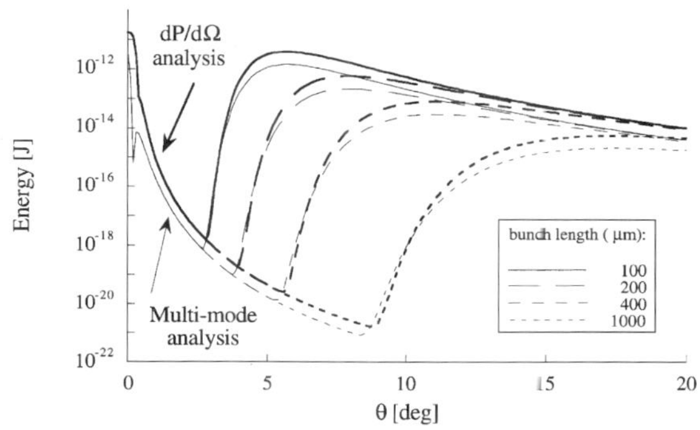

Fig. 4. The effect of bunch length on COUR. $\gamma = 400$, $a_w = 1.44$, $L = 0.5057$ m. The $dP/d\Omega$ analysis gives the total energy radiated. For the multi-mode analysis, the energy in the lowest-order mode is calculated, and $z_R = 1$ m.

where $\hat{n} = \cos\theta\hat{z} + \sin\theta\hat{x}$ is the direction of observation (see Fig. 3). $\boldsymbol{\beta}$ is given by Eqs. (12) (without the primes).

The total power is given by $P = (1/r^2)(dP/d\Omega)\Delta A$, where ΔA is the area of the detector and r is the distance of the detector from the wiggler, and the total energy radiated W is given by integrating with respect to time.

2.4. N electrons

The energy deposited in a mode by a bunch of N electrons includes an expression which is related to the bunch shape [20,21]:

$$W_{\text{tot}} = W_{1e^-}[N + N(N-1)f(\omega)], \tag{16}$$

where $f(\omega)$ is the bunch form factor and is related to the function that represents the electron bunch density:

$$f(\omega) = \left|\iint dy\,dz S(y)S(z)e^{-iky\sin\theta + ikz\cos\theta}\right|^2. \tag{17}$$

$S(y)$ and $S(z)$ represent the normalized transverse and longitudinal electron bunch densities, respectively. Note that for small angles, $f(\omega)$ is the square of the Fourier transform of the function which represents the longitudinal density. The result shows that the structure of the bunch, and not only the bunch size (as described before), affects the intensity of the radiation.

Calculations have been performed with the code OFFAX to predict the outcome of an experiment. In Fig. 4, the energy radiated as a function of angle is shown for several bunch lengths. Note the close agreement between the multi-mode analysis and the $dP/d\Omega$ analysis. Furthermore, the results show that a large fraction of the total energy is radiated into the lowest-order optical mode.

3. Experiment

The experimental goals are as follows:

1. To observe COUR and to compare with calculations.
2. To use data from the emitted radiation to infer the structure of an electron bunch.
3. To evaluate this technique in relation to other techniques.

We plan to perform the COUR experiment at the Source Development Lab (SDL) at Brookhaven National Laboratory. The SDL is dedicated to developing sources and experiments for the next generation of synchrotron radiation research. The newly constructed linear accelerator is scheduled to produce a 200 MeV electron beam by October 1998. Electron bunches will range from 20 μm to 1 mm in length [23].

COUR will be produced on this beamline from a short wiggler which is a prototype for the 10 m NISUS wiggler. This 0.5 m NISUS prototype consists of 13 periods, each 3.89 cm long, a peak field of 5.6 kG, and a wiggler parameter of up to 1.44 [24].

Radiation will exit the beamline via a mirror and a diamond window, and will be detected by a liquid-helium-cooled bolometer. The extremely sensitive bolometer has a responsivity of 10^5 V/W, a noise level of 10 nV/Hz$^{1/2}$, and response time of 1 ms [25]. The mirror in the beamline may be moved so that radiation can be viewed at various angles.

By measuring the energy radiated at various angles, we hope to verify the plot shown in Fig. 4. Note that there have been no attempts so far to observe COUR in experiments.

4. Conclusion

We have shown that when an electron passes through a short wiggler, coherent radiation may be emitted at angles to the beam axis. The way this radiation becomes coherent as the angle of observation changes is determined by the size and structure of the electron bunch. Thus, it may be possible to characterize an electron bunch by observing COUR. Producing COUR form a short wiggler disturbs the electron bunch negligibly. Thus, this diagnostic may be useful in cases where an electron bunch is to be characterized in real time while it is being used for another purpose. Specifically, this electron beam diagnostic would be useful for SASE FELs.

We have shown that the COUR effects are independent of beam energy. Thus this electron beam diagnostic may be used without exact knowledge of the beam energy or in cases where the beam energy is not stable.

We plan to conduct an experiment to demonstrate COUR at the Source Development Lab at Brookhaven National Laboratory.

Acknowledgements

The authors wish to thank William Graves, Erik Johnson, Lawrence Carr, and James Murphy of Brookhaven National Laboratory and Henry Freund of Science Applications International Corporation for helpful discussions pertaining to this project. This work was supported by the U.S. Army AASERT Program. The computational work was supported by the North Carolina Supercomputing Center.

References

[1] J. Schwinger, Phys. Rev. 75 (1949) 1912.
[2] J. Nodvick, D. Saxon, Phys. Rev. 96 (1954) 180.
[3] G.P. Gallerano et al., Nucl. Instr. and Meth. A 358 (1995) 78.
[4] J. Denard, AIP Conference Proceedings, 333 (1995) 224.
[5] E. Nishimura, Nucl. Instr. and Meth. A 358 (1995) ABS 89.
[6] T. Nakazato, Phys. Rev. Lett. 63 (1989) 1245.
[7] Y. Shibata et al., Phys. Rev. E 57 (1998) 1061.
[8] J.H. Brownell, J. Walsh et al., Nucl. Instr. and Meth. A 393 (1997) 323.
[9] H. Motz, J. Appl. Phys. 22 (1951) 527.
[10] J.M.J. Madey, J. Appl. Phys. 42 (1971) 1906.
[11] J.M.J. Madey, personal communication.
[12] C. Pellegrini, AIP Conference Proceedings 367 (1995) 3.
[13] P.G. O'Shea, Nucl. Instr. and Meth. A 358 (1995) 36.
[14] P.G. O'Shea, C.P. Neuman, J.M.J. Madey, H.P. Freund, Nucl. Instr. and Meth. A 393 (1997) 129.
[15] K.-J. Kim, M. Xie, Nucl. Instr. and Meth. A 331 (1993) 359.
[16] P.G. O'Shea, Phys. Rev. E 57 1081.
[17] H.P. Freund, P.G. O'Shea, Phys. Rev. Lett. 80 (1998) 520.
[18] P. Luchini, H. Motz, Undulators and Free-Electron Lasers, Oxford University Press, New York, 1990.
[19] J.M.J. Madey, Physics of Free Electron Lasers, Course at Duke University, Spring 1997.
[20] E.B. Blum, U. Happek, A.J. Sievers, Nucl. Instr. and Meth. A 307 (1991) 568.
[21] R. Lai, A.J. Sievers, AIP Conference Proceedings 367 (1995) 312.
[22] J.D. Jackson, Classical Electrodynamics, Wiley, New York, 1975, p. 663.
[23] Conversation with William Graves, Brookhaven National Laboratory, June 1998.
[24] D.C. Quimby et al., Nucl. Instr. and Meth. A 285 (1989) 281.
[25] Conversation with Lawrence Carr, Brookhaven National Laboratory, June 1998.

Nuclear Instruments and Methods in Physics Research A 429 (1999) 293–298

NUCLEAR INSTRUMENTS & METHODS IN PHYSICS RESEARCH Section A

www.elsevier.nl/locate/nima

Potential diagnostics for the next-generation light sources☆

A.H. Lumpkin*, B.X. Yang

Advanced Photon Source, Argonne National Laboratory, 9700 South Cass Avenue, Argonne, IL 60439, USA

Abstract

There is continued interest in developing a diffraction-limited soft or hard X-ray source. Candidate paths include the storage-ring-based free-electron laser (FEL) and the linac-based self-amplified spontaneous emission (SASE) FEL for the two regimes, respectively. As previously discussed, target beam parameters are $\sigma_{x,y} \sim 10$ μm, $\sigma_{x',y'} = 1$μrad, and $\sigma_t \sim 1$ ps (0.1 ps for the linac case). We report the use of few- to sub-angstrom radiation emitted by a 7-GeV beam transitting a 198-period diagnostics undulator and a bending magnet to characterize the particle beam. A particle beam divergence as low as $\sigma_{y'} = 3.3$ μrad was measured using an X-ray monochromator. Additionally, a particle beam size of $\sigma_y < 45$ μm and a bunch length of $\sigma_t = 28$ ps with 4-ps resolution were measured using an X-ray pinhole camera with a unique synchroscan and dual-sweep X-ray streak camera as the detector. The adjustable pinhole aperture was varied by more than a factor of 100 to assess spatial resolution and dynamic range issues in the system. These diagnostics demonstrations should scale to next-generation applications. © 1999 Published by Elsevier Science B.V. All rights reserved.

PACS: 41.60.Cr; 41.60.Ap; 07.85.Qe

Keywords: Diagnostics; SASE; FEL; Spatial resolution; X-ray streak camera; Undulator

1. Introduction

At the Rome FEL conference, it was suggested that the pursuit of next-generation light sources would involve the development of certain diagnostic capabilities as well [1]. In particular, targeted beam parameters for a diffraction-limited hard X-ray source are $\sigma_{x,y} \sim 10$ μm, $\sigma_{x',y'} \sim 1$ μrad, and $\sigma_t \sim 1$ and 0.1 ps for the storage-ring-based case and linac-based case, respectively. We report here measurements on the path to these objectives based on detecting few- to sub-angstrom X-ray synchrotron radiation (XSR) emitted by a 7-GeV e^+ beam as it transitted a 198-period diagnostics undulator as well as a separate dipole magnet. From the first source a particle beam divergence of $\sigma_{y'} = 3.3$ μrad was measured using an X-ray monochromator. From the second source a particle beam size of tens of microns was measured, and a bunch length of tens of ps with few-ps resolution was measured using a unique synchroscan and dual-sweep X-ray streak camera as the detector in an X-ray pinhole system.

2. Experimental background

The Advanced Photon Source (APS) is a third-generation hard X-ray synchrotron radiation research facility. Its main particle beam parameters

*Corresponding author. Tel.: + 1-630-252-4879; fax: + 1-630-252-4732.

☆Work supported by the U.S. Department of Energy, Office of Basic Energy Sciences, under Contract No. W-31-109-ENG-38.

PII: S0168-9002(99)00130-8

include a 7-GeV beam energy, 100-mA stored beam current, and a 8.2×10^{-9} mrad natural emittance. The measurement and monitoring of particle-beam quality is one of the operational requirements. In support of these tasks one of the 40 sectors in the ring has been dedicated to particle-beam diagnostics using both optical synchrotron radiation (OSR) and XSR. The push towards lower vertical coupling to enhance X-ray beam brilliance has also resulted in providing a test bed for techniques for measuring beam parameters in one plane relevant to next-generation sources. As shown in Table 1, vertical coupling of 1% implies few-μrad divergence and 0.1% coupling implies 1 μrad. The corresponding vertical beam sizes are 29 and 9 μm. As shown schematically in Fig. 1, our techniques include an X-ray pinhole system on one of the dipole sources in the ring and an X-ray monochromator on the diagnostics undulator. The adjustable pinhole based on a remotely controlled four-jaw aperture is located 10.1 m from the source resulting in a magnification of 1.67 at the photocathode (PC) of the X-ray streak camera. An important practical aspect of the set-up was the remote-controlled positioning of the camera using three stacked translation stages. These stages were under the camera mainframe and provided x-, y-, and $z(t)$-axis motion. This facilitated the alignment of the strip photocathode (6 mm (H) × 50 μm (V)) to the X-ray pinhole image field.

The Au photocathode provides sensitivity to radiation from at least 10 eV to 10 keV and has some efficiency at energies up to 20–25 keV. The soft X-rays are strongly attenuated by the Be windows in the transport. The X-ray tube is housed in a mainframe compatible with the plug-in units of the Hamamatsu C5680 series [2]. For these experiments the synchroscan plug-in provided the

Table 1
APS storage ring parameters in the straight section (for natural emittance $\varepsilon = 8.2$ nm rad)

Coupling	Parameter	Horizontal	Vertical
	Beta (m)	14.2	10.1
10%	$\sigma_{x,y}$ (μm)	325	87.0
10%	$\sigma_{x',y'}$ (μrad)	23	8.6
1%	$\sigma_{x,y}$ (μm)	340	29.0
1%	$\sigma_{x',y'}$ (μrad)	24	2.8
0.1%	$\sigma_{x,y}$ (μm)	341	9.0
0.1%	$\sigma_{x',y'}$ (μrad)	24	0.9

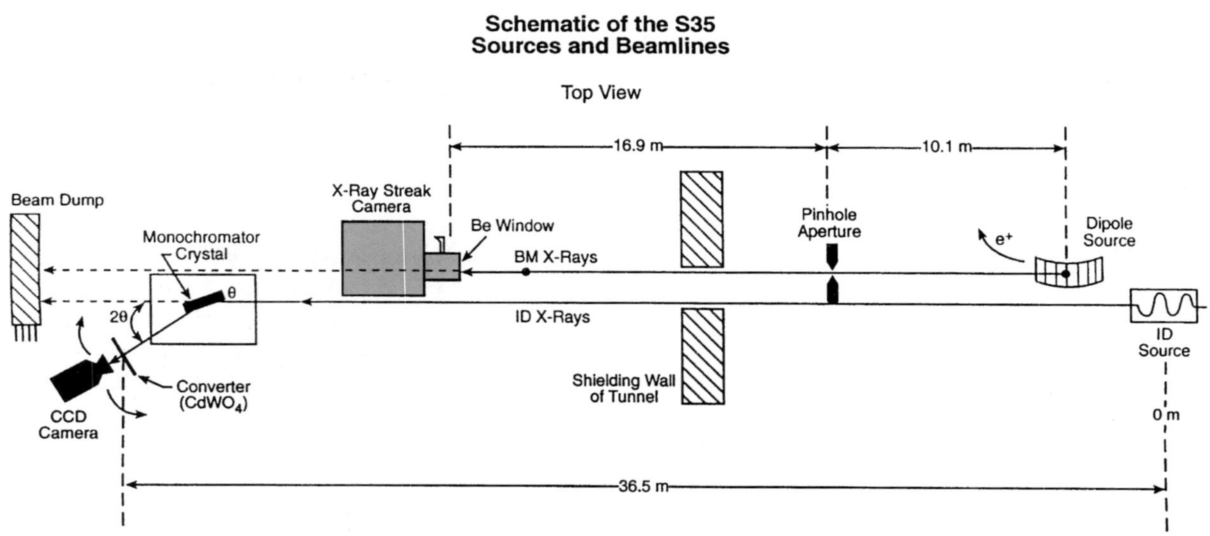

Fig. 1. A schematic diagram of the APS diagnostics beamline. The X-ray pinhole and X-ray streak camera in the dipole source line are indicated as well as the monochromator on the undulator beamline.

vertical (fast time) sweep and the M5679 unit the horizontal (slow time) sweep.

The insertion device (ID), or diagnostics undulator in this case, is a 198- period permanent magnet undulator specified by APS and fabricated by STI Optronics, Inc. [3]. It has a period $\lambda_u = 1.8$ cm and length $L = 3.56$ m. With a 7-GeV particle beam, the on-axis fundamental $\lambda \sim 0.5$ Å can be determined from Eq. (1) where λ_n is the harmonic wave number, λ_u is the undulator period, γ is the Lorentz factor, K is the undulator deflection parameter, and θ is the angle with respect to the beam axis:

$$\lambda_n(\theta) = \frac{\lambda_u}{n2\gamma^2}\left(1 + \frac{K^2}{2} + \gamma^2\theta^2\right). \quad (1)$$

The projected on-axis cone angle σ_{p0} from Eq. (2) is 2.6 μrad in the fundamental compared to the nominal 8-μrad divergence of the APS stored e^+ beam at the baseline 10% vertical coupling and for $\beta_y = 10.1$ m in the ID straight section:

$$\sigma_{p0} \cong \sqrt{\frac{\lambda}{2L}}. \quad (2)$$

Beam divergences were measured using an in-air monochromator based on a Huber goniometer, a Si (2 2 0) or Ge (4 4 0) crystal, a $CdWO_4$ converter crystal, and a charge-coupled device (CCD) camera. The converter crystal was located 36.5 m from the ID source point so that the divergence of the particle beam was the largest contributor to the ~ 26 keV X-ray beam image's transverse dimensions. The other contributions are the beam transverse size and the undulator radiation cone. The beta functions were determined independently, and the analysis of the data is similar to that done by Tarazona and Ellaume [4].

3. Experimental results

In this section we discuss results on beam transverse size, bunch length, and beam divergence measurements whose values are relevant to future diffraction-limited sources.

Initial reports of our first direct measurements of the beam transverse size and beam bunch length using a unique synchroscan and a dual-sweep X-ray streak camera have been reported in Ref. [5]. However, the relevance to future fourth-generation light sources based on an FEL mechanism warrants reporting directly to this community as well. In Fig. 2, the focus mode of the X-ray camera is used to display (a) the Hg light (UV) profile, (b) the pinhole X-rays illuminating the Au PC, and (c) the image from the "off-axis" photo cathode profile. The respective vertical full-width at half-maximum (FWHM) sizes are 4.5, 9.1, and 22.5 channels (ch). The conversion factor in transverse dimensions is

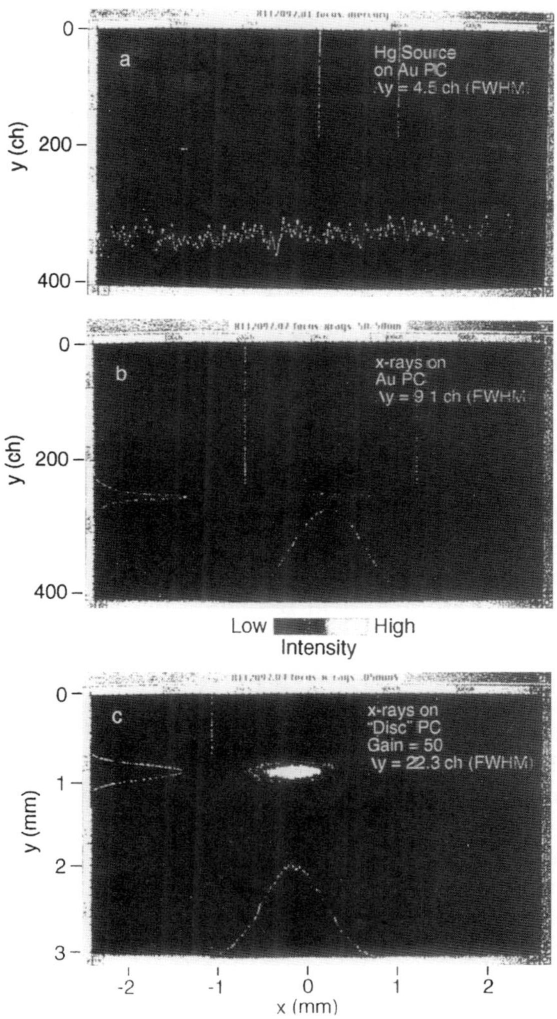

Fig. 2. Focus mode images from the X-ray streak camera for (a) Hg source illumination, (b) X-rays on the Au PC, and (c) X-rays on the "off-axis disk PC".

Table 2
Effects of pinhole aperture size on the observed vertical profile size

Vertical aperture height[a] (μm)	Vertical profile FWHM (μm)	Observed vertical profile σ_{est} (μm)	Actual vertical profile σ_{est} (μm)[b]
20	102	43	40.3
30	104	44.3	40.3
40	109	46.2	40.7
60	124	53	43.6

[a]The horizontal aperture size was held constant at 20 μm. The MCP gain was reduced as needed to compensate for the increased signal strength.
[b]A camera contribution of about $\sigma_{cam} = 12$ μm to the limiting resolution was estimated, which was subtracted in quadrature with the aperture contribution.

about 8 μm/ch, and in the temporal domain on the fastest range it is 0.3 ps/ch. The first two values illustrate the effects of the larger photo-electron energy spread on the static spread function for X-rays (Fig. 2b) compared to UV light (Fig. 2a). The third value is attributed to imaging the full vertical profile of $\sigma_y \sim 45$ μm at the source. We postulate that the metal disk that holds the Au slit PC acted as a PC, albeit with 100 times less sensitivity than the Au. Its vertical dimension is sufficient to detect the full image. The aperture was at 50×50 μm^2.

More recent data have been taken that address the spatial resolution limits by tracking the observed size as one varied the effective pinhole aperture from 20×20 to 20×60 μm^2. The contribution to the vertical system resolution by the aperture size was thus qualitatively assessed as shown in Table 2. By estimating the aperture's rms contribution to the observed size, and by subtracting both it and the estimated camera limiting resolution of 12 μm in quadrature, the actual beam size of about $\sigma_y = 40$ μm was determined. As shown in column 4, the actual beam size is the same after addressing these contributions to resolution.

In Fig. 3 we show an example of (a) a focus mode image and (b) a synchroscan image with observed bunch length of 65 (FWHM) or 28 ps ($\sigma_{est.}$). The pinhole aperture was set at 100 (H) $\times$ 100 μm (V) with a magnification of about 1.6:1 for these data. On this streak deflection range the camera resolution is determined mostly from the focus mode image size to be $\sigma_{res} \sim 4$ ps. The fastest range available is about four times faster and implies 1–1.5 ps (σ) resolution. This same tube was previously tested with a UV laser at 2480 Å and $\sigma_{res} \cong 0.6$ ps was determined [6]. These two resolution numbers for

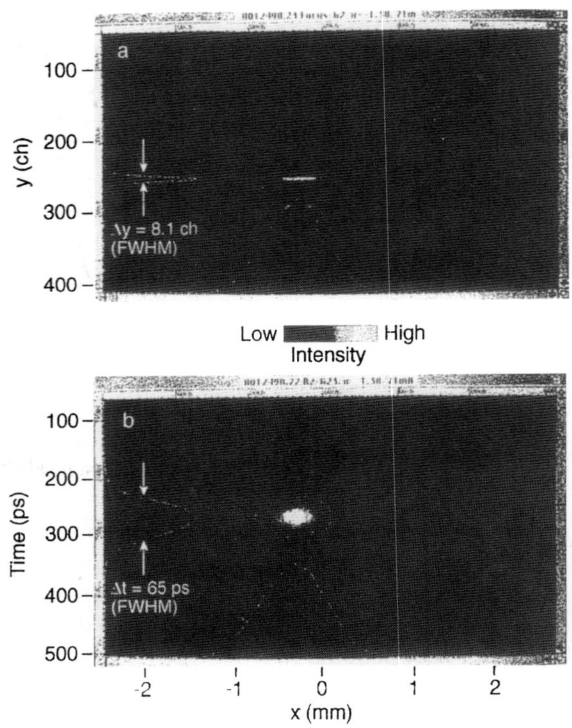

Fig. 3. X-ray streak camera images of the X-ray pinhole field in (a) focus mode and (b) synchroscan mode. The latter shows the time profile of 65 (FWHM) or 28 ps (σ) on the vertical display axis.

the fastest sweep range bound the expected resolution performance for incident photon wavelengths from 2480 to ~ 1 Å. This trend reflects the increasing contribution of the photo-electron energy spread with the higher energy incident photons. Streak images were also taken with the $10 \times 10\ \mu m^2$ aperture at 80-mA stored beam, so this implies that with the aperture as in Fig. 3, a circulating single bunch involving current < 1 mA could be measured although single turn data from the dipole source seem unlikely. The dual-sweep feature was also demonstrated and data are reported in Ref. [5].

Analysis of the projected beam image from the diagnostics undulator in Fig. 4 leads to an implied vertical divergence of only 3.3 μrad with the

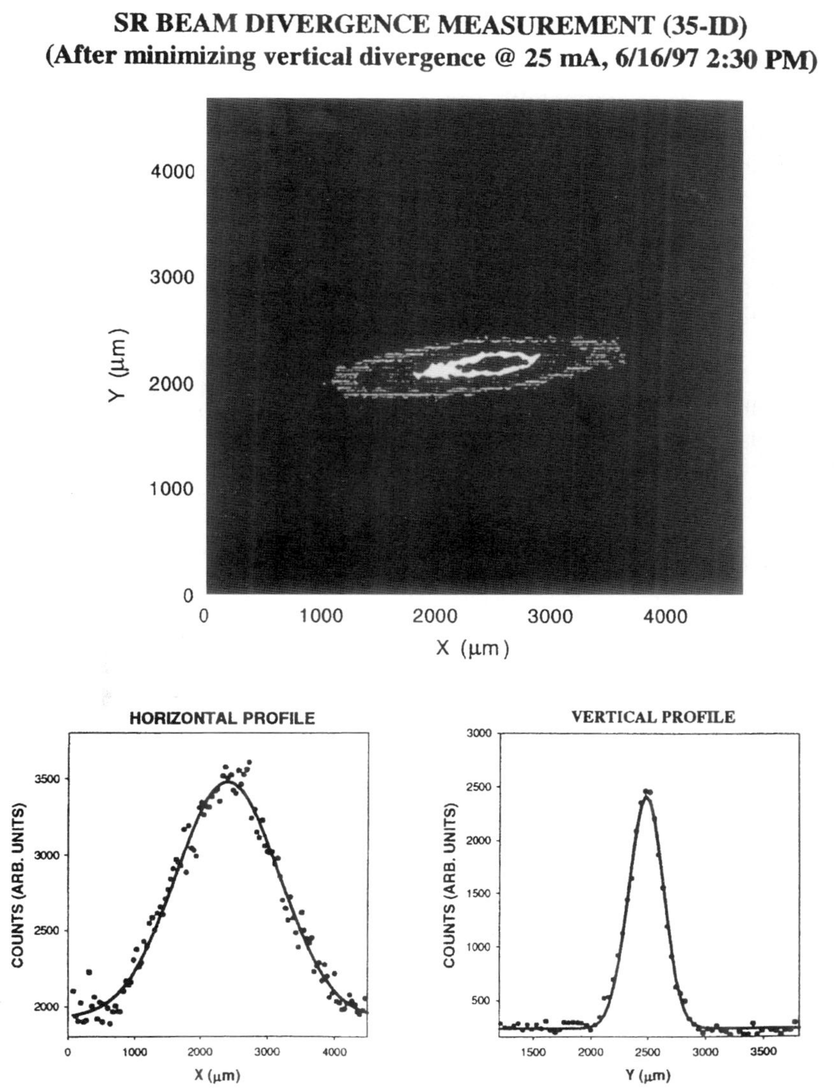

Fig. 4. Positron beam divergence data from the monochromator on the diagnostics undulator. The image (top) and the x- and y-profiles (bottom) imply $\sigma_{x'} = 22$ μrad and $\sigma_{y'} = 3.3$ μrad.

horizontal divergence of 22 μrad. The vertical coupling was evaluated at 1.6% with a total natural emittance of 7.1 ± 0.9 nmrad. The 3.3-μrad value is one of the lowest divergences directly measured to date on a storage ring. If this diagnostics undulator were used with a 15-GeV beam, the fundamental would be at 110 keV with an on-axis cone angle of about 1.2 μrad. It thus seems realistic to propose that one leg on the linac-based FEL array of beamlines should have such a diagnostics undulator. It is noted that these measurements were done with a 30-ms integration time in the CCD camera corresponding to an integrated charge of 2.4 mC. However, the ID gap was at 34 mm resulting in only 0.2 W in the fundamental. Scaling issues are readily addressed by closing the gap to 14 mm (6 mm) to gain a factor of 10^3 (10^4) increase in power for the source, using cryocooled or image-intensified CCDs to gain 10^2–10^3 in sensitivity, and using a low horizontal emittance as well. The combined factors of 10^6–10^7 in signal and detector options imply sensitivity to a single pulse (train) with $Q \leq 1$ nC total charge. This value is close to nominal fourth-generation light source (4GLS) designs. Further studies are planned with an in-vacuum monochromator recently installed in the beamline and eventually with a cryocooled monochromator.

4. Summary

In summary, we have used low vertical coupling in the APS storage ring and a unique set of diagnostics devices to generate and measure some beam parameters near prototypical of a diffraction-limited hard X-ray source or 4GLS. The unique X-ray streak camera and the diagnostics undulator are key, scalable devices that may ultimately be in the complement of diagnostics for an eventual next-generation light source facility.

Acknowledgements

The authors acknowledge the support of John Galayda and Glenn Decker of the APS Accelerator Systems Division, interactions with W. Cieslik of Hamamatsu Photonic Systems on the X-ray streak camera, and discussions with K. Robinson and D. Quimby of STI Optronics, Inc. on the design of the diagnostic undulator.

References

[1] A.H. Lumpkin, Nucl. Instr. and Meth. A 393 (1997) 147.
[2] Hamamatsu Photonics, C5680 Universal Streak Camera Data Sheet, May 1993.
[3] B.X. Yang, A.H. Lumpkin, G.A. Goeppner, S. Sharma, E. Rotela, A.C. Sheng, E. Moog, Status of the APS diagnostics undulator beamline, Proceedings of the 1997 Particle Accelerator Conference, May 12–16, 1997, Vancouver, BC, Canada, submitted for publication.
[4] E. Tarazona, P. Ellaume, Rev. Sci. Instr. 66 (1995) 2.
[5] A.H. Lumpkin, B.X. Yang, First multi-GeV particle-beam measurements using a synchroscan and dual-sweep X-ray streak camera, Proceedings of BIW'98, SLAC, May 4–7, 1998, AIP Conference Proc. No. 451, 1998, p. 214.
[6] A. Lumpkin, B. Yang, W. Gai, W. Cieslik, Initial tests of the dual-sweep streak camera system planned for APS particle-beam diagnostics, Proceedings of the 1995 PAC, vol. 4, Dallas, Texas, May 1–5, 1995 (1996) p. 2476.

ELSEVIER

Nuclear Instruments and Methods in Physics Research A 429 (1999) 299–303

NUCLEAR INSTRUMENTS & METHODS IN PHYSICS RESEARCH Section A

www.elsevier.nl/locate/nima

Simulation and experimental study on tungsten field emission cathode

Yoshiaki Tsunawaki[a,*], Yoshiko Tokura[a,1], Mitsuhiro Kusaba[a], Nobuhisa Ohigashi[b], Kunioki Mima[c], Masayuki Fujita[d], Kazuo Imasaki[d], Sadao Nakai[c], Makoto Shiho[e]

[a]*Department of Electrical Engineering and Electronics, Osaka Sangyo University, 3-chome, Naka-gaito, Daito, Osaka 574-8530, Japan*
[b]*Department of Physics, Kansai University, Yamate-cho, Suita, Osaka 564-8680, Japan*
[c]*Institute of Laser Engineering, Osaka University, Yamada-oka, Suita, Osaka 565-0871, Japan*
[d]*Institute for Laser Technology, Yamada-oka, Suita, Osaka 565-0871, Japan*
[e]*Fusion Research Establishment, Japan Atomic Energy Research Institute, Tokai-mura, Naka, Ibaraki 319-1195, Japan*

Abstract

We have been developing a tungsten field emission cathode for a Cherenkov FEL or Smith–Purcell FEL. The properties of the electron beam from a tungsten tip have been investigated using numerical simulation and some experiments. Although the 2nd gate reduced the emission current due to a decrease of the effective field at the tungsten tip, it focused and enlarged the electron beam current in a certain potential range of the 2nd gate depending on the cathode voltage and the distance between 1st and 2nd gates. The electron beam diameter was reduced to $\sim$ 2 mm on the fluorescent screen at the anode.

Keywords: Field emission; Tungsten cathode; Electron beam

1. Introduction

Since the 1930s, field emission from a very fine needle has been researched to study the phenomenon itself, the surface structure/phenomena of material, and to apply it as an electron beam source [1]. Recent studies give the expectation that it would be a good electron beam source for a free electron laser (FEL) [2,3], gyrotron [4], monotron [5], etc. because of its very small energy spread and very high brightness. Some researchers have investigated semiconductor field emission arrays for an FEL beam source by numerical simulation [6,7] and experiment [8,9]. We have also been studying a field emission cathode of tungsten (W) wire for a Cherenkov or Smith–Purcell FEL which are expected to be sufficiently small to be used in a typical laboratory unscreened from X-ray radiation [10,11].

* Corresponding author. Tel.: + 81-720-75-3001; fax: + 81-720-70-8189.

E-mail address: ytsuna@elec.osaka-sandai.ac.jp (Y. Tsunawaki)

[1] Present address: Tabai Espec Co. Ltd., 5-6-3, 3-chome, Tenjimbashi, Kita, Osaka 530-8550, Japan

PII: S0168-9002(99)00132-1

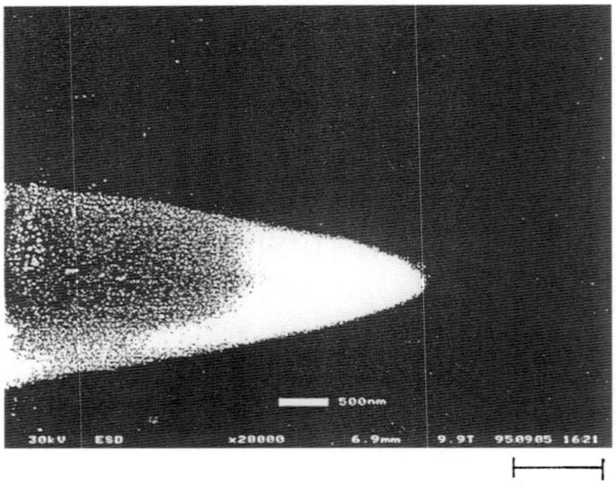

Fig. 1. SEM micrograph of a tungsten tip.

In this study, we have investigated the detailed properties of the field emission current from a tungsten cathode and its dependence on the voltage of electrodes by means of EGUN simulation [12] and experiment whose elements consisted of a W needle, two electrodes and an anode. The existence of the optimum 2nd electrode potential was found to depend especially on the cathode voltage and the distance between 1st and 2nd electrodes.

2. Simulation and experiment

The tungsten needle was made of a normal W wire with a diameter of 100 μm by an electrolytic etching method in an electrolyte of KOH solution. Fig. 1 shows the SEM micrograph of the etched W tip with a curvature of 200 –300 nm.

The field emission experiment was carried out in a chamber (inner diameter: 146 mm, height: 170 mm) evacuated to 6.5×10^{-9} Torr. While the electron beam was being emitted, the pressure of the chamber was $\sim 3 \times 10^{-8}$ Torr. The tip was set on a stem which is used for an electron microscope filament, because it was aged for about 15 min at $\sim 500°C$. The apparatus contains, besides the W tip cathode, the first electrode (1st gate) to draw a field emission, the second electrode (2nd gate) to focus the electron beam and a Faraday cup (anode) to collect the electrons. The 1st gate was grounded. The apertures of the molybdenum gates are 1 mm and their widths are 0.3 mm. The distance between the 1st gate and the anode is about 30 mm. The tip was set in such a way that its end was inside the aperture of the 1st gate. The distance between 1st and 2nd gates was adjusted by ceramics with a proper thickness.

The simulation was performed after the procedures of the simulation were verified to be accurate by comparison with the Fowler–Nordheim (F–N) relation between 1st gate voltage and anode current for a system without 2nd gate. As is well known, the whole system cannot be simulated by one EGUN procedure in order to estimate the detailed field distribution around the cathode tip. Therefore, a factor of 10 enlargement was repeated 5 times in this work. According to the experiment mentioned above, it was assumed that curvature of a cathode tip was 250 nm and the distance between 1st gate and anode was 31.4 mm. However, the radial distance was only 1 or 2 mm because the number of meshes for calculation was limited. We did not simulate at the cathode voltage so high that space-charge effect had to be considered, because such a high voltage could not be applied to the cathode in the experiment due to formation of a discharge.

3. Results and discussion

3.1. Simulation

In order to calculate the electron beam current, we must decide the positions on the surface of W tip where electrons are emitted. The extreme position in them was assumed to be the point where the current reduces to one-tenth of that at the center of the tip surface. It was, furthermore, assumed that electrons are emitted from 60 points on the surface and the distances between neighboring points of them are equal.

With no 2nd gate it was observed as expected that the electron beam diverged and the current increased with the cathode voltage in accordance with the F–N equation. The addition of the 2nd

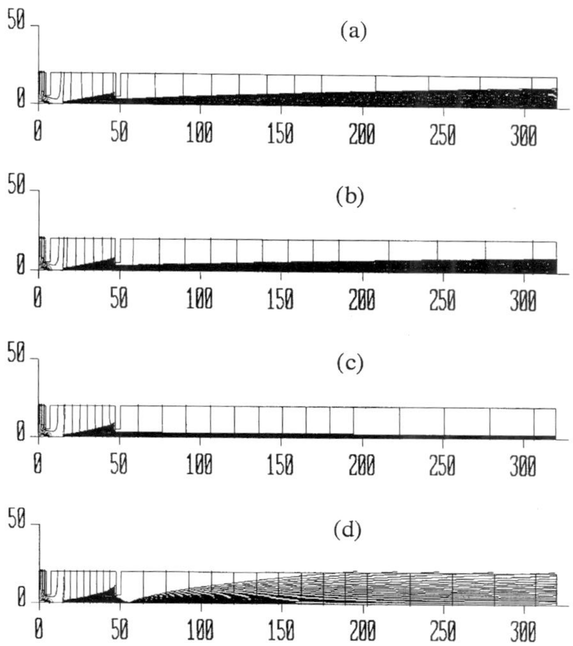

Fig. 2. Electron beam trajectories depending on the 2nd gate voltage for a cathode voltage of − 4 kV and 2nd gate voltage of (a) − 2.5, (b) − 3, (c) − 3.5 and (d) − 4 kV.

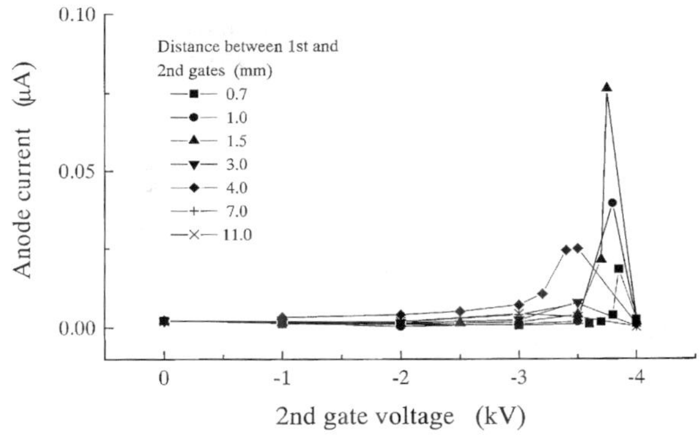

Fig. 3. Anode current depending on the 2nd gate voltage when cathode and anode voltages are − 4 and 4 kV, respectively.

gate with negative voltage made the electron beam focus. However, it diverged after converging if the 2nd gate voltage was around same as that of cathode.

Fig. 2 shows, for example, the electron beam trajectories dependence on the 2nd gate voltage under the condition that the distance between the 1st and 2nd gates (Δl) is 4 mm and the voltages of cathode and anode are − 4 and 4 kV, respectively. It is seen that the focusing of the electron beam is controlled only by a small change of the 2nd gate voltage, which is in a certain narrow range. Fig. 3 shows the current at the anode with diameter of 1 mm depending on the 2nd gate voltage for various distances of Δl under the voltages of cathode and anode same as mentioned above. The decrease of the 2nd gate potential gives the following tendencies; (1) at first, the anode current decreases due to the reduction of effective field at the tip, and then (2) it increases to a peak because of the enhancement of focusing power of 2nd gate, and after that (3) it drops abruptly because the electron beam diverges with crossing trajectories due to overfocussing by the 2nd gate as shown in Fig. 2(d). After the peak current increases with increase of Δl, it reduces due to the increase of electrons interrupted by the 2nd gate surface. It was furthermore seen that the addition of higher voltage to the anode enhanced the anode current because of a focusing effect. Similar results to those mentioned above were also obtained at higher cathode voltage.

3.2. Experiment

With no 2nd gate, the anode current increased with both cathode and anode voltages. The dependence of anode current on cathode voltage satisfied the F–N relation within an error of ± 1.5%. The maximum current was ~ 20 μA when the voltages at cathode and anode were − 6 and 50 kV, respectively. These phenomena are qualitatively in good agreement with the results obtained from the simulation.

The addition of the 2nd gate decreased the anode current by one half to one-fifth lower than the value expected from the simulation. The experimental results for Δl of 4 mm as a typical example are shown in Figs. 4–6. Fig. 4 shows the anode current dependence on the cathode voltage when the 2nd gate voltage is − 4 kV. The cathode voltage threshold is observed around − 3.5 kV for every anode voltage. However, when the anode voltage was

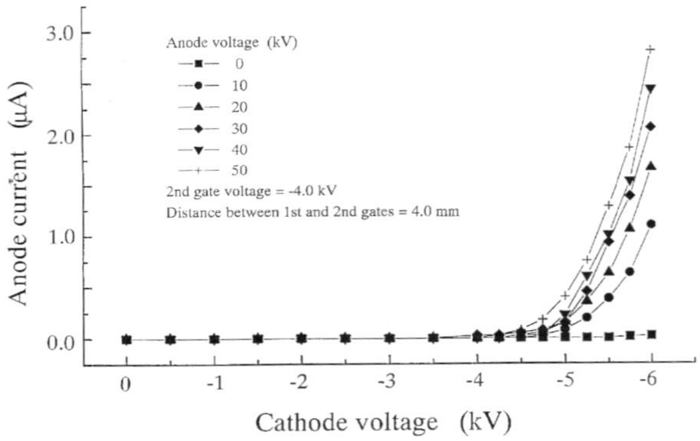

Fig. 4. Experimental anode current depending on the cathode voltage, when the 2nd gate voltage is -4 kV and the distance between 1st and 2nd gates is 4 mm.

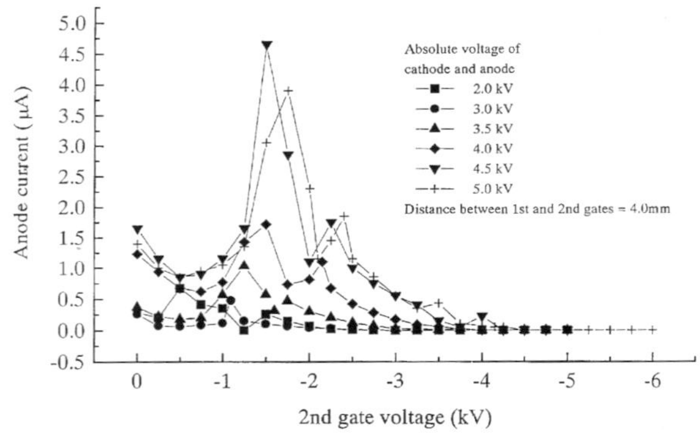

Fig. 5. Experimental anode current depending on the 2nd gate voltage, when the distance between the 1st and 2nd gates is 4 mm.

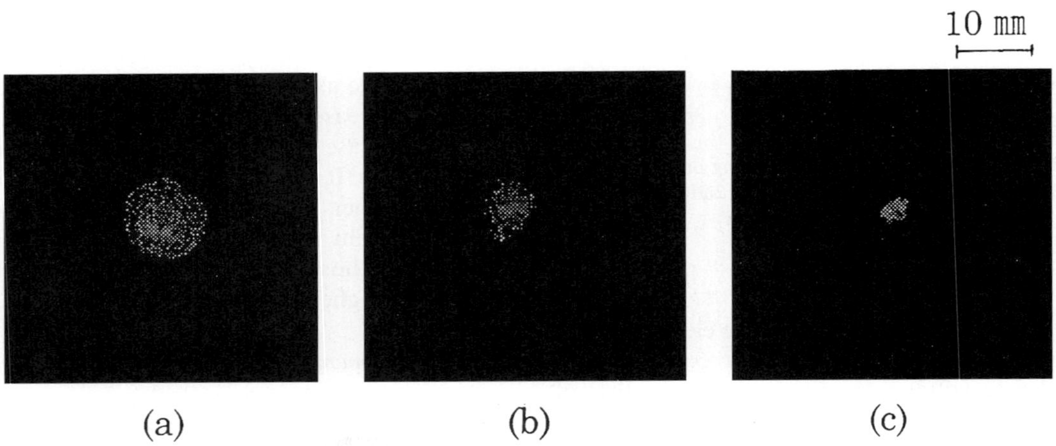

Fig. 6. Beam images for 2nd gate voltage of (a) 0, (b) -1 and (c) -2 kV, respectively, under the condition that the absolute voltages of cathode and anode are 5 kV.

0 kV, noticeable increment of the current was not observed because most of the emitted electrons flew onto the 1st and 2nd gates. Although the anode current generally does not depend on the anode voltage, it is seen that increment ratio of the current to the anode voltage is larger as anode voltage is higher. This phenomenon seems to be attributed to some focusing effect of the anode as mentioned in the simulation.

Fig. 5 shows the dependence of anode current on the 2nd gate voltage under the absolute potential of 4 kV for both cathode and anode. It is seen that the anode current decreases, on the whole, with decreasing negative voltage of the 2nd gate, and it builds up in a certain voltage range of the 2nd gate depending on the cathode potential and the distance of Δl. The result with cathode voltage of -4 kV corresponds to the simulation shown in Fig. 3. Such variation of the current is qualitatively explained by the results of the simulation. However, it is seen that the 2nd gate voltage giving the peak current is higher than that of the simulation.

ELSEVIER Nuclear Instruments and Methods in Physics Research A 429 (1999) 304–309

NUCLEAR INSTRUMENTS & METHODS IN PHYSICS RESEARCH Section A

www.elsevier.nl/locate/nima

Enhancement of an electron beam buncher for a CW FEM

A.I. Al-Shamma'a*, A. Shaw, R.A. Stuart, J. Lucas

The University of Liverpool, Department of Electrical Engineering and Electronics, Brownlow Hill, Liverpool L69 3GJ, UK

Abstract

We present the design, construction and experimental results of an X-band waveguide electron beam buncher. The effect of the voltages across the buncher on the output power is reported. The electron beam used in this experiment is produced by thermonic electron gun which was successfully operated up to 60 kV and 5 mA. The nominal beam diameter was 1 mm passing through an X-band cavity resonator and guided by magnetic field of 170 G.

1. Introduction

At the university of Liverpool we have developed several pre-punched FEMs [1–3] operating at voltages ≥ 80 kV and frequencies ranging from 8.17 to 10.2 GHz. These devices produced results, which demonstrated coherent emission and gain with a beam current ≥ 5 mA. The buncher section in these FEMs is part of the high voltage (50 kV) section and is followed by a drift distance of 65 cm in order to achieve fully bunched electrons. In this paper we introduce further developments with the buncher to maximise the power coupled out.

2. FEM system arrangements

The FEM system, shown in Fig. 1, consists of four main components: the electron gun, the microwave buncher, the wiggler magnet, and the cavity resonator. These components can be described briefly as follows:

2.1. Electron gun characteristics

A triode electron gun arrangement has been used, which incorporates a cathode, an anode, and a grid [4]. the application of a negative biasing voltage to the grid electrode is used to effectively define the cathode surface emitting area. Therefore, the grid behaves like a variable beam aperture. The cathode is designed from different sources including tungsten (≤ 3 mA), thoriated tungsten (≤ 5 mA), a single LaB_6 crystal (≤ 20 mA), and an oxide coated filament. The latter is under development for higher current emission > 200 mA. These sources were capable of delivering a compressed beam with a nominal 2 mm diameter, for fundamental research applications. The gun is capable of operating at voltages up to 100 kV.

2.2. Microwave buncher cavity

The buncher section is placed directly after the gun and about 5 mm from the anode aperture.

* Corresponding author.

0168-9002/99/$ - see front matter
PII: S0168-9002(99)00157-6

The electron beam image at the anode was observed using a fluorescent screen and a CCD camera as shown in Fig. 6 where cathode and anode voltages were -5 and 5 kV, respectively, and the potential of the 2nd gate was changed to be 0, -1 and -2 kV. It was observed that the electron beam converged to the diameter of 11 to ~ 2 mm corresponding to the decrease of the 2nd gate voltage and diminished at its voltage lower than -2.5 kV. The brightness of the screen was greatest at the 2nd gate voltage around -2 kV as expected from the simulation and observation of current.

Thus, though the simulation might qualitatively explain these experimental results, further quantitative explanation seems to require the enlargement of the radial distance in the simulation model to approximate the experimental dimension.

4. Summary

Field emission from a tungsten cathode has been simulated and its experiment has been carried out using a system consisting of W tip, 1st and 2nd gates and anode.

When there was no 2nd gate, the current increased up to ~ 20 μA with the absolute voltage of cathode according to the Fowler–Nordheim equation. The addition of the 2nd gate decreased the current to one half to one-fifth lower than the value expected from the simulation. However, it focused and enlarged the current of the electron beam in a certain voltage range of 2nd gate depending on the cathode potential and the distance between 1st and 2nd gates. The reduction of the electron beam diameter to ~ 2 mm was also observed by a fluorescent screen. Detailed simulations supported these phenomena qualitatively.

Acknowledgements

The authors would like to thank Prof. M. Boussoukaya of C.E. Saclay for his significant suggestion. This work is supported in part by a grant-in-aid for Scientific Research from the Ministry of Education, Science, Sports and Culture of Japan.

References

[1] W.P. Dyke, W.W. Dolan, Adv. Electron. Electron Phys. 8 (1956) 90.
[2] C.A. Brau, Nucl. Instr. and Meth. A 393 (1997) 426.
[3] C.A. Brau, Nucl. Instr. and Meth. A 407 (1998) 1.
[4] M. Garven, SN Spark and ADR Phelps, Conference Digest of 19th International Conference on Infrared and Millimeter Waves, Sendai, Japan, 1994, p. 472.
[5] K. Yokoo, T. Ishihara, Int. J. Infrared Millimeter Waves 18 (1997) 1151.
[6] C.M. Tang, A.C. Ting, T. Swyden, Nucl. Instr. and Meth. A 318 (1992) 353.
[7] C.M. Tang, M. Goldstein, T.A. Swyden, J.E. Walsh, Nucl. Instr. and Meth. A 358 (1995) 7.
[8] H. Ishizuka, S. Kawasaki, H. Kubo, A. Watanabe, M. Shiho, Jpn. Appl. Phys. 35 (1996) 5471.
[9] H. Ishizuka, S. Kawasaki, M. Arai, H. Shimawaki, K. Yokoo, H. Kubo, A. Watanabe, M. Shiho, J. Itoh, Nuc. Instr. and Meth. A 393 (1997) 479.
[10] K. Mima, T. Taguchi, N. Ohigashi, Y. Tsunawaki, M. Shiho, S. Kuruma, M. Fujita, K. Imasaki, C. Yamanaka, S. Nakai, Nucl. Instr. and Meth. A 341 (1994) ABS103.
[11] Y. Tokura, Y. Tsunawaki, M. Kusaba, N. Ohigashi, K. Mima, M. Fujita, K. Imasaki, M. Shiho, Proceedings of the 19th International FEL Conference, Beijing, China, 1997, p. II-101.
[12] W.B. Herrmannsfeld, R. Becker, I. Brodie, A. Rosengreen, C.A. Spindt, Nucl. Instr. and Meth. A 298 (1990) 39.

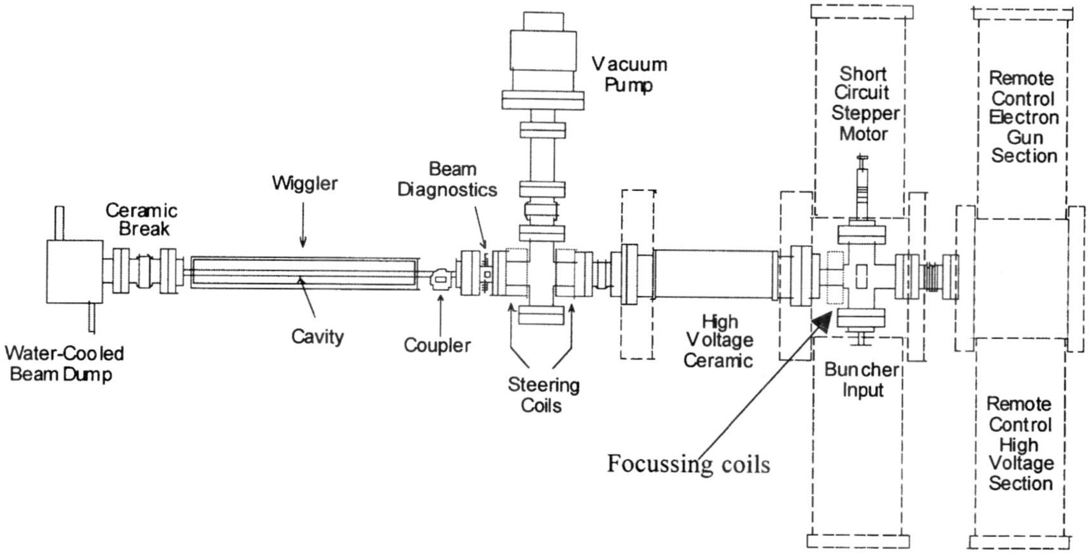

Fig. 1. Compact FEM system for industrial applications.

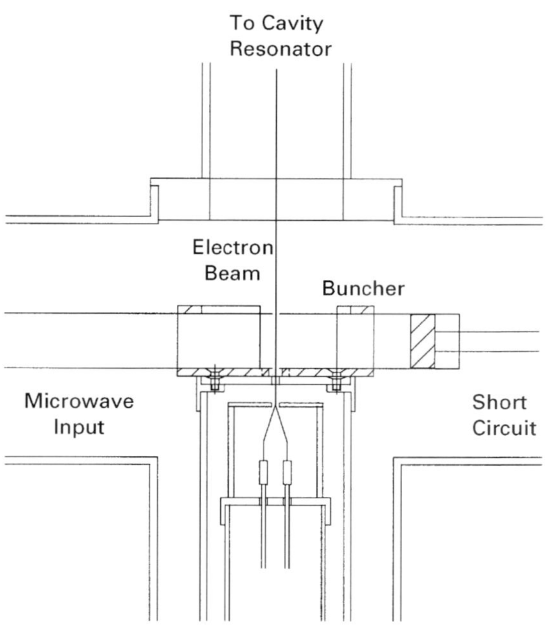

Fig. 2. Microwave buncher cavity.

This buncher is designed using a standard X-band waveguide, WG16/WR90, with internal dimensions 22.86 mm and 10.16 mm, as shown in Fig. 2. Microwave power (up to 1W) is fed into one end of the buncher; the other end is terminated by a tuneable short circuit. The short circuit section is driven by a stepper motor and can tune for frequencies from 8 to 9.3 GHz. The frequency limitation is due to the length of the buncher, which is about 50 mm. Two 3 mm diameter apertures allow the electron beam to pass through the narrow dimension of the buncher. To maximize the buncher electrical field, an inductive window with a 5 mm aperture was placed 20 mm from the electron beam apertures. The buncher cavity resonances have been measured with an average of Q factor of 6000.

2.3. Wiggler magnet

The efficiency and the gain of an FEM system are largely dependent on the wiggler design parameters such as the wiggler period, the magnetic field, and the number of periods. In a specific design, it is desirable for the wiggler to provide a magnetic field, which has uniform amplitude oscillations over its length. The wiggler magnet is designed from NdFeB magnet bars having a remnant field of 1.1 T and consists of 25 periods. The 19 mm wiggler period is formed by two 8 mm wide magnet bars with 1.5 mm spacing. Two magnet bars per period are used to satisfy the synchronous condition of an

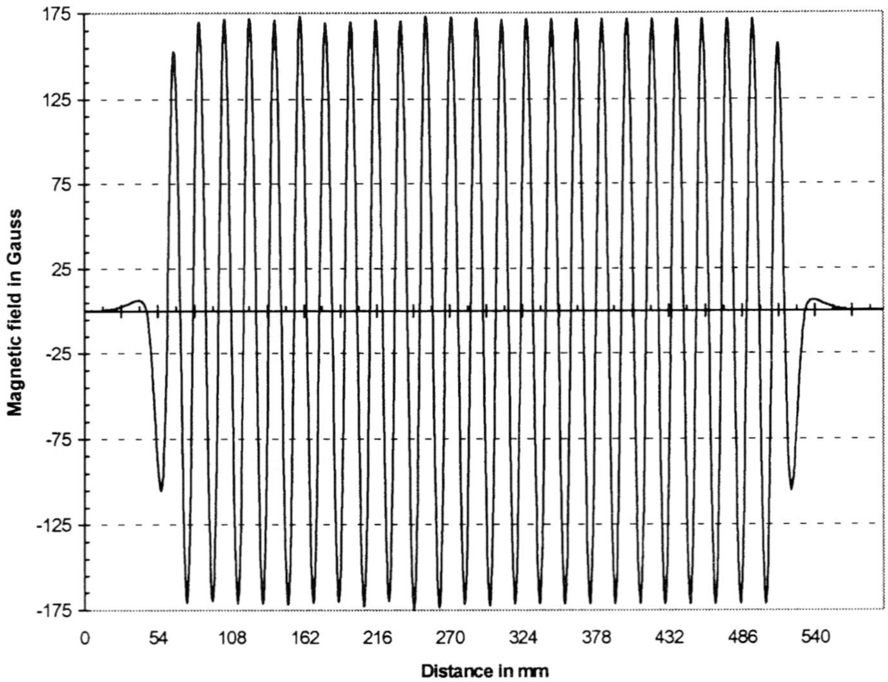

Fig. 3. Wiggler magnetic field distribution.

FEM operating at voltages > 80 kV. The average magnetic field, measured at the centre of the 25 mm wiggler gap, is 170 G, as shown in Fig. 3.

2.4. Wiggler cavity resonator

Standard X-band copper rectangular waveguide has been used to construct the wiggler cavity. A layout of the wiggler cavity is shown in Fig. 4. The total length of the cavity is 60 cm and it is short-circuited at each end by two 7 mm thick irises with a 9 mm diameter aperture. The wiggler cavity resonances formed by these irises are shown in Fig. 5 with an average measured Q factor of 2000. A 24 dB cross coupler was placed at about $1/4\lambda_g$ from the entrance section of the cavity and used for coupling out the required power. At the exit of the cavity the electron beam was collected using either a water-cooled beam dump or by an energy recovery section. Table 1 shows the experimental parameters of the FEM system.

Table 1
FEM system parameters

Operating frequency	8.0–9.3 GHz
Electron beam	
Voltage range	40–76 kV (through wiggler)
Diameter	1 mm
Current	1 mA
Pre-buncher	
Type	Tuneable waveguide cavity
Dimensions	X-band (WG16/WR90)
Frequency	8.0–9.3 GHz
Microwave input power	1 W max.
Wiggler	
Type	PM (NdFeB)
Period	19 mm
Period-to-gap ratio	0.75
Magnetic field on axis	170 G
Number of periods	25
Wiggler waveguide (Standard X-band)	
Internal a-dimension	22.86 mm
Internal b-dimension	10.16 mm
Length	600 mm

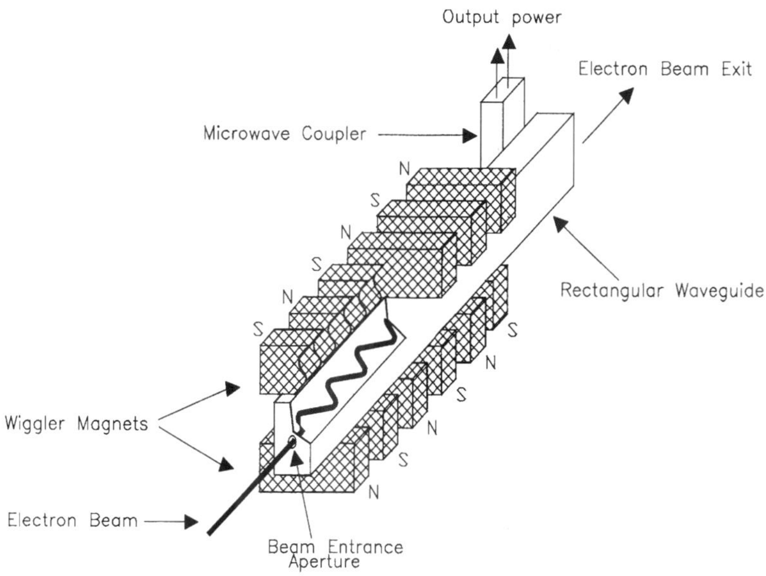

Fig. 4. Schematic diagram of the Wiggler cavity resonator.

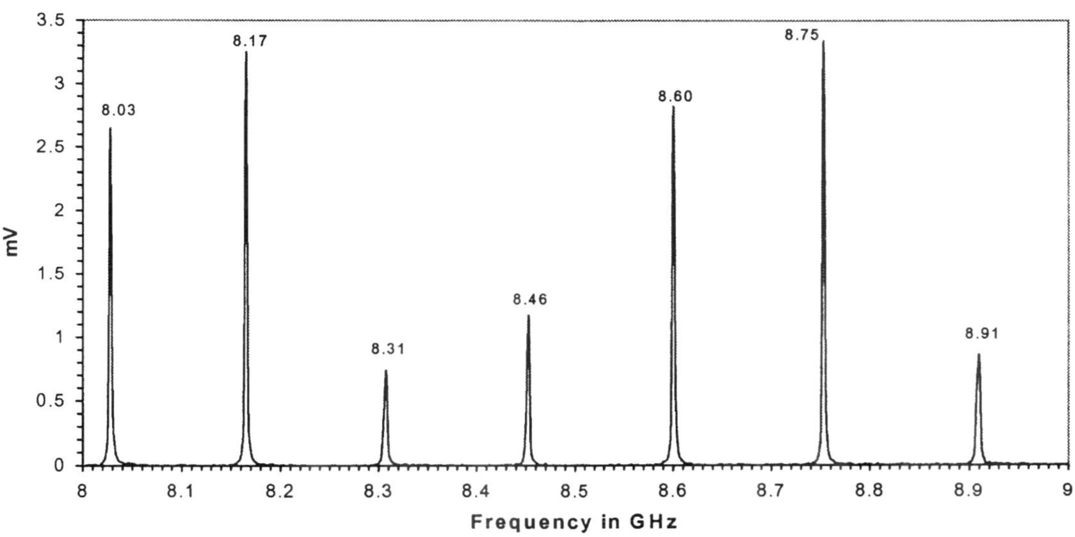

Fig. 5. Wiggler cavity resonances.

3. Experimental results

A series of experiments have been performed and the results are listed below

3.1. Output power versus beam current

The buncher frequency and the wiggler cavity frequency were set at one particular resonance, in

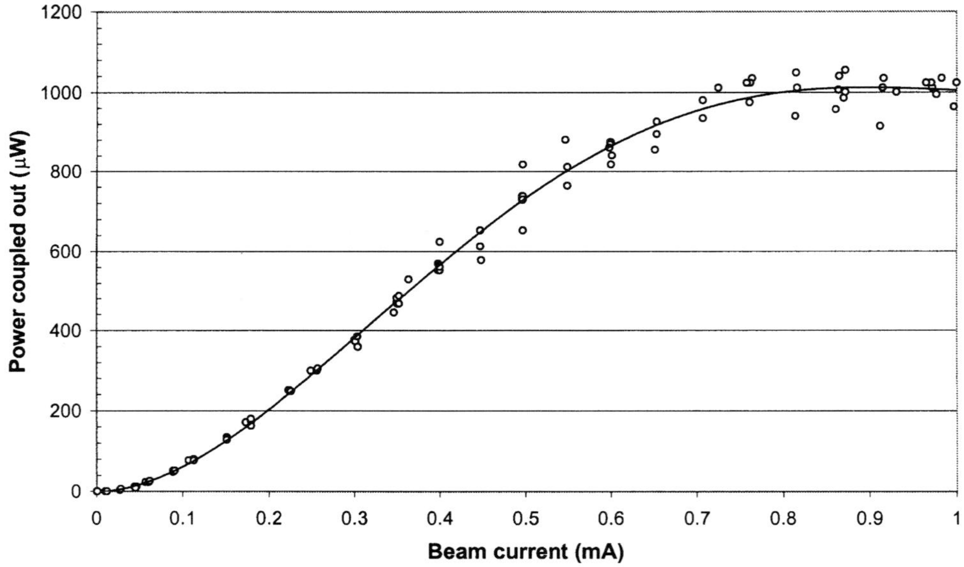

Fig. 6. Power coupled out through a 24 dB coupler versus FEM beam current.

this case 8.6 GHz. The buncher voltage, which could be varied from 5 to 11 kV, was set to 11 kV. The input power of the buncher cavity was 0.3 W. To satisfy the synchronous condition at 8.6 GHz, the operating voltage was adjusted remotely until it achieved the interacting value of 46.8 kV. Fig. 6 shows the relationship between the electron beam current and the power that was coupled out of the 24 dB cross coupler. This indicated that the maximum power to be coupled out was 0.4% of the power inside the wiggler cavity. The solid line in Fig. 6 represents a theoretical model of the 4th order polynomial equation. The electron beam current was limited to 1 mA in order to run the FEM system continuously for a longer time period. With a beam current of 1 mA the coupled power was about 1.1 mW. It was found that if we compared these results with [1], the power coupled out is higher by a factor of 27 and if compared with Refs. [2,3] by a factor of 9.

3.2. *Spontaneous emission (output power versus beam voltage)*

For this experiment, the electron beam was set to 1 mA (nominal) and the voltage across the buncher cavity was varied from 7 to 11 kV. The operating frequency was 8.6 GHz. Fig. 7 shows the spontaneous emission of the FEM system operating at 46.8 kV at different voltages across the microwave buncher. It can be seen that the maximum power, which could be coupled out, was 240, 520 and 1200 μW at 7, 8 and 11 kV, respectively. Further increasing the voltage across the buncher, i.e. exceeding 11 kV, however, resulted in a reduction in the power coupled out. This indicated that at 11 kV the electrons were fully bunched but above 11 kV the electrons were over bunched. Further tests were carried out at 8.17, 8.31, 8.45, 8.75, and 8.89 GHz to prove the tunability of the FEM system.

4. Conclusions and future work

The results presented in this paper show that the power coupled out of the 24 dB coupler is at least 9 times higher than that in the previous systems [1–3]. They are also in good agreement with the theoretical analysis [5], indicating that FEM operation can be easily observed and predicted by simple models at low beam voltages (< 100 kV). The effect of the focusing coils on the output power

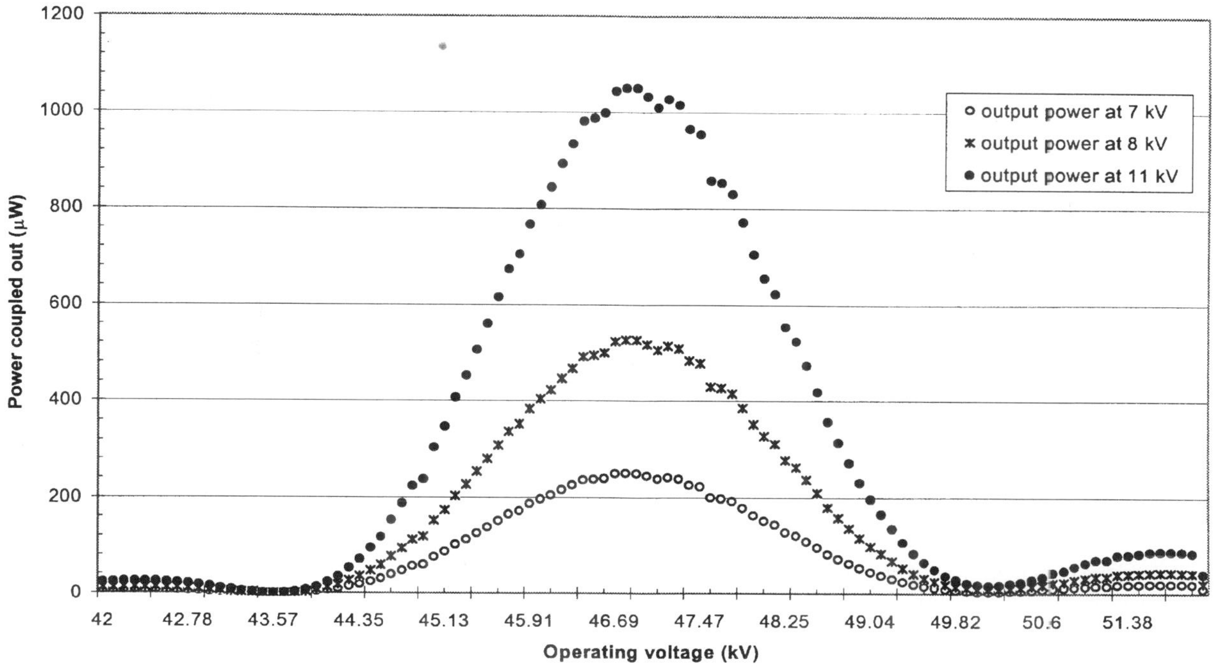

Fig. 7. Spontaneous emission of the FEM system.

was monitored and 91% of beam transportation was achieved. Experiments are in progress to measure the FEM gain at different wiggler cavity resonances and testing will soon begin on a new buncher cavity that has been designed to maximize the bunching factor and hence the power output.

Acknowledgements

The authors would like to thank the EPSRC for their financial support.

References

[1] G. Dearden et al., Nucl. Instr. and Meth. A 318 (1992) 230.
[2] G. Dearden et al., Nucl. Instr. and Meth. A 341 (1994) 80.
[3] A.I. Al-Shamma'a et al., Nucl. Instr. and Meth. A 393 (1997) II 83.
[4] V.E. Cosslett, Oxford University Press, Oxford, 1946.
[5] J. Lucas et al., Proceedings of the 20th Free Electron Laser Conference, Williamsburg, VA, Elsevier Science B.V., Amsterdam, 1999, p. II-95.

ELSEVIER
Nuclear Instruments and Methods in Physics Research A 429 (1999) 310–314

NUCLEAR INSTRUMENTS & METHODS IN PHYSICS RESEARCH Section A

www.elsevier.nl/locate/nima

Self-consistent simulation of the CSR effect on beam emittance☆

R. Li

Thomas Jefferson National Accelerator Facility, 12000 Jefferson Ave., Newport News, VA 23606, USA

Abstract

When a microbunch with high charge traverses a curved trajectory, the curvature-induced Coherent Synchrotron Radiation (CSR) and space-charge forces may cause serious emittance degradation. Earlier analyses based on rigid-line charge model are helpful in understanding the mechanism of this curvature-induced bunch self-interaction. In reality, however, the bunch has finite transverse size and its dynamics respond to the CSR force. In this paper, we present the first self-consistent simulation for the study of the impact of CSR on beam optics. With the bunch represented by a set of macroparticles, the dynamics of the bunch under the influence of the CSR force are simulated, where the CSR force in turn depends on the history of bunch charge distribution and current density in accordance to causality. This simulation is bench-marked with previous analytical results for a rigid-line bunch. The algorithm applied in the simulation will be presented, along with the simulation results obtained for bending systems in the Jefferson Lab FEL lattice.

Keywords: Microbunch; Curved trajectory; Space-charge forces

1. Introduction

When a short bunch with high charge is transported through a magnetic bending system, curvature-induced coherent synchrotron radiation and space-charge forces set a wakefield across the bunch, inducing energy spread and causing emittance growth. This phenomenon has raised considerable concern in the design of free-electron-laser drivers containing bunch-compression chicanes and recirculation arcs. Circumventing this deleterious effect demands a thorough understanding of the physics involved as well as computational tools for the prediction of the CSR effect in lattice designs.

Earlier analyses of the CSR-induced wakefield were largely based on the rigid-line-charge model. By studying the longitudinal CSR wakefield for a periodic circular orbit as well as a transient trajectory both in free space and with shielding [1–5], these analytical works help us to understand the mechanism of the curvature-induced bunch self-interaction. In reality, however, the bunch has finite transverse size and its dynamics responds to the CSR interaction. So, a self-consistent simulation is needed to study the actual dynamical system. The feedback of the CSR-induced wakefields on bunch emittance was first simulated in DESY [3], where the wakefields were computed from the rigid-line-charge model using designed bunch dynamics, and its effect on the bunch emittance degradation was

☆This work was supported by the U.S. Dept. of Energy under Contract No. DE-AC05-84ER40150.

0168-9002/99/$ - see front matter
PII: S0168-9002(99)00135-7

consequently obtained. This scheme works when the CSR-induced change in the bunch dynamics is small.

In this paper we present a self-consistent simulation for the CSR effect on beam dynamics in free space. The dynamics of the bunch, under the influence of the CSR wakefield, is simulated by the motion of a set of macroparticles. This CSR wakefield in turn depends on the history of bunch dynamics in accordance with causality. The simulation is benchmarked against analytical results obtained for a rigid-line charge.

2. Simulation algorithm

2.1. General problem

First we outline the problem to be solved in the simulation. The dynamics for an electron in the bunch is governed by

$$\mathrm{d}(\gamma m_e \dot{\boldsymbol{v}})/\mathrm{d}t = e(\boldsymbol{E} + \boldsymbol{\beta} \times \boldsymbol{B}) \tag{1}$$

with $\boldsymbol{\beta} = \boldsymbol{v}/c$, $\boldsymbol{E} = \boldsymbol{E}^{\text{ext}} + \boldsymbol{E}^{\text{self}}$, and $\boldsymbol{B} = \boldsymbol{B}^{\text{ext}} + \boldsymbol{B}^{\text{self}}$. Here $\boldsymbol{E}^{\text{ext}}$ and $\boldsymbol{B}^{\text{ext}}$ are the external designed electromagnetic (EM) fields, and $\boldsymbol{E}^{\text{self}}$ and $\boldsymbol{B}^{\text{self}}$ are the EM fields from bunch self-interaction, which in turn depends on the history of the bunch charge distribution ρ and current density $\boldsymbol{J}$ via the scalar and vector potentials ϕ and $\boldsymbol{A}$:

$$\boldsymbol{E}^{\text{self}} = -\nabla\phi - \partial\boldsymbol{A}/c\partial t, \qquad \boldsymbol{B}^{\text{self}} = \nabla \times \boldsymbol{A} \tag{2}$$

with

$$\phi(\boldsymbol{r},t) = \int \frac{\mathrm{d}\boldsymbol{r}'}{|\boldsymbol{r}-\boldsymbol{r}'|}\rho(\boldsymbol{r}',t'),$$

$$\boldsymbol{A}(\boldsymbol{r},t) = \frac{1}{c}\int \frac{\mathrm{d}\boldsymbol{r}'}{|\boldsymbol{r}-\boldsymbol{r}'|}\boldsymbol{J}(\boldsymbol{r}',t'). \tag{3}$$

Here the retarded time t' is defined as

$$t' = t - |\boldsymbol{r} - \boldsymbol{r}'|/c. \tag{4}$$

For an ultrarelativistic bunch on a circular orbit, the self EM fields are dominated by CSR fields.

2.2. Macroparticle model

A straightforward way to simulate the bunch self-interaction is to use a macroparticle model.

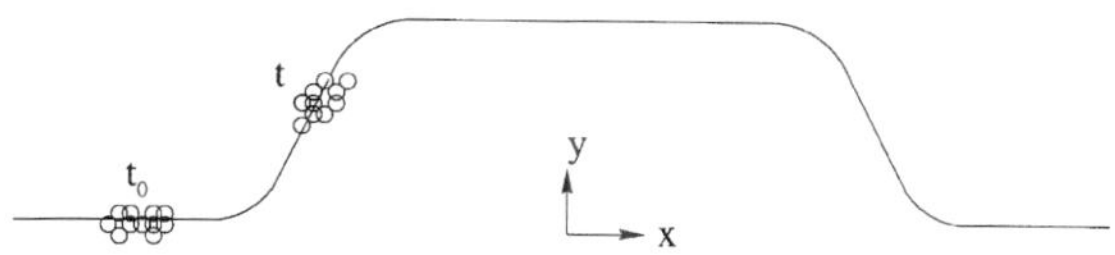

Fig. 1. Macroparticles in longitudinal–horizontal plane.

Since CSR mainly couples the bunch length effect into the bend plane (typically horizontal) dynamics, at this stage we only simulate the longitudinal-horizontal dynamics in the $z = 0$ plane (z stands for vertical offset from design orbit), with the bunch distribution simulated by a set of round Gaussian discs in this longitudinal–horizontal plane with zero vertical extent. An example configuration is illustrated in Fig. 1.

The single macroparticle density distribution at $\boldsymbol{r} = \{x,y\}$ at time t is given by

$$n_{\mathrm{m}}(\boldsymbol{r} - \boldsymbol{r}_0(t)) = \frac{1}{2\pi\sigma_{\mathrm{m}}^2} \mathrm{e}^{-(x-x_0(t))^2 + (y-y_0(t))^2/2\sigma_{\mathrm{m}}^2} \tag{5}$$

where $\boldsymbol{r}_0(t) = \{x_0(t), y_0(t)\}$ is the centroid of the macro-particle at time t, and σ_{m} is its rms size. Knowing the centroid location $\boldsymbol{r}_0^{(i)}(t)$ and velocity $\boldsymbol{\beta}_0^{(i)}(t)$ for each macroparticle, one can construct the charge distribution $\rho(\boldsymbol{r},t)$ and the current density $\boldsymbol{J}(\boldsymbol{r},t)$ of the whole bunch:

$$\rho(\boldsymbol{r},t) = q\sum_{i=1}^{N} n_{\mathrm{m}}(\boldsymbol{r} - \boldsymbol{r}_0^{(i)}(t)) \tag{6}$$

$$\boldsymbol{J}(\boldsymbol{r},t) = q\sum_{i=1}^{N} \boldsymbol{\beta}_0^{(i)}(t) n_{\mathrm{m}}(\boldsymbol{r} - \boldsymbol{r}_0^{(i)}(t)) \tag{7}$$

with i the index of the macroparticle and N the total number of macroparticle in a bunch. For a bunch with total charge Q, the charge per macroparticle is $q = Q/N$.

It can be shown that if the centroids in Eq. (6) have a Gaussian distribution with bunch length σ_{s}, and each macroparticle has a Gaussian distribution as in Eq. (5), then the effective bunch length simulated by Eq. (6) is $\sigma_{\mathrm{eff}} = \sqrt{\sigma_{\mathrm{s}}^2 + \sigma_{\mathrm{m}}^2}$. In general, one should choose $\sigma_{\mathrm{m}} \ll \sigma_{\mathrm{s}}$ in order to have $\sigma_{\mathrm{eff}} \simeq \sigma_{\mathrm{s}}$. The number of macroparticle N should be chosen to ensure the overlap of macroparticles for the suppression of shot noise.

2.3. CSR Wakefields in macroparticle model

The computation of the CSR wakefield $\boldsymbol{E}^{\text{self}}$ and $\boldsymbol{B}^{\text{self}}$ in Eq. (1) is the core of the simulation. After applying the macroparticle model as described in Eqs. (6) and (7) to ϕ and $\boldsymbol{A}$ in Eq. (3), one differentiates over the potentials and gets from Eq. (2) the CSR wakefield on $(\boldsymbol{r},t)$ generated by the whole bunch, which is the superposition of contributions from individual macroparticles:

$$\boldsymbol{E}^{\text{self}}(\boldsymbol{r},t) = \sum_{i=1}^{N} \boldsymbol{E}^{(i)}(\boldsymbol{r},t)$$

$$B_z^{\text{self}}(\boldsymbol{r},t) = \sum_{i=1}^{N} B_z^{(i)}(\boldsymbol{r},t) \tag{8}$$

where the single-particle-generated CSR wakefields are 2D integrals over the area surrounding the source macroparticle's previous path:

$$\boldsymbol{E}^{(i)}(\boldsymbol{r},t) = q\int \frac{\mathrm{d}^2\boldsymbol{r}'}{|\boldsymbol{r}-\boldsymbol{r}'|} n_{\mathrm{m}}(\boldsymbol{r}'-\boldsymbol{r}_0^{(i)}(t')) \times \left\{ \frac{\boldsymbol{r}'-\boldsymbol{r}_0^{(i)}(t')}{\sigma_{\mathrm{m}}} - \frac{\dot{\boldsymbol{\beta}}_0^{(i)}(t')}{c}\sigma_{\mathrm{m}} - \boldsymbol{\beta}_0^{(i)}(t') \times \left[\frac{\boldsymbol{r}'-\boldsymbol{r}_0^{(i)}(t')}{\sigma_{\mathrm{m}}} \cdot \boldsymbol{\beta}_0^{(i)}(t') \right] \right\} \tag{9}$$

$$B_z^{(i)}(\boldsymbol{r},t) = q\int \frac{\mathrm{d}^2\boldsymbol{r}'}{|\boldsymbol{r}-\boldsymbol{r}'|} n_{\mathrm{m}}(\boldsymbol{r}'-\boldsymbol{r}_0^{(i)}(t')) \times \left[\boldsymbol{\beta}_0^{(i)}(t') \times \frac{\boldsymbol{r}'-\boldsymbol{r}_0^{(i)}(t')}{\sigma_{\mathrm{m}}} \right] \tag{10}$$

with t' given in Eq. (4). To evaluate the integrand of the above integrals at $\boldsymbol{r}'$, one needs to interpolate the phase parameters $\boldsymbol{r}_0^{(i)}$ and $\boldsymbol{\beta}_0^{(i)}$ for the centroid at retarded time t', using the history of the phase parameters at discrete timesteps $t^{(k)}$ obtained from the leap-frog scheme in Section 2.4.

The 2D integral representation of single-macroparticle-generated CSR wakefields can be understood from the retardation nature of radiation: if we divide the ith macroparticle into grids, the wakefield observed at $(\boldsymbol{r},t)$ is the sum of fields emitted by each grid at its own retarded location and time. In our simulation the integrals in Eqs. (9) and (10) are numerically implemented by applying the 2D Simpson's Rule, in which we divide a neighborhood of the previous path of the bunch into a grid, and sum up the integrand on the grids using appropriate coefficients. This procedure converges fast with respect to the number of grid points. The singularities in Eqs. (9) and (10) are taken care of by a singularity removal technique.

2.4. Leap-frog scheme

With the EM field obtained in Section 2.3, we integrate the equation of motion in Eq. (1) numerically using a leap-frog scheme. In 2D cylindrical coordinates, Eq. (1) is written as

$$\begin{aligned} &(\gamma m_e \dot{r})/\mathrm{d}t - \gamma m_e r\dot{\theta}^2 = e(E_r + r\dot{\theta}B_z/c) \\ &d(\gamma m_e r\dot{\theta})/\mathrm{d}t + \gamma m_e \dot{r}\dot{\theta} = e(E_\theta - \dot{r}B_z/c). \end{aligned} \tag{11}$$

Consider a circular orbit with design radius r_0 and design energy $\gamma_0 m_e c^2$. For a macroparticle with its centroid at (r,θ) and energy $\gamma m_e c^2$, we let $s = r_0\theta$ be its longitudinal coordinate and $x = r - r_0$ be its horizontal offset from the design orbit. With k denoting the index of the timestep, and Δt the increment of time at each time step, the discrete form of Eq. (11) in a leap-frog scheme for the centroid of the macroparticle is then

$$[(\gamma\beta_x)_{k+1} - (\gamma\beta_x)_k]/c\Delta t = [\tilde{E}_x - \beta_s(\gamma\beta_s/r - \gamma_0\beta_0/r_0 + \tilde{B}_z)]_{k+1/2} \tag{12}$$

$$[(\gamma\beta_s)_{k+1} - (\gamma\beta_s)_k]/c\Delta t = [\tilde{E}_s + \beta_x(\gamma\beta_s/r - \gamma_0\beta_0/r_0 + \tilde{B}_z)]_{k+1/2}. \tag{13}$$

Here $(\beta_x,\beta_s) = (\dot{r}, r\dot{\theta})/c$, $\gamma^2 = (\gamma\beta_x)^2 + (\gamma\beta_s)^2 + 1$, and $\tilde{E}_{x,s} = (e/m_e c^2)E_{x,s}^{\text{self}}$, $\tilde{B}_z = (e/m_e c^2)B_z^{\text{self}}$. Since $(\gamma\beta_{x,s})_{k+1/2}$ on the right-hand side of Eqs. (12) and (13) are defined as

$$(\gamma\beta_{x,s})_{k+1/2} = [(\gamma\beta_{x,s})_k + (\gamma\beta_{x,s})_{k+1}]/2 \tag{14}$$

Eqs. (12) and (13) are two coupled nonlinear equations to be solved simultaneously in order to obtain the reduced momentum $(\gamma\beta)_{x,s}$ at each time step. Consequently, the longitudinal–horizontal coordinates for the macroparticle centroids at the next half-time step yield

$$x_{k+3/2} = x_{k+1/2} + (\beta_x)_{k+1} c\Delta t$$

$$s_{k+3/2} = s_{k+1/2} + \left(\frac{r_0 \beta_s}{r}\right)_{k+1} c\Delta t. \quad (15)$$

By setting $r_0 \to \infty$, the above equations naturally reduce to equations in Cartesian coordinates, which are used for the straight sections of a design orbit. Notice that from Eqs. (12)–(14), it is straightforward to derive the energy equation:

$$(\gamma_{k+1} - \gamma_k)/c\Delta t = (\tilde{\boldsymbol{E}} \cdot \boldsymbol{\beta})_{k+1/2}. \quad (16)$$

This implicit energy conservation is the reason we chose the leap-frog scheme for the numerical integration of Eq. (1) in our simulation. Note that at each time step, only the centroids' dynamics gets advanced, and the macroparticles move translationally without rotation about their centroids. Therefore the speed of light is not exceeded in this simulation.

3. Simulation results

3.1. Benchmark

First, we turn off the CSR force by setting $E^{\text{self}}_{x,s} = B^{\text{self}}_z = 0$ in Eqs. (12) and(13) and make sure the single-particle dynamics agrees with results from other optics codes, such as DIMAD in the ultrarelativistic case. Next, we benchmark the simulated CSR wakefield with earlier analytical results for a simple beamline: straight-bend-straight, using LCLS bend parameters [7]: $\theta = 11.4$ deg, $r_0 = 25.13$ m, $\sigma_s = 50$ μm and $Q = 1$ nC. To imitate the rigid-line Gaussian bunch used in the analysis, we let the macroparticle centroids have a Gaussian distribution along the design orbit, and keep the bunch rigid by not responding to the CSR interaction. The simulation results of the CSR wakefield and their comparison with analysis are shown in Fig. 2 for the bunch at steady state. It is clear that the simulation results agree perfectly with the analytical result obtained for the effective bunch length σ_{eff}, which gets closer to the analytical result for actual bunch length σ_s as σ_m gets smaller. Fig. 2 also shows that the radial field $E_x - \beta B_z$ is an order of magnitude smaller than the longitudinal field, as was observed in Refs. [3,6]. The transient longitudinal wakefield at the entrance of a bend also agrees well with analysis.

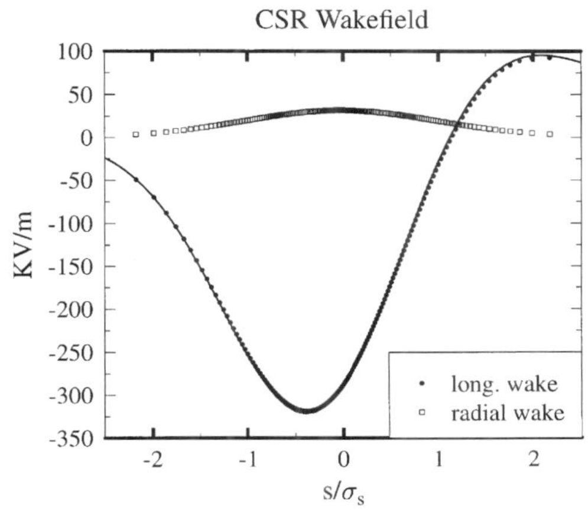

Fig. 2. CSR wakefield on the bunch at steady state; the dots are the simulated wakefields with the rms of the macroparticle centroid distribution being σ_s, and the solid curve is the analytical longitudinal wakefield for the effective bunch length σ_s^{eff}. Here $\sigma_m/\sigma_s = 0.4$.

3.2. Results for JLab FEL chicane

The goal of our simulation is to compute the CSR-induced emittance growth generated from the bending systems in the Jefferson Lab FEL lattice. Of particular interest is the effect of the optical chicane in front of the wiggler. To proceed, we first generate a macroparticle phase space distribution using the design beam parameters upstream of the optical chicane obtained from PARMELA. Then the CSR wakefield is computed at each time step, which is used to advance the macroparticle dynamics to the next time step. The transient radiation interaction after the bunch exits from a bend is also computed. Due to the short drift length between magnets for the JLab FEL chicane, the simulation shows nontrivial coupling between adjacent bends. With all these effects included, our simulation yields a 15% emittance growth in the horizontal phase space as a result of the CSR effect in the chicane. This is to be compared with the JLab FEL experiment. Recently the JLab FEL successfully lased at 150 W infrared average power. More measurements of CSR effect will be carried out in the near future for the comparison of experiment and simulation.

Acknowledgements

The author thanks C.L. Bohn and J.J. Bisognano for many helpful discussions, and acknowledges the information provided by P. Emma for bench-marking, the useful discussions with P. Emma, B. Yunn, D. Douglas and H. Liu, and the continuing technical support from M. Baddourah in NERSC. This work was supported by the U.S. Dept. of Energy under Contract No. DE-AC05-84ER40150.

References

[1] Y.S. Derbenev, J. Rossbach, E.L. Saldin, V.D. Shiltsev, DESY Report No. TESLA-FEL-95-05, 1995.
[2] B. Murphy, S. Krinsky, R.L. Gluckstern, BNL-63090, 1996.
[3] M. Dohlus, T. Limberg, Proceedings of the 1996 FEL Conference, Rome, Italy, 26–31 August 1996. NL-63090, 1996
[4] E.L. Saldin, E.A. Schneidmiller, M.V. Yurkov, DESY-TESLA-FEL-96-14, 1996.
[5] R. Li, C.L. Bohn, J.J. Bisognano, Proceedings of the SPIE, San Diego, 1997.
[6] Y. Derbenev, V. Shiltsev, Fermilab-TM-1974, May 1996.
[7] Private communication with P. Emma of SLAC, 1998.

ELSEVIER

Nuclear Instruments and Methods in Physics Research A 429 (1999) 315–319

NUCLEAR INSTRUMENTS & METHODS IN PHYSICS RESEARCH Section A

www.elsevier.nl/locate/nima

Numerical analysis of shielded coherent radiation and noninertial space-charge force with 3-D particle tracking

Ryoichi Hajima*, Eisuke Ikeno

Department of Quantum Engineering and Systems Science, University of Tokyo, Hongo 7-3-1, Bunkyo-ku, Tokyo 113-8656, Japan

Abstract

A 3-D particle tracking code is developed to estimate emittance growth in bunch-compressors. The simulation code enables one to calculate the electron beam distribution in six-dimensional phase space along a given beam line and estimate emittance growth caused by shielded coherent radiation and noninertial space-charge forces as well as the usual space-charge force and higher-order aberrations. We present some examples of calculations for unshielded and shielded coherent synchrotron radiation, including a comparison with 1-D theoretical prediction. The relation between emittance growth and the complex radiation resistance of conducting walls is also introduced. © 1999 Elsevier Science B.V. All rights reserved.

1. Introduction

In recent years several studies have shown that emittance growth due to the self-field of electrons traveling through circular motion may become a severe problem in the design of X-ray free-electron lasers, where an electron beam of extremely low emittance generated by a photo-cathode RF-gun is accelerated in a linear accelerator without dilution of emittance and compressed by magnetic bunch compressors to attain high peak current. This self-field in circular motion is called the coherent synchrotron radiation (CSR) force and the noninertial space-charge force.

Coherent synchrotron radiation has been studied since the 1940s when high-energy accelerators were invented [1]. The early studies were conducted mainly to estimate the amount of energy loss caused by radiation which was relevant to the construction of a high-energy accelerator. In the present, CSR is an important issue regarding emittance growth in a bunch compressor for X-ray FELs.

There are several studies on this subject: one-dimensional analysis of the CSR force on short electron bunches [2], transient analysis including finite length of magnets [3], and CSR in the motion of small periods such as in an undulator [4]. As for numerical studies, it has been shown that the Liénard–Wiechert potential together with particle tracking can be applied to the calculation of CSR and noninertial space charge forces [5]. Emittance growth by shielded CSR has also been discussed [6,7].

* Corresponding author. Tel.: + 81-3-5689-7283; fax: + 81-3-3818-3455.

E-mail address: hajima@q.t.u-tokyo.ac.jp (R. Hajima)

0168-9002/99/$ - see front matter
PII: S0168-9002(99)00070-4

In the present study, we extend Carlsten's method to a geometry with two parallel conducting walls and numerical results are compared with theoretical prediction. The relation between complex radiation resistance and emittance growth due to the shielded CSR force is introduced.

2. Space-charge force with Liénard–Wiechert potential

Radiative and static fields of relativistic electrons can be obtained in the form of a retarded potential, which is reduced to Liénard–Wiechert potential form in the case of a small line charge of density λ traveling through a circular trajectory of radius R, and the azimuthal electric field in the static frame is expressed as [5]

$$E_\theta = \frac{\lambda}{4\pi\varepsilon_0} \frac{1}{r_{\rm ret} - \boldsymbol{r}_{\rm ret} \cdot \boldsymbol{u}_{\rm ret}/c} \times \left(\frac{1}{\gamma^2} - \beta^2 \frac{x}{R} + \frac{r}{R} \beta^2 [1 - \cos(\zeta')] \right) \Bigg|_{\zeta_r}^{\zeta_f} \tag{1}$$

where the first term is the usual space charge force proportional to γ^{-2}, which is often neglected for high-energy electron beams, the second term is the noninertial space-charge force which does not appear in a one-dimensional analysis, and the third term represents the CSR force. The retarded position of the source particle, ζ', is obtained from geometrical considerations and found as the solution of a transcendental equation:

$$\beta^2 R^2 (\zeta' - \zeta)^2 = \rho^2 + 2R(R + x)(1 - \cos \zeta'). \tag{2}$$

A 3-D particle tracking code JPP [8] has been modified to calculate the self-field caused by CSR and by noninertial space-charge effects, where the numerical routine for the self-field is based on Carlsten's method.

While transient analysis is necessary to estimate the CSR force affecting the electrons traveling through magnets of finite length [3], the transient effect is not included in the calculations through the present study. This is because the transient analysis of the CSR force in six-dimensional phase space requires quite heavy calculation and useful results such as the relationship between emittance growth and the complex impedance of a beam duct can be obtained even if the transient effect is neglected.

3. Coherent synchrotron radiation force without shielding

The change of electron energy caused by the unshielded CSR force is calculated and compared with theoretical predictions [2] in this section. We consider an electron beam transport through a two-dipole dog-leg as shown in Fig. 1. Parameters are chosen as: radius of circular path $R = 2.292$ m, path length in each magnet $L_{\rm d} = 20$ cm, energy of electrons $E = 400$ MeV. The electron bunch is assumed to be a Gaussian shape in the longitudinal direction, $\sigma_z = 0.4$ mm, uniform in transverse direction and to have 1 nC charge. In the calculations, the noninertial space-charge term is neglected and only the CSR term is taken into account to compare the results with one-dimensional theory where the noninertial space charge does not appear.

Fig. 2 shows the calculated longitudinal phase plot after the dog-leg. The results of the one-dimensional theory are plotted as a solid line. The numerical result in the case of an electron beam having a relatively small radius, $r_{\rm b} = 0.1$ mm, shows good agreement with the theoretical one. The result for $r_{\rm b} = 1$ mm, however, shows a deviation from the theoretical curve. This is because the retarded position in a Liénard–Wiechert potential depends on the transverse position of the source and observer particles as shown in Eq. (2) and a one-dimensional treatment does not give a good approximation for an electron bunch of relatively large transverse size. This limitation of the one-dimensional treatment

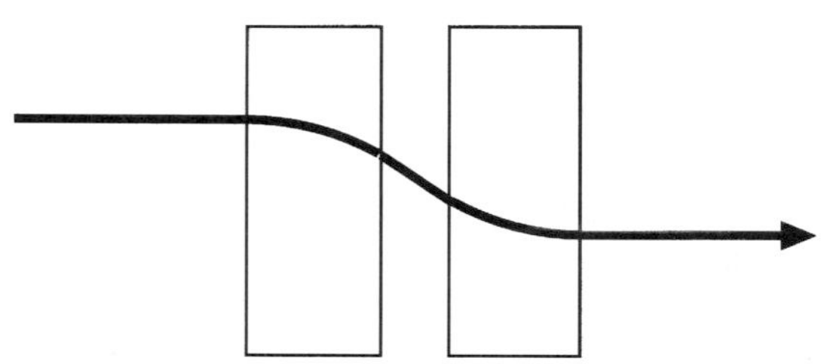

Fig. 1. Two-dipole dog-leg: achromatic beam line consisting of two rectangular magnets.

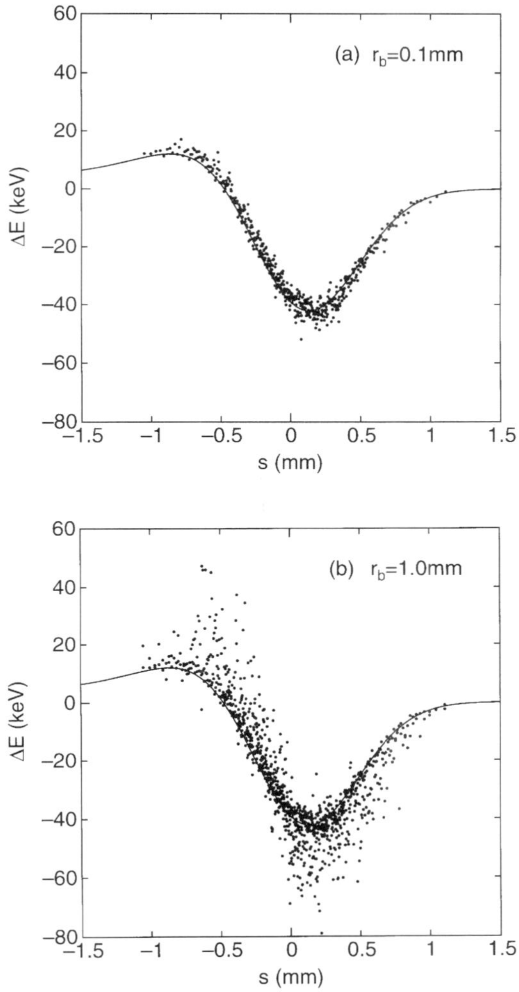

Fig. 2. Longitudinal phase plot of electron bunch after the dog-leg. The transverse size of the bunch is (a) $r_b = 0.1$ mm and (b) $r_b = 1$ mm.

should be considered carefully in the analysis of bunch compressors where an electron bunch spreads in the horizontal direction due to dispersion of the beam line.

4. Coherent synchrotron radiation force shielded by conducting walls

In this section, the shielded CSR force is calculated by the JPP code and compared with the results of impedance analysis.

The energy loss of an electron bunch by the CSR force in a metallic beam pipe can be expressed as

$$\left(\frac{\mathrm{d}W}{\mathrm{d}t}\right)_{\mathrm{CSR}} = -(qc\beta/R)^2 \sum_{n=-\infty}^{\infty} |f_n|^2 \mathrm{Re}[Z_n] = -(qc\beta/2\pi R)^2 \mathscr{R} \qquad (3)$$

where q is charge of the bunch, Z_n is complex impedance of the beam pipe, $\mathscr{R}$ is radiation resistance and f_n is a longitudinal Fourier component of an electron bunch defined as

$$f_n = \frac{1}{2\pi}\int_0^{2\pi} \exp(-\mathrm{i}n\theta)\lambda(\theta)\,\mathrm{d}\theta \qquad (4)$$

where $\lambda(\theta)$ is the longitudinal charge distribution of the bunch.

For two infinite metallic plates placed parallel to the orbit plane, the impedance can be expressed as [1]

$$\frac{Z_n}{n} = \frac{2\pi^2 Z_0}{\beta}\frac{R}{h}\sum_{j=1,3,\ldots}\left[\beta^2 J_n'(J_n' + \mathrm{i}Y_n') + \frac{(j\pi R/h)^2}{(n\beta)^2 - (j\pi R/h)^2} J_n(J_n + \mathrm{i}Y_n)\right] \qquad (5)$$

where h is the distance between two plates, the argument of Bessel functions is $\sqrt{(n\beta)^2 - (j\pi R/h)^2}$ and we assume perfect conductivity of the plates.

The effect of conducting plates on the self-field in a particle tracking code can be completely simulated by the image charge method. Although an infinite number of image charges are required to fulfill the boundary condition on the two parallel plates, we can truncate the number of image charges without degradation of accuracy in the simulation, regarding the field dependency on the distance between source and observer. The number of image charges in the following calculations is carefully determined not to introduce numerical error arising from this truncation. In the calculation of retarded position, a tentative solution with a "pencil beam" approximation is refined through an iterative algorithm. The approximation is, however, not appropriate for image particles far from the central axis of the trajectory. A special approximation, therefore, is introduced to obtain a

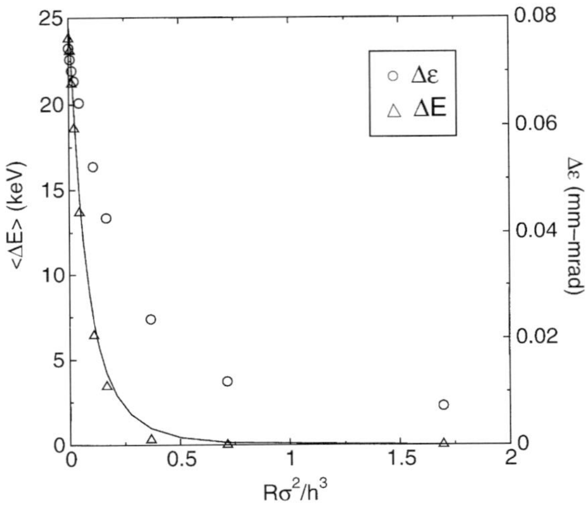

Fig. 3. Averaged energy loss of an electron and emittance growth as a function of dimensionless shielding parameter: energy loss (△) and emittance growth (○) calculated by JPP code, energy loss calculated from impedance analysis (solid line).

tentative retarded position for the image particles [8].

We consider a two-dipole dog-leg having a beam line surrounded by two infinite parallel plates to study the shielding effect on coherent radiation and emittance growth. The energy loss of electrons and normalized emittance growth are calculated for various distances between the two plates. Fig. 3 shows the calculated energy loss and emittance growth as a function of the dimensionless shielding parameter $R\sigma^2/h^3$, where the simulation parameters are chosen as: $E = 400$ MeV, $r_b = 0.1$ mm, $q = 1$ nC, $\sigma_z = 0.4$ mm, $R = 2.292$ m, $L_d = 0.2$ m.

It can be seen that the energy loss obtained by the JPP code agrees with the result of impedance analysis, where energy loss can be neglected for a large shielding parameter as predicted before [9].

It seems that the reduction of emittance growth by shielding of metallic walls is not as drastic as the energy loss. While the energy loss due to coherent radiation for $h = 10$ mm ($R\sigma^2/h^3 = 0.37$) is less than 1 keV, $\frac{1}{30}$ of the energy loss without shielding, the emittance growth is about $\frac{1}{3}$ of that for an open duct. This different dependence of the shielding effect between energy loss and emittance growth can be explained by a longitudinal phase plot after the dog-leg as shown in Fig. 4. It shows that relatively large energy modulation still exists, while the total energy loss of the electron bunch is almost zero. This energy modulation originating in the chicane magnets results in emittance growth.

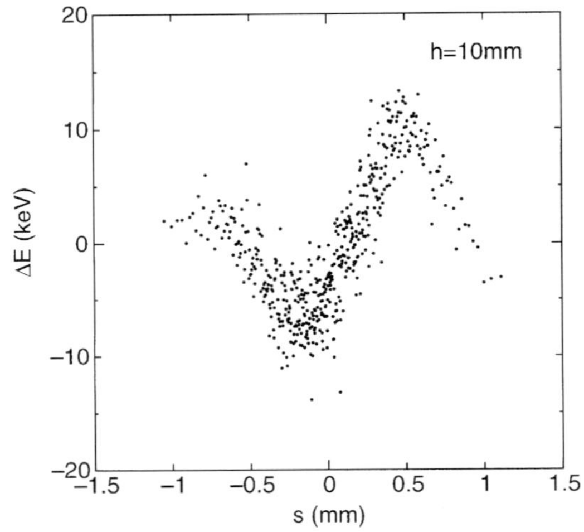

Fig. 4. Longitudinal phase plot of electron bunch after the dog-leg.

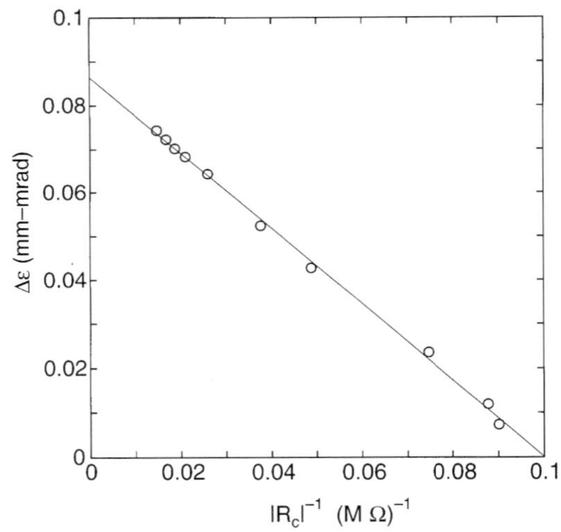

Fig. 5. Obtained emittance growth by JPP code as a function of inverse of magnitude of complex radiation resistance.

The estimation of emittance growth, therefore, requires an amount of energy modulation of the electron bunch. It means that we should consider the imaginary part of impedance in the emittance analysis as well as the real part which is directly related with total CSR power. We introduce a concept of complex radiation resistance to include the imaginary part of the impedance:

$$\mathscr{R}_c = (2\pi)^2 \sum_{n=-\infty}^{\infty} |f_n|^2(\mathrm{Re}[Z_n] + \mathrm{i}\,\mathrm{Im}[Z_n]). \quad (6)$$

Fig. 5 shows the emittance growth obtained by the JPP code as a function of the inverse of magnitude of the complex radiation resistance calculated from Eq. (5). It seems that the emittance growth of an electron bunch due to shielded coherent synchrotron radiation can be expressed as a function of complex radiation resistance.

5. Conclusion

In the present study we described a simulation code which enables one to calculate the noninertial space charge and coherent synchrotron radiation forces, which should be considered in the design of bunch compressors for X-ray FELs. Averaged energy loss of an electron due to the CSR force is calculated with the simulation code and found to be consistent with one-dimensional analytical results. In the case of the shielded CSR force, the energy loss obtained agrees with the result of impedance analysis. It is found that the emittance growth of an electron bunch caused by the shielded CSR force can be expressed as a function of the complex radiation resistance of the conducting walls.

Acknowledgements

We wish to thank Katsuya Nomura for his contribution to refining the numerical routine for image charges.

References

[1] J. Schwinger, On radiation by electrons in a betatron, unpublished, 1945.
[2] Ya Derbenev et al., Microbunch radiative tail–head interaction, DESY-print TESLA-FEL 95-05, 1995, unpublished.
[3] E.L. Saldin et al., On the coherent radiation of an electron bunch moving in an arc of a circle, DESY-print TESLA-FEL 96-14, 1996, unpublished.
[4] E.L. Saldin et al., Radiative interaction of electrons in a bunch moving in an undulator, DESY-print TESLA-FEL 97-08, 1997, and Proceedings of this Conference.
[5] B.E. Carlsten, Phys. Rev. E 54 (1996) 838.
[6] R. Li et al., Shielded transient self-interaction of a bunch entering a circle from a straight path, Proc. SPIE, San Diego, 1997.
[7] M. Dohlus et al., Uncorrelated emittance growth in the TTF-FEL bunch compression sections due to coherent synchrotron radiation and space charge effects, Proceedings of EPAC Conference, 1998, Stockholm, and DESY-M-98-060.
[8] R. Hajima, Proceedings of the International Computational Accelerator Physics Conference (ICAP98).
[9] S.A. Kheifests, B. Zotter, Shielding effects on coherent synchrotron radiation, CERN-SL-95-22-AP, 1995, unpublished.

ELSEVIER

Nuclear Instruments and Methods in Physics Research A 429 (1999) 320–323

NUCLEAR INSTRUMENTS & METHODS IN PHYSICS RESEARCH Section A

www.elsevier.nl/locate/nima

Reduction of bend-plane emittance growth in a chicane pulse compressor☆

David H. Dowell*

Boeing Information, Space and Defense Systems, Seattle, WA 98124-2499, USA

Abstract

Emittance preservation in beam bending systems is vitally important in the production of bright, high-current electron microbunches. Generally, the emittance increase occurs in the bend plane and results from changes in the microbunch energy distribution as the beam transits the bend. This redistribution of electron energies increases the beam's divergence, and hence the emittance, by spoiling the achromatic transport of the bending system. In this paper we investigate the correlated emittance growth in a three dipole chicane compressor due to coherent synchrotron radiation (CSR). Breaking the symmetry of the chicane partially cancels the CSR-induced correlation thereby reducing the bend plane emittance growth. The consequences of this emittance compensation scheme are discussed. © 1999 Elsevier Science B.V. All rights reserved.

Keywords: Microbunch compression; Emittance growth; Space charge; Coherent synchrotron radiation

1. Introduction

Emittance growth of high charge electron beams in transport systems typically occurs in the bend plane as a result of changes in the microbunch energy distribution. While there are many effects capable of producing this redistribution, the most common are the space charge force, wake fields and coherent synchrotron radiation (CSR). These three phenomena are similar in their effect upon the bend plane emittance. They all change the microbunch energy causing a combination of correlated and uncorrelated emittance growth. This paper demonstrates that small modifications to the bend transport can significantly reduce the correlated emittance. Only the growth due to CSR is described and corrected, however the basic principles could be applied to any one or combination of these three effects.

2. Comparison of computed CSR with experiment

In earlier papers[1] [1,2], we modified the CSR overtake potential given by Debenev et al. [3] to account for the startup of CSR radiation in the

☆Work supported by USA/SSDC under Contract DASG60-97-C-0105.

*Tel.: + 1-206-544-5526.

E-mail address: david.h.dowell@boeing.com (D.H. Dowell)

[1] The sign of the pole face rotations given in Ref. [1] are in error. The correct angles for the center dipole are " + 19.5°", as shown in Fig. 2.

PII: S0168-9002(99)00071-6

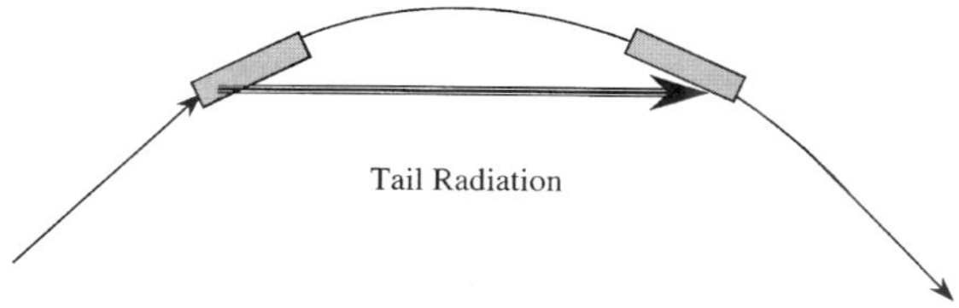

Fig. 1. Coherent synchrotron radiation (CSR) transfers energy from the tail to the head of the microbunch.

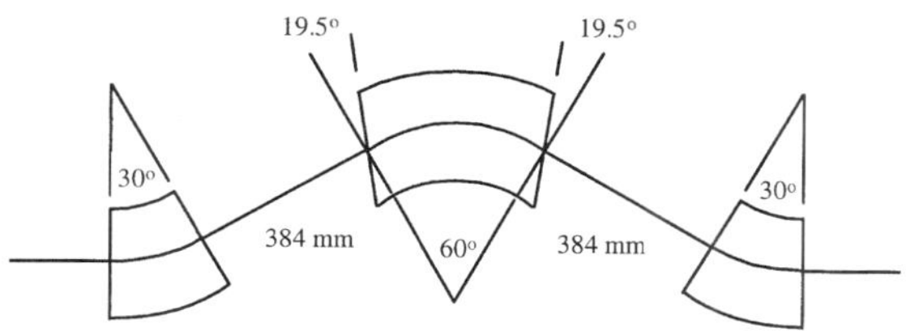

Fig. 2. Boeing chicane composed of three $n = \frac{1}{2}$ dipoles: the outer two being sector magnets, i.e. no pole face rotations, while the center dipole has 19.5° rotations on both the entrance and exit pole faces. These rotations on the center dipole make the chicane doubly achromatic. The chicane is non-isochronous with $R_{56} = 3$ mm/%.

catch-up length and used it to compute the emittance growth in specific bend designs (Fig. 1). This paper continues this work by showing that modifications to the bend parameters can significantly reduce the correlated emittance growth. We begin by comparing experimental emittance with those computed by PARMELA and our formulation of CSR effects. The chicane compressor studied here is shown in Fig. 2.

In the experiment, 2.7 nC microbunches were compressed from 7 to 1.0 mm (rms), boosting the peak current from 50 to 270 A [1,4]. The experimental and computed bend plane emittances are given in Fig. 3. The PARMELA calculation began with the experimental emittance of 8.8 π mm-mrad (rms) before compressing and includes only space charge forces. Combining this emittance in quadrature with the CSR emittance gives good agreement with the experiment, and verifies we have accurately modeled CSR effects.

3. Reduction of CSR-induced correlated emittance

The energy distributions produced by space charge forces, wakefields and CSR in a bending

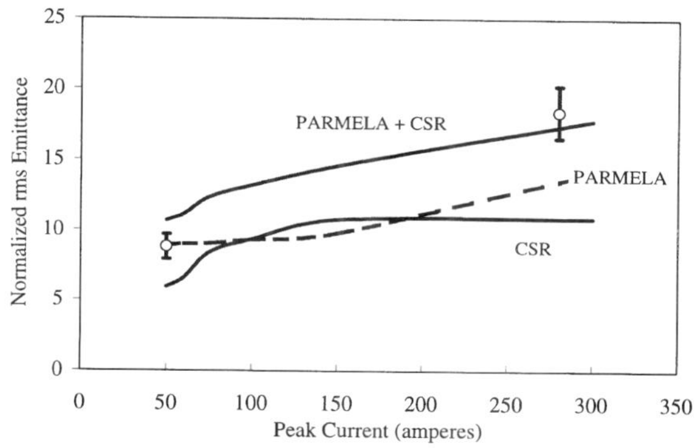

Fig. 3. Comparison of experiment with PARMELA and CSR computed emittance growth.

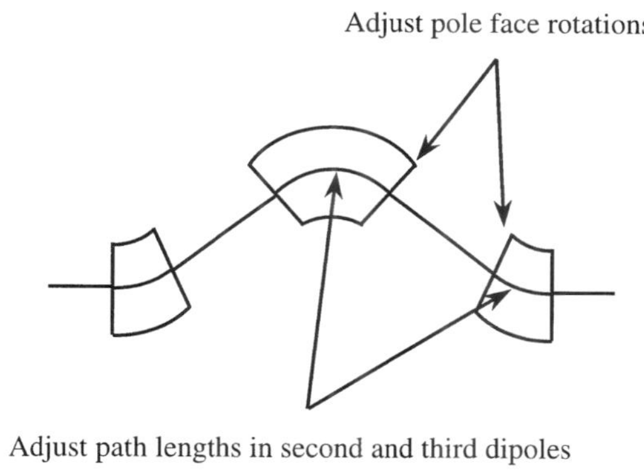

Fig. 4. Modifying the path lengths and pole face rotations of the second and third dipoles reduces the correlated bend plane emittance growth due to CSR.

system typically generate correlations of electron position and angle with the longitudinal position in the microbunch. Uncorrected, these correlations become scrambled in the transport following the compressor and contribute to the random, statistical emittance. However, as we will demonstrate, by making the compressor transport asymmetric the emittance growth is significantly reduced. This is done in the Boeing chicane by modifying the path lengths and pole face rotations of the second and third dipoles from the achromat design values (Fig. 4).

We demonstrate this type of emittance compensation with two examples. The first is similar to the

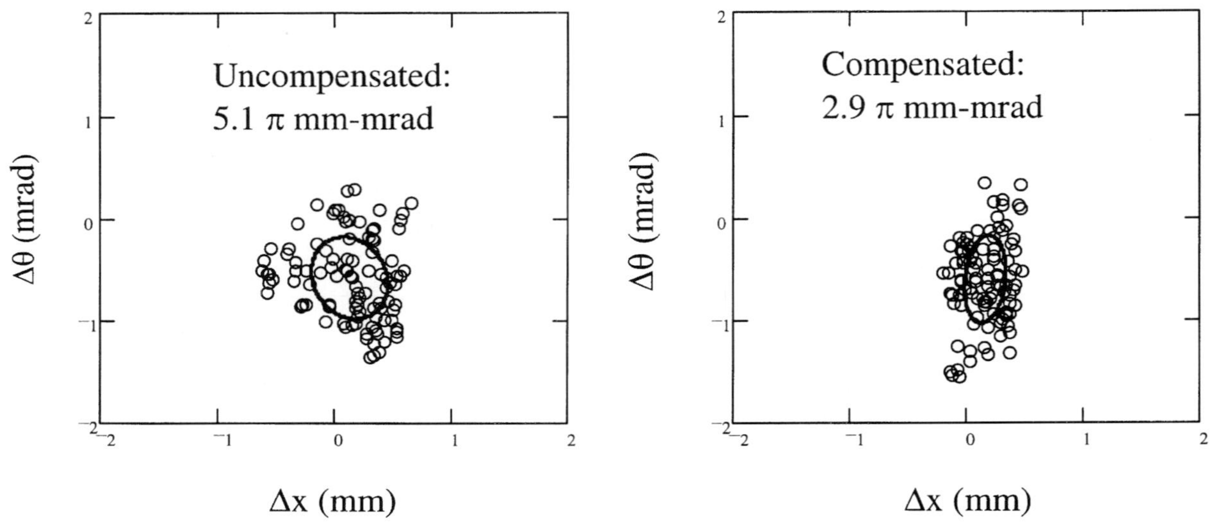

Fig. 5. Bend-plane phase spaces for a microbunch being compressed to 313 A. Increasing the path length in the center dipole by 0.8%, compensates for the CSR-induced emittance growth. The ellipse represents the rms. emittance.

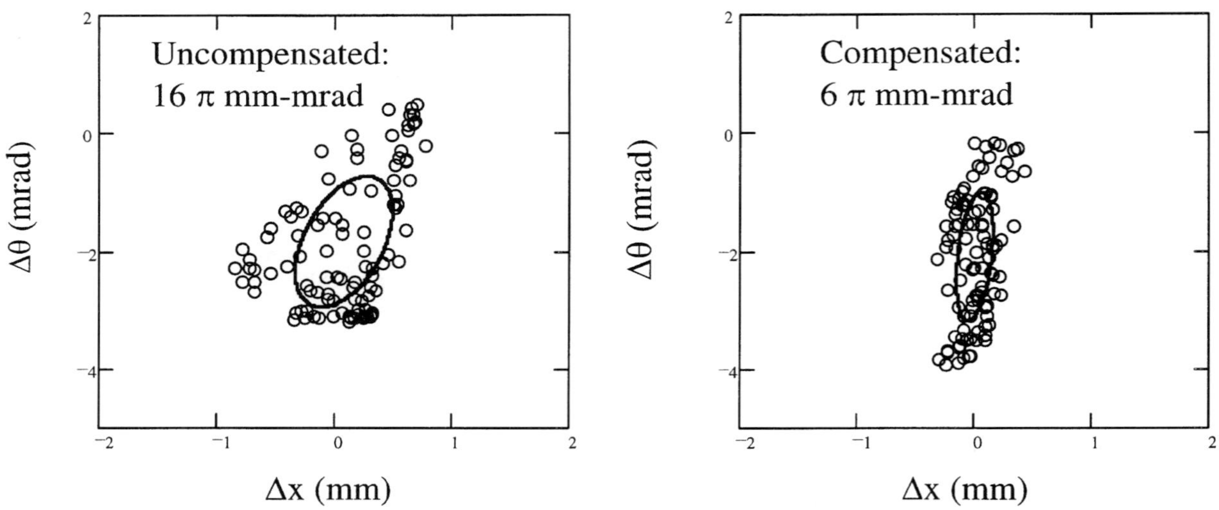

Fig. 6. Bend-plane phase spaces corresponding to microbunch compression from 50 to 1000 A. Increasing the last dipole path length 4% and changing the entrance pole face rotation from 0 to − 15.6° significantly reduces CSR emittance growth.

experiment just described. In this case, a 1 π mm-mrad, 2.7 nC microbunch is compressed from 45 to 313 A of peak current, and the emittance grows to 5.1 π mm-mrad. However, increasing the path length in the center dipole by 0.8% reduces the final emittance to 2.9 π mm-mrad. Fig. 5 shows the uncompensated and compensated phase spaces.

The second example represents a more extreme case where a 3 nC microbunch is compressed from 50 to 1000 A. The uncompensated emittance grows to 16 π mm-mrad. However, in this case, by increasing the third dipole path length 4% and rotating its entrance pole face to − 15.6° reduces this to 6 π mm-mrad. The relevant phase spaces are shown in Fig. 6.

4. Summary and conclusions

The energy distributions produced by space charge forces, wakefields and CSR in a bending system typically lead to correlations between electron angle and longitudinal position in the microbunch. The result is a component of the emittance growth in bends, which is correlated to various orders of the z-position in the microbunch. This work shows it is possible to cancel this correlation by breaking the symmetry of the bend. In one case considered here, the CSR-induced emittance growth is reduced by more than a factor of two.

References

[1] D.H. Dowell, P.G. O'Shea, Coherent Synchrotron Radiation Induced Emittance Growth in a Chicane Buncher, Proceedings of Particle Accelerator Conference, Vancouver, BC, 1997, submitted for publication.

[2] D.H. Dowell, Compensation of Bend-Plane Emittance Growth in Pulse Compressors, Proceedings of Particle Accelerator Conference, Vancouver, BC, 1997, submitted for publication.

[3] Y.S. Derbenev et al., DESY TESLA-FEL Technical note 95-05 (1995).

[4] D.H. Dowell et al., Nucl. Instr. and Meth. A 393 (1997) 184.

ELSEVIER

Nuclear Instruments and Methods in Physics Research A 429 (1999) 324–331

NUCLEAR INSTRUMENTS & METHODS IN PHYSICS RESEARCH Section A

www.elsevier.nl/locate/nima

The design for the LCLS RF photoinjector☆

R. Alley[a], V. Bharadwaj[a], J. Clendenin[a], P. Emma[a], A. Fisher[a], J. Frisch[a], T. Kotseroglou[a,*], R.H. Miller[a], D.T. Palmer[a], J. Schmerge[a], J.C. Sheppard[a], M. Woodley[a], A.D. Yeremian[a], J. Rosenzweig[b], D.D. Meyerhofer[c], L. Serafini[d]

[a] *Stanford Linear Accelerator Center, Stanford University, Stanford, CA 94309, USA*

[b] *University of California, Los Angeles, CA, USA*

[c] *Department of Physics and Astronomy, Department of Mechanical Engineering, University of Rochester, Rochester, NY 14627, USA*

[d] *INFN, University of Milan, Milan, Italy*

Abstract

We report on the design of the RF photoinjector of the Linac Coherent Light Source. The RF photoinjector is required to produce a single 150 MeV bunch of ~ 1 nC and ~ 100 A peak current at a repetition rate of 120 Hz with a normalized rms transverse emittance of $\sim 1\pi$ mm-mrad. The design employs a 1.6-cell S-band RF gun with an optical spot size at the cathode of a radius of ~ 1 mm and a pulse duration with an rms sigma of ~ 3 ps. The peak RF field at the cathode is 150 MV/m with extraction 57° ahead of the RF peak. A solenoidal field near the cathode allows the compensation of the initial emittance growth by the end of the injection linac. Spatial and temporal shaping of the laser pulse striking the cathode will reduce the compensated emittance even further. Also, to minimize the contribution of the thermal emittance from the cathode surface, while at the same time optimizing the quantum efficiency, the laser wavelength for a Cu cathode should be tunable around 260 nm. Following the injection linac the geometric emittance simply damps linearly with energy growth. PARMELA simulations show that this design will produce the desired normalized emittance, which is about a factor of two lower than has been achieved to date in other systems. In addition to low emittance, we also aim for laser amplitude stability of 1% in the UV and a timing jitter in the electron beam of 0.5 ps rms, which will lead to less than 10% beam intensity fluctuation after the electron bunch is compressed in the main linac.

Keywords: RF photoinjector; Linac coherent light source

1. Introduction

The Linac Coherent Light Source (LCLS) [1] is a future free electron laser (FEL) in the 1 Å wavelength regime, in which a low-emittance electron beam is generated with an RF photoinjector, accelerated to 15 GeV using existing sectors of the Stanford Linear Accelerator Center (SLAC) 3-km linac, and compressed at intermediate energies by two magnetic chicanes to a length of 20 μm. Subsequently, the beam is injected into a 100 m undulator. FEL lasing is produced in a single pass of the electron beam through the

☆Work supported by Department of Energy contract DE-AC03-76SF00515.

* Corresponding author.

E-mail address: theo@slac.stanford.edu (T. Kotseroglou)

0168-9002/99/$ - see front matter

PII: S0168-9002(99)00072-8

undulator by Self-Amplified Spontaneous Emission (SASE).

The electron source for the LCLS consists of a high-gradient 1.6-cell S-band RF gun with a copper photocathode illuminated by intense optical pulses at 260 nm provided by an Nd:YAG-pumped Ti:Sapphire laser system whose fundamental frequency is tripled. The source must produce a nominal 1 nC pulse with low transverse and longitudinal emittance at a pulse repetition rate of 120 Hz. The required 1 σ rms emittance for the nominal charge is $< 1\pi$ mm-mrad. A significant part of the overall emittance produced by the source is expected to be contained in the spatial and temporal halo of the pulse, which can be removed downstream of the photoinjector, if desired, with scrapers. To meet the emittance requirements, the optical spot size at the cathode has a radius of ~ 1 mm, the pulse duration has an rms sigma of ~ 3 ps, and the RF field at the cathode is 130–150 MV/m.

2. RF photocathode gun

The RF gun design is based on the 1.6-cell gun developed by a BNL/SLAC/UCLA collaboration [2]. The parameters of the gun as applied to the LCLS are listed in Table 1. To minimize emittance growth due to the E_z component of the TM110 mode, the gun has been symmetrized. The original BNL zero-mode-suppressed side coupling has been replaced by side-coupling RF into the full cell only. In consequence the cell-to-cell coupling between the two cells has been improved by increasing the iris size, which also increases the mode separation between the zero and π-modes. This arrangement allows for more precise field balancing during tuning. To provide more RF focusing and decrease the peak field on the cell-to-cell iris, the half-cell length has been slightly increased.

A schematic of the gun is shown in Fig. 1. There are two identical coupling ports in the full cell, located 180° apart. The RF waveguide is connected to one port and a vacuum pump to the second. For diagnostic purposes, an RF monitoring loop is also located in the second port. The photocathode is located at the geometric center of the end plate of the half-cell. The end plate is removable to facilitate installation of cathode material other than Cu by implantation or by using an insert. The laser beam can be brought to the cathode either along the axis of the gun or at grazing incidence through the 72°

Table 1
Photoinjector gun parameters. A uniform temporal laser pulse profile is assumed

Parameter	Value
Cathode material	Cu (or possibly Mg)
Usable diameter of cathode	12 mm
Quantum efficiency	10^{-5} at 260 nm
Nominal extraction field	140 MV/m
Charge per bunch	~ 1 nC
Peak current	~ 100 A
Beam energy at gun exit	7 MeV
Energy spread at exit	0.2% rms
RF frequency	2856 MHz
Bunch rep rate	120 Hz
RF pulse duration	3–4 μs
RF peak power	15 MW
Number of cells	1.6
Length	0.168 m

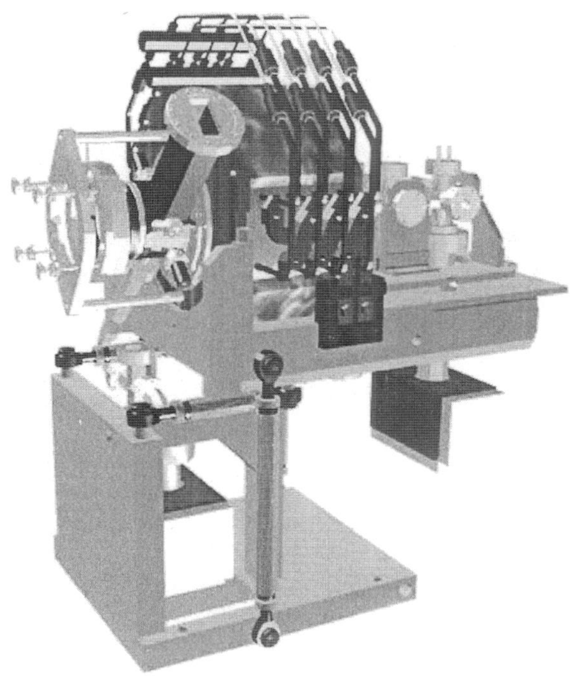

Fig. 1. Perspective of the gun assembly. The beam exits to the right (figure by C. Stelmach and S. Pjerov, NSLS, BNL).

side port (not shown in the figure). The peak electric field at the cathode is 150 MV/m and the laser pulse is injected 57° ahead of the peak.

The 15 MW in RF power required for the gun will be provided by the first of the two klystron stations in Linac 0, the first accelerator section after the RF gun and up to a chicane (see Section 5). Phase and amplitude for the gun can be set independently of the downstream injector accelerating sections. A diagnostic section of 0.75 m will separate the exit of the gun from the entrance of Linac 0. Apart from including vacuum isolation valves, this drift space will contain a beam position monitor (BPM), current monitor, Faraday cup, profile screen, and steering magnets.

3. Photocathode

A metal photocathode is chosen for the RF gun. The quantum efficiency (QE) for Cu illuminated with UV light depends somewhat on surface preparation, but a QE of 10^{-5} at 260 nm seems reasonably conservative for normal incidence [3–5]. A gain in QE by a factor of two to four can be achieved by illuminating the cathode at a grazing angle. At 260 nm an optical pulse of $\sim$ 500 μJ on the cathode is required to produce 1 nC of charge when the QE is 10^{-5}.

The principal advantage of a metal cathode is that its QE is relatively impervious to exposure to atmospheric air or the vacuum environment of the operating gun. Thus special cathode preparation and installation (e.g. load-lock) techniques are less critical.

The photoelectric response of metal cathodes is on the sub-picosecond level, thus imposing no limitation on any desired temporal pulse shaping. Finally, the use of Cu as the cathode allows the entire end plate of the half-cell to be formed in the standard manner of Cu RF cavities, permitting operation at the highest field values. The use of a cathode plug or insert in an S-band gun has so far limited the cathode field to about 80 MeV/m, whereas simulations indicate the transverse emittance drops with increasing field up to about 140 MeV/m. Nonetheless, since the QE of Mg cathodes is several times higher than for Cu, we are searching for a way to use an Mg cathode with fields on the order of 150 MV/m.

4. Laser

The laser system [6] is required to deliver a 500 μJ pulse of UV photons to the cathode at a repetition rate of 120 Hz. To meet the emittance requirements of the source, the laser must have a pulse length in the range of 3–5 ps rms and radius of $\sim$ 1 mm hard edge on the cathode in order to balance the effects of space charge and RF on emittance growth in the gun itself. Pulse shaping is required to optimize both the transverse and longitudinal emittance. Stability of the laser pulses of better than 0.5 ps rms in time and 1% in energy in the UV is crucial for the bunch compressors and FEL. The laser system is shown in Fig. 2. A CW, frequency doubled, diode pumped Nd:YAG laser provides energy in the green (532 nm) to pump the CW mode-locked Ti:Sapphire oscillator, which delivers a continuous train of 10 nJ, 100 fs FWHM pulses that repeat at 119 MHz. By using this frequency, which is the 24th sub-harmonic of the linac's 2856 MHz RF, the timing of the laser pulse can be locked to the phase of the RF in the linac and gun. The timing system is shown in Fig. 3 and described in more detail in Ref. [1]. The wavelength of the oscillator is tuned to 780 nm, which after amplification is frequency tripled to 260 nm. With the possible exception of a timing stability of 0.5 ps rms, oscillators of this type are commercially available. A Pockels cell and polarizer are used to gate single pulses, at 120 Hz, from the 119 MHz pulsetrain. The selected pulses are then amplified by two multipass Ti:Sapphire crystals, both configured as 4-pass "bow-tie" amplifiers. Both are pumped by a single Q-switched, doubled Nd:YAG laser that can produce a 120 Hz train of 3 ns pulses, which is also a commercial system. The pump output is stabilized by "trimming" a few percent of the main pump pulse by the use of a Pockels cell.

The system uses chirped pulse amplification (CPA) [7] to avoid damage and nonlinearities in the amplifiers. The large bandwidth of Ti:Sapphire enables us to use pulse shaping in time at the 0.1 ps level in order to optimize the emittance of the

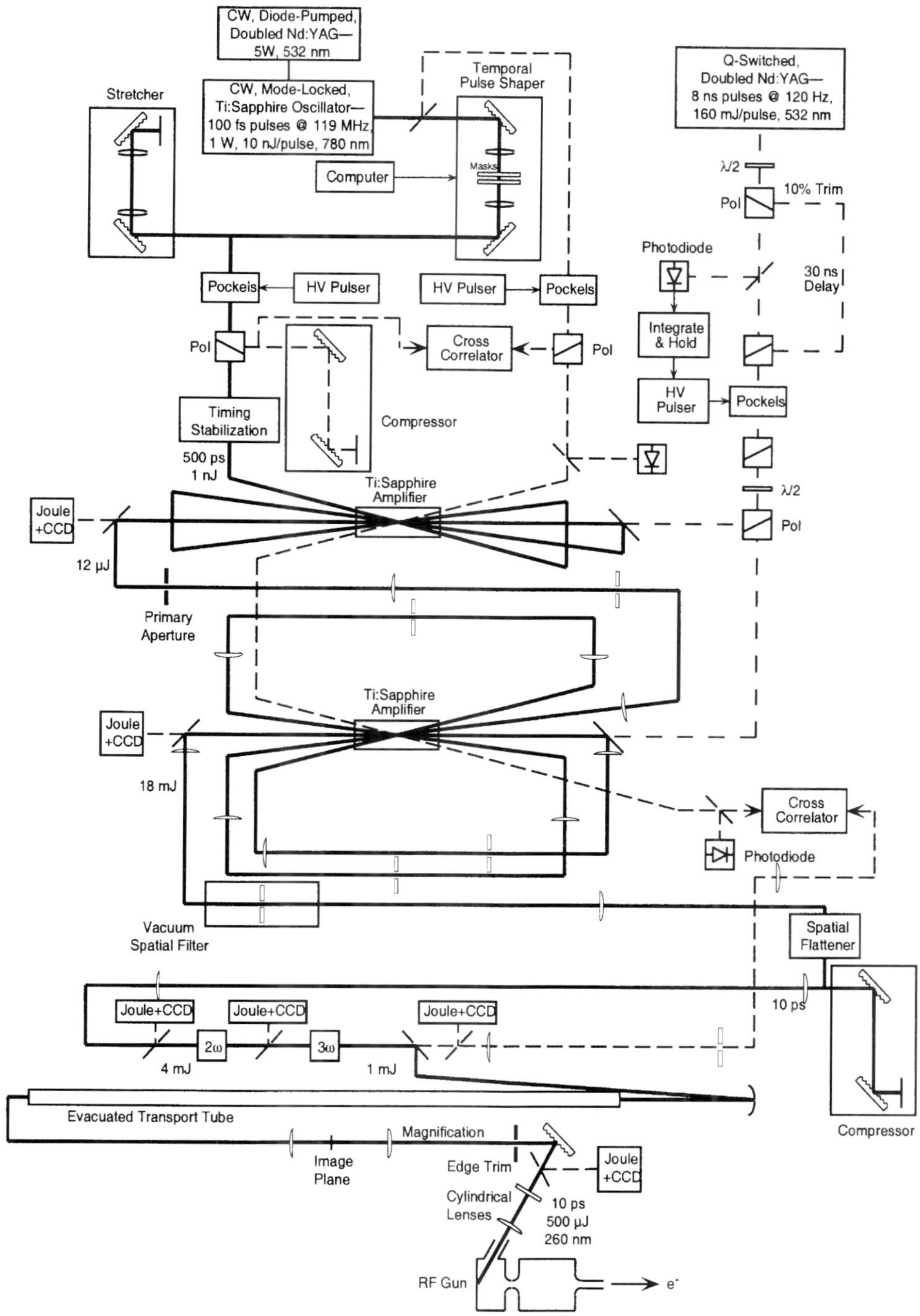

Fig. 2. The drive laser for the RF photocathode electron gun for the LCLS. The thick lines show the main beam path, the closely spaced, dashed lines indicate diagnostic beams, and the widely spaced, dashed lines are pump beams.

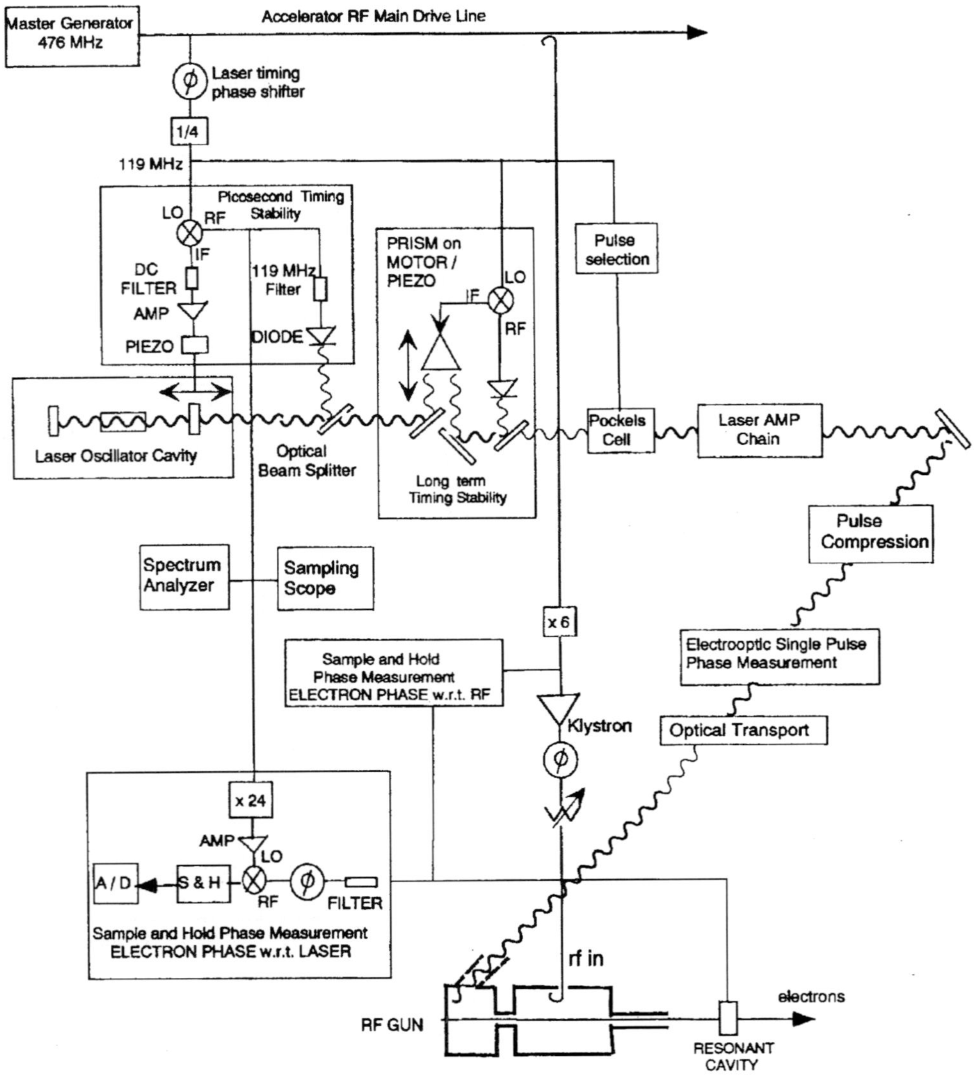

Fig. 3. Timing stabilization scheme. The laser oscillator cavity length is stabilized by the use of a piezo on the cavity end mirror. An external optical delay line corrects for slow drifts of the phase of the laser w.r.t the electrons produced.

electron beam in the gun and injector. This is done by a computer controlled liquid crystal display (LCD) mask, which is positioned in the Fourier plane of two gratings just after the oscillator.

After the second amplifier, the transverse shape of the pulse is modified from Gaussian to uniform to better match the requirements for obtaining a low emittance from the gun. The laser pulse is then tripled in frequency and transported to the gun by means of Fourier imaging that also includes compensation for the temporal and spatial distortion caused by grazing incidence onto the photocathode.

The energy management of the system is as follows: transmission through the spatial flattener $\sim 50\%$, through the compressor $\sim 50\%$, through the tripling stage ~ 25% and through the optical transport to the gun ~ 50%. Consequently starting from 18 mJ in the IR after the second amplifier, the required 500 μJ of UV is delivered to the cathode. Table 2 is a summary of the laser parameters.

5. Linac 0

The design of Linac 0 was originally described in Ref. [8]. Some additional details can be found in Ref. [1]. Linac 0 consists of four SLAC-type 3 m S-band accelerator sections, associated solenoid, and necessary drift sections with diagnostics. It is preceded by the RF gun and diagnostic section and followed by a second diagnostic section and an achromatic bend system (low-energy dog-leg) that brings the beam into the 15 GeV linac. The layout of the photoinjector system is shown in Fig. 4. The overall length from the photocathode to the last accelerator section output flange is 15.8 m, including four drift sections of 75, 47, 105 and 105 cm, respectively. The gun solenoid is 20 cm in length and is located 19.8 cm downstream of the photocathode. Two linac solenoids, each 70 cm in length are located at the end of accelerating sections 2 and 3, respectively.

Table 2
Laser system requirements

Parameter	Value
Operating wavelength	260–280 nm
Pulse rep rate	120 Hz
Number of micropulse per pulse	1 (for LCLS-I), 40 (for LCLS-II)
Pulse energy on cathode	> 500 μJ
Pulse radius on cathode	~ 1 mm HWHM
Pulse risetime	1.0 ps
Pulse length	10 ps (tunable)
Longitudinal pulse form	Various
Transverse pulse form	Uniform
Homogeneity on cathode	10% ptp
Pulse-to-pulse energy jitter	< 1%
Laser to RF stability	< 0.5 ps rms
Spot diameter jitter on cathode	1% ptp
Pointing stability	< 1% ptp

Two SLAC-type 5045 klystrons are required, one to power the gun and the first two 3 m sections and the other for the last two 3 m sections. They will be operated in an un-SLEDed mode [9]. As shown in Fig. 4 high-power RF phase shifters and attenuators will allow the independent control of the gun and first two 3 m sections.

In Linac 0, after each accelerator section there will be a beam position monitor (BPM), a current monitor, and a screen. Between the second and third sections, an X-band cavity will be placed alongside a ceramic gap in the beamline. The charge-normalized RF intensity generated in this cavity as an electron bunch passes can be used to minimize the bunchlength. Transverse emittance diagnostics are included in the dog-leg system.

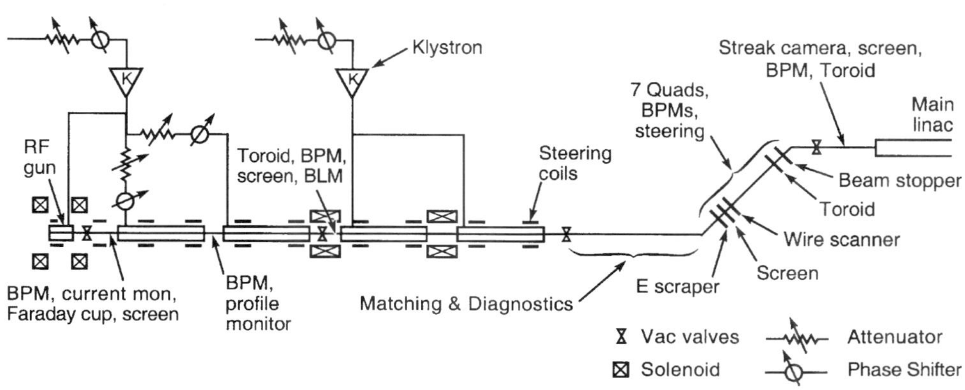

Fig. 4. Overall layout of the LCLS photoinjector showing the RF gun, Linac 0, and the low-energy dog-leg, with drift sections and diagnostics indicated, see Ref. [8].

6. Emittance compensation

In 1989 Carlsten [10] pointed out that by simply placing a solenoidal field at the proper location at the gun one can reverse the evolution of phase space such that at a latter point downstream the individual phase spaces of the various slices of the beam will momentarily be aligned. Emittance compensation in the RF gun is done by using a solenoid with precisely defined field symmetry and positioned at the gun's exit to an accuracy of 25 μm. An identical compensation magnet with current flow in the opposite sense will be used to null the magnetic field at the cathode. We base our design on recent experiments at the ATF at BNL [11] that have shown encouraging results on emittance compensation, as shown in Fig. 5.

Emittance compensation in Linac 0 was studied extensively using PARMELA as originally described in Ref. [8]. In the first 3-m section, where the gradient is 7 MV/m, a 1.5% energy spread for the full beam is introduced to help in emittance compensation by placing the centroid of the beam $\sim 5^\circ$ ahead of the RF crest. This energy spread is removed in the last two 3 m sections, where the gradient is 17 MeV/m, by placing the RF such that the beam is slightly behind the crest. The magnetic field in the emittance compensation solenoid at the gun is slightly less than 3 kG with an effective length of $\sim$ 20 cm, while the solenoid between the accelerating sections is slightly greater than 3 kG with an effective length of 70 cm.

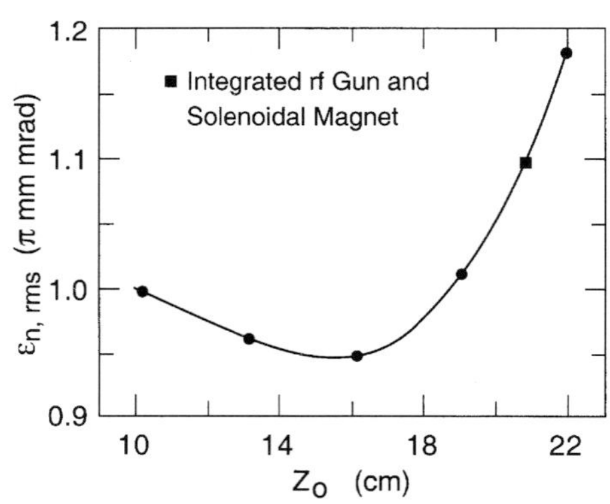

Fig. 5. PARMELA simulation of compensated normalized emittance of ATF gun and beamline as a function of position (defined as the center) of the emittance compensating solenoid. The actual position of the solenoid during the experiments is shown by the square data point.

At the entrance to the third 3-m section, the beam energy is $\sim$ 50 MeV so that space charge effects are much reduced. The RF field gradient though, is more than double in this section as in the previous two, so the beam radius is purposely reduced by using a solenoid between the second and third sections. The transverse space charge distribution at the cathode is assumed to be uniform and the temporal shape Gaussian with an rms sigma of 4.4 ps, truncated at $\pm$ 2 sigma, resulting in overall rms length of 3.8 ps. Table 3 and Figs. 6 and 7 show the resulting electron beam parameters at the end of Linac 0.

Table 3
Electron beam parameters at end of Linac 0, see Ref. [8]

Parameter	Value
Energy	150.5 MeV
Bunch charge	1.0 nc
rms pulse length	2.3(0.69) ps(mm)
Halo population	7.7%
rms relative energy spread	0.13%
Normalized rms core emittance	1.1π mm mrad

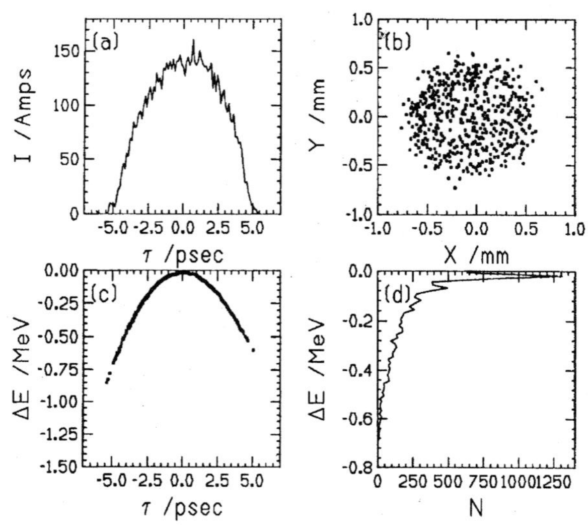

Fig. 6. PARMELA results for bunch at 150 MeV showing: (top left) temporal distribution; (top right) x–y space; (bottom left) longitudinal phase space; and (bottom right) energy distribution, see Ref. [8].

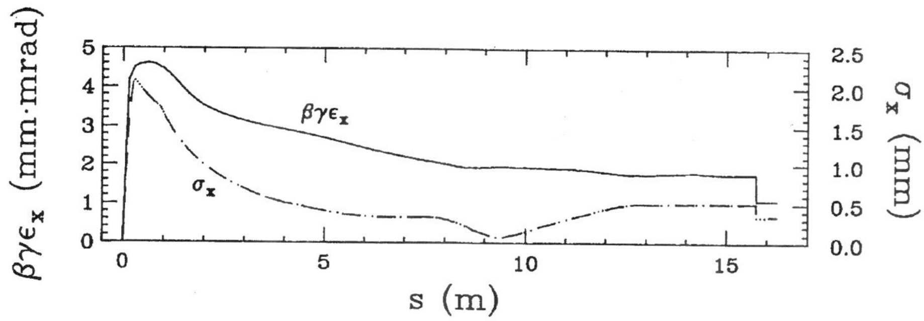

Fig. 7. Normalized emittance (solid line) and rms beam size (dashed line) along the beamline from the cathode ($s = 0$) to 150 MeV at the end of Linac 0. The step at $s \sim 16$ m is a 7.7% halo cut (in simulation only), see Ref. [8].

7. Summary

We have presented the design of the RF photoinjector of the LCLS. The RF photoinjector is required to produce a single 150 MeV bunch of ~ 1 nC and ~ 100 A peak current at a repetition rate of 120 Hz with a normalized rms transverse emittance of $\sim 1\pi$ mm-mrad. The design employs a copper cathode, 1.6-cell S-band RF gun, which has been symmetrized, a CPA, Ti:Sapphire laser system that is pulse-shaped, frequency tripled and stabilized both in timing and intensity and a linac 0 with four 3-m SLAC-type accelerating sections. PARMELA simulations show that this design will produce a beam with the desired characteristics at 150 MeV.

References

[1] LCLS Design Study Report, SLAC-R-521, UC-414, April 1998.
[2] D.T. Palmer et al., SPIE 2522 (1995) 514.
[3] T. Srinivasan-Rao et al., J. Appl. Phys. 69 (1991) 3291.
[4] P. Davis et al., Proceedings of the 1993 Particle Accelerator Conference, p. 2976.
[5] E. Chevallay et al., Nucl. Instr. and Meth. A 340 (1994) 146.
[6] T. Kotseroglou et al., The LCLS source laser, SLAC-PUB-7755, paper presented at the International Conference on Lasers '97, New Orleans, LA, December 15–19, 1997.
[7] P. Maine et al., IEEE J. Quantum Electron. QE- 24 (1988) 398.
[8] A.D. Yeremian et al., Proposed injector for the LCLS Linac, IEEE Proceedings of the Particle Accelerator Conference, vol. 3, Vancouver BC, 1997, p. 2855.
[9] Z.D. Farkas et al., SLED: A method of doubling SLAC's energy, Proceedings of the Ninth International Conference on High Energy Accelerators, SLAC, CA, May 2–7, 1974, p. 576 and SLAC-PUB-1453.
[10] B.E. Carlsten, Nucl. Instr. and Meth. A 285 (1989) 313.
[11] D.T. Palmer et al., Experimental results of a single emittance compensation solenoidal magnet, SLAC-PUB-7421, presented at the seventeenth IEEE Particle Accelerator Conference, Vancouver, B.C., Canada, May 12–16, 1997.

ELSEVIER

Nuclear Instruments and Methods in Physics Research A 429 (1999) 332–335

NUCLEAR INSTRUMENTS & METHODS IN PHYSICS RESEARCH Section A

www.elsevier.nl/locate/nima

The on axis coupled structure type RF gun

Fumihiko Oda[a,*], Minoru Yokoyama[a], Akihiro Nakayama[a], Eiji Tanabe[b]

[a]*Kawasaki Heavy Industries Ltd., 118 Futatsuzuka, Noda, Chiba 278, Japan*
[b]*AET Associates, Inc., 1-3-4 Kamiasao, Asaoku, Kawasaki, Kanagawa 215, Japan*

Abstract

The fundamental design of this newly developed RF gun with a thermionic cathode is the $\pi/2$ mode standing wave structure. It has two accelerating cells and a coupling cell located on the beam axis, a so-called on axis coupled structure (OCS). This structure offers a stable operation for high beam current, owing to high group velocity and wide bandwidth. It is important to reduce damage onto the cathode caused by back bombardment, especially for long macropulse operation, such as in an FEL injector. Back bombardment, as well as output beam profile was simulated by using the electromagnetic field analytical codes 'EMSYS'(2D) and 'MAFIA'(3D). The cavity shape was optimized to reduce back bombardment power without sacrificing beam emittance.

1. Introduction

Today, RF guns are commonly used as injectors which obtain a higher brightness beam than conventional guns [1]. Laser-driven RF guns can emit extremely high brightness beams, but the lifetimes of cathodes are too short to operate stably and mode-locked drive lasers are considerably expensive. Although RF guns with thermionic cathodes are compact and low-cost injectors, back bombardment is the critical disadvantage especially for long macropulse operation.

Kawasaki Heavy Industries (KHI) is constructing an RF-linac based compact FEL system of which the wavelength range is 5–16 μm [2]. A combination of an S-band RF gun with a thermionic cathode and an alpha-magnet was selected as a compact injector which can generate a sufficiently brilliant beam for FEL lasing.

We chose the standing wave $\pi/2$ mode as the accelerating mode which is advantageous for the field stability over the beam loading or structural imperfection. The coupling cavity is located on the beam axis, a so-called on axis coupled structure (OCS) [3,4]. The cavity shape was designed by using electromagnetic field analytical codes to minimize the back bombardment power without sacrificing beam emittance. We will describe the details of the design and the results of the beam simulations.

2. Outline of design

The design goals established for the gun operating as an injector of the FEL system are: (1) the usable charge (i.e. electrons with energy $\geq 95\%$ of the peak energy) included in each bunch should be more than 0.15 nC, (2) the peak momenta should be about 2 MeV, (3) linear, monotonic dependence of energy on time for 10–20 ps during a bunch to

* Corresponding author.

0168-9002/99/$ - see front matter
PII: S0168-9002(99)00074-1

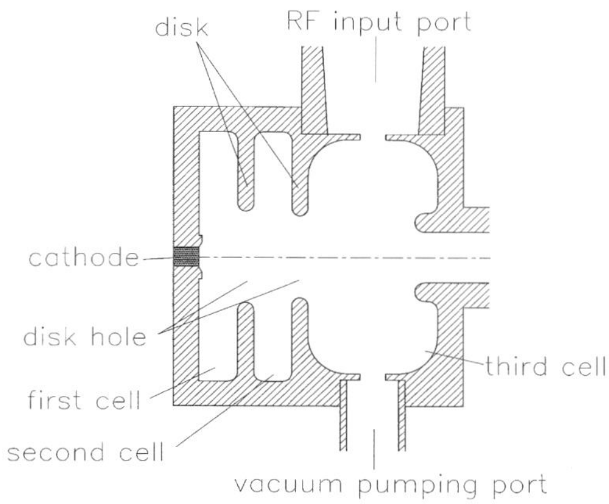

Fig. 1. Cross-sectional view of the OCS type RF gun.

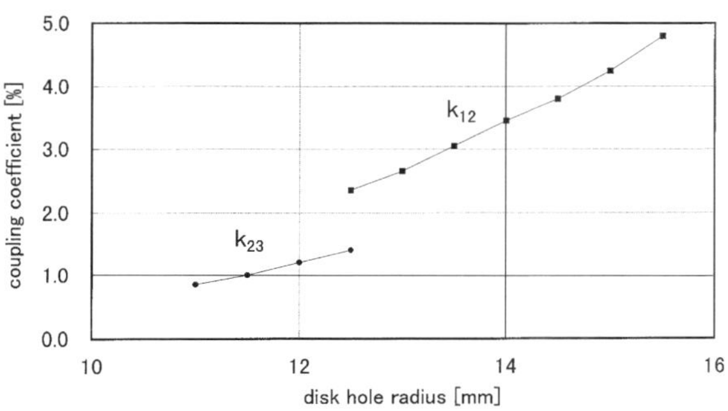

Fig. 2. Coupling coefficients (k_{12}, k_{23}) as functions of disk hole radii.

allow bunch compression by an alpha magnet, (4) the normalized emittance of the output beam should be less than 20 πmm-mrad, (5) minimized back bombardment power onto the cathode.

Fig. 1 is a cross-sectional view of the OCS type RF gun. The standing wave $\pi/2$ mode was selected as the accelerating mode and ideally the accelerating field does not exist in the second cell. This mode has an advantage in field stability over beam loading or frequency errors of the individual cells [5]. In the third cavity, there is an RF input port which feeds RF power to the cavity. A vacuum pumping port is located at the axially opposite position of the RF input port in order to minimize the asymmetric field.

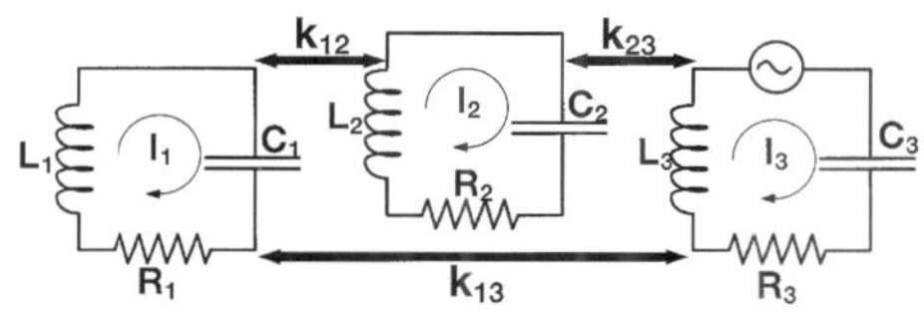

Fig. 3. Coupled resonator model for the OCS type RF gun with nearest neighbor and second nearest-neighbor coupling.

Each cell is electrically coupled with other cells at the disk hole on the beam axis. The coupling coefficient between neighboring cells depends on the radius of the disk hole, as the disk thickness is kept constant (5 mm). Fig. 2 shows the relation between the radius of the disk hole and the coupling coefficient.

Fig. 3 shows a coupled resonator model for the OCS type RF gun with nearest and second nearest-neighbor coupling. For the lossless case (high Q), the coupled equations in matrix format generated from the model above are [3,6]

$$\begin{pmatrix} 2\delta_1 & \frac{k_{12}}{2} & \frac{k_{13}}{2} \\ \frac{k_{12}}{2} & 2\delta_2 & \frac{k_{23}}{2} \\ \frac{k_{13}}{2} & \frac{k_{23}}{2} & 2\delta_3 \end{pmatrix} \begin{pmatrix} X_1 \\ X_2 \\ X_3 \end{pmatrix} = \begin{pmatrix} I_1 \\ I_2 \\ I_3 \end{pmatrix},$$

where

$$\delta_n = \frac{\omega - \omega_n}{\omega} \quad (n = 1,2,3). \tag{1}$$

The quantities I_n ($n = 1,2,3$) are forcing terms, X_n are amplitudes and ω_n are resonant frequencies for individual cells. There are three solutions to the homogenous equations ($I_n = 0$) corresponding to the $0, \pi/2$ and π modes. Assuming the $\pi/2$ mode, Eq. (1) gives the amplitude ratio of the first and third cells (α) as follows:

$$\frac{k_{12}}{k_{23}} = \frac{|X_3|}{|X_1|} \quad (\equiv \alpha). \tag{2}$$

That means we can design the field balance between the first and third cells by slightly changing the radii of the disk holes. The field ratio is the important parameter for emittance of the output beam and the back bombardment power [7]. We determined the field ratio so as to minimize the back bombardment power without sacrificing emittance by using the beam simulations described in the next section.

3. Simulation and design optimization

The electric field profiles were calculated by using the field analytic modules of EMSYS and MAFIA, and beam dynamics was calculated by the full relativistic and electromagnetic particle-in-cell analytic module of EMSYS. Electric field properties were calculated by SUPERFISH and the results were consistent with those given by EMSYS and MAFIA.

In order to meet design criterion (1), current density of 16 A/cm^2 and cathode diameter of 6 mm were determined assuming that the gun will employ an impregnated-tungsten dispenser cathode. We determined the coupling coefficient of the input coupler to obtain maximum accelerating field under beam-loading. The electrode shape of the third cell was designed to obtain proper focusing, while maintaining a high shunt impedance.

As described above, we can vary α by slightly changing the radii of disk holds. To satisfy design criterion (3) we also varied the length of the first cell as α varied. A larger α offers lower back bombardment power, but aggravates the emittance of the output beam because the accelerating field on the cathode goes smaller. Fig. 4 shows calculated emittance of the output beam and back bombardment power as functions of α. We can see a monotonic decrease of back bombardment power and emittance aggravation at $\alpha > 2.6$. We determined $\alpha = 2.6$ from this result. Final parameters of the RF gun are listed in Table 1.

Fig. 5(a) gives a calculated energy spectrum of the output beam. Energy spread (FWHM) is less than 30 keV. Fig. 5(b) shows the energy of the output beam as a function of time. The dependence of energy on time is sufficiently linear for about 10 ps from the head of the beam, and the charge contained during 10 ps is about 0.17 nC. Average

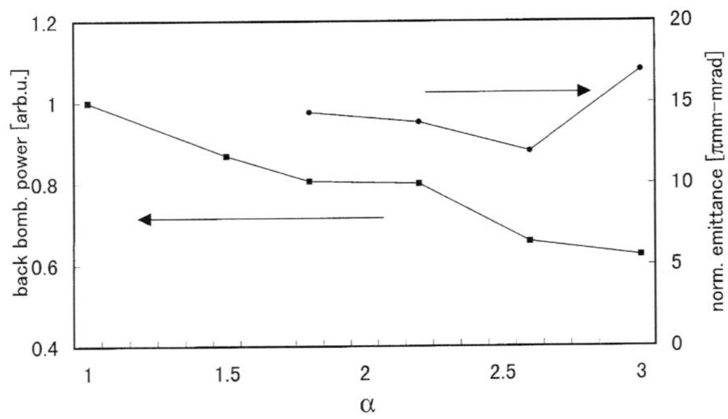

Fig. 4. Emittance of output beam and back bombardment power as functions of α calculated by EMSYS.

Table 1
Optimized parameters of the RF gun

Drive frequency	f_0	2856 MHz
Coupling coefficient of the input coupler	β	3
Coupling coefficient between cells	k_{12}	1.4%
	k_{23}	3.6%
Max. accelerating field (first cavity)	E_{p1}	30 MV/m
Max. accelerating field (third cavity)	E_{p3}	65 MV/m
Max. surface field	E_s	140 MV/m

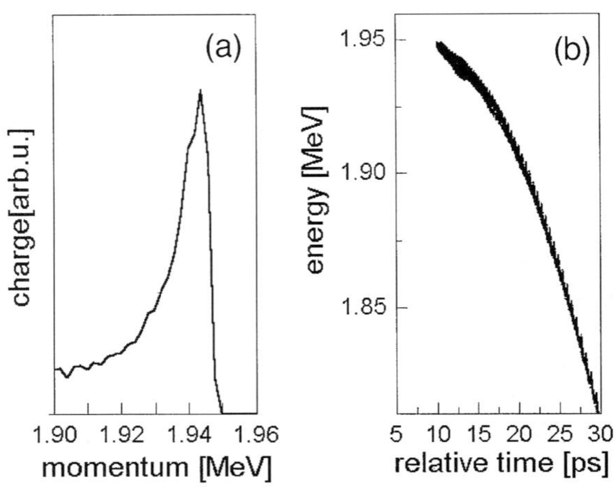

Fig. 5. (a) Energy spectrum, (b) time dependence of energy of output beam calculated by EMSYS.

Table 2
Parameters of output beam and back bombardment calculated by EMSYS

Peak momentum of output beam	p_p	1.9 MeV
Norm. rms emittance	ε_n	11 πmm-mrad
Energy spread (FWHM)	ΔE	25 keV
Bunch length	Δt	5 ps
Average back bombardment power	P_{bb}	1.1 W

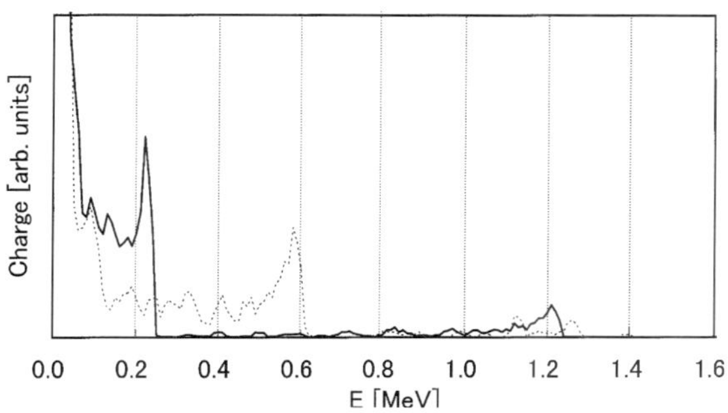

Fig. 6. Calculated energy spectra of back bombarding beam. Solid line represents the optimized cavity ($\alpha = 2.6$). Dashed line represents smaller α ($= 1.0$) case.

back bombardment power is about 1.1 W, assuming 5 μs of macropulse duration time and 10 pps of macropulse repetition rate. Parameters of output beam and back bombardment are listed in Table 2.

4. Profiles of back bombarding beam

Fig. 6 shows the calculated energy spectra of back bombarding beam for an optimized cavity ($\alpha = 2.6$) and a cavity of smaller α ($= 1.0$) of which $E_{p1} = 50$ MV/m and $E_{p2} = 40$ MV/m. Average back bombardment power of the optimized cavity is about 60% of the $\alpha = 1.0$ case. We can see peaks in the energy spectra at 0.2 MeV for the optimized cavity and 0.6 MeV for the $\alpha = 1.0$ case which represent the electrons back from the end of the first cell. For larger α (i.e. smaller E_{p1}), the energy of the peak is smaller. This is a major occasion of the monotonic decrease of the back bombardment power as a function of α. Fig. 7 shows a radial

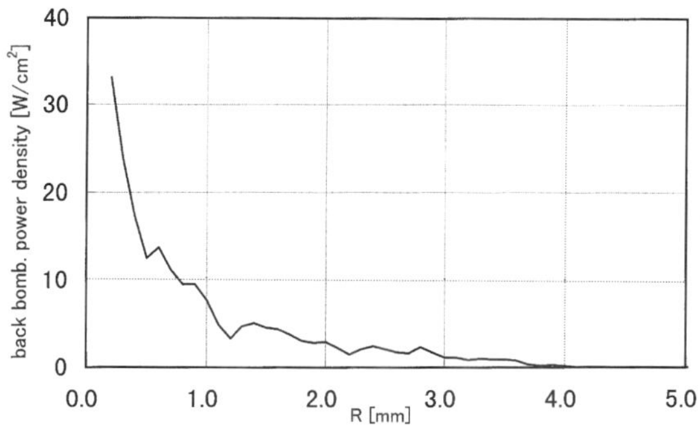

Fig. 7. Calculated radial distribution of back bombardment power density.

distribution of back bombardment power density for the optimized cavity. High power density around the center of the cathode is caused by high momentum electrons accelerated toward the center of the cathode. This means effective back bombardment may be sharply reduced by using an annulus-shaped cathode which has a small beam dumper built in the center.

5. Summary

A new type of thermionic RF gun (OCS type) was developed by using field analytical codes. The cavity was designed to satisfy the design criteria established for the gun operating as an injector of the KHI compact FEL system. The annulus-shaped cathode may sharply reduce the effective back bombardment.

References

[1] C. Travier, Nucl. Instr. and Meth. A 304 (1990) 285.
[2] M. Yokoyama, Nucl. Instr. and Meth. A 429 (1999) 269.
[3] E.A. Knapp, B.C. Knapp, J.M. Potter, Rev. Sci. Instr. 39 (1968) 979.
[4] T. Nishikawa, S. Giordano, D. Carter, Rev. Sci. Instr. 37 (1966) 652.
[5] M. Foley, T. Jurgens, Proceedings of the Part. Accel. Conference, 1993, p. 852.
[6] Y. Yamazaki, Proceedings of the Linear Accel. Conference, 1992, p. 580.
[7] E. Tanabe, M. Borland, M.C. Green, R.H. Miller, L.V. Nelson, J.N. Weaver, H. Wiedemann, Proceedings of the Linear Accel. Conference, 1989, p. 106.

ELSEVIER

Nuclear Instruments and Methods in Physics Research A 429 (1999) 336–340

NUCLEAR INSTRUMENTS & METHODS IN PHYSICS RESEARCH Section A

www.elsevier.nl/locate/nima

Optical techniques for electron-beam characterizations on the APS SASE FEL project[☆]

A.H. Lumpkin*, B.X. Yang, W.J. Berg, M. White, J.W. Lewellen, S.V. Milton

Advanced Photon Source, Argonne National Laboratory, Argonne, IL 60439, USA

Abstract

At the Advanced Photon Source (APS) the injector linac's DC thermionic gun is being supplemented by a low-emittance RF thermionic gun that will support the SASE FEL project. To address the anticipated smaller beam sizes, the standard Chromox beam-profiling screens are being complemented by optical transition radiation (OTR) and Ce-doped YAG single-crystal converters. Direct comparisons of the effective conversion efficiency, spatial resolution, and time response of the three converter screen types have been performed using the DC thermionic gun's beam accelerated to 400–650 MeV. An apparent blurring of observed beam size with increasing incident charge areal density in the YAG crystal was observed for the first time. Only the OTR was prompt enough for the few-ps domain micropulse bunch length measurements performed with a streak camera. Initial beam images of the RF-thermionic gun beam have also been obtained.

PACS: 41.60.Cr; 41.85.Ew

Keywords: APS; OTR; YAG crystal; RF thermionic gun; Beam size; Bunch length

1. Introduction

The higher brightness beams needed for a self-amplified spontaneous emission (SASE) free-electron laser (FEL) injector linac require enhanced characterization capability as well [1]. At the Advanced Photon Source (APS) we are addressing the issue with adjustments to the optical techniques used for electron beam characterizations on the injector linac. The original DC thermionic gun is being supplemented by a low-emittance RF thermionic gun [2]. The standard Chromox beam profiling screens are being complemented by the addition of optical transition radiation (OTR) and Ce-doped YAG single-crystal converters at selected positions. Both of these latter converters show better spatial resolution and temporal response than the Chromox screens. However, only the OTR mechanism is prompt enough to support streak camera measurements on the few-ps micropulse bunch length. In the course of these direct screen comparisons with OTR as the common reference, we observed for the first time an apparent "blurring" of beam image size in the YAG converter with increasing charge areal density. These conditions were achieved with a 30-ns-long macropulse involving only a few nC from the DC thermionic gun

☆Work supported by the U.S. Department of Energy, Office of Basic Energy Sciences, under Contract No. W-31-109-ENG-38.

*Corresponding author. Tel.: + 1-630-252-4879; fax: + 1-630-252-4732.

0168-9002/99/$ - see front matter
PII: S0168-9002(99)00075-3

accelerated to 600 MeV. Since we had an elongated beam focus with a 10 to 1 aspect ratio, the charge areal density was reduced for a given incident charge in these experiments. These results are relevant to emerging SASE FEL projects where electron beam quality is critical.

2. Experimental background

The APS facility's injector system uses a 250-MeV S-band electron linac and an in-line S-band 450-MeV positron linac. The primary electron gun is a conventional DC thermionic gun. For one alternate configuration, an RF thermionic gun, designed to generate low-emittance beams ($< 5\pi$ mm mrad) and configured with an α magnet, injects beam just after the first linac accelerating section [3]. Then both in-line linacs can be phased to produce 100–650-MeV electron beams when the positron converter target is retracted.

The RF gun's projected, normalized emittance at higher peak current is about an order of magnitude lower than that expected of the DC gun, and correspondingly smaller beam spot sizes should result. The standard intercepting screens are based on Chromox of 0.25-mm thickness (rotated 45° to the beam) and with a 300-ms decay time. Previous experiences on the Los Alamos linac-driven FEL with a low-emittance photoelectric injector (PEI) support the applicability of optical transition radiation (OTR) screens oriented at 45° to the beam direction for profiling small beams [4]. In addition, we have installed a Ce-doped YAG single crystal of 0.5 mm thickness normal to the beam direction with a metal mirror at 45° to the beam just behind it. The OTR screen and the YAG-mirror assembly are on two separate actuators on orthogonal faces of a beam-line cubic-shaped cross. The light is transported out of the tunnel via two 150-mm diameter achromat lenses to an optics table. The OTR or YAG light is viewed by a charge-coupled device (CCD) camera; a charge-injection device (CID) camera; a gated, intensified CCD camera; and a Hamamatsu C5680 dual-sweep streak camera. The most useful vertical sweep plug-in has been a synchroscan unit phase-locked to 119.0 MHz, the 24th subharmonic of the 2856 MHz linac frequency. A low-jitter countdown circuit has been built using Motorola ECLIN PS logic to generate the 24th subharmonic. It has been tested with a 0.7-ps (rms) jitter pulse generator, and the total jitter was observed to be 1.1 ps. Bandpass filters on the output result in a clean 119.0-MHz sine wave to be used with the synchroscan unit [5].

The initial tests of the optical diagnostics have been done with a beam from the DC gun. These were done mostly in the 500–600-MeV region with 2–5 nC in the macropulse comprised of 86 micropulses. Recently, the RF gun beam was imaged on the upstream beamline locations and at the end of the linac.

3. Experimental results

We will report results on both transverse and longitudinal aspects of the electron beam.

3.1. Transverse measurements

In our preliminary results using an in-tunnel camera, we initially compared the smallest beam sizes we could observe on a half Chromox/half Ti OTR foil [6]. From these measurements we deduced the Chromox limiting resolution of $\sigma \approx 200$ μm since the OTR screen provided much better resolution.

In this paper we focus our attention more on the comparisons of the YAG-mirror assembly and the OTR screen. There have been previous reports on the achievement of high spatial resolution comparable to OTR screens, with about 1000 times more visible light emitted and with a response time of 80 ns (FWHM) [7]. Those measurements were done with a PEI beam accelerated to 50 MeV using about 0.1 nC of charge in a single micropulse. In our studies we have used 500–600-MeV beams and a 30-ns-long macropulse comprised of 86 micropulses. A macropulse average current of 8 mA corresponds to about 250 pC. We have taken data at macropulse currents up to 625 mA. However, in Fig. 1 we show an OTR image at 100 mA (~ 3 nC) with a horizontal profile size 180 μm (FWHM), or ~ 76 μm (σ), and an aspect ratio (H/V) of $\frac{1}{3}$. In Fig. 2 we show an image from the YAG screen with additional neutral density filter and a bandpass

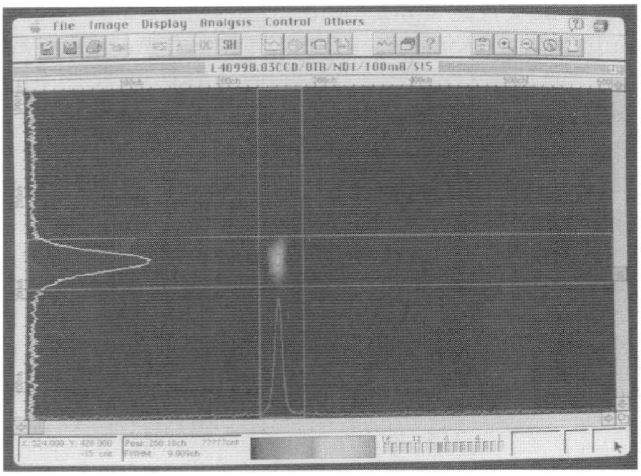

Fig. 1. CCD camera beam image at 600 MeV and 100 mA using the OTR screen. The observed horizontal profile size is 180 μm (FWHM) with a spot aspect ratio (H/V) of 1/3.

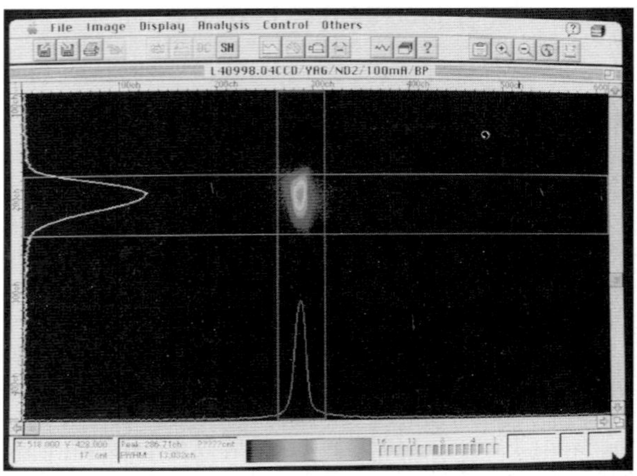

Fig. 2. CCD camera beam image at 600 MeV and 100 mA using the YAG : Ce single-crystal converter. The observed horizontal size of 286 μm (FWHM) is larger than that of Fig. 1. Filters were added to attenuate the signal relative to the OTR signal.

filter to compensate for the YAG brightness. In this case, the best focus we could obtain horizontally was 45% larger at 286 μm (FWHM) and an aspect ration (H/V) of $\frac{1}{3}$.

Subsequently, we have done a more controlled series of measurements as a function of beam current. In this case the aspect ratio (H/V) of $\frac{1}{10}$ was used. We carefully adjusted the camera position on a rail to compensate for the displacement of the

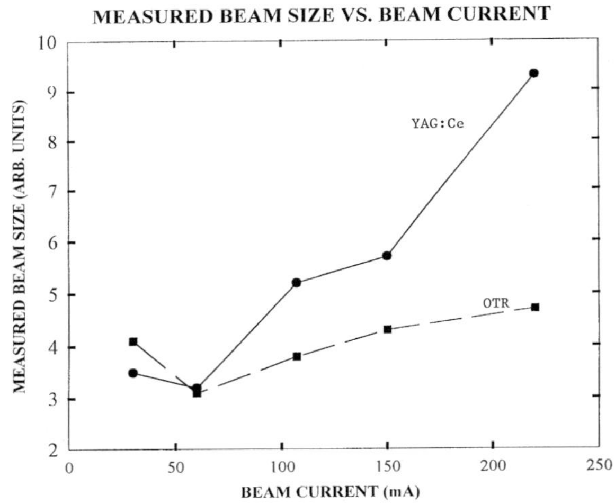

Fig. 3. Plot of the observed beam size versus macropulse average current for the YAG : Ce (solid circles) and the OTR (solid squares) converter screens. An image size growth for the YAG data relative to the OTR data is seen with increasing current beyond 60 mA.

YAG crystal from the OTR 45° plane in the object distances for each current. We also adjusted the neutral density filter used between YAG and OTR for the increasing current. As shown in Table 1 and Fig. 3, the horizontal image sizes of 70–80 μm (FWHM), or 30–35 μm (σ), were achieved at 30 and

Table 1
Comparison of beam spot sizes using the YAG and OTR converter screens (June 1998 data). The aspect ratio H/V was $\sim\frac{1}{10}$. ND = 0.3 is a factor of 2 in attenuation

Converter	ND filter	I_B (mA)	X-size (FWHM) (μm)
OTR	0.0	30	90 ± 7
YAG	2.0	30	77 ± 7
OTR	0.5	60	68
YAG	2.5	60	70
OTR	0.8	107	84
YAG	2.8	107	114
OTR	0.8	150	95
YAG	2.8	150	125
OTR	1.0	220	103
YAG	2.8	220	205

60 mA for both screens, but at 100 mA and above, the observed beam image sizes increased much faster with current using the YAG crystal than using the OTR screen. These data suggest a saturation or blooming phenomenon in the YAG crystal at higher incident areal charge densities. The 33% increase in observed beam size from the YAG:Ce at 150 mA (or 4.7 nC) occurred when a $40 \times 400\ \mu m^2$ spot size was seen on the OTR screen. At 625 MeV, the charge areal density of about 0.3 pC per μm^2 appears to be sufficient for the effect to be detected, and this would indicate reason for concern for 0.5 nC focused into a $20 \times 20\ \mu m^2$ spot. In an initial beam size measurement with 170 mA from the RF gun at a beam energy of 400 MeV and at the $140 \times 140\ \mu m^2$ spot size (as observed with OTR), the YAG:Ce image also was larger. We still need to test this phenomenon further with lower-emittance beams and search for alternative explanations.

3.2. *Longitudinal measurements*

The evaluation of DC gun micropulse bunch length was also done using the streak camera. Previous measurements have been done with an RF phasing technique [8].

In Fig. 4 we show the synchroscan image of the 4 micropulses from the 86 micropulses in the macropulse. This synchronous sum has low jitter (< 1 ps), and by using the bandpass filter any chromatic temporal dispersion effects were also reduced. The observed bunch length was about 4.4 (FWHM), or 1.9 ± 0.2 ps (σ), in good agreement with the 2.2 ± 0.1 ps of Ref. [8]. The streak camera resolution is about 0.6 ps (σ). In Fig. 5 a very graphic example of the slower response time of the YAG crystal is shown. The vertical band of light from the scintillator's decay completely obscures the few-ps time domain.

In Fig. 6 a dual-sweep streak image is shown at a macropulse average beam current of 625 mA. The 100-ms horizontal time axis displays separately the macropulses arriving at 30-ms intervals. The micropulse bunch length of 4.9 ps (FWHM) at 0.22 nC/micropulse implies a peak current of about 45 A.

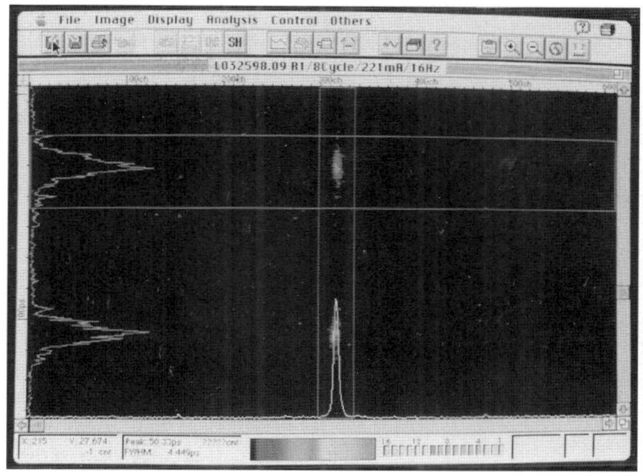

Fig. 4. Synchroscan streak camera image of the linac micropulses using the OTR mechanism. The RF phase delay is such that some micropulses are seen on the up sweep and some on the down sweep of the 119.0 MHz sinusoidal deflection. The observed bunch length is about 4.4 ps (FWHM) including the about 1.6 ps (FHWM) limiting resolution contribution.

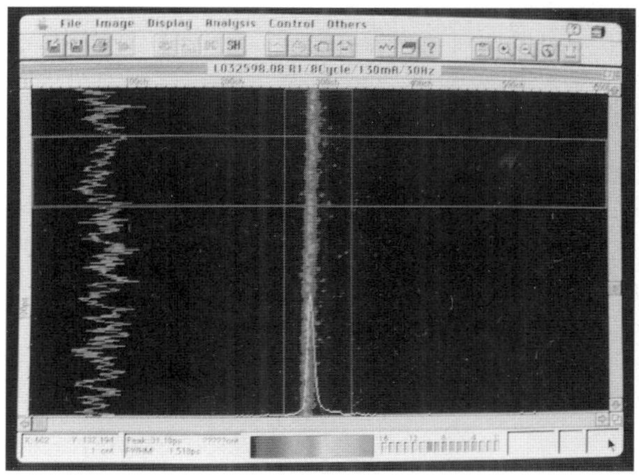

Fig. 5. Synchroscan streak camera image of the linac micropulse using the YAG single crystal. All micropulse bunch length detail is obscured by the 80-ns response time of the crystal. This results in the vertical band filling the fast time axis that only spans 120 ps.

3.3. *RF gun accelerated beam*

In August 1998, initial images were obtained of the RF gun beam at 50, 150, and 400-MeV stations. Preliminary measurements using the streak cameral for the beam at 400 MeV indicated

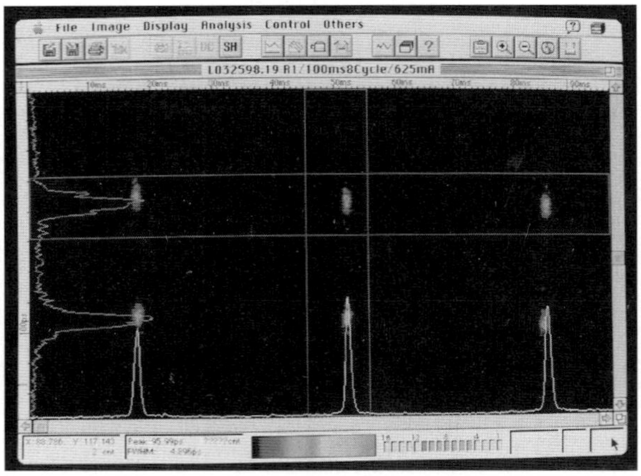

Fig. 6. Dual-sweep synchroscan images of the linac pulse. The horizontal axis is 100 ms and shows the 30-ns separation of the macropulses. The micropulse average bunch length is 4.9 ps (FWHM).

a bunch length of the un-optimized beam of 8–10 ps (FWHM).

4. Summary

In summary, optical techniques have been developed for electron beam measurements within the APS SASE project. The OTR and Ce-doped YAG crystals provide a trade on conversion efficiency, response time, and spatial resolution that cover the near-term needs of the project. The techniques have been validated in the DC thermionic gun beam, and initial measurements on the RF thermionic gun beam have been performed.

References

[1] A.H. Lumpkin, Nucl. Instr. and Meth. A 393 (1997) 147.

[2] M. Borland, An improved thermionic microwave gun and emittance preserving transport line, Proceedings of the 1993 Particle Accelerator Conference, Washington, DC, May 17–20, 1993, pp. 3015–3017.

[3] S.V. Milton et al., The advanced photon source low-energy undulator test line, Proceedings of the 1997 Particle Accelerator Conference, Vancouver, B.C., May 12–16, 1997, submitted for publication.

[4] A.H. Lumpkin et al., Nucl. Instr. and Meth. A 296 (1990) 769.

[5] R. Laird, F. Lenkszus, Argonne National Laboratory, private communication, April 1997.

[6] A.H. Lumpkin, W.J. Berg, B.X. Yang, Planned optical diagnostics for the APS low-energy undulator test line, Proceedings of the 1997 Particle Accelerator Conference, Vancouver, B.C., May 12–16, 1997, submitted for publication.

[7] W.S. Graves, R.D. Johnson, P.G. O'Shea, A high resolution electron beam profile monitor, Proceedings of the 1997 Particle Accelerator Conference, Vancouver, B.C., May 12–16, 1997, submitted for publication.

[8] N. Sereno et al., Bunch length measurements at the Advanced Photon Source (APS) linear accelerator, Proceedings of the 1995 Particle Accelerator Conference, vol. 2, 1996, pp. 1070–1072.

ELSEVIER
Nuclear Instruments and Methods in Physics Research A 429 (1999) 341–346
NUCLEAR INSTRUMENTS & METHODS IN PHYSICS RESEARCH Section A
www.elsevier.nl/locate/nima

Transverse emittance measurements from a photocathode RF gun with variable laser pulse length

D.A. Reis[a], M. Hernandez[b,*], J.F. Schmerge[b], H. Winick[b], M.J. Hogan[c]

[a] *University of Rochester, Dept. of Physics and Astronomy, Rochester, NY 14627, USA*
[b] *Stanford Synchrotron Radiation Laboratory, SLAC Box 4349, Stanford, CA 94309, USA*
[c] *Stanford Linear Accelerator Center, Box 4349, Stanford, CA 94309, USA*

Abstract

The gun test facility (GTF) at SSRL was started in 1996 to develop an appropriate injector for the proposed linac coherent light source (LCLS) at SLAC. The LCLS design requires the injector to produce a beam with at least 1 nC of charge in a 10 ps or shorter pulse with no greater than 1π mm mrad normalized rms emittance. The photoinjector at the GTF is 1.6 cell S-band symmetrized gun and emittance compensation solenoid. Emittance measurements, reported here, were made as function of laser pulse width using Gaussian longitudinal pulses. The lowest achieved emittance to date with 1 nC of charge is 5.6π mm mrad and was obtained with a pulse width of 5 ps (FWHM) and is in agreement with simulation. There are indications that the accelerator settings for these results may not have been optimal. Simulations also indicate that a normalized emittance meeting the LCLS requirement can be obtained using appropriately shaped transverse and temporal laser/electron beam pulses. Work has begun on producing temporal flat top laser pulses which combined with transverse clipping of the laser is expected to lower the emittance to approximately 1π mm mrad for 1 nC with optimal accelerator settings.

Keywords: GTF; Photocathode; RF gun; LCLS

1. Introduction and motivation

The proposed linac coherent light source (LCLS) at the Stanford Linear Accelerator Center (SLAC) is a single-pass, self-amplified spontaneous emission, free-electron laser (FEL) utilizing the last third of the 3 km SLAC linac [1] and operating at a wavelength of 1.5 Å. A critical component of the LCLS is a high brightness electron source. The source must be able to produce a 1 nC beam with $\leq 1\pi$ mm mrad normalized rms transverse emittance and 10 ps pulse length so that the FEL can saturate in a 100 m long wiggler (assuming the pulse is compressed to 70 fs rms and the emittance grows in the linac and bunch compressors to 1.5π mm mrad at the wiggler entrance).

The gun test facility (GTF) was constructed to develop an appropriate injector for the LCLS and is located in the Stanford Synchrotron Radiation Lab (SSRL) injector vault. The GTF consists of a photo-cathode drive laser, RF power stations, 3 m

* Corresponding author.

0168-9002/99/$ - see front matter
PII: S0168-9002(99)00079-0

SLAC linac section, RF gun and emittance compensating solenoid as well as electron beam diagnostics to characterize the beam [2]. The first gun being characterized at the GTF is the result of a collaboration between Brookhaven National Laboratory (BNL), SLAC and the University of California at Los Angeles. The prototype 1.6 cell gun and emittance compensation solenoid was characterized at BNL [3]. Emittance measurements reported here are made as a function of laser pulse length. The GTF expects to produce lower emittance beams by controlling the laser (and thus the electron beam) pulse length and shape.

2. Drive laser

The photocathode drive laser is a chirped-pulse amplification based Nd:glass system. The oscillator is a passively mode-locked diode pumped glass laser capable of generating < 200 fs transform limited pulses [4]. The repetition rate of the oscillator is set by the cavity length and is roughly 119 MHz, the 24th subharmonic of the accelerating RF. Low-timing jitter (< 3 ps rms) has been achieved by using a slow photo-diode (< 150 MHz) to sample the pulse train from the laser oscillator and frequency multiplying the signal in a phase-locked multiplier to generate the 2856 MHz master RF.

The pulse train from the oscillator is chirped to ~ 300 ps/nm in a grating pair expander, and then a single pulse is selected out at 1.25 or 2.5 Hz for amplification in a Nd : glass regenerative amplifier. The ~ 2 mJ, 1054 nm pulse is then compressed to as short as 1.1 ps (FWHM) using a grating pair with the opposite chirp of the expander. The pulse is frequency doubled and quadrupled in a 10 mm KD*P and 5 mm BBO crystals, respectively, to generate 200–300 μJ of 263 nm light. The data reported here was obtained with variable Gaussian temporal pulse widths by adjusting the chirp in the optical compressor. Pulses as short as 4.6 ps (FWHM) in the UV have been measured using a sub-picosecond resolution streak camera [5] and include ~ 1 ps spatial chirp across the beam. Fig. 1 shows a typical measurement of the 3 UV pulse widths used in the experiments. The UV pulse is typically imaged onto the cathode at near-normal incidence using an 8 m 1:1 telescope. Grazing incidence is also possible and was used for the cleaning of the cathode. Transverse shaping can be accomplished by imaging an aperture, after the quadrupling crystal, onto the cathode.

3. Timing jitter measurements

Typical measurements of the timing jitter at the GTF were conducted using a double-balanced

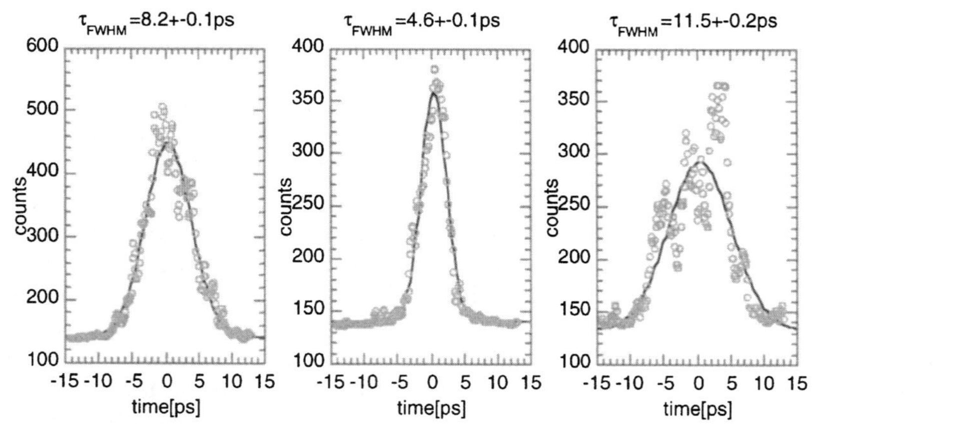

Fig. 1. Streak camera output of three different UV pulses with Gaussian fits. The different pulses lengths are obtained by varying the chirp in the compression stage.

mixer as a phase bridge to measure shot to shot variations in the phase of different RF signals relative to the 2856 MHz master clock derived from the laser. It was found that the total jitter between the RF gun and this master clock was about 0.5 ps rms over a few minutes with negligible contributions from the klystron and its drive. It is believed that this jitter comes from a shot to shot change in the difference between the drive frequency and the gun resonance and could be produced from either ~ 100 Hz noise in the laser oscillator or $\sim 0.1°$C change in the gun temperature.

The jitter between the 2856 MHz master clock and the 119 MHz from the laser was measured with a fast sampling scope and found to be stable to < 2 ps rms where the upper limit is set by the trigger jitter of the scope. The amplified laser pulse was found to be stable to < 2 ps rms by phase comparing the bandpass filtered output of a photodiode to the 119 MHz again using a double-balanced mixer. While no direct measurement of the jitter in arrival time of the laser at the cathode have been made, we can estimate the jitter to be < 3 ps rms (assuming the above measurements are uncorrelated). However, the long-term drift between the laser and the gun can be on the order of tens of ps in the course of a day. In principle, this could be eliminated with a slow feedback system on the laser cavity length, the gun temperature, or on the laser injection phase.

4. Electron beam measurements

Because of the low quantum efficiency ($\sim 5 \times 10^{-6}$ electron/photon) and large scale ($\sim 100\%$) non-uniformity of our copper cathode, we elected to clean the cathode in situ. A relatively high intensity ($\sim 2 \times 10^9$ W/cm^2) laser pulse was incident on the cathode in the presence of ~ 100 MV/m electric field after attempts to clean the cathode without the presence RF power failed. During the cleaning process ~ 1 μC of charge is extracted from the cathode in ~ 50 ns depleting the stored energy in the gun. The enhanced emission during this process appears to be the same as previously reported [6] and was achieved by focusing the laser onto the cathode at grazing incidence to roughly $1 \times 1/3$ mm^2. The cathode was "cleaned" by continuously scanning the laser across an area of roughly 5×3 mm^2 along the short axis of the laser at a rate of 22 shots/mm while maintaining the enhanced emission throughout each pass and with an overlap of $\frac{1}{3}$ mm between passes. The cleaning resulted in a larger quantum efficiency (typically $2\text{–}3 \times 10^{-5}$ at our normal operating conditions) and a more uniform cathode (large scale). However, the dark current (field emission) from the cathode also increased by roughly an order of magnitude. The cleaning process has qualitatively been seen as a micro-roughening of the cathode surface and may explain the increase in dark current by creating areas of increased localized field. It is believed that the increase in the quantum efficiency may be from removal of copper oxide from the cathode surface in addition to the localized field enhancement.

The electron yield/photon was measured as a function of the laser phase (Schottky scan) on a Faraday cup just after the emittance compensation solenoid. Assuming a 100% collection efficiency this is equivalent to the quantum efficiency. A typical scan is shown in Fig. 2. The data was obtained at a single solenoid setting for phases below 100°, but due to a significantly decreased energy and increased energy spread, the solenoid had to be varied for phases above 100° to optimize charge collection on the Faraday cup. The scan is used to determine the operating phase by observing where emission first takes place. Due to the finite duration of the laser pulse, the error in determining this phase is on the order of the pulse duration. For the emittance scans, the zero phase was taken to be a few degrees into emission. The Schottky scan can also be used in determining the thermal emittance of the electron beam. Following the formalization given in Ref. [7] we can estimate the thermal emittance to be $\sim 0.2\pi$ mm mrad (from the data presented in Fig. 2 and assuming a 1 mm radius spot at the cathode). Because of uncertainties in the charge collection, only data taken at the constant solenoid fields are used in the calculation.

The horizontal emittance of the electron beam was measured using the standard quadrupole scan technique. While measurements were made at peak currents of up to 200 A at an energy of 35 MeV, simulations using the envelope equation with space

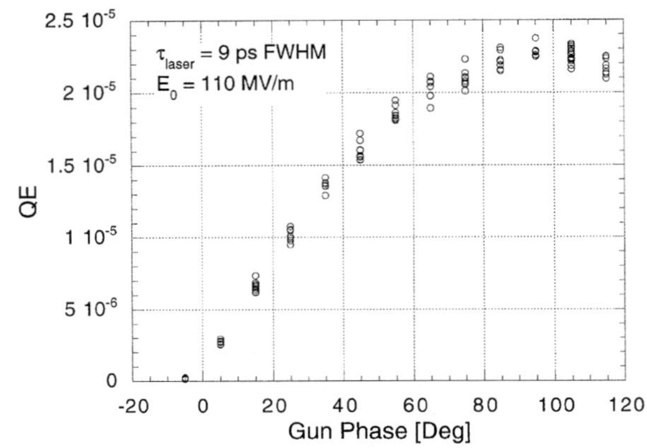

Fig. 2. A typical Schottky scan from which the laser phase relative to the gun and the thermal emittance can be estimated. The quantum efficiency is determined from the charge measured on the Faraday cup (with the average dark-current background subtracted) and is normalized to the number of laser photons.

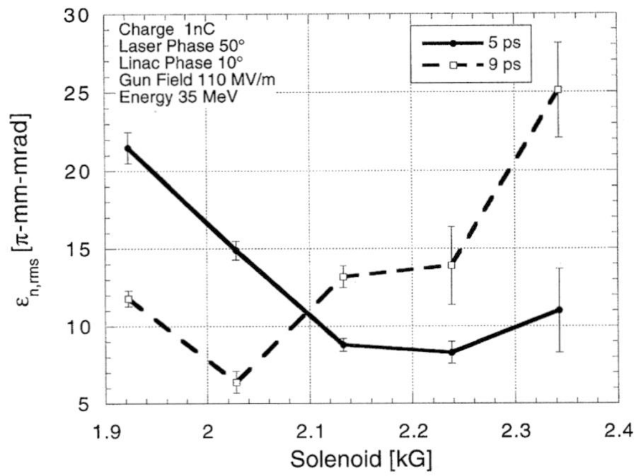

Fig. 3. Emittance as a function of solenoid field with 1 nC of charge for two different laser pulse lengths. The laser phase is referenced to the zero-crossing in the cathode-cell of the gun, and the linac phase is referenced to the crest in the linac. The gun field is the maximum on axis accelerating field.

charge [8] have indicated no appreciable increase in the calculated emittance. The screen material used was SLAC chromate [9]. The image was acquired with an 8-bit CCD Pulnix camera and DataTranslation frame grabber both synchronized to the electron beam. The beam was focused to the smallest possible spot and an image was acquired. The image was examined for saturation of the camera which was controlled by using a set of crossed polarizers between the lens and the camera. This allowed intensity control without changing the depth of field of the system. The lens used was a microscope objective with an adjustable zoom. A set of 127 μm diameter wires on the screen were used for calibration purposes. The lens was set to give a calibration of 8 μm per pixel to allow a compromise between capturing the full variation of the beam size and maximizing the resolution of smallest spot size. Once it was determined that the scan would be free of saturation, the quadrupole was standardized. Five images of the beam and a background shot (taken with the RF on, but the laser off) were taken at each quadrupole setting. The following signals were logged for each shot: the field probe in the full cell of the gun, a Joule meter sampling the laser energy incident on the cathode, the phases of the laser and the linac, a toroid upstream of the screen, and the image of the electron beam. Quad scans were performed at various solenoid settings to try to find the magnetic field which produced the best emittance compensation [10] for a given set of running conditions. The background subtracted images of the beam were analyzed off line and the rms widths of the beam profiles were found using the peak in the distribution as the center.

Emittance measurements were made with 1 nC of charge using a transverse Gaussian pulse with sigmas of 1.3 and 0.8 mm for x and y, respectively, at a laser phase of 50° and at 35 MeV. Lengthening the laser pulse from 5 (FWHM) to 9 ps (FWHM) reduced the emittance from $8.3 \pm 0.7\pi$ mm mrad at a solenoid field of 2.2 kG to $6.4 \pm 0.7\pi$ mm mrad at a solenoid field of 2.0 kG, respectively, as shown in Fig. 3. This is believed to be due to the reduction of space-charge forces by lengthening the electron bunch. Measurements were also made with a laser pulse length of 11 ps (FWHM) using a transverse Gaussian pulse with sigmas of 1.3 and 0.8 mm for x and y, respectively, at a laser phase of 50° and at 35 MeV which produced $6.6 \pm 0.2\pi$ mm mrad at a solenoid field of 2.0 kG. However, only 0.7 nC of charge was available, so this data was not included in Fig. 3. Conversely, for a lower charge of 0.4 nC with a transversely clipped beam (1 sigma) the 5 ps laser pulse produced an emittance of $6.3 \pm 0.4\pi$ mm mrad at a solenoid field of 2.0 kG while the 9 ps bunch produced $8.6 \pm 0.4\pi$ mm mrad at a solenoid field of 1.9 kG both at 35 MeV. These measurements were only performed once and will be repeated during the next experimental run. However, individual quad scans which were repeated were consistent within the uncertainty of the measurement process.

The operating parameters for the data reported here were not optimum from theoretical considerations but were considered necessary from an operational point of view. Optimal emittance compensation is anticipated for fields around 140 MV/m [11]. However, the field in the gun was limited to 110 MV/m to avoid occasional arcs which tripped off the RF system. The laser phase which produces the maximum charge emitted from the gun (see Fig. 2) is not at the same phase which produces a bunch with the maximum energy (for fixed field level in the gun and laser power on the cathode). Since we were interested in producing a high charge, in most cases the laser phase was chosen closer to the peak in the Schottky scan than is ideal. Future measurements of the effect of laser phase on emittance are planned. PARMELA [12] simulations indicate that, for the present distance between the gun and linac, more efficient emittance compensation is achieved for a lower linac gradient yielding a final beam energy of 26 MeV. This is believed to be due to better matching of the RF focusing in the linac at the lower gradient. In fact, the lowest emittance measured for 1 nC of charge was $5.6 \pm 0.2\,\pi$ mm mrad at 26 MeV at a laser phase at 60° and solenoid field of 2.0 kG. However, it was found that for a higher gradient, the energy spread of the electron beam was smaller, which made the determination of the phase of the bunch with respect to the RF in the linac more precise and enabled greater consistency in tuning the beam. Therefore, the energy of the beam was chosen to be 35 MeV for most of the measurements despite the expected lower emittance at reduced linac gradients.

According to PARMELA simulations, altering the temporal laser profile from a Gaussian pulse shape to a flat-top-pulse shape will reduce the emittance to the level required by the LCLS [11]. Work is currently underway to produce temporal flat-top laser pulses by use of a Michelson interferometer-based pulsed stacker (similar in design to Ref. [13]) or by the use of a mask in the Fourier plane of the optical compressor.

Acknowledgements

The authors would especially like to acknowledge the work of the late Jim Weaver. Jim was instrumental in building the GTF RF gun and other components of the GTF. In addition, we greatly appreciate the loan of the GLX-200 Nd : glass laser oscillator from Steve Milton at the APS, the Nd:glass regenerative amplifier head and Pockels' cells from David Meyerhofer and Adrian Melissinos at the University of Rochester and the Hamamatsu FESCA-500 streak camera from Bob Siemann at SLAC. The RF gun currently under test at the GTF was part of the thesis work of Dennis Palmer at SLAC with contributions from Roger Miller at SLAC, X.-J. Wang at BNL and others. Additional assistance with the laser system was provided by David Meyerhofer. The authors would also like to acknowledge the assistance of the entire SSRL staff in the construction of the GTF. Finally, we acknowledge Max Cornacchia at SSRL and Jym Clendenin, Alan Fisher, Theo Kostseroglou, and Dennis Palmer at SLAC for useful discussions. This work was supported by the Department of Energy under contract number DE-AC03-76SF00515.

References

[1] M. Cornacchia, Performance and design concepts of a free electron laser operating in the X-ray region, SPIE, Free-Electron Laser Challenges, vol. 2988, 1997, pp. 90–96, San Jose.

[2] J.F. Schmerge et al., SLAC RF photocathode gun test facility, SPIE, Free Electron Laser Challenges, vol. 2988, 1997, pp. 2–14, San Jose.

[3] D.T. Palmer et al., Emittance Studies of the BNL/SLAC/UCLA 1.6 cell Photocathode RF Gun, IEEE Particle Accelerator Conference, 1997.

[4] Time-Bandwidth Products, model GLX-200, Zurich.

[5] Hamamatsu, model FESCA-500, streak camera.

[6] X.J. Wang et al., J. Appl. Phys. 72 (3) (1992) 888.
[7] J.E. Clendenin, G.A. Mulhollan, High quantum yield, low-emittance electron sources, 15th ICFA Advanced Beam Dynamics Workshop, 1998.
[8] M. Reiser, Theory and Design of Charged Particle Beams, Wiley, New York, 1994.
[9] S. Yencho, D.R. Walz, IEEE Trans. 32 (1985) 2009.
[10] B.E. Carlsten, Nucl. Instr. Meth. A 285 (1989) 313.
[11] M. Hernandez et al., Emittance measurements for the SLAC gun test facility, IEEE Particle Accelerator Conference, 1997.
[12] PARMELA, Lloyd Young, Los Alamos National Lab.
[13] C.W. Siders, A.J. Taylor, High-energy pulse train and shaped pulse generator using a "100%" throughput 2^n-pulse Michelson interferometer, Ultrafast Opt. Conf., Monterey, CA, 1997.

ELSEVIER

Nuclear Instruments and Methods in Physics Research A 429 (1999) 347–351

NUCLEAR INSTRUMENTS & METHODS IN PHYSICS RESEARCH Section A

www.elsevier.nl/locate/nima

A numerical study of emittance growths in RF guns

Kai Masuda[a,*], Takashi Inamasu[b], Masaaki Sobajima[a], Jiro Kitagaki[a], Masami Ohnishi[a], Hisayuki Toku[a], Kiyoshi Yoshikawa[a]

[a]*Institute of Advanced Energy, Kyoto University, Gokasho, Uji, Kyoto 611-0011, Japan*
[b]*Sharp Corporation, shinjo-cho, Kita-katsuragigunn, Nara 639-2198, Japan*

Abstract

A beam with greatly reduced emittance is required for further improvements of FELs, in particular, for FELs of shorter wavelengths, and of narrower bandwidths. From this viewpoint, the BNL/SLAC/UCLA 1.6-cell S-band photocathode RF gun performance characteristics were calculated, first in order to evaluate what may contribute to the emittance growths in photocathode RF guns. We developed an RF gun to produce an electron beam with an extremely low emittance, by using a 2-D simulation code. It is found that, by optimizing the laser injection phase, the drive laser spot radius and the cavity shape around the laser spot, the beam emittance by the 1.6-cell RF gun can be greatly reduced to 2.1 πmm mrad, from the previous 4.4 πmm mrad of the original shape.

1. Introduction

The objectives of this study are, first, to evaluate the contributions to the beam emittance in a present RF gun, and, then, with their improvements, to design the optimal gun configuration and operation scheme to produce an extremely low emittance beam.

As the prototype (original) RF gun, the BNL/SLAC/UCLA 1.6-cell photocathode RF gun [1–3] was calculated, which has shown experimentally a good beam performance of 1 nC, 4πmm mrad [3]. The interest of this study is to clarify quantitatively what factors may result in emittance growths in the gun, and how much the emittance can be reduced with the original cavity configuration as well. The relationship between the laser injection phase and the emittance, the space charge effects, and the effects of the initial thermal emittance at the cathode were investigated. In addition to achieving a much lower beam emittance, design refinements of the laser spot size, and the cavity shape around the laser spot were carried out.

2. Modeling of the RF gun

A set of 2-D codes [4–6], which we have developed, was used to model the RF gun. Relativistic electron motions are calculated in the time-domain, taking into account space charge effects by solving Maxwell's equations self-consistently. Also, cavity fields are calculated, and taken into account in the trajectory calculations.

Fig. 1 shows the upper-half of the configuration of the prototype 1.6-cell RF gun with a calculated eigenmode pattern. The RF frequency is 2856

* Corresponding author. Tel.: + 81-774-38-3442; fax: + 81-774-38-3449.

E-mail address: masuda@iae.kyoto-u.ac.jp (K. Masuda)

0168-9002/99/$ - see front matter
PII: S0168-9002(99)00085-6

MHz, and the maximum electric field on the axis is taken to be 124 MV/m as a reference. A typical simulation result is presented in Fig. 2 showing snapshots of electron trajectories (upper) and their corresponding energies (lower). Electrons are produced at $z = 0$ by the drive laser incident onto the copper cavity wall. The incident electron beam parameters are chosen as; 1 nC charge, longitudinally Gaussian (21 ps FWHM) pulse shape within 15 ps pulse length with a ramp edge, and radially flat top within 1 mm radius with a ramp edge. In the simulations, 50 particles are injected at each time step ($dt = 0.24$ ps). The injection phase ϕ of the drive laser was varied from 0° to 55° with respect to the incidence of the 15 ps laser pulse.

3. Emittances with the original design

Fig. 3 shows the normalized rms emittances $\varepsilon_{n,rms}$ at $z = 1$ and 50 cm from the cathode as functions of ϕ. The emittance $\varepsilon_{n,rms}$ at $z = 50$ cm for $\phi \geq 10°$ was calculated to be $4.3 < \varepsilon_{n,rms} < 4.4\,\pi$mm mrad, which is found to agree well with the experimentally measured value of 4.74 ± 0.24 πmm mrad [3].

As is shown in Fig. 3, it is found that the emittances tend to take minima for $\phi = 0$. Furthermore, the emittance growth from $z = 1$ to 50 cm tends to increase as ϕ increases. The sliced normalized rms emittance $\varepsilon_{n,rms}$ is found to be decreasing as ϕ increases ($\varepsilon_{n,rms} = 1.7, 1.0$, and $0.8\,\pi$mm mrad for $\phi = 5°, 25°$, and 45° at $z = 50$ cm). It is defined by the normalized rms emittance of the particles injected at each time step, and is mainly varied by nonlinearity of the radial cavity fields. This is mainly due to increased bunch length (see Fig. 4), which

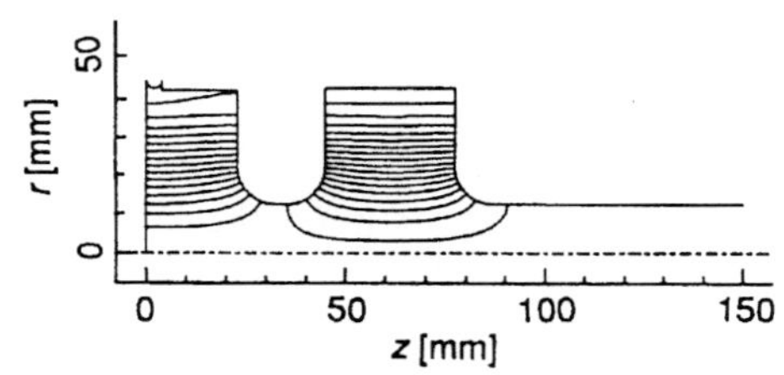

Fig. 1. Calculated cavity fields in the 1.6-cell RF gun.

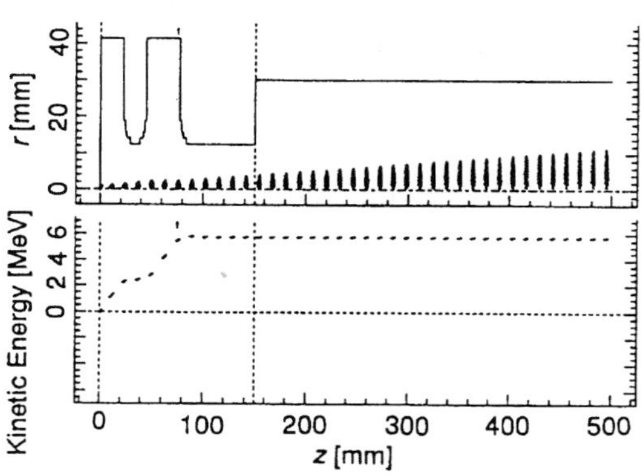

Fig. 2. Simulation results, showing snapshots of electron bunches, and corresponding kinetic energies.

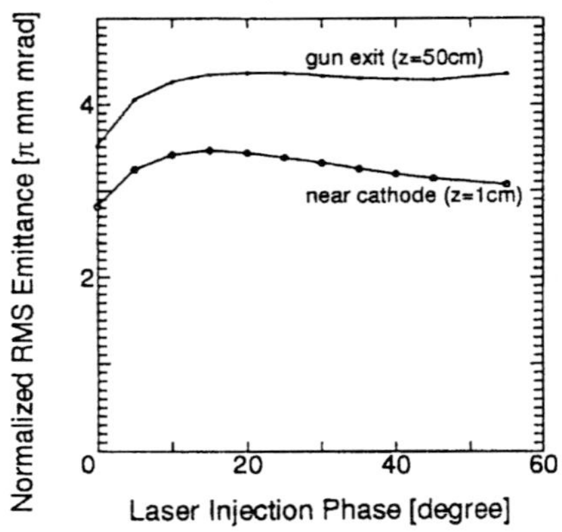

Fig. 3. Normalized rms emittances as functions of laser injection phase.

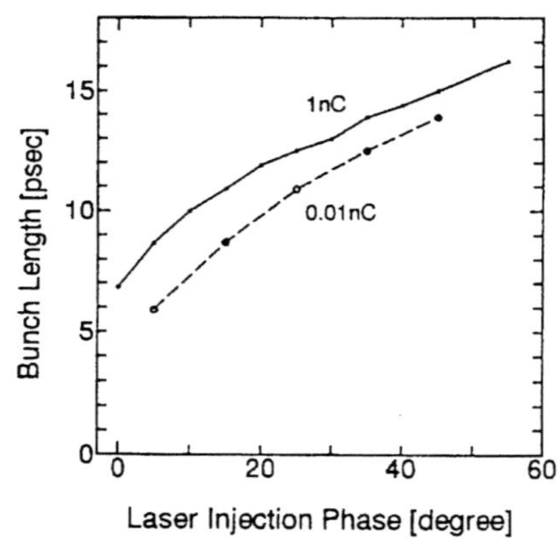

Fig. 4. Bunch lengths as functions of laser injection phase.

eventually increases the RF contribution to the emittance growth [3,7] (see Fig. 5) as ϕ increases. On the other hand, for the $\varepsilon_{n,rms}$ at $z = 1$ cm for $\phi \geq 15$, it is found to decrease as ϕ increases. This is due to decreased transverse space charge effects, leading to a small beam radius R. As is shown in Fig. 6, the RF contribution is found to scale as $\varepsilon_{n,rms} \propto R^2$.

To evaluate and compare the space charge effects on the emittance growth, beams with 1, 0.1, and 0.01 nC were chosen for $\phi = 15°$. In Fig. 7, it is clearly seen that the contribution due to the space charge is significant. The $\varepsilon_{n,rms}$ at $z = 50$ cm for 0.01 nC, i.e., in the case of non-space charge effects only, was calculated to be 0.43 πmm mrad, which may be regarded as the theoretically achievable minimal emittance with the pulse length of 15 ps.

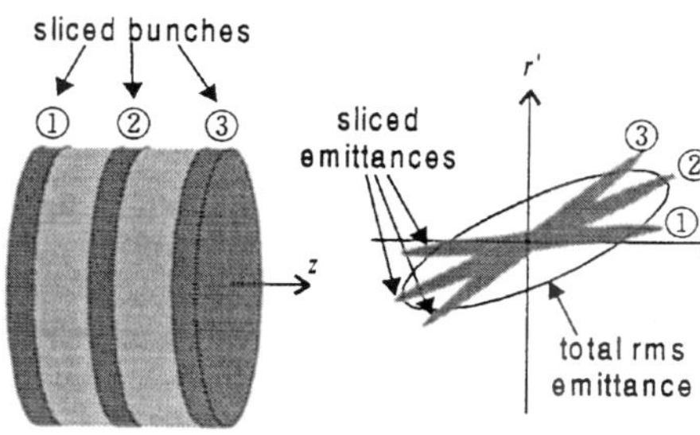

Fig. 5. Schematics of the RF contribution to the emittance. Each sliced bunch has a low sliced emittance with $r'/r \approx$ constant. However, since the sliced bunches have different r'/r with respect to their incident phases due to time-varying RF fields, the total emittance is found comparatively high.

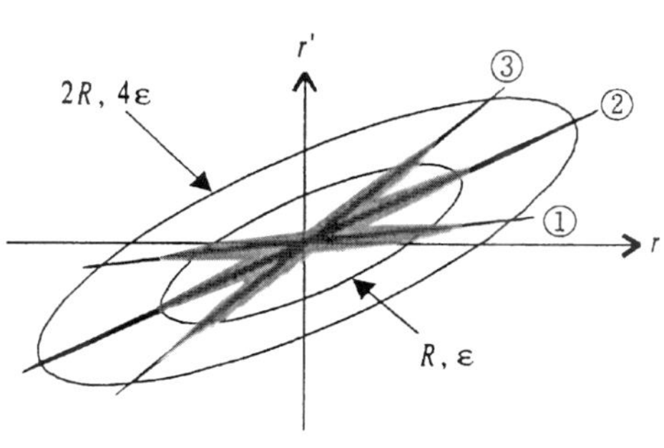

Fig. 6. Schematics of relation between beam radius and the RF contribution. Even with the same sliced emittances, the total emittance becomes higher for a larger beam radius, due to the RF contribution to the total emittance.

Also, the longitudinal space charge forces are found to result in bunch lengthening as shown in Fig. 4, which consequently increases the RF contribution to the emittance growth as well.

Also, the contribution to the emittance due to the initial thermal emittance was evaluated, by comparing the emittances with or without an initial thermal emittance of $\varepsilon_{0,n,rms} = 0.4\,\pi$mm mrad. As presented in Table 1, for the 0.01 nC beam, the contribution is found to be $\Delta\varepsilon_{n,rms} = 0.13\,\pi$mm mrad, which is much smaller than the thermal emittance of $\varepsilon_{0,n,rms} = 0.4\,\pi$mm mrad. In addition, for the 1 nC beam, the contribution is found to be much smaller, $\Delta\varepsilon_{n,rms} = 0.01\,\pi$mm mrad. This indicates that the thermal emittance contribution is much smaller than other contributions due to RF, and space charge forces. Especially, for the 1 nC beam, the thermal emittance contribution is found to be negligible, $\Delta\varepsilon_{n,rms}/\varepsilon_{n,rms} = 0.003$.

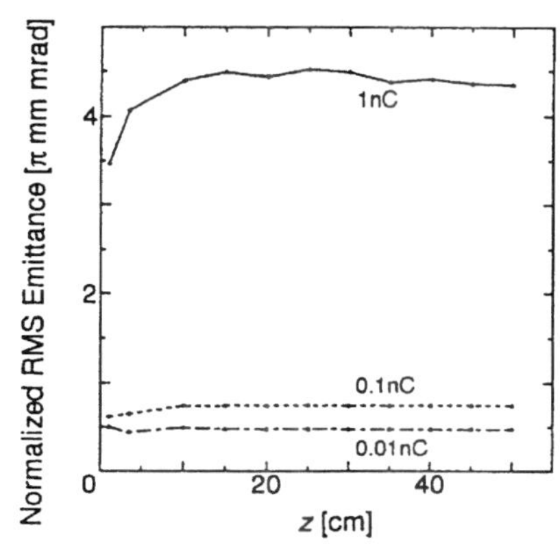

Fig. 7. Emittance evolutions of 1, 0.1, and 0.01 nC bunches along the axis.

Table 1
Thermal emittance contribution to the total emittance, for beam charges of 0.01 and 1 nC

Charge (nC)	$\varepsilon_{n,rms}$[πmm mrad]		$\Delta\varepsilon_{n,rms}$ (πmm mrad)
	$\varepsilon_{0,n,rms} = 0$	$\varepsilon_{0,n,rms} = 0.4$	
0.01	0.61	0.48	0.13
1	4.37	4.38	0.01

4. Design refinements

From the above discussions (1) a short bunch length, and (2) a small beam radius are found essential, in order to reduce the RF contribution to make a low emittance beam of 1 nC. The former can be achieved by optimizing the laser injection phase ϕ as is shown in Fig. 3, and/or by use of a shorter pulse laser with higher laser power. As for the latter to achieve a smaller beam radius, and a consequent improved emittance, the incident beam radius R_0 (laser spot radius), and the cavity shape were optimized in this study with the following results.

Fig. 8 shows $\varepsilon_{n,rms}$ as a function of R_0, for $\phi = 5°$. The $\varepsilon_{n,rms}$ is found to take the minimum of 3.2 πmm mrad for $R_0 = 2.2$ mm. For $R_0 < 2.2$ mm, due to transverse space charge forces, the beam radius tends to become rather large in the gun cavity, which consequently causes a higher emittance at the gun exit.

To compensate the transverse space charge forces, and, as a consequence, to produce a smaller beam radius, the cavity shape around the cathode plane had to be improved. Repeated trial-and-error design refinements eventually resulted in the cavity shape shown in Fig. 9. The cavity wall around the laser spot is set $\alpha = 15°$ off the laser spot plane to produce transverse focusing fields. With a larger $\alpha > 15°$, the electron beam was found to be too focused, which eventually result in a rather large beam radius due to increased space charge forces, and the consequent higher emittance at the gun exit.

With this refined cavity shape, for $\phi = 5°$, the $\varepsilon_{n,rms}$ was calculated as a function of R_0 as is shown in Fig. 10, where the circle corresponds to the cavity shape and the laser spot size of Fig. 9. This design refinement is found efficient in offsetting the transverse space charge forces, even for small R_0. As a consequence, with the original R_0 of 1 mm, the $\varepsilon_{n,rms}$ is found to be greatly reduced to 2.1 πmm mrad, compared with 4.1 πmm mrad with the original cavity shape for $\phi = 5°$.

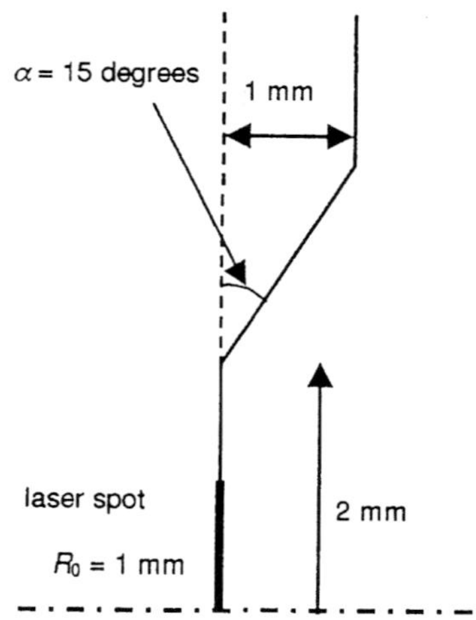

Fig. 9. Design refinements of the cavity shape around the laser spot, to compensate the transverse space charge forces.

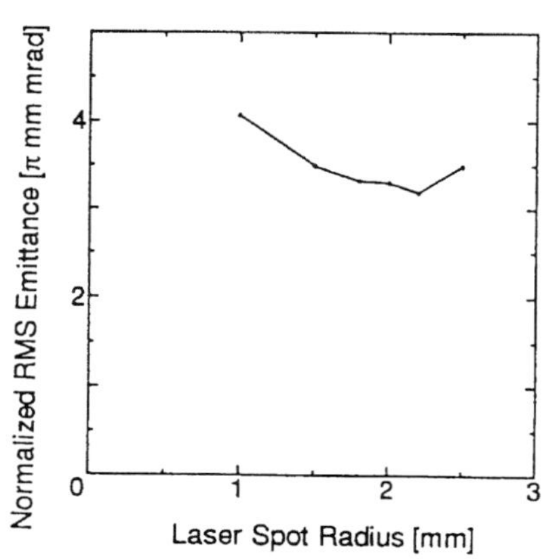

Fig. 8. Emittance as a function of laser spot radius.

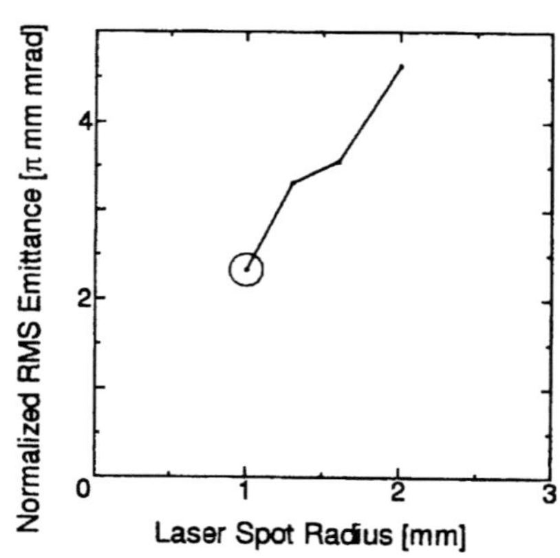

Fig. 10. Emittance as a function of laser spot radius, with the refined cavity shape shown in Fig. 9. The circle corresponds to the laser spot radius shown in Fig. 9.

5. Conclusions

In case of a 1 nC electron beam, it has been found that the contribution to the beam emittance due to space charge effects is dominant through the interaction with RF fields, compared with initial thermal emittance at the cathode. For reduction of emittance, it has been found essential to achieve electron beams with short bunch lengths, and small beam radii in order to reduce the RF contribution, i.e., unfavorable interaction with time-varying RF fields, which tends to enhance beam emittance. Through numerical evaluations, it is found that the bunch length can be reduced by adjusting the laser injection phase, while the beam radius can be reduced by refining the cavity shape around the laser spot. With these improvements adopted, the beam emittance consequently has been found to be reduced down to 2.1 πmm mrad, from the original 4.4 πmm mrad.

References

[1] D.T. Palmer, X.J. Wang, I. Ben-Zvi, R.H. Miller, SLAC-PUB-7420, 1997.

[2] D.T. Palmer, X.J. Wang, I. Ben-Zvi, R.H. Miller, J. Skaritka, SLAC-PUB-7421, 1997.

[3] D.T. Palmer, X.J. Wang, R.H. Miller, M. Babzien, I. Ben-Zvi, C. Pellegrini, J. Sheehan, J. Skaritka, H. Winick, M. Woodle, V. Yakimenko, SLAC-PUB-7422, 1997.

[4] K. Masuda, Ph.D. Thesis, Dept. of Electrical Engineering, Kyoto University, 1997.

[5] K. Masuda, K. Yoshikawa, M. Ohnishi, Y. Yamamoto, H. Toku, M. Sobajima, J. Kitagaki, Development of an improved two-dimensional finite element code for cylindrically symmetric eigenmodes, IEEE Trans. Microwave Theory Tech., in press.

[6] Y. Yamamoto, T. Inamasu, K. Masuda, M. Sobajima, M. Ohnishi, K. Yoshikawa, H. Toku, E. Tanabe, Nucl. Instr. and Meth. A 393 (1997) 443.

[7] L. Serafini, J.B. Rosenzweig, Phys. Rev. E 55 (6) (1997) 7565.

ELSEVIER

Nuclear Instruments and Methods in Physics Research A 429 (1999) 352–357

NUCLEAR INSTRUMENTS & METHODS IN PHYSICS RESEARCH Section A

www.elsevier.nl/locate/nima

High average current 2-MeV electron accelerator for a high-power free-electron laser

Byung Cheol Lee[a,*], Young Uk Jeong[a], Sung Oh Cho[a], Jongmin Lee[a], Sergey Miginsky[b], Gennady Kulipanov[b]

[a] *Laboratory for Quantum Optics, Korea Atomic Energy Research Institute, P.O. Box 105, Yusong, Taejon 305-600, South Korea*
[b] *Budker Institute of Nuclear Physics, Novosibirsk, Russia*

Abstract

A high average current CW RF electron accelerator has been developed for the free electron laser programs at KAERI. The accelerator is composed of a 300-keV electron gun, one RF bunching cavity, and two RF acceleration cavities. The kinetic energy of the electron beam is 1.5 MeV nominally and 2 MeV at maximum. The duration of a pulse is 350 ps and its repetition rate is variable from a single pulse to 22.5 MHz. The peak current is 6 A, and the average current, at the maximum repetition rate, is 45 mA. The resonant frequency of the RF cavities is 180 MHz. The energy gain of an RF acceleration cavity is 0.6 MeV nominally and 0.85 MeV at maximum. The total RF power supplied into the RF cavities is 262 kW. A millimeter-wave FEL driven by the 2-MeV accelerator is under design. The 2-MeV accelerator will be used as an injector of a high-average-power infrared FEL.

PACS: 41.60.Cr; 41.75.Ht

Keywords: Free electron laser; Electron accelerator; Electron gun; RF cavity

1. Introduction

In order to drive a high average power free-electron laser (FEL), a high average current electron accelerator is indispensable. A recirculating radio frequency (RF) accelerator with energy recovery seems to be the best choice for this purpose [1–3]. With the energy recovery method, the energy of the "used" electron beam is converted into RF power, and the energy of the "dumped" electron beam is lower than that of the "injected" electron beam. Thus the RF power consumption and the radiation hazard are reduced dramatically. With the multiple-turn recirculation concept, the system dimension, the power consumption, and thus the system cost are also reduced very much.

One of the critical issues, in the development of the recirculating accelerator with energy recovery, is the generation of high-average-current electron beam with proper energy for injection and with beam quality good enough for recirculation and energy recovery. The choice of accelerator structure, especially the type and the frequency of the RF cavity, is very important. Two high-average-current accelerators, using super-conducting acceleration cavities, have been in successful operation

* Corresponding author. Tel.: + 82-42-868-8378; fax: + 82-42-861-8292.

E-mail address: bcleey@nanum.kaeri.re.kr (B.C. Lee)

0168-9002/99/$ - see front matter
PII: S0168-9002(99)00086-8

[4,5]. Recently, KAERI has finished installation and commissioning of a 2-MeV, CW, normal-conducting in cooperation with BINP. The average current of the electron beam from the accelerator is 45 mA at maximum. The accelerator will be used as the injector for a high-energy recirculating accelerator for a high power FEL. In our short-term plan, the accelerator will also be used for a high power millimeter-wave FEL and a tunable X-ray generator.

2. The electron accelerator system

Fig. 1 shows a schematic of the whole accelerator facility including the RF generator and the control system. Fig. 2 shows a schematic of the electron accelerator, and Fig. 3 shows a photograph of the accelerator. The accelerator is composed of a 300-keV electron gun, one RF bunching cavity, two RF acceleration cavities, a 180-degree bending magnet, and a beam dump. Main parameters of the electron

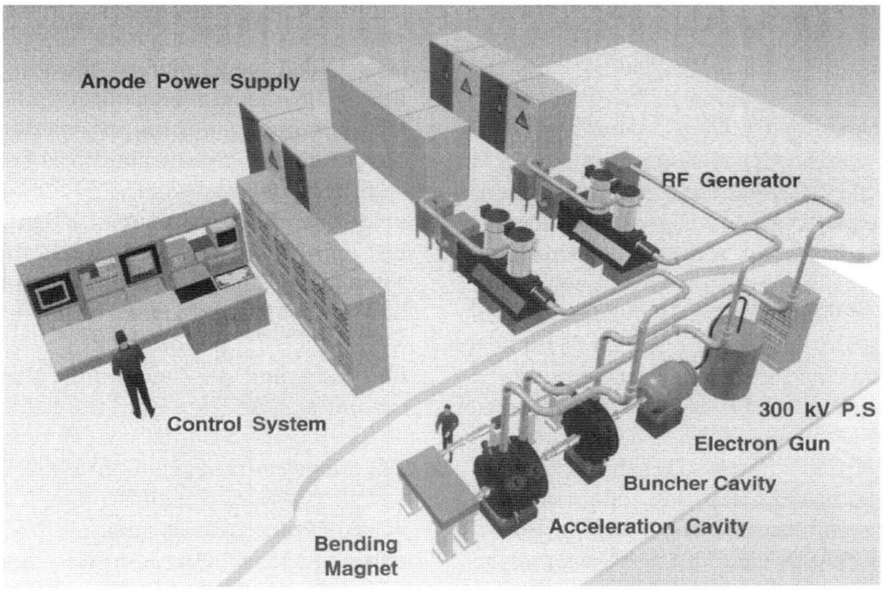

Fig. 1. Schematic of the high-average-current 2-MeV electron accelerator facility.

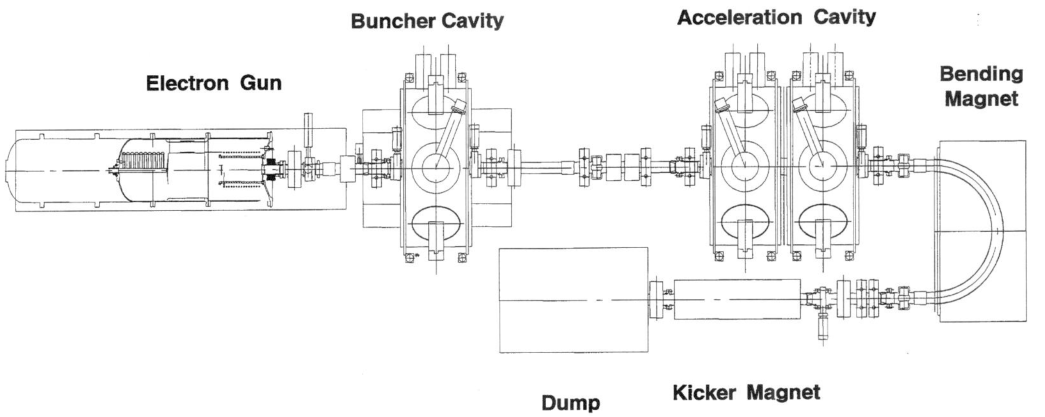

Fig. 2. Schematic of the electron beam accelerator.

Fig. 3. Photograph of the accelerator.

Table 1
Main parameters of the 2-MeV electron accelerator

Electron energy (kinetic)	1.5 MeV (Nominal) 2.0 MeV (Maximum)
Current	6 A (Peak) 45 mA (Average)
Emittance	40π μm rad
Repetition rate	0...22.5 MHz
Pulse duration	350 ps
RF frequency	180 MHz
Operation mode	CW/Single pulse

accelerator are listed in Table 1. The resonance frequency of the RF cavity is 180 MHz. The maximum repetition rate of the electron beam should be equal to the RF frequency divided by the number of turns, that is $\approx 180/8 = 22.5$ MHz. The normalized emittance should be $\varepsilon_n \sim 150\,\pi\,\mu$m rad in order to enable lasing at $\sim 5\,\mu$m wavelength. The bunch length should not exceed 300 ps (transformed to the energy spread $\Delta E/E \sim 2\times10^{-3}$ in the undulator). To obtain peak current ~ 100 A in the undulator it is necessary to get ~ 6 A peak current and $\Delta E/E \sim 10^{-3}$ from the injector.

2.1. The electron gun

The electron gun is composed of a cathode-grid unit controlled by a modulator, a static linear accelerating tube placed in a high-pressure vacuum vessel with insulating SF_6 gas, a 300-kV DC power supply with control electronics, a power supply for the high-voltage part of the gun with an isolation transformer, and control electronics with optical signal transmission to and from the high-voltage part. Fig. 4 shows a schematic of the electron gun. The parameters of the electron gun are listed in Table 2.

In order to get a very short pulse electron beam (1.5 ns), we use a tetrode-type thermionic cathode. The cathode-grid unit operates in the space-charge-limited emission mode. The gap and voltage between the cathode and the grid electrode are 0.1 mm and 100 V, respectively. The voltages on the grids are optimized to get minimum emittance of the electron beam. The emittance calculated by uisng a simulation code does not exceed $20\,\pi\mu$m

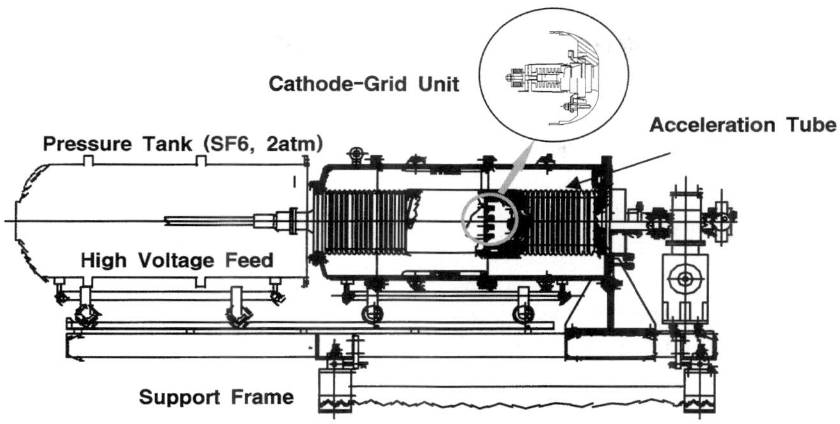

Fig. 4. Schematic of the 300-keV, 45-mA electron gun.

Table 2
Parameters of the 300-keV electron gun

Electron energy (kinetic)	300 keV
Current:	1.25 A (Peak) 45 mA (Average)
Emittance	160 π μm rad
Repetition rate	0...22.5 MHz
Pulse duration	1.6 ns
Cathode material	BaO:CaO:MgO
Operation mode	CW/Single pulse

rad. The pulse duration of 1.5 ns corresponds to an $\approx 100°$ phase interval at the main accelerating (and bunching) frequency of 180 MHz. If the duration is larger, a significant number of electrons are lost during bunching.

2.2. The RF cavities

The resonance frequency of the bunching RF cavity is 180 MHz [6]. The cavity has copper clad stainless-steel walls. The Q value of the cavity is 42 000, and the shunt impedance is 8.5 MΩ. The tuning range of the resonant frequency is 320 kHz, and the tuning speed is 5 kHz/s. The heat loss at a cavity voltage of 850 kV is 100 kW.

The RF power supplied into the bunching cavity is 2 kW, and the corresponding RF-voltage in the cavity is 110 kV. For the best bunching of the electron beam, the RF phase is chosen so that the energy loss of the electron bunch is 30 keV after passing the bunching cavity. Since the power lost by electrons is comparable to the RF power, the phase and the amplitude of RF in the cavity should be controlled very carefully.

The main accelerating section is composed of two independently-controlled RF cavities. The accelerating cavities have almost the same structure and parameters as those of the bunching cavity. The energy gain from each accelerating cavity is 600 keV nominally, and 800 keV at maximum. The distance between the bunching cavity and the first accelerating cavity is chosen such that the bunching process occurs optimally.

2.3. The transport beamline and the beam measurement system

The transport beamline is intended for guiding a beam through the vacuum chamber, minimizing emittance at the output of the injector, and fitting parameters of the beam to the recirculating main accelerator. It consists of sets of magnetic correctors, solenoidal lenses, quadrupole lenses, and a 180° bending magnet.

The beam measurement system is intended for measuring all the significant parameters of the electron beam, namely: position, average current, average energy, time structure, vertical and horizontal emittances, and energy spread. It consists of a beam position monitor; a DC current monitor; a pulse current monitor; and a beam profile monitor module with a kicker magnet. The pulse current monitor measures the image charge due to pulse electron beam, and its time resolution is 100 ps.

3. Beam dynamics

One of the critical issues in the design and the operation of the accelerator is the question: is the pulse duration of 1.5 ns short enough to capture all of the electron beam? The pulse duration of 1.5 ns is almost the state of the art limit we can get by using an electrical modulator. Fig. 5 shows the result of the simulation of beam trajectories starting from the cathode-grid unit and ending at the exit of the second accelerating RF cavity. The space charge effect and the influence of the cavities, solenoids, and vacuum chamber are taken into account. The vertical axis in the figure is the energy of electron beam in keV, while the horizontal axis is longitudinal position expressed in picoseconds. The simulations show that most part of a 1.5-ns pulse can be bunched to < 100 ps duration with an energy spread of less than 20 keV. The "tails" laying out of 1.5 ns duration hardly can be bunched, and will be lost after passing the bending magnet due to large energy spread. Fig. 6 shows the measured pulse waveforms of an electron bunch (a) at the exit of the second RF cavity and (b) at the exit of the electron gun.

Another important issue is the beam instability at a high average current of electron beam. With up to 1 MHz of repetition rate, the percent of beam transport to the dump is more than 95%. At 4 MHz, the transport decreases to 75%, which seems to be related to the beam instability at

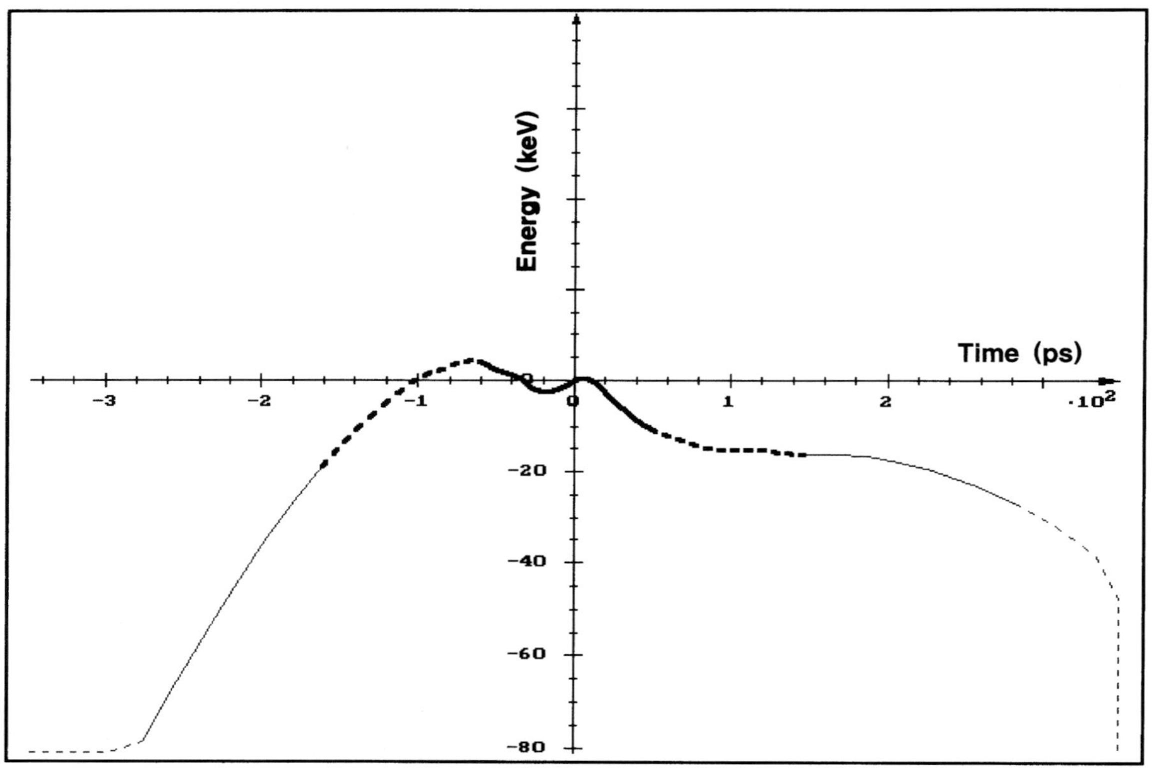

Fig. 5. Phase space portrait of an electron bunch at the exit of the second RF cavity. The pulse duration of the bunch before entering the bunching cavity is 1.5 ns.

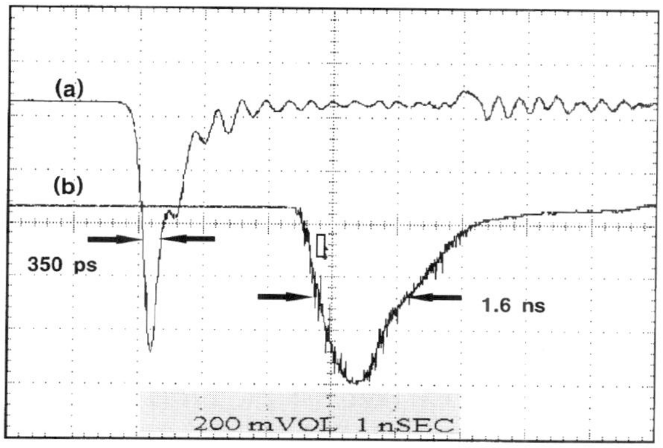

Fig. 6. Measured pulse forms of a bunch (a) at the exit of the second RF cavity and (b) at the exit of the electron gun.

high-average current. The analysis of beam instability and the improvement of beam transport is under investigation.

4. Conclusion

A high-average-current CW electron accelerator using normal-conducting RF cavities has been developed and commissioned. The energy and the average current of the electron beam are 2 MeV and 45 mA, respectively. The accelerator will be used as an injector of a higher energy accelerator for high-power infrared FEL. In the short-term plan, the accelerator will be used for a high-power millimeter-wave FEL, which is now under design [7].

References

[1] N.A. Vinokurov et al., Nucl. Instr. and Meth. A 375 (1996) 403.
[2] N.A. Vinokurov et al., LBNL-40081 UC414, March 1997.
[3] K.W. Berryman et al., SPIE 2376 (1995) 53.
[4] E. Minehara et al., Proceedings of 19th International Conference on Free Electron Laser, II-45, Elsevier, Amsterdam, 1998.
[5] S.V. Benson et al., Nucl. Instr. and Meth. A 407 (1998) 401.
[6] V.S. Arbuzov et al., Proceedings of the 1993 Particle Accelerator Conference, vol. 2, 1993, p. 1226.
[7] Sun Kook Kim et al., High-average-power millimeter-wave FEL driven by a 2-MeV CW electron accelerator, Abstract Book of the 20th International FEL Conference, 1998.

ELSEVIER

Nuclear Instruments and Methods in Physics Research A 429 (1999) 358–364

NUCLEAR INSTRUMENTS & METHODS IN PHYSICS RESEARCH Section A

www.elsevier.nl/locate/nima

The magnetic and diagnostics systems for the Advanced Photon Source self-amplified spontaneously emitting FEL☆

E. Gluskin[a], C. Benson[a], R.J. Dejus[a], P.K. Den Hartog[a], B.N. Deriy[a], O.A. Makarov[a], S.V. Milton[a], E.R. Moog[a,*], V.I. Ogurtsov[a], E.M. Trakhtenberg[a], K.E. Robinson[b], I.B. Vasserman[a], N.A. Vinokurov[c], S. Xu[a]

[a]*Advanced Photon Source, Argonne National Laboratory, XFD-401, 9700 S. Cass. Avenue, Argonne IL 60439, USA*
[b]*STI Optronics, 2755 Northup Way, Bellevue WA 98004, USA*
[c]*Budker Institute of Nuclear Physics, 11 Ac. Lavrentyev Prosp., 630090 Novosibirsk, Russia*

Abstract

A self-amplified spontaneously emitting (SASE) free-electron laser (FEL) for the visible-to-ultraviolet spectral range is under construction at the Advanced Photon Source at Argonne National Laboratory. The amplifier part of the FEL consists of twelve identical 2.7-m-long sections. Each section includes a 2.4-m-long, 33-mm-period hybrid undulator, a quadrupole lens, and a set of electron beam and radiation diagnostics equipment. The undulators will operate at a fixed magnetic gap (approx. 9.3 mm) with $K = 3.1$. The electron beam position will be monitored using capacitive beam position monitors, YAG scintillators with imaging optics, and secondary emission detectors. The spatial distribution of the photon beam will be monitored by position sensitive detectors equipped with narrow-band filters. A high-resolution spectrograph will be used to observe the spectral distribution of the FEL radiation.

Keywords: Free-electron laser

1. Introduction

The SASE FEL now under construction [1,2] at the Advanced Photon Source (APS) will consist of two major parts: the APS injection system 600-MeV linac coupled to a small-emittance electron gun, and a set of 12 undulators. Initial operations will be at a lower energy in order to produce visible 532-nm light. The parameters for initial operations are given in Table 1. Further information about the particle beam production and characteristics are given in Refs. [1,2]. After experience is gained with visible light, the linac energy will be increased, and the FEL will produce 120-nm ultraviolet light.

Simulations of the beam bunching in an FEL were performed [3], showing that the undulator line can have separated undulators without significant deleterious effects on the bunching. The drift spaces between undulators can then accommodate

☆Supported by the US Department of Energy, BES-Material Sciences, under Contract No. W-31-109-Eng-38.

* Corresponding author. Tel.: + 1-630-252-5926; fax: + 1-630-252-9303.

E-mail address: moog@anl.gov (E.R. Moog)

0168-9002/99/$ - see front matter
PII: S0168-9002(99)00087-X

diagnostics and quadrupoles for horizontal beam focusing. Since the horizontal focusing can be separate from the undulators, the design for the undulators becomes much simpler, and the approach used for the design and construction of the undulators on the APS storage ring can be directly applied to the FEL. Also, the insertion device magnetic measurement facility of the APS and the magnetic tuning techniques developed for the APS undulators are directly applicable to the measurement and tuning of the undulators for the FEL. A drift space also allows much greater flexibility in the choice of diagnostics for both the particle and light beams.

Table 1
FEL parameters

Wavelength	532 nm
Beam energy	220 MeV
Normalized emittance	5π mm-mrad
Peak current	150 A
Energy spread	0.1%
Focusing	Separate quadrupoles
Undulator period	33 mm
Undulator parameter K	3.1
Undulator effective field	10.061 kG
Nominal magnetic gap	9.3 mm (fixed)
Undulator length	2.4 m
Cell length	2.7265 m
Number of cells	12
Gain length	~ 0.8 m

These advantages have led to the choice of a separated undulator design for the APS FEL.

The FEL line will consist of a series of twelve identical cells, where each cell includes an undulator, a diagnostics section, and a quadrupole singlet. The choice of the lattice was made with the help of the FEL simulation codes [3]. Configurations with a single quadrupole, a doublet, or a triplet placed in the drift section were considered. The singlet was found to give the best particle beam bunching. The codes have also been used to optimize the quadrupole strength. The beta function for an undulator cell is shown in Fig. 1 for a quadrupole strength of 7.22 kG (for a focal length of 1 m) and an energy of 220 MeV. Vertical beam focusing is intrinsic to the undulator field; no additional vertical focusing between the undulators will be added. The horizontal focusing will match the vertical focusing for equal two-plane focusing.

The length of the drift space between successive undulators must be carefully chosen so as to maintain the proper phasing between undulators [4]. The phasing is affected by the strength of the undulator magnetic field and by the end fields. Measurements of the end field on existing magnetic structures and a calculation of their effect on the phasing have led to choosing a drift length of 326.5 mm from the end of one undulator to the beginning of the next undulator, for an overall cell length of 2.7265 m.

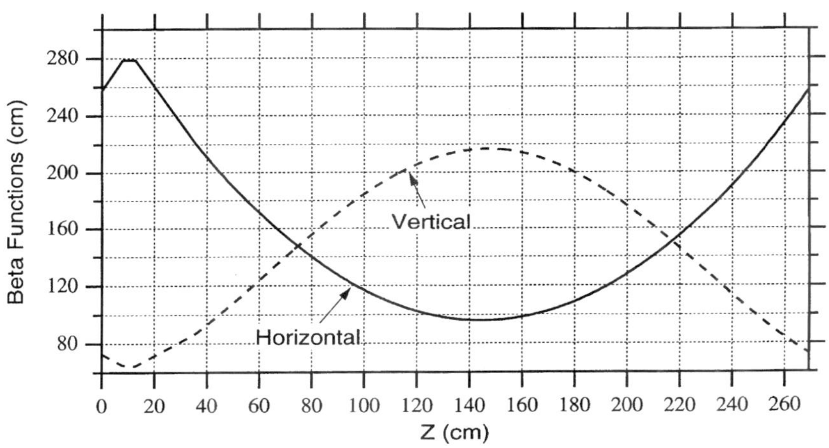

Fig. 1. The beta function for a cell of the FEL lattice. The cell consists of: a 7.785-cm drift length, followed by a 5-cm quadrupole, then a 22.665-cm drift length for optical diagnostics, and finally a 233.7-cm undulator (not including the end poles).

The functions of beam steering and horizontal beam focusing have been combined in the quadrupole magnet. Separate windings on the quadrupole poles allow it to also serve as a dipole correction magnet, steering the beam vertically and horizontally.

2. Characteristics of the FEL undulator

The period length of the undulators to be used for the FEL is 33 mm. Simulations of the expected gain have been performed using period lengths as short as 27 mm, and the results showed very little sensitivity to changes in the period. (Changing the period length from 27 mm to 33 resulted in a gain length change between + 20% and − 7% depending on whether the undulator K parameter was kept constant or allowed to increase as it normally would.) Therefore, since the well-understood and standard APS undulator A is a 33-mm-period device, the decision was made to proceed with that period length, and, in fact, with the same design for the magnetic structure. Since horizontal focusing is separate from the undulator, there is no need to cant the undulator poles [4], and the standard storage-ring undulator design can be used. However, the criteria for magnetic tuning for the single-pass, fixed-gap FEL are somewhat different than for a variable-gap undulator A in the storage ring, as described below. STI Optronics, of Bellevue, WA, designed, built and tuned the undulators A; they are also building and tuning the FEL magnetic structures. The final tuning of the ends to match the phasing [4] to the drift length will be done at APS.

The magnetic structures will be held at a fixed magnetic gap when they are installed in the FEL tunnel. For convenience in tuning, however, a variable-gap support and drive system will be used in the measurement room. Once the magnetic structure is tuned, it will be mounted on the fixed-gap system and its magnetic characteristics will be confirmed. In both the fixed-gap and variable-gap systems, the supports holding the magnetic structures apart are located at the same longitudinal positions along the undulator to minimize differences in the strongback deflections.

Some of the magnetic tuning requirements for FEL undulators are more demanding than for a storage ring undulator. For an FEL, it is critically important that the particle beam trajectory coincide with the axis of the emitted radiation, and that the coincidence extend over many gain lengths. The most convenient means of achieving this is to keep the trajectory of the particle beam straight through the undulator end regions as well as through the full-field regions. This leads to the requirement that the second field integral (averaged over each period) remain less than 3300 G-cm^2 through the entire length of the undulators, including the end sections. (Note that this is much less than the second integral of the vertical component of the earth's field over the length of an undulator, so that differences in the ambient field between the measurement room and the installation site need to be taken into account.) For a beam energy of 220 MeV, the requirement corresponds to a trajectory displacement of 45 μm. (The corresponding requirement for a storage ring undulator is that the second field integral through the full-field region be below 10^5 G-cm^2 for all gaps, with no special requirement for the ends.) The requirement that we be able to confirm the trajectory straightness in the vertical direction to this accuracy means that the horizontal field component must be measured accurately, despite the planar Hall effect. The work done to develop this capability is reported elsewhere [5].

The effective magnetic field strengths for each undulator must be matched so that the light produced by one undulator is at the resonant wavelength for the next. Simulations were carried out in which a K of 3.10 was assigned to some undulators and a K of 3.11 to others, corresponding to a change in undulator field of 32 G. The electron bunch peak current density dropped by ~ 35% compared to the ideal case at one position along the FEL, but began to recover further downstream. This change in field strength is larger than the 8 G difference that would make the wavelengths from different undulators different by 5% of the width of the first harmonic peak from one gain length of undulator. The undulators will be tuned to place all the effective fields within a 15-G-wide range. A field strength change of 15 G could be caused by a change in magnetic gap of 16 μm. The undulator gap will be near 9.3 mm, but the gap of each undulator will be adjusted individually to compensate for undulator-to-undulator variations.

The requirement that the field strengths for the undulators be nearly identical leads to a temperature uniformity requirement for the FEL line. The permanent magnets are made of Nd–Fe–B, which loses strength reversibly with a temperature coefficient of 0.09%/°C. There is also a change in gap, since the spacer blocks that hold the magnetic structures apart will expand. For a 1°C increase in temperature, the loss in magnet strength will result in a field decrease of 9 Gauss, and the thermal expansion of brass spacers will result in a field decrease of about 0.2 Gauss. To keep a < 15 Gauss variation requirement, the temperature of any undulator must be the same as the temperature of any other undulator to within 1.5°C (or ± 0.75°C).

The vertical focusing of the particle beam by the undulator is the result of variation in the undulator field with vertical position, so variations in the midplane from pole to pole can affect the focusing. We sought to determine the vertical center under each pole with an accuracy better than the 50 μm tolerance on the overall vertical position [6]. The preferred technique, given the drift in the Hall probe zero, was to measure $B_y(y)$ under each pole (y is vertical position). The minimum value of B_y is the center height. The results of the measurements are shown in Fig. 2. Most of the scatter in the magnetic center heights is not due to measurement error. Instead, it arises from a variety of factors inherent in the assembly of undulators. The permanent magnet blocks vary in strength; for this device the rms error in the magnet strength was 0.43%. There can be errors in the mechanical height of the poles. The shims, especially the skew shims, used for tuning the device can affect the magnetic center height, as can other sources of magnetic errors such as the presence of ambient magnetic material. This scatter is inconsequential, however, since it is on a much shorter scale than the β-function of the particle beam.

Other FEL requirements are less demanding than the corresponding requirements for storage-ring undulators. Since the FEL undulators will operate at a single fixed gap, magnetic tuning only needs to be done at that one gap. Also, a small phase error is important for a storage ring undulator to ensure high brilliance in higher order harmonics, whereas FEL operation only relies on a brilliant first harmonic. Because the brilliance of the first harmonic is much less affected by phase errors than the brilliance of higher harmonics, the rms phase error requirement is less demanding for an FEL undulator. The criterion used for the FEL is that the first harmonic intensity should not decrease by more than 5% due to phase errors, which leads to the requirement that the rms phase error be less than 10°.

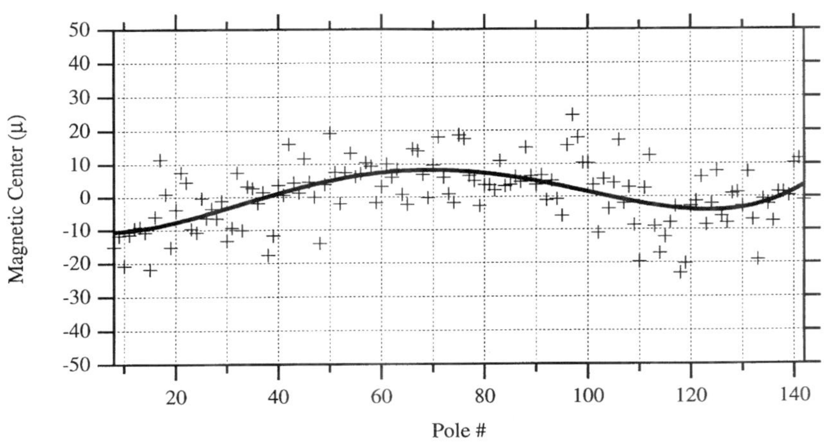

Fig. 2. The height of the magnetic center for each pole along an undulator at a gap of 9.3 mm. The rms error is 23 μm.

Simulations [6] have also been used to determine tolerances for the alignment of the undulators to the beam and to adjacent undulators. The calculated tolerances given in Table 2 are based on requiring that the power output does not change more than approximately 10% for a given parameter. The focal length of the quadrupole assumed for these calculations was 2.4 m. The simulations of the effect of the quadrupole strength on the beam bunching found that while a focal length of 1 m gave the best beam bunching, it also resulted in tighter tolerances. This somewhat longer focal length may be a better compromise.

Table 2
Acceptable tolerances

Parameter	Tolerance
Longitudinal undulator displacement	1 mm
Vertical undulator displacement	50 μm
Lateral undulator displacement	1 mm
Horizontal alpha function, α_x	0.20
Vertical alpha function, α_y	0.20
Horizontal beta function, β_x	0.50 m
Vertical beta function, β_y	0.20 m
Horizontal incident beam coordinate, x_0	200 μm
Vertical incident beam coordinate, y_0	50 μm
Horizontal incident beam angle, x'_0	100 μrad
Vertical incident beam angle, y'_0	50 μrad

3. Beam diagnostics

The diagnostics serve two purposes: one is to monitor and maintain the coincidence between the particle beam and the undulator radiation, the second is to evaluate the characteristics of the light that is produced by the FEL.

A schematic of the diagnostics section that will be located between the undulators is shown in Fig. 3. Since it is critical that the particle beam and the axis of the emitted light beam coincide through the entire series of undulators, three different and complementary monitors of the particle beam position have been included. The capacitive button BPM, or beam position monitor, is the same as the BPMs

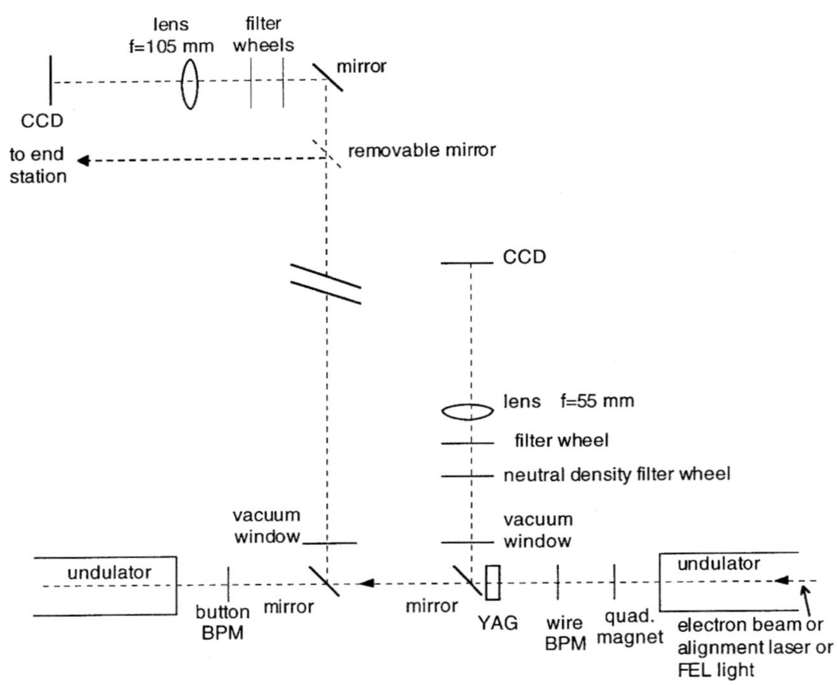

Fig. 3. Schematic of the diagnostics section (not to scale).

used at the ends of the insertion device straight sections in the APS storage ring. The relative positions of the buttons are different than in the storage ring, however – since the FEL vacuum chamber has a smaller vertical aperture than the usual storage ring ID vacuum chamber, the buttons will be vertically closer. They will also be closer transversely in order to improve their sensitivity [7]. The secondary-emission wire BPM is an absolute position monitor that consists of two perpendicular sets of four parallel wires. The spacing between the 15-μm wires is, in order, 0.5, 1, and 0.5 mm. The current to individual wires is monitored as the particle beam is steered to strike the wires. The beam can be centered vertically and horizontally by steering it to determine where it hits the wires on opposite sides of the beam centerline, then splitting the difference. During normal operation the beam will not strike the wires because the spacing between the central wires will be a few times the beam size. The third beam position monitor is the CCD image of the YAG scintillator crystal. The optical system will be designed to make the size of a CCD pixel comparable to the 10-μm resolution reported [8] for the YAG crystal itself.

Upstream of the undulators, there will be a chicane for the particle beam [9]. The synchrotron radiation produced at its bends will be monitored as a means of characterizing the particle beam, and it will also provide a place for an alignment laser to be inserted. The alignment laser will be directed down the inside of the vacuum chamber and will be used to define the desired straight-line beam path. Since the alignment laser light will travel through the same optical systems as the FEL light and the light from the YAG crystal, the desired position of the light on the CCD arrays can be defined.

The lens and CCD in the upper left of Fig. 3 will be used to check the distribution in angle of the incoming light as well as its position, by varying the focus from infinity to the downstream end of the nearest undulator. When the focus is at infinity, i.e., the distance between the CCD and the lens is the focal distance, all the light incident on the lens parallel to a particular angle will be imaged to the same point on the CCD. In this configuration, all position information about the incoming light is lost and the image on the CCD will reflect the distribution in angle of the incoming light. A deflection between undulators or a trajectory kick within an undulator will appear as a displacement in the CCD image. A bandpass filter will limit the angular spread of red-shifted light that reaches the lens. With the focus adjusted to lie at closer distances, such as within the undulator, the positions of the emitted light along the length of the undulator will be monitored.

One of the filter wheels in the upper left of Fig. 3 will carry bandpass filters; the other will have neutral-density filters so that the light levels can be adjusted to suit the CCD. Of the bandpass filters that have been selected, one will pass the on-axis first-harmonic FEL light. Another will pass red-shifted light, which will be off-axis and in the shape of a cone around the axis, with the angle between the cone and the axis depending on the wavelength transmitted. Using the red-shifted light to guide adjustments of the relative trajectories through two consecutive undulators may allow more accurate adjustments. The red-shifted light appears as a ring, and two rings are easier to align than two spots. Also, the width of the annulus of red-shifted light is smaller than the size of the on-axis spot, so the difference is between aligning two sharp rings as opposed to two broader spots.

As shown in Fig. 3, a mirror is inserted into the particle (and light) beam path in order to reflect the FEL or alignment laser light into the optics at the upper left of the figure. This mirror will have three positions: one where the mirror is removed from the beam path, one where the mirror completely blocks the beam and reflects all the light, and one with a small hole to allow the particle beam to pass, unperturbed, while still reflecting much of the light into the optical system. Demanding requirements have been placed on the motion of this mirror so that the position of the light on the CCD is repeatable to within a pixel despite the approximately 1-m-long distance between this mirror and the next mirror in the light path. In order to more readily achieve this repeatability, the direction of motion of the mirror between its different positions is parallel to the plane of the mirror face.

Another use for the optics in the upper left of Fig. 3 is as a diagnostic for the light produced by the FEL. Each set of these optics will be calibrated to the same intensity standard. They will then be used to measure the intensity from each undulator individually, as follows. The mirror in the particle beam path after the first undulator will be positioned so that the hole allows the particle beam to pass unperturbed. A small fraction of the undulator light will also pass through the hole, but most of it will be reflected into the optics where the absolute intensity of the light from the first undulator will be measured. The small amount of light that passes through the hole is still all the light from the first undulator that would interact with the particle beam in the second undulator to induce bunching. When the light is viewed after the second undulator, the contribution from the first undulator will be a small portion of the total intensity. Since the undulators are longer than a gain length, almost all the intensity will be from the second undulator. The light intensity from the different undulators can then be compared. If no beam bunching is occurring, then the absolute intensity seen after each undulator will be the same. Intensity measurements can be made after every one of the undulators for a single incident beam pulse, so that the intensity growth can be studied without variations introduced by changes in the incident beam.

A second diagnostic of the FEL light will be located in an end station. A Paschen–Runge-type spectrograph will be placed on the low-radiation side of a shielding wall at the downstream end of the undulator line. A schematic of the spectrograph is shown in Fig. 4. It will be used for high-resolution spectral measurements near the first harmonic, and since the goal is to measure the spectral structure in the SASE light, each pulse from the linac will be individually measurable. Light sent to this station will have been picked off after any one of the undulators (including after the last undulator), using the removable mirror shown in the upper left of Fig. 3, and then passed through a hole in the shielding wall. It will go through a bandpass filter

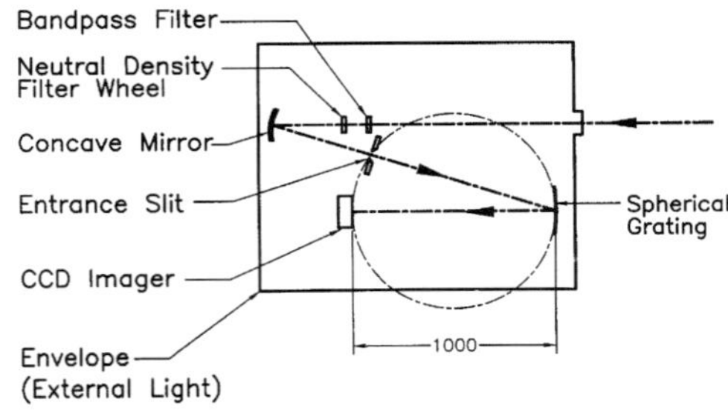

Fig. 4. A top view of the Paschen–Runge-type spectrometer that will analyze the light from the SASE FEL.

and, if necessary, a neutral density filter before being reflected and focused onto the slit by a concave mirror. The slit, the spherical grating, and the detector all lie on a Rowland circle. The CCD will be cooled to reduce the dark current and improve the signal-to-noise ratio. It is expected that the light from a single undulator with no FEL amplification and from a single incident bunch will be readily measurable.

References

[1] S.V. Milton et al., Nucl. Instr. and Meth. A 407 (1998) 210.

[2] S.V. Milton, J.N. Galayda, E. Gluskin, The advanced photon source low-energy undulator test line, Proceedings of PAC97, held 12–16 May 1997, Vancouver, BC, Canada.

[3] R.J. Dejus, O.A. Shevchenko, N.A. Vinokurov, Nucl. Instr. and Meth. A 429 (1999) 225.

[4] K.E. Robinson, D.C. Quimby, J.M. Slater, IEEE J. Quantum Electron. QE-23 (1987) 1497.

[5] I. Vasserman, Argonne National Laboratory Report No. ANL/APS/TB-32, 1998.

[6] R.J. Dejus, I.B. Vasserman, Argonne National Laboratory Report No. APS/IN/LEUTL/98-1, 1998.

[7] Glenn Decker, Argonne National Laboratory, private communication.

[8] J. Safranek, P.M. Stefan, Proceedings of EPAC '96, the Fifth European Particle Accelerator Conference, p. 1573.

[9] B. Yang, Argonne National Laboratory, private communication.

ELSEVIER

Nuclear Instruments and Methods in Physics Research A 429 (1999) 365–370

NUCLEAR INSTRUMENTS & METHODS IN PHYSICS RESEARCH Section A

www.elsevier.nl/locate/nima

Up-grading a 4.7-cm-period plane electromagnetic undulator

V.A. Bogachenkov, K.V. Kondratyev, V.A. Papadichev*

P.N. Lebedev Physical Institute, 53 Leninsky Prospect, 117924 Moscow, Russia

Abstract

Electromagnetic undulators have a number of advantages over permanent-magnet undulators. They are less expensive to fabricate and their field is easily regulated by changing the current, without requiring a complex and expensive precision system for changing the undulator gap. Their main weaknesses are that they require a large power supply and that the field is limited due to yoke saturation mainly because of large axial stray fluxes, particularly in simple constructions.

Modernization of a 4.7-cm-period, 20-period long plane electromagnetic undulator of simple design is described. Samarium-cobalt permanent magnets were used to increase the field and decrease power consumption. They were placed between adjacent rings (with opposite sign of field) and while increasing the working field they decreased saturation of the iron yokes. Small lateral displacements of permanent magnets were used to equalize field amplitudes in half periods of the undulator with 0.1% accuracy. Matched input and output to and from the undulator, respectively, were formed by means of auxiliary permanent magnets and special magnetic screens

Keywords: Electromagnetic undulator; Yoke saturation

1. Introduction

Electromagnetic undulators were the first to be employed for generating undulator radiation at storage rings [1] and in FELs [2] on various accelerators. Both helical and planar undulators were developed, the former being primarily pulsed (if not superconducting as in [2]) and the latter DC. Large power consumption dictates using iron, which allows a saving of nearly 50% in ampere-turns and 75% in power consumption (iron is also beneficial in facilitating electron beam focusing in both transverse directions). However, power consumption remains large and a problem, the practical limit of magnetic field amplitude being set by iron saturation. The periodic structure of an undulator with magnetic field of opposite signs in neighboring gaps, in contradistinction to dipole magnets, is unavoidably connected with noticeable parasitic axial stray fluxes between axially adjacent poles. These stray fluxes are especially large in simple designs. Placing the currents near the pole surface results in smaller stray fluxes, but this hampers considerably the manufacturing of a water-cooled coil of meander shape.

Some combined (EM + PM) schemes to counterbalance early saturation of yokes were proposed in Ref. [3]. A design with two sets of yokes, each having a common coil, and auxiliary

* Corresponding author.

0168-9002/99/$ - see front matter
PII: S0168-9002(99)00088-1

permanent magnets (as in a classical hybrid undulator) was tested with a 7-period (8-cm period, 3-cm gap) mock-up [4,5] yielding 5.8 kG peak field strength.

Two other prototypes based on a modified section of the electromagnetic undulator of the Paladin FEL experiment [5] used permanent magnets placed on the poles either near the gap, or on their sides. The results were also presented in several other publications [6–8]. Some increase of undulator field was obtained, but no results of further studies were published, and it seems that not all possible advantages and drawbacks of the proposed scheme were revealed.

A similar idea (with two separate sources of magnetic field) was used in a hybrid undulator [9]. The iron yokes had two air gaps: one centered on the undulator axis and the other on the opposite side (farthest from the axis). Three sets of permanent magnets were employed: one placed in the axial gaps between adjacent poles as in a classical hybrid undulator and two others in the side gaps of right and left arrays of yokes (instead of coils). The device yielded 7.4 kG for a gap/period ratio of $\frac{1}{3}$. It should be noted that the yokes in this device are less saturated due to the side air gaps (where PMs are placed). The higher field compared to Ref. [3] was obtained because the working gap was shunted by higher magnetic resistance due to the side air gap. So performance of such undulators is different from a combined EM + PM undulator having single gap near the undulator axis. Another difference is that the field amplitude is not varied in this PM + PM undulator. Therefore, the study of PM assisted electromagnetic undulators of various geometry is of some interest.

We use PM-added side-flux suppression in a simple design and have not yet employed water cooling of the coil. An undulator deflection parameter $K \approx 1$ in a 4.7-cm-period undulator was obtained. Further increase requires design optimization and a water-cooled coil.

Focusing in the orbit plane was ensured (as in the initial design) by a wedge-like cross-section form of the poles. Field errors were minimized to a 0.1% level by small displacements of permanent magnets in the transverse plane. Two methods of matched undulator input formation were realized employing an iron shield and supplementary permanent magnets.

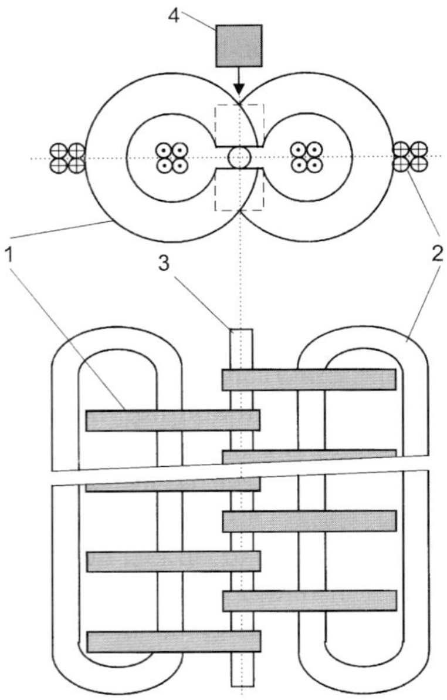

Fig. 1. Schematic drawing of a plane electromagnetic undulator. (1) C-shaped yokes, (2) coil, (3) beam transport tube, (4) permanent magnet (SmCo) insertion (placed into dashed squares in up-graded undulator).

2. Up-grading a 4.7-cm-period undulator

A 20-period undulator was fabricated six years ago[1] for a Lebedev Physical Institute FEL project [10] emphasizing simple design and inexpensive manufacturing. Two sections of a projected (not yet realized) water-cooled coil were to excite two sets of yokes (odd and even). The gaps of C-shaped yokes were along the undulator axis with odd yokes on one side of the axis and even ones on the other (see Fig. 1). Steel with 0.2% carbon was used instead of expensive Armco or Permendur as pole and yoke material. The rings were machined to have a wedge-like cross-section (slight cones) and gaps of 1.6 cm were then cut. Aluminum spacers ensured precision assembly of the undulator.

[1] The undulator was designed by Zakharov [10].

Preliminary tests with a simple uncooled coil showed that the maximum 1.4 kG field amplitude ($K = 0.6$ at 30 A) was limited by saturation of steel yokes. This was due to low saturation induction in steel and also non-optimal ratio of pole width to axial gap between adjacent poles. The maximum amplitude was slightly increased to 1.8 kG by placing one set of gaps 1.5 cm from the other in the lateral direction, sacrificing nearly a half of the aperture for transporting electron beam (the ring width is 3.4 cm, axial thickness – 1.2 cm).

Since the limiting field amplitude was primarily due to saturation of yokes, it was decided to insert SmCo permanent magnets between odd and even poles in the intersection region of their axial projections, with magnetic moments parallel or anti-parallel to the undulator axis as in the well-known hybrid undulator scheme (position 4 in Fig. 1). Gap fields excited by the coil and PMs are summed and yoke fluxes subtracted. A pair of magnets had $1 \times 4 \times 4\ cm^3$ overall dimensions. The remaining air gap of nearly 0.15 cm was filled with 0.35 mm thick transformer iron plates. Magnetic moments were measured preliminarily and the magnets grouped in pairs to yield minimal 3% rms deviation (6% maximal). Such a variation in moments was permissible since it could be compensated by a small lateral displacement of PMs. The magnetic field was measured by means of a Hall probe.

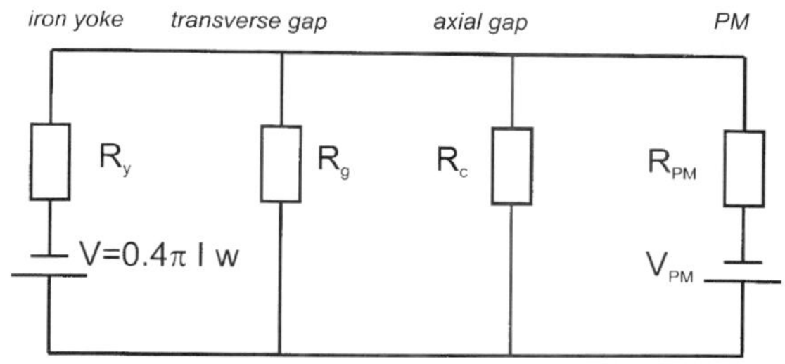

Fig. 3. Undulator magnetostatic circuit: R_y – magnetic resistance of iron yoke, R_g – resistance of a transverse air gap between upper pole and lower one, R_{PM} – resistance of an axial gap occupied by PM insertion, R_c – resistance of a remaining axial gap in projection intersection area (partial coupling resistance).

The measured dependencies of field amplitude on undulator current are shown in Fig. 2. Light circles (curve 1) and triangles (curve 2) correspond to the initial design for zero and 1.5 cm mutual lateral displacement of odd and even poles, respectively. Solid diamonds (curve 3) correspond to inserted PMs. Light squares represent an excitation curve without PMs when both sets of gaps (and yokes) have the same sign (dipole regime).

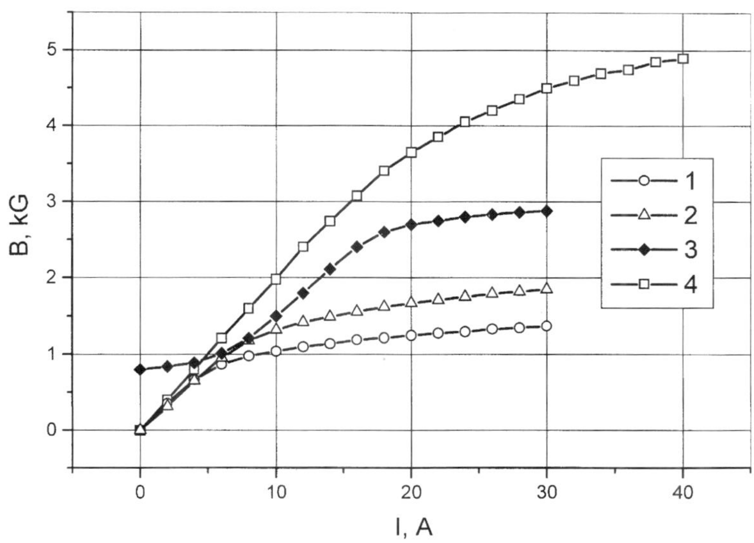

Fig. 2. Field amplitude dependencies on undulator current: (1 and 2, and 4) without permanent magnets; (1) zero mutual lateral displacement of sets of odd and even gaps, (2) 1.5 cm displacement; (3) with permanent magnets and without displacement; (4) without permanent magnets in dipole regime (same sign of field in all gaps).

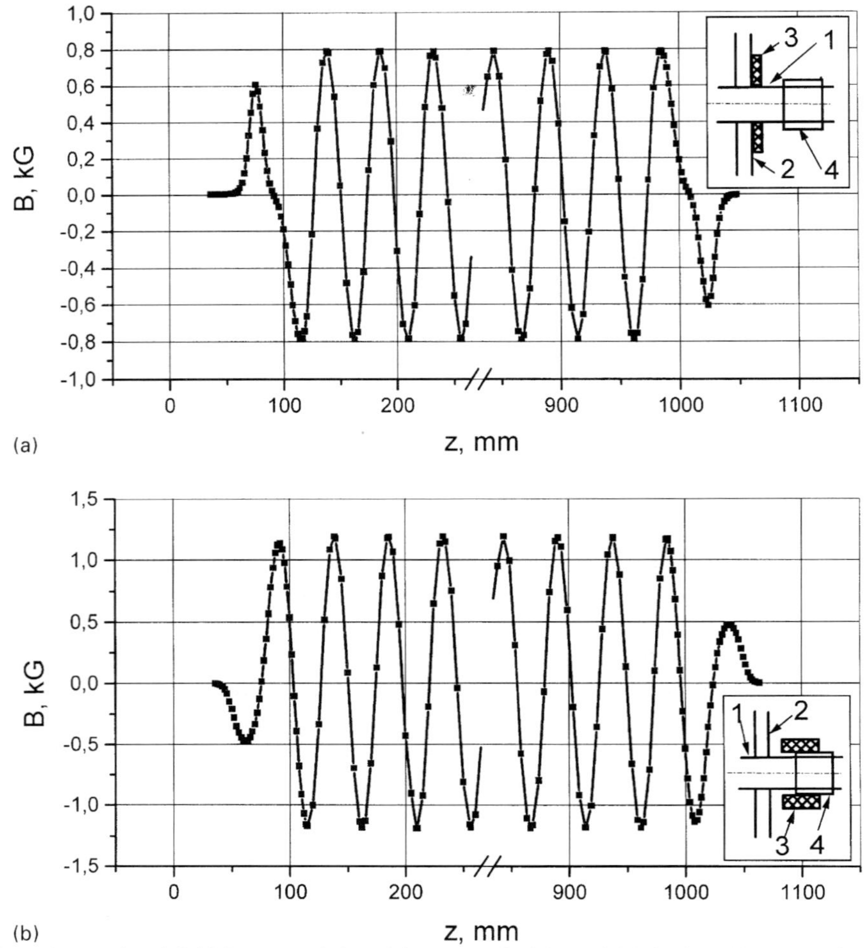

Fig. 4. Axial dependence of end field in up-graded undulator matched for axial injection: (a) zero excitation current, auxiliary permanent magnets are attached to end yokes, (b) excitation current $I = 8$ A, auxiliary magnets generating field parallel to undulator field are placed near the axis apart from end yokes. 1 – Beam transporting tube, 2 – end yoke, 3 – permanent magnet, 4 – magnetic screen.

Table 1

	Initial model	Up-graded model	Possible improvements
Undulator period, l_0, cm	4.7	4.7	4.7
Number of periods, N	20	20	20
Undulator length, L, m	0.94	0.94	0.94
Magnetic gap, h, cm	1.6	1.6	1.6
Specific power, P, kW/m	6	6	$\sim$10
Deflection parameter, K	0.6	1.3	$\sim$2.6
Field errors (RMS values), $\Delta B/B$,%	$\geqslant$1	0.2–0.4	0.2–0.4

A simple magnetostatic circuit can be drawn to assess various possible improvements (Fig. 3). It is possible to use experimental data of Fig. 2 and magnetization curves of yoke material for better fitting circuit parameters and predicting undulator field variation with changes in circuit elements. One can determine the value of saturation induction for steel yokes by measuring the axial dependence of field in a dipole regime and taking into account transverse stray fluxes. It was approximately 10 kG, which is less than one-half of that for Vanadium Permendur (VP). So with VP one can attain 5.6 kG even in this non-optimized geometry (ampere-turns and PM moments should be doubled, of course).

A further way to increase the undulator field is to optimize yoke and pole widths. The former could be made up to four times wider, the latter – nearly 1.5 times thinner. The optimum is a compromise between direct field increase due to PM flux through the undulator median plane, which gets larger with yoke saturation, and higher field excited by ampere-turns, when saturation is less influential because of PM flux of opposite sign in the yokes. Rough estimates based on the above magnetic circuit show a possible increase of 1.4 times up to 6–7 kG. The power consumption is not high: 1.6 kW/m for this model with $K = 1$, and 10 kW/m for possible further improvements with $K = 2$ and even, perhaps, $K \approx 3$ (See Table 1).

To control field amplitude with high precision when adjusting the lateral position of PMs, the 100 Hz current component was suppressed up to 0.2%, resulting in 0.04% alternating field component. DC current was stabilized with 0.1% accuracy.

RMS error is 0.1% (three half-periods at each end of the undulator were excluded from error calculation since they were included in fringe field integrals). After adjusting amplitudes at $I = 8$ A with 0.1% accuracy, one can vary the current in an 8–16 A interval (up to $K = 1$) with rms errors not exceeding 0.4%. This accuracy is sufficient, since calculations of permissible spectral line width give an estimate of 2% (gamma = 13.5), corresponding to 2 mm lateral beam displacement [11].

The simplest method of beam injection in an undulator is to form its fringe field so that electrons moving parallel to the axis acquire the necessary transverse velocity corresponding to the orbit angle at zero field points (so called matched input). End field formation in such a combined undulator differs from both parent prototypes: a classical plane electromagnetic undulator with iron and a PM hybrid one. Two methods were tried in this study, both employing a magnetic shield (iron tube) to cut excessive magnetic flux and auxiliary SmCo permanent magnets to add missing flux. Field axial dependencies for both cases are presented in Fig. 4 with schematic drawings of iron tube and auxiliary magnets. Fig. 4a is for zero excitation current, b – for $I = 8$ A. The second method (b), when auxiliary PMs are placed apart from end yokes, is more universal and can be employed for any undulator current (and in any plane undulator).

3. Conclusion

The up-graded electromagnetic undulator (combined with SmCo permanent magnets) has some advantages over pure electromagnetic and PM undulators:

- Axial stray fluxes may be greatly reduced and attainable field level limited by yoke saturation increased;
- power consumption per unit length of undulator is lowered by several times, being 1.6 kW at $K = 1$ for this device;
- no correction coils are necessary to adjust field amplitude homogeneity: it can be done with 0.1% precision by small lateral movements of permanent magnets;
- no complex device for precision movement of one half of the undulator relative to the other to change field amplitude (as in PM undulators) is needed;
- the simple design means less expensive manufacturing, less material consumption etc., especially for manufacturing inexpensive microundulators;
- using yokes of high-saturation material (like Vanadium Permendur) and optimizing axial dimensions of yokes and poles, it seems possible to obtain a 6–7 kG undulator field for a moderate specific power consumption (~ 10 kW/m), though requiring a water-cooled coil;
- employing a water-cooled coil might permit even higher fields with moderate power consumption.

References

[1] K. Halbach, Nucl. Instr. and Meth. A 246 (1986) 77.
[2] D.A.G. Deacon, L.R. Elias, J.M.J./ Madey, G.J. Ramian, H.A. Swettman, T.I. Smith, Phys. Rev. Lett. 38 (1977) 892.

[3] K. Halbach, Nucl. Instr. and Meth. A 250 (1986) 115.
[4] M.J. Burns, G.A. Deis, R.H. Holmes, R.D. Van Maren, K. Halbach, IEEE Trans. Magn. 24 (1988) 978.
[5] G.A. Deis, M.J. Burns, T.C. Christensen, F.E. Coffield, B. Kulke, D. Prosnitz, E.T. Scharlemann, K. Halbach, IEEE Trans. Magn. 24 (1988) 986.
[6] G.A. Deis, A.R. Harvey, C.D. Parkison, D. Prosnitz, J. Rego, E.T. Scharlemann, K. Halbach, IEEE Trans. Magn. 24 (1988) 1090.
[7] T.C. Chritensen, M.J. Burns, G.A. Deis, C.D. Parkison, D. Prosnitz, K. Halbach, IEEE Trans. Magn. 24 (1988) 1094.
[8] R.D. Schlueter, G.A. Deis, Nucl. Instr. and Meth. A 331 (1993) 711.
[9] A.A. Varfolomeev, S.N. Ivanchenkov, A.S. Khlebnikov, C. Pellegrini, G.A. Baranov, V.I. Michailov, Nucl. Instr. and Meth. A 318 (1992) 813.
[10] A. Agafonov et al., Nucl. Instr. and Meth. A 331 (1993) 186.
[11] V.A. Papadichev, Nucl. Instr. and Meth. A 358 (1995) 429.

ELSEVIER

Nuclear Instruments and Methods in Physics Research A 429 (1999) 371–376

NUCLEAR INSTRUMENTS & METHODS IN PHYSICS RESEARCH Section A

www.elsevier.nl/locate/nima

A simple method of forming a matched input in helical undulators

V.A. Papadichev*, I.V. Sinilshchikova, O.A. Smith

P.N. Lebedev Physical Institute RAS, 53 Leninski Prospect, 117924 Moscow, Russia

Abstract

A helical undulator was used in the initial experiment to realize a free electron laser [D.A.G. Deacon et al., Phys. Rev. Lett. 38 (1977) 892]. Since then it has been widely used in FELs with pulsed electron beams. To inject a beam, an adiabatic input, 5–10 periods long, is generally used. Such a section is also often used at the undulator output. In all, this lengthens the undulator by 30–50%. An alternative method of forming a matched fringe field with a short winding no more than a period long, in accordance with a special law, is effective but difficult to realize. A new, simpler method of matching the undulator input, buttressed by calculations and experimentally realized, is proposed. In this case, the fringe field is formed by means of conductor segments in the form of straight pieces and circumference arcs. This enables one to accurately meet the geometric tolerances. The axial length of this section is about $\frac{1}{6}$ of an undulator period. Experimental results obtained on a 4.8 cm-period undulator gave good agreement with calculations. © 1999 Elsevier Science B.V. All rights reserved.

Keywords: Helical undulators; FEL; Electron beams

1. Introduction

The motion of particles of a beam of finite emittance in an ideal field of a free electron laser (FEL) undulator can be represented as betatron oscillations of electrons about an ideal orbit of the undulator (in the presence of errors in the field, about the distorted orbit) [1–4]. Minimal shift and width of the radiated spectrum and maximum FEL amplification are attained for the smallest deviation of the beam from the undulator axis (or median plane). This can be accomplished by imparting by means of external devices to central particles of the beam a deviation and angle (transverse velocity) equal to the coordinate and angle of the orbit at the undulator input as usual in transport systems of storage rings and large RF linear accelerators, microtrons, etc.

However, a different method that is more convenient and simpler is widely used, especially in amplifiers and single-pass FELs on high-current accelerators, X-ray FELs, etc. The beam is directed into the undulator along its axis and matching of initial conditions at injection is achieved by deflection of the beam in a specially formed undulator fringe field. The setup is simpler and the overall FEL length is shorter. Moreover, while changing particle energy and the undulator field amplitude,

* Corresponding author. Tel.: 7-95-132-6300; fax: 7-95-938-2251.

E-mail address: papadich@sci.lpi.ac.ru (V.A. Papadichev)

0168-9002/99/$ - see front matter
PII: S0168-9002(99)00089-3

matched injection conditions are maintained automatically.

If orbit distortion is not large (corrected if necessary), one injects an electron beam on the ideal orbit, i.e., with zero transverse angle at field maxima. The orbit amplitude is usually much smaller than beam radius and betatron amplitudes, so one can neglect orbit displacement (but this can be easily taken into account if necessary). The fringe field length is usually not large (of the order of winding diameter) compared to betatron period $l_\beta = l_0\gamma\sqrt{2}/K$, so electrons do not change their transverse position, experiencing only a transverse kick. The deflection angle is proportional to field integral along the axis and should be zero at the first regular field maximum.

It is necessary to integrate the equation of motion and correct the initially calculated (integrated) angle and field curve if the betatron period is comparable to the undulator period.

In a plane undulator with almost two-dimensional field, formation of a matched fringe field is relatively simple: a half-period with approximately one-half the amplitude of the field is added at the undulator input. The integral along the axis of this field (taking into account scattering fringe field) to the first maximum of regular undulator field (in the second or third half-period) should be equal to zero.

Less often, for matching injection in a plane undulator, a field having the same period and gradually increasing amplitude, a so-called adiabatic input [5,6], is used.

Helical electromagnetic undulators with a pulsed feed are simpler and considerably cheaper to fabricate than other types of undulators. They are, therefore, widely used in FELs with pulsed-periodic beams providing beam focusing in both transverse directions and when necessary permitting the use of a longitudinal magnetic field. Such a field is often used, for example, for generation and transport of intense electron beams. However, it is much more difficult to produce a matched input in a helical undulator than in a planar one due to the three dimensionality of the fringe field.

It should be noted that for injection of particles along the axis, for example, in a helical undulator with "abrupt" break of the periodic part of the winding, the initial angle of betatron oscillations $\beta_\beta/\beta_\parallel$ for one of the coordinates is approximately equal to the angle of the ideal orbit $\beta_{\perp 0}/\beta_\parallel$, and the amplitude a_β due to this oscillation is $\gamma\sqrt{2}/K$ times the amplitude of the orbit a_0. Here, γ is the total energy of particles in units of rest energy, $K = eB_{\perp 0}l_0/(2\pi m_0 c^2) = 0.934 B_{\perp 0}\,(\mathrm{T})\, l_0\,(\mathrm{cm})$ is the parameter of undulator deflection, $\beta_\parallel = v_\parallel/c$, $\beta_\beta = v_\beta/c$, $\beta_{\perp 0} = v_{\perp 0}/c$, where $v_\parallel$, v_β and $v_{\perp 0}$ are axial, betatron and orbital velocities of electrons, respectively, e and m_0 are the electron charge and rest mass respectively, c is the speed of light in vacuum, and l_0 the period of the undulator.

Most often, helical undulators have an adiabatically rising field [7–12]. This is achieved in various ways: winding of the input part on a cone tapering down to the cylinder of the main part of the undulator [8,9], shunting current of the input turns of the undulator [8,10–12], branching of the total current of a multi-turn winding with the current increasing to the main part of the undulator [7,12], etc.

For injection along the axis, the amplitude of betatron oscillations in an undulator with adiabatic input is $1/2\pi N_{\mathrm{inp}}$ of that in an undulator with an "abrupt" edge [13]. However, in certain cases it is necessary for N_{inp} to be large to decrease the spread of longitudinal velocities in the beam, which lengthens the undulator. A similar section with gradual decreasing field is often used at the undulator output, significantly increasing FEL length. Another shortcoming is the decrease of adiabatic input effectiveness near gyroresonace [13,14]. Moreover, fringe fields due to an unavoidable sharp "break" at the adiabatic input winding leads to considerable transverse deviations of the beam and requires additional measures to reduce them [9].

The method of forming a matched input with a short-section winding following a special law $z(\varphi)$, where z is the axial coordinate and φ the azimuthal angle, was proposed by Gaskevich [15] and successfully realized [16,17]. The length of the input section in this case varies from about a half period to a full period, depending on the variant chosen. This makes it possible to significantly decrease the overall length of the undulator as compared to the variant with adiabatic input and output. The shortcoming of the method is the difficulty in making the grooves and laying the wire in the profiled part with the required accuracy, which is quite high.

We propose a new method, which is simpler in fabrication, to form a fringe field providing matched injection of a beam in a helical undulator with initial motion of the beam along the undulator axis. The fringe field is formed by means of conductor segments in the form of straight elements and arcs, which makes it possible to keep within the stringent geometric tolerances. The extent of this section is about $\frac{1}{6}$ of an undulator period, which practically does not increase the overall length of the undulator. Such an input automatically maintains matching of injection conditions during changing amplitude of undulator field and the energy of particles in the absence of an axial magnetic field. Experimental results obtained on a 4.8 cm-period undulator showed good agreement with calculations.

2. A new method of forming fringe field: calculations and experiment

The magnetic field of a helical undulator was calculated numerically by the Biot–Savart–Laplace law. A helical undulator consists of a bifilar winding: the second winding is displaced by $\frac{1}{2}$ period relative to the first, with currents flowing in opposing directions. Calculations were performed for undulators with various ratios of winding diameter to period. Experimental investigations were performed on an undulator having the following parameters: undulator period $l_0 = 4.8$ cm, radius of spiral winding $R = 1.2$ cm, undulator length = 20 periods and spiral current $I = 1$ kA. The undulator is schematically shown in Fig. 1.

The dependences of magnetic field components along the undulator axis (B_x, B_y) on coordinate z for "abrupt break" of winding (20 periods without input and output section) are shown in Fig. 2 (a and b). Integrals along the axis from this field to the maxima of the regular periodic field of the undulator were calculated P_x, P_y.

In calculating integrals, the upper limit corresponded to the axial coordinate five periods distant from the edge of the undulator. Thus, calculations on forming a matched input are valid to a high accuracy for the case when the period of betatron oscillations is 10–15 times greater than the undulator period, which is usually the case for most FEL experiments ($l_\beta = l_0\sqrt{2}/\beta_{\perp 0}$). When the ratio of the period of betatron oscillations to the undulator period is less, it is necessary to integrate the equation of motion in the fringe field (forming section) and 3–5 periods within the undulator and introduce corrections to the parameters of the forming section.

In a matched fringe field, these integrals should equal zero. In the case considered, for abrupt edge, P_x and P_y differ from zero. For quantitative evaluation of the required correction to these values, integrals are calculated from a component of regular, periodic magnetic field for a quarter of a period M_x, M_y, which determine the components of the maximum transverse velocity of orbital motion. Of course, $M_x = M_y$ for a helical undulator of circular cross-section. Calculations showed that P_x exceeds

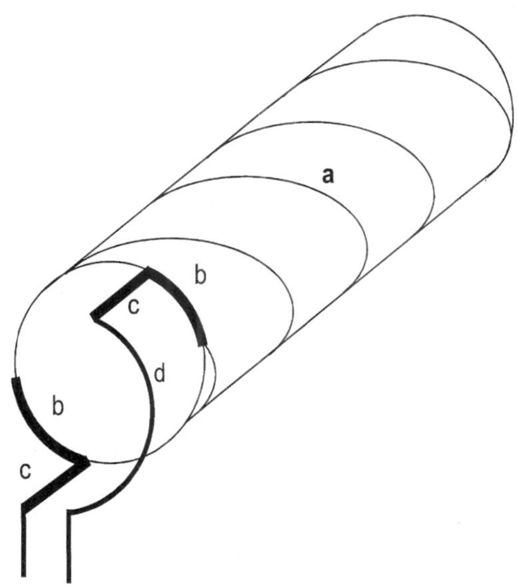

Fig. 1. Schematic design of a helical undulator: (a) regular (periodic) part with an abrupt break; (b, c) corrector: (b) two arc pieces, (c) straight pieces, (d) semicircle arc to place both feed conductors on the same side, minimizing stray flux.

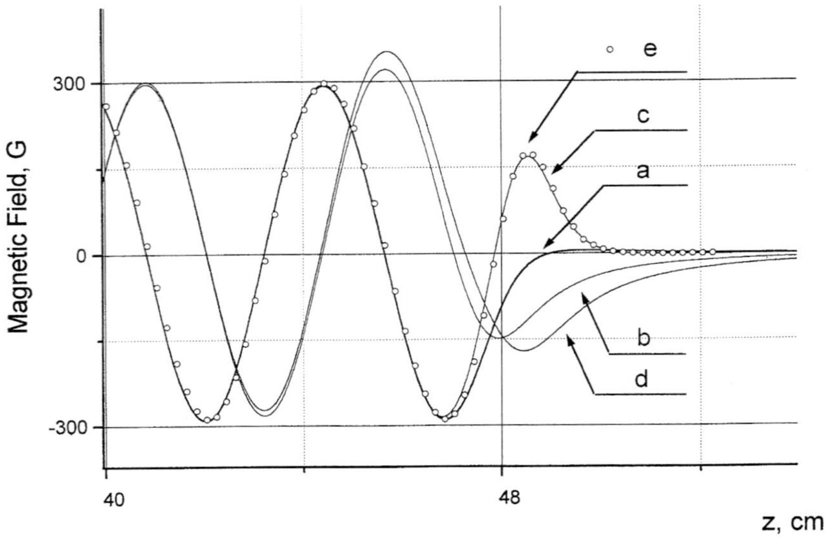

Fig. 2. Helical undulator fringe fields: (a, b) B_x, B_y for abrupt break; (c, d) B_x, B_y for proposed matched input; (e) experiment (open circles).

M_x by about 15% and P_y is approximately only 6% of M_y (integrating up to $3l_0$ within undulator).

To form a matched fringe magnetic field, segments of wire in the form of straight pieces and arcs were used. Two straight wires are placed along generatrices of the cylinder on which the undulator spiral is wound. These generatrices are on diametrically opposite sides of the cylinder. The wires then take the form of a circumference arc and, finally, pass over to the undulator winding. Power feed conductors are twisted or go alongside each other to minimize stray flux; then one performs a semicircle to reach the straight piece (Fig. 1).

The field at the undulator axis from a current of generatrix and circumference arc was calculated analytically, which made it possible to select their parameters. The total field of the input device together with the undulator fringe field was calculated numerically. The field B_y at the undulator axis from a generatrix segment at radius r and length $z = 2a$, i.e., $(+a, -a)$, is

$$B_y(z) = \frac{I}{cR}\left(\frac{z+a}{\sqrt{R^2+(z+a)^2}} - \frac{z-a}{\sqrt{R^2+(z-a)^2}}\right).$$

It can be seen that $B_y(-z) = B_y(z)$. The deviation of particles in the x direction is determined by $\int_{-\infty}^{\infty} B_y(z)\,\mathrm{d}z = 4Ia/cR$. The deviation in the field of a conductor pair is twice this: $\int_{-\infty}^{\infty} B_y(z)\,\mathrm{d}z = 8Ia/cR$.

The field at the axis from arc $\varphi_2 - \varphi_1$ at radius R with current I is

$$B(z) = \frac{IRz}{c\,(R^2+z^2)^{3/2}}(\sin\varphi_2 - \sin\varphi_1)$$

$$B(z) = -\frac{IRz}{c\,(R^2+z^2)^{3/2}}(\cos\varphi_2 - \cos\varphi_1).$$

Both components are odd functions of z and upon integration yield zero, providing no contribution to deflection of particles. Arcs serve to locate the straight sections at the required azimuth (A), permitting correction simultaneously of B_x and B_y.

The angle (azimuth) at which it is necessary to turn the pair of anti-parallel currents is found from conditions: $I_c \sin A = -P_x$, $I_c \cos A = -P_y$, i.e., $\tan A = P_x/P_y$.

The length of the generatrices is

$$2a = -P_x\frac{cR}{4I\sin A}.$$

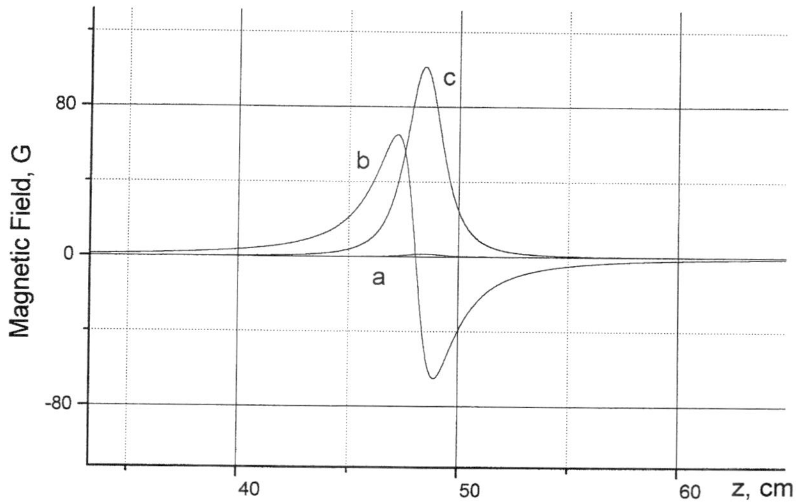

Fig. 3. Field curves: (a) B_y of a pair generatrix currents; (b) B_y of a pair of arcs; (c) B_x of a pair of generatrix currents.

Thus, the magnetic field of anti-parallel currents turned by angle A corrects both components B_x and B_y of an undulator with an "abrupt edge", providing matching of injection conditions. Curves of the magnetic field on axis, created individually by a pair of generatrix segments and a pair of arcs carrying current, are given in Fig. 3 (a–c). The calculated values of parameters of straight segments are: $L_0 - 0.788$ cm and $A = 90.9°$.

The magnetic field of an undulator taking into account the considered correction elements and connecting semicircle are shown in Fig. 2 (c and d). Measured values of magnetic field of the described undulator are also indicated in the figure. Satisfactory agreement exists between calculated and experimental data.

Calculation of the integral of the fringe magnetic field showed that its value does not exceed 5% of the integral of the regular field over a quarter of a period. Parameters of the input section for several ratios of winding diameter to undulator period are shown in Fig. 4.

Finally, the effect of the bend in the conductor in passing over from the straight segment to the arc has been analyzed. This bend can be represented as a quarter of a circle with a certain radius. (The arc was replaced by a chord in calculating the field.) Accounting for the obtained field led to shortening the length of the straight segments by $2r$, r being the bend radius. The azimuthal angle A decreased by 1.3°. The arc length depends also on wire diameter and its position in the groove. This was taken into account in fabricating undulator correcting elements.

3. Conclusions

1. Adiabatic input, used most widely in helical undulators, has a number of shortcomings.
2. The method of forming by means of a short section with winding according to a given $z(\varphi)$ law is effective since the matching section and field have small extent and do not create undesirable fringe fields. However, making the groove and placing the conductor in it with the required high accuracy are difficult to realize.
3. Our proposed method of forming a matched input has the following advantages:

(a) The extent of the input section is least of the three considered methods;
(b) Simplicity of fabrication;
(c) The possibility of using axial fields larger than with adiabatic input;
(d) Simplicity in realization and effectiveness of the method are reasons for optimism regarding its

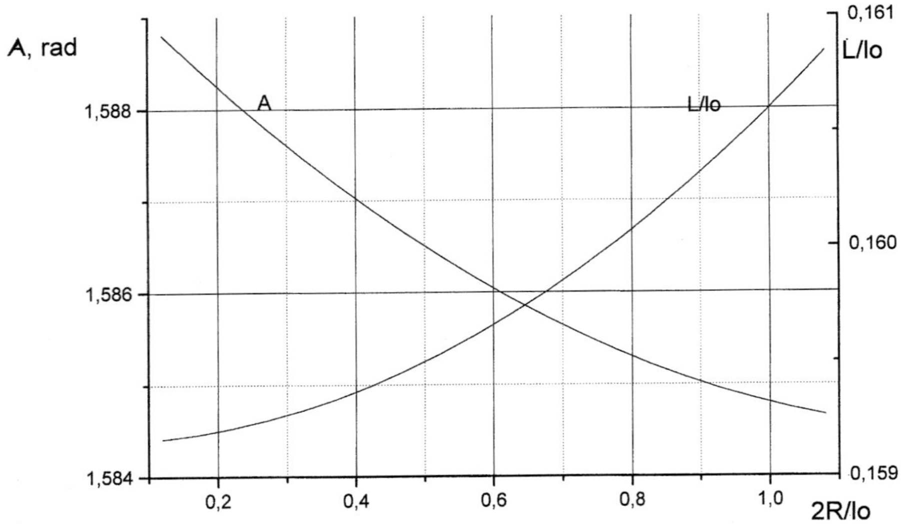

Fig. 4. Angle A (rad) of arcs and length of straight pieces in units of undulator period (L/l_0) as functions of the ratio of winding diameter to the period $2R/l_0$.

wide use in FELs. Calculation of parameters of the input section for several ratios of winding diameter to undulator period show weak dependence on this ratio.

References

[1] J.P. Blewett, R. Chasman, J. Appl. Phys. 48 (1977) 2692.
[2] E.T. Scharlemann, J. Appl. Phys. 58 (1985) 2154.
[3] A.N. Lebedev, G.V. Martirosjan, JTF 56 (1986) 1130.
[4] V.A. Papadichev, Nucl. Instr. and Meth. A 358 (1995) 429.
[5] M. Shiho, K. Sabamoto, S. Maebara et al., Nucl. Instr. and Meth. A 304 (1991) 141.
[6] Y.C. Huang, H.C. Wang, R.H. Pantell, J. Feinstein, J. Harris, Nucl. Instr. and Meth. A 341 (1994) 431.
[7] L.R. Elias, J.M. Madey, Rev. Sci. Instr. 50 (1979) 1335.
[8] J. Fajans, J. Appl. Phys. 55 (1984) 43.
[9] V.A. Papadichev, O.A. Smith, Nucl. Instr. and Meth. A 318 (1992) 803.
[10] H. Bottollier-Curtet et al., Nucl. Instr. and Meth. A 304 (1991) 197.
[11] H. Bottollier-Curtet et al., Nucl. Instr. and Meth. A 318 (1992) 131.
[12] A.K. Kaminsky, A.A. Kaminsky, S.N. Sedykh, A.P. Sergeev, Field distribution investigation in few types of helical undulators, Presented at the 18th International FEL Conference, Free Electron Lasers 1996, Amsterdam.
[13] V.A. Papadichev, Nucl. Instr. and Meth. A 304 (1991) 749.
[14] V.A. Papadichev, Nucl. Instr. and Meth. A 375 (1996) 474.
[15] E.B. Gaskevich, Trudy FIAN 214 (1993) 164.
[16] A.I. Bukin, E.B. Gaskevich, V.G. Kurakin, O.V. Savushkin, A helical undulator for the far infrared free electron laser, Presented at the 18th International FEL Conference, Free Electron Lasers 1996, Amsterdam.
[17] V.A. Papadichev, I.V. Sinilshchikova, O.A. Smith, Methods of precise measurements of magnetic field and its correction in helical undulators, Presented at the 18th International FEL Conference, Free Electron Lasers 1996, Amsterdam.

ELSEVIER

Nuclear Instruments and Methods in Physics Research A 429 (1999) 377–385

NUCLEAR INSTRUMENTS & METHODS IN PHYSICS RESEARCH Section A

www.elsevier.nl/locate/nima

An electrostatic undulator with single-polarity feed

V.A. Papadichev*

P.N.Lebedev Physical Institute RAS, 53 Leninski Prospect, Moscow 117924, Russia

Abstract

The analytical formulae for field and potential in a plane electrostatic undulator with single-polarity feed were derived using an original method of summing infinite series. The system is placed inside a zero-potential metallic housing to simulate its real use in FELs. Such a geometry allows a precise calculation of the transverse velocity of electrons and their change of energy when moving in the undulator. A single-polarity feed can be advantageous in permitting higher field amplitudes, since electrical breakdown inside the undulator is excluded, in contradistinction to opposite-polarity $(+V_0, -V_0)$ feed of adjacent conductors in usual designs. Also the housing can be placed at a distance much greater than undulator transverse dimensions, thus hampering external discharge. Some peculiarities of particle dynamics and regimes of FEL with such an undulator were analysed. © 1999 Elsevier Science B.V. All rights reserved.

Keywords: Electrostatic undulator; Single-polarity feed; Free electron lasers

1. Introduction

One of the main trends in FEL research and development is the elaboration of a compact device for laboratory use. The greatest gain in this regard is given by reducing accelerator energy and length correspondingly. For the same radiation wavelength, this means reducing the undulator period l_0 proportional to gamma squared ($\gamma = E/m_0c^2$), i.e., employing microundulators with millimeter and submillimeter periods if the initial gamma exceeds 10. Unfortunately, the field amplitude $B_\perp$ falls sharply with reduced period because of technical problems: in electromagnetic undulators due to current density limitation, in permanent magnet (PM) and hybrid ones due to difficulties in fabricating high-quality thin plates of brittle PM. Undulator parameter $K = (eB_\perp l_0)/(2\pi m_0 c^2) = 0.934 B_\perp(\mathrm{T}) l_0(\mathrm{cm})$ falls faster ($\propto l_0^p$), where $p = 1.5$–2, resulting, for example, in $K = 0.13$ for 8-mm-period and 2 kG field [1].

As will be shown in Section 4, electrostatic undulators after some vacuum treatment and conditioning can withstand fields up to 0.5–1 MV/cm (equivalent to 1.5–3 kG for relativistic electron velocities). Thus, their use becomes expedient for 0.1–10 mm undulator periods. Much simpler manufacturing compared to magnetic undulators and the small amount of inexpensive materials required make their share in total expenses for an FEL negligible.

* Tel.: + 7-95-132-6300; fax: + 7-95-938-2251.
E-mail address: papadich@sci.lpi.ac.ru (V.A. Papadichev)

0168-9002/99/$ - see front matter
PII: S0168-9002(99)00091-1

Periodic electrostatic systems of various types have been proposed for focusing electron beams in RF generators (and later in linear accelerators) since the late 1930s [2–8]. A method of obtaining Doppler-shifted high-frequency radiation of relativistic electrons performing transverse oscillations was suggested in 1947 [9] and further developed for magnetostatic and electrostatic undulators [10]. Application of relativistic particle radiation in electrostatic undulators for precise measurement of its velocity was considered later [11,12]. A proposal to use electrostatic undulators with large and variable period in storage rings for obtaining X-ray radiation was published in 1989 [13,14], and a scheme of a multichannel electrostatic undulator was published in 1993 [15]. Field parameters were calculated in some works using conformal mapping, but more often the electric field was presented only as the fundamental harmonic, and its amplitude was not expressed through the potential of undulator electrodes of finite dimensions, which is necessary for practical applications. Detailed calculations of the potential and the electric field in electrostatic undulators of various types with bipolar feed were performed using conformal mapping, allowing one to obtain results in a simple and compact form [16–18]. A single-polarity-feed undulator treated below may have some advantages over traditional bipolar schemes (see Section 4).

Lowering the undulator parameter K and energy γ requires increasing the number of periods N in order to obtain the same gain per pass $G \approx j_b K^2 N^3 \lambda^2 \gamma$ (λ is the radiation wavelength). This is possible if the beam emittance ε is small: the maximum number of periods $N \leqslant (1 + K^2/2)\beta\beta_{||}/(\varepsilon K \gamma k_0 \sqrt{2})$, where $\beta = (1-\gamma^2)^{-1/2}$, $\beta_{||} = (1 - 1/\gamma^2 - \beta_{\perp}^2)^{1/2}$, $k_0 = 2\pi/l_0$. Electron beams with small emittance $\varepsilon \leqslant 3 \times 10^{-7}\pi$ m·rad are obtained on cold-emission sources for microFELs [19–23] and electrostatic accelerators [1,24], so that electrostatic undulators could be first tried there.

A periodic system of electrodes (wires, foils) with opposite polarities of adjacent elements is used in typical undulator schemes [2–18]. The highest field values in optimal designs are on the inward surface of electrodes (facing the median plane). Beam motion can be distorted, leading to lower FEL gain and even total loss of the beam if electrical discharge occurs between electrodes of opposite polarity. A single-polarity feed excludes such a possibility because no discharge can occur in the beam region, and one can suppress external breakdown by increasing the vacuum gap between undulator and zero-potential housing. The axial magnetic field used for beam transport may be also beneficial for exclusion of external discharges.

The design and field calculation method for a single-polarity-feed undulator differ from those of a typical bipolar undulator. First, it is necessary to shift one layer of wires by a half-period relative to the other in order to have nonzero transverse field in the undulator median plane. Second, simple (plane) surfaces of zero potential are outside the working region, in contradistinction to the bipolar-feed case [16–18]. Therefore, it is not possible to isolate a single conductor in a zero-potential box of simple form (orthogonal parallelepiped) and then apply conformal mapping by means of well-known trigonometric and homographic functions. An original method of summing infinite series of electric fields of a single conductor was used, resulting in simple analytic expressions for field and potential. They are calculated for the entire region within the zero-potential housing. Formulae were obtained for the field on the electrode surface and in the median plane, which allow to optimize undulator design, calculate transverse particle velocities and oscillation of their energy, and assess focusing properties of the undulator. The expressions derived are valid for circular-cross-section conductors with arbitrary (but periodic) relative position.

2. Electrostatic field and potential calculation

Calculation is carried out in two-dimensional geometry using a complex potential and field (Fig. 1). The electric potential $w(z)$ and field $\boldsymbol{E} = \overline{-\mathrm{i}\,\mathrm{d}w/\mathrm{d}z}$ of a single infinitely thin and infinitely long rectilinear conductor piercing perpendicularly the xy-plane at a point $z_0 = x_0 + y_0$ and with charge q per unit length are

$$\begin{aligned} w &= -2q\mathrm{i}\ln(z - z_0) \\ \boldsymbol{E} &= 2q/(\bar{z} - \bar{z}_0) \end{aligned} \qquad (1)$$

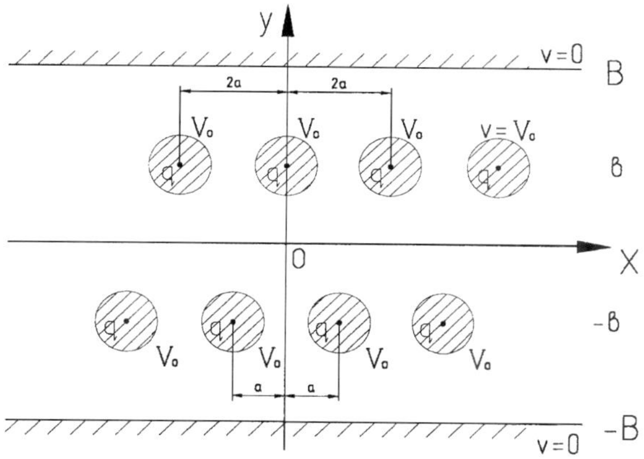

Fig. 1. Geometry of an undulator with single-polarity feed.

where $z = x + \mathrm{i}y$ is the point of observation, $\mathrm{i} = \sqrt{-1}$, $\bar{z} = x - \mathrm{i}y$. The electric field of a system consisting of two arrays of charged conductors in planes $y_1 = b$ and $y_2 = -b$ at points $z_{1k} = 2ak + \mathrm{i}b$ and $z_{2k} = 2ak + a - \mathrm{i}b$, respectively, (Fig. 1) can be written as

$$\boldsymbol{E} = 2q \sum_{k=-\infty}^{\infty} \left[\frac{1}{\bar{z} - 2ak + \mathrm{i}b} + \frac{1}{\bar{z} - 2ak - a - \mathrm{i}b} \right]. \tag{2}$$

Expanding $\cot(\pi t)$ [25] and transforming it, one obtains

$$\cot(\pi t) = \frac{1}{\pi t} + \frac{1}{\pi} \sum_{k=1}^{\infty} \left(\frac{1}{t-k} + \frac{1}{t+k} \right) = \frac{1}{\pi t} + \frac{1}{\pi} \sum_{\substack{k=-\infty \\ k \neq 0}}^{\infty} \frac{1}{t-k} = \frac{1}{\pi} \sum_{k=-\infty}^{\infty} \frac{1}{t-k}. \tag{3}$$

The following expressions can be obtained after simple transformations:

$$\boldsymbol{E} = \frac{\pi q}{a} \left[\cot\left(\frac{\pi}{2a} (\bar{z} + b\mathrm{i}) \right) + \cot\left(\frac{\pi}{2a} (\bar{z} - a - b\mathrm{i}) \right) \right] = \frac{2\pi q}{a} \frac{\cos(\pi \bar{z}/a)}{\sin(\pi \bar{z}/a) + \mathrm{i} \sinh(\pi b/a)};$$

$$w(z) = -\mathrm{i} \int \bar{\boldsymbol{E}} \, \mathrm{d}z = -2q\mathrm{i} \ln[\sin(\pi z/a) - \mathrm{i} \sinh(\pi b/a)] + \text{const.} \tag{4}$$

The constant is found from the boundary conditions: since a small-period undulator should always be placed inside a vacuum chamber or other device at ground potential, we let $v = 0$ at planes $y = B$ and $y = -B$. This means mathematically summing fields of an infinite number of conductor layers (reflections relative to planes $\pm B$ with charges $-q$ at $y = \pm(2B \pm b + 4Bk)$ and $+q$ at $y = \pm(4B \pm b + 4Bk)$, where $k = 1,2,3,\ldots$. Periodic components of the potential and field evanesce exponentially with distance. It is sufficient therefore to add only fields of the two nearest pairs of layers and require separately that $v_{\mathrm{dc}} = 0$ at $y = \pm B$ for a direct component of the potential.

One needs to obtain the inverse function $z(w)$, then put $v = v_0$ and use parametric dependencies $x(u,v_0)$ and $y(u,v_0)$ on u to draw equipotentials. The problem can be simplified keeping in mind the more rapid attenuation of higher harmonics with distance from a layer: It is sufficient that only the direct component and the first harmonic be zero at $y = \pm B$. The potential in this case is

$$w = -2q\mathrm{i} \ln \left[\frac{\sin(\pi z/a) - \mathrm{i} \sinh(\pi b/a)}{A[\mathrm{i} + \alpha \sin(\pi z/a)]} \right] \tag{5}$$

where α represents the sum of small linear terms of $\sin(\pi z/a)$ from reflections gathered in the denominator in order to get a linear equation for $\sin(\pi z/a) = f(w)$ and then $z = (a/\pi)\arcsin(f(w))$:

$$z = \frac{a}{\pi} \arcsin \left[\mathrm{i} \frac{A e^{\mathrm{i}w/2q} + \sinh(\pi b/a)}{1 - \alpha A e^{\mathrm{i}w/2q}} \right]. \tag{6}$$

A and α are derived by putting $v = 0$ at $y = B$ and solving a pair of coupled equations (the first-order expansion term of the minor root of α was taken):

$$\alpha = 2\sinh(\pi b/a)/[\cosh(2\pi B/a) + \cosh(2\pi b/a) - 1];$$

$$A = \sqrt{[\cosh(2\pi B/a) + \cosh(2\pi b/a) - 1]/[2 + \alpha^2 \cosh(2\pi B/a)]}. \tag{7}$$

It is convenient for practical use of the obtained formulae to put the potential $v = V_0$ at the conductor surface (at the point $x = 0,\ y = y_c < b$) and find an expression for q for substituting it in $w(z)$ and $E(z)$:

$$q = -(V_0/2)/\ln\left[\frac{\sinh(\pi b/a) - \sinh(\pi y_c/a)}{A(1 + \alpha\sinh(\pi y_c/a)}\right]$$

$$w(z) = V_0 \mathrm{i}\ln\left[\frac{\sin(\pi z/a) - \mathrm{i}\sinh(\pi b/a)}{A(1 + \alpha\sinh(\pi z/a)}\right] \Bigg/ \ln\left[\frac{\sinh(\pi b/a) - \sinh(\pi y_c/a)}{A(1 + \alpha\sinh(\pi y_c/a)}\right]. \qquad (8)$$

Equipotentials are shown in Figs. 2(I) and (II) and 3 for $b = a$, $B = 2b$ and $1.5b$, respectively. The correction factor inside the undulator is $1 - \cosh^2(\pi b/a)/\cosh^2(\pi B/a) \approx 1$, and one can let $\alpha = 0$ in most practically important cases when $B > 2b$. Electric

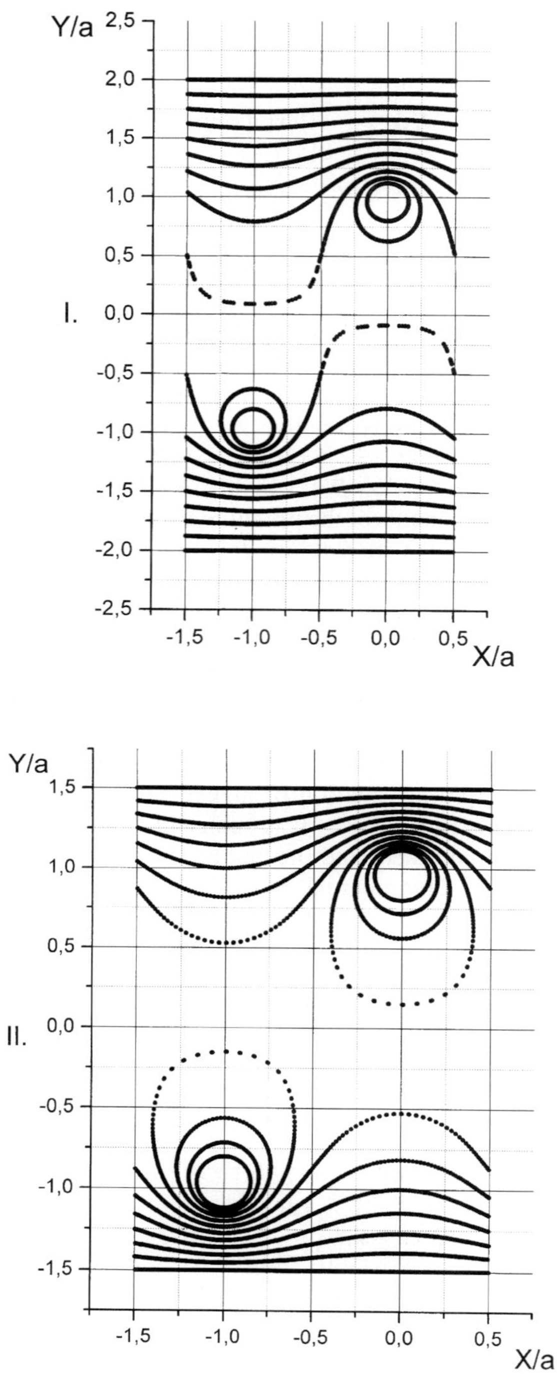

Fig. 2. Equipotentials: (I) for $b = a$, $B = 2b$, $V_0 = 1$ (conductor potential), $\Delta v = 0.1V_0$; (II) for $b = a$, $B = 1.5b$, $V_0 = 1$, $\Delta v = 0.1V_0$.

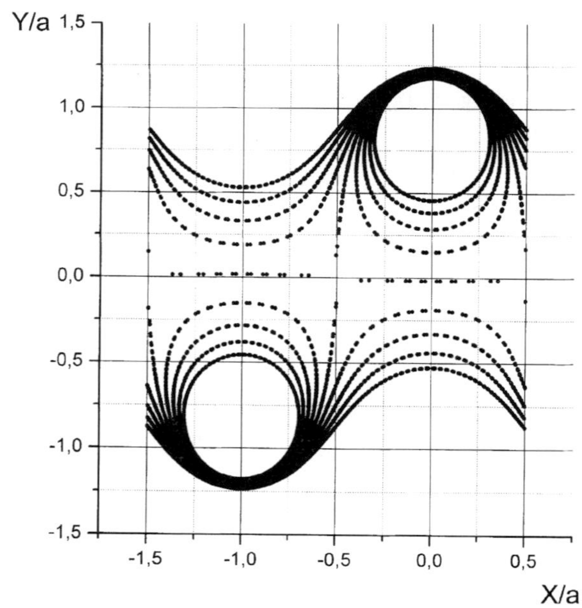

Fig. 3. Magnified map of equipotentials near the median plane: $V_0 = 1$ at conductors, the equipotential nearest to the conductor is at $v = 0.6V_0$, step $\Delta v = 0.02V_0$, $b = a$, $B = 1.5b$ (as in Fig. 2-II).

field in the median plane is

$$E = E_0\left[\frac{\cos(\pi x/a)}{\sin(\pi x/a)/\sinh(\pi b/a) + \mathrm{i}} - \frac{\alpha\sinh(\pi b/a)\cos(\pi x/a)}{\alpha\sin(\pi x/a) + \mathrm{i}}\right]$$

$$E_0 = -\frac{\pi V_0}{a\sinh(\pi b/a)} \Bigg/ \ln\left[\frac{\sinh(\pi b/a) - \sinh(\pi y_c/a)}{A(1 + \alpha\sinh(\pi y_c/a))}\right]. \quad (9)$$

Harmonic content (normalized to the fundamental) is the same, if $\alpha \approx 0$, as in the bipolar-feed undulator [3].

The field on the electrode surface can be calculated by putting $v = V_0$ or $v = 0$ (housing) in $E(\bar{w})$. Calculation of the absolute value $|E| = \sqrt{E\bar{E}}$ reveals the most stressed points. The field value can be lowered then by reshaping the undulator. Fig. 4 gives an example of the $|E|$ distribution on electrodes.

3. Beam focusing and energy spread

Orbital motion in a plane electrostatic undulator, as opposed to a magnetic one, is in the plane of the undulator field. Betatron oscillations are also in the same plane. They are due to field increase when nearing a conductor and orbit undulation. Since orbital velocities are not large, $\beta_{\perp 0} = v_{\perp 0}/c \approx 0.01$, the natural undulator focusing may be insufficient (the betatron oscillation period $l_\beta = \sqrt{2l_0/\beta_{\perp 0}}$) and external focusing may be necessary. Both an axial magnetic field and transverse magnetic quadrupole/sextupole focusing are compatible with an electrostatic undulator. When $B_{||} \neq 0$, electrons move along field lines performing slow azimuthal drift in the transverse plane [26]. Betatron oscillations occur in both transverse directions if $B_{||} = 0$ and focusing along both transverse coordinates is ensured.

Transverse movement along undulator electric field is connected with particle energy change. Orbit amplitude $a_0 = \beta_{\perp 0}l_0/(2\pi\beta_{||})$ is very small (micrometers for $l_0 = 1$ mm, $\beta_{\perp 0} = 0.01$), so energy oscillations have a negligible amplitude of about 5 eV. It should be noted that even much greater

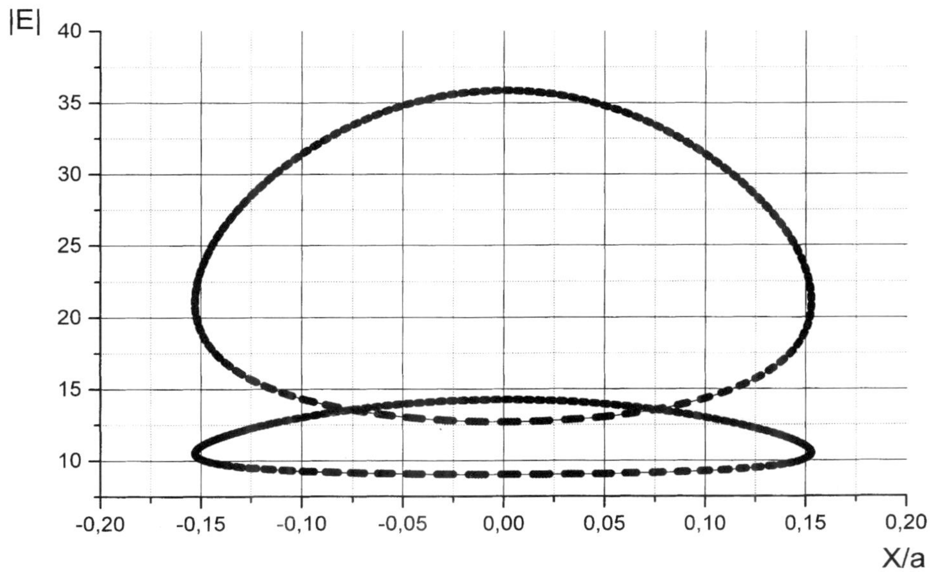

Fig. 4. Absolute value of electric field (normalized to field amplitude in the median plane) on the conductor surface ($v = V_0$) and inner surface of the housing ($v = 0$).

amplitudes could not affect FEL performance, since they do not cause slow oscillation of $\beta_{||}$ (only a DC component and the second harmonic of the fundamental are present).

Slow 'betatron' oscillations could be more harmful. Change of energy in this case can be calculated using a magnified map of equipotentials near the median plane (Fig. 3). It can be written as $\gamma = \gamma_0 + \Delta\gamma \cos^2(k_\beta z)$, where $k_\beta = 2\pi/l_\beta$ (l_β is the slow oscillation period). The example in Fig. 3 shows that energy variation $\Delta E = 0.5$ keV and $\Delta\gamma/\gamma = 8 \times 10^{-4}$ for an electron energy of 100 keV and 3×10^{-4} for 1 MeV (undulator period 1 mm, applied voltage $V_0 = 10$ kV). The longitudinal velocity spread in a beam occupying a half of the vertical aperture $\Delta\beta_{||}/\beta_{||} = (\Delta\gamma/\gamma)/(\beta_{||}^2\gamma^2)$ is $\approx 1.4 \times 10^{-3}$ and 4.2×10^{-5} for the above electron energies, which seems to be acceptable. One obtains $N \leqslant 125$ for the permissible number of periods when $E = 100$ keV and $N \leqslant 660$ when $E = 1$ MeV. Thus the fastest electron does not outstrip the synchronous (slowest) ones in the beam more than $\lambda/2$ at the end of the undulator, and they cannot absorb energy from light. Using such 'long' (125 and 660 mm) undulators more than compensates for the small K. The spread of $\beta_{||}$ in the beam caused by transverse velocity of betatron oscillations is also not large: $\Delta\beta_{||}/\beta_{||} < 2 \times 10^{-4}$ for a beam diameter equal to a half of a geometrical aperture.

It should be noted that electrons acquire or lose energy entering the undulator (approximately by eV_0) depending on voltage sign, but this can be easily taken into account.

4. Comparison of electrostatic and magnetic microundulators

Reducing the period of a magnetic undulator leads, unfortunately, to lowering its field amplitude. It is connected with current density limitation in coils (windings) of electromagnetic undulators: in scaled devices $B_\perp \propto Iw/g \propto jS_w/g \propto jl_0^2/l_0 \propto jl_0$, where Iw is ampere-turns, g the gap, j the current density, S_w the coil cross-section area, and l_0 the undulator period. This is true for DC (superconducting or not) and pulsed undulators. The field limit is about 0.7 T for 8-mm-period SC devices with iron (one should expect 0.2 T for a 3-mm period) [27,28]. 1 T was achieved in an ironless pulsed undulator with a 5-mm period [29–31]. It functioned near the creep limit of a current-carrying conductor caused by nonuniform current distribution and local overheating. Thus, the repetition rate and total number of pulses were limited and some difficulties arose with field amplitude reproducibility [31].

One could expect that scaled decrease of dimensions of a permanent-magnet undulator should not change its field amplitude since the equivalent ampere-turns of a PM are proportional to the period l_0 (not l_0^2 as for an electromagnetic one). But due to the inferior quality of thin PM plates, the undulator field amplitude in fact decreases with lower undulator period [1,32–34].

The only type of magnetic undulator in which the field amplitude should be independent of its period is one proposed in Ref. [35], a so-called solenoid-derived type. Axial field created by a solenoid is redistributed by a periodic structure of small iron blocks with a staggered array, and the magnetic field topography should be the same independent of device scale. 1 T transverse field was attained in a 1-cm-period device. The precision of manufacturing must be very high in small-period undulators; possible corrections of field distortions will be hampered by the small size of parts. The technology of manufacturing such microundulators is yet to be developed. One of the drawbacks of such undulators is that the resulting transverse magnetic field is tightly tied with the axial, and one can change the former independently of the latter only by precisely varying the gap. A pulsed undulator, based on the same principle but with smaller field 0.3 T for $l_0 = 6$ and 2 mm gap was realized [36]. The lower field was partly due to a necessity (because of skin-effect) to use laminated blocks made of transformer steel, which has worse magnetic parameters than Vanadium Permendur used for the DC case.

A similar principle is used when homogeneous axial field is restructured by eddy currents in metallic periodic systems [37,38]. The magnetic field is pulsed in this case, and the pulse duration should be proportional to l_0^2 in scaled devices. The ratio $B_\perp/B_{||}$ can be changed within some limits by varying the pulse length. Usually, ratios attained are 0.2–0.3 (much less than for iron structures in DC field).

Thus, extrapolating the existing experimental data on most widely used magnetic undulators (excluding the solenoid-derived type) to a 1-mm period, one could expect field amplitudes of about 1 kG.

Electrostatic undulators are much simpler and inexpensive in design and manufacturing. One can employ widely used and elaborate techniques of optical mechanics, so undulator periods up to dozens of micrometers look realizable. The main problem, it is generally believed, is the very low attainable field amplitude. Careful analysis of the recent state of vacuum insulation shows that this is not the case.

Electrostatic undulators are usually considered being fed by bipolar voltage. DC and pulsed regimes can be used, and both are discussed below. The merits and demerits of the single-polarity-feed undulator are also analysed for both regimes.

The theoretical breakdown limit of a vacuum gap is set by the Fowler–Nordheim law as the electric field strength when sharp rise of field emission current begins. It is about 7×10^9 V/m for a 'perfect' vacuum gap, i.e., with very even and clean surface of electrodes made of specific metals, heated and conditioned in ultra-high (10^{-8}–10^{-9} mbar) vacuum [39]. Prebreakdown current starts usually at microtips where local microfield is several times (sometimes dozens) greater than average macrofield in the gap.

The best materials for electrodes are stainless steel, titanium and molybdenum; the worst – copper and nickel. Breakdown is usually initiated at microtips and surface nonhomogeneities (dielectric and foreign metal inclusions, microparticles on the surface), so electrodes should be conditioned (treated in situ) after polishing, washing and outgassing by baking in situ at UHV. It is aimed to 'burn out' nonhomogeneities and remove or blunt microtips in a mild manner so as not to create new ones. Several procedures are used [39,40]:

- current conditioning: current is increased in small steps and prebreakdown current is allowed to stabilize at each stage before proceeding to the next;
- glow-discharge conditioning: AC discharge is induced in He at 10^{-3}–10^{-2} mbar pressure;
- gas conditioning: Ar at 10^{-3} mbar fills a vacuum chamber and prebreakdown current (i.e. 50 μA) flows at high voltage (20 kV);
- Spark conditioning: electric polishing by means of short pulses of high voltage and current, etc.

DC field strength of (1–2)10^8 V/m (0.3–0.7 T) can be attained with small surface electrodes after conditioning [39,40]. A breakdown threshold of 1.3×10^8 MV/m was attained experimentally with steel electrodes and 0.25-mm gap at 10^{-6} mbar vacuum with oil pumping; 10^8 V/m was sustained during more than 1 h without breakdown [41]. Magnetic insulation cannot be employed directly (by applying axial magnetic field) in a bipolar-feed undulator because the electric field on electrodes is in all directions in the axial plane perpendicular to the conductors, and both fields would be parallel in the axial gap between adjacent conductors. Pulsed electric fields of the same order can be applied, since breakdown time for small gaps is rather short (less than 1 μs). Thus, 10^8 V/m can be taken as a breakdown limit for bipolar-feed electrostatic undulators which have inevitably small gaps $\approx l_0/3 - l_0/2$.

The single-polarity-feed undulator has four advantages over the bipolar one: (1) no breakdown in working region can occur, (2) a large gap between conductors under high voltage and a grounded shield can be used, (3) a beneficial effect of the sign of the electrode voltage with higher electric field can be exploited and (4) magnetic insulation can be employed by applying an axial magnetic field.

Since breakdown is usually initiated at microtips and also influenced by the total voltage effect, the breakdown voltage V_b is not proportional to the electrode separation d, but more likely to $d^{0.6}$. Nevertheless, increasing the distance from 1 to 100 mm (as in the case of a single-polarity-feed undulator), one could gain a factor of 10 in V_b.

If the electric field is nonhomogeneous, a positive voltage on the electrode with higher field results in higher breakdown voltage than with negative. For a point-plane system the difference can reach a factor of three. An anode field of up to 10^9 V/m does not influence breakdown voltage, occurring at a nearly constant cathode field of 10^9 V/m [40]. The electric field of a small-diameter, short undulator structure falls as r^{-2} (like that of a sphere) or as r^{-1} in a coaxial geometry. Thus, the field on the chamber walls (cathode) may be 10 –30 times less than that on the undulator.

Magnetic isolation can be applied to prevent prebreakdown currents to evolve to full breakdown (with plasma formation, photo- and secondary electron and ion emission, etc.). It was studied up to now in a pulsed regime [42,43]. A 2×10^8 V/m macroscopic electric field was achieved in a 5-mm gap during 100 ns in the first experiment with large-area electrodes (more than 1 m^2). Insulation was insured by current self-magnetic field. Gap closure was caused by plasma jets. 5×10^8 V/m was obtained in the second experiment [43], where additional external magnetic isolation was used. One can hope that for much larger gaps, much smaller electrode area and closed drift orbits of electrons plasma formation will be considerably hampered, and its expansion with much lower density and velocity will not be harmful, resulting in much longer voltage pulses (or even suppressed breakdown in the DC regime).

Taking the above into account, one can hope to reach on the electrodes an electric field strength of (3–6) 10^8 V/m (equivalent to 1–2 T for relativistic electrons) for a single-polarity-feed undulator. It is better to feed it by positive polarity, placed in the center of a medium-diameter (20–40 cm) vacuum chamber and, perhaps, immersed in an axial magnetic field. Pulsed regimes may be preferable, allowing higher fields. The energy required is small due to small capacitance of undulator structure relative to the grounded chamber ($\approx$ 10 pF). Of course the pulses should have the same repetition rate and be phased with electron beam macropulses.

Calculated maximal field strength for a single-polarity-feed undulator with a 1-mm period gives $E_0 = 0.3$ MV/cm (1 kG) for the median plane, $E_c = 0.36$ MV/cm at zero-potential chamber walls and $E_{V0} = 1$ MV/cm at charged conductor external surface. The following parameters were used: $B = 2a = l_0 = 1$ mm, $V_0 = 36$ kV, $b = a/3 = 0.17$ mm, aperture $2y_c = a/3 = 0.17$ mm. Increasing the distance to chamber walls would lead to higher voltage in the planar case: $B = 50$ mm would require $V_0 \approx 1.8$ MV to retain the same field strength in the median plane. Small transverse dimensions of the undulator ($\approx l_0$) insures rapid field decrease ($\propto r^{-1}$, like in the coaxial geometry) and a lower applied voltage and chamber-wall field compared to the planar case: $V_0 = 170$ kV (still too high), $E_0 = 0.3$ MV/cm, $E_c \approx 8$ kV/cm, $E_{V0} = 1$ MV/cm. The helical undulator corresponds to lower $B \approx 2b - y_c = a/2$ and requires 3 times lower voltage. Proportionally lower voltage can be used for smaller periods if 0.3 MV/cm (1 kG) is enough for an FEL. Otherwise, one can have up to 1 T (breakdown limit) for smaller period and higher voltage.

Of course, experimental studies are necessary in a geometry more suitable for FEL undulator application, as well as optimization of undulator parameters.

5. Conclusion

- Electrostatic undulators can be of simple design and inexpensive to manufacture. They may have a higher field than magnetic undulators for millimeter and submillimeter periods. Decrease of the undulator parameter K can be compensated (to have the same gain) by increasing the number of periods up to 10^3 if beams with small emittance and energy spread are used. The total undulator length does not exceed 300–600 mm. Electrostatic accelerators and cold emission microsources are the first devices to be tried with electrostatic undulators.
- Electrostatic undulators permit using axial magnetic field which is often employed for beam generation and transport.
- External focusing by transverse magnetic field can be used if necessary.
- Single-polarity-feed undulators could allow higher field amplitudes since there is no discharge inside the undulator, and external breakdown to zero-potential housing can be prevented by using a larger gap, auxiliary magnetic field, etc.
- Schemes to get higher FEL effectiveness can be more easily realized: compensation of beam energy loss at the end of the undulator, field amplitude profiling, partial conditioning, etc.

- Longitudinal velocity spread in the beam due to transverse inhomogeneity of potential and field is small and should not have any influence on FEL performance.
- The analytical methods developed in this report can be applied for calculation of any electrostatic undulator with conductors of circular cross-section (single-polarity or bipolar, multichannel, etc.)

Acknowledgements

The author is grateful to A.N. Lebedev for useful discussions and G.V. Rybalchenko and O.A. Smith for their assistance in preparing the manuscript.

References

[1] L.R. Elias, I. Kimel, D. Larson et al., Nucl. Instr. and Meth. A 304 (1991) 219.
[2] W.C. Hahn, G.F. Metcalf, Proc. IRE 27 (1939) 106.
[3] L.P. Smith, P.L. Hartman, J. Appl. Phys. 11 (1940) 220.
[4] J.B. Blewett, Phys. Rev. 88 (1952) 1197.
[5] A.M. Clogston, H. Heffner, J. Appl. Phys. 25 (1954) 436.
[6] P.K. Tien, J. Appl. Phys. 25 (1954) 1281.
[7] J.S. Cook et al., Proc. IRE 45 (1957) 1517.
[8] J.S. Cook et al., J. Appl. Phys. 29 (1958) 583.
[9] V.L Ginsburg, Izv. Acad. Nauk SSSR (Bull. Acad. Sci. USSR), Ser. Phys. 11 (1947) 165.
[10] H. Motz, J. Appl. Phys. 22 (1951) 527.
[11] N.A. Korkhmazyan, Izv. Acad. Sci. Armen. SSR, Fiz. 5 (1970) 287, 418.
[12] V.L. Ginsburg, Pis'ma JETP 16 (1972) 501.
[13] R. Tatchyn, J. Appl. Phys. 65 (1989) 4107.
[14] R. Tatchyn, Rev. Sci. Instr. 60 (1989) 2571.
[15] A. Murai et al., Nucl. Instr. and Meth. A 331 (1993) 680.
[16] V.A. Papadichev, Nucl. Instr. and Meth. A 393 (1997) 403.
[17] V.A. Papadichev, Nucl. Instr. and Meth. A 393 (1997) 409.
[18] V.A. Papadichev, G.V. Rybalchenko, Nucl. Instr. and Meth. A 407 (1998) 428.
[19] J.E. Walsh, J.B. Murphy, IEEE J. Quantum Electron. QE-18 (1982) 1259.
[20] M. Shiho et al., Proceedings of the Forty Seventh Annual Meeting of Physics Society of Japan, 1992, p. 193.
[21] K. Mima, S Nakai, T. Taguchi et al., Nucl. Instr. and Meth. A 331 (1993) 550.
[22] H. Ishizuka, Y. Nakahara, S. Kawasaki et al., Nucl. Instr. and Meth. A 331 (1993) 577.
[23] C.M. Tang, M. Goldstein, T.A. Swyden, J.E. Walsh, Nucl. Instr. and Meth. A 358 (1995) 7.
[24] G. Ramian, J. Kaminski, S.J. Allen, Nucl. Instr. and Meth. A 393 (1997) 220.
[25] I.S. Gradstein, I.M. Ryzhik, Tables of Integrals, Sums, Series and Products, Nauka, Moscow, 1971.
[26] V.A. Papadichev, Nucl. Instr. and Meth. A 375 (1996) 479.
[27] X. Zhang, I. Ben-Zvi, G. Ingold, S. Krinsky, L.H. Yu, Nucl. Instr. and Meth. A 331 (1993) 689.
[28] V.A. Papadichev, V.S. Vysotsky, V.N. Tsikhon, V.T. Eremichev, Nucl. Instr. and Meth. A 331 (1993) 748.
[29] R.W. Warren, Nucl. Instr. and Meth. A 318 (1992) 789.
[30] R.W. Warren, C.M. Fortgang, Nucl. Instr. and Meth. A 331 (1993) 706.
[31] C.M. Fortgang, R.W. Warren, Nucl. Instr. and Meth. A 341 (1994) 436.
[32] R. Tatchyn, P. Csonka, J. Phys. D 20 (1987) 394.
[33] G. Ramian, L.R. Elias, I. Kimel, Nucl. Instr. and Meth. A 250 (1986) 125.
[34] V.A. Papadichev, G.V. Rybalchenko, Nucl. Instr. and Meth. A 407 (1998) 419.
[35] A.H. Ho, R.B. Pantell, J. Feinstein, B. Tice, Nucl. Instr. and Meth. A 296 (1990) 631.
[36] V.L. Bratman, Yu.D. Grom, A.V. Savilov, V.A. Papadichev, G.V. Rybalchenko, Proceedings of the Twentieth International FEL Conference, Williamburg, Elsevier Science B.V., Amsterdam, 1999, p. II.15.
[37] V.A. Bogachenkov, V.L. Bratman, G.L. Denisov et al., Kratk. Soobshch. Fiz. 6 (1983) 38, in Russian.
[38] V.A. Bogachenkov, V.A. Papadichev, Sinilshchikova, O.A. Smith, in: C. Yamanaka (Ed.), Proceedings of the Sixth International Conference on High-Power Electron and Ion Beams, Kobe, Japan, 1986, p. 548.
[39] R.V. Latham, High Voltage Vacuum Insulation: the Physical Basis, Academic Press, London, 1981.
[40] I.N. Slivkov, Electrical insulation and Breakdown in Vacuum, Atomizdat, Moscow, 1972.
[41] A.S. Denholm, Can. J. Appl. Phys. 36 (1958) 476.
[42] R.W. Stinnett, M.A. Palmer, R.B. Spielman, R. Bengtson, IEEE Trans. PS-11 (1983) 216.
[43] A.A. Kolomensky, E.G. Krastelev, Trans. on Electr. Insul. EI-20 (1985) 775.

ELSEVIER

Nuclear Instruments and Methods in Physics Research A 429 (1999) 386–391

NUCLEAR INSTRUMENTS & METHODS IN PHYSICS RESEARCH Section A

www.elsevier.nl/locate/nima

Field fine tuning by pole height adjustment for the undulator of the TTF–FEL

J. Pflüger*, H. Lu[1], T. Teichmann

Hamburger Synchrotronstrahlungslabor HASYLAB, at Deutsches Elektronen-Synchrotron, DESY, Notkestr 85, 22603 Hamburg, Germany

Abstract

The field of the undulator for the VUV–FEL at the TESLA Test Facility has to meet very tough tolerances in order to guarantee a close overlap between the electron beam and the laser field. Consequently the undulator was designed to have height-adjustable poles in order to allow for fine tuning of the vertical undulator field in such a way that the trajectory is straightened. The signature of local pole height and gap changes on the field distribution was investigated. It was seen that changes are not restricted to the pole itself. Its effect can be seen up to the next eight neighboring poles. In this contribution we describe an algorithm in detail, which allows the prediction of required pole height changes in order to correct for field errors. As input data field errors deduced from precise magnetic field measurements are used together with the signatures of pole movements. A band diagonal system of linear equations has to be solved to obtain the pole height corrections. For demonstration of the method the field of the 0.9 m long prototype structure was optimized to have a straight trajectory. Since only a sparse band diagonal system of equations has to be solved, the method has the potential to be used in very long undulators having 600 –1000 poles. © 1999 Elsevier Science B.V. All rights reserved.

Keywords: Undulator; Electron beam; Laser field

1. Introduction

At DESY in Hamburg a Free Electron Laser (FEL) for the VUV spectral range down to 6.4 nm using the principle of self amplified spontaneous emission (SASE) [1,2] is under construction. It will use the electron beam generated by the TESLA test facility (TTF) [3,4] and will be built in two stages which are described in detail in Refs. [3,5]. The complete undulator system has a maximum length of about 30 m. It is a fixed gap structure and is described in detail in Refs. [6–9]. Table 1 reproduces its magnetic parameters. The device is a combined function undulator which integrates two functions. First, it provides the sinusoidally-shaped undulator field so that the FEL process can take place. Second, an alternating gradient field of about ± 20 T/m for the FODO lattice which is superimposed to the undulator field is generated.

The magnetic design was chosen to combine the following properties:

1. It is a completely planar structure, which allows for very good access to the field region at the

* Corresponding author. Fax.: 49-40-8998-4475.

E-mail address: pflueger@desy.de (J. Pflüger)

[1] On leave from FEL laboratory at Institute for High Energy Physics (IHEP), P.O. Box 2732 Beijing 100080, Peoples Republic of China.

PII: S0168-9002(99)00112-6

Table 1
Undulator parameters for the undulator for the VUV–FEL at the TESLA Test Facility

Gap (fixed) (mm)	12
Period length (mm)	27.3
Undulator peak field (T)	0.5
K-parameter	1.27
Design gradient (T/m)	18.3
Number of poles per undulator module	327
Total length per module (mm)	4492.2
Length of FODO quad section (mm)	136.5
FODO period length (m)	0.9555
Number of FODO periods per module	5
Separation between undulator modules (m)	0.2853

beam position allowing for high accuracy field measurements as well as an easy installation of the vacuum chamber without breaking of any magnetic circuits.
2. The gradient can be as large as ≈ 20 T/m.
3. The exact value and the precise location of the quadrupole axis is fine tunable.
4. Undulator and focusing fields are decoupled. This means that on the quadrupole axis the sign and magnitude of the field gradient has no influence on the undulator field and vice versa.

The magnetic fields of undulators for SASE FELs have to meet tough specifications [10]. Careful optimization and fine tuning of an undulator is required after it has been assembled. Two steps are planned. First the "naked" undulator, i.e., the undulator without the focusing magnets will be measured and optimized in such a way as to obtain a straight trajectory in the horizontal and vertical plane. In the horizontal plane the height adjustable poles will be used to obtain an optimum B_y field distribution. In the vertical plane field errors are not expected to be serious. Good sorting of the magnets using simulated annealing will minimize residual error fields [11] in any case. Small remaining field errors can be treated using very few suitable shims.

In a second step the focusing magnets will be attached and their strength as well as the exact position of quad centers will be fine tuned as described in Ref. [12].

This contribution deals with the tuning of the B_y field of the "naked" undulator. A numerical procedure which allows the prediction of pole adjustments to obtain a straight trajectory from precise magnetic measurements was developed and is described in detail.

2. Magnetic measurements, experimental

The new 12 m long bench was used to characterize the magnetic performance of the prototype structure. It provides sufficient mechanical accuracy for the magnetic measurements of the combined function undulator for the FEL at the TTF. Measurements could be performed with both high spatial and high-field resolution. For the field measurements the resolution is given by $\Delta B/B = 5 \times 10^{-4}$, the spatial resolution of the encoder system is 1 μm.

The magnetic measurements presented in this contribution were made on a 0.9 m long prototype of the undulator for the VUV–FEL with the same parameters as in Table 1. The pole heights of this structure can be adjusted by tuning hex screws by an estimated $\pm$ 0.5 mm. The screws having a 0.7 mm pitch act on the poles under an angle of 60° so that there results a 3.9 μm pole height change for a rotation angle of 1° on the tuning screws. The tuning screws were actuated manually using special hex keys with degree scales. Due to backlash between hex keys and the screws as well as to stick effects and elasticity in the 150 mm long key it was estimated that the screws could be adjusted with an accuracy of about $\pm$ 5–10° leading to a relative inaccuracy of the pole heights of about 20–40 μm.

3. Description of the method

The idea for the error control by pole height adjustment is simple: Error fields are determined by means of precise magnetic measurements. Precise knowledge of the influence of pole height changes on the field distribution, the so-called signature, can be obtained from magnetic measurements. Then an algorithm may be used to calculate the required pole shifts, which minimize these errors. Similar approaches has been reported by Stoner and Bekefi [13] who calculated shunt impedances to optimize the peak field homogeneity of a pulsed electromagnetic short period undulator and Ramian et al. [14,15] who describe a robotic system which is able to fine tune the peak field of PM undulator systems. Due to the measurement techniques applied, only the peak field was optimized in Refs. [13–15]. In contrast, the goal of this work was to optimize the complete trajectory. The treatment needs some assumptions and conventions, which are described below. In a first step the error field on each pole is determined by the "deviation" from the ideal field. Each pole is characterized by one number, the mean field deviation, which is the deviation averaged over the half period length of that pole at its nominal position. We tried two different ways to determine the deviation from the ideal field:

1. Applying a λ filter, which means convoluting the magnetic field data with a square function of width λ, the period length. In this way all the information with periodicity λ/n ($n = 1,2,\ldots$) is removed from the data and only nonperiodic perturbations are left over. The convolution however smoothes the data and smears out spatial information over one period length.
2. Alternatively, we tried various harmonic fits to the data up to order 11 to determine the periodic content and subtracted it from the field to obtain the deviation. This method gives noisier data, but preserves spatial information.

Both methods were compared and gave comparable results. Fig. 1 demonstrates the steps in the case of the application of a λ filter. Fig. 1(a) shows the field distribution of the upper structure half of a 0.9 m long prototype structure. One pole, #37, has been detuned on purpose to demonstrate how it shows up in the analysis. In Fig. 1(b) the λ filter has been applied to the data of Fig. 1(a). The perturbation applied on pole #37 is now very clearly visible. The large excursion at the ends is an artifact of the λ filter. This region therefore has been excluded from the analysis. Fig. 1(c) finally shows the assignment of the field errors to the poles. The endpole region has been excluded. Spatial information is smeared out and the perturbation of pole #37 has now been split up on the next nearest-neighboring poles as well. This is the input data to calculate the pole height adjustment.

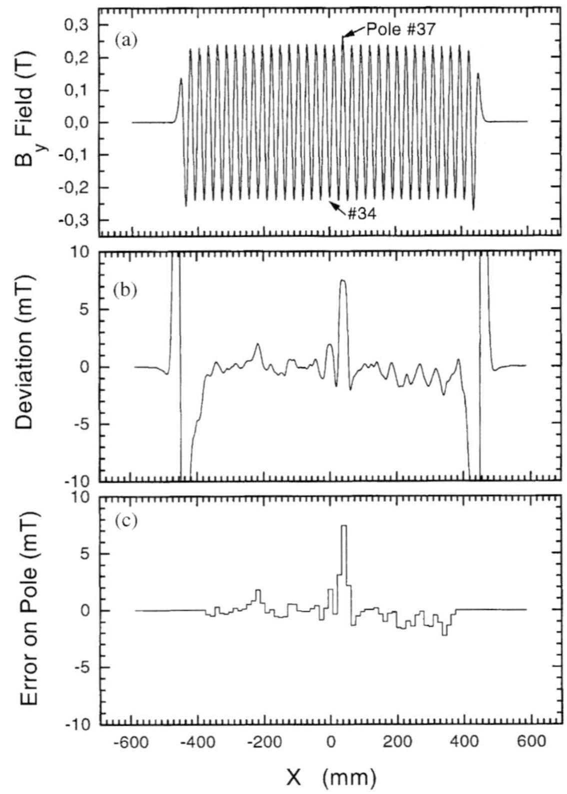

Fig. 1. Example for generation of the error fields on the poles for the upper structure half. The field was measured 6 mm below the poles of the upper structure half at the nominal beam position at 12 mm gap. (a) Field data, pole #37 has been detuned on purpose to demonstrate the effect. (b) After application of the λ-filter. The large end excursion at the ends is an artifact of the λ-filter. (c) Average error field at the poles. Due to the convolution, the adjustment is now smeared out over next nearest-neighbors. This is clearly seen at pole #37. The end region has been excluded.

In the second step we assume a linear relationship between the movement of a pole and the perturbation induced at any other pole. This is an assumption which only holds for small pole shifts of a few tenth of a millimeter. We assume an undulator consisting of N identical poles. Poles near the ends again have to be excluded, since due to the truncation they cannot be considered as identical with poles well inside the undulator. Excluding the outermost 4–5 poles was found to be sufficient to neglect both, artifacts from the λ filter and truncation effects as well.

We are interested in the field change S_j on a pole with index j if another pole with index i is moved by an amount p_i. In order to get the total field change on pole j one has to sum over all contributions from poles i. The S_j may be interpreted as minus the field change on pole j which is required to obtain a perfect undulator field.

These considerations lead to the following linear system of equations:

$$S_1 = a_0 \cdot p_1 + a_1 \cdot p_2 + a_2 \cdot p_3 + \cdots + a_{i-1} \cdot p_i + \cdots + a_{N-1} \cdot p_N$$

$$S_2 = a_{-1} \cdot p_1 + a_0 \cdot p_2 + a_1 \cdot p_3 + \cdots + a_{i-2} \cdot p_i + \cdots + a_{N-2} \cdot p_N$$

$$S_3 = a_{-2} \cdot p_1 + a_{-1} \cdot p_2 + a_0 \cdot p_3 + \cdots + a_{i-3} \cdot p_i + \cdots + a_{N-3} \cdot p_N$$

$$\cdot = \quad \cdot \quad \cdot$$

$$\cdot = \quad \cdot \quad \cdot$$

$$\cdot = \quad \cdot \quad \cdot$$

$$S_j = a_{1-j} \cdot p_1 + a_{2-j} \cdot p_2 + a_{3-j} \cdot p_3 + \cdots + a_{i-j} \cdot p_i + \cdots + a_{N-j} \cdot p_N$$

$$\cdot = \quad \cdot \quad \cdot$$

$$\cdot = \quad \cdot \quad \cdot$$

$$\cdot = \quad \cdot \quad \cdot$$

$$S_N = a_{1-N} \cdot p_1 + a_{2-N} \cdot p_2 + a_{3-N} \cdot p_3 + \cdots + a_{i-N} \cdot p_i + \cdots + a_0 \cdot p_N.$$

The strength of the "interaction" is characterized by the coefficients a_m where $m = i - j$. These a_m coefficients are the response of a pole shift p_i on the field change S_j on pole j normalized by the pole shift p_i. Their dimension is therefore [T/mm]. The assumption of identical poles is reflected in the fact that they depend only on the difference $i - j$ and are assumed to be the same for all poles. The coefficients a_m have been determined experimentally by measuring the response of the magnetic field to a known pole shift (0.5 mm) of a known pole (pole #34, the central pole at $X = 0$.) of the upper structure half of the prototype structure. In Fig. 2, the effect of this pole shift is shown (dotted line). The full line shows the average field on the poles in the vicinity of pole #34. The index m is indicated. Under the pole itself there is the strongest effect, but up to the next four neighbors there is a measurable effect with opposite sign. Integrated over all neighbors this sign change eats up a considerable amount of the correction of the pole itself. The shape of the dotted curve is that of a dipole layer of magnetic surface charges brought on the pole.

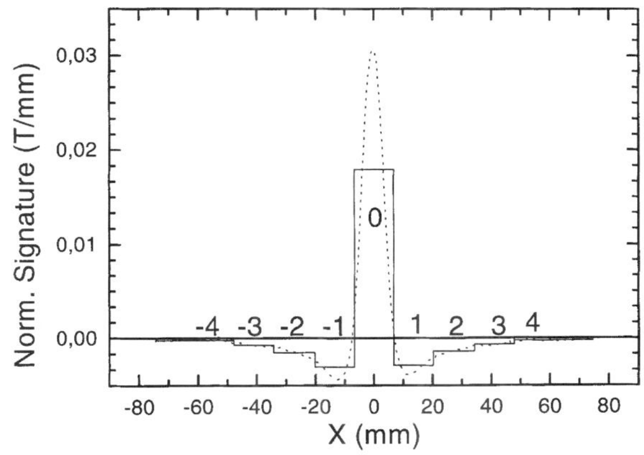

Fig. 2. Normalized response of a pole shift of the central pole # 34 of the upper magnet structure. Dotted line: Measured difference. Full line: Average field on the poles.

It is seen that the interaction on both sides are very similar so that the assumption $a_m = a_{-m}$ holds. We assume the same interaction for all poles so that the coefficient generated from pole #34 are also valid for all other poles. It is also seen that it is sufficient to treat only the four next nearest-neighbors. For larger distances the correction becomes negligibly small. The linear system of equations is therefore greatly simplified. Below and above the main diagonal there are only four side diagonals which have to be considered. Using a specialized equation solver for band diagonal systems of equations reduces numerical effort and also minimizes memory requirements dramatically. Large pole numbers can be treated at moderate effort on a PC. Up to 1000 poles have been tried in simulations. Even larger numbers may be possible.

There is no fundamental difference if a structure half, i.e. the top or bottom structure of an undulator or a full magnet structure with closed gap is optimized. Only the a_m coefficients are different and have to be determined for both cases in the proper way as described above. This means that for structure halves the signature of a single pole movement has to be evaluated and in the case of a closed gap the poles in the upper and lower structure have to be moved simultaneously resulting in a local gap change.

For the prototype structure it was found very helpful to separately optimize the two structure halves first and to do the final fine tuning with the gap closed to the nominal position in a second step. This second step needs only minor additional adjustments.

4. Results

The 0.9 m long prototype structure was used to test the optimization procedure. Again only the upper structure was used. The field was measured 6 mm below the poles corresponding to the nominal electron beam location with respect to the upper structure at a 12 mm gap. The resulting second field integral distribution from four iteration steps are shown in Fig. 3. The resulting electron beam excursion can be calculated by

$$z\ (\mu\text{m}) = 587/\gamma \cdot I_2\ (\text{Tmm}^2)$$
$$= 0.3.I_2\ (\text{Tmm}^2)/E_{\text{Kin}}\ (\text{GeV})$$

where γ is the kinetic electron energy divided by the electron's rest mass energy. Emphasis was put on a straight trajectory in the undulator itself. It is seen in Fig. 3 that the endpoles are not adjusted properly. This is of minor importance in this context since they are easily adjustable and they can be tuned to result in a straight trajectory before and after the undulator. It is seen that within a few Tmm^2 no deviations can be seen corresponding to sub micrometer beam excursions. A full structure with 12 mm gap results in twice the field. The field integral deviations in this case therefore will be twice as large. For each iteration step the error fields were calculated and the resulting pole height adjustments were calculated. A computer printout was generated containing the pole's number and the amount of adjustment needed on that pole. The

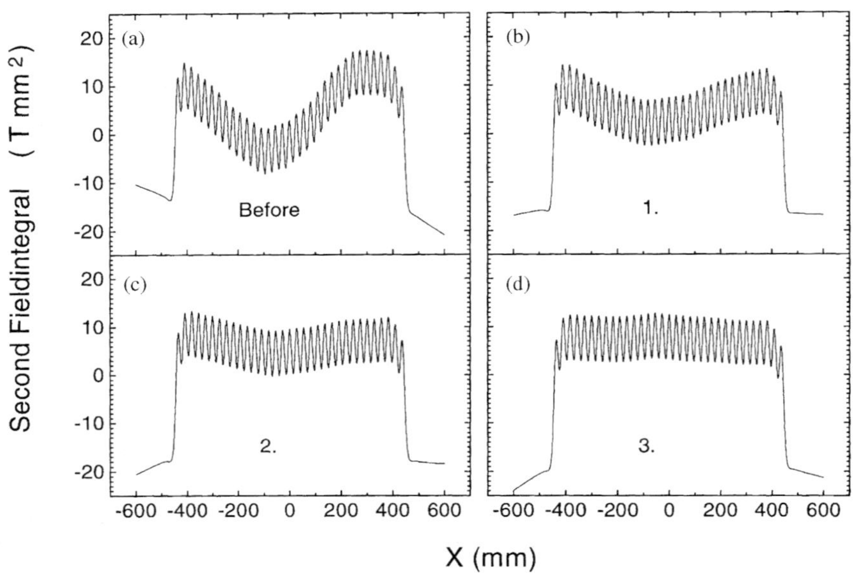

Fig. 3. Demonstration of the optimization of the upper structure half in three iteration steps. The second field integral is shown which is proportional to the beam excursion. At 300 MeV which will be used for Phase I 10 Tmm^2 corresponds to 10 μm beam excursion.

amount of adjustments was much larger in the first optimization. For the last iteration step needed only very little correction was needed.

5. Limitations, accuracy

The quality of the optimization is influenced by three factors

1. the accuracy of the field measurements,
2. the accuracy of the coefficients a_m,
3. the accuracy of the pole height adjustment.

The measurements accuracy was conservatively estimated to be: $\Delta B/B = 5 \times 10^{-4}$. In practice it may even be better.

Inaccurate coefficients a_m influence the convergence of the procedure more than the accuracy of the result, which means that more iterations are needed to get the optimized result.

The main source of error is the modest accuracy of local pole height adjustment. $\pm$ 5–10° of angular accuracy leading to an estimated inaccuracy of the pole height of $\pm$ 20–40 μm which in turn leads to a field error of about 0.3–0.6%. This is by far the dominant contribution. To improve the accuracy of pole height adjustment one has to measure these changes directly using a suitable dial gauge with micrometer resolution instead of just counting the tuning screw angle. This improvement is underway. With this moderate effort it should be possible to reduce the pole height adjustment error to 5–10 μm or better so that this contribution amounts to less than 0.075–0.15%.

References

[1] A.M. Kondratenko, E.L. Saldin, Part. Accel. 10 (1980) 207.
[2] R. Bonifacio, C. Pellegrini, L. Narducci, Opt. Commun. 50 (1984) 373.
[3] A VUV Free Electron Laser at the TESLA Test Facility at DESY, Conceptual Design Report, TESLA-FEL Report 95-03, DESY, Hamburg, April 1995.
[4] D.A. Edward (Ed.), TESLA Test Facility linac design report, TESLA Report 95-01, Hamburg, March 1995.
[5] W. Brefeld, B. Faatz, Y.M. Nikitina, J. Pflüger, J. Rossbach, E.L. Saldin, E.A. Schneidmiller, M.V. Yurkov, Nucl. Instr. and Meth. A 393 (1997) 119.
[6] Y.M. Nikitina, J. Pflüger, Nucl. Instr. and Meth. A 375 (1996) 325.
[7] J. Pflüger, Y.M. Nikitina, TESLA-FEL 96-02, 1996.
[8] Y.M. Nikitina, J. Pflüger, TESLA-FEL 96-03, 1996.
[9] J. Pflüger, Y.M. Nikitina, Nucl. Instr. and Meth. A 381 (1996) 554.
[10] B. Faatz, J. Pflüger, Y. Nikitina, Nucl. Instr. and Meth. A 393 (1997) 380.
[11] B. Faatz, J.Pflüger, Sorting strategy for the TTF–FEL undulator magnets TESLA-FEL Report, in preparation.
[12] J. Pflüger, H. Lu, T. Teichmann, Nucl. Instr. and Meth. A 407 (1998) 386.
[13] R. Stoner, G. Bekefi, IEEE J. Quantum Electron. 31 (6) (1995) 1158.
[14] G. Ramian, Nucl. Instr. and Meth. A 318 (1992) 225.
[15] G. Ramian, J. Kaminski, S.J. Allen, Nucl. Instr. and Meth. A 393 (1997) 220.

ELSEVIER

Nuclear Instruments and Methods in Physics Research A 429 (1999) 392–396

NUCLEAR INSTRUMENTS & METHODS IN PHYSICS RESEARCH Section A

www.elsevier.nl/locate/nima

Development of solenoid-induced helical wiggler with four poles per period

N. Ohigashi[a,*], Y. Tsunawaki[b], M. Kiyochi[a], N. Nakao[c], M. Fujita[d], K. Imasaki[d], S. Nakai[c], K. Mima[c]

[a]*Department of Physics, Kansai University, Suita, Osaka 564-8680, Japan*

[b]*Department of Electrical Engineering and Electronics, Osaka Sangyo University, Daito, Osaka 574-8530, Japan*

[c]*Institute of Laser Engineering, Osaka University, Suita, Osaka 565-0871, Japan*

[d]*Institute for Laser Technology, Suita, Osaka 565-0871, Japan*

Abstract

A new type of helical wiggler consisting of two staggered-iron arrays inserted into a solenoid field has been developed. The field measured by a test wiggler showed linear increment with the period. It was seen that 24% of the solenoid field contributed to the induced wiggler field when the gap length and the period of the wiggler were 16 and 24 mm, respectively. This wiggler would be useful for an FEL with a low-energy electron beam propagating in a strong axial guiding field.

1. Introduction

For a free electron laser (FEL) using a low-energy electron beam, we have studied a planar wiggler in which a staggered-iron array was put into a solenoid field. This kind of wiggler has also been experimentally studied by using an array of ferromagnetic material [1,2]. The electron beam in a planar wiggler with a solenoid field diverges in the direction of $\nabla B_w \times \boldsymbol{B}_0$, where B_w is the wiggler field and $\boldsymbol{B}_0$ is the solenoid field vector. This phenomenon introduces electron beam loss into the drift tube wall [3,4]. It occurs more remarkably when the electron is closer to the magnetic pole face because of the high gradient of the field of B_w. Therefore, a sheet electron beam is favourable to prevent such drift motion. There is, otherwise, another idea that two parallel currents at both sides of the electron beam flow along the wiggler axis. These currents work as an equivalent quadrupole magnet to focus the electron beam. If a helical wiggler is adopted, these currents will not be required because of its focusing power on an electron beam.

In this work, we have investigated a helical wiggler consisting of two staggered-iron arrays inserted perpendicular to each other in a solenoid field. A pair of staggered iron blocks is arranged spirally along the axis of the solenoid field. The field is induced between them and it rotates by an angle of 90° every quarter period along the wiggler axis. At first glance one might think that this kind of helical wiggler would not create a strong wiggler field because of the leakage field between neighboring iron blocks which is higher than that for a planar wiggler. Therefore, we constructed and

* Corresponding author. Fax: + 81-6-6330-3770.

E-mail address: ohigasi@ipcku.kansai-u.ac.jp (N. Ohigashi)

0168-9002/99/$ - see front matter
PII: S0168-9002(99)00113-8

compared experimentally both helical and planar wigglers using staggered-iron blocks. The ratio (σ_{wo}) of induced wiggler field to solenoid field increased with the period (λ_w) for both wigglers. Although the field of the helical wiggler was less than that of the planar one with increasing periods, σ_{wo} was almost same for both wigglers with a period of about 20 mm and the field was also adequately strong. These results suggest the use of this design as a practical helical wiggler for a microwave-FEL (FEM).

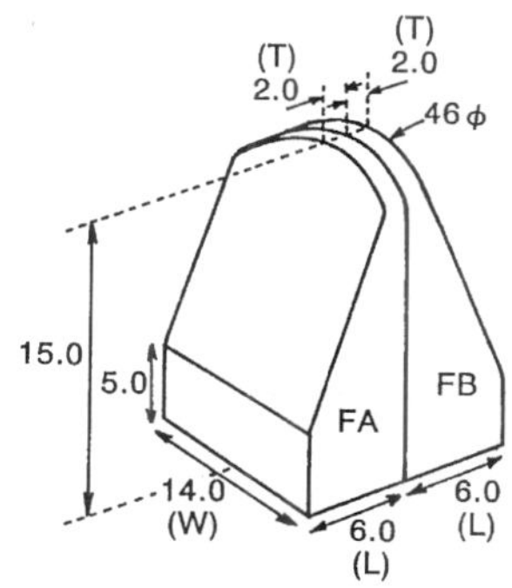

Fig. 2. A pair of iron-blocks: FA and FB (unit: mm).

2. Structure of the wiggler

The structure of the wiggler is shown in Fig. 1. The axial field B_0 is created by a copper solenoid coil wound (3371 turns) on a bobbin of a copper tube (diameter = 50 mm, length = 1270 mm). The designed field is 3 kG for the coil current if 100 A. The staggered-iron array is constructed by iron blocks (FA and FB) and nonmagnetic spacers (SA) which are installed between the drift tube (SUS304) and brass retainer frame as shown in Fig. 1(a). Two sets of them are aligned on x–z and y–z planes, respectively, as shown in Fig. 1(b) and (c). The phases between these arrays differ by a quarter of a period. The detailed dimensions of an iron block is shown in Fig. 2. The helical wiggler field is induced between every pair of staggered iron

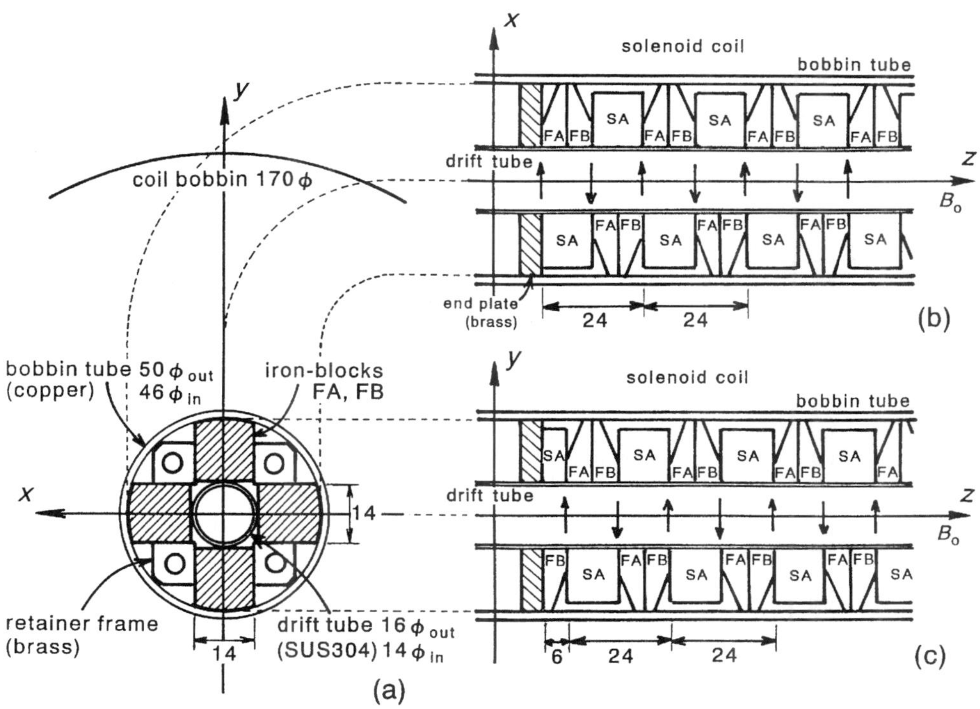

Fig. 1. Solenoid induced helical wiggler: cross sections in (a) xy-plane, (b) xz-plane, and (c) yz-plane.

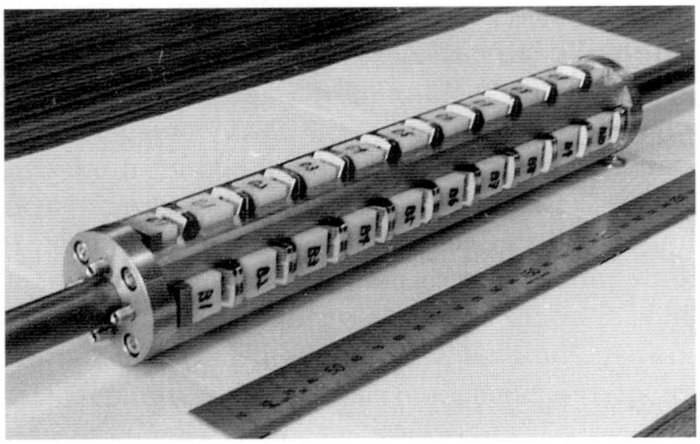

Fig. 3. Test helical wiggler consisting of two staggered-iron-arrays.

blocks. Typical period and gap length in this work are 24 and 16 mm, respectively. The period is, furthermore, changed by the insertion of iron, and nonmagnetic spacers. In this work, we selected periods of 24, 36 and 48 mm. Fig. 3 shows a test wiggler with the period $\lambda_w = 24$ mm and period number $N_w = 10$.

3. Field measurement of the wiggler

The magnetic field was measured using a Hall element with an active region of 1.8 mm $\varnothing$. In order to analyze the precise distribution of the

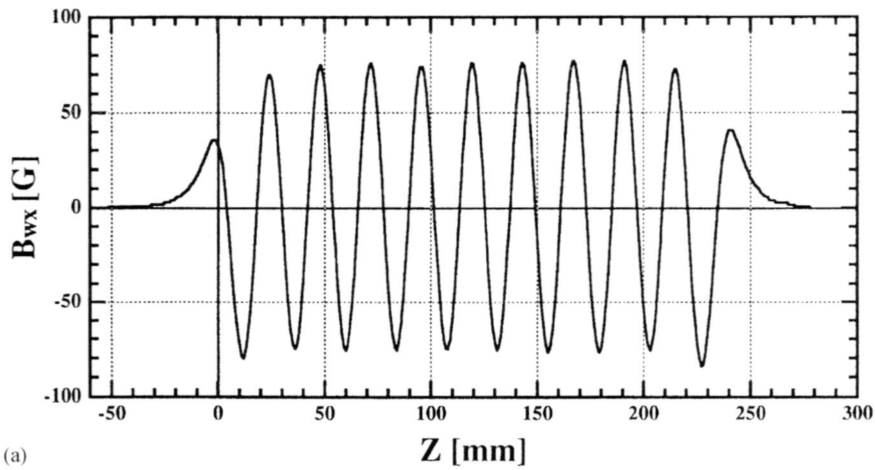

(a)

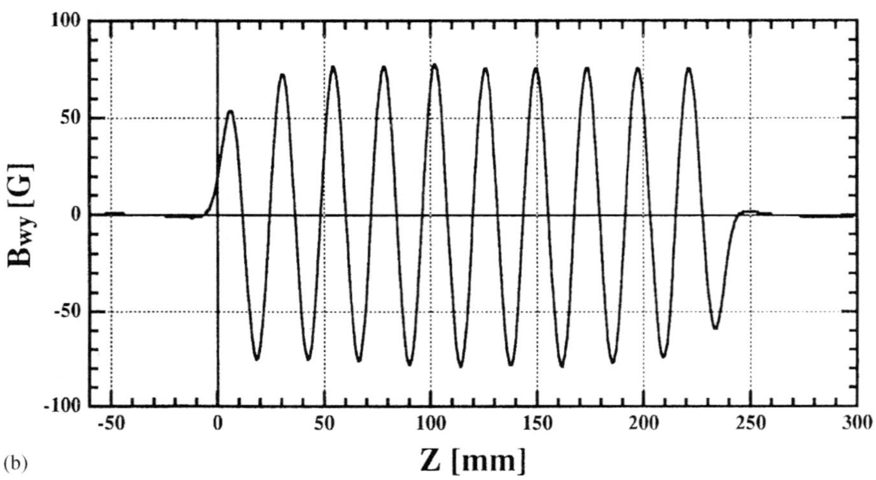

(b)

Fig. 4. Induced wiggler field distributions measured for the wiggler (gap length $l_G = 16$ mm, $\lambda_w = 24$ mm, $N_w = 10$).

wiggler field (B_{wx} and B_{wy}), the Hall element was set at the center of a cylindrical retainer made of brass block whose diameter was slightly less than that of the drift tube. The retainer was moved smoothly with the Hall element inside the drift tube along the wiggler axis.

When the coil current was 10 A, the axial field B_{00} of 324 G was reduced to $B_0 = 320$ G upon the installation of staggered-iron arrays. The induced wiggler field distributions of B_{wx} and B_{wy} are shown in Fig. 4(a) and (b), respectively, for the wiggler with a period of 24 mm. $z = 0$ means the starting position where the initial iron blocks were arranged. It is seen that the field forms a homogeneous helix. The dependence of the wiggler field on the period is shown in Fig. 5. Solid and dotted lines show the results for helical and planar wigglers, respectively. It should be noted that the size of iron blocks in the planar wiggler is different from Fig. 2; $W = 20$, $L = 5.5$, and $T = 1.5$ mm. It is observed that the field and $\sigma_{w0}(= B_w/B_{00})$ increase almost linearly with period of the wiggler. This phenomenon is explained by the magnetic resistance which increases linearly between the neighboring iron blocks in z direction with increasing period. The field of the helical wiggler is lower than that of the planar wiggler due to the higher leakage field. However, for both the wigglers with a short period to ~ 20 mm, the fields are almost the same. If the coil current of the solenoid is increased to 100 A, one expects to obtain an axial field of $B_0 = 3.2$ kG ($B_{00} = 3.24$ kG) and helical wiggler field of $B_w = 760$ G ($\sigma_{w0} = 24\%$) for the period of 24 mm. These fields seem adequate for an FEL using an electron beam with an energy lower than 1 MeV.

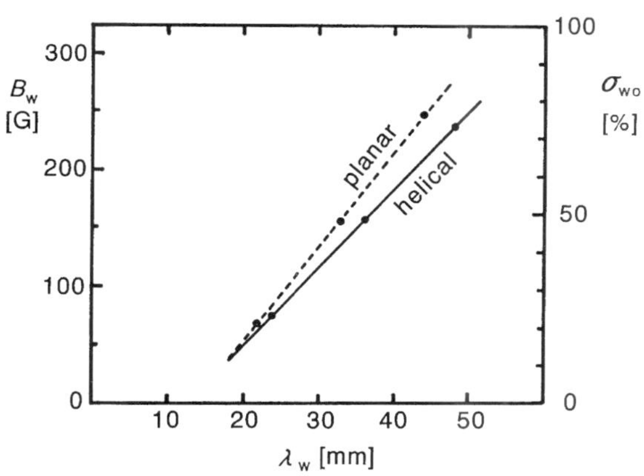

Fig. 5. Dependencies of wiggler field B_w and induction ratio σ_{w0} on period λ_w. ($B_{00} = 324$ G).

4. Parameter survey for FEM

As is well known, the electromagnetic wave of an FEL grows through the coupling between the modes of the slow space charge wave and an electromagnetic waveguide [5–7]. The axial velocity of an electron is estimated from the constant axial velocity solution to give steady state orbits for an electron [3].

In this study, the behavior of an FEM was analyzed under the conditions of $B_0 = 3$ kG, $B_w = 750$ G, $\lambda_w = 24$ mm, $N_w = 30$ for the wiggler and $I_b = 100$ mA, $r_b = 3$mm, $E_b = 305$–400 keV, $\Delta\gamma/\gamma = 0.01$ for the electron beam. The frequency of the FEM was estimated to vary from 19.6 to 44.2 GHz depending on E_b, by the 3D theory [8] for orbit I and TE_{11} waveguide mode. The small signal gain was calculated as 12.8% for an electron beam with $E_b = 400$ keV (power $P_b = 40$ kW) from the ordinary FEL gain formula [9]. The saturation power is, therefore, 330 W according to the relation of $P_{sat} = P_b/4N_w$. If the noise power is assumed to be $P_0 = 1$ μW, this saturation power will be accomplished in 163 round tips of scattered radiation between two resonator mirrors.

5. Conclusions

A solenoid-induced helical wiggler with four poles per period has been developed. From the magnetic field measurement using a short test wiggler, an axial field of 3.2 kG and a wiggler field of 750 G will be expected when the staggered-iron arrays with period of 24 mm are inserted into a solenoid field of 3.24 kG. This type of helical wiggler is considered to be useful for an FEM using a low-energy electron beam.

References

[1] A.H. Ho, R.H. Pantell, J. Feinstein, B. Tice, Nucl. Instr. and Meth. A 296 (1990) 631.

[2] Y.C. Huang, H.C. Wang, R.H. Pantell, J. Feinstein, J. Harris, Nucl. Instr. and Meth. A 341 (1994) 431.

[3] H.P. Freund, T.M. Antonsen Jr., Principles of Free-Electron Laser, Chapman & Hall, New York, 1992, pp. 41, 66 and 67.

[4] J.A. Pasour, F. Mako, C.W. Roberson, J. Appl. Phys. 53 (1982) 7174.

[5] H.P. Freund, P. Sprangle, D. Dillenburg, E.H. da Jornada, R.S. Schneider, B. Liberman, Phys. Rev. A 26 (1982) 2004.

[6] S.H. Gold, W.M. Black, H.P. Freund, V.L. Granatstein, A.K. Kinkead, Phys. Fluids 27 (1984) 746.

[7] J. Fajans, G. Bekefi, Y.Z. Yin, B. Lax, PFC/JA-84-41, Plasma Fusion Center, MIT, December 1984, preprint.

[8] H.P. Freund, S. Johnston, P. Sprangle, IEEE J. Quantum Electron. QE-19 (1983) 322.

[9] C.A. Brau, Free-Electron Lasers, Academic Press, New York, 1993, pp. 100–110.

ELSEVIER

Nuclear Instruments and Methods in Physics Research A 429 (1999) 397–406

NUCLEAR INSTRUMENTS & METHODS IN PHYSICS RESEARCH Section A

www.elsevier.nl/locate/nima

X-ray optics design studies for the SLAC 1.5–15 Å Linac Coherent Light Source (LCLS)

R. Tatchyn[a,*], J. Arthur[a], R. Boyce[a], A. Fasso[a], J. Montgomery[a], V. Vylet[a], D. Walz[a], R. Yotam[a], A.K. Freund[b], M. Howells[c]

[a]*Stanford Synchrotron Radiation Laboratory, Stanford Linear Accelerator Center, Stanford, CA 94305, USA*

[b]*European Synchrotron Radiation Facility, B.P. 220, 38043 Grenoble CEDEX, France*

[c]*Advanced Light Source, Lawrence Berkeley National Laboratory, Berkeley, CA 94720, USA*

Abstract

In recent years, a number of systematic studies have been carried out on the design and R&D aspects of X-ray free-electron laser (XRFEL) schemes based on driving highly compressed electron bunches from a multi-GeV linac through long (30 m – 100 + m) undulators. These sources, when operated in the self-amplified spontaneous emission (SASE) mode, feature singularly high peak output power densities and frequently unprecedented combinations of phase-space and output-parameter values. This has led to correspondingly pivotal design challenges and opportunities for the optical materials, systems, components, and experimental configurations for transporting and utilizing this radiation. In this paper we summarize the design and R&D status of the X-ray optics section of the SLAC Linac Coherent Light Source (LCLS), a 1.5 Angstrom SASE FEL driven by the last kilometer of the SLAC 3-kilometer S-band linac.

Keywords: X-ray optics design studies; Linac coherent light source; SASE; XRFEL

1. Introduction

In recent years a notable direction in FEL source development has been toward ever-decreasing X-ray wavelengths [1,2]. Proposed or developing designs have included amplifiers and oscillators based both on storage rings and microtrons, as well as single-pass amplifiers driven by linacs. Of all these, the latter currently appear to offer the greatest flexibility and range in temporal pulse parameters, as well as the highest attainable peak powers. To date, a number of design and R&D studies have been conducted or are in progress [3–8] on linac-driven XRFEL schemes based on driving low emittance, high current electron bunches through long undulators to attain gain saturation in either the self-amplified spontaneous emission (SASE) or coherent amplifier (CA) modes [9–13]. The output parameters of these devices feature peak powers in the 10–100 GW range, full transverse and low-to-moderate longitudinal coherence, pulse durations

* Corresponding author. Tel.: 650-926-2731; fax: 650-926-4100.

E-mail address: tatchyn@ssrl.slac.stanford.edu (R. Tatchyn)

0168-9002/99/$ - see front matter
PII: S0168-9002(99)00115-1

in the 50–500 fs range, pulse repetition rates ranging from $\sim 10^2$ to as high as $\sim 10^8$ Hz, broad spontaneous continua with integrated powers comparable to or greater than the coherent output, potentially flexible polarization parameters, and intense bremsstrahlung and secondary-particle spectra. These unique (and often unprecedented) combinations of optical source properties have, in turn, highlighted correspondingly demanding requirements, as well as innovative design opportunities, on optical systems, instrumentation, and components for processing these pulses for scientific or technological applications [14].

Over the period 1996–1998, a directed design study of the X-ray optical system for the SLAC LCLS, a proposed 1.5–15 Å FEL driven by the last kilometer of the SLAC 3 kilometer S-band linac, has been carried out at the Stanford Linear Accelerator Center. This effort, part of a more general, funded study of a complete LCLS R&D facility [7], has included modeling and theoretical investigations of the source properties of the LCLS, development of designs for its overall optical system and beam line layouts, and the initiation of directed R&D studies of selected optical components, instrumentation, and ultra-short-pulse radiation/matter interactions. In this paper we present an overview and summary of the current status of these activities.

2. Optical transport and experimental facility layouts

The radiation emanating from the LCLS consists of three basic types: (1) the coherent spectral lines (the FEL fundamental and its harmonics); (2) the ordinary spontaneous insertion device spectrum; and (3) a very broad bremsstrahlung and secondary-particle spectrum consisting of gamma rays, neutrons, muons, and other particles with energies bounded by the maximum operating energy of the LCLS (5–15 GeV). All three types of radiation emerge more or less co-terminously with the high-energy electron bunch that drives their production. The primary function of the undulator-to-experimental-area section of the LCLS is to deflect the electron beam away from the (predominantly neutral) radiation, dump it, and then pass the radiation on through a system of spectral-angular filters whose function is to transport either the spontaneous or coherent undulator photons to the experimental end stations while suppressing, as much as possible, the transmission of the bremsstrahlung component and any secondary noise generated by it.

Apart from the performance of these critical functions [15], the current design of the optical system and its component layouts was predicated on a number of additional factors. These include: (1) handling of ultrashort X-ray pulses of unprecedented peak power density; (2) a spectral range of the FEL coherent lines (0.9–8.5 keV in the fundamental and up to 25 keV in the 3rd harmonic) that requires the use of both specular (for the full spectral range) and crystal (for $\lambda < 3$–5 Å) optics; (3) the consequent necessity of unobstructed high-vacuum transport through to the experimental areas, with ultra-high vacuum (UHV) environments for selected specular instrumentation; (4) the requirement of separating the spontaneous synchrotron radiation (SR) spectrum (extending out to 1 + MeV at 15 GeV) from the coherent lines, and (5) the provision of a basic set of spectral-angular filtering functions. A plan-view layout of the LCLS undulator-to-experimental-area transport system, which has been designed to address or incorporate these basic requirements [16] is shown in Fig. 1. A side-view layout detailing the currently-assigned locations of the major components is shown in Figs. 1 and 2.

In approaching the question of mitigating potential damage to the optical components, experimental samples, and instrumentation, two approaches were explored. The first, incorporated into the depicted design, is a gas (or liquid) attenuation cell. Its purpose is: (1) to attenuate the LCLS power to levels manageable by ordinary optics, allowing for the initial characterization of the LCLS output with conventional instrumentation; and (2) to provide a continuous transition to power densities at which meaningful research on the interaction of LCLS-type pulses with matter can proceed. In this regard, the cell itself constitutes an appropriate physical system for performing initial studies of this type. The second approach, a long beam line (up to ~ 800 m), has also been assessed

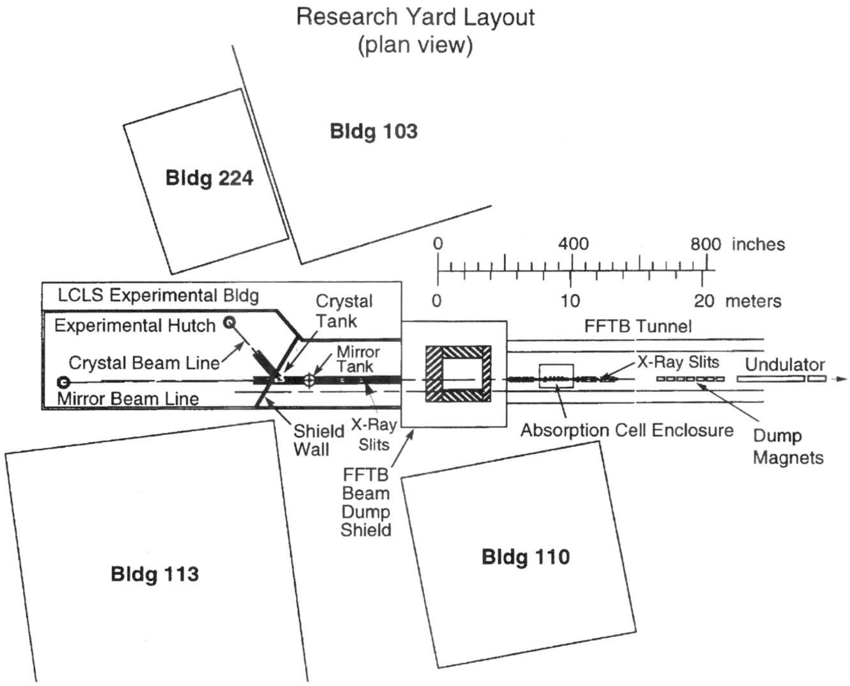

Fig. 1. LCLS undulator-to-experimental area layout (plan view) in the SLAC Research Yard.

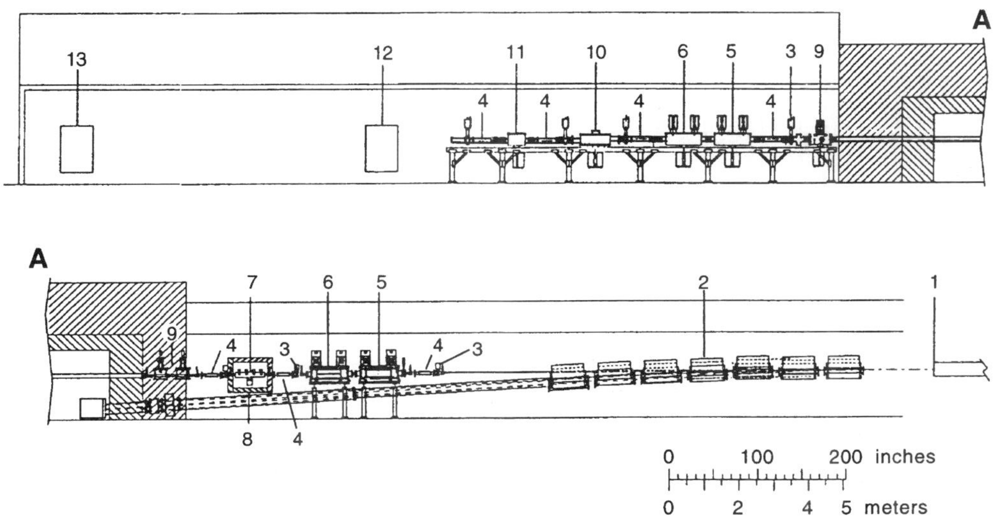

Fig. 2. Detailed layout (elevation view) of electron beam dump line and the X-ray optics system components in the LCLS undulator-to-experimental-area. (1) Undulator exit; (2) electron beam dump magnets; (3) vacuum valve; (4) Differential Pumping Section (DPS); (5) vertical X-ray slits; (6) horizontal X-ray slits; (7) absorption cell; (8) absorption cell radiation enclosure; (9) beam shutter; (10) mirror tank; (11) crystal tank; (12) crystal beam line end station; (13) specular beam line end station.

and may be incorporated into a future phase of the LCLS R&D or user facility design.

3. LCLS X-ray optics design studies: overview

An overview of an initial block representation of the LCLS X-ray optics system depicting some of the basic features that have driven the present design and selected R&D studies is shown in Fig. 3. These efforts have centered on a number of subject areas, including: (1) source modeling and description of the coherent, spontaneous, and bremsstrahlung output radiation; (2) transport of the coherent and spontaneous source properties from the undulator to the experimental stations; (3) transport and suppression of bremsstrahlung and bremsstrahlung-generated noise; (4) peak power density damage, and damage mitigation issues; (4) instrumentation principles and design, including short-pulse effects; (5) novel optics; and (6) experimental design.

4. Source and radiation transport studies

A basic goal of the LCLS X-ray optics design and R&D efforts is to provide a detailed and reliable description of the source radiation. Although a large body of literature exists wherein idealized SASE or CA systems have been analyzed or simulated [9–13,17–23], it appears evident that further work based on more realistic electron beam and insertion device modeling will be required. For example, the peak output characteristics of the two limiting cases of the LCLS listed in Table 1 are based on idealized beam distributions and optimal transport and field error assumptions. In fact, more detailed studies have indicated that the inclusion of

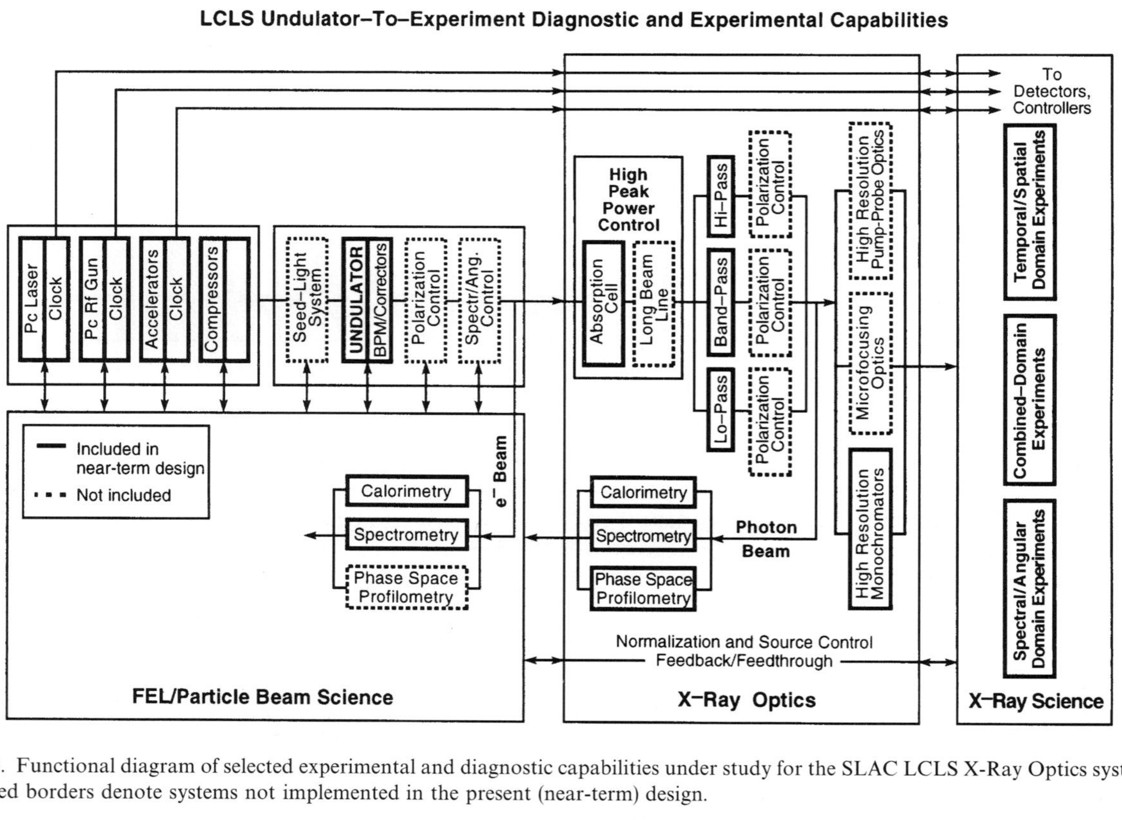

Fig. 3. Functional diagram of selected experimental and diagnostic capabilities under study for the SLAC LCLS X-Ray Optics system. Dashed borders denote systems not implemented in the present (near-term) design.

Table 1
Optical and source parameters of the LCLS. Undulator $K = 3.67$. $N_u = 3300$ periods. Undulator period $\lambda_u = 3$ cm

Radiation wavelength (Å)	1.5	15
Norm. emitt. $\gamma\varepsilon$ (mm-mrad)	1.5	2.0
Electron energy (GeV)	14. 35	4.54
Peak current (A)	3400	3400
Bunch duration (f_s, FWHM)	277	277
Peak spontaneous power (GW)	81	4.9
Peak coherent power[a] (GW)	9	11
Average coherent power[b] (W)	0.31	0.35
Energy/pulse (mJ)	2.5	0.64
Coherent photons/pulse ($\times 10^{12}$)	1.9	23
Approx. Bandwidth (BW) (%)	0.1	0.1
Peak brightness[c] ($\times 10^{32}$)	12	1.48
Peak degeneracy parameter ($\times 10^9$)	3.3	412
Average brightness[c] ($\times 10^{21}$)	40	4.9
Transverse size (μm, FWHM)[c]	78	93
Divergence angle (μrad, FWHM)[c]	1	8
Spontaneous fundamental opening angle (μrad, FWHM)	4.9	15.5
Spontaneous fundamental transverse size (μm, FWHM)	82	131
Peak power density (W/mm²)[d] ($\times 10^{12}$)	**1.88**	**1.62**
Peak field (V/m)[d]($\times 10^{10}$)	**3.8**	**3.5**

[a]Output fully transversely coherent
[b]At 120 Hz pulse rep rate
[c]Photons/s/mm²/mrad²/0.1%Bw
[d]At undulator exit

more realistic versions of these and other effects [7,24] could substantially affect the listed peak brightness, B, and other associated parameters. To illustrate, we recall the definition of B as the photon energy density in a phase space with dimensions x, $x'(=\mathrm{d}x/\mathrm{d}z)$, y, $y'(=\mathrm{d}y/\mathrm{d}z)$, f, and τ. This space contains two intersecting distributions of electrons and photons, which are related through an association between each mode of the radiation field and the electron (or electron structure) generating it. As a result, the phase space statistics of the radiation can be characterized by the statistics of the electron distribution concatenated with the corresponding mode parameters. A unique aspect of the SLAC LCLS is that the radiation field is partitioned into a large number of causally unconnected longitudinal regions ("slippage lengths"), each of which contains a further aggregate of independent modes, of an average length on the scale of the "cooperation length" ($\sim N_u\lambda/10$, where N_u is the number of undulator periods). This partitioning precludes the self-consistent treatment of the complete electron and radiation distributions as two statistically homogeneous interacting systems. The photon distribution is in fact a longitudinal array of a large number of independent sub-distributions, each of which must be assigned an independent set of statistical parameters for consistency. Thus, indexing the slippage lengths by i and the modes within each slippage length by j, one is led to a specification of the brightness more detailed than the one typically presented, viz.,

$$B = \frac{\langle N_{\mathrm{phot}(ij)}\rangle/\langle\Delta\tau_{ij}\rangle}{\langle\eta_{ij}^3\rangle\,\langle\Delta x_{ij}\rangle\,\langle\Delta x'_{ij}\rangle\,\langle\Delta y_{ij}\rangle\,\langle\Delta y'_{ij}\rangle(\langle\Delta f_{ij}\rangle/\langle f_{ij}\rangle)}. \tag{1}$$

Here the brightness is interpreted as an average brightness corresponding to the radiation distribution projected onto an observation plane. In this picture the probability density functions (pdfs) of the projected distribution are each the convolution of the corresponding indexed pdfs along the bunch (the phase space (say, $x - x'$) area of an indexed region is given by $\eta_{ij}\Delta x_{ij}\Delta x'_{ij}$, where, say, Δx is the standard deviation of the concatenated x-positions of the electrons and photons). This phase space description also provides a natural formalism for describing the variation, or correlation, of the distributions' statistical moments vs. z (or t), information which is expected to be of critical importance to instrumentation designers and users of LCLS radiation. Given the description represented by Eq. (1), a number of effects that can substantially dilute the brightness cited in Table 1 can be identified (e.g., wakefield-induced gradients in f_{ij}, corresponding "chirps" induced in the other phase-space variables by chromatic electron optics, mode splitting, etc.). In view of this, the projected performance charts of the LCLS (see Fig. 4) explicitly indicate an expected range of variation in the brightness as a consequence of some of these effects.

For purposes of the design study, a fairly comprehensive set of calculations of the far-field spontaneous spectra and normal-incidence power density distributions from a zero-emittance beam in an idealized (error-free) LCLS undulator were performed (e.g., cf Figs. 5 and 6). These idealizations were justified by the purpose of the calculation,

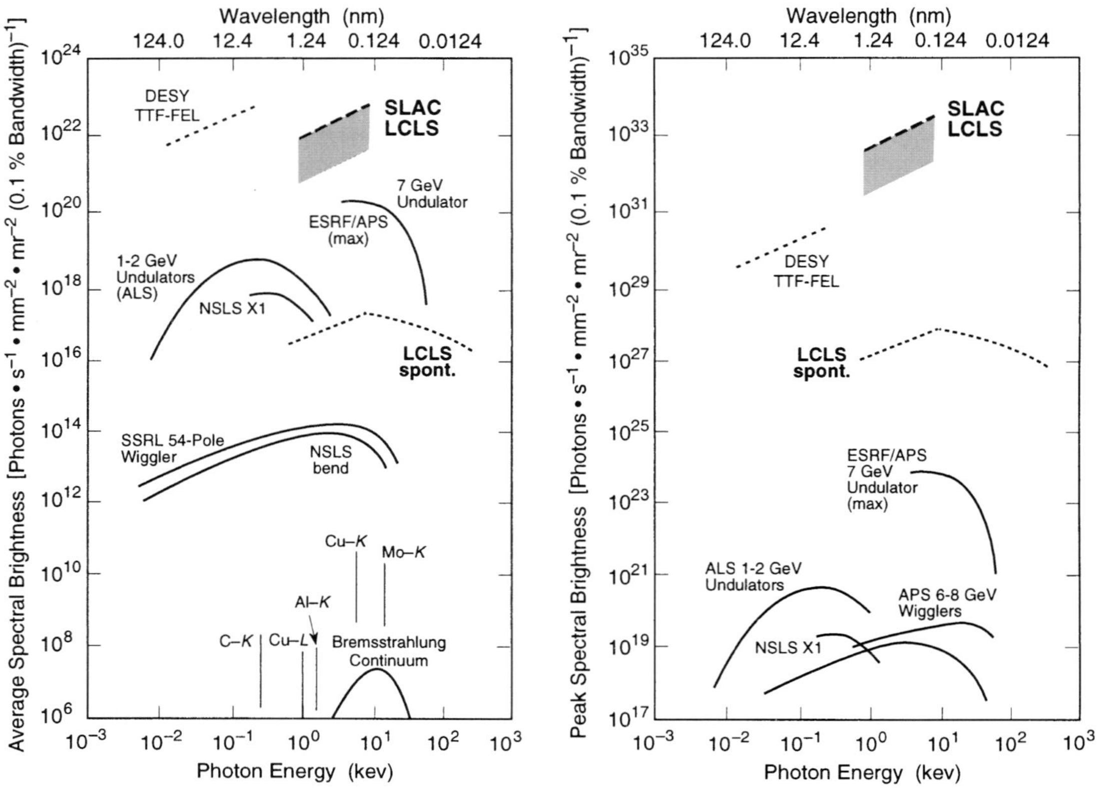

Fig. 4. Comparative peak and average brightness curves contrasting the LCLS with alternative existing or funded quasi-coherent sources.

viz., to provide maximal estimates of peak power loading on the LCLS optical elements. Nonetheless, more realistic calculations [8,25,26] will be required for more refined simulations of power loading, as well as for instrumentation design and experimental applications. In particular, the present location of the LCLS optics is well within the near-field radiation zone of the undulator, which transmits spectral-angular distributions substantially different from those of the far field [27].

Analytical estimates and numerical simulations of bremsstrahlung generation along the linac-through-undulator system and its transport through the X-ray optics into the experimental hall were carried out [7,28–32]. These calculations have been used to verify that adequate personnel and equipment radiation safety should be implementable using conventional design practices. With regard to the potential impact of the thermal neutron flux generated by the bremsstrahlung pulses on experimental data acquisition, more detailed modeling, particularly in the time domain, will need to be performed. We note that, as a result of preliminary analytical estimates of this effect, a thermal neutron shield wall has been incorporated into the design of the experimental hall (see Fig. 1).

In view of the strong interest of the scientific community in the coherence of the LCLS, a long-term goal of the R&D and design efforts is to quantitatively describe the propagation of its coherence parameters through the X-ray optics system. Although formalisms exist that are valid for assessing weak-field coherence propagation [33–36], non-linear and other effects associated with the interactions of the ultra-short LCLS pulses with matter still remain to be systematically assessed.

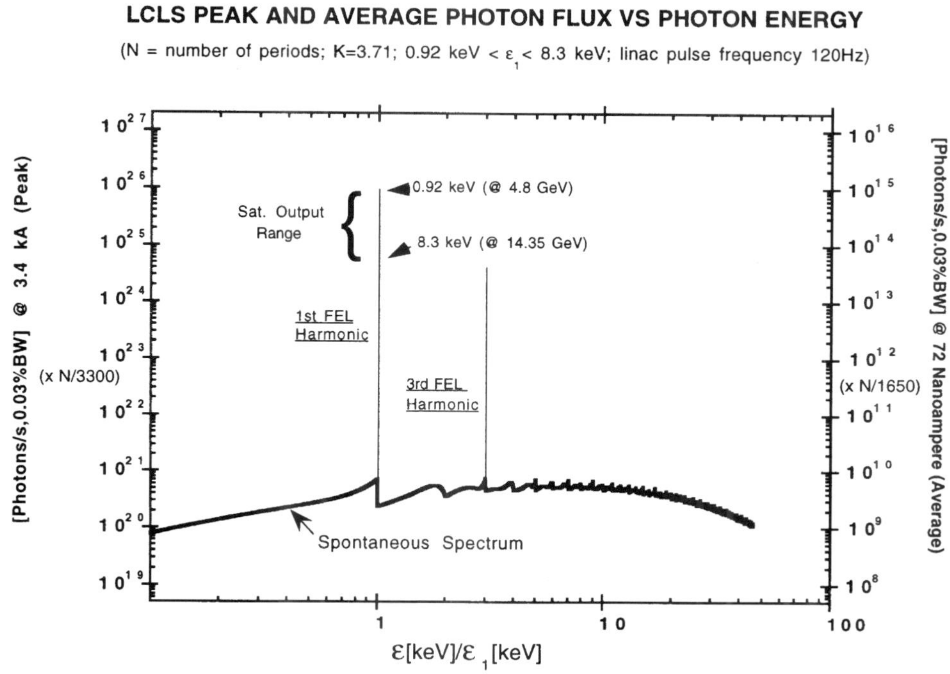

Fig. 5. LCLS angle-integrated spontaneous and coherent spectral flux densities. Peak (left ordinate) vs. time-averaged (right ordinate) values are based on practical operating parameters of SLAC's 3-km S-band linac.

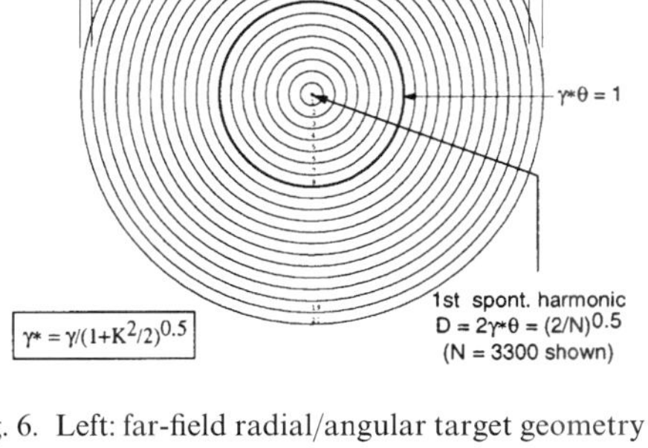

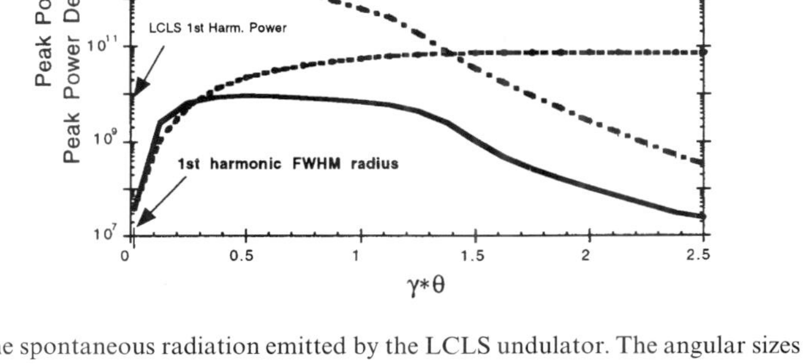

Fig. 6. Left: far-field radial/angular target geometry for the spontaneous radiation emitted by the LCLS undulator. The angular sizes of the spontaneous and coherent fundamental lines are of the same order. The $\theta = 1/\gamma^*$ contour contains $\sim 85\%$ of the spontaneous power. Right: peak power, angle-integrated peak power, and peak power density distributions averaged over the annular regions vs. $\gamma^*\theta$ at an LCLS energy of 15 GeV. The peak power density of the coherent LCLS fundamental is marked on the ordinate.

5. Peak power density studies

The full parameter regime of the LCLS has not, to our knowledge, been experimentally precreated or explored at any prior or existing alternative source. Even theoretical and numerical studies of the interactions of ultrashort, high-intensity radiation pulses with matter have heretofore tended to be restricted to the IR/visible/UV spectral range [37]. In view of the fact that additional strong focusing of the raw LCLS beam is being planned for scientific applications such as, e.g., microstructure imaging and non-linear physics, corresponding interaction and damage issues associated with fields with upper limits in the 10^{15}–10^{16} V/m range are expected to become physically relevant [14]. It is anticipated that continuing theoretical, numerical, and experimental studies will be undertaken throughout the course of the X-ray optics R&D program to investigate these and related issues [38,39]. In particular, studies of high-peak-power effects in the materials utilized in the LCLS optical system design (viz., gases, crystals, metals, and dielectrics) are expected to be critical to the longer-term development of the LCLS as a user facility.

6. Instrumentation design studies

The design and successful use of optical instrumentation for diagnosing the LCLS radiation or for processing and transporting it to the experimental stations will depend on the extent to which peak power damage effects can be mitigated or circumvented. However, even with these issues resolved, care will still need to be exercised in factoring in the temporal-profile effects of the ultra-short radiation pulses on the performance of conventional instrumentation [14,40]. A substantial depth of experience in handling both peak and average power density problems on beam lines has been accumulated at 3rd and earlier-generation storage rings sources [41–50], and this has been both drawn upon [8] and is expected to continue serving as an invaluable guide to the ongoing design program. Apart from the performance of conventional optical functions, a longer-term goal of the LCLS optics R&D program is to extend the output parameter ranges of the raw (unprocessed) LCLS beam by substantial margins or factors, in many instances to enable the execution of new classes of scientific experiments. Thus, optics will need to be developed to perform functions such as, e.g., microfocusing [51–54], polarization control, temporal and angular pulse compression and dilation [14], and beam splitting. Implementing these capabilities will, in turn, necessitate the development of diagnostic instruments, detectors, and techniques (e.g., interferometry) with correspondingly extended ranges. The use of novel materials, multi-phase optics, new modes of operation (e.g., use of dynamic, or "disposable" optics) and the development of new instrumentation concepts and configurations are all expected to play key roles in achieving these ends.

7. Experimental design

Selected possibilities for novel scientific and technological applications opened up by a 1.5 Å LCLS have been noted in prior literature [55]. Since 1995, a number of workshops covering various scientific or technological applications in a more systematic fashion have been held [56–58], with a resulting strong positive response from the international scientific community. In the short term, initial experiments on the LCLS will in all likelihood explore various aspects of the interaction of LCLS pulses with matter. Concurrently, selected investigations not requiring the full per-pulse flux (e.g., imaging, structural studies, microfocusing, etc.) will also be possible. The development of both general and specialized instrumentation for these applications is expected to be an interactive process, undertaken in conjunction with the experimental groups carrying out the research. As the LCLS evolves into a user facility, attention will be given to the development and standardization of instrumentation capable of supporting the broadest anticipated range of user requirements.

8. Summary

An overview of the LCLS X-ray optical system and selected activities of the associated R&D program has been presented. A common theme underlying the design process has been, and remains, a concern over the potentially deleterious effects of high peak power densities impinging on the system components. In the initial stages of LCLS operation, the design proposes to implement a gas absorption cell to permit continuous variation of peak power density from levels at which damage phenomena can be studied to those at which damage is not a concern. Additionally, it will enable the use of conventional instrumentation to perform critical diagnostics of the LCLS beam at turn-on, as well as the execution of initial scientific experiments. In the longer term, an ultra-long beam line may be implemented. The nearer-term evolution of the optics to the point where the majority of experiments that require unattenuated radiation can be supported is envisaged to be an interactive R&D process. Scattering and damage studies on initial samples (perhaps using the absorption cell itself), coupled with theoretical and numerical studies, will provide much-needed knowledge and experience on which the design of reliable LCLS instrumentation can be based. A parallel R&D program will be the continuing study of the source and coherence properties of the LCLS radiation. Here the modeling and representation of the spontaneous and coherent outputs described in the sections above will be extended. The ultimate goal of these studies will be a reliable quantitative description of the LCLS radiation and its propagation through the complete optical system. Research on the development of LCLS optics and instrumentation, including theoretical, numerical and experimental studies of short pulse effects, will continue. These programs will be carried out in collaboration with various laboratories and research groups (e.g., LLNL, LANL, LBNL, BNL, APS, ESRF, DESY, etc.) that have the expertise, facilities, and interest to participate in the required studies.

Acknowledgements

We thank Herman Winick, Claudio Pellegrini, Kwang-Je Kim, Gerd Materlik, Art Bienenstock, Karl van Bibber, Dick More, John Goldstein, Brian Newnam, Paul Csonka, and numerous other colleagues engaged in LCLS or alternative FEL and X-ray laser research for stimulating inputs and conversations. This work was supported in part by the Department of Energy Offices of Basic Energy Sciences and High Energy and Nuclear Physics, and Department of Energy Contract DE-AC03-76SF00515. Other portions of this work were supported by Department of Energy CRADA SLAC-9302. Selected calculations of the LCLS spontaneous spectrum utilized resources of the National Energy Research Scientific Computing Center, which is supported by the Office of Energy Research of the US Department of Energy.

References

[1] W.B. Colson, Nucl. Instr. and Meth. A 375 (1996) 669.
[2] M.E. Couprie, Storage ring FEL applications and related source requirements in LURE, in: H. Kitamura and Spring-8 Project Team (eds.), Report of the International Workshop on 30-m Long Straight Sections, JAERI, 1996, pp. 137–148.
[3] E.D. Johnson, Nucl. Instr. and Meth. A 358 (1995) 544.
[4] R. Tatchyn et al., Nucl. Instr. and Meth. A 375 (1996) 274.
[5] R. Brinkmann, G. Materlik, J. Rossbach, A. Wagner (eds.), Conceptual Design of a 500 GeV e + e − Linear Collider with Integrated X-Ray Laser Facility, DESY 97-048, Hamburg, 1997.
[6] M. Cornacchia et al., SPIE Proc. 2988 (1997) 2.
[7] LCLS Design Study Group, Linac Coherent Light Source (LCLS) Design Study Report, SLAC-R-521, UC-414, 1998.
[8] R. Tatchyn, J. Arthur, R. Boyce, T. Cremer, A. Fasso, J. Montgomery, V. Vylet, D. Walz, R. Yotam, A.K. Freund, M.R. Howells, SPIE Proc. 3154 (1998) 174.
[9] K.-J. Kim, Phys. Rev. Lett. 57 (13) (1986) 1871.
[10] J.B. Murphy, C. Pellegrini, Introduction to the physics of the free electron laser, in: M. Month, S. Turner, H. Araki, et al. (Eds.), Frontiers of Particle Beams, Lecture Notes in Physics No. 296, Springer, Berlin, 1988, pp. 163–212.
[11] Y.H. Chin, K.-J. Kim, M. Xie, Phys. Rev. A 46 (10) (1992) 6662.
[12] R. Bonifacio, L. DeSalvo, P. Pierini, N. Piovella, C. Pellegrini, Phys. Rev. Lett. 73 (1) (1994) 70.
[13] W.M. Fawley, SPIE Proc. 2988 (1997) 98.
[14] R. Tatchyn, LCLS Optics: technological issues and scientific opportunities, Proceedings of the Workshop on Scientific Applications of Short Wavelength Coherent Light Sources, SLAC Report 414; SLAC-PUB 6064, March 1993.
[15] S.H. Rokni, E.C. Benson, D.L. Burke, T.M. Jenkins, J.C. Liu, G. Nelson, W.R. Nelson, H.E. Smith, P. Tenenbaum, V. Vylet, D.R. Walz, Health Physics 71 (5) (1996) 786 SLAC-PUB 6784.

[16] H. Winick, Synchrotron Radiation Sources, Research Facilities, and Instrumentation, in: H. Winick, S. Doniach (Eds.), Synchrotron Radiation Research, Plenum Press, New York, 1980, p. 27.
[17] R. Bonifacio, F. Casagrande, Nucl. Instr. and Meth. A 237 (12) (1985) 168.
[18] R.L. Gluckstern, H. Okamoto, S. Krinsky, Saturation of a high gain FEL, Proceedings of the 1993 Particle Accelerator Conference, IEEE Cat. No.93CH3279-7, vol. 2, 1993, p. 1545.
[19] H.P. Freund, Phys. Rev. E 52(5), Part B (1995) 5401.
[20] L.H. Yu, C.M. Hung, D. Li, S. Krinsky, Phys. Rev. E 51 (1) (1995) 813.
[21] M. Xie, Design optimization for an X-ray free electron laser driven by SLAC linac, Proceedings of the 1995 Particle Accelerator Conference, IEEE Cat. No.95CH35843, vol. 1, 1995, p. 183.
[22] Nuhn, H.-D. et al., Alignment and magnet error tolerances for the LCLS X-ray FEL Proceedings of the 1995 Particle Accelerator Conference, IEEE Cat. No.95CH35843, vol. 1, (1995), p. 231.
[23] E.L. Saldin, E.A. Schneidemiller, M.V. Yurkov, Statistical properties of radiation from VUV and X-ray free electron laser, DESY Technical Note TESLA-FEL 97-02, April 1997.
[24] R. Tatchyn, J. Arthur, Phase space and coherence properties of an X-ray FEL source, presented at the 1998 SPIE Annual Meeting, August 20–24, 1998, San Diego, CA, SPIE Proc. 3451, 1998, p. 193.
[25] R. Tatchyn, A.D. Cox, S. Qadri, SPIE Proc. 582 (1986) 47.
[26] R. Coisson, R. Tatchyn, Introduction to optical properties of insertion devices excited by synchrotron sources, SPIE Tutorial T1, International Conference on Insertion Devices for Synchrotron Sources, October 27–30, 1985.
[27] R. Walker, Nucl. Instr. and Meth. A 267 (1988) 537.
[28] A. Fassò et al, in: A. Palounek (Ed.), FLUKA92, Proceedings of the SARE-1, Santa Fe Jan. 11–15, 1993, Los Alamos LA–12835–C 1994, p. 134.
[29] A. Ferrari, M. Pelliccioni, P.R. Sala, Nucl. Instr. and Meth. B 83 (4) (1993) 518.
[30] W.R. Nelson, Y. Namito, Computer code MUON83.MORTRAN, SLAC Radiation Physics Dept., 1981.
[31] A.H. Sullivan, A Guide to Radiation and Radioactivity Levels near High Energy Particle Accelerators, Nucl. Tech. Publishing, Ashford, Kent, UK, 1992.
[32] L.P. Keller, Muon background in a 1.0 TeV Linear Collider, SLAC-PUB 6385, 1993.
[33] J.W. Goodman, Statistical Optics, Wiley, New York, 1985.
[34] L. Mandel, E. Wolf, Optical Coherence and Quantum Optics, Cambridge University Press, Cambridge, 1995 chapter 4.
[35] K.-J. Kim, SPIE Proc. 582 (1986) 2.
[36] R. Coisson, Some Remarks on Coherence and Degeneracy, Proceedings of the Workshop on PEP as a Synchrotron Radiation Source, SSRL Report 88/06.
[37] R. Tatchyn, G. Materlik, A. Freund, J. Arthur (Eds.), Proceedings of the SLAC/DESY International Workshop on the Interactions of Intense Sub-Picosecond X-Ray Pulses with Matter, SLAC, Stanford, CA, Jan. 23–24, 1997, SLAC-WP-12.
[38] K.C. Kulander, M. Lewenstein, Multiphoton and Strong-Field Processes, in: G.W.F. Drake (Ed.), Atomic, Molecular, and Optical Physics Handbook, American Institute of Physics, Woodbury, New York, 1996, p. 828.
[39] R. More, Interaction Experiments with Advanced Photon Sources, presented at the SLAC/DESY International Workshop on the Interactions of Intense Sub-Picosecond X-Ray Pulses with Matter, SLAC, Stanford, CA, Jan. 23–24, 1997.
[40] S.A. Akhmanov, V.A. Vysloukh, A.S. Chirkin, in: Y. Atanov, (Tr.), Optics of Femtosecond Laser Pulses, American Institute of Physics, New York, 1992.
[41] A.K. Freund, Opt. Eng. 34 (1995) 432.
[42] S. Lindaas, M. Howells, B. Calef, D. Pinkas, C. Jacobsen, A. Kalinovsky, X-Ray Gabor Holography, presented at the SLAC/DESY International Workshop on the Interactions of Intense Sub-Picosecond X-Ray Pulses with Matter, SLAC, Stanford, CA, 1/23-24/97.
[43] A.K. Freund, Rev. Sci. Instr. 67 (1996) 4 (CD ROM), Paper B01.
[44] J. Susini, Opt. Eng. 34 (1995) 388.
[45] L.E. Berman, Rev. Sci. Instr. 66 (2), Part 2 (1995) 2041.
[46] J. Arthur, Opt. Eng. 34 (2) (1995) 441.
[47] G. Grubel, J. Als-Nielsen, A.K. Freund, J. Phys. IV, Colloq. (France) 4 (C9) (1994) 27.
[48] P.L. Cowan, S. Brennan, Rev. Sci. Instr. 60 (1989) 1987.
[49] S. Brauer, G.B. Stephenson, M. Sutton, J. Synchrotron Radiat. 2 (1995) 163.
[50] A.K. Freund, Crystal optics for the LCLS, presented at the SLAC/DESY International Workshop on the Interactions of Intense Sub-Picosecond X-Ray Pulses with Matter, SLAC, Stanford, CA, 1/23-24/97.
[51] R. Tatchyn, P. Csonka, H. Kilic, H. Watanabe, A. Fuller, M. Beck, A. Toor, J. Underwood, R. Catura, SPIE Proc. 733 (1986) 368.
[52] E. Ziegler, O. Hignette, M. Lingham, A. Souvorov, X-ray focusing using elliptically bent multilayers, SPIE Proc. 2856, 1996, p. 61.
[53] A. Snigirev, V. Kohn, I. Snigireva, A. Lengeler, Nature 384 (6604) (1996) 49.
[54] Ya. Hartman, A.K. Freund, I. Snigireva, A. Souvorov, A. Snigirev, Nucl. Instr. and Meth. A 3 (85) (1997) 371.
[55] J. Arthur, G. Materlik, R. Tatchyn, H. Winick, Rev. Sci. Instr. 66 (2), Part 2 (1995) 1987.
[56] J. Arthur, G. Materlik, H. Winick (Eds.), Proceedings of the Workshop on Scientific Applications of Coherent X-Rays, SLAC-437.
[57] J.-L. Laclare (Ed.), Proceedings of the 10th ICFA Beam Dynamics Panel Workshop on 4th Generation Light Sources, ESRF, Grenoble, 1/22-25/96.
[58] B.H. Wiik et al, International Workshop on X-Ray Free Electron Laser Applications, (Copies of Transparencies), DESY, Hamburg, Germany, Sept. 16–17, 1996.

ELSEVIER

Nuclear Instruments and Methods in Physics Research A 429 (1999) 407–413

NUCLEAR INSTRUMENTS & METHODS IN PHYSICS RESEARCH Section A

www.elsevier.nl/locate/nima

Beam-based alignment for the LCLS FEL undulator☆

P. Emma*, R. Carr, H.-D. Nuhn

Stanford Linear Accelerator Center, Stanford, CA 94309, USA

Abstract

The linac coherent light source (LCLS) (LCLS Design Study Report, SLAC-R-521, 1998) is an X-ray FEL based on a long, permanent magnet undulator comprised of multiple segments which provide space for electron beam position monitors (BPMs) and quadrupole focusing magnets. The 15 GeV electron trajectory within the undulator must be straight to ~ 5 μm over a gain length of 11 m so that the photon beam overlaps efficiently with the electron beam. This alignment precision is not achievable with available mechanical survey methods. For this reason, an empirical beam-based alignment technique is developed that uses BPM readings in conjuntion with large, deliberate variations in the electron energy. We describe here a detailed alignment simulation including BPM and magnet-mover calibration errors, quadrupole gradient errors, and dipole field errors which demonstrate that the alignment can be achieved with high confidence.

1. Introduction

The LCLS uses a 112-m long, planar-type, permanent magnet undulator which includes a discrete quadrupole focusing array to keep the electron beam well focussed. The strong gradient and precise beam overlap requirements create tight alignment tolerances not readily achievable with traditional alignment techniques. The inclusion of high-resolution BPMs and magnet movers in the design, however, provides the ability to measure and correct both BPM and quadrupole misalignments by taking advantage of their sensitivity to beam energy.

The undulator is composed of 52 1.92-m sections, each including 128 dipoles. The section is followed by a 24-cm long gap which includes a ~ 1-μm resolution RF-cavity BPM and a 12-cm long, 45-T/m gradient permanent magnet quadrupole. Each quadrupole is mounted on a remotely controlled *x*-and-*y*-mover to provide trajectory corrections. The beam-based alignment method uses a full set of BPM readings taken at three different electron beam energies to deduce the quadrupole magnet and BPM misalignments which are then corrected with magnet movers and software offsets, respectively.

2. Trajectory analysis

The readback of the *i*th BPM which measures the centroid of the transverse position of the electron bunch at location s_i along the undulator can be written as

$$m_i = \sum_{j=1}^{i} \theta_j C_{ij} - b_i \tag{1}$$

☆Work supported by the Department of Energy, contract DE-AC03-76SF00515.

* Corresponding author.

0168-9002/99/$ - see front matter
PII: S0168-9002(99)00117-5

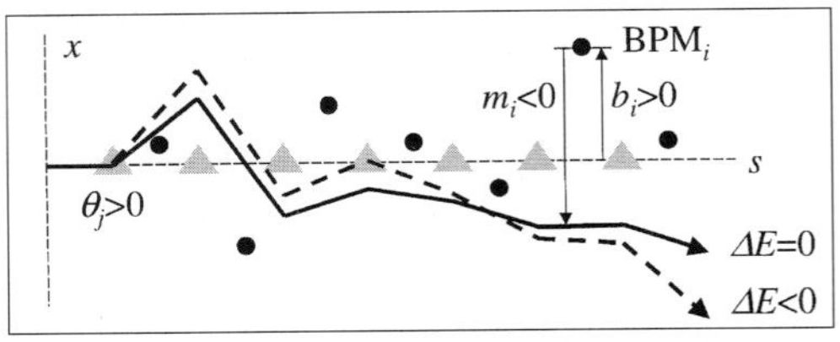

Fig. 1. Schematic diagram of e^- trajectory for nominal energy ($\Delta E = 0$) and for a lower energy ($\Delta E < 0$). BPMs are represented as dots and kick sources as triangles.

where θ_j is the kick angle at location s_j ($s_j < s_i$) due to a transversely misaligned quadrupole magnet or dipole field error upstream of BPM-i, C_{ij} is the transfer coefficient which maps a beam angle at point j to a position at point i, and b_i is the readback offset (mechanical and/or electrical) of BPM-i. The initial launch conditions are temporarily ignored here for clarity (more on this below). Eq. (1) is shown graphically in Fig. 1.

The kick angles in the figure are represented as dipoles, however they are completely equivalent to either misaligned quadrupoles and/or field errors of the undulator dipoles (or other sources). The quadrupole focusing is not explicitly shown in the figure, but is represented mathematically in the transfer coefficient, C_{ij}.

Since the kick angles, θ_j, are inversely proportional to beam momentum, p, whereas the BPM offsets, b_i, are independent of momentum, variations of beam energy (momentum) can be used to measure both parameters simultaneously. This is clear if θ_j is replaced by a dipole field error, ΔB_j, (or quadrupole misalignment) to explicitly show the momentum dependence of Eq. (1):

$$m_{ik} = \frac{1}{p_k}\sum_{j=1}^{i} e\Delta B_j \ell C_{ij}(p_k) - b_i\,. \tag{2}$$

Here ℓ is the dipole (or quadrupole) length, e is the electron charge and the subscript, k, on momentum and BPM readback indicates the different values of beam momentum ($\geqslant 2$ required). The transfer coefficients, $C_{ij}(p_k)$, are, in general, also momentum dependent.

A matrix expression for this linear system is given in Eq. (3) with N BPMs, N kicks (misaligned quadrupoles) and two different momenta ($k = 1{,}2$). Here $-\boldsymbol{I}$ is an $N \times N$ negative-unity matrix and the elements $P_{ij}(k) \equiv eC_{ij}/p_k$ are the scaled momentum-dependent transfer coefficients which map the jth field error (kick) to the ith BPM Eq. (3):

$$\begin{bmatrix} m_{11} \\ m_{21} \\ \vdots \\ m_{N1} \\ m_{12} \\ m_{22} \\ \vdots \\ m_{N2} \end{bmatrix} = \begin{bmatrix} & P_{11}(1) & 0 & \cdots & 0 \\ & P_{21}(1) & P_{22}(1) & \cdots & 0 \\ -\boldsymbol{I} & \vdots & \vdots & \ddots & \vdots \\ & P_{N1}(1) & P_{N2}(1) & \cdots & P_{NN}(1) \\ & P_{11}(2) & 0 & \cdots & 0 \\ -\boldsymbol{I} & P_{21}(2) & P_{22}(2) & \cdots & 0 \\ & \vdots & \vdots & \ddots & \vdots \\ & P_{N1}(2) & P_{N2}(2) & \cdots & P_{NN}(2) \end{bmatrix} \cdot \begin{bmatrix} b_1 \\ b_2 \\ \vdots \\ b_N \\ \Delta B_1\ell \\ \Delta B_2\ell \\ \vdots \\ \Delta B_N\ell \end{bmatrix}. \tag{3}$$

There are many undulator dipoles (potential field errors) per BPM (128) which cannot all be solved. However, the quadrupole alignment corrections tend to cancel all net dipole field errors (undulator dipole errors, quadrupole offsets and other fields). The quality of this cancellation is examined in the simulations described below.

To explicitly write Eq. (3) in terms of misalignments, $\Delta B_j\ell$ is replaced by the quadrupole magnet misalignment, Δx_j, and $P_{ij}(k)$ is replaced by

$$P_{ij}(k) \rightarrow [1 - Q^j_{11}(k)]R^{ji}_{11}(k) - Q^j_{21}(k)R^{ji}_{12}(k) \tag{4}$$

where $Q^j_{11}(k)$ and $Q^j_{21}(k)$ are the 'thick-lens' transfer matrix elements across the jth quadrupole evaluated at the kth momentum, and $R^{ji}_{11}(k)$ and $R^{ji}_{12}(k)(R^{ji}_{12} = C_{ij})$ are the position-to-position and angle-to-position, respectively, transfer matrix elements from the exit of the jth quadrupole to the ith BPM, also evaluated at the kth momentum.

In practice, the linear system of Eq. (3) is solved by imposing 'soft-constraints' on the solutions to stabilize the system. This is equivalent to including the additional information that quadrupole and BPM offsets are zero to within ~ 1 mm. These constraints weight the normalized residual fit error (χ^2) such that the solutions do not wander out to large values over long distances (see Ref. [1]).

By ignoring the initial launch conditions of the beam at the undulator entrance in Eqs. (1)–(3) we define the reference line for the determined BPM and quadrupole misalignments as the incoming position and angle of the beam. At a point when Eq. (3) has been solved and the N solutions of b and Δx are available, we then extract a best line-fit (initial position and angle) from these data and accordingly adjust the launch conditions (with steering dipoles prior to the undulator) so that the quadrupole and BPM offset corrections will not systematically follow these initial erroneous launch conditions. On the next application of the alignment procedure, the launch conditions will be much closer to the axis defined by the initial BPM and quadrupole positions (initial tunnel survey).

3. Operational procedure

The full alignment procedure used in initial machine commissioning is given in Table 1.

Table 1
Beam-based alignment procedure

Step	Description	~ Time/h
1	Rough correct incoming trajectory launch using 1st few und. BPMs	0.2
2	Apply weighted steering to minimize BPM readings and mover changes	0.2
3	Save ~ 100 sets of BPM readings for each of 5, 10 & 15 GeV	1.5
4	Run all BPM data through analysis	0.1
5	Correct launch position and angle	0.1
6	Set quadrupole movers to new positions and correct BPM offsets	0.1
7	Steer BPM readings to ~ 0 using a minimum number of magnet movers	0.2
8	Repeat steps 3–7 until peak of BPM readings at 5 GeV is < ~ 50 μm	2/iter.

Steps 1 and 2 are rough steering steps used only in initial commissioning when the trajectory amplitude is potentially very large. Step-2 involves a weighted steering procedure where both the absolute readings of the BPMs and the magnet mover changes are simultaneously minimized with respect to a 50 μm weight. This is used because an exact 1-to-1 steering can require mover changes of > 1 mm in order to exactly zero each BPM reading; an undesirable steering accuracy at this early stage. Step-5 is a precise launch correction based on the BPM and quadrupole offsets determined in step-4 (see below). Step-7 is a minor cleanup to remove remaining small oscillations which are due to mover errors. The procedure is iterated as necessary.

4. Simulation results

Simulations have been run for the entire beam-based alignment procedure, from initial rough steering through final precision alignment. The simulations use the LCLS undulator parameters with one BPM (reads both x and y) located at the upstream face of each quadrupole. A conservative set of statistical and systematic errors are included in the simulations as summarized in Table 2.

The BPM resolution requirement of 1 μm is an rms value averaged over ~ 100 beam pulses. Thermal stability of the BPM to ~ 1 μm over ~ 2 h is,

Table 2
Errors used in alignment simulation

Description (Gaussian rms errors)	Value	Unit
BPM rms resolution (multi-pulse res.)	1	μm
BPM rms offsets (uncorrelated)	50	μm
BPM rms offsets (correlated)	300	μm
BPM mean calibration error (systematic)	10	%
BPM rms calibration error (random)	3	%
Quad. rms offsets (uncorrelated)	50	μm
Quad. rms offsets (correlated)	300	μm
Quad. mean gradient error (systematic)	0.3	%
Quad. rms gradient error (random)	0.3	%
Undulator rms dipole error (x, random)	0.1	%
Mover mean calibration error (systematic)	5	%
Mover rms calibration error (random)	3	%
Mover reproducibility (digitization error)	±1	μm
Incoming trajectory bias (beam size units)	10	σ
Incoming rms orbit jitter (beam size units)	0–0.1	σ

however, required. The 'correlated' offsets (misalignments) listed in Table 2 describe a random walk where the expectation value of the square of the misalignment, $\langle x^2 \rangle$, is related linearly to its distance from the undulator entrance. This treatment approximates the long undulator survey 'wander' error arising over 100 m. The BPMs and quadrupoles both follow the same random walk path plus an additional 50 μm rms 'uncorrelated' component applied to both BPMs and quadrupoles separately. Two adjacent elements are therefore misaligned with respect to each other by $\sqrt{2} \times 50$ μm ≈ 70 μm rms. Over longer distances the relative misalignment increases.

Errors in 3300 dipole pairs, x-plane only, are also included. The 'calibration error' in the table implies that the BPMs (magnet movers) are mis-scaled so that an actual displacement of 1 mm will read back as (move by) 1.1 mm (1.05 mm). The 'mover reproducibility' simulates a random digitization error such that the final position of the mover achieves its desired position plus a uniformly distributed random component of up to ± 1 μm. The 'incoming trajectory bias' is a static beam launch error which is ten times that of the rms beam size in both position and angle (10×30 μm and 10×1.7 μrad). The 'incoming orbit jitter' is a randomly varying launch position and angle error which occurs during the energy-scan data acquisition. The simulation shown here includes no orbit jitter, but in fact the results are insensitive to launch jitter up to 10% of the nominal beam size (net rms after averaging ~ 100 pulses). The effective jitter can, in practice, easily be reduced even further by acquiring ~ 100 orbits and using the linac BPMs to select only those 70–80 orbits which produce the most constant mean trajectory.

Fig. 2 shows the quadrupole and BPM horizontal misalignments used in this simulation with respect to a line defined by the upstream linac axis (as are all figures except Fig 7). The vertical plane is also fully simulated but not shown here.

The resultant, un-steered horizontal e^- beam trajectory at 15 GeV through the undulator is shown in Fig. 3 (prior to step-1). The BPM *readbacks* (diamonds in all plots) are, in practice, the

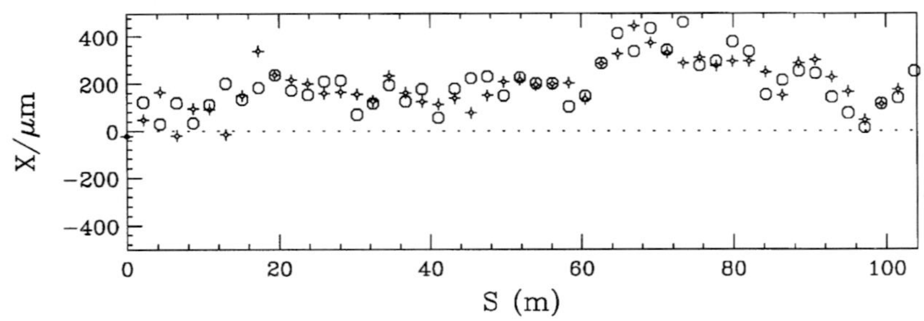

Fig. 2. Quadrupole (cross) and BPM (circle) absolute misalignments used in this simulation, with respect to the line defined by the upstream linac axis.

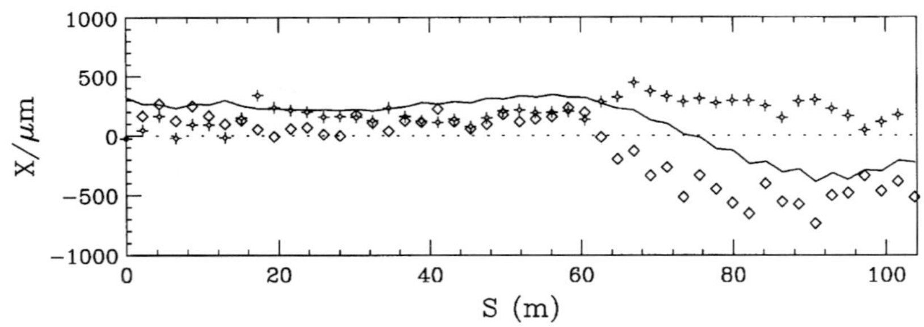

Fig. 3. Trajectory (solid) through undulator before step-1 including an incoming trajectory bias, BPM readbacks (diamond) and absolute quadrupole positions (cross).

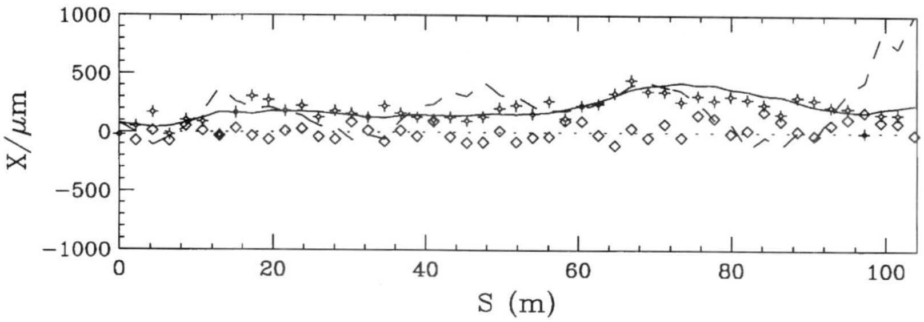

Fig. 4. Trajectory at 15 GeV (solid) after steering (after steps 1 and 2) including BPM readbacks (diamond) and absolute quadrupole positons (cross). The trajectory produced at 5 GeV (dash) is also overlaid.

only known quantities at this point. The absolute quadrupole positions are shown as crosses " + " in all plots.

Fig. 4 shows the 15 GeV e^- trajectory after the rough launch correction and steering of steps 1 and 2. The 5-GeV orbit is also shown (without the 5-GeV BPM readbacks) where peak trajectory amplitudes of ~ 1 mm can arise.

The 15 GeV BPM readback data of Fig. 4 are saved to disk and the energy is then lowered to 10 GeV by switching off klystrons in the linac. The fields of the linac magnets are scaled to the new energy and any detected beam position differences upstream of the undulator are manually corrected until the undulator entrance launch position at 10 GeV is within a few microns of that at 15 GeV. Any beam *angle* difference at the undulator entrance is not detectable with the linac BPMs. However, such an angle is eventually incorporated into the repositioning of the first undulator quadrupole(s). No changes are made to the undulator components during the energy scan. Only the trajectory launch is adjusted, if necessary, to maintain a constant beam position at the entrance. Finally, the e^- energy is lowered again to 5 GeV (dashed in Fig. 4) and the BPM readbacks are again saved.

Since the fields of the magnets upstream of the undulator are scaled to the lower energies, the beta functions at the undulator entrance will be constant during the procedure. With permanent magnet undulator quadrupoles, however, these entrance beta functions will not be matched to the undulator focussing at 5 GeV causing a worst case 80 μm beam size beat through the undulator as compared to the nominal matched 45 μm maximum size at 5 GeV. This worst case beat should have no significant effect on the alignment procedure.

With the multi-energy BPM readback data saved, the analysis program calculates BPM and quadrupole offsets. Fig. 5 shows the calculated quadrupole offsets (results of step-4) as well as the true offsets (used in this simulation).

The fine structure of the calculated offsets agree well with the true offsets. The values differ, however, by a straight line which is due to (1) the initial launch bias which has only been roughly corrected, and (2) the correlated component of the BPM and quadrupole offsets. One line (per plane) is then fitted simultaneously to both the calculated quadrupole and BPM (not shown) offsets (see Fig. 5). The slope and offset of the line-fit is used to readjust the initial beam launch angle and position, respectively, at the undulator entrance. In this way the large linear component of the quadrupole and BPM offsets is not incorporated into their corrections.

Fig. 6 shows the new electron trajectory after the launch conditions, the quadrupole magnet movers, and the BPM offsets are corrected, and the minor steering of step-7 is finished. The offset corrections which are applied have the line-fit of Fig. 5 removed. At each stage of the simulation the magnet mover limitations (calibration errors and reproducibility errors) and BPM errors (calibration and resolution) are incorporated.

The linear component which remains in the Fig. 6 trajectory is due to the correlated quadrupole and BPM offsets (random walk of initial tunnel survey). Since these offsets are due mainly to the

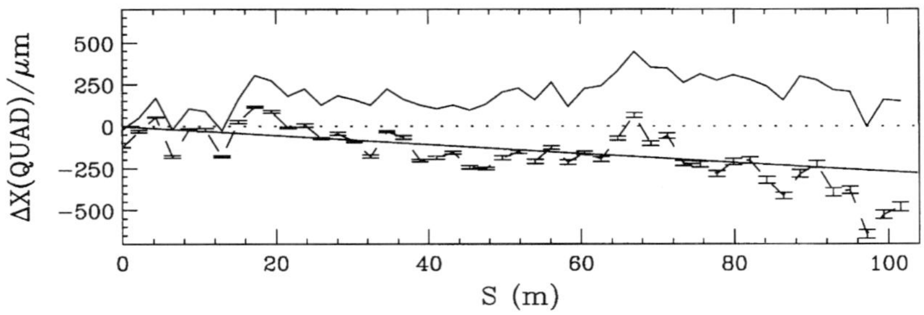

Fig. 5. True (solid) and calculated (dash) quadrupole offsets determined in step-4. The best line-fit to the calculated offsets (to be used in step-5) is also overlaid.

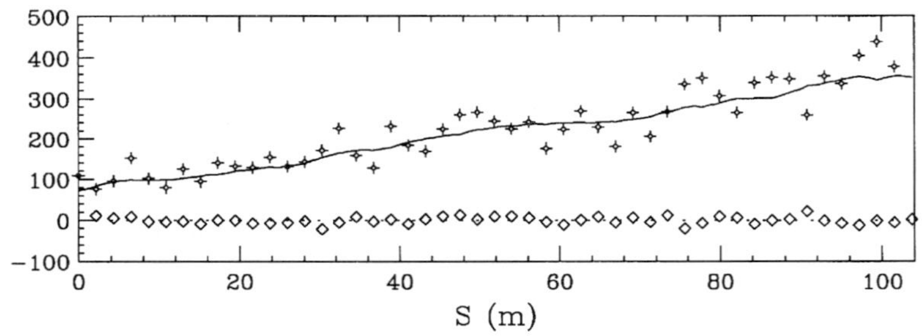

Fig. 6. Trajectory (solid) at 15 GeV after one iteration of steps 3–7. BPM readbacks (diamond) and absolute quadrupole positions (cross) are also shown.

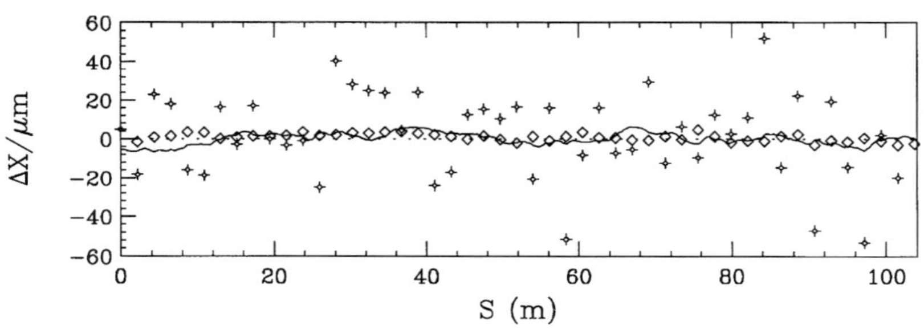

Fig. 7. Trajectory (solid) at 15 GeV (linear component removed for clarity) after *two* iterations of steps 3–7, with BPM readbacks (diamond) and absolute quadrupole positions (cross). The rms of the trajectory is 3 μm.

difference between the line defined by the linac axis and the slightly different line established by the undulator survey, the procedure inevitably launches the electrons straight down the undulator vacuum chamber which presumably follows these correlated initial tunnel survey errors. The trajectory shown in Fig. 6 is then actually the most desired trajectory where a slight change in beam position and angle at the undulator entrance is used to launch the electron beam down the center of the undulator. The tiny dispersion generated by this slight position and angle change is completely negligible. After one iteration of steps 3–7, the trajectory rms, with respect to a straight line over the length of the undulator is ~ 20 μm.

Fig. 7 shows the final trajectories after a second iteration of steps 3–7 is applied. In this case the linear component of Fig. 6 is removed from the plot for clarity. The rms of the electron trajectory over the length of the undulator, with respect to a straight line, achieves a value of 3 μm. The final alignment requires changes to the magnet movers of ~ 100 μm rms. Many random seeds have been tested with similar results. Doubling the uncorrelated quadrupole and BPM offsets to 100 μm, or the gradient errors to 0.6% still produces acceptable results.

5. Summary

A final electron trajectory of 3 μm rms with respect to a straight line over the length of the LCLS undulator is achievable using energy scanned beam-based alignment. Prior to the alignment procedure, BPM and quadrupole uncorrelated misalignments of 100 μm rms plus 300 μm rms correlated misalignments are tolerable for the initial survey. Effects such as BPM and magnet mover calibration errors, quadrupole field gradient errors, and dipole field errors have been included in a detailed simulation which demonstrates this level can be achieved realistically with a high confidence level. A single pass of the alignment procedure will be reapplied approximately monthly with a stretched wire feedback system and occasional weighted steering maintaining daily operations.

References

[1] P. Emma, Beam Based Alignment of Sector-1 of the SLC Linac, SLAC-PUB-5787, March 1992.

ELSEVIER

Nuclear Instruments and Methods in Physics Research A 429 (1999) 414–418

NUCLEAR INSTRUMENTS & METHODS IN PHYSICS RESEARCH Section A

www.elsevier.nl/locate/nima

Numerical study of effects of the beam tube on laser fields with a three-dimensional simulation code using the finite element method

Masaaki Sobajima[a,*], Yonggui Li[b], Tetsuo Yamazaki[a], Kiyoshi Yoshikawa[a], Masami Ohnishi[a], Hisayuki Toku[a], Kai Masuda[a], Jiro Kitagaki[a], Takeshi Nakamura[c]

[a]*Institute of Advanced Energy, Kyoto University, Gokasho, Uji, Kyoto 611-0011, Japan*
[b]*Institute of High Energy Physics, Chinese Academy of Sciences, P.O. Box 918, Beijing 100039, People's Republic of China*
[c]*SPring-8, Kamigori-cho, Ako-gun, Hyogo 678-12, Japan*

Abstract

In January 1997, the Beijing FEL observed large laser amplification at 8–18 μm. However, through the collaborative work, it was found from both experiments and numerical simulations that the laser loss on the beam tube wall was not negligible, and that the saturation was not seen in the relatively long wavelength range because of this loss. This calls for further investigation on the effects of the beam tube of finite size. In order to include such effects self-consistently, we have developed a new three-dimensional code that can solve equations with the boundary conditions of the beam tube by using the Finite Element Method. Results show that the beam tube effects are dominant in deriving higher laser modes in the tube, compared with the optical guiding effects, and consequently reduced gain especially in the longer wavelength range, where the beam tube effects are greatly emphasized. It is also found that TEM_{02} mode is the most dominant higher mode in the beam tube, and is also the main cause of gain reduction.

PACS: 02.60.Cb; 41.60.Cr; 42.60. − v

Keywords: Simulation; Vacuum tube; Beijing FEL

1. Introduction

In collaboration with the Beijing FEL (BFEL) group, we have been analyzing BFEL experimental results by numerical simulations since September 1996. Through our collaboration, it was found from both experiments and numerical simulations that the effects of the relatively narrow BFEL beam tube on the laser field cannot be ignored [1].

We have thus investigated by our three-dimensional FEL code previously developed [2,3], in particular, the diffractive loss which occurs when the laser field goes through the beam tube entrance.

* Corresponding author. Tel.: + 81-774-38-3443; fax: + 81-774-38-3449.

E-mail address: sobajima@iae.kyoto-u.ac.jp (M. Sobajima)

0168-9002/99/$ - see front matter
PII: S0168-9002(99)00119-9

On the other hand, the effects of the beam tube size can also create higher-order transverse modes in the laser field which could reduce the degree of interaction between the electron beam and the laser field.

Since our previously developed simulation code, aiming at the evaluation of optical guiding effects, imposes a periodic boundary condition in the transverse plane in the infinite, not finite beam tube because of the use of Fourier transformation, the code cannot treat the beam tube effects self-consistently.

We have thus developed a new three-dimensional code that solves equations with boundary conditions of the beam tube by using the Finite Element Method, in order to deal with the effects associated with the finite beam tube, as well as optical guiding effects.

Through the numerical simulations, we have studied the evolution of the laser profiles, and evaluated the effects of the beam tube of finite size by comparing the results of the old code and the new one, and also by decomposing the transverse laser fields into Gaussian beam modes we observed the generation of higher order modes.

2. Simulation code and boundary conditions

In our new code, the vector potential of laser field, $\boldsymbol{A}_{\rm L}$ is assumed to have the form,

$$\boldsymbol{A}_{\rm L}(x,y,z)=\boldsymbol{x}|A_{\rm L}(x,y,z)|\cos\{k_{\rm L}z-\omega_{\rm L}t+\phi_{\rm L}(x,y,z)\} \quad (1)$$

$$A_{\rm L}(x,y,z)=|A_{\rm L}(x,y,z)|\exp\{{\rm i}\phi_{\rm L}(x,y,z)\} \quad (2)$$

where $\phi_{\rm L}$, $k_{\rm L}$ and $\omega_{\rm L}$ stand for the phase, the wave number and the angular frequency of laser, respectively. Substitution of this vector potential into Maxwell's equations, applying SVAP approximation and averaging over a laser period, gives the following equations,

$$\left(-\frac{\rm i}{2k_{\rm L}}\nabla_\perp^2+\frac{\partial}{\partial z}\right)a_{\rm L}(x,y,z)=s(x,y,z) \quad (3)$$

$$s(x,y,z)=\sum_{j=1}^{N_{\rm e}} s_j\delta(x-x_j)\delta(y-y_j) \quad (4)$$

$$s_j=-\frac{e^2}{4\pi\varepsilon_0 mc^2\gamma_j}\{|a_{uj}|(J_0(\xi_j)-J_1(\xi_j))\exp(-{\rm i}\zeta_j)+{\rm i}a_{\rm L}\} \quad (5)$$

$$\xi_j=|a_{uj}|^2/\{2(1+|a_{uj}|^2)\} \quad (6)$$

$$a_{\rm L}=\frac{eA_{\rm L}}{\sqrt{2}mc}\quad a_{uj}=\frac{eA_{uj}}{\sqrt{2}mc} \quad (7)$$

where $N_{\rm e}$ is the number of electrons in the laser wavelength of $\lambda_{\rm L}$, J_n is the nth order Bessel function, and $\boldsymbol{a}_{\rm L}$ and $\boldsymbol{a}_{uj}$ are, respectively, the dimensionless amplitudes of the vector potentials of the laser field, and the undulator field at the position of the jth electron. Furthermore, x_j, y_j, γ_j, and ζ_j are, respectively, the x, y coordinate, Lorentz factor, and the phase in the ponderomotive potential of the jth electron averaged over an undulator period.

The new code solves these equations by using the Finite Element Method with the boundary condition which is given by the tangential component on the wall $a_{\rm L\parallel}=0$, assuming a beam tube wall made of a perfect conductor. The periodic boundary condition is imposed for the laser propagation from the mirrors to the start of the beam tube in order to solve equations analytically using Fourier transformation as is given in the old code. And the laser field outside the opening of the beam tube entrance is lost.

In the simulations, $\rm TEM_{00}$ Gaussian mode is given analytically at the entrance of the beam tube. This could, however, inevitably induce components of other higher modes because of the finite x, y-plane in the calculations.

3. Results and discussions

Simulations for laser evolution are made for the reference FEL parameters shown in Table 1, characteristic of BFEL, and the cross section of the beam tube in the calculation is discretized by a finite number of triangular elements as shown in Fig. 1.

Evolutions of the laser profiles in the beam tube on the x, and y-axis are shown, respectively, in Figs. 2 and 3 for the wavelengths of 8 and 18 μm. It is obvious in Fig. 3, that the transverse higher-order

Table 1
Beijing FEL parameters

Parameter	Value
Optical signal	
Wavelength, λ_L	8–18 μm
Injected power, P_{in}	1 W
Rayleigh range, Z_R	0.65 m
Cavity length, L_m	2.52 m
Electron beam	
Peak current, I_p	15 A
Radium, σ_x, σ_y	0.64 mm
Normalized emittance, ε_{nx}, ε_{ny}	26 π mm mrad
Energy, γ	35–55
Energy spread, σ_γ	0.5%
Macro-bunch	4 μs (230 pass)
Undulator	
Period, λ_u	3.0 cm
Length, L_u	1.5 m
Undulator parameter, K	1.0
Beam tube	
Length, L_d	2.2 m
Cross section ($x \times y$)	15 mm × 10 mm

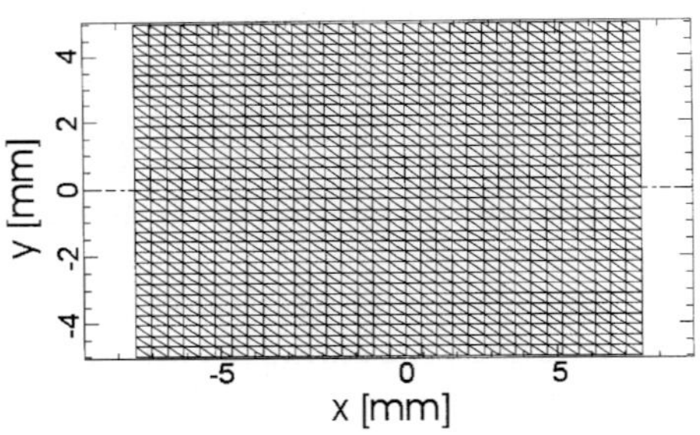

Fig. 1. Mesh of calculation region.

modes for 18 μm are strongly affected by the beam tube.

The gain as a function of the laser wavelength (solid curve) and without the beam tube effects for comparison(dotted curve) calculated by the old code is shown in Fig. 4. It is seen that the gain decreases by about 1% at 18 μm due to the effects of the beam tube, and the difference becomes smaller as the wavelength and the spot size of the laser decreases. It is to be noted that since the macro-bunch length of the BFEL is only 4 μs, the gain itself becomes very crucial to achieve the saturation and lasing for the BFEL.

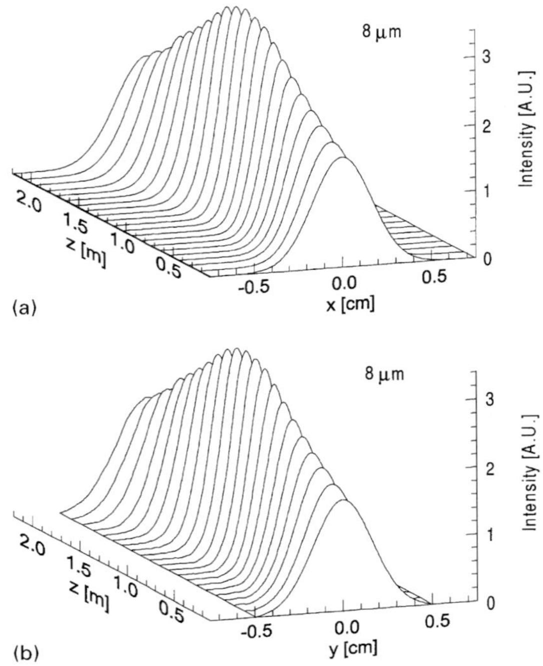

Fig. 2. Evolution of laser profiles on (a) x-axis, and (b) y-axis for the wavelengths of 8 μm. (a) x-axis; (b) y-axis.

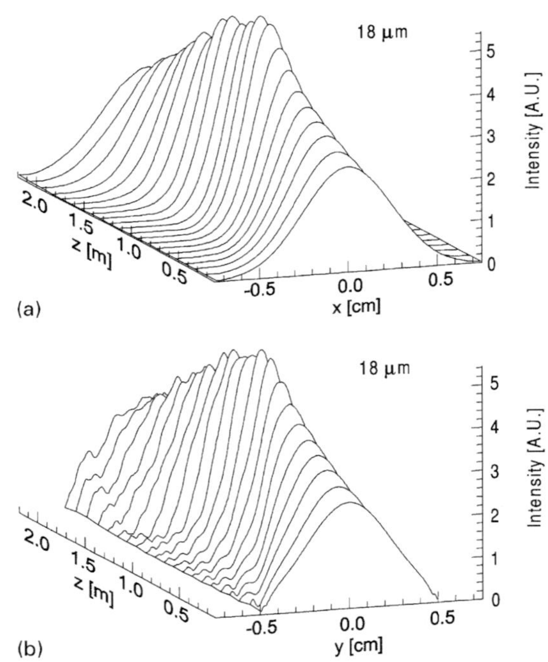

Fig. 3. Evolution of laser profiles on (a) x-axis, and (b) y-axis for the wavelengths of 18 μm. (a) x-axis; (b) y-axis.

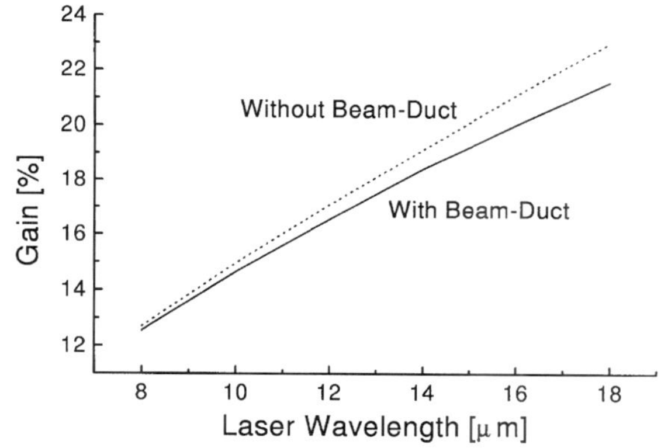

Fig. 4. Gain with beam tube effects (solid line) and without them (dotted line) as a function of laser wavelength.

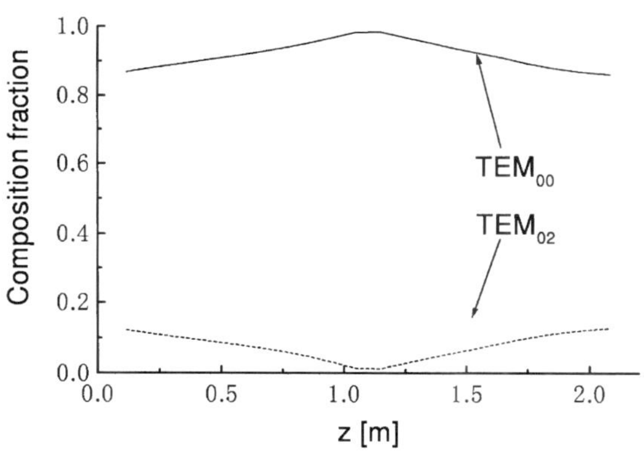

Fig. 5. Evolution of the mode composition fraction in beam tube.

Among the higher modes for 18 μm in Fig. 4, the TEM_{02} is dominant. Its evolution along the beam tube is shown in Fig. 5, together with the fundamental mode TEM_{00}. It is also to be noted that the evolution of laser propagation mode in the beam tube without interaction with electron beams is found to be almost the same as that in Fig. 5.

Other higher modes are found to be almost negligible as can be seen in Fig. 5.

It is seen that the fraction of the most dominant mode, TEM_{00}, increases near the center of the beam tube, while the second most dominant mode, TEM_{02}, decreases inversely, indicating energy transfer between modes.

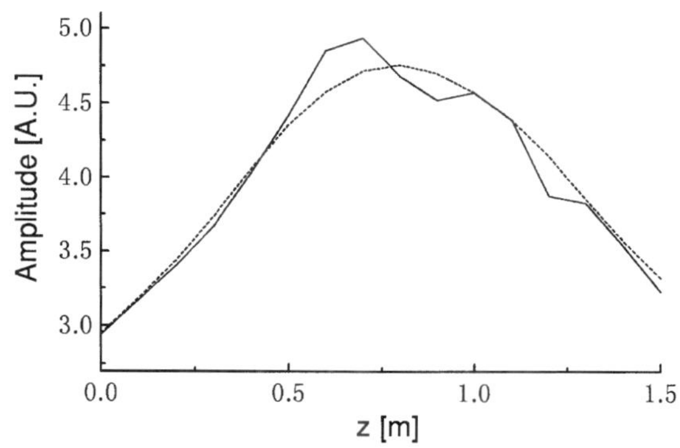

Fig. 6. Evolution of laser amplitude on the z-axis for finite (solid line) and infinite size (dotted line).

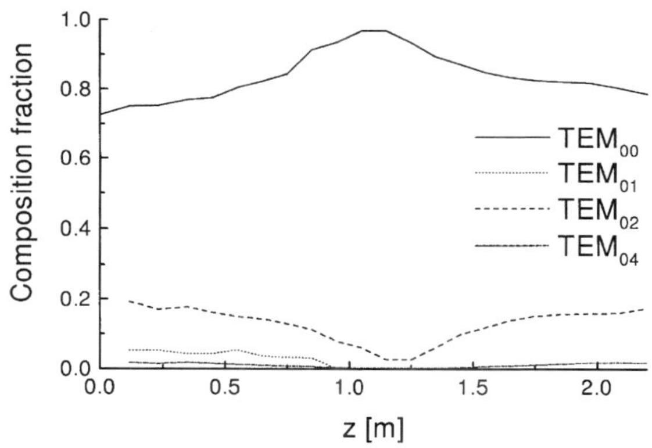

Fig. 7. Evolution of the mode composition fraction in beam tube at the 230th pass (18 μm).

This energy transfer between modes occurs because of the laser field reflection on the beam tube wall.

Unlike in the case of an infinite beam tube where higher modes can freely propagate and diffuse, in the beam tube of finite size, since all the modes are confined and forced to interact with the electron beam, more unfavorable effects by the higher-order modes from the view point of gain evolution are expected in this case.

Actually, from the evolution of the laser amplitude on the z-axis for both finite (solid line) and infinite size (dotted line) as shown in Fig. 6, it can be seen that the averaged laser amplitude simulated with the beam tube is less than that without the beam tube effects.

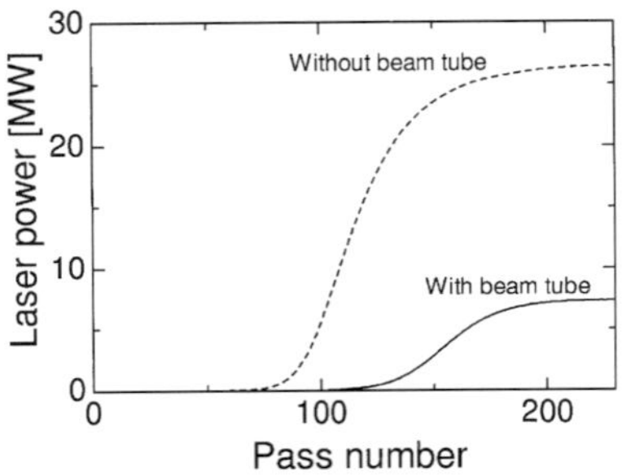

Fig. 8. Laser power evolution calculated by new code (solid line) and old code (dashed line) plotted as a function of pass number.

The realistic FEL modes for the wavelength of 18 μm in the cavity can be obtained by simulations of the laser that resonates in the cavity. Evolutions of the mode composition fractions on the *y*-axis at the 230th pass in the beam tube are shown in Fig. 7. It can be seen that the fraction of higher modes becomes larger than that at the first pass through the propagation between the cavity mirrors and the undulator, that is, higher modes are mostly generated by the interception that occurs when the laser goes into the beam tube.

Evolutions of the laser power in the optical cavity for 18 μm were calculated by both the new and the old code as is shown in Fig. 8. The laser power calculated by the new code is shown to be about 27% of that calculated by the old code because it includes the diffraction loss and the higher mode effects.

Throughout the simulations, it is found that the evolution of the higher modes is mainly dominated by the beam tube effects, not as much by optical guiding effects in the case of IR–FELs (18 μm) of the present BFEL.

4. Conclusions

We have developed a new FEL code to deal with the beam tube effects self-consistently with practical boundary condition by using the Finite Element Method and the simulated BFEL experimental results with this code. Results show that the beam tube effects are dominant in deriving higher laser modes in the tube, compared with the optical guiding effects, and consequently reduce the gain especially in the longer wavelength range, where the beam tube effects are much emphasized.

It is also found that the TEM_{02} mode is the most dominant higher mode in the beam tube, and is the main cause of gain reduction.

The realistic FEL modes in the cavity are obtained, and they are found to include higher modes, such as TEM_{01} and TEM_{04} in addition to TEM_{02}. And the laser power at saturation calculated by the new code is about 27% of that calculated by the old code due to the beam tube loss.

References

[1] M. Sobajima et al., Nucl. Instr. and Meth. A 407 (1998) 121.

[2] M. Sobajima et al., Nucl. Instr. and Meth. A 393 (1997) 280.

[3] T. Nakamura, Proceedings of the fourteenth Linear Accelerator Meeting in Japan, 1989, p. 236 (in Japanese).

ELSEVIER

Nuclear Instruments and Methods in Physics Research A 429 (1999) 419–423

NUCLEAR INSTRUMENTS & METHODS IN PHYSICS RESEARCH Section A

www.elsevier.nl/locate/nima

Helical magnetized wiggler for synchrotron radiation laser

Mei Wang[a,*], S.Y. Park[a,b,c], J.L. Hirshfield[a,c]

[a]*Department of Physics, Yale University, New Haven, CT 06520-8120, USA*
[b]*Department of Physics, POSTECH, Pohang, South Korea*
[c]*Omega-P, Inc., New Haven, CT 06520-2008, USA*

Abstract

A helical magnetized iron wiggler has been built for a novel infrared synchrotron radiation laser (SRL) experiment. The wiggler consists of four periods of helical iron structure immersed in a solenoid field. This wiggler is to impart transverse velocity to a prebunched 6 MeV electron beam, and thus to obtain a desired high orbit pitch ratio for the SRL. Field tapering at beam entrance is considered and tested on a similar wiggler. Analytic and simulated characteristics of wigglers of this type are discussed and the performance of the fabricated wigglers is demonstrated experimentally. A 4.7 kG peak field was measured for a 6.4 mm air gap and a 5.4 cm wiggler period at a 20 kG solenoid field. The measured helical fields compare favorably with the analytical solution. This type of helical iron wigglers has the potential to be scaled to small periods with strong field amplitude.

Keywords: Synchrotron radiation; Laser; Iron wiggler

1. Introduction

A recent development in free-electron laser technology to achieve high strength, short period wiggler fields is to use high saturation magnetization of an iron alloy structure immersed in a strong applied field. A planar "Staggered-Array Wiggler" system utilizing shifted alternating ferromagnetic and nonferromagnetic bars immersed in an axial field was investigated by a group at Stanford University in 1994 [1]. A coaxial arrangement of alternating rings with the central portion of the coax shifted by one half period (CHI wiggler) providing cylindrically symmetric periodic fields has been studied by a group at Naval Research Laboratory [2,3]. It has been pointed out that an iron helical structure can provide helical wiggler fields based on the same principle [4]. This paper describes a helical magnetized iron wiggler of this type that has been studied analytically and experimentally. To test this idea, a helical ferromagnetic iron structure is positioned in the room temperature bore of a super-conducting magnet. The high permeability helical iron structure deflects the solenoid field to form helical alternating poles and thus to form helical transverse fields. As shown in Fig. 1, the iron wiggler immersed in the first field region is to impart transverse velocity to a prebunched electron beam from a 6 MeV RF gun for a synchrotron radiation laser (SRL) experiment. Interaction for the SRL is to be in a quasi-optical cavity in the second field region, with a field uniformity of $\pm 0.1\%$. The desired high transverse

* Corresponding author. Tel.: +1-203-432-5167; fax +1-203-432-6926.

E-mail address: mei.wang@yale.edu (M. Wang)

0168-9002/99/$ - see front matter
PII: S0168-9002(99)00121-7

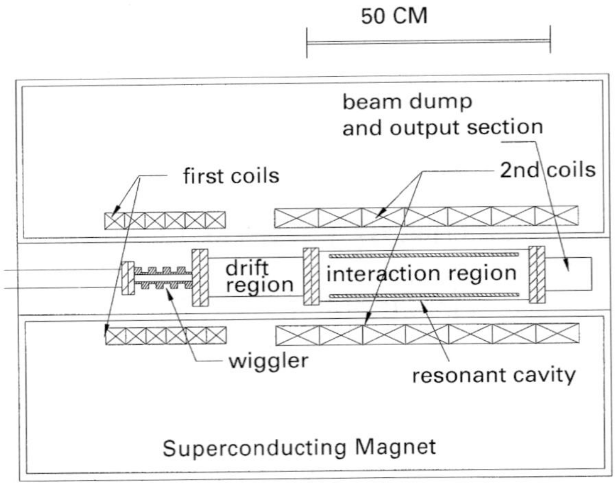

Fig. 1. Synchrotron radiation laser experiment setup.

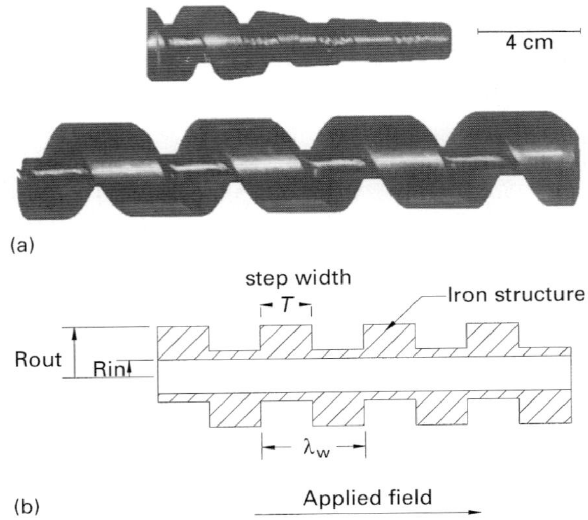

Fig. 2. Helical magnetized iron wiggler configuration. (a) Pictures of the fabricated wigglers; the top wiggler is tapered.

velocity (close to c) can be tuned by adjusting strengths of the axial fields in the two regions. SRL is a device based on a high synchrotron harmonic gain interaction between relativistic gyrating electrons and the fields of a TEM_{00q} mode in an optical resonator [5,6]. This concept has the potential of providing voltage and/or magnetic field tunable, multi-kilowatt, coherent infrared power at wavelengths in the range of 50–1500 μm.

2. Analytic field approximation

An analytic solution for the field near a helical magnetized structure within a solenoid coil can be obtained by solving Laplace's equation with appropriate boundary conditions. To approximate, it is useful to solve for the on-axis field, and to treat the ferromagnetic structure as having a uniform magnetization $\boldsymbol{M}$, which is independent of applied field. (This is approximately true after the ferromagnetic structure is saturated.) Magnetic induction $\boldsymbol{B}$ outside a local magnetic dipole moment distribution can be obtained by integrating over the distribution. In this approximation, the magnetic induction $\boldsymbol{B}$ is given by

$$\boldsymbol{B}(\boldsymbol{r}) = \int_{V'} \frac{3\hat{\boldsymbol{n}}\,(\hat{\boldsymbol{n}}\cdot\boldsymbol{M}\,\mathrm{d}V') - \boldsymbol{M}\,\mathrm{d}V'}{|\boldsymbol{r}-\boldsymbol{r}'|} \tag{1}$$

where $\hat{\boldsymbol{n}}$ is a unit vector in the direction $\boldsymbol{r}-\boldsymbol{r}'$.

For an infinitely long helical structure as in Fig. 2, one just needs to calculate on-axis field at $z=0$, and fields at other locations can be deduced by helical symmetry. $\boldsymbol{r}' = a\cos\phi'\,\hat{\boldsymbol{e}}_x + a\sin\phi'\,\hat{\boldsymbol{e}}_y + z'\,\hat{\boldsymbol{e}}_z$, where a is the radius, $z' = z'_0 + \mathrm{d}z'$, $\phi' = k_w z'_0$ is the helix angle; $k_w = 2\pi/\lambda_w$, and $\mathrm{d}V' = a\,\mathrm{d}z'\,\mathrm{d}a\,\mathrm{d}\phi'$. Substituting $\boldsymbol{r}'$, $\mathrm{d}V'$ in Eq. (1), we have

$$B_x(0,0,z) = B_w \sin k_w z$$
$$B_y(0,0,z) = -B_w \cos k_w z$$
$$B_z(0,0,z) = 0 \tag{2}$$

where

$$B_w = 4M \sin\left(\frac{k_w\,\tau}{2}\right) k_w^2 \int_{R_{in}}^{R_{out}} aK_1(k_w a)\,\mathrm{d}a \tag{3}$$

where τ is the step width shown in Fig. 2, and K_1 is the first-order modified Bessel function.

Using the published saturation field of 2.4 T for vanadium permendur, Fig. 3 shows the predicted characteristics from Eq. (3) of a helical permendur wiggler for variable inner radius (air gap radius) and outer radius. The step width τ is taken to be half the wiggler period in all the following cases if not stated otherwise. In Fig. 3, the outer radius is taken to be infinity in order not to play a role. It

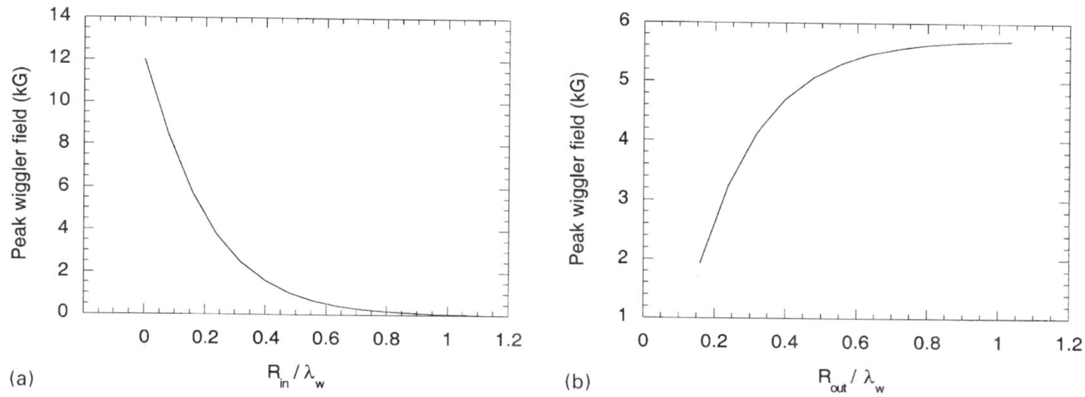

Fig. 3. Variation of helical wiggler field with parameters. (a) Peak wiggler field as a function of normalized inner radius (R_{out} = infinity); (b) peak wiggler field as a function of normalized outer radius ($R_{in}/\lambda_w = 0.16$).

can be seen in Fig. 3b that when R_{out} reaches one period in length, the field begins to saturate. It is shown in Fig. 3a that a transverse field approaching 1.2 T can be achieved with this type of wiggler, i.e., roughly half the saturation magnetization. The general variations of wiggler field magnitudes shown in Fig. 3 are similar to results in Refs. [1–3].

3. Comparisons between analytic solution and POISSON simulation

The 2D code POISSON was used to study characteristics of the wiggler, and results were compared with the 3D analytical solution. In POISSON, an internal *B–H* table is built in for 1010 low carbon steel, which has a saturation magnetization of 2.05 T. To compare, a saturating field was applied in the POISSON simulation and the 2.05 T saturation magnetization was used in the analytical solution Eq. (3) for the examples shown here. Fig. 4 shows normalized wiggler field strengths as functions of the variables. It should be noted that the normalization factor used for the POISSON simulation results as approximately 1.4 times the normalization factor for the analytical solution results as, for the same geometry, the POISSON simulation always gives a higher field. This is possibly due to the different dimensionality used in these two models. Nevertheless, there are discrepancies between the POISSON simulations and the analytic solution. In Fig. 4a, both show a fast drop as the inner radius increases. The POISSON simulations and the analytical solution agree well in Fig. 4b. In Fig. 4c, when the step width to period ratio is bigger than 0.5, the POISSON simulations agree well with the analytical solution, but have a big difference when the ratio is smaller than 0.5. In the POISSON simulations, the optimal step width to period ratio is 0.55, while in the analytical solution, the optimal is 0.5. These discrepancies are not surprising considering the different dimensionality and the very simplified model used in the analytical solution.

4. Measurements

Two helical wigglers of this type have been fabricated from cold rolled steel, as shown in Fig. 2a; one is tapered and the other is uniform. Measurements of the solenoid field used to magnetize the wiggler show a field uniformity of $\pm 0.1\%$ for a range of 30 cm. A transverse Hall probe was moved along the axis of the wiggler to measure the field strengths. The x, y components of transverse field were measured by rotating the probe by 90°.

Figs. 5 and 6 show the measurements made on the longer uniform wiggler, which has four periods with a period length of 5.4 cm. The step width τ is taken to be half the period length. An important difference between the actual wigglers and the

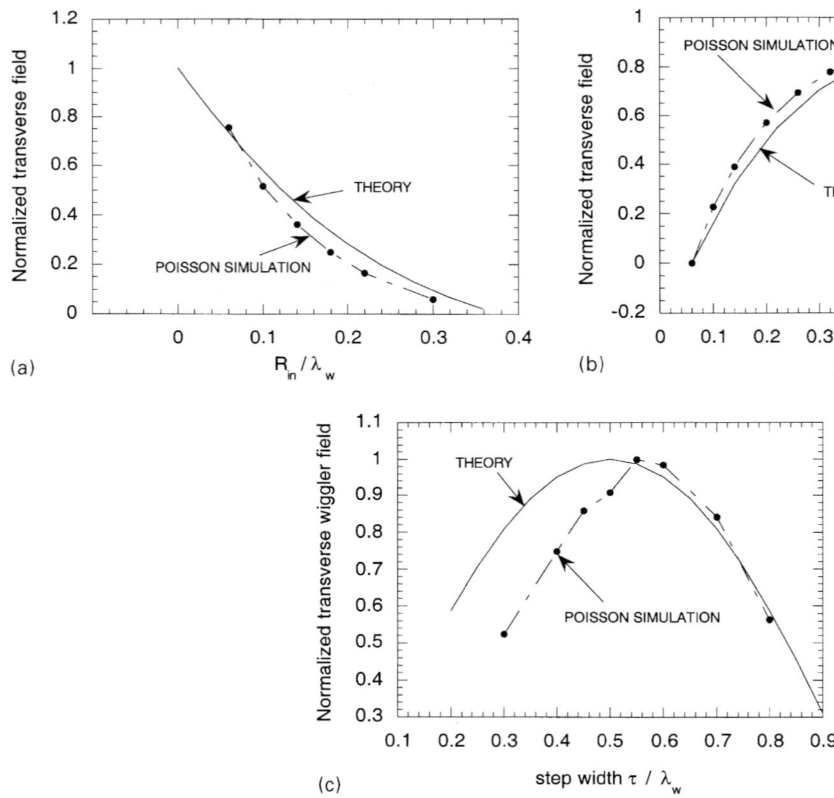

Fig. 4. Comparisons between the 2D code POISSON and the 3D analytic approximation results. The axial field is 20 kG. (a) Normalized transverse field as a function of inner radius. $R_{out}/\lambda_w = 0.38$; (b) normalized transverse field as a function of outer radius. $R_{in}/\lambda_w = 0.16$; (c) normalized transverse field as a function of step width τ.

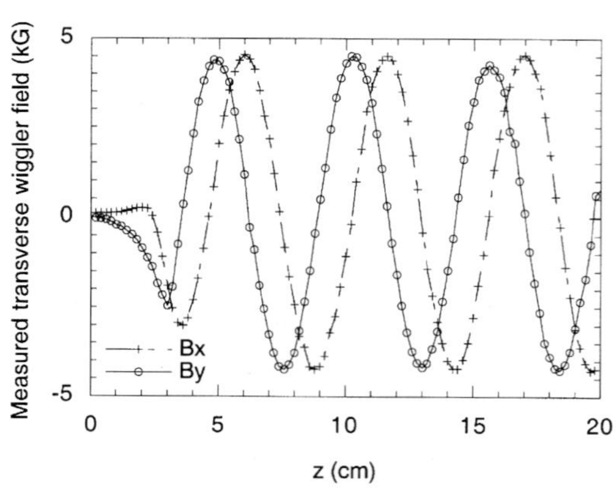

Fig. 5. Measured transverse field profile along the axis. $\lambda_w = 5.4$ cm, $R_{in} = 3.2$ mm, $R_{out} = 19$ mm, Pipe wall thickness = 2.5 mm, step width $\tau = 2.7$ cm, axial field 15 kG.

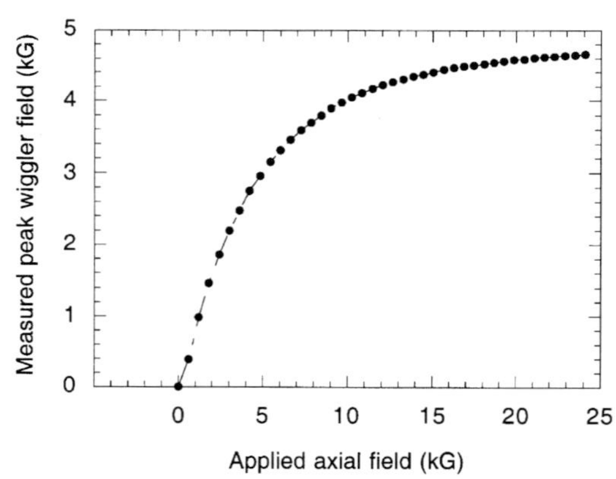

Fig. 6. Measured peak wiggler field as a function of axial field. Wiggler parameters are the same as in Fig. 5.

assumed analytical model is that the helical slots in the iron pipe are not cut through, but a thin inner wall 0.25 cm thick remains to provide rigid mechanical support. The POISSON code has been modified for this change, but to save space, the results are not shown here. An effective inner radius instead of the actual radius was used in analytic solution Eq. (3) to approximate the influence of the uncut wall. Fig. 5 shows the measured field profile as a function of axial distance with an applied axial field of 1.5 T, and Fig. 6 shows the measured peak transverse field as a function of applied axial field. In Fig. 6, as the axial field reaches 1.5 T, the iron begins to saturate, and the wiggler field is thereafter insensitive to axial field strength. As shown in Fig. 6, a helical field of about 4.7 kG was achieved with this wiggler. This value compares favorable with the analytic prediction of 4.4 kG for the measured 1.6 T cold-rolled steel saturation, when an effective inner radius of 4.0 mm instead of the actual 3.2 mm is taken. For a similar wiggler made out of vanadium permendur, one can expect 50% more field due to a higher saturation for that material.

Orbit considerations in SRL dictate that the wiggler should be tapered gradually to a desired value and terminated abruptly. A shorter wiggler of 3 cm period length was fabricated to test field-tapering. The outer radius of this shorter wiggler was tapered from 0.7 cm to a full radius of 1.5 cm in three periods. Fig. 7 shows the measured transverse fields of the tapered test wiggler as a function of axial position. It is seen that a reasonably symmetric gentle taper can be achieved.

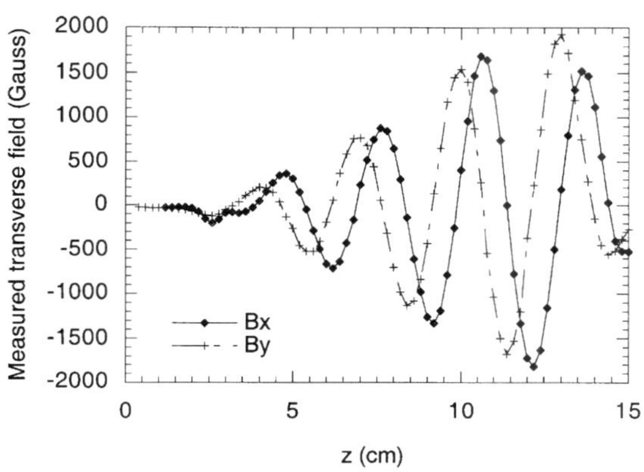

Fig. 7. Measured tapered wiggler field strength as a function of axial distance. $\lambda_w = 3.0$ cm, $R_{in} = 5.0$ mm, $R_{out} = 7$–15 mm, pipe wall thickness = 2 mm, step width $\tau = 1.5$ cm, axial field = 6 kG.

5. Conclusions

Helical magnetized iron wigglers for the synchrotron radiation laser experiment have been fabricated and tested. A handy analytical formula provides a simple, quick approach to the preliminary design of wigglers of this type. Measurements compared favorably with the analytical solution. A peak helical wiggler field of 4.7 kG has been measured for a cold-rolled steel structure with a period of 5.4 cm, in a solenoid field of 2 T. The effects of tapering the structure were demonstrated and showed that a reasonably symmetric smooth taper is achievable. This configuration of helical iron wiggler has the advantages of easy fabrication, low cost, field tunability and high-field strengths for short periods.

Acknowledgements

The authors would like to thank Dr. Arne Fliflet and Dr. Steven Gold of NRL for loan of the superconducting magnet; Dr. M.A. LaPointe of Omega-P Inc. for tremendous help in setting up the magnet and good discussion; Dr. P. Yang at University of California at Santa Barbara for measuring the saturation magnetization of our iron sample. This work is supported in part by US Department of Energy.

References

[1] Y.C. Huang, H.C. Wang, R.H. Pantell, J. Feinstein, J.W. Lewellen, IEEE J. Quantum Electron 30 (5) (1994) 1289.
[2] R.H. Jackson, H.P. Freund, D.E. Pershing, J.M. Taccetti, Nucl. Instr. and Meth. A 375 (1996) 496.
[3] R.H. Jackson, H.P. Freund, D.E. Pershing, J.M. Taccetti, Nucl. Instr. and Meth. A 341 (1994) 454.
[4] H.A. Leupold, E. Potenziani II, ARL-TR-946, 1996, pp. 56–59, Unpublished Report.
[5] J.L. Hirshfield, Phys. Rev. Lett. 68 (1992) 792.
[6] J.L. Hirshfield, G.S. Park, Int. J. Infrared and Millimeter Waves 12 (3) (1991) 275.

ELSEVIER

Nuclear Instruments and Methods in Physics Research A 429 (1999) 424–428

NUCLEAR INSTRUMENTS & METHODS IN PHYSICS RESEARCH Section A

www.elsevier.nl/locate/nima

Regenerative FEL amplifier at the TESLA test facility at DESY

B. Faatz[a], J. Feldhaus[a], J. Krzywinski[b], E.L. Saldin[c], E.A. Schneidmiller[c], M.V. Yurkov[d,*]

[a]*Hamburger Synchrotronstrahlungslabor at Deutsches Elektronen-Synchrotron (DESY), Notkestrasse 85, D-22607 Hamburg, Germany*
[b]*Institute of Physics of the Polish Academy of Sciences, 02688 Warszawa, Poland*
[c]*Automatic Systems Corporation, 443050 Samara, Russia*
[d]*Particle Physics Laboratory (LSUE), Joint Institute for Nuclear Research, Dubna, 141980 Moscow Region, Russia*

Abstract

This paper presents a conceptual design of a regenerative FEL amplifier (RAFEL) as an extension of the single-pass free electron laser project at the TESLA test facility (TTF) at DESY. The proposed scheme requires the additional installation of only two optical components for a narrow-band feedback system and is fully compatible with the present design and the infrastructure developed for the TTF FEL project. It would allow to construct a tunable VUV laser with a minimum wavelength around 60 nm, a pulse duration of about 1 ps, a peak power of about 300 MW and an average power of about 25 W. The output radiation of the regenerative FEL amplifier would possess all the features which are usually associated with laser radiation: full transverse and longitudinal coherence and shot-to-shot stability of the output power. The degeneracy parameter of the output radiation would be about 10^{14} and thus have the same order of magnitude as that of a quantum laser operating in the visible.

PACS: 41.60.Cr; 52.75.Ms; 42.25.Kb

Keywords: FEL Amplifier; TESLA test facility; DESY

1. Introduction

Free-electron laser techniques provide the possibility to extend the energy range of lasers into the vacuum ultraviolet (VUV) and X-ray regime using a single-pass free electron laser (FEL) amplifier based on the self-amplification of spontaneous emission (SASE). A characteristic feature of a single-pass SASE FEL is its rather large amplification bandwidth. When the process of amplification starts from noise, it produces a relatively wide spectrum of output radiation with only a short longitudinal coherence length. The relative bandwidth at VUV wavelengths would be of the order of one percent. Each radiation pulse consists of many independent wavepackets. The length of each wavepacket is much less than the radiation pulse length and there is no phase correlation between them.

For many applications this behavior is a serious limitation because, for example, one would observe large intensity fluctuations behind a narrow-band monochromator. Therefore, the improvement of

* Corresponding author. + 7-09621-62154; fax: + 7-09621-65767.

E-mail address: yurkov@sunse.jinr.ru (M.V. Yurkov)

0168-9002/99/$ - see front matter
PII: S0168-9002(99)00123-0

the longitudinal coherence of VUV and X-ray FELs is of great practical importance. In order to obtain full longitudinal coherence it is necessary to seed the FEL amplifier with a sufficiently narrow band of radiation at a power level well above the effective power of shot noise in the electron beam. The optimum bandwidth of the seeding radiation, $\Delta\lambda/\lambda$, is related to the pulse length σ_z by $\Delta\lambda/\lambda = \lambda/\pi\sigma_z$. In this paper we present the seeding scheme based on the concept of regenerative FEL amplifier (RAFEL) [1] which will be incorporated into the FEL project at the TESLA test facility (TTF) at DESY [2].

2. Properties of the TTF FEL without seeding

The RAFEL will be installed at the end of the first stage (Phase I) of the TTF FEL project. During this phase the linear accelerator will operate at electron beam energies up to 390 MeV and the FEL will produce radiation with a wavelength down to $\sim$ 40 nm. Phase I is intended to test various novel hardware components and to prove the SASE principle for the first time at short wavelengths. In Phase II of the project the accelerator will be extended to increase the electron beam energy to more than 1 GeV and to drive a FEL facility down to wavelengths of a few nanometers.

The radiation from a high-gain SASE FEL is spatially (or transversely) coherent. The temporal (or longitudinal) coherence, however, is poor due to the start-up from noise. Consequently, also the frequency spectrum exhibits a similar structure (Fig. 1). It is interesting to note that the characteristic line width in the frequency domain which is determined by the bunch length, is close to the natural line width of atomic core levels given by the lifetime of the excited states. If one would select such a narrow band of radiation by a monochromator behind the FEL, the output intensity would change from pulse to pulse with a probability distribution close to the negative exponential distribution with the normalized standard deviation equal to 1 [3]. On the other hand, if seeding were possible, the spectral brilliance would be much higher and intensity fluctuations could be reduced to small level even for the smallest possible bandwidth given by the bunch length, providing ideal conditions for high-resolution spectroscopy.

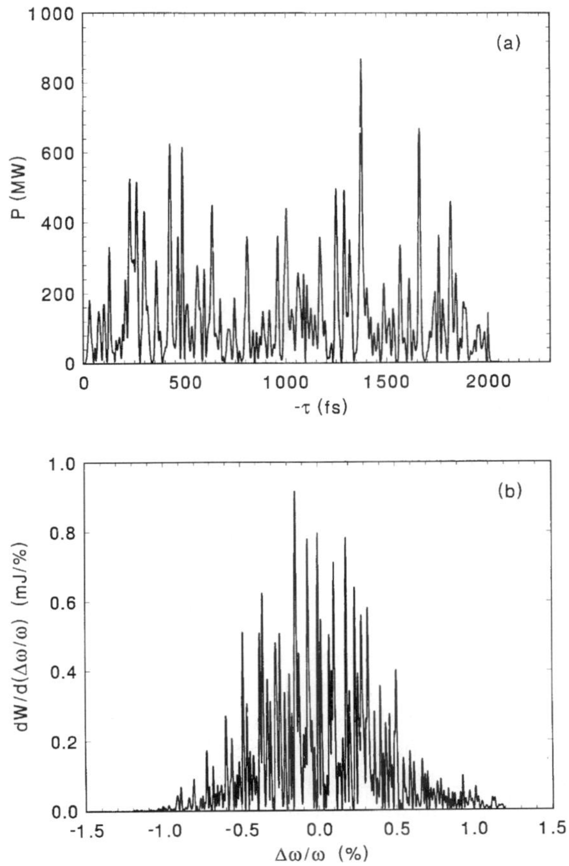

Fig. 1. Typical temporal structure (a) and spectrum of the radiation pulse (b) at the exit of 70 nm SASE FEL.

3. Regenerative FEL amplifier at the TTF

The technical parameters of the VUV FEL at DESY are such that it can be easily extended by a narrow-band optical feedback system turning the single-pass SASE FEL starting from noise into a regenerative FEL amplifier [4] which could provide fully coherent, powerful VUV laser radiation continuously tunable between approximately 200 and 60 nm. The installation of the feedback is greatly facilitated by the fact that there is free space

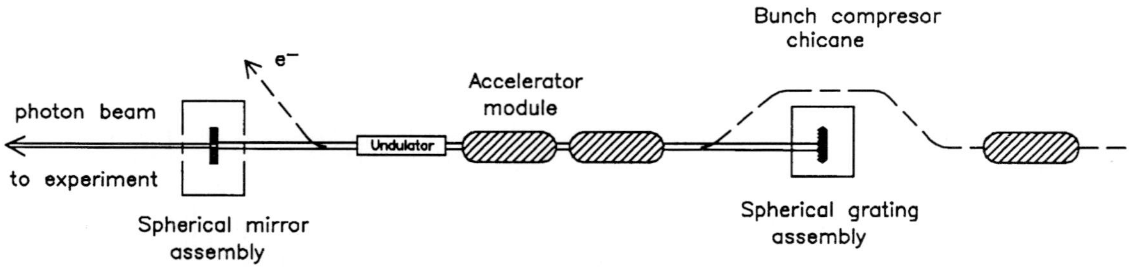

Fig. 2. Layout of regenerative FEL amplifier at the TESLA test facility.

available for the optical components at exactly twice the distance between two electron bunches when the accelerator is operated in a 9 MHz multi-bunch mode. The schematic layout of the feedback system is shown in Fig. 2. The optical system consists of only a mirror and a grating, because one can use optical components with good reflectivity near normal incidence. SiC appears to be particularly well suited for the energy range of the Phase I facility (see Fig. 3). This material has excellent thermal properties such that surface distortions by the average absorbed power are negligible. A simple, far too pessimistic estimate shows that the high peak power should also not pose a serious problem. If the energy absorbed per pulse were transformed into heat within the same absorption volume, the temperature would increase by about 1000 K – far too little to cause laser ablation.

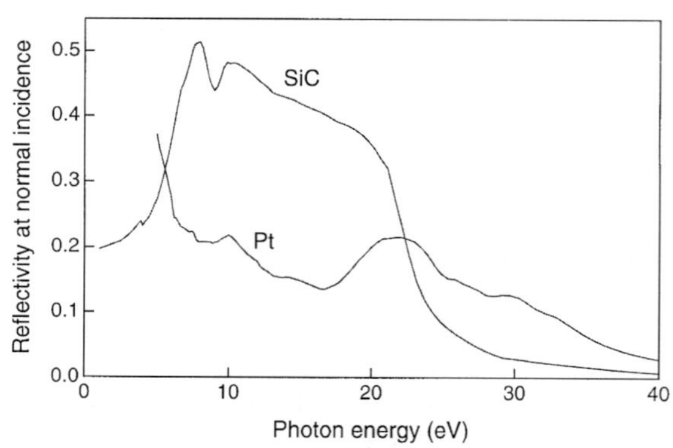

Fig. 3. Reflectivity of SiC and Pt at normal incidence calculated from the optical constants using REFLEC [5].

Parameters of RAFEL are presented in Table 1. The RAFEL at the TESLA test facility operates as follows. The first bunch in a train of up to 7200 bunches amplifies shot noise and produces intense, but wide-band radiation as shown in Fig. 1.

Table 1
Parameters of the RAFEL at the TESLA test facility

Parameters of the electron beam	
Energy	180–325 MeV
Charge per bunch	1 nC
Peak current	500 A
Bunch length (RMS)	250 μm
Bunch width (RMS)	70 μm
Energy spread (RMS)	500 keV
Normalized emittance (RMS)	2 mm mrad
Number of bunches per train	7200
Bunch separation	111 ns
Repetition rate	10 Hz
Parameters of the planar undulator	
Period	2.73 cm
Magnetic gap	12 mm
Peak magnetic field	0.497 T
Length of undulator module	4.5 m
Number of modules	3
Parameters of the feedback system	
Distance between mirrors	66.4 m
Monochromator resolution	5×10^{-5}
Total transmission	5×10^{-4}
Parameters of the radiation	
Wavelength	60–200 nm
Micropulse duration (RMS)	500 fs
Energy in the radiation pulse	0.36 mJ
Peak output power	300 MW
Average power	25 W

A fraction of the radiation is back-reflected by a spherical mirror using a magnification of the order of one. The spherical grating in Littrow mounting which is installed in a straight section in the electron bunch compression area between the first and the second accelerator module, disperses the light and focuses a narrow band of radiation back on the entrance of the undulator.

Calculations show that the aberrations in the optical system can be reduced down to the required level [4]. An alignment accuracy of about 2 μrad is sufficient for the reliable operation of the optical feedback. The alignment of the system can be made by conventional surveying techniques reducing the number and the values of free parameters to a minimum. This should allow one to achieve the optimum overlap between photon and electron bunches within 10 to 20 steps in either direction of the grating.

The bandwidth of the feedback is chosen to produce a photon pulse length about four times as long as the electron bunch length in order to avoid effects from a ± 1 ps time jitter. This requires a resolving power $\lambda/\Delta\lambda = 2\times 10^4$ at $\lambda = 120\,\mathrm{nm}$. If the distance between mirror and grating is properly adjusted, the monochromatic photon pulse coming from the first electron bunch in the train travels together with the fifth electron bunch. At the entrance of the undulator the overlap is optimum and the power density of the monochromatic light exceeds that of the shot noise by about three orders of magnitude such that it is amplified to saturation on its way through the undulator. Hence, the fifth and all following pulses exhibit a narrow bandwidth determined by the pulse length at a nearly constant, saturated power level with very little dependence on the input power of the seeding pulse. Accordingly, the time evolution of the light pulse will be smooth and the radiation will be fully coherent. Since the total saturated pulse energy is independent of the bandwidth of the seeding radiation provided it is smaller than the amplifier bandwidth, the spectral brilliance of the RAFEL is approximately two orders of magnitude larger than that of the SASE FEL.

The optical system shown in Fig. 2 is only intended to test the seeding at short wavelengths and to gain experience with the FEL operating in the deep saturation regime. In order to minimize effort and risk, the optical system has been simplified as much as possible, particularly in order to facilitate the alignment and stability of the system at a distance of 66.4 m between mirror and grating. The characterization of the radiation will be done by measuring the intensity and the spectral distribution of the photons scattered from a thin probing wire in the beam.

If one wanted to use the output radiation for experiments, one would drill a central hole into the mirror M passing up to 90% of the radiation to the experiment. However, the hole in the mirror causes diffraction which dominates the propagation of the light and thus has to be taken into account in the design of the optical system. During a preliminary study [4] we have investigated a similar system with a plane mirror reflecting $\sim 90\%$ of the radiation to an experiment and passing the central part of the beam through a 1 mm hole to a spherical mirror. In this study the spherical grating was replaced by a plane grating in a combination with a spherical mirror in order to improve the optical performance of the system over a wider spectral range.

4. Present status

At present the conceptual design of the RAFEL project is finished and the technical design of the feedback system is under way. The feedback system will be installed at the TTF site in 1999. We expect that the experiment will start in the beginning of the year 2000.

Acknowledgements

We wish to thank P. Gürtler, T. Limberg, G. Materlik, T. Möller, U.-K. Müller, C. Pagani, J. Pflüger, J. Roßbach, J.R. Schneider, S. Schreiber, B. Sonntag and H. Weise for useful discussions. We are grateful to U. Hahn, J. Halik, A. Kabel and W. Sobala for technical support.

References

[1] J. Goldstein, D. Nguyen, R. Sheffield, Nucl. Instr. and Meth. A 393 (1997) 137.
[2] A VUV free electron laser at the TESLA test facility: conceptual design report, DESY Print TESLA-FEL 95-03, Hamburg, DESY, 1995.
[3] E.L. Saldin, E.A. Schneidmiller, M.V. Yurkov, Opt. Commun. 148 (1998) 383.
[4] B. Faatz, J. Feldhaus, J. Krzywinski, E.L. Saldin, E.A. Schneidmiller, M.V. Yurkov, Regenerative FEL amplifier at the DESY TESLA test facility as a fully coherent VUV laser, DESY Print TESLA-FEL 97-07, Hamburg, DESY, 1997.
[5] F. Schäfers, M. Krumrey, BESSY Technical Report TB 201/96, 1996.

ELSEVIER

Nuclear Instruments and Methods in Physics Research A 429 (1999) 429–434

NUCLEAR INSTRUMENTS & METHODS IN PHYSICS RESEARCH Section A

www.elsevier.nl/locate/nima

The FEL and IFEL design study for the proposed NTHU photon–electron dynamics laboratory

Y.C. Huang[a,*], C.S. Hsue[b], R.H. Pantell[c], T.I. Smith[d]

[a]*Department of Atomic Science, National Tsinghua University, Hsinchu 30043, Taiwan*
[b]*Department of Physics, National Tsinghua University, Hsinchu 30043, Taiwan*
[c]*Department of Electrical Engineering, Stanford University, Stanford, CA 94305, USA*
[d]*Department of Physics, Stanford University, Stanford, CA 94305, USA*

Abstract

A photon–electron dynamics laboratory is being proposed at National Tsinghua University, Taiwan. The mission of the proposed laboratory is to conduct advanced research on free-electron laser, inverse free-electron laser, and laser-driven particle acceleration. We discuss in this paper the design study of the Stanford high efficiency FEL and the reconfigured Stanford/TRW wiggler in the proposed laboratory. The Stanford high efficiency FEL is to demonstrate a 19% beam-to-laser efficiency. The reconfigured, 1-m Stanford/TRW wiggler may serve as an inverse FEL that provides 1.5-μm bunched electrons for various kinds of laser-driven acceleration experiments.

1. Introduction

A photon–electron dynamics laboratory is being proposed at National Tsinghua University (NTHU), Taiwan. The goal of the proposed laboratory is to study photon–electron interactions, including free-electron laser (FEL) and laser-driven particle acceleration. Continuation of the Stanford high efficiency FEL (HEFEL) project [1] and resuscitation of the Stanford/TRW FEL [2] are being discussed. The Stanford HEFEL employs a 1.35-m, staggered-array wiggler, which was designed to serve at the same time as a microwave re-acceleration section for high efficiency operation. The 4.8-m Stanford/TRW FEL was driven by the Stanford super-conducting accelerator and operated both at the visible and IR wavelengths in the past. When using a 25 MeV electron beam, both FELs could generate laser radiation at a ~ 10 μm wavelength, if designed properly. We discuss in this paper the design of the two FELs driven by a single S-band RF linac in the proposed laboratory.

Free electron lasers, operated under the self-amplified spontaneous emission (SASE) mode, are promising for generating coherent, short wavelength radiation in the future. Alternatively, if there exits an accelerator that produces ultra-short electron bunches, efficient coherent radiation at short wavelengths could be generated from a compact

* Corresponding author. Tel.: 886-3-5715131; fax: 886-3-5718649.

E-mail address: ychuang@faculty.nthu.edu.tw (Y.C. Huang)

0168-9002/99/$ - see front matter
PII: S0168-9002(99)00124-2

undulator. Therefore, laser-driven particle accelerators, operating at optical wavelengths, have been actively studied recently [3]. In an effort to develop this type of accelerator, experimental study in laser-driven particle acceleration is one of the major goals of the proposed NTHU photon–electron dynamics laboratory. When functioning as an inverse FEL (IFEL) at a 10 μm wavelength, the Stanford/TRW wiggler could generate bunched electrons that are suitable for testing various kinds of laser-driven acceleration schemes.

Fig. 1 shows the equipment layout of the proposed NTHU photon–electron dynamics laboratory, where a single S-band linac provides electrons to both the HEFEL and the Stanford/TRW FEL/IFEL. The electron source is a microwave thermionic gun, producing macropulses of a few μs at a 10 Hz rate. When the Stanford/TRW wiggler is configured as an IFEL, a CO_2 laser modulates and bunches the electrons inside the wiggler. The bunched electrons are then used for CO_2 laser driven particle acceleration experiments in the downstream beam line.

We discuss in this paper the design issues of the HEFEL and the Stanford/TRW FEL/IFEL. In particular, recent studies on the IFEL bunching mechanism are presented.

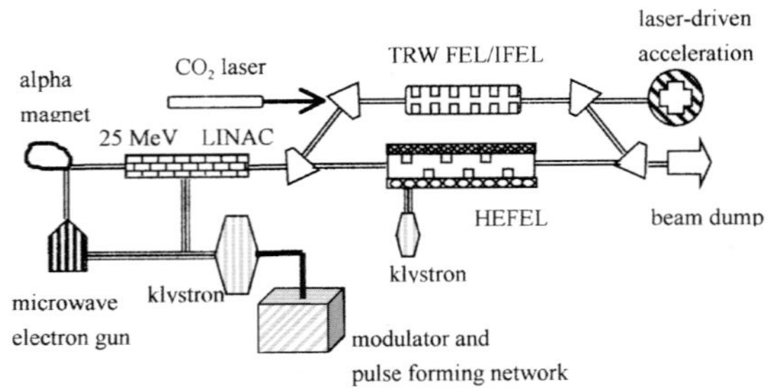

Fig. 1. The equipment layout of the proposed NTHU photon–electron dynamics laboratory. An S-band, RF thermionic gun provides electrons to a 25 MeV linac, which in turn drives both the Stanford high efficiency FEL (HEFEL) and the Stanford/TRW FEL/IFEL. The synchronism wavelength is 10 μm for both wigglers. The bunched electrons from the IFEL are suitable for various kinds of laser-driven particle acceleration experiments in the down-stream beamline.

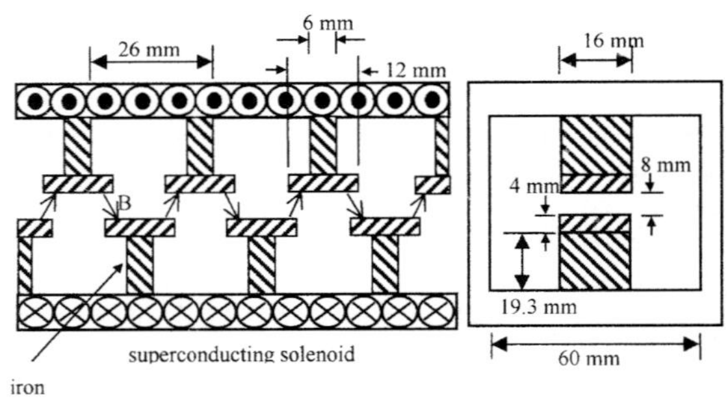

Fig. 2. The wiggler/linac of the Stanford high efficiency FEL. The wiggler field is generated from a staggered iron array inside a superconducting solenoid. The structure not only provides the alternating magnetic field for FEL operation, but also functions as a slow-wave microwave accelerator to maintain electron synchronism in the optical field.

2. The high efficiency FEL

In a FEL, electrons fall off synchronism with the optical wave after giving away their energies to the optical fields. This desynchronization limits the FEL efficiency. It is possible to synchronize the electron and the optical field over a longer distance by tapering the wiggler field. Unfortunately, tapering the wiggler field has the disadvantage of reducing small signal gain at FEL buildup.

A novel scheme to enhance the FEL efficiency is to compensate the electron energy loss by microwave re-acceleration at FEL saturation. The electron beam essentially acts as a catalyst to convert the microwave energy to the optical energy. By adjusting the buildup time of the microwave acceleration field, one can dynamically achieve high conversion efficiency at FEL saturation while having high gain at FEL buildup. This novel scheme imposes the need of designing a combined wiggler/accelerator section for the FEL. The FEL wiggler not only serves as a magnetic device but also functions as a slow-wave structure for particle acceleration.

The Stanford HEFEL wiggler is a 1.35 m, staggered-array wiggler, in which two iron arrays, displaced by a half-wiggler wavelength with respect to each other, are inserted inside a super-conducting solenoid, as Fig. 2 shows. The iron poles deflect the solenoid field periodically in the transverse direction, providing the wiggler field for FEL operation.

The longitudinal magnetic field in the solenoid also helps to confine the electron beam. With the dimensions shown in the figure, POISSON simulation predicts a peak wiggler field of 5.1 kG over a 1.35 m wiggler length at a 15 kG solenoid field. The HEFEL was designed to radiate at a 10 μm fundamental wavelength when driven by a 25-MeV electron beam. The small signal gain is about 200% for typical beam parameters from a S-band linac [4].

The dimensions of the Stanford HEFEL wiggler were at the same time optimized by the computer code MAFIA for particle acceleration. The rectangular iron head design is optimized for obtaining a high wiggler field and a good acceleration field. The FEL efficiency of the proposed microwave re-acceleration scheme is given by

$$\eta = \frac{f_t E_m L}{K_e + E_m L} \tag{1}$$

where f_t is the electron trapping fraction, E_m is the peak microwave field on the axis in volt/m, L is the wiggler/linac length in meters, and K_e is the electron kinetic energy in electron volts. The numerator $f_t E_m L$ is the energy that the electrons obtain from microwave acceleration or equivalently that the electrons give to the optical fields. With our design parameters [1,4]. $f_t = 84\%$ after a 30 cm prebuncher, $E_m = 5.1$ MV/m, $L = 1.35$ m, and $K_e = 24$ MV, the FEL efficiency from (1) is expected to be 19%.

3. The Stanford/TRW FEL/IFEL

The 4.8 m Stanford/TRW wiggler consists of nine equal-length sections with a wiggler period $\lambda_w = 3.56$ cm. Each section has 15 wiggler periods and a length of 53.4 cm. The wiggler consists of two linear arrays of $SmCo_5$ permanent magnets. The peak wiggler field is $B_{w,peak} = 2.9$ kG, and the wiggler parameter a_w, defined as $a_w = 0.093 \times B_{w,rms}(\mathrm{kG}) \times \lambda_c(\mathrm{cm})$, is $a_w = 0.68$. The wiggler gap is about 1 cm wide.

With a 25-MeV beam, the radiation wavelength is 10 μm. At this wavelength, the wiggler length is limited by diffraction. For example, if the laser radius need not exceed 2 mm at either end of the wiggler, the maximum length of the wiggler is limited to ~ 1 m or 30 wiggler periods for a confocal resonator configuration. Since the electron linac to be used is of the same type as the one driving the Mark III infrared free-electron laser [5], one may estimate the small signal gain of the 1-m Stanford/TRW FEL from appropriate gain scaling. When the wiggler length is equal to twice the optical Rayleigh length, the small signal gain scaling is given by [6]

$$G_0 - 1 \propto \frac{I_e a_w^2 L_w^2}{\gamma} [J_0(\xi) - J_1(\xi)]^2 \tag{2}$$

where I_e is the electron peak current, L_w is the wiggler length, γ is the electron energy in units of the electron rest mass, and the variable ξ is defined as $\xi \equiv a_w^2/(2(1 + a_w^2))$. For the Mark III FEL, the parameters $\gamma \approx 85$, $a_w = 1.0$, and $L_w = 108$ cm gave a 28% small signal gain in experiments. Therefore, the small signal gain of a 1-m Stanford/TRW wiggler driven by a 25 MeV linac is about 21%, according to the scaling law (2). The gain scaling for the Stanford/TRW FEL could be conservative, because the beam quality requirements for a long wavelength FEL are more forgiving.

It is possible to avoid the diffraction problem by installing a waveguide in the Stanford/TRW wiggler. However, when configured as an IFEL, the Stanford/TRW wiggler has a favorable length of 1 m, as will be shown below.

In our design, the 1-m Stanford/TRW wiggler may function as an IFEL whenever there is a need to generate bunched electron beams for laser-driven particle acceleration experiments. The bunching mechanism inside an IFEL can be best understood from the pendulum equation. The highest possible energy modulation for the trapped electrons is the bucket height in the longitudinal phase space. Electron bunching may not occur efficiently, if the electron energy modulation is smaller than electron intrinsic energy spread. To avoid over bunching, the wiggler length must be less than a quarter the synchrotron wavelength, which can be seen from a typical phase space diagram. The

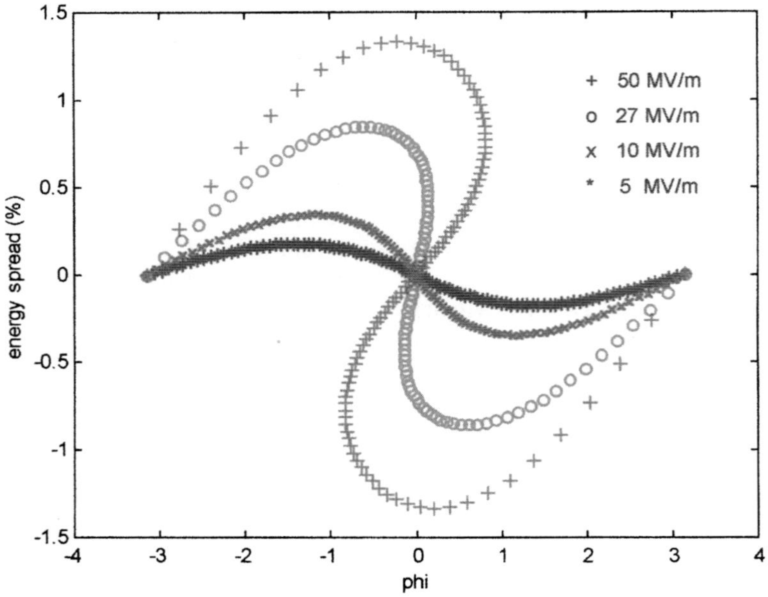

Fig. 3. Electron phase-space distribution for various laser field strengths. In our simulation, 100 monochromatic electrons enter the 1 m Stanford/TRW wiggler at the synchronism energy and exit with energy modulation from the laser fields. The optimal bunching occurs at a 27 MV/m peak laser field.

synchrotron wavelength is inversely proportional to the square root of the laser field strength, whereas the bucket height is proportional to the square root of the laser field strength [7]. Although the electrons only need small laser power to oscillate a quarter synchrotron wavelength in a long wiggler, the induced energy modulation from a long wiggler could be relatively small compared to electron intrinsic energy spread. In fact, it is straightforward to show that

$$\left(\frac{\Delta\gamma}{\gamma}\right)_{\max} = \frac{2\lambda_w}{\lambda_s} \tag{3}$$

where $(\Delta\gamma/\gamma)_{\max}$ is the bucket height, and λ_s is the synchrotron wavelength. Therefore, to obtain a large energy modulation, it is desirable to have a short synchrotron wavelength or equivalently a short wiggler length at the cost of a strong laser field.

Fig. 3 shows the phase space simulations for various laser field strengths in the 1 m IFEL. For each simulation curve, 100 monochromatic electrons are injected at the synchronism energy and propagate to the end of the 1 m Stanford/TRW wiggler. The vertical axis is the percentage energy change from the synchronism energy and the horizontal axis is the phase of the ponderomotive wave. It is seen that the electrons are under-bunched at a 5 MV/m laser field and become over-bunched as the laser field gradually increases beyond 27 MV/m.

At the optimal laser field strength 27 MV/m, we simulate 2000 electrons with 0.5% peak-to-peak energy spread in the IFEL and plot in Fig. 4a the electron distribution versus distance at the wiggler exit. The intrinsic energy spread indeed broadens the electron bunch length. The histogram in Fig. 4b shows that more than 50% electrons are bunched into a 1.5 μm distance at the exit of the 1 m Stanford/TRW wiggler. Future studies will include particle debunching due to emittance.

For confocal laser focusing, the mean laser area is $\sim 7\,\mathrm{mm}^2$ for a 1 m long wiggler. Thus the 27 MV/m laser gradient corresponds to 6.8 MW CO_2 laser power, which is fairly modest.

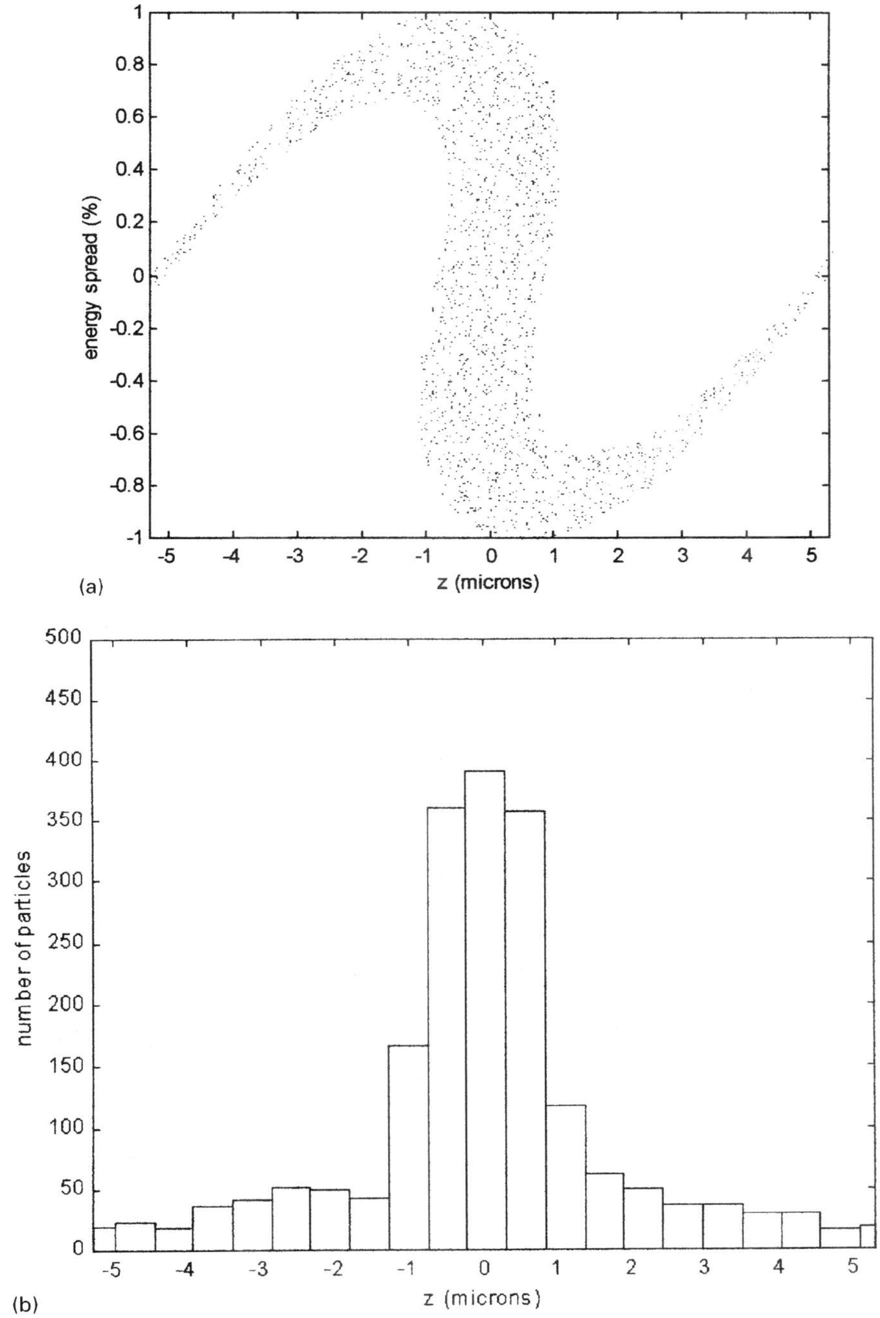

Fig. 4. (a) Electron energy modulation versus distance at the wiggler exit for 2000 electrons with 0.5% intrinsic energy spread. The intrinsic energy spread has the effect of broadening the electron pulse. (b) More than 50% electrons are bunched within a 1.5 μm distance at the wiggler exit.

4. Conclusions

A photon–electron dynamics laboratory at National Tsinghua University, Taiwan, is being proposed. The mission of the proposed laboratory is to conduct advanced research in novel FEL concepts and laser-driven particle accelerators. We study in this paper the two FEL programs to be carried out in the proposed NTHU photon–electron dynamics laboratory.

The Stanford high efficiency FEL employs a super-conducting staggered-array wiggler/linac, wherein the electron beam serves as a catalyst to convert microwave power to optical power. With typical S-band linac beam parameters at a 25 MeV energy, our calculation showed 19% beam-to-laser efficiency at a 10 μm laser wavelength. The 4.8 m Stanford/TRW wiggler, originally driven by 66 MeV electron beams from the Stanford super-conducting accelerator, can be reduced in length for FEL and IFEL operation driven by a 25 MeV S-band linac. With wiggler length = 1 m, a 21% small signal gain for the Stanford/TRW FEL is predicted. When configured as an IFEL, the 1 m Stanford/TRW wiggler generates 1.5 μm electron bunches at a laser field gradient of 27 MV/m. The short electron bunches are suitable for various kinds of laser-driven particle acceleration experiments and for generating harmonics of the bunching wavelengths.

Acknowledgements

Y.C. Huang thank D. Quimby of STI Optronics for helping the electron bunching calculations, and H.A. Schwettman for loaning the TRW wiggler to the proposed laboratory. This work is supported in part by a gift from Simpatico Industries Co., Ltd.

References

[1] J.F. Schmerge, J.W. Lewellen, J. Harris, Y.C. Huang, J. Feinstein, R.H. Pantell, Nucl. Instr. and Meth. A 341 (1994) 335.
[2] G.R. Neil, J.A. Edighoffer, S.W. Fornaca, C.E. Hess, T.I. Smith, H.A. Schwettman, Nucl. Instr. and Meth. A 237 (1985) 199.
[3] Y.C. Huang, T. Plettner, R.L. Byer, R.H. Pantell, R.L. Swent, T.I. Smith, J.E. Spencer, R.H. Siemann, H. Wiedemann, Nucl. Instr. and Meth. A 407 (1998) 316. See also in: Swapan Chattopadhyay (Ed.), Proceedings Advanced Accelerator Concepts Workshop, Lake Tohoe, CA, October 1996, and in: Victor Granatstein (Ed.), Proceedings Advanced Accelerator Concept Workshop, Baltimore, MA, July 1998.
[4] J.F. Schmerge, R.H. Pantell, Y.C. Huang, J. Harris, J. Feinstein, L. Zitelli, Z.Q. Yan, J.W. Lewellen, Nucl. Instr. and Meth. A 331 (1993) 558.
[5] S.V. Benson, M.J. Madey, J. Schultz, M. Marc, W. Wadensweiler, G.A. Westenskow, M. Velghe, Nucl. Instr. and Meth. A 250 (1986) 39.
[6] C. Brau, Free-Electron Lasers, Academic Press, New York, 1990, p. 278.
[7] C. Brau, Free-Electron Lasers, Academic Press, New York, 1990, pp. 114–115.

ELSEVIER

Nuclear Instruments and Methods in Physics Research A 429 (1999) 435–439

NUCLEAR INSTRUMENTS & METHODS IN PHYSICS RESEARCH Section A

www.elsevier.nl/locate/nima

Transverse velocity modulated e-beam propagating through an undulator as a source of coherent spontaneous radiation☆

A.A. Varfolomeev*, T.V. Yarovoi

Coherent Radiation Laboratory, RRC "Kurchatov Institute", Moscow 123182, Russia

Abstract

A new kind of e-beam bunching enabling the production of coherent spontaneous emission (CSE) is proposed and considered. It is shown that an e-beam in passing through an undulator will induce CSE along with incoherent spontaneous radiation on the same wavelength if the e-beam transverse velocity is modulated with a spatial period twice as long as the radiation wavelength. Such angular modulated beams can be regarded as alternatives to conventionally density modulated beams for CSE production. Requirements for e-beams are similar in both cases.

1. Introduction

It is believed that only a density-modulated e-beam can produce coherent radiation on wavelengths corresponding to the spatial phase period of the beam. It is true for all kinds of e.b. radiation devices including conventional FELs based on using initially unmodulated beams. Bunching in this case is created by the forces of the radiated fields, thus a self-bunching process develops. An induced momentum modulation transforms after some distance into an e.b. phase or density modulation. Initial bunching or prebunching of the e-beam provides very strong enhancement of the FEL radiation (see, for example Ref. [1]). If the problem of short wavelength bunching were resolved, very strong coherent radiation sources based on undulators [2] would be possible. Keeping in mind that any new approach to the e.b. bunching process at small wavelengths is desirable we have considered a new type of modulation, namely transverse e.b. velocity modulation, which can be pure transverse or mixed with longitudinal phase modulation. Regarding superradiance as a test for cooperative properties of these new beams we have calculated the spontaneous emission of a transversely modulated beam. We show that radiation with characteristics typical for CSE [2] can really be induced by beams with no density modulation, and we describe the main features of it as well as requirements for e-beam parameters.

2. Superradiance of transversely modulated beam. Qualitative approach

Let us consider an electron beam propagating through an undulator at small angle α to the z-axis

☆Presented at the 20th International FEL Conference, August 16–21, 1988, Williamsburg, VA, USA.

* Corresponding author.

E-mail address: varfol@dnuc.polyn.kiae.su (A.A. Varfolomeev)

0168-9002/99/$ - see front matter
PII: S0168-9002(99)00128-X

of the undulator. The average longitudinal and transverse electron velocity components at the injection point are, respectively, $\beta_{0\parallel} = \beta_0 \cos\alpha$ and $\beta_{0\perp} = \beta_0 \sin\alpha$ in c units. We will assume that the conventional planar undulator provides electron velocity oscillations along the x-direction

$$\boldsymbol{\beta}_j = \boldsymbol{\beta}_{0j} + \boldsymbol{e}_x (a_w\sqrt{2}/\gamma)\cos k_w z \tag{1}$$

where β is velocity normalized to speed of light, $k_w = 2\pi/\lambda_w$, λ_w is the spatial undulator period, $a_w = eH_{0w}/2^{3/2}\pi mc^2$ is the rms undulator field parameter and γ is the electron energy in units of mc^2. In a common case the initial velocities of electrons β_{oj} are not equal but have some statistical distribution and can vary with the time of entrance into the undulator t_0. Some bunching parameters can be introduced for non-uniform beams

$$b_{\parallel}(\omega) = \frac{1}{T}\int_0^T e^{i(\omega/c)t_0}\beta_{0\parallel}(t_0)\,dt_0,$$

$$b_{\perp}(\omega) = \frac{1}{T}\int_0^T e^{i(\omega/c)t_0}\beta_{\perp 0}(t_0)\,dt_0.$$

Here $b_{\parallel}(\omega)$ is the longitudinal bunching parameter and $b_{\perp}(\omega)$ is the transverse bunching parameter. We will be interested mostly in the case when $b_{\perp} \neq 0$ for some frequency.

It is known [3] that an electron travelling along the undulating trajectory (1) radiates spontaneously with maximum intensity at the resonance wavelength λ_s depending on the observation angle. For the z-axis direction this wavelength is given by

$$\lambda_s = \lambda_w(1 + a_w^2 + \gamma^2\alpha^2)/2\gamma^2, \quad \Delta\lambda_s/\lambda_s = 1/N_w. \tag{2}$$

Here $\alpha = \beta_{\perp 0}/\beta_0$ is the noncollinearity angle, $\Delta\lambda_s$ is the spectral width and N_w is the number of undulator periods. Spontaneous radiation of a given frequency has a limited angular distribution with angles

$$\theta < 1/\gamma N^{1/2}. \tag{3}$$

Suppose that the bunching parameter is $b_{\perp}(\omega_b) \neq 0$. This means that $\beta_{\perp 0}$ and hence α has some nonvanishing harmonic component at frequency ω_b. As a result, radiation at zero angle to the undulator axis will have an oscillating wavelength (2) depending on the arrival time of the electron at the undulator entrance and consequently on the phase of the radiation wave. This means that conditions for cooperative electron interaction and coherent spontaneous emission at some frequency will be provided. We would like to underline that in this case coherence is provided because the radiation waves do not interfere constructively if they are not of the same frequency and do not propagate in the same direction. This is a key point of this new mechanism of CSE. So we can imagine that the transverse modulated beam consists of some density modulated angular modes having their own direction of propagation defined by average angles α of the modes. Any collinear ($\alpha = 0$) or noncollinear ($\alpha \neq 0$) mode can be considered as a partial density modulated electron beam consisting of a train of microbunches of respective wavelength. With this approach we can easily make a conclusion about the main features of the radiation emitted by the transversely bunched electron beam.

For sake of simplicity we consider a thin quasi monochromatic e-beam with $\Delta\gamma/\gamma \ll 1/N_w$ and diameter $d_e < (\lambda_s N_w \lambda_w)^{1/2}$, and we analyze spontaneous radiation emitted along the z-axis (at zero angle with respect to average e.b. velocity). For this "collinear geometry" the effect of transverse bunching on CSE should be more evident.

From plain physical analysis it follows that CSE of a transversely bunched beam should have the following characteristics:

1. The total radiation intensity is proportional to $b^2(\omega_b)N_e^2$ where N_e is the total number of electrons in the entire bunch length L_b and $b(\omega_b)$ is the bunching Fourier component.
2. If the undulator is tuned to ω_s then the maximum of CSE intensity in the forward direction is reached at a bunching frequency $\omega_b = \omega_s/2$.
3. The frequency spread of CSE is of the order

$$\Delta\omega_s/\omega_s = 1/(N_W + n_b - 1) \tag{4}$$

where $n_b = L_b\omega_b/\pi\beta_{\parallel}$ is the number of microbunches in the e.b. bunch.

4. The electron beam angular spread should be limited for the CSE to be intense. This limit can

be found from Eqs. (2) and (4)

$$\alpha_e < [(1 + a_W^2)/(N_W + n_b - 1)]^{1/2}/\gamma. \quad (5)$$

Note that for $n_b \gg N_w$ limit (5) is lower than that following from Eq. (3) for short bunches.

5. A strong decrease of the CSE effect is expected also if the angular bunching amplitude $a_{\perp 0}$ exceeds the limit in Eq. (5).

One more comment is needed on the problem of production of the transversely modulated beams. From dynamics relations $dp_\perp = m\gamma dv_\perp$ and $dp_\| = m\gamma^3\, dv$ one can see that the transverse velocity can be changed by the weaker forces than the longitudinal one. In the optical wavelength range, transverse modulation can be provided, for example, by the fields of a moderate intensity laser [4].

3. Radiation of transversely bunched beams. Analytical approach and numerical results

To demonstrate that the transverse modulation of e.b. is really efficient for CSE production we will make more precise calculations. For the above-mentioned routine planar undulator the electron trajectory can be written in the form

$$x_j = A_x \sin \omega_W(t - t_0) + \beta_{x0j}c(t - t_0) + x_{0j},$$

$$y_j = \beta_{y0j}c(t - t_0) + y_{0j},$$

$$z_j = B_z c(t - t_0) - A_z \sin 2\omega_W(t - t_0) - A_x\beta_{0xj}c \sin \omega_W(t - t_0) \quad (6)$$

where $A_x = \sqrt{2}a_W c/\gamma\omega_W$, $A_z = a_W^2 c/4\gamma\omega_w$, $B_z = 1 - (1 + a_w^2)/2\gamma^2 - \beta_{\perp 0}^2/2$, $\beta_{\perp 0}^2 = \beta_{x0}^2 + \beta_{y0}^2$, ω_w is $2\pi\beta_\| c/\lambda_w$ and t_0 is the time of electron j entering the undulator at the point (x_{0j}, y_{0j}). The H_ω component of the radiation field can be calculated using the classical formulae [5,6]

$$\boldsymbol{H}_\omega = (e\omega/R_0c^2)\int dt[\boldsymbol{n} \times \boldsymbol{v}_j(t)] \exp[i\omega(t - \boldsymbol{r}_j(t)\cdot \boldsymbol{n}/c)] \quad (7)$$

where $\boldsymbol{n}(\sin\theta\cos\varphi, \sin\theta\sin\varphi, \cos\theta)$ is the unit vector $\boldsymbol{k}/|\boldsymbol{k}|$, $\boldsymbol{k}$ is a wave vector, $\boldsymbol{v}_j(t)$ is the velocity of the electron defined by Eq. (6) and R_0 is the distance to a radiation detector. For small transverse electron deviations and the small detection angle defined by the inequalities $a_w/\gamma \ll 1$, $\beta_{0\perp} \ll 1$, $\gamma\theta \ll 1$ the field harmonic (7) takes the form

$$\boldsymbol{H}_\omega = \boldsymbol{h}\frac{e\omega}{R_0c^2} 2A_x \exp i\varphi_{oj} \sum_{m,n=-\infty}^{\infty} (J_m(\xi_1) - J_{m+1}(\xi_1))J_n(\xi_2)\cdot\frac{[\exp 2\pi N_W i\eta - 1]}{i\eta} \quad (8)$$

where $\boldsymbol{h} = \boldsymbol{e}_x\cos\theta - \boldsymbol{e}_z \sin\theta\sin\varphi$, $\varphi_{0j} = \omega(t_0 - r_{0j}\sin\theta/c)$, $J_m(\xi)$ is an mth order Bessel function, $\xi_1 = A_z\omega\cos\theta/c\gamma$, $\xi_2 = (A_x\omega/c)(\beta_{x0} - \sin\theta\cos\varphi)$, $\eta = (\omega/2\gamma^2\omega_w)[1 + a_w^2 + \gamma^2\alpha^2 - (2m + n + 1)]$, α is a noncollinearity angle, which for small observation angles θ takes the form $\alpha = [\beta_{\perp 0}^2 - 2\theta(\beta_{0x}\cos\varphi + \beta_{0y}\sin\varphi) + \theta^2]^{1/2}$. The formulae (8) enables definition of the radiation field for any initial conditions relating to electron primary coordinates and velocities at the undulator entrance. The field induced by the beam bunch can be found by the respective summation of the electron fields (8) with taking into account the initial transverse conditions and phase φ_0 for each electron.

With the purpose of getting more detailed information on the transverse bunching effect we will introduce a model. According to this model the transverse velocities of all electrons are modulated on the same frequency so that at the undulator entrance they are described as

$$\boldsymbol{\beta}_{\perp 0j} = \boldsymbol{b}_{\perp 0j} + \boldsymbol{a}_{\perp 0}\sin\omega_b t_0 \quad (9)$$

where $\boldsymbol{b}_{\perp 0j}$ are primary electron velocities (statistically distributed in the common case according to beam emittance) and $\boldsymbol{a}_{\perp 0}$ is the amplitude of regular angular modulation. At small angular divergence of the e-beam providing $|\beta_{\perp 0}| \ll 1/\gamma\sqrt{N_W}$ sensitivity to the transverse bunching is very high. It can be seen from the formulae obtained for the peak intensity of radiation (per unit frequency range and unit solid angle near resonant frequency ω_s) emitted at the angle $\theta = 0$

$$\frac{dI}{d\omega\, d\Omega} = \frac{e^2}{8c}\left(\frac{\omega}{\omega_W}\right)^2 \frac{a_W^2}{\gamma^2}[J_0(\xi_1) - J_1(\xi_1)]N_w^2 N_e 1 \left[1 + \frac{\pi^2 N_W^2 N_e \gamma^4 a_{0\perp}^4}{16\cdot(1 + a_W^2 + \gamma^2\beta_{o\perp}^2)^2}\right]. \quad (10)$$

It is seen that one part proportional to N_e^2 is presented if $a_{\perp 0} \neq 0$.

For more detailed analysis numerical calculations were made. Results are presented in the figures. The following parameter values were used for the calculations:

$N_w = 100$, $N_e = 10^9$, $\lambda_w = 5$ cm, $\gamma = 100$,

$a_w = 1$, $\lambda_s = 5\times 10^{-4}$ cm, $\omega_b = \pi c/\lambda_s$.

The density function of the beam was taken in the form valid for the $\beta_{\|}ct_0$ interval $-\sigma \div \sigma$

$$s(t_0) = N_e[\cos(\pi\beta_{\|}t_0/\sigma)/\sigma + 1]/2\sigma \quad (11)$$

where σ is a length parameter of the bunch. Parameters $a_{\perp 0}$ and σ are shown in the figures. Some calculations (Figs. 1 and 2) were made for a well collimated e-beam with $\boldsymbol{b}_{\perp 0j}\cong 0$ for all electrons of the bunch. Estimation of the e-beam angular spread influence on CSE (Fig. 4) was found by averaging radiation field strength (8) over $\boldsymbol{b}_{\perp 0j}(b_{x0}, b_{y0})$ parameters within the limits shown. This averaging as well as averaging over t_0 was made by the respective integration of the field (8) over these parameters.

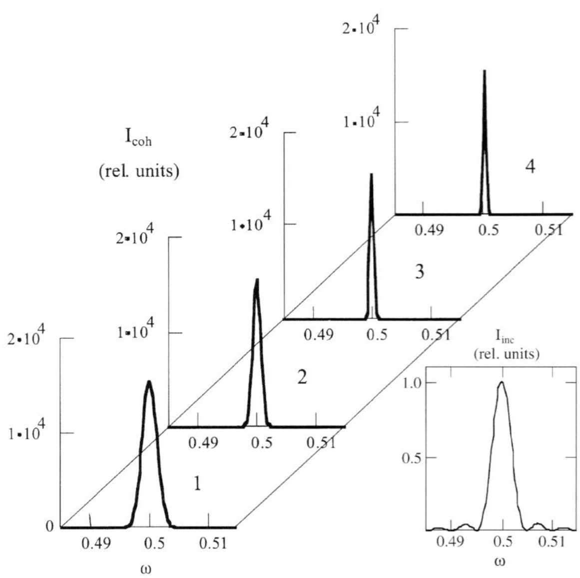

Fig. 1. Frequency spectra of coherent (I_{coh}) and incoherent (I_{inc}) radiation versus bunch length σ. Here as well as in all other figures below the intensity is normalized to the maximum of spontaneous radiation and frequency is expressed in $\lambda_w/2\gamma^2\lambda_s$ units. 1 – $\sigma = 100\lambda_s$; 2 – $\sigma = 200\lambda_s$; 3 – $\sigma = 400\lambda_s$; 4 – $\sigma = 600\lambda_s$.

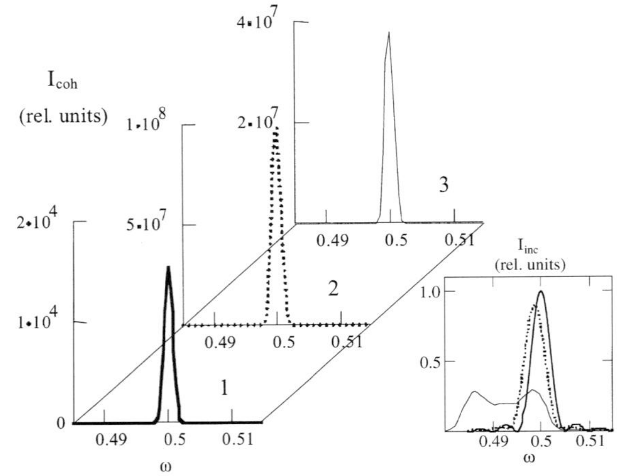

Fig. 2. Radiation frequency spectra versus transverse velocity modulation amplitude for $\sigma = 0.1$ cm 1 – $a_{\perp 0} = 0.01/\gamma$; 2 – $a_{\perp 0} = 0.1/\gamma$; 3 – $a_{\perp 0} = 0.25/\gamma$.

Fig. 1 presents frequency spectra of coherent and incoherent spontaneous radiation emitted at zero angle $\theta = 0$ with different bunch lengths σ and a well collimated e-beam ($b_{\perp 0j} = 0$). The same modulation amplitude $a_{\perp 0} = 0.01/\gamma$ was used. All intensities are expressed in units of maximal value of the incoherent intensity. It is seen that the frequency width decreases with increasing σ in accordance with Eq. (4). Incoherent spontaneous spectra are all of the same shape (does not depend on bunch length). The maximum I_{coh} is also the same for all these cases and equal to 1.54×10^4 in agreement with the above formulae (10).

Fig. 2 shows the dependence on the bunching amplitude $a_{\perp 0}$ for a fixed bunch length $\sigma = 0.1$ cm. It is seen that the peak value of the CSE is decreasing with increasing $a_{\perp 0}$. This can be explained by more intense radiation at larger angles θ when $a_{\perp 0}$ increases. The frequency width of the CSE is still small as Eq. (4) predicts. The frequency spread

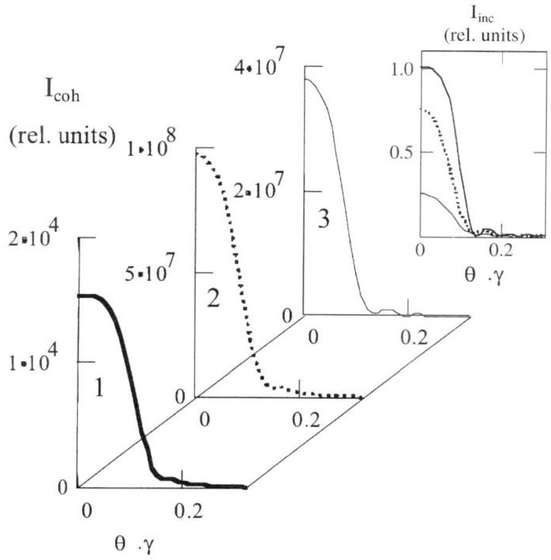

Fig. 3. Angular distribution for coherent and incoherent radiation versus modulation amplitude: 1 – $a_{\perp 0} = 0.01/\gamma$; 2 – $a_{\perp 0} = 0.1/\gamma$; 3 – $a_{\perp 0} = 0.25/\gamma$.

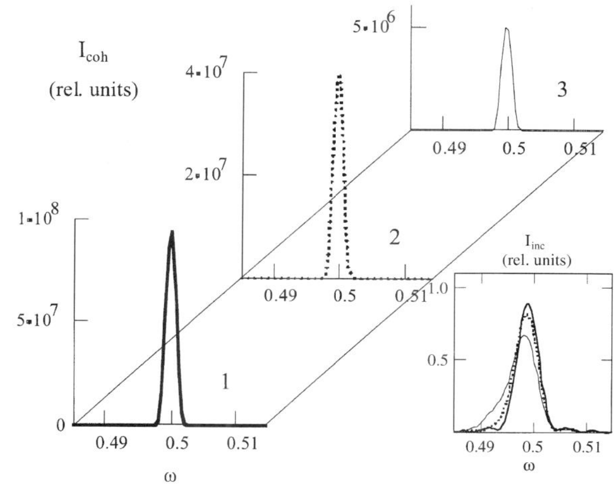

Fig. 4. Frequency spectra versus amplitudes of beam angular spread: 1 – $b_{\perp 0} = 0.01/\gamma$; 2 – $b_{\perp 0} = 0.1/\gamma$; 3 – $b_{\perp 0} = 0.25/\gamma$.

of the incoherent radiation becomes larger due to modulation of the noncollinearity angle in Eq. (2) defining the resonance frequency.

In Fig. 3 the angular distributions of the radiation at the resonance frequency are given. One can see that both coherent and incoherent radiation have equal angular sizes independent of the transverse modulation amplitude in full accordance with Eq. (3).

Fig. 4 presents radiation intensities corresponding to the fields averaged over random distributions of the initial transverse velocities $b_{\perp 0}$. The modulation amplitude is $a_{\perp 0} = 0.01/\gamma$. It is seen that the CSE is not destroyed if the angular beam spread does not exceed the modulation angles ($b_{\perp 0} < a_{\perp 0}$).

4. Discussion and conclusion

The analytical and numerical results presented above clearly show that transversely bunched electron beams induce both routine spontaneous radiation and coherent radiation with high intensity, low frequency spread and small angular divergence typical for CSE. This radiation is adequate to CSE of density modulated beams. It means that transverse velocity modulated beams can be used for efficient production of CSE in undulators or prebunched FEL devices. The respective requirements for the electron beam quality are not more severe than those for the case of using a density modulated beam or for conventional FEL devices. Rather small transverse bunching amplitudes are required ($a_\perp < 1/\gamma\sqrt{N}$). Being in this sense an alternative to the usual density modulated beams transverse modulated beams can be more easily produced in real experiments and can be considered as a good prospect for creation of high intensity superradiant sources.

References

[1] I. Schnitzer, A. Gover, Nucl. Instr. and Meth. A 237 (1985) 124.

[2] D.F. Alferov, Yu.A. Bashmakov, E.G. Bessonov, J. Tech. Phys. 48 (1978) 1592; 1598 (in Russian).

[3] A.A. Varfolomeev, Yu.Yu. Lachin, Nucl. Instr. and Meth. A 296 (1990) 411.

[4] A.A. Varfolomeev, T.V. Yarovoi, Possible schemes for transverse modulation of e-beams and its applications, Proceedings of the 16th ICFA Beam Dynamics Workshop on Nonlinear and Collective Phenomena in Beam Physics, 1–5 September 1998, Arcidosso, Italy, submitted for publication.

[5] L.D. Landau, E.M. Lifshiz, Theory of Field, Nauka, Moscow, 1973 (in Russian).

[6] J.D. Jackson, Classical Electrodynamics, second ed., Wiley, New york, 1975.

ELSEVIER

Nuclear Instruments and Methods in Physics Research A 429 (1999) 440–444

NUCLEAR INSTRUMENTS & METHODS IN PHYSICS RESEARCH Section A

www.elsevier.nl/locate/nima

Optical-wavelength bunching of the low-voltage electron beam in the resonant gas

I.V. Smetanin*, A.N. Oraevsky

P.N. Lebedev Physics Institute, Academy Science of Russia, Moscow, Leninskii Prospect 53, 117924 Moscow, Russia

Abstract

We propose in this paper the use of the dissipative instability of an electron beam in a gas with resonant dielectric susceptibility as a possible mechanism for beam bunching at optical wavelengths. © 1999 Elsevier Science B.V. All rights reserved.

Keywords: Beam bunching; Resotron; Resonant gas

One of the proposed ways to enhance the efficiency of FEL operation is the use of pre bunched beams. In this paper we intend to consider a new concept for short-wavelength bunching of an electron beam which is based on the dissipative instability of a beam in a gas at a resonant absorption frequency of the gas [1–3].

The problem in question is also interesting for gas-loaded FEL physics. The key principle of gas-loaded FELs is that a dielectric medium is introduced into the interaction region (undulator) to alter the synchronism condition. As a result, the eigenwave of the electrodynamic system is slowed down with a coefficient of $\sim \sqrt{\varepsilon}$. In this conventional scheme the non-resonant part of the dielectric susceptibility is used. However, the non-resonant gas dielectric susceptibility is usually close to unity, so, to have a reasonable effect, one must use a gas at rather high pressure, of the order of one atmosphere or higher [4]. This necessity leads to considerable problems in beam transport [5,6]. On the other hand, the resonant part of the dielectric susceptibility can be very high even at rather low pressures ($\leqslant 1$ Torr), which are optimal with respect to beam transport. However, the eigen electromagnetic mode of the system becomes rapidly damped due to absorption, and lasing in the conventional gas-loaded FEL schemes becomes impossible.

An alternative approach can be developed on the basis of the beam dissipative instability which arises at resonant frequencies in a gas. The instability is absent for transparent media, and the growth rate is determined by the imaginary part of the gas dielectric susceptibility. The possibility of optical-wavelength bunching of the electron beam propagating in the gaseous medium was suggested for the first time in Ref. [1]. The new concept of a high-power short-wavelength FEL oscillator

* Corresponding author. Fax: + 7-095-938-2251.

E-mail address: smetanin@neur.lpi.msk.su (I.V. Smetanin)

PII: S0168-9002(99)00129-1

(resotron), which is based on the effect of amplification of electromagnetic wave by the electron beam near the surface of resonantly absorbing material, was developed in Ref. [2,3].

In this paper, the gas-loaded resotron concept [2,3] is discussed. In this concept, the magnetically guided electron beam moves in a waveguide filled with a resonant gas. We show that in such a system a bunching of the electron beam along with an amplification of an electromagnetic wave at optical (from far infrared to visible) frequencies, corresponding to the gas absorption line, are possible. The amplification mechanism is the radiative dissipative instability. It will be shown that rather high gain coefficient can be realized. The oscillation frequency and the bunching period, are determined mainly by the gas medium's resonant frequency and do not depend strongly on the beam electron energy. As a result, oscillation and bunching at optical wavelengths are possible using high-current low-voltage electron beams. The use of low-energy beams is not crucial in the scheme proposed. However, the instability growth rate and the gain scales with the beam energy as $\sim\gamma^{-3/2}$, and thus drops rather rapidly with electron energy.

Let us consider the process of co-propagation of the relativistic electron beam and an electromagnetic wave in a waveguide, which is filled with a gas of dielectric susceptibility $\varepsilon=\varepsilon_1+\mathrm{i}\varepsilon_2$, $\varepsilon_2>0$. We intend to find conditions under which the amplification of the signal electromagnetic wave with a frequency corresponding to the gas absorption line becomes possible. To be precise, we will consider the case of TM_{0n} wave in the cylindrical waveguide of radius a, so the wave fields components are $(E_z,E_r,H_\phi)\exp[\mathrm{i}(hz-\omega t)]$. The relativistic electron beam of initial density n_0, velocity v_0 and energy γ penetrates in the z-direction. To provide the coupling between the beam and the wave, a guiding magnetic field H_0 is applied along the waveguide (z-)axis.

We will consider the small-signal operation regime. To the first order with respect to perturbations caused by the signal wave, we have for high-frequency perturbations of electron velocity, $(\delta v_r,\delta v_\phi,\delta v_z)\exp[\mathrm{i}(hz-\omega t)]$, the following relations:

$$\frac{\delta v_r}{c}=-\mathrm{i}\frac{e}{m\gamma c}\left[(E_r-\beta_0H_\phi)-\mathrm{i}\frac{\omega_c}{hv_0-\omega}\times(E_\phi+\beta_0H_r)\right]\left(1-\frac{\omega_c^2}{(hv_0-\omega)^2}\right)^{-1}$$

$$\frac{\delta v_\phi}{c}=-\mathrm{i}\frac{e}{m\gamma c}\left[(E_\phi+\beta_0H_r)+\mathrm{i}\frac{\omega_c}{hv_0-\omega}\times(E_r-\beta_0H_\phi)\right]\left(1-\frac{\omega_c^2}{(hv_0-\omega)^2}\right)^{-1}$$

$$\frac{\delta v_z}{c}=-\mathrm{i}\frac{eE_z}{m\gamma c}\frac{1}{hv_0-\omega}. \tag{1}$$

Here $\omega_c=eH_0/m\gamma c$ is the cyclotron frequency. To get simple analytical qualitative results, we will restrict our consideration to the limiting case of an infinite guiding magnetic field, emerging under the condition

$$\left|\frac{\omega_c}{hv_0-\omega}\right|\gg 1 \tag{2}$$

under which transverse components of the electron velocity vanish. It will be shown below that the solution is of the order $hv_0-\omega\sim\Omega_b$, where $\Omega_b=4\pi n_0e^2/m\gamma^3c$ is the beam plasma frequency. Then the condition (2) takes the form $H_0\gg 4\pi j(\beta\gamma^2c)^{-1}$, where j is the beam current density. Assuming for example $\gamma=3$, $j=5\ \mathrm{kA/cm^2}$, we have $H_0\gg 700$ G, a rather reasonable value.

In the cold beam limit (hydrodynamic approximation), the current and density perturbations at signal frequency are

$$\delta j_z=-\frac{e\omega n_0}{hv_0-\omega}\delta v_z,\quad \delta n=-\frac{hn_0}{hv_0-\omega}\delta v_z \tag{3}$$

and, after substitution in Maxwell equations, we derive by the conventional procedure the equations for the high-frequency field components

$$\frac{1}{r}\frac{\mathrm{d}}{\mathrm{d}r}\left(r\frac{\mathrm{d}E_z}{\mathrm{d}r}\right)+\kappa^2E_z=0$$

$$E_r=-\mathrm{i}\frac{h}{h^2-\varepsilon\omega^2/c^2}\frac{\mathrm{d}E_z}{\mathrm{d}r},\quad H_\phi=\frac{\varepsilon\omega}{hc}E_r.$$

Here the transverse wave number κ is

$$\kappa^2=\left(\varepsilon\frac{\omega^2}{c^2}-h^2\right)\left(1-\frac{\Omega_b^2}{\varepsilon(hv_0-\omega)^2}\right). \tag{4}$$

Thus, the transverse dependence of the longitudinal component of the wave electric field amplitude is guided by the usual formula, $E_z \sim J_0(\kappa r)$, where $J_0(x)$ is the zero-order Bessel function. Taking into account the boundary conditions on the waveguide walls (which assumed to be perfect), we find the dispersion equation

$$\kappa = \frac{\alpha_n}{a} \tag{5}$$

where α_n is the nth root of the Bessel function, $J_0(\alpha_n) = 0$.

This dispersion equation has two pairs of solutions. It is easy to understand it from the consideration of the limiting case of infinite waveguide radius, $\alpha_n/a \to 0$ [1]

$$\left(h^2 - \varepsilon\frac{\omega^2}{c^2}\right)\left(1 - \frac{\Omega_b^2}{\varepsilon(hv_0 - \omega)^2}\right) = 0.$$

The first pair of solutions, $h^2 = \varepsilon\omega^2/c^2$, corresponds to the free electromagnetic wave in a gas. If the wave frequency coincides with the gas resonant absorption line, these modes become rapidly damped within a few wavelengths, $\mathrm{Im}\, h \sim (\omega/c)\sqrt{(|\varepsilon| - \varepsilon_1)/2} \sim (1/2)(\varepsilon_2/\varepsilon_1)^{1/2}\omega/c)$. The second pair of solutions satisfy the following dispersion equation [1]:

$$(\omega - hv_0)^2 = \Omega_b^2/\varepsilon \tag{6}$$

The solution to this equation is $h = (\omega/v_0) \pm \Omega_b/(v_0\sqrt{\varepsilon})$, which represents two waves, fast and slow with respect to the electron beam, one of which (slow) is growing $\mathrm{Im}\, h \approx (\Omega_b/v_0)\sqrt{(|\varepsilon| - \varepsilon_1)/2}$. It is easy to see that these are the potential waves, corresponding to space charge waves.

In the more general case of finite waveguide size, the first pair of solutions of Eq. (5) is near the points $h \sim \sqrt{\varepsilon\omega^2/c^2 - \alpha_n^2/a^2}$, which corresponds to the rapidly damped eigenwaves of the waveguide in the absence of an electron beam. To provide coupling of these modes with the beam, one should use rather high gas densities to match their phase velocities with the electron velocity (Cherenkov interaction). However, the lasing or banching become possible only when the Cherenkov gain exceeds $\mathrm{Im}\, h$, that seems to be very hard to realize.

The most interesting is the second pair of solutions, which, in analogy with Eq. (6), is expected to be near the point $hv_0 - \omega \sim \Omega_b \ll \omega$. As even in the far infrared region ω exceeds Ω_b by a few orders of magnitude, one can assume $h = \omega/v_0 + \delta h$, $\delta h \ll \omega/v_0\gamma^2$, and we have

$$\delta h = \pm\frac{1}{\sqrt{\varepsilon}}\frac{\Omega_b}{v_0}\left(1 - \frac{\alpha_n^2\beta^2}{k_0^2a^2(\varepsilon\beta^2 - 1)}\right)^{-1/2}. \tag{7}$$

Here $k_0 = \omega/c$ is the vacuum wave number, and $\beta = v_0/c$. These are the fast and the slow (with respect to the beam electrons) waves, the phase velocities of which are close to the electron velocity v, and thus the energy exchange between the waves and the electron beam becomes very effective. As a result, fast wave attenuation and slow wave amplification occurs, with the following increment

(a) in the case of rather low gas pressure $((\varepsilon_1 - 1), \varepsilon_2 \ll \gamma^{-2})$

$$\mathrm{Im}\,\delta h \sim \frac{\Omega_b}{2v_0}\left(1 + \frac{1}{2}\left(\frac{\alpha_n\beta\gamma}{k_0a}\right)^2\right)\mathrm{Im}\,\varepsilon \tag{8}$$

(b) in the opposite case of rather high electron energy $((\varepsilon_1 - 1), \varepsilon_2 \gg \gamma^{-2})$,

$$\mathrm{Im}\,\delta h \sim \frac{\Omega_b}{2v_0}\left(1 + \left(\frac{\alpha_n}{k_0a}\right)^2\frac{1}{|\varepsilon| - 1}\right)\mathrm{Im}\,\varepsilon. \tag{9}$$

Note that for a waveguide of radius a few millimeters, and for an IR signal, the inequality $k_0a \gg \alpha_n\gamma$ easily holds for low-energy relativistic beams.

The bunching period is thus $\sim 2\pi(\mathrm{Re}\, h)^{-1}$, and, assuming the beam frequency is small compared to ω, is of the order of the wavelength $2\pi c/\omega$.

Relations (8) and (9) show the amplification in the scheme discussed is fundamentally determined by the absorption properties of the medium at the frequency of the signal. Really, the instability is absent for the transparent ($\varepsilon_2 \to 0$) medium. The physical origin of this amplification mechanism is in the electron beam dissipative instability [2,3]. The above solutions correspond to the fast and slow space charge waves, the slow one is the wave of negative energy and thus is unstable due to the presence of dissipation (i.e., absorbing medium) in the system.

It is worth stressing that the resotron scheme under discussion differs drastically from conventional Cherenkov devices (see details in Refs. [2,3]). In the conventional schemes, the amplification occurs at the resonance condition, that is, the velocity of electrons is close to the phase velocity of the eigen wave of the electrodynamic system. In contrast, in the scheme under discussion the eigenwaves of the system without the electron beam are rapidly damped. The mode amplified is the combination of the radiated electromagnetic wave and the potential wave caused by the longitudinal density wave, so the resonance with the beam is an intrinsic property of the resotron. Note that the beam-signal synchronism arises automatically. Contrary to the Cherenkov mechanism, the effect under discussion has no threshold with respect to the beam velocity (i.e., beam energy).

The above consideration was carried out in the case of monoenergetic electron beam. To determine the limit for this cold beam approximation the kinetic Vlasov equation should be analyzed. Assuming the perturbation of electron distribution function δf driven by the signal amplified to be small in comparison with its unperturbed value f_0 the solution of Vlasov equation results in the following relations (instead of relations (3)) for amplitudes of beam current and density perturbations at signal frequency

$$\delta j_z = \mathrm{i}\frac{e^2 E_z}{m\gamma^3}\int \mathrm{d}v f_0(v)\frac{\omega}{(hv-\omega)^2} \times [1-(1+\mathrm{i}\delta hz)\exp(-\mathrm{i}\delta hz)]$$

$$\delta n = \mathrm{i}\frac{e^2 E_z}{m\gamma^3}\int \mathrm{d}v f_0(v)\frac{h}{(hv-\omega)^2} \times \left[1-\left(1+\mathrm{i}\frac{\delta hz\omega}{hv}\right)\exp(-\mathrm{i}\delta hz)\right]. \quad (10)$$

At the regime of high gain (or long interaction distances L), we have $\mathrm{Im}(\delta hz) \gg 1$ throughout almost the entire amplification length, and exponents on the right-hand sides of Eqs. (10) are vanishing. The restriction for the electron energy spread is thus independent on the length of the device and found to be of the standard form [7]

$$\frac{\Delta\gamma}{\gamma} < \frac{\mathrm{Im}\,\delta h}{h}. \quad (11)$$

Note, that in the opposite limit of low gain (or short distance L) $\mathrm{Im}\,\delta hL < 1$ the energy spread restriction becomes dependent on L, and this dependence is rather complex.

The linear (small-signal) approximation accepted in the present analysis is valid as long as the perturbations of variables are small enough, $\delta j_z/j_z \ll 1$, $\delta n/n_0 \ll 1$ that results in an estimation of wave field strength limiting amplitude as

$$E_z \sim \frac{m\gamma^3}{e^2\omega}(hv-\omega)^2. \quad (12)$$

When the electron beam propagates in a gas, the electrons suffer collisions that lead to the changes in their energy and beam emittance. The optimum gas pressure is thus restricted by the violation of the condition (11) due to collisions. To estimate this effect, one can use relations [5,6] for the electron energy losses

$$\frac{\mathrm{d}\gamma}{\mathrm{d}z} = -\frac{\gamma}{\lambda_\mathrm{r}} - 4\pi N_0 Z\left(\frac{e^2}{mc^2}\right)^2 \ln\left(\frac{\pi mc^2\gamma^{3/2}}{ZW_\mathrm{r}}\right) \quad (13)$$

and emittance growth

$$\frac{\mathrm{d}Q}{\mathrm{d}z} = \frac{4\pi R^2 c^2}{\lambda_\mathrm{r}}\frac{\hbar c}{e^2}. \quad (14)$$

Here R is the beam radius, N_0 the gas density, Z the effective nuclear charge of gas molecules, $W_\mathrm{r} \sim 13$ eV, and λ_r the electron Bremsstrahlung radiation length,

$$\lambda_\mathrm{r} = 4N_0 Z(Z+1)\frac{e^2}{\hbar c}\left(\frac{e^2}{mc^2}\right)^2 \ln(183Z^{-1/3}). \quad (15)$$

To make estimates, one should specify the gas medium. Let us consider the molecular hydrogen which is one of the best investigated systems. In the far infrared region, hydrogen has a number of well resolved rotation transitions. The most intense rotation transition is determined (in the pure rotation spectra) by the equation

$$(2J+1)(J+1) = 3T/(hcB) \quad (16)$$

where J is the rotation quantum number, T the gas temperature, and B the molecule rotation constant. At room temperature, the most intense transitions in hydrogen are at 587 cm^{-1} (ortho-hydrogen) and

814 cm^{-1} (para-hydrogen). The line center is situated at $2B(2J+3)$. The maximum value of imaginary part of gas dielectric susceptibility is given by the relation

$$(\varepsilon_2)_{max} \approx \frac{4\pi\mu^2 N_0}{3\hbar\gamma_g}(J+1)\frac{\hbar\omega}{T}\frac{hcB}{T} \times \exp(-hcBJ(J+1)/T). \quad (17)$$

Here γ_g is the gas line width, $\varepsilon - 1 \sim (\omega^2 + \gamma_g^2)^{-1}$. At gas pressures $\leqslant 10^{-2}$–10^{-3} Torr and up to few atmospheres, the line width is determined by intermolecular collisions and can be estimated as $\gamma_g \sim 2.7 \times 10^7$ Hz/Torr. As a result, in this pressure region the resonant value of dielectric susceptibility becomes independent of gas pressure. Making the estimate for hydrogen, one can find $\varepsilon_{2max} \sim 0.1$ at the frequencies mentioned above and at room temperature. Thus, it is possible to operate at the optimum beam transport pressures of a few Torr [4,5]. At these pressures characteristic lengths, determined by relations (13) and (14) are a few meters and the scattering processes do not violate the beam quality condition (11). The estimate of the gain at $\gamma \sim 5$ current density $j_b \sim 5$ kA/cm^2, waveguide radius $a = 3$ mm, and the above hydrogen parameters is thus $\mathrm{Im}\,\delta h \sim 0.05$ cm^{-1}.

This research is supported by the Russian Basic Research Foundation.

References

[1] A.N. Oraevsky, Kvantovaya Elektronika 7 (1980) 495 (in Russian).

[2] A.N. Oraevsky, I.V. Smetanin, JETP Lett. 62 (1995) 258.

[3] A.N. Oraevsky, I.V. Smetanin, Laser Phys. 7 (1997) 155.

[4] M.B. Reid, J. Feinstein, R.H. Pantell, A.S. Fisher, Nucl. Instr. and Meth. A 272 (1988) 268.

[5] E.P. Lee, R.K. Cooper, Part. Accel. 7 (1976) 83.

[6] W. Sharp, M. Lampe, Phys. Fluids 23 (1980) 2383.

[7] M.V. Kuzelev, A.A. Rukhadze, Electrodynamics of Dense Electron Beams in a Plasma, (in Russian), Nauka, Moscow, 1990.

ELSEVIER Nuclear Instruments and Methods in Physics Research A 429 (1999) 445–450

NUCLEAR INSTRUMENTS & METHODS IN PHYSICS RESEARCH Section A

www.elsevier.nl/locate/nima

Free-electron lasers driven by supercurrent

I.V. Smetanin*

P.N. Lebedev Physics Institute, Academy Science of Russia, Moscow, Leninskii Prospect 53, 117924 Moscow, Russia

Abstract

We propose new concepts of free-electron lasers which are based on using the DC current in superconductors as a driver. It is found that these FEL-type devices are very promising due to the high critical current density and quantum coherence of the condensate in superconductors. Cherenkov and resotron (based on the dissipative instability) schemes driven by supercurrent are considered, the small-signal exponential gain is calculated.

The key problem in conventional short-wavelength FEL design is to enhance current density of an electron beam at sufficiently low-energy spread and emittance. The limiting beam current density is of about 10–20 kA/cm^2 due to the beam quality requirements. Another type of electron current – supercurrent – can provide much higher electron densities: typical critical current densities in thin superconductor films exceed ~ 100 MA/cm^2. The problem of electron beam emittance and energy spread, crucial for FEL operation, is absent in the case of supercurrent due to the quantum coherence of Cooper pairs in the condensate. These features makes the FEL-type device driven by supercurrent very promising.

In this paper we propose the supercurrent as a driver for short-wavelength FEL-type devices. It is considered as two schemes, namely, the Cherenkov FEL and the dissipative instability-based FEL (which was named resotron in Refs. [1,2]). In the Cherenkov scheme, we assume a thin superconductor film, carrying DC current, is situated within a slow-wave waveguide structure (see Fig. 1a). It is found that the Cherenkov resonance arises when the phase velocity of electromagnetic wave is equal to the characteristic velocity of the diffusion of phase of the order parameter. In this case, the strong coupling between the supercurrent and electromagnetic wave exists, and an amplification of this wave becomes possible. In the resotron scheme, thin superconductor film is placed in a waveguide which is filled by absorbing dielectric (see Fig. 1b), which is characterized by the complex dielectric susceptibility $\varepsilon = \varepsilon' + i\varepsilon''$. In this case dissipative instability of the supercurrent is developed and amplification of an electromagnetic signal at frequencies corresponding to the dielectric absorption line becomes possible.

Let us first consider the Cherenkov scheme. To get straightforward and descriptive analytical results, we restrict our consideration by the simplest case. We assume that the superconductor film, carrying DC current, is thin, i.e., film thickness is less than both the coherence length (i.e., the size of Cooper pair) and the field penetration depth. In this case, the dependence of the order parameter on

* Fax: + 7-095-938-2251.
E-mail address: smetanin@neur.lpi.msk.su (I.V. Smetanin)

0168-9002/99/$ - see front matter
PII: S0168-9002(99)00131-X

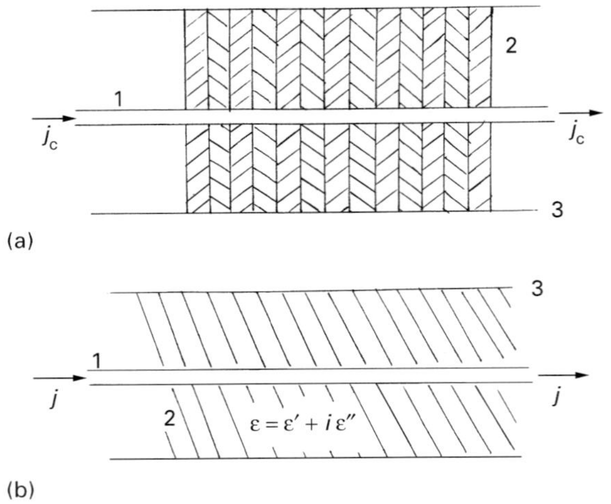

Fig. 1. The geometry of FELs driven by supercurrent. (a) Cherenkov oscillator: 1 – thin superconductor film, carrying DC current j, 2 – dielectric superlattice, which is formed by alternate layers of two materials of different dielectric susceptibilities, d is the superlattice period, 3 – external waveguide. (b) The dissipative instability ("resotron") scheme: 1 – thin superconductor film, carrying DC current j, 2 – absorbing dielectric medium of resonant susceptibility $\varepsilon = \varepsilon' + i\varepsilon''$, 3 – external waveguide (metal walls).

transverse coordinates is insignificant, and the problem becomes one dimensional. The waveguide is filled by the medium of periodically changing refractive index. It may be the dielectric superlattice, for example, which is formed by alternate layers of two transparent (at signal frequency) materials of different dielectric susceptibilities ε_1 and ε_2. The electromagnetic field in such a superlattice, due to Flouqet theorem, can be represented as a series of space harmonics [3]

$$E = \sum_{n=-\infty}^{+\infty} E_n(x,y,t)\exp(-i\omega t)\exp\left[i\left(kz + \frac{2\pi}{d}n\right)\right],$$

$$n = 0, \pm 1, \pm 2, \ldots \quad (1)$$

Here d is the period of dielectric superlattice and k is the so-called Bloch wave number, which is, in a definite sense [3], a mean value of signal wave numbers in corresponding layers, $k_1 = (\omega/c)\varepsilon_1$, $k_2 = (\omega/c)\varepsilon_2$, $k_1 < k < k_2$. When the superlattice period is small with respect to the signal wavelength $d \ll \lambda = 2\pi c/\omega$ (the modern technology can provide superlattices with the period less than 10 nm at a sufficient quality), phase velocities of even the lowest modes in the solution (1) are much less than the speed of light and thus may become close to the characteristic propagation velocities of condensate perturbations. Thus the conditions of an effective energy exchange between the supercurrent and the signal wave arises. It is the situation of our particular interest.

To describe the dynamics of the superconductor order parameter, we will use the generalized nonstationary Ginzburg–Landau equation. There exists a number of such generalizations in literature [4–10]. For definiteness, we will use Gor'kov–Eliashberg equation for the parameter Δ (it has been derived by using microscopic approach for the case of gapless superconductor, see Ref. [4]

$$\frac{\partial \Delta}{\partial t} + \frac{\tau_s}{3}\left[\pi^2(T^2 - T_c^2) + \frac{1}{2}|\Delta|^2\right]\Delta + D\left(-i\nabla - \frac{2e}{c}A\right)^2 \Delta = 0. \quad (2)$$

The current density is defined by the usual way

$$j = \frac{N\tau_1\tau_s}{2m}\left[ie(\Delta\nabla\Delta^* - \Delta^*\nabla\Delta) - \frac{4e^2}{c}A\right]. \quad (3)$$

Here $D = \frac{1}{3}\tau_1 v_F^2$ is the phase diffusion coefficient, v_F is the Fermi velocity, T and T_c are the current and critical temperatures, respectively (measured in s^{-1}), N is the density of superconductive condensate, and $\tau_{1,s}$ are the characteristic spin revolution times [4]. It is assumed the Lorentz gauge for the electromagnetic field vector potential div $A = 0$.

When the high-frequency field is absent, we represent the unperturbed parameter Δ in the form $\Delta = S_0\exp(ik_0 z)$, where the value k_0 is determined by the DC current density. In this gauge, the vector potential A_0 of the self-induced magnetic field of thin film DC current is small even at current densities close to the critical one, $(2e/\hbar c)A_0 \ll k_0$, and can be omitted. By substitution of the unperturbed Δ in Eq. (2), we find

$$S_0^2 = -\frac{6Dk_0^2}{\tau_s} + 2\pi^2(T_c^2 - T^2). \quad (4)$$

When the current density is close to the critical one,

$$(S_0^2)_{\max} = \frac{12}{\tau_s}(Dk_0^2)_{\max} = \frac{4}{3}\pi^2(T_c^2 - T^2). \tag{5}$$

We are interested in the small-signal quasi-stationary amplification regime, when the signal pulse length $c\tau_p$ is much greater than the length of the interaction region L, $c\tau_p \gg L$. The coupling between the electromagnetic wave and supercurrent exists only in the case of unvanishing longitudinal component of the high-frequency vector potential, that is provided by the presence of waveguide. As in the small-signal approximation the problem becomes linear with respect to the electromagnetic field, one can consider the only spatial harmonic

$$A_z = a(z)\exp(\mathrm{i}[hz - \omega t]) + \text{c.c.} \tag{6}$$

where $a(z)$ is the slow varying function, $\mathrm{d}a/\mathrm{d}z \ll ha$, and, according to Eq. (1), the wave number is $h = k + 2\pi n/d$.

It is convenient to seek the solution of Eq. (2) in the form

$$\Delta = S_0 \exp(\mathrm{i}[k_0 z + f(z,t)]) \tag{7}$$

$$\mathrm{Re} f(z,t) = g_1(z)\exp(\mathrm{i}[hz - \omega t]) + \text{c.c.}$$

$$\mathrm{Im} f(z,t) = g_2(z)\exp(\mathrm{i}[hz - \omega t]) + \text{c.c.} \tag{8}$$

where $f(z,t)$ is the small perturbation of phase of the order parameter caused by the high-frequency signal wave (6), and $g_{1,2}$ are slow varying functions. The perturbation is small when

$$|f| \ll 1, \quad \left|\frac{\partial f}{\partial z}\right| \ll k_0, \quad \frac{2e}{\hbar c}a(z) \ll k_0. \tag{9}$$

After linearization, in the first order of perturbation theory, Eq. (2) is reduced to the system of coupled equations

$$(\mathrm{i}\Omega - 1)g_1 + 2\mathrm{i}\frac{\partial g_1}{\partial x} + \frac{\partial^2 g_1}{\partial x^2} = 2\xi\frac{\partial g_2}{\partial x} + 2\mathrm{i}\xi g_2$$

$$(-\mathrm{i}\Omega + 1 + \partial)g_2 - 2i\frac{\partial g_2}{\partial x} - \frac{\partial^2 g_2}{\partial x^2}$$

$$= -U + 2\xi\frac{\partial g_1}{\partial x} + 2\mathrm{i}\xi g_1. \tag{10}$$

Here the following dimensionless parameters, $\Omega = \omega/(Dh^2)$, $\xi = k_0/h$, $\delta = \tau_s S_0^2/(3Dh^2)$, $U = 4ek_0 a/(ch^2)$, and new dimensionless variable $x = hz$ are introduced.

It is easy to calculate the eigenvalues of the system (10) and find the characteristic damping coefficient, Γ, $\Gamma = \frac{1}{\sqrt{2}}([\Omega\delta/2 + (\sqrt{\Omega\delta/2} \mp \Omega)^2]^{1/2} \pm \sqrt{\Omega\delta/2})^{1/2}$. As a result, within the small distance l_d

$$l_d \sim (3/2\omega\tau_1)^{-1} v_F \tau_1, \quad \Omega \gg \delta$$

$$l_d \sim (\omega\tau_1)^{-1/4}(v_F\tau_1/k_0)^{1/2}, \quad \Omega \ll \delta \tag{11}$$

functions g_1,g_2 reach their constant values

$$g_1 = \frac{2\mathrm{i}\xi U}{(1 + \delta - \mathrm{i}\Omega)(1 - \mathrm{i}\Omega) - 4\xi^2}$$

$$g_2 = \frac{U(1 - \mathrm{i}\Omega)}{(1 + \delta - \mathrm{i}\Omega)(1 - \mathrm{i}\Omega) - 4\xi^2}. \tag{12}$$

Evolution of the electromagnetic wave amplitude is guided by the wave equation

$$2\mathrm{i}h\frac{\partial a}{\partial z}\exp(\mathrm{i}[hz - \omega t]) = -\frac{4\pi}{c}\mathrm{Im}\, j_z \tag{13}$$

where the longitudinal high-frequency component of supercurrent is determined by Eq. (3).

Finally, we find the exponential gain coefficient

$$\alpha = \frac{1}{a}\frac{\partial a}{\partial z} = 16\pi N \tau_s \tau_1 r_0 S_0^2 \frac{k_0^2}{h^3} \times \sigma \frac{(1 + \delta - 4\xi^2 - \Omega^2)\Omega}{(1 + \delta - 4\xi^2 - \Omega^2)^2 + (1 + \delta)^2\Omega^2}. \tag{14}$$

Here r_0 is the classical radius of an electron, and σ is the geometrical factor corresponding to the space overlap between the optical mode and the supercurrent, which is of the order of the ratio of film thickness to the transverse size of the waveguide.

The amplification ($\alpha > 0$) of signal wave emerges under the condition

$$v_{\mathrm{ph}} = \frac{\omega}{h} < v_0 = (\mathrm{D}h)[1 + \delta - 4\xi^2]^{1/2}. \tag{15}$$

It is easy to see that this inequality has the typical form of Cherenkov condition: the phase velocity of the signal amplified must be less than the definite characteristic velocity v_0. Note, that when the current in the film is close to the critical one, we have

$\delta \approx 4\xi^2$, and velocity v_0 coincides with the characteristic diffusion velocity, Dh, of phase of the order parameter which perturbations have a scale length $\sim h^{-1}$. Amplification of the electromagnetic wave leads to the decrease in the DC current.

Let us make some estimates at typical values of gapless superconductor parameters, $\tau_1 \sim \tau_s \sim 10^{-12}$ s^{-1}, $T_c \sim 10^{-3}$ eV (10 K), $v_F \sim 10^8$ cm/s, $N \sim 10^{23}$ cm^{-3}. When the current is close to its critical value, one can find, according to relation (5), $k_0 \approx (\pi/\sqrt{3})T_c/v_F \sim 3\times 10^4$ cm^{-1}. When the signal frequency is close to the critical temperature, $\omega \sim T_c$ ($\sim 1.5\times 10^{12}$ s^{-1}). Assuming $h \sim k_0$ (i.e., the superlattice period is $d \sim 2$ μm) and that the real value of geometrical factor is $\sigma \sim 10^{-3}$–10^{-4}, we have the signal phase velocity is $\omega/h \approx 5\times 10^7$ cm/s and exponential gain coefficient is rather high $\alpha \approx 10^7\ \sigma$ cm^{-1}.

Let us consider now the resotron scheme (see Fig. 1b) in which the mechanism of signal amplification is the dissipative instability of supercurrent. We assume the thin superconductor film, carrying DC current, is situated within a waveguide which is filled by absorbing dielectric. In this case, the situation is more complex with respect to the above consideration as the transverse structure of signal amplified must be accounted for. As the dielectric is absorbing at the signal frequency ($\mathrm{Im}\,\varepsilon > 0$), one should expect solutions of the surface wave type. This fact implies that our previous assumption of transverse-independent order parameter of the condensate falls. One should solve in this situation the Gor'kov–Eliashberg Eq. (2) for the order parameter joint with the boundary condition

$$n\left(-\mathrm{i}\nabla\Delta - \frac{2e}{c}A\Delta\right) = 0 \tag{16}$$

where n is the vector, normal to the superconductor boundary. That is, the transverse current density must be equal to zero on the superconductor–dielectric interface. In the case of our interest, when the superconductor film is thin (i.e., its thickness is less than both the coherence length and the field penetration depth), one can assume that the condition (16) is fulfilled not only on the interface but within all the superconductor volume. With this assumption, only the derivatives with respect to longitudinal coordinate z survive in Eq. (2), and the dependence of the order parameter on transverse coordinate becomes parametrical. Thus we reduce the problem of the order parameter to the above solution (Eqs. (7), (8) and (13)). However, one should keep in mind that the parameters S_0, k_0, $g_{1,2}$ are now transverse-dependent functions, and this dependence is determined by Eq. (16). As we are interested in the small-signal amplification regime, one can expect that this dependence will be rather weak to be omitted in the lowest order of perturbation theory.

As in the Cherenkov scheme, we are interested in the amplification of quasi-continuous signal wave (6) of TM type. Solving by conventional procedure Maxwell equations, one can easily find that the transverse characteristic wave numbers are $\kappa_d^2 = \varepsilon\omega^2/c^2 - h^2$ in the dielectric region, $E_z(x) \sim \exp(\pm \mathrm{i}\kappa_d x)$, and $\kappa_s^2 = (\omega^2/c^2 - h^2)(1 - \mathrm{i}4\pi\alpha/\omega)$ within the superconductor film, $E_z(x) \sim \exp(\pm\kappa_s x)$. Here α is the high-frequency complex conductivity of the superconductor film determined by Eqs. (3), (8) and (12), $j_z = \alpha E_z$, and x is the transverse coordinate. Finally, using the electromagnetic field boundary conditions at the dielectric-superconductor interface ($x = \pm d$) and at the metal walls of the waveguide ($x = \pm a$), we derive the following dispersion equation:

$$\left(\left[\frac{\kappa_s\kappa_d - \varepsilon(h^2 - (\omega^2/c^2))\tan(\kappa_d(a-d))}{\kappa_s\kappa_d + \varepsilon(h^2 - (\omega^2/c^2))\tan(\kappa_d(a-d))}\right]\exp(2\kappa_s d)\right)^2 = 1. \tag{17}$$

This dispersion equation contains two different types of solutions, which correspond to different choice of the sign after extracting of square root. To understand the nature of these solutions, it is convenient to investigate dispersion equation (17) in the limiting case of infinitely thick dielectric ($a \to \infty$), i.e., when we have the pure surface wave solution. Assuming also that the superconducting film is thin, $\kappa_s d \ll 1$, we have two roots of Eq. (17), the first one is

$$\frac{\varepsilon}{\kappa_d}\left(h^2 - \frac{\omega^2}{c^2}\right)d = \mathrm{i}. \tag{18}$$

This equation corresponds to the eigen mode of the electrodynamic system, which is rapidly damping with longitudinal coordinate z due to the resonant absorption in the dielectric. One can easily see it in the limit $d = 0$, in which the only solution is $h^2 = \varepsilon\omega^2/c^2$. The second root of Eq. (17) leads to the following dispersion equation:

$$\mathrm{i}\frac{\kappa_{\mathrm{d}}d}{\varepsilon}\left(1 - \frac{4\pi\alpha}{\omega}\right) = 1 \tag{19}$$

which corresponds to the dissipative mode and is of our particular interest. This mode exist only in the presence of supercurrent and vanishes in the limit $d \to 0$. Let us consider now this equation in more details. It is evident that the solution scalings are quite different depending on the relations between parameters of the problem at hand, which determines the high-frequency current complex response (12), Ω, ξ, δ.

The case of very slow wave is realized when $h \gg S_0/v_{\mathrm{F}} \sim \sqrt{2}\pi T_{\mathrm{c}}/v_{\mathrm{F}} \sim 4\times10^5\ \mathrm{cm}^{-1}$ and $h \gg \sqrt{\omega/D} = \sqrt{3\omega/\tau_{\mathrm{s}}}v_{\mathrm{F}}^{-1} \sim 5\times10^4\ \mathrm{cm}^{-1}$ (we assume here the above typical superconductor parameters, $T_{\mathrm{c}} \sim \omega \sim 10^{13}\ \mathrm{s}^{-1}$, Fermi velocity $v_{\mathrm{F}} \sim 10^8$ cm/s, and $\tau_{\mathrm{s}} \sim 10^{-12}$ s). This situation becomes possible, for example, when the current density is close to the critical one. In this limit, Eq. (19) is reduced to the form

$$h^3 = \frac{\mathrm{i}}{\varepsilon}\frac{\Lambda d}{D^2\omega} \tag{20}$$

where the new parameter Λ is $\Lambda = 32\pi N\tau_1\tau_{\mathrm{s}}(e^2/m)S_0^2Dk_0^2 = 32\pi ej_0Dk_0$, and j_0 is the DC current density.

One can easily see from this equation that the instability ($\mathrm{Re}\,h > 0$, $\mathrm{Im}\,h < 0$) arises under the only condition

$$\mathrm{Re}\ \varepsilon < 0 \tag{21}$$

i.e., can be realized only at frequencies of the material resonant absorption line. Instability is absent for the transparent medium. The instability growth rate ($\mathrm{Im}\,h$) is of the same order of magnitude as the characteristic wave number $h_0 \sim (\Lambda d/D^2\omega|\varepsilon|)^{1/3}$. Assuming for estimates the critical current density is $j_0 \sim 100\ \mathrm{MA/cm^2}$, $|\varepsilon| \sim 20$, and film thickness $d \sim 10$ nm, we have $h_0 \sim 1.5\times10^6\ \mathrm{cm}^{-1}$.

In an intermediate case, $h \ll S_0/v_{\mathrm{F}} \sim \sqrt{2}\pi T_{\mathrm{c}}/v_{\mathrm{F}}$, $h \gg \sqrt{\omega/D} = \sqrt{3\omega/\tau_{\mathrm{s}}}v_{\mathrm{F}}^{-1}$, the dispersion equation leads to the following solution:

$$h = \frac{\mathrm{i}}{\varepsilon}\frac{3\Lambda d}{\tau_{\mathrm{s}}S_0^2D\omega} \tag{22}$$

and, finally, when the parameters are that holds $h \ll \sqrt{\omega/D} = \sqrt{3\omega/\tau_{\mathrm{s}}}v_{\mathrm{F}}^{-1}$, we have

$$h = -\varepsilon\frac{S_0^2\tau_{\mathrm{s}}\omega^2}{3\Lambda d}. \tag{23}$$

In both these cases, the instability condition ($\mathrm{Re}\,h > 0$, $\mathrm{Im}\,h < 0$) holds only within the resonant material absorption line, at $\mathrm{Re}\,\varepsilon < 0$.

It is interesting to note also, that in these three cases the solution scales with DC current, signal frequency, and superconductor film thickness by quite different ways, $\sim (j_0d/\omega)^{1/3}$, $\sim j_0d/\omega$, and $\sim \omega^2/j_0d$, respectively.

In conclusion, we considered in this paper the new FEL concepts, in which the supercurrent is used as a driver instead of relativistic electron beam. Both the schemes analyzed, based on Cherenkov and dissipative instabilities, are found to be very promising due to the high critical current densities and quantum coherence in superconductors. Estimates show that the devices proposed can become the very compact (of centimeter size) high-gain sources of coherent radiation in a rather broad spectral region.

Acknowledgements

This work is supported in part by Russian Basic Research Foundation, and is initiated by Profs. D.F. Zaretsky and E.A. Nersesov.

References

[1] A.N. Oraevsky, I.V. Smetanin, JETP Lett. 62 (1995) 258.
[2] A.N. Oraevsky, I.V. Smetanin, Laser Phys. 7 (1997) 155.
[3] F.G. Bass, A.A. Bulgakov, A.P. Tetervov, High-Frequency Properties of Semiconductors With Superlattices, Nauka, Moscow, 1989, in Russian.

[4] L.P. Gor'kov, G.M. Eliashberg, Zh. Eksp. Teor. Fiz. 54 (1968) 612. [Sov. Phys. JETP 27 (1968) 328].
[5] A.M. Gulian, G.F. Zharkov, Superconductors in External Fields, Nauka, Moscow, 1989, in Russian.
[6] B.T. Heilikman, V.Z. Kresin, Kinetic and Non-Stationary Effects in Superconductors, Nauka, Moscow, 1972, in Russian.
[7] A.A. Abrikosov, Essentials of the Theory of Metals. Nauka, Moscow, 1987, in Russian.
[8] V.F. Elesin, Yu.V. Kopaev, Usp. Fiz. Nauk 133 (1981) 259. [Sov. Phys. Usp. 24 (1981) 116].
[9] A.N. Oraevsky, Opt. Acoust. Rev. 1 (1990) 63.
[10] A.N. Oraevsky, JETP 76 (1993) 480.

ELSEVIER

Nuclear Instruments and Methods in Physics Research A 429 (1999) 451–456

NUCLEAR INSTRUMENTS & METHODS IN PHYSICS RESEARCH Section A

www.elsevier.nl/locate/nima

Electron bunch length, optical pulse duration and bandwidth considerations for the far-infrared CLIO-2 FEL project

F. Glotin*, C. Rippon, R. Prazeres, J.-M. Ortega

LURE, Bât. 209D, Université Paris-Sud, B.P. 34, 91898 Orsay Cedex, France

Abstract

The CLIO-2 or ETOILES project is the envisioned extension of the CLIO facility, aimed at covering the far-infrared range between 50 μm and 1 or 2 mm. It should consist of a new independent linac-based FEL, operating in S- or L-band at energies between 4 and 16 MeV. Synchronized with the actual CLIO FEL and Optical Parametric Oscillators, it would allow covering the whole infrared spectral range and performing multi-color experiments.

Keywords: CLIO project; Electron bunch length; Optical pulse duration; Bandwidth consideration

1. Introduction

The CLIO infrared free-electron laser (FEL) has operated as a user facility since 1993 [1]. About 2400 h of laser beam time are now produced annually, of which 800 are dedicated to FEL physics and optimization and 1600 for laser users. The CLIO spectral range spans from 3 to 50 μm and its peak power is several MW in (sub-)picosecond pulses. With such features and the corresponding Fourier limitation on the minimum laser linewidth $\Delta\lambda/\lambda \approx 0.2\%$, one understands that this laser is dedicated to fast kinetics and nonlinear optics-based experiments rather than high-resolution spectroscopy. This relatively large bandwidth and the corresponding short pulses originate mainly from the short electron bunches (10 ps FWHM) delivered by the RF-linacs.

As the infrared range below $\lambda = 8$ μm is now well covered by optical parametric oscillators (OPO) also available at the CLIO facility, we have upgraded the spectral range of CLIO from 2–17 μm to its actual range, and we plan to develop further in the far-infrared. However, the actual machine is mainly limited to shorter wavelengths by the size of the vacuum chamber inside the undulator, which clips the optical mode. The size of the chamber cannot be increased without limiting the undulator gap range and then the spectral range. The use of a waveguide and cylindrical mirrors for the far-infrared will not be appropriate at shorter wavelengths, because of the consequent bad matching of the TEM mode with the electron beam, and because frequent changes of the mirrors would not be convenient for facility operation. Adding

* Corresponding author. Tel.: + 33-1-6446-8017; fax: + 33-1-64-46-41-48.

E-mail address: glotin@lure.u-psud.fr (F. Glotin)

PII: S0168-9002(99)00136-9

Table 1
Main parameters of the CLIO-2 laser beam

Spectral range	50 μm to $\geqslant$ 1 mm
Min. spectral width	0.2–1%
Min. pulse length	2–50 ps
Peak power	1–10 MW
Average power	1–10 W
Micro-pulses repetition rate	31.25, 62.5 and 125 MHz
Macro-pulses repetition rate	6.25, 25 and 50 Hz

a second beamline on the same accelerator would also be difficult due to the lack of space. This is why we envision building another independent machine, CLIO-2, to achieve our spectral extension. The new machine should be installed near the other one and synchronized with it, allowing a smooth joint machine management and multi-color experiments. The main optical parameters of CLIO-2 are summarized in Table 1.

2. Accelerator

The electron energy is determined by the desired spectral range, considering a maximum value $K = 2.25$ of the peak deflection parameter for a permanent magnet undulator.

The synchronization of CLIO-2 with CLIO leads to a requirement on the accelerator frequency. Either both machines operate in S-band at the same frequency (2998 MHz), or CLIO-2 operates in the L-band at $f = 1311.5$ MHz, in order to be able to match the lowest micropulse repetition rate of CLIO at $f/42 = 31.25$ MHz.

The L-band solution allows one to work with long pulses (30 ps) and consequently a higher bunch charge (2–3 nC) while keeping a modest energy spread ($< 1\%$). However, RF components are more expensive and less common than in S-band, and our frequency is slightly different than the standard one. This is why we prefer to envision for now an S-band solution and consider as a reference the performance achievable with the actual CLIO injector [2–4], composed of a 90 kV thermionic gun, a 500 MHz prebuncher cavity and a 4 MeV/3 GHz standing-wave buncher. A thermionic gun is

Table 2
Main parameters for an S-band linac

Frequency	2998 MHz
Total electron energy	4–16 MeV
Micro-pulses repetition rate	31.25, 62.5 and 125 MHz
Macro-pulses repetition rate	6.25, 25 and 50 Hz
Bunch charge	1.5 nC
Macropulse length	15 μs
Micropulse length	10–30 ps (FWHM)
Current (macropulse)	180 mA
Peak current (micropulse)	25–80 A
Max. average power	2 kW
RF power	10 MW
Energy spread	1–3% (FWHM)
Emittance $\varepsilon = 4\pi\beta\gamma\sigma\sigma'$	$\approx 100\ \pi$ mm mrad

likely to be chosen rather than a better performing model based on a photo-cathode, because of its simplicity and reliability: an important issue for a user-facility. A 1 m long travelling-wave section should be sufficient to achieve the required energy. The main parameters of such a linac are summarized in Table 2.

The main difficulty here consists in limiting the energy spread, whereas one looks for relatively long electron bunches, 10–30 ps FWHM. While accelerated in the travelling-wave section, top-hat electron bunches will acquire a corresponding energy spread from 0.3% to 2.2% simply due to the sine shape of the wave and this finite length. A further spread of 100 keV is induced by the buncher and must also be accounted for; see Fig. 1. This will correspond to a spread of 2.5% at the lowest energies, but to only 0.6% at 16 MeV. Therefore, we estimate the maximum energy spread between 2.3 and 3.2%, depending on the energy. A minimal spread $\approx 1\%$ will be achieved for $E > 10$ MeV and a 10 ps long bunch.

As displayed in Fig. 1 and according to PARMELA simulations [5], it is possible to switch between 10 and 30 ps long pulses by slightly adjusting the buncher phase and field, keeping the energy spread essentially the same. According to the simulations and our experience with CLIO [4], the peak current should vary between 25 and 80 A, depending on the bunch length.

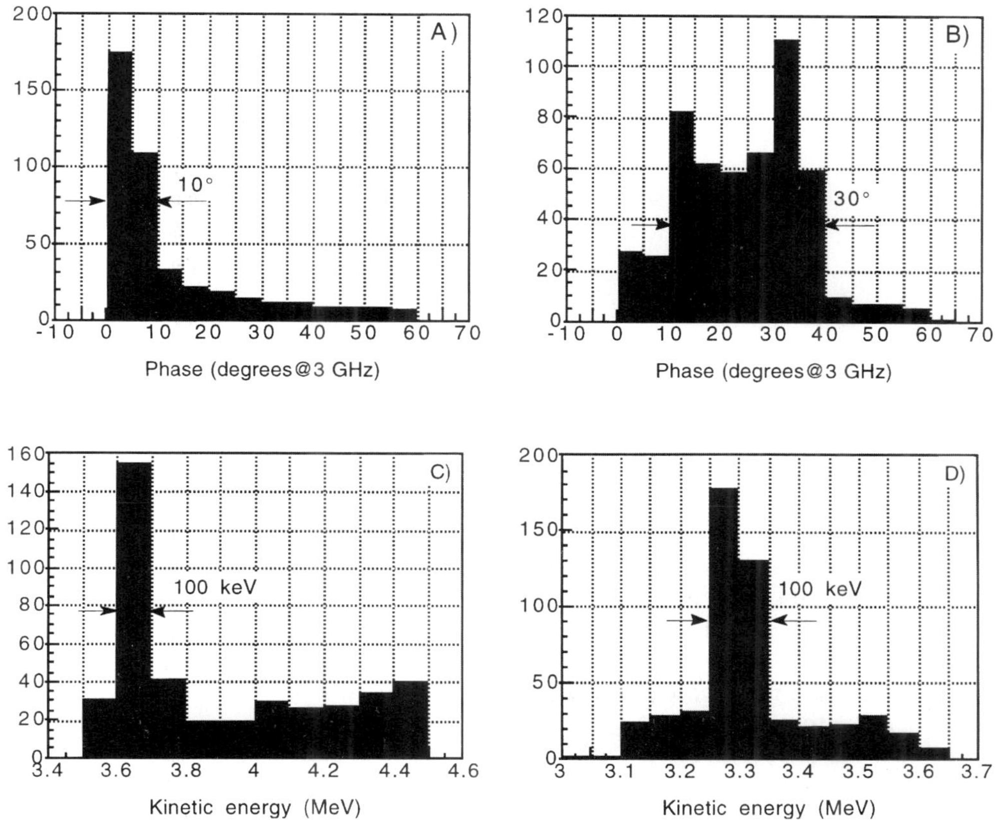

Fig. 1. PARMELA phase and energy histograms at the buncher exit for short (A, C) and long (B, D) electron bunches with different tunings of the buncher (1° at 3 GHz = 0.9 ps).

We have also investigated the effect of the velocity mismatch between the low-energy bunches and the accelerating wave in the section, and found there is no increase of the energy spread when the final energy of the bunch is maximized.

Transport of the beam from the section to and through the undulator has still not been fully studied. Achromatism is of course required, but a strict isochronism is not mandatory since we can tolerate and even look for a bunch lengthening to decrease the laser linewidth at long wavelength; see Table 3. The undulator will be installed after a deviation and a slit, used to clean the beam of particles with out-of-range energy. The overall length will be as short as possible, to limit the space charge effects at low energy.

3. Optical cavity and undulator

3.1. Optical cavity

Both CLIO and CLIO-2 will have a 4.8 m long optical cavity (to match their repetition rates), metallic mirrors, and a hole-coupling extracting scheme.

At the long wavelengths planned, diffraction losses become large for a freely propagating mode, and use of a waveguide is mandatory. Usually, it is only necessary to guide the wave in the plane of the undulator magnetic field. The FEL wavelength is then

$$\lambda = \lambda_{\mathrm{v}} \frac{1+\beta_z}{1 \pm \sqrt{1-(1-\beta_z^2)[1+\lambda_0/\lambda_{\mathrm{s}}]}}, \tag{1}$$

Table 3
Minimum spectral width vs. wavelength for $N = 40$, $p = 5\%$

λ(μm)	L_e/c(ps)	L_s/c(ps)	m	δ	L_p/c(ps)	L/c(ps)	$\Delta\lambda/\lambda$(%)
50	10	7	5	0.8	11	28	0.30
50	30	7	5	0.8	11	48	0.17
500	10	67	0.5	0.8	11	88	0.95
500	30	67	1	1.7	23	119	0.70
1000	30	133	0.5	1.7	23	186	0.90

where $\lambda_v = \lambda_0/2\gamma^2(1 + K^2/2)$ is the usual resonant wavelength without waveguide, K the undulator peak deflecting parameter, β_z the reduced longitudinal velocity of the electrons, λ_0 the undulator period, a the waveguide width and $\lambda_s = 4a^2/\lambda_0$ the "zero slippage wavelength" [6,7]. If $\lambda_v = \lambda_s$, the waveguide prevents the slippage between light and electrons inside the undulator [8]. In our case, we are interested in long pulses and wish to keep a normal slippage length $L_s = N\lambda$ where N is the number of periods. We can then choose $a = 10$ mm and $\lambda_0 = 50$ mm leading to $\lambda s = 8$ mm $\gg \lambda_v$ and we can neglect the effect of the waveguide on λ and L_s.

3.2. Undulator

With this value of $\lambda_0 = 50$ mm, we can achieve a maximum value of $K \geqslant 2$ with an undulator gap of 15 mm and a Halbach-type undulator made with permanent $SmCo_5$ magnets. There is enough space in the cavity for a 2 m long undulator, with 40 periods. However, the two-section undulator of CLIO has proven to be very useful and has provided new insights in FEL physics [9,10]. Therefore, we envision building again an undulator with two or more sections, to investigate multicolor operation in the far-infrared, efficiency dependence with N and super-radiance.

3.3. Gain

The small-signal gain for a full 40-period undulator is calculated with the above parameters and the usual formulae including inhomogeneous effects [11]. To estimate the filling factor, one considers a constant transverse size for the electron beam with $\sigma = 0.6$ mm, and cylindrical mirrors with $R = 3$ m for the cavity. Results are summarized in Fig. 2.

In Fig. 2, the gain is expressed in %/A. With the estimated peak currents of 25 and 80 A for bunch lengths of 30 and 10 ps, respectively, high values of gain are achieved, up to 480%. This will allow to operate and reach saturation with a half or even shorter undulator, a requirement needed for multicolor operation of the FEL.

4. Optical pulse length and spectral width

Some users of CLIO are interested in short pulses for fast kinetic studies, and some others in as small as possible spectral width (0.2–0.4%) for nonlinear spectroscopy or pump-probe measurements at modest time resolution. The same will apply for CLIO-2, so we must be attentive to this latter parameter when operation of a linac-based FEL would favor a short pulse and broadband regime.

For the longest and shortest laser pulses produced through cavity length variation, we can assume we operate in a nearly Fourier-limited regime. This is the case for an infrared FEL like FELIX or CLIO, and seems especially likely when the slippage distance, represented as L_s here, is non-negligible compared to the electron bunch length. The Fourier limit is usually expressed as $\Delta\nu \cdot \Delta\tau = \alpha$, where $\Delta\nu$ and $\Delta\tau$ are, respectively, the width of the laser frequency spectrum and the pulse length (both FWHM), and $\alpha \approx 0.5$, a constant depending on the pulse shape ($\alpha = 0.414$ for a Gaussian) [12]. Another useful expression is $\Delta\lambda/\lambda = \alpha/n$ where n is the

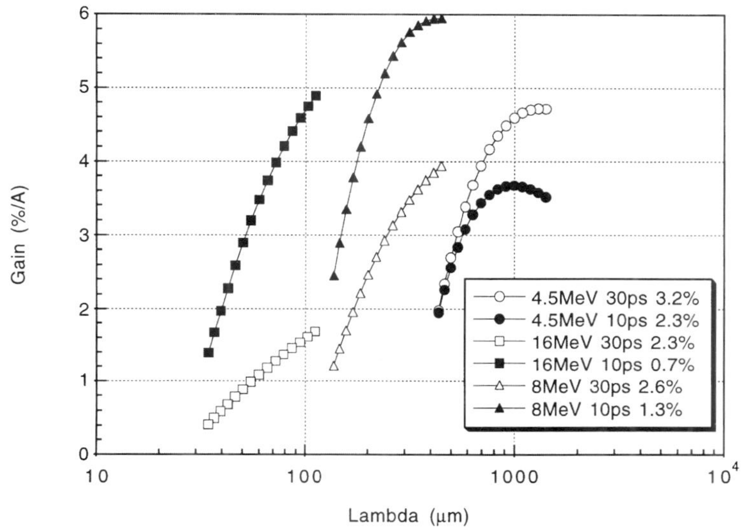

Fig. 2. Small-signal gain for various total energies, electron bunch lengths (FWHM), and energy spreads (FWHM).

number of optical periods within $\Delta\tau$ and $\Delta\lambda/\lambda$ the relative spectral width (FWHM).

4.1. Longest pulses

We can estimate the rough length of the longest pulses we can hope to achieve by $L = L_e + L_s + L_p$.

- L_e is the electron bunch length, or at least the useful part of the bunch where the current is large and then the optical gain is higher than the cavity losses.
- $L_s = N\lambda$ is the well-known slippage length [8], due to the mismatch between the electron velocity and the phase velocity of the wave. N is the number of undulator periods and λ the laser wavelength. Note that this definition of L_s is still valid for us although we operate with a waveguide, as mentioned above.
- L_p is the length increase due to the optical cavity length detuning. If the cavity is shortened by $\delta = m\lambda$ relatively to the exact length for synchronism between electrons and photons, the light is pushed forward by the same amount from one round-trip to the other. The advance of its front edge is eventually limited by the cavity losses p and if one considers an attenuation by a factor 2 for this latter, we can define L_p by $L_p = (-\ln(2) \ / \ \ln(1-p))\delta$. Unless the gain is high enough such that saturation is achieved in one or very few passes, it is clear that one must keep $\delta \ll L_e$ to preserve a minimum of gain.

Table 3 presents an estimate of the achievable linewidth for different wavelengths, detunings and bunch lengths. As appears here, the minimum spectral width should be kept below 1%, with the smaller linewidths for the shorter wavelengths as L_e becomes relatively less important for greater wavelengths.

Note that this simple model is in good agreement with experimental results on FELs like FELIX and CLIO [1,13], with $\Delta\lambda/\lambda = 0.2$–$0.4\%$ and $m \geqslant 5$.

4.2. Shortest pulses

The question of the shortest pulses achievable with an FEL such as CLIO-2 is still open. One could think it is limited by L_e or L_s, but experiments on similar machines have denied this [14,15]. To overcome the apparent limitation by

L_e, one can remark that in practice the electron bunches are not top-hat profiled but present a peaked and asymmetric structure [4]. Therefore, it is possible to use the combined dependence of the gain with both the current and the desynchronism δ in a way such as to limit the effective part of the bunch to a length shorter than L_e. To limit the laser growth on a length smaller than L_s, one must consider an unusual mechanism: as studied by Jaroszynski et al. [16], superradiance can explain the very short pulses already observed on existing machines and allows hope for the same effects to happen on CLIO-2. However, it is difficult to evaluate how much shorter than L_s our laser pulses will be, even through simulations, as the process should strongly depend on the detail of the electron bunch shape.

To provide an estimate of the shortest bunch length for different wavelengths, we can consider a rather conservative value of $L_s/3$, based on our experience. For $N = 40$ undulator periods, it corresponds to 13 optical cycles long pulses with a Fourier-limited spectrum 7.5% wide. At 50 μm, short pulses about 2 ps long should be produced, and up to 44 ps at $\lambda = 1$ mm. However, nothing prevents us from equaling or beating the record of six optical cycles obtained at FELIX [15], especially as one can guess that the shortest pulses are produced when $L_s \approx L_e$, i.e. when the different parts of the micropulse undergo different values of gain due to the slippage.

5. Conclusion

In conjunction with CLIO and OPOs, the proposed new independent FEL CLIO-2 allows us to cover the whole infrared spectral range, with synchronized and tunable lasers. Even operated with an S-band short pulse linac, the FEL linewidth can be kept $< 1\%$ in the far infrared.

First applications envisioned in the far-infrared concern the fields of semiconductor and quantum wells, chemistry and biochemistry of systems with hydrogen bonds, studies of small high-T_c superconductive monocrystals, and nonlinear spectroscopy of interfaces by sum-frequency generation [17].

References

[1] J.-M. Ortega et al., Nucl. Instr. and Meth. A 375 (1996) 618.
[2] J.C. Bourdon et al., Nucl. Instr. and Meth. A 304 (1991) 322.
[3] R. Chaput et al., Nucl. Instr. and Meth. A 267 (1993) 265.
[4] F. Glotin, J.M. Berset, R. Chaput, D. Jaroszynski, J.M. Ortega, R. Prazeres, Nucl. Instr. and Meth. A 341 (1994) 49.
[5] The PARMELA Program (version 5.00)', B. Mouton, LAL/SERA (Orsay)
[6] A. Doria, G.P. Gallerano, A. Renieri, Opt. Commun. 80 (1991) 417.
[7] S.K. Ride, R.H. Pantell, J. Feinstein, Appl. Phys. Lett. 57 (1990) 1283.
[8] W.B. Colson, C. Pellegrini, A. Renieri (Eds.), Laser Handbook, vol. 6, Elsevier, Amsterdam, 1990.
[9] D. Jaroszynski, R. Prazeres, F. Glotin, J.M. Ortega, Phys. Rev. Lett. 72 (1994) 2387.
[10] D. Jaroszynski, R. Prazeres, F. Glotin, J.M. Ortega, D. Oepts, A. van der Meer, G. Knippels, W. van Amersfoort, Phys. Rev. Lett. 74 (1995) 2224.
[11] G. Dattoli, T. Letardi, J.M.J. Madey, A. Renieri, IEEE J. Quantune. Electron. QE-20 (6) (1984) 637.
[12] E.P. Ippen, C.V. Schank, in: S.L. Shapiro (Ed.), Ultrashort Light Pulses Springer, New York, 1977.
[13] G. Knippels, Ph. D. Thesis, University of Amsterdam, 1996.
[14] F. Glotin et al., Nucl. Instr. and Meth. A 375 (1996) 558.
[15] G. Knippels et al., Phys. Rev. Lett. 75 (1995) 1755.
[16] D. Jaroszynski et al., Phys. Rev. Lett. 78 (1997) 1699.
[17] A. Tadjeddine, A. Peremans, J. Electroanal. Chem. 409 (1996) 115.

ELSEVIER

Nuclear Instruments and Methods in Physics Research A 429 (1999) 457–461

NUCLEAR INSTRUMENTS & METHODS IN PHYSICS RESEARCH
Section A

www.elsevier.nl/locate/nima

A new far infrared free-electron laser

J.E. Walsh*, J.H. Brownell, J.C. Swartz, J. Urata, M.F. Kimmitt

Department of Physics and Astronomy, Dartmouth College, Hanover, NH 03755-3528, USA

Abstract

The operation of a new ultra compact diffraction grating coupled free-electron laser (FEL) has been demonstrated. The basic elements of the device which is termed a grating coupled oscillator (GCO) are the beam in a scanning electron microscope (SEM) and a diffraction grating which is mounted in the e-beam focal region of the SEM. The e-beam is controlled by the SEM's electron optical system and distributed feed back is provided by the grating itself. Recent experimental results are presented and techniques for extending the wavelength and power coverage are discussed.

Keywords: SEM; Grating; Electrons

1. Introduction

Observations of radiation produced by electrons moving near the surface of a diffraction grating were first described by Smith and Purcell in 1953 [1] and suggestions that this radiative mechanism could be the basis of a useful radiation source were made even earlier [2]. Over the intervening near half-century, interest has been sustained. At visible and near visible wavelengths experiments similar to the one described in Ref. [1] were carried out by a number of researchers. In the millimeter range of the spectrum coherent sources [3–6] which used gratings as coupling elements, were developed. At short wavelengths (the visible and near infrared range) the limited brightness of the electron beams employed in the experiments limited the degree of coherence. However, this is not a fundamental constraint. Recently [7], the electron beam in a scanning electron microscope (SEM) and a diffraction grating mounted in the e-beam focal region have been used to produce coherent radiation at far infrared (FIR) wavelengths. The characteristics of this device will be described and conditions required for extension toward operation at shorter wavelengths will be discussed.

2. The SEM-based FEL

A schematic cross-section of the SEM-FEL is shown in Fig. 1. The beam is formed on a Tungsten "hairpin" cathode and it is focused and positioned over the grating with the SEM's internal electron optical system. When used as a microscope very small beam currents are typically employed. However, the present system is capable of running in the low milliamp range provided care is taken with heat sinking. This current level is more than sufficient for operating in the super radiant regime.

* Corresponding author.

0168-9002/99/$ - see front matter
PII: S0168-9002(99)00140-0

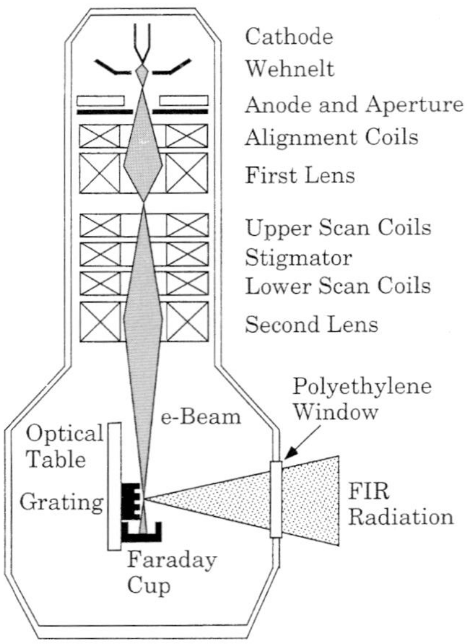

Fig. 1. Schematic diagram of the SEM-GCO.

Various short pulse schemes would yield still higher currents. The beam voltage may be varied up to 40 kV. Again this is a convenient limit for the present apparatus but not a fundamental constraint. Operation at 1.5 THz (200 μm wavelength) has been achieved with a 25 kV beam. Although run in a regime well above its intended design limit the SEM beam retains the tight focusing and high brightness that is characteristic of these devices. At present the smallest beam waist radius that can be easily attained is approximately 12 μm. This yields peak beam current densities in the 200 A/cm^2 range. The beam emittance can be measured in situ with either a 3-wire profilometer or knife edges. Normalized emittance values are typically in the range at $10^{-2}\pi$mm mrad, a range that is two orders of magnitude smaller than the standard high-quality RF linac. Thus although the current in the SEM is much lower than typical linac currents beam brightness may be comparable.

The gratings used in the present experiment are 2–4 mm wide and 10–15 mm long. The gratings are mounted on a miniature optical table that has been drilled with holes on $\frac{1}{4}''$ centers. Other components including Faraday cups and the profilometer are mounted on the table. The complete assembly is mounted on the microscope stage which brings flexibility to the task of precise alignment.

3. Experiments

The SEM in its original role as a microscope can be used to align the grating. Peak signal is obtained when the beam axis and grating surface are parallel. When the beam current is at the low end of its range and/or it is not tightly focused, signal power increases linearly with beam current. In this regime, coupling conditions are comparable to those encountered in the early experiments of Smith and Purcell. Energy is transferred to the radiation field via a velocity synchronous coupling of the beam electron and one of the slow space harmonic components of the field. Space harmonic components propagating at the same velocity as the beam electrons "evanesce" away from the surface with a characteristic scale length given by the expression

$$\lambda_e = \lambda\beta\gamma/2\pi$$

where λ is the operating wavelength and β and γ are the velocity of a beam electron relative to the speed of light and the relative energy. In order to couple efficiently λ_e must be comparable to the position of the beam centroid above the grating. In the experiments carried out by Smith and Purcell and in similar early work the beam diameter was much greater than λ_e. When this is the case the emission from individual electrons add incoherently and power increases linearly with beam current. This is the spontaneous emission or shot noise regime.

Each mode on the grating consists of a set of space harmonics. Space harmonic components with phase velocities which fall within the light cone on a dispersion plane produce the outgoing radiation. Space harmonic components which have velocities which fall outside of the light cone are trapped on the surface. In addition to providing beam-grating coupling these field components provide a distributed feed back. This is always the case but when the beam current density is sufficient the system enters a super linear regime. Output power then increases in proportion to a substantial power of the current.

The details of this super radiant operating regime are still under investigation but certain general

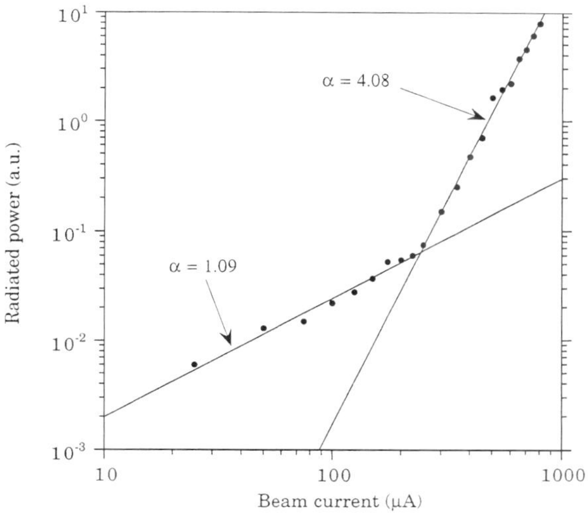

Fig. 2. Typical power versus beam current relation.

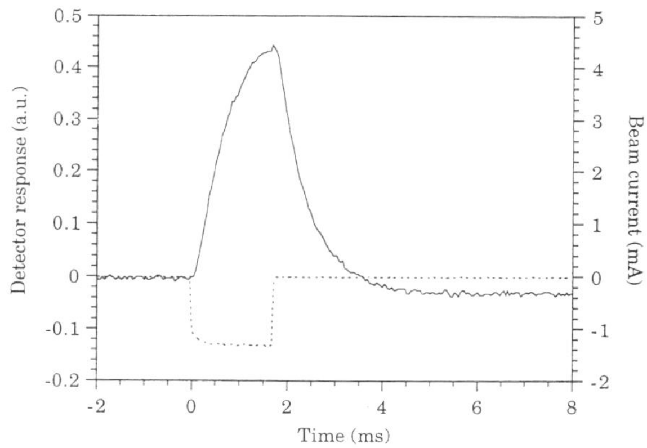

Fig. 3. A single unaveraged FIR output pulse (solid) for an e-beam pulse (dotted) near the high end of the SEM current range. The wavelength is approximately 500 μm (0.6 THz), the grating period was 168 μm, and the accelerating potential 25 kV.

trends have emerged. The threshold currents increase with decreasing beam density. Threshold currents below 100 μA have been observed but typical values are in the 200–400 μA range. Since the threshold decreases with increasing beam current density but decreasing total current, it is assumed that limits set by the "surface resonator" characteristics of the grating are a dominant factor. A typical output power versus beam current curve is shown in Fig. 2. In the lower power linear regime the measured power, 100s of pW in the threshold region, are in accord with predictions [8] of the theory governing the spontaneous emission. Above threshold peak powers of 100s on nW are easily obtained and power levels in the μW range (at the detector) have been observed. At high levels the signal is observed directly at the output of the silicon composite bolometer. A typical pulse is shown in Fig. 3.

The intrinsic saturation limit for the system has not yet been established. If it is assumed that the basic constraints governing e-beams and gratings are similar to those which apply to devices such as Cherenkov free-electron lasers, ultimate electronic efficiencies in the 0.1–1.0% range can be expected from an optimized system. This would provide milliwatt level powers from an SEM driven device. A pulsed system may produce greater power. The present power levels are already useful for spectroscopic investigations and for driving heterodyne mixers with low-power requirements.

4. Tuning

The operating wavelength is determined by the expression

$$\lambda = \frac{\ell}{|n|}\left(\frac{1}{\beta} - \cos\theta\right) \qquad (2)$$

which is often referred to as the Smith–Purcell formula. In Eq. (2), ℓ is the grating period, $|n|$ is the order of the diffraction and θ is the angle of emission relative to the beam axis. The optical system in the present device collects only the normal emission. Wavelengths have been measured with an FIR grating spectrometer, polarizing and non-polarizing Michelson interferometers and cutoff filters. A compilation of the interferometer data is shown in Fig. 4.

Operation on orders other than the first is possible. By choosing the depth of the slot in order to reinforce the efficiency on a particular order, emission on that order will be dominant. A typical spectrum from the Martin–Puplett FTIR spectrometer is shown in Fig. 5 and the data from

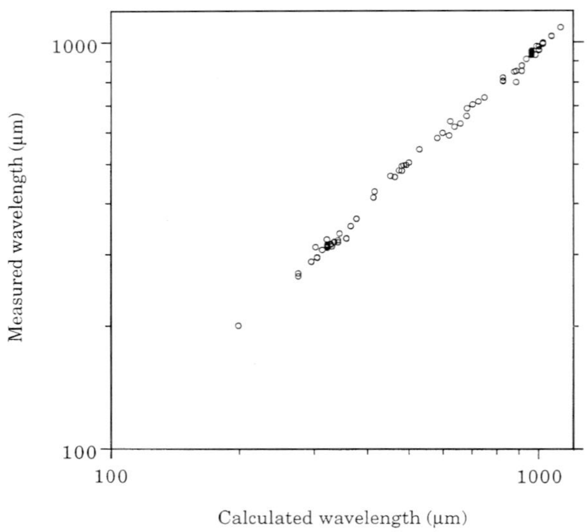

Fig. 4. Compilation of measured versus predicted wavelengths for emission in the normal direction.

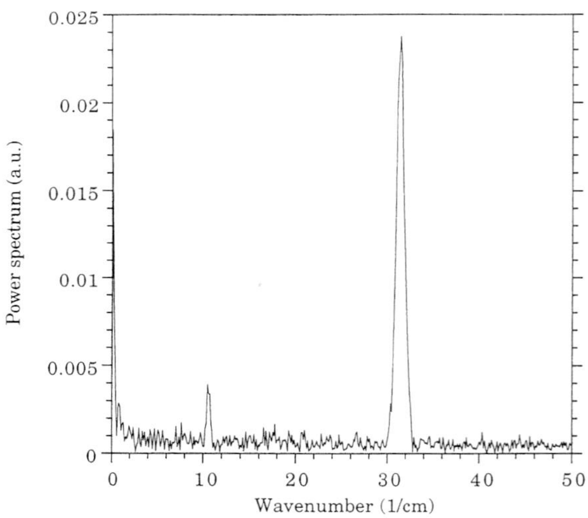

Fig. 5. A typical spectrum produced by the Martin–Puplett interferometer. The line width is comparable to the resolving power.

Fig. 4 is re-plotted as a function of the beam velocity in Fig. 6. Grouping in various orders is evident. Power generated on the higher order is comparable to the output that can be obtained from a grating which has the same period but which has been blazed for the first order.

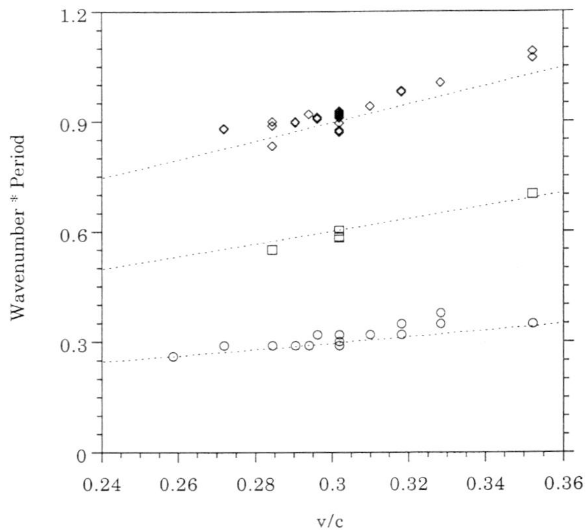

Fig. 6. Measured wave numbers in units of the grating period(s) versus relative beam velocity. Three orders are shown.

5. Conclusion

To date operation of the GCO has been extended down to 200 μm wavelength. Although the details of a theoretical picture are still unfolding, evidence suggests that device operation relies on a basic three-wave coupling. The high current density beam supports fast and slow (relative to the beam velocity) space charge waves. Coupling of these beam modes with a co-synchronous phase velocity space harmonic on the grating leads to a growing mode that is similar to a Cherenkov instability [9]. On the basis of this analogy, and presuming that e-beams representative of the electron optical state of the art are employed, extensions of GCO operation down into the 10s of μm wavelength range is a realistic expectation. With improved emittance this extension can be accomplished without a dramatic change in the beam current and voltage. The operational limit of this device is primarily a question of brightness. If pulsed field emitters or photo assisted field emission cathodes are adapted to GCO operation, and the beam quality maintained at higher (mildly relativistic) beam energies, then the wavelength would be further reduced and output power increased. Ultimately, when extending the range of any electron beam driven coherent source, beam quality and beam energy are key parameters. A fundamental conclusion that may be drawn from the present experiments is that substantial improvements in beam emittance and energy spread can reduce the need for high beam energy.

Acknowledgements

The support of ARO grant DAAH04-95-0640, DOD/AF DURIP Contract F49620-97-1-0287 and Vermont Photonics Inc. is gratefully acknowledged.

References

[1] S.J. Smith, E.M. Purcell, Phys. Rev. 92 (1953) 1069.
[2] W.W. Salisbury, Super High Frequency Electromagnetic Wave Generator, US Patent 2,634,372, Applied – October 26, 1949, Awarded April 7, 1953.
[3] F. Rusin, G. Bogomolov, Proc. IEEE 720 (1969).
[4] K. Mizuno, S. Ono, Y. Shibato, IEEE Trans. Electron. Dev. ED- 20 (1973) 749.
[5] D. Wortmann, H. Dropkin, R. Leavitt, IEEE J. Quant. Electron. QE- 17 (1981) 1341.
[6] E. Price, J.E. Walsh, Appl. Phys. Lett. 61 (1992) 252.
[7] J. Urata, M. Goldstein, M.F. Kimmitt, A. Naumov, C. Platt, J.E. Walsh, Phys. Rev. Lett. 80 (1998) 516.
[8] M. Goldstein, J.E. Walsh, M.F. Kimmitt, J. Urata, C. Platt, Appl. Phys. Lett. 71 (1997) 452.
[9] E. Fisch, J.E. Walsh, Appl. Phys. Lett. 60 (1992) 1298.

ELSEVIER Nuclear Instruments and Methods in Physics Research A 429 (1999) 462–470

NUCLEAR INSTRUMENTS & METHODS IN PHYSICS RESEARCH Section A

www.elsevier.nl/locate/nima

Principles of high-contrast energy modulation and microbunching of electron beams

R. Tatchyn*

Stanford Synchrotron Radiation Laboratory, Stanford University, Stanford, CA 94309, USA

Abstract

Recent advancements in ultra-short-pulse terawatt IR/visible/UV laser technology have made it possible to consider particle-beam energy modulation schemes in which the practical field amplitudes of the laser and undulator fields in an ultrarelativistic electron bunch's average rest frame can attain comparable magnitudes. This parameter regime, well outside the radiation field strengths attainable by conventional free electron lasers, makes possible the exploitation of "relativistic interferometry", viz., the phenomenon of high-contrast interference, to isolate regions in the longitudinal space of an electron bunch on the order of one laser wavelength. In this paper we review selected requirements on the laser and undulator fields to generate highly efficient microbunching followed by coherent radiation at a substantially reduced wavelength. Conditions on the electron beam quality, as well as selected possibilities for new modes of operation, are summarized. © 1999 Elsevier Science B.V. All rights reserved.

Keywords: Microbunching; High-contrast modulation; Free electron lasers; Relativistic interferometry

1. Introduction

In practically every free electron laser (FEL) heretofore designed or operated, the radiation field amplitude in the rest frame of the electron bunch is overwhelmingly dominated by the field amplitude of the insertion device. For example, in a gain-saturated 1.5 Å SASE LCLS driven by a 15 GeV beam with a 1 mm mrad normalized emittance [1], an output power of 100 GW and an undulator field of 1.3 T in the laboratory frame are transformed, respectively, into bunch-frame amplitudes of ~0.0400 T versus ~13000 T. In prior work, the relatively recent advent of ultrashort-pulse IR/visible/UV lasers in the terawatt range [2] was adopted as a premise for considering the preparation of electron bunch frames in which the practical field amplitudes and wavelengths of the laser and insertion device could be made equal, leading to the possibility of inducing high-contrast (HC) interference effects, in particular on the spontaneous radiation and particle energy [3] distributions. A basic motivation in these initial studies was – and remains – the generation of ultrashort (subfemtosecond) radiation pulses. In subsequent papers [4,5], the energy modulation of single particles in this regime was analyzed and confirmed the initiation of conditions for rapid bunching, which, if attained, could be used to generate coherently

* Tel.: + 1-650-926-2731; fax: + 1-650-926-4100.
E-mail address: tatchyn@slac.stanford.edu (R. Tatchyn)

PII: S0168-9002(99)00142-4

enhanced radiation at substantially reduced wavelengths. More recently [6], these studies were extended to preliminary simulations of collective dynamics within a cold bunch. In the present paper the basic principles of HC systems are reviewed with the aim of: (1) clarifying the basic physics involved, and (2) deriving simple expressions for the initial evaluation or design of practical configurations. Selected possibilities for further developing the HC modulation technique are discussed and some recent preliminary simulations of warm-beam effects on the bunching process are presented.

2. Physical principles: single-electron dynamics

A layout of the basic constituents of a HC modulation (HCM) and bunching system is schematized in Fig. 1. A high-intensity radiation (e.g., laser) pulse (or synchronized group of superimposed pulses) of wavelength λ_r and normalized vector potential K_r intercepts a high-quality, high-density electron beam at angle(s) θ along the axis of a static or dynamic field synthesizer (FS) [7].[1] The FWHM bunch length is $\sim\sqrt{2\pi}\sigma_B$ and the FS parameters are λ_u and K_u. We adopt, as a practical working definition of HCM, the relation $O(K_r) \cong O(K_u)$. The polarizations of the field sources can be arbitrary, but each one will be assumed linear for the present discourse. An essential attribute of the FS is its ability to generate arbitrary near-axis field distributions, and in particular accurately-scaled replications of the superimposed laser-field components. Inside the FS, the laser imparts an energy modulation profile to a region of the bunch. This modulation induces a (ballistic) bunching drift inside the beam which can be further modified with suitably designed dispersion regions, which can also comprise sections of a FS. The conditions for maximum-contrast interference of the radiation and FS fields in the rest frame of the electron bunch are straightforwardly derived to be [8]

$$\lambda_r = \lambda_u(1 - \beta^* \cos\theta) \tag{1}$$

and

$$B_r \cong B_u/(1 - \beta^* \cos\theta) \tag{2}$$

where $\beta^* = \sqrt{1 - \gamma^{*-2}}$, $\gamma^* = \gamma/\sqrt{1 + K_u^2/2}$, and γ is an (incoming) electron's Lorentz factor. For $\theta \ll 1$, Eqs. (1) and (2) reduce to $\lambda_r \cong (\lambda_u/2)((\gamma^*)^{-2} + \theta^2)$ and $B_r \cong 2B_u/((\gamma^*)^{-2} + \theta^2)$. Upon compression of the modulated region down to a (microbunched) length Δ_{min}, the beam, or a suitably filtered portion thereof, is sent through a radiator of period $\lambda_R \geq 4\gamma^{*2}\Delta_{min}$, from which a coherently enhanced radiation pulse of wavelength $\lambda_R \ll \lambda_r$ is emitted. The maximum number of radiator periods usable for this purpose will in principle be limited by the time it takes for the bunching to relax, diffuse, or scatter back to the condition $\Delta_{min} > (\lambda_R/2\gamma^{*2})$. It should be evident from these descriptions that the basic elements and selected processes of the HC system have their counterparts in the optical klystron (OK) [9–11] and free electron laser (FEL) [12]. Indeed, the basic single-electron dynamics underlying the energy modulation can be described by the same equations. However, as will be outlined below, significant differences in the detailed physics can arise from the highly disparate parameter regime of HC modulation, as well as from the basically different phenomena involved in the collective (bunching) dynamics.

Our single-particle analysis will apply explicitly to linearly polarized fields, following the approach of Colson [13]. The equations of motion for the transverse (y) and longitudinal (z) trajectory components can be written as

$$\frac{d(mc\gamma\beta_y)}{dt} = qE_y + cq\beta_z B_x \tag{3}$$

and

$$\frac{d(mc\gamma\beta_z)}{dt} = -cq\beta_y B_x \tag{4}$$

along with the energy-transfer equation

$$\dot{\gamma} = \frac{q}{cm}\beta_y E_y. \tag{5}$$

[1] Although in the present article we focus primarily on ultrashort laser pulses (down to the order of a few wavelengths long) and static magnetic-field FSs, the general HC modulation and bunching system need not be restricted to these conditions.

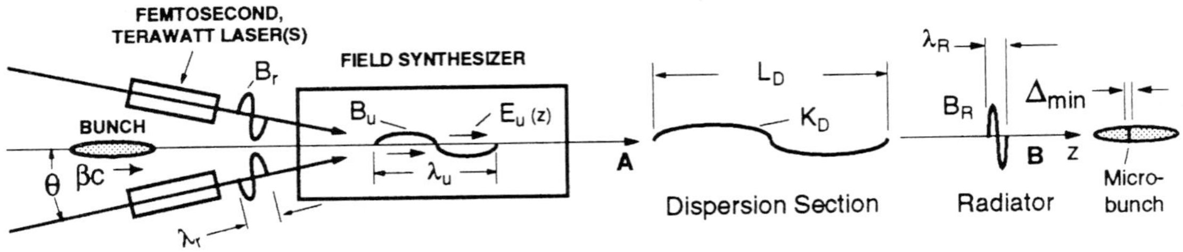

Fig. 1. High contrast modulation (HCM) and dispersion-assisted bunching (A) followed by coherence-enhanced radiation (B). For ballistic bunching, $K_D = 0$. The FS can, in general, generate arbitrary static or dynamic field distributions with components along all three axes.

Taking $E_y = E_r \sin^{[1]}(2\pi(ct-z)/\lambda_r + \phi_r)$ and $B_x = B_u \sin^{[1]}(2\pi z/\lambda_u + \phi_u) - (E_r/c)\sin^{[1]}(2\pi(ct-z)/\lambda_r + \phi_r)$, with $\theta = 0$ rad, and the bracketed exponents denoting the number of integral sinusoidal cycles, the following expressions for the transverse c-normalized velocity and longitudinal acceleration are derived:

$$\beta_y = -\frac{K_r}{\gamma}\cos^{[1]}\left(\frac{2\pi}{\lambda_r}(ct-z) + \phi_r\right) - \frac{K_u}{\gamma}\cos^{[1]}\left(\frac{2\pi}{\lambda_u}z + \phi_u\right) \tag{6}$$

and

$$\dot{\beta}_z = \frac{2\pi c K_r K_u}{\lambda_u \gamma^2} \left\{\sin^{[1]}\left(2\pi\left[\frac{(ct-z)}{\lambda_r} - \frac{z}{\lambda_u}\right] + \phi_u - \phi_r\right) + \frac{1}{2}\left[\sin^{[2]}\left(\frac{4\pi}{\lambda_r}(ct-z) + 2\phi_r\right) - \sin^{[2]}\left(\frac{4\pi z}{\lambda_u} + 2\phi_u\right)\right]\right\} \tag{7}$$

where the indexed ϕ's determine the phases of the laser and undulator fields with respect to the electron in question.[2] Although this does not allow for a fully general source decomposition, it suffices to explore the basic physics of HCM. We note the explicit dependence of both motion components on z. In general, the functional dependence of z on t is complicated, and can ordinarily be expressed as a Fourier series (or, in our limit, a Fourier integral) in t. The significance of higher-order spectral terms is dependent on the amplitude of the coefficient of Eq. (7), which is seen to be directly proportional to $K_r K_u$ and inversely proportional to $\lambda_u \gamma^2$. Although the investigation of Eqs. (6) and (7) for small-through-large values of both these factors is of legitimate concern for HCM, we restrict ourselves in this paper to the ranges $\lambda_u \gamma^2 \gg 1$ m and $0.005 < K_r \sim K_u < 0.25$, the lower bound of the latter constraint stemming from present-day technological limitations on implementing long-period insertion devices with overly weak fields. In this regime, $z \cong \beta^* ct$, the z-velocity in the bunch frame, v'_z, is non-relativistic, and the visualization of the dynamics contained in Eqs. (6) and (7) is correspondingly simplified.

Given the above approximations and assumptions, a number of basic results follow. For an initial transverse velocity of zero, the fields impart no net transverse kick to the electron, irrespective of their phases, so long as their 1st field integrals are zero (Eq. (6)). The initial transverse position, however, proportional to the difference of the 2nd field integrals, need not remain constant [4]. For the condition on λ_u versus λ_r expressed in Eq. (1), the upper expression in square brackets in Eq. (7) approaches zero, and the time-dependent parts of the arguments in the lower bracketed expressions approach equality. Thus, the magnitude and sign of the longitudinal acceleration imparted to a bunch

[2] These field distributions are, strictly speaking, unphysical. The approximation is, notwithstanding, reasonable in that: (1) it allows the systematic study of HC modulation by a progressive increase in the number of "single" cycles, and (2) it can in fact provide insight into the evolution of an actual HC bunching distribution, which, due to the highly non-linear dynamics, can be made to develop a single dominant bunched peak. In the general case more realistic field distributions can of course be employed using numerical simulations.

electron are determined primarily by its phase relative to the laser and FS fields. In this regard we note that the direction of electron motion in z is always directed toward the nearest minimum of $\dot{\beta}_z$, indicating the onset of dynamic bunching.

We can now derive a number of useful expressions for assessing HC modulation systems in this parameter regime. Using the Einstein velocity-addition formula [8] and the above equations, the maximum normalized energy and velocity kicks (i.e., for $|\phi_r - \phi_u| = \pi/2$) imparted to a bunch electron in the lab frame are given, respectively, by: (1) $\Delta\gamma \cong 2\pi\gamma K_r K_u$, and (2) $\Delta\beta_z \cong 2\pi K_r K_u/\gamma^2$. The corresponding absolute energy and velocity kicks in the bunch frame are: (1) $\Delta E' \cong 2\pi^2 m_e c^2 (K_r K_u)^2$ J, and (2) $\Delta v'_z \cong 2\pi c K_r K_u$ m/s. We can now compare selected features of HCM versus OK (or FEL) systems running at similar wavelengths. In the latter, $K_r \sim O(10^{-5}\text{–}10^{-7})$ at saturation, and 4–5 orders of magnitude smaller than that at the FEL entrance. Thus, even for a high-K_u FEL the net longitudinal kinetic energy extracted from a bunch electron will be considerably smaller than in a typical HCM system with, e.g., $K_r \sim K_u \sim 0.1$. Furthermore, the net energy exchange in an OK or FEL accrues over many periods N_u – typically of the order of 10^3, whereas the net gain (or loss) in HCM is attained over a substantially smaller number of periods (e.g., for the case being considered, $N_u = 1$). Thus, the average relative energy transfer per period in a typical FEL is of the order of $1/(N_u)^2$, or $\sim 10^{-6}$, justifying its characterization as "adiabatic", whereas the HCM loss (or gain) per period, typically $O(0.1 - 0.01)$, could be more aptly termed as "high-action", or "rapid". Based on these observations, a relevant criterion for distinguishing a HCM system versus a FEL or OK modulating at a similar wavelength can be expressed as $(K_r K_u)_{HCM} \gg (K_r K_u)_{FEL,OK}$.

3. Physical principles: collective dynamics

As opposed to a FEL or OK, in which the radiation scattered off the electrons and the inter-particle Coulomb forces both act, in general, as essential factors in the energy modulation and bunching processes, the scattered radiation in an HCM system plays an initially minor, if not negligible, role. Upon modulation by the (exogenous) radiation field, the bunching in the electrons' rest frame is mediated, in the non-relativistic limit[3] by an exchange between the electrons' acquired kinetic energy and the electrostatic potential energy associated with the gradually evolving particle density. Just as in an FEL, however, the dynamics is still governed by the self-consistent interaction among all the particles moving in the combined undulator, Coulomb, and radiation fields.

The combined collective modulation and bunching in the general HCM system can be described by an unrestricted Vlasov equation [14]. If the overall process can be approximated as a sequence of two independent sub-processes, viz., modulation followed by ballistic bunching,[4] the (dissipation-free) Vlasov equation governing the dynamics of the bunching phase can be written as

$$\frac{\partial f}{\partial t} + \boldsymbol{v} \cdot \frac{\partial f}{\partial \boldsymbol{r}} = \frac{e^2}{4\pi\varepsilon_0} \frac{\partial f}{\partial \boldsymbol{v}} \cdot \int \frac{\partial}{\partial \boldsymbol{r}} \frac{1}{|\boldsymbol{r} - \boldsymbol{r}'|} f(\boldsymbol{r},\boldsymbol{v},t)\, d\boldsymbol{r}'\, d\boldsymbol{v} \tag{8}$$

where the term on the right describes the interparticle Coulomb interactions, and f, the particle distribution, is assumed to carry the fields' velocity modulation imprint, $\boldsymbol{v}(\boldsymbol{r}(0))$, as an initial condition. We point out, in passing, that for arbitrary HCM fields and distributions f, Eq. (8) cannot in general be linearized and must be solved either numerically or approached with alternative analytical techniques.

Prior to performing numerical simulations based on Eq. (8), it is instructive to first assess the bunching process for an idealized case. In our treatment, we reduce the dimensionality of the problem from 3-D to 1-D by first partitioning the bunch longitudinally into infinitesimal slices of thickness δ_z. In view of the finite bunch radius, r_b, we assume the

[3] In this limit we can disregard interactions stemming from the particles' magnetic fields.

[4] This approximation (equivalent to taking $K_r \sim K_u \ll 1$), although admittedly inexact, is justifiable for the systems analyzed in this paper. Since $\Delta v'_z \ll c$, and the modulation employs single-cycle fields, the assumption is that the particles have moved over only a small fraction of $\lambda'_u (= \lambda_u/\gamma^*)$ by the time the two fields have passed through.

force between any two such slices to vary inversely with $(1 + (\Delta z'/r_b)^2)$, where $\Delta z'$ is the inter-slice distance in the bunch frame. We next estimate the potential energy of a collapsed sheet containing all the charge from a linearly modulated region of extent λ'_u, by integrating work over pairs of charge sheets. We thus find that the peak velocity required (assuming linear modulation) to compress the given region into an infinitesimal sheet (at which point the total kinetic energy will vanish) is approximable by

$$\Delta v'_{z(CR)} \cong \sqrt{\frac{q r_b \rho_B \lambda_u}{m_e \gamma^2 \varepsilon_0} \tan^{-1}\left(\frac{\lambda_u}{2\gamma r_b}\right)} \text{ m/s} \tag{9}$$

where q and m_e are the electronic charge and mass, ρ_B is the lab-frame charge density, and ε_0 is the permittivity of free space, all in MKS units. We note that for any set $(\rho_B, \lambda_u, r_b, \gamma)$ there are evidently three relevant modulation regimes, viz.: I. $\Delta v'_z < \Delta v'_{z(CR)}$ (undermodulation); II. $\Delta v'_z > \Delta v'_{z(CR)}$ (overmodulation); III. $\Delta v'_z \cong \Delta v'_{z(CR)}$ (critical modulation); and that $\Delta v'_{z(CR)}$ varies inversely with a power of γ higher than unity.

Using the 1-D model approximation to Eq. (8), several numerical simulations of a typical HCM systems's bunching dynamics using different energy modulation distributions of Gaussian beams [6] were performed. Recent extensions of the model code to include energy spread within the bunch were also calculated. Here we present results for a linear modulation profile, $\Delta v'_z = (2\gamma^* |\Delta v'_{z(MAX)}|/\lambda_u) z'$, with $z' = 0$ the position midway between the approaching fields. In the top line of Fig. 2 we plot the evolution of bunching for regimes I. (a), II. (b), and III. (c) in a cold 10 nC beam using velocity modulation amplitudes, $|\Delta v'_{z(MAX)}|$, of 0.1c, 0.15c, and 0.12c, respectively. As expected, undermodulation leads to the longest bunching time and results in underbunching, while overmodulation induces the quickest bunching with only a transitory attainment of the bunching peak. In the second line of Fig. 2 we present histograms of the evolving macroparticle distribution in regime III. (corresponding to Fig. 2(c)) for bunch frame times of 0 ps (d), 3.5 ps (e), and 5.25 ps (f). The last plot (f) shows only a coarse approximation to the actual theoretical maximum of 1000 since: (1) no search for the precise time of the maximum peak was made, and (2) the electric field fluctuations associated with the model distribution's (relatively large) density fluctuations would in any case tend to inhibit the attainment of a perfectly compressed peak. In Fig. 2(g) the effect of an 0.1% 1-σ energy spread on the bunching process of (d–f) is displayed. In Figs. 2(h) and (i) the evolution of the electric field distributions in the bunch frame for the regime III case with 0.0% versus 0.1% energy spreads are shown. The graphs reveal the basic features of the bunching process, supporting the results derived analytically in the above-cited work. As our numerical simulation code has not yet been extended to incorporate the modulation interaction or the action of additional dispersion fields into the bunching dynamics, analytical estimates for the minimum attainable bunch length Δ_{min} in a dispersion section of length L_D in terms of the electron beam's emittance parameters (assuming a net linear modulation) have been worked out [5]. The basic results can be summarized by the ratio

$$\frac{\Delta_{min}}{\lambda_r} \approx \frac{2\gamma}{\Delta\gamma_{1N}} \sqrt{2\pi\sigma_\varepsilon^2 + (\pi(\sigma_V'^2 + \sigma_H'^2))^2 \gamma_D^{*4}} \tag{10}$$

where $\Delta\gamma_{1N}$ is the total modulation increment, σ_ε is the standard deviation of the normalized random energy spread, σ'_V and σ'_H are the standard deviations of the beam's transverse emittance angles, and $\gamma_D^* = \gamma/\sqrt{1 + (K_D^2/2)}$, where K_D is the normalized vector potential of the dispersion region. Eq. (10) can be used (with $K_D = 0$) as an approximate check on the peak bunching reduction displayed in Fig. 2(g).

4. Radiation performance

Estimates of the radiative output characteristics of the microbunched beam in a radiator field fulfilling the condition $\lambda_R \geq 4\gamma^{*2}\Delta_{min}$ are readily derived. Here we restrict ourselves to assessments of the total emitted energy and output power in the lab frame. The maximum possible spontaneous emission power can be expressed in MKS units as $P_{SP} \cong N_e[\gamma^4 q^2 \langle \dot{v}_y^2 \rangle_t / (6\pi\varepsilon_0 c^3)]$, where N_e is the number of bunch electrons and the quantity in brackets

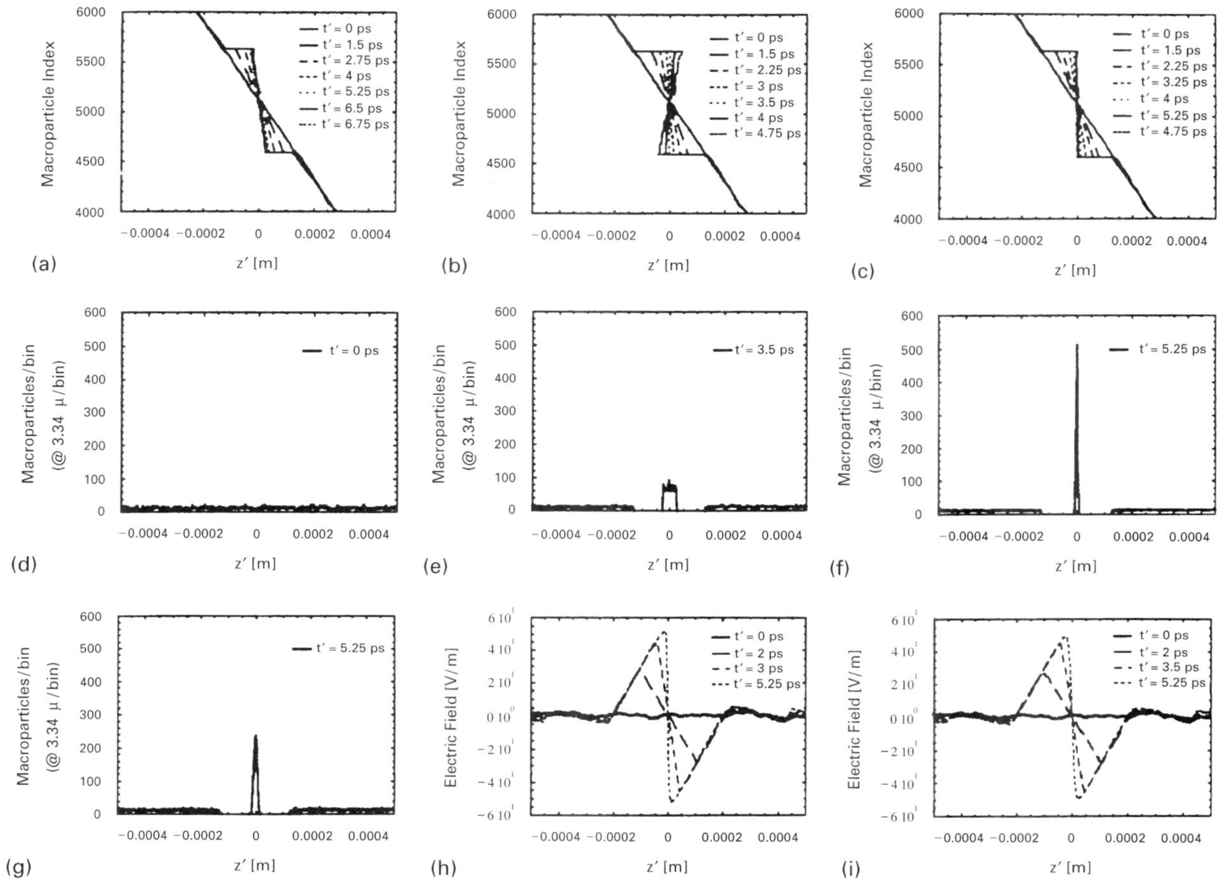

Fig. 2. Bunching and electric field distributions in the rest frame of a cylindrical electron beam with longitudinally Gaussian and transversely uniform density profiles. Beam parameters: $\gamma = 300$; $\sqrt{2\pi}\sigma_B = 0.5$ mm; $r_b = 100\,\mu$m; $N_e = 10^{11}$; number of macroparticles $\cong 10^4$ (contained in a region of length $10\lambda_u/\gamma^*$ at the center of the bunch); electrons /macroparticle $\cong 8.9\times10^4$. To enhance the resolution only the central 2000 macroparticles containing the modulated region are displayed in a–c; in d–i, the central 3700. For graphs a–f, and h, $\sigma_\varepsilon = 0.0$; for g and i, $\sigma_\varepsilon = 0.001$.

is an approximation to the relativistic Larmor formula [15]. The maximum coherent emission power is given by a similar expression, but with the factor N_e replaced by $(N_e\lambda_r/\sqrt{2\pi}\sigma_B)^2$. The ratio η^{Ph}_{HCM} of the total number of coherent photons, N_{COH}, emitted by the microbunch to the spontaneous number N_{SP} emitted by the entire bunch is consequently $\eta^{Ph}_{HCM} \equiv (N_{COH}/N_{SP}) \cong N_e(\lambda_r/\sqrt{2\pi}\sigma_B)^2$. For broad ranges of readily implemented linac beam parameters, it is readily verified that $\eta^{Ph}_{HCM} \gg 1$. In practical terms, the number of spontaneous photons emitted by the N_e bunch electrons passing through an N_R-period radiator can be expressed as $N_{SP} \cong 7.61\times10^{-3}N_eK_R^2N_R$, and an upper limit on the emitted coherent power by

$$P_{COH}[W] \cong 1.585\times10^{-14}\gamma^4\eta^{Ph}_{HCM}N_e(B_R[T])^2. \quad (11)$$

A set of parameters for a representative HCM buncher/radiator case is listed in Table 1 (cf. Fig. 1 and Eq. (10)). We observe, in Eq. (11), the scaling of P_{COH} with $(B_R)^2$. This indicates that in cases where K_R is initially small the coherent in-band power from a single microbunch could be substantially increased – perhaps well toward the multi-gW level – by substantially increasing the radiator field. Fields up to the order of 5 T (versus the ~ 1 T assumed in the table) could be readily generated using a pulsed FS.

Table 1
System and radiation output parameters of a HCM ultra-short pulse buncher/radiator. Longitudinal beam profile Gaussian; transverse profile flat

Ultra-short pulse generator parameters ($\gamma = 300$; $r_b = 100\,\mu m$; $\rho_B = 10.6\,C/m^3$)	
Bunch charge	1 nC
$\sqrt{2\pi}\sigma_B$	3000 μm
λ_r	4800 Å
λ_u	8 cm
σ_e	0.1%
σ'_V	0.1 mr
σ'_H	0.1 mr
K_D	1
L_D	0.355 m
$\Delta\gamma/\gamma$	0.05
Δ_{min}	240 Å
K_R	~ 1
Compression ratio (Δ_{min}/λ_r)	6%
480 Å coherent photons/pulse	7.0×10^9
Peak coherent output power	0.18 GW

5. Selected aspects of high contrast modulation and bunching

Perhaps the most critical aspect of prior and current studies of HCM has been the emphasis on ultra-short pulse generation. This has facilitated the tractability of our exploratory analyses of both modulation and bunching using single-cycle fields. In actuality, of course, an ultra-short pulse may span more than one cycle and will always have non-instantaneously decaying edge derivatives. If the FS replicates this distribution then it is clear that the modulated region will also extend over more than twice one period. We can note, however, that the non-linearity of the bunching and radiation process will tend to give a strong preference to the global maxima of the velocity interference pattern. Consequently, by using, e.g., Fourier transform techniques to, say, sharply peak the center of a laser pulse containing a small number of cycles, it should be possible to generate a micro-bunched pulse train in which the radiation from a single microbunch is overridingly dominant. In the general case, many other interesting possibilities arise. For example, for sufficiently small K_r and K_u the modulation pattern will approximate to the convolution of the laser and FS pulse trains. Thus, two fields with flat-top envelopes will generate a pattern with a pyramid-shaped envelope. This hints at the wide range of possibilities – and challenges – for imposing arbitrary bunch structures on arbitrarily-distributed beams, e.g., in the event that both K_r and K_u are appreciable, it should be evident that for N_r,$N_{FS} \gg 1$ the dynamics of the system will no longer be separable into independent modulation and bunching phases, the general form of the Vlasov equation will need to be employed in computer simulations, and the modulation/bunching patterns will no longer be approximable, in general, by a straightforward convolution of the system fields.

A second critical point concerns our assumption of $\theta = 0$ rad for the analyses and calculations described in this paper. By varying this parameter (in conjunction with the k-vector(s) and polarizations of the radiation and FS fields), substantial reductions in the scaling of the required laser field strength and period in relation to those of the FS could be realized, and a broad range of novel variations in the energy modulation and bunching distributions would become accessible.

An interesting possibility for extending the present HCM scheme would be to post-accelerate the beam following an initial modulation phase. Eq. (9) reveals that the potential barrier to bunching is lowered inversely with a power of γ greater than 2. Thus, tailoring the initial modulation and post-accelerating field profiles could lead to the preparation of optimally microbunched beams at substantially higher energies. Another interesting extension could involve the restriction of FS and laser fields to the combinations of one or more higher multipole distributions for tailoring the energy modulation in directions transverse to the beam axis. An extension of this technique would be to prepare non-zero field integrals of selected multipole components (including the dipole). This would allow the design of transverse velocity modulation profiles, which could be used for phase-space filtering of the modulated/bunched region(s) independently of the overall bunch.

A potentially important application of HCM would be to attempt the longitudinal compression of a whole electron bunch at a relatively low value

Table 2
Fundamental differences between "single-cycle" HCM/bunching and multiple-period FEL (or OK) systems

FEL/OK	HCM
Modulation and bunching adiabatic	Modulation and bunching rapid
Relative energy change/period $\sim O(10^{-6})$	Relative energy change/period $\sim O(10^{-2})$
Electrons interact over "co-operation length" $\sim N_u\lambda_r/10$	Electrons interact over O(1) period
Self-consistent interactions through radiation + Coulomb fields essential to modulation/bunching	Electron-scattered radiation plays an initially minor role. Self-consistent interactions through Coulomb fields dominate
Dynamics can be linearized	Dynamics strongly non-linear
Modulation/bunching process periodic (quasi-periodic (chirped) in a tapered FEL)	Modulation/bunching patterns less restricted
Modulated bunch typically $\gg N_u\lambda_r/10$	Modulated bunch region typically $\sim O(\lambda_r)$

of gamma, followed by acceleration. In this case $\lambda_r > \sqrt{8\pi\sigma_B}$, implying the use of high-intensity mm-wavelength fields. Due to the small (and "soft") radiation losses in such a system, the technique, if successfully developed, could perhaps help to circumvent potential limits set by coherent synchrotron radiation (CSR) effects in high-energy chromatic compressors [16,17].

Finally, throughout this paper we have emphasized the relationship of HCM to OK and FEL phenomena. We conclude this section with a summarization of some of the basic distinctions between these two physical regimes in Table 2.

6. Summary

We have summarized recent studies of HCM systems and pointed out a number of potential applications for particle-beam and radiation source physics. It is perhaps not superfluous to note that HCM, as analyzed in this article, could be considered practicable for beam energies up to the 2–3 GeV range, provided FS structures of up to 2–4 m in length and with fields controllable down to the order of a few gauss could be implemented.[5] These requirements could of course be substantially eased with the advent of reliable UV lasers approaching the petawatt range, or, alternatively, by operating at $\theta > 1/\gamma^*$ with lower-power sources set in axisymmetric configurations. In the shorter term, it appears that the still-limited availability of femtosecond, terawatt lasers, is in part compensated by the fact that such devices are likely to be employed anyway at FEL facilities driven by photocathode RF guns, and there is a reason to expect that these may start becoming more prevalent. Moreover, collateral improvements in laser performance and parameter values [18], coupled with reductions in cost, can be expected to continue. Apart from this, the other basic constituents of an HCM and bunching/radiating system, viz., a high-quality, high current electron beam and a FS insertion device(s) are either in the process of being developed at a number of laboratories [7,19], or can be approximated to with existing resources. For example, even with the substantially poorer compression expected for a storage ring beam, the impressive output predicted by Eq. (11) – coupled with, say, a high-K_R radiator for harmonic generation – strongly suggests that HCM could be productively applied to recirculating beams as well.[6] Given the apparent richness of HCM physics, we

[5] A basic limit on the modulating FS length will be set by the Rayleigh range and waist size of the laser pulse (assuming no additional optics in the interaction region).

[6] For example, for a 2 GeV beam with 3×10^{11} particles/bunch and a bunch length of 3 cm, coherent output powers could approach the TW range. We note, however, that the higher relative energy spread of the circulating beam, which would inhibit the attainment of large wavelength reductions (as predicted by Eq. (10)), would make the attainment of even higher in-band powers substantially more difficult.

believe that undertaking such efforts is likely to further the understanding, control, and metrology of particle beams, and contribute to the development of FEL, OK, and other 4th generation radiation sources.

Acknowledgements

This work was supported in part by the Department of Energy Offices of Basic Energy Sciences and High Energy and Nuclear Physics, and Department of Energy Contract DE-AC03-76SF00515. Selected calculations utilized resources of the National Energy Research Scientific Computing Center, which is supported by the Office of Energy Research of the U.S. Department of Energy.

References

[1] R. Tatchyn et al., Nucl. Instr. and Meth. A 375 (1996) 274.
[2] C.P.J. Barty, T. Guo, C. LeBlanc, F. Raksi, C. Rose-Petruck, J.A. Squier, B. Walker, K.R. Wilson, V.V. Yakovlev, K. Yamakawa, Sub-20-fs multiterawatt lasers, Proceedings of Ultrafast Phenomena 1997, OSA Technical Digest Series vol. 8, 1996, p. 162.
[3] R. Tatchyn, Quantum limited temporal pulse generation, in: M. Cornacchia, H. Winick (Eds.), Proceedings of the Workshop on Fourth Generation Light Sources, SSRL Report 92/02, p. 482.
[4] R. Tatchyn, Nucl. Instr. and Meth. A 358 (1995) 56.
[5] R. Tatchyn, SPIE Proc. 2988 (1997) 112.
[6] R. Tatchyn, High-contrast energy modulation of electron beams for improved microbunching and coherence of free-electron radiation sources, presented at the SPIE 1998 Annual Meeting, San Diego Ca, July 19–24, 1998; SPIE Proc. (1998) 3451, in press.
[7] R. Tatchyn, Fourth generation insertion devices: new conceptual directions, applications, and technologies, in: M. Cornacchia, H. Winick (Eds.), Proceedings of the Workshop on Fourth Generation Light Sources, SSRL Report 92/02, p. 417.
[8] J. Aharoni, The Special Theory of Relativity, 2nd ed., Clarendon Press, Oxford, 1965.
[9] P.L. Csonka, Part. Accel. 11 (1980) 45.
[10] N.A. Vinokurov, A.N. Skrinsky, Institute of Nuclear Physics, Novosibirsk, USSR, Preprint 77/67.
[11] R. Bonifacio, R. Corsini, P. Pierini, Phys. Rev. A 45 (6) (1992) 4091.
[12] J.B. Murphy, C. Pellegrini, Introduction to the physics of the free electron laser, in: M. Month, S. Turner (Eds.), Lecture Notes in Physics vol. 296, Springer, Berlin, 1988, pp. 163–219.
[13] W.B. Colson, Free electron laser theory, Ph. D. Dissertation, Department of Physics, Stanford University, 1977.
[14] T.-Y. Wu, Kinetic Equations of Gases and Plasmas, Addison-Wesley, Palo Alto, 1966.
[15] J.D. Jackson, Classical Electrodynamics, Wiley, New York, 1975 (Chapter 14).
[16] B.E. Carlsten, J.C. Goldstein, Nucl. Instr. and Meth. A 393 (1977) 490.
[17] The LCLS Design Study Group, Linac Coherent Light source (LCLS) Design Study Report, Section 7, SLAC-R-521, 1998.
[18] C. Le Blanc, E. Baubeau, F. Salin, J.A. Squier, C.P.J. Barty, C. Spielmann, IEEE J. Selected Topics Quantum Electronics 4 (2) (1998) 407.
[19] J.F. Schmerge, D.A. Reis, M. Hernandez, D.D. Meyerhofer, R.H. Miller, A.D.T. Palmer, J.N. Weaver, H. Winick, D. Yeremian, Nucl. Instr. and Meth. A 407 (1988).

ELSEVIER

Nuclear Instruments and Methods in Physics Research A 429 (1999) 471–475

NUCLEAR INSTRUMENTS & METHODS IN PHYSICS RESEARCH
Section A

www.elsevier.nl/locate/nima

Measurement of far-infrared subpicosecond coherent radiation for pulse radiolysis

T. Kozawa*, Y. Mizutani, K. Yokoyama, S. Okuda, Y. Yoshida, S. Tagawa

The Institute of Scientific and Industrial Research, Osaka University, 8-1 Mihogaoka, Ibaraki, Osaka 567-0047, Japan

Abstract

Using a magnetic bunch compression method, a 26.5 MeV subpicosecond electron single bunch was generated with the L-band linac of Osaka University. The coherent transition radiation emitted from the subpicosecond single bunch was observed at wavelengths from 100 to 700 μm. The intensity was 7.9×10^9 times higher than that of the incoherent transition radiation obtained by calculation. The length of the compressed electron bunch was evaluated to be roughly 50 fs (rms) from the analysis of the spectra of the transition radiation. The coherent transition radiation has high enough intensity to be applied to pulse radiolysis as a pulsed light source.

PACS: 29.17; 41.60; 41.75; 82.50

Keywords: Coherent transition radiation; Subpicosecond electron bunch; Far-infrared; Pulse radiolysis; Pulsed light source

1. Introduction

The radiation emitted from an electron bunch is coherent at wavelengths longer than the length of the electron bunch. The intensity of the coherent radiation is enhanced by the factor of the number of electrons in a bunch. Coherent radiation is expected to be a new pulsed light source in the far-infrared region. Coherent radiation was first observed for synchrotron radiation [1]. For the other radiation processes, coherency was also observed at wavelengths longer than the bunch length [2,3].

In the field of radiation chemistry, the pulse radiolysis method is used to observe ultrafast phenomena. In particular, a stroboscopic pulse radiolysis method is available for time-resolved absorption spectroscopy in the picosecond time region. Investigation on ultrafast phenomena in the picosecond time region has been started using a picosecond electron bunch generated with a linac and the Cherenkov radiation emitted from them [4–6]. For the absorption spectroscopy, the electron bunch and the Cherenkov radiation were used as an irradiation source and a probe light, respectively. Cherenkov radiation was available as a pulsed light source in the visible region. Pulse radiolysis is one of the most powerful methods to investigate radiation-induced reactions. Using the pulse radiolysis method, ultrafast phenomena such as excitation, ionization, relaxation, solvation, geminate ion recombination and so on have been studied in the radiation chemistry, physics, biology and applied fields such as material science [4–8].

* Corresponding author. Tel.: + 81-6-879-8511; fax: + 81-6-875-4346.

E-mail address: kozawa@sanken.osaka-u.ac.jp (T. Kozawa)

0168-9002/99/$ - see front matter
PII: S0168-9002(99)00147-3

A new time-resolved absorption spectroscopy system using a femtosecond laser as an analyzing light instead of the Cherenkov radiation was developed at the Institute of Scientific and Industrial Research (ISIR), Osaka University [9–12]. The synchronization of a femtosecond laser with a picosecond linac enabled the measurement of ultrafast phenomena induced by radiation with a wide wavelength range from 250 nm to 1.5 μm. The investigation of ultrafast phenomena in the picosecond time region has been started on several subjects such as organic compounds, σ-conjugated polymers and lithographic materials [13]. Now, to obtain more information such as vibrational and rotational relaxation of molecules, a new pulsed light source in the far-infrared region is required.

In this paper, using an ultrashort electron bunch generated by a magnetic bunch compression method, the far-infrared coherent transition radiation emitted from the subpicosecond electron bunch was investigated.

2. Experimental setup

The ISIR L-band linac consists of a thermionic electron gun, two 108 MHz ($\frac{1}{12}$ of the main accelerating microwave frequency of 1300 MHz) subharmonic bunchers (SHB), a 216 MHz ($\frac{1}{6}$ of the main accelerating frequency) SHB, a 1300 MHz travelling wave-type prebuncher, a 1300 MHz travelling wave-type buncher, a 1300 MHz travelling wave-type accelerating tube and a focusing system [14]. The accelerating potential of the electron gun is provided by 95 kV DC.

A bunch compressor was installed at the ISIR L-band linac as shown in Fig. 1. The achromatic bunch compressor consists of two 45° sector magnets, four quadrupole magnets and a vertical beam slit. The energy of an electron bunch is modulated for the bunch compression by adjusting the RF phase in the accelerating tube. By translating the energy dispersion into the difference of trajectory lengths in the magnetic field of the bunch compressor, the electron bunch is compressed. The length of the compressed bunch was evaluated by measuring the Cherenkov radiation emitted by the compressed bunch in air at the end of the beam line. The Cherenkov radiation was measured by using a picosecond streak camera (Hamamatsu Photonics Co. Ltd.) which has a time resolution of 2 ps. An optical band pass filter, which is centered at 461.5 nm and has a half-width of 10.7 nm, was used to avoid pulse broadening due to optical dispersion in the convex lenses used in the measurement.

Fig. 2 shows the experimental setup for measuring the transition radiation emitted from the compressed single bunches. The transition radiation was generated with an A1 foil at the end of the beam line. Spherical and plane mirrors were used for focusing and reflecting the light, respectively. The intensity of light was measured with a liquid-He cooled silicon bolometer. The spectrum of light was measured with a grating monochromator at wavelengths of 100–700 μm. The absolute sensitivity of the detection system including the monochromator was calibrated with a high pressure Hg lamp of 250 W which was supposed to be a black-body radiation source of 5000 K.

3. Results and discussions

The RF phase in the accelerating tube was tuned so as to accelerate electrons in the early phase of the bunch more than those in the later phase of the bunch. The peak electric field in the accelerating tube was adjusted to 10 MV/m. The phase of the accelerating electric field was 70°. The peak energy of the accelerated bunch was 26.5 MeV. The energy spread after the modulation was 9.4% (FWHM). The magnetic fields of the bunch compressor were adjusted so as to make the shortest bunch. The compressed bunch length could not be evaluated by using the streak camera because its length was shorter than the time resolution of the streak camera. The horizontal and vertical beam sizes (full width) of the compressed bunch were 1.0 and 1.0 mm, respectively. The charge was 1.27 nC/bunch.

Fig. 3 shows the spectrum of the transition radiation emitted from the compressed single bunch. The spectral energy of the transition radiation emitted from an electron per unit frequency interval

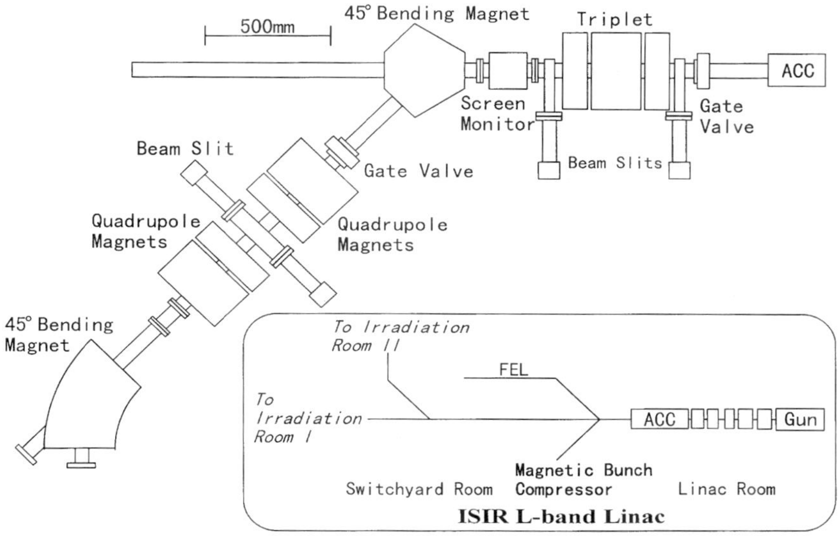

Fig. 1. Magnetic bunch compressor.

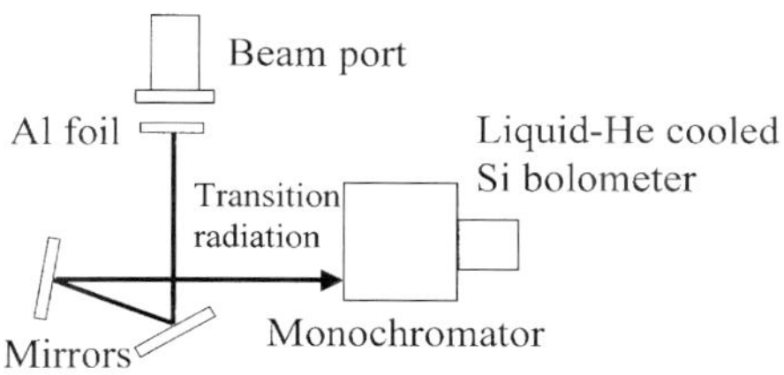

Fig. 2. Measurement system of transition radiation.

can be expressed [3] as

$$E(\omega) = \frac{e^2}{\pi c}\left(\ln\frac{2}{1-\beta} - 1\right) \tag{1}$$

where ω is the angular frequency of the radiation, e an elementary electric charge, c the speed of light, v the electron velocity, and $\beta = v/c$. The broken line in Fig. 3 shows the calculated spectrum of the incoherent transition radiation.

When the wavelength of radiation emitted from an electron bunch is longer than the length of the electron bunch, the radiation becomes coherent. When the cross section of the bunch is small and the observation point is far from the emitting point, the intensity of coherent radiation $[P_{\mathrm{coh}}(\lambda)]$ is expressed [15] as

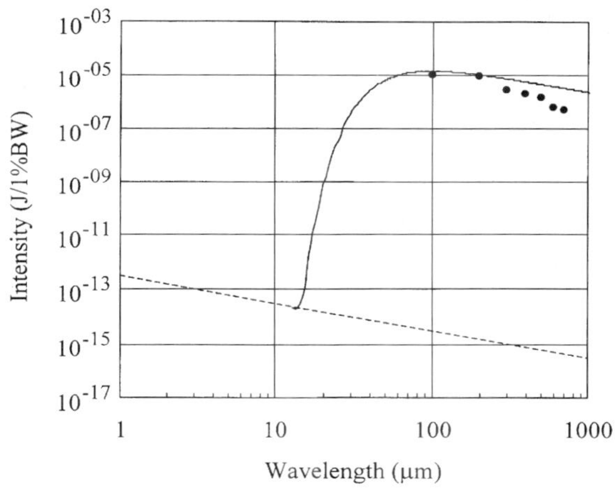

Fig. 3. Spectrum of the transition radiation. The solid points represent the experimental data, the solid line shows the spectral intensity expected for a Gaussian distribution, and the dashed line shows the calculated spectral intensity of the incoherent transition radiation.

$$P_{\mathrm{coh}}(\lambda) = P_{\mathrm{incoh}}(\lambda)[1 + (N-1)f(\lambda)] \tag{2}$$

where λ is the wavelength, N the number of electrons in a bunch, $P_{\mathrm{incoh}}(\lambda)$ the intensity of the incoherent radiation emitted from N electrons, and $f(\lambda)$ the bunch form factor. Assuming that the bunch shape is a Gaussian distribution, the bunch form factor is expressed [15] as

$$f(\lambda) = \exp(-\alpha^2/2), \quad \alpha = 2\pi\sigma/\lambda \tag{3}$$

where σ is the longitudinal bunch length. The solid line in Fig. 3 shows the spectrum of the coherent transition radiation calculated by the factor of $\sigma = 15.9$ μm (53 fs). The bunch length was chosen to fit the peak wavelength of the calculated spectrum to 100 μm. The bunch length is thought to be shorter than 15.9 μm because the peak could not be observed with the present detection system. The experimental data are in agreement with the calculated intensities. The observed intensities at wavelengths from 100 to 700 μm are 7.9×10^9 (number of electrons in a bunch) times higher than those of the incoherent transition radiation.

The observed intensity of the coherent transition radiation is high enough for the probe light in pulse radiolysis. We are planning new pulse radiolysis for far-infrared absorption spectroscopy to investigate the phenomena of rotational relaxation of molecules. The available pulsed light sources at ISIR are shown in Fig. 4. The free electron laser is also one of the promising light sources for the pulse radiolysis.

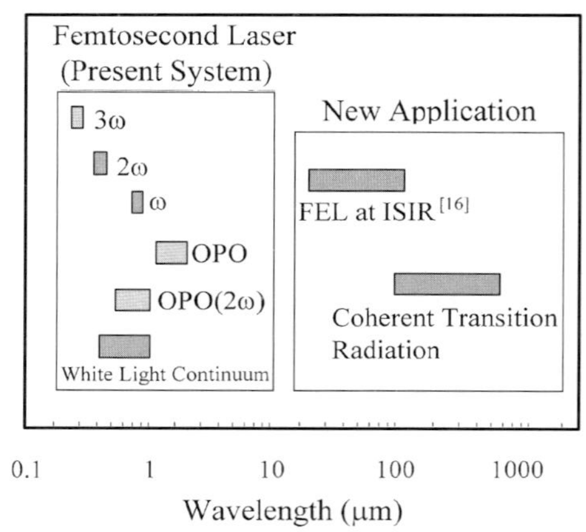

Fig. 4. Wavelengths of available pulsed light sources for pulse radiolysis at ISIR. ω, 2ω, 3ω and OPO mean fundamental, second harmonic, third harmonic and optical parametric oscillation, respectively. (see also Ref. [16]).

New studies on the relaxation dynamics of three important excited states (electronic, vibrational and rotational excited states) in the primary processes of radiation chemistry will be started by using the various light sources which cover the wide wavelength region from 250 nm (UV) to 700 μm (far-infrared).

4. Conclusion

Using the magnetic bunch compression method, a subpicosecond electron single bunch with a bunch length of roughly 50 fs (rms) and a charge of 1.27 nC was generated. The bunch length was evaluated from the analysis of the spectra of the transition radiation. The observed intensity of the coherent transition radiation at wavelengths from 100 to 700 μm was 7.9×10^9 times higher than that of the incoherent transition radiation. The coherent transition radiation is a new promising pulsed light source for the stroboscopic pulse radiolysis method.

References

[1] T. Nakazato et al., Phys. Rev. Lett. 63 (1989) 1245.
[2] J. Ohkuma, S. Okuda, K. Tsumori, Phys. Rev. Lett. 66 (1991) 1967.
[3] U. Happek, A.J. Sievers, E.B. Blum, Phys. Rev. Lett. 67 (1991) 2962.
[4] M.J. Bronskill, W.B. Taylor, R.K. Wolff, J.W. Hunt, Rev. Sci. Instr. 41 (1970) 333.
[5] C.D. Jonah, Rev. Sci. Instr. 46 (1975) 62.
[6] H. Kobayashi, Y. Tabata, Nucl. Instr. Meth. B 10/11 (1985) 1004.
[7] Y. Yoshida, S. Tagawa, M. Washio, H. Kobayashi, Y. Tabata, Radiat. Phys. Chem. 34 (1989) 493.
[8] S. Tagawa, N. Hayashi, Y. Yoshida, M. Washio, Y. Tabata, Radiat. Phys. Chem. 34 (1989) 503.
[9] Y. Yoshida, S. Tagawa, Proceedings of the International Workshop Femtosecond Tech., Tsukuba, Japan, 1995, p. 63.
[10] S. Tagawa, Y. Yoshida, M. Miki, T. Yamamoto, K. Ushida, Y. Izumi, Proc. Int. Workshop Femtosecond Tech., Tsukuba, Japan, 1996, p. 31.
[11] Y. Yoshida, Y. Mizutani, T. Kozawa, M. Miki, S. Seki, T. Yamamoto, K. Hori, S. Tagawa, Y. Izumi, K. Ushida, Proceedings of the Asian Particle Accelerator Conference, Tsukuba, Japan, 1998, in press.

[12] T. Kozawa, K. Hori, Y. Mizutani, T. Yamamoto, S. Suemine, Y. Yoshida, S. Tagawa, Proceedings of the Asian Particle Accelerator Conference, Tsukuba, Japan, 1998, in press.
[13] T. Kozawa, S. Nagahara, Y. Yoshida, S. Tagawa, T. Watanabe, Y. Yamashita, J. Vac. Sci. Technol. B 15 (1997) 2582.
[14] S. Okuda, Y. Honda, N. Kimura, J. Ohkuma, T. Yamamoto, S. Suemine, T. Okada, S. Takeda, K. Tsumori, T. Hori, Nucl. Instr. and Meth. A 358 (1995) 248.
[15] Y. Shibata et al., Nucl. Instr. and Meth. A 301 (1991) 161.
[16] R. Kato, S. Okuda, G. Kondo, S. Kondo, H. Kobayashi, T. Iog, S. Suemine, G. Isoyama, Nucl. Instr. and Meth., in press.

ELSEVIER

Nuclear Instruments and Methods in Physics Research A 429 (1999) 476–480

NUCLEAR INSTRUMENTS & METHODS IN PHYSICS RESEARCH Section A

www.elsevier.nl/locate/nima

An FEL based high-intensity gamma source at the TESLA Test Facility at DESY

C. Pagani[a], E.L. Saldin[b], E.A. Schneidmiller[b], M.V. Yurkov[c,*]

[a]*INFN Milano-LASA, Via Cervi, 201, 20090 Segrate (MI), Italy*

[b]*Automatic Systems Corporation, 443050 Samara, Russia*

[c]*Joint Institute for Nuclear Research, Particle Physics Laboratory (LSVE), Dubna, 141980 Moscow Region, Russia*

Abstract

One possible extension of the FEL activity at DESY is connected with the installation of an additional FEL beamline providing tunable UV radiation with peak and average power of 220 GW and 7 kW, respectively. This report presents the feasibility study of a high-intensity, polarized, monochromatic gamma source at the TESLA Test Facility. Gamma quanta are produced in the process of Compton backscattering of the UV FEL radiation on 1 GeV electrons of the TTF accelerator. The ultimate intensity of the gamma source can reach a value up to 10^{12} gamma quanta per second with a maximum energy of about 100 MeV. The energy resolution of the gamma source can be reduced down to a value of about 0.2%. Potential applications of the intense gamma source at the TESLA Test Facility are discussed as well.

PACS: 41.60.Cr; 41.75.Ht; 25.20.Dc

Keywords: Free electron laser; Compton backscattering; Gamma source

1. Introduction

Compton backscattering of laser light on high-energy electrons is now considered as the most promising way to obtain high-intensity monochromatic gamma rays [1]. Recently several investigations have been performed showing the possibility of constructing gamma sources with the energy of gamma quanta of the order of tens to hundreds of MeV with the ultimate intensity 10^{10}–10^{14} s^{-1} [2,3]. These proposals suggest using the scattering of free-electron laser (FEL) radiation on the electron bunches in a storage ring.

In this paper we point to the possibility of constructing a high-intensity gamma source at a high-power linear accelerator. A 1 GeV superconducting linear accelerator under construction at the TESLA Test Facility (TTF) at DESY will produce an electron beam with high average and peak power, and low energy spread and emittance. The main practical application of this accelerator is driving a soft

* Corresponding author. Tel: + 7-09621-62154; fax: + 7-09621-65767.

E-mail address: Yurkov@sunse.jinr.ru (M.V. Yurkov)

0168-9002/99/$ - see front matter
PII: S0168-9002(99)00148-5

X-ray FEL [4]. The construction of a high-power UV FEL at the TTF has also been proposed [5]. The radiation from a low-power master oscillator will be amplified in the FEL amplifier with a tapered undulator providing peak and average output power up to 220 GW and 7 kW, respectively. The Compton backscattering of this radiation on the TTF electron beam allows one to reach a gamma-quanta intensity of up to 10^{12} s^{-1} with the maximum energy of about 100 MeV. This unique gamma-ray source could be used for investigations in nuclear physics as well as for testing the technical solutions of positron sources and gamma–gamma options of future linear colliders.

2. High-intensity gamma source

A powerful UV FEL at the TTF [5] can be used for construction of a high-intensity, polarized, monochromatic gamma source. The scheme of the gamma source is presented in Fig. 1. Gamma quanta are produced in the process of Compton backscattering of laser photons on incoming electrons. The FEL interaction induces an energy spread in the electron bunch. If the laser radiation is scattered from the bunches involved in the FEL process, this will result in a decrease of the energy resolution of the gamma source. To avoid this harmful effect, the master laser should operate at half the repetition rate of the electron bunches. The laser radiation amplified by an electron bunch is focused on the next, unperturbed electron bunch.

The maximum gamma-quanta energy is defined by the frequency of the primary photons ω and the energy of the electron $\mathscr{E}$: $(\hbar\omega_\gamma)_{\max} = \mathscr{E}\chi/(1+\chi)$. Here $\chi = 4\gamma\hbar\omega/m_e c^2$, m_e is the electron mass and $\gamma = \mathscr{E}/m_e c^2$ is the relativistic factor. The maximum energy of the gamma quanta is about 86 MeV at a primary photon wavelength of 200 nm and an electron energy of 1 GeV. In principle, the energy of the TTF accelerator can be upgraded to a value of 2 GeV. In this case the maximum available gamma energy will be about 320 MeV. The general parameters of the gamma source at the TTF are presented in Table 1.

Table 1
Parameters of the gamma source at the TESLA Test Facility

General parameters	
Maximal energy of gamma quanta	85 MeV
Yield of gamma quanta	10^{12} s^{-1}
Energy resolution	1%
Angular divergence	0.5 mrad
Repetition rate	10 pps
Macropulse duration	800 μs
Number of pulses per macropulse	7200
Micropulse repetition rate	9 MHz
Micropulse duration (FWHM)	300 fs
Parameters of the electron beam	
Energy	1 GeV
Charge per bunch	1 nC
Energy spread (RMS)	1 MeV
Normalized emittance (RMS)	2 mm mrad
Beam size at the conversion point	14 μm
Bunch length (RMS)	50 μm
Parameters of the FEL	
Wavelength	208 nm
Micropulse duration	230 fs
Energy in the radiation pulse	50 mJ
Peak output power	220 GW
Average power	3.6 kW
Parameters of the focusing system	
Focuse distance of the mirror	3 m
Radiation spot size on the mirror	2.5 cm
Incident radiation flux on the mirror	350 W/cm^2
Reflectivity of the mirror	99%
Radiation spot size at the conversion point	8 μm
Rayleigh length	1 mm
Conversion efficiency	0.23%

The frequencies of the incident and scattered photons, ω and ω_γ, are connected by the relation

$$\hbar\omega_\gamma = \frac{\mathscr{E}\chi}{1+\chi+\gamma^2\theta^2} \tag{1}$$

where θ is the scattering angle. It is seen from Eq. (1) that there is a strict correlation between the energy of the backscattered photon and the scattering angle. This allows one to use angular collimation to select the gamma quanta with the required energy. In the case of monoenergetic and laminar electron beam this technique allows one, in principle, to achieve any required energy resolution of gamma quanta. In practice an electron beam has a finite energy spread which imposes a lower limit on energy resolution of the gamma source. The influence of the angular spread in the electron beam on the energy resolution of the gamma source can be eliminated by an appropriate choice of the beta function at the conversion point. The value of the rms angle spread in the electron beam is $\sigma_\theta = \sqrt{\varepsilon/\beta}$, where ε is the emittance of the electron beam and β is the beta function at the conversion point. Minimal energy resolution of the gamma source due to the angle spread is given by the expression

$$(\Delta\hbar\omega_\gamma)/\hbar\omega_\gamma \simeq (\gamma\sigma_\theta)^2/2. \tag{2}$$

It should also be noted that there is strict correlation between the energy of the gamma quantum and its polarization, so the angular selection provides not only monochromatization, but also determines the polarization of gamma quanta.

To obtain a higher yield of gamma quanta one should provide maximum spatial density of the primary photons and of the electron beam in the conversion region. We assume the laser beam to be close to Gaussian. The power flux density at the waist of a Gaussian laser beam is given by the expression $\mathrm{d}P(r)/\mathrm{d}S \propto \exp(-2r^2/w^2)$, where w is the size of the Gaussian beam waist. The Rayleigh length of the laser beam, $Z_\mathrm{R} = \pi w^2/\lambda$, and the beta function of the electron beam focusing system, β, are assumed to be large with respect to the length of the laser and the electron pulse. Thus, we can neglect the variation of the transverse size of the photon and electron pulses during the collision. As a result, the yield of gamma quanta is [5]

$$\frac{\mathrm{d}N_\gamma}{\mathrm{d}t} = fN_\mathrm{e}\delta, \quad \delta = \frac{A\sigma_\mathrm{c}}{2\pi\hbar\omega(\varepsilon\beta + w^2/4)} \tag{3}$$

where A is the energy in the radiation pulse, f is the repetition rate of collisions, N_e is the number of electrons per bunch, δ is the conversion efficiency (number of gamma quanta produced by one electron) and

$$\sigma_\mathrm{c} = 2\pi r_\mathrm{e}^2\left[\frac{1}{\chi}\ln(1+\chi) - \frac{8+4\chi}{\chi^3}\ln(1+\chi) + \frac{8}{\chi^2} + \frac{2+\chi}{2(1+\chi)^2}\right] \tag{4}$$

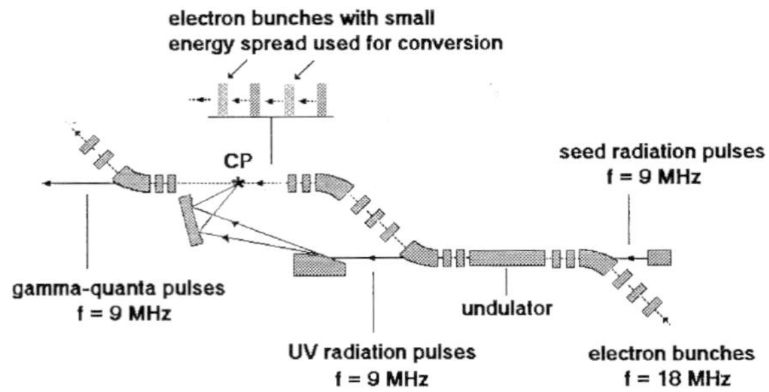

Fig. 1. Scheme of the gamma source at the TESLA Test Facility.

is the total Compton cross section on unpolarized electrons, where $r_\mathrm{e} = e^2/m_\mathrm{e}c^2$ is the classical radius of the electron.

The ultimate energy resolution of the gamma source presented in Fig. 1 is limited by the energy spread in the electron beam induced by the SASE FEL process. Even in the case when there is no seed signal at the entrance of the FEL amplifier, the process of the amplification of the radiation is triggered by the shot noise in the electron beam. Simulation of this process shows that the induced energy spread due to the SASE FEL process is about 5 MeV at the exit of the undulator [5] which imposes a limit on ultimate energy resolution of the gamma source of 1%. Taking into account the angular spread in the electron beam, the beta function of the electron beam at the conversion point should be about 20 cm in order to provide this energy resolution.

The radiation spot size at the conversion point has been set to $w \simeq 8$ μm (which corresponds to a Rayleigh length of 1 mm). In this case there are no fluctuations of the gamma-source intensity due to the time jitter of the electron pulse of about 1 ps. According to Eq. (3), at chosen parameters of the colliding photon and electron bunches conversion efficiency is about $\delta \simeq 0.23\%$. At a pulse repetition rate of the accelerator equal to 18 MHz the yield of gamma quanta is about 10^{12} s^{-1}.

There exists another option for organization of the conversion region prior to the entrance of the electron beam into the undulator. It provides the possibility to double the yield of gamma quanta, because each electron bunch can be used for producing gamma quanta. Due to a small value of the conversion efficiency this will not destroy the electron beam and it can be used for the amplification of the radiation in the undulator. Besides, an ultimate energy resolution of the gamma source of about 0.2% can be achieved. This limit is defined by the energy spread in the electron beam of 1 MeV. The value of the beta function at the conversion point should be increased up to 1 m. The yield of the gamma source with ultimate energy resolution of 0.2% will be about of $2 \times 10^{11}\ s^{-1}$.

3. Possible applications for nuclear physics

The gamma source at the TTF is a unique one providing an extremely high flux of polarized, monochromatic gamma quanta. The proposed gamma facility is an extremely selective probe for investigations of not only conventional nuclear physics, but of exotic nuclear states (similar to those which occur inside the core of neutron stars or the quark-gluon plasma) and reaction mechanisms. Up to 200 MeV we are dealing with the internucleon distance 7 about 1 fm (which is of the order of the size of nucleon). Nowadays the investigation of microscopic nuclear structure with resolution $\gtrsim 200$ MeV is an unexplored field [6]. The idea is to work in the kinematical region where the processes on free nucleons are forbidden (the cumulative particle production or the underthreshold particle production). This intermediate energy region is covered by nonperturbative QCD effects. Investigations of this transition energy region, from the perturbative to nonperturbative QCD, are the most attractive direction for relativistic nuclear physics to understand the role of the nonnucleonic (pion, Δ, etc., and few-nucleon configurations) and quark-gluon degrees of freedom.

Another interesting topic of these investigations is a possibility to extract additional information about the hadron or the nuclear structure which is usually hidden in the spin-averaged analysis [7]. Polarization observables have a promise of opening a new field in the photoproduction of pions from nucleons and nuclear targets [8,9], in the processes of photodisintegration of the lightest nuclei [10,11] and others. Many problems in photonucleon and photonulear physics are not resolved till now due to the lack of high-quality photon beams.

4. Test facility for the positron source

The projects of the next generation linear colliders plan to produce positrons by gamma quanta in a thin target [12,13]. Gamma quanta are produced by the spent electron beam in a long wiggler or undulator. For instance, in the TESLA project the 250 GeV electron beam passes a wiggler of 35 m length producing a powerful gamma beam [12]. The mean energy of gamma quanta is about 25 MeV. Positrons produced in a thin target have large energy and angle spread. After the target they should be captured and accelerated. Such a complicated system requires experimental verification prior to construction of the full-scale facility. The test facility can be constructed on the base of the high-intensity gamma source described above. The energy spectrum of the gamma source is similar to that used in a full scale facility. The intensity of the gamma source described above has been limited by the requirement of the energy resolution. For the positron source test facility the value of the beta function at the conversion point can be reduced down to a value of about 1 cm (the angle divergence in the beam is about $1/\gamma$ in this case). As a result, the yield of gamma quanta reaches the value of $10^{13}\ s^{-1}$. Remembering that the conversion efficiency of gamma quanta into electron–positron pairs is about 0.5 at a target thickness of about one radiation length, we estimate the yield of the positrons to be about $5 \times 10^{12}\ s^{-1}$. Such a positron yield is sufficient for operation of the test facility. Taking into account that the TTF accelerator has the same time structure as the TESLA accelerator, this facility will allow experimental verification of all the systems of the positron source (optimization of the target, capture efficiency, etc.).

The test facility for the positron source could reveal a unique opportunity for experimental investigation of the polarized positron source proposed in Ref. [14]. In this case, the FEL amplifier should be equipped with a helical undulator.

5. Test facility for the gamma–gamma collider

An option of a gamma–gamma collider is included in several projects of future generation linear colliders [12,13]. The scheme for realization of gamma–gamma collisions can be explained as follows. Two electron bunches are focused into the interaction point. Prior to the collision each electron bunch is irradiated by a powerful laser pulse. High-energy gamma quanta are produced in the process of Compton backscattering and, following the electron trajectories, collide at the interaction region. The time structure of the laser pulses should be identical to the time structure of the electron bunches. This means that the laser should have a capability of precise synchronization with the electron bunches and should provide a high repetition rate. The last parameter is extremely severe for the TESLA collider requiring a sub-terawatt level peak power laser with a repetition rate of about 1 MHz. In this case there is no reliable technical solution for a conventional laser system with the required parameters, and an FEL system is considered now as a candidate for the laser [12,15]. The parameters of the high-power UV FEL at the TTF and of the optical system of the gamma source are very close to those used in the project of the gamma–gamma collider. Indeed, the peak and average laser power of 220 GW and 7 kW, respectively, are close to those required for the gamma collider. The time structure of the laser pulses is similar to the time structure of the TESLA collider. The problems connected with alignment of the mirrors, possible radiation damage of the mirrors, time jitter and the problems of focusing the laser beam to a tiny spot are similar to those to be met at a full-scale facility. So the successful realization of this project will serve as an experimental verification of the technical feasibility of the gamma–gamma option at TESLA.

Acknowledgements

We wish to thank R. Brinkmann, S. Choroba, B. Dwersteg, J. Feldhaus, K. Flöttman, A. Gamp, P. Gürtler, J. Krzywinski, G. Materlik, T. Möller, J. Pflüger, J. Roßbach, J.R. Schneider, S. Schreiber, B. Sonntag, D. Trines, I. Will and J. Weisend II for many useful discussions.

References

[1] A. D'Angelo, Review of Compton scattering projects, presented at the EPAC-98 Conference, Stockholm.

[2] E.L. Saldin et al., Nucl. Instr. and Meth. A 375 (1996) 606.

[3] V.N. Litvinenko, J.M.J. Madey, Nucl. Instr. and Meth. A 375 (1996) 580.

[4] J. Rossbach, Nucl. Instr. and Meth. A 375 (1996) 269.

[5] C. Pagani, E.L. Saldin, E.A. Schneidmiller, M.V. Yurkov, Nucl. Instr. and Meth. A, 423 (1999) 190.

[6] L. Frankfurt, SLAC Workshop on High Energy Electroproduction and Spin Physics, SLAC-392 Conf-920266 UC-414, 1992.

[7] R.P. Feynman, Photon–hadron Interactions, W.A. Benjamin, Inc., New York, 1972.

[8] A.A. Chumbalov et al., Phys. Lett. B 321 (1994) 317.

[9] D. Drechsel, MKPH-T-93-9, Mainz, 1993.

[10] V. Isbert et al., DAPNIA/SPhN 93 64, C.E. Saclay, 1993.

[11] H. Arenhövel, M. Sanzone, Photodisintegration of the Deuteron, Springer, New York, 1991.

[12] R. Brinkmann, G. Materlik, J. Rossbach, A. Wagner (Eds.), Conceptual design of a 500 GeV e + e − linear collider with integrated X-ray laser facility, DESY 97-048, Hamburg, 1997.

[13] Zeroth-order design report for the next linear collider, LBNL-PUB-5424, SLAC Report 474, UCRL-ID-124161, May 1996.

[14] V.E. Balakin, A.A. Mikhailichenko, Proceedings of the Twelth International Conference on High Energy Accelerators, Batavia, 1983, p. 127; Preprint INP 79-85, Novosibirsk, 1979.

[15] E.L. Saldin et al., Nucl. Instr. and Meth. A 361 (1995) 101.

ELSEVIER

Nuclear Instruments and Methods in Physics Research A 429 (1999) 481–484

NUCLEAR INSTRUMENTS & METHODS IN PHYSICS RESEARCH Section A

www.elsevier.nl/locate/nima

Adsorption of pyridine on gold, studied by difference frequency generation (DFG) using the CLIO-FEL

A. Tadjeddine*, A. Le Rille, O. Pluchery, P. Hébert, W.Q. Zheng, T. Marin

Centre Universitaire Paris Sud, LURE-CNRS, Batiment 209D, BP 34, F-92898 Orsay Cédex, France

Abstract

The geometry of the adsorption of pyridine on gold single-crystal electrode has been investigated by in situ difference frequency generation (DFG) in aromatic ring spectral range, by using the CLIO free electron laser. We found that pyridine adsorbs in an end-on configuration, via the nitrogen lone pair, the molecular plane being tilted. The potential dependence of the vibrational modes showed evidences of a reorientation of the molecule around the potential of zero charge.

1. Introduction

Pyridine has been extensively used as a model compound for the study of bonding and coordination with metal surfaces. This molecule contains a delocalized π-electron system and a lone pair located on the N atom, both of which participate in bonding. As early as 1971, Chao et al. inferred from ellipsometric measurements a change of the pyridine bonding on a polycrystalline gold electrode, from a flat lying orientation at negative potential to a vertical orientation at positive potential [1]. The potential dependence of the adsorption geometry of pyridine on Au(*hkl*) electrodes has been studied by Stolberg et al. [2–4]. They found that the molecules adsorb in a flat orientation at negatively charged surfaces, and in an N-bonded vertical orientation at positively charged surfaces. Therefore, pyridine undergoes a phase transition at the potential of zero charge (pzc). Pettinger et al. have also investigated this transition, using second harmonic generation [5,6]. Brolo et al. have been able to record SERS spectra of pyridine adsorbed on a Au(210) single-crystal electrode [7]. However, the signal from the bulk molecules can hide that of the adsorbed molecules and careful data treatment is needed to extract the interfacial spectrum from the overall signal. On the other hand, non-linear second-order processes such as sum (SFG) and difference (DFG) frequency generation are inherently interface sensitive between centrosymmetric media where bulk generation is forbidden in the electric dipole approximation [8]. Direct measurement of vibrational spectra of adsorbed species is thus possible, without using the modulation or difference methods required in infrared spectroscopy. We present the first investigation of the pyridine–gold electrochemical interface by DFG using the CLIO-FEL laser in the aromatic ring spectral range, around 10 μm.

* Corresponding author. Tel.: + 33-1-64-46-80-02; fax: + 33-1-64-46-41-48..

0168-9002/99/$ - see front matter
PII: S0168-9002(99)00151-5

2. Experimental

The optical set-up used in this experiment has been described elsewhere [9]. An actively modelocked frequency-doubled YAG laser is synchronised with the IR CLIO FEL. The temporal profiles of both lasers consist of micropulses at a repetition rate of 62 MHz emitted in 10 μs bunches at 25 Hz. The micropulse duration is 70 ps for the YAG and 1 ps for CLIO. The spectral width of the IR pulses is 10 cm^{-1}. The green (30 mW) and the IR (150 mW) beams were focussed on the sample at incidence angles of 55° and 65°, respectively. This geometry allows generation of a DFG signal, which, unlike SFG, is not possible in any configuration, due to phase matching conditions [10]. The DFG signal is detected by a photomultiplier after spatial and spectral filtering and normalised to a SFG reference generated in the bulk of a ZnS crystal. Measurements were performed in a spectroelectrochemical cell equipped with a 60° BaF_2 prismatic window to prevent strong refraction of the laser beams. The working electrode was a Au(111) single-crystal disc polished and annealed following the standard electrochemical procedure. The counter electrode was a large area platinum wire and the reference potential was a standard calomel electrode (SCE) from which all potentials are referred. All solutions were prepared from ultra-pure water and reagents. During the experiment, the solution was continuously purged by an argon stream. The electrode was lightly pressed against the cell window to minimize absorption of the IR light by the electrolyte solution. DFG spectra were recorded at a fixed potential between − 0.9 and + 0.7 V.

3. Results and discussion

Fig. 1 shows the DFG spectrum recorded for Au(111) in contact with a solution of 0.1 M $NaClO_4$ + 0.002 M pyridine, at a potential of + 0.3 V, between 900 and 1150 cm^{-1}. The experimental data were best fitted with five resonances at 981, 1008, 1028, 1050, and 1092 cm^{-1} [11].

The pyridine molecule belongs to the C_{2v} point group with two orthogonal mirror planes contain-

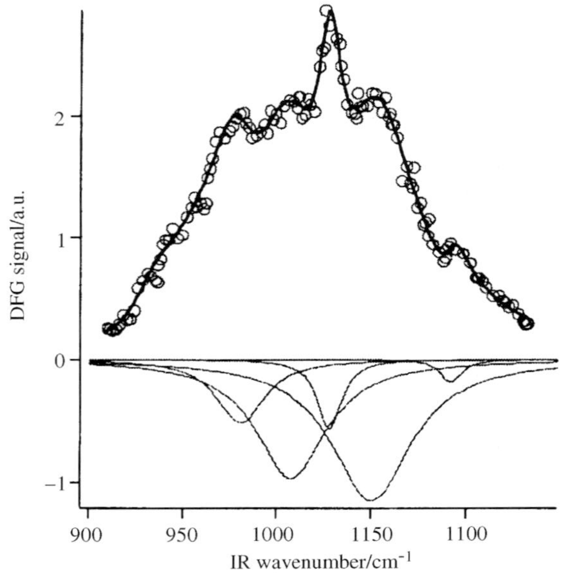

Fig. 1. DFG spectrum of Au(111) electrode in contact with (0.1 M $NaClO_4$ + 0.002 M pyridine) at E = + 0.3 V/SCE. Also shown are the six bands obtained after deconvolution of the DFG spectrum.

ing the C_2 axis. The 27 fundamental modes of vibration of liquid pyridine taken from and labelled following Corrsin et al. [12] consist of 10 A_1, 3 A_2, 9 B_1, and 5 B_2 modes. In the gas phase, the A_1, B_1, and B_2 modes are both Raman an IR active, the A_2 modes being only Raman active. The A_1 and B_1 modes account for the in-plane vibrations, whereas the A_2 and B_2 modes represent the out-of-plane vibrations. The frequencies of the bands observed for adsorbed molecules are likely to be different from those of the free molecules, due to their electronic interactions with the metal.

The bands at 1008 and 1028 cm^{-1} are in good agreement with those obtained by Brolo et al. on Au(210) [7]. They are assigned to the symmetric ring breathing mode 1 and the asymmetric ring breathing mode 12, respectively. The bands at 1050 and 1092 cm^{-1} are assigned to the modes 18a and 18b, respectively. These two modes have the same frequency of 1068 cm^{-1} in liquid pyridine. The band at 981 cm^{-1} can be assigned to the A_2 mode 17a which is IR inactive in the gas phase. However, it has a weak IR activity in the liquid phase [12,13] which explains the observed DFG spectrum.

The symmetry of the modes observed in the vibrational spectra can give information on the adsorption geometry of the molecule. For a given geometry, a mode will be observed if it gives rise to an IR dynamic dipole normal to the surface. The atomic displacements have been given for each mode of vibration of pyridine by Long and Thomas [14]. Pyridine can be bound perpendicular to the surface via the N atom, in a way retaining both symmetry planes. In this case only A_1 modes will be observed. The molecule can lie flat on the surface, the mirror plane perpendicular to the molecular plane being retained, in which case only out-of-plane modes will be observed. Finally, the molecule can be tilted, due to simultaneous N and π-bonding. If the mirror plane perpendicular to the molecular plane is retained, B_1 modes will not be observed. If the molecule is tilted and rotated around the C_2 axis, all active modes will be observed. Our spectra show that the pyridine molecule is bound via the N atom and is tilted towards the surface and rotated around the C_2 axis (Fig. 2). This geometry has been inferred, from AFM observations, to the orientation of pyridine adsorbed on heulandite (010) in aqueous solutions [15].

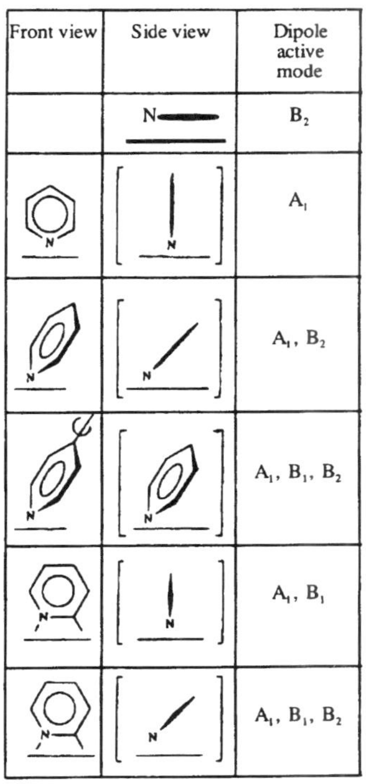

Fig. 2. Front and side views of the different adsorption configurations of pyridine, with their vibrational dipole active mode.

No significant frequency shift of the ring breathing mode 1 was detected within our spectral resolution, while a shift of 10 cm^{-1} V^{-1} was observed for mode 18b. These results are in agreement with the shifts of mode 1 reported for pyridine adsorbed on silver (6 cm^{-1} V^{-1}) [16], on copper (10 cm^{-1} V^{-1}) and Au(210) (no significant shift) [7]. The weak potential dependence of the mode vibrational frequency is consistent with an N-bonded adsorption geometry.

Fig. 3 shows the evolution of the oscillator strength of the 1008 and 1092 cm^{-1} bands, as a function of the applied potential. These bands correspond to the ring breathing mode 1 (A_1) and to the C–H in-plane bend mode 18b (B_1), respectively. The oscillator strength of both the A_1 and the B_1 mode is constant at negative potential and increases at potential positive of + 0.05 V, close to the pzc. This behavior indicates that a reorientation occurs around the pzc. Moreover, the activity of a B1 mode is possible only if the molecule is

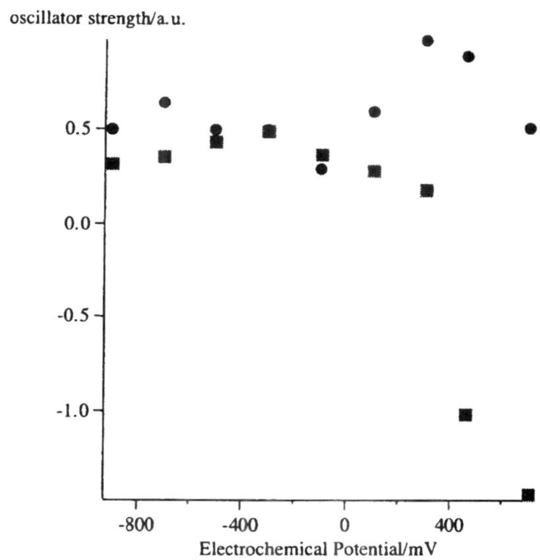

Fig. 3. Potential dependence of the oscillator strength of the vibrational bands at ca. 1008 cm^{-1} (circles) and at ca. 1090 cm^{-1} (squares).

rotated around the C_2 axis. This conclusion is further supported by the observed change of the sign of oscillator strength of mode B1, indicative of a change of orientation of the dipole of the vibrational mode.

4. Conclusion

In situ vibrational spectra of pyridine adsorbed on a gold single-crystal electrode have been measured by DFG as function of the electrode potential, in the 900–1150 cm^{-1} spectral range, using the CLIO-FEL. Pyridine was found to be adsorbed via the nitrogen lone pair, in an end-on configuration, the molecular plane being tilted. Our results give spectroscopic evidence of the reorientation of pyridine around the potential of zero charge. It demonstrates the effectiveness of CLIO to probe the electrochemical interface in high IR wavelength.

Acknowledgements

We thank the CLIO staff for their technical support throughout this experiment.

References

[1] F. Chao, M. Costa, A. Tadjeddine, Bull. Soc. Chim. France 7 (1971) 2465.
[2] L. Stolberg, S. Morin, J. Lipkowski, D.E. Irish, J. Electroanal. Chem. 307 (1991) 241.
[3] L. Stolberg, J. Lipkowski, D.E. Irish, J. Electroanal. Chem. 238 (1987) 333.
[4] L. Stolberg, J. Lipkowski, D.E. Irish, J. Electroanal. Chem. 296 (1990) 171.
[5] B. Pettinger, S. Mirwald, J. Lipkowski, 397 (1993) 395.
[6] B. Pettinger, J. Lipkowski, S. Mirwald, A.J. Friedrich, J. Electroanal. Chem. 329 (1992) 289.
[7] A.G. Brolo, D.E. Irish, J. Lipkowski, J. Phys. Chem B 101 (1997) 3906.
[8] Y.R. Shen, Nature 337 (1989) 519.
[9] A. Peremans, A. Tadjeddine, P. Guyot-Sionnest, Nucl. Instr. and Meth. A 341 (1994) 146.
[10] A. Tadjeddine, A. Le Rille, in: A. Wieckowski (Ed.), Interfacial Electrochemistry: Principles and Applications, Marcel Dekker, New York, in press.
[11] P. Hébert, A. Le Rille, W.Q. Zheng, A. Tadjeddine, J. Electroanal. Chem. 447 (1998) 5.
[12] L. Corrsin, B.J. Fax, R.C. Lord, J. Chem. Phys. 21 (1953) 1770.
[13] D.P. Di Lella, H.D. Stidham, J. Raman Spectrosc. 9 (1980) 90.
[14] D.A. Long, E.L. Thomas, Trans. Faraday Soc. 59 (1963) 783.
[15] M. Komiyama, T. Shimaguchi, T. Koyama, M.M. Guu, J. Phys. Chem. 100 (1996) 15 198.
[16] R. Kötz, E. Yeager, J. Electroanal. Chem. 123 (1981) 335.

ELSEVIER

Nuclear Instruments and Methods in Physics Research A 429 (1999) 485–488

NUCLEAR INSTRUMENTS & METHODS IN PHYSICS RESEARCH Section A

www.elsevier.nl/locate/nima

FEL multiphoton dissociation and isotope separation of boron

Masaki Hashida[a,*], Masayori Matsuoka[b], Yasukazu Izawa[b], Yukinori Nagaya[c], Muneo Ayabe[d]

[a]*Institute for Laser Technology, 2-6 Yamadaoka, Suita Osaka 565, Japan*
[b]*Institute of Laser Engineering, Osaka University, 2-6 Yamadaoka, Suita Osaka 565, Japan*
[c]*The Kansai Electric Power Co., Inc., 3-22, 3-Chome, Nakanoshima, Kitaku Osakan 530, Japan*
[d]*Ishikawajima-Harima Heavy Industries Co., Ltd., 1 Shin-Nakahara-Cho, Isogu-Ku, Yokohama 235, Japan*

Abstract

The dependencies of dissociation yield and isotope selectivity on FEL frequency for boron trichloride were measured. The measurements have been performed by using metallic cell (38 cm^3) filled with a mixture of BCl_3 (2 Torr) and H_2S (10 Torr). The maximum dissociation yield was obtained at a laser frequency shifted by 10 cm^{-1} toward the red side from the vibrational resonance of BCl_3. The maximum selectivity was $\sim$ 1.5, which increased gradually with decreasing the laser frequency toward the red side from the resonance.

Keywords: Micropulse energy; Isotope separation; FEL frequency; Multiphoton dissociation

1. Introduction

Dissociation of polyatomic molecules in a strong infrared laser occurs through multiphoton absorption of radiation by the molecular vibration levels. The dissociation process is isotopically selective when the laser pulse duration is much shorter than the relaxation (V–V transfer) time. Isotopically selective dissociation of polyatomic molecules in a strong IR laser field has been reported by many authors since the 1970s and knowledge of the dissociation process seems to be abundant. However, target molecules have been limited by discretely oscillating IR laser frequencies. Free electron laser (FEL) light has continuous tunability in a wide frequency region, short micro pulse duration, and relatively high-peak (micropulse) power. The FEL will greatly increase the range of molecules suitable for isotope separation and improve the isotope selectivity. We have demonstrated selective dissociation experiments with the FEL for the first time. The present paper reports the dissociation yield and isotope selectivity for boron trichloride.

Naturally occurring boron consists of two isotopic components (^{10}B: 19.0% and ^{11}B: 81.0%). Boron compounds are used as neutron absorbers in nuclear power plants and also in the medical field due to the relatively large neutron absorption cross section of ^{10}B. By the use of ^{10}B enriched compounds, the amounts of boron compound added into the cooling water of nuclear power reactor and injected into the human blood will be reduced.

* Corresponding author.

0168-9002/99/$ - see front matter
PII: S0168-9002(99)00152-7

Multiphoton dissociation of BCl_3 and isotope separation of ^{10}B and ^{11}B by using TEA CO_2 laser have been reported by several authors [1–9]. However, the isotope selectivity was less than 2 because of the long laser pulse width compared to the V–V transfer relaxation time [10]. The FEL micropulse duration (10 ps) is much shorter than V–V transfer relaxation time ($\tau \sim 250$ ns at 2.0 Torr) of BCl_3 for our experimental condition. The FEL will be a useful tool for improving the isotope selectivity of BCl_3. BCl_3, BF_3, and BBr_3 have a strong ν_3 absorption band in the 10, 7, and 12 μm region, respectively [11,12]. The isotope shift for ^{10}B and ^{11}B in BCl_3, BF_3, and BBr_3 is about 40 cm^{-1}. These wavelength regions are covered by the FEL. Thus BCl_3, BF_3, and BBr_3 are good candidates for FEL laser isotope separation. For TEA CO_2 laser, the wavelength oscillates discretely in the range of 9–10 μm. For the first experiment of multiphoton dissociation and isotope separation experiment using FEL, we selected BCl_3 as a target molecule. The dissociation yield and the isotope selectivity are compared for FEL and CO_2 laser irradiation.

2. Experiments

The multiphoton dissociation process is described by

$$BCl_3 + nh\nu \leftrightarrow BCl_2 + Cl \quad (1)$$

In order to prevent inverse reactions by recombination, H_2S was added as a scavenger gas, which induces the following reaction [5,13]:

$$BCl_2 + Cl + H_2S \rightarrow HSBCl_2 + HCl \quad (2)$$

Finally, the dissociated molecules are collected as $HSBCl_2$ in liquid phase.

For the dissociation experiments of BCl_3, we used the FEL1 at the Free Electron Laser Research Institute (FELI), Japan. The oscillation characteristics of FEL1 are summarized in Table 1. The FEL time structure consists of two components (macropulse and micropulse). The macropulse is composed of a train of 268 micropulses.

The mixture of BCl_3 and H_2S was filled in a stainless steel cell with four windows. Initial pressures of BCl_3 and H_2S were 2.0 and 10.0 Torr, respectively. FEL irradiation was focused with a ZnSe lens ($f = 30$ mm) into the cell horizontally through KCl windows. At the focal position, the FEL beam was nearly circular with a diameter of 400 μm. The maximum intensity at the focal point was 1.5 GW/cm^2. During the laser irradiation experiment, the changes in concentrations of $^{10}BCl_3$ and $^{11}BCl_3$ were monitored by measuring the absorption spectrum by FTIR spectrometer through the two KCl windows. The measured FTIR absorption spectra were compared to the initial absorption spectrum to determine the fractional dissociation. The FEL macropulse energy was monitored at the entrance and exit of the cell with a calorimeter. For CO_2 irradiation experiments, the energy of laser was varied from 50 to 600 mJ with a polarizer.

Table 1
Oscillation characteristics of FEL1 at FELI

Wavelength tuning range	5.5–12.5 μm
Micropulse	
energy	18 μJ
Pulse width	10 ps
Repetition	22.3 MHz
Macropulse	
energy	5 mJ
Pulse width	12 μs
Repetition rate	10 Hz
Average output power	50 mW
Line width ($\Delta\nu/\nu$)	0.6%

3. Results and discussion

Fig. 1 shows the decrease in the concentrations of $^{10}BCl_3$ and $^{11}BCl_3$ at the laser frequency of 944 cm^{-1}. The concentration N was normalized by the initial concentration N_0. The solid and open squares are the experimental data for $^{10}BCl_3$ and $^{11}BCl_3$, respectively. The solid curves are obtained by using the following equation:

$$N = N_0 \exp(-d_x S)$$

where S is the number of laser shots, and d_x is the apparent dissociation yield averaged over the

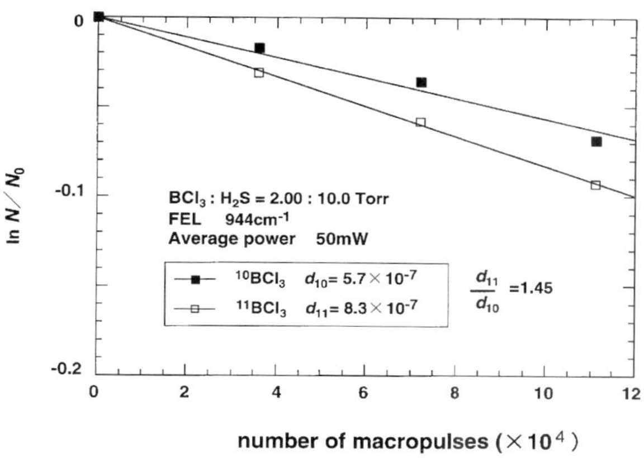

Fig. 1. Multiphoton dissociation of BCl_3 by the FEL at the frequency of 944 cm^{-1}.

whole cell volume, and slope of line corresponds to the dissociation yield. In this experiment the dissociation yields for one macropulse were 5.7×10^{-7} and 8.3×10^{-7} for $^{10}BCl_3$ and $^{11}BCl_3$, respectively. The isotope selectivity defined by d_{11}/d_{10} was 1.45.

Fig. 2 shows the laser frequency dependence of the dissociation yield. The absorption spectrum of BCl_3 is shown as a solid line. The FEL line width is 6 cm^{-1}, which is much less than the isotope shift between $^{10}BCl_3$ and $^{11}BCl_3$. The maximum dissociation yield was obtained for the frequency shifted by 10 cm^{-1} towards the red side from the vibrational resonance of BCl_3. Similar results had been reported earlier for BCl_3 molecule [10,14], and have been attributed to shift of spectrum of transitions in the quasi-continuum of high vibrational states because of the anharmonicity of target molecules. At 985 cm^{-1}, the dissociation of $^{10}BCl_3$ was slightly faster than $^{11}BCl_3$.

In Fig. 3, the dissociation yields by FEL at 944 cm^{-1} are compared to those obtained with a TEA CO_2 laser. Open and closed circles are the dissociation yield for 100 ns CO_2 laser pulse width. Open and closed squares are the FEL data. The horizontal axis shows the laser intensity at the focal point. Solid and dashed curves are the calculated dissociation yields for CO_2 laser and FEL, respectively, under the assumption that in the region where the laser intensity exceeds the dissociation threshold all the BCl_3 molecules dissociate for one CO_2 laser pulse or for one macropulse of FEL. The dissociation threshold for BCl_3 with CO_2 laser has been reported to be in the range of 10^8–10^9 W/cm^2 by several researchers [1,14,15]. In this calculation, the dissociation threshold of 540 MW/cm^2 was used. This curve shows that below a dissociation threshold no dissociation occurs, and with increasing laser intensity, the volume in the gas where the laser intensity exceeds the dissociation threshold increases according to the plot. The experimental data for CO_2 laser fit the calculation very well. It is concluded that for CO_2 irradiation with intensities above its dissociation threshold is 100%. In the case of FEL, the experimental data shows a smaller dissociation yield than calculated. However, at an intensity of 1 GW/cm^2 the decrease in BCl_3

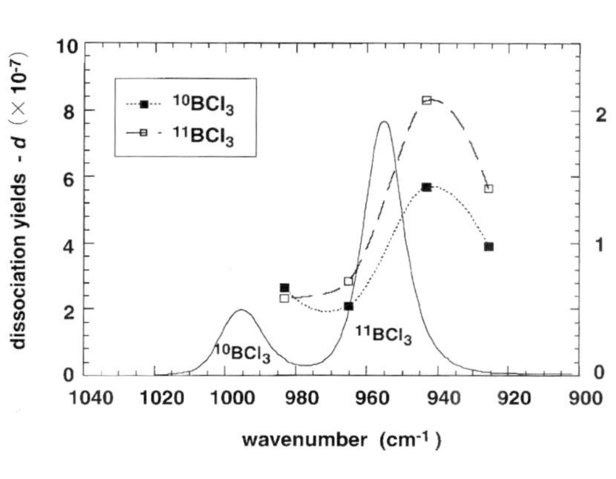

Fig. 2. Laser frequency dependence of the dissociation yield.

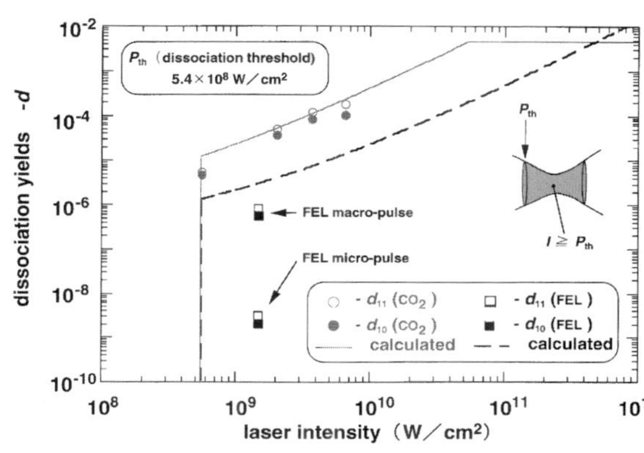

Fig. 3. Dissociation yields depend on laser intensity of FEL and CO_2 laser at the frequency of 944 cm^{-1}.

concentration was too low to determine the dissociation yield. Thus the dissociation threshold for FEL may be larger than that for CO_2 laser, because the pulse energy of the FEL is much lower than that of a CO_2 laser. If the dissociation threshold for FEL is $\sim 1\ GW/cm^2$, the experimentally obtained dissociation yield for FEL becomes very close to the calculation. The dissociation yield for a micropulse shown in the figure is a lower limit, because it was impossible to determine the yield for one micropulse.

Since the micropulse duration (10 ps) is much shorter than V–V transfer relaxation time ($\tau \sim 250$ ns) of BCl_3 at 2.0 Torr, the isotope selectivity will be improved when BCl_3 molecules are dissociated perfectly with one micropulse irradiation. However, the isotope selectivity for the FEL was ~ 1.5. This value is nearly same as for CO_2 laser irradiation. This fact suggests that the successive excitation of BCl_3 by the train of FEL micropulses is the important process to lead the dissociation, in other words, BCl_3 molecules are selectively excited in highly vibrational states by one micropulse of irradiation. In order to improve the isotope selectivity, the dissociation should be finished before the next micropulse irradiates the BCl_3 molecule. For this reason we have a proposal to perform the experiment by two-color laser irradiation scheme, where the precisely tuned FEL micropulse excites BCl_3 up to the 3–4 vibrational states followed by a CO_2 laser pulse leading to dissociation.

4. Summary

Multiphoton dissociation of BCl_3 using an FEL was demonstrated for the first time. It was found that the dissociation yield in the laser irradiation region where the laser intensity exceeds the dissociation threshold is not 1. The maximum dissociation yield was obtained at the laser frequency shifted by 10 cm^{-1} towards the red side from the vibrational resonance of BCl_3. At this frequency, an isotope selectivity of ~ 1.5 was found. The low value of isotope selectivity suggests that the successive excitation of BCl_3 by the train of FEL micropulses is the dominant process in the multiphoton dissociation. In order to improve the isotope selectivity, a two-color irradiation by FEL and CO_2 laser scheme will be suitable.

Acknowledgements

The FEL irradiation was carried out at FELI. The authors are grateful to the staffs of FELI for their comments and technical helps.

References

[1] R.V. Ambartsumyan, V.S. Letokhov et al., JETP Lett. 20 (1974) 273.
[2] J.J. Comeford, S. Abramowitz, J. Chem. Phys. 61 (1974) 3571.
[3] N.V. Krlov, Appl. Opt. 13 (1974) 301.
[4] R.V. Ambartsumyan et al., Sov. Phys. JETP 42 (1975) 36.
[5] S.M. Freund, J.J. Ritter, Chem. Phys. Lett. 32 (1975) 255.
[6] C.T. Lin, T.D.Z. Atvers et al., J. Appl. Phys. 48 (1977) 1720.
[7] V.S. Letokov, Phys. Today (1977) 23.
[8] R.V. Ambartsumyan, V.S. Letokhov, Acc. Chem. Res. 10 (1977) 61.
[9] C.T. Lin, T.D.Z. Atvers, J. Chem. Phys. 68 (1978) 4233.
[10] P.L. Houston, A.V. Nowak et al., J. Chem. Phys. 58 (1973) 3373.
[11] I.W. Levin, S. Abramowittz, J. Chem. Phys. 43 (1965) 4213.
[12] G. Hertzberg, Molecular spectra and Molecular Structure, II Infrared and raman Spectra of Polyatomic Molecules, Van Nostrand, New York, 1945.
[13] J. Bouix, H. Hillel, Can. J. Chem. 51 (1973) 292.
[14] Yu.R. Kolomiiskii, E.A. Ryabov, Sov. J. Quantum Electron. 8 (1978) 375.
[15] Y. Ishikawa, O. Kurihara, R. Nakane, S. Arai, Chem. Phys. 52 (1980) 143.

ELSEVIER

Nuclear Instruments and Methods in Physics Research A 429 (1999) 489–496

NUCLEAR INSTRUMENTS & METHODS IN PHYSICS RESEARCH Section A

www.elsevier.nl/locate/nima

Applications of UV-storage ring free electron lasers: the case of super-ACO

L. Nahon[a,b,*], E. Renault[a,b], M.E. Couprie[a,b], F. Mérola[a], P. Dumas[a], M. Marsi[c], A. Taleb-Ibrahimi[a], D. Nutarelli[a,b], R. Roux[a], M. Billardon[a,d]

[a] *LURE, bat 209d, Université Paris-Sud, BP 34, 91898 Orsay Cedex, France*

[b] *SPAM/DRECAM/DSM, bat 522, CE de Saclay, 91191 Gif sur Yvette Cedex, France*

[c] *Sincrotrone Trieste, 34012 Trieste, Italy*

[d] *ESPCI, 10 rue Vauquelin, 75231 Paris Cedex, France*

Abstract

The potential of UV-storage ring free electron lasers (SRFELs) for the performance of original application experiments is shown with a special emphasis concerning their combination with the naturally synchronized synchrotron radiation (SR). The first two-color FEL + SR experiment, performed in surface science at Super-ACO is reported. The experimental parameters found to be the most important as gathered from the acquired experience, are underlined and discussed. Finally, future prospects for the scientific program of the Super-ACO FEL are presented with two-color experiments combining the FEL with SR undulator-based XUV and VUV beamlines as well as with a SR white light bending magnet beamline emiting in the IR-UV (20 μm–0.25 μm). © 1999 Elsevier Science B.V. All rights reserved.

PACS: 07.85.Qe; 41.60.Cr; 42.62.Fi; 71.20.Mq

Keywords: Synchrotron radiation; Free electron laser; Pump/probe experiment; Surface photovoltage effect; Absorption spectroscopy

1. Opportunities offered by UV-SRFELs sources

As it is obviously appearing in their names, UV-storage ring free electron lasers (UV-SRFELs) happen to emit coherent light in the UV range and are based on storage rings ! These features are potentially extremely interesting in the sense that they open two important directions for applications, which are not directly possible to follow with the more and more challenging conventional lasers.

1.1. Emission in the UV

In recent years, as the quality of the electron beams available for storage ring FELs improved, photons have been produced in the UV, for instance at Super-ACO since 1991 with wavelengths around 350 nm [1], around 239 nm at UVSOR since 1996 [2] and more recently around 234 nm at NIJI-4 [3] and around 240 nm at Duke [4], and very likely in the future in the VUV range (below 180 nm). This makes the UV-SRFEL a very interesting and sometimes unique tool for the performance of one-photon time-resolved and/or frequency-resolved experiments in photochemistry and photobiology, since most of organic and

* Corresponding author. Tel.: + 33-1-64-46-88-60; fax: + 33-1-64-46-41-48.

E-mail address: nahon@lure.u-psud.fr (L. Nahon)

PII: S0168-9002(99)00153-9

inorganic molecules have their first electronically excited states lying in the 380 – 200 nm range. For instance the time-resolved polarized fluorescence decay of a biological molecule, the NADH coenzyme, has been studied with the Super-ACO FEL leading to very interesting results [5]. In such experiments, beside the spectral range, one takes advantage of the linear polarization, the pulse-to-pulse stability and the high repetition rate of SRFELs which is very suitable for time-correlated photon-counting-type experiments. One can set up the same experimental environment around SRFELs in terms of detection technique and sample preparation (e.g. a UHV chamber or a molecular beam), as for conventional table top lasers. One can also consider the coupling of a SRFEL with synchronized external lasers for the performance of two-color experiments where one of these colors has to be tunable in the UV range, for instance in resonance enhanced multiphoton ionization (REMPI) experiments.

1.2. The natural synchronization with the synchrotron radiation

SRFEL and synchrotron radiation (SR) pulses have the same origin, i.e. the electron bunches travelling in the storage ring, leading to a natural pulse-to-pulse synchronization between the two sources. Although several groups have been successful in the synchronization of mode-locked lasers with the SR [6–10], as listed in Table 1, these synchronisation set-ups, usually based on an electronic locking between the RF cavity of the ring and of the mode-locker RF, can present some instabilities and drifts. In addition, the non-unity repetition rate ratio can be a disadvantage for the exploration of long-time dynamics in time-resolved experiments. Furthermore, and this is the main point, the UV range covered by SRFELs, which is not accessible with a reasonable power by mode-locked conventional lasers, makes SRFELs a unique tool for the performance of pump/probe two-color experiments using the complementary features of FEL and SR: high average power and spectral/temporal resolution for the former and large spectral range and easy tunability for the latter.

Two-color experiments can be classified in different categories:

(i) Frequency-resolved experiments, i.e. with a fixed pump-to-probe delay, in which one takes advantage of the synchronization for achieving a high temporal confinement of the two pulses, which strongly increases the two-photon pumping scheme efficiency via a short-lived intermediate state. This gain factor, determined by the temporal overlap of the two sources on the time-scale of the sample excited state lifetime, as compared with the use of a cw laser in combination with the SR, depends of course on the duty factor of the SR source with typical values of 100 (second generation SR source) to 10^4 (third generation SR source) for the same laser average power.

(ii) Time-resolved experiments, in which the pump-to-probe delay is scanned in order to study the relaxation dynamics of the excited state, as it has been performed for decades in the visible with pulsed laser [13,14]. The probe-photon then samples the temporal decay curve of the excited state while allowing a cw detection scheme, since all the dynamical information resides in the pump-to-probe delay. The time range accessible for such time-resolved FEL + SR studies is usually limited on the short-time side by the SR pulse width ΔT_{SR} and on the long-time side by the repetition period of the SR pulses (T_{SR}) (see Table 1 for characteristic values). As it can be seen in Fig. 1, this range corresponds to the dynamic timescale of many relaxation processes either in the gas phase or in condensed matter.

(iii) A combination of (i) and (ii) when for instance, for a series of given pump-to-probe delays another variable of the experiment belonging to the frequency domain, such as a photon or an electron energy, is scanned, so that informations on a 2D (time and energy) level is gained. The experiment on semi-conductor interfaces presented in Section 2 belongs to this last category.

In contrast to the challenging conventional laser-based UV sources, usually produced by non-linear processes requiring high-energy pulses at low repetition rates, SR-FELs are pulsed sources with a high repetition rate, in the MHz range, producing low-energy pulses. This can be a very interesting property since the small number of events per shot allows the performance of coincidence experiments

Table 1
Temporal and spectral characteristics of pulsed lasers and SR used in combination for the performance of two-color experiments

SR spectral range	SR pulse width FWHM ΔT_{SR} (ps)	SR rep.rate $1/T_{SR}$ (MHz)	Laser type	Laser spectral range (nm)	Laser pulse width FWHM ΔT_{laser} (ps)	Laser rep. rate $1/T_{SR}$ (MHz)	Laser Max. average power (W)	Location
UV	460	90	Mode-Locked Nd:YAG SHG[a]	532	70	90	1.6	UVSOR Japan [6]
IR and XUV	400	52.9	Mode-Locked Nd:YAG SHG[a]	532	100	105.8	0.5	NSLS USA [7]
VUV	700	8.32	Mode-Locked Ar^+	514	300	74.9	1.7	Super-ACO France [8]
XUV	400	4.80	Ti:Saph SHG[a]	390–423	8	83.3	0.1	BESSY Germany [9]
VUV	400	90.1	Ti:Saph THG[b]	243–280	2	90.1	0.05	UVSOR Japan [10]
XUV	600	8.32	SRFEL in operation	345–355	20–80	8.32	0.3	Super-ACO France [11]
IR to XUV	40	3.57	Future SRFEL	180-visible	< 5	3.57	10–80	SOLEIL France [12]

[a] Second harmonic generation.
[b] Third harmonic generation.

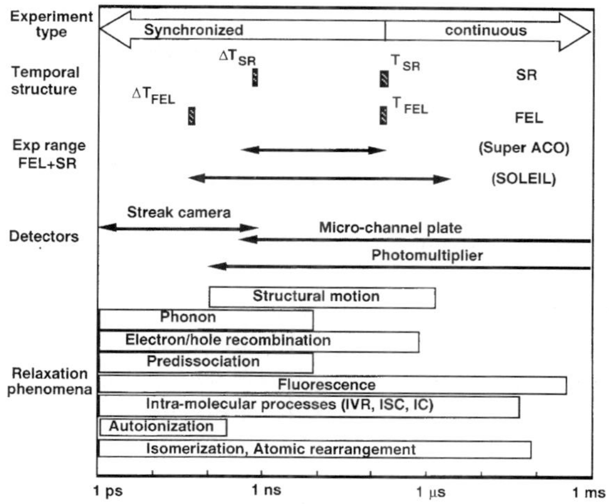

Fig. 1. Temporal range accessible for the SRFEL + SR combination together with the timescale of several relaxation processes dynamics.

without too many false coincidences. Also, in the case of photoemission experiments, widely used as a probe technique, the small number of electrons created per shot insures a negligible space charging effect, allowing the performance of high-resolution photoelectron spectroscopy.

Finally, and this is a major point, SR-FELs are usually part of a synchrotron radiation facility gathering many scientists from different areas. It is then a very favorable place for developing tight collaborations at both the instrumental and scientific level, leading to the performance of original two-color experiments.

2. The first two-color experiment FEL + SR at super-ACO

2.1. Presentation of the surface photo voltage experiments

In this section, we present briefly the first two-color FEL + SR experiment ever performed. It started at Super-ACO in 1994 and is relevant to the surface sciences, concerning the time-resolved study of the surface photo voltage effect (SPV) induced by FEL UV photons (at 350 nm) on semiconductors surfaces and interfaces such as Si(1 1 1) 2 × 1 and Ag/GaAs [11,15]. The SPV is due to the motion of FEL-created electron/hole pairs, leading to modifications of the band bending. One then directly measures the SPV by shifts of the core photoelectron peaks generated by the SR from the SU3 undulator beamline. Owing to the natural synchronization between the FEL and SR pulses, effects due to the nanosecond and sub-nanosecond timescale dynamics of the process have been observed by scanning the FEL-to-SR delay with an optical delay line on the FEL beam path.

Fig. 2 shows the SPV intensity fluctuations as a function of the delay between the two pulses for three different surface qualities of cleaved Si(1 1 1)2 × 1. The long-time behavior is due to the steady-state regime of photocarrier relaxation as observed in the bulk. In contrast, the strongly oscillating transient regime, observed during the first half-nanosecond, is attributed to the π (bonding) and π^* (anti-bonding) electronic surface states [11]. One can then understand the very different behavior observed on a surface showing many defects known to accelerate the electronic relaxation (curve c in the figure). These results, as well as recent data obtained on other silicon interfaces [16], are very important for the understanding of surface recombination phenomena and their role in the dynamics of photon-induced electron/hole pairs.

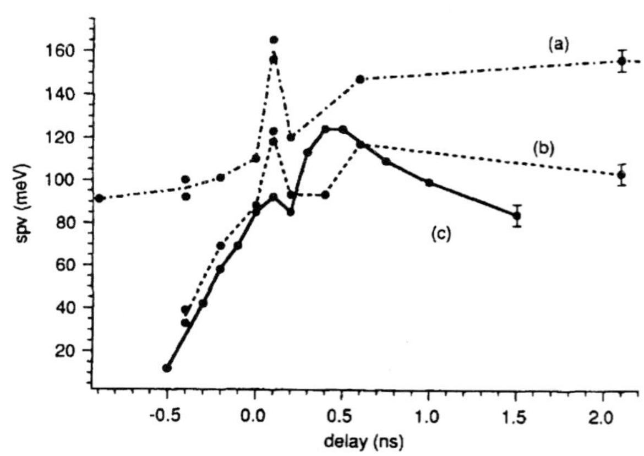

Fig. 2. SPV intensity as a function of the FEL/SR delay for 3 Si(1 1 1) surfaces: (a) Si(1 1 1)2 × 1 freshly cleaved surface; (b) partially contaminated surface; (c) surface with a high defect density.

Such studies allow one to track the sub-nanosecond electrostatic charge density evolution of off-equilibrium systems, with an application in semi-conductor physics related to the transient behavior of Schottky diodes. Besides, there are also interesting high "photon density" effects as observed in spectro-microscopy experiments on semiconductor surfaces (photoelectron analysis on a micron size volume) performed with a very brilliant SR beam as available on third generation SR sources. Inducing with the FEL such high-flux density effects, which could be linked to a SPV effect, permits their precise and quantitative evaluation.

2.2. *What can we learn from the SPV experiment on the experimental procedure*

During the performance of the SPV experiment, several points appeared to be critical for obtaining reliable data, which are of general importance:

FEL operation lifetime: A complex photoemission experiment, like many other endstations encountered around SR sources, requires some time to become operational (warming up of the mirrors carrying the SR, checking procedure for the electron analyzer...), even though the sample has been prepared previously to the FEL operation. The typical acquisition time for a given temporal delay is in the range of 10 min. Therefore, the investigation of several temporal delay values for a given sample, with the possible intermediate fine tuning of the FEL pointing, requires a minimum of several hours of satisfactory operation of both the SR and the FEL. This was the case at Super-ACO, where, depending on the operation of the FEL (with or without the use of the 500 MHz harmonic RF cavity [17]) the lifetime of the FEL in useful experimental conditions (power, stability) is usually in the range of 3–6 h. This is definitely one of the greatest advantages of operating the FEL at the rather high energy of 800 MeV, the nominal energy of Super-ACO, as compared to other SRFELs.

Flux: Because of the thermal load deposited of the FEL cavity front mirror, exposed to the intense photon flux of the undulator, and because of the high-order coherent synchrotron oscillations which can be triggered at high current, the FEL operation requires a moderate stored current, typically below

80–100 mA at Super-ACO (in 2 bunches) instead of a maximum stored current of 400 (resp. 200) mA for the standard SR operation in the multi-bunches (resp. 2 bunches) mode. Nevertheless, the Super-ACO FEL has the highest operating current of any SRFEL. Besides, because of the high repetition rate of SR-FEL leading to low-energy pulses, the electronic transitions are usually non-saturated making two-photon transitions of second order. For all these reasons, the flux factor of merit, usually given by the product of the FEL and SR flux, has to be optimized, for instance by operating the ring with a high energy. Also, the SR should be produced when possible by insertion devices instead of bending magnets. For most experiments, it is not reasonable to perform these kind of experiments below 30 mA at 800 MeV or 80 mA at 500 MeV. The FEL flux should be maximized by adjusting the transmission of the mirror through which the photons are extracted. Recently a usable power of 300 mW has been achieved at Super-ACO owing to an increased transmission of one of the cavity mirrors [17,18].

FEL/SR spatial overlap: In the case of the SPV experiment, because of the high target density leading to a reasonably intense two-color signal, one could directly find and optimize the spatial overlap of the two beams onto the sample by maximizing the SPV signal. More generally, and especially in the case of dilute species for which the two-color signal might be very weak, the spatial overlap can be checked independently with a small area photodiode located at the sample position, showing a signal due to the contribution of the two beams. For this procedure, the SR monochromator has to be set to the zero-order position. A precise calibration then requires a high quality mechanism for the monochromator motion, such that the diffraction zero-order and the dispersed orders coincide spatially.

FEL/SR synchronization: The absolute synchronization, i.e. the "zero SR-to-FEL delay" is an important parameter to determine. Its measurement is performed with a small area fast photodiode connected to a fast oscilloscope, providing an absolute calibration precision in the range of $\pm$ 100 ps. For such a measurements, the SR monochromator is set to zero-order so that all the SR light, including the visible and near-UV part, is sent onto the photodiode, which otherwise would not be sensitive enough in the dispersed orders. This condition is important and cannot always be fulfilled on all beamlines, for instance on some undulator beamlines.

Overall temporal jitter: This is an important issue, especially in the case of short-living excited states and when sharp features are expected in the temporal decay curve, which is the case of the SPV experiment. This jitter is linked to the SR pulse-to-pulse stability, limited by the jittering of the electron bunches and to the FEL/SR pulse-to-pulse synchronization, limited by the jitter of the FEL micropulses with respect to electronic longitudinal distribution. Owing to the good longitudinal stability of the electron beam [19] and to a longitudinal feedback on the laser [20], the overall jitter is less than 10 ps at Super-ACO.

Intensity fluctuations: This is a major point for the SPV experiment, as well as for many others, since in the case of important fluctuations, the transient regime might lead to some parasitic effects, such as the broadening of the spectra, which could lead to unreliable data. In order to achieve a macro-temporal high stability (around 1%) over a long time (several hours), the longitudinal feedback was constantly used in order to lock the laser micro-pulses at perfect synchronism with the electron bunches leading to a cw macro-temporal structure [21].

3. Future prospects at super-ACO

As it is clear from the previous section, two-color FEL + SR experiments are not straightforward and require a careful and optimized operation of both the FEL and the SR. This imposes some severe constraints on the choice and type of beamline to be used. According to the spectral domain to be covered by the SR, one can use different types of beamline, listed here by increasing wavelength:

(a) X-ray: for this photon energy range, a beamline based on a two-crystal monochromator, could be used for the time-resolved study of structural motion on FEL-excited samples probed by photodiffraction. To our knowledge, such an experiment has not yet been performed, nor is planned.

(b) Soft X-ray: this spectral range allows the investigation of electronic properties of matter at the inner shell level, such as electron correlations. It is typically the spectral domain of grazing incidence grating monochromators, allowing a resolving power $\lambda/\Delta\lambda$ in the 10^4 range for the best beamlines with up to 10^{12} ph/s on third generation sources, and typically a few 10^9 ph/s for a 3000 resolving power (case of SU3) on a second generation machine such as Super-ACO. Considering the small ultimate slit width required for the high resolution operation, an emittance as low as possible and a high brilliance as delivered by an undulator is preferable. Such a grazing incidence beamline has allowed the performance of the SPV experiment for which future prospectives in surfaces sciences are two-fold: (i) continuation of the experiments on other interfaces such as Au/Si in order to have a better insight into the role of the surface states in the relaxation dynamics; (ii) performance of two-color two-photon photoemission experiments via the conduction band in order to probe of the empty electronic states. In the gas phase, inner-shell electronic relaxation of FEL-produced radicals could be investigated, as it has been done with a cw visible Ar ion laser, in the specific case of halogen atoms requiring only visible light to be produced by laser-induced dissociation of the precursor halogen molecule [22].

(c) VUV: this spectral domain is covered by normal incidence monochromators for which, due to the minimization of optical aberrations, a resolving power of up to 10^5 with a flux in the 10^{10} ph/s range can be obtained, which is strived for the new SU5 beamline at Super-ACO presently under construction [23]. This beamline, whose primary scientific goals are in the fields of gas-phase photochemistry and photon-induced processes involving the valence shells, should be used in the future in combination with the FEL for the performance of two-photon photoionization experiments on rare gases and small molecules as well as for the temporal study of intra-molecular dynamics by time-resolved photoelectron spectroscopy with SR on FEL-excited molecules. If in addition one considers the exotic polarization capabilities of the SU5 beamline, allowing the study of the anisotropic properties of matter owing to a versatile electromagnetic crossed undulator [24], one can foresee on the long-range a strong program of the UV + VUV combination, including for instance the study of circular dichroism on FEL-produced and aligned photofragments.

(d) UV-IR range: this is a spectral range which has not yet been widely used, but the SR emitted from an bending magnet can cover the UV and visible range reaching the far-IR region. At Super-ACO, a dedicated new beamline, named SA5 and depicted in Fig. 3, is currently at its final stage of construction, aimed at extracting and collimating the white light continuum of a bending magnet in the 0.25–20 μm range. Such a beamline should be operational on the short range for the performance of transient absorption spectroscopy of FEL-excited samples. The vertical collection angle of 15 mrad should allow the extraction of more than 20% of the light emitted at 20 μm while the horizontal

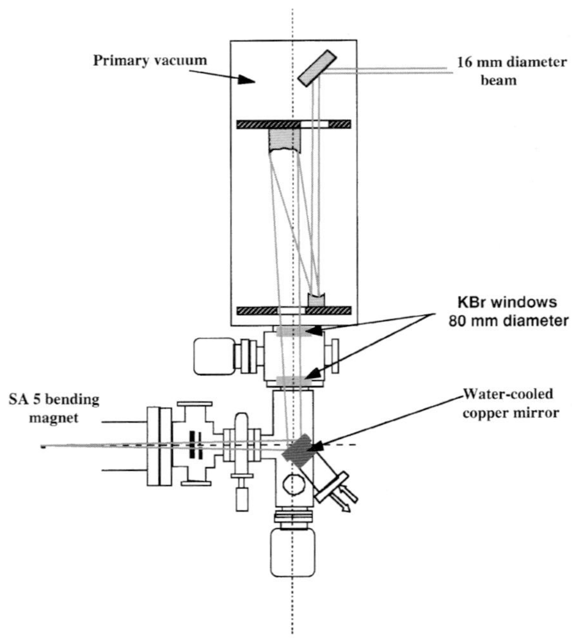

Fig. 3. Schematic of the optical layout of the SA5 beamline; a water-cooled Al-coated copper mirror collects and deflects the beam towards a focusing optical element made of two-spherical conjugated off-axis mirrors, which collimate and demagnify the beam for an easy transportation towards the experimental hutch via simple flat mirrors.

aperture of 45 mrad should allow one to obtain a high photon flux, sufficient in all cases to saturate most of the absorption lines. The geometry of the beamline has been especially studied so that a relative delay of -2 ns up to 8 ns would be easily tuned by a computer-controlled optical delay line.

The scientific case as well as the experimental set-up of the SA5 beamline is two-fold according to the part of the SR spectrum to be used:

- in the IR part of the SR spectrum, up to 20 μm, absorption spectroscopy will be carried out through a Fourier-transform spectrometer allowing high resolution and sensitivity. Note that in the IR spectral range, the SR is the only available pulsed and white source with no equivalent in world of lasers. The UV + IR combination is potentially very fruitful since it implies both the nuclear and electronic motions and their possible couplings, which can be investigated by IR (pump) + UV (probe) experiments requiring high-power IR lasers such as IR-FELs [25] or, as here, in the UV (pump) + IR (probe) sequence. In materials sciences, such an experiment can bring new insights on the photocarriers density [26] in large gap semi-conductors, with possible applications to UV photodetectors. Time-resolved modifications of thin films and interfaces under UV irradiation such as the photon-induced polymerisation of adsorbed molecules or their photodesorption could also be studied. In the gas phase, the ns temporal confinement of the pump and probe pulses should allow the performance of experiments in a gas cell at pressures up to a few Torr in a collisionless regime. Then, the ro-vibrationnal energy content of molecular fragments produced by FEL-induced photodissociation could be mapped out. Reciprocally, the FEL could be used to produce, by laser-induced dissociation from a suitable molecular precursor, radicals whose ro-vibronic structure is interesting to study. Finally, the dynamics of intramolecular relaxation processes, due to electronic state coupling, such as inter-system coupling should be investigated.
- in the visible and UV (down to 250 nm) part of the SR spectrum, the absorption profiles of the sample in solution will be obtained via a 750 mm focal length monochromator equipped with a CCD detection. The scientific program will be devoted to the field of time-resolved photobiology in which both the photophysics and photochemistry of natural chromophores, such as nucleic acids, coenzymes or neurotransmitters are studied in order to be able to understand and control the chemical reactions involved in the metabolism, including, for instance, the action of certain drugs used in anti-tumor therapy. Along the temporal evolution of the photochemical reaction triggered by the FEL photon, reaction intermediates such as radicals or ions, as well as the final products will be analyzed owing to their spectral signature on the absorption spectrum.

4. Conclusion

The development of the source and of the scientific applications program is pursued in parallel at Super-ACO, since both activities are mutually benefiting from this procedure. The progress on the source is often driven by the application requirements in terms of stability, extracted power, spectral range ..etc.., while reciprocally it is the achievement of such progress which allows the performances of these experiments. The development of the FEL-based experimental program, mainly devoted to the study of off-equilibrium matter, will certainly be many-fold in the near future. One direction to be followed will be the decrease of the emission wavelength towards the deep UV and the VUV (on the long-run) leading to new potential applications with the opening of the first ionization channels. Another interesting property which could be used in the future lies in the possibility of producing circularly-polarized UV light with helical undulator-based SR-FEL, as projected on third generation machine such as ELETTRA and SOLEIL. Such a polarization feature could make UV (and VUV) SR-FELs a very powerful tool for the investigation of the chiral and anisotropic properties of matter.

References

[1] M.E. Couprie, D. Garzella, M. Billardon, Europhys. Lett. 21 (1993) 909.

[2] H. Hama, K. Kimura, H. Hosaka, J. Yamazaki, T. Kinoshita, Proceeding of AFEL 97 Free Electron Laser

and its applications in Asia, Ionics Publishing Co, Tokyo, 1997, p. 17–24.

[3] K. Yamada, N. Sei, T. Yamazaki, H. Ohgaki, V.N. Litvinenko, T. Mikado, M. Kawai, M. Yokoyama, Nucl. Instr. and Meth., A 429 (1999) 159.

[4] V.N. Litvinenko, Y. Wu, S.H. Park, I.V. Pinayev, Nucl. Instr. and Meth., A 429 (1999) 151.

[5] M.E. Couprie, F. Mérola, P. Tauc, D. Garzella, A. Delboulbé, T. Hara, M. Billardon, Rev. Sci. Instrum. 65 (1994) 1485.

[6] T. Mitani, H. Okamoto, Y. Takagi, M. Watanabe, K. Fukui, S. Koshihara, C. Ito, Rev. Sci. Instrum. 60 (1989) 1569.

[7] D.L. Ederer, J.E. Rubensson, D.R. Muller, R. Shuker, W.L. O'Brien, J. Lai, Q.Y. Dong, T.A. Callcott, G.L. Carr, G.P. Williams, C.J. Hirshmugl, S. Etemad, A. Inam, D.B. Tanner, Nucl. Instr. and Meth. A 319 (1992) 250.

[8] J. Lacoursière, M. Meyer, L. Nahon, P. Morin, M. Larzillière, Nucl. Instr. and Meth. A 351 (1994) 545.

[9] J. Gatzke, R. Bellmann, I. Hertel, M. Wedowski, K. Godehusen, P. Zimmermann, T. Dohrmann, A.v.d. Borne, B. Sonntag, Nucl. Instr. and Meth. A 365 (1995) 603.

[10] M. Mizutani, M. Tokeshi, A. Hiraya, K. Mitsuke, J. Synchrotron, Radiat. 4 (1997) 6.

[11] M. Marsi, M.E. Couprie, L. Nahon, D. Garzella, A. Delboulbé, T. Hara, R. Bakker, G. Indlekofer, M. Billardon, A. Taleb-Ibrahimi, Appl. Phys. Lett. 70 (1997) 895.

[12] M.E. Couprie, D. Nutarelli, M. Billardon, Nucl. Instr. and Meth. B 144 (1998) 66.

[13] A.H. Zewail, Science 242 (1988) 1645.

[14] L. Nahon, M. Meyer, J. Lacoursière, M. Gisselbrecht, P. Morin, M. Larzillière, in: A. Yagishita, T. Sasaki (Eds.), Frontiers Sciences Series Atomic and molecular photoionization, Academic Press, Tokyo, 1996, pp. 1–10.

[15] M. Marsi, L. Nahon, M.E. Couprie, D. Garzella, T. Hara, R. Bakker, M. Billardon, A. Delboulbé, G. Indlekofer, A. Taleb-Ibrahimi, J. Elec. Spectr. 94 (1998) 149.

[16] M. Marsi et al., Nucl. Instr. and Meth., 1999, These Proceedings.

[17] M-E. Couprie, R. Roux, D. Nutarelli, E. Renault, M. Billardon, Nucl. Instr. and Meth. A 429 (1999) 165.

[18] D. Nutarelli, M-E. Couprie, E. Renault, R. Roux, L. Nahon, A. Delboulbé, M. Billardon, Proceedings 20th Free Electron Laser Conference, Williamsburg, VA, 1998, Elsevier Science B.V., Amsterdam, 1999, p. II.63.

[19] R. Roux, M.E. Couprie, R.J. Bakker, D. Garzella, D. Nutarelli, L. Nahon, M. Billardon, Phys. Rev. E 58 (1998) 6584.

[20] M.E. Couprie, D. Garzella, T. Hara, J. Codarbox, M. Billardon, Nucl. Instr. and Meth. A 358 (1995) 374.

[21] D. Garzella, M.E. Couprie, A. Delboulbé, T. Hara, M. Billardon, Nucl. Instr. and Meth. A 341 (1994) 24.

[22] L. Nahon, P. Morin, Phys. Rev. A 45 (1992) 2887.

[23] L. Nahon, B. Lagarde, F. Polack, C. Alcaraz, O. Dutuit, M. Vervloet, K. Ito, Nucl. Instr. and Meth. A 404 (1998) 418.

[24] L. Nahon, M. Corlier, P. Peaupardin, F. Marteau, O. Marcouillé, P. Brunelle, C. Alcaraz, P. Thiry, Nucl. Instr. and Meth. A 396 (1997) 237.

[25] "Recent report of the Workshop on scientific opportunities for Infrared Free Electron Lasers", Lawrence Berkele Laboratory, Berkeley, PUB 26783 (1989) and "Combustion Dynamics Facility", Lawrence Berkeley Laboratory and Sandia National Laboratory, Berkeley, PUB-5284, 1990.

[26] G.L. Carr, J. Reichman, D. DiMarzio, M.B. Lee, D.L. Ederer, K.E. Miyano, D.R. Mueller, A. Vasilakis, W.L. O'Brien, Semicond. Sci. Technol. 8 (1993) 922.

ELSEVIER

Nuclear Instruments and Methods in Physics Research A 429 (1999) 497–501

NUCLEAR INSTRUMENTS & METHODS IN PHYSICS RESEARCH Section A

www.elsevier.nl/locate/nima

The free electron laser: a system capable of determining the gold standard in laser vision correction

W. Craig Fowler[a,b,*], John G. Rose[a], Daniel H. Chang[a,b], Alan D. Proia[c]

[a]*Department of Ophthalmology, Duke University Medical Center, Durham, NC 27710, USA*
[b]*Free Electron Laser Laboratory, Duke University Medical Center, Durham, NC 27710, USA*
[c]*Department of Pathology, Duke University Medical Center, Durham, NC 27710, USA*

Abstract

Introduction. In laser vision correction surgery, lasers are generally utilized based on their beam-tissue interactions and corneal absorption characteristics. Therefore, the free electron laser, with its ability to provide broad wavelength tunability, is a unique research tool for investigating wavelengths of possible corneal ablation. Methods. Mark III free electron laser wavelengths between 2.94 and 6.7 μm were delivered in serial 0.1 μm intervals to corneas of freshly enucleated porcine globes. Collateral damage, ablation depth, and ablation diameter were measured in histologic sections. Results. The least collateral damage (12–13 μm) was demonstrated at three wavelengths: 6.0, 6.1 (amide I), and 6.3 μm. Minimal collateral damage (15 μm) was noted at 2.94 μm (OH-stretch) and at 6.2 μm. Slightly greater collateral damage was noted at 6.45 μm (amide II), as well as at the 5.5–5.7 μm range, but this was still substantially less than the collateral damage noted at the other wavelengths tested. Conclusions. Our results suggest that select mid-infrared wavelengths have potential for keratorefractive surgery and warrant additional study. Further, the free electron laser's ability to allow parameter adjustment in the farultraviolet spectrum may provide unprecedented insights toward establishing the gold-standard parameters for laser vision correction surgery.

PACS: 87.50.Hj; 41.60.Cr; 87.90. + y

Keywords: Free electron laser; Infrared laser; Refractive surgery; Cornea; Laser vision correction; Collateral damage

1. Introduction

The free electron laser represents a new and unique research tool in science and medicine because it allows broad-wavelength tunability while providing excellent optical-beam quality [1]. It affords investigators the unprecedented opportunity to evaluate laser-tissue interactions serially over multiple wavelengths by "making use of a simple and elegant gain medium – an electron beam in a magnetic field" [1].

In laser vision correction surgery, lasers are generally utilized based on their beam-tissue interactions and corneal absorption characteristics, with principal regard to their corneal cutting potential. The greatest absorption of laser light in

* Corresponding author. W. Craig Fowler, MD, Medical Director, Duke Free Electron Laser Laboratory, Box 90319, Durham, NC 27708, USA. Tel.: +1-919-660-2655; Fax: +1-919-681-7416.

PII: S0168-9002(99)00154-0

the cornea is by water and biomolecules in the far-ultraviolet region of the electromagnetic spectrum (<300 nm), the mid-infrared region (with peaks at about 3 and 6 μm), and the far-infrared regions (>10 μm). The cornea is virtually transparent to light at wavelengths from 0.3 to 1.3 μm [2–4].

Infrared investigations for possible usage in corneal surgery began with the original report by Fine in 1967 utilizing the continuous wave carbon dioxide laser, operating at a wavelength of 10.6 μm [5]. Significant thermal damage of adjacent tissue was noted, and to improve heat dissipation, pulsed CO_2 laser emissions were investigated [6]. To further reduce the amount of energy deposition, Q-switching was implemented, yielding an observed decrease in collateral damage to only 25–30 μm [7]. Despite these advances, the far-infrared carbon dioxide laser was abandoned in favor of the mid-infrared hydrogen fluoride (HF, 3.0 μm), Raman-shifted neodynium: YAG (Nd:YAG, 2.80 and 2.92 μm), and subsequently the erbium-YAG (Er:YAG, 2.94 μm) lasers because of their higher corneal absorption characteristics, which allowed for more defined corneal incisions and decreased thermal damage [2,3,8–12].

In this study, we sought to investigate the corneal cutting characteristics of serial mid-infrared

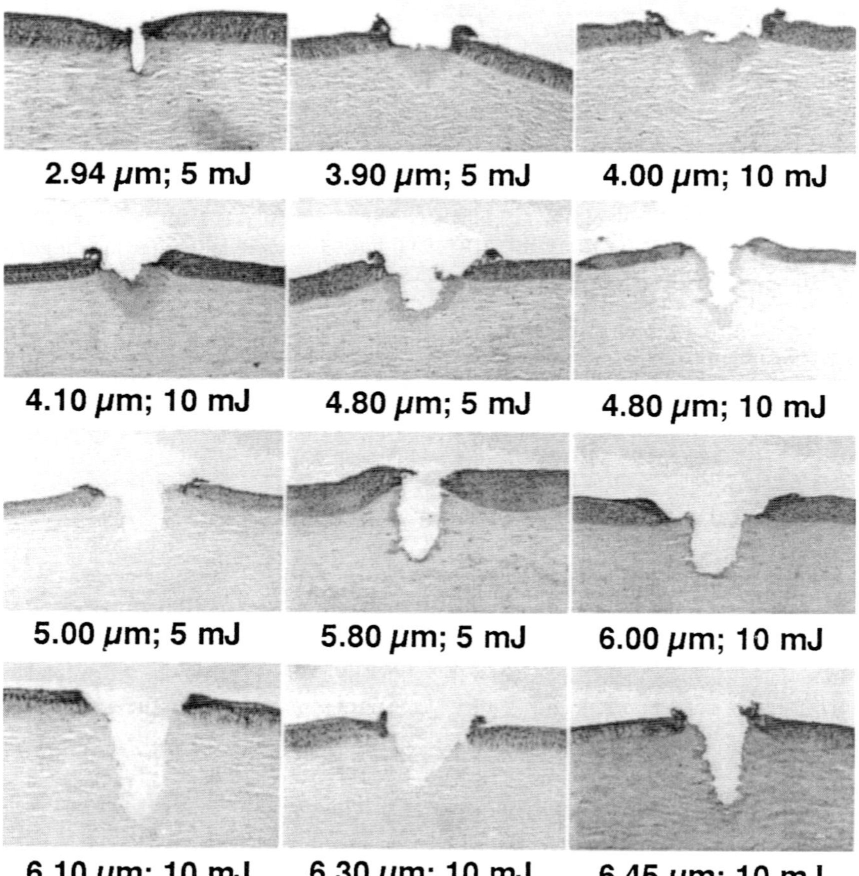

Fig. 1. Representative photomicrographs demonstrating the variability in cutting and collateral damage at serial wavelengths and differing energies. Note the intensely staining (hypereosinophilic) regions of collateral damage surrounding the ablation zone.

wavelengths as generated by the free electron laser at Duke University. Dimensions of ablation and extent of collateral damage were systematically evaluated in an ex vivo model.

2. Methods

Mark III FEL generated wavelengths between 2.94 and 6.7 μm were delivered in serial 0.1 μm

Table 1
Table comparing diameter and depth of ablation to collateral damage for serial infrared wavelengths

Wavelength (μm)	Ablation diameter (μm)	Ablation depth (μm)	Collateral damage (μm)
(A) Energy = 5 mJ			
2.94	27	141	15
3.90	130	56	135
4.00	193	63	24
4.40	81	299	52
4.50	94	220	25
4.80	159	165	49
4.90	106	40	117
5.00	299	262	44
5.10	90	232	30
5.30	92	207	25
5.40	79	276	20
5.50	237	199	26
5.60	95	209	13
5.80	94	61	34
6.00	85	42	13
6.10	86	144	13
6.20	198	102	15
6.30	122	228	13
6.45	208	118	18
6.70	162	142	18
(B) Energy = 10 mJ			
3.10	105	111	30
3.20	86	384	34
3.90	159	44	120
4.00	234	59	104
4.40	93	208	73
4.50	146	280	36
4.80	124	340	116
4.90	146	174	74
5.00	180	208	44
5.10	179	213	42
5.20	126	216	35
5.30	226	295	69
5.40	244	222	45
5.50	86	237	17
5.60	74	167	13
5.70	184	154	16
5.80	92	149	40
6.00	256	121	12
6.10	416	214	13
6.20	419	86	15
6.30	192	216	12
6.45	277	113	31

intervals to the corneas of freshly enucleated porcine globes. All globes were used within 6 h of harvesting and were stored in saline-moistened chambers to retard epithelial desiccation. Laser parameters included macropulse energies of 5 and 10 mJ per pulse (as measured by a Molectron™ pyro-electric energy detector) with duration of 4.0 μs, frequency of 10 Hz; micropulse duration was 1.0 ps, with a frequency of 2.8 GHz. The beam was delivered through a calcium fluoride (CaF_2) lens, and the tissue was positioned at the focus of the normally incident beam with a three-dimensional translation stage, yielding an 80 μm spot diameter. Number of macropulses delivered equaled 4 in all data presented, and a minimum of 5 ablations were performed at each wavelength.

Treated globes were immediately fixed in 3.7% phosphate-buffered neutral formaldehyde for at least 48 h. Corneas were then excised and embedded in paraffin for standard hematoxylin and eosin stained sections. Collateral damage was calculated as the mean linear distance of histologically abnormal corneal stroma surrounding ablation treatment areas. The abnormal stroma was hypereosinophilic (demonstrated by increased staining – see Fig. 1) and appeared smudged when compared with the stroma more peripheral to the wound. This appearance is typical of that resulting from thermal damage in vivo. A confocal microscope (Meridian™) with a calibrated on-screen histologic micrometer system was used to perform all measurements.

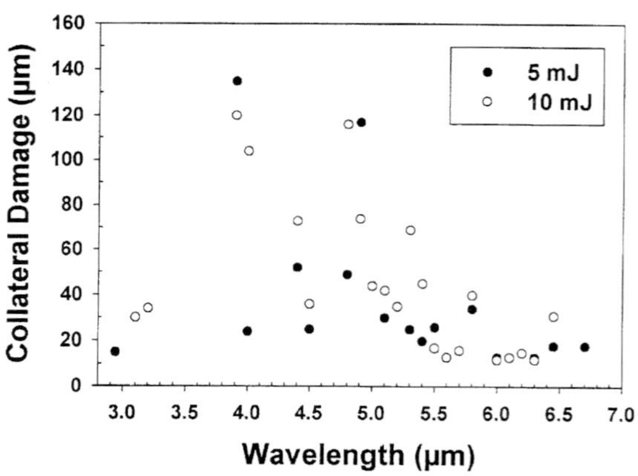

Fig. 2. Collateral damage as a function of serial free electron laser mid-infrared wavelength (2.94–6.7 μm) and energy (5 and 10 mJ). The least collateral damage, 12–13 μm, was demonstrated at three wavelengths: 6.0, 6.1 (amide I stretch), and 6.3 μm. Minimal collateral damage of only 15 μm was also noted at 2.94 (OH stretch) and 6.2 μm. Slightly greater collateral damage was noted at 6.45 μm (amide II), as well as the 5.5–5.7 μm range, but this was still substantially less than the collateral damage noted at the other wavelengths tested. Single measurements of the most representative histological sections for each wavelength are presented.

3. Results

Representative photomicrographs that demonstrate the variability in cutting and the degree of collateral damage resulting from mid-infrared free electron laser ablation are presented in Fig. 1. Quantification of ablation zone dimensions and collateral damage as a function of wavelength is given in Table 1 and Fig. 2.

4. Discussion

Our results demonstrate that the free electron laser operating in the mid-infrared range is capable of cutting porcine corneas, although the depth of ablation and the degree of collateral damage varies widely depending on wavelength. In general, the best cutting and the lowest levels of collateral damage occurred at wavelengths near and corresponding to known resonance frequencies of protein bonds (6.1–6.45 μm) and of water (3.0 μm). The least collateral damage corresponded to 6.0, 6.1 (amide I stretch), and 6.3 μm. Limited collateral effects were also noted at 6.2 and 2.94 μm (OH stretch). Slightly greater collateral damage was noted at 6.45 μm (amide II), as well as at 5.5–5.7 μm, but this was still less than that noted at other wavelengths. These results are consistent with those obtained by analyzing these investigational ablations three-dimensionally using confocal microscopy [13].

With the development of widespread commercial interest in laser systems for keratorefractive surgery, resources and attention have largely been diverted from basic science investigations towards the development of clinical delivery systems suitable for human application. Yet, many technical questions still remain, including the determination of optimal fluence, optimal wavelength, and optimal profile of the newly ablated corneal surface. The free electron laser now provides the ideal research laser system to answer these questions.

Previous free electron laser corneal studies have provided significant information with gelatin [14], ex vivo porcine corneas [13,15–19], and cadaveric human corneas [20]. However, it is important to stress that loss of corneal tear film, alterations in corneal hydration, absence of dynamic physiology, inconstant intraocular pressure, disrupted aqueous metabolism, and lack of dynamic wound healing processes may lead to results significantly different from in vivo corneal investigations. This ex vivo tissue-dependent variability was particularly evidenced by some of the higher energy levels in this study, yielding some unanticipated ablation depths not predicted by theoretical considerations. Therefore, future laser-corneal studies ideally should be performed on in vivo systems, using an animal with a cornea that histologically resembles that of the human.

Our results suggest that select mid-infrared wavelengths have potential for keratorefractive surgery and warrant additional study. Furthermore, the free electron laser's ability to allow parameter adjustment in the far-ultraviolet spectrum, where current clinical systems operate, may provide unprecedented insights toward establishing the gold-standard parameters for laser vision correction surgery.

Acknowledgements

The authors would like to thank Genny A. Barnett, Ph.D., Joel R. Ross, Ph.D., and David M. Howell, MD., Ph.D., John Madey, Ph.D., Karl David Straub, M.D., Ph.D., and Brett Hooper, Ph.D. We also thank the many other scientists and staff at the Duke Free Electron Laser Laboratory. This work was supported in part by the Office of Naval Research through the medical arm of the Free Electron Laser Program Grant Number N 00014-94-1-0818.

References

[1] C. Brau, Free Electron Lasers, Academic Press, Boston, 1990.
[2] H. Loertscher, S. Mandelbaum, R.K. Parrish, J.M. Parel, Am. J. Ophthalmol. 102 (1986) 217.
[3] T. Seiler, J. Marshall, S. Rothery, J. Wollensak, Lasers Ophthalmol. 1 (1986) 49.
[4] D.S. Gartry, The development of excimer laser corneal surgery: Beam tissue interactions, in: C.N.J. McGhee, H.R. Taylor, D.S. Gartry, S.L. Trokel (Eds.), Excimer Lasers in Ophthalmology, Butterworth-Heinemann, Boston, 1997.
[5] B.S. Fine, S. Fine, G.R. Peacock, W.J. Geeraets, E. Klein, Am. J. Ophthalmol. 64 (1967) 209.
[6] H. Beckman, A. Rota, R. Barraco, H.S. Sugar, E. Gaynes, Am. J. Ophthalmol. 71 (1971) 1277.
[7] R.H. Keates, L.S. Pedrotti, H. Weichel, W.H. Possel, Ophthalmic Surg. 12 (1981) 117.
[8] D. Stern, C.A. Puliafito, E.T. Dobi, W.T. Reidy, Ophthalmology 95 (1988) 1434.
[9] T. Bende, M. Kriegerowski, T. Seiler, Laser Light Ophthalmol. 2 (1989) 263.
[10] T. Seiler, H. Schmidt-Petersen, R. Leiacker, R. Steiner, J. Wollensak, Am. J. Ophthalmol. 120 (1995) 668.
[11] M. Mrochen, T. Seiler, Int. Ophthalmol. Clin. 36 (1996) 73.
[12] G.A. Peyman, R.M. Badaro, B. Khoobehi, Ophthalmology 96 (1989) 1160.
[13] J. Kim, J. Rose, W.C. Fowler et al., Invest Ophthalmol. Vis. Sci. 37 (1996) s64.
[14] B. Jean, T. Bende, J. Refract Corneal Surg. 10 (1994) 433.
[15] W.C. Fowler, S.R. Wehrly, N.R. Imami et al., Invest Ophthalmol. Vis. Sci. 35 (1994) 2027.
[16] N.R. Imami, W.C. Fowler, S.R. Wehrly et al., Invest Ophthalmol. Vis. Sci. 35 (1994) 2027.
[17] E.C. Vietorisz, J.G. Rose, W.C. Fowler et al., Invest Ophthalmol. Vis. Sci. 37 (1996) s64.
[18] S.R. Wehrly, W.C. Fowler, N.R. Imami et al., Invest Ophthalmol. Vis. Sci. 35 (1994) 2027.
[19] T. Bende, R. Walker, B. Jean, J. Refract Surg. 11 (1995) 129.
[20] G. Edwards, R. Logan, M. Copeland et al., Nature 371 (1994) 416.

ELSEVIER

Nuclear Instruments and Methods in Physics Research A 429 (1999) 502–506

NUCLEAR INSTRUMENTS & METHODS IN PHYSICS RESEARCH Section A

www.elsevier.nl/locate/nima

Systematic infrared ablations with the mark III FEL at 2.94 μm in the chicken cornea

W. Craig Fowler[a,b,*], Daniel H. Chang[a,b], Alan D. Proia[c]

[a]Department of Ophthalmology, Duke University Medical Center, Durham, NC 27710, USA
[b]Free Electron Laser Laboratory, Duke University Medical Center, Durham, NC 27710, USA
[c]Department of Pathology, Duke University Medical Center, Durham, NC 27710, USA

Abstract

Introduction. Previous studies with the free electron laser have only assessed the corneal tissue response in *ex vivo* models. This is the first free electron laser investigation that has examined the clinical and histologic response of the cornea *in vivo*. Methods. The Mark III free electron laser was tuned to 2.94 μm for a total of 212 separate *in vivo* 1 mm spot ablations at varied fluences (0.4–2.04 J/cm^2), repetition rates (1–15 Hz), and number of pulses (1–32). Also, 22 separate *in vivo* central scanning patterns were performed. Results. For the *in vivo* chicken cornea, the threshold of ablation is approximately 0.64 J/cm^2 at a repetition rate of 10 Hz. Clinical biomicroscopy and light microscopy revealed parameter-dependent collateral damage (10–40 μm) induced by laser energy. Wound healing following free electron laser ablation exhibits patterns similar to those seen in excimer laser anterior keratectomy, including increased epithelial thickness, new collagen deposition, increased numbers of keratocytes, and patterns congruent with attempted profiles of myopic ablation. Conclusions. With further study, continued parameter adjustment, and refined delivery schemes, the potential for marked improvement and possible application appears promising.

PACS: 87.50.Hj; 41.60.Cr; 87.90. + y

Keywords: Free electron laser; Infrared laser; Photorefractive keratectomy; Cornea; Collateral damage; Wound healing

1. Introduction

Depth of tissue penetration is an important characteristic of lasers that may be suited for vision correction surgery. Recently, most attention has focused on the FDA-approved argon–fluoride excimer laser, which employs an ultraviolet wavelength of 193 nm and has a penetration depth of 2–3 μm [1]. However, the infrared erbium : YAG (Er: YAG) laser, at 2.94 μm, has a penetration depth of only 0.75 μm [1]. Laser corneal ablation at 2.94 μm has therefore been investigated because of its potential accuracy for tissue removal and the possibility that it may minimize detrimental aspects of wound healing, such as corneal haze [2–6].

Previous studies with the free electron laser have only assessed the corneal tissue response in ex vivo models [7–13]. This is the first free electron laser

* Corresponding author. Duke Free Electron Laser Laboratory, Box 90319, Durham, NC 27708, USA. Tel.: + 1-919-684-3799; fax: + 1-919-681-3572.

E-mail address: fowle006@mc.duke.edu (W.C. Fowler)

0168-9002/99/$ - see front matter
PII: S0168-9002(99)00155-2

investigation that has examined the clinical and histologic response of the cornea in vivo.

2. Methods

The Mark III free electron laser at Duke University was tuned to the mid-infrared wavelength of 2.94 μm for all laser applications in this study and was operated in accordance with University radiation safety guidelines. White leghorn chickens (*Gallus gallus*) were used for the laser experiments and were anesthetized with an intramuscular injection of 2 mg/kg ketamine/xylazine (advanced to 6 mg/kg as needed based on clinical observations) and topical instillation of 0.5% proparacaine hydrochloride. Immediately prior to each laser treatment, the corneal epithelium of the right eyes were mechanically debrided over the intended treatment zone, and a total of 212 separate in vivo 1 mm spot ablations were performed at varied fluences (0.4–2.04 J/cm^2), repetition rates (1–15 Hz), and number of pulses (1–32). Average macropulse duration was 1.3 μs (range 0.5–1.6 μs) and consisted of a train of 1.0 ps micropulses at a 2.9 GHz repetition rate. Additionally, 22 separate in vivo central corneal scanning ablations (Fig. 1) were performed. These refractive scanning ablations were designed to simulate the excisional flattening profile used clinically to correct myopia by removing the greatest amount of tissue over the central treatment zone. Wound healing was studied through clinical biomicroscopic assessments from 0 to 6 months post-operatively and documented photographically. For histologic analysis, the chickens were euthanized post-operatively by CO_2 asphyxiation or by intravenous injection of Euthasol (1 ml/10 lb). The globes were immediately enucleated and fixed in 3.7% phosphate-buffered neutral formaldehyde. All animals were handled and housed in accordance with animal care and use guidelines, as approved by the Duke University Institutional Animal Care and Use Committee and the North Carolina State University Division of Poultry Science.

3. Results

For the in vivo chicken cornea, linear regression analysis demonstrated the approximate threshold of ablation to be 0.64 J/cm^2 at a macropulse

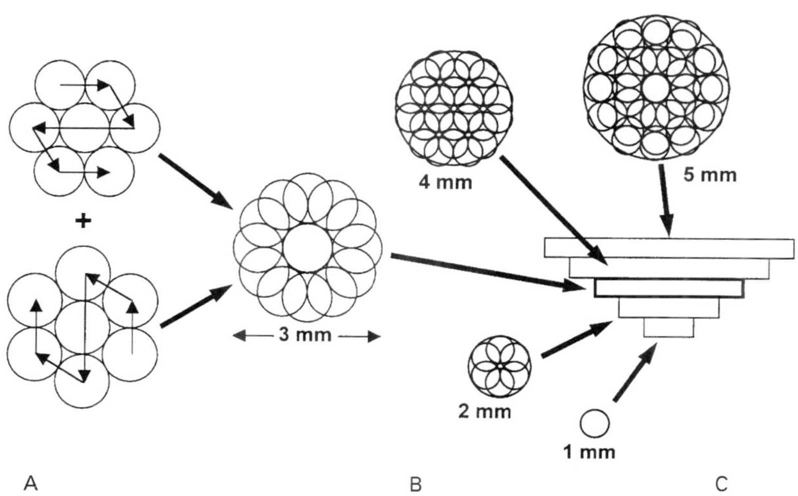

Fig. 1. Scanning algorithm to generate a 5 mm refractive ablation zone. Myopic correction is simulated by removing more tissue centrally than peripherally. (A) The 1 mm laser spot with 15 Hz frequency is translated at 15 mm/s in the two patterns shown by the arrows. (B) The resultant 3 mm zone is combined with similar zones of varied diameters (5, 4, 2, and 1 mm) to create the tapered profile, as shown in (C).

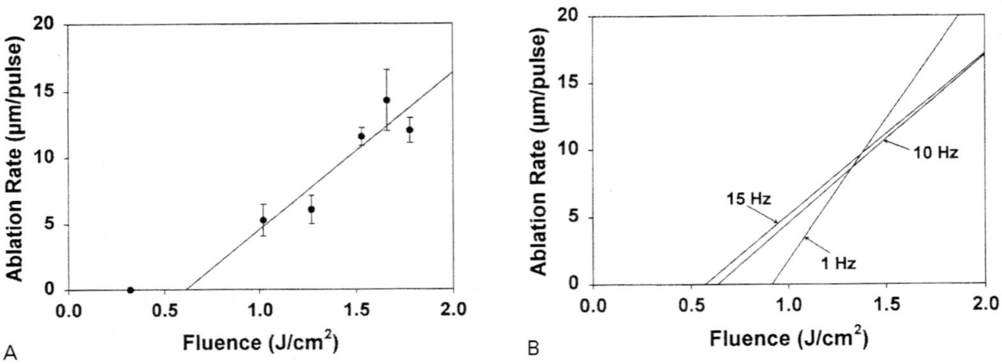

Fig. 2. (A) Ablation rate as a function of fluence for the free electron laser at 2.94 μm and 10 Hz. Linear regression analysis yields an approximate ablation threshold of 0.64 J/cm^2. Error bars indicate standard error of the mean. (B) Diagrammatic representation depicting the relationship of frequency to ablation rate as a function of fluence. The linear regressions were calculated from 8, 20, and 12 values for 1, 10, and 15 Hz, respectively.

repetition rate of 10 Hz (Fig. 2A). The rate of ablation as a function of fluence for various macropulse repetition rates is shown in Fig. 2B. Our results suggested possible decreased ablation threshold with increased repetition rates.

Clinical biomicroscopy revealed increased laser-related collateral damage with increasing depth of ablation (increased number of pulses), pulse duration, fluence > 1.2 J/cm^2, and repetition rates >10 Hz. Histologic analysis confirmed the observed collateral damage (Fig. 3). The collateral damage was manifested histologically as hypereosinophilic stroma with loss of the normal lamellar architecture. The appearance of the collateral damage resembled thermal damage observed in connective tissue in other body sites. On average, the collateral damage approximation ranged from 10 to 40 μm. The greatest collateral damage was noted at the most superficial regions of the ablation site, and the least collateral damage was noted at the base.

Average time to reepithelialization after laser keratectomy was 3 to 4 days. Post-operative clinical biomicroscopic haze peaked between 2 and 4 weeks, and marked clearing was observed by 4–6 months (Fig. 4). Histologic examination at 2 months revealed increased epithelial thickness over the site of laser surgery, new collagen deposition, increased numbers of keratocytes, and patterns congruent with the attempted profiles of myopic ablation (Fig. 5).

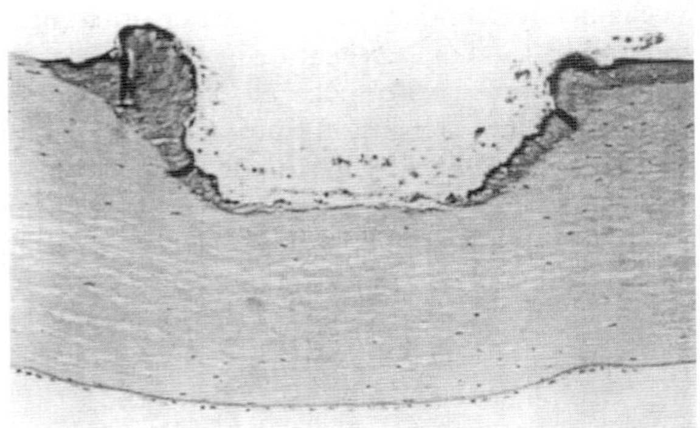

Fig. 3. Histologic section of in vivo corneal tissue following varied depths free electron laser ablation at 2.94 μm. Hypereosinophilic regions demonstrate collateral damage around ablation zones. Note decreased collateral damage at base relative to superficial regions. Fluence = 1.53 J/cm^2, frequency = 10 Hz, number of pulses = 16; original magnification × 130.

4. Discussion

This study reports the first in vivo investigation of the free electron laser for refractive corneal surgery application. The infrared wavelength of 2.94 μm was targeted for this investigation because of several important associated advantages. First of all, the cornea has an absorption peak at this wavelength. Secondly, there is a very limited depth of penetration affording the possibility of

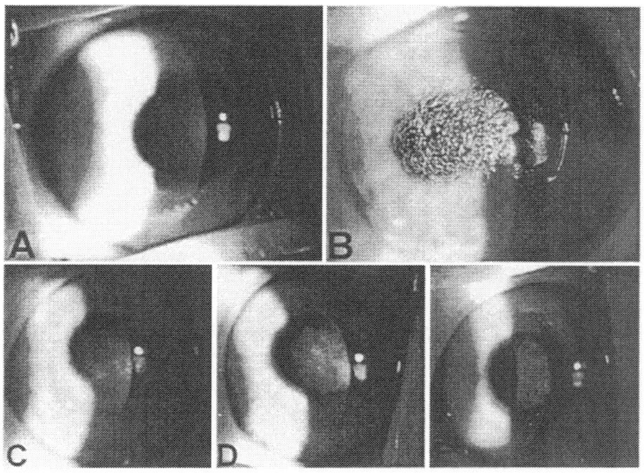

Fig. 4. Serial time-point photographs depicting clinical biomicroscopic haze following treatment by the free electron laser at 2.94 μm using a myopic ablation profile. (A) Normal (untreated) chicken cornea. (B) Cornea 2 hours post-operative. Note central scanning pattern in deepithelialized zone. (C–E). Clinical biomicroscopic haze at 2, 4, and 10 weeks, respectively. Note central corneal haze development and regression over time.

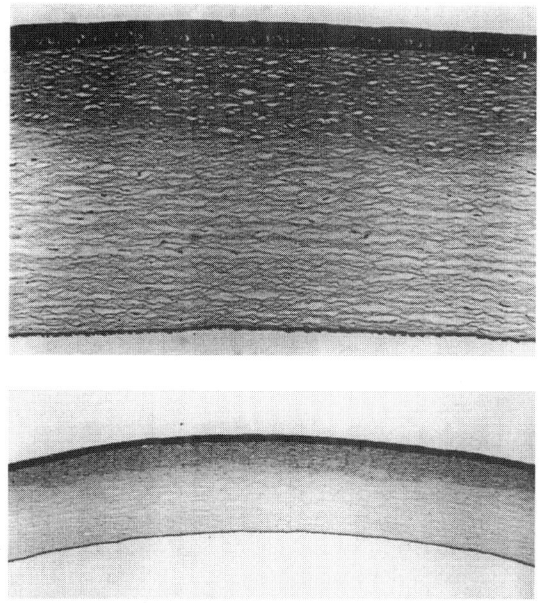

Fig. 5. Histologic cross sections of corneal tissue following a myopic refractive scanning ablation with 2.94 μm free electron laser. (A) Note anterior stromal collagen deposition and hypercellularity (increased numbers of keratocytes). (B) Lower magnification view demonstrating tapered profile of ablation and wound healing. Section taken 2 months post-operative. Original magnification of A: × 170, B: × 52.

controlled excision. In addition, infrared wavelengths are transmissible through fiber optics and lack the mutagenic potential associated with ultraviolet radiation [1,14]. Finally, the Er:YAG laser at 2.94 μm has been suggested as a potential alternative to the excimer laser for photorefractive keratectomy because it produces a similar corneal wound healing response [6,15].

Our results demonstrated that the Mark III free electron laser can be utilized at 2.94 μm to ablate the chicken cornea and that the amount of biomicroscopic corneal haze compares favorably with the corneal haze noted with early excimer laser systems. Those early systems lacked ideal beam homogenization, and their refractive delivery algorithms were crude [1]. Furthermore, it is not surprising that the amount of biomicroscopic corneal haze created by this first refractive application of the free electron laser is greater than that created by the current FDA-approved excimer laser systems in this model [unpublished data, manuscript in preparation].

Continued parameter refinement will be essential for future investigations. Based on our results, these should be optimized with fluences slightly above threshold (0.8–1.0 J/cm^2), with frequencies from 1 to 10 Hz, and with shortened pulse duration ($\leqslant$500 ns), where there is less biomicroscopic haze and decreased thermal-related changes.

Additionally, the corneal reprofiling algorithms for the refractive scanning ablations were a significant limitation in this study. The treatment transition zones were too abrupt, and the regions of overlapping ablations were not optimally blended. This resulted in excess collateral damage and a very uneven bed of refractive ablation. Considerable refinement should improve the smoothness of the treated corneal surface and decrease clinical haze associated with the related corneal wound healing response.

In conclusion, with further study, continued parameter adjustment, and refined delivery schemes, the potential for marked improvement and possible application appears promising.

Acknowledgements

The authors would like to thank Susanne A. Hauger, Ph.D., Julianne H. Thompson, CVT, Ramona R. Nelson, HTL/ASCP, Bruce C. Roberts, BS, B.D. Guenther, Ph.D., Vladimir, Litvinenko, and Ying Wu, Ph.D., for their invaluable contributions. We also thank the many other scientists and staff at the Duke Free Electron Laser Laboratory. This study was supported in part by the US Office of Naval Research under Grant No. N00014-94-1-0818.

References

[1] S.L. Trokel, History of the excimer and other ophthalmic refractive lasers, in: C.N.J. McGhee, H.R. Taylor, D.S. Gartry, S.L. Trokel (Eds.), Excimer Lasers in Ophthalmology, Butterworth-Heinemann, Boston, 1[illegible]7.

[2] T. Bende, B. Jean, M. Matallana, Klin Monatsbl Augenheilkd 202 (1993) 52.

[3] T. Bende, M. Kriegerowski, T. Seiler, Laser Light Ophthalmol. 2 (1989) 263.

[4] M. Mrochen, T. Seiler, Int. Ophthalmol. Clin. 36 (1996) 73.

[5] M. Mrochen, K. Vogler, N. Glaeske, Laser Optoelektronik 28 (1996) 42.

[6] T. Seiler, H. Schmidt-Petersen, R. Leiacker et al., Am. J. Ophthalmol. 120 (1995) 668.

[7] J. Kim, J. Rose, W.C. Fowler et al., Invest Ophthalmol. Vis. Sci. 37 (1996) s64.

[8] W.C. Fowler, S.R. Wehrly, N.R. Imami et al., Invest. Ophthalmol. Vis. Sci. 35 (1994) 2027.

[9] N.R. Imami, W.C. Fowler, S.R. Wehrly et al., Invest. Ophthalmol. Vis. Sci. 35 (1994) 2027.

[10] E.C. Vietorisz, J.G. Rose, W.C. Fowler et al., Invest. Ophthalmol. Vis. Sci. 37 (1996) s64.

[11] S.R. Wehrly, W.C. Fowler, N.R. Imami et al., Invest. Ophthalmol. Vis. Sci. 35 (1994) 2027.

[12] T. Bende, R. Walker, B. Jean, J. Refract. Surg. 11 (1995) 129.

[13] G. Edwards, R. Logan, M. Copeland et al., Nature 371 (1994) 416.

[14] T. Seiler, T. Bende, K. Winckler, J. Wollensak, Graefes Arch. Clin. Exp. Ophthalmol. 226 (1988) 273.

[15] G.A. Peyman, R.M. Badaro, B. Khoobehi, Ophthalmol. 96 (1989) 1160.

FREE ELECTRON LASERS 1998
G.R. Neil and S.V. Benson (Eds.)
1999 Elsevier Science B.V.

The Effect of a Short Circuit Waveguide on the Coupled Power of a Pre-Bunched CW FEM

A.I. Al-Shamma'a*, R.A. Stuart, A. Shaw, S.R. Wylie and J. Lucas

*The University of Liverpool, Department of Electrical Engineering and Electronics, Brownlow Hill, Liverpool L69 3GJ, U.K.

Abstract

The possibility of highly efficient parametric amplifications of the power coupled out of an X-Band microwave cavity resonator is discussed. This paper presents the effect of using a short circuit waveguide, connected into 24dB coupler, on the output power. The variation of the spontaneous emission output power, with and without the short circuit section, at various operating beam voltages, ranging from 40kV to 60kV are reported.

1. Introduction

The FEM research program, for the production of industrial FEM's, has been going for several years at the University of Liverpool. Compact, low voltage ≤ 80 kV, FEM prototypes have been developed as, an amplifier using a microwave pre-buncher technique [1,2] in CW mode, and as an oscillator (under construction) in pulsed mode. The operating frequency is in the range between 8 to 12GHz, and is dependent upon the wiggler cavity resonator. The design target for these FEM's is the generation of high power [3] for industrial applications. The CW pre-buncher FEM is being constructed and tested in order to examine the physics of a high Q factor for an X-band microwave buncher and to act as a test bed for new ideas and techniques. The complete schematic diagram of this system is shown in Fig. 1. The thermionic electron gun is mounted inside the high voltage terminal with a deliverable current ≥ 5mA. The wiggler magnet consists of 30 periods of NdFeB bars with a 19mm wiggler period. The electron beam, which has a 2mm (nominal) diameter, is transported down the wiggler magnet through a cavity resonator. This cavity is designed from an X-band waveguide with internal dimensions, a=22.86mm and b=10.16mm. The total length of the cavity is 600mm and it is short-circuited at each end by two 7mm thick irises, with a 9mm diameter aperture. A 24dB cross coupler was placed at about 1/4 λ_g from the entrance section of the cavity. The required power was coupled out from one end of this coupler with the other end connected to a short circuit, or terminated by a matched load. At the exit of the wiggler cavity the electron beam was collected using either a water-cooled beam dump or an energy recovery section.

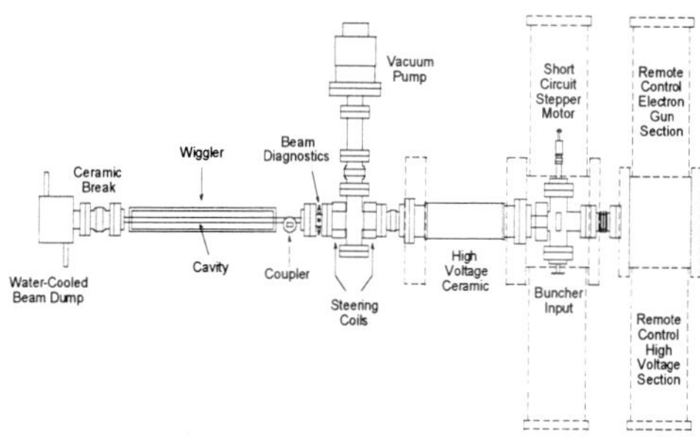

Fig. 1: A Compact FEM system for industrial applications.

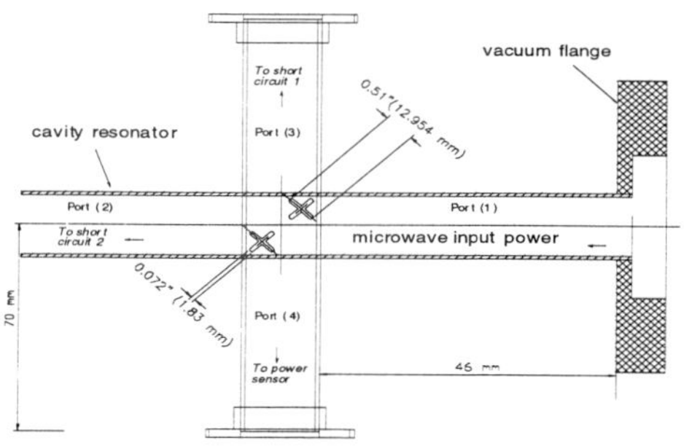

Fig. 2: An X-band cross coupler

In this paper, we discuss the X-band cross coupler, and the effect the connection of an X-band short-circuited section has on the output power.

2. Cavity resonator with a cross coupler

Figs. 1 and 2 show the set up of the cavity and the cross coupler within the FEM system layout. For this test, port 2 of the cross coupler was connected to a short circuit, port 3 was connected to a resonant cavity, and port 4 was connected to a matched power sensor.

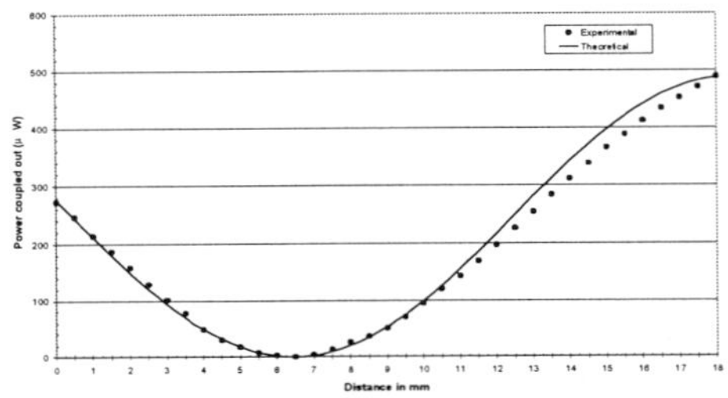

Fig. 3: Power coupled out of an FEM.

Fig. 3 shows an experimental measurement of the power coupled out by the 24dB coupler vs. the short circuit plunger position at 8.32GHz, with a synchronous voltage of 47.5kV and a 0.5mA beam current. The power data was remotely collected for every 0.5mm movement of the plunger. Once again the results were in good agreement with the theory. Several tests at different FEM operating settings (beam voltage, beam current and frequency) have been carried out and similar results are achieved. Further investigations, using a matched load instead of a short circuit, were performed, and the results can be explained as follows:

(a)To obtain a clear comparison between the use of a short circuit or a matched load, an experimental set up to measure the spontaneous power has been created, with an FEM operating at a voltage of 50.5 kV at 9.07 GHz with a beam current of 0.5mA. For the first test, a short circuit was connected to the cross coupler, then the test was repeated with a matched load (a section of an X-band cavity with a tapered ferrite inside).

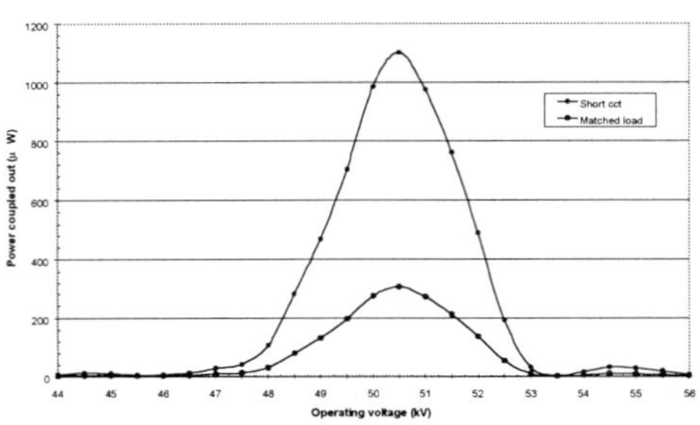

Fig. 4: Spontaneous power vs. FEM beam

Fig. 4 shows some interesting results for the FEM operating voltages, varying from 44 to 56kV. The effects of adding a short circuit to the FEM cavity arrangements is clearly shown, particularly at 50.5kV, where the maximum power coupled was 1100μW. By comparison, when the matched load was used, the maximum power coupled out was 330μW.

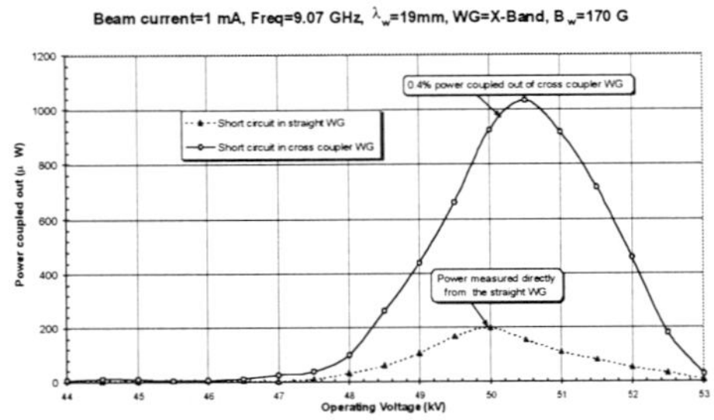

Fig. 5: Spontaneous power vs. FEM operating voltage with straight WG and WG with a cross coupler.

(b) The same settings of (a) were used but this time the output power was measured against the beam current. The results obtained and show once again that the power coupled out, in the case of the short circuit, was much higher than when the matched load was used, 3.4 times higher, for example, at 1mA beam current.

3. Cavity resonator without a cross coupler

A section of a straight cavity, without a cross coupler, which had a resonance the same as that in the buncher cavity, was used. Fig. 5 shows the spontaneous emission power using the previous operating parameters, which was 5 times less than when the cross coupler cavity was used.

4. Conclusions

This paper clearly demonstrates the benefit of using a short circuit section, connected to a cross coupler on a cavity resonator. The results show that the power coupled out when a short circuit was used, was 3.4 times higher than for the matched load case, and about 5 times higher than when a straight cavity resonator was used.

5. References

[1] G. Dearden et al., Nucl. Instr. and Meth. A 341 (1994) 80.

[2] A.I. Al-Shamma'a et al., Nucl. Instr. and Meth. A 393 (1997) II 83.

[3] J. Lucas et al, these proceedings (20th Int. Free Electron Laser Conf., Virginia, USA, 1998).

FREE ELECTRON LASERS 1998
G.R. Neil and S.V. Benson (Eds.)
1999 Elsevier Science B.V.

Effect of the sidebands on the efficiency of the 1.0 kW FEL☆

Claudio G. Parazzoli*

Boeing Laser & Electro-Optical Systems Organization, P.O. Box 3999 MC 87-85, Seattle, WA 98124-2499, USA

1. Introduction

This paper describes the effects of the onset of the sidebands on the wiggler extraction and laser efficiency, for the 1 kW visible FEL that is being built at the Boeing FEL laboratory in Seattle [1,2].

A state-of-the-art, three dimensional, time dependent code has been used to simulate in detail the effects of sidebands on the performance of the 1 kW visible FEL. The results of the calculations indicate that sideband suppression greatly increases the extraction efficiency at high wiggler energy tapers.

The calculations have been performed with the Boeing version of the LANL-developed FELEX, a fully three-dimensional time dependent code [3]. This is the first time that the sidebands effects have been numerically investigated in detail with a fully three-dimensional and time dependent code.

We use the electron beam parameters for the Boeing 1 kW visible FEL (see Section 2). When the wiggler energy taper is set at 8%, the pass-averaged laser extraction efficiency η_L, is 4.64%, if we neglect finite electron pulse length effects and the onset of sidebands. The finite pulse length effects reduce η_L to 2.40%. If the sidebands are not suppressed (free running case) by the use of an appropriate optical element, η_L is further reduced to 1.0 %. When the wiggler taper is reduced from 8 to 5%, η_L is 1.21% for a free running case. This indicates that in the parameter region of our design, for a free running case, a weaker taper yields a better performance.

☆Work supported by USA/SSDC under Contract DASG60-97-C-0105.

* Tel. +1 253 773 8299,
e-mail: claudio.g.parazzoli@boeing.com

Key words: FEL, sidebands, and simulation.

In the rest of this paper we will distinguish between the laser extraction efficiency η_L and wiggler extraction efficiency η_W. The relationship between η_L and η_w is:

$$\eta_L = \frac{1-R}{1-R(1-L_T)}\eta_w ,$$

Here L_T is the total resonator loss and R the output-coupler reflectivity.

The added diffraction losses at the leading and trailing edges of the optical pulse, due to lack of gain guidance of the optical beam, increase L_T thus decreasing the η_L/η_w ratio. In this paper the laser and wiggler extraction efficiency are averaged over the total number of micropulses for which the simulation has been performed. In this way the start up effects are appropriately accounted for.

2. Simulation results

Free electron laser sidebands are generated through the modulation of the optical field by synchrotron oscillations [4,5]. They have been widely analyzed in the literature in the absence of diffraction effects; we will not review all the existing work here.

The current pulse is a super-Gaussian of exponent 6 with 10. ps FWHM. The electron pulse peak current is 250 A. The rms normalized x-emittance and y-emittance were 15.0 π and 6.25 π mm-mrad respectively. The e-beam normalized energy is γ = 186.91. The optical cavity mirror losses are 4.5% and output-coupling fraction is 25%. The Rayleigh range is 500 cm, the small signal gain of the order of 10, the laser wavelength is 0.80 μm The wiggler peak field is 8.689kG and the wiggler pitch is 2.18 cm. The wiggler energy taper varies between 8 and 5%.

In Fig. 1 the results of a run with 5% energy taper and free running are compared to 8% energy taper, with sidebands control and free running. The lower energy taper provides higher η_w (1.91%) than the higher taper when the laser is free running (1.58%). Both cases show a decrease in η_w with increasing pass number as the pulse width broadens. The 5% energy taper has similar micropulse wiggler extraction efficiency to the 8% case, but with more constant output as function of the pass number and a more rapid start up. In the presence of sidebands control the 8% taper yields $\eta_w = 2.89$ %.

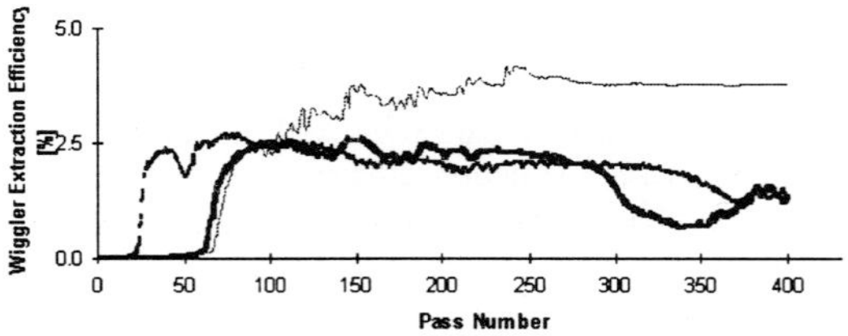

Figure 1. Micropulse wiggler extraction efficiency as function of the pass number. Free running, 5.0% energy taper, pass average efficiency 1.91 % is medium heavy dashed line (--). Sidebands control 8% energy taper, pass average efficiency 2.89% is light solid line (–). Free running, 8% energy taper, pass average efficiency 1.58 % is heavy solid line (–).

Finally, in Figure 2 η_L is computed from the data of Figure 1. The pass averaged laser efficiency η_L is 2.40% with 8% wiggler energy taper and sidebands control (down from η_w = 2.89%) and 1.21 % with 5% wiggler energy taper and free running (down from η_w = 1.91%).

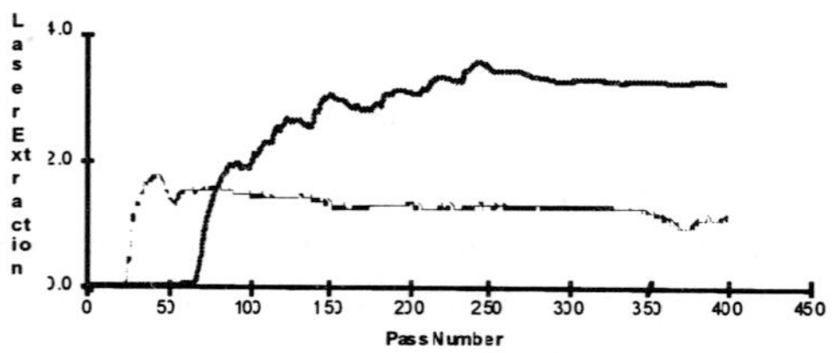

Figure 2. Micropulse laser extraction efficiency as function of the pass number. Free running, 5.0% energy taper, pass average efficiency 1.21 % is light dashed line (--). Sidebands control 8.0% energy taper, pass averaged efficiency 2.40% is heavy solid line (–).

References

[1] J.L. Adamski et al., Nucl. Instr. And Meth. A 375 (1996) 288

[2] D.H. Dowell et al., Nucl. Instr. And Meth. A 393 (1997) 184

[3] B. D. McVey, Nucl. Instr. And Meth. A250 (1986) 449.

[4] 'Free Electron Lasers' , C A Brau, Academic Press, 1990.

[5] 'Classical Free electron laser theory' W B Colson in 'Laser Handbook, Free Electron Lasers'. Eds. W B Colson, c Pellegrini, A Ranieri, North Holland 1990

FREE ELECTRON LASERS 1998
G.R. Neil and S.V. Benson (Eds.)
1999 Elsevier Science B.V.

Simulations of the Proposed TJNAF 20 kW Free Electron Laser

R. E. LeGear, R. B. Steele, R. D. McGinnis, and W. B. Colson

Physics Department, Naval Postgraduate School, Monterey, CA 93943 USA

Abstract

In support of the Navy's requirement for an anti-ship cruise missile defense system, Thomas Jefferson National Accelerator Facility (TJNAF) is constructing the first kW FEL with a goal of improving the design to 20 kW. The 20 kW design proposes to use a klystron undulator to improve gain in weak optical fields. Simulations compare the performance of the conventional undulator with a variety of klystron undulators.

The US Navy is continuing to conduct research to improve surface ship self-defense against anti-ship cruise missiles (ASCM's). One method currently under consideration uses a High Power Directed Energy Beam produced by a Free Electron Laser (FEL) [1].

To support this objective Thomas Jefferson National Accelerator Facility (TJNAF) is constructing the first kilowatt FEL, with a goal of improving the performance to 20 kW. TJNAF's design includes the use of a klystron undulator to improve gain in weak optical fields. This paper discusses the results from simulations comparing the performance of the conventional undulator with a variety of klystron undulators.

Two of the goals of the TJNAF FEL are to achieve an output power of 20 kW at the target while maintaining a final energy spread of 6 percent or less to ensure proper recirculation of the electrons through the accelerator. The simulations varied dispersive section strengths D of the klystron undulator, and varied desynchronism values d between the optical and electron pulses, in order to compare gain G, final power P, and final electron energy spread $\Delta\gamma/\gamma$ [3].

Gain: Simulations were conducted for values of dispersive strength of D = 0, 1, 2, and 3, with D = 0 representing the conventional undulator. The desynchronism between the sequence of electron pulses and the stored optical pulse was varied from d = 0.0 to 0.3 with a step size of 0.02. Figure (1) shows the results of the simulations. The conventional undulator achieved a maximum gain of G = 0.2, occurring at a desynchronism value of d = 0.03. A maximum steady-state gain of G = 0.67 was observed with the klystron undulator using a dispersive strength of D = 2 and a desynchronism value of d = 0.11.

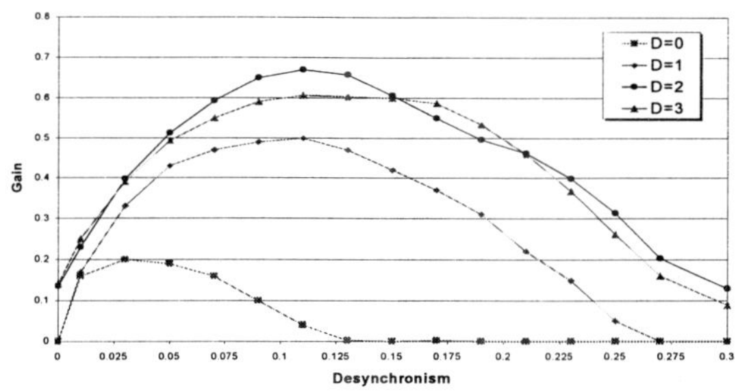

Figure 1. Steady State Gain

Power: The resulting final average power for the klystron undulator are shown in figure (2). Though not included on this graph, the conventional undulator using a desynchronism value of d = 0.007 achieved a final average power 14 times greater than the maximum klystron undulator power with a dispersive strength of D = 1. Much lower final power levels were observed for the klystron undulator using stronger dispersive strengths of D = 2 and 3.

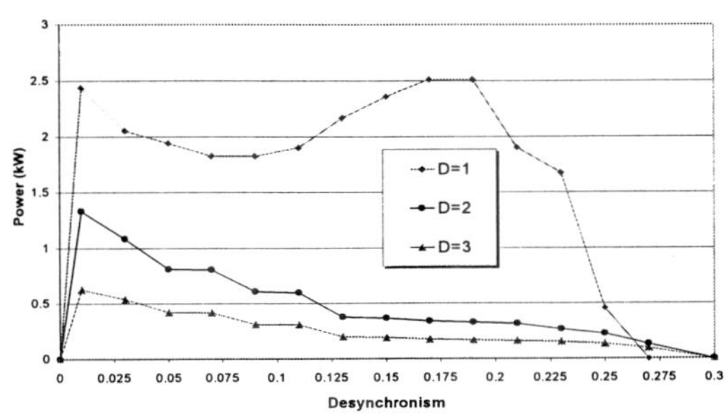

Figure 2. Final Power for D = 1, 2, and 3.

Energy Spread: It is estimated that the TJNAF 20 kW FEL design can tolerate a final energy spread of 6 % in order to recirculate the electron beam. Figure (3) shows a plot of the fractional energy spread of the electrons versus the desynchronism. For the conventional undulator, the energy spread peaks at $\Delta\gamma/\gamma = 8.6$ % at $d = 0.007$, which also corresponds to the point of maximum gain for this undulator. The energy spread decreases with increased desynchronism, reaching the desired value of 6% at $d = 0.03$, and remains below the 6 % level for all values of d greater than 0.03. This corresponds to a final power level of 20 kW for the conventional undulator design. For the klystron undulator design the energy spread is maximum at $\Delta\gamma/\gamma = 5$ % at $d = 0.003$. In klystrons of strength $D = 2$ and 3, the energy spread is approximately a constant 2 %.

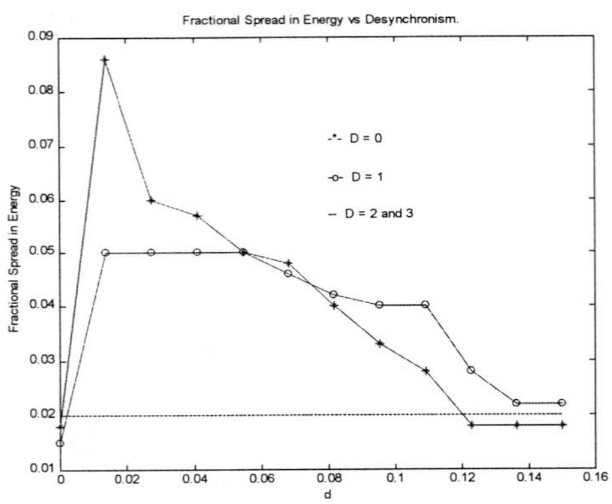

Figure 3. Energy Spread vs. d.

Conclusions: The purpose of these simulations was to quantify the behaviors of steady-state gain, final average power, and induced energy spread as functions of desynchronism d and klystron dispersive strength D, for the proposed 20 kW klystron free electron laser. The two major design goals are an average optical power of 20 kW while maintaining a final energy spread of 6% or less.

Gain is maximum at $d = 0.03$ in the conventional undulator, and at $d = 0.11$ for klystron undulators of strength $D = 1$, 2, and 3. The gain increases with the dispersive strength up to an optimal value of $D = 2$. Further increases in D result in diminishing gain because of beam quality. A klystron with strength $D = 2$ provides 30 % more gain than $D = 1$, and 70 % more than a conventional undulator.

The conventional undulator provides 14 times as much power as a klystron with $D = 1$. For $D = 1$, the peak power occurs at $d = 0.17$, and falls off rapidly to zero at $d = 0.25$. For $D = 2$ and 3, the FEL's power peaks at $d = 0.01$, and then gradually declines to zero at $d = 0.3$. The resultant energy spread is 2 % for $D = 2$ and 3 for all values of desynchronism. The energy spread exceeds the design goal of 6 % for the conventional undulator only within the range of $d = 0.01$ to $d = 0.03$.

Based on the above simulation results it appears that the optimum configuration is to use a conventional undulator with a desynchronism value of $d = 0.03$. The final power with this configuration of 19.2 kW is close to the 20 kW objective and corresponds to a final gain of 19 percent.

Acknowledgements:

The authors are grateful for support from the Naval Postgraduate School and the SPAWAR Directed Energy Office.

References:

[1] A. M. M. Todd, W. B. Colson, and G. R. Neil, "*Megawatt-Class Free Electron Laser Concept for Shipboard Self-Defense*", SPIE 2988, 176 (1997).

[2] E. Anderson, R. Lyon, W. B. Colson, G. Neil and S. Benson, "*Exploring the Feasibility of an FEL Naval Weapon System*", Nuclear Instruments and Methods in Physics Research A393, II-149 (1997).

[3] W. B. Colson, C. Pellegrini, and A. Renieri, editors for the "*Free Electron Laser Handbook*", North-Holland Physics, Elsevier Science Publishing Co. Inc., The Netherlands (1990).

FREE ELECTRON LASERS 1998
G.R. Neil and S.V. Benson (Eds.)
1999 Elsevier Science B.V.

FEL pulse shape measurements with 100-fs temporal resolution using a 10-fs Ti:Sapphire laser and differential optical gating

X. Yan*, A.M. MacLeod, W.A. Gillespie
Division of applied physics and materials, School of Engineering, University of Abertay Dundee, Bell Street, Dundee DD1 1HG, U.K.

G.M.H. Knippels, M.J. van de Pol, A.F.G. van der Meer
FOM-Institute for plasma physics 'Rijnhuizen', P.O. Box 1207, 3430 BE, Nieuwegein, The Netherlands

C.W. Rella
FOM-Institute for atomic and molecular physics, Kruislaan 407, 1098 SJ, Amsterdam, The Netherlands

Characterization of the ultrashort optical and electron pulses produced by free-electron lasers (FELs) is important both for the further development of the FEL and for its operation as a research tool. Conventional methods of characterizing short pulses are not generally applicable in the far infrared. Second or higher order autocorrelation techniques become increasingly impractical at longer wavelengths due to a lack of suitable non-linear materials, and although a far-infrared streak camera has recently been demonstrated on a FEL with sub-picosecond resolution, this is an expensive solution which is not generally available. Furthermore, autocorrelation methods suffer from the drawback that the pulse shape is difficult to resolve, and cross correlation of the FEL pulse with an external laser is beset with synchronization difficulties which generally require that the measurement be obtained in a single shot requiring arrays of fast, sensitive long-wavelength detectors.

Recently, a simple technique called differential optical gating (DOG) has been developed, in which these synchronization problems are considerably reduced [1]. It relies on a simultaneous measurement of the sample pulse intensity at two closely spaced moments in time by cross-correlating it with two pulses (one is delayed with respect to another) from an external ultrafast laser. Provided the time delay between the two pulses is known, each two-pulse measurement allows the FEL pulse intensity $I(t_0)$ and its time derivative $I'(t_0)$ to be calculated at the time of measurement, t_0, which is in principle unknown because of the presumed jitter between the two lasers. Subsequent measurements produce further pairs $(I(t), I'(t))$ at different times, t. These accumulate —either as the result of jitter or by altering the delay between the two lasers — to provide a map of I′ as a function of I, i.e. $I'=f(I)$. Integration of this equation allows $I(t)$ to be recovered except at zeros of $f(I)$. Clearly, this technique can only be used under the assumption that the sample pulse does not change during the repeatative measurements.

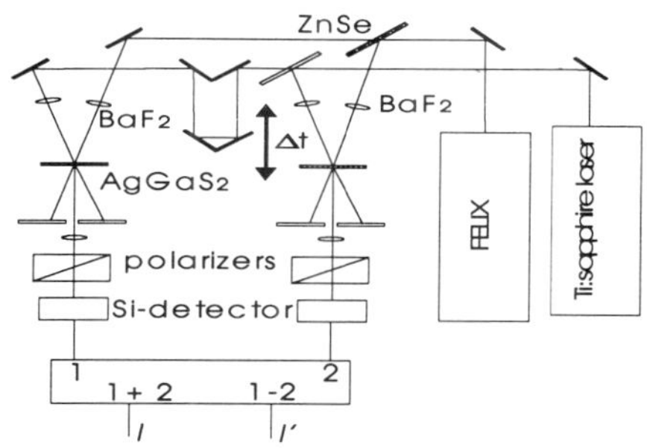

Fig.1 DOG experimental setup

In this contribution we present measurements of different FEL pulse shapes obtained on FELIX using the DOG technique with a synchronized 10-fs Ti:Sapphire laser (FemtoSource Pro, Femto Lasers, Vienna, Austria). The measurements were made at a wavelength of 9 μm using the sum-frequency generated by the FEL radiation and the 800 nm radiation from the Ti:Sapphire laser in a 100-μm thick $AgGaS_2$ crystal (MolTech, Berlin, Germany). We have achieved a temporal resolution of the order of 100 fs, a factor of five improvement over previous results [1]. The resolution could be improved even further by optimizing the design of the focussing optics for the Ti:Sapphire laser and using thinner, less dispersive, sum-frequency mixing crystals. See Figure 1 for the experimental setup.

We have studied the FEL optical pulse shape in two important regimes of operation. First, when the laser cavity is detuned away from perfect synchronism, relatively

* Corresponding author: Tel. +31-30-6096999, Fax. +31-30-6031204, email:lynn@rijnh.nl

long FEL pulses are generated. A typical DOG picture is shown in Fig. 2(A), and the reconstructed pulse shape in Fig. 2(B). The leading edge fits well to an exponential, whereas the trailing edge is fitted by a Gaussian. These results are taken at the early part of the macropulse, before the formation of a subpulse sets in. Similar results were observed at the Stanford FEL [1]. The reconstructed pulse is also in good agreement with separate measurements of the intensity autocorrelation function (1.33 ps fwhm) and the optical spectrum (time-bandwidth product of 0.35).

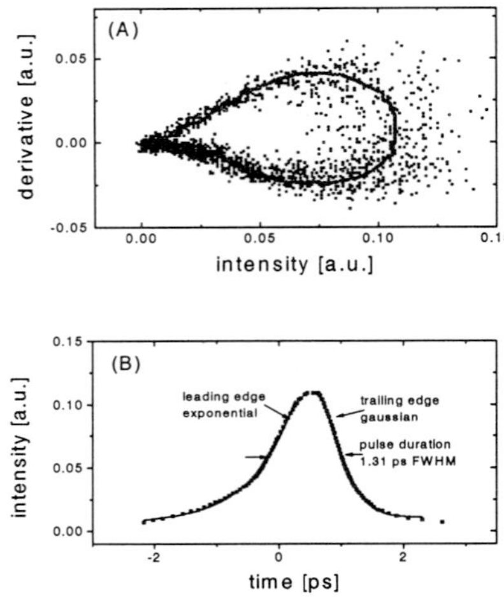

Fig. 2 Raw DOG data (A) and reconstructed FEL pulse (B) for ΔL=-22μm, λ=9μm.

Under certain conditions the formation of subpulses, which have a clear signature in the DOG data, is seen. An example is given in Fig. 3, with a (preliminary) reconstructed micropulse shape shown as an inset.

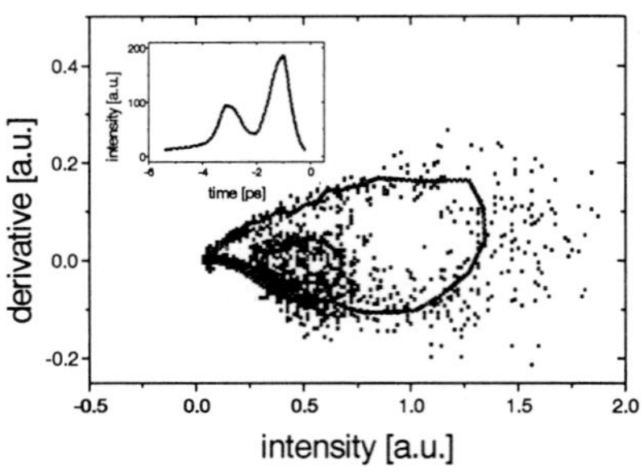

Fig. 3 DOG data with a subloop, indicating the formation of subpulses on the leading edge. A preliminary reconstruction of the pulse is shown as an inset.

In a second series of measurements we focused on the shortest possible FEL pulses to demonstrate the improved time-resolution of approximately 100 fs that we obtained. The data presented here represent the first high-resolution pulse shape measurements of these ultrashort FEL pulses, and should provide a basis for more stringent tests of FEL theory in this operating regime than were previously possible. The DOG data shown in Fig. 4 correspond to an FEL pulse of approximately 200 fs fwhm in duration. Although our current time resolution is perhaps still not good enough to resolve the details of the pulse shape fully, we have fitted it for illustration purposes with the the first supermode [2], which is an analytic expression for the FEL pulse shape at small values for the cavity detuning:

$$|A_1(\xi)|^2 \approx \nu \; exp(3^{1/2}\,(2/\nu)^{1/3}\,(\xi-1))\sin^2(\pi\,\xi).$$

In this formula, ν and ξ are the (normalized) cavity detuning and z coordinate respectively. Our fit yields a value for ν of 0.09, whereas the calculated value yields ν = 0.03.

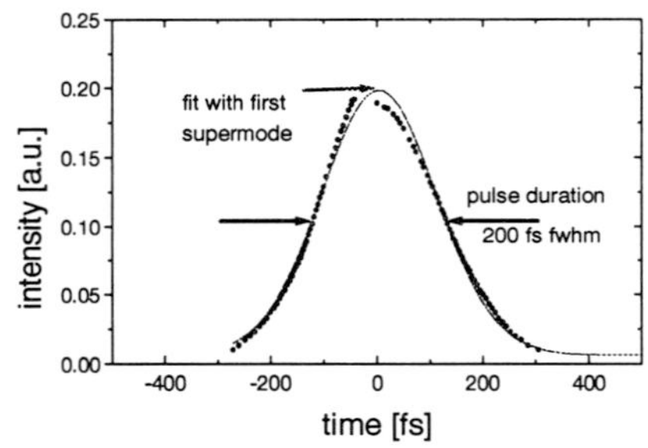

Fig. 4 Reconstruction of a short FEL pulse taken at ΔL=-10μm and λ=9μm. The solid line corresponds to the first supermode (see text).

Acknowledgements

This work is part of the research program of the 'Stichting voor Fundamenteel Onderzoek der Materie (FOM)', which is financially supported by the 'Nederlandse Organisatie voor Wetenschappelijk Onderzoek (NWO)'.

References

[1] C.W. Rella, G.M.H. Knippels, D.V. Palanker, T.I. Smith, H.A. Schwettman, to be published in the proceedings of the 19th Int. FEL. Conf., Beijing, China, 1997; C.W. Rella, G.M.H. Knippels, D.V. Palanker, H.A. Schwettman, Submitted to Opt. Commun.

[2] N.Piovella, P.Chaix, G.Shvets, D.Jaroszynski, proceedings of the 17th International FEL Conference, New York, USA, 1995, p.156.

FREE ELECTRON LASERS 1998
G.R. Neil and S.V. Benson (Eds.)
1999 Elsevier Science B.V.

Short Micropulse Generation in an FEL using Modulated Desynchronism

Todd I. Smith, Takuji Kimura

Stanford Picosecond FEL Center, Hansen Experimental Physics Laboratory
Stanford University, Stanford, CA 94305-4085

1. Introduction

During the past few years there has been much interest in the topic of generating ultra short FEL pulses [1]–[5]. In this paper we present a method of producing short pulses using modulated desynchronism in an FEL. At the Beijing FEL Conference we reported observing that the micropulse length of our mid-infrared FEL decreased from greater than one picosecond to about 350 femtoseconds when the electron beam energy was modulated by a few tens of KeV at 40 KHz [6]. These observations were made at a wavelength of 4.5 μm with a time averaging autocorrelator. It seems quite likely that they arise from modulation by a few microns of the desynchronism parameter of the FEL. This modulation is expected from coupling between temporal dispersion in magnetic dipoles in the electron beam line and the 40 KHz energy modulation. Here we report the first results of time synchronized measurement of the micropulse length, the optical power spectrum, and the optical power as a function of the phase of the modulating signal.

2. Experiment

The FEL beam line at the SCA/FEL consists mainly of four accelerator structures and two FEL undulators. The first two structures accelerate the beam to about 20 MeV. After passing through the magnetic chicanes of the far-infrared FEL, the beam undergoes further acceleration by two more structures before entering the 72 period mid-infrared FEL undulator ($\lambda_w = 3.1$ cm). The rms wiggler parameter is $K = eB_w\lambda_w/2\pi mc = 0.83$. During our typical FEL operation, the wavelength of the mid-infrared beam is stabilized through a feedback loop on the first accelerating structure. By opening this loop and applying signals from a function generator, the energy of the electron beam can be modulated.

The autocorrelation signal and the optical power signal are recorded as a function of time and are shown in Fig. 1. Each of the three plots corresponds to a different modulation level. The flat signals taken without modulation are shown for comparison. As the modulation level increases, the maximum of the autocorrelation becomes much higher than the normal level, while the change in power is less prominent. This is an indication of very short micropulses.

By gating the signals from the autocorrelator, the spectrum analyzer, and the power detector, respectively, simultaneous measurements of the micropulse length, the spectral width, and the optical power as a function of the phase of the modulating signal can be made. Fig. 2 shows one set of measurement at 5.0 μm wavelength. The measured micropulse length ranged from 300 fs to 800 fs (FWHM) while the spectral width varied from 2.3% to 0.9% (FWHM). The product remained transform limited throughout the entire cycle.

3. Discussion

The magnetic chicanes in the FEL beam line are non-isochronous, i.e., higher energy electrons pass through them quicker than lower energy ones. The effect is calculated to be 0.03 ps/KeV. Thus modulation of the beam energy is translated into modulation of the electron bunch repetition frequency. If we assume that the energy modulation takes the form $(\Delta E/2)\sin(\omega_m t)$, where ΔE is the peak-to-peak energy modulation in KeV, and ω_m is the modulation frequency in rad/sec, then the fractional change in the electron bunch repetition frequency is given by,

$$\frac{\Delta f}{f_{FEL}} = 1.5 \times 10^{-14}\ \omega_m\ \Delta E\ \cos(\omega_m t) \qquad (1)$$

where f_{FEL} is the nominal value of the repetition frequency. Since a change in repetition frequency has the same effect as the cavity length detuning $\delta L = L_{cav}(\Delta f/f_{FEL})$, thus an equivalent cavity length de-

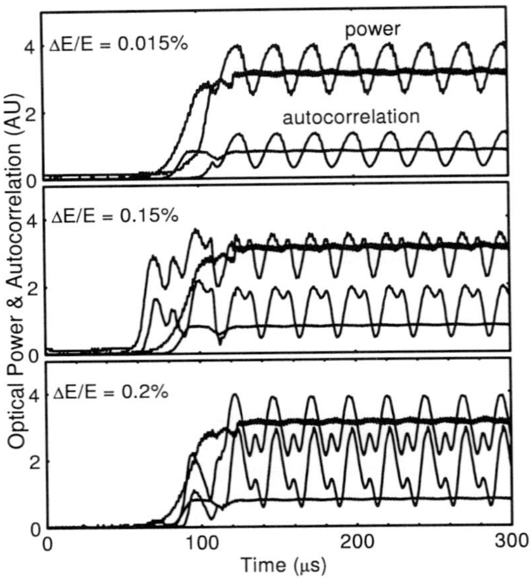

Fig. 1. Plots of autocorrelation and optical power signals as a function of time with different levels of beam energy modulation. The flat curves are for situations without the energy modulation.

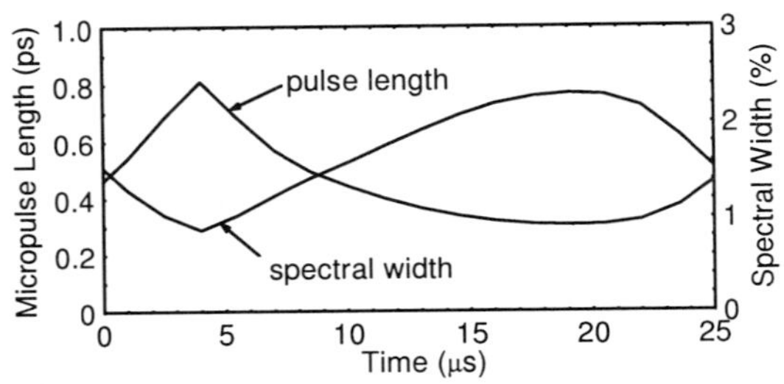

Fig. 2. Plot of micropulse length (FWHM) and spectral width (FWHM) as a function of time when the beam energy is modulated at 40 KHz.

tuning can be calculated from Eq. (1). In our experiment, the magnitude of the energy modulation is monitored with an electric field probe inside the first accelerator structure. Measurement shows the maximum energy modulation saturates at between 50 KeV and 100 KeV ($\Delta E/E = 0.15\% - 0.3\%$). From Eq. (1), that corresponds to a variation in the repetition frequency of $5 - 10$ Hz out of 11.8 MHz, or is equivalent to a change in the cavity length by $5 - 10\,\mu$m out of a total length of 12.7 m. This change is a significant fraction of the power vs. detuning curve. However an energy modulation of 0.3% is well within the energy acceptance range of the FEL since the Fourier transform limited bandwidth of a 350 fs micropulse at 5 μm is 2%.

In an FEL, the length of the micropulse is determined by the temporal overlap of the optical field and the electrons and can be adjusted by changing the cavity desynchronism, which is defined as the difference between the round trip time of the optical pulse and the electron bunch repetition period. To counteract the "laser lethargy" effect, a cavity length that is slightly shorter than perfect synchronism is usually chosen. As cavity desynchronism approaches zero, the pulse length becomes shorter, while the optical power grows. However the gain also decreases, and optical power starts to decay once the gain becomes less than the cavity loss. The decay time of our cavity is about 2 μs. We believe that by modulating the beam energy we are able to operate briefly in the small desynchronism, short pulse, high efficiency region which is otherwise unstable. Simulation based on the model provided in [4] produced results for the optical power signal similar to those in Fig. 2, including the higher harmonic components.

4. Conclusion

Time synchronized measurements of the micropulse length and spectral width show that micropulses as short as 300 fs are achieved when electron beam energy is modulated at about 40 KHz, and the beam stays transform-limited at all time. There is strong evidence that the coupling between temporal dispersion in the magnetic dipoles and the energy modulation causes the cavity desynchronism to oscillate between the high power, short pulse regime and the low power, long pulse operating regime.

The authors wish to thank Seiji Yamamoto for his help in assembling the electronics necessary for the gated measurement.

REFERENCES

1. N. Piovella *et al.*, Phys. Rev. E **52**, 5470 (1995).
2. D. A. Jaroszynski *et al.*, Phys. Rev. Lett. **70**, 3412 (1993).
3. R. J. Bakker *et al.*, Phys. Rev. E **48**, 3256 (1993).
4. N. Piovella, Phys. Rev. E **51**, 5147 (1995).
5. G. Shvets, *et al.*, Phys. Rev. E **56**, 3606 (1997).
6. T. I. Smith, "Beating the Fourier Transform Limit with an FEL" in *Proceedings of the Beijing FEL Conference*, August 1997.

FREE ELECTRON LASERS 1998
G.R. Neil and S.V. Benson (Eds.)
1999 Elsevier Science B.V.

5 ~ 15 Hz FEL macropulse operation at MIR-FEL facility of the FELI

Y.Kanazawa, A.Zako, E.Oshita, T.Takii, A.Nagai, T.Tomimasu

Free Electron Laser Research Institute, Inc.(FELI)
2-9-5,Tsuda-Yamate,Hirakata,Osaka 573-0128,Japan

Abstract

A klystron modulator used for three FEL facilities (FEL-1, FEL-2 and FEL-4) was reinforced to increase the repetition rate. This paper describes some results on a test operation of 5 ~ 15 Hz FEL macropulse operation and two-color lasings at 6.87 μm and 7.03 μm using a stepwise 24-μs rf pulse.

1. Introduction

The usual FEL beam consists of a train of 3 picoseconds pulse (micropulse) repeating at 22.3125 MHz or at 89.25 MHz and the FEL micropulse train continues for 18 μs (macropulse) and the macropulse repeats at 10 Hz at MIR-FEL facility. The FEL peak power per micropulse is 10 MW at the exit of the cavity mirror and the maximum average power is 100 mW at 22.3125-MHz micropulse repetition rate, and 300 mW at 89.25 MHz at the user rooms. In order to meet recent FEL user's demands [1], the klystron modulator used for three FEL facilities (FEL-1, FEL-2 and FEL-4) was reinforced to increase the macropulse repetition rate from 10 Hz up to 20 Hz in Sept. 1996.

This paper describes recent results and problems on 5~15-Hz FEL macropulse operation at MIR-FEL facility (FEL-1).

2. Rf power sources and rf waveguides

Fig. 1 shows a schematic layout of a thermionic gun, a 714-MHz prebuncher, a 2856-MHz buncher, seven 2.93-m accelerating waveguides, a 714-MHz klystron and two 2856-MHz klystrons, rf waveguides, four S-type BT lines and four FEL facilities. The prebuncher, the buncher and the seven 2.93-m accelerating waveguides are maintained to be their designed temperature of 40℃ by a closed-loop, temperature control system.

The rf waveguides from the klystrons to the injector and seven accelerating waveguides are filled with pressurized SF_6 gas and are also maintained to keep their temperature constant with the closed-loop, temperature control system. The first klystron therefore has been always used for all FEL facilities but the second one has been only used for FEL-3 at a 10-Hz macropulse repetition rate.

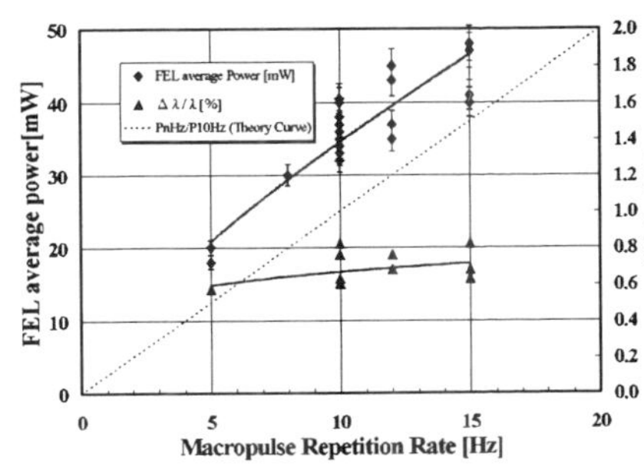

Fig. 2. Macropulse repetition rate dependence of 7.1-μm FEL average power and its spectral spread.

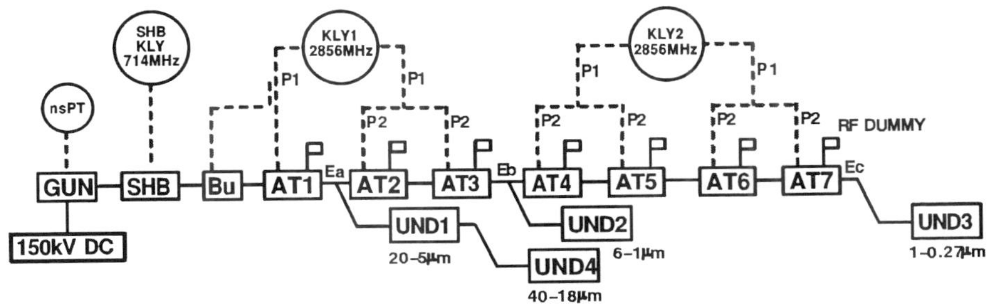

Fig. 1. The schematic layout of a thermionic gun, a 714-MHz prebuncher, a 2856-MHz buncher, seven 2.93-m accelerating waveguides, a 714-MHz klystron and two 2856-MHz klystrons, rf waveguides, four S-type BT lines and four FEL facilities

3. 5~15-Hz FEL macropulse operation at MIR-FEL facility

The 7.1-μm FEL was obtained from FEL-1 of a 20.7 mm undulator gap (K=0.838) using a 28.6-MeV electron beam. Fig. 2 shows repetition rate dependences of 7.1-μm FEL average power measured at the user's station-1 and its spectral spread ($\Delta\lambda/\lambda$, FWHM) measured at the monitor room, where a quarter of the FEL beam from FEL-1 (1.5-mmφ mirror aperture of FEL-1 cavity mirror) is delivered to a beam diagnostic station using a fan-shaped mirror with a 90° opening angle and three quarters of the FEL beam is delivered to the user station-1.

The results shown in Fig. 2 were obtained at the optimum oscillation condition for each macropulse repetition rate. The average power is approximately proportional to the repetition rate and the spectral spread is almost independent of the repetition rate.

However, whenever the repetition rate is changed, the average power is smaller than that being proportional to the repetition rate and the spectral spread becomes broader than that of the optimum oscillation. Mostly, a change of the repetition rate stopped lasing or sometimes induced two-color lasing. The reason is due to a distortion of the 24-μs rf pulse shape induced by the change of repetition rate. We therefore had to always adjust all parameters to reach the optimum condition at each repetition rate by remotely adjusting distributed inductor's inductance of the pfn network [2]. Fine adjustments of the rf phase for the prebuncher, the buncher and the seven 2.93-m accelerating waveguides including the rf waveguides from the klystrons are also essentially needed, because of slight changes in their equilibrium temperature and in SF_6 pressure filled in the rf waveguides.

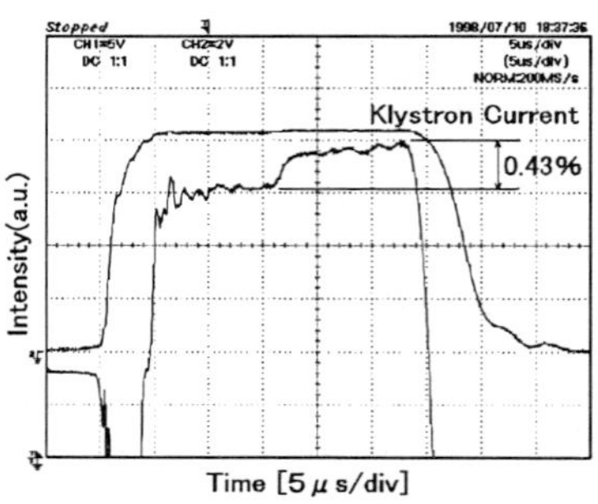

Fig. 3. A typical stepwise 24-μs klystron current shape

Fig. 3 shows an example of a stepwise 24-μs klystron current pulse shape. This shape was easily obtained by remotely adjusting distributed inductor's inductance of the pfn network.

4. Two-color lasing

Fig. 4 shows a typical two-color FEL spectra of 6.87 and 7.03 μm induced intentionally at a 10-Hz macropulse operation. A 10-μs front part of the klystron current pulse shown in Fig. 3 contributes to the 7.03 μm and a 12-μs long back part does to the 6.87 μm. A relative wavelength difference $\Delta\lambda/\lambda = 0.16\ \mu m / 7\ \mu m = 2.3\ \%$ is found in Fig. 3.

The 0.43 % step of the klystron current pulse shape induces a 0.36 % step in an accelerated electron spectrum, because the energy difference is 5/6 times of the 0.43 % step.

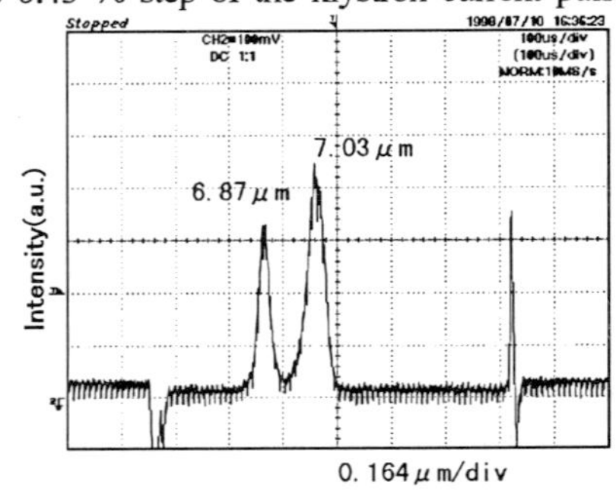

Fig. 4. A typical two-color FEL spectra

Furthermore, the 0.36 % step in the electron spectrum induces a 0.72 % step in an FEL spectrum because the FEL wavelength is inversely proportional to the square of electron energy. The estimated value of 0.72 % can explain only one third of the relative wavelength difference of 2.3 %. Two-third seems to be due to phase adjustments for the buncher and the first accelerating waveguide.

References

[1] T.Tomimasu, et al., CP413, "Towards X-Ray Free Electron lasers" edited by R.Bonifacio and W.A. Barletta (1997,AIP) pp.127-141.

[2] E.Oshita, et al., IEEE Proc. PAC'95 (Dallas,May1-5,1995) pp.1608~1610.

FREE ELECTRON LASERS 1998
G.R. Neil and S.V. Benson (Eds.)
1999 Elsevier Science B.V.

Cyclotron autoresonance maser in the regime of trapping and adiabatic deceleration of electrons

V.L.Bratman[1], Yu.D.Grom[1], A.W.Cross[2], Yu.K.Kalynov[1], A.D.R.Phelps[2], S.V.Samsonov[1], and A.V.Savilov[1]

[1] *Institute of Applied Physics, Russian Academy of Sciences 46 Ulyanov St., Nizhny Novgorod, 603600, Russian Federation*

[2] *University of Strathclyde, Dept. of Physics and Applied Physics John Anderson Building, 107 Rottenrow, Glasgow G4 0NG, Scotland, UK*

1. Introduction

A natural method for efficiency enhancement of Free-Electron Lasers (FELs) is providing a mechanism of prolonged resonance during the process of electron-wave energy exchange. For various types of FELs (in particular, for CARMs), one such mechanism is the regime of trapping of electrons by the RF wave, and of adiabatic deceleration of the trapped particles by smoothly profiling the parameters of the system (a similar regime was first proposed for TWTs [1]) [2-6]. According to the theory [1-6], this regime can provide a considerable increase in efficiency with a weak sensitivity to spread in electron velocity, as compared with the "traditional" regime of compact electron bunching [7]. Just the realization of the trapping regime allowed the achievement of a high electron efficiency for the first time in a FEL-amplifier experiment [8]. However, in oscillators the realization of this regime is more complicated due to several reasons. Evidently, the main ones are difficulties in providing single-frequency operation, and in the excitation of the long, profiled system, especially for a relatively short electron pulse. In order to avoid these difficulties, one can develop the idea of sectioning (see e.g. [9]) and study a FEL-oscillator of the twystron type with the microwave system consisting of the first, self-exciting section, being a relatively short cavity, and the second prolonged section, where trapping is provided [10,11].

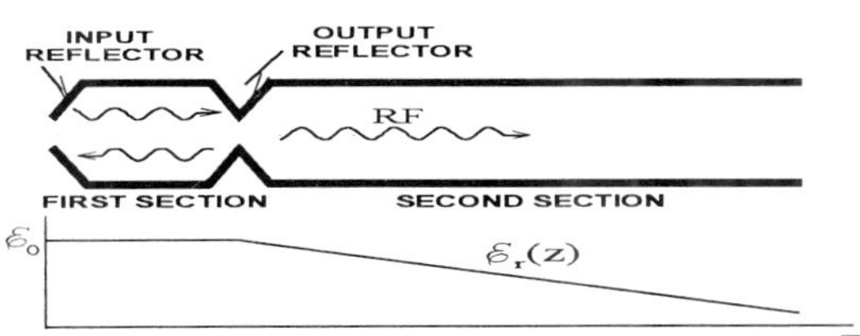

Figure 1. Schematic of the sectioned oscillator.

2. Sectioned FEL-oscillator

In the simplest model of a sectioned FEL-oscillator (Fig.1), the self-exciting section is a relatively short segment of a waveguide with RF feedback, which is provided by very short reflectors at the input and output of this section, with the reflection coefficients 100% and R respectively. This section is excited due to the "traditional" mechanism of compact electron bunching. The RF field passes through the output reflector and comes into the second section, being a long waveguide without RF feedback This section represents an amplifier; the input signals are both the RF wave, having passed through the output reflector of the first section, and perturbations in the electron density due to electron bunching inside the first section. In the second section, the energy of the exact resonance of electrons with the wave, $\mathcal{E}_r(z)$, decreases with the longitudinal coordinate (Fig.1) due to profiling the parameters of the system (for instance, profiling the guiding magnetic field in the CARM, or the undulator field in the ubitron). A part of electrons is trapped by the potential well formed by the RF field. Energies of the trapped particles decrease with the decrease of $\mathcal{E}_r(z)$, energies of the other particles practically do not vary. Thus, the efficiency is approximately defined by the formula $\eta = K[\gamma_0 - \gamma_r(z)]/[\gamma_0 - 1]$, where K is the trapping coefficient, γ is the electron Lorentz factor.

According to calculations [10], for smooth profiling the trapping coefficient is defined basically by the parameters of the self-exciting section; namely, by the reflection coefficient at the output of the first section. If it is small enough, $R = 0.1 - 0.3$, then almost total trapping, $K = 0.9 - 0.95$, can be achieved in a very wide region of parameter space. The increase of R leads to a significant decrease of the trapping coefficient because in this case the system contains a non-adiabatic de-trapping factor. This is a fast change of the complex RF amplitude in the region of the transition from the first section into the second one, because of the strong reflection of the RF wave by the output reflector.

This difficulty can be avoided by using a prolonged output reflector: the self-exciting section can be a cavity with the distributed feedback provided by two reflectors: a short input reflector, and a Bragg reflector with a length approximately equal to the total length of the section (Fig. 2) [11]. In this case, the distribution of the RF field within the region of the transition between the sections is smoother. This should provide more accurate trapping of electrons at the beginning of the second section.

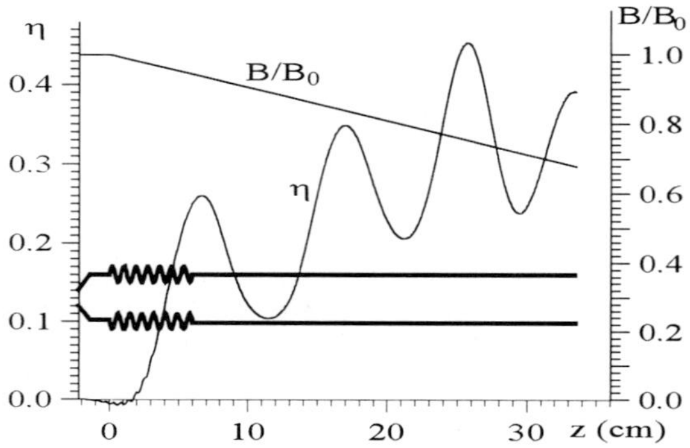

Figure 2. Design and calculations for the CARM at the fundamental cyclotron resonance. Electron efficiency and guiding magnetic field versus the longitudinal coordinate.

3. Preliminary experiment

The proposed scheme can be effectively used for various FELs, including ubitrons, CARMs and Cherenkov devices. As for the CARM, our goal is a significant enhancement of their efficiency in the experiments at the fundamental [12] and the second [13] cyclotron harmonics. As calculations show (Fig.2), a high electron efficiency (about of 40 - 50 %) can be achieved.

Advantages of this scheme were confirmed by a preliminary experiment under the realization of the CARM-twystron in the regime of trapping (the fundamental cyclotron harmonic, 8 mm wavelength, 100 A / 500 kV electron beam). The main disadvantage of this experiment were a quite bad quality of the electron beam (tens per cent of velocity spread). Due to this fact, for the traditional (uniform) scheme of the oscillator, the power was essentially lower than in the experiment [12] (3.6 MW with the electron efficiency of 7 %). In addition, tapering of the magnetic field was a non-ideal, and the length of the interaction region was significantly shorter (Fig. 3) as compared with the preliminary design (Fig.2). However, even in this situation, the experiment demonstrated the possibility of a significant increase of the output power due to the use of the regime of trapping: the power of 5.2 MW with the electron efficiency of 10.5% was achieved.

Acknowledgments: This work was supported by the Russian Foundation for Basic Research, Grants No. 96-02-18971 and 98-02-17068, and by the United Kingdom DERA.

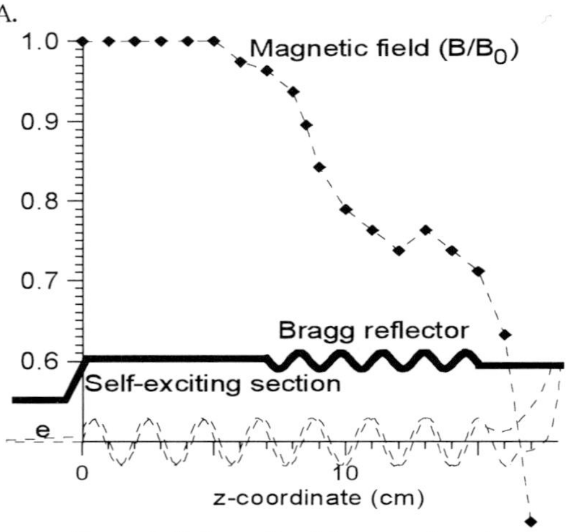

Figure 3. Schematic of the experiment

References

[1] E.D.Belyavsky, Radiotekh. i Electron.16 (1971) 208.
[2] N.M.Kroll, P.L.Morton, M.N.Rosenbluth, Phys. Q. Electron. 7 (1980) 113.
[3] P.Sprangle, C.-M.Tang, W.N.Manheimer, Phys. Rev. Lett. A. 21 (1980) 302.
[4] N.S.Ginzburg, I.A.Man'kin et al., Relativ. HF Electron. 5 (1988) 37.
[5] V.L.Bratman, N.S.Ginzburg, A.V.Savilov, Relativ. HF Electron. 7 (1992) 22.
[6] G.S.Nusinovich, Phys. Fluids B 4 (1992) 1989.
[7] V.L.Bratman, N.S.Ginzburg, M.I.Petelin, Opt. Commun. 30 (1979) 409.
[8] T.Orzechowski, B.Anderson, J.Clark et al., Phys. Rev. Lett. 57 (1986) 2172.
[9] E.B.Abubakirov, A.V.Smorgonsky, Radiotekh. i Electron. 35 (1990) 2644.
[10] A.V.Savilov, IEEE Trans. on Plasma Sci. 26 (1998) 36.
[11] A.V.Savilov, V.L.Bratman, A.D.R.Phelps, S.V.Samsonov, Nucl. Instr. Meth. A407 (1998) 480.
[12] V.L.Bratman, G.G.Denisov, B.D.Kol'chugin, S.V.Samsonov and A.B.Volkov, Phys. Rev. Lett. 75 (1995) 3102.
[13] S.J.Cooke, A.W.Cross, W.He and A.D.R.Phelps, Phys. Rev. Lett. 77 (1996) 4836.

FREE ELECTRON LASERS 1998
G.R. Neil and S.V. Benson (Eds.)
1999 Elsevier Science B.V.

FEM with pulsed short-period undulator and low-energy electron beam

V.L.Bratman, Yu.D.Grom, A.V.Savilov

Institute of Applied Physics RAS, 46 Ulyanov St., Nizhny Novgorod, Russia 603600

V.A.Papadichev*, G.V.Rybalchenko

P.N.Lebedev Physical Institute RAS, 53 Leninski Prospect, Moscow, Russia 117924

Abstract

Main elements of an oscillator with a planar 6-mm-period undulator in the form of a periodic system of ferromagnetic bodies immersed in the pulsed solenoid have been studied. At a solenoid field of 1.0 T, the undulator provides a 0.3-0.2 T transverse magnetic field for the 2-3 mm gaps. The microwave system is a rectangular waveguide of 2 × 10 mm cross-section terminated by input and output reflectors. An electron beam with a 50 A current, 500 kV voltage, 20 ns pulse duration, and 1.8 mm diameter has been obtained. Simulations predict a MW level of output power at a 2 mm wavelength with 5-7% electronic efficiency.

1. Introduction

Realization of an effective FEM with a short-period undulator is still an unsolved but very attractive problem. The use of undulators with millimeter periods and low-energy electron beams produced by high-current accelerators with "cold" emitters opens in particular one of the simplest ways for producting powerful coherent radiation at millimeter and submillimeter wavelengths. Such short-period undulators can provide a sufficiently large transverse component of electron velocity, which allows obtaining a relatively high electron efficiency of the FEM. This work shows the possibility for realization of a FEM-oscillator with a planar 6-mm-period undulator, which is formed by a periodic system of ferromagnetic bodies immersed in a solenoid [1,2], pulsed in our case. Such a system redistributes the external magnetic field so that a significant periodic transverse component of the field is provided.

2. Undulator

Axial magnetic field $B_{||}$ is used for high-current, pulsed, electron beam generation and transport. The most suitable undulator is one which modifies axial magnetic field to transverse periodic. Such devices were used for many years with various methods of field formation: ferromagnetic insertions for dc fields [1,2] and metallic ones with pulsed fields. Skin-effect (eddy currents) is used in the latter case. The most effective way to obtain high transverse field in our case, where pulsed magnetic field of a solenoid is employed, is to make insertions fabricated of laminated transformer steel to avoid eddy current effect (parasitic in this scheme).

Upper and lower halves of the undulator are arrays of 4 × 10 × 25 mm blocks which are sets of 0.35-mm-thick plates of transformer steel. The blocks are placed with 6-mm period. The halves can be assembled with various gaps, thus changing transverse to longitudinal field ratios. 2 and 3 mm gaps were used for field measurements. Longitudinal shift of one half relative to the other changes transverse field amplitude in the median plane as well as harmonic content. Maximum amplitude of the fundamental is attained with a half-period shift. This geometry was used in the experiments. At a solenoid field of 1.0 T the measured transverse field amplitude amounted to 0.3 T for a 2 mm gap and 0.2 T for a 3 mm. Its dependence on axial magnetic field for a 3-mm gap is shown in Fig. 1.

Adiabatic entry often used in such undulators is too long, so matched fringe field formation was chosen to have a shorter undulator. The end block was moved outwards by several millimeters relative to the others. This allows to balance the entry field integral, so that electrons are placed exactly on orbit (field integral should be zero at field maxima).

*Corresponding author. Tel.: (095) 132 6300; fax: (095) 938 2251; e-mail: papadich@sci.lpi.ac.ru

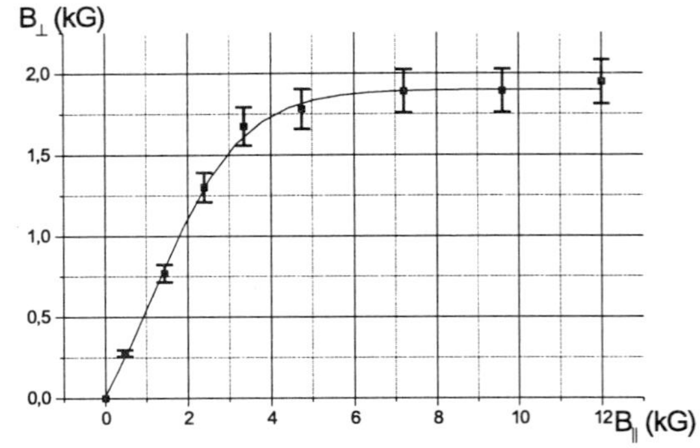

Fig. 1. Transverse undulator field, $B_{\perp}$, versus solenoid field, $B_{\parallel}$, for a 3-mm gap.

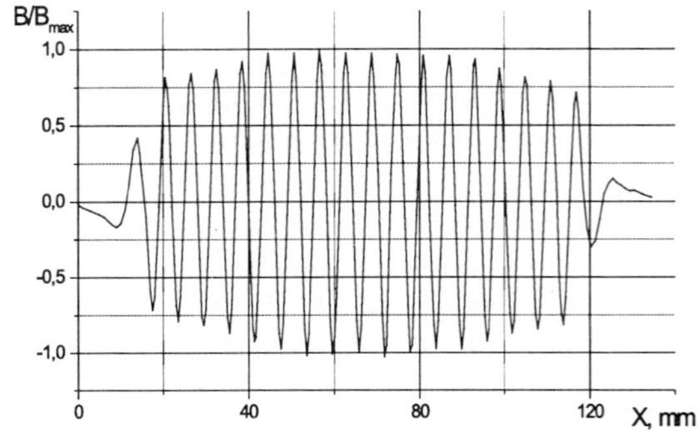

Fig. 2. Axial dependence of the undulator field.

Axial dependence of undulator field is presented in Fig. 2.

Field amplitude modulation along the axis is due to the small number of periods and can be excluded by adjusting the transverse position of each block. Assessment of its influence on particle dynamics shows that this is not necessary.

It is worth noting that the studied undulator has a shorter period 6 mm, than previous ones 10 mm periods [2-4] and even 25 mm [1]. Laminated iron insertions in pulsed magnetic field and matched undulator entry were not employed earlier and proved to be effective for field formation and reducing undulator length considerably compared to devices with an adiabatic entry.

3. Electron beam and microwave system

As an electron source for a FEM with the described undulator, one can use any injector forming thin high-current electron beams. For instance, such beams are produced by explosive emitters in high-current direct-action accelerators. At the Institute of Applied Physics (N.Novgorod, Russia), one of such accelerators, the SINUS-6, allows obtaining electron beams with an energy of 300-700 keV, pulse duration 20 ns, and a total current of several kA. The use of a quasi-plane magnetic-insulated diode with a narrow outlet provides good selection of "the most rectilinear" electrons. This results in forming a small-diameter (1.5-2.0 mm) high-quality electron beam with an operating current of 20-100 A [5].

In the design of the FEM-oscillator, it is planned to use a cavity of 2 × 10 mm cross-section as the microwave system. It is formed by a rectangular operating waveguide terminated by two (input and output) selective reflectors. The operating magnetic field of the pulsed solenoid is assumed to be 1.0 T. The amplitude of the transverse field component would be about 0.2 - 0.3 T.

According to simulations, for 15 - 20 undulator periods, 0.85-0.90 cavity feedback coefficient and nearly 500 kV accelerating voltage, the use of a 20-50 A beam allows obtaining a MW level of output power with 5-7% electronic efficiency. The frequency of the output radiation can be varied from 100 to 240 GHz by small changes of the accelerating voltage with corresponding adjustment in parameters of the cavity reflectors (Fig.3).

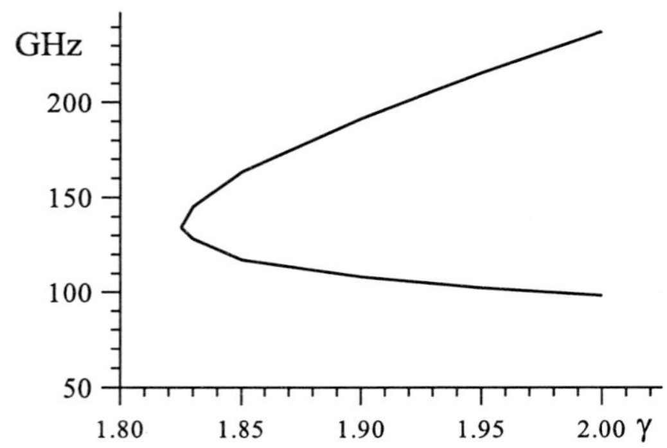

Fig. 3. Upper and lower resonant frequencies versus the electron Lorentz factor (γ = 1.82 corresponds to the regime of grazing).

References

[1] A.H. Ho, R.B. Pantell, J. Feinstein and B. Tice, Nucl. Instr. and Meth. A296 (1990) 631.

[2] A.A. Varfolomeev, S.N. Ivanchenkov, A.S. Khlebnikov and N.S. Osmanov, Nucl. Instr. and Meth. A318 (1992) 809

[3] Y.C.Huang, H.C.Wang, R.H.Pantell et al., Nucl. Instr. and Meth. A341 (1994) 431

[4] J.M.Tacceti, R.H.Jackson, H.P.Frenud et al., Nucl. Instr. and Meth. A358 (1995) 193

[5] V.L. Bratman et al., Phys. Rev. Lett. 75, 3102, (1995)

FREE ELECTRON LASERS 1998
G.R. Neil and S.V. Benson (Eds.)
1999 Elsevier Science B.V.

A far-infrared light source by using the coherent radiation from high intensity single-bunch beams of the L-band linac at ISIR

K. Yokoyama[a], Y. Miyauchi[a], S. Okuda[a,*], R. Kato[a], T. Takahashi[b]

[a]*The Institute of Scientific and Industrial Research, Osaka University 8-1 Mihogaoka, ibaraki, Osaka 567-0047, Japan*

[b]*Research Reactor Institute, Kyoto University, Kumatori, Osaka 590-0494, Japan*

1. Introduction

The radiation from a high-energy electron bunch becomes coherent and highly intense at wavelengths about or longer than the length of the bunch. This phenomenon was first observed for synchrotron radiation [1] and after that, for various kinds of radiation. Radiation such as synchrotron radiation, transition radiation and Cherenkov radiation has a continuous spectrum. From these features the coherent radiation is expected to be applied to spectroscopy in a submillimeter to millimeter range as a new intense pulsed light source.

At The Institute of Scientific and Industrial Research (ISIR) in Osaka university, self-amplified spontaneous emission [2] in the radiation process from a wiggler and the coherent radiation [3] are being observed by using the high intensity single-bunch electron beams generated with the 38-MeV L-band linear accelerator (linac). The purpose of the present work is to establish a new light source by using the coherent radiation.

2. Characteristics of the coherent radiation

For an electron bunch sufficiently narrow around the beam axis, the intensity of radiation from the bunch at a wavelength of λ is expressed as follows:

$$P(\lambda) = p(\lambda)\, N\, [\, 1+ (N-1) f(\lambda)\,], \qquad (1)$$

where $p(\lambda)$ is the intensity of radiation from one electron, N the number of electrons in the bunch and $f(\lambda)$ the bunch form factor which is determined from the shape of the bunch. At wavelengths enough longer than the bunch length $f(\lambda)$ is nearly equal to 1 and hence, $P(\lambda)$ is N times the intensity of the ordinary incoherent radiation. For the single-bunch beam of the ISIR linac N is 5×10^{11} in maximum. The length of the bunch is from a few to a few tens of picoseconds (about one to ten millimeters) in FWHM. The wavelengths of the coherent radiation is in a submillimeter to millimeter range. Another feature of the radiation is polarization, which depends on the radiation process.

In most experiments a multibunch electron beam forming a bunch train is used. The intensity of radiation from the beam is $P(\lambda)\, G(\lambda)$, where

$$G(\lambda)= (\, \sin(\pi n d/\lambda) \,/\, \sin(\pi d/\lambda)\,)^2. \qquad (2)$$

In this equation d is the interval between the bunches. This factor gives the oscillation on the spectrum due to the interference between the light pulses. When the radiation is applied to spectroscopy this fact determines the lowest value of the resolution in wavelength: it is typically 1%.

The characteristics of the coherent radiation show that the ideal beams used for a new light source are high-intensity single-bunch beams. In the experiments performed so far single-bunch beams were used in Osaka University and Cornell University [4].

*Corresponding author. Tel.: +81-6-879-8511; fax: +81-6-875-4346; e-mail: s-okuda@sanken.osaka-u.ac.jp.

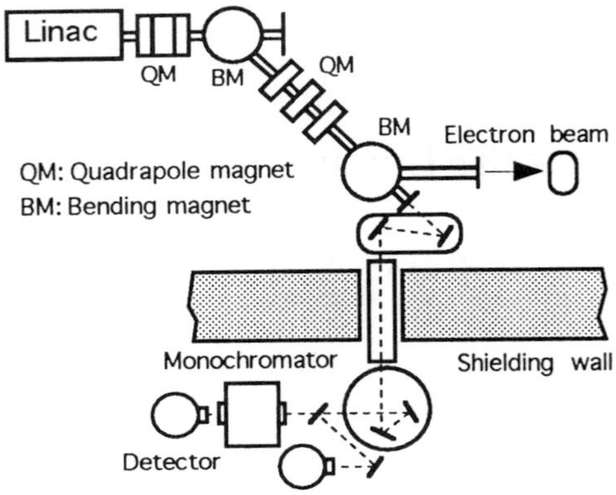

Fig. 1. Schematic diagram of the experimental setup for the measurements of the coherent synchrotron radiation.

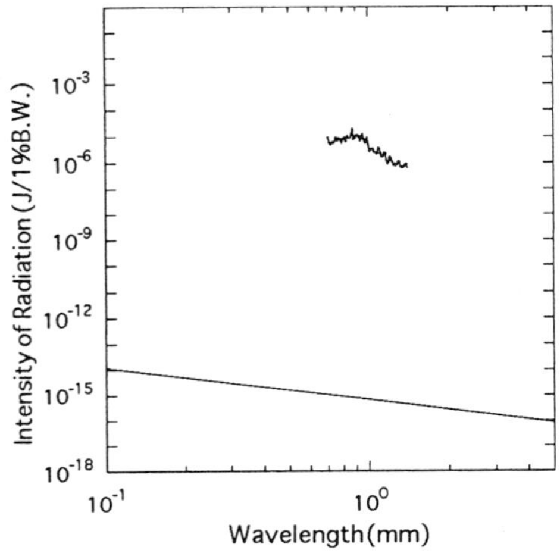

Fig. 2. Spectrum of the coherent synchrotron radiation from an electron bunch of the single-bunch beam at 30 nC: the intensity of the incoherent radiation given by calculation is indicated with a solid line.

3. Experimental

The ISIR linac has a specific subharmonic buncher system to generate the high-intensity single-bunch beams. The maximum charge of electrons in a bunch is 73 nC. The operational conditions of the linac are determined so as to obtain high intensity radiation. The length of the bunch can be controlled from 10 to 50 ps.

Fig. 1 shows the setup for the present experiments. The synchrotron radiation emitted from a bending magnet at a solid angle of 0.01 sr is transported. Most path of radiation is in vacuum. The spectrum of the radiation is measured by using a grating spectrometer and a liquid-He cooled Si-bolometer. The resolution in wavelength of the spectrometer is about 1% which is determined from the size of the grating. The radiation is partly splitted to another detector to monitor the intensity of the incident light.

4. Results and discussion

Fig.2 shows the spectrum of the coherent radiation measured at wavelengths from 0.7 to 1.4 mm. The energy of the beam is 28 MeV and the charge of electrons in a bunch is about 30 nC. The spectrum of the incoherent radiation given by calculation is indicated with a solid line. The intensity of radiation is 10 orders of magnitude enhanced according to the coherence effects, as shown in this figure. Structure on the spectrum is attributed to the reflection in the vacuum window on the path of radiation and to the shape of the electron bunch. The total energy of radiation at this range of wavelengths is 4 x 10^{-4} J. This is more than 7 orders of magnitudes higher than that of the other far-infrared light sources having continuous spectra. The degree of polarization is 0.62 for the wavelength range from 1 to 2mm. The stability of the intensity of radiation averaged over 60 pulses is below $\pm 1\%$.

By using the light source the absorption spectra of N_2O gas are being measured. This will be applied to the researches in various fields.

References

[1] T. Nakazato, M. Oyamada, N. Niimura, S. Urasawa, O. Konno, A. Kagaya, R. Kato, T. Kamiyama, Y. Torizuka, T. Nanba, Y. Kondo, Y. Shibata, K. Ishi, T. Ohsaka and M. Ikezawa, Phys. Rev. Lett. 63 (1989) 1245.

[2] J.Ohkuma, S.Okuda, and K.Tsumori, Phys. Rev. Lett. 63 (1991) 1967.

[3] S. Okuda, J. Ohkuma, N. Kimura, Y. Honda, T. Okada, S. Takamuku, T. Yamamoto and K. Tsumori, Nucl. Instr. and Meth. A331 (1993) 76.

[4] U. Happek, A. J. Sievers, and E. B. Blum, Phys. Rev. Lett. 67 (1991) 2962.

FREE ELECTRON LASERS 1998
G.R. Neil and S.V. Benson (Eds.)

Realisation of a High Power Microwave Source

J. Lucas, A.I. Al-Shamma'a, and R.A. Stuart

The University of Liverpool, Department of Electrical Engineering and Electronics, Brownlow Hill, Liverpool L69 3GJ, U.K.

Abstract

Microwaves have already found a large number of applications in manufacturing industry for processing. The majority of these applications are mainly concerned with heating or drying using existing magnetron devices, which produce kW powers at low frequencies 0.9GHz and 2.46GHz. A possible replacement device for the magnetron at higher frequencies up to 100GHz, capable of producing kW powers, is the free electron microwave source (FEM). In order to consider the potential of such devices, a simulation of a tuneable FEM device operating up to saturation conditions has been given. It has been shown for an X band device operating at 8.7 GHz with a voltage of 90 kV and a beam current of 300 mA, that 3.7 kW of microwave power is available for industrial applications. Doubling the device current to 600 mA would increase the power available to more than 10 kW. By introducing cylindrical waveguides, frequencies up to 100 GHz are readily attainable with comparable powers. The Liverpool FEM is currently being upgraded to produce such power outputs by enhancing the electron beam current. The operating frequency is up to Ku band (12 to 18 GHz).

1. Introduction

The magnetron is able to produce a significant amount of microwave power (30 kW at 0.9 GHz and 8 kW at 2.46 GHz) with an efficiency of 30%. Based upon these attributes, new applications of microwaves are currently being explored for manufacturing industries. Four of the more recent applications are:

1. Microwave Plasma Torch for the welding or cutting of metals and ceramics [1].
2. Microwave vitrification of radioactive waste for environmentally friendly, long-term storage [2].
3. Microwave chemistry for the optimum cleaning of contaminated catalytic surfaces.
4. UV light sources for waste water management.

Although manufacturing industry currently employs magnetrons for microwave-based processes, such as drying and baking, there are many industrial processes that could use microwaves but are restricted by the limitation of the magnetron. However, the industry has a requirement for shorter wavelength microwave sources having a range of tuneable frequencies and power outputs at the kW level. Existing magnetron devices can not satisfy these requirements but they readily fall within the capabilities and potential of the Free Electron Microwave source (FEM).

2. The Free Electron Microwave Source

At the University of Liverpool, a compact, low voltage, prototype free electron microwave (FEM) source has been developed to examine the physics of pre-bunched FEM operation and to act as a test bed for new ideas and techniques. The operational parameters of this low current demonstration device have been published previously [3], together with experimental results [4,5,6]. The FEM which is currently operating using an X band waveguide (8-12) GHz device employing a 55 kV, 1 mA electron beam together with a permanent magnet wiggler. The electron beam is pre-bunched at 5 kV on exiting the electron gun, prior to entering an electrostatic accelerator as shown in Fig. 1. This technique is necessary to achieve adequate gain with the low current capability of the present power supply. **Table 1** gives the specifications of the prototype FEM.

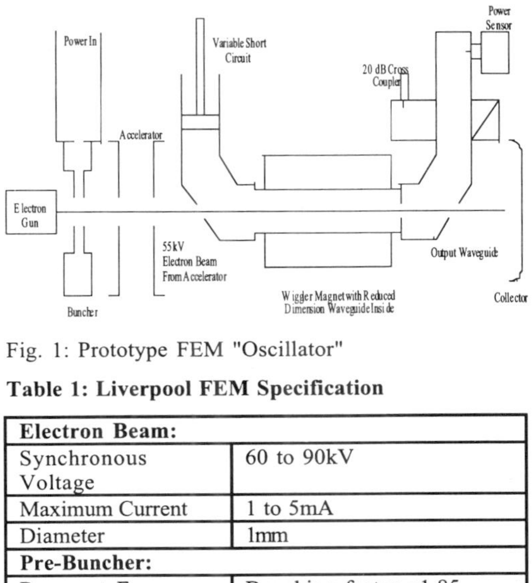

Fig. 1: Prototype FEM "Oscillator"

Table 1: Liverpool FEM Specification

Electron Beam:	
Synchronous Voltage	60 to 90kV
Maximum Current	1 to 5mA
Diameter	1mm
Pre-Buncher:	
Resonant Frequency	Bunching factor =1.85
Waveguide Resonator:	
Dimensions	12.5 × 25mm
Cut-off Frequency	7.5GHz
Resonant Frequency	8 to 12GHz
Quality Factor	Length =500mm
Wiggler Magnet	
Material	NdFeB
Period	25mm
Gap	24mm
Number of Periods	20
Peak Field	500 Gauss

The experimental results given previously [4,5,6] are of the general form and can be quantitatively explained by a linear theory. However for higher powers non linear interaction occur and there is no quantitative explanation for the amount of power capable of being produced as well as giving a reasonable estimate of the saturated output.

3. Simulation results of the FEM interaction

The physics of the interaction between the electron beam and electromagnetic wave within a cavity resonator whilst under the influence of the wiggler magnetic field may be described either using a two-wave model or a single particle model. A single particle code has been created to examine how an electron interacts with the electromagnetic wave as it travels down the wiggler magnet. Using this model, the rate of electron energy may be calculated for different values of field strength (E_o), initial phase angle (Φ), and the initial electron energy (eV) for various FEM system parameters and the simulation parameters are summarised in **Table 1**.

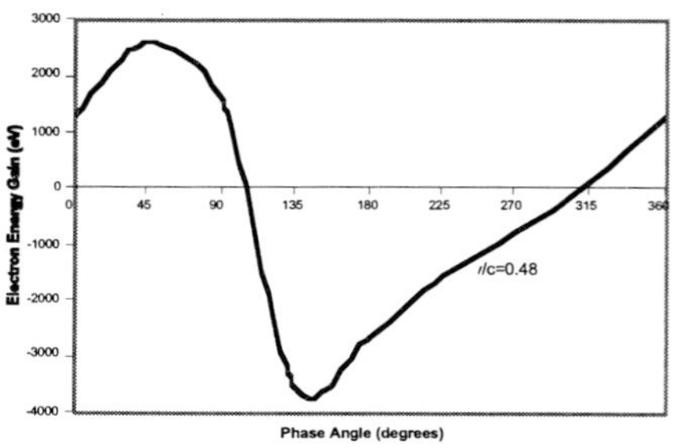

Fig. 2: Electron interaction with the EM wave as a function of the initial phase angle

The results of the simulation are shown in Fig. 2, which examines the synchronous interaction as the phase angle between an electron, and the electromagnetic wave is varied. The results in Fig. 2 are specifically for $E_0 = 80$ kV/m and (v_z/c) varying between 0.47 and 0.49. Detailed results have been obtained for E_o from 5 to 1280 kV/m and the results for $E_o = 80$ kV/m illustrate the general trends. At relatively small signal levels there is a sinusoidal variation in the interaction strength with phase angle. For a continuous stream of electrons some of the electrons will gain energy whilst others will lose a corresponding amount of energy. For a certain phase angle there will be a maximum energy transfer (eV) from the electrons to the EM wave. At approximately 180° extra phase shift, the process is reversed and a similar amount of energy is transferred from the EM wave to the electron. For a continuous electron beam the net energy transfer is close to zero. However, if the electron current is pre-bunched it is possible for the electrons to enter the wiggler with only a small range of phase shift angles and hence interaction at the optimum conditions can occur resulting in a large transfer of energy from the electron to the EM wave. The amount of energy transferred to the EM wave by the electron depends not only on the phase angle but also on the electron velocity (v/c).

Fig. 3 shows the energy transfer, maximised for phase angle, as a function of (v/c) and clearly shows a resonance effect. As E_o increases, the energy transfer curve becomes less symmetrical with a rapid fall off at the higher (v/c) values. For high values of E_0=1280 kV/m the maximum energy lost by the electron is 28,000 eV for an electron beam velocity (v/c) = 0.525, which corresponds to an electron beam energy of 90 keV. Under these high-energy transfer conditions it is important to observe the rate of energy transfer while the electron is travelling through the wiggler. As shown in Fig. 4, for E_o = 160 kV/m the electron energy loss variation is approximately linear with distance along the wiggler although at the end of the wiggler there is evidence of saturation occurring. For E_o = 1280 kV/m there exists a maximum energy transfer occurring at 50% along the wiggler and at 100% of the distance along the wiggler the electron has recovered all the energy it previously had transferred to the EM wave. We therefore conclude that under such high field conditions the length of the wiggler has to be matched to the value of E_o. This is advantageous as the Q of the waveguide will be higher and will result in less cavity losses.

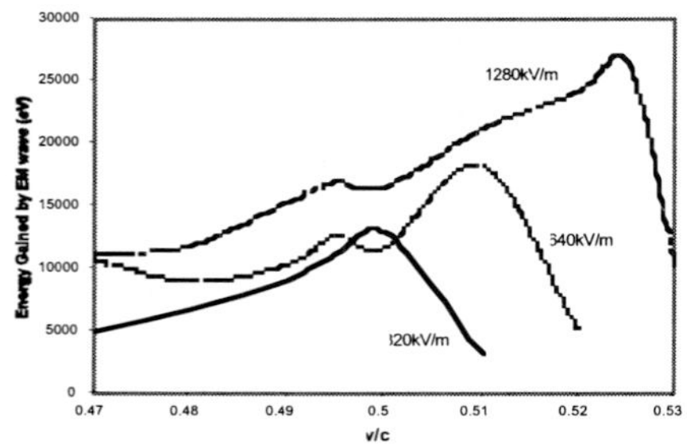

Fig. 3: The loss of energy by the electron as it traverses the cavity

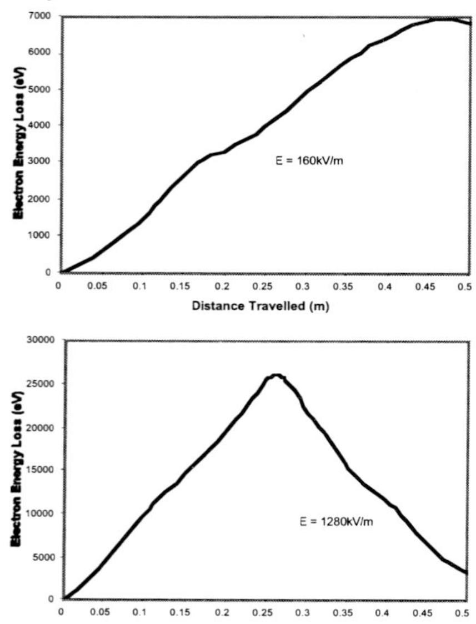

Fig. 4: The loss of energy by the electron as it traverses the cavity

4. FEM Output Power Capability

In order to evaluate the output power capability of the FEM it is necessary to equate the energy transferred from the electron beam to the EM wave with the energy lost by the EM wave within the cavity and by applications. The energy transferred is $eV_I\ I$, where V_I is the interaction voltage and I is the beam current. Both the cavity losses and the useful energy available for applications represent the energy loss. The balance equation is:

$$eV_I I = P2\alpha L + kP = k^1 P \qquad (1)$$

Where L is the length of the cavity, α is the attenuation coefficient and k is the fraction of cavity power as useful output. However, the mean power within a rectangular waveguide is

$$P = \frac{V^2}{4Z_0}\left(\frac{\lambda}{\lambda_g}\right)^{ab} \qquad (2)$$

Then substitution from (1) gives

$$eV_I I = k^1 \frac{E_o^2}{4Z_o}\frac{\lambda . ab}{\lambda_g} \qquad (3)$$

If $k^l = 2\alpha L$ then all the power generated in the cavity is used to satisfy the Ohmic losses of the cavity and the beam current is I^l. All the extra current $(I-I^l)$ can then generate useful power for application purposes. The results of the single particle analysis have been summarised in Table 2. The Liverpool FEM is capable of being operated with three electron guns giving beam currents of 2 mA, 15 mA and 50 mA. The latter has the capability of being enhanced to produce a beam current of 300 mA. For the 300 mA gun currents, useful output power can be produced of the order of 3.7 kW. At these operating conditions, the beam current required to offset the cavity losses is 159 mA so that only half of the beam current (~ 141 mA) is producing useful power. It is not unusual to have Klystron with 1 A electron beams which would have the capability to produce 21 kW of useful power. For the lower gun currents of 50 mA, 15 mA and 2 mA the useful power capabilities are 162 W, 14 W and 400 mW respectively.

5. Conclusion

A free electron microwave (FEM) source has been simulated for the X band range of frequencies (8–12 GHz). It has been shown for a 2 mA electron beam that 407 mW of microwave beam power can be produced whilst for a 300 mA beam 3666 W of power is produced. These high output energies are produced because the electron beam has initially been pre-bunched at the required frequency. A series of confirming experiments are currently being undertaken with the aim of using the FEM device on industrially based projects.

6. Acknowledgements

The authors wish to thank the EC funding program for their financial support towards Liverpool FEM Thematic Network programme.

7. References

[1]. Internal Report, The Welding Institute, 1995.
[2]. Internal Report, British Nuclear Fuel, 1996
[3]. G. Dearden et al., Nucl. Instr. and Meth. A 318 (1992) 230.
[4]. G. Dearden et al., Nucl. Instr. and Meth. A 341 (1994) 80.
[5]. A.I. Al-Shamma'a et al., Nucl. Instr. and Meth. A 393 (1997) II 83.
[6]. S. Mayhew et al., Nucl. Instr. and Meth. A 393 (1997) 365.

Table 2: The minimum current (I^1) necessary to maintain the electric field strength E_o within the cavity

Simulation Results λ=35 mm			Cavity Losses		Useful power produced by the cavity (W)			
P cavity kW	E_o (kV/m)	V_I (V)	I^1 mA	$V_I I^1$ (W)	2mA gun	15mA gun	50mA gun	300mA gun
72526	1280	26000	159	4134	-	-	-	3666
43456	640	18300	135	2477	-	-	-	3020
13579	320	12600	61	774	-	-	-	3011
3386	160	6900	28	193	-	-	152	1877
847	80	4200	11.5	48.3	-	14.7	162	1212
53.0	20	1154	2.62	3.02	-	14.3	54.6	343
13.3	10	581	1.30	0.756	0.407	7.96	28.3	173
3.32	5	291	0.65	0.189	0.392	4.18	14.4	87

FREE ELECTRON LASERS 1998
G.R. Neil and S.V. Benson (Eds.)
1999 Elsevier Science B.V.

A Path Towards the 1kW or Higher FEL Light Output at the JAERI Superconducting rf Linac based FEL

N. Nishimori, E. Minehara, M. Sawamura, R. Nagai, N. Kikuzawa, M. Sugimoto and T. Yamauchi
Free Electron Laser Laboratory, Advanced Photon Research Center, Kansai Research Establishment, Japan Atomic Energy Research Institute, (JAERI), Tokai, Naka, Ibaraki 319-1195, Japan

The first lasing of the JAERI FEL has been successfully achieved in February 26th, 1998. As a next step, it is planned to extract average FEL light output of 1kW or higher. A possible path is discussed to improve the existing JAERI FEL performances.

1. Introduction

The superconducting rf linear accelerator (rf linac) is the best option for a high average power free electron laser (FEL) because of the negligible rf losses on the accelerating cavity. Especially, the cavity with the lower frequencies than the S and L bands can accept the larger amount of electrons per bunch for acceleration. As the accelerating frequency of the JAERI superconducting cavity is 499.8 MHz and several times larger longitudinal and transversal acceptances are available, the JAERI FEL should become one of the most powerful FEL systems for high power operation. The JAERI FEL has been designed and constructed to be a quasi-CW (Continuous-Wave) far infrared laser with a 1 ms long macropulse and a 10 Hz repetition rate [1], because of poor radiation shield, small capacity of power supply and low cooling capacity of the liquid He recondensing refrigerator system. After the first lasing [2], we plan to make the average output well higher than 1 kW for prototyping industrial superconducting FEL applications and also test platform of CW operation in the future.

2. JAERI FEL system

The electron gun was originally designed to produce the electron beam of 4 ns micropulse width and about 100 mA peak current. After some improvements of the grid pulser, an FWHM of the width is below 3 ns, the estimated peak current to be around 150 mA, the fluctuation of micropulses below 3 %. To increase the beam current from the electron gun, we need some improvements for the grid pulser to shorten the micropulse width and injection line length, to minimize SHB voltage and energy spread of the beam, and to maximize the peak current. To transport almost 100 % electron beam from the electron gun, and to reduce beam losses in the injection line, a high quality electron beam like a Stanford SCA/FEL [3] is required to be realized in the JAERI FEL. The normalized emittance from the electron gun at the Stanford is estimated to be 3π mm-mrad and micropulse charge is 21pC/3ps with 11.8 MHz. This is equivalent to the emittance of 8π mm-mrad and micropulse charge of 448pC/9ps with 10.4125 MHz in the JAERI low frequency system. In this case the beam current is 4.6 mA. If the micropulse of four times longer width is assumed to be accepted in longitudinal direction with an almost the same quality beam, the beam current of 18.4 mA will be available.

An improvement of the rf feed back system is one of the most important for the JAERIs' first lasing achievement. Errors of the rf amplitude inside the cavities during the beam loading had been typically kept to be below 0.5 % for the pre-accelerators and 0.1 % for the main accelerators in peak to peak sense. The error of the rf phase is below 1 degree in peak to peak for all the accelerators. As an example, cavity field and phase fluctuation due to 2 mA beam loading for a pre-accelerator is shown in Fig. 1. Thus, the rf feedback system has worked well for each accelerating module. The maximum rf power capacity fed by the rf power supply is 12 kW for the pre-accelerators and 100 kW for the main. Under the present configuration, the pre-accelerator power supply restricts the maximum beam current to 6 mA for 2 MV operation. We need to replace power supplies for pre-accelerators for the higher beam current operation.

The optical resonator length is 14.4 m to match the repetition rate of 10.4125 MHz of the electron beam. The main characteristic of the optical cavity and undulator is listed in ref [4,5]. The hole coupling at the downstream mirror is estimated to be 0.15 % and used to align the optical cavity mirrors and to monitor the optical field. The Q-value of optical resonator is 91 determined by the decayed time constant of the FEL light. The obtained optical cavity loss is 1.1 %. The extraction of the laser power is made by a Au coated and movable scraper mirror of 20 mm diameter supported by a rod of 1 mm

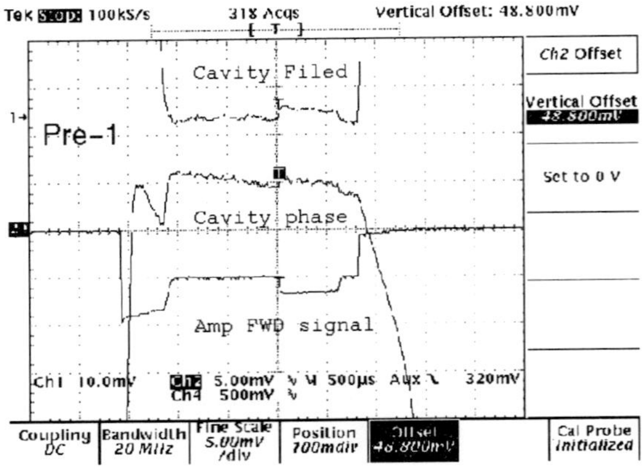

Fig. 1. The cavity amplitude and phase fluctuation due to 2 mA beam loading for a pre-accelerator with macropulse width of 600 μs. As the full scale of the cavity field is 200 mV, the fluctuation in a macropulse is below 0.5 %. The phase fluctuation is below 1 degree. An amplifier feed forward signal (Amp FWD) shows the power provided by a power supply. The figure shows that two times larger power can be transmitted to the cavity.

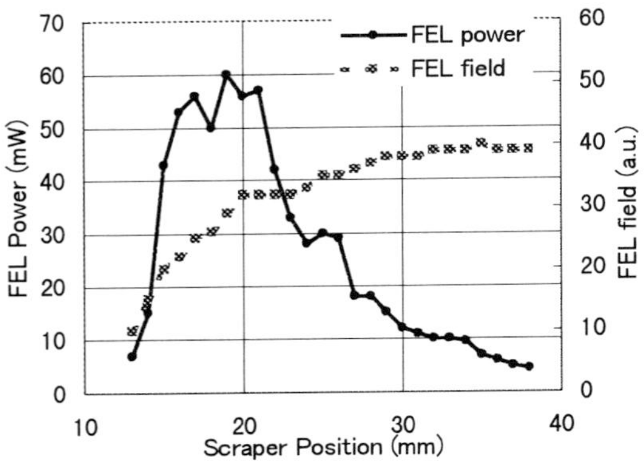

Fig. 2. The solid line shows FEL power outcoupled by a movable scraper mirror, and measured by a thermal detector (Spectra-Physics 407A). The dotted line shows FEL field outcoupled thorough a hole coupling, and measured by a Ge-Cu detector. Both data were obtained with macropulse duration of 150 μs and 10 Hz at the same time. The FEL field at maximum outcoupling decreased to 70 % of its maximum.

diameter. At the same time the intensity of FEL field in the resonator cavity was measured through the 2 mm diameter hole coupling. At the highest power extraction by the scraper mirror, the FEL field decreased to about 70 % of its maximum as shown in Fig. 2.

3. A path towards 1 kW oscillation and possible improvements

In a case of the electron current 2 mA and energy 15.5 MV, the total beam power is 31 kW. The extraction efficiency is estimated to be 0.375 % from the simple formula [6]. The scraper mirror coupling is 5.2 % from a calculation and the measured cavity loss is 1.1 %. The KRS5 window loss is estimated to be 70 % for 25 μm FEL. Finally the average FEL power is calculated to be 47 W. As the measured power is scattered around 40 W, they are almost consistent with each other.

In the JAERI FEL, maximum output of the rf power supply for the pre-accelerators is 12 kW, and that for the main accelerators 100 kW. Total power of 112 kW can be transferred to the electron beam power. In the beginning stage, we plan to concentrate to maximize theFEL power extraction from the electron beam power. As a typical case, we assume the electron beam current 6 mA and the energy 18 MV. In its case the gain is calculated as 19.7 % compared with the present case of 7.9 % from the one-dimensional approximation. The outcoupling is estimated to be 13.6 %. Finally the calculated average power is estimated to be 183 W. In the later stage, we plan to introduce a new undulator having 26 wiggler periods in order to increase the extraction efficiency twice. However, it is still not enough to reach the 1 kW FEL power.

Therefore, we have to increase the electron beam power by increasing either the current or voltage or both of them. The required current to reach 1 kW FEL power is 18 mA. The 18 mA operation in the JAERI FEL would be possible in our configuration by using familiar technologies. In this case, two 20 kW power supplies for the two pre-accelerators and the energy recovery system for the main accelerators are required to feed and then to achieve the 1 kW FEL light output.

References

[1] E. J. Minehara et al., Nucl. Instr. and Meth. A 331 (1993) 276.
[2] E. J. Minehara et al., these proceedings.
[3] K.W. Berryman, Ph.D. Thesis, Stanford University, 1995.
[4] R. Nagai et al., Nucl. Instr. and Meth. A 358 (1995) 403.
[5] R. Kato et al., Nucl. Instr. and Meth. A 358 (1995) abs77.
[6] C. A. Brau, Free Electron Lasers, (Academic Press, Boston, 1990), 236.

FREE ELECTRON LASERS 1998
G.R. Neil and S.V. Benson (Eds.)
1999 Elsevier Science B.V.

A 36.4 GHz FEL-amplifier driven by a linear induction accelerator

N.Yu. Peskov*, G.G. Denisov, N.S. Ginzburg and D.A. Lukovnikov

Institute of Applied Physics, Russian Academy of Sciences, 603600 Nizhny Novgorod, Russia

A.K. Kaminsky, A.A. Kaminsky, S.N. Sedykh and A.P. Sergeev

Joint Institute for Nuclear Research, 141980 Dubna, Russia

Abstract

This paper is devoted to progress in an experimental study of an FEM-amplifier with a guide magnetic field driven by the LIU-3000 linac (1 MeV / 200 A / 200 ns). A new wideband RF-input scheme based on the effect of microwave beam multiplication is suggested for the experiment. So far amplification of 20 dB has been obtained at the frequency of 36.4 GHz.

The JINR-IAP FEM-oscillator with guide magnetic field has been studied on the basis of using the LIU-3000 linac [1, 2]. In the oscillator with a two-mirror Bragg resonator an output power of 37 MW and an efficiency of 26%, the highest for a millimeter wavelength band FEM-oscillator, were achieved at the frequency of 31 GHz under reversed guide field orientation.

In recent experiments an FEM amplifier scheme has been developed. We suggest a new broadband RF-input (Fig.1). The operating principle of the RF-input is based on the effect of microwave beam multiplication [4]. In accordance with this effect, a narrowly directed monochromatic wave with cross-section a, when injected into a wide waveguide with its cross-section $A > a$, excites a set of eigenmodes with certain definite amplitudes and phases such that at a distance $L = A^2/p\lambda$ from the input of the wide waveguide the field looks like a set of p split wave beams with amplitude profiles identical to the injected beam. In principle, the new unit makes it possible to separate microwave and electron beams and to provide a high (up to 100%) transformation of the input signal into the operating wave of an oversized waveguide without any obstacles on the electron beam aperture. Moreover, this efficient transformation is possible in a rather wide (about 10%) frequency range.

For this project the RF-input was designed with an 85% input signal transformation into the operating $TE_{1,1}$ mode of a circular waveguide in the frequency region from 32 to 37 GHz. Results of "cold" microwave measurements are presented in Fig.2. Channel 1 corresponds to transformation into the interaction region and channel 2 (the remaining

*Corresponding author. Tel. +7 8312 384 575, fax +7 8312 362 061, e-mail peskov@appl.sci-nnov.ru

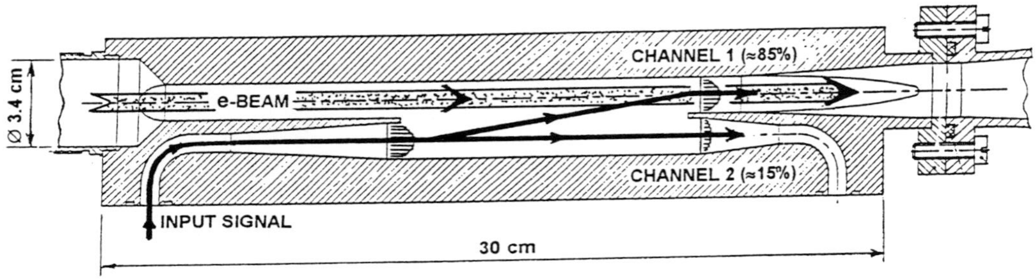

Fig.1 The RF-input scheme based on the effect of microwave beam multiplication.

15% of the input signal) is designed and used for on-line monitoring.

An experimental study of the amplifier is carried out on the LIU-3000 linac (JINR, Dubna) which generated electron energies of 0.8 MeV and a 200 A beam current pulse, of duration ~200 ns, at a repetition rate up to 1 Hz. The helical wiggler of period 6 cm and transverse magnetic field amplitude of up to 3.5 kG is used to drive the electron transverse oscillations. The wiggler was immersed in a uniform axial magnetic field generated by a solenoid. The strength of this field could be varied up to 7 kG. The reversed guide magnetic field regime is chosen for FEM operation.

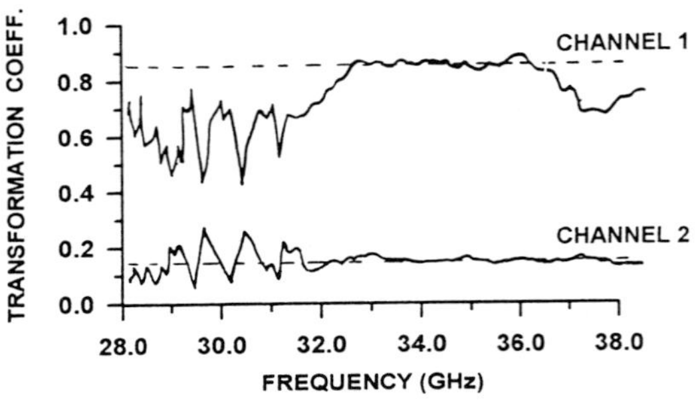

Fig.2 Results of "cold" microwave measurements of the RF-input: transformation coefficient versus frequency.

A magnetron with a frequency of 36.4 GHz was used to drive the amplifier. At the present stage of the experiment an amplification of 20 dB and an output power of 5 MW were obtained when the beam of current of 110 A was utilized. We expect that at the next stage an improvement in the beam transport through the new unit and the use of the full beam current produced by the accelerator (about 180 - 200 A) will increase the amplification and output power.

This work is supported by grants 97-02-16643 and 97-02-17379 of Russian Foundation for Basic Research.

References

[1] N.S.Ginzburg, A.K.Kaminsky, A.A.Kaminsky, N.Yu.Peskov, S.N.Sedykh, A.P.Sergeev, A.S.Sergeev, Proc. of 17th Int. FEL Conf., New York, USA, 1995, p.215.

[2] N.S.Ginzburg, A.A.Kaminsky, A.K.Kaminsky, N.Yu.Peskov, S.N.Sedykh, A.P.Sergeev, A.S.Sergeev, IEEE Trans. on Plasma Sci., 1998, v.26, no.3, p.541.

[3] G.G.Denisov, D.A.Lukovnikov, M.Y.Shmelyov, Proc. of SPIE, Edit. by J.R.Birch and T.J.Parker, 1993, vol.2104, p.485.

FREE ELECTRON LASERS 1998
G.R. Neil and S.V. Benson (Eds.)
1999 Elsevier Science B.V.

Status of the Microwave Inverse FEL Experiment

R. B. Yoder,[a] T. B. Zhang,[c] T. C. Marshall,[b] Mei Wang,[a] and J. L. Hirshfield[a,c]

[a]*Physics Dept., Yale University, PO Box 208120, New Haven, CT 06520-8120*
[b]*Dept. of Applied Physics, Columbia University, New York, NY 10027*
[c]*Omega-P, Inc., 202008 Yale Station, New Haven, CT 06520-2008*

A proof-of-principle experiment to accelerate electrons via the inverse FEL interaction at microwave frequencies has been in development for several years at Yale University and Omega-P, Inc. Since inverse FEL acceleration using a transverse wiggler field was proposed by Palmer in 1972 [1] it has been extensively studied theoretically [2]; groups at Columbia University and at Brookhaven have demonstrated some acceleration at optical wavelengths using the IFEL mechanism [3,4], but neither attempted to exhibit high trapping fractions or large energy gain. The Yale accelerator, known as the Microwave Inverse FEL Accelerator or MIFELA, is intended to demonstrate the IFEL principle with strong trapping [5], and will give the opportunity to explore the possibilities of this mechanism with regard to phase stability and beam quality. Simulation results and details of the experimental hardware will be presented here; operation and data collection are expected to begin in the very near future.

The acceleration structure of MIFELA is a cylindrical waveguide into which RF power at 2.856 GHz enters via two feeds 90° apart, creating a circularly polarized traveling wave. The waveguide is operated in the TE_{11} mode and is near cutoff ($n = \omega / ck = 0.2$) to minimize the magnet bore size and maximize acceleration gradient. The undulator field is provided by a bifilar helical winding which is pulsed at high current and which is tapered in pitch for maximum acceleration gradient; the undulator parameter a_w varies from 2.4 to 2.75 at an initial period of 11.75 cm. An axial magnetic field, also tapered, is used for orbital stability and guiding. Profiles of the fields, which have been optimized through simulation studies, are shown in Fig. 1. The particle source for the experiment is a 2-1/2 cell RF gun, which produces 5–10 ps bunches of up to 10^9 electrons at 6 MeV with low emittance. Up to 25 MW of RF power is available from a SLAC-type klystron and is divided between the gun and accelerator, insuring that beam and RF are in phase. Beam focusing and energy selection are carried out by a 19-element beamline and achromat before injection into the accelerator. Numerical studies, discussed further below, indicate that with 15 MW of RF power, such a beam will be accelerated to 6.7 MeV over 95 cm, and that this result can be scaled to higher energies. Beam loading is expected to be negligible.

Since the long wiggler period gives rise to relatively large-diameter beam orbits, the beam travels first through an injection region (occupying the first 5 wiggler periods) in which the wiggler and guiding fields are gradually ramped up from zero to their acceleration values, so that the beam can be 'spun up' to its final gyration radius while ensuring that the orbits remain centered on the axis [6]; little energy change occurs in this region in computations (See Fig. 1). Maximal orbit stability is obtained with a nonlinear wiggler field up-taper, as shown. The beam's transverse velocity is converted to axial velocity by tapering the fields down to zero again at the end of the device.

Simulation results

Numerical simulations of acceleration in the MIFELA have been computed using a fully nonlinear, three-dimensional, slow time-scale FEL simulation code written by Freund and Ganguly and known as ARACHNE, which has been benchmarked extensively against FEL experiments [7]. The injected beam for these model calculations is assumed to be evenly spread in azimuth and radius, with an outer radius of 0.7 mm, and centered on axis with no energy spread. Injection in the correct phase window for acceleration is essential [8]; note, however, that at these long RF wavelengths, the phase window being used is almost negligibly small. A conservative estimate of $\pi/10$ (i.e., a 17 ps bunch) was used for these simulations.

Fig. 1 shows the average value of the energy parameter γ over the MIFELA. In the entry and exit regions, there is little change in energy. The gradient in the acceleration region is essentially constant at 0.81 MeV/m. After extraction, the beam transverse velocity has nearly vanished ($\beta_\perp = 0.03$) and the beam may be analyzed in a spectrometer. Output energy spread is 2% for all particles in the simulation.

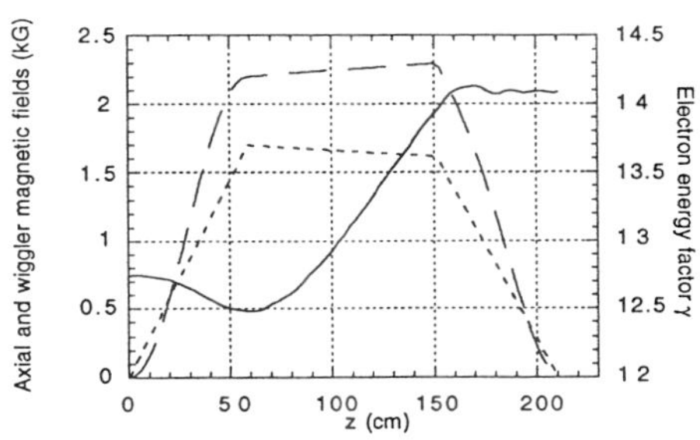

Fig. 1. Axial (short dashes) and wiggler (long dashes) magnetic field profiles in MIFELA, together with average computed electron energy factor γ (solid line).

TABLE 1. Experimental parameters for MIFELA.

Entry region $(0 < z < 58.75$ cm)	
Electron beam energy	$\gamma = 12.8$
Electron beam radius	$r_b = 0.7$ mm
Wiggler field	$B_w = 0$–2.2 kG, sine-squared ramp
Axial magnetic field	$B_z = 0$–1.7 kG, linear
Wiggler period	$\lambda_w = 11.75$ cm, constant
Wiggler radius	$r_w = 3.84$ cm
Peak electron beam current	$I_b = \leq 0.1$ A
Macropulse length	$\tau_p = 2$ μs
Acceleration region $(58.75$ cm $< z < 151.5$ cm)	
Waveguide radius	$R = 3.14$ cm
Free-space RF wavelength	$\lambda_s = 10.5$ cm
Waveguide index	$n = 0.2$
Input RF power level	$P_{in} = 15$ MW
Normalized RF field strength	$a_s = 0.14$, circularly polarized
Wiggler period	$\lambda_w = 11.75$–12.32 cm, linear ramp
Wiggler coil radius	$r_w = 3.84$ cm
Wiggler current	$I_w \leq 60$ kA
Wiggler field strength	$B_w = 2.2$–2.3 kG
Axial magnetic field	$B_z = 1.7$–1.61 kG

Technical design

A schematic drawing of the structure, consisting of the acceleration waveguide and couplers for RF input and output, is shown in Fig. 2. The 15-period wiggler is powered by a capacitor bank capable of delivering over 50 kA with a pulse length of about 40 μsec. The heavy-gauge copper windings of the wiggler are wound directly onto the outside of the waveguide, which is also the vacuum vessel, over a layer of insulator. Because the beam dynamics depend sensitively on the wiggler field profile at injection, the wiggler field is brought up from zero to its initial value with a nonlinear taper that is calculated to give optimum orbit stability; this up-taper is achieved with a series of resistive shunts between the two wiggler windings, which are spaced so as to give the correct current profile. An insulating spacer of varying thickness is used to control the winding period accurately as it is tapered along the structure. The axial field is provided by a series of 18 coils, independently controlled by computer, which can produce the desired profile.

Initial testing of the accelerator systems has included diagnostics on the beam, performance tests of the beamline, and wiggler field measurements. The Yale 2-1/2 cell RF gun has been operated at beam energies from 3 to 7 MeV and has produced bunches of up to 10^9 electrons. Use of the beamline achromat to remove the initial beam energy spread has produced beam spot sizes on the order of the 0.7 mm radius used in the simulations discussed above. A pulsed undulator field of up to 2.5 kG has been demonstrated to be attainable.

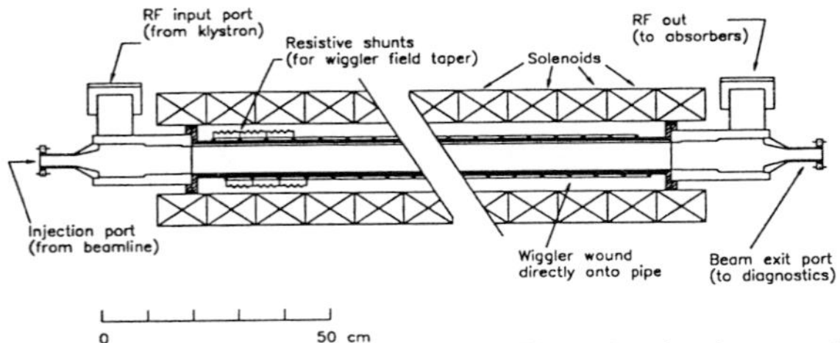

Fig. 2. Schematic diagram of the MIFELA, showing input and output couplers for RF on either end of the acceleration waveguide itself, the wiggler helix wound directly onto the beam pipe, and the axial field coils. The resistive shunts which are used to create the initial field up-taper are shown symbolically.

Future prospects

The construction of MIFELA is essentially complete, and operation is expected to begin very shortly. The experiment is intended to investigate IFEL dynamics in detail, and also to explore the phase response of the IFEL and its applications to beam bunchers of interest to accelerator physicists. While this experiment operates at low power, a MIFELA scaled to 150 MW of 34 GHz RF power is estimated to have an acceleration gradient of at least 40 MV/m, which could make it a useful high-energy accelerator component.

Acknowledgements

The authors acknowledge the collaboration of M. A. LaPointe (Omega-P), S. Y. Park (Postech, Korea), and M. Shapiro (MIT). This work was supported by the US Department of Energy, Division of High Energy Physics.

References

[1] R. B. Palmer, J. Appl. Phys. 43 (1972) 3014.
[2] See e.g., P. Sprangle and C. M. Tang, IEEE Trans. Nucl. Sci. NS-28 (1981) 3346; P. A. Sprangle and A. C. Ting, Part. Accel. 22 (1987) 149; E. D. Courant, C. Pellegrini and W. Zakowicz, Phys. Rev. A 32 (1985) 2813; and references cited in these papers.
[3] I. Wernick and T. C. Marshall, Phys. Rev. A 46 (1992) 3566.
[4] A. van Steenbergen, J. Gallardo, J. Sandweiss, and J. M. Fang, Phys. Rev. Lett. 77 (1996) 2690; Y. Liu, et al., Phys. Rev. Lett. 80 (1998) 4418.
[5] J. L. Hirshfield, T. C. Marshall, T. B. Zhang, A. K. Ganguly, and P. A. Sprangle, Nucl. Instr. and Meth. A 358 (1995) 129; R. B. Yoder, T. B. Zhang, T. C. Marshall, and J. L. Hirshfield, *Proceedings of the 1996 Advanced Accelerator Concepts Workshop,* (AIP Conf. Proc. 398), p. 629.
[6] T. B. Zhang and T. C. Marshall, Nucl. Instr. and Meth. A. 375 (1996) 515.
[7] A. K. Ganguly and H. P. Freund, Phys. Rev. A 32, (1985) 2275; H. P. Freund and A. K. Ganguly, Phys. Rev. A 34 (1986) 1242; A. K. Ganguly and H. P. Freund, IEEE Trans. Plasma Sci. PS-20 (1992) 245, and references therein.
[8] T. B. Zhang and T. C. Marshall, Phys. Rev. E 50 (1994) 1491.

FREE ELECTRON LASERS 1998
G.R. Neil and S.V. Benson (Eds.)
1999 Elsevier Science B.V.

Initial Optical Measurements on the IR Demo*

Dick Oepts, Stephen V. Benson, George R. Neil, and Michelle D. Shinn,

Thomas Jefferson National Accelerator Facility, 12000 Jefferson Ave., Newport News, VA23606, USA

Abstract

The initial measurements from lasing of the IR Demo laser are presented. Early operation was around 5 microns. Both pulsed and CW operation were achieved. Spectra at various detunings and outcouplings are shown. Up to 311 Watts of CW power was produced.

1. Introduction

The IR Demo laser was operated with an electron energy of 38 MeV and 60 pC of charge per bunch. For other parameters see ref [1]. The electron bunch length was minimized using the coherent transition radiation from an aluminum foil, and estimated to be in the order of one picosecond FWHM [2].
The normal micropulse repetition frequency, for one optical pulse circulating in the laser cavity, is 18.7125 MHz. The cavity was surveyed during its assembly to be set at the synchronous length for this frequency to within 100 μm, and an adjustment by some 60 μm sufficed to obtain first lasing.
A mirror with 98% reflection was used as the outcoupler in the initial runs, to facilitate start-up of the laser. This outcoupler was later replaced by one with 90% reflection, for more output power.
To check the effect of larger outcoupling, or lower effective gain, the electron bunch repetition frequency was reduced by successive factors of two, so that the optical pulse experienced gain only in ½, ¼, etc., of its roundtrips.

2. Small outcoupling

In Fig. 1 we show the cavity detuning curve under standard conditions with the 98% reflective outcoupler. Lasing was insensitive to beam parameters and optical alignment. The width of the curve and the strongly increasing power near zero detuning also show that the laser operates far above threshold.
Some spectra of the output radiation are illustrated in Figs.2-4. In Fig.2, the cavity is shortened by 20 μm from its synchronous length, leading to a narrow spectrum. In this large detuning case, the optical pulse develops an exponential slope at its leading edge, with a length determined by the cavity loss and by the amount of detuning [3]. The observed full width at half maximum of about 4 nm for the main peak in Fig. 2 corresponds to a 3 ps 1/e-time for the pulse slope. This would indicate a cavity loss of about 4% rather than the 2.8% total mirror loss expected at this wavelength.

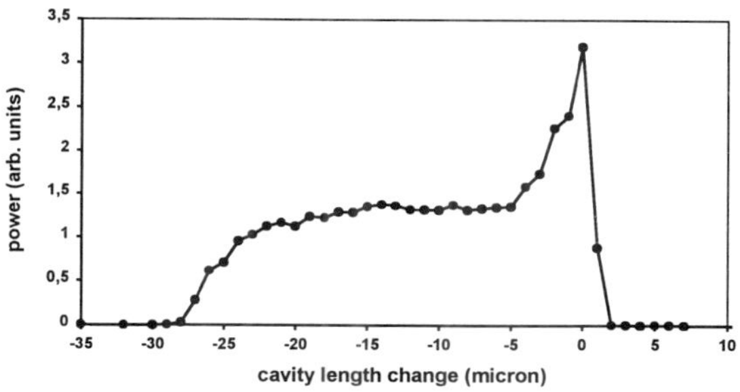

Fig. 1 Cavity detuning curve, 2% outcoupling

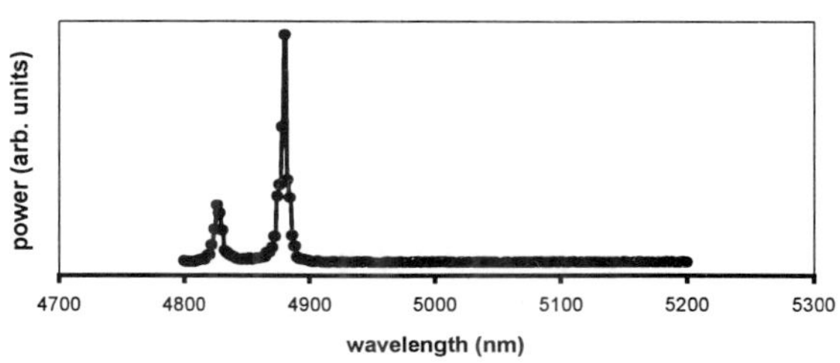

Fig. 2. Spectrum at 20 μm cavity detuning

*Work supported by the U.S. Department of Energy under contract DE-AC05-84-ER40150, the Office of Naval Research, the Commonwealth of Virginia, and the Laser Processing Consortium.

The origin of the smaller peak, shifted to shorter wavelengths by 53 nm, or 1.1%, is not yet explained.

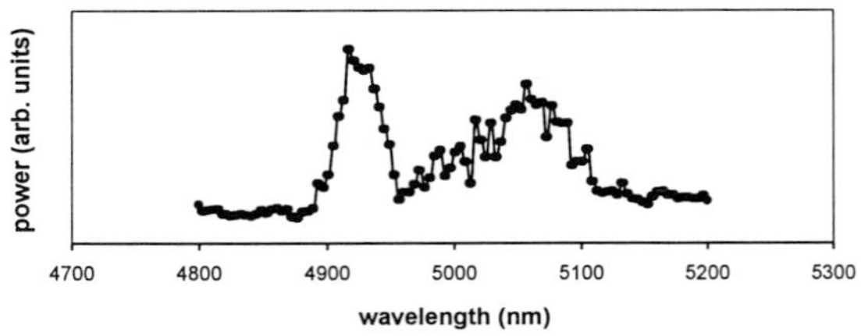

Fig. 3. Spectrum at 2 micron cavity detuning

A prominent sideband is seen in Fig. 3, at 2μm detuning. This indicates an intracavity power high enough to excite a sideband instability. At a slightly longer cavity length, as shown in Fig. 4, the measured spectrum suggests that the intracavity power has reached the regime of chaotic electron dynamics.

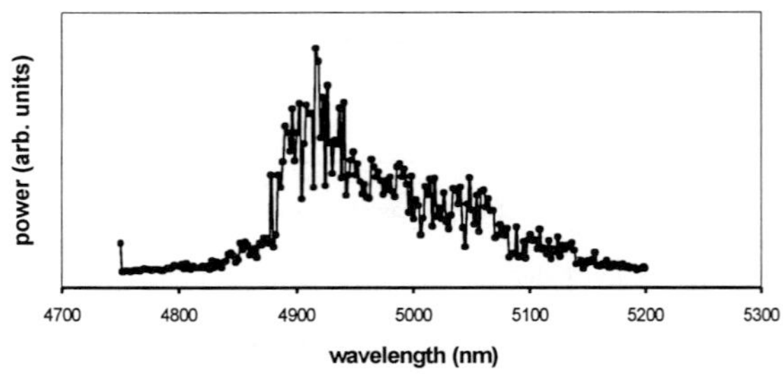

Fig. 4. Spectrum at near-zero cavity detuning

Lasing could still be maintained with a reduction of the electron bunch frequency by a factor of 8, which means that the gain is at least 25% per pass.

The results described so far were obtained with an electron beam consisting of macropulses of 200 μs duration, repeated at 2 Hz. Next, the laser was operated in the CW mode, and a stable output of 155 Watts was obtained.

3. Large outcoupling

In a second experimental run, an outcoupler with 90% nominal reflection was used. Again, detuning curves and spectra were measured for different micropulse repetition rates and also for a series of different bunch charges. The saturated output power at small detuning is reduced roughly by the same factor as the average current in both cases. The appearance of the detuning curves is different, however, because the effective detuning increases when the optical pulse has to make more roundtrips before meeting an electron bunch again.

An example of the spectrum at small detuning, under otherwise standard conditions, is shown in Fig. 5. The fine-structure is due to absorption by atmospheric water vapor. The presence of the sideband at the long wavelength side shows that the regime of nonlinear saturation is still reached with the outcoupling of 11% at this particular wavelength. While the output power is increased, the intracavity power is lower than in the small-outcoupling case, which reduces the load on the cavity mirrors.

In CW operation, an output power of 311 Watts was achieved with the nominally 10% outcoupling mirror.

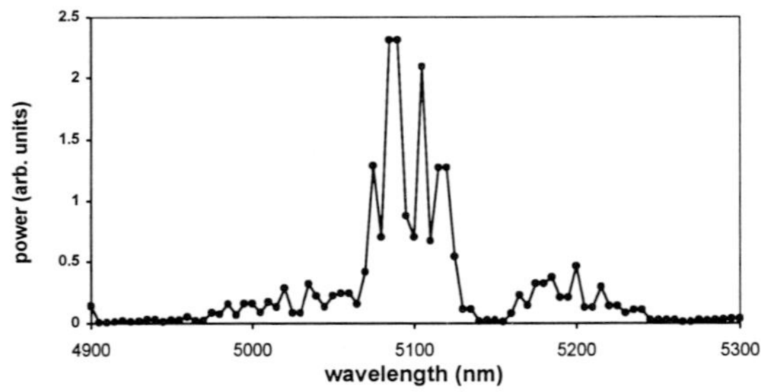

Fig.5. Spectrum at 1 micron cavity detuning, 11% outcoupling

4. Conclusion

The first measurements on the laser output show that the IR Demo basically lases as expected, although some details merit further study.

References

[1] S.V. Benson, et.al. "First Lasing of the Jefferson Lab IR Demo FEL", *these proceedings*

[2] C.L. Bohn, et al. "Performance of the Accelerator Driver for Jefferson Laboratory's Free Electron Laser", *these proceedings*

[3] W.B. Colson, in: *Laser Handbook,* Vol 6, W.B. Colson, C. Pellegrini, and A. Renieri, Eds, North-Holland, Amsterdam 1990, p.179.

FREE ELECTRON LASERS 1998
G.R. Neil and S.V. Benson (Eds.)
1999 Elsevier Science B.V.

Progress of the FELICITA I Free-Electron-Laser Experiment at DELTA[1]

Matthias Hirsch, Dirk Nölle, Henrich Quick, Thomas Schmidt[2], DELTA Machine Group
Institute for Accelerator Physics and Synchrotron Radiation, University of Dortmund, GER

Abstract

This paper presents the progress of the FELICITA I experiment, under commissioning at the storage ring DELTA. This Free-Electron Laser is based on an electromagnetic undulator. This undulator cannot only be operated as a conventional undulator, but also in the optical klystron configuration. Due to the higher gain, this operation mode is chosen for the commissioning.
During the commissioning [1], the machine was characterised at different energies between 300—800 MeV. Due to the experience made during this process, the commissioning energy was fixed to 450 MeV. The start-up wavelength is fixed to 470 nm operating the undulator at a K of 2. During the last runs the optical cavity was commissioned with beam, and losses as low as 1.5% have been demonstrated. Using this value and the measured electron beam characteristics, the threshold average current was calculated to be about 7 mA.

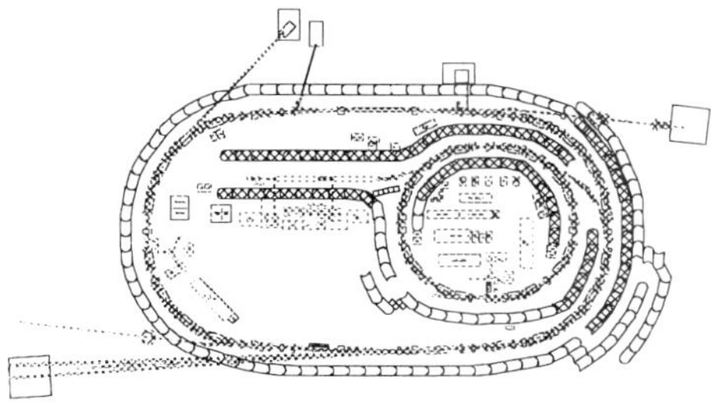

Fig. 1: Layout of DELTA. FELICITA I is located in the northern superstraight. Two FEL beamlines are available to probe the radiation coupled out through upstream and downsteam mirror.

1 INTRODUCTION

Free-Electron-Lasers (FEL) have been one of the main goals of the DELTA facility just from the start of the project. This idea had a serious impact on the design of the storage ring. To guarantee a low impedance, the vacuum system was designed under the prerequisite to avoid all changes of the cross section of the vacuum chamber. This implied the design of new kicker magnets [2] and other low impedance vacuum components [3]. Due to the combination of all these measures, the overall impedance of the vacuum system was calculated to be Z/n = 0.4 Ω [4].

In order to start a long term FEL program aiming for operation within the regime of 100 nm, a first milestone was defined with the FELICITA I project addressing the visible and near UV [5].

2 SETUP OF THE FELICITA I EXPERIMENT

The first FEL designed for the DELTA facility is called FELICITA I. With a undulator period of 25 cm and a K value below 3, it is possible to operate this FEL in the visible part of the spectrum. Due to the long period length the undulator is constructed as an electromagnetic device, featuring two operation modes. The magnet can either be operated as a conventional FEL or in the optical klystron configuration usual for storage ring FELs.

As shown in Fig. 1, this design allows for an optical cavity of 14.4 m mirror distance, corresponding to 4 bunch operation of DELTA.

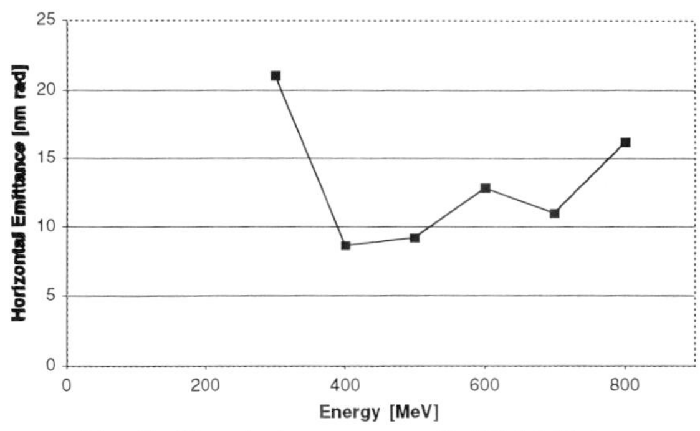

Fig. 2: The start-up energy of 450 MeV is a good compromise between emittance, stability and FEL gain.

3 STATUS OF THE COMMISSIONING

During fall 97, the storage ring was characterised at different energies including the operation of the undulator, especially in the low energy regime, in order to figure out the best energy for FEL operation. It was decided to start the commissioning of FELICITA I at 450 MeV. This energy was chosen as a compromise between machine stability and FEL gain. Furthermore, the

---

[1] This work was supported by the BMBF under contract 05 3PEAAI

[2] Now a member of the SLS Team, Villingen, Switzerland

measurements indicate a minimum of the emittance at this energy (Fig. 2).

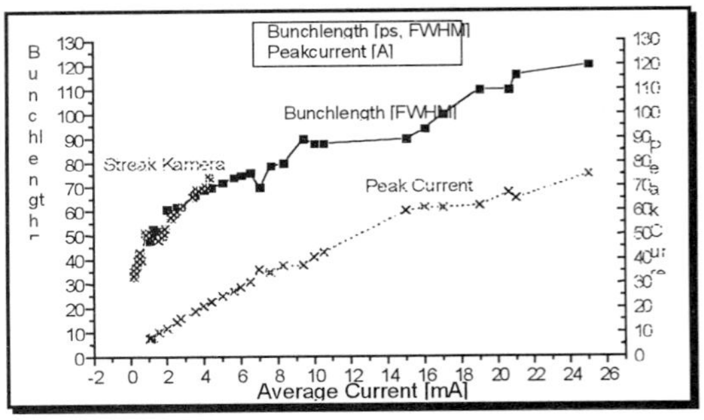

Fig. 3: Measurement of length and peak current with a fast photodiode and a streak camera, respectively.

The commissioning wavelength is fixed to 470 nm, to stay well in the visible. At 450 MeV this wavelength corresponds to an undulator K of 1.98. Furthermore, it was decided to start using the optical klystron configuration with the strongest possible dispersive section, in order to work at the highest possible gain.

During the 450 MeV runs currents up to 24 mA in single bunch mode and 30 mA in 4 bunch operation have been achieved. At present, the 4 bunch operation cannot be used for the FEL, as strong synchrotron oscillations disturb the accuracy of the spacing between the 4 bunches. For single bunch operation a simple feedback system, acting on the RF phase of the transmitter, was used to suppress the longitudinal oscillations in the RF bucket. Fig. 3 shows the bunch length and peak current of DELTA, measured with a fast photodiode and a streak camera, respectively[3].

The optical cavity could be tuned to losses as low as 1.5 % per pass, resulting in about 5 % losses per electron pass for single bunch operation

Collecting all data taken from the ring and from the optical cavity, the expected threshold current for lasing can calculated to about 7 mA for single bunch operation (Table 1). This is well in the operation regime of DELTA.

Up to now, two periods of dedicated FEL runs have been performed. During the first period in winter 97/98 the electron beam had the required performance, but the alignment of the optical cavity and the matching of the electron beam to the optical cavity was not sufficient. This could be improved in a second run period during spring 98, but due to a shortcut in an internal sextupole magnet, DELTA could not provide currents high enough to reach the threshold. The next beam time is scheduled for fall/winter 98.

[3] The measurements with the streak camera have been recorded during a demonstration of Hamamatsu, Germany at DELTA.

4 CONCLUSION

The FELICITA I experiment at DELTA is under commissioning. Both the required electron beam quality as well as the operation of the undulator and the optical cavity have been demonstrated. Therefore, the successful operation of this device is expected in the near future.

ACKNOWLEDGEMENT

The DELTA FEL group would like to thank all the people supporting the FEL activities at DELTA, especially A. Renieri and the Super ACO FEL team, for their continuous support.

e⁻ Beam Parameter		FEL	Op. Klyst.
Number of Bunches		1	1
Bunchlength (FWHM)	ps	90	90
Energy Spread		7,500E-04	7,500E-04
natural Emittance	m rad	1,940E-08	1,940E-08
Beam Energy	MeV	456	453
Undulatorparameter			
K—Value		2,000	1,980
Periodlength	mm	250	250
Number of Periods		19	7
Slippage Parameter Nd			1,138E+02
Optical Cavity			
Wavelength	nm	470	470
Mirror Refl. per Pass		98,676%	98,676%
Mirror Refl. per e- Pass		94,806%	94,806%
Cavity Length	m	14,4	14,4
Radius of Curvature	m	8,0	8,0
Beam Waist	m	5,992E-04	5,992E-04
Raileigh Length	m	2,40	2,40
FEL Performance Data			
JJ Faktor		6,529E-01	6,553E-01
Filling Factor		2,784E-01	3,755E-01
Current Density * FF	A/m^2	7,669E+07	6,615E+07
Colson Current per mA	1/mA	1,6030E-02	8,046E-04
1 + Nd/Nu			1,726E+01
Energy Spread Gain Deg.			5,229E-01
Gain and Currents			
Gain per mA	*1/mA*	*0,22%*	*0,73%*
Laser Threshold	*%*	*5,19%*	*5,19%*
Theshold Current	*mA*	*24,00*	*7,15*

REFERENCES

[1] U. Berges et al., Status of the Dortmund Electron Test Accelerator Facility, Proc. of the EPAC 98, Stockholm, Sweden

[2] G. Blokesch et al., A Slotted-Pipe Kicker for High-Current Storage Rings, Nucl. Instr. & Meth. A338, 1994, p151

[3] B. Hippert et al., The DELTA Vacuum System, Proc. of the EPAC 96, Sitges, Spain

[4] M. Negrazus, Ph.D. Thesis, University of Dortmund, 1994

[5] D. Nölle et al., FEL Projects at DELTA, Nucl. Instr. & Meth. A296, 1990, p263

FREE ELECTRON LASERS 1998
G.R. Neil and S.V. Benson (Eds.)
1999 Elsevier Science B.V.

Coherent Harmonic Generation in the NIJI-IV FEL

K.Yamada[a], N.Sei[a], H.Ohgaki[a], T.Mikado[a] T.Yamazaki[b]

[a] *Electrotechnical Laboratory, 1-1-4 Umezono, Tsukuba, Ibaraki 3058568 Japan*
[b] *Institute of Advanced Energy, Kyoto University, Gokasho, Uji, Kyoto 611, Japan*

Abstract

Coherent harmonic generation (CHG) is an attractive method to obtain tunable coherent light at the vacuum ultraviolet (VUV) range in a low-gain free electron laser (FEL) system. To study the feasibility of the CHG in the NIJI-IV FEL, third harmonic of a pulsed Nd:YAG laser was introduced onto the stored beam in the NIJI-IV as a source for electron bunching. A small sub-peak reproducibly appeared on the fifth harmonic line with the YAG laser on, which can correspond to the coherent harmonic photons at 71 nm.

1. Introduction

At the electrotechnical laboratory (ETL) efforts to shorten the FEL wavelength toward the vacuum ultraviolet (VUV) are being made[1,2] in the NIJI-IV FEL. To obtain a direct lasing in the VUV, it is essential to achieve sufficient FEL gain to overcome a large optical-cavity loss. Coherent harmonic generation (CHG) is another attractive method to obtain tunable coherent light especially at the wavelengths shorter than ~ 150 nm where optical-cavity loss will be much larger. We already investigated the spectral characteristics of higher harmonics[3] from the NIJI-IV FEL system, aiming at the generation of coherent VUV photons through CHG process. Recently we started to introduce the third harmonic of a pulsed Nd:YAG laser to realize the CHG in the NIJI-IV FEL. Here we report our CHG system and some preliminary experimental results.

2. Experiment

The experimental setup is shown in Fig.1. The third harmonic at 355 nm of a pulsed Nd:YAG laser was focused onto the electron beam in the 6.3-m optical klystron (OK) as an external source for electron micro-bunching. The beam energy of the NIJI-IV and the deflection parameter K of the OK were set at 309 MeV and 2.27 to obtain the fundamental undulator radiation of 355 nm. The NIJI-IV was operated in single bunch mode to suppress coupled bunch instability and to obtain higher peak beam current with smaller beam energy spread. The YAG laser can generate the pulse energy of 400 mJ at 355 nm by nonlinear frequency up-conversion with its rating frequency of 10 Hz. Considering the pulse width of ~5 ns, the laser peak power is estimated to be ~ 80 MW. Equipped with an injection seeder, this laser can operate with a single longitudinal mode.

Coincidence between the electron bunch and the YAG laser is most delicate and important point in the CHG experiment. To operate the YAG laser at 10 Hz synchronously with the electron bunch with the revolution frequency of 10.1 MHz, a careful trigger system for the YAG laser based on the ring RF signal was assembled. As a result, the YAG laser was synchronized with the electron bunch with a long-term drift of $\pm$ 2 ns which is in acceptable level for this experiment.

The YAG laser was focused on to the electron bunch in the modulator undulator of the OK with a quartz lens. The focal

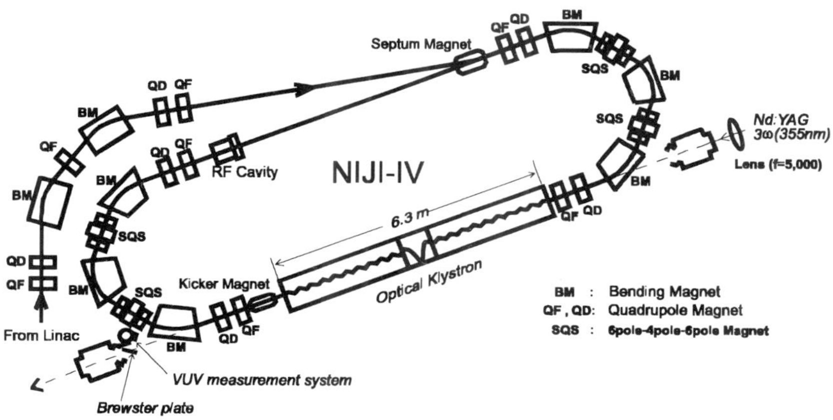

Fig.1 Experimental setup for CHG on the storage ring NIJI-IV.

length of the lens and the Rayleigh range of the focusing laser beam was 5 m and 0.7 m, respectively. Since the focal spot size is estimated to be 0.06 cm in diameter, the laser intensity can reach 3 x 10^{10} W/cm^2 at its focal region.

The output light from the OK was partly reflected by a quartz flat placed with the Brewster's angle for 355-nm light and sent to the spectrum measurement system. To observe the harmonics in the VUV, a real-time spectrum measurement system based on a ultra-high vacuum 0.2-m monochronator was used. This monochromator was equipped with a three-stage microchannel plate (MCP) and a phosphor as a highly sensitive VUV detector. By transferring the spectrum on the phosphor to the CCD camera with imaging optics, we can make a real-time observation of the VUV spectrum with very high sensitivity. In this experiment, the coherent VUV photons should be emitted with 10 Hz, the YAG repetition frequency, while the incoherent photons are generated with 10.1 MHz, the electron beam revolution frequency. Therefor, the first stage of the MCP was gated by a 1-kV and 200-ns signal to improve the signal-to-noise ratio.

3. Preliminary results and discussion

Harmonic radiation from the OK includes coherent and incoherent components, according to the degree of micro-bunching. From our previous estimation[3], we can expect a larger ratio of the coherent component to the incoherent component at smaller N_d of the OK. Therefor, N_d was set at a rather small value (~ 5) in this experiment. Figure 2 shows a spectrum of the fundamental OK radiation. A coarse modulation with only one or two peaks is found, due to a small N_d. The YAG peak at 355 nm is seen as a sharp spike on the spectrum. Although light amplification is not necessary in the CHG process, the OK condition was tuned so that the YAG peak lay at the wavelength where the light amplification is expected, to identify the CHG photons in the VUV range.

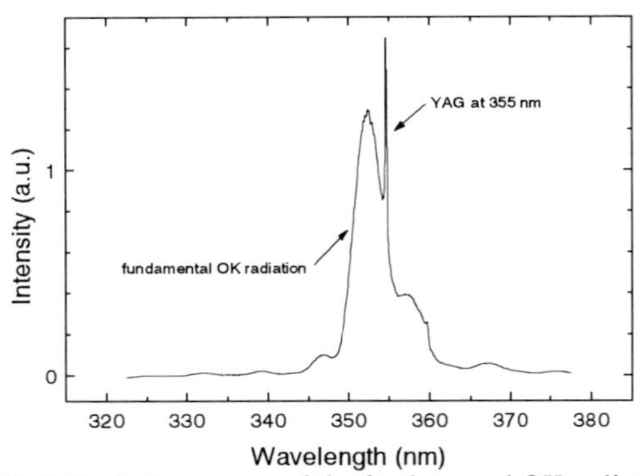

Fig.2 Typical spectrum of the fundamental OK radiation. YAG peak at 355 nm is also indicated.

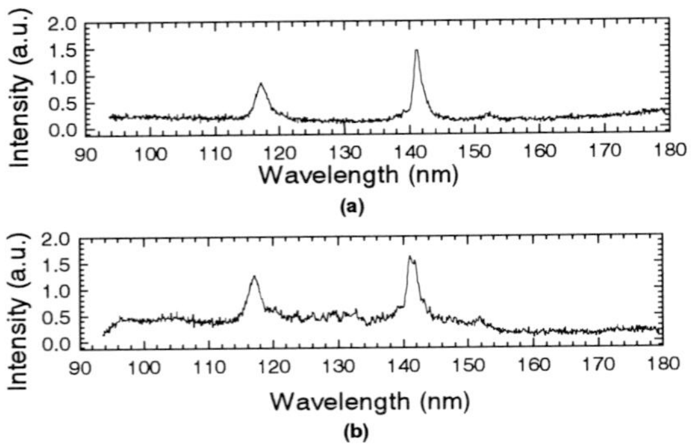

Fig.3 Typical harmonic spectra around 140 nm obtained with YAG laser off (a) and on (b). The average beam current was ~ 8 mA.

Figure 3 (a) and (b) show the typical VUV spectrum around 140 nm obtained with YAG laser off and on, respectively. The peaks at ~118 nm and ~142 nm correspond to the third harmonic, 3ω, and the second order diffraction of fifth harmonic, 5ω. So, the peak observed at 142 nm lies at 71 nm actually. The peak at 3ω is overlapped by the second order diffraction of 6ω [3] and the third order diffraction of 9ω. In this experiment, we found that a small sub-peak reproducibly appeared on the 5ω line when the YAG laser was on, as shown in Fig.3 (b). This sub-peak can correspond to the coherent fifth harmonic photons at 71 nm. Side peak was not observed on the 3ω in this experiment. This is probably because the spectral structure on the 3ω was disturbed by overlapping with the 6ω and 9ω.

4. Conclusion

CHG experiment has been started on the NIJI-IV FEL system by introducing the third harmonic of a pulsed Nd:YAG laser. In a preliminary experiment, a small sub-peak was reproducibly observed on 5ω line with YAG laser on. This sub-peak can correspond to the coherent fifth harmonic photons at 71 nm. Further experiment will be necessary to confirm the generation of coherent harmonics

References

[1] K. Yamada, N.Sei, T.Yamazaki, H.Ohgaki, V.N. Litvinenko, T.Mikado, S.Sugiyama, M.Kawai, M.Yokoyama, a paper in this proceedings.

[2] K.Yamada et al., *Proc. SPIE, eds. P.G.O'shea and H.E.Bennet, San Jose, 1997* (SPIE, Bellingham, 1997), Vol. 2988, p.134.

[3] K.Yamada, T.Yamazaki, N.Sei, H.Ohgaki, T. Mikado, S. Sugiyama, M.Kawai, M.Yokoyama, Nucl. Instr. and Meth. A407 (1998) 193.

FREE ELECTRON LASERS 1998
G.R. Neil and S.V. Benson (Eds.)
1999 Elsevier Science B.V.

FEL Trajectory Analysis for the VISA Experiment*

P. Emma, H.-D. Nuhn

Stanford Linear Accelerator Center, Stanford University, Stanford, CA 94309-0210, USA

1. System Description

The Visible to Infrared SASE Amplifier (VISA) experiment [1], proposed by a collaboration of several institutions in the United States, is designed to provide a complete test of the Self-Amplified-Spontaneous-Emission (SASE) Free-Electron-Laser (FEL) theory in the wavelength region between 600 to 800 nm. The experiment will be carried out at the Accelerator Test Facility (ATF) at the Brookhaven National Laboratory (BNL). The ATF linac will be upgraded for this experiment to reach the required electron energies of up to 85 MeV. The present schedule for the experiment aims to obtain a complete set of data by Spring 1999.

The VISA undulator comprises four 99-cm of arrays of pure permanent magnet blocks (segments) with a period length of 1.8 cm. Four FODO focusing cells per 99-cm segment are superposed on the beam by means of permanent magnets in the gap alongside the beam to maintain an average β-function of 30 cm at 600 nm and 27 cm at 800 nm. The FODO cell length is 24.75 cm (one quarter of a segment length), a compromise between a requirement for small β-funtions and practical considerations. The magnet positions are not adjustable.

Each segment has two beam position monitors and two sets of x-y dipole correctors. The trajectory walk-off in each segment is reduced to a value smaller than the rms beam radius by means of magnet sorting, precise fabrication, and post-fabrication shimming and trim magnets.

Mechanical and magnetic errors of the undulator that remain after sorting and installation will be corrected by magnet measurements and shimming. The magnet measurements will use a pulsed stretched wire system [2], which reads the electron trajectory directly. It is expected that the random walkoff of the trajectory over the length of two segments can be kept below the rms beam radius. However, this leaves possible inter-segment alignment errors. It is the main subject of this paper to determine tolerances for this alignment process. During operation, the beam position is measured with intercepting YAG crystal beam position monitors, which are located at the 25 and 75 cm points of each segment. The beam position is corrected with a pair of trim coils in each plane. Kick angles of up to about 4 mrad are achievable.

2. Correction Strategy

Trajectory correction studies have been performed to determine the number of BPMs required per quadrupole and to evaluate the effectiveness of various steering methods. Simulated e^- trajectories are generated which include 50 μm rms random quadrupole magnet and BPM misalignments for the cases of one BPM per quadrupole, one BPM per 2 quadrupoles, and one BPM per 4 quadrupoles. A weighted steering is performed using a dipole corrector well upstream of each BPM. The trajectories are then used in the FRED3D code to test their impact on FEL gain (see description below). Only the trajectories produced with the sparse case of one BPM per 4 quadrupoles show significant impact on FEL gain. The first two cases are superior, but they are also technically impractical, requiring 32 and 16 total BPMs, respectively, over the 4-m undulator. The third, more practical, case of 8 total BPMs is only acceptable if post-fabrication shimming of the undulator magnets is done based on alignment measurements of the quadrupole centers using a pulsed stretched wire [2]. With this method, alignment of the quadrupoles can be achieved to a level of 5-10 μm which justifies the chosen layout of one BPM per 4 quadrupoles (2 BPMs per 99-cm undulator segment).

In this configuration, however, beam-based alignment, as proposed for the LCLS undulator [3], is difficult to apply to the VISA undulator. Its sparse number of BPMs per quadrupole, finite BPM resolution, and the limited energy stability of its focusing array negate the usefulness of this method.

Table 1: Basic VISA FEL Parameters.

Parameter	Value
Electron Beam Parameters:	
Electron Energy	72.6-83.8 MeV
Norm. Electron Beam Emittance	2 mm mrad
Peak Current	200 A
RMS Bunch Length	428 μm
RMS Beam Radius	62-60 μm
Uncorrelated Energy Spread	0.18-0.15 %
Undulator Parameters:	
Undulator Period	1.8 cm
Peak Magnetic Field	0.75 T
Resonant Wavelength	600-800 nm
FEL Parameters:	
FEL Parameter ρ (1D)	88-77 x 10^{-4}
Rayleigh Length	30-38 mm
Power Gain Length	17.1-19.1 cm
Error Free Saturation Length	3.4-3.8 m
Quadrupole Focusing (FODO Lattice):	
Cell Length	24.75 cm
Quadrupole Length	9 cm
Quadrupole Gradient	33.3 T/m
Ave β-function	0.27-0.30 m
Electron Trajectory Correction:	
Steering Coil Separation (Center)	0.5 m
Number of Steering Coils each plane	2 /m
Max. Correction Field	50 G
Effective field length	20 cm
Max. Kick Angle	4.1-3.6 mrad
Number of Crystal Monitors (BPMS)	2 /m

*Work supported by the Department of Energy (Contract DE-AC03-76SF00515), Office of Basic Energy Sciences, Division of Material Sciences.

3. Trajectory Simulations

To test the impact of various alignment errors, a computer code was written which generates electron trajectories through the undulator. The code includes misalignments of the BPMs and undulator segments, 5-10 μm misalignments of the quadrupoles within the segment, and a weighted steering algorithm. The simulated errors included in the code are summarized below. Each effect is applied independently to the horizontal and vertical planes.

INTERNAL SEGMENT ERRORS: Each 99-cm undulator segment includes misalignments of its 8 quadrupole magnets which approximately replicates the trajectory produced by the pulsed wire alignment. The e^- trajectory is randomly distorted but bounded by ± 50 μm.

EXTERNAL SEGMENT ERRORS: Each rigid segment is randomly misaligned (gaussian) at both ends by 50-150 μm rms as described below.

RANDOM BPM MISALIGNMENT: Each BPM is randomly misaligned (gaussian) by 50 μm rms.

WEIGHTED STEERING ALGORITHM: BPM readbacks and associated corrector changes are simultaneously minimized, in a least squares sense, so that unnecessarily large corrections are not applied.

No incoming trajectory launch errors are included here. It is expected that such effects are correctable by using steering elements and mutually aligned BPMs which are upstream of the undulator. Several different rms misalignments and random seeds are used to create various steered e^- trajectories. Each simulated e^- trajectory is converted to a set of undulator dipole errors which replicates both the horizontal and vertical trajectories. In this conversion, the β-function through the undulator is taken as a constant since FRED3D requires a constant focusing gradient. A file describing the set of relative dipole errors is read into FRED3D where the original simulated trajectory is reproduced.

4. FEL Simulations

FRED3D [4] is a monochromatic FEL simulation code that simulates the interaction between the electron beam and the optical field in the wiggler of an FEL amplifier. In each half-period, a transverse momentum increment corresponding to the magnetic field error at that magnetic pole is added to the motion of each particle. Recently, the code has been modified by its author [5] to accept a list of pairs of pole errors from data files.

Three different random seeds are used per amplitude of misalignment. External segment misalignments of 50, 100, and 150 μm rms plus a constant random BPM error of 50 μm rms are used, for a total of 9 different tested trajectories. Fig. 1 shows an example horizontal and vertical trajectory, both before and after the weighted steering is applied, for random external segment and BPM misalignments of 50 μm rms. The small, bounded internal quadrupole misalignments are also included. The simulations use a 4.95-m undulator (i.e. one additional segment) so that the saturation length can be calculated well past the 4-m point.

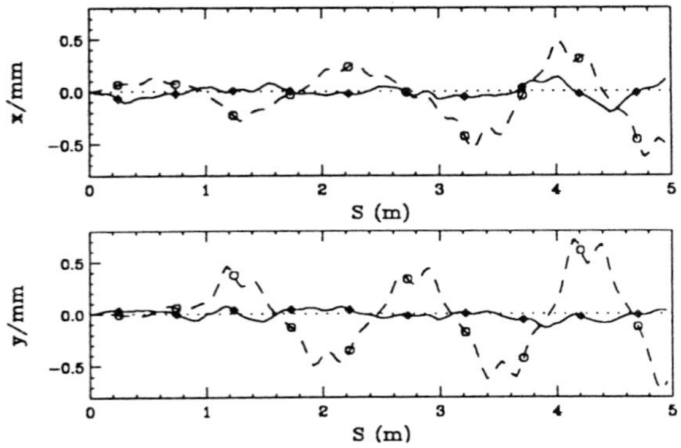

Figure 1: Horizontal and vertical e^- trajectory through the VISA undulator before (*dashed*, $\sigma_x = 238$, $\sigma_y = 331$ μm) and after (*solid*, $\sigma_x = 59$, $\sigma_y = 45$ μm) steering for 50 μm BPM and external segment misalignments. (Plot symbols indicate BPM locations.)

Each trajectory is used in the FRED3D FEL code to determine the new saturation length. Fig. 2 shows the saturation length and saturation power for the 9 trajectories, plotted versus the net horizontal and vertical rms of each trajectory, $\sigma_{x,y}$.

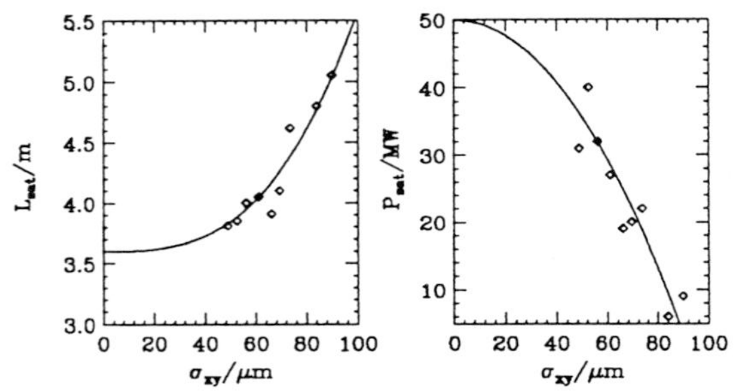

Figure 2: Saturation length, L_{sat}, and power, P_{sat}, versus combined x and y rms amplitude of e^- trajectories. The solid line is a polynomial fit to the data.

The data are fitted to empirical polynomials which are forced through $L_0 = 3.6$ m and $P_0 = 50$ MW, respectively, at $\sigma_{x,y} = 0$. The results show that random external segment and BPM misalignments should both be <50 μm rms in order that saturation occurs at a length less than 4 m.

5. References

[1] L. Bertolini, R. Carr, M. Cornacchia, E. Johnson, M. Libkind, S. Lidia, H.-D. Nuhn, C. Pellegrini, G. Rakowsky, J. Rosenzweig, and R. Ruland, these proceedings.

[2] R. Warren in *NIM*, vol. A272, p. 257, 1988.

[3] P. Emma, R. Carr, and H.-D. Nuhn, these proceedings.

[4] E.T. Scharlemann *J. Appl. Phys.*, vol. 58(6), pp. 2154–2161, 1985.

[5] E.T. Scharlemann, 1998. Private Communication.

FREE ELECTRON LASERS 1998
G.R. Neil and S.V. Benson (Eds.)
1999 Elsevier Science B.V.

Upgrade of the Simulation Code TDA3D

S. Reiche and B. Faatz

Deutsches Elektronen Synchrotron DESY, Notkestraße 85, 22603 Hamburg, Germany

Abstract

The well-known Free Electron Laser (FEL) amplifier simulation code TDA3D has been upgraded to include most of the important features needed for present and future design studies of VUV and X-ray single-pass FELs. A number of bugs have been removed and I/O has been changed and extended to make it more flexible. To features such as wiggler errors, quadrupole misalignment has been added and corrector coils to adjust the beam trajectory. In addition, multi-section undulators can now be simulated with arbitrary (integrated or separated) focusing structures.

1. Introduction

With the design and construction of Free Electron Lasers (FEL), many codes have been developed over the years in order to describe the physics taking place in different regimes (see for example [1]). One such code was a code developed by T.-M. Tran and J.S. Wurtele at MIT, a Three Dimensional Axi-symmetric simulation code (TDA). The physics in this code has been well described as well as suggestions for updates, such as the inclusion of space charge [2]. The code was extended to include non-axisymmetric modes, space charge modes and ion channel focusing [3]. The version of TDA decribed in this paper is based on this code.

There are several new features of importance for future VUV and X-ray FELs. The undulator can be described as modules with driftspaces in between. In addition to the standard FODO lattice, an arbitrary focusing structure can be given as an additional input file. In order to simulate the influence of wiggler errors more realistically, random displacement of quadrupoles has been added. Correction stations, with values and position given in the same input file, are included. Furthermore, the possible parameters that can be given in the output file has been extended and made more flexible.

2. Simulation results

Table 1. Undulator and optics parameters for the TTF FEL (Phase I at 230 MeV).

Electron beam	
Peak current	500 A
Normalized rms emittance	2π mm mrad
rms energy spread	0.2 %
average beam size	77 μm
undulator	
number of modules	3
length of module	4.5 m
period length	27.3 mm
undulator peak field	0.497 T
length of quadrupoles	136.5 mm
number of quads./module	10
distance between quads.	341 mm
quad. gradient	12.5 T/m

As an example, the TTF-FEL parameters have been used to show some typical output. Parame-

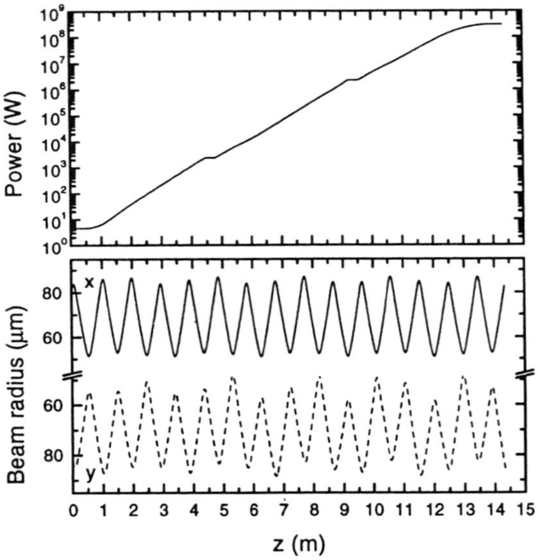

Fig. 1. *Power (top figure) and electron beam radius (bottom figure) along the TTF undulator. Relevant parameters are given in Table 1.*

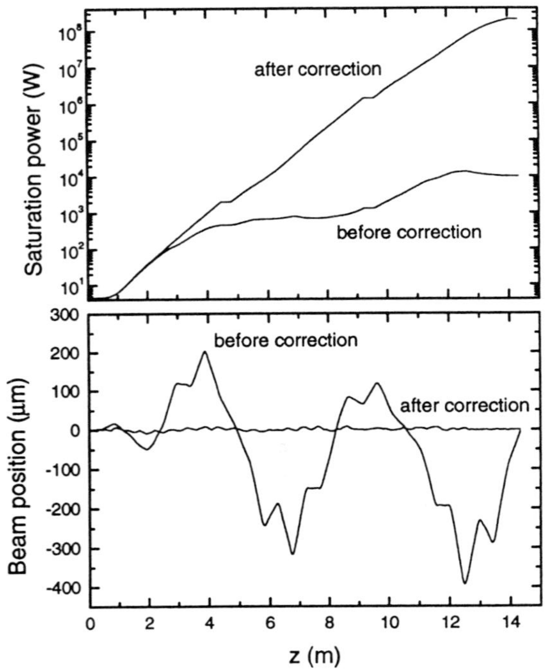

Fig. 2. *Power along the TTF-FEL undulator including undulator dipole field errors (rms value of 0.2%) and quadrupole misalignment (maximum misalignment is* $\pm 20\,\mu m$.*)*

ters are given in Table 1. The undulator consists of multiple modules. In Fig. 1, the power growth and the beam radii in both directions are shown. The FODO structure has been integrated into the undulator, with drift section field-free. As can be seen, the variation in the electron beam radius is rather large for this FODO lattice.

The possibility for calculating wiggler errors and quadrupole misalignment, both without and including correction, is also shown. Comparing the corrected beam with the result in Fig. 1 shows that the power has been reduced by less than a factor of two, but the saturation length stays virtually the same.

Remarks: The Fortran 77 source code as well as the manual can be found on the website:

http://www.desy.de/~tda3d

Acknowledgement: The authors would like to thank P. Pierini and G. Travish for there feedback and usefull discussions.

References

[1] G.A. Travish, *Where Do We Stand with High Gain FEL Simulations?*, Towards X-Ray Free Electron Lasers, AIP Conference proceedings 413 Eds. R. Bonifacio and W.A. Barletta, June 1997, Gargnano, Italy.

[2] T.-M. Tran and J.S. Wurtele, *TDA - A Three-Dimensional Axisymmetric Code For Free-Electron-Laser (Fel) Simulation*, Computer Physics Comm. 54 (1989) 263;

[3] P. Jha and J.S. Wurtele, *Three-Dimensional Simulation of a Free-Electron Laser Amplifier*, Nucl. Instr. Meth. **A331** (1993) 477.

FREE ELECTRON LASERS 1998
G.R. Neil and S.V. Benson (Eds.)
1999 Elsevier Science B.V.

Numerical Study of the 3-D Effect on FEL Performance and Its Application to the APS LEUTL FEL *

Yong-Chul Chae
Argonne National Laboratory, 9700 So. Cass Ave., Argonne, Illinois 60439, U.S.A.

I. Introduction

A Low-Energy Undulator Test Line (LEUTL) is under construction at the Advanced Photon Source (APS) [1]. In LEUTL periodic focusing is provided by external quadrupoles. This results in an elliptical beam with its betatron oscillation envelope varying along the undulators. The free-electron laser (FEL) interaction with such a beam will exhibit truly 3-D effects. Thus the investigation of 3-D effects is important in optimizing the FEL performance. The programs GINGER [2] and TDA3D [3], coupled with theoretically known facts, have been used for this purpose. Both programs are fully 3-D in moving the particle, but model the interaction between particles and axially symmetric electromagnetic waves. Even though TDA3D can include a few azimuthal modes in the interaction, it is still not a fully 3-D FEL code. However, we show that these 2-D programs can still be used for an elliptical beam whose aspect ratio is within certain limits.

We present numerical results of FEL performance for the circular beam, the elliptical beam, and finally for the beam in the realistic LEUTL lattice.

II. Matched, Circular Beam

In this section we assume a long planar undulator with symmetric x-y focusing whose strength is half of the natural undulator focusing. We also assume that the electron beam is matched, namely the envelope of betatron oscillation is constant along the undulator. With these assumptions we present in Table 1 a set of parameters of the LEUTL FEL that can be considered as nominal values for this study.

Table 1
Nominal Parameters for LEUTL FEL

Energy	E	220.0	MeV
Energy Spread	dE	0.1	%
Norm. Emittance (rms)	ϵ_n	5.0	mm-mrad
Undulator Period	λ_w	3.3	cm
Undulator Parameter (Peak)	K	3.1	
Matched β	β_0	1.46	m
Wavelength	λ_r	516.75	nm

We used the parameterization developed by M. Xie [4] in order to estimate the FEL performance and compare with the result from the GINGER simulation operated in SASE mode. The agreements in the saturation power and length are excellent, as shown in Figure 1. Also shown are the results from TDA3D and GINGER operating in single-frequency FRED mode. The gain lengths from the simulations and theory are similar to within two percent.

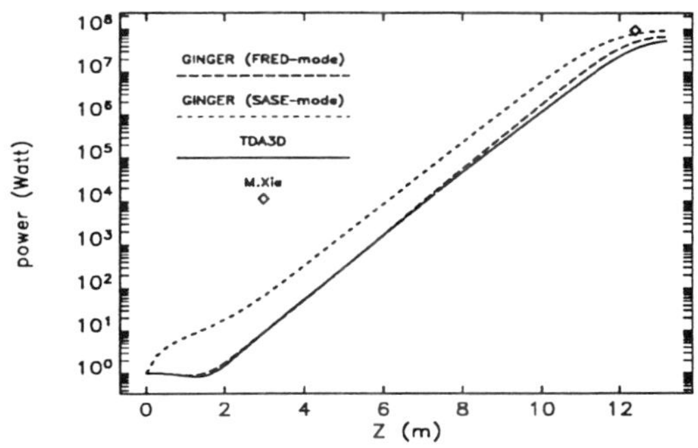

Figure 1
Comparison of results from GINGER (SASE and FRED) and TDA3D and M. Xie's formula.

III. Matched, Elliptical Beam

For design optimization we considered betatron focusing, which is asymmetric in the x and y directions. The matched beam will have elliptical cross section. We used a three-dimensional FEL theory generalized to this case [5]. There it was shown that the growth rate of the fundamental guided mode can be expressed by using six dimensionless scaling parameters

$$\frac{Re(q)}{k_w D} = F\left(2k_1\epsilon_x, 2k_1\epsilon_y, \frac{\sigma_\gamma}{D}, \frac{k_{\beta x}}{k_w D}, \frac{k_{\beta y}}{k_w D}, \frac{k-k_1}{k_1 D}\right), \quad (1)$$

where $Re(q)$ is the growth rate in the exponential growth regime. The growth rate is related to the power gain length L_g as $Re(q) = 1/2L_g$.

We solved the dispersion relation, specifically Eq. (39) in Ref. [5], to obtain the growth rate $Re(q)$ for a Gaussian beam. One of the results is shown in Figure 2, where we varied the aspect ratio σ_x/σ_y while keeping the cross-sectional area the same as the circular beam considered in the previous section. The growth rate is normalized by the circular beam results.

Based on the above results, we can make the aspect ratio of the beam in the LEUTL FEL less than 2 and only suffer a reduction in the growth rate of less than 5% compared with the matched, circular beam case.

Also presented in the figure are the results from TDA3D. These results were generated using only circular optical mode, whereas the results from theory are generated using the correctly matched elliptical optical mode. From these results we see that the growth rates of the two fundamental

* Work supported by U.S. Department of Energy, Office of Basic Energy Sciences under Contract No. W-31-109-ENG-38.

guide mode types, circular or elliptical, are very close to one another up to an aspect ratio of 3. This has practical implication in that the beam aspect ratio along the LEUTL lattice will be less than 2. We see that we still can use the 2-D program in estimating the performance of the LEUTL FEL.

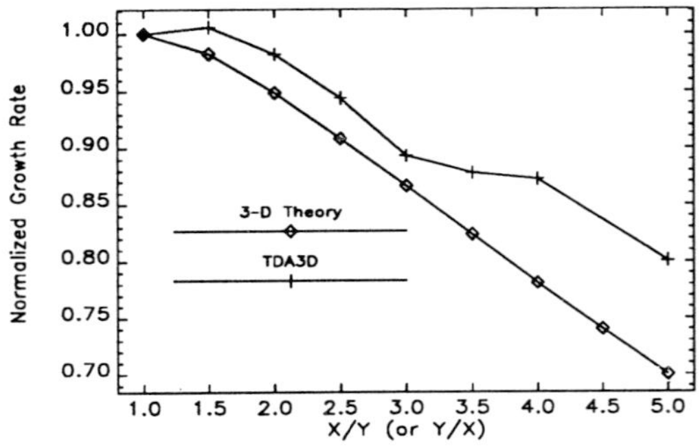

Figure 2
Growth rate as functions of beam aspect ratios.

IV. Beam in LEUTL Lattice

The undulators in LEUTL are separated by a drift where a horizontally focusing quadrupole is located. Each planar undulator provides the necessary vertical focusing to the beam. The lattice then is FOFO in the horizontal plane and FODO in the vertical plane. The horizontal/vertical phase advance per sector is $110^0/120^0$, respectively. The maximum β-function beat occurs in the middle of the undulator where $\beta_y/\beta_x \sim 2.0$. The periodicity of the electron beam envelope, an output generated by the TDA3D simulation, is shown in Figure 3. Also shown is the radiation beam size along the LEUTL. This clearly shows the gain-guiding in the undulators and the diffraction in the drift spaces.

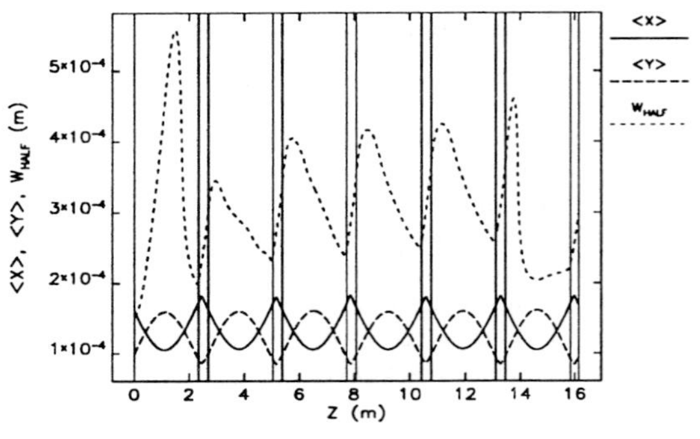

Figure 3
Electron and radiation beam size variation along LEUTL FEL.

Using TDA3D simulation, the exponential gain in the LEUTL FEL is shown in Figure 4.

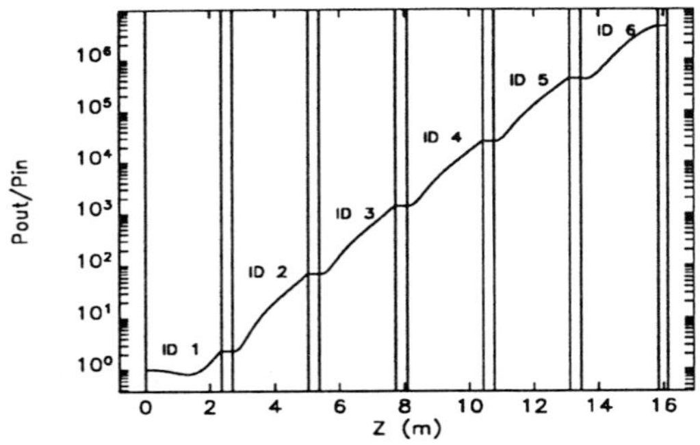

Figure 4
Gain along the LEUTL FEL.

Finally, we show a result for the undulators with an error in K. We consider three undulators of which the middle one has an error ΔK. Figure 5 shows the results from TDA3D for $\Delta K/K = 0$, $\pm 0.5\%$. For $\Delta K > 0$ the gain is considerably reduced, but the growth rate becomes larger than ΔK=0 in the third undulator and consequently ends with better performance. This shows the effect of tapering and an optimization on the saturation length is underway.

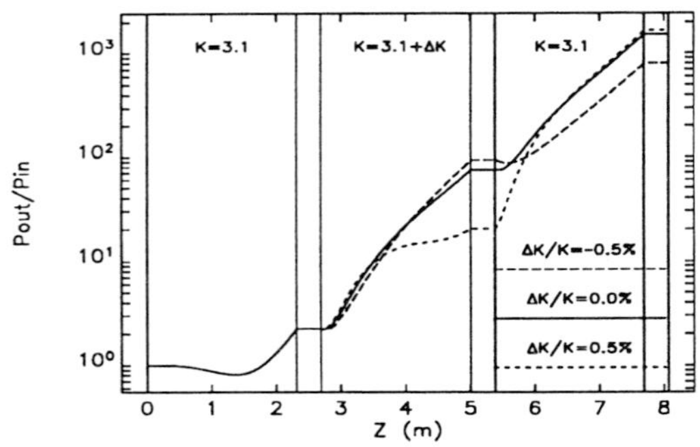

Figure 5
Effects of gap variation (tapering).

V. Acknowledgments

I appreciate the encouragement from and discussions with Lee Teng, Steve Milton, Ed Crosbie, and Kwang-Je Kim. I thank Gil Travish for obtaining and using the new version of TDA3D.

VI. References

[1] S.V. Milton et al., *Nucl. Inst. Meth.* A407, 210, 1998.
[2] W.M. Fawley, *private communication.*
[3] T.M. Tran and J.S. Wurtele, *Comp. Phys.Comm.* 54, 263, 1989.
[4] M. Xie, Proc. 1995 Particle Accel. Conf., 183, 1995.
[5] Y.H. Chin et al., *Nucl. Inst. Meth.* A331, 429, 1993.

FREE ELECTRON LASERS 1998
G.R. Neil and S.V. Benson (Eds.)
1999 Elsevier Science B.V.

SIMULATION CODE DEVELOPMENT OF SASE FEL USING SDE METHOD

Michio Goto * Sin-ichiro Kuruma. ** , Kunioki Mima.*, Kazuo Imasaki**, Nobuhisa Ohigashi***, Chiyoe Yamanaka.**

* Institute of Laser Engineering, 2-6 Yamadaoka,Suita Osaka 565 Japan
** Institute for Laser Technology, 2-6 Yamadaoka,Suita Osaka 565 Japan
***Kansai Univ. 3-3-350 Yamate-cho, Suita Osaka 565 Japan

Abstract

To analyze an amplification process of short wavelength or high power FELs, we have developed a 3-D SASE simulation code. We calculate the evolution of some radiation properties for the parameters of Free Electron Laser Research Institute, Inc. (FELI). We confirm that the electron beam traps the radiation during the amplification process.

Introduction

In the short wavelength (ultra violet ~X-ray) or ultra-high power FEL, it is not possible to use a Fabry-Perot type resonator either because highly reflective mirrors do not exist, or because mirror damage limits the power. Therefore, it is necessary to amplify the radiation in a very long wiggler where the radiation is trapped in the electron beam and reaches saturation. This amplification scheme is called by Self Amplified Spontaneous Emission(SASE).

To analyze this amplification process, we adopt the Source Dependent Expansion (SDE) method [2-3] to describe the radiation field. In the SDE, the configuration is assumed to be cylindrically symmetric, so Gauss-Laguerre functions are used, where the normalization of radial coordinate depends on the axial position z.

In this paper, we calculate the evolution of the radiation amplitude and radial profile for the parameters of FELI.

Model and configuration

In SDE, the radiation fields are expanded by the wave modes as [3]

$$a(r,z) = \sum_m a_m(z) L_m(\xi) e^{-\{1-i\alpha(z)\}\frac{\xi}{2}}, \tag{1}$$

$$\xi = \frac{2r^2}{r_s^2(z)}, \tag{2}$$

where a_m is m-th radiation mode amplitude, r_s is 0th mode radiation radius, L_m is the m-th Gauss-Laguerre function.

The radiation radius r_s, curvature radius α and radiation amplitude a_m are described by following equations [3],

$$\frac{d\hat{r}}{d\hat{z}} = \frac{2\alpha}{\hat{\omega}\hat{r}_s} - \frac{2}{\hat{\omega}\hat{r}_s} C(z)\langle \sin\psi \rangle \quad , \tag{3}$$

$$\frac{d\alpha}{d\hat{z}} = \frac{2(1+\alpha^2)}{\hat{\omega}\hat{r}_s^2} - \frac{4}{\hat{\omega}\hat{r}_s^2} C(z)\left(\langle\cos\psi\rangle + \alpha\langle\sin\psi\rangle\right), \tag{4}$$

$$\frac{d\hat{a}_m}{d\hat{z}} = -\hat{A}_m\hat{a}_m - i\left\{\hat{F}_m - m\hat{B}\hat{a}_{m-1} - (m+1)\hat{B}^*\hat{a}_{m+1}\right\}, \tag{5}$$

where,

$$\hat{A}_m = \frac{2}{\hat{\omega}\hat{r}_s^2}\{\alpha + i(2m+1) - C(z)[\langle\sin\psi\rangle + i(2m+1)\langle\cos\psi\rangle]\}, \tag{6}$$

$$\hat{B} = \frac{2}{\hat{\omega}\hat{r}_s^2}\left\{\left(\alpha^2 - \alpha - C(z)\langle\cos\psi\rangle\right) + iC(z)\langle\sin\psi\rangle\right\}, \tag{7}$$

$$\hat{F}_m = -4\frac{\nu}{\hat{\omega}} K \frac{a_0}{|a_0|}\left\langle \frac{e^{-i\psi}}{\gamma} \right\rangle \frac{\left(\hat{r}_s^2 - \hat{r}_b^2\right)^m}{\left(\hat{r}_s^2 + \hat{r}_b^2\right)^{m+1}}, \tag{8}$$

$$C(z) = \frac{2\nu}{\gamma} \frac{\left(\hat{r}_s^2 - \hat{r}_b^2\right)\hat{r}_s^2}{\left(\hat{r}_b^2 + \hat{r}_s^2\right)^2} \frac{K}{|a_0|}. \tag{9}$$

where r_s and r_b, are the normalized radii of the radiation and electron beams.

Numerical analysis

Using this code, we calculated the evolution of radiation power, spectrum and amplitude for SASE FEL. The simulation parameters are listed in Table 1.

Table 1 Simulation parameters

Electron beam		
	Beam energy	E_{b0}=150 Mev
	Beam current	I_b=200 A
	Beam radius	r_{b0}=100 μm
Laser		
	Wavelength	λ_{s0}=65 nm
	Input power	P_0=1 W
	Laser radius	r_{s0}=120 μm
	Mode number (Laguerre)	m=5
Wiggler		
	Wiggler period	λ_w=1 cm
	Wiggler field	B_w=0.535 T
	Wiggler number	N_w=1000

Fig.1 shows radiation power for each wavelength. At the wavelength of 64.90nm, it is found that the resonance condition is satisfied well and the radiation power in the MW range can be attained. As shown in Fig.2, the corresponding radiation radius is nearly maintained at the initial radius and it is clear that the radiation is trapped by the electron beam until the power saturates, namely, z~9.0m.

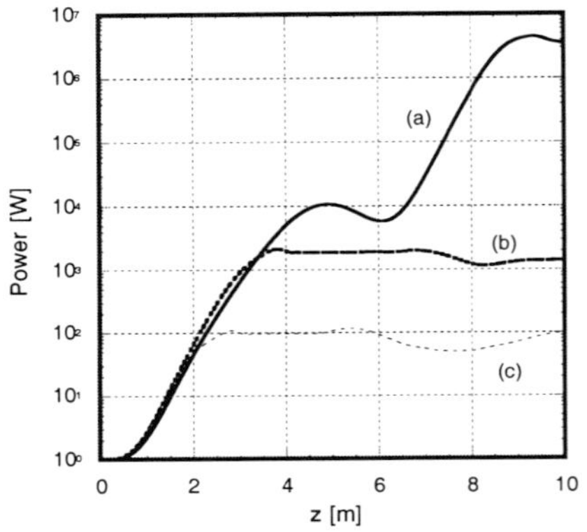

Fig. 1 The evolution of the radiation power along the z-axis. λs=(a) 64.90 nm, (b) 64.95 nm, (c) 65.0 nm

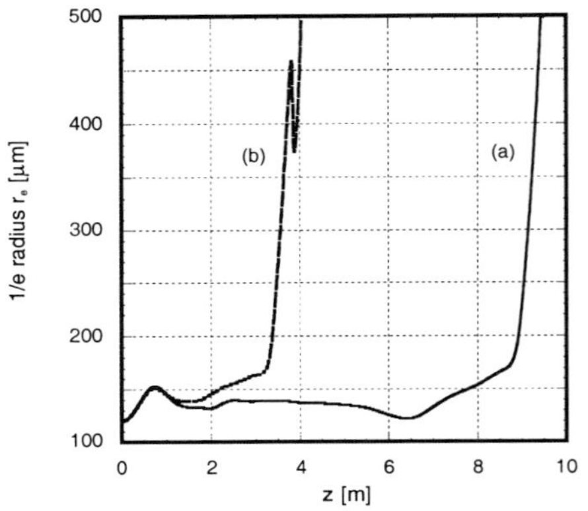

Fig. 2. The evolution of the spot size of the radiation beam for λs=(a) 64.90 nm, (b) 64.95 nm

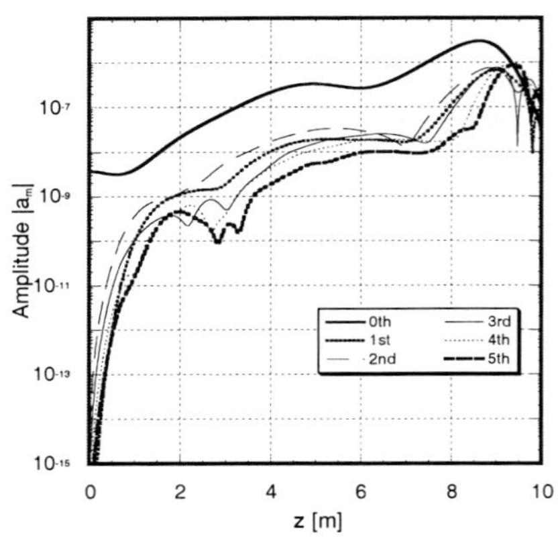

Fig. 3 The radiation amplitude evolution for each Gauss-Laguerre mode along the z-axis.

Fig.3 shows the evolution of each radiation mode amplitude at the wavelength 64.90nm. Due to the growth of the higher order Gauss-Laguerre modes, the radiation radius can be controlled so as not to diverge. Fig.4 shows the radial profile of the radiation at the points z=(a) 0.75 m, (b) 6.5 m, (c) 9.0 m. In the case of (b), it is found that the radius is a minimum and therefore the radiation gets more gain.

In Fig.1, an initial saturation in the case of λ_s =64.90 nm is evident. We consider this detail due to the frequency spectrum narrowing in the exponential gain regime [2] or to the electron's overshooting resonance.

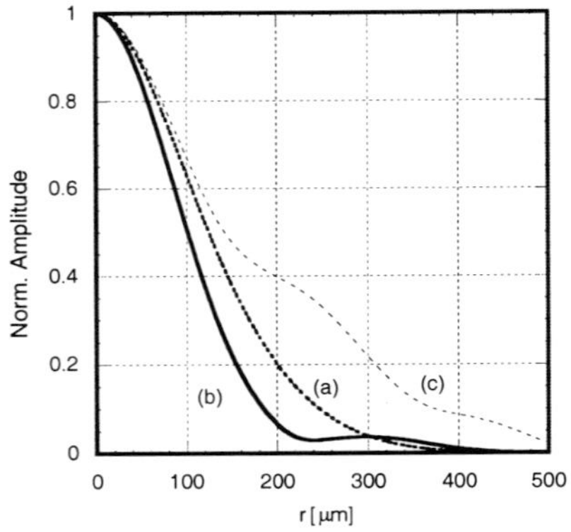

Fig. 4 The distribution of normalized radial profile for z=(a) 0.75 m, (b) 6.5 m, (c) 9.0 m.

Summary

To analyze the amplification characteristics of SASE FEL, we developed the 3-D simulation code using the SDE method. Using this code, we calculated the evolution of radiation power, amplitude for each mode and radiation amplitude profile for the parameters of a FEL at FELI, Osaka, Japan. And it is confirmed that the radiation is partially trapped by the electron beam in the amplification process. In the future work, we will find the optimal parameters of SASE in the X-ray regime, where the radiation is trapped in an electron beam.

REFERENCES

[1]C.A. Brau, Free Electron Laser (Academic Press, 1990)

[2]K. J. Kim, Phys. Rev. Lett. 57(1986),1871-1874

[3]P.Sprangle A.Ting and C.M.Tang Phys.Rev. A36 (1987), 2773-2781

[4]B.Hafizi, P.Sprangle, and A. Ting, Phys.Rev. A36 (1987), 1739-1746

[5]E.T. Scharlemann, J. Appl. phys. 58 (1985), 2154-2161

FREE ELECTRON LASERS 1998
G.R. Neil and S.V. Benson (Eds.)
1999 Elsevier Science B.V.

Performance of the Accelerator Driver of Jefferson Laboratory's Free-Electron Laser*

C. L. Bohn, on behalf of the Jefferson Lab FEL Team
Thomas Jefferson National Accelerator Facility
12000 Jefferson Avenue, Newport News, VA 23606 (USA)

Abstract

The driver for Jefferson Lab's infrared free-electron laser is a superconducting, recirculating accelerator that recovers about 75% of the electron-beam energy and converts it to radiofrequency power. It is designed to lase continuous-wave at 3-6 μm at kW-level power. In achieving first light, the accelerator operated "straight-ahead" to deliver 38 MeV, 1.1 mA cw current through the wiggler for lasing at wavelengths in the vicinity of 5 μm. The waste beam was then sent directly to a dump, bypassing the recirculation loop. Stable operation at power levels up to 311 W cw have thus far been achieved in this mode. The accelerator has recently recirculated up to 0.6 mA cw current with energy recovery. In this mode it has lased pulsed and cw at low power. It remains to clean up the transport for high-power cw lasing.

1 Introduction

Thomas Jefferson National Accelerator Facility (Jefferson Lab) recently constructed a cw, kW-level, 3-6 μm free-electron laser (hereafter called the IR Demo) [1] and this summer has been engaged in an intensive commissioning effort. The IR Demo incorporates a recirculating superconducting linac for energy recovery. First lasing [2] was achieved on 15 Jun 98, with 155 W cw power at 5 μm wavelength reached just two days later. This was done without energy recovery; however, on 29 Jul 98 lasing was accomplished with a recirculated beam consisting of 200 μs macropulses delivered at 2 Hz repetition rate while recovering energy from the macropulses. That same day, the IR Demo delivered 311 W cw power without energy recovery. To date, lasing with pulsed beam has taken place at currents up to 2 mA averaged over a macropulse. Up to 0.6 mA cw has been recirculated with energy recovery. The machine has also lased cw at low power (6 W) with recirculation. Plans for the immediate future are to optimize the performance of the machine and begin supporting user experiments. This paper outlines the commissioning sequence and accelerator performance leading to these achievements.

2 Overview of the Machine

The IR Demo, pictured in Figure 1, incorporates a recirculating accelerator comprising a 10 MeV injector and a 32 MeV linac to produce a 42 MeV electron beam for cw kW-level lasing. It is designed to achieve the top-level electron-beam requirements listed in Table 1 of Ref. [1] while transforming 75% of the beam power back into rf power. The design of the machine is also discussed in more detail in Ref. [1]. Beam parameters originally thought to be required for first light differ from those needed for kW power, however, and they are listed in Table 1 below.

First lasing involved operating the machine in the "straight-ahead" mode, in which the beam is deposited in the "42 MeV dump" depicted in Figure 1. Doing so allowed us to achieve the first-lasing milestone before construction of the recirculation loop had been fully completed. Subsequently, as mentioned earlier, the machine was operated in the recirculation mode with pulsed beam, first without lasing, then with lasing, and with energy recovery from the pulses. In this mode, the beam is deposited in the "10 MeV dump" after deceleration through the cryomodule. What follows is a brief description of the commissioning sequence that led to achieving the first-light electron beam.

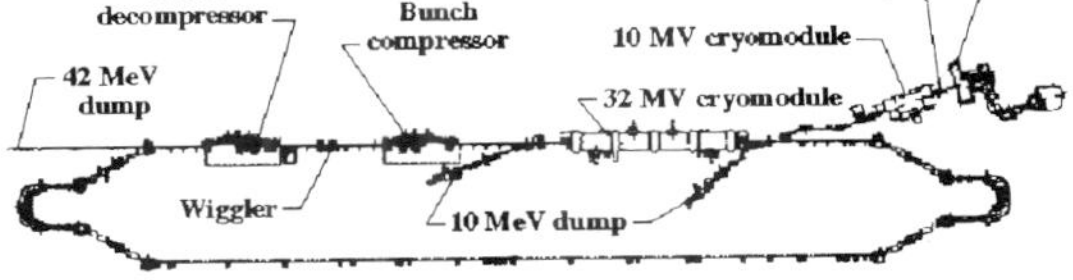

Figure 1. Schematic of IR Demo.

*Work supported by the U. S. Department of Energy under contract DE-AC05-84-ER40150, the Office of Naval Research, the Commonwealth of Virginia, and the Laser Processing Consortium.

Table 1: Beam Requirements at Wiggler for First Lasing.

Parameter	Required	Measured
Kinetic Energy	38 MeV	38.0±0.2 MeV
Average current	1.1 mA	1.10±0.05 mA
Bunch charge	60 pC	60±2 pC
Bunch length (rms)	<1 ps	0.4±0.1 ps
Peak current	22 A	60±15 A
Trans. emittance (rms)	<8.7 mm-mr	7.5±1.5 mm-mr
Long. emittance (rms)	33 keV-deg	26±7 keV-deg
Pulse repetition frequency	18.7 MHz	18.7 MHz

3 Photocathode Gun Performance

Testing the gun off-line ultimately yielded bunch charges from 0-120 pC with phase-space parameters that looked to agree reasonably well with PARMELA [3,4]. Consequently the gun was installed in the FEL injection line. The small space between the gun and the cryounit excludes diagnostics, so reasonable confidence in the gun's performance had to be established prior to its installation, and the cited set of experiments sufficed.

To date the gun has been operated in the FEL to a maximum bunch charge of 60 pC in view of the first-light requirements in Table 1 as well as the desire to preserve cathode lifetime. Experience in running the cited gun experiments hinted that cathode lifetime is significantly shorter at higher bunch charges. Experience to date is that the e-folding lifetime of the GaAs cathode is ~10-20 hours at 60 pC, 1.1 mA average current. Cathode lifetime is seen to depend sensitively on the quality of the ambient vacuum, which in turn may be influenced by beam operations via ionization of residual gas and back-bombardment of ions onto the cathode. Available data is too sparse to support a more quantitative statement.

Recently a decision was made to anodize the outer extremities of the cathode wafer for suppression of possible electron emission and beam halo. Although the benefit has been hard to ascertain conclusively, subsequent operation leading to first light proceeded with easily achievable beam transmission to the straight-ahead dump at 1.1 mA cw, something that had been more difficult to achieve prior to anodization.

4 Key Diagnostics and Accelerator Performance

Diagnostics for the IR Demo include [5]: suites of beam-position monitors, optical-transition-radiation viewers, and beam-loss monitors; two interferometric bunch-length monitors, one (BL1) at the entrance to the linac cryomodule and the other (BL2) just after the wiggler; two multislit transverse-emittance monitors, one (MS1) after the injector cryounit and the other (MS2) at the entrance to the linac cryomodule; two beam-current monitors; and picoammeters at the beam dumps. Key diagnostics that ultimately led to the decision to install the wiggler and try for first light were BL2, a multimonitor emittance measurement using five viewers in the wiggler region, and an energy spread measurement using the dipole magnets and viewer in the second optical chicane. Cleanup of the electron beam proceeded systematically and led to gradual improvement in the six-dimensional properties of the beam. Beam parameters measured at the wiggler location shortly before the 13 Jun 98 wiggler installation were as listed in Table 1. All of them agree with PARMELA to within 10% except the energy spread, for which the measured value was a factor of two higher, and correspondingly so was the longitudinal emittance. The accelerator is capable of generating 48+ MeV beam.

Despite good beam quality at the wiggler, the injector has yet to be fully optimized. Its present setup produces a total beam energy of 9.5±0.1 MeV as inferred from the injection-line dipole strengths, close to the desired 9.65 MeV. Measurements with MS1 yielded a normalized rms transverse emittance of 5.5±0.6 mm-mr, ~30% higher than PARMELA [6]. The beam at MS2 is off-nominal enough that good measurements with MS2 or BL1 have yet to be possible, but the bunch compression, inferred by measuring the M_{55} ($=\partial\phi_{in}/\partial\phi_{out}$) transfer function using a pickup cavity, is close to PARMELA.

5 Status and Plans

The IR Demo achieved first light within six hours from turn-on after inserting the wiggler, a testimony to the fine quality of the electron beam and the FEL systems. It routinely operates with 1.1 mA average current in the straight-ahead mode. The first attempt to take beam around the recirculation loop occurred on 28 Jul 98. In the ten days of machine operations that followed prior to this Conference, we had recirculated up to about 0.6 mA cw with energy recovery and had lased cw at low power in the recirculation mode. Immediate plans are to clean up the beam transport to support kW-level lasing, to begin user experiments, and to measure emittance growth from coherent synchrotron radiation in magnetic bending systems [7].

References

[1] C. L. Bohn, in Proc. 1997 Part. Accel. Conf. (in press).
[2] S. V. Benson, et al., these Proceedings.
[3] D. Engwall, et. al, in Proc 1997 Part. Accel. Conf. (in press).
[4] D. Kehne, et al., in Proc. 1997 FEL Conf. (in press).
[5] G. Krafft, et. al., in Proc. 1997 Part. Accel. Conf. (in press).
[6] P. Piot, et. al., in Proc. 1998 EPAC Conf. (in press).
[7] C. L. Bohn, R. Li, J. J. Bisognano, Proc. 1996 FEL Conf., ed. by G. Dattoli and A. Renieri, (Elsevier, Amsterdam, 1997), II-139.

FREE ELECTRON LASERS 1998
G.R. Neil and S.V. Benson (Eds.)
1999 Elsevier Science B.V.

Direct Measurement of the Shape of Short Electron Bunches

D. Oepts* and G.M.H. Knippels

FOM-instituut voor Plasmafysica "Rijnhuizen", P.O. Box 1207, 3430 BE Nieuwegein, The Netherlands

Abstract
We describe a technique to measure the longitudinal density distribution of electron bunches with sub-picosecond resolution by means of electro-optic sampling of the Coulomb field of the electrons.

1. Introduction

Recently, techniques of high-speed electronic measurements have been extended into the THz frequency domain. The availability of ultrashort Ti:Sapphire laser pulses has enabled the measurement of electric fields with sub-picosecond time resolution. This opens up new possibilities for the study of short relativistic electron bunches such as used in free-electron lasers. A direct measurement of the longitudinal shape of the electron bunches now becomes feasible, using methods developed for THz electric field probing.

2. Electro-Optic sampling

In an Electro-Optic (EO) sampling set-up, the field-induced optical activity (Pockels effect) in a sample crystal is measured with an ultra-short probe laser pulse. The change in the polarization state of the probe beam can be detected with high sensitivity, and the response time of the crystal can be very short. Therefore, EO sampling can be used for time resolved probing of both local fields and free-space propagating pulses in the THz range [1].
We intend to apply this technique to the measurement of the Coulomb field of the electron bunches in the FELIX free electron laser. The set-up is sketched in Fig.1. The electron beam passes close to a ZnTe sample crystal, and the EO effect is sensed by a Ti:Sapphire probe pulse traversing the crystal in parallel. A photodiode detects the signal caused by the polarization change.

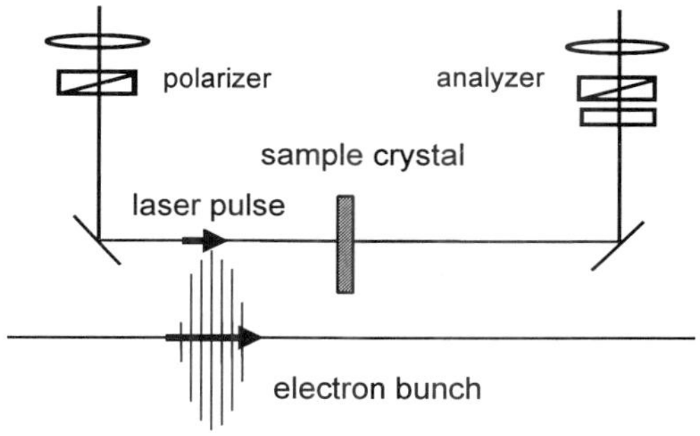

Fig.1 Diagram of the electro-optic sampling set-up

3. Resolution

The Coulomb field of an electron moving at relativistic speed in a straight line is concentrated in the direction perpendicular to its trajectory, within an angle of order $1/\gamma$ [2]. The instantaneous field in a point P at a distance R from a pencil electron beam is mainly determined by the electrons in a range $\pm R/\gamma$ of the bunch length, so that the time resolution obtained by probing the field at P could, ideally, be $\tau \approx 2R/\gamma c$). As an example, for R=10 mm and $\gamma=100$, $\tau \approx 0.7$ ps; this improves for higher energies. The resolution degrades for a beam with finite transverse emittance, but this is a minor effect when the normalized emittance is smaller than R. When the probe pulse and the THz field pulse traverse the sample with the same speed, the thickness of the crystal adds to the magnitude of the EO effect without reducing the resolution. In reality, there is a velocity mismatch

* Corresponding author. Present address: Jefferson Lab., Newport News VA 23606, USA
Tel. +01 757 269 5849, Fax +01 757 269 7658, e-mail oepts@jlab.org

due to the dispersion in the material, which limits the useable thickness. An even more important aspect of the dispersion is the response time of the sample's electric polarization. ZnTe, for example, has an LO-phonon resonance at 5.3 THz, which means that field pulses shorter than about 200 fs will be distorted and attenuated. It has been shown, however, that frequencies up to 37 THz can still be measured with ZnTe [1] and that the effects of dispersion and absorption can be accurately modeled [3].

4. Sensitivity

The x-component of the electric field at P(R,0,0), caused by an electron with charge q at (0,0,z(t)), moving along the z-axis, is [2]

$$E_x = \frac{2\gamma q}{4\pi\varepsilon_0 \left(R^2 + \gamma^2 z^2\right)^{3/2}}$$

For a narrow beam of electrons with a longitudinal density distribution $\rho(z)$ varying little over a distance R/γ, the integrated field is

$$E_x = \frac{2\rho(z)}{4\pi\varepsilon_0 R}$$

Taking an average $\rho(z)$ equal to $\rho = Q/l_b$, where Q is the total charge of the bunch and l_b its effective length, we have for the average field produced by the bunch:

$$E_{av} = \frac{2Q}{4\pi\varepsilon_0 R l_b}$$

For example, with Q=200 pC, as in FELIX, R=10 mm, and l_b=1 mm, we have $E_{av} \approx$ 350 kV/m. For a probe wavelength of 800 nm and a ZnTe sample of 0.1 mm thick with a Pockels coefficient $r_{14}=4\times10^{-12}$ m/V, this gives a retardation in the order of 4×10^{-3} wavelengths, which should give an easily measurable intensity variation at the detector.
The sensitivity can be considerably improved by a differential detection scheme using a pair of balanced detectors measuring the difference of the signals at orthogonal polarizations [1].

5. Synchronization

In the above, it was tacitly assumed that the probe pulse can reproducibly be made coincident with the field of a given slice of the electron bunch. In practice, synchronization to within the required time resolution is not easily achieved. It has been shown, however, that the jitter between the probe pulses and the output pulses of FELIX can be reduced to 400 fs (rms) through active control of the cavity length of the probe laser [4]. For a better time resolution, the remaining jitter can be overcome by using the differential optical gating (DOG) technique [5].
In principle, it is also possible to perform single-shot measurements, e.g. by using a chirped probe pulse and a spectrometer to separate the signals from different time slices [6]. Alternatively, one could use an intentional velocity mismatch combined with space resolved detection [7]. In these cases, synchronization is not a problem, provided that the sensitivity is high enough to obtain reliable results from a single pulse.

6. Conclusion

Application of the EO sampling technique to the measurement of the Coulomb field of short electron bunches promises to be a valuable method to determine the longitudinal density distribution of the bunches on a subpicosecond timescale. Asymmetric bunch shapes present no special problem, and there is no time reversal ambiguity as in more common methods based on autocorrelation.

Acknowledgment

This work was performed as part of the research program of FOM with financial support from NWO.

References

[1] Q. Wu, and X.-C. Zhang, Appl. Phys. Lett. **68** (1996) 1604-1606

[2] J.D. Jackson, in "Classical Electrodynamics", Wiley, New York, 2nd ed.1975, p. 555

[3] H.J. Bakker et al., J. Opt. Soc. Am. B. **15** (1998), 1795-1801

[4] G.M.H. Knippels et al, this conference, WE4-02

[5] X. Yan et al., this conference, WE1-02

[6] Jiping Jiang and X.-C. Zhang, Appl. Phys. Lett **72** (1998) 1945-1947

[7] R. Thatchyn, APS Particle Accelerator Conference, Vancouver 12-16 May 1997, 5P.66

FREE ELECTRON LASERS 1998
G.R. Neil and S.V. Benson (Eds.)
1999 Elsevier Science B.V.

Performance of an Electron Gun Based on Gated Field-emitter Arrays

H. Ishizuka[a], H. Yamazaki[a], H. Shimawaki[b], K. Yokoo[b], S. Kawasaki[c], A. Watanabe[d] and M. Shiho[d]

a Faculty of Engineering, Fukuoka Institute of Technology, Higashi-ku, Fukuoka 811-0295, Japan
b Research Institute of Electrical Communication, Tohoku University, Katahira, Aoba-ku, Sendai 980-8577, Japan
c Department of Physics, Faculty of Science, Saitama University, Shin-ohkubo, Urawa. Saitama 338-8570, Japan
d Japan Atomic Energy Research Institute, Fusion Research Establishment, Ibaraki 311-0193, Japan

Abstract

Electron beams with currents up to 20 mA and 100 μA were generated by a triode gun using single-gated and double-gated field-emitter arrays (FEA), respectively, as the cathode. Beam emittance was significantly lowered when individual beamlets emerging from the tips were collimated by the focusing lens of double-gated FEAs. Further improvements in collimation capability and reliability of FEAs are the key for practical application of this type of gun to free electron micro-lasers.

1. Introduction

Typical parameters of a driving beam for free electron micro-lasers are [1-3]: energy ≥ 100 keV, spot radius ~ 10 μm, rms emittance ~ 10^{-8} π-m-rad and current 100 μA–10 mA. Theoretical studies have indicated that such a beam can be produced using an FEA as the source, but the technical feasibility has not been established yet. This paper is a continuation of our previous work [4].

2. Gun structure and electron optics

The electron gun consisted of an FEA cathode mounted on a TO-5 header, an anode at 15 mm and a wehnelt. A wide variety of FEAs were tried, the number of tips ranging from 1 to 10,000 and tip material being either molybdenum or silicon. Most of them had a single gate for electron extraction, but some were also provided with second gate (focus electrode) for collimation of individual beamlets emerging from the tips. The gun, followed by a lens system, was installed in a 1 m long test stand [5] and voltages of up to – 10 kV were applied to the cathode.

Electron optical properties of FEA and the gun were studied using a computer code EGN2W. Figure 1 shows electron trajectories through the gun and einzel lenses. Here a beamlet started at 0.1 mm from the axis with a half divergence angle of 50 mrad at 70 eV. In Fig. 1(b) the lens system is operated to form a magnified image of the cathode surface for diagnosis of the FEA. On the other hand, the operating mode in Fig. 1(c) is for spot formation. Simulations indicate that the beam should be collimated at the initial stage of generation for good quality to be assured. This was verified experimentally; the beam profile was quite sensitive to the wehmelt voltage as well as to the voltage V_F applied to the focus electrode of FEA with respect to the base.

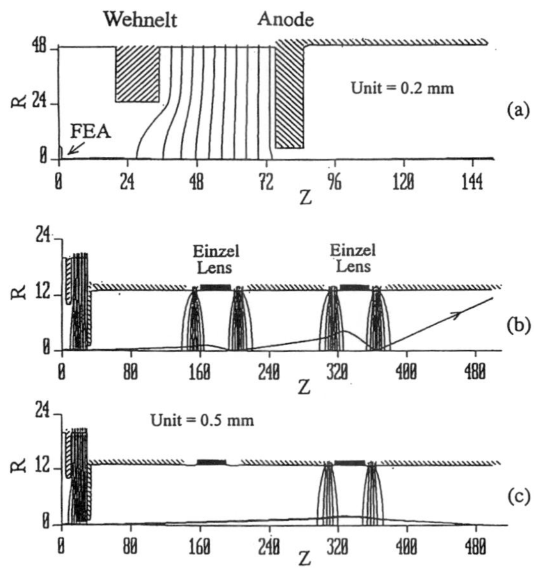

Fig. 1. Electrode configuration, equipotential plot and electron trajectories. (a) Triode gun. (b) Lens system succeeding the gun operated as an electrostatic microscope to examine the performance of FEA. (c) The system operated in the probe forming mode.

3. Emission and collimation characteristics of FEA

FEAs were operated cw when observing the beam cross section by means of fluorescent images, and pulsed when measuring the peak electron yield. Beam current of up to 20 mA were obtained from a 10,000 tip single-gated FEA in a 1 μs pulse, and 100 μA dc was reached with a double-gated FEA having 600 tips. Fraction of working tips was on the order of 10 % for large FEAs like those [6], but each and every tip of small arrays having less than 10 tips used to work if the gate voltage V_G was sufficiently high..

The measured beam current from a 1 x 10 array is shown in Fig. 2 as a function of V_F, where the gate voltage V_G is the parameter. Calculated half-angle of divergence at 100 eV is also shown.. As V_F was lowered, both the emission and the divergence angle were reduced. The emission was recovered by increasing V_G, but full collimation was not achieved because breakdown failure of the FEA happened between the gate and the focus electrode when $V_F \leq 10$ V.

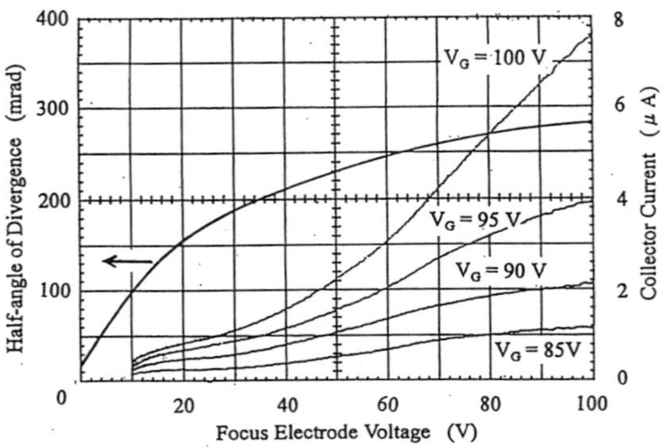

Fig. 2. Focus electrode voltage dependence of emission current (measured for 1 x 10 array) and divergence (calculated at 100 eV).

Beams from single-gated FEAs were seriously affected by aberrations unless scraped with an aperture stop. With double-gated FEAs, the whole beam was transported and the beamlets formed separate images on the screen (Fig. 3).

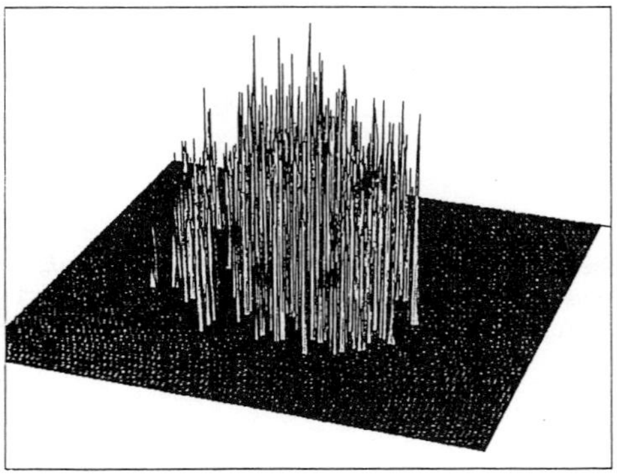

Fig. 3. 3-D profile of fluorescence produced by the beam. The terrain was smooth when the beam was not collimated at FEA.

Fig. 4. Cross section of beamlets vertically narrowed to the minimum width by a quadrupole lens. Beam energy is 5 keV.

4. Emittance of beams and beamlets

Beam emittance was lowered by beamlet collimation at the FEA; for a 600-tip FEA of 0.1 mm radius, when the focusing lens was operated, the normalized emittance was a few $10^{-7}\pi$-m-rad independently of the beam current within the measured range of up to 100 μA.

Figure 4 shows fluorescence produced by beamlets whose cross sections were narrowed to the minimum by a quadrupole lens. The cathode is a 600 tip double-gated FEA. Half-width of each pattern is less than 0.15 mm. The beamlets were aligned when the strength of the quadrupole lens was increased to a certain value. This alignment indicates that the beamlets were ejected parallel to each other from the FEA. That the alignment did not occur at the minimum beamlet width gives a proof of insufficiency of collimation (half-angle of divergence ≈ 0.1 rad at 100 eV), consistently with the operating condition of FEA as mentioned in the previous section. From the dependence of image width on the strength of quadrupole lens, the normalized edge emittance of beamlets carrying ≤ 1μA was estimated to be within $10^{-8} - 10^{-9}$ π-m-rad .

4. Conclusions

Beam current from FEAs suffices the requirement of low current FELs. Double-gated FEAs proved to decrease the beam divergence at very low energies. Still, the beam brightness is not high enough; the beam emittance need to be lowered by one order of magnitude or more. The largest problem in attaining the goal appears to be in FEA itself.

The promising aspect of FEA-based electron gun is that FEAs are making steady progress in regard to packing density, operating voltage, vacuum environment, etc. Further improvements of FEAs in collimation capability and reliability are the key to generate electron beams meeting the conditions of free electron micro-lasers.

References

[1] T. Taguchi and K. Mima, Nucl. Instr. and Meth. A331 (1993) 597.
[2] C. M. Tang et al, Nucl. Instr. and Meth. A358 (1993) 7.
[3] C.A. Brau, Nucl. Instr. and Meth. A393 (1997) 426.
[4] H. Ishizuka et al, Nucl. Instr. and Meth. A393 (1997) 479.
[5] H. Ishizuka et al, Jpn. J. Appl. Phys. 35 (1996) 5471.
[6] C. Constancias and R. Bapist, Technical Digest of IVMC '97 (1997) 215.

FREE ELECTRON LASERS 1998
G.R. Neil and S.V. Benson (Eds.)
1999 Elsevier Science B.V.

Design study of a short-undulator based bunch length monitor for UTNL subpico-second linac

Ryoichi Hajima [1], Katsumi Takahashi

Department of Quantum Engineering and Systems Science, University of Tokyo, Hongo 7-3-1, Bunkyo-ku, Tokyo 113-5378, Japan

1. Introduction

A photocathode rf-gun and a chicane based magnetic bunch-compressor have been installed on the S-band linac of UTNL (Nuclear Engineering Research Laboratory, University of Tokyo), where electron bunches as short as 400 femto-second are available. Both streak camera and coherent transition radiation interferometer are used for the measurement of electron bunch length [1]. These measurements, however, requires a screen target as a radiation emitter on the beam line, which causes fatal destruction of the electron beam. Since the bunch length is one of important parameters for experiments in a linac, a non-destructive monitor which enable one to measure bunch length without interruption of experiments is desired. We design a non-destructive bunch length measurement system utilizing off-axis radiation from a short undulator for UTNL linac.

2. Design of a bunch length monitor

In recent years a bunch length monitor utilizing off-axis radiation from a short undulator has been proposed [2]. This monitor is based on a principle that transition of incoherent radiation into coherent radiation occurs for electron bunch with longitudinal size close to the radiation wavelength, and radiation power is sensitive to bunch length. We optimize design parameters of a bunch length monitor for UTNL linac in this section. In the following design study, we assume that electron bunch has energy of 20MeV, charge of 0.8nC, reputation rate of 10Hz. An existing undulator originally designed for an FEL experiment is converted into a diagnosis use: $K = 1.4$, $N_u = 10$ and $\lambda_u = 4.0$cm.

Fundamental wavelength of undulator radiation for off-axis angle, θ, is obtained by

$$\lambda_r = \frac{\lambda_u}{2\gamma^2}\left(1 + 2\gamma^2 + K^2 - 2\gamma^2\cos\theta\right) \ ,$$

where λ_u is undulator period, K is undulator parameter. For relativistic electron beam, Eq(1) can be reduced into

$$\lambda_r \sim \lambda_u(1 - \cos\theta) \ .$$

If the wavelength is close to the order of the bunch length, the radiation is enhanced by transition of incoherent undulator radiation into coherent radiation [3]. While radiation power is proportional to the number of electrons in a bunch, N, for incoherent radiation, coherent radiation power is proportional to N^2. Radiation power, therefore, shows strong dependency on bunch length, if the fundamental radiation wavelength is the same order of bunch length to be measured.

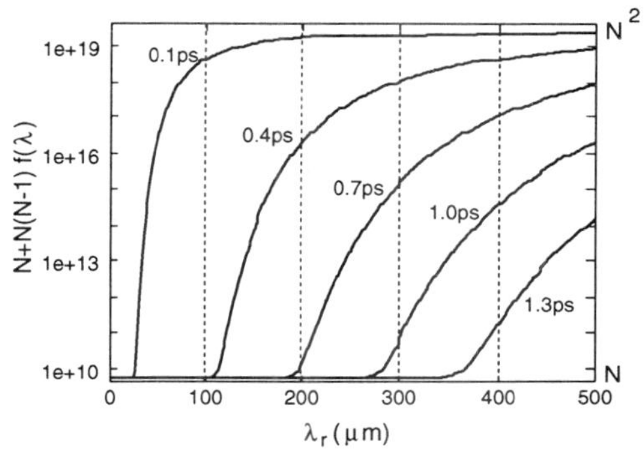

Figure 1. Power enhancement by transition of incoherent radiation into coherent radiation.

[1] Corresponding author. Tel: +81-3-5689-7283;
FAX: +81-3-3818-3455; e-mail: hajima@q.t.u-tokyo.ac.jp

Figure 1 shows the enhancement of radiation power caused by transition into coherent regime, where the radiation power is plotted as a function of wavelength for bunch length between σ =0.1ps and 1.3ps. It can be seen that the radiation power between 200μm and 300μm changes drastically with bunch length in subpico-second region, which is the bunch length to be measured. The observation off-axis angle has been determined at 6 degree so that the fundamental radiation wavelength is 256μm. Designed configuration of measurement system is shown in figure 2, where off-axis radiation is extracted through polyethylene window of 1.2mm square and detected with Golay cell, which is a broadband detector in sub-millimeter region.

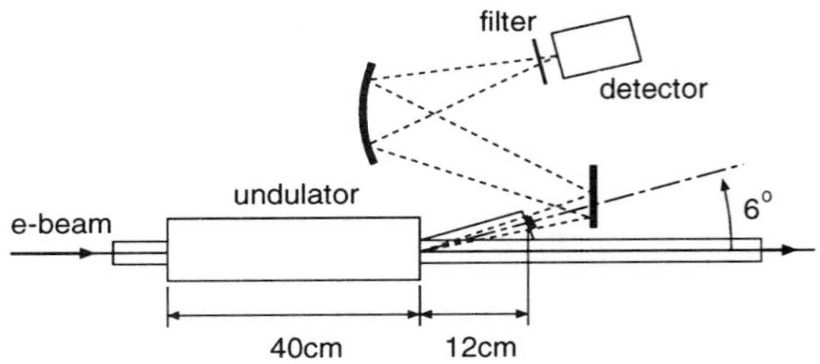

Figure 2. Designed configuration of a measurement system.

Since undulator radiation has higher harmonics and a Golay cell has broad band sensitivity, a bandpass filter is required to detect only wavelength around 256μm, where the radiation power largely depend on bunch length. A bandpass filter in sub-millimeter region can be realized with metal grid. If we choose geometry of metal grid so that grid-interval is 230 μm and the ratio of grid strip width to grid-interval is 0.5, performance of the filter becomes: maximum transmittance is 0.78 at the central wavelength 258.06μm and band width is about 60μm (FWHM). Calculated radiation spectrum after the bandpass filter is plotted in figure 3.

Figure 4 shows estimated signal to noise ratio of the detector, where the noise power is calculated from specification data of Golay cell OAD-7. It is shown that the detector signal larger than noise level can be obtained for bunch length below σ = 0.55ps, that is 1.2ps(FWHM), and the signal increases more than 10 times for the change of bunch length of 0.1ps. This system, therefore, can be used as a bunch length monitor for subpico-second electron bunches. Electron bunch longer than 1.2ps is also measureable, if we use a polyethylene window of larger size.

In figure 4, signal to noise ratio for multi pulse mode with a thermionic gun, which continues 4.5 μsec with 2ns intervals, is also plotted.

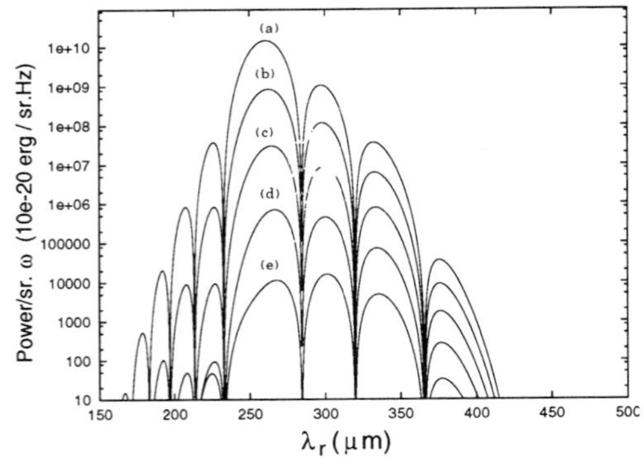

Figure 3. Calculated off-axis radiation spectrum after the bandpass filter. (a) σ_z = 0.5ps, (b) σ_z = 0.6ps, (c) σ_z = 0.7ps, (d) σ_z = 0.8ps, (e) σ_z = 0.9ps.

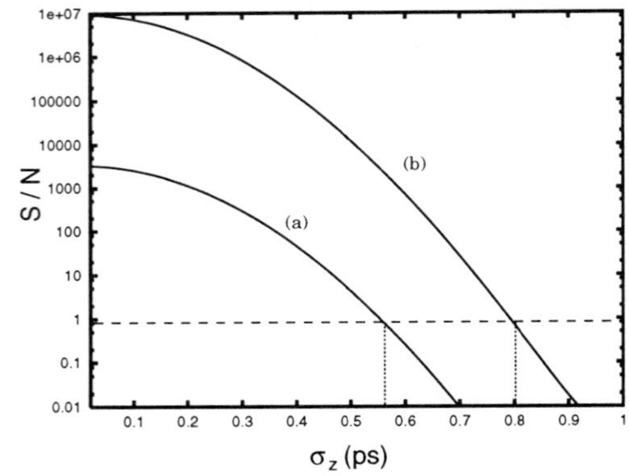

Figure 4. Estimated signal to noise ratio. (a) single pulse from photocathode rf-gun, (b) multi pulse from thermionic gun.

3. Conclusion

In the present study, a bunch length measurement system utilizing off-axis radiation from a short undulator has been designed. We have optimized off-axis angle for the measurement and parameters for a bandpass filter. It is shown that measurement of subpico-second electron bunch is available with the system.

References

[1] J. Sugahara et al., in Proc. EPAC-98.

[2] M.L.Ponds et al., Nucl. Inst. Meth. A375 (1996), 136-139.

[3] G.A.Krafft et al., Proc. 1995 Part. Acel. Conf., 2601-2603.

FREE ELECTRON LASERS 1998
G.R. Neil and S.V. Benson (Eds.)
1999 Elsevier Science B.V.

Beam Position Monitors for the Linear Coherent Light Source

Roger Carr , Robert Hettel, Don Martin, Clive Field
Stanford Linear Accelerator Center

The Linear Coherent Light Source (LCLS) FEL requires that the radiation created by the electron beam in an undulator continuously bathe the electron beam, in order to have efficient FEL gain. Beam position monitors will be used both to establish the initial alignment of the electron beam, and to provide runtime feedback to counteract drifts. There are tight tolerance and stability requirements on the beam position monitors. There are also restrictions on impedence that would cause unwanted wakefield effects, and bandwidth because of the short, 250 fsec electron pulses. To select a BPM technology, we examined the known BPM techniques, including detection of synchrotron radiation, transition radiation, diffraction radiation, fluoresence, photoemission or bremsstrahlung from thin wires, Compton scattering from laser wires, and image currents from the electron beam.

The LCLS will produce intense pulses of coherent x-rays in the 15-1.5 Å range generated by self-amplified spontaneous emission (SASE) from a 5-15 GeV electron beam passing through a 112 m long undulator [1] . The undulator has 52 segments, each 1.92 m long, separated by 24 cm gaps containing vacuum pumps, quadrupole focusing magnets and diagnostics. The focusing magnets are equipped with transverse movers that are used for initial beam based alignment, and for feedback response to steering errors detected by beam position monitors. The undulator gap is 6 mm, and the vacuum chamber within has a 5 mm ID.

It is possible to align BPM's mechanically to perform these functions at 5 GeV, where the overlap tolerance is relaxed, but not possible at 15 GeV, so we developed a method of beam based alignment [2] By recording the readings of roughly aligned BPMs as a function of beam energy (varied between 5 and 15 GeV) and by fitting a model of the undulator electron transport optics to those readings, offset errors for quadrupoles, BPMs, and incoming beam trajectory can be calculated and corrected. When this process is repeated 2-3 times (which may take a few hours), simulations indicate that BPM offsets and electron beam straightness in the 112 m long undulator can be established and maintained with < 5 μm rms accuracy.

We will also install insertable intercepting monitors that provide a means to measure rough positions and measure beam size.

Micron resolution and stability is needed only in a bandwidth comparable to thermal drift frequencies (<< 1 Hz, over periods of days), implying that BPM readings from many beam pulses can be averaged for higher resolution. Single shot resolution of order 1 μm for a 1 nC bunch is desired to detect 120 Hz pulse-pulse trajectory instability. A dynamic range of 40 dB is needed for low and high intensity operation.

The total longitudinal beam impedance of 52 BPMs must be kept well below a loss factor of ~1 kV/pC to keep the correlated energy spread of the electron bunch below 0.1%; otherwise the FEL saturation length would increase beyond 112 m.

We have investigated several intercepting and non-intercepting beam position monitor technologies that might meet the performance needs for the LCLS undulator. Among intercepting monitors, YAG crystals offer superior performance to phosphosrs, because they do not bloom, but any intercepting monitor will be struck by the combined electron and photon beams, confusing the signal source. One can imagine filters, like 100 μm tungsten foils, to strip off the x-rays. Transition radiation foils could be used, but also suffer from combined beam effects.

Wire scanners are used successfully at SLAC to measure micron or smaller rms beam sizes. [3] Overlap between the electron beam and a precisely positioned carbon wire is detected downstream of the undulator by measuring either bremsstrahlung gamma rays or energy degraded electrons. By stepping a wire across the beam, pulse by pulse for 10-20 pulses, using a linear motion stage or by steering the beam across the wire, a profile of the beam can be measured. The beam shape is fitted on-line, with a typical uncertainty of 2% of the width, and the center position obtained within 1-2 μm with respect to an external fiducial on the motion stage.

A distinct advantage of the carbon wire monitor is that it can be used for simultaneous measurement of electron and undulator photon beam position. While the impinging electron beam generates bremsstrahlung, the undulator will diffract from the wire in a powder diffraction pattern. An experiment at SSRL using 7 mm amorphous carbon wire filaments and 1.5 Å x-rays showed that an intensity maximum for Bragg scattering occurs at 25.8 degrees from the beam axis.

We conclude that the preferred intercepting monitor for the LCLS is the wire scanner because of its ability to measure both electron and photon position at high operating intensities and because of its proven micron-level performance.

Candidates for non-intercepting position monitors include diffraction radiation monitors, laser wire (or spot) monitor, and more commonly used RF BPMs.

A diffraction radiation (DR) monitor [4] having a 2 mm radius aperture within the 2.5 mm radius undulator vacuum chamber would produce micron wavelength radiation that could be observed with a simple camera system to determine beam size and position. While the measured radiation pattern is sensitive to the transverse displacement of the electron beam from the center of the aperture, we are not convinced this monitor would have the micron position resolution required. Furthermore, the monitor also has a high impedance (~75 V/pC loss factor), implying that only 10 monitors could be inserted during FEL operation.

The success of the laser wire monitor for measuring micron beams at the SLAC Linear Collider Final Focus [5] prompted us to investigate a method of measuring 1.06 µm light from a high powered laser (e.g. a 100 Compton scattering from a 1 µm x 10 µm laser "spot" MW peak pulsed YAG laser). Because of the large background expected from bremsstrahlung and high energy undulator photons, a measurement of degraded energy electrons at the end of the undulator might offer better performance. A principal problem with the laser spot monitor is that, due to possible changes in laser optical components over time caused by the high pulsed laser power and radiation environment, the absolute stability of the laser spot position is uncertain and there is no clear method for monitoring it. Another drawback is that if the electron beam is off the laser spot, there is no indication of which way to steer.

Uncertainties in performance of the DR and laser spot monitors led us to concentrate on specifying an appropriate non-intercepting RF BPM pickup and processing system for the LCLS undulator. The properties of these monitors are summarized below:

Monitor Type	Parameters	Center Ac'cy	Resolution	Oper. Freq.	Issues
Wall Current	z=6mm, R_B=2W	100 mm	0.7 mm/nC	> 1 GHz	Ferrite saturation
Stripline	z=9mm, Z_o=40W	100 mm	0.2 mm/nC	2–5 GHz	Strips on ceramic cylinders
Microwave Aperture	3.0x1.5mm slot to waveguide	100 mm	0.1 mm/nC	> 50 GHz	Operating freq >chamber cutoff.; HOM errors
Cavity	f_{ID}=7mm,z=2.8mm	50 mm	1 mm/nC	~ 32 GHz	f_o ~ cutoff; low Q
Stripline	z=40mm,Z_o=50W	50 mm	0.2 mm/nC	0.5–2 GHz	Technical maturity
Cavity	f_{ID}=60mm,z=5mm	5 mm	0.2 mm/nC	~ 6 GHz	Robust;TM_{010} mode

Of those investigated, the Cavity BPM best meets the design requirements. Because of the natural symmetry of circular machining and the availability of ultra-precision diamond lathes, micron level absolute mechanical and electrical center accuracy can be achieved. Button type BPM's are also strong candidates, if we can work around their poor absolute center calibration.

We conclude that non-intercepting RF cavity electron BPMs, together with a beam based alignment system, are best suited for this application. We also plan for insertable intercepting wire monitors for rough alignment, for beam size measurements, and for simultaneous measurement of electron and photon beam position.

ACKNOWLEDGEMENTS

The authors wish to thank the following people for contributions to this work: J. Arthur, S. Brennan, M. Cornacchia, H.-D. Nuhn, J. Sebek (SSRL); P. Emma, S. Smith, S. Wagner (SLAC); G. Lambertson, W. Barry (LBL); A. Lumpkin (APS); I. Ben-Zvi (BNL); and R. Fiorito (NSWC).. This work was supported by the U.S. Department of Energy, Office of Basic Energy Sciences, under contract number DE-AC03-76SF00515.

REFERENCES

[1] LCLS Design Study Report, SLAC Pub. SLAC-R-521, April 1998

[2] P. Emma, R. Carr, H-D Nuhn, these Proceedings.

[3] C.Field, Nucl. Instr. & Meth. A360 (1995) 467

[4] D. Rule, R. Fiorito, W. Kimura, "Non-Interceptive Beam Diagnostics Based on Diffraction Radiation", *Proc. of the 7th Beam Instrumentation Workshop*, AIP 390 (1996) 510

[5] M. Ross et al., "A Laser-Based Beam Profile Monitor for the SLC/SLD Interaction Region", *Proc. of the 7th Beam Instrumentation Workshop*, AIP 390 (1996) 281.

FREE ELECTRON LASERS 1998
G.R. Neil and S.V. Benson (Eds.)
1999 Elsevier Science B.V.

Electron Gun Control for a CW-FEM

A. Shaw*, A.I. Al-Shamma'a, S.R. Wylie, R.A. Stuart, J. Lucas

*The University of Liverpool, Department of Electrical Engineering and Electronics, Brownlow Hill, Liverpool L69 3GJ, UK.

Abstract

At Liverpool University, a low cost computerised control system has been established for the operation of a pre-bunched CW-FEM electron gun. This paper presents the design of the automated electron gun control system.

1. Introduction

A series of single user, low cost computerised control systems have been developed [1] for the control of the CW-FEMs built at the University of Liverpool. This paper outlines the control system used in the electron gun section for the latest generation of pre-bunched CW-FEM.

There are two levels of control required for the present pre-bunched CW-FEM [2], one is an extensive hands on programme which has every bit of information displayed on the screen for interpretation by the research team. The second is a simple graphical display with minimal user input to enable a single unskilled operator to control the FEM, but will not be discussed in this paper.

2. Pre-bunched Electron Gun Assembly

The pre-bunched electron gun is split into 3 sections (fig.1). Electron beam generation, bunching and acceleration.

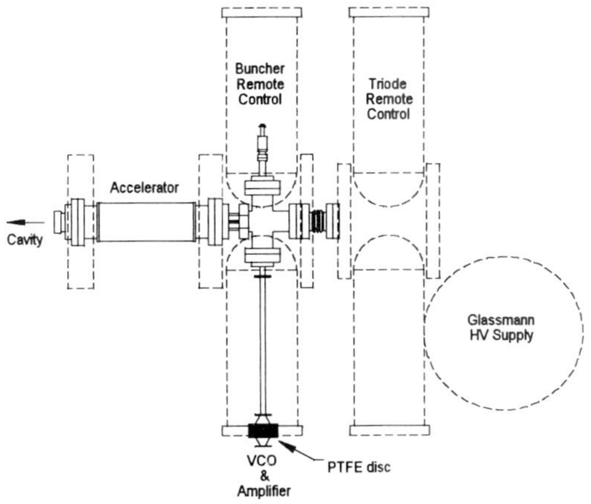

Fig.1: Pre-bunched electron gun layout.

2.1 Electron Beam Generation

The first stage of the pre-bunched electron gun consists of a simple planar triode (fig.2). The triode is capable of delivering a 3-10keV, 20mA electron beam, depending upon the cathode material used. The triode is designed to allow a tungsten wire (≤ 3mA), thoriated tungsten wire (≤ 5mA) or a single LaB_6 crystal (≤ 20mA) to be used depending on the experiment being performed.

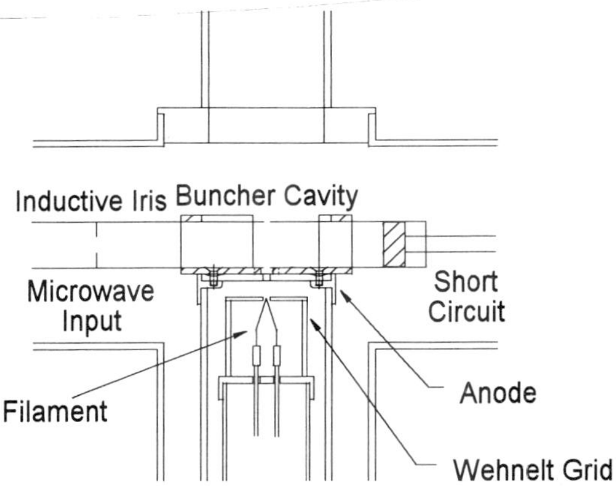

Fig.2: Schematic of triode and buncher cavity.

2.2 Electron Beam Bunching

The second stage of the electron gun consists of an X-band waveguide bolted to the anode of the triode. A 3mm diameter hole in the centre of the 'b' dimension allows the electron beam to pass through. The X-band cavity has a moveable short circuit at one end and an inductive iris at the other. 1W of microwave power (between 8 and 9.3 GHz) is passed into the cavity via a quartz window and tuned to maximise the electric field across the centre of the cavity using the moveable short circuit. Once the beam exits the buncher, a magnetic lens re-focuses the beam to the focal point of the electrostatic immersion lens in the acceleration stage. A saddle coil is placed after the lens to compensate the earth's magnetic field and any misalignment in the triode assembly.

2.3 Electron Beam Acceleration

The final stage of the electron gun consists of two concentric cylinders arranged as an electrostatic immersion lens that accelerates the bunched electrons to the desired energy for interaction in the FEM (≤200keV).

3 Electron Gun Computer Control System

The electron gun control system uses a combination of direct and remote control. A PC is used as the direct control system with a series of Amplicon Liveline A/D and D/A interface cards attached to each device. The remote control is performed by a low cost micro-controller network [3]. The micro-controllers are connected in a star arrangement with the direct control PC as the central node (fig.3), each node is connected to the PC via a duplex fibre optic link. The electron gun consists of 3 sections: triode, bunching and acceleration, each at a different potential.

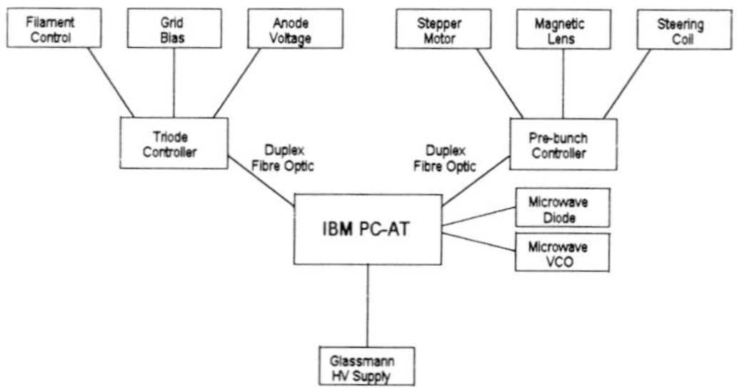

Fig.3: Block diagram of control scheme.

3.1 Triode.

The triode consists of a 0-1kV negative DC power supply, a 0-5A DC constant current source and a 0-10kV positive DC power supply. The output of the 0-1kV power supply is used to bias the Wehnelt grid of the triode, to control the amount of electron emission from the cathode. The 0-5A constant current source is used to heat the filament to its particular operating temperature and finally, the 0-10kV supply applies the potential difference between the anode and the cathode to produce the electron beam. Each aspect of these supplies is controlled via a single micro-controller allowing the triode operating parameters to be set and monitored by the PC.

3.2 Electron Bunching

The second stage consists of a 1.8° stepper motor, a 0.5A DC constant current source and a ±0.2A constant current source. The stepper motor is connected to the moveable short circuit via a vacuum rotary drive, which allows the short circuit to be positioned to within 5μm. The 0.5A DC constant current source supplies the magnetic lens that focuses the electron beam. The ±0.2A constant current source drives the saddle type steering coil. These are both controlled by a single micro-controller and form the second node of the micro-controller network.

The electron bunching section also consists of a voltage controlled oscillator, feeding a 40dB microwave amplifier. The amplifier delivers 1W of microwave power into the bunching cavity, through the high voltage chamber, using two horns either side of a 5mm thick PTFE disc. The microwave power is then passed into the bunching cavity via a vacuum-sealed quartz window. The PC can directly control the VCO and by measuring the voltage across the microwave diode on the cross coupler, the short circuit can be moved to maximise the resonance at the required frequency.

3.3 Electron Accelerating

The final stage consists of a 200kV, 18mA DC Glassmann high voltage power supply. This is controlled by two 0-10V signals, one to adjust the operating voltage, the other to set the current limit. Two 0-10V feedback signals are monitored by an A/D interface card, one measuring the voltage and the other measuring the current supplied.

4. Conclusion

It has been found that, although its possible to control the whole of a CW-FEM via a network of low cost micro-controllers [3] the time taken to respond to any change in the system is long (≈2 seconds). The combination of a distributed micro-controller network and direct control via a PC allows a fast response to any system changes and therefore means a reliable closed loop control can be achieved. This control scheme has been successfully implemented in the current CW-FEM at the University of Liverpool [2].

5. References.

[1] C. Balfour et al., Nucl. Instr. and Meth. in Phys., A375 (1996), 430.

[2] A.I. Al-Shamma'a et al., Nucl. Instr. and Meth. in Phys., Submitted at 1998 conference, USA.

[3] C. Balfour et al., Nucl. Instr. and Meth. in Phys., A393 (1997), II-73.

FREE ELECTRON LASERS 1998
G.R. Neil and S.V. Benson (Eds.)
1999 Elsevier Science B.V.

Transition Metal Carbide Field Emitter Cathodes for FEL Applications

Cha-Mei Tang[a,*] and William Mackie[b]

[a] *Creatv MicroTech, Inc., Potomac, MD 20854-1221, USA*
[b] *Applied Physics Technologies, Inc., McMinnville, OR 97128, USA*

The cathode requirements for free electron lasers (FELs) are high brightness and high peak current. Cold field emission cathodes fabricated from appropriate material have the potential of delivering the required current for a range of FELs. The electron beam emitted from the sharp tip of a single field emitter is by nature very bright. For applications requiring collimated, high brightness beams with very high peak currents approaching one Ampere, single field emitter tips may be the best cathode choice.

Historically, tungsten, W, and molybdenum, Mo, field emitters are used and are capable of relatively high field emission currents, but are very sensitive to contaminants and require ultra-high vacuum for stable operation. Silicon or gallium arsenide emitters are much more tolerant of the operating environment, but have a much lower field emission current limit.

Field emission cathodes made from refractory transition metal carbides have many desirable properties for this high-brightness and high-current application. The two most promising materials are hafnium carbide, HfC, and zirconium carbide, ZrC. These carbides have low chemical reactivity, are very stable thermally, and are extremely resistant to sputtering and diffusion. The work functions for these carbides are slightly more than 1 eV lower than for Mo or W. Field emitters have been made from single carbide crystals [1,2] and, in addition, these materials have also been used as thin-film coatings on Mo field emitters and Mo FEAs to improve emission property [3]. The following table summarizes relevant experimental data (the radius of curvature reported is based on SEM measurements).

Table I: Summary of Emission Measurements from HfC and ZrC

- Stable DC currents of 3 mA to 5 mA have been routinely obtained from single crystal ZrC and HfC tips at 1500 K with radius of curvature about 60 nm.
- Pulsed current of 49 mA has been obtained from a single crystal ZrC tip at 300 K with radius of curvature about 90 nm with pulse length of 10 μs and duty factors of 0.01 to 0.001.
- Stable DC currents about 0.5 mA have been obtained from single crystal Mo tips coated with ZrC at 300 K with radius of curvature about 30 nm to 60 nm.
- Calculated emission current densities are in the 10^8 A/cm^2 range at the tip.
- Lifetime >> 3000 hours at 5 mA in ultra-high vacuum has been demonstrated. Test was terminated not due to tip failure.
- Emitters have been operated in the 100-500 μA range while at pressures as high as 5×10^{-5} torr of air.

While DC field emission currents from a single carbide field-emitter have routinely been measured in the mA range, higher currents can be obtained from tips with a larger tip radius. In addition, lasers could be used to significantly increase the emission current. Photoelectric field emission or high field photoemission may be able to deliver a peak current from a single tip as high as one Ampere. These techniques are currently under investigation at Vanderbilt University [4-6] and at Brookhaven National Laboratory [7,8].

No experimental measurement of normalized emittance from a single gated tip has been reported, but indirect experimental data indicate that the normalized emittance may be better than $10^{-2}\pi$ mm-mrad [9,10]. An estimate of normalized beam brightness from a single tip is on the order of 10^{15} A/(m^2-steradian) at current I=10 mA [4,5].

Using designs obtained from 2D and 3D simulation codes, we have the ability to use extraction apertures and integrated focusing elements to provide beam confinement and efficient electron extraction for pulsed applications. We are currently working to develop electron guns using these materials and designed for FEL applications, for micro FELs and for the Infrared FEL at Thomas Jefferson National Accelerator Facility.

Acknowledgments

The work is supported by DoE SBIR program, Contract No. DE-FG02-98ER82560.

References

[1] W. A. Mackie, R. L. Hartman and Paul R. Davis, "High Current Density field-emission from transition Metal Carbides," Applied Surface Science 67 (1993) 29.

[2] W. A. Makie, R. L. Hartman, M. A. Anderson and P. R. Davis, "Transition Metal Carbides for Use as Field-Emission Cathodes," J. Vac. Sci. Technol. B 12 (1994) 722.

[3] T. Xie, W. A. Mackie and P. R. Davis, "Field Emission from ZrC films on Si and Mo Single Emitters and Emitter Arrays," J. Vac. Sci. Technol. B 14 (1996) 2090.

[4] C. A. Brau, "High-Brightness Electron Beams – Small Free-Electron Lasers", Proc. of the 19th Intl. Free Electron Laser Conf., Beijing, China, Nuclear Instrum. and Methods A407, (1998) 1-7.

[5] C. A. Brau, "Needle Cathodes as Sources of High-Brightness Electron Beams," AIP Conf. Proc. No. 413, Towards X-Ray FELs, eds. R. Bonafacio and W. A. Barletta, (1998) 277.

[6] C. H. Garcia and C. A. Brau, "Photoelectric Field Emission from Needle Cathodes," Abstract Book 20th Intl. FEL98 Conf., Williamsburg, VA, August 16-21, 1998, TU-1-03, manuscript to be published in this proceedings.

[7] T. Srinivasan-Rao, I. Ben-Zvi, J. Smedley, X. J. Wang, M. Woodle, D. T. Palmer, R. H. Miller, "Performance of Megnesium Cathode in the S-Band RF Gun," to be published in the Proc. of the 1997 Particle Accelerator Conference, Vancouver, Canada, May 1997.

[8] Triveni Srinivasan-Rao, personal communication.

[9] C. M. Tang and T. A. Swyden, "Beam collimation from field-emitter arrays with linear planar lenses," Society for Information Display Intl. Symposium Digest of Technical papers, Vol. XXVIII, (1997) 115.

[10] H. Ishizuka, H. Yarnazaki, H. Shimawaki, K. Yokoo, S. Kawasaki, A. Wantanabe and M. Shiho, "Performance of an Electron Gun Based on Gated Field-Emitter Arrays," Abstract Book 20th Intl. FEL98 Conf., Williamsburg, VA, August 16-21, 1998, TU-3-01.

FREE ELECTRON LASERS 1998
G.R. Neil and S.V. Benson (Eds.)
1999 Elsevier Science B.V.

Small-Angle Approximation for the Impedance due to Wall Surface Roughness

G. V. Stupakov
Stanford Linear Accelerator Center, Stanford University, Stanford, CA 94309

1 INTRODUCTION

The design of Linac Coherent Light Source (LCLS) at SLAC requires the energy spread of the beam to be less then 0.1% [1]. Longitudinal wakefields in the accelerator tend to redistribute the bunch energy and, if large enough, can cause degradation of the beam performance. It has been pointed out by Bane, Chao and Ng [2] , that one of the major sources of wakefields for the LCLS might be the wall surface roughness in the undulator. The model developed in Ref. [2] assumes that a rough surface can be represented as a collection of bumps of relatively simple shapes (hemisheres, half cubes, etc.), and the impedance is calculated a sum of impedances for each shape with account of density of the bumps per unit area.

In this paper, we develop a theory of impedance due to the surface roughness of a perfect conductor, using a small-angle approximation for the wall surface. Our final result is the expression for the impedance in terms of the spectral function of the surface profile. The formula represents the contribution of different scales, and can be used for estimation of the impedance based on statistical properties of the surface. A simple model of a fractal random surface is considered, and it is shown that the impedance is proportional to the rms square of the bumps height divided by the correlation length of the bumps. The result is compared to the Bane-Chao-Ng model.

2 ASSUMPTIONS AND STATISTICAL PROPERTIES OF A ROUGH SURFACE

The detailed derivation of the impedance of round pipe with rough surface can be found elsewhere [3]. Here we outline the main assumptions and present the final formula for the longitudinal impedance.

Our approach is based on the assumption of *small-angle approximation*. If we assume that the rough surface is given by equation $y = h(x, z)$, where x, y and z are the cartesian coordinates, and h is the local height of the surface, then the small-angle approximation means that the angle between the normal to the surface and the vertical direction (y axes) is much smaller than unity, or

$$|\nabla h| \ll 1 \,. \tag{1}$$

This assumption allows to develop a rather general theory of impedance, which with a good accuracy works even when ($|\nabla h| \sim 1$).

In addition to Eq. (1), we also require that the h[illegible] of the bumps and their characteristic size g be small compared to the radius of the pipe b_0,

$$g,\ |h| \ll b_0 \,. \tag{2}$$

Evidently, this inequality is easily satisfied for realistic values of g, h and b_0.

There is one more condition that simplifies the consideration. Typically, the size of the surface bumps g is on the order of microns, and the bunch length σ_z in FEL is of the order of at least tens of microns. This means that the characteristic frequency of interest $\omega \sim c/\sigma_z$ is small compared to c/g,

$$\omega \ll c/g \,. \tag{3}$$

To describe the statistical properties of a rough surface, we introduce the correlation function $K(x, y)$ such that

$$K(x - x', z - z') = \langle h(x', z')h(x, z)\rangle \,, \tag{4}$$

where the angular brackets denote averaging over possible random profiles $h(x, z)$. Eq. (4) assumes that statistical properties of the surface do not depend on the position, which is true due to the macroscopic nature of the roughness. An important statistical characteristic of the roughness is the *spectral density* (or *spectrum*) $R(\kappa_z, \kappa_x)$, defined as a Fourier transform of the correlation function,

$$R(\kappa_x, \kappa_z) = \frac{1}{(2\pi)^2} \int dx\, dz\, K(x, z) e^{-i\kappa_x x - i\kappa_z z} \,. \tag{5}$$

If the surface is statistically isotropic (all direction in the $x - y$ plane are statistically equivalent), the spectrum R depends only on the absolute value of the wave number $\kappa = \sqrt{\kappa_x^2 + \kappa_z^2}$, $R = R(\kappa)$.

The main result of Ref. [3] is that the longitudinal impedance of a circular pipe of radius b_0 with a rough perfectly conducting surface characterized by the spectral function $R(\kappa_x, \kappa_z)$ in the frequency range limited by the condition (3) is given by the following equation:

$$Z(\omega) = -\frac{ikZ_0}{2\pi b_0} \int d\kappa_z\, d\kappa_x\, R(\kappa_x, \kappa_z) \frac{\kappa_z^2}{\kappa} \,, \tag{6}$$

where now the z-axes is directed along the pipe axes, and the x axes is locally directed along the azimuthal coordinate θ. We note again, that due to assumed smallness of the surface structures, we can use the local Cartesian coordinate system x, y and z in Eqs. (4) – (6) instead of the global cylindrical coordinate system θ, r and z.

Eq. (6) tells that the contribution to Z of roughness in longitudinal (z) and azimuthal directions are not equal: the presence of the factor κ_z^2 in the integrand means that bellow-type variations on the surface are more dangerous than ridges on the surface going in the longitudinal direction.

3 SURFACE MODELS

As a model of a rough surface, we consider here a power spectrum, limited at small wavelengths, $R(\kappa) = A\kappa^{-q}$ for $\kappa > \kappa_0$, and $R(\kappa) = 0$ for $\kappa < \kappa_0$, where κ_0 is the minimal value of spectrum, $q > 0$ is a power factor, and A defines the amplitude of the roughness. The parameter κ_0 can be expressed in terms of the characteristic correlation length, l_c, of the random profile, $\kappa_0 \sim \pi/l_c$. We can also relate the factor A to the rms height d of the roughness, using the relation

$$d^2 = 2\pi \int_0^{\infty} \kappa \, d\kappa \, R(\kappa) = \frac{2\pi A}{q-2} \kappa_0^{2-q} . \qquad (7)$$

For the convergence of the integral we have to require $q > 2$. The shape of the surface for two different values of q

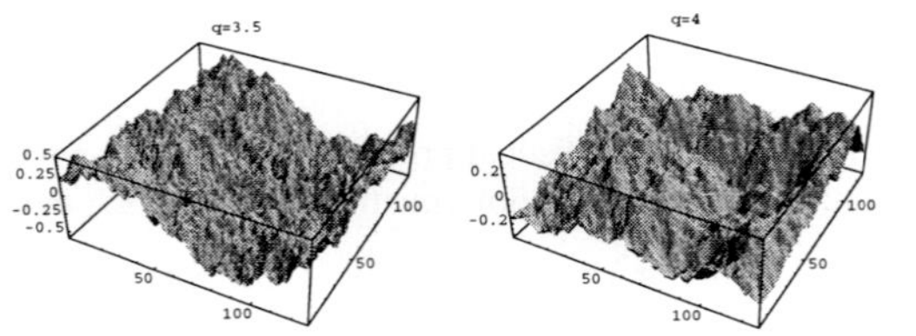

Figure 1: Fractal surfaces for $q = 3.5$ and $q = 4$. Smaller values of q give more "spiky" profiles.

obtained with a help of computer code described in [4] is shown in Fig. 1. It turns out, that increasing the value of q makes the surface smoother. Using Eq. (6) we can calculated the impedance of such surface,

$$Z(\omega) = -\frac{ikZ_0}{4\pi b_0} \frac{q-2}{q-3} d^2 \kappa_0 . \qquad (8)$$

Again, for convergence, we need to require that $q > 3$, otherwise the integral diverges as $\kappa \to \infty$. This requirement is stronger than the convergence condition for Eq. (7), and is due to a relatively slow decay of the spectrum at large κ.

4 COMPARISON WITH BANE-NG-CHAO MODEL

To compare our result with Ref. [1], we write down here the impedance from [1]

$$Z(\omega) = -\frac{ikf\alpha Z_0}{3\pi^2 b_0} a , \qquad (9)$$

where a is the height of the bumps, f is a form-factor that depends on the choice of specific shape modeling the bumps (f varies from about 5 to 20 for different shapes [2]), and α is the filling factor characterizing the number of bumps per unit area. For numerical estimate in [1] it was assumed that $f = 5$ and $\alpha = 0.5$. To compare this result to our model, we have to express the rms height d in terms of a; a simple calculation gives $d = \sqrt{\alpha/2}d$.

Returning to the fractal model, we will chose $q = 4$ as reasonable approximation for a real surface profile. A choice of the correlation length l_c that would be compatible with [1] requires that l_c be of the order of the bump's height, that is $\kappa_0 \sim \pi/d$. This reduces Eq. (8) to

$$Z(\omega) = -\frac{ikZ_0}{2b_0} d . \qquad (10)$$

Comparing Eqs. (9) and (10), we see that they have same scaling, but different numerical factors. Eq. (10) gives about three times larger impedance (for $\kappa_0 \sim \pi/d$) than quoted in [1].

We want to emphasize here that the right choice of the correlation length l_c is critical for the estimation of the impedance (8). Although we do not have detailed data for the roughness spectrum of a real surface (which, of course, depends on the particular manufacturing process involved), there are some indication, that in many cases l_c may be 10-30 times larger than assumed above [5]. If that assumption is correct, then the impedance would be about an order of magnitude smaller then estimated above.

5 CONCLUSIONS

We have developed a theory of impedance of perfectly conducting rough surface in small-angle approximation. The effect of finite conductivity is independent of the geometrical wake, and is additive to the one found in this paper. Using as an example a statistically fractal surface with a power spectrum, we calculated the longitudinal impedance as a function of statistical characteristics of the surface.

6 ACKNOWLEDGMENT

The author is thankful to K. Bane, M. Cornacchia and P. Emma for useful discussions.

This work was supported by Department of Energy contract DE-AC03-76SF00515.

7 REFERENCES

[1] Linac Coherent Light Source (LCLS) Design Study Report. SLAC-R-521, Apr 1998. 381pp.

[2] K.L.F. Bane, C.K. Ng and A.W. Chao. SLAC-PUB-7514 (1997).

[3] G. V. Stupakov, Preprint SLAC–PUB–7908, 1998.

[4] R. E. Maeder. *The Mathematica Programmer II*, Academic Press, San Diego, 1996.

[5] D. J. Whitehose. *Handbook of Surface Metrology*, IOP Publishing, 1994.

FREE ELECTRON LASERS 1998
G.R. Neil and S.V. Benson (Eds.)
1999 Elsevier Science B.V.

Rectifier for RF Power Recovery

M. A. LaPointe,*# Changbiao Wang,# and J.L. Hirshfield*#

*Omega-P, Inc., 202008 Yale Station, New Haven, CT 06520-2008
#Daprtment of Physics, Yale University, P.O. Box 208124, New Haven, CT 06520-8124

One method to increase the efficiency of free-electron lasers is to recover some of the spent beam power by converting it to rf power as the beam excites a superconducting cavity, as in the CEBAF FEL. The rf power must then be rectified to dc or pulsed power which can be re-circulated to decrease the required wall-plug power. A method to achieve this rectification is to accelerate a low voltage, high current beam and then to use a single-stage depressed collector to deliver the beam power to an external circuit. This paper describes experiments underway at Yale University to demonstrate such a rectifier.

A cyclotron autoresonance accelerator (CARA) currently in use at Yale has demonstrated the transfer of rf power to electron beam energy with very high efficiency[1]. The proposed rectifier under construction uses a 2.856 GHz CARA as the low voltage, high current accelerating structure, and a single-stage depressed collector is being developed to recover energy from the accelerated beam. It should be noted that the 2.856 GHz frequency used for these experiments is due to the availability of a former SLAC XK-5 klystron installed at Yale. Scaling to the CEBAF frequency of 1.5 GHz is straightforward.

RF absorption

CARA was run at rf levels of 0.23, 0.8 and 1.2 MW to determine the viability of CARA at absorbing relatively low power levels. Prior experiments had been performed at 3-8 MW of input rf power. Fig. 1 shows the 1.23 MW case where 97.4% of the incident power was absorbed by the 78.5 kV, 22 A beam. 32 kW of power was reflected from the input coupler (2.5%), which gave a net 95% absorption of the incident rf. For this case the final beam voltage was 131 kV. The other cases, not shown here, also had overall rf-to-beam efficiencies in excess of 90%

One consideration for an installed system is the manner in which it responds to variations in input power. CARA absorption was optimized for 1.2 MW of input rf power and then, without reoptimizing the beam or magnetic field parameters, the input rf was varied by ± 20%. As is seen in Fig. 2 the efficiency drops by only 2% (from 94% to 92%) for a ± 10% variation in input power. As the power level is changed by ± 20% the absorption drops to about 80%.

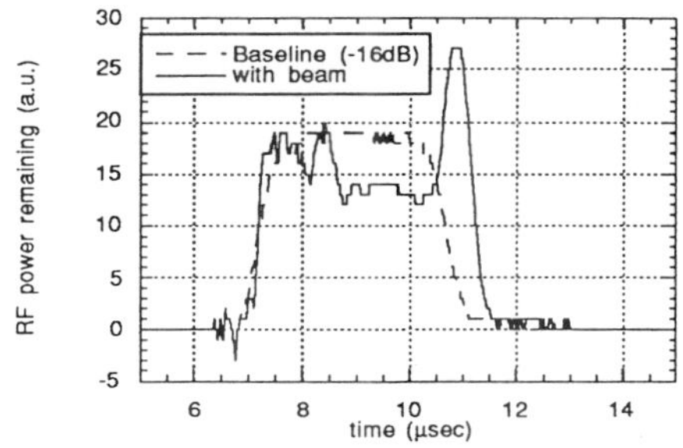

Fig.1. Absorbtion measurement at the end of CARA, showing in excess of 97% of the power incident on the beam to be absorbed.

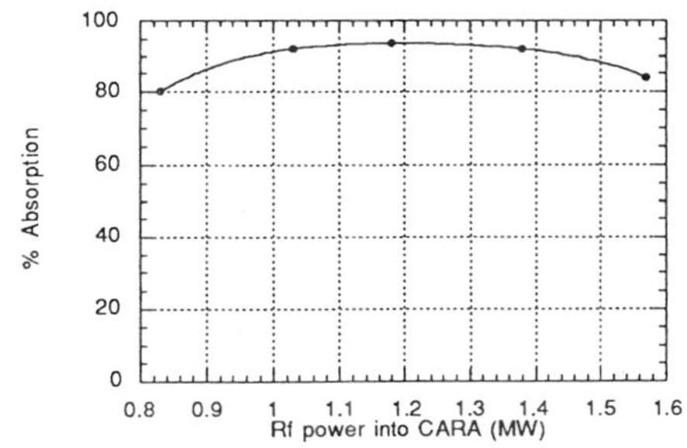

Fig. 2. Absorption efficiency as input rf is changed, showing the relative insensitivity of CARA to ±10% changes in rf conditions.

Conversion efficiency

The overall rectifier efficiency η_{rf} for converting rf to dc is given by

$$\eta_{rf} = \frac{P_{coll} - P_{bi}}{P_{rf}} \tag{1}$$

where P_{coll} is the power recovered on the depressed collector, P_{bi} is the initial beam power and P_{rf} is the incident rf power to be rectified. If one defines the collector efficiency as $\eta_{coll} = P_{coll}/P_{bf}$ and the final beam power as $P_{bf} = P_{bi} + \eta_{CARA} P_{rf}$ where η_{CARA} is the efficiency of CARA, then one can obtain the following relation for the overall collector efficiency

$$\eta_{rf} = \eta_{coll}\eta_{CARA} - (1 - \eta_{coll})\frac{P_{bi}}{P_{rf}} \qquad (2)$$

From Eq. 2 it is seen that the maximum efficiency is given, as expected, by the product of the CARA and collector efficiencies with a correction due to the use of the electron beam. What is also evident is that to maximize the overall efficiency the ratio of beam power to rf power should be minimized.

Simulation results

Simulations were performed using both cold and warm beams to model both CARA and the depressed collector. For a cold beam the collector efficiency is near 100% and the overall rf conversion efficiency is limited by the CARA efficiency as seen in Eq 2. Simulations were also run for a warm beam with initial rms axial velocity spreads on the electron beam ranging from 0.4% to 1%. For all the simulations the initial beam was taken as 10A, 46.4 kV

CARA detuning	CARA γ_{min} (kV)	CARA end $\Delta\gamma$ (kV)	CARA end $<\gamma>$ (kV)	η_{CARA} (%)	depressed voltage (kV)	η_{rf} (%)
0	121.2	40.40	146.2	99.75	118.2	71.39
-0.03	131.6	21.02	145.2	98.82	128.6	81.75
-0.04	134.8	15.67	144.0	97.55	131.8	84.94
-0.05	134.1	12.56	141.7	95.34	131.1	84.24

Table I: Rectifier simulation using a 0.5% rms axial beam spread on the initial 10A, 46.4 kV beam.

Table I shows the simulation results for an axial velocity spread of 0.5% for the initial beam. It has been shown previously that slight detunings of the magnetic field away from exact resonance in CARA can greatly limit the velocity spread growth during the acceleration process with only slight degradation in the absorption efficiency.[2] As seen in Table I, for a detuning parameter of -0.04, a maximum rf conversion efficiency of nearly 85% can be achieved. For initial axial velocity spreads of 0.4% and 1.0%, the maximum rf efficiencies are 86% and 79% respectively, all with a detuning parameter of -0.04.

The simulations of the collector have led to a preliminary design indicated in Fig. 3. The simulations shown are for a collector in which primary electrons require a minimum of three reflections to reverse their direction and escape into the retarding region The solid angle for low energy secondary electrons to escape into the retarding region, where they would be accelerated back toward CARA, is small. The final design will minimize the effect of secondary electrons.

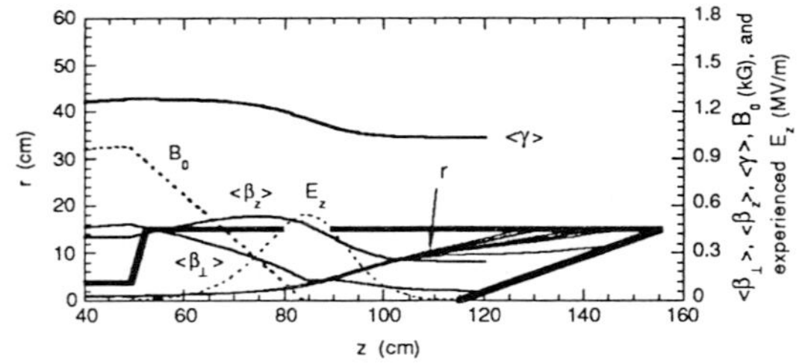

Fig. 3. Preliminary design of the single stage depressed collector showing the particle trajectories, energies, axial magnetic field and retarding potential.

Summary

It has been demonstrated experimentally that the expected 1 MW of rf power generated from the slowing down of the spent CEBAF FEL beam could be absorbed by CARA with efficiencies up to 95%. Simulations have indicated that the accelerated beam from CARA could be rectified using a single stage depressed collector with rf–to–dc efficiencies as high as 85%.

Acknowledgement

This work sponsored by the Department of Energy, Division of High Energy and Nuclear Physics.

References

[1] J. L. Hirshfield, M. A. LaPointe, A. K. Ganguly, R. B. Yoder and Changbiao Wang, *Phys. Plasmas*, **3**, 2163 (1996); M. A. LaPointe, R. B. Yoder, Changbiao Wang, A. K. Ganguly and J. L. Hirshfield, *Phys. Rev. Lett.* **76**, 2718 (1996).

[2] J.L. Hirshfield, Changbiao Wang, and A.K. Ganguly, *IEEE Trans. on Plasma Sci.*, **24**, 825 (1996)

FREE ELECTRON LASERS 1998
G.R. Neil and S.V. Benson (Eds.)
1999 Elsevier Science B.V.

Electron Bunch Profile Measurements with 300 Femtosecond Resolution

K. N. Ricci, T. I. Smith, E. R. Crosson

Stanford Picosecond FEL Center*
W. W. Hansen Experimental Physics Laboratory
Stanford University
Stanford, CA 94305-4085, USA

Abstract

Using the longitudinal dispersion of a magnetic chicane, and off-phase acceleration in a radio frequency electric field, it is possible to transform a highly relativistic electron bunch such that a temporal slice from the initial phase-energy space is converted into an energy slice of the final phase space. An energy spectrometer can be used to analyze the final energy distribution, revealing the longitudinal density profile of the initial electron bunch. We estimate the limits of resolution attainable with this technique, and we present bunch profile measurements from the Stanford Superconducting Accelerator front end with a resolution better than 300 femtoseconds.

1. Introduction

Extremely short relativistic electron bunches are usually measured in the frequency domain using spectral analysis of coherent radiation,[1] or they are measured in the time domain using a streak camera or radio frequency (RF) accelerator phasing techniques.[2] Streak cameras have only recently achieved truly sub-picosecond resolution.[3] The expense and effort of installing such a streak camera have led many facilities to prefer off-phase RF acceleration because the required instruments—accelerator phase and amplitude controls, and a beam energy spectrometer—are already in operation in most accelerator systems. In its simplest form, the off-phase acceleration (OPA) technique requires running a section of the RF accelerator at such a phase that the electric field is ramping up or down rapidly while a relativistic electron bunch passes through. Electrons entering the OPA section receive an energy increment that varies in proportion to their time of arrival. The final energy distribution, measured in a spectrometer, should resemble the temporal profile of the initial bunch if the energy spread imparted by the OPA is much greater than the initial energy spread. However, for many linear accelerators the initial energy spread of the beam is large enough that sub-picosecond bunch profiles cannot be measured this way.

* Work supported in part by ONR Grant N000140-94-1-1024.

Here we demonstrate a modification of the OPA method, suggested three years ago,[4] that can significantly improve temporal resolution. The beam passes first through a longitudinally dispersive magnetic chicane and then into the OPA and spectrometer. The dispersion of the chicane is set so that each time slice of the initial beam phase space will map to a unique energy in the final phase space, regardless of its initial energy spread. This has enabled us to measure electron bunch profiles at the Stanford Superconducting Accelerator (SCA) with a resolution better than 300 femtoseconds.

2. Theory of the Technique

The longitudinal motions of electrons within a bunch will be described here using the coordinates t (time) and E (energy) relative to the central position and mean energy of the bunch. As the bunch passes through the chicane and OPA, its coordinates transform according to the linear transport matrices in equation (1):

$$\underset{\text{OPA}}{\begin{bmatrix}1 & 0\\ \alpha & 1\end{bmatrix}}\underset{\text{chicane}}{\begin{bmatrix}1 & \delta\\ 0 & 1\end{bmatrix}}\underset{\text{initial}}{\begin{bmatrix}t\\ E\end{bmatrix}}=\underset{\text{final}}{\begin{bmatrix}t+\delta E\\ \alpha t+(1+\alpha\delta)E\end{bmatrix}}, \tag{1}$$

where δ is the longitudinal dispersion of the chicane in picoseconds/keV, and α is the rate of change of the OPA in keV/picosecond. Notice that if the chicane and OPA are adjusted so that $\alpha\delta = -1$, then the final energy coordinate will be $E_{final} = \alpha t_{initial}$, and a measurement of the final energy distribution will yield the temporal distribution of the initial bunch. In effect, the chicane disperses the initial beam in just such a way that the OPA removes the initial energy spread from each temporal slice of the phase space. To first order, such a profile measurement has resolution

$$\Delta t = \Delta E_{spec} / \alpha \,, \qquad (2)$$

where ΔE_{spec} is the energy resolution of the spectrometer. In conventional OPA measurements without the chicane, $\delta = 0$ in equation (1) and the final energy distribution contains the temporal profile convolved with the initial energy spread ΔE_{beam}. Then the profile measurement is degraded to a resolution $\Delta E_{beam}/\alpha$ at best, if $\Delta E_{beam} >> \Delta E_{spec}$ as is usually the case. Note that adding even a small amount of longitudinal dispersion before the OPA can improve this situation by changing the effective beam energy spread to $(1+\alpha\delta)\Delta E_{beam}$.

3. Measurements at the SCA

Figure 1 shows electron bunch profiles measured with this method. The SCA front end accelerated a beam with $4.2\text{x}10^7$ electrons per bunch to 19 MeV, with an initial energy spread of 130 keV FWHM (full width at half maximum) in the spectrometer. The beam is typically bunched to a few picoseconds or less in the SCA front end to provide high current density for the operation of a free electron laser (FEL). Magnetic chicanes provided –0.012 ps/keV longitudinal dispersion at 19 MeV. The final sections of the 1.3 GHz RF accelerator were run off-phase to provide an acceleration that ramped up at 84 keV/ps.

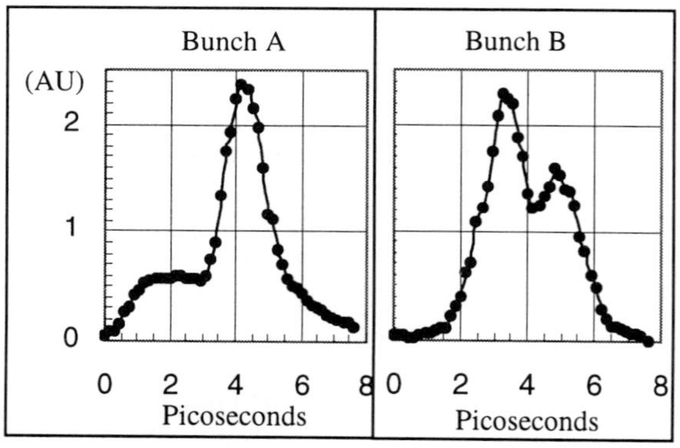

Figure 1. Profiles of electron bunches from the SCA front end. Bunch A was generated with the SCA configured for FEL operation. Bunch B was then produced by adjusting the phase and amplitude of the buncher and front end accelerator structures.

At these settings, the final beam energy was 27 MeV and the profile resolution in a 0.08% energy spectrometer was approximately 260 fs, from equation (2).

Bunch A of Figure 1 was produced with the SCA in a near-optimal state for FEL operation. Notice that most of the current is bunched tightly in the central peak with a FWHM of 1.4 ps, but a significant amount remains in a 3 ps long lobe at the front of the pulse. Bunch B was produced by monitoring the pulse shape in the spectrometer while repeatedly adjusting the SCA front end RF phases and amplitudes in an effort to create a structured bunch. A wide variety of interesting pulse shapes can be obtained this way.

4. Discussion

With higher acceleration gradients and better spectrometer resolution, this measurement could be extended to much shorter time scales. For example, consider a 100 MeV beam passing through a chicane and into a 10 MeV section of a 10 GHz linear accelerator running 90° off-phase. With α=630 keV/ps, and a 0.01% energy spectrometer, equation (2) gives a resolution of 16 fs. On these time scales, care must be taken to consider second order coupling from transverse phase space and nonlinear longitudinal dispersion. These calculations have been presented elsewhere.[5] At the SCA, the nonlinear chicane dispersion is about 30 fs, while transverse focusing produces errors on the order of 90 fs. This transverse focusing limit may be difficult to overcome with the typical SCA emittances of 10 π mm mrad. However, in newer accelerators with an order of magnitude less emittance, these terms will be much smaller, and bunch profiles with less than 30 fs resolution may be achieved.

We expect that this bunch profile diagnostic technique will help in optimizing bunch length and shape in accelerators where ultra short electron bunches are required. In the near future we will use this method at the SCA to evaluate and to supplement our coherent radiation diagnostics for bunch length and longitudinal phase space.

References

[1] Gi. Schneider, et al, NIM-A 396 (1997) pp. 283-292.
[2] D. X. Wang, et al, Phys. Rev. E 57 (1998) pp. 2283-2286.
[3] A. Takahashi, et al, "New Femtosecond Streak Camera," in *Proceedings of SPIE* vol. 2116 (1994) pp. 275-284.
[4] E.R. Crosson, et al, "Technique for Measuring an Electron Beam's Longitudinal Phase Space with Sub-picosecond Resolution," in *AIP Conf. Proc.* vol. 367 (1996) p. 400.
[5] K.N. Ricci, T.I. Smith and E.R. Crosson, "Sub-picosecond Electron Bunch Profile Measurement Using Magnetic Longitudinal Dispersion and Off-Phase RF Acceleration," submitted to *Advanced Accelerator Concepts Workshop, AIP Conf. Proc* (1998).

FREE ELECTRON LASERS 1998
G.R. Neil and S.V. Benson (Eds.)
1999 Elsevier Science B.V.

Complete characterization of UV dielectric multilayer mirrors for performance improvement of Free Electron Lasers

D. Nutarelli* (a,b), M-E Couprie(a,b), E. Renault(a,b), R. Roux(b), L. Nahon(a,b), A. Delboulbé(b), C. Boccara(c), M. Billardon (b,c)

a) *Service Photons, Atomes et Molécules; CEA/DSM/DRECAM Bat.522 91191 Gif sur Yvette France.*
b) *Laboratoire pour L'Utilisation du Rayonnement Electromagnétique (LURE), Bât. 209D, BP 34 91898 Orsay Cédex, France.*
c) *Ecole Supérieure de Physique et Chimie Industrielle (ESPCI), 10 rue Vauquelin 75231 Paris Cédex, France.*

Abstract

The Free Electron Laser operation in the UV for user applications at short wavelengths requires dielectric multilayer mirrors with very low initial loss. Complete characterization of UV mirrors is described and new Super-ACO FEL performance is reported.

1.Introduction

Ultra-violet Free-electron Laser operation requires high reflectivity dielectric multi layer mirrors.

The total losses (P) for such a mirror are given by the relation (1):

$$P = T + A + D \qquad (1)$$

with T the transmission, A the absorption and D the scattering losses.

The choice of the number of layers determines the mirror transmission, which can thus optimize the FEL extracted power de-pending on the gain and total losses. Low values of the transmission maximize the intracavity laser power for Coherent Harmonic Generation produced by the FEL.

2. Complete characterization system

In order to accurately characterize the initial loss and mirror performance each component of eq. (1) should be measured separately. At Super-ACO, the transmission is measured with a CARY spectrometer, and the total losses are measured with an accuracy of about 5% using the well-known (since 1980) phase shift method [1-2]. More recently (1997), the absorption has been directly measured (down to 10 ppm), with a system based on the mirage effect [3, 4]. The roughness, which is responsible for the scattering losses, has first been measured (down to 20 pm) with a uni-dimensional interferometric profilometer (1992) [5]. More recently (1998), a two-dimensional profilometer system has been developed so two-dimensional roughness pro-files can be acquired over a surface area of about 1 mm^2. Direct measurements of the total scattered light on substrates have also been made in collaboration with the optical laboratory of the "Ecole Nationale Superieure de Physique de Marseille" [3, 6]. Nevertheless, at Super-ACO, the scattering losses are mainly evaluated from the roughness measurements.

2.1 Surface roughness and scattering losses

According to the "Scalar Theory" from Bennet and Porteus [7], the following relation gives the scattering losses:

$$D = \left(\frac{4\pi}{\lambda}\delta\right)^2 \qquad (2)$$

where λ is the wavelength of the incident light and δ is the roughness defined as RMS value of the mirror surface profile. Nevertheless, for a dielectric multilayer mirror, eq. (2) cannot be simply applied using the surface roughness because the multilayer structure leads to multiple scattering at the interfaces. Assuming an interface roughness δ, and transmission T_{in}, the total scattering of the N multilayers (with high n_h and low n_l indices) is:

$$D = \left(\frac{4\pi}{\lambda}\delta\right)^2 \frac{1 - T_{in}^N}{1 - T_{in}} \qquad (3)$$

with $T_{in} = 4n_h \cdot n_l / (n_h + n_l)^2$ In fact, the scattering losses increase with shorter wavelengths, and depend also on the number of layers and on the layer materials.

* Corresponding author : D. Nutarelli. Tel 33 1 64 46 81 42 fax 33 1 64 46 41 48 e-mail "nutarelli@lure.u-psud.fr"

The scattering losses of an Al_2O_3/SiO_2 mirror is thus 7 times larger than that of a Ta_2O_5/SiO_2 mirror of identical transmission.

3. Experimental results

Figure 1 illustrates the relation between the scattering losses and the roughness. The scattering losses have been estimated experimentally, from the independent measurements of the total losses, transmission and absorption ($D = P - (T + A)$), and calculated by using eq. (3). The scattering loss follows a parabolic variation as a function of roughness, $D = a \cdot \delta^2$, with $a = 2.5 \cdot 10^{-4}$ $[Å]^{-2}$. This study gives a good criterion to select the substrates before layer deposition.

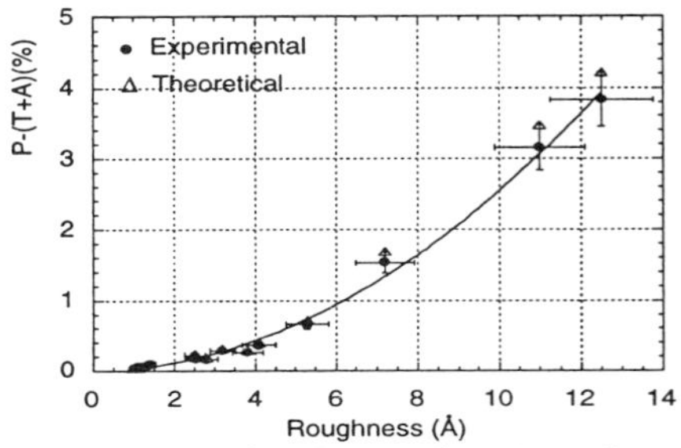

Figure 1. Scattering losses as a function of roughness for a set of Ta_2O_5/SiO_2 dielectric multilayers mirrors at the wavelength of 350 nm. (•): Experimentally estimated. (Δ): Calculated from the roughness values.

4. Consequences for the FEL

The tunability range and output power of the FEL depend on the difference between maximum laser gain and optical cavity loss. Good optics are thus very important for FEL performance. In addition, for a given value of the transmission, the output laser power dramatically decreases if A and D increase.

A 500 MHz RF harmonic cavity has been installed in the Super-ACO storage ring providing an enhancement of the laser gain by a factor 1.5 to 3 (depending on the 500 MHz RF voltage) with respect to that obtained with the 100 MHz cavity. So mirrors with a transmission of 0.1 % have been installed providing an output power of 300 mW at 70 mA of stored current, the highest output power of any UV FEL. Figure 2 shows the enhancement of the output power due to the gain and transmission optimization. Considering the estimated gain with the 500 MHz cavity and the value of the useless losses (A+D), an operation with the optimized value of the transmission may provide an output power up to about 800 mW available for user applications.

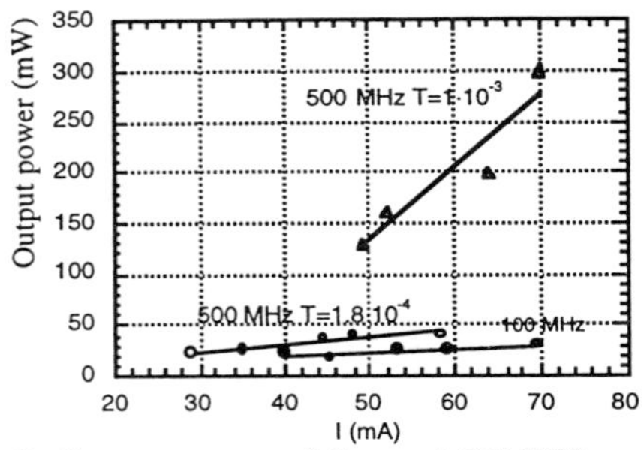

Figure 2. Output power of Super-ACO FEL versus the total stored current. The values are measured for different operation: (•) the 100 MHz RF cavity, (o) 100 and 500 MHz RF cavities for the same set of mirrors (transmission of $1.8 \cdot 10^{-4}$), (Δ) 100 and 500 MHz RF cavities with a set of more transparent mirrors (transmission of $1 \cdot 10^{-3}$).

References

[1] J.M. Herbelin, J. A. Mckay, M. A. Kwok, R. H. Ueunten, D. S. Urevig, D. J. Spencer and D. J. Benard, *Appl. Opt.*, **19**, 144-147, (1980)

[2] M. Billardon, M.E.Couprie, J.M. Ortéga et M.Velghe*Applied Optics* **vol 30**, No. 3, 344-351, 1991.

[3] D. Garzella, PhD Thesis at the "University Paris Sud" April 1996.

[4] J.P. Roger, F. Charbonnier, D. Fournier, C. Boccara, P. Robert, Proceedings of the International Conference on Lasers 1988, Mc Lean Editor, STS Press, VA, USA (1989)

[5] P. Gleyzes, Ph.D. thesis, University of Paris Sud (1993).

[6] C. Amra, C. Grèses-Besset, S. Maure, D. Torricini, Optical Interference Coatings, F. Abelès Editor, Proc. SPIE 2253, pp. 1184-1200 (SPIE, Washington, 1994).

[7] H.E. Bennet and J.O. Porteus, J. Opt. Soc. Am. 51 (1961) 123.

FREE ELECTRON LASERS 1998
G.R. Neil and S.V. Benson (Eds.)
1999 Elsevier Science B.V.

A FELI FEL beam profile and beam sharing system

M. Yasumoto[a], N. Umesaki[a], T. Tomimasu[b], Y. Kanazawa[b], A. Zako[b]

[a]*Osaka National Research Institute, AIST, 1-8-31 Midorigaoka, Ikeda Osaka 563-8577, Japan*
[b]*Free Electron Laser Research Institute, Inc. , 2-9-5 Tsuda-yamate, Hirakata Osaka 573-0128, Japan*

Abstract

A new beam sharing system with a 90°-opening angle fan-shaped mirror has been developed to allow for simultaneous FEL experiments at two or more different stations and to allocate the FEL machine time effectively. Through the FEL profile measurement the possibility of the simultaneous experiment with the system is shown.

1. Introduction

Free Electron Laser (FEL) has so attractive properties such as wavelength flexibility that FEL users have been increasing in Japan. The machine time thus should be allocated effectively. The FELI has been open for about twenty user groups belonging to company, university and national laboratory, and is operating four FEL facilities (FEL-1 ~ FEL-4), which cover the spectral range from 0.27μm to 40μm [1]. There are four user rooms (Lab.1~Lab.4) and twenty-one user stations. The infrared FEL is transported to the user stations with two vacuumed optical lines. The user stations are ranked in the optical line for the series, therefore, by monopolizing one oscillator on each optical line, since it is correspondent to one oscillator, only one user will use it. An FEL beam sharing system with a fan shaped mirror has been carried out as a trial attempt [2]. In this paper (1) to propose a new FEL beam sharing system with a fan shaped mirror and (2) to prove the possibility of simultaneous experiments with the system are written through the discussion of the FEL mode and profile measurements.

2. Beam transport and sharing system

The infrared FEL beam come out from a 1.5mm pinhole of a center of a hole coupling optical cavity (FEL-1) and 50~80m long transported in the optical pipe. The extracted FEL beams are modified to a near-parallel light with a parallel beam modification unit composed of a spherical mirror in order to prevent to lose it in the optical line. A monitor room (MR) and four user stations in user rooms (Lab.4, Lab.3, Lab.2 and Lab.1) are located at 53m, 54m, 61m, 69m and 82m distance from the FEL-1, respectively. The beam sharing system is presently installed on the No.1 optical line at the MR. The sharing mirror is made of a 90° part of 100mm diameter copper disk coated with gold on the reflection surface. At present a quarter of the FEL reflected with the mirror is delivery for a diagnostic station or an ablation experimental station [3]. At the while, three quarters of the FEL beam passed through the system is then delivered to the user stations (Lab.1~4).

3. FEL beam profile and GCR measurement

The FEL beam profile at the user station depends on an optical transverse mode (profile) in the coupling optical cavity [4] and on diffraction at the pinhole of the upstream mirror. Although there are many reports of the optical profile analysis in the coupling optical cavity [5,6], there is few reports of the laser property at the user station. The beam profiles at user stations are complicated, however it is essential for the FEL users to estimate the irradiated FEL power density. It is reported from an ordinary laser beam-mode analysis that Gaussian beam has a basic profile which is lowest order mode (TEM_{00}, Gaussian shape). The higher mode laser such as TEM_{01}, TEM_{10} has more complicated profiles. In this paper we thus define the term "Gaussian component ratio (GCR)" to refer to Gaussian profile correlation-rate comparing with the perfect Gaussian beam. The quantitative correlation (A) can be defined in the equation of $A_{min}=\sum_{x}\sum_{y}(Z_{xy}-S_{xy})^2$, where

Z_{xy} = amplitude of the pixel data at (x, y) and S_{xy} = amplitude of fitted Gaussian surface at (x, y). The GCR means a relative value for how well the profile matches the fitted Gaussian surface. In other word, the higher value of the GCR means a better Gaussian shape than the smaller GCR. The lower-order mode of the optical transverse mode matches the Gaussian fit and means the highest value of the GCR.

In order to analyze the FEL transverse mode, three dimensional (3D)-profiles of the FEL beam were measured with a 3D - pyroelectric detector system (Spiricon Pyrocam II & LBA200) at the MR and the user rooms. The detector has 32×32 elements in 2-inch (51.2mm) square detectable area and 12bit resolution dynamic range. Fig.1 shows 3D-profiles of the FEL, which are measured at the Lab.4, when the FEL wavelength is at (a)6.4μm (E=28.6MeV, K=0.668) and (b) 8.8μm (E=28.6MeV, K=1.166). Fig.1 (a) and (b) have 69% and 75% at the GCR, respectively. The higher GCR FEL (Fig.1 (b)) shows more concentric profile than the lower one (Fig.1 (a)). The higher GCR FEL forms an Gaussian profile, which is suited for high power density irradiation experiments. It is because that the Gaussian profile is most concentrated of the power at the peak compared with other profile FEL. On the contrary, the lower GCR FEL resembles the top-hat beam laser in the profile. The GCR FEL is good for the uniform irradiation experiment.

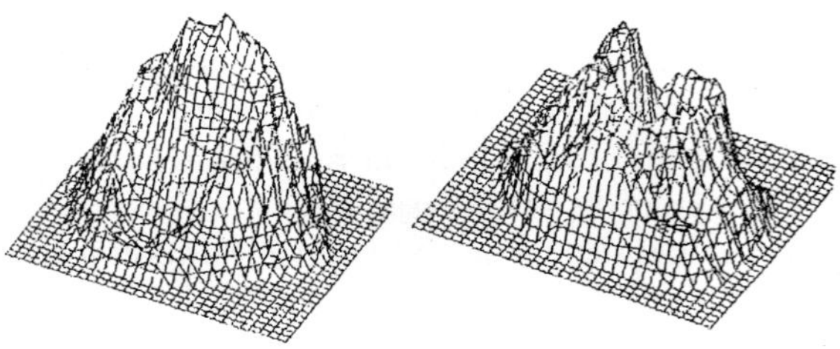

(a) 6.4μm (b) 8.8μm

Fig.1. FEL 3D-beam profiles.

In order to estimate the sharing effect on the beam, the GCR of the 270° shared and normal FEL were measured and plotted in Fig.2. A solid and dotted line indicate the GCR of the shared and normal 9.5μm FEL as a function of the distance from the FEL-1, respectively. The sharing system is located 52m apart from the FEL-1 facility. Although the normal beam GCR gradually decrease 82% to 52%, the shared beam GCR also decrease 65% to 52%. The spherical aberration due to the spherical mirror at the parallel beam modification unit is gradually appeared on the normal beam profiles according to the distance. After about 30m flight path from the sharing system, which means 82m apart from the FEL-1 facility, the shared beam is the same value as that of the normal beam in view of the GCR. This means the shared beam becomes the same shape as the normal beam after 30m-flight path.

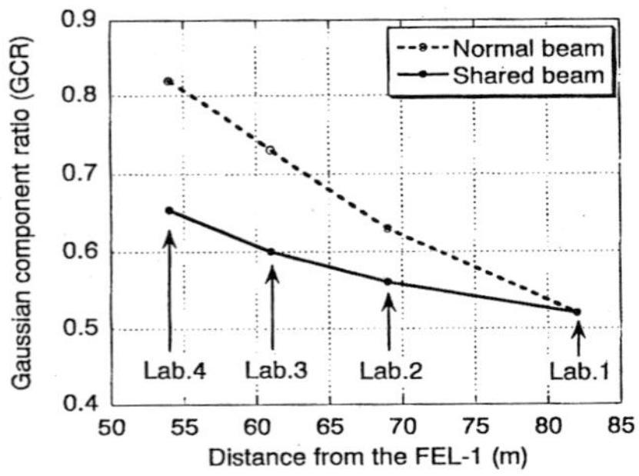

Fig.2. Gaussian component ratio (GCR) for the normal and 270°shared beams as a function of distance from the FEL-1 facility.

4. Conclusion

The beam profile of the shared beam recovers after 30m-flight path as the normal beam in view of the GCR. The sharing system therefore can be adapted for every experiment which does not need the GCR at any place and for other experiments needing the GCR at any place after 30m flight path.

References

[1] T. Tomimasu et al., Nucl. Instr. and Meth. A375 (1996) 626.

[2] S. Okuma et al., Nucl. Instr. and Meth. A375 (1996) 654.

[3] M.Yasumoto et al., Nucl. Instr. and Meth. B144 (1998) 176.

[4] K. Saeki et al., Nucl. Instr. and Meth. A 375 (1996) 12.

[5] B. Faatz et al., Nucl. Instr. and Meth. A331 (1993) 636.

[6] R. Prazeres et al., Nucl. Instr. and Meth. A341 (1994) 54.

FREE ELECTRON LASERS 1998
G.R. Neil and S.V. Benson (Eds.)
1999 Elsevier Science B.V.

Construction and adjustment of a wiggler for Storage ring FEL

Liu Bo, Ma Youwu, Lu Yuzhu, Zhang Zhaoming,
Zhang Fengming, Zhang Lijie, Chui Zhongwei, Hu Jinquan, Pu Dexiu
China Institute of Atomic Energy, Beijing, China

1. Introduction

Owing to the good beam qualities, the synchrotron radiation storage ring has become a drive source for shorter wavelength FEL. For the researches on this aspect the National Laboratory of Synchrotron Radiation (NLSR) of University of Science and Technology of China (USTC) has had a project to make experiments utilizing a straight section with limited length on the storage ring [1]. The key device for the project is the wiggler or optical klystron. This wiggler has three sections: modulation section, dispersive section and gain section. Modulation section and gain section has same structure while the dispersive section between them has higher peak field. In the modulation section energy modulation is taken place owing to the interaction between electron beam and laser field. After the electrons pass through the dispersive section, energy modulation is changed to density modulation, that is bunching, the coherent radiation can enhanced, so the gain is increased.

The whole construction the wiggler including field distribution design, structure building and field adjustment was taken in the linac laboratory of China Institute of Atomic Energy (CIAE). During the construction we consider the compatibility between SR-FEL research and SR operation. We design the device have two working modes: one is the optical klystron and another is as normal insert SR undulator. The working modes can be switched easily.

2. Basic requirements

According to the physics parameters of the electron beam in the storage ring, the following basic requirements were given for this wiggler:

Configuration type	Halbach type with permanent magnet
Dimension of magnet unit	18mm×18mm×100mm
Peak Fields	B_0=0.3T (gap = 40mm, for modulation and gain section), B_0=05~0.7T (gap=40 for dispersive section)
Period Length	λ=72mm (for modulation and gain section) λ=216mm (for dispersive section)
Number of period	N=12 (for modulation and gain section), N=1 (for dispersive section)
Gap	gap=25mm~150mm,variable
Errors for Peak Field	±1.0% (when gap=40mm)
The 1st integral at the exit	≤0.005 T·cm
The 2nd integral at the exit	≤0.05 T·cm^2

3. Fields distribution design

For Halbach type wiggler the peak field at the central axis is determined as follows:

$$B_0 = 1.8B_r(1-e^{-2\pi h/\lambda})e^{-\pi g/\lambda}$$

Based on the requirements above the material for the magnet unit must have enough Br and Hc. We selected NdFeB (N35 type) as the magnet material. For this material its remanence field Br=(1.17~1.20) T, the coercive force Hc=(11.0~11.5) KOe. The dimension of each magnet pole is 18mm×18mm×100mm. Because of the production reason, the pole must consist of two smaller magnet block. By means of gluing technique, the block units for magnet poles are formed.

And then we measured the field distribution for each block unit along its transverse central line, 20mm apart from its surface. From these measurements we found that among the all used magnet block units the diversity is ±3% for Br and is ±3° for the magnetization direction. We used the PANDIRA code to do the simulation calculation for the fields distribution of the wiggler [2]. The calculation results show that if the diversity for Br and magnetization direction is respectively ±3% and ±3°, it will be hard for the peak fields to reach 0.3T in modulation and gain section. So we must do optimization for the arrangement of each magnet block unit to decrease the effect of the diversity of Br and magnetization direction. Our optimization procedure includes two steps. In the first step, we arrange manually the magnet block according to these criteria: 1) for the magnet poles where the peak field is located, we select the block units which enable the magnetization direction errors of up and down poles will have same sign and errors of nearby peak poles will have different sign after assembly, 2) for the magnet poles where the zero field is located we select those block units which enable the magnetization direction errors of up, down and nearby poles have relative different sign. On the basis of the step one, we used simulated annealing optimization technique in step 2 to determine the position of each magnet block unit. Through the simulated calculation, we can see obviously the effect of the optimization.

4. Mechanical structure

The mechanical structure of the wiggler is C type. Each magnet block is glued in an aluminum slot to form a magnet block unit. All magnet units are fixed to the up and down main girder. The basement of wiggler is frame with the dimension of 2000mm×400mm×900mm. The three sections, modulation section, dispersive section and gain section, have their own supporter. The position of each supporter can be adjusted separately. We used the ball leading screw mechanism to change the working gap of the wiggler. For modulation and gain section, two sets of ball leading screws and three sideways are used. For the shorter dispersive section, we used one ball leading screw and three sideways. By this kind of structure the mechanical central plane can be guaranteed during the adjustment of the gap. The allowed gap adjustment arrange is 25mm~150mm. We have designed a special mechanical structure by which the dispersive section can be easily switched to another section that has same structure as modulation and gain section. After the change, this wiggler will serve as a normal SR undulator in storage ring.

5. Fields adjustment

After assembly of whole wiggler we had made the measurement and adjustment on the fields distribution. The apparatus we used is automatic magnetic fields measurement system. In this system, the magnetic field is measured by the hall probe and the data acquiring and processing is controlled by the computer [3]. Because the longest distance can be measured by the system is 1700mm while the range need to be measured of this wiggler is 2100mm, so the measurement and adjustment of the wiggler is carried out in two steps. We divided the wiggler in two ranges for the measurement and adjustment. The magnetic fields in the first range is measured and adjusted in the first step. After completion of measurement for the first range, the wiggler was move along its longitudinal axis and we began the measurement and adjustment for the second range. We used the field data in the first step as the basis for the adjustment in second step. The measurement step is 1mm, the measuring precision of the system $\leq 3\times10^{-4}$.

We set the gap of modulation section and gain section to 40mm and the gap of dispersive section to 55mm and we have done the following measurements:

1) At the central line (0, 0, Z) of wiggler, the distribution of vertical component By via Z is measured. Where, Z=0~2100mm. From this we can get the 1st and 2nd integral of the field. These integrals will reflect the angular deflection and relative orbital displacement of the electron beam.

2) Apart form the central line but still on the central plane (X, 0, Z), the distribution of vertical component By via Z is measured. Where Z=0~2100, X=-10mm~10mm.

3) We selected several peak positions in modulation and gain section and peak position of the dispersive section to measure the change of peak field Bp via Gap. Where Gap=40mm~100mm.

4) Setting Gap=40mm, we measured the change of peak fields in three sections via transverse position X. From this we can get the field homogeneous region.

We adopted two methods for the adjustment of the wiggler field. In one method, we can slightly change the gap of each magnetic block unit and in another method, we can change the local field distribution by means of shimming. The material for shimming is 20mm×100mm steel plate with the thickness 0.3mm. We found the placed position of the shimming plate has close relation to change of local field distribution. The following figure shows the 2nd integral of the fields at the central line.

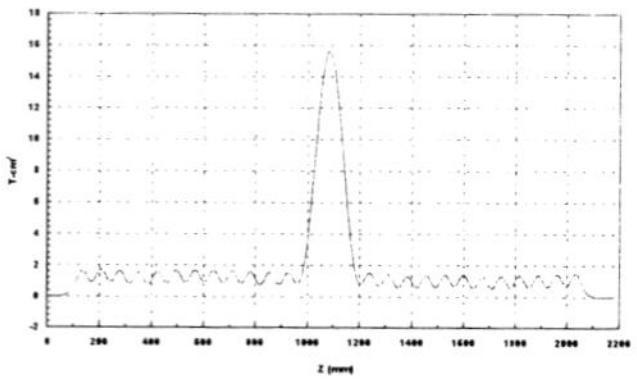

Fig. The 2nd integral of the field (at the central line, gap=40mm in modulation and gain section and gap=55 in dispersive section)

The final features of the wiggler is as following:

Peak Fields	B_0=0.296T (gap = 40mm, for modulation and gain section), B_0=0.707T (gap=55mm for dispersive section)
Errors for Peak Field	≤1.0% (when gap=40mm)
The 1st integral at the exit	0.004 T·cm
The 2nd integral at the exit	0.05 T·cm2
Field uniformity in transverse direction	0.26% (X= -10mm ~ +10mm)

Reference

[1] Jia Qika, Coherent harmonic FEL using non-symmetrical optical klystron, High Power Laser and Particle Beam,(in Chinese), 1994, 6(3),450

[2] Ma Youwu etc. Research of a hybrid Undulator, China Nuclear Science and Technology Report, CNIC-00994, 1995

[3] Liu Bo etc. A Magnetic Field Measurement System for Undulator China Nuclear Science and Technology Report, CNIC-00785, 1993

FREE ELECTRON LASERS 1998
G.R. Neil and S.V. Benson (Eds.)
1999 Elsevier Science B.V.

Enhanced Harmonic Spontaneous Radiation Using a Novel Undulator

T.C. Marshall and Yichen Shao
Department of Applied Physics, Columbia University, New York City 10027
and
Zohreh Parsa
Brookhaven National Laboratory, Upton, New York 11973

In a typical free electron laser (FEL), the electron beam interacts with a "dipole" undulator that has a sinusoidal magnetic field variation; the electron motion is sinusoidal in the plane transverse to this field and emits odd-numbered harmonic radiations along the axis. However, one need not limit the choice of undulator field profile to the sinusoid, providing other profiles result in significant advantages. In connection with the IFEL accelerator, in the past we have pointed out [1] that the use of an undulator profile that approximates a "square wave" will result in an enhanced acceleration gradient, by as much as a factor of two (equivalent in effect to an increase of laser drive intensity by a factor of four). This improvement (essentially at the fundamental FEL resonance) results largely from the fact that, for a given peak undulator field amplitude, the rms electron acceleration obtained from the square wave undulator is larger than that from the sinusoid; furthermore, the electron orbit is stable as well. In this paper, we find additional advantages that should result particularly at harmonic numbers f >1 if the undulator field profile is nearly "square wave": namely, a large enhancement of the harmonic spontaneous power radiated, together with enhanced FEL gain. The modification of undulators to enhance FEL gain has been examined in the past[2,3,4,5], usually with a particular design in mind, but with similar conclusions.

A purely square-wave undulator field profile is unphysical, however it is possible to develop a strong nonlinearity of the sinusoidal pattern in an electromagnet undulator containing ferromagnetic material by operating the device at a field where the material becomes saturated. As saturation progresses, the square wave profile could become a limiting case. Our analysis consists in a calculation of the spontaneous (single-electron) radiation in an undulator having N periods, where the axial magnetic field profile is approximated by the first few Fourier components of a square wave (we have computed the cases n = 1, 3, 5, and 7, but present here only the example where we use just the first three Fourier components; Fig. 1). This provides a more physical approximation to the undulator field that actually could be produced; but more importantly, it permits us to extend a straightforward calculation originally made by Colson [6] which expands the electron orbit in harmonics of the undulator period and gives an expression for the radiated power in terms of a series of Bessel functions. We retain the "undulator approximation", namely that not only is the amplitude of the motion K/γ small but also $K< 1$: then the radiation will have sharp lines at the harmonics on the axis since the radiation cone (width ~ $1/\gamma$) overlaps the orbit.

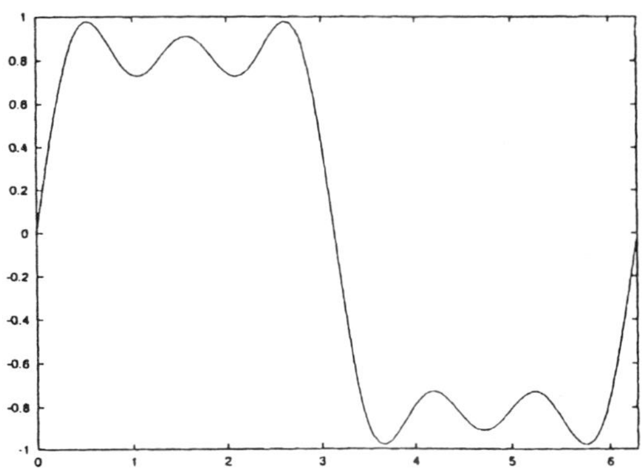

Fig. 1: Representation of a "square-wave" undulator field by the first three Fourier components. Ordinate, normalized magnetic field; abscissa, axial distance measured in radians over one period (2π).

In a long undulator, the spectrum becomes sharply peaked at frequencies satisfying $\omega_{rf} = \omega_0 f / (1 - \beta_0 \cos\theta) = f\omega_{r1}$, where $\omega_0 = k_0 c$ (the undulator wavenumber times the speed of light), f is the harmonic number, θ is the angle from the axis of motion along the undulator, ω_{r1} is the FEL resonance frequency for f=1, and ω_{rf} is the resonance frequency for the fth harmonic. The resonance line widths are all $\sim \omega_{r1}/N$. K is the normalized magnetic vector potential, eB_0/k_0mc^2, and B_0 is the peak undulator field. The undulator is the planar dipole type, and we have computed only radiation directed along the axis. We point out that taking higher n in the undulator expansion will smooth the top of the square wave, but will also increase the slope of the jump. The overall plan of the computation has been presented in detail in [3,6] and due to lack of space here, we will only present certain results.

The spontaneous power was computed numerically, and in Fig. 2 we show a typical result where we have taken K = 0.64 (note that $\omega_B/\omega_0 = K/\gamma \sim \beta_x$) and $\gamma = 80$ (40MeV). Only the peak power emission data point is plotted at each harmonic, all intermediate points are close to zero and are not plotted, and we compare the sinusoidal undulator with two approximations to the square wave undulator, where we include respectively only the n = 1 and 3 components, or the n= 1,3, and 5 components. The striking feature is the very substantial enhancement of spontaneous power emitted at the higher harmonics. That there should be *some* enhancement of radiation is apparant from the fact that electron radiation depends on the square of the electron speed, and the latter is proportional to the integral of the undulator field. The ratio of emission from the "square-wave" undulator to the "sinusoid" is 2.0 for f=1; this is the ratio of the mean square motion of the electron in these two different undulators that have equal peak field amplitudes. For the "approximate" square-wave undulator here, this ratio is about 1.3. However, discounting this factor of 1.3, there is still a remaining factor ~ 10 in enhanced radiation at the fifth harmonic, and larger enhancements (~100) at the higher harmonics. The cause for the power enhancement is in the axial oscillation motion, where the component of the motion driven by the large n=1 term of the field mixes with the components driven by the weaker n harmonics; this arises from the relation $\beta_z \approx \beta_0 - \beta_x^2/4\beta_0$. Harmonic enhancement would be clearly identifiable in a simple experiment.

The enhancement of harmonic spontaneous emission using the "square-wave" undulator profile has further implications. From Madey's theorem[7], the FEL gain will be proportional to the derivative of the spontaneous spectrum. But, the linewidth (~1/N) of this radiation does not depend on the details of the undulator field, just the number of periods. ; hence the FEL gain should be proportional to the spontaneous power emitted at the various harmonics. The enhanced gain has been the subject of previous work[2-5] where specially-devised undulators, not of this type, are described.

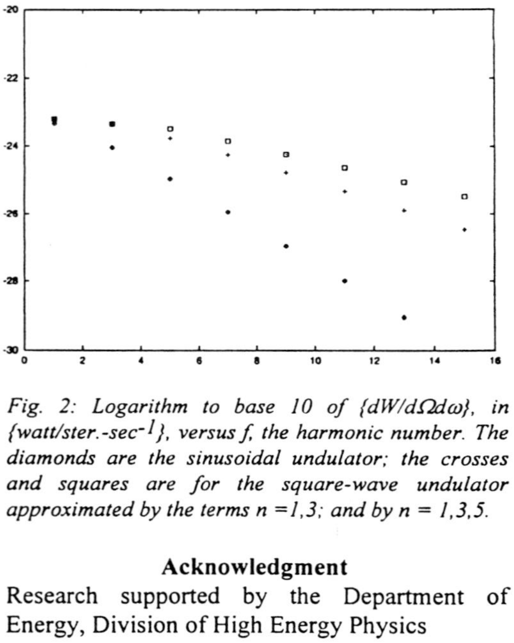

Fig. 2: Logarithm to base 10 of {dW/dΩdω}, in {watt/ster.-sec^{-1}}, versus f, the harmonic number. The diamonds are the sinusoidal undulator; the crosses and squares are for the square-wave undulator approximated by the terms n =1,3; and by n = 1,3,5.

Acknowledgment

Research supported by the Department of Energy, Division of High Energy Physics

References

[1] "Enhanced IFEL Performance Using a Novel Wiggler", Z. Parsa and T.C. Marshall, May 1997 Particle Accelerator Conference, Vancouver, Canada. BNL report #64491 and CAP - 168-ATF-97C; to be publ., Conference Proceedings [IEEE].

[2] M.J. Schmitt and C.J. Elliott, IEEE J. Quantum Electronics **QE-23**, 1552 (1987)

[3] M. Asakawa et al, Nucl. Instrum. Meth. in Phys. Res. **A318**, 539 (1992)

[4] D. Iracane, D. Touati, and P. Chaix, Nucl. Instrum. Meth. in Phys. Res. **A341**, 220 (1994)

[5] M. Asakawa et al, Nucl. Instrum. Meth. in Phys. Res. **A375**, 416 (1996)

[6] W.B. Colson, IEEE J. Quantum Electronics **QE-17**, 1417 (1981)

[7] J.M.J. Madey, Nuov. Cim. **50B**, 64 (1979).

FREE ELECTRON LASERS 1998
G.R. Neil and S.V. Benson (Eds.)
1999 Elsevier Science B.V.

Design and Construction of a Low-Cost Electromagnetic Undulator

Till Grübler[1], Dirk Nölle, Christian Piel, Henrich Quick, Thomas Schmidt[2], Thomas Weis
Institute for Acceleratorphysics and Synchrotronradiation, University of Dortmund,
D-44221 Dortmund, GER

Abstract

This paper deals with the design and construction of a low-cost electromagnetic undulator. The conception for this device was developed during the design study for a compact and cheap FEL [1] in the mid and far IR.
The undulator yoke is constructed out of only a few identical iron pieces that can be simply milled in a mechanical workshop. Only two main coils are needed to excite the field in this magnet. The device, built at the University of Dortmund in the DELTA workshop, has a period length of 25 mm and a gap of 8 mm. The k-value that can be reached with a 400 A 20 V power supply is about 0.4. As this device is suitable for LINAC applications only, also a pulsed operation might be possible. This could be a good alternative to overcome cooling problems.
This paper presents the design, field calculations, the construction and field measurements of the magnet using hall-probes and the pulsed wire method. Furthermore a rough cost estimate is given.

1 INTRODUCTION

Free Electron Lasers, needing less than 20 m^2 in space and available with a limited budget would be a very interesting device for scientific and industrial applications. One important component for such a device is a small and cheap undulator magnet: It has to be simple to build, easy to assemble, and must be adjustable without difficult mechanics. For a compact IR light source, electron beam energies of 5 to 10 MeV are a good choice, as the shielding is reduced, as these energies are below the neutron production thresholds. The emitted radiation will be tuneable in the mid and far infrared (40 to 150 μm) to be well away from commercial tabletop gas lasers.

2 DESIGN-CRITERIA

For a sufficient gain using a typical LINAC as driver and for the reason of limited space, the magnet must fulfil several criteria:

i) The magnet must be small, e.g. less than one meter in length.
ii) Fast and simple tune-ability is compulsory.
iii) The magnetic field must be stable within $\Delta B/B \leq 0.5$ % for ± 2 mm around axis.
iv) The gap-size must be 8 mm or more to include the wave-guide and the vacuum tube.
v) The material must be cheap, and machining should be easily possible.

Taking all these points into account, an electromagnetic undulator with a period length of 25 mm and a maximum field of 0.17 T (k = 0.4) seems to be adequate due to the experience at DELTA with electromagnets. Using 62 poles, the total length is less than 0.8 m.

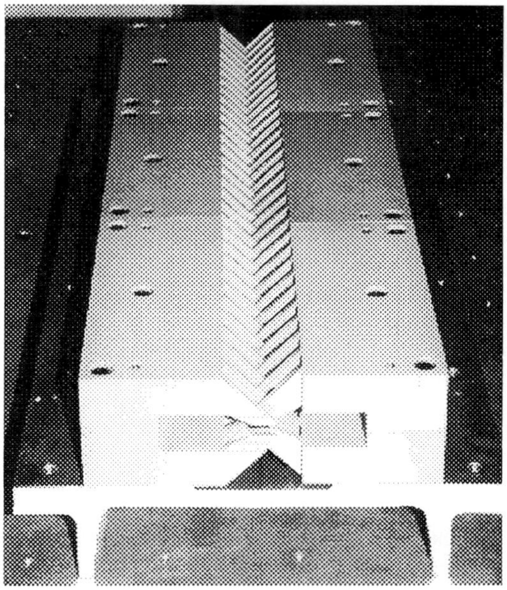

Fig. 1: The two C-shaped yokes are shown. The magnet consists of twelve segments, fixed by bolts and screws.

To avoid problems with individual coils for each pole, a design was chosen using only two main coils for the whole device. This leads to an unconventional shape [2] of the poles and the yoke: Two iron blocks in C-shape are the basic components of this undulator. The poles fit to each other like the cogs of a cogwheel. There are two main coils (one for each block) generating the magnetic field. This

[1] Now at the IWF, University of Braunschweig, GER
[2] Now at the SLS, Paul-Scherrer-Institute, Villingen, CH

offers the opportunity to use water-cooled copper tubes for the coils. Properly designed poles and homogenous iron give a very low distortion of the magnetic field. The two C-shaped yokes of magnetic iron can be machined. The shape of the poles is shown in Fig. 1 and 2.

During the design-phase, simulations were performed to optimise the pole-shape and the gap-size. The above mentioned criteria are fulfilled for a gap-size of 8 mm and a pole-width of 18 mm. The pole-thickness is 6 mm which is a compromise: optimal matching with a sinusoidal field is reached at less than 6 mm, but the stability during machining is increased for 6 mm pole-thickness, and the distortion of the magnetic field is small, e. g. less than 0.25 %. Reducing the thickness of the first and last pole by 50 % gives a 1/2—1/2 adaptation, which is important to obtain a straight trajectory.

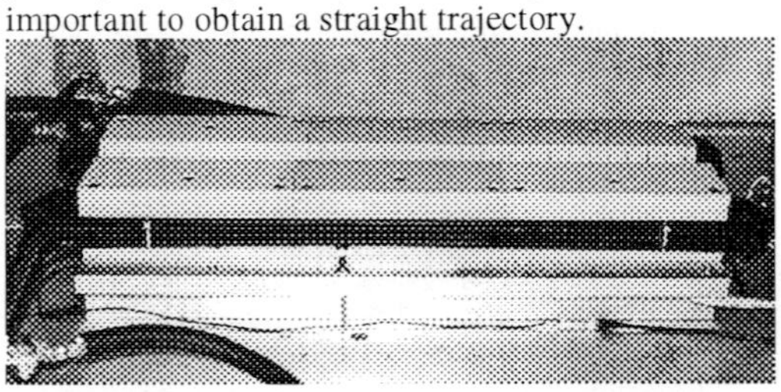

Fig. 2: The two coils, consisting of nine turns each, excite the magnet. They are made from conventional copper tubes. Using water-cooling offers the opportunity to drive the magnet with currents up to 400 A.

The gap–size and the pole to pole distance are in the same order of magnitude, therefore the whole device is strayfield dominated. This stray effect and the C-shape shift the peak of the magnetic field towards the side of the yoke. To match the peak of the magnetic field and the electron beam, a small step was introduced. A 0.5 mm step 5.5 mm from the centre of each pole shifts the peak of the magnetic field to the position of the electron beam which is in the middle of the pole. In addition, the homogeneity of the magnetic field is increased by a factor of two. The maximum field remains the same.

The magnet design consists of twelve more or less identical pieces, to allow easier machining, transport and assembly. Simulations with MAFIA show that the distortion of the magnetic field due to the slits and the screws are negligible. The eight segments in the middle are identical. Another set of drawings is needed for two end-segments. The remaining two end-segments are symmetrical to the other end-segments after a mirror-transformation.

3 PERFORMANCE OF THE MAGNET

Measurements of the magnetic performance were made using a hall-probe and the pulsed-wire-technique [3]. Hall-probe measurements of several poles were done to check the height of the field, its uniformity and deviations from the simulations. Every pole in the central section showed almost perfect matching with the simulations (Fig. 3 and 4). Deviations from the simulations occurred at the end-poles only: The magnetic field in the end-poles is stronger than calculated due to the unsymmetrical distribution of the iron: In the central section, the flux through each pole is weakened by the flux through the neighbour poles. The lack of the neighbouring pole at the end gives a higher field than calculated. This means, the electrons will experience an angular kick during their passage through the undulator. However, the correction can be done easily using two trim coils at the entrance and the exit of the undulator.

Including the measurements using the pulsed-wire-technique to determine the values of the first and second field integral, four trim coils were mounted. The position of these coils is not critical, therefore a

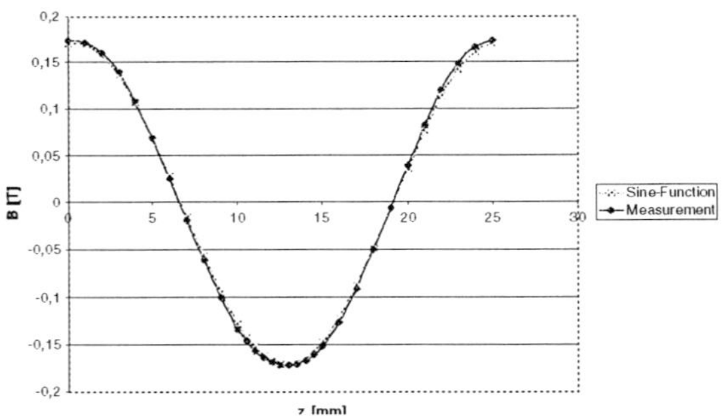

Fig. 3: Hall-probe measurements show the good agreement of the magnetic field and a sine-function of corresponding amplitude.

regular distribution was chosen. The desired current is about 20 A for each trim coil, compared to 400 A in the main coils.

The use of a computer controlled power supply offers the opportunity to change the magnetic field easily. Data tables for the trim coils help to keep the field quality as desired.
With 240 A in the main coils, a peak field of 0.173 T is reached. The field quality (without end-poles) is $\Delta B/B \leq 0.43$ % for $\pm$ 2 mm. The maximum current of 400 A through each coil gives a magnetic field of 0.186 T. The field quality is still within the order of $\Delta B/B = 0.5$ %.

MAGNET DATA	*VALUE*	*REMARKS*
Excitation	electro-magnetical	low costs, good tuneability, use of standard components
Yoke Material	iron	use of old magnetic material possible
Period-Length	25 mm	
Number of Poles	62	provides sufficient gain
Peak-Field (@ 240 A)	0.173 T	
Uniformity of the Field (@240A)	0.43 %	for 4 mm cross-section of electron beam
Deviation from Sinusoidal Field	0.24 %	
K-Value	0.4	
Adaptation	1/2–1/2	by reduction of pole-thickness
Gap-Size	8 mm	sufficient space for wave-guide
Pole-Thickness	6 mm	good field-uniformity and stability
Pole-Length	18 mm	integrated step for best field performance
Magnet-Length	0.766 m	compact light source possible
Number of Segments	12	simple machining, easy transport

Table 1: The performance data of the undulator.

4 COSTS

The total costs of the magnet and its power supplies are lower than 30,000 US$. A brief overview about the amount of money necessary for each component is given in Table 2. The magnetic iron was free, due to the fact that parts of an old spectrometer magnet of the University of Bonn, that was also used to construct the FELICITA I undulator [4], were still available. Less than two weeks of machining in the DELTA-workshop by one technician must be added.

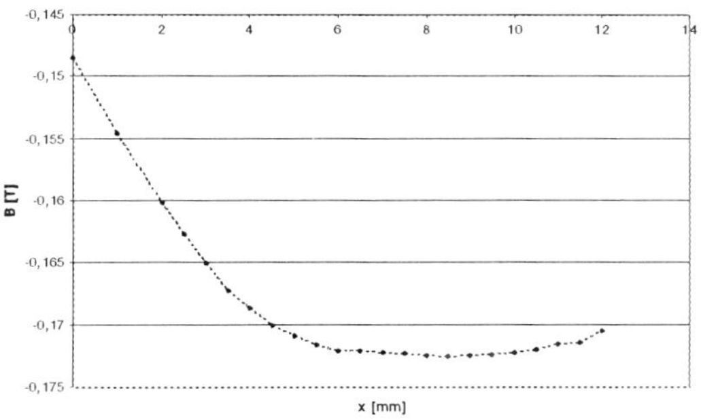

Fig. 4: The homogeneity of the field is better than 0.43 %. In this example, the centre of the beam is located at x = 8 mm.

A student, during his thesis work, did the design, the simulations, and the tests of the magnetic performance. One power supply (400 A, 20 V) delivers the current for the main coils. The price is about 18,000 US$, which is more than 50 % of the device's costs. Two additional small power supplies for the trim coils (20 A, 20 V) cost less than 2,000 US$. One is needed to match the angle of the electron beam, and one is necessary to match the first and second field integral. The tubes for the main coils, the electrical insulation and their water supply sum up to 1,000 US$. All components were made from standard materials with well-known technologies.

5 CONCLUSIONS

A small and simple magnet for a Free-Electron Laser in the mid and far Infrared was designed and build. Its magnetic performance is sufficient for an average LINAC with pulse compression as driver. For less than 30,000 US$ (without iron) such a compact undulator can be build without any special technique.

REFERENCES

[1] D. Nölle et al., IR-LICHT, Abstr. of the FEL Conf. 96, Rome, Italy, p. TU C06 P18

[2] T. Grübler, Diplomarbeit, University of Dortmund, 1998 (unpublished)

[3] R. Warren, Nucl. Instr. & Meth. **A272**, 1988, p. 257

[4] A. Geisler et al., Nucl. Instr. & Meth. **A375**, 1996, p. 445

PART OF THE MAGNET	*COSTS [US$]*
Iron	free (material of an old magnet re-used)
Design, Construction, Measurements	12 month thesis work, 1 student
Fabrication of the Yoke	2 weeks machine-time, 1 technician
1 Main Power Supply (400 A, 20 V)	18,000
2 Trim Coil Power Supplies (20A, 20 V)	2,000
Copper Tubes, Insulation	1,000

Table 2: A rough estimation of the magnet's costs.

FREE ELECTRON LASERS 1998
G.R. Neil and S.V. Benson (Eds.)
1999 Elsevier Science B.V.

Field performance studies of a split-shunt-plate scheme for strong-field LCLS undulator tapering*

R. Tatchyn†

Stanford Synchrotron Radiation Laboratory, Stanford University, Stanford, CA 94309 USA
†*Corresponding author: Tel.: 650-926-2731; FAX: 650-926-4100; E-mail: tatchyn@slac.stanford.edu*

Abstract

In recent years, design studies at the Stanford Linear Accelerator Center (SLAC) have advanced and demonstrated the feasibility of using simple hybrid/permanent magnet (PM) configurations consisting of cuboid permeable and PM pieces for attaining the high fundamental field amplitudes necessary for Linac Coherent Light Source (LCLS) applications. Included in this work were recent field studies of a gap-variable scheme for shunt-plate tuning of the hybrid/PM field structure, which revealed a desirable range of sensitivity of the fundamental field strength to an easily-prepared range of shunt plate parameters. In this paper we explore an alternative variation of the initial shunt plate configuration that may offer mechanical advantages for implementation and operation.

1. Introduction

In recent work, a simple hybrid/PM undulator design consisting of cuboid pieces of permeable and magnetic material has been assessed for short-wavelength Free Electron Laser (FEL) and short-period synchrotron radiation (SR) applications [1]. The simplicity and high degree of symmetry of the structure make it possible to add critical functions such as superimposed focusing and dipole field tuning in an essentially independent fashion. For the latter option, for example, a scheme for tuning the dipole filed amplitude over a limited range with a pair of shunt plates with a variable gap placed above and below the undulator structure (see Fig. 1) was analyzed. Simulations with the 3D electromagnetic modeling package AMPERES [2] established that this approach could be used to tune the field with acceptable sensitivity over a range of several percent, allowing for independent tapering of the undulator field to correct for SR-induced energy loss or to investigate the effects of tapering on gain/bandwidth products in FEL systems.

Due to the importance of these two functions, an alternative variant of the shunt plate tuning technique with advantages for mechanical implementation and field control has been investigated and a brief summary of the basic results is presented in this paper.

2. Split shunt plate tuning

A dimensioned layout of the top half of the new variant is depicted in Fig. 2. The shunt structure, consisting of two identical plates of transverse dimensions $t_S x w_S$, is centered at a fixed distance d over the top of the undulator. The undulator field is tuned by varying the gap g_S. The method allows for the strong vertical force component Fz to be controlled with a brace

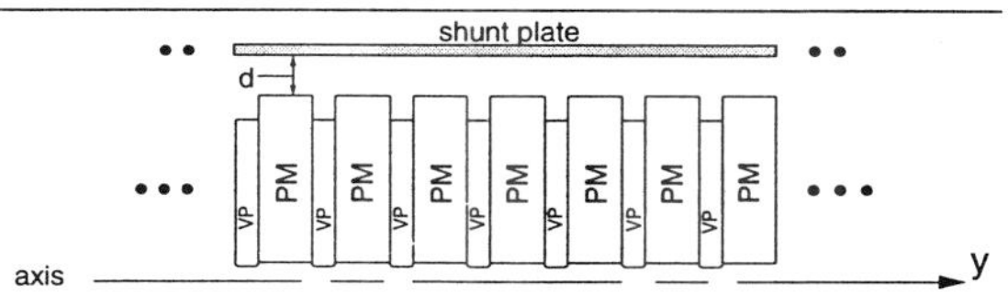

Figure 1. Top half of hybrid/PM LCLS undulator with shunt plate tuning (side view).

with a fixed vertical gap and the tuning motion to coincide with the direction of the weaker force component Fx. A coarse initial estimate of this force component (per unit length in the y direction) can be obtained by estimating the energy density of the field's z

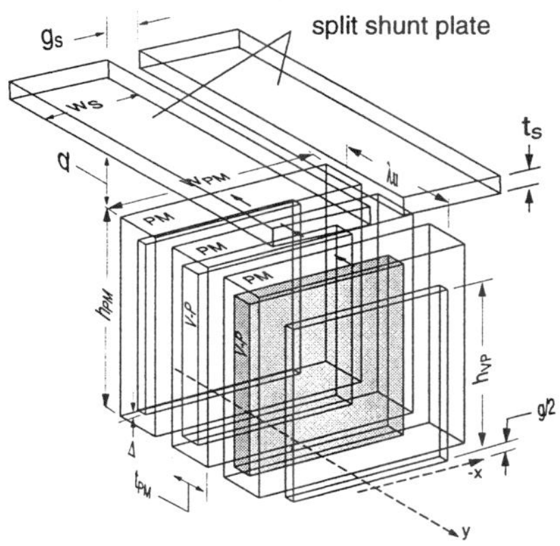

Figure 2. Split-shunt-plate tuning configuration.

component in the gap region immediately above the undulator and then multiplying it by t_S. An analogous estimate can be performed for the Fz component. These reveal the approximate proportionality of Fx to t_S and of Fz to w_S, indicating two of the primary design factors for

*Portions of this work were supported in part by the Department of Energy Offices of Basic Energy Sciences and High Energy and Nuclear Physics, and Department of Energy Contract DE-AC03-76SF00515.

controlling the magnitude of forces in the tuning structure. Of course the reduction of these is limited by the point at which the shunt plates: 1) start approaching saturation, and 2) become too narrow to significantly load the gap field.

3. Performance studies

Studies of the structure in Fig. 2 were performed with the AMPERES modeling package. Here we present results for the following set of fixed parameters: λ_u=3 cm, t_{PM}=10.5 mm, h_{PM}=34 mm, w_{PM}=4 cm, t_{VP}(=(λ_u-2t_{PM})/2)=4.5 mm, h_{VP}=3 cm, w_{VP}=3 cm, Δ=0 mm, g/2=3 mm, t_s=5 mm, w_s=20 mm, and d=100 μ. The electromagnetic model consists of the configuration shown in Fig. 2, extended by symmetry reflections in the x-y and x-z planes, leading to a total model length of approximately 10 cm. The pole material is Vanadium Permendur and B_r=1.25T (H_c~11900 Oersted) for the selected PM material (Nd/Fe/B).

In Fig. 3, the dependence of the on-axis field amplitude Bz on g_s is plotted. For the particular dimension of d it is seen that a total tuning range of approximately 2.5% can be attained over a reasonably broad range of transverse motion. In Fig. 4 the force

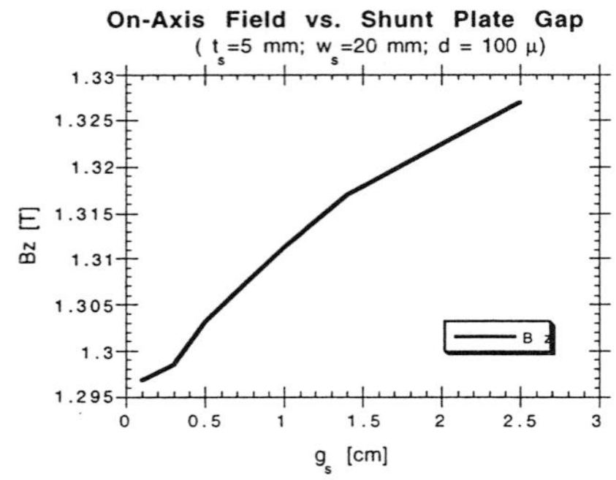

Figure 3. On-axis undulator field vs shunt plate gap.

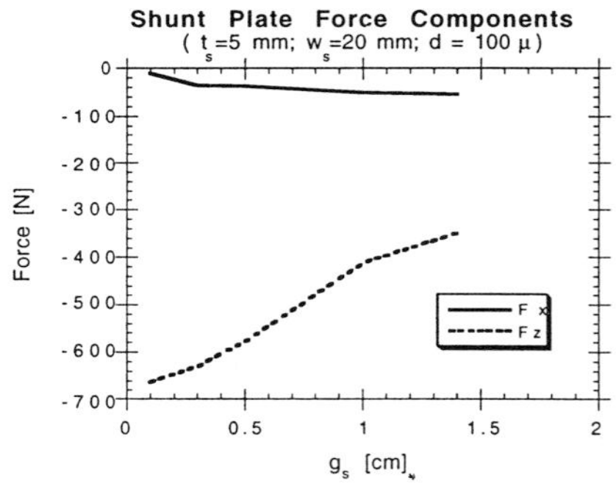

Figure 4. Fx and Fz vs. gap (shunt plate length 10 cm).

components Fx and Fz are plotted as functions of g_s for the same range. As expected, the force in the direction of tuning is relatively small in comparison to the vertical force, indicating the expectation of a relatively straightforward mechanical implementation.

4. Summary

A fixed-vertical-gap variant of a simple shunt plate tuning scheme applicable to a cuboid high-field hybrid/PM LCLS undulator design developed recently at SSRL [3,4] has been investigated. Initial modeling studies indicate that the proposed variant appears promising for a straightforward mechanical implementation. The relatively weak influence of the shunt structure on Bz implies additional favorable properties for implementation. These include a less stringent dimensional tolerance and a reduced sensitivity of the undulator field's multipole components (including those of any focusing lattice in the undulator gap) to misalignments and tuning motion tolerances. Of course, in an actual implementation of the scheme, these sensitivities will need to be quantitatively established with further simulations.

5. References

[1] R. Tatchyn, "Field design studies of a planar hybrid/permanent magnet undulator with strong planar permanent magnet focusing," presentad at the International FEL97 Conference, Beijing, China, Aug. 18-21, 1977; to appear in Nuclear Instruments and Methods, 1998.

[2] Integrated Engineering Software,Inc., Winnipeg, Manitoba.

[3] S. Caspi, R. Schlueter, R. Tatchyn, "High-Field Strong-Focusing Undulator Designs for X-Ray Linac Coherent Light Source (LCLS) Applications," Proc. IEEE PAC95, Dallas, TX, May 1-5, 1995; SLAC-PUB-95-6885.

[4] R. Tatchyn et al, NIM A 375, 274(1996).

FREE ELECTRON LASERS 1998
G.R. Neil and S.V. Benson (Eds.)
1999 Elsevier Science B.V.

The LCLS FEL Undulator

Roger Carr
Stanford Linear Accelerator Center

The LCLS undulator design has been in process for several years, and has evolved to a candidate design that was reviewed and found to be technically sound. [1,2,3] The undulator system comprises a hybrid permanent magnet undulator in 52 segments each with 64 periods of 30 mm length. The main design challenge is to provide a magnetic lattice of sufficient mechanical and magnetic precision that the electron beam trajectory can be held to a walkoff of less than 5 microns per field gain length, while maintaining the emittance and energy spread. If this is done, the undulator should saturate at 1.5 Å wavelength in a total length of 112 m. It is important that we build a saturating device, because the output will be much more stable than it would be on the exponential part of the gain curve. Though the permanent magnet hybrid is the present candidate design, we are reviewing other options, such as a superconducting bifilar helix, which would saturate at shorter length and allow a larger diameter beampipe.

The LCLS is designed for electron energies of 5-15 GeV, a beam of 233 fsec FWHM pulse duration and 1.5π mm-mrad normalized emittance. The beam diameter is 65 μm rms and the field gain length at 15 GeV is 11.7 m. The magnetic gap is 6 mm. The hybrid undulator is planar, and yields vertically plane polarized radiation.

The hybrid permanent magnet design calls for conventional NdFeB magnets alternating with vanadium permendur pole pieces. Reasonably tight tolerances on the magnet materials, computerized sorting through a very large population of magnets, and magnetic measurements and shimming after the final assembly are required to produce a field with random errors less than 0.1%. The gap will have to have some slight variability, by segment, for two reasons. First, it is possible to sort a population of magnets to an average value, but it is difficult to 'pull' that average to a desired value. Second, the desired rms average values of the magnetic field will taper very slightly over the length of the undulator, to make up for energy losses by spontaneous radiation and resistive wakefields.

Decreasing the average beta function below the natural focusing value (70 m for two-plane focusing) is necessary to minimize the saturation length. To do this we interpose a strong focusing FODO lattice of permanent magnet quadupoles between the undulator segments. The optimal average beta function is 18 m. The focusing magnets will be 12 cm long, with 45 T/m gradients.

For our undulator parameters, radiation of any wavelength will advance by one wavelength relative to the electron beam in a 24 cm drift space. We put such spaces between undulator segments to provide space for beam position monitors, vacuum ports, and alternately focusing and defocusing quadrupole magnets, as shown in Figure 1.

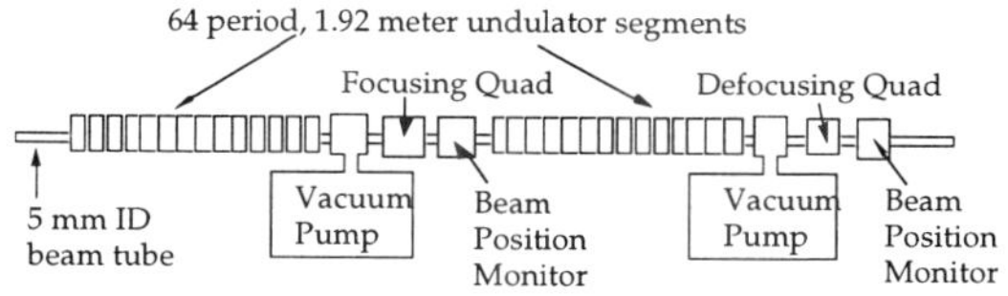

FIGURE 1. A schematic side view of the undulator structure, showing the FODO lattice with separations between 2m undulator sections for diagnostics, focusing correctors, and vacuum ports.

The focusing magnets are mounted on 2D transverse movers, for two purposes. First, the beam based alignment of the electron trajectory is done by finding the quadrupole position where the trajectory varies least with energy. [4,5] This alignment strategy solved the major problem in the LCLS design; it could not have been done with mechanical alignment methods. Second, the quadrupoles are moved at run-time to offset trajectory errors detected by the beam position monitors. The resolution of the motions must be about 1 μm.

The beam position monitors will be either RF cavity devices, or a four-quadrant button design. This choice was arrived at after an examination of all available BPM technologies [6]. The BPMs must have 1 μm sensitivity, high stability (possibly achieved through frequent self-calibration), and low impedence so as not to cause energy degradation or emittance growth in the electron beam. We plan to install about ten carbon wire beam position monitors as well, for initial alignment and beam size measurements. We have also developed a strategy which would allow us to use carbon wires as simultaneous electron-photon beam position monitors, but detecting diffracted x-rays from the photon beam, and bremsstrahlung created by the electron beam. Initial mechanical alignment, to trajectory tolerances of about 50 μm, is sufficient to see FEL gain to saturation at 15 Å (5 GeV)

The beam position monitors will be mounted on the same girders as the undulator segments, and the position of the girder will be monitored by measuring offsets with respect to two suspended wires that run the full length of the undulator. The wires are very stable, so the system can detect slow thermal and geophysical drifts of the magnets. These drifts are expected to have magnitudes of about 100 μm.[7] Feedback control will then move the girders to counteract drifts, using a cam mover system that is capable of submicron resolution. [8] The undulator structure will also be stabilized by temperature controlled water flowing in channels in the girders.

The very small 6 mm gap required for the hybrid undulator causes some problems. NdFeB is not very radiation hard, and it can be damaged by stray beams. Collimators and interlocks will be necessary to prevent this damage. Good thermal stabilization of the magnetic material is also essential to minimize radiation vulnerability.

Even worse than radiation is the problem of wakefield effects. With a 5 mm ID stainless steel beampipe, internally copper plated, we have resistive wall wakefield effects on energy of $\Delta E/E = 0.0003$ and on emittance of $\Delta\varepsilon/\varepsilon = 3\%$. which are tolerable. However, beamppe roughness is potentially more serious. Some calculations show significant wakefield effects for surface roughness as small as 40 nm. We do not consider the calculations to be conclusive, but they do serve as a warning. Fortunately, the semiconductor industry has developed gas handling tubing with internal roughness on the order of 100 nm, and we hope to improve upon this with electropolishing and other techniques. This is an area of active theoretical and experimental work.

We have briefly described above some of challenges that arose in the design of an x-ray FEL at SLAC. Among the problems that any such design must address are: 1) tight mechanical tolerances, 2) geophysical and thermal environmental problems 3) beam position monitoring 4) initial alignment strategy 5) stability of alignmnent 6) radiation dose managment and 7) wakefield effects. Of these, we consider the initial alignment the most challenging, because we have no precedent for it. Also, beampipe roughness could be a severe problem. The other issues have been addressed in existing the SLC and FFTB machines at SLAC, and in work at other laboratories.

ACKNOWLEDGEMENTS

The author is pleased to thank the many contributors to the LCLS undulator design study, including John Arthur, Richard Boyce, Jeff Corbett, Max Cornacchia, Robert Hettel, Don Martin, Heinz-Dieter Nuhn, Jim Sebek, Roman Tatchyn,(SSRL), Karl Bane, Vinod Bharadwaj, Gordon Bowden, Paul Emma, Alberto Fasso Clive Field, Cho Ng, Pantaleo Raimondi, Sayed Rokni, Robert Ruland John Sheppard, Vaclav Vylet, DieterWalz (SLAC), Klaus Halbach, Kwang-je Kim, Steve Lidia, Ross Schlueter, Ming Xie, (LBNL) Lou Bertolini, Lou Griffith, Marcus Libkind, (LLNL) Ilan Ben-Zvi,(BNL), and Claudio Pellegrini (UCLA). This work was supported by the US Department of Energy, Office of Basic Energy Sciences under contract number DE-AC03-76SF00515.

REFERENCES

[1] R. Carr, in *Towards X-Ray Free Electron Lasers*, AIP Proceedings 413, (1997), p. 143

[2] *LCLS Design Study Report*, SLAC Pub. R-521, (1998) Chapter 8.

[3] *LCLS Design Review Report*, J. Bisognano (ed.), (1997)

[4] P. Emma, R. Carr, H-D Nuhn, these Proceedings

[5] P. Castro: TESLA-FEL 97-04 (Aug. 97) and Proc. EPAC 98 Stockholm

[6] R. Carr, R. Hettel, D. Martin, C. Field, these Proceedings

[7] *Zero Order Design Report for the Next Linear Collider*, Appendix C, SLAC Report 474

[8] G. Bowden, P. Holik, S.R. Wagner, G. Heimlinger, & R. Settles, Nucl. Inst & Methods A368 (1996), p. 579

FREE ELECTRON LASERS 1998
G.R. Neil and S.V. Benson (Eds.)
1999 Elsevier Science B.V.

The VISA FEL Undulator

Roger Carr [1], Ilan Ben-Zvi, [2] Lou Bertolini, [3] Max Cornacchia, [1] Paul Emma, [1] Pedro Frigola, [5] Erik Johnson, [2] Marcus Libkind, [4] Steve Lidia, [3] Heinz-Dieter Nuhn, [1] Claudio Pellegrini, [5] George Rakowsky, [2] James Rosenzweig, [5] Robert Ruland [1]

1) Stanford Linear Accelerator Center, 2) Brookhaven National Laboratory, 3) Lawrence Berkeley Laboratory, 4) Lawrence Livermore National Laboratory , 5) University of California at Los Angeles

The Visible-Infrared SASE Amplifier (VISA) FEL is an experimental device designed to show Self Amplified Spontaneous Emission (SASE) to saturation in the 800 - 600 nm range, where silicon detectors may be used to characterize the optical properties of the FEL radiation. VISA is the first SASE FEL designed to saturate, and its diagnostics will provide important checks of theory [1].

VISA has a 4 m pure permanent magnet undulator comprising four 99 cm segments, each with 55 periods of 18 mm length. The saturation length is calculated numerically to be 3.4 (3.8) m at 71 (85) MeV for an ideal undulator with 18 mm period, and a maximum magnetic field of 0.75 T. These parameters can be achieved with a pure Halbach permanent magnet design with no permeable materials, NdFeB magnets with B_r = 1.25 T, and a fixed gap of 6 mm [2]. Figure 1 shows the magnetic scheme for the VISA undulator.

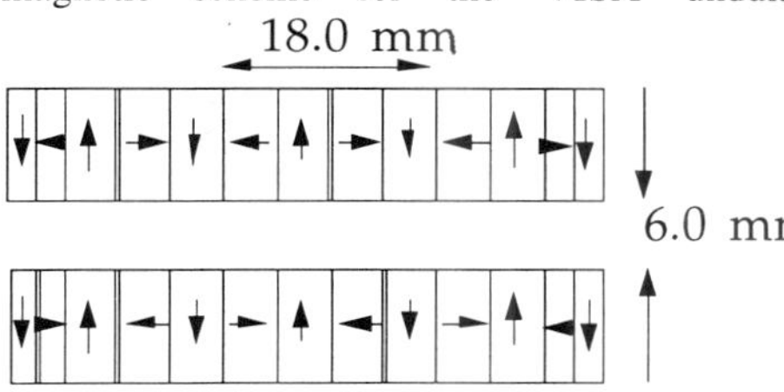

Figure 1: Schematic side view of two periods of the VISA undulator structure, showing a symmetric two-half-block termination scheme. Arrows within the magnet blocks indicate the direction of magnetization.

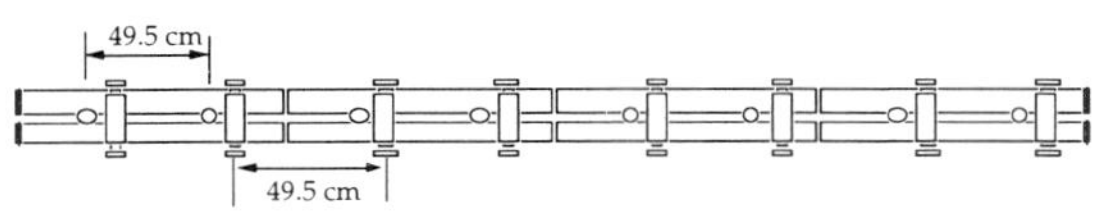

Figure 2: Schematic side view of the VISA undulator, with end terminating magnets on both ends, pop-in diagnostic ports (circles) and steering trim coils (rectangle groups).

Figure 2 shows the entire four-segment undulator. The electron beam in VISA has an rms diameter of 120 µm, and numerical simulations show that saturation length is adversely increased if the trajectory walks off a straight line by more than 50 µm per field gain length of 34 cm (38 cm) at 71 MeV (85 MeV). A magnetic field with rms errors of less than 0.4% is required in order to achieve this trajectory walkoff tolerance.

Natural focusing is too weak for a 4 m undulator to saturate at these wavelengths, so we added strong focusing to the undulator. To achieve an average beta function of 27(30) cm at 71 (85) MeV we put a FODO lattice with four cells per segment into the undulator. This will be done by placing rows of paired magnets alongside the beam, as shown in Figure 3 [3].

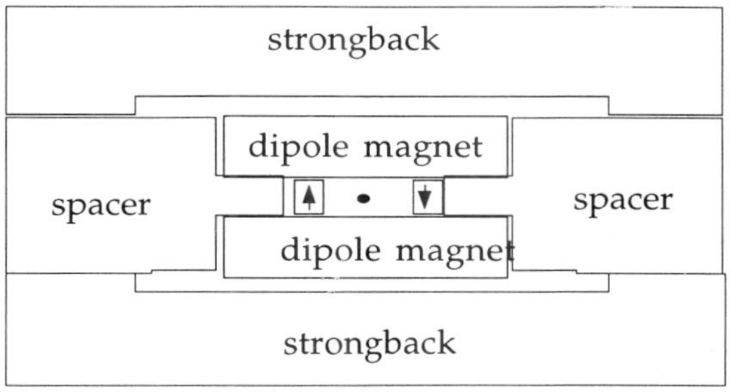

Figure 3: Schematic end view of the VISA undulator. Focusing magnets are shown as rectangles on either side of the central beam.

The focusing assemblies are 100 mm long with a gap of about 10.75 mm between them. They generate a gradient of 33 T/m on-axis.

Magnetic field errors of a pure permanent magnet undulator can be controlled in several ways. First, magnet material is chosen within a certain tolerance on its net magnetic moment and the direction of its magnetization. We specified NdFeB material with Br = 1.25 T, moment errors of no more than 1.5% of B_r in each of the three principal axes of the rectangular blocks, and direction errors of no more than 1.5°. After Helmholtz coil measurements are made of all the blocks, a sorting algorithm is executed on the data [4]. The errors in a randomly assembled undulator can be reduced to 0.4% by the action of the algorithm.

After sorting and assembly, the undulator assemblies will have magnetic errors from magnet block measurements and mechanical imperfections, so we employ magnetic shimming to correct the trajectory in each segment. The NSLS magnetic measurements group at BNL will do this work using a pulsed wire technique [5]. With these magnet measurements and magnet shimming, we expect to be able to control trajectory walkoff to less than ±50 μm per field gain length. Run time corrections may also be done with trim coils. Our shim magnets are small blocks of NdFeB that are used in fours, as shown in figure 3. The blocks are mounted in pairs on movers so the strength of the shim effect can be controlled.

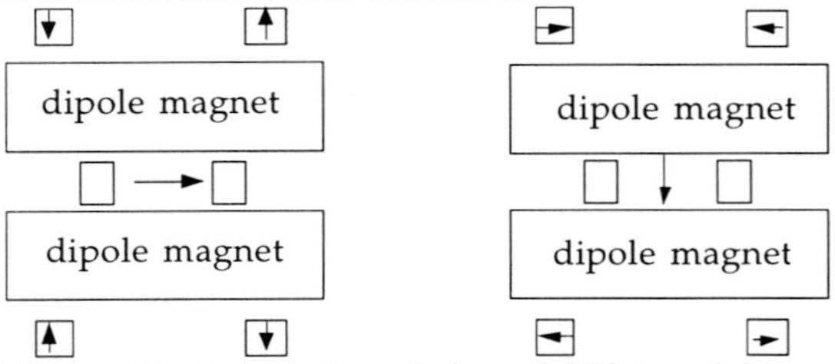

Figure 3: Schematic end view of VISA undulator, showing placement of shim magnets in groups of four. With magnetizations arranged as at left, the shim magnets create a net horizontal field on axis, and on the right they create a vertical field.

We will also use the pulsed wire to find the axis of the quadrupole focusing magnets; this position is then transferred to external fiducials that are used to align the segments together. After shimming individual segments, we will set up pairs of segments, butted together, and shim the trajectory across the joint. When this is finished, the segments are placed on kinematic mounts inside a long vacuum box; the mounts are attached through bellows to x-z slides on an external support beam, to isolate the undulator from the box.

Optical survey alignment should bring the axis to straightness within about 200 μm. A fixture with slits, referenced to tooling balls and previously calibrated to the magnetic axis, will be used to align the entrance and exit of the undulator to a beamline reference laser, which will also be used to calibrate the diagnostic pop-ins. Two laser straightness interferometers, also aligned parallel to the beamline reference laser, will then be used to achieve an alignment tolerance of 20 μm to the magnetic axes. One interferometer can be moved on a path level with the axis horizontally, and the other is vertically above the axis.

We chose Ce:YAG crystals for beam position monitors because when the electron beam strikes 0.5 mm thick slice of this material, it causes the crystal to fluoresce with negligible blooming.[6] The fluorescent light reflects from two 45 degree mirrors into a CCD camera, as shown in figure 4:

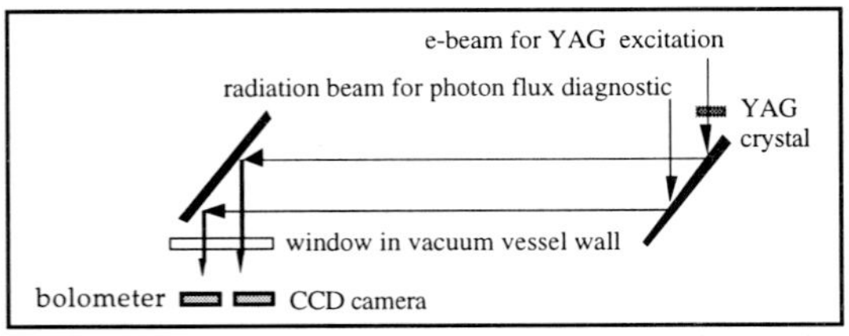

Figure 4: Schematic view of pop-in diagnostic. The mirrors and YAG crystal are translated into two positions that intercept the combined photon and electron beams. The diagram reflects two positions of the periscope; the beams are in the same place with respect to the undulator.

The diagnostic pop-ins are mounted on the vacuum vessel, but they are just periscopes that bring light out to CCD cameras. The cameras are mounted on the same support beam as the undulator, so that exact position repeatability for the pop-in is unnecessary. The resolution of the BPMs should be about 20 μm. The pop-ins also carry a mirror that can be used to bring FEL output into a bolometer for gain curve measurements.

There are approximately two betatron oscillation periods in the length of the VISA undulator, and the diagnostics and steering trim coils are placed at intervals of roughly $\pi/2$ phase advance. Simulations have been performed to show that errors as large as 200 μm can be reduced to 50 μm, but can require kicks up to 1 mrad.

This work was supported by the United States Department of Energy, Office of Basic Energy Sciences under contract No. DE-AC03-76SF00515. The authors are pleased to acknowledge valuable contributions from Lorraine Solomon, John Skaritka, Jeff Aspenleiter, and Michael Lehecka, all of Brookhaven National Laboratory.

REFERENCES

1 R. Bonifacio, C. Pellegrini, and L.M. Narducci, Opt. Commun. 50, (1984) p. 373.

2 K. Halbach, Nucl. Inst. & Methods 187 (1981) p. 109

3 A.A. Varfolomeev and A.H. Hairetdinov, Nucl. Inst. & Methods A341 (1994) p. 462

4 S. Lidia and R. Carr, Rev. Sci. Inst. 66 (1995) p. 1865

5 R. Warren and C. Fortgang, Nucl. Inst. & Methods A341 (1994) p. 444

6 E. D. Johnson, W. S. Graves, K.E. Robinson, Proceedings of the 8th Beam Instrumentation Workshop (1998)

FREE ELECTRON LASERS 1998
G.R. Neil and S.V. Benson (Eds.)
1999 Elsevier Science B.V.

PM Wiggler for a Pulsed-FEM

S.R. Wylie*, C.C. Wright, R.A. Stuart, A.I. Al-Shamma'a, and J. Lucas

**The University of Liverpool, Department of Electrical Engineering and Electronics, Brownlow Hill, Liverpool L69 3GJ, U.K.*

Abstract

We present the design of a permanent magnet wiggler for a pulsed 60kV FEM using two NdFeB bars per period. The wiggler period is 19mm and the maximum magnetic field measured in the gap centre was 320 Gauss. A comprehensive simulation has been carried out to study the effect of the end poles on the magnetic field distribution and the actual electron beam trajectory measurements are discussed and compared with the simulation results.

1. Introduction

The Free Electron Laser (FEL) research group is currently in the process of designing a new pulsed FEM system as a continuation of the research to produce a compact, low cost system, for industrial applications. An oscillator prototype FEM, i.e. without pre-bunching, is currently under construction and will have an output power of 50 watts at X-band frequencies. Unlike the previous systems, this FEM will use a higher current, ≥300mA, and a 55 to 80kV pulsed electron beam. The electron accelerator design is based around a Pierce type gun acquired from EEV. A new pure permanent magnet wiggler has been designed for this system, using a specially developed simulation program. The wiggler uses a new end section design that greatly reduces the mechanical complexity, and therefore the cost, compared to previous wigglers [1,2].

2. Wiggler Simulation Software

Two programs have been developed to simulate the effect of different wiggler designs on the electron beam [3]. MagSim calculates and displays the magnetic field profile of wiggler magnet and TrajectSim calculates the electron beam trajectory through the wiggler.

2.1. Magnetic Field Simulation Program

MagSim provides a realistic representation of a physical magnet, allowing the size, the position, the orientation and the remanence to be adjusted.

2.2. Simulation of Electron Beam Trajectories

Calculations of the electron beam trajectory were performed by simulating an electron as a particle with a rest mass, a relativistic mass, a charge, an energy, a position and a velocity direction. TrajectSim is used to calculate and store the path of the electron through the simulated magnetic field produced by the MagSim program.

3. The Wiggler Magnet Design

The wiggler magnet was designed with thirty 19mm periods, using two NdFeB permanent magnets, with a residual field of 12500 Gauss (1.25 Tesla), per period. The MagSim program was used to simulate the effect of the wiggler, with the assumption that all the magnets were perfectly matched. To display the design approach more clearly, the simulation results have been produced for a scaled wiggler. The magnet and gap dimensions have been doubled and the number of wiggler periods has been reduced to eight and half. In each case a 60kV electron has been fired towards the wiggler, starting at (0,0,-300) heading left to right along the Z-axis.

3.1. Full magnet bar at each end

The initial design began with all the wiggler magnets having uniform dimensions,

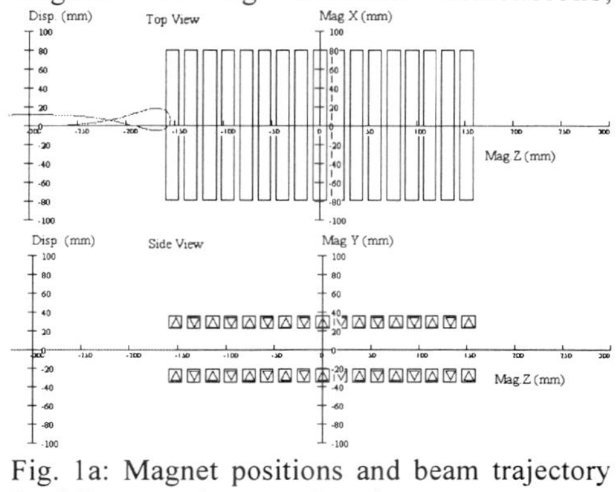

Fig. 1a: Magnet positions and beam trajectory for full magnet bar at each end.

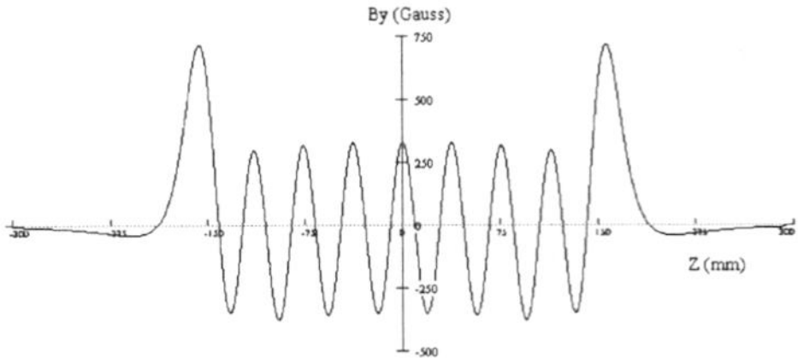

Fig. 1b: Magnetic field distribution for full magnet bar at each end.

as shown in Fig.1a. The trajectory, as seen from the top view of the wiggler magnet, shows that electrons at 60kV cannot enter the wiggler. This is due to the large "rabbit ears" in the magnetic field distribution (B_y), shown in Fig.1b, where the field is strong enough to turn the electrons away.

3.2. Tuneable half magnet bar at each end

To reduce the magnetic field offset at each end, the half-width magnets was used and gradually moved towards their adjacent full magnet. As the gap was reduced the offset decreased, and this trend continued until the gap was −1mm, i.e. the half-width magnets had overlapped the adjacent magnet, as shown in Fig.2a.

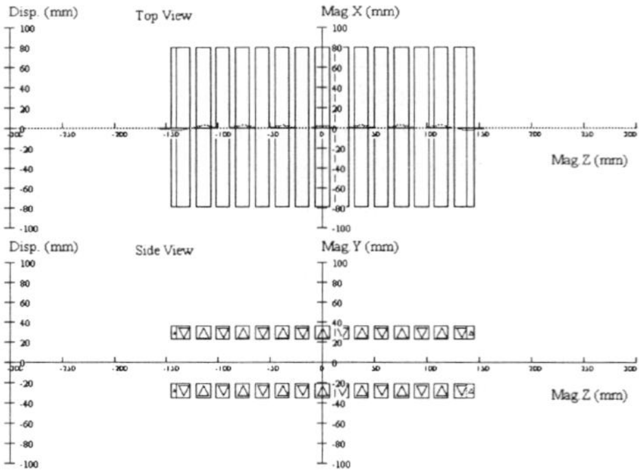

Fig. 2a: Magnet positions and beam trajectory for overlapped half magnet bar at each end.

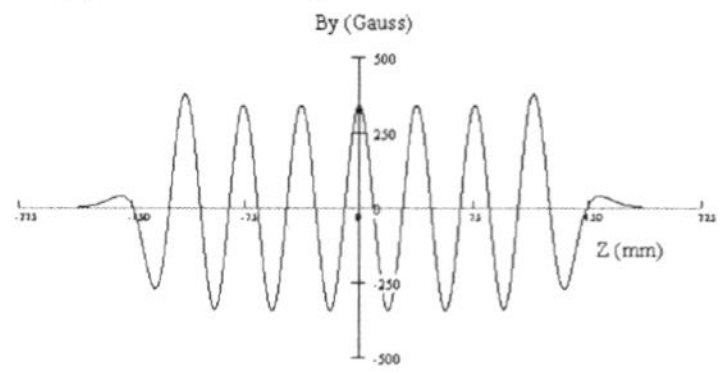

Fig. 2b: Magnetic field distribution for overlapped half magnet bar at each end.

It can be seen from Fig.2b that the "rabbit ears" effect has almost disappeared, and as a result, the electrons travel much closer to the axis inside the wiggler. Notice also that there is now no visible external field when |Z|>200.

3.3. Final Design

At first glance, the previous design does not present a practical solution, but, as superposition applies, the two overlapping magnets can be substituted, as shown in Fig.3, by two magnets and an air gap, by effectively removing the overlapping volume.

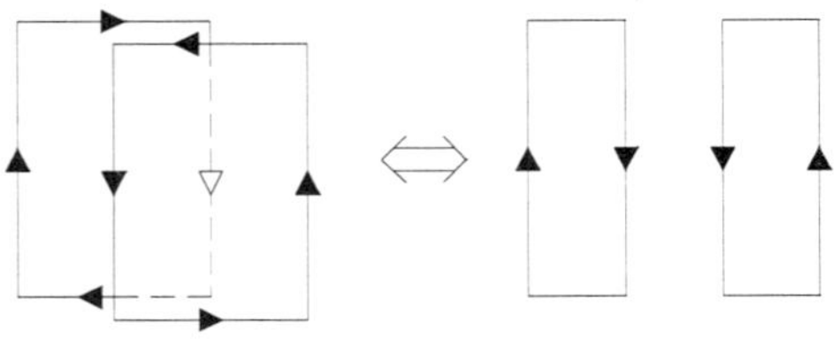

Fig. 3: Superposition of magnets.

For the actual wiggler design, the first and last full magnets were reduced to a width of 6mm, and the half magnets' width was reduced to 2.5mm. Fig.4 show how these modifications, produced an identical magnetic field distribution and beam trajectory as in Fig.2a and Fig.2b.

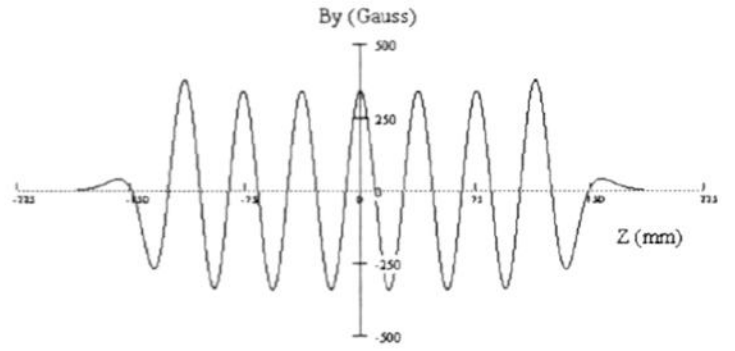

Fig. 4: Magnetic field distribution for final design.

4. Actual Wiggler Measurements

Having produced a suitable theoretical design, a NdFeB magnet wiggler was built with the specifications listed in Table 1. The magnetic field measurements have been made using a Hall probe with an effective cross-sectional area of approximately 0.04mm^2, and an accuracy of ±1.3 Gauss.

The distribution of the wiggler magnetic field has been obtained and is shown in Fig.5, the average magnetic field along the gap centre is 320 Gauss (0.032 Tesla). The electron beam trajectory was calculated from a double integration of the magnetic field distribution, and the results showed that the wiggler would produce a 1mm off-axis shift, which was in agreement with the simulated results.

Table 1: Wiggler magnet parameters

Remanent field Br	1.25 Tesla
Permanent magnet material	NdFeB
Wiggler period (λw)	19mm
Wiggler gap (g)	22mm
Average magnetic field	320 Gauss
Number of magnets per period	2
Total No. of periods	30
Overall length	570mm

5. Conclusions and Future Work

MagSim and TrajectSim are two simulation programs, which have been developed as a design tool for a pure permanent magnet wiggler, for a new pulsed FEM. The software has enabled the wiggler to be constructed with simplified end sections, and so reduced the overall cost. Magnetic field distribution measurements for the wiggler are in good agreement with the simulated results.

Further simulations are in progress to investigate the ability of side magnets [4] to correct the effects of poor magnet matching in the wiggler's central section, on the electron beam trajectory. Preliminary results suggest that the use of side magnets could correct the trajectory of a centrally fired electron even with ±5% errors in the magnets' residual fields, and give an acceptance angle at the wiggler entry point of about 10 degrees. This would be of great benefit in terms of the speed of wiggler construction.

6. References

[1] A.I. Al-Shamma'a et al., Nucl. Instr. and Meth. in Phys. Res., A 375 (1996) 424.

[2] A.I. Al-Shamma'a et al., Nucl. Instr. and Meth. in Phys. Res., A 393 (1997) II 87.

[3] C.C. Wright, Internal Report, University of Liverpool, 1998.

[4] M. Cohen et al, Phys. Rev. Lett. 74 (1995) 3812.

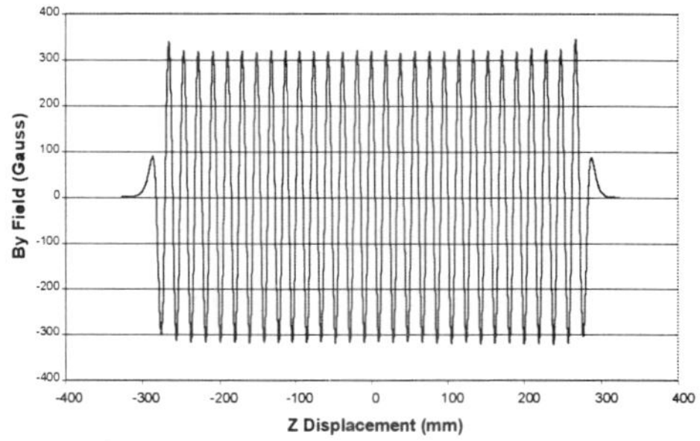

Fig. 6: Wiggler Magnetic Field Distribution for 30 periods.

FREE ELECTRON LASERS 1998
G.R. Neil and S.V. Benson (Eds.)
1999 Elsevier Science B.V.

A STRONG FOCUSING UNDULATOR SCHEME*

A. S. Khlebnikov a), N. S. Osmanov, A. V. Smirnov, and R. Tatchyn†
Russian Research Center 'Kurchatov Institute', 123182 Moscow, Russia
†Stanford Synchrotron Radiation Laboratory, Stanford Linear Accelerator Center, Stanford, CA 94305, USA
a)Corresponding author. Tel./Fax: 7(095)196 7749; E-mail: khlebnik@ismain.isssph.kiae.ru

Abstract

Improvements of two limiting cases of a novel hybrid/permanent magnet (hybrid/PM) undulator design are considered. For the short-period (< ~1cm) case the magnetic structure doesn't contain any small-scale PM pieces or permeable blocks. At a 4.5 mm vertical gap and a 9 mm period an on-axis field amplitude of ~3.1 kGauss and a period-averaged focusing gradient of up to 500 T/m can be attained. For the case emcompassing period lengths > ~1 cm the magnetic field amplitude can be increased up to 5.1 kGauss for the same gap/period ratio by installing an additional set of bias PM pieces. 3D-simulation results of selected magnetic field optimizations are discussed.

1. Introduction

The recent introduction of X-ray FEL (XRFEL) projects based on Self-Amplified Spontaneous Emission (SASE) has led to the development of specialized high-field undulator designs with strong superimposed focusing. In a number of approaches proposed or introduced in recent years period-averaged focusing gradients of up to ~100 T/m have been reported for superimposed permanent magnet (PM) lattices [1,2,3], and up to ~50 T/m with shaped/canted pole hybrid/PM designs [4], at undulator gaps and periods down to ~5 mm and ~3 cm, respectively. Continuing in this direction, a novel design of a strong-focusing hybrid/PM undulator with even higher focusing gradients and shorter attainable periods has recently been advanced by us and our collaborators [5,6]. For periods down to the sub-centimeter range the magnetic structure doesn't contain any small-scale PM pieces or permeable blocks, the undulator field being channeled through four steel monoblocks with machined periodic profiles. The steel geometry provides for a straightforward mechanism of leakage-field suppression and the PM monoblocks that generate the field are substantially displaced from the electron beam axis, a favorable configuration for high-bremsstrahlung environments. Experimental and numerical studies of selected variations of the basic structure indicated the attainability of averaged focusing gradients in the ~300-500 T/m range. At these focusing strengths, the undulator's dipole field amplitude can attain up to ~0.29 T for a gap/period ratio of 1/2, which is comparable to or greater than the dipole fields attained with previously reported pure-PM monoblock structures featuring similar ratios [7,8,9]. Despite these results, the importance of increasing the maximum mid-plane field of our design even further, up to the regime typical of the Halbach ("Type 3") configuration [10], was clearly recognized by us from the outset. In the present paper we describe three improvements on our original scheme that lead to increased on-axis field amplitudes. Two of these can be used for sub-centimeter period lengths (Case 1), while the third design, which includes added PM material in closer proximity to the gap, is more applicable to periods in excess of ~1cm (Case 2).

*Portions of this work were supported in part by the Department of Energy Offices of Basic Energy Sciences and High Energy and Nuclear Physics, and Department of Energy Contract DE-AC03-76SF00515.

2. Strong-Focusing Design Variations

2.1. Case 1

A schematic view of the top half of the improved undulator design is shown in Fig.1. The uniform part of

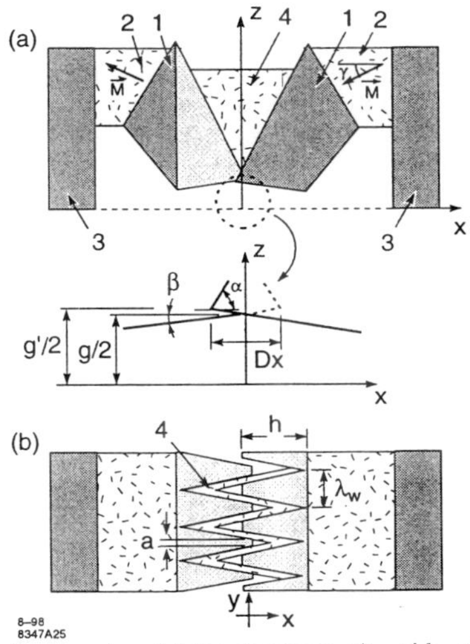

Figure 1. Improved undulator structure: a) - side view of the top part of the structure; b) - top view. 1 - steel yokes; 2 - PM material; 3 - steel plates; 4 - bias PM pieces (Case 2 only).

the magnetic structure consists of four steel blocks (1) and two additional steel plates (3) which form right and left yokes with a magnetic gap (g) and four PM blocks (2) with magnetization vectors inclined at an angle γ. No bias PMs are employed. Each steel block (1) has a periodic structure (with period λ_W) machined into the poles. The poles have a wedge shape with dimensions h and a in the XY-plane and an acute angle $\alpha+\beta$ in the XZ-plane. The right and left parts of the assembly are shifted relative to each other along the Y-axis by the value $\lambda_W/2$ and along the X-axis by Dx. The undulator's magnetic field is created by the superposition of the modulated magnetic fluxes of both yokes. The average value of the on-axis magnetic field is equal to 0. Both the displacement Dx and the specific steel yoke profile provide some amount of leakage flux suppression between the yokes' poles.

2.2 Case 2

When the undulator period is larger than ~1 cm one can add a set of PM bias magnets (Fig. 1). These (4) are affixed in the slots between poles in orientations similar to those of Ref. [4], and can be machined either as single pieces or as serrated profiles of a larger PM monoblock. In the simulations we used λ_W = 9 mm, g = 4.5 mm, Vanadium-Permendur yokes, and NdFeB permanent magnets (Br=1.2 T). The angle γ between the magnetization vectors of the PM blocks (2) and the X-axis was ± 20°. The parameter h was equal to 10 mm.

The scheme without the added PM bias pieces can provide an enhanced magnetic field amplitude of up to 0.305 T for the parameters a=2.2 mm, α=80°, Dx=-2.0 mm (parallel gap, β=0°), and up to 0.317 T for the same a, α, and Dx=-1.8 mm (canted gap, β=-7°, g'=4.72 mm).

In the design with the bias magnets the magnetic field amplitude can reach up to 0.513 T for the parameters a=1.0 mm, α=80°, Dx=-2.7 mm (parallel gap, β=0°). Figure 2 shows the mid-plane field and its gradient as functions of Dx for these parameters. One can see that the respective maxima of the undulator field and gradient are attained for different values of Dx.

The dependence of the optimal values of the field amplitude B_Z (1) and the parameter Dx (2) on a is displayed in Figure 3. It is evident that small values of a (for which the approaching similarity of our structure to

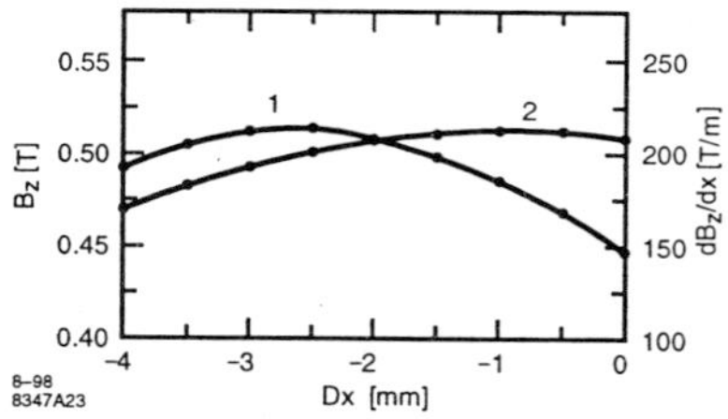

Figure 2. Undulator field (1) and its focusing gradient (2).

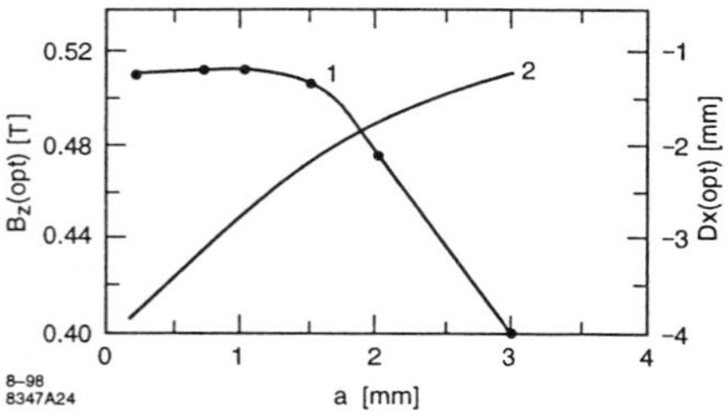

Figure 3. Optimization curves for Case 2 (bias PM blocks included).

the Halbach "Type 3" field configuration becomes increasingly evident) are preferable. However, even the original scheme [5] (with a=3 mm) is seen to generate a substantial field increase of ~33% when the PM bias magnets are employed.

Thus, our work indicates that the described structures, once suitably optimized, can provide strong-focusing average gradients of ~200 T/m or more, together with dipole field strengths approaching those of an optimal Halbach configuration.

3. References

[1]. R. Tatchyn, Nucl. Instr. and Meth. A341 (1994) 449.

[2]. A. A. Varfolomeev, V. V. Gubankov, A. A. Hairetdinov, S. N. Ivanchenkov, A. S. Khlebnikov, N. S. Osmanov and S. V. Tolmachev, Nucl. Instr. and Meth. A358 (1995) 70.

[3]. Y. M. Nikitina, J. Pfluger. Nucl. Instr. and Meth. A375 (1996) 295.

[4]. R. D. Schlueter, Nucl. Instr. and Meth. A358 (1995) 44.

[5]. M. Caplan, V.V.Gubankov, A.S.Khlebnikov, N. S. Osmanov, A. V. Smirnov, R. Tatchyn, L. M. Tkachenko, S. V. Tolmachev and N. S. Yarimov. A short-period strong focusing undulator scheme. Presented at 19-th International Free Electron Laser Conference, August 18-22 August, Beijing, China. Preprint of Russian Research Center 'Kurchatov Institute', IAE-6096/14 (1998), Moscow.

[6]. A. S. Khlebnikov, A. V. Smirnov, R. Tatchyn. A short-period strong focusing undulator scheme. Presented at the 6-th European Particle Accelerator Conference, 22-26 June, 1998, Stockholm.

[7] G. Ramian, L. Elias and I. Kimel, Nucl. Instr. and Meth. A250 (1986) 125.

[8] R. Tatchyn, A. Toor, J. Hunter, R. Hornady, D. Whelan, G. Westenskow, P. Csonka, T. Cremer and E. Kallne, J. of X-ray Science and Technology 1, (1989)79.

[9] A. Toor, P. Csonka, R. Tatchyn, Rev. Sci. Instrum. 60(7), 1439(1989).

[10] M. J. Burns, G. A. Deis, R. H. Holmes, R. D. Van Maren, K. Halbach, IEEE Trans. Magnetics 24(2), 978(1988).

FREE ELECTRON LASERS 1998
G.R. Neil and S.V. Benson (Eds.)
1999 Elsevier Science B.V.

Low-energy beam focusing in various types of undulators

Yoshikazu Miyahara
SPring-8, Mihara, Mikazuki, Sayo-gun, Hyogo 679-5198, Japan

In a low energy FEL, electron beam focusing is important to get a higher FEL gain. It has been thought that a helical field produced by a double helix coil is very suitable for this purpose. Since a superconducting coil for a high peak field is not easy to construct, other methods such as additional quadrupole magnets, a parabolic pole undulator and an angled edge undulator have been investigated [1-3]. Meanwhile, several kinds of helical field undulators composed of permanent magnets have been recently proposed and constructed; a crossed array undulator [4], a double array undulator [5] and a canted array undulator [6]. In the present paper, orbit trajectories of a low energy electron beam in these undulators are numerically investigated and focal lengths of the beam are obtained against the beam energy, and compared with that of a planer undulator, which has a vertical focusing.

The field of Halbach type *planer undulator* is given by

$$B_x=-B_0(k_x/k_y)\sin(k_x x)\sinh(k_y y)\cos(k_z z)$$

$$B_y= B_0\cos(k_x x)\cosh(k_y y)\cos(k_z z)$$

$$B_z=-B_0(k_z/k_y)\cos(k_x x)\sinh(k_y y)\sin(k_z z)$$

where B_0 is the peak field, $k_y{}^2-k_x{}^2=k_z{}^2$ and $k_z=2\pi/\lambda_0$ with λ_0 the undulator period.

The pole face of *parabolic pole undulator* is concaved to introduce a sextupole field, and the field is obtained by replacing k_x with ik_x in the field for the planer undulator.

Crossed array undulator is composed of horizontal and vertical planer magnet arrays. The field of horizontal array is given above. The field of vertical array is given by

$$B_x'= B_0\cos(k_y'y)\cosh(k_x'x)\cos(k_z z+\pi/2)$$

$$B_y'=-B_0(k_y'/k_x')\sin(k_y'y)\sinh(k_x'x)\cos(k_z z+\pi/2)$$

$$B_z'=-B_0(k_z/k_x')\cos(k_y'y)\sinh(k_x'x)\sin(k_z z+\pi/2)$$

where $k_x'^2-k_y'^2=k_z{}^2$. The total field is given by the sum with that for the planer undulator.

The field of *double array undulator*, composed of two Halbach types of undulators, can be expressed as

$$B_x= B_0(K_x/K_y))\sin(\phi/2)\cos(K_x x)\cosh(K_y y)\sin(k_z z)$$

$$B_y= B_0\cos(\phi/2)\cosh(k_y y)\cos(k_z z)$$
$$+B_0\sin(\phi/2)\sinh(K_x x)\sin(K_y y)\sin(k_z z)$$

$$B_z=-B_0(k_z/k_y)\cos(\phi/2)\sinh(k_y y)\sin(k_z z)$$
$$+B_0(k_z/K_y)\sin(\phi/2)\sin(K_x x)\cosh(k_y y)\cos(k_z z)$$

where $K_x=\pi(10/g)^{1/2}$, $K_y=(K_x{}^2+k_z{}^2)^{1/2}$ with g the magnet gap height, and ϕ is phase difference of the magnet arrays. This formula is valid in the range $K_x x\leq\pi/2$. Circularly polarized radiation is produced when $\tan(\phi/2)=K_y/K_x$ is satisfied.

Canted array undulator is basically composed of a Halbach type undulator, the top and bottom magnet arrays of which are canted by θ against the beam axis in the horizontal plane. The field is expressed as

$$B_x=-(B_0/2)\sin\theta\{\exp(k_z y)\sin[k_z f(z,+,x)]$$
$$+\exp(-k_z y)\sin[k_z f(z,-,x)]\}$$

$$B_y= (B_0/2)\{\exp(k_z y)\cos\{k_z f(z,+,x)]$$
$$+\exp(-k_z y)\cos[k_z f(z,-,x)]\}$$

$$B_z=-(B_0/2)\cos\theta\{\exp(k_z y)\sin[k_z f(z,+,x)]$$
$$-\exp(-k_z y)\sin[k_z f(z,-,x)]\}$$

Table 1 Parameters and conditions for numerical calculation of orbit trajectories.

Beam energy	E(MeV)	5–40
Undulator period	λ_0(mm)	60
Peak field	B_0(T)	0.3
Planer undulator	$k_x=0$	
Parabolic undulator	$k_x=k_y=k_z/\sqrt{2}$	
Crossed array undulator	$k_x=k_y'=0$	
	$k_y=k_x'=k_z$	
Double array undulator	$k_x=0$	
	g=30 mm	
	ϕ=129 deg	
Canted array undulator	λ_{eff}=60 mm	
	$\theta=\pi/4$	

where $f(z,\pm,x)=z\cos\theta\pm x\sin\theta$, and $k_x=0$ and $k_y=k_z$ are assumed. The undulator produces an elliptical field with the ratio $\sin\theta$ between the horizontal and vertical peak fields. The undulator has an effective period $\lambda_{eff}=\lambda_0/\cos\theta$, with λ_0 the period before canted.

The field in *double helix coil* is approximately given by

$$B_x \cong -B_0\{[1+(1/8)k_z^2(3x^2+y^2)]\sin(k_z z) - (1/4)k_z^2 xy\cos(k_z z)\}$$

$$B_y \cong B_0\{[1+(1/8)k_z^2(x^2+3y^2)]\cos(k_z z) - (1/4)k_z^2 xy\sin(k_z z)\}$$

$$B_z \cong -B_0[1+(1/8)k_z^2(x^2+y^2)][k_z x\cos(k_z z) + k_z y\sin(k_z z)]$$

Orbit trajectory of an electron beam in the undulators was calculated numerically for the above fields with the equation, $m\gamma d^2\mathbf{r}/dt^2=-e\mathbf{v}\times\mathbf{B}$, where m and e are the electron mass and charge, respectively, γ is the relativistic beam energy and **v** is the velocity. Field correction at the entrance of undulators was not considered. Instead, beam injection angle into the undulators was adjusted to reduce the orbit amplitude. The focal length is defined as 1/8 of one period of average orbit oscillation. Table 1 shows the parameters and conditions for the numerical calculation.

Fig.1 shows the beam energy dependence of the vertical focal lengths obtained from the orbit trajectories. Vertically the helix coil and the crossed undulator have the same focal length as that of the planer undulator. The canted undulator has a shorter focal length. The double array undulator and the parabolic undulator have 1.5 times longer focal length than that of the planer undulator. Horizontally, the helix coil, the crossed undulator and the parabolic undulator have the same focal lengths as those in the vertical plane. There is no focusing in the canted undulator and the double array undulator, although these undulators produce helical fields. These results suggest that all the focusing force in the undulators originally come from the vertical edge focusing. The double array undulator and the canted undulator, basically composed of plane magnet arrays in the top and bottom, have no focusing force horizontally. In the parabolic undulator, the original edge focusing is divided into the horizontal and vertical planes by the sextupole field in the undulator. The same thing is realized by changing the edge angle in a planer undulator. Since the focal lengths are all proportional to the beam energy, as shown in the figure, they can be scaled at a higher beam energy.

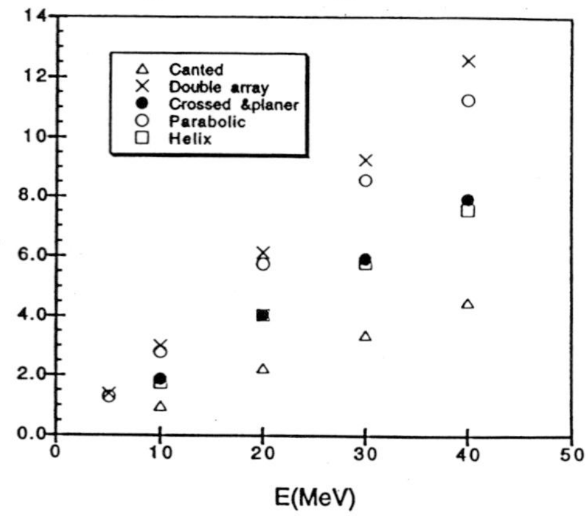

Fig.1 Energy dependence of vertical focal length in various type undulators. Horizontal focal lengths are the same except for planer, double array and canted array undulators, which have no horizontal focusing.

[1] E.T.Scharlemann, J.Appl.Phys. 58 (1985) 2154-2161.

[2] Y.Tsunawaki et al., Nucl.Instr. and Meth. A 304 (1991) 753-758.

[3] K.Mima et al., Nucl.Instr. and Meth. A 318 (1992) 834-838.

[4] H.Onuki, Nucl.Instr. and Meth. A 246 (1986) 94.

[5] S.Sasaki et al., Jpn.J.Appl.Phys. 31(1992) L1794-1796.

[6] Y.Miyahara and H.Kobayashi, Nucl. Instr. and Meth. A 410 (1998) 297-303.

FREE ELECTRON LASERS 1998
G.R. Neil and S.V. Benson (Eds.)
1999 Elsevier Science B.V.

Nonlinear gain and electron acceleration in the down-conversion IR FEL

I.V. Smetanin
P.N.Lebedev Physics Institute, Ac.Sci.of Russia, Moscow, Leninskii prospect 53, 117924 Russia

Abstract

The unconventional scheme of an optical undulator IR free-electron laser that achieves Doppler frequency down-conversion is investigated. Analytical non-perturbative theory of this down-conversion FEL in the nonlinear strong-signal low-gain Compton regime is developed. The electromagnetic field distribution and electron beam evolution inside the interaction region are investigated and FEL saturation intensity is determined..

The optical undulator FEL is considered now as a promising coherent and rather compact x-ray source. In the IR spectral domain, the new unconventional "down-conversion" scheme of the optical undulator FEL is suggested [1,2]. In the "down-conversion" FEL a high-power electromagnetic pump wave (of wavelength λ_i) and the REB propagate in the same direction, in contrast with the conventional FEL concepts. The signal amplified is then counter-propagating and its wavelength (λ_s) is determined by the reverse of the conventional relation ($\gamma >> 1$) $\lambda s \approx 4\gamma^2 \lambda_i$. Assuming a high-power Nd-laser (λ_i=1.06 μm) radiation as a pump and $\gamma \approx 5$ we have $\lambda_s \approx$ 100μm. Thus, it is possible to cover all the IR spectral domain using low-voltage high current electron accelerators. That is the clear advantage of the scheme.

The linear theory of a down-conversion IR FEL in the low-gain Compton regime was given in Ref. [2]. It was found that this scheme reveals some specific features that are quite different from those of conventional FELs. As a result of lasing, in the down-conversion FEL the beam electrons have to be accelerated instead of decelerating in the conventional devices, because the energy of pumping wave is shared between the amplified signal wave and electron beam. Beam quality restrictions are substantially less severe than those in conventional FELs [2]

$$\langle \Delta\Theta^2 \rangle / 2, \Delta\gamma / \gamma < 4\gamma^2 / 2N \ , \qquad (1)$$

here $\Delta\gamma$ is the beam energy spread, $\langle\Delta\Theta^2\rangle$ is the mean angle spread due to the beam emittance, and N is the number of undulator periods.

In this paper, we develop a non-perturbative strong-signal 1D nonlinear theory of the down-conversion FEL in the low-gain Compton regime. A homogeneous REB is propagating in the positive *z*-axis direction at a velocity βc. The electromagnetic field potential is assumed to be a superposition of pumping (i) and signal (s) plane waves of circular polarization and slowly varying amplitudes $A_{i,s}(z)$. The pumping wave propagates in the beam direction, the signal is a counter-propagating wave.

Under the assumptions above, the dynamics of electrons are described by a one-dimensional kinetic equation, which can be reduced by a well-known procedure to a set of quasi-Bloch equations [3]

$$\frac{\partial f_0}{\partial z} = -\frac{ke^2}{pc^2}\frac{\partial R_1}{\partial p}$$

$$\frac{\partial R_1}{\partial z} + \mu R_2 = -2\frac{ke^2}{pc^2} A_i^2 A_s^2 \frac{\partial f_0}{\partial p} \qquad (2)$$

$$\frac{\partial R_2}{\partial z} - \mu R_1 = 0$$

Here f_0 is the steady-state part of the electron distribution function, f_1 is the part which oscillates at the frequency of ponderomotive potential, $iR_1 = A_i^* A_s f_1 - A_s^* A_i f_1^*$ and $R_2 = A_i^* A_s f_1 + A_i A_s^* f_1^*$, $\mu = m\gamma\omega/pc - k$ - is the detuning parameter of an electron from exact resonance with the ponderomotive wave of the frequency $\omega = \omega_i - \omega_s$ and wavenumber $k = k_i + k_s$, $p = m\gamma\beta c$ is the momentum of an electron. In analogy with conventional Bloch equations, the functions f_0, $R_{1,2}$ can be regarded as the medium excitation and real and imaginary parts of medium polarization, respectively. The boundary condition for equation (3) is the steady state Gaussian momentum distribution at the entrance of undulator.

The wave equations are then reduced to the form

$$\frac{\partial A_{i,s}^2}{\partial z} = -\frac{2\pi n_e e^2}{mc^2 k_{i,s}} \int R_1 \frac{dp}{\gamma} \qquad (3)$$

In the small-signal approximation, the set of equations (2),(3) can be solved using perturbation theory, which leads to the results of Ref.[2].

In the strong-signal nonlinear operation regime, perturbations of the electron distribution function become large and are to be accounted exactly. Assuming the amplitudes to be slow varying functions, we derive from (2) the equation for R_l [5]

Assuming for estimates γ=3.3, λ_l=1.06 μm, $I_i=10^{14}$ W/cm^2, $n_e=10^{12}$ cm^{-3} (i.e., ≈5 kA/cm^2), and L=2 the undulator length, we have the signal saturation intensity is $I_s \approx 0.33$ MW/cm^2.

To describe the evolution of the electron beam distribution function, one shoul integrate the first of the quasi-Bloch equations (2). As a result, we find

$$f_0(x,p) = f_0(0,p) - \frac{4\sigma}{\sqrt{\pi}\sigma_0}\exp\left[-\frac{y^2}{4} - \frac{y_0^2}{4\varepsilon_+^2}\right] \bullet$$

$$\sum_{n=0}^{\infty}\frac{1}{2^n n!}\frac{\varepsilon_-^{n-1}}{\varepsilon_+^{n-1}}\left[\frac{1}{2}H_{n+1}\left(\frac{y}{\sqrt{2}}\right) - nH_{n-1}\left(\frac{y}{\sqrt{2}}\right)\right] \bullet$$

$$\left[\frac{\varepsilon_-^2}{2\varepsilon_+^2}H_{n+1}\left(\frac{y_0}{\sqrt{2\varepsilon_+\varepsilon_-}}\right) - nH_{n-1}\left(\frac{y_0}{\sqrt{2\varepsilon_+\varepsilon_-}}\right)\right] \bullet$$

$$\frac{\sin^2(\lambda_n x/2)}{\lambda_n^2}$$

This solution can determines the broadening of the electron distribution and the gain in energy of beam electrons which result from the lasing in the down-conversion FEL. The acceleration of electrons in the down-conversion FEL is the direct consequence of energy conservation law in an elementary Compton event. Numerical calculations [5] show that, at the signal intensity of ≈1/4 saturation intensity (9), the characteristic width of the distribution function increases in ~3 times at the exit of undulator, and the mean energy gain per electron $\delta\gamma/\gamma \sim 2\gamma^2/kl \sim 0.3\%$.

Acknowledgements. This research is supported by Russian Basic Research Foundation.

References

[1]. A. Gover. Opt.Commun., **45** (1983) 281.
[2]. I. V. Smetanin. JETP, **79** (1994) 856 .
[3]. F. A. Hopf, P. Meystre, M. O. Scully, and W. H. Louisell. Phys. Rev. Lett. **37** (1976) 1342 .
[4]. H. Bateman, A. Erdelyi. Higher Transcendental Functions. Vol.2. McGraw-Hill, 1953.
[5]. I.V.Smetanin. Laser Phys., **8** (1998) 136.

FREE ELECTRON LASERS 1998
G.R. Neil and S.V. Benson (Eds.)
1999 Elsevier Science B.V.

The FEL Projects at the Rossendorf Radiation Source ELBE

F. Gabriel, P. Gippner, E. Grosse*, D. Janssen, P. Michel, H. Prade, A. Schamlott, W. Seidel, U. Steegmüller, M. Wenzel, A. Wolf, R. Wünsch and the ELBE-crew
Forschungszentrum Rossendorf, Postfach 510119, 01314 Dresden, Germany

The Forschungszentrum Rossendorf (FZR) is building a superconducting ELectron accelerator [1] with high Brilliance and low Emittance (ELBE) which can deliver a cw beam of 40 MeV and 1 mA. ELBE will be equipped with a free electron laser system for the production of infrared light in the range of 5 to 150 μm. The electron beam will also be used to generate X-rays, bremsstrahlung, positrons and neutrons.

Electron gun: The electron gun is designed to deliver microbunches of 85 pC at 11.8 MHz yielding an average current of 1 mA. To ensure good gain for the infrared radiation (IR) production in the free electron laser (FEL) special care is taken to keep the beam emittance low. The normalized transverse emittance is expected to be below 7 mm mrad (rms). For the thermionic gun with a subharmonic/fundamental (0.26/1.3 GHz) buncher system the longitudinal emittance will be well below 100 keV·ps (rms) with a rather large rms energy spread of up to 100 keV imposed by the necessarily quite high bunching voltage.

We therefore started the development of a superconducting photo-electron gun [2] which would allow us to reduce the longitudinal emittance by about a factor of 4 and especially the energy spread to below 30 keV. This gun is presently being built in collaboration with the Budker Institute at Nowosibirsk, where the surrounding acceleration cavity ($E_{z,max} \approx 20$ MV/m) has been designed. Robust Cs_2Te-photocathodes for 5 eV photon energy will be used inside of the superconducting cavity. A 11.8 MHz repetition rate laser system with rather large average power (1 W) at a wavelength of 262 nm will be developed by the Max-Born-Institut für Nichtlineare Optik und Kurzzeitspektroskopie at Berlin.

Accelerator: The maximum electron energy of about 40 MeV will be reached by using 4 superconducting 1.3 GHz Nb-cavities with 9 cells each. These cavities are identical to the ones developed for the DESY TESLA-Test-Facility (TTF), where they are supposed to reach at least 25 MV/m when used at low duty cycle, whereas for ELBE a cw-operation with a minimum of 10 MV/m is envisaged. After chemical treatment and tests of the cavities performed at the DESY site, the cavities will be arranged pairwise in two cryostats of 3.2 m length. Between them there will be enough space to allow the installation of a small superconducting energy modulator cavity and a corresponding magnetic separator system such that subsequent beam bunches of different energy can be produced [3]. This will allow pump-probe experiments with independently tunable wavelengths from two parallel driven FELs. The design of the cryostats and the accelerator set-up was performed in close contact with the Hansen Experimental Physics Laboratory (HEPL), where a similar arrangement is planned for the upcoming upgrade of the Stanford Picosecond FEL Center. A floor plan of the ELBE project is shown in fig. 1. The beam line system will lead the electrons to a number of experimental caves, which have to be shielded heavily by concrete walls due to the rather high beam power. Target stations to generate positrons, neutrons and bremsstrahlung for investigations related to nuclear physics and technology are planned and special emphasis is put on channeling and parametric X-ray production. The rather good quality of that radiation [4] makes it a candidate for its use in independent experiments as well as in correlation to the IR-studies to be performed with the FELs for which two caves are reserved.

FEL-undulators: For the far-infrared wavelengths an electromagnetic undulator is being developed and built at the FZR. It will have a period of 90 mm and rms K-values from 0.6 to 1.2 are to be realized by adjusting the current in the magnetic coils. Varying the electron energy from 10 to 40 MeV will result in FEL radiation between 150 and somewhat below 25 μm. This undulator will first be installed downstream from the accelerator and the beam line leading to it will be equipped with a chicane for beam compression.

The double dog leg beam line leading to the mid-infrared FEL cave has been especially designed to allow control of the bunch length as well as of the transverse phase space. Thus an optimum matching to the undulator can be realized. To obtain bunch compression the buncher and accelerator have to be phased such that high momentum particles enter the beam line first - in contrast to the situation when a chicane is used.

In the initial phase of ELBE use will be made of two existing permanent magnet undulators for the mid-infrared region which we envisage to install for some period of time. The 50 mm period undulator with 45 elements built at ENEA/Frascati [5] is housed in a large vacuum chamber and its widely variable gap should result in K-values ranging from about 0.5 to 2.0. With 40 MeV electrons the shortest wavelength to be reached will be about 6 μm and a small–signal gain above 20% can be reached for wavelengths up to 50 μm.

---

*Corresponding author, also at TU Dresden

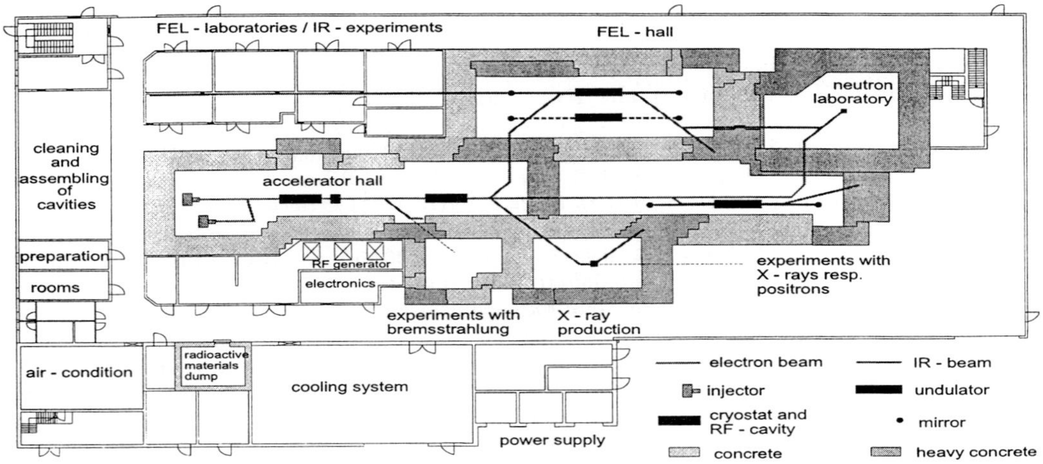

FIG. 1. Floor plan of the radiation source ELBE

For a better coverage of shorter wavelengths a smaller undulator period is desirable. For the TTF at DESY a 15 m long wiggler [6] with 27 mm period is under construction and plans have been made to temporarily install at ELBE two undulator units of 33 segments each which are not integrable in the long DESY device. The undulator has to be set up such that gap (and K-value) can be varied. A reasonable small signal gain can be obtained over a wide range in K allowing wavelengths down to 3 μm to be reached at 40 MeV. The gain will be nearly a factor of 2 higher when low energy-spread beam-bunches, as obtainable with a photo-electron-gun, are injected.

Optical resonator and beam line: Corresponding to the electron beam repetition rate of 11.8 MHz the length of the optical resonator will be 12.7 m. The projected Rayleigh length of $\approx$1 m is a compromise between gain considerations, stability criteria, mode size and diffraction losses at mirror and beam line components. At present the consequence of single and multi-hole resonator outcoupling to the transverse mode structure of the beam is numerically investigated. The IR beam will be examined with respect to its power, wavelength, bandwidth, pulse duration and transverse dimensions in a control cave before it is directed to the user caves. According to FZR's concept of becoming a user facility ample laboratory space is arranged for. We are presently investigating the advantage of installing a multi-user pump-probe set-up as well as a cavity ring down experiment; a target station for gaseous probes, eventually windowfree, is under consideration.

Outlook: After completion of the approval procedure, which was complicated by the extended radiation-hazard studies, the groundbreaking took place this summer, such that the accelerator installation can start in the spring of 1999. We expect the first electron beam before the end of that year and the first infrared light not too much later. As indicated in this report, the conditions for high power laser emission are under special study at ELBE.

We have increasing contacts to possible users of our facility from other groups in the Forschungszentrum Rossendorf as well as from outside. We have held user's meetings for applications in biomedicine and also for investigations in semiconductor research and physical chemistry. In addition to the wide wavelength coverage important for material science and biomedical or physicochemical investigations a high average radiation intensity is aimed for to facilitate IR-induced modifications of materials and surfaces. Under consideration is also to deliver radiation produced by other processes for a simultaneous use together with the FEL infrared light.

Acknowledgment: We are greatful to our collaboration partners at HEPL, DESY, ENEA, JLab, Budker-Institut, FELIX, and last but not least TU Darmstadt.

[1] ELBE Design Report 1998 (in preparation)
[2] D. Janssen et al. Conf. Proc. of Particle Acc. Conf., Vancouver, Canada, (1997)
[3] E.R. Crosson, G.E. James, H.A. Schwettman, T.I. Smith, R.L. Swent, *Nucl. Instr. Meth.* **A 393** (1997) II-33
[4] H. Genz, L. Groening, P. Hoffmann-Stascheck, A. Richter, M. Höfer, J. Hormes, U. Nething, J.P.F. Sellschop, C. Toepffer, M. Weber, *Phys. Rev.* **B 53** (1996) 8922
[5] F. Ciocci, E. Fiorentino, G.P. Gallerano, A. Renieri, E. Salia, *Nucl. Instr. Meth.* **A 250** (1997) 380-384
[6] B. Faatz, J. Pflüger, Y. M. Nikitina, *Nucl. Instr. Meth.* **A 375** (1996) 618-625

FREE ELECTRON LASERS 1998
G.R. Neil and S.V. Benson (Eds.)
1999 Elsevier Science B.V.

Spontaneous Emission and Gain in Free-Electron Laser With Circularly Polarized Wiggler and Ion-Channel Guiding.

Punit Kumar, Kamna Pande and Pallavi Jha.
Department of Physics, University of Lucknow, Lucknow - 226007, India.

A relativistic electron beam (REB) injected into an underdense ionized plasma channel ejects plasma electrons, leaving behind a positive ion core which attracts and confines the beam electrons [1]. For an electron beam perfectly injected into a free-electron laser (FEL) with a circularly polarized wiggler, the presence of an ion channel leads to significant amplification of a co-propagating radiation field at the upshifted wiggler frequency [2,3]. In this paper, we show that if the electron is axially injected into the interaction region of such an FEL configuration, its transverse trajectory becomes a complicated function of the wiggler and ion-channel betatron frequencies. Axial oscillations arise in the electron motion and indicate the possibility of harmonic generation. The trajectories obtained for axial injection have been used in evaluating the spontaneous emission spectrum at the upshifted wiggler frequency and its harmonics. Finally, gain in radiation has been obtained for the same frequencies in the low-gain-per-pass limit.

Consider a relativistic electron (charge $-e$, rest mass m) injected axially with energy $\gamma_o mc^2$ and velocity $\beta_o c$ into the interaction region of an FEL having a circularly polarized wiggler magnetic field of amplitude B_w, wavenumber $k_w (= 2\pi/\lambda_w)$, described by $\vec{B_w} = B_w(cosk_w z, sink_w z, 0)$, in the presence of a pre-ionized plasma channel whose axis is coincident with the wiggler axis. Once the plasma electrons are expelled, the transverse electrostatic field generated by the ion channel is given by $\vec{E_i} = 2\pi e n_i(x, y, 0)$, n_i being the density of positive ions having charge e. For an electron propagating in the ion-focussed regime, routine instabilities that arise in transport get damped [1]. On solving the relativistic Lorentz force equations perturbatively in orders of the wiggler magnetic field, the electron trajectories may be obtained. The transverse electron trajectory is found to be a superposition of two helices, one characterized by the wiggler frequency ω_w, and the other by the betatron frequency ω_i. This gives rise to the oscillations in the axial direction.

The intensity of spontaneous emission per unit solid angle per unit frequency interval emitted by the electron (along the wiggler axis) in the presence of the combined ion-channel and wiggler fields, at a frequency $\omega = \frac{f\beta_o\omega_w}{(1-\bar{\beta}_{zo})}$ is found to be

$$\frac{d^2W}{d\Omega d\omega} = \frac{e^2 K_i^2 N_w^2 f^2 \alpha_s^2}{4\gamma_o^2 c(1-\bar{\beta}_{zo})^2} \frac{sin^2(\frac{\pi N_w \nu_f}{\beta_o\omega_w})}{(\frac{\pi N_w \nu_f}{\beta_o \omega_w})^2}. \qquad (1)$$

where $\nu_f = (1-\bar{\beta}_{zo})\omega - f\beta_o\omega_w, \alpha_s^2 = P_X^2 + P_Y^2$,

$$P_X = \sum_{l=-\infty}^{\infty}\sum_{m=-\infty}^{\infty}\sum_{n=-\infty}^{\infty} J_l(a_1)J_m(a_2)J_n(a_3) [\delta_{m,l-f}\delta_{n,\frac{-(l+m-1)}{2}} + \delta_{m,l-f}\delta_{n,\frac{-(l+m+1)}{2}} - \delta_{m,l-f-1}\delta_{n,\frac{-(l+m)}{2}} - \delta_{m,l-f+1} \cdot \delta_{n,\frac{-(l+m)}{2}}],$$

$$P_Y = \sum_{l=-\infty}^{\infty}\sum_{m=-\infty}^{\infty}\sum_{n=-\infty}^{\infty} J_l(a_1)J_m(a_2)J_n(a_3) [X_i\delta_{m,l-f}\delta_{n,\frac{-(l+m-1)}{2}} - X_i\delta_{m,l-f} \cdot \delta_{n,\frac{-(l+m+1)}{2}} - \delta_{m,l-f-1}\delta_{n,\frac{-(l+m)}{2}} + \delta_{m,l-f+1}\delta_{n,\frac{-(l+m)}{2}}],$$

$a_1 = \frac{K_i^2\omega}{2\beta_o^2\gamma_o^2\omega_w}(\frac{X_i}{\gamma_o^2} - 1)(\frac{X_i-1}{X_i+1}), a_2 = \frac{K_i^2\omega}{2\beta_o^2\gamma_o^2\omega_w}(\frac{X_i}{\gamma_o^2} + 1)(\frac{X_i+1}{X_i-1})$,
$a_3 = \frac{K_i^2\omega}{8\gamma_o^2\omega_w}(\frac{X_i^2-1}{X_i})$, N_w is the number of wiggler periods and $J_l(a_1), J_m(a_2), J_n(a_3)$ are Bessel functions of the first kind.

The spectrum will be sharply peaked around the frequency $\beta_o\omega_w/(1-\bar{\beta}_{zo})$ and its odd harmonics. The line broadening for every harmonic depends only on the length of the interaction region $L(=\frac{2\pi N_w c}{\omega_w})$ whereas the intensity depends on f as well as the electron beam, wiggler and ion-channel parameters.

Let an electromagnetic radiation of frequency $f\omega_r(=cfk_r=\frac{2\pi fc}{\lambda_r})$, amplitude E_r and phase $\xi(=fk_r z-f\omega_r t+\phi)$, represented by $\vec{B_r}=E_r(sin\xi,cos\xi,0),\vec{E_r}=E_r(cos\xi,-sin\xi,0)$, co-propagate with the electron in the FEL interaction region. For relativistic electrons, the transverse force due to the radiation field is nearly zero, hence transverse trajectories have nearly the same form as in the absence of radiation. The axial velocity is obtained using the energy conservation relation. Once the trajectories are known, we find the rate of exchange of the electron's energy with the radiation field and eliminate the fast-oscillatory terms to get

$$\dot{\gamma}=\frac{eK_iE_r}{2mc\gamma}P_J cos(\zeta_w+\phi), \quad (2)$$

where $\zeta_w=fk_r\bar{z}-f(\omega_r-\beta_o\omega_w)t$ is a slowly evolving dimensionless parameter,

$$P_J = \sum_{l=-\infty}^{\infty}\sum_{m=-\infty}^{\infty}\sum_{n=-\infty}^{\infty} J_l(A)J_m(B)J_n(C) \left[(1+X_i)\delta_{m,l-f}\delta_{n,\frac{-(l+m+1)}{2}}+(1-X_i).\delta_{m,l-f}\delta_{n,\frac{-(l+m-1)}{2}}-2\delta_{m,l-f+1}\delta_{n,\frac{-(l+m)}{2}}\right],$$

$A=\frac{K_i^2f\omega_r}{2\beta_o\gamma_o^2\omega_w}(\frac{1-X_i}{X_i+1})$, $B=\frac{K_i^2f\omega_r}{2\beta_o\gamma_o^2\omega_w}(\frac{X_i+1}{X_i-1})$, and $C=\frac{K_i^2f\omega_r}{8\beta_o\gamma_o^2\omega_w}(\frac{X_i^2-1}{X_i})$. It may be seen that the coupling factor P_J determines the energy transfer and is non-zero only for odd-integral values of f.

Differentiating ζ_w twice with respect to time, we get the pendulum equation

$$\ddot{\zeta}_w=\frac{eK_iE_rf\omega_r(1+\eta K_i^2)}{2mc\gamma^4}P_J cos(\zeta_w+\phi). \quad (3)$$

where $\eta=2+(\frac{X_i^2-1}{2})$

Following Colson [4], the FEL gain is evaluated as

$$G=\frac{\pi e^2N_bK_i^2L^3f\omega_r(1+\eta K_i^2)P_J^2}{2m\beta_o^3c^3\gamma_o^5}F(\Delta\omega T), \quad (4)$$

where $F(\Delta\omega T)=\frac{1}{(\Delta\omega T)^3}[2-2cos\Delta\omega T-\Delta\omega T sin\Delta\omega T]$, $\Delta\omega=f\beta_o\omega_w-(1-\bar{\beta}_z)f\omega_r$ is the resonance parameter, $T(\simeq L/c\beta_o)$ is the time taken by the electron to traverse the interaction length L and N_b is the electron beam density.

The gain is maximum when $\Delta\omega T=2.6$ at a frequency

$$f\omega_r=\frac{f\beta_o\omega_w}{1-\bar{\beta}_z}(1-\frac{2.6}{Lk_w})\simeq\frac{2\gamma_o^2f\beta_o\omega_w}{1+\eta K_i^2}. \quad (5)$$

Since f is an odd integer, gain is obtained at the fundamental upshifted wiggler frequency and its odd harmonics.

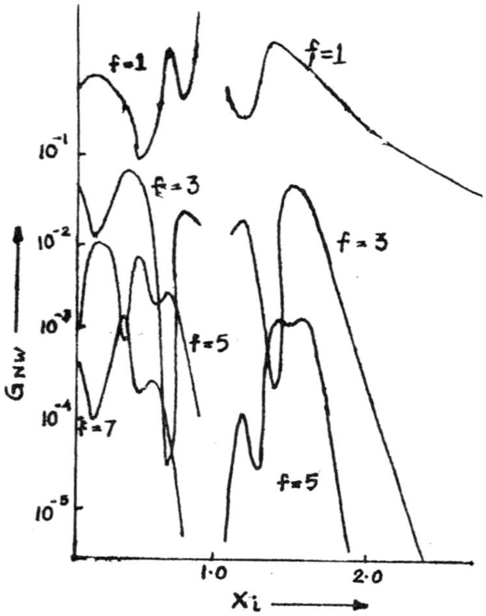

The above figure shows the variation of maximum gain G_{Nw} (normalized by $\frac{0.14\pi e^2N_b\omega_wL^3}{mc^3\beta_o^2\gamma_o^3}$) with X_i for $K(=\frac{eB_w}{mc\omega_w})$ fixed at 1.0 and f=1,3,5 and 7. For all values of f, the gain is seen to be an oscillatory function of X_i. This is due to the complicated form of the coupling coefficient P_J.Gain at the fundamental increases as the wiggler frequency tends to the betatron frequency. On the contrary, the gain at the harmonic frequencies falls as X_i approaches 1. For $X_i<1$, gain at the 3rd and 7th harmonics is maximum at $X_i\simeq0.4$, whereas gain at the 5th harmonic is maximum at $X_i\simeq0.2$. For $X_i>1$, gain at all frequencies maximizes close to $X_i\simeq1.5$ and then falls rapidly.

References

1. D. H. Whittum, Ph.D. dissertation, University of California at Berkeley (1990).
2. P. Jha and P. Kumar, IEEE Trans. Plasma Sci. 24 (1996) 1359.
3. P. Jha and P. Kumar, Phys. Rev. E 57 (1998) 2256.
4. W. B. Colson, Ph. D. dissertation, Stanford University (1977).

FREE ELECTRON LASERS 1998
G.R. Neil and S.V. Benson (Eds.)
1999 Elsevier Science B.V.

Thematic Network Project for Microwave Applications

J. Lucas *, R.A. Stuart, and A.I. Al-Shamma'a

* *FEL Laboratory, Dept. of Electrical Engineering & Electronics, The University of Liverpool, P.O. Box 147, Liverpool L69 3BX, U.K.*

Abstract

This paper is specifically concerned with the generation of microwaves, by such devices as magnetrons and industrial Free Electron Lasers (FELs), and their industrial applications. This paper covers the European dimensions and partnership in this project and summarises the setting up of the Network to allow the promotion of microwave sources and applications to be achieved within the EC.

1. Introduction

This proposal is specifically concerned with the generation of microwaves, by such devices as magnetrons and industrial free electron lasers (FEL's), and their industrial applications. Microwaves are long wavelength electromagnetic waves which have wavelengths ranging from L Band (150-300mm) through the S, C, X and K Bands to the Millimetre Band (1-7.5mm).

Magnetrons are capable of producing high output powers for the longer wavelength microwaves. To extend the high output power capabilities into the higher frequency bands required the development of a new device, termed the industrial Free Electron Laser (FEL). The FEL may be regarded as operating in a similar manner to the Klystron and the TWT, but having the capability to produce the large microwave powers over the entire microwave spectrum, up to millimetre waves. FEL devices possess the unique property of being electronically tuneable and are able to produce the exact wavelength needed to match the requirements of the industrial application.

The leading group in the EC for research and development of microwave industrial FEL's is The University of Liverpool (UK). This work is underpinned by strong FEL activities at the University of Twente (Holland) and ENEA Frascati (Italy), who have developed the capability of producing high pulsed powers and shorter microwave frequencies (millimetre band) than the Liverpool group.

A programme of research, mainly funded by the UK DTI, was started at Liverpool in 1990 and a tuneable X band IFEL was demonstrated in 1993/4 producing 1W of output power. During 1994/5 the system efficiency and power capability were enhanced to 20% and 100W respectively. The targets for 1996/8 are 30% efficiency and 1kW output with further power increases in subsequent years. During 1997/8 FEL's operating in the Ku, K and Ka bands (12GHz to 40GHz) will be demonstrated producing powers in excess of 100W.

Microwaves produced by magnetrons have found a large number of well established applications in manufacturing industry for the processing of both solids and liquids.

During the exploratory phase of the Network it was revealed that many more important applications could be undertaken by the availability of tuneable microwave devices such as the industrial FEL over a wider frequency spectrum.

2. European Dimension and Partnership

To undertake the Implementation Phase, a partnership of Universities, Research Organisations and Industry have been brought together into the Network. They are currently researching either FEL systems or FEL components or manufacturing microwave sources and systems or developing microwave applications. Therefore a group of twenty-five partners from five EC countries have agreed to participate in the Implementation Phase.

On the microwave applications side there are a multitude of benefits to be attained. It is a great opportunity for the Universities, Research Organisations and Industries to interact to each other. The Universities and Research Organisations have necessary detailed knowledge concerning microwaves in direct contrast to Industry, which possesses detailed knowledge of manufacturing processes. The Network will allow the Universities and Research Organisations to educate on the theory and components required for microwave generation and propagation. Another benefit of the Network is that microwave education will be given to EC industry. The University Partners are from five EC countries and plan to jointly produce a common set of documents on microwave sources and applications which will form the basis of seminars and short education courses for industry. For a nominal fee SME's will visit the University for such a course whilst for the larger industries, the University will visit the factory to put on an industrial course.

To summarise the setting up of the Network will allow the promotion of microwave sources and applications to be achieved within the EC. In addition it will facilitate the setting up of an EC IFEL based industry in the near future which will have a lead over the USA and Japanese based companies.

3. Potential Applications

Microwaves have already found a large number of applications in manufacturing industry for the processing of both solids and liquids. The majority of these applications are concerned with mainly heating or drying activities. This is because existing microwave sources are based upon the magnetron devices, which only produce low operation frequencies 0.90 GHz, and 2.46 GHz. However magnetrons are reasonably compact and have high output powers (35 kW at 0.9 GHz and 8kW at 2.46 GHz respectively) with a wall plug efficiency of 30%.

The applications of microwaves have not been fully explored within manufacturing industries and many more important applications exist. The availability of a powerful, tuneable microwave source, covering the wide microwave spectrum, will allow many more applications to be identified. These include such activities as plasma material processing, vitrification, waste water treatment, adhesive curing, fashion cloth marking, communications, medical applications. The partners within the industrial FEL project will investigate these applications.

4. Role and Tasks of the Partners

The project co-ordination is by the University of Liverpool (UK). There are 26 partners whose aim is to promote Microwave applications of Industrial Free Electron Lasers. To obtain this objective the consortium of partners have been classified in two main group activities, A (Industrial Free Electron Lasers) and B(Microwave Applications). These are listed in table 1 and the main activity of each partner has been specified.

5. Acknowledgements

The authors wish to thank the EC Thematic program for their financial support towards Liverpool FEM Thematic Network project.

Table 1: Group Activities

Organisation	Activity	Country	Group
The Univ. of Liverpool	Overall FEL Design and communication	UK	A
Univ. of Milan	Electron sources	Italy	B
ENEA, Frascati	Overall FEL design	Italy	A
Comex Technologies	Insitu surface treatment	France	B
Sincrotrone Trieste	Wiggler design and construction	Italy	A
The Univ. of Twente (NCLR)	Overall FEL design	Holland	A
Coaxial Power Supply	RF sources and magnetrons	UK	A
The Univ., of Essex	Dielectric waveguide for medical applications	UK	B
British Nuclear Fuel (BNFL)	Vitrification for nuclear waste	UK	B
Cybernetix	Microwave sub sea communications	France	B
The Welding Institute (TWI)	Plasma material processing using FEL	UK	A
UMIST	Microwave pre-punchers for FEL design	UK	A
The Univ. of Thrace	Radiation monitoring, control, and safety of microwaves	Greece	A
Product Innovation	High voltage ceramics and electron source	UK	A
The Univ. of Gent (INTEC)	Sub sea communications	Belgium	B
Leybold Vakuum	Vacuum systems	Germany	A
Magnet developments Limited	Wiggler magnets	UK	A
IFR	Microwave components	UK	B
WITHDRAWN	Sterilisation of surgical implements	France	B
ABC	Catalytic cleaning for automobiles	UK	B
Broideries-Leveaux	Fashion cloth	France	B
Titanas	Plasma torch material processing	Greece	B
Surface Transforms	Catalytic cleaning for gas exhaust and fuel cells	UK	B
CLAIRE	Microwave plasma / Laser welding	France	B
Univ. of Limerick	Fibre optic sensor systems	Ireland	B

FREE ELECTRON LASERS 1998
G.R. Neil and S.V. Benson (Eds.)

Sub-picosecond synchronization of a 10-fs-Ti:Sapphire laser to FELIX

G.M.H. Knippels,* M.J. van de Pol, A.F.G. van der Meer

FOM-Institute for Plasma Physics 'Rijnhuizen', P.O. Box 1207, 3430 BE, Nieuwegein, The Netherlands

In this contribution we report the first successful sub-picosecond synchronization of a Kerr-lens mode-locked 10-fs Ti:Sapphire laser system (Femto Source Pro, Femto Lasers, Vienna, Austria) to an FEL. Previous attempts to synchronize a different FEL with a regeneratively mode-locked Ti:Sapphire laser have led to several picoseconds of jitter [1]. The Ti:Sapphire laser is pumped by 5 W of 532 nm from an intra-cavity frequency-doubled Nd:YVO_4 laser (Millenia V, Spectra Physics, Mountain View, CA USA), and employs so-called 'chirped-mirror' technology to achieve reliable and stable production of the ultrashort optical pulses [2]. We modified the cavity by mounting a tiny (5 mm diameter, 1 mm thick) high-reflector in the short arm of the cavity on a PZT with 10 μm range. This piezo-controlled mirror is used for active cavity stabilization. A fast Si-photodiode (-3 dB at 1.5 GHz) is used to monitor the 100-MHz optical pulse train. The 1 GHz component of this signal and a reference signal of the 1 GHz clock of the FEL are combined in a double-balanced mixer, the output of which serves as the error signal for driving the PZT. The unity-gain bandwidth of the loop is 1 kHz, determined primarily by the PZT drive electronics and the loop filter. A schematic layout of the feedback loop is given in Fig. 1. The use of the 10th harmonic of the 100 MHz roundtrip frequency in the feedback loop provides better locking because of the reduced influence of amplitude-to-phase-noise in the loop [3]. Figure 2 shows the phase-noise side bands of the Ti:Sapphire laser at the 10th harmonic around 1 GHz. Trace A is taken with the Ti:Sapphire laser in free-running mode, while trace B is taken when the feedback loop is activated and stabilizes the cavity length to match the clock-frequency. The noise peaks around 200 Hz are traced back to the times-four phase-locked loop. The width of the central peak is determined by the frequency stability of the 250 MHz signal generator.

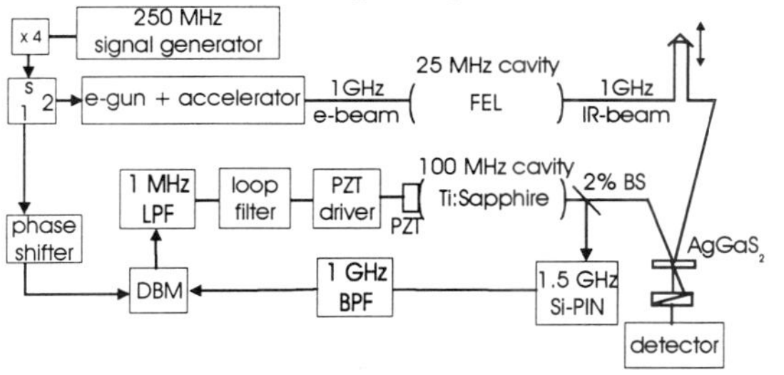

Fig. 1. Schematic layout of synchronization electronics and optics. An ultra-stable rf-clock (SMG 801.0001.52, Rohde & Schwarz, Munich, Germany) running at 250 MHz is multiplied by four to yield 1 GHz. An rf-splitter splits the 1 GHz signal in two. One arm serves as the reference clock input for the double-balanced mixer (DBM), while the other part drives the electron gun and accelerating structures. A fast silicon-diode detects the 100 MHz Ti:Sapphire pulse train and filters out the 1 GHz component, which is mixed with the 1 GHz clock in the double-balanced mixer to yield the phase difference. After passing through a 1-MHz low-pass filter and a loop filter (unity gain at 1 kHz) the error signal is send to a PZT-driver to adjust the cavity length. A phase shifter is used to control temporal overlap.

When the FEL and the Ti:Sapphire laser are frequency-locked to each other, the temporal overlap is

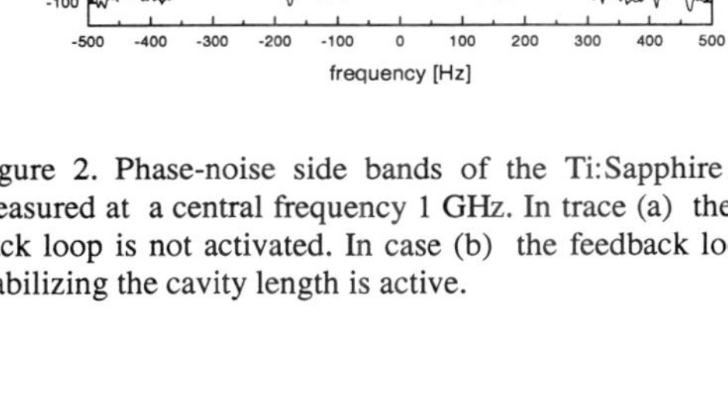

Figure 2. Phase-noise side bands of the Ti:Sapphire laser, measured at a central frequency 1 GHz. In trace (a) the feedback loop is not activated. In case (b) the feedback loop for stabilizing the cavity length is active.

* Corresponding author: Tel. +31-30-6096999, Fax. +31-30-6031204, email: knippels@rijnh.nl

found by simply scanning a phase-shifter. Type-I sum-frequency generation in a 100-μm thick $AgGaS_2$ crystal (MolTech, Berlin, Germany) has been used at an FEL wavelength of 9 μm to measure the degree of jitter that is left between the two lasers in a direct optical cross-correlation experiment. The sum-frequency is detected through a calcite polarizer that blocks the Ti:Sapphire background (see Fig. 1). In Fig. 3, a delay line was scanned to vary the optical delay between the two laser pulses, and individual data points correspond to a fifty-shot-averaged measurement over successive macropulses. The duration of the Ti:Sapphire laser pulses was measured separately with an autocorrelator close to the $AgGaS_2$ sample and was found to be 10 fs fwhm (see inset in Fig. 3 at the right-hand side). The infrared FEL pulse duration was similarly measured with a CdTe-based autocorrelator to be 435 fs fwhm in duration (see inset on left-hand side). The jitter from the cross-correlation signal can thus be estimated to be 400 fs rms (~900 fs fwhm, assuming un-correlated gaussian noise). The achieved degree of synchronization is present over many minutes and the resulting system is very competitive compared to other commercially available systems that allow locking of a Ti:Sapphire laser to an external clock. For example Spectra-Physics and Coherent typically specify a jitter of less than 3 ps rms (~6 ps fwhm) on a 60-second time scale or shorter for table-top-systems. In our case the FEL is located 40 meters away from the Ti:Sapphire laser, and the excellent synchronization achieved demonstrates the intrinsically stable cavity design of the Ti:Sapphire laser, as well as the tight locking of the FEL to its 1 GHz clock. The large separation between the lasers leads to a slow thermal drift in synchronization of about 1 ps per half-hour.

Although we demonstrated the synchronization of the Ti:Sapphire laser to the FEL at an infrared wavelength of 9 μm, this is only limited by the availability of suitable sum-frequency mixing crystals (and has been verified between 7 and 13 μm wavelength with the same $AgGaS_2$ crystals). The synchronization is present over the whole FELIX wavelength (4.2-300 μm) range since we directly locked to the FEL's internal RF-clock. Furthermore, we also are able to synchronize the Ti:Sapphire laser to FELIX when it runs at 25 or 50 MHz repetition rate instead of 1 GHz. This is achieved with a slight modification of the feedback loop so that first locking to the 25/50 MHz is obtained, before the 1 GHz lock is activated. Identical synchronization performance is observed in these modes since the 25/50 MHz signals are just necessary in the beginning to select the appropriate 1 GHz cycle, but do not influence the loop characteristics (the 25/50 MHz signal can even be removed, once locking is obtained).

Conclusions

We have successfully synchronized a passively-modelocked 10-fs Ti:Sapphire laser to FELIX with a measured jitter of only 400 fs rms. The system allows for a whole series of new two-color experiments, especially where high peak powers in the infrared are required or tunability beyond the range covered with OPA/OPG's systems.

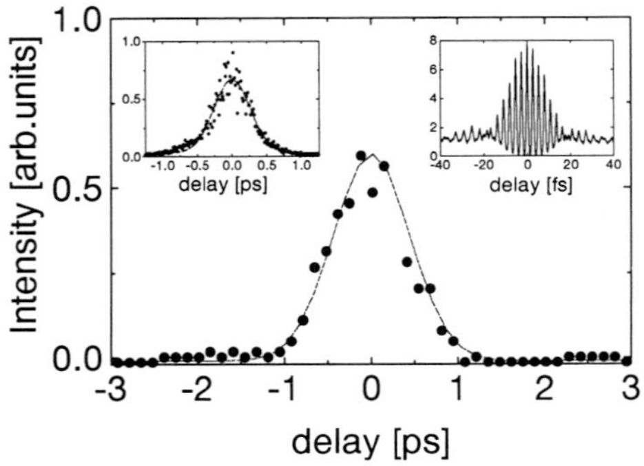

Figure 3. Measured optical cross-correlation between the FEL running at 9 μm and the Ti:Sapphire laser at 800 nm in a 100-μm thick $AgGaS_2$ crystal. The inset on the left-hand side shows the background-free autocorrelation measurement of the 435-fs FEL pulse with a home-built autocorrelator based on CdTe, and the inset on the right-hand side shows the 10-fs fringe-resolved autocorrelation function of the Ti:Sapphire pulse. The derived jitter is 400 fs rms (~900 fs fwhm). The scan took approximately two minutes to record.

Acknowledgements

This work is part of the research programme of the 'Stichting voor Fundamenteel Onderzoek der Materie (FOM)', which is financially supported by the 'Nederlandse Organisatie voor Wetenschappelijk Onderzoek (NWO)'.

References

[1] R.J. Stanley, R.L. Swent, T.I. Smith, Opt. Commun. **115**, 87-92 (1995).

[2] L. Xu, G. Tempea, A. Poppe, M. Lenzner, C. Spielmann, F. Krausz, A. Stingl, K. Ferencz, Appl Phys. B **65**, 151-159 (1997).

[3] D. von der Linde, Appl. Phys. B **39**, 201-217 (1986).

Author index